The HUTCHINSON Dictionary *of* SCIENTIFIC BIOGRAPHY

VOLUME I

The **HUTCHINSON** Dictionary *of* SCIENTIFIC BIOGRAPHY

VOLUME I

(Abbe to Leavitt)

Consultant Editors
Roy Porter
Marilyn Ogilvie

Helicon

First published in Great Britain in six volumes as
The Biographical Dictionary of Scientists:
Biologists, Chemists 1983;
Astronomers, Physicists 1984;
Engineers and Inventors, Mathematicians 1985

Second edition published 1994
Third edition published 2000

Helicon Publishing Limited
42 Hythe Bridge Street
Oxford OX1 2EP
email: admin@helicon.co.uk
Web site: www.helicon.co.uk

Printed and bound in Slovenia by DELO tiskarna
by arrangement with Korotan Ljubljana

ISBN 1-85986-304-3

British Library Cataloguing in Publication Data
A catalogue for this book is available
from the British Library

CONTENTS

Volume I

Volume II

CONTRIBUTORS

Neil Ardley, Gareth Ashurst, Jim Bailey, Mary Basham, Alan Bishop, Anthony Burton, William Cooksey, Christopher Cooper, David Cowey, Michael Darton, Keld Fenwick, Lorraine Ferguson, Judy Garlick, Richard Gulliver, Ian Harvey, Maggy Hendry, Keith Hutton, Nicholas Law, Lloyd Lindo, Robert Matthews, Nigel Morrison, Robert Mortimer, Patricia Nash, Valarie Neal, Jon and Lucia Osborne, Adam Ostaszewski, Caroline Overy, David Pimm, Roy Porter, Helen Rapson, Peter Rodgers, Mary Sanders, Martin Sherwood, Robert Smith, Robert Stewart, E. M. Tansey, Fred Taylor, Christopher Tunney, Martin Walters, Zusa Vrbova, David Ward, Edward Young

EDITORS

Consultant Editors
Roy Porter
Marilyn Ogilvie

Editorial Director
Hilary McGlynn

Managing Editor
Elena Softley

Technical Editor
Tracey Auden

Project Editor
Catherine Thompson

Researcher and Contributor
Maggy Hendry

Text Editors
Malgorzata Colquhoun
Sara Jenkins-Jones
Karen Lines

Bibliographies
Andrew Colquhoun

Production Director
Tony Ballsdon

Page Make-up
TechType

Art and Design Manager
Terence Caven

PREFACE TO THIRD EDITION

This new, revised, and updated edition of *The Hutchinson Dictionary of Scientific Biography* contains over 80 new biographies, focusing particularly on contemporary scientists, such as Stanley Prusiner, the US winner of the 1997 Nobel Prize for Physiology or Medicine, and on women scientists not included in the previous edition, such as Rosalind Franklin, the British biophysicist, and the Russian mathematician Sofia Kovalevskaya. In total this third edition includes over 1,280 in-depth biographies of scientists throughout the ages and across the major scientific disciplines, ranging from around 500 words to mini-essays of over 1,200 words.

Also added for this edition are the selected bibliographies, chronologies, quotations, and tables of scientific discoveries and of Nobel prizewinners. The historical reviews of the major sciences have been brought up to date, reflecting the increasing speed of change in many scientific disciplines, and an enlarged glossary of scientific terms is also included, enabling this to be used as a stand-alone reference book for students and the general reader alike.

INTRODUCTION TO SECOND EDITION

Science is sometimes viewed as impersonal: a method, a system, a technique for generating knowledge. But it is highly personal. The story of science is the story of the individuals who have discovered its truths.

This volume details the lives of the men and women responsible for the world's greatest scientific achievements. Its form – a biographical dictionary, alphabetically arranged – may seem straightforward and common sensical. But not so. For it inevitably raises the question of the role of the individual in history, and that has been fiercely disputed.

The idea that a single mind can produce a superlative novel, opera, poem or painting is uncontroversial – all apart from the most radical literary critics are happy to talk of Jane Austen, Richard Wagner, John Keats, or Pablo Picasso and their output. Yet debate rages about the comparable notion in science. The Great Man (or Woman) theory of scientific creativity is widely contested, and has been dismissed as outmoded. There are many historians and sociologists who would say that the discovery of the law of gravity (to take one signal instance) should not particularly be credited to Sir Isaac Newton. Many would say that the work of precursors should be credited – it was, after all, Newton himself who likened himself to a dwarf standing on a giant's shoulders. Marxists might insist that the theory of gravity should be seen less as an idea created by some lightning-flash in Newton's head than a necessary response to the industrial and technological needs of early capitalism – if Newton hadn't formulated the law, someone else would have done. Others would argue that it was an expression of the *Zeitgeist* or spirit of the age. Social and institutional historians would see that scientific breakthrough as the collective product of the shared labours of the late seventeenth-century scientific community or of the Fellows of the Royal Society. Many historians nowadays are of the opinion that to attribute creativity in science to individual genius is the fallacious and obsolete legacy of nineteenth century individualism and Romanticism.

The biographical arrangement of this dictionary is not intended as a reassertion of the individualistic theory of creativity. Rather it is designed to spell out and make accessible the multi-dimensional nature of science as thought and activity: meditation, observation, experimentation, exploration, debate, synthesis. Some of these are mainly solitary activities; others are deeply social, cooperative, and competitive. Abundant attention is given, of course, to the experimental and theoretical breakthroughs made by the men and women whose careers are charted in the almost thirteen hundred concise, factually well-documented biographies that follow in this work, from Abbe to Zworykin. Entries explore the logic of their trains of thought, their flights of imagination, their demolition of traditional patterns of argument. But coverage is also given to the collective social lives of these scientists and doctors, thinkers and tinkers. The accounts below explore their positions in national and religious cultures, their education, apprenticeships and training, the societies and institutions which they founded or to which they belonged, their colleagues, friends – and foes! Distinguished scientists are seen to be children of their times, but also the progenitors of new schools of thought, in a continuum stretching from past to present. Some achieved greatness in their own lifetimes. Others, like Semmelweis for instance, led lives of comparative obscurity or were ridiculed in their own day, and have won fame only posthumously.

The first function of a dictionary like this is, of course, to provide basic information, clearly, accurately, helpfully. But it is also to point to relationships, to indicate new connections. It must contain the information the reader was seeking. But it must also whet the appetite by revealing new and unexpected knowledge. This Dictionary eschews narrowness and avoids the mistake of talking about individuals merely as *scientists* – alter all, that technical professional term dates only from the 1830s. Entries show how such-and-such a person was not merely (say) a pre-eminent physicist or a celebrated chemist, but also politically active, or a major poet, a deep humanitarian, a man or woman of profound religious convictions – or, in some cases, a dabbler in magic or an industrial magnate.

The individuals biographized in the following pages all have something in common: – all made major contributions to the various scientific disciplines, astronomy, botany, biology, chemistry, cosmology. engineering, exploration, geology, mathematics, physics, physiology, and so forth. But a browse through the pages will equally highlight the differences, the wide range of skills and outlooks former scientists have possessed. Science has been the work both of men and women – in former centuries mainly a male preserve, but more latterly women have been more fully represented in its ranks (and hence in this dictionary). Scientific leaders have come from the Old and the New Worlds; they have been professionals and amateurs, full-timers and part-timers. In origins they have been princes and paupers, professors, obscure surveyors and lens-grinders. Science has been pursued for the love of Truth, the love of God, the solution of technical problems, a passion for beauty, the desire for money or fame. Many have been driven by some inner demon. Many of the scientists represented in the pages below have attempted, not just to expand the frontiers of science but also to reflect – philosophically, autobiographically, or even poetically – upon scientific thinking, the scientific process, and the scientific personality.

Nowadays some fear that science has become rigid, formal, bureaucratized; nothing but a mechanized process for churning out findings and results and solving technical problems. The lives of the scientists documented below would prove an antidote to such pessimism. They afford abundant evidence of the visions and goals that have driven the pursuit of science in history.

Roy Porter
Professor of the History of Medicine
Wellcome Institute for the History of Medicine

Historical Reviews of the Sciences

ASTRONOMY

ASTRONOMY IS THE oldest of all sciences, its origins dating back at least several thousands of years. The word astronomy comes from the two Greek words *astron* 'star' and *nomos* 'law' and perhaps the first attempt at understanding the laws that governed the stars and their apparent movement across the night sky was prompted by the need to produce an accurate calendar.

In order to predict the flooding of the River Nile, and hence the time when the surrounding lands would be fertile enough for crops to be planted, the Egyptians made observations of the brightest star in the night sky, Sirius. It was discovered that the date when this star (called Sothis by the Egyptians) could first be seen in the dawn sky (the heliacal rising) enabled the date of the flooding to be calculated. This also enabled the length of the year to be calculated quite accurately; by 2780 BC the Egyptians knew that the time between successive heliacal risings was about 365 days. More accurate observations enabled them to show that the year was about 365 1/4 days long, with a slight difference of 20 minutes between the sidereal year (the time between successive appearances of a star in the same position in the sky) and the tropical year (the time between successive appearances of the Sun on the vernal equinox).

Although evidence suggests that the grouping of stars into constellations was done before a fairly accurate calendar was drawn up, this latter achievement was probably the first scientific act carried out in the field of astronomy.

The prediction of phenomena, which is the fundamental activity of any science, was also being carried out by other ancient civilizations, such as the Chinese. Some historians suggest that the existence in Europe and elsewhere of megalithic sites such as Stonehenge in England and Carnac in France (some of which date back to almost 3000 BC) shows that even minor early civilizations could calculate their calendars and, possibly, predict basic astronomical events such as eclipses.

Certainly by 500 BC the prediction of eclipses had become quite accurate. Thales of Miletus (624–547 BC), a Greek philosopher, ended a war between the Medes and the Lydians by accurately predicting the occurrence of an eclipse of the Sun on 28 May 585 BC. His prediction was probably based on the Saros, a period of about 18 years, after which a particular sequence of solar (and lunar) eclipses recurs. This interval was known to Babylonian and Chaldean astronomers long before Thales' time.

The Greeks and early astronomy

The next 800 years of astronomy were dominated by the Greeks. Ionian natural philosopher Anaximander (611–547 BC), a pupil of Thales, helped bring the knowledge of the ancient Egyptians and others to Greece and introduced the sundial as a timekeeping device. He also pictured the sky as a sphere with the Earth floating in space at the centre.

Anaxagoras (*c.* 500–428 BC) made great advances in astronomical thought, with his correct explanation of the cause of lunar and solar eclipses. In addition, he considered the Moon to be illuminated by reflected light and all material in the heavens to be composed of the same material; a rocky meteorite falling on the Aegean coast in 468 BC may have brought him to this conclusion. All these thoughts are now known to be broadly correct.

The Greek philosopher Plato (*c.* 420–340 BC) effectively cancelled out these advances with his insistence on the perfection of the heavens, which according to him implied that all heavenly bodies must follow the perfect curve (the circle) across the sky. This dogma cast its dark shadow over astronomical thought for the next 2,000 years. Both Eudoxus (406–355 BC) and Callippus (370–300 BC) tried to convert observations into proof for this idea: each planet was set into a sphere, so that the universe took on an onion-ring appearance, with the Earth at the centre. More spheres had to be added to make the theory even approximate to the observations, and by the time of Callippus there were 34 such spheres. Even so, the theory did not match observations.

It was Greek philosopher and astronomer Heraklides of Pontus (388–315 BC) who noted that the apparent motion of the stars during one night might be the result of the daily rotation of the Earth, not the stars, on its axis. He also maintained that Mercury and Venus revolved around the Sun, but he held to the geocentric belief that the Earth was the centre of the universe.

The mathematician and astronomer Aristarchos of Samos (*c.* 320–240 BC) took these ideas one step further. In *c.* 260 BC he put forward the heliocentric theory of the Solar System, which puts the Sun at the centre of the system of the six planets then known, with the stars infinitely distant. This latter conclusion was based on the belief that the stars were motionless, their apparent motion resulting simply from the Earth's daily rotation. Aristarchos still maintained that the planets moved in Platonically perfect circles, however.

By noting that when the Moon was exactly half-illuminated it must lie at the right angle of a triangle formed by lines joining the Earth, Sun, and Moon, Aristarchos was able to make estimates of the relative sizes and distances of the Sun and Moon. Unfortunately, although his theory was correct, the instruments he needed were not available, and his results were highly inaccurate. They were good enough, however, to show that the Sun was more distant and much larger than the Moon. Although this provided indirect support for the heliocentric theory, since it seemed logical for the small Earth to orbit the vast Sun, the geocentric dogma prevailed for another 1,800 years.

Greek astronomy had its last success between 240 BC, when the scholar and polymath Eratosthenes (*c.* 276–*c.* 195 BC) made his calculation of the Earth's size, and *c.* AD 170, when Ptolemy died. In between those two events, Hipparchus (*c.* 146–127 BC) replaced the Eudoxian theory of concentric spheres with an even more contrived arrangement based on the ideas of Apollonius. The Sun and planets were all considered to revolve upon a small wheel, or epicycle, the centre of which revolved around the Earth, (the centre of the universe, on a larger circle (the deferent)). Using this theory, Ptolemy (or Claudius Ptolemaeus) (*c.* AD 2nd century) was able to predict the position of the planets to within 1°, for example, about two lunar diameters.

Having begun so well, Greek astronomy ended somewhat dismally, preferring dogma to observable evidence and so failing to come to terms with the heliocentric theory of Aristarchos.

It was not until some 600 years after the death of Ptolemy that astronomy started once again to move forward. The lead was taken by the Arabs. Their great mathematical skill and ingenuity with instruments enabled them to refine the observations and theories of the Greeks and produce better star maps, which were becoming increasingly useful for navigation, one of the spurs to astronomical research for many centuries to come. During the Middle Ages, nevertheless, European astronomers did little more than tinker with Ptolemy's epicycles. The Roman Catholic Church decided that the geocentric theory was the only one compatible with the Scriptures, thus making anyone who attempted to put the Sun at the centre of the Solar System guilty of heresy, which was punishable by death.

It was, however, the publication of one book that made Europe the centre of astronomical development, after the science had effectively stood still for many centuries.

The heliocentric theory

In 1543 the Polish astronomer Nicolaus Copernicus (1473–1543) published a book entitled *De Revolutionibus Orbium Coelestium/On the Revolutions of the Heavenly Spheres,* in which he demonstrated that by placing the Sun at the centre of the Solar System, with the planets orbiting about it, it was possible to account for the apparent motion of the planets in the sky much more neatly than by geocentric theory. To explain the phenomenon of retrogression, for example, cosmologists had been compelled to add complication upon complication to Ptolemy's theory. With the heliocentric theory it could be explained simply as the result of a planet's distances from the Sun.

The 'Copernican revolution' was far from a complete break with the past, nevertheless. Copernicus continued to believe that all celestial motion must be circular, so that his model retained the epicycles and deferents of Ptolemy. But whereas Aristarchos' heliocentric belief had been left to moulder for centuries, Copernicus' ideas were picked up by other cosmologists who, by the end of the 16th century, had far more accurate observations at their disposal.

The supplier of these more accurate observations was the Danish astronomer Tycho Brahe (1546–1601), who, using only the naked eye, was able to make observations accurate to two minutes of arc, five times more precise than those of Ptolemy. The effect was enormous. Calendar reform took place, with the Gregorian calendar, now used throughout the Western world, being instituted in 1582. More important for astronomy, Brahe's observations were used by his German assistant, Johannes Kepler (1571–1630), to establish the heliocentric theory of the Solar System once and for all. Out went the complicated systems of epicycles and the dogma of the perfect circle; for Kepler's idea that the planets followed elliptical paths around the Sun, which itself sat at one focus of the ellipses, accorded nicely with the observations of Brahe.

In *Astronomia Nova/New Astronomy*, published in 1609, Kepler enunciated the first two laws of planetary motion; the first stated that each planet moved in an elliptical orbit; the second that it did so in such a way that the line joining it to the Sun sweeps out equal areas in equal times. Kepler's third law, that the cube of the distances of a planet from the Sun is proportional to the time required for it to complete one orbit, was announced in 1618.

At the same time as Kepler was making these theoretical breakthroughs, the invention of the telescope by the Dutch optician Hans Lippershey (1587–1619), in 1608 was effecting a revolution in observation. Italian physicist and astronomer Galileo Galilei (1564–1642) quickly put the instrument to use. In the years 1609 to 1610 he discovered the phases of Venus (showing that planet to be orbiting inside the path of the Earth) and identified four satellites of Jupiter; he also established the stellar nature of the Milky Way. As if to underline the imperfect nature of the heavens, contrary to the Platonic dogma that had crippled astronomy for so long, Galileo also discovered spots on the Sun and mountains on the 'perfect' sphere of the Moon.

The next major breakthrough was again a theoretical one, the discovery by Isaac Newton (1642–1727) of the law of universal gravitation in 1665. Gravity is the single most important force in astronomy and Newton's discovery enabled him to deduce all three of Kepler's laws.

By nature a somewhat reticent man, Newton did not publish his discoveries, in the *Philosophiae Naturalis Principia Mathematica/Mathematical Principles of Natural Philosophy,* until 1687, after the prompting of his friend the English mathematician, physicist, and astronomer Edmund Halley (1656–1742). Newton showed that his law could account even for small effects such as the precession of the equinoxes discovered by Hipparchus and that the slight deviations of the planets from their Keplerian orbits were the result of their mutual gravitational attraction. By applying this perturbations theory to the Earth–Sun–Moon system, Newton was able to solve problems about the various motions that had baffled Kepler and his predecessors.

Newton also made a significant contribution to observational astronomy in 1668, when he built the first reflecting telescope, with optics that were free of some of the defects of the refractors then in use.

Seven years later, in 1675, Charles II founded the Royal Observatory at Greenwich, essentially to find an accurate way of determining longitude for British ships involved in overseas exploration and colonization. The line of zero longitude was set at Greenwich, and the first Astronomer Royal, John Flamsteed (1646–1719), drew up a new star catalogue with positions accurate to 20 seconds of arc. Published in 1725, it was the first map of the telescopic age.

Careful observer as Flamsteed was, he failed to notice anything strange about the star which he noted in his catalogue as 34 Tauri. Its true nature was discovered by German-born English astronomer William Herschel (1738–1822), using the best reflector then in existence, in 1781. The 'star' was in fact, a previously undiscovered planet, and was named Uranus. Its discovery doubled the dimensions of the known Solar System.

Beginnings of astrophysics

The opening of the 19th century marked the beginning of one of the most important branches of astronomical observation: spectroscopy. In 1802 English chemist and physicist William Wollaston (1766–1828) discovered that the Sun's spectrum was crossed by a number of dark lines, and in 1815 German physicist and optician Joseph Fraunhofer (1787–1826) made a detailed map of these lines. Fraunhofer noticed that the spectra of stars were slightly different from that of the Sun, but the enormous significance of this observation was not grasped until half a century later by German physicist Gustav Kirchhoff (1824–1887). In the meantime, Fraunhofer's skill as a telescope-maker enabled German astronomer Friedrich Bessel (1784–1846) to determine the distance to the star, 61 Cygni; he found it to lie at a distance of about six light years, a term that had then became commonly used as a measure of stellar distance.

Although Kepler's laws enabled the calculation of the relative sizes of the planetary orbits to be obtained, a figure in absolute terms for the mean Earth–Sun separation was still needed. In the 19th century much effort was expended on this task, using the method suggested by Kepler of observing transits of Venus from widely separated places in order to take parallax observations. By the middle of the century a figure within 2% of the correct value had been obtained.

The year 1846 saw another triumph for Newton's theory of gravitation, when English astronomer John Adams (1819–1892) and French astronomer Urbain Le Verrier (1811–1877) predicted the position of an as-yet-unseen planet. Their prediction was based on the observed discrepancies in the motion of Uranus – discrepancies which the two astronomers took to be caused by another, massive planet orbiting beyond the path of Uranus. By calculating the new planet's orbit and its position at certain times, Johann Galle (1812–1910) spotted it on 23 September 1846, less than two lunar diameters from the predicted spot. The new planet was later named Neptune.

Fraunhofer's study of the dark lines in the solar spectrum bore fruit in 1859 with Kirchoff's explanation of them. The lines were absorption lines and were the result of the presence in the Sun of certain chemical elements. Kirchhoff's discovery made it possible to determine the chemical composition of the Sun from Earth-bound observations. His work was extended by English astronomer William Huggins (1824–1910), who was able to show that the stars were made of similar elements to those found on the Earth, thus supporting the 2,300-year-old contention of Anaxagoras. Huggins used the new invention of photography to record the spectra. He also made the first measurement of stellar red shift, to determine the relative motion of stars towards or away from the Earth. These developments were crucial to the future development of astronomy and astrophysics.

By the end of the 19th century, photography had begun to take over from naked-eye observations of the universe. Stars, comets, nebulae, and the Andromeda galaxy had all been photographed by 1900. Spectroscopy had also been used on all these objects, and the composition, motion, and distance of many of them determined. This latter achievement was possible because of the discovery by Danish astronomer and physicist Ejnar Hertzsprung (1873–1967) of a relationship between the spectrum of a star and its intrinsic luminosity. The relation was also found by US astronomer Henry Russell (1877–1957) in the USA and published nine years after Hertzsprung in 1914. As a result, the diagram plotting the luminosities of stars against their spectra is called the Hertzsprung–Russell diagram. Its importance lies in its ability to show how stars evolve and how distant a star of a given spectrum and apparent brightness is.

Hertzsprung was able, by means of his diagram, to calculate the distance of a Cepheid variable star. As US astronomer Henrietta Leavitt (1868–1921) discovered in 1904, variable stars exhibit a relationship between their intrinsic luminosity and their period of variability, so that a measurement of the period above can give the Cepheid's distance. As such a measurement could be carried out even on Cepheids in other galaxies, the distances of these galaxies could now be calculated by means of Leavitt's period-luminosity law, published in 1912.

The origins of the universe

In 1916 German-born US theoretical physicist Albert Einstein (1879–1955) published his 'Foundation of the General Theory of Relativity' in *Annalen der Physik.* Essentially a theory of gravitation, it marked the greatest theoretical advance in our understanding of the universe since Newton's *Principia* and like Newton's theory it had far-reaching implications for astronomy.

Einstein's theory immediately cleared up a long-standing problem concerning the orbit of Mercury, which was slowly rotating at the rate of about 43 seconds of arc per century. Einstein's theory showed this to be a result of effects arising from the high orbital velocity of Mercury and the intense gravitational field of the Sun. The theory also made two predictions. First, that in the presence of an intense gravitational field, light should be red-shifted to longer wavelengths as it struggled to escape the field. In 1925 this was found by US astronomer Walter Adams (1876–1956) to be true of the spectrum of the white-dwarf companion of Sirius. Second, that according to General Relativity, light should be bent by the space-time curvature pictured in the theory as being the cause of gravitation. Observations of a solar eclipse in 1919 showed the light of stars close to the position of the Sun was indeed bent by the amount predicted by Einstein.

When he applied his theory to the entire universe, in an attempt to reach a universal understanding of dynamics, Einstein was dismayed to discover that in its pure form it would only apply to a universe that is in overall motion. Such a prediction was contrary to the contemporary belief that, apart from the individual motions of the stars within galaxies, the universe was static.

However, by applying spectroscopy to the light of distant galaxies, US astronomer Vesto Slipher (1875–1969) was able to show that the galaxies were in a state of recession. Surprisingly, it was not until 1924 that proof that the galaxies were star-systems separate from our own galaxy was given by US astronomer Edwin Hubble (1889–1953). Armed with this knowledge, Hubble was then able to show that the universe as a whole was expanding, and that the rate at which a galaxy appeared to be moving away from Earth was proportional to its distance from us (as determined by the Cepheid law). Thus, Einstein's theory was correct in predicting an expanding universe.

By the end of the 1920s the idea that the universe was born in a 'Big Bang', as proposed by Abbé George Lemaître (1894–1966) in 1927, had become, as it still remains, the established dogma of cosmology. Estimates based on Hubble's law currently set the date of the creation at about 15,000 million years ago.

Will the universe expand forever? This depends upon the mean density of matter in the universe. If the density exceeds a certain critical value (roughly equal to three protons per 1 cu m/35 cu ft) the expansion will halt sometime in the future and the universe will collapse to a 'big crunch'. With this in mind, astronomers have tried to estimate the mean density of matter, but have been hampered by the fact that most of the material in the universe is invisible; this conclusion was reached in the 1980s by considering the gravitational pulls on galaxies. The hunt for the invisible 'dark matter' continues in the 1990s. Modern theories of particle physics have thrown up a wealth of possible dark matter candidates. These theories also allow cosmologists to discuss events in the early Big Bang, possibly as far back as 10^{-43} seconds after the initial event. One theory dealing with the very early universe was developed by US cosmologist Alan Guth, in which the universe is supposed to have undergone a brief phase of accelerated expansion (called inflation) around 10^{-35} seconds after its birth.

The year 1930 saw the discovery by photographic means of a ninth planet in the Solar System, Pluto. After a painstaking search involving millions of star images, US astronomer Clyde Tombaugh (1906–1997) detected the tiny speck of light on plates taken at the Lowell observatory in Arizona.

In 1938 the long-standing problem of the power source of the Sun and stars was finally solved by German-born US physicist and astronomer Hans Albrecht Bethe (1906–) and German theoretical physicist Carl Friedrich von Weizsäcker (1912–). They found that the vast outpourings of energy were attributable to the fusing of hydrogen nuclei deep within stars, the process being so efficient that luminosity could be sustained for thousands of millions of years.

The universe in a new light

Experiments in America by US radio engineer Karl Jansky (1905–1950) and US radioastronomer Grote Reber (1911–) in the 1930s heralded the start of a new era in astronomical observation, marked by the use of wavelengths other than light, in particular radio waves. Solar radio emission was detected in 1942 and following a theoretical prediction by Dutch astronomer Hendrik van de Hulst (1918–) in the Netherlands in 1944, interstellar hydrogen radio emission was detected and this was used to produce a map of our Galaxy which was not limited to those regions not obscured by light-absorbing dust.

A radically different theory of the universe, which was free of the Big Bang and its attendant difficulties was proposed by Austrian-born US astronomer and physicist Thomas Gold (1920–), Austrian-born British scientist Hermann Bondi (1919–), and English cosmologist and astrophysicist Fred Hoyle (1915–) in 1948. Called the steady-state theory, it pictured the universe as being in a constant state of expansion, with new matter being created to compensate for the dilution caused by the Hubble recession. The theory aroused much criticism, but it was not until 1965 that the theory was considered by many to have been finally disproved.

In that year, experiments by German-born US radio engineer Arno Penzias (1933–) and US radioastronomer Robert Wilson (1936–) in the USA resulted in the discovery that the universe appears to contain an isotropic background of microwave radiation. On the Big Bang picture, that can be interpreted as the red-shifted remnant of the radiation generated in the original Big Bang. Since it is difficult to reproduce the properties of this background on the steady-state theory, the theory was abandoned by most astronomers.

Radioastronomy, which had progressed far in attempts to discover the true nature of the universe by observations of distant galaxies, had two major successes in the 1960s. The first was the discovery of quasi-stellar objects, and their identification as extremely remote yet very powerful sources by Dutch astronomer Maarten Schmidt (1929–) in 1963. Their power source remains enigmatic, but the current belief is that they are galactic centres which contain massive black holes, sucking material into themselves. The second discovery was made by Northern Irish astronomer Jocelyn Bell Burnell (1943–) at Cambridge in 1967. Rapid bursts of radio energy at extremely regular intervals were picked up and interpreted as being generated by a rapidly rotating magnetized neutron star, or pulsar. Such an object may be the result of a supernova explosion. In February 1987, the supernova SN1987A exploded in the Small Magellanic Cloud, the first naked-eye supernova since 1604. Observations of SN1987A allowed refinement of supernova theory. In particular, observation of neutrinos from the explosion confirmed that neutrinos carry off much of the energy released in the explosion.

Techniques have been developed in which the outputs from two or more radio telescopes are combined to allow better resolution than is possible with a single dish. In aperture synthesis, several dishes are linked together to simulate the performance of one very large single dish. This technique was pioneered by English astronomer Martin Ryle (1918–1984) at Cambridge, England. Very long baseline interferometry uses radio telescopes spread across the world to resolve minute details in radio sources.

Orbiting observatories

Earth-bound observations, which have always been hampered by the interference of a turbulent and polluted atmosphere, are now being supplemented by orbiting observatories operating at a wide range of wavelengths. The first X-ray observatory, UHURU, was launched in 1971, and it detected many sources of X-rays both within and beyond our galaxy. Successors include Einstein, EXOSAT, ROSAT, Chandrasekhar, and XMM; the latest of these have sensitivities which allow measurements of distant objects which record the state of the universe at early times.

Infrared missions have made similar advances, including the Cosmic Background Explorer (COBE), which measured the diffuse infrared and cosmic microwave background radiation. In 1992 COBE detected slight variations in the strength of the background radiation, believed to mark the first stage in the formation of galaxies. The National Aeronautics and Space Administration (NASA) is developing the Microwave Anistropy Probe (MAP) mission and the European Space Agency (ESA) the Planck mission as improved versions of COBE. ESA's Far Infrared Space Telescope (FIRST) will be able to detect the highly luminous infrared emission associated with the high rates of early star formation in dust-enshrouded galaxies.

Following the success of the Hubble Space Telescope, NASA is planning the Next-Generation Space Telescope (NGST) to address the question of galaxy formation at high redshift, for example during the early period of galaxy formation.

Planetary missions

The late 1960s and early 1970s were marked by the advent of the crewed exploration of the Moon by the US Apollo mission, beginning in 1969 with the expedition of *Apollo 11*. Throughout the 1960s and 1970s, uncrewed probes to the planets revealed more about them than had been discovered in all the previous centuries of study combined. The missions were either fly-bys, beginning with the US *Mariner 2* probe of Venus in 1962, or landings, such as the descent of the Russian *Venera 4* on to the surface of Venus in 1967.

Detailed maps of the four terrestrial planets, Mercury, Venus, Earth, and Mars, have now been made, and US Pioneer and Voyager probes passed by Jupiter and Saturn, taking in much of their satellite systems, in the 1970s and early 1980s. *Voyager 2* flew past Uranus in 1986, and Neptune in 1989.

Other highlights of planetary astronomy include the 1986 rendezvous of the *Giotto* probe with Halley's comet, the 1990 launch of the Hubble Space Telescope, and the 1990 arrival of *Magellan* at Venus. In 1991, the probe *Galileo* flew to within 1,600 km/994 mi of the asteroid Gaspra. In 1992 *Giotto* flew at a speed of 14 km/8.7 mi per second to within 200 km/128 mi of comet Grigg-Skjellerup, at a distance of 240 million km/150 million mi from Earth (13 light minutes away). On 25 September 1992 the *Mars Observer* was launched from Cape Canaveral, the first US mission to Mars for 17 years. Unfortunately, the craft was lost in 1993 as it was about to go into orbit around the planet. NASA re-flew its payload on *Mars Global Surveyor* (launched 7 November 1996) and *Mars Climate Orbiter* (launched 12 December 1998), and on 4 July 1997 delivered *Mars Pathfinder* to the surface of Mars. *Pathfinder* and its small *Sojourner* rover explored a region know as Ares Vallis, where water is believed to have flowed on the Martian surface in the distant past.

On 28 August 1993 *Galileo* flew past the asteroid Ida, and went on to orbit Jupiter on 7 December 1995. On the same day, the *Galileo* entry probe plunged into Jupiter's atmosphere and made measurements down to a pressure of over 20 bars. The orbiter completed its nominal two-year mission, and a two-year extension called the Galileo Europa mission, gathering huge amounts of data about Jupiter, its satellites and environment, including evidence for a subsurface ocean on Europa. An international mission to Saturn, the *Cassini* spacecraft, left Earth in 1997 and will land a probe on Saturn's largest moon, Titan, in 2004.

In 1996 astronomers from the San Francisco State University claimed the discovery of two planets orbiting stars 35 light years away that could support life. The two planets were detected by sophisticated techniques that analysed 'wobbles' in the period of the parent stars, 70 Virginis (in the constellation Virgo) and 47 UMa (in Ursa Major). Using similar techniques, astronomers had already claimed two other planet discoveries in other constellations, neither of which could sustain life. One planet is orbiting a pulsar or neutron star; the other, closely orbiting the star 51 Pegasi in the constellation of Pegasus, was spotted in 1995 by Swiss astronomers.

New telescopes

The 1990s saw the installation of an array of outstanding telescopes, incorporating radical new technology. There are currently nine telescopes plus one antenna of the Very Long Baseline Array in full operation on Mauna Kea, Hawaii, including the Subaru (Japan National Large Telescope), the Keck I and II observatories, the Submillimetre Array, the James Clerk Maxwell Telescope, the California Institute of

Technology 10.4-m/34-ft Submillimetre Telescope (CSO), and the Gemini Northern 8-m/26-ft telescope. The European Southern Observatories (ESO) at La Silla, Chile, have a 15-m/49-ft diameter parabolic antenna devoted to radio observations, two 3.6-m/11.8-ft optical New Technology Telescopes, and others ranging in aperture from 2.2 m/7.2ft to 0.5 m/1.6 ft. The Very Large Telescope (VLT) project at Cerro Paranal consists of four 8.2-m/27-ft telescopes, working independently or in combined mode. When completed, the total light collecting power of the telescope will equal that of a 16-m/52-ft single telescope. The ESO VLT will then be the world's largest and most advanced optical telescope.

IRAM (Institut de Radio Astronomie Millimétrique) is the world's largest telescope operating at wavelengths between 0.8 mm/0.03 in and 3.5 mm/0.14 in. Founded in 1979, it is mainly the responsibility of France, Germany, and Spain, and operates two major facilities: a 30-m/98-ft diameter telescope on Pico Veleta in the Sierra Nevada (southern Spain) and an array of five 15-m/49-ft diameter telescopes on the Plateau de Bure in the French Alps.

The US National Radio Astronomy Observatory's Very Large Array (VLA) in New Mexico consists of 27 antennas in a huge Y pattern. Each antenna is 25 m/81 ft in diameter and weighs 230 tonnes; they are combined electronically to give the resolution of an antenna 36 km/22 mi across, with the sensitivity of a dish 130 m/422 ft in diameter.

The Department of Astronomy at the University of Arizona and its associated research division, Steward Observatory, operate a 10-m/32.8-ft Submillimetre Telescope and the 1.8-m/5.9-ft Lennon reflector (on Mount Graham); a 4.5-m/14.8-ft equivalent aperture MMT reflector (on Mount Hopkins); a 2.3-m/7.5-ft Bok reflector and a 0.9-m/3-ft reflector (on Kitt Peak); a 1.6-m/5.2-ft Bigelow reflector and a 0.4-m/1.3-ft Schmidt telescope (on the Caralina site); a 1.5-m/4.9-ft NASA reflector and a 1-m/3.3-ft reflector (on the Mount Lemmon site); and a 0.5-m/1.6-ft reflector on campus.

In the 21st century, new observing systems are likely to lead to a greatly improved understanding of the nature of the universe, and its evolution. Even more excitingly, observatories will be realized in the next decade or two which will not only detect the presence of planets around distant stars, but will also have the spatial resolution and optical precision to obtain spectra of the radiation reflected and emitted from the planet, with tolerable levels of contamination from the much brighter parent star nearby. Once this is possible, suitable instrumentation will be able to analyse the spectrum of the planet in an effort to establish its composition and state, including, eventually, surface temperature and pressure, and detection of such key molecules as water vapour and ozone (as a proxy for the more difficult oxygen). In short, we may learn in the next century not only about the existence of Earthlike planets elsewhere in the universe, but also whether life is present on them.

BIOLOGY

LIKE SO MUCH else, the systematic study of living things began with the Greeks. Earlier cultures such as those in Egypt and Babylon in the Near East, and the early Indian and Chinese civilizations in Asia, had their own approaches to the study of Nature and its products. But it was the Greeks who fostered the attitudes of mind and identified the basic biological problems and methods from which modern biology has grown. Enquiry begun by Greek philosophers such as Alcmaeon (born *c.* 535 BC) and Empedocles (*c.* 492–*c.* 432 BC) culminated in the biological work of Greek polymath Aristotle (384–323 BC) and the medical writings of the Greek physician Hippocrates (*c.* 460–377 BC) and his followers.

Natural philosophy

Aristotle was an original thinker of enormous power and energy who wrote on physics, cosmology, logic, ethics, politics, and many other branches of knowledge. He also wrote several biological works which laid the foundations for comparative anatomy, taxonomy (classification), and embryology. He was particularly fascinated by sea creatures; he dissected many of these as well as studying them in their natural habitats. Aristotle's approach to anatomy was functional: he believed that questions about structure and function always go together and that each biological part has its own special uses. Nature, he insisted, does nothing in vain. He therefore thought it legitimate to enquire about the ultimate purposes of things. This teleological approach has persisted in biological work until the 20th century. In addition, Aristotle studied reproduction and embryological development, and he established many criteria by which animals could be classified. He believed that animals could be placed on a vertical, hierarchical scale ('scale of being'), extending from humans down through quadrupeds, birds, snakes, and fishes to insects, molluscs, and sponges. Hierarchical thinking (as reflected in the terms 'higher' and 'lower' organisms) is still present in biology. One of Aristotle's pupils, the Greek philosopher and scientist Theophrastus (*c.* 372–287 BC), founded for botany many of the fundamentals that Aristotle had established for zoology.

What Aristotle achieved for biology, Hippocrates and his followers contributed for medicine: they established a naturalistic framework for thinking about health and disease. Unlike earlier priests and doctors, they did not regard illness as the result of sin, or as a divine punishment for misdeeds. They were keen observers whose most influential explanatory framework saw disease as the result of an imbalance in one of the four physiologically active humours (blood, phlegm, black bile, and yellow bile). The humours were schematically related to the four elements (earth, air, fire, and water) of Greek natural philosophy. Each person was supposed to have his or her own dominant humour, although different humours tended to predominate at different times of life (such as youth and old age) or seasons of the year. The therapy of Hippocrates aimed at restoring the ideal balance through diet, drugs, exercise, change of life style, and so on.

Galen's influence on medicine

After the classical period of Greek thought, the most important biomedical thinker was Greek physician, anatomist and physiologist Galen (*c.* 129–*c.* 199), who combined Hippocratic humoralism with Aristotle's tendency to think about the ultimate purposes of the parts of the body. Galen should not be blamed for the fact that later doctors thought that he had discovered everything, so that they had no need to look at biology and medicine for themselves. In fact, Galen was a shrewd anatomist and the most brilliant experimental physiologist of antiquity. For most of that time, human dissection was prohibited and

Galen learnt his anatomy from other animals such as pigs, elephants, and apes. It took more than a thousand years before it was discovered that some of the structures which he had accurately described in other animals (such as the five-lobed liver and the rete mirabile network of veins in the brain) were not present in the human body. Like that of Aristotle, Galen's anatomy was functional, and often his tendency to speculate went further than sound observation would have permitted, as when he postulated invisible pores in the septum of the heart which were supposed to allow some blood to seep from the right ventricle to the left.

Following Galen's death at the end of the 2nd century and the collapse of the Roman Empire, biology and medicine remained stagnant for a thousand years. Most Classical writings were lost to the European West, to be preserved and extended in Constantinople and other parts of the Islamic Empire. From the 12th century these texts began to be rediscovered in southern Europe – particularly in Italy, where universities were also established. For a while scholars were content merely to translate and comment on the works of men such as Aristotle and Galen, but eventually an independent spirit of enquiry arose in European biology and medicine. Human dissections were routinely performed from the 14th century and anatomy emerged as a mature science from the fervent activity of Belgian physician Andreas Vesalius (1514–1564), whose *De Humani Corporis Fabrica/On the Fabric of the Human Body* (1543) is one of the masterpieces of the Scientific Revolution. His achievement was to examine the body itself rather than relying simply on Galen; the illustrations in his work are simultaneously objects of scientific originality and of artistic beauty. The rediscovery of the beauty of the human body by Renaissance artists encouraged the study of anatomy by geniuses such as Italian, artist, inventor and scientist Leonardo da Vinci (1452–1519). Shortly afterwards, the English physician William Harvey (1578–1657) discovered the circulation of the blood and established physiology on a scientific footing. His little book *De Motu Cordis/On the Motion of the Heart* (1628) was the first great work on experimental physiology since the time of Galen. The eccentric wandering Swiss physician and chemist Paracelsus (1493–1541) had also deliberately set aside the teachings of Galen and other Ancients in favour of a fresh approach to Nature and medicine and to the search for new remedies for disease.

The age of discovery

While these achievements were happening in medicine, anatomy, and physiology, other areas of biology were not stagnating. Voyages of exploration alerted naturalists to the existence of many previously unknown plants and animals and encouraged them to establish sound principles of classification, to create order out of the apparently haphazard profusion of Nature. Zoological and botanical gardens began to be established so that the curious could view wonderful creatures like the rhinoceros and the giraffe. And just as the Great World (Macrocosm) was revealing plants and animals unknown in Europe only a short time before, so the invention of the microscope in the 17th century gave scientists the opportunity of exploring the secrets of the Little World (Microcosm). The microscope permitted Dutch microscopist Anton van Leeuwenhoek (1632–1723) to see bacteria, protozoa, and other tiny organisms; it enabled English physicist Robert Hooke (1635–1703) to observe in a thin slice of cork regular structures which he called 'cells'. And it aided Italian physician Marcello Malpighi (1628–1664) to complete the circle of Harvey's concept of the circulation of the blood by first seeing it flowing through capillaries, the tiny vessels that connect the arterial and venous systems. Many of these microscopical discoveries were communicated to the Royal Society of London, one of several scientific societies established during the mid-17th century.

The full potential of the microscope as a biological tool had to wait for technical improvements effected in the early 19th century. But it also led scientists along some blind alleys of theory. Observations of sperm 'swimming' in seminal fluid provided some presumed evidence for a theory that was much debated during the 18th century, concerning the nature of embryological development. Aristotle had thought that the body's organs (heart, liver, stomach, and so on) only gradually appear once conception has initiated the growth of the embryo. Later scientists, including William Harvey, extended Aristotle's theory with new observations. But now the visualization of moving sperm suggested that some miniature, but fully formed organism was already present in the reproductive fluids of the male or female. The tiny *homunculi* were thought to be stimulated to growth by fertilization. If the homunculus was always there, it followed that its own reproductive parts contained all of its future offspring, which in turn contained its future offspring and so on, back to Adam and Eve (depending on whether the male or the female was postulated as the carrier). This doctrine, called preformationalism, was held by most 18th-century biologists, including Swiss physiologist Albrecht von Haller (1708–1777) and Italian physiologist Lazzaro Spallanzani (1729–1799), two of the century's greatest scientists. Both men, like virtually all scientists of the period, were devout Christians, and preformationalism did not conflict with their belief that God established regular, uniform laws which governed the development and functions of living things. They did not believe that inert matter could join together by accident to make a living organism. They rejected, for instance, the possibility of spontaneous generation, and Spallanzani devised some ingenious experiments designed to show that maggots found in rotting meat or the teeming life discoverable after jars of water are left to stand did not spontaneously generate. Haller, Spallanzani, Scottish anatomist John Hunter (1728–1793), and most other great 18th-century experimentalists held that the actions of living things could not be understood simply in terms of the laws of physics and chemistry. They were Vitalists, who believed that unique characteristics separated living from non-living matter. The special attributes of humans were often ascribed to the soul, and lower animals and plants were thought to possess more primitive animal and vegetable souls which gave them basic biological capacities such as reproduction, digestion, movement, and so on.

The systematic use of improved microscopes revolutionized the way in which biologists conceived organisms. In the closing years of the 18th century a French physician named Xavier Bichat (1771–1802), aided only with a hand lens, developed the idea that organs such as the heart and liver are not the ultimate functional units of animals. He postulated that the body can be divided into different kinds of tissues (such as nervous, fibrous, serous, and muscular tissue) which make up the organs. Increasingly, biologists and doctors began thinking in terms of smaller functional units, and microscopists, such as Scottish botanist Robert Brown (1773–1858), began noticing regular structures within these units, which we now recognize as cells. Brown called attention to the nucleus in the cells of plants in 1831 and by the end of the decade the German botanist Matthias Schleiden (1804–1881) and German physiologist Theodor Schwann (1810–1882) systematically developed the idea that all plants and animals are composed of cells. The cell theory was quickly established for adult organisms, but in certain situations – such as the earliest stages of embryological development or in the appearance of 'pus' cells in tissues after inflammation or injury – it appeared that new cells were actually crystallized out of an amorphous fluid which Schwann called the 'blastema'. The notion of continuity of cells was enlarged upon by the German pathologist Rudolf Virchow (1821–1902), who summarized it in his famous slogan 'All cells from cells'. The cell theory gave biologists and physicians a new insight into the architecture and functions of the body in health and disease.

Microorganisms and disease

Concern with single-celled organisms also lay behind the work of French chemist and microbiologist Louis Pasteur (1822–1895), which helped to establish the germ theory of disease. Pasteur trained as a chemist, but his researches into everyday processes such as the souring of milk and the fermentation of beer and wine opened for him a new understanding of the importance of yeast, bacteria, and other microorganisms in our daily lives. It was Pasteur who finally convinced scientists that animals do not spontaneously generate on rotting meat or in infusions of straw; our skin, the air, and everything we come into contact with can be a source of these tiny creatures. After reading about Pasteur's work, English surgeon Joseph Lister (1827–1912) first conceived the idea that by keeping away these germs (as they were eventually called) from the wounds made during surgical operations, healing would be much faster and post-operative infection would be much less common. When in 1867 Lister published the first results using his new technique, antiseptic surgery was born. He spent much time developing the methods, which were taken up by other surgeons who soon realized that it was better to prevent infection altogether (asepsis) by carefully sterilizing their hands, instruments, and dressings. By the time Lister died, he was world famous and surgeons were performing operations that would have been impossible without his work.

After Lister drew attention to the importance of Pasteur's discoveries for medicine and surgery, Pasteur himself showed the way in which germs cause not only wound infections, but also many diseases. He first studied a disease of silkworms which was threatening the French silk industry; he then turned his attention to other diseases of farm animals and human beings. In the course of this research, he discovered that under certain conditions an organism could be grown which, instead of causing a disease, actually prevented it. He publicly demonstrated these discoveries for anthrax, then a common disease of sheep, goats, and cattle which sometimes also affected human beings. He proposed to call this process of protection 'vaccination', in honour of English biologist Edward Jenner (1749–1823), who in 1796 had shown how inoculating a person with cowpox (vaccinia) can protect against the deadly smallpox. Pasteur's most dramatic success came with a vaccine against rabies, a much-dreaded disease occasionally contracted after a bite from an infected animal.

By the 1870s, other scientists were investigating the role played by germs in causing disease. Perhaps the most important of them was German bacteriologist Robert Koch (1843–1910), who devised many key techniques for growing and studying bacteria, and who showed that tuberculosis and cholera – prevalent diseases of the time – were caused by bacteria. Immunology, the study of the body's natural defence mechanisms against invasion by foreign cells, was pioneered by another German bacteriologist, Paul Ehrlich (1854–1915), who also began looking for drugs that would kill disease-causing organisms without being too dangerous for the patient. His first success, a drug named Salvarsan, was effective in the treatment of syphilis. Ehrlich's hopes in this area were not fully realized immediately, and it was not until the 1930s that the synthetic sulfa drugs, also effective against some bacterial diseases, were developed by German bacteriologist Gerhard Domagk (1895–1964) and others. Slightly earlier Scottish bacteriologist Alexander Fleming (1881–1955) had noticed that a mould called *Penicillium* inhibited the growth of bacteria on cultures. Fleming's observation was investigated during World War II by Australian-born British bacteriologist Howard Florey (1898–1968) and German-born British biochemist Ernst Chain (1906–1979), and since then many other antibiotics have been discovered or synthesized. But antibiotics are not effective against diseases caused by viruses; such infections can, however, often be prevented using vaccines. An example is poliomyelitis, vaccines against which were developed in the 1950s by Russian-born US viriologist Albert Sabin (1906–1993) and US microbiologist Jonas Salk (1914–1995).

Many of these advances in modern medical science are a direct continuation of discoveries made in the 19th century, although of course we now know much more about bacteria and other pathogenic microorganisms than did Pasteur and Koch.

Modern biological sciences

The rapidly developing discipline of biochemistry became more and more dominant in 20th-century biology. Molecular biologists and chemists have been concerned with determining the structures of many large biological molecules, such as the muscle protein myoglobin by English biochemist John Kendrew (1917–1997) and Austrian-born British molecular biologist Max Perutz (1914–). In their researches they often interpreted the diffraction patterns produced when X-rays pass through these complex molecules, a technique pioneered by English chemist and crystallographer Dorothy Hodgkin (1910–1994).

Molecular biology is only one of several new biological disciplines to be developed during the past century. The oldest of these, biochemistry, was established in Britain by English biochemist Frederick Gowland Hopkins (1861–1947) who, along with Polish-born US biochemist Casimir Funk (1884–1967) and US biochemist and nutritionist Elmer McCollum (1879–1967), is remembered for his fundamental work in the discovery of vitamins, substances that help to regulate many complex bodily processes. Other biochemists such as Czech-born US biochemists Carl Cori (1896–1984) and Gerty Cori (1896–1957) have studied the ways in which organisms make use of the energy gained when food is broken down. Many of these internal processes are also moderated by the action of hormones, one important example of which is insulin, discovered in the 1920s by Canadian physiologist Frederick Banting (1891–1941) and others.

Modern biologists also often use physics in their work, and biophysics is now an important discipline in its own right. English physiologist Archibald Hill (1886–1977) and German-born US biochemist Otto Meyerhof (1884–1951) pioneered in this area with their work on the release of heat when muscles contract. More recently, German-born British physiologist Bernhard Katz (1911–) has used biophysical techniques in studying the events at the junctions between muscles and nerves, and at the junctions between pairs of nerves (synapses). The events at synapses are initiated by the release of chemical substances such as adrenaline and acetylcholine, as was demonstrated by English physiologist Henry Dale (1875–1968) and German pharmacologist Otto Loewi (1873–1961). The way in which nerve impulses move along the nerve axon has been investigated by English physiologists Alan Hodgkin (1914–1998) and Andrew Huxley (1917–). For this work, they made use of the giant axon of the squid, an experimental preparation whose importance for biology was first shown by English zoologist John Young (1907–1997). The complicated way in which the nervous system operates as a whole was first rigorously investigated by English neurologist Charles Sherrington (1857–1952).

Another area of fundamental importance in modern biology and medicine is immunology. For instance, the discovery by Austrian-born US physician Karl Landsteiner (1868–1943) of the major human blood group system (A, B, and O) permitted safe blood transfusions. The development of the immune system – and the way in which the body recognizes foreign substances ('self' and 'not-self') has been investigated by such scientists as British immunologist Peter Medawar (1915–1987) and Australian immunologist Frank Macfarlane Burnet (1899–1985). Much of this knowledge has been important to transplant surgery, pioneered for kidneys by English surgeon Roy Calne (1930–) and for hearts by South African caardiothoracic surgeon Christiaan Barnard (1922–).

Taxonomy and classification

Experimental biology was well established in the 18th century; another great area of 18th-century activity was classification. Again inspired by Aristotle, and drawing on the work of previous biologists such as the British naturalist John Ray (1627–1705). The Swedish botanist Carolus Linnaeus (1707–1778) spent a lifetime trying to bring order to the ever increasing number of plants and animals uncovered by continued exploration of the Earth and its oceans. His *Systema Naturae/System of Nature* (1735) was the first of many books in which he elaborated a philosophy of taxonomy and established the convention of binomial nomenclature still followed today. In this convention, all organisms are identified by their genus and species; thus human beings in the Linnaean system are *Homo sapiens.* Depending on the nature of the characteristics examined, however, plants and animals could be placed in a variety of groups, ranging from the kingdom at the highest level through phyla, classes, orders and families, and so on beyond the species to the variety and, finally, the individual. Naturalists had traditionally accepted that the species was the most significant taxonomic category, Christian doctrine generally holding that God had specially created each individual species. It was also assumed that the number of species existing was fixed during the Creation, as described in the Book of Genesis – no new species having been created and none becoming extinct. Linnaeus, however, believed that God had created genera and that it was possible that new species had emerged during the time since the original Creation.

Although still based on the Linnaean system, modern taxonomists use a range of methods in describing and classifying species, and these may now include biochemical and genetic comparison, in addition to the more traditional consideration of morphological characteristics. The extent to which classification can, or even should, seek to reflect evolutionary relatedness continues to the present day. Classification is necessarily an abstraction, being a 'snapshot' in an evolutionary process, but is nevertheless essential as a tool for describing and studying organisms. With increasing pressure on habitats, and a rising rate of human-induced extinctions, the need to build up a detailed inventory of species is ever more urgent.

Palaeontology and evolution

Some 18th-century naturalists such as French naturalist Georges Buffon (1707–1788) began to suggest that the Earth and its inhabitants were far older than the 6,000 or so years inferred from the Bible. General acceptance of a vastly increased age of the Earth, and of the reality of biological extinction, awaited the work of early 19th-century scientists such as French zoologist Georges Cuvier (1769–1832), whose reconstructions of the fossil remains of large vertebrates like the mastodon and dinosaurs found in the Paris basin and elsewhere so stirred both the popular and scientific imaginations of his day. Despite Cuvier's work on the existence of life on Earth for perhaps millions of years, he firmly opposed the notion that these extinct creatures might be the ancestors of animals alive today. Rather he believed that the extent to which any species might change (variability) was fixed and that species themselves could not change much over time. His contemporary and scientific opponent, French naturalist Jean Baptiste Lamarck (1744–1829), argued however that species do change over time. He insisted that species never become extinct; instead they are capable of change as new environmental conditions and new needs arise. According to this argument, the ancestors of the giraffe need not have had such a long neck, which instead might have slowly developed as earlier giraffelike creatures stretched their necks to feed on higher leaves. Lamarck believed that physical characteristics and habits acquired after birth could – particularly if repeated from generation to generation – become inherited and thus inborn in the organism's offspring. We still call the doctrine of the inheritance of acquired characteristics 'Lamarckianism', although most naturalists before Lamarck had already believed it. It continued to be generally accepted (for instance, by English naturalist Charles Darwin (1809–1882)) until late in the 19th century. The debates between Cuvier and Lamarck were part of the new possibilities opened up by the revolution in thinking about the age of the Earth and of life on it.

Charles Darwin was not the first to suggest that biological species can change over time, but his book *The Origin of Species* (1859) first presented the idea in a scientifically plausible form. As a young man, Darwin spent five years (1831–36) on HMS *Beagle*, during which he studied fossils, animals, and geology in many parts of the world, particularly in South America. By 1837 he had come to believe the fact of evolution; in 1838 he hit upon its mechanism: natural selection. This principle makes use of the fact that organisms produce more offspring than can survive to maturity. In this struggle for existence, those offspring with characteristics best suited to their particular environment will tend to survive. In this way, Nature can work on the normal variation which plants and animals show and, under changing environmental conditions, significant change can occur through selective survival.

Darwin knew that his ideas would be controversial so he initially imparted them to only a few close friends, such as the Scottish geologist Charles Lyell (1797–1875) and the English botanist Joseph Hooker (1817–1911). For 20 years he continued quietly to collect evidence favouring the notion of evolution by natural selection, until in 1858 he was surprised to receive a short essay from Welsh naturalist Alfred Russel Wallace (1823–1913), then in Malaya, perfectly describing natural selection. Friends arranged a joint Darwin–Wallace publication, and then Darwin abandoned a larger book he was writing on the subject to prepare instead *The Origin of Species.* In it he marshalled evidence from many sources, including palaeontology, embryology, geographical distribution, ecology (a word coined only later), and hereditary variation. Darwin did not have a very clear idea of how variations occur, but his work convinced a number of scientists, including English scientist and humanist Thomas Huxley (1825–1895), German zoologist Ernst Haeckel (1834–1919) and English scientist, inventor, and explorer Francis Galton (1822–1911), Darwin's cousin. Huxley became Darwin's chief publicist in Britain, Haeckel championed Darwin's ideas in Germany, and Galton quietly absorbed the evolutionary perspective into his own work in psychology, physical anthropology, and the use of statistics and other forms of mathematics in the life sciences.

Genetics

Meanwhile, unknown to Darwin (and largely unrecognized during his lifetime), an Austrian monk named Gregor Mendel (1822–1884) was elucidating the laws of modern genetics through his studies of inheritance patterns in pea plants and other common organisms. Mendel's work on the inheritance patterns of peas was published in 1866, but had little influence on biological thinking until 1900. Then it was scrutinized by Dutch botanist and geneticist Hugo de Vries (1848–1935), German biologist Karl Erich Correns (1864–1933), and by the Austrian scientist Erich von Tschermak-Seysenegg (1871–1962). Their work signalled the start of serious research into heredity. Mendel studied characteristics that were inherited as a unit; this enabled scientists to understand such phenomena as dominance and recessiveness in these units, called 'genes' in 1909 by the Danish biologist W L Johannsen (1859–1927). By 1900, when de Vries, English geneticist William Bateson (1861–1926), and others were recognizing the importance of Mendel's pioneering work, much more was known about the microscopic appearances of cells both during adult division (mitosis) and reduction division (meiosis). In addition, German zoologist August Weismann (1834–1914) had developed notions of the continuity of the inherited material (which he called the 'germ plasm') from generation to generation, thus suggesting that acquired characteristics are not inherited. Only a few scientists in the 20th century, such as the Soviet botanist Trofim Lysenko (1898–1976), have continued to believe in Lamarckianism, for modern genetics has accumulated overwhelming evidence that characteristics such as the loss of an arm or internal muscular development do not change the make-up of reproductive cells. In 1906, Bateson introduced the word 'genetics'. He demonstrated that certain inherited traits tend to be transmitted as a group, thus establishing the concept of genetic linkage. In 1909 he used the term 'allele' for alternate forms of a gene.

It is now believed that new inheritable variations occur when genes mutate. The study of this process and of the factors (such as X-rays and certain chemicals) that can make the occurrence of mutations likely was pioneered by geneticists such as US geneticists Thomas Hunt Morgan (1866–1945) and Herman Muller (1890–1967). They did much of their work with fruit flies (*Drosophila*). While these and other scientists were showing that genes are located on chromosomes – strands of darkstaining material in the nuclei of cells – other researchers were trying to determine the exact nature of the hereditary substance itself. Originally it was thought to be a protein, but in 1953 English molecular biologist Francis Crick (1916–) and US biologist James Watson (1928–) were able to show that it is dioxyribonucleic acid (DNA). Their work was an early triumph of molecular biology, a branch of the science that has grown enormously since the 1950s.

Scientists now know a great deal about how DNA works. Among those who have contributed are US biologist George Beadle (1903–1989), US microbiologist Edward Tatum (1909–1975), French biochemist Jacques Monod (1910–1976), US geneticist Joshua Lederberg (1925–), and New Zealand-born British biochemist Maurice Wilkins (1916–). In the mid-20th century the chemical link between the gene and the expression of that gene in the organism was beginning to be unravelled. This link proved to be the production by the gene of a specific protein, called an enzyme, which has a particular effect in the cell in which it is produced. In 1941, Beadle and Tatum established the gene-enzyme hypothesis, showing that one gene was responsible for the production of one specific enzyme (known famously as 'one gene, one enzyme').

Genetic engineering

Genetic engineering consists of a collection of methods used to manipulate genes. The first of these techniques dates back to 1952 when Joshua Lederberg found that bacteria exchange genetic material contained in a body he called a plasmid. The next year it was established that plasmids were rings of DNA free from the main DNA in the chromosome of the bacteria. The next step was taken by Werner Arber (1929–), who studied viruses (called bacteriophages) which infect bacteria. He found that bacteria resist phages by splitting the phage DNA using enzymes. By 1968 Arber had discovered the enzymes produced by bacteria that split DNA at specific locations. In addition, he found that different genes that have been split at the same location by one of the restriction enzymes, as they are called, will recombine when placed together in the absence of the enzyme. The resulting product is called recombinant DNA.

In 1973 Stanley H Cohen (1922–) and US organic chemist Herbert W Brown (1912–) combined restriction enzymes with plasmids in the first genetic engineering experiment. They cut a chunk out of a plasmid found in the bacterium *Eschericia coli* and inserted in the gap a gene created from a different bacterium. In 1976 Indian chemist Har Gobind Khorana (1922–) and co-workers constructed the first artificial gene to function naturally when inserted into a bacterial cell.

In 1995 a team at the Institute for Genomic Research in Gaitherburg, Maryland, USA, unveiled the first complete genetic blueprint for a free-living organism – a bacterium *Haemophilus influenza.* In theory, the blueprint, consisting of 1.8 million genetic instructions, would allow scientists to construct the bacterium from scratch. The achievement demonstrates the speed of the DNA decoding techniques used by genetic researchers.

A logical progression from such research resulted in the production of useful genetically engineered products. In 1977, the US biochemist Herbert Boyer (1936–), of the firm Genentech, fused a segment of human DNA (deoxyribonucleic acid) into the bacterium *Escherichia coli,* allowing it to produce a human protein (in this case somatostatin). This was the first commercially produced genetically engineered product.

Now this technology has moved on, to the point where genetically engineered bacteria and yeasts are used almost routinely to produce useful products such as human growth hormone and insulin. These essential drugs were amongst the first genetically engineered products to go on sale.

In recent years it has become possible to use genetic engineering techniques to alter the genetic makeup of plants. This offers the possibility of creating new varieties of species, notably crops, with specific new characteristics. For example, resistance to low temperatures can be introduced to crops from warm climates, so they may be grown successfully elsewhere, as can resistance to insect pests and virus diseases. Work is under way with cassava (a staple tropical crop) to introduce disease resistance, and to increase its protein content. Genetically modified (GM) crops are now produced in many parts of the world, and are beginning to appear in shops alongside their more 'natural' relatives.

The application of genetic engineering to medicine has progressed rapidly. The Human Genome Organization was established 1988 in Washington, DC, with the aim of mapping the complete sequence of human DNA. In 1985 the first human cancer gene was isolated by US researchers. In 1993 the gene for Huntington's disease was discovered.

The first person to undergo gene therapy (in September 1990) was a four-year-old girl suffering from a rare enzyme (ADA) deficiency that cripples the immune system. A healthy ADA gene was inserted into a virus that had been rendered harmless. The virus was inserted into a blood-forming cell taken from the child's bone marrow. This cell reproduced, creating millions of cells containing the missing gene. Finally these cells were infused into the child's bloodstream, to be carried to the bone marrow where they produced healthy blood cells complete with the ADA gene. This was only partially successful in that the child still required additional treatment with synthetic ADA.

The final years of the 20th century saw further developments in genetic engineering, many of which have important implications for medicine. It is now possible to produce what are known as transgenic animals, by injecting genes from one animal into the fertilized egg of another. This technique was first perfected in 1981 by scientists at Ohio University, USA, who injected genes from one animal into the fertilized egg of a mouse. The resulting transgenic mouse then has the foreign gene in many of its cells, and the gene is passed on to its offspring, creating permanently altered (transgenic) animals. In 1982, for example, a gene controlling growth was transferred from a rat to a mouse, producing a transgenic mouse which grew to double its normal size.

Such transgenic techniques promise to benefit medicine, and in particular transplant surgery. One of the major problems in surgery has been in transplanting organs – the main difficulties being the supply of suitable organs, and the process of rejection by the host. Transgenic pigs have now been bred which incorporate some human genes which reduce the risk of tissue rejection when the pigs' organs are transplanted into human patients, in a process called xenotransplantation.

Parallel research has established that it may soon be possible to grow kidneys which are derived from embryo kidney cells. If successful, these techniques could simultaneously address the twin problems of supply and rejection of human organs. Nevertheless, such research has raised difficult ethical problems which have yet to be fully resolved.

Many of the rapid advances in animal genetic engineering are of direct medical benefit to people. Examples of products now able to be manufactured from engineered animals are: human blood clotting factors, from sheep and goats; human growth hormone and beta-interferon from cattle; and humanized organs for transplant from pigs. Beta-interferon is used to treat multiple sclerosis.

Human evolution

The 20th century saw major discoveries of human fossils, which helped to shed light on the evolution of our own species. Around the turn of the 20th century, fossils of a species called *Homo erectus* (Java Man) were discovered. This humanlike creature (hominid) lived between 0.5 and 1.5 million years ago. In the 1920s and 1930s, an even older form was discovered, and named in a separate genus *Australopithecus.* This hominid lived between about 1 and 2 million years ago.

Since the 1950s, even older forms of *Australopithecus* have come to light, in Africa. More recently, in 1994, *Australopithecus ramidus* has been dated at 4.4 million years old. This form was probably mainly vegetarian and lived in a woodland habitat. Like humans, it walked upright on its hind legs.

In the 1960s and 70s, a form regarded as intermediate between modern humans (*Homo sapiens)* and *Australopithecus* was found in Africa. This was named in 1964, by British archaeologist, anthropologist, and palaeontologist Louis Leakey (1903–1972), South African anatomist and physical anthropologist Philip Tobias (1925–), and others, *Homo habilis* (Handy Man), after its apparent use of stone tools. The fossils came from the now famous Olduvai Gorge in Tanzania, and have been dated at between 1.6 and 2 million years ago.

Various further finds of *Homo erectus* (named for its humanlike upright stance) have been made, in Africa and in Europe, mainly in northern Kenya, but also in Algeria, Morocco, Ethiopia, and South Africa. The most likely story emerging is that this form originated in Africa, then spread to Europe and Asia. *Homo erectus* probably used tools made of wood, such as handaxes, and were hunters and gatherers, with a mixed diet.

The techniques of genetics and biochemistry were also applied to physical anthropology, and analyses of protein and DNA from fossils and living humans and apes were undertaken. These studies revealed that modern humans are more closely related to chimpanzees and pygmy chimpanzees (bonobos) than either are to gorillas, gibbons, and orangutans. They also helped shed light on the evolutionary pathways on the fossil hominid line.

Analysis of DNA and fossils suggests that modern humans are only a few hundred thousand years old, and that the evolutionary split between humans and apes occurred between 5 and 8 million years ago, probably in Africa. Our own species, *Homo sapiens* (Wise Man) emerged as recently as around 400,000 years ago. All living people are closely related, and probably share a common ancestor in Africa.

A form of human, *Homo neanderthalensis,* known as Neanderthal Man (from the Neander valley in Germany where it was first discovered) evolved in Europe around 200,000 years ago. A genetic analysis carried out on DNA extracted from fossil Neanderthal bones indicated in 1997 that Neanderthals shared a common ancestor with modern humans no later than 600,000 years ago, and proved conclusively that they were not our direct ancestors but represent a separate evolutionary line.

Extinction and the discovery of new species

More than 99.9% of all evolutionary lines that once existed on Earth have become extinct, so extinction has loomed large in the history of life on Earth, and continues to feature in the natural world today. On the other hand, our knowledge of the natural world is still so sparse that species are still being discovered and described.

Discoveries by palaeontologists and geologists have helped to explain the apparently catastrophic extinctions which seem to have occurred from time to time in the distant past. The most famous of these was the relatively sudden demise of the dinosaurs towards the end of the Cretaceous period.

Analysis of a clay layer in deposits dated at the boundary of the Cretaceous and Tertiary periods by US physicists Walter and Luis Alvarez in 1980 revealed unusually rich traces of the heavy metal iridium. It was then found that this enriched iridium layer was present in similar aged deposits worldwide. The Alvarezes proposed the theory that this was caused by the impact on the Earth of a large asteroid, which might have had other ecological effects which explain the sudden extinction of many creatures, including the dinosaurs.

US palaeontologists looked at other mass extinctions of the past, and found that these were periodic, happening around every 26 million years. They developed the theory that something periodically disturbed the cloud of comets (known as the Oort cloud) at the edge of our Solar System, causing some of the comets to fall towards the Sun, occasionally hitting the Earth or other planets.

As we reach the end of the 20th century, new species of animals and plants are still being discovered and described. Expeditions to tropical rainforests routinely find undescribed insects, notably beetles, but there have also been recent discoveries of new vertebrates, even some quite large mammals. For example, a new species of whale from Chile was described in 1996, and a new species of muntjac deer from Vietnam in 1997 and another in 1999.

In 1995 a new phylum (a group of creatures that share a distinct body plan) was discovered. The phylum, Cycliophora, is only the 36th ever described for all the 1.5 million or so named organisms. Most other phyla were described in the 1800s. A single species has so far been assigned to the phylum. *Symbion pandora* is a tiny creature found clinging to the mouthparts of Norwegian lobsters by zoologists at the University of Copenhagen.

CHEMISTRY

CHEMISTRY SEEMS TO have originated in Egypt and Mesopotamia several thousand years before Christ. Certainly by about 3000 BC the Egyptians had produced the copper-tin alloy known as bronze, by heating the ores of copper and tin together, and this new material was soon common enough to be made into tools, ornaments, armour, and weapons. The Ancient Egyptians were also skilled at extracting juices and infusions from plants, and pigments from minerals, which they used in the embalming and preserving of their dead. By 600 BC the Greeks were also becoming a settled and prosperous people with leisure time in which to think. They began to turn their attention to the nature of the universe and to the structure of its materials. They were thus the first to study the subject we now call chemical theory. The Greek philosopher Aristotle (384–322 BC) proposed that there were four elements – earth, air, fire, and water – and that everything was a combination of these four. They were thought to possess the following properties: earth was cold and dry, air was hot and moist, fire was hot and dry, and water was cold and moist. The idea of the four elements persisted for 2,000 years. The Greeks also worked out, at least hypothetically, that matter ultimately consisted of small indivisible particles, *atomos* – the origin of our word 'atom'.

From the Egyptians and the Greeks comes *khemeia,* alchemy and eventually chemistry as we know it today. The source of the word *khemeia* is debatable, but it is certainly the origin of the word chemistry. It may derive from the Egyptians' word for their country *Khem,* 'the black land'. It may come from the Greek word *khumos* (the juice of a plant), so that *khemeia* is 'the art of extracting juices'; or from the Greek *cheo* 'pour or cast', which refers to the activities of the metal workers. Whatever its origin, the art of *khemeia* soon became akin to magic and was feared by the ordinary people. One of the greatest aims of the subject involved the attempts to transform base metals such as lead and copper into silver or gold. From the four-element theory, it seemed that it should be possible to perform any such change, if only the proper technique could be found.

The Arabs and alchemy

With the decline of the Greek empire *khemeia* was not pursued and little new was added to the subject until it was embraced by the increasingly powerful Arabs in the 7th century AD. Then for five centuries *al-kimiya,* or alchemy, was in their hands. The Arabs drew many ideas from the *khemeia* of the Greeks, but they were also in contact with the Chinese – for example, the idea that gold possessed healing powers came from China. They believed that 'medicine' had to be added to base metals to produce gold, and it was this medicine that was to become the philosopher's stone of the later European alchemists. The idea that not only could the philosopher's stone heal 'sick' or base metals, but that it could also act as the elixir of life, was also originally Chinese. The Arab alchemists discovered new classes of chemicals such as the caustic alkalis (from the Arabic *al-qalíy)* and they improved technical procedures such as distillation.

Western Europe had its first contact with the Islamic world as a result of the Crusades. Gradually the works of the Arabs – handed down from the Greeks – were translated into Latin and made available to European scholars in the 12th and 13th centuries. Many people spent their lives trying in vain to change base metals into gold; and many alchemists lost their heads for failing to supply the promised gold.

A new era in chemistry began with the researches of Irish chemist Robert Boyle (1627–1691), who carried out many experiments on air. These experiments were the beginning of a long struggle to find out what air had to do with burning and breathing. From Boyle's time onwards, alchemy became chemistry and it was realized that there was more to the subject than the search for the philosopher's stone.

Chemistry as an experimental science

During the 1700s the phlogiston theory gained popularity. It went back to the alchemists' idea that combustible bodies lost something when they burned. Metals were thought to be composed of a calx (different for each) combined with phlogiston, which was the same in all metals. When a candle burned in air, phlogiston was given off. It was believed that combustible objects were rich in phlogiston and what was left after combustion possessed no phlogiston and would therefore not burn. Thus wood possessed phlogiston but ash did not; when metals rusted, it was considered that the metals contained phlogiston but that its rust or calx did not. By 1780 this theory was almost universally accepted by chemists. English chemist Joseph Priestley (1733–1804) was a supporter of the theory and in 1774 he had succeeded in obtaining from mercuric oxide a new gas which was five or six times purer than ordinary air. It was, of course, oxygen but Priestley called it 'dephlogisticated air' because a smouldering splint of wood thrust into an atmosphere of this new gas burst into flames much more readily than it did in an ordinary atmosphere. He took this to mean that the gas must be without the usual content of phlogiston, and was therefore eager to accept a new supply.

It was French chemist Antoine Lavoisier (1743–1794) who put an end to the phlogiston theory by working out what was really happening in combustion. He repeated Priestley's experiments in 1775 and named the dephlogisticated air oxygen. He realized that air was not a single substance but a mixture of gases, made up of two different gases in the proportion of 1 to 4. He deduced that one-fifth of the air was Priestley's dephlogisticated air (oxygen), and that it was this part only that combined with rusting or burning materials and was essential to life. Oxygen means 'acid-producer' and Lavoisier thought, erroneously, that oxygen was an essential part of all acids. He was a careful experimenter and user of the balance, and from his time onwards experimental chemistry was concerned only with materials that could be weighed or otherwise measured. All the 'mystery' disappeared and Lavoisier went on to work out a logical system of chemical nomenclature, much of which has survived to the present day.

Early in the 19th century many well-known chemists were active. French chemist Claude Berthollet (1748–1822) worked on chemical change and composition, and French chemist Joseph Gay-Lussac (1778–1850) studied the volumes of gases that take part in chemical reactions. Others included Berzelius, Cannizzaro, Avogadro, Davy, Dumas, Kolbe, Wöhler, and Kekulé. The era of modern chemistry was beginning.

Atomic theory and new elements

An English chemist, John Dalton (1766–1844), founded the atomic theory in 1803 and in so doing finally crushed the belief that the transmutation to gold was possible. He realized that the same two elements can combine with each other in more than one set of proportions, and that the variation in combining proportions gives rise to different compounds with different properties. For example, he determined that one part (by weight) of hydrogen combined with eight parts of oxygen to form water, and if it was assumed (incorrectly) that a molecule of water consisted of one atom of hydrogen and one atom of oxygen, then it was possible to set the mass of the hydrogen atom arbitrarily at 1 and call the mass of oxygen 8 (on the same scale). In this way Dalton set up the first table of atomic weights (now called relative atomic masses), and although this was probably his most important achievement, it contained many incorrect assumptions. These errors and anomalies were researched by Swedish chemist Jöns Berzelius (1779–1848), who found that for many elements the atomic weights were not simple multiples of that of hydrogen. For many years, oxygen was made the standard and set at 16.000 until the mid-20th century, when carbon (= 12.000) was adopted. Berzelius suggested representing each element by a symbol consisting of the first one or two letters of the name of the element (sometimes in Latin) and these became the chemical symbols of the elements as still used today.

At about the same time, in 1808, English chemist Humphry Davy (1778–1829) was using an electric current to obtain from their oxides elements that had proved to be unisolatable by chemical means:

potassium, sodium, magnesium, barium, and calcium. His assistant, English physicist Michael Faraday (1791–1867), was to become even better known in connection with this technique, electrolysis. By 1830, more than 50 elements had been isolated; chemistry had moved a long way from the four elements of the ancient Greeks, but their properties seemed to be random. In 1829 the German chemist Johann Döbereiner (1780–1849) thought that he had observed some slight degree of order. He wondered if it was just coincidence that the properties of the element bromine seemed to lie between those of chlorine and iodine, but he went on to notice a similar gradation of properties in the triplets calcium, strontium, and barium and with sulphur, selenium, and tellurium. In all of these examples, the atomic weight of the element in the middle of the set was about half-way between the atomic weights of the other two elements. He called these groups 'triads', but because he was unable to find any other such groups, most chemists remained unimpressed by his discovery. Then in 1864 English chemist John Newlands (1837–1898) arranged the elements in order of their increasing atomic weights and found that if he wrote them in horizontal rows, and started a new row with every eighth element, similar elements tended to fall in the same vertical columns. Döbereiner's three sets of triads were among them. Newlands called this his 'Law of Octaves' by analogy with the repeating octaves in music. Unfortunately there were many places in his chart where obviously dissimilar elements fell together and so it was generally felt that Newland's similarities were not significant but probably only coincidental. He did not have his work published.

In 1862 a German chemist, Julius Lothar Meyer (1830–1895), looked at the volumes of certain fixed weights of elements, and talked of atomic volumes. He plotted the values of these for each element against its atomic weight, and found that there were sharp peaks in the graph at the alkali metals – sodium, potassium, rubidium, and caesium. Each part of the graph between the peaks corresponded to a 'period' or horizontal row in the table of the elements, and it became obvious where Newlands had gone wrong. He had assumed that each period contained only seven elements; in fact the later periods had to be longer than the earlier ones. By the time Meyer published his findings, he had been anticipated by the Russian chemist Dmitri Mendeleyev (1834–1907), who in 1869 published his version of the periodic table, which was more or less as we have it today. He had the insight to leave gaps in his table for three elements which he postulated had not yet been discovered, and was even able to predict what their properties would be. Chemists were sceptical, but within 15 years all three of the 'missing' elements had been discovered and their properties were found to agree with Mendeleyev's predictions.

The beginnings of physical chemistry

Until the beginning of the 19th century, the areas covered by the subjects of chemistry and physics seemed well defined and quite distinct. Chemistry studied changes where the molecular bonding structure of a substance was altered, and physics studied phenomena in which no such change occurred. Then in 1840 physics and chemistry merged in the work of Swiss-born Russian chemist Germain Hess (1802–1850). It had been realized that heat – a physical phenomenon – was produced by chemical reactions such as the burning of wood, coal, and oil, and it was gradually becoming clear that all chemical reactions involved some sort of heat transfer. Hess showed that the quantity of heat produced or absorbed when one substance was changed into another was the same no matter by which chemical route the change occurred, and it seemed likely that the law of conservation of energy was equally applicable to chemistry and physics. Thermochemistry had been founded and work was able to begin on thermodynamics. Most of this research was done in Germany and it was Latvian-born German chemist Wilhelm Ostwald (1853–1932), towards the end of the 19th century, who was responsible for physical chemistry developing into a discipline in its own right. He worked on chemical kinetics and catalysis in particular, but was the last important scientist to refuse to accept that atoms were real – there was at that time still no direct evidence to prove that they existed. Other contemporary chemists working in the new field of physical chemistry included Dutch physical chemist Jacobus van't Hoff (1852–1911) and Swedish physical chemist Svante Arrhenius (1859–1927). Van't Hoff studied solutions and showed that molecules of dissolved substances behaved according to rules analogous to those that describe the behaviour of gases.

Arrhenius carried on the work which had been begun by Davy and Faraday on solutions that could carry an electric current. Faraday had called the current-carrying particles 'ions', but nobody had worked out what they were. Arrhenius suggested that they were atoms or groups of atoms which bore either a positive or a negative electric charge. His theory of ionic dissociation was used to explain many of the phenomena in electrochemistry.

Towards the end of the 19th century, mainly as a result of the increasing interest in the physical side of chemistry, gases came under fresh scrutiny and some errors were found in the law that had been proposed three centuries earlier by Robert Boyle. German-born French physical chemist Henri Regnault (1810–1878), Scottish physicist James Clerk Maxwell (1831–1879) and Austrian theoretical physicist Ludwig Boltzmann (1844–1906) had all worked on the behaviour of gases, and the kinetic theory of gases had been derived. Taking all their findings into account, Dutch physicist Johannes van der Waals (1837–1923) arrived at an equation that related pressure, volume, and temperature of gases and made due allowance for the sizes of the different gas molecules and the attractions between them. By the end of the century Scottish chemist William Ramsay (1852–1916) had begun to discover a special group of gases – the inert or rare gases – which have a valency (oxidation state) of zero and which fit neatly into the periodic table between the halogens and the alkali metals.

Organic chemistry becomes a separate discipline

Meanwhile the separate branches of chemistry were emerging and organic substances were being distinguished from inorganic ones. In 1807 Berzelius had proposed that substances such as olive oil and sugar, which were products of living organisms, should be called organic, whereas sulphuric acid and salt should be termed inorganic. Chemists at that time had realized that organic substances were easily converted into inorganic substances by heating or in other ways, but it was thought to be impossible to reverse the process and convert inorganic substances into organic ones. They believed in Vitalism – that somehow life did not obey the same laws as did inanimate objects and that some special influence, a 'vital force', was needed to convert inorganic substances into organic ones. Then in 1828 German chemist Friedrich Wöhler (1800–1882) succeeded in converting ammonium cyanate (an inorganic compound) into urea. In 1845 German chemist Adolf Kolbe (1818–1884) synthesized acetic acid, squashing the Vitalism theory forever. By the middle of the 19th century organic compounds were being synthesized in profusion; a new definition of organic compounds was clearly needed, and most organic chemists were working by trial and error. Nevertheless there was a teenage assistant of German organic chemist August von Hofmann (1818–1892), the English chemist William Perkin (1838–1907), who was able to retire at the age of only 35 because of a brilliant chance discovery. In 1856 he treated aniline with potassium chromate, added alcohol, and obtained a beautiful purple colour, which he suspected might be a dye (later called aniline purple or mauve). He left school and founded what became the synthetic dyestuffs industry.

Then in 1861 the German chemist Friedrich Kekulé (1829–1886) defined organic chemistry as the chemistry of carbon compounds and this definition has remained, although there are a few carbon compounds (such as carbonates) which are considered to be part of inorganic chemistry. Kekulé suggested that carbon had a valency of four, and proceeded to work out the structures of simple organic compounds on this basis. These representations of the structural formulae showed how organic molecules were generally larger and more complex than inorganic molecules. There was still the problem of the structure of the simple hydrocarbon benzene, $C_6H_{6,}$ until 1865 when Kekulé suggested that rings of carbon atoms might be just as possible as straight chains. The idea that molecules might be three-dimensional came in 1874 when van't Hoff suggested that the four bonds of the carbon atom were arranged tetrahedrally. If these four bonds are connected to four different types of groups, the carbon atom is said to be asymmetric and the compound shows optical activity – its crystals or solutions rotate the plane of polarized light. German organic chemist Viktor Meyer (1848–1897) proposed that certain types of optical isomerism could be explained by bonds of nitrogen atoms. French-born Swiss chemist Alfred Werner (1866–1919) went on to demonstrate that this principle also applied to metals such as cobalt, chromium, and rhodium, and succeeded in working out the necessary theory of molecular structure, known as coordination theory. This new approach allowed there to be structural relationships within certain fairly complex inorganic molecules, which were not restricted to bonds involving ordinary valencies. It was to be another 50 years before enough was known about valency for both Kekulé's theory and Werner's to be fully understood, but by 1900 the idea was universally accepted that molecular structure could be represented satisfactorily in three dimensions.

Modern synthetic organic chemistry

Kekulé's work gave the organic chemist scope to alter a structural formula stage by stage, to convert one molecule into another, and modern synthetic organic chemistry began. Practical techniques for the synthesis of organic compounds were developed. French chemists Paul Sabatier (1854–1941) and Jean Senderens (1856–1936) discovered the Sabatier–Senderens reduction, Grignard and Gilman reagents were developed by French chemist Victor Grignard (1871–1935) and US chemist Henry Gilman (1893–1986) respectively, and German organic chemists Kurt Alder (1902–1958) and Otto Diels (1976–1954) found a method to synthesize cyclic carbon compounds, an essential step in drug development. New advances continued to be made throughout the 19th century.

In 1954, the Wittig reaction, a route to produce unsaturated hydrocarbons, was developed by German chemist George Wittig (1897–1987). In the 1960s US chemist Charles Pedersen (1904–1990), French chemist Jean Marie Lehn (1939–), and US chemist Donald Cram (1919–) discovered and developed crown ethers, cryptands and crytates, versatile organic reagents with broad applications in biochemistry and organic synthesis. US chemist Elias J Corey (1928–) developed retrosynthesis, a powerful tool for building complex molecules from smaller, cheaper, and more readily available ones. Retrosynthesis can be used to picture a molecule like a jigsaw, working backwards to find reactive components to complete the puzzle. Modern chemists use retrosynthesis to design everything from insect repellents to better drugs.

Significant advances also occurred in the field of structural analysis. In 1909 German physicist Max von Laue (1879–1960) began a series of brilliant experiments. He established that crystals consist of atoms arranged in a geometric structure of regularly repeating layers, and that these layers scatter X-rays in a set pattern. In so doing, he had set the scene for X-ray crystallography to be used to help to work out the structures of large molecules for which chemists had not been able to determine formulae. This field was advanced by scientists such as English physicist Lawrence Bragg (1890–1971), English chemist and X-ray crystallographer Rosalind Franklin (1920–1958), and English biochemist Dorothy Hodgkin (1910–1994) who developed the technique to allow the determination of a wide range of crystal structures, from common salt to DNA and insulin.

German organic chemist Richard Willstätter (1872–1942) was able to work out the structure of chlorophyll and another German organic chemist, Heinrich Wieland (1877–1957), determined the structures of steroids. Russian-born Swiss chemist Paul Karrer (1889–1971) elucidated the structures of the carotenoids and other vitamins and English organic chemist Robert Robinson (1886–1975) tackled the alkaloids – he worked out the structures of morphine and strychnine. The alkaloids have found medical use as drugs, as have many other organic compounds. The treatment of disease by the use of specific chemicals is known as chemotherapy and was founded by the German bacteriologist Paul Ehrlich (1854–1915). The first anti-bacterial drug, protosil red, was discovered by German chemist Gerhard Domagk (1895–1964). A series of drugs known as the sulpha drugs were developed from his discovery. The need for drugs to combat disease and infection during World War II spurred on research, and by 1945 the antibiotic penicillin, first isolated by Australian-born British bacteriologist Howard Florey (1898–1968) and German-born British biochemist Ernst Chain (1906–1979), was being produced in quantity. Other antibiotics such as streptomycin and the tetracyclines soon followed.

In 1912, Polish-born US biochemist Casimir Funk (1884–1967) isolated vitamin B from yeast. A year later US biochemist Elmer McCollum (1884–1967) discovered vitamin A and found vitamin D in 1920.

Hungarian-born US biochemist Albert Szent-Györgyi (1893–1986) isolated vitamin C from cabbages. Vitamin E was isolated by US chemists soon afterwards. Once the vitamins had been isolated in pure form, their structure could be determined. The structure of vitamin B_1 was determined in 1934 and, by the 1940s, most of the vitamins we know today had been found, isolated, and synthesized in laboratories.

Some organic molecules contain thousands of atoms; some, such as rubber, are polymers and others, such as haemoglobin, are proteins. German organic chemist Hermann Staudinger (1881–1965) pioneered the concept of macromolecules and his theories formed the foundation of polymer science. Synthetic polymers have been made which closely resemble natural rubber; the leader in this field was US organic chemist Wallace Carothers (1896–1937), who also invented nylon. German organic chemist Karl Ziegler (1898–1973) and Italian chemist Giulio Natta (1903–1979) worked out how to prevent branching during polymerization, so that plastics, films, and fibres can now be made more or less to order. Work on the make-up of proteins had to wait for the development of chemical techniques such as chromatography (by Italian-born Russian botanist Mikhail Tswett (1872–1919) and by English biochemists Archer Martin (1910–) and Richard Synge (1914–1994)) and electrophoresis (by Swedish chemist Arne Tiselius (1902–1971)). In the forefront of molecular biological research are English biochemists Frederick Sanger (1918–), John Kendrew (1917–1997), and Austrian-born British molecular biologist Max Perutz (1914–).

The rise of the chemical industry

The 20th century was a time of advancement and discovery for the chemical industry. German chemists Fritz Haber (1868–1934) and Carl Bosch (1874–1940) developed industrial techniques using high pressures, catalysts, and high temperatures to manufacture chemicals that could not be produced economically in the 19th century, notably nitrogen fixation which lead the development of modern artificial fertilisers and explosives. The 19th century chemical industry had been based on the conversion of coal to chemicals. This was superseded by developments in the USA in the 20th century. Pioneers such as US chemist William Burton (1865–1964), French-born US inventor Eugene Houdry (1892–1962),and US chemical engineer Warren K Lewis (1882–1975) developed techniques for the conversion of petroleum oil and natural gas into chemicals. Compounds produced in this way are called petrochemicals. This route allowed chemicals to be produced cheaper and in greater quantities than ever before. Petroleum also provided the chemical industry with a variety of previously unavailable feedstocks which lead to the development of plastics, synthetic rubber, and synthetic fibres. Materials such as nylon, Teflon, Lycra, and neoprene rubber were developed from the 1930s onwards.

Modern atomic theory

In 1897 English physicist J J Thomson (1856–1940) proved the existence of the first subatomic particle, the negatively charged electron. New Zealand-born British physicist Ernest Rutherford (1871–1937) deduced that the unit of positive charge was a particle quite different from the electron, which was the unit of negative charge, and in 1920 he suggested that this fundamental positive particle be called the proton. In 1895 German physicist Wilhelm Röntgen (1845–1923) discovered X-rays, but other known radiation components – alpha and beta rays – were found to be made up of protons and electrons.

In 1911, Rutherford evolved his theory of the nuclear atom, which suggested that sub-atomic particles made up the atom. Rutherford's model of the atom had most of its mass located in a small positively charged core called a nucleus surrounded by a mist of electrons which occupied almost all of the space. Why the negatively charged electrons were not drawn into the nucleus was explained by Danish physicist Niels Bohr (1885–1962), who postulated that the electrons orbited the nucleus in stable orbits called shells.

In about 1902 it was proved, contrary to all previous ideas, that radioactive elements changed into other elements, and by 1912 the complicated series of changes of these elements had been worked out. In the course of this research, English chemist Frederick Soddy (1877–1956) realized that there could be several atoms differing in mass but having the same properties. They were called isotopes and we now know that they differ in the number of neutrons which they possess, although the neutron was not to be discovered until 1932, by the English physicist James Chadwick (1891–1974). The question now was, how did the nuclear atom of one element differ from that of another?

In 1913 the young English scientist Henry Moseley (1887–1915) found that there were characteristic X-rays for each element and that there was an inverse relationship between the wavelength of the X-ray and the atomic weight of the element. This relationship depended on the size of the positive charge on the nucleus of the atom, and the size of this nuclear charge is called the atomic number. Mendeleyev had arranged his periodic table, by considering the valencies of the elements, in sequence of their atomic weights, but the proper periodic classification is by atomic numbers. Scientists used this discovery to update the periodic table into the form it is used in today. It was now possible to predict exactly how many elements were still to be discovered. Since the proton is the only positively charged particle in the nucleus, the atomic number is equal to the number of protons; the neutrons contribute to the mass but not to the charge. For example, a sodium atom, with an atomic number of 11 and an atomic weight (relative atomic mass) of 23, has 11 protons and 12 neutrons in its nucleus.

Isotopes and biochemistry

The new electronic atom was also of great interest to organic chemists. It enabled theoreticians such as English organic chemist Christopher Ingold (1893–1970) to try to interpret organic reactions in terms of the movements of electrons from one point to another within a molecule. Physical chemical methods were being used in organic chemistry, founding physical organic chemistry as a separate discipline. US theoretical chemist and biologist Linus Pauling (1911–1994), who was to suggest in the 1950s that proteins and nucleic acids possessed a helical shape, worked on the wave properties of electrons, and established the theory of resonance. This idea was very useful in establishing that the structure of the benzene molecule possessed 'smeared out' electrons and was a resonance hybrid of the two alternating double bond/single bond structures. The concept of isotopes was clarified by English chemist and physicist Francis Aston (1877–1945) with the mass spectrograph. This instrument used electric and magnetic fields to deflect ions of identical charge by an extent that depended on their mass – the greater the mass of the ion, the less it was deflected. He found for instance that there were two kinds of neon atoms, one of mass 20 and one of mass 22. The neon-20 was ten times as common as the neon-22, and so it seemed reasonable that the atomic weight of the element was 20.2 – a weighted average of the individual atoms and not necessarily a whole number. In some cases, the weighted average (atomic weight) of a particular atom may be larger than that for an atom of higher atomic number. This explains the relative positions of iodine and tellurium in the periodic table, which Mendeleyev had placed correctly without knowing why.

In 1931 US chemist Harold Urey (1893–1981) discovered that hydrogen was made up of a pair of isotopes, and he named hydrogen-2 deuterium. In 1934 it occurred to the Italian-born US physicist Enrico Fermi (1901–1954) to bombard uranium (element number 92, the highest atomic number known at that time) to see whether he could produce any elements of higher atomic numbers. This approach was pursued by US nuclear chemist Glenn Seaborg (1912–1999) and the transuranium elements were discovered, going up from element 94 but becoming increasingly difficult to form and decomposing again more rapidly with increasing atomic number.

In November 1994, researchers working at the GSI heavy-ion cyclotron at Darmstadt, Germany, produced element 110. The element, atomic mass 269, was produced when atoms of lead were bombarded with atoms of nickel. As is usual for super heavy atoms, the new element has a very short half-life; it decayed in less than a millisecond. A second element was discovered in December 1994. Three atoms of element 111, atomic mass 272, were detected when bismuth-209 was bombarded with nickel atoms. It decayed into two previously unknown isotopes of elements 109 and 107 after about a millisecond. In February 1996 element 112 was discovered by the same team. Elements 114, 116, and 118 were created in 1999.

The boundaries between chemistry and other sciences

The area between physics and chemistry has been replaced by a common ground where atoms and molecules are studied together with the forces that influence them. A good example is the discovery in the early 1990s of a new form of carbon, with molecules called buckyballs, consisting of 60 carbon atoms arranged in 12 pentagons and 20 hexagons to form a perfect sphere.

The boundary between chemistry and biology has also become less well defined and is now a scene of intense activity, with the techniques of chemistry being applied successfully to biological problems. Electron diffraction, chromatography, and radioactive tracers have all been used to help discover what living matter is composed of, although it is possible that these investigations in biology are only now at the stage that atomic physics was at the beginning of this century. It was Lavoisier who said that life is a chemical function, and perhaps the most important advance of all is towards understanding the chemistry of the cell. Biochemical successes of recent years include the synthesis of human hormones, the development of genetic fingerprinting, and the use of enzymes in synthesis. The entire field of genetic engineering is essentially biochemistry.

The advances in chemistry have come with a price. The sheer scale of the chemical industry has led to global pollution, environmental damage, and the development of chemical and biological weapons. As most chemicals are manufactured from petroleum, huge quantities of crude oil have to be transported all over the world. Accidents such as the *Exxon Valdez* oil spill in Alaska's Prince William Sound or deliberate releases such as during the Gulf War by the Iraqi forces, cause catastrophic environmental damage. Waste products from chemical manufacture have polluted water supplies and the widespread use of aerosols and chlorinated solvents was shown by US chemist F Sherwood Roland (1927–), Mexican chemist Mario Molina (1943–), and Dutch chemist Paul Crutzen (1933–) to accelerate ozone depletion in the upper atmosphere. Greenhouse gas production and acid rain are consequences of the dependence on fossil fuels to provide industry with energy. The misuse of drugs, especially antibiotics has lead to the emergence of resistant strains of bacteria, and the overuse of fertilisers has introduced levels of nitrates into the environment, poisoning water supplies and killing aquatic life. The indiscriminate use of insecticides such as DDT, highlighted in US science writer Rachel Carson's (1907–1964) influential book *Silent Spring*, have had adverse effects, such as the emergence of chemical-resistant insects and the killing of beneficial insect species. A particularly dark aspect of industrial production is the availability of poisonous chemical agents in sufficient quantities for creating a viable weapon in warfare. Widespread use of chemical weapons during World War 1 and subsequently in various countries around the world has shown that this development is here to stay despite international pressure.

Chemistry in the future

The chemist of the future will have two primary roles, to continue to develop and produce the chemicals that society requires and to provide solutions to address the problems of pollution. There is still much research to do in the field of chemotherapy. Less toxic anticancer agents, antiviral drugs, and the next generation of antibiotics are all still to be developed. Already companies are developing alternatives to CFCs and other ozone-depleting chemicals. Biodegradable plastics are being developed and the understanding of the environmental impact of chemicals has become a science in its own right, environmental chemistry. Alternative sources of industrial raw materials to replace petroleum, such as biomass, are being developed to eliminate the risk of oil spillages, and operational procedures and safeguards at chemical factories are continually being improved to increase the safety of chemical production. Species-specific insecticides and pesticides are a possible development which would minimize the impact of chemicals in agricultural production and more careful management of artificial fertilisers is also an area of study. Economic ways to reduce greenhouse gas levels, such as chemically combining carbon dioxide to form inert compounds which can be safely stored, is an area of intense study. In short, the chemist of the future will be a global scientist, assessing the impact of his or her products on the environment and finding ways in which their manufacture and use can be made safer.

EARTH SCIENCE

THE GEOLOGICAL SCIENCES have arisen in part from the need for further utilization of Earth's resources as well as from the ever-present desire to understand our origins.

The concept of time has been central to development of the geological sciences. The history of the development of geology, or earth sciences as it is more commonly known in recent times, is in most ways inseparable from historical progress in the other physical sciences, including chemistry, physics, and the allied field of biology. But it is the colossal time (in human terms) that initially confounded our understanding of Earth and it is this same element of time that often requires unique scientific perspectives.

Antiquity

Scientific thinking about the Earth grew out of traditions of thought which took shape in the Middle East and the Eastern Mediterranean. Early civilization needed to adapt to the seasons, to deserts and mountains, volcanoes and earthquakes. Yet inhabitants of Mesopotamia, the Nile Valley, and the Mediterranean littoral had experience of only a fraction of the Earth. Beyond lay *terra incognita.* Hence legendary alternative worlds were conjured up in myths of burning tropics, lost continents, and unknown realms where the gods lived.

The first Greek philosopher about whom much is known was Thales of Miletus (*c.* 640–546 BC). He postulated water as the primary ingredient of material nature. Thales' follower, Anaximander, believed the universe began as a seed which grew; and living things were generated by the interaction of moisture and the Sun. Xenophanes (*c.* 570–475 BC) is credited with a cyclic worldview: eventually the Earth would disintegrate, returning to a watery state.

Like many other Greek philosophers, Empedocles (*c.* 500–*c.* 430 BC) was concerned with change and stability, order and disorder, unity and plurality. The terrestrial order was dominated by strife. In the beginning, the Earth had brought forth living structures more or less at random. Some had died out. The survivors became the progenitors of modern species.

The greatest Greek thinker was Aristotle. He considered the world was eternal. Aristotle drew attention to natural processes continually changing its surface features. Earthquakes and volcanoes were due to the wind coursing about in underground caves. Rivers took their origin from rain. Fossils indicated that parts of the Earth had once been covered by water.

In the 2nd century AD, Ptolemy composed a geography that summed up the Ancients' learning. Ptolemy accepted that the equatorial zone was too torrid to support life, but he postulated an unknown land mass to the south, the *terra australis incognita.* Antiquity advanced a 'geocentric' and 'anthropocentric' view. The planet had been designed as a habitat for humans. A parallel may be seen in the Judaeo-Christian cosmogony.

The centuries from Antiquity to the Renaissance accumulated knowledge on minerals, gems, fossils, metals, crystals, useful chemicals and medicaments, expounded in encyclopedic natural histories by Pliny (AD 23–79) and Isidore of Seville (AD 560–636). The great Renaissance naturalists were still working within this 'encyclopedic' tradition. The most eminent was Konrad Gesner, whose *On Fossil Objects* was published in 1565, with superb illustrations. Gesner saw resemblances between 'fossil objects' and living sea creatures.

At the same time, comprehensive philosophies of the Earth were being elaborated, influenced by the Christian revelation of Creation as set out in 'Genesis'. This saw the Earth as recently created. Bishop Ussher (1581–1656) in his *Sacred Chronology* (1660), arrived at a creation date for the Earth of 4004 BC.

In Christian eyes, time was directional, not cyclical. God had made the Earth perfect but, in response to Original Sin, he had been forced to send Noah's Flood to punish people by depositing them in a harsh environment, characterized by the niggardliness of Nature. This physical decline would continue until God had completed his purposes with humans.

The 16th and 17th centuries brought the discovery of the New World, massive European expansion and technological development. Scientific study of the Earth underwent significant change. Copernican astronomy sabotaged the old notion that the Earth was the centre of the system. The new mechanical philosophy (Descartes, Gassendi, Hobbes, Boyle, and Hooke) rejected traditional macrocosm–microcosm analogies and the idea that the Earth was alive. Christian scholars adopted a more rationalist stance on the relations between Scripture and scientific truth. The possibility that the Earth was extremely old arose in the work of 'savants' like English physicist Robert Hooke (1635–1703). For Enlightenment naturalists, the Earth came to be viewed as a machine, operating according to fundamental laws.

The old quarrel as to the nature of fossils was settled. Renaissance philosophies had stressed the living aspects of Nature. Similarities between fossils and living beings seemed to prove that the Earth was capable of growth. Exponents of the mechanical philosophy denied these generative powers. Fossils were petrified remains, rather like Roman coins, relics of the past, argued Hooke. Such views chimed with Hooke's concept of major terrestrial transformations and of a succession of faunas and floras now perished. Some species had been made extinct in great catastrophes.

This integrating of evidence from fossils and strata is evident in the work of Danish naturalist Nicolaus Steno (1638–1686). He was struck by the similarity between shark's teeth and fossil *glossopetrae.* He concluded that the stones were petrified teeth. On this basis, he posited six successive periods of Earth's history. Steno's work is one of the earliest 'directional' accounts of the Earth's development that integrated the history of the globe and of life. Steno treated fossils as evidence for the origin of rocks.

The Enlightenment

Mining schools developed in Germany. German mineralogists sought an understanding of the order of rock formations which would be serviceable for prospecting purposes. Johann Gottlob Lehmann (1719–1776) set out his view that there were fundamental distinctions between the various *Ganggebergen* (masses formed of stratified rock). These distinctions represented different modes of origin, strata being found in historical sequence. Older strata had been chemically precipitated out of water, whereas more recent strata had been mechanically deposited.

German geologist Abraham Gottlob Werner (1749–1817) was appointed in 1775 to the Freiberg Akademie. He was the most influential teacher in the history of geology. Werner established a well-ordered, clear, practical, physically based stratigraphy. He proposed a succession of the laying down of rocks, beginning with 'primary rocks' (precipitated from the water of a universal ocean), then passing through 'transition', 'flútz' (sedimentary), and finally 'recent' and 'volcanic'. The oldest rocks had been chemically deposited; they were therefore crystalline and without fossils. Later rocks had been mechanically deposited. Werner's approach linked strata to Earth history.

Thanks to the German school, but also to French observers like botanist, palaeontologist, and stratigrapher Jean-Etienne Guettard, (1715–1786), chemist Antoine Lavoisier (1743–1794), and geologist Déodat Dolomieu (1750–1801), and to Swedish scientists like chemist and physicist Torbern Bergman (1735–1784), stratigraphy was beginning to emerge in the 18th century.

Of course, there were many rival classifications and all were controversial. In particular, battle raged over the nature of basalt: was it of aqueous or igneous origin? The Wernerian, or Neptunist, school saw the Earth's crust precipitated out of aqueous solution. The other, culminating in Hutton, asserted the formation of rock types from the Earth's central heat.

A pioneer of this school was French naturalist Georges-Louis Buffon (1707–1778). He stressed ceaseless transfigurations of the Earth's crust produced by exclusively 'natural' causes. In his *Epochs of Nature* (1779) he emphasized that the Earth had begun as a fragment thrown off the Sun by a collision with a comet. Buffon believed the Earth had taken at least 70,000 years to reach its present state. Extinction was a fact, caused by gradual cooling. The seven stages of the Earth explained successive forms of life, beginning with gigantic forms, now extinct, and ending with humans.

Though a critic of Buffon, Scottish natural philosopher James Hutton (1726–1797) shared his ambitions. Hutton was a scion of the Scottish Enlightenment, being friendly with Adam Smith and James Watt. In his 'Theory of the Earth' (1795), Hutton demonstrated a steady-state Earth, in which natural causes had always been of the same kind as a present, acting with precisely the same intensity ('uniformitarianism'). There was 'no vestige of a beginning, no prospect of an end'. All continents were gradually eroded by rivers and weather. Debris accumulated on the sea bed, to be consolidated into strata and thrust upwards by the central heat to form new continents. Hutton thus postulated an eternal balance between uplift and erosion. All the Earth's processes were gradual. The Earth was incalculably old. His maxim was that 'the past is the key to the present'.

Hutton's theory was much attacked in its own day. Following the outbreak of the French Revolution in 1789, conservatives saw all challenges to the authority of the Bible as socially subversive. Their writings led to ferocious 'Genesis versus Geology' controversies in England.

The 19th century

New ideas about the Earth brought momentous social, cultural, and economic reverberations. Geology clashed with traditional religious dogma about Creation. Modern state-funded scientific education and research organizations emerged. German universities pioneered scientific education. The Geological Survey of Great Britain was founded, after English geologist Henry De la Beche (1796–1855) obtained state finance for a geological map of southwest England. De la Beche's career culminated in the establishment of a Mines Record Office and the opening in 1851 of the Museum of Practical Geology and the School of Mines in London.

Specialized societies were founded. The Geological Society of London dates from 1807. In the USA, the government promoted science, and various states established geological surveys, New York's being particularly productive. In 1870, Congress appointed US geologist John Wesley Powell (1834–1902) to lead a survey of the natural resources of the Utah, Colorado, and Arizona area. Powell would later become the second director of the US Geological Survey, which was founded in 1879.

Building on Werner, the great achievement of early 19th-century geology lay in the stratigraphical column. After 1800, it was perceived that mineralogy was not the master key. Fossils became regarded as the indices enabling rocks of comparable age to be identified. Correlation of information from different areas would permit tabulation of sequences of rock formations, thereby displaying a comprehensive picture of previous geological epochs.

In Britain the pioneer was English geologist William 'Strata' Smith (1769–1839). Smith received little formal education and became a canal surveyor and mining prospector. By 1799 he set out a list of the secondary strata of England. This led him to the construction of geological maps. In 1815 he brought out *A Delineation of the Strata of England and Wales,* using a scale of five miles to the inch. Between 1816 and 1824 he published *Strata Identified by Organized Fossils,* which displayed the fossils characteristic of each formation.

Far more sophisticated were the Frenchmen zoologist and palaeontologist Georges Cuvier (1769–1832) and naturalist and geologist Alexandre Brongniart (1770–1837), who worked on the Paris basin. Cuvier's contribution lay in systematizing the laws of comparative anatomy and applying them to fossil vertebrates. He divided invertebrates into three phyla and conducted notable investigations into fish and molluscs. In *Researches on the Fossil Bones of Quadrupeds* (1812), he reconstructed such extinct fossil quadrupeds as the mastodon, applying the principles of comparative anatomy. Cuvier was the most influential paleontologist of the 19th century.

Fossils, in Cuvier's and Brongniart's eyes, were the key to the identification of strata and Earth history. Cuvier argued for occasional wholesale extinctions caused by geological catastrophes, after which new flora and fauna appeared by migration or creation. Cuvier's *Discours sur les révolutions de la surface du globe* (1812) became the foundation text for catastrophist views.

Classification of older rock types was achieved by English geologist Adam Sedgwick (1785–1873) and Scottish geologist Roderick Murchison (1792–1871). Sedgwick unravelled the stratigraphic sequence of fossil-bearing rocks in North Wales, naming the oldest of them the Cambrian period (now dated at 500–570 million years ago). Further south, Murchison delineated the Silurian system amongst the *grauwacke.* Above the Silurian, the Devonian was framed by Sedgwick, Murchison and De la Beche. Shortly afterwards, English geologist Charles Lapworth (1842–1920) developed the Ordovician.

Uniformitarianism

Werner's retreating-ocean theory was quickly abandoned, as evidence accumulated that mountains had arisen not by evaporation of the ocean, but through processes causing elevation and depression of the surface. This posed the question of the rise and fall of continents. Supporters of 'catastrophes' argued that terrestrial upheavals had been sudden and violent. Opposing these views, Scottish geologist Charles Lyell (1797–1875) advocated a revised version of Hutton's gradualism. Lyellian uniformitarianism argued that both uplift and erosion occurred by natural forces. Expansion of fieldwork undermined traditional theories based upon restricted local knowledge. The retreating-ocean theory collapsed as Werner's students travelled to terrains where proof of uplift was self-evident.

Geologists had to determine the Earth movements that had uplifted mountain chains. Chemical theories of uplift yielded to the notion that the Earth's core was intensely hot, by consequence of the planet commencing as a molten ball. Many hypotheses were advanced. In 1829, *Researches on Some of the Revolutions of the Globe* linked a cooling Earth to sudden uplift: each major mountain chain represented a unique episode in the systematic crumpling of the crust. The Earth was like an apple whose skin wrinkled as the interior shrank through moisture loss. The idea of horizontal (lateral) folding was applied in the USA by US mineralogist, crystallographer, and geologist James Dwight Dana (1813–1895) to explain the complicated structure of the Appalachians. Such views were challenged by Lyell in his bid to prove a steady-state theory. His classic *Principles of Geology* (1830–33) revived Hutton's vision of a uniform Earth that precluded cumulative, directional change in overall environment; Earth history proceeded like a cycle, not like an arrow. In *Principles of Geology*, Lyell thus attacked diluvialism and catastrophism by resuscitating Hutton's vision of an Earth subject only to changes currently discernible. Time replaced violence as the key to geomorphology.

Lyell discounted Cuvier's apparent evidence for the catastrophic destruction of fauna and flora populations. For over 30 years he opposed the transmutation of species, reluctantly conceding the point at last only in deference to his friend, English naturalist Charles Darwin(1809–1882), and the cogency of Darwin's *Origin of Species* (1859).

Ice ages

Landforms presented a further critical difficulty. Geologists had long been baffled by beds of gravel and 'erratic boulders' strewn over much of Northern Europe and North America. Bold new theories in the 1830s attributed these phenomena to extended glaciation. Swiss palaeontologist Louis Agassiz (1807–1873) and others contended that the 'diluvium' had been moved by vast ice sheets covering Europe during an 'ice age'. Agassiz's *Studies on Glaciers* (1840) postulated a catastrophic temperature drop, covering much of Europe with a thick covering of ice that had annihilated all terrestrial life.

The ice-age hypothesis met opposition but eventually found acceptance through Scottish geologist James Geikie (1839–1915), Scottish physicist and geologist James Croll (1821–1890), and German geographer and geologist Albrecht Penck (1858–1945). Syntheses were required. The most impressive unifying attempt came from Austrian geologist Eduard Suess (1831–1914). His *The Face of the Earth* (1885–1909) was a massive work devoted to analysing the physical agencies contributing to the Earth's geographical evolution. Suess offered an encyclopedic view of crustal movement, the structure and grouping of mountain chains, of sunken continents, and the history of the oceans. He made significant contributions to structural geology.

Suess disputed whether the division of the Earth's relief into continents and oceans was permanent, thus clearing the path for the theory of *continental drift.* Around 1900, the US geologist and cosmologist Thomas C Chamberlin (1843–1928) proposed a different synthesis: the Earth did not contract; its continents were permanent. Continents, Chamberlin argued, were gradually filling the oceans and thereby permitting the sea to overrun the land.

The 20th century

At the start of the 20th century the tools for significant advancements in the geological sciences were being developed. Geophysics had already emerged as a distinct discipline in the late 19th century. Study of the Earth's magnetic and gravitational fields came to early prominence as did the application of chemistry to geological problems. Studies of the propagation of seismic waves through the Earth revealed the nature of the planet's interior. In 1919 the American Geophysical Union was formed, and 1957 was designated the International Geophysical Year. The modern term 'earth sciences', replacing geology as the name of the discipline, marks the triumph of geophysics. Application of breakthroughs in all of these subdisciplines gave rise to the development of plate tectonics, representing one of the great scientific achievements of the century.

In the year 1900 the Earth was estimated to be about 90 million years old by comparing the amount of salt in the oceans to the rate at which salt was delivered to the oceans by rivers. In 1907 US chemist Bertram Boltwood (1870–1927) used the recent discovery that some forms of lead were the products of radioactive decay of the element uranium to demonstrate that some rocks were as old as 2,200 million years. The Earth was apparently much older than had been previously contemplated. In *The Age of the Earth* (1913), English geologist Arthur Holmes (1890–1965) pioneered the use of this new tool of radioactive decay methods for rock-dating.

Mapping Earth's magnetic field was a major goal of the geological sciences in the early 20th century. Deployment of the research sailing vessel *Carnegie* in 1909 was a manifestation of the desire for detailed geomagnetic surveys. The ship, operated by the Carnegie Institution of Washington, was built of nonmagnetic materials for the purpose of measuring Earth's magnetic field as it sailed the oceans. It logged over 342,000 miles in this endeavour before being destroyed by fire in 1929.

By 1922 distortions in Earth's gravitational field were being used to locate salt domes in the Gulf of Mexico and the oil trapped adjacent them. Geophysics was proving to be an important tool for prospecting of economic resources.

In 1906 it was found that something lay at the centre of the Earth that blocked the path of seismic waves (sound waves transmitted through the Earth). By 1914 it was established that the object incapable of transmitting certain kinds of seismic waves was a molten core. Over the decades abrupt changes in the velocities of seismic waves were used to deduce further details of Earth's layered structure. Changes in density with depth as revealed by seismologists allowed physical chemists and mineral physicists to show that minerals have an entirely different structure at very high pressures in the Earth and in other terrestrial planets.

Toward the close of the 20th century three-dimensional imaging of Earth's structure was made possible through the use of global seismic tomography. Developed by geophysicists at Harvard University, this method is analogous to CAT scans of the human brain. With seismic tomography it is possible to follow the fate of great slabs of Earth's crust as they descend into the mantle, and it is also possible to find the sources of sustained volcanism such as that that created the Hawaiian Islands in the Pacific Ocean.

In order to explain the structure of Earth and other rocky bodies of the Solar System, scientists began to study the behaviour of materials at extreme pressures and temperatures. Rock deep in a planet is subjected to crushing pressures of many thousands of atmospheres. At the turn of the century the maximum pressure that could be achieved in the laboratory was about 2,000 atmospheres. In 1910, Nobel laureate and US physicist Percy Bridgman (1882–1961) invented a device that allowed him to squeeze all manner of materials to pressures of 20,000 atmospheres. This enabled scientists to examine the nature of rock, as well as other materials like water, to pressures corresponding to the base of Earth's crust. Toward the close of the century it is now possible to investigate materials at pressures of several millions of atmospheres, corresponding to the centre of our planet, using diamond anvils. The advent of the laser has allowed earth scientists to heat materials to extreme temperatures even as they are being squeezed to crushing pressures. In so doing it is possible to simulate conditions that exist in Earth's core. Fundamental information about Earth's deep interior are still being investigated with these methods. The temperatures of the core is one such problem that has thus far eluded precise determination.

With the capability for investigating matter at very high pressures and temperature has come a better understanding of not only Earth, Mars, and the other rocky planets, but also the structures of the gaseous, hydrogen-rich giant planets Jupiter and Saturn and the icy planets Uranus and Neptune as well. It is in these bodies of the Solar System that materials that we think of as gases turn to metals and ultimately to exotic materials that we can not yet classify as solids, gases, or liquids. Indeed the disciplines of geology and stellar physics are beginning to overlap as mineral physicists contemplate the nature of matter in objects that span the transition from giant planet to star.

The origin of life

The latter half of the 20th century saw renewed focus on the issue of how life began on Earth. Traditionally, fossils comprised the contribution of the geological sciences to the question of the origin of life. Widespread acceptance of the existence of ancient pre-animal (pre-metazoan) life came from the discovery in 1954 of numerous kinds of microscopic fossil organisms in the Gunflint rocks along the shores of Lake Superior in the USA. The Gunflint find showed that life was prevalent 2,000 million years ago.

Through their studies of fossil life forms, earth scientists have contributed to the concepts of evolution. Mass extinctions, in which large proportions of all of Earth's species die out, have been a particular source of intrigue in the 1980s and 1990s. This intrigue was stimulated in 1980 by the suggestion of Nobel prizewinning physicist Luis Alvarez (1911–1988) and his colleagues that the dinosaurs and 70% of all other species of the Earth were killed off 65 million years ago by the impact of an asteroid or comet. Others suggested that profuse volcanism in India at that time was responsible for the so-called 'KT' mass extinction. Testing of these hypotheses has promoted dialogue between meteoriticists, biologists, geochemists, and cosmochemists, and the boundaries of what is called earth science has been once again blurred toward the close of the century.

The suggestion in 1996 by the US National Aeronautics and Space Administration (NASA) researchers that the SNC meteorite ALH84001, thought to be a piece of Mars, contains evidence for bacterial life is arguably the principal motivation for the renewed interest in the origin of life (SNC is the name given to a class of meteorites named after the original members Shergotty, Nakhla, and Chassigny; ALH reflects the discovery of the sample in the Alan Hills of Antarctica). Indeed, it has even been suggested by some scientists that life on Earth could have originated on Mars and spread here by exchange of planetary fragments like the SNC meteorites. But another motivation is the realization that the organic and inorganic worlds are not as separate and distinct as was once believed. It has been demonstrated recently that the amount of biological material in ocean surface water is partly controlled by the amount of the element iron, an inorganic metal. Metals are now recognized as important to the functioning of cells in living things and earth scientists are beginning to study the movements of these important life-giving inorganic substances within Earth's hydrosphere, atmosphere, and rocky regolith.

The chemistry and global cycle of nitrogen is also proving to be instrumental in understanding the possible origins of life. Nitrogen is an essential component of amino acids (the constituents of proteins) and nucleic acids (DNA, RNA), the very building blocks of life. Earth scientists, including mineral physicists and biogeochemists, are studying how the element nitrogen could have found its way from inert forms, such as the nitrogen in our atmosphere, to more reactive states suitable for forming amino acids and nucleic acids. Results suggest the critical reactions occurred on the ocean floor along the great mid-ocean ridges where ocean water is heated by underlying volcanic magma. To simulate the pressures exerted by the water in the deep ocean basins, mineral physicists employ some of the same tools once used for studying the behaviour of rocks and minerals deep in Earth's crust.

Continental drift

Several scientists of the 19th and 20th centuries had suggested that the continents were once joined and then drifted apart, but German geologist Alfred Wegener (1880–1930) went further and declared that continental rafts might actually slither horizontally across the Earth's face. Wegener published his ideas of 'Die Verschiebun der Kontinente', or continental displacement, from 1913 to 1924. His ideas became known as continental drift. Empirical evidence for continental drift lay, he thought, in the close jigsaw-fit between coastlines on either side of the Atlantic, and notably in palaeontological similarities between Brazil and Africa. Wegener was also convinced that geophysical factors would corroborate wandering continents.

Wegener supposed that a united supercontinent, Pangaea, had existed in the Mesozoic. This had developed numerous fractures and had drifted apart, some 200 million years ago. During the Cretaceous, South America and Africa had largely been split, but not until the end of the Quaternary had North America and Europe finally separated. Australia had been severed from Antarctica during the Eocene.

What had caused continental drift? Wegener offered a choice of possibilities. One was a westwards tidal force caused by the Moon. The other involved a centrifugal effect propelling continents away from the poles towards the equator (the 'flight from the pole'). In its early years, drift theory won few champions, and in the English-speaking world reactions were especially hostile. A few geologists were intrigued by drift, especially the South African, Alexander Du Toit (1878–1948), who adumbrated the similarities in the geologies of South America and South Africa, suggesting they had once been contiguous. In *Our Wandering Continents* (1937), Du Toit maintained that the southern continents had formed the supercontinent of Gondwanaland.

The most ingenious support for drift came, however, from the English geophysicist Arthur Holmes. Assuming radioactivity produced vast quantities of heat, Holmes argued for convection currents within the crust, in 1929. Radioactive heating caused molten magma to rise to the surface, which then spread out in a horizontal current before descending back into the depths when chilled. Such currents provided a new mechanism for drift.

The real breakthrough required diverse kinds of evidence accumulated during the 1940s, especially through studies in oceanography and palaeomagnetism. Sonar, developed as part of the World War II war effort, became available for mapping the topography of the ocean floor. The result was discovery of the semi-continuous mid-ocean ridge, a mountain chain 50,000 km/31,250 mi in length beneath the oceans. In addition US geologist William Maurice Ewing (1906–1974) ascertained that the crust under the ocean is much thinner than the continental shelf.

The US geologist Harry Hess (1906–1969), in an unpublished manuscript in 1959, suggested that new crust was produced in mid-ocean ridges, whereas deep-ocean trenches marked the sites where old crust descended into the depths of the mantle. Hess's concept, eventually published in 1962, was that the ocean crust was like a giant conveyor belt. This concept was termed 'seafloor spreading' and was shortly thereafter confirmed by palaeomagnetic studies of the ocean floor. In 1963, English geophysicists Fred Vine (1939–) and Drummond Matthews (1931–), showed that reversal in Earth's magnetic field had left a record of seafloor spreading in the form of stripes on the ocean floor. The stripes were defined by the direction of magnetism locked in to the rocks as they were formed at the ridges.

Seafloor spreading constituted the great breakthrough in understanding the dynamics of the Earth. Continents were seen to be gliding across the surface as passive passengers on great plates bounded by mid-ocean ridges and deep oceanic trenches. Constantly being formed and destroyed, the ocean floors were young; only continents – too light to be drawn down by the current – would preserve testimony of the remote geological past. Support came from J Tuzo Wilson (1908–1993), a Canadian geologist, who provided backing for the seafloor spreading hypothesis.

Seafloor spreading led to the formal development of the theory of 'plate tectonics' in the late 1960s by English geophysicist Dan McKenzie (1942–) and French geophysicist Xavier Le Pichon (1937–), then at the Lamont Observatory in New York, USA.

The majority of Earth scientists accepted the new plate tectonics model with remarkable rapidity. In the mid-1960s, a full account of plate tectonics was expounded. The Earth's surface was divided into six major plates, the borders of which could be explained by way of the convection-current theory. Deep earthquakes were produced where one section of crust was driven beneath another, the same process also causing volcanic activity in zones like the Andes. Mountains on the western edge of the North and South American continents arose from the fact that the continental 'raft' is the leading edge of a plate, having to face the oncoming material from other plates being forced beneath them. The Alps and Himalayas are the outcome of collisions of continental areas, each driven by a different plate system.

Geologists of the late 1960s and 1970s undertook immense reinterpretation of their traditional doctrines. Well-established stratigraphical and geomorphological data had to be redefined in terms of the new forces operating in the crust. Tuzo Wilson's *A Revolution in Earth Science* (1967) was a persuasive account of the plate tectonics revolution.

Geology is remarkable for having undergone such a dramatic and comprehensive conceptual revolution within recent decades. The fact that the most compelling evidence for the new theory originated from the new discipline of ocean-based geophysics has involved considerable reassessment of skills and priorities within the profession. Above all, the ocean floor now appears to be the key to understanding the Earth's crust, in a way that Wegener never appreciated.

Satellite observations

In recent years Earth observation satellites have measured continental movements with unprecedented accuracy. The surface of the Earth can be measured using global positioning geodesy (detecting signals from satellites by Earth-based receivers), satellite laser ranging (in which satellites reflect signals from ground transmitters back to ground receivers), and very long-baseline interferometry, which compares signals received at ground-based receivers from distant extraterrestrial bodies. These techniques can measure distances of thousands of kilometres to accuracies of less than a centimetre. Movements of faults can be measured, as can the growth of tectonic plates. Previously, such speeds were calculated by averaging displacements measured over decades or centuries. The results show that in the oceanic crust, plate growth is steady: from 12 mm/0.05 in per year across the Mid-Atlantic Ridge to 160 mm/6.5 in per year across the East Pacific Rise. The major continental faults seem to be very irregular in their movement; the Great Rift Valley has remained stationary for 20 years, when long-term averages suggest that it would have opened up about 100 mm/4 in in that time.

The future

At the start of the new millennium Earth sciences is a field in transition. Increasingly, the field of geology, that was once focused on studies of Earth's near-surface, is giving way to a broader discipline in which the Earth is studied in the context of its interplanetary environs. New methods are being devised to study not only Earth's unseen, deep interior, but also to examine rocky bodies throughout the Solar System. Discovery of extra-solar giant planets in the past several years has hastened the desire to understand how planets like Earth might form around other stars, and if life is an inevitable consequence of such terrestrial planet formation. Scientists now consider that the best way to gain answers to these questions is to understand the phenomena of planets and life in our Solar System as a whole. Thus, where earth scientists once studied the structure of our own planet, they now compare Earth to Mars and the other planets of the inner Solar System. Where they once concerned themselves with the evolution of life on Earth, now they steer their research towards understanding the origin of life in general, wherever and however it may have started. An example is the state of water on other planets. In 1999 geologists are contemplating the role that glaciers may have had on the form of the largest volcano in the Solar System, Olympus Mons on Mars.

ENGINEERING AND TECHNOLOGY

OUR EXISTENCE TODAY is powerful evidence of our ability to invent. Were it not for the invention of simple tools made from sharp-edged stones some 2 million years ago, it is doubtful whether our relatively weak and slow ancestors would have survived for long. Such tools enabled early humans to fight off predators and to hunt for food.

As well as being surrounded by potentially hostile animals, early humans were at the mercy of the climate. It was the second of the major inventions of prehistory, a means of creating fire, that enabled them to survive the Ice Ages. *Homo erectus* was using this to live through the second Ice Age some 400,000 years ago.

These two inventions served our ancestors well for an extremely long time. Not until the foundation of Jericho, the world's first walled town, around 7000 BC does another major invention reveal itself, in the development of pottery. Behind the foundation of Jericho, were the beginnings of organized agriculture, and it was this that provided the stimulus for the development of increasingly sophisticated tools, such as the plough and the sickle. Copper appeared at around this time and the alloy made from copper and tin called bronze, appeared around 5000 BC.

That most famous and significant of inventions of the ancient world, the wheel, first appeared around 3000 BC, in what is now the region covered by Iran and Iraq. These early wheels were solid, wooden and fixed to sleds which had previously been dragged across the ground. Although its first use was in transporting heavy loads, the wheel and axle combination later became a feature of milling devices, and irrigation systems. Sumerian and Assyrian engineers used wheel-driven water drawing devices in irrigation networks which are still in use today.

Greek influences

For several thousand years, these early inventions were enlarged and improved upon, without any major advances being made. By the time mathematician and physicist Archimedes (*c.* 287–*c.* 212 BC) was investigating the principle of the lever and producing his famous helical screw, the scene of the most significant developments had shifted from the Middle East to Greece. The measurement of time in particular remained a continuing challenge, and by the 2nd century BC, Ctesibius of Alexandria had developed the Egyptian *clepsydra* (water-clock) to give accuracy not surpassed until well into the Middle Ages.

Hero of Alexandria (*c.* AD 60) was the last of the Greek technologists, his most famous invention being of the *aeolipile,* a primitive steam turbine that gave a hint, as early as the 1st century BC, of the potency of steam as a power-source. One of the most important inventions of the classical world was the water wheel. In its simplest form, the Norse wheel, it was just a horizontal scoop wheel set in a fast-moving stream, driving the stones of a grain mill. The vertical wheel, first described in the 1st century BC by the Roman Pollio Vitruvius was far more efficient, and by the use of gearing could drive all kinds of machinery from water pumps to the hammers of a forge. It remained the main source of industrial power until the invention of the rotative steam engine at the end of the 18th century.

Roman and Chinese influences

Despite their astonishing ability in geometry, physics, and mathematics, the Greeks were unable to make the advance which transformed architecture and civil engineering: the arch. The pre-Roman Etruscans used the semicircular arch as an architectural feature, but it was the Romans who put the arch to full use. Because of its ability to spread imposed stresses more evenly, the arch allowed greater spans in buildings than the Greeks' simple pillar-and-beam arrangements. Aqueducts comprising 6 m/20 ft wide arches were possible as early as 142 BC, as evidenced by the Pons Aemilius in Italy.

Combining the structural economy of the arch with the availability of good cement the Romans set up an infrastructure and communications network that gave them a standard of living hitherto unprecedented in history. It also enabled them to extend that standard, and maintain it by its armies, over a similarly unprecedented expanse of the world.

Animals such as oxen had been used for ploughing by the ancient Egyptians and the Greeks. However, it was the Chinese who first produced an efficient animal harness, enabling animals to be used in tandem to haul great loads. The harness did not reach the West, however, until the 9th century, well after the fall of the Roman Empire. China was the birthplace of a number of other major inventions: of paper from pulp (by Ts'ai Lun around AD 100), of the magnetic compass, and of gunpowder, around 500 and 850 respectively.

The development of Western technology

It was also around this time that north-western Europe began its climb to ascendency in technology that it has held onto for centuries since. The poorer climate of this region, combined with the need to develop a new form of agriculture, was responsible for the emergence around the 8th century of the crop rotation methods still used today. The wind was put to use in both sea-going vessels and land-based mills. The region grew more populous and, by the 11th century, northern Europeans were moving their influence into the Mediterranean and Middle East.

As the societies grew, becoming more complex, the need for metals for housing, tools, equipment, and coinage increased concomitantly. This caused a renewed interest in the extraction and treatment of ores, which were also in increasing demand by rulers anxious to develop weapons capable of keeping their rivals at bay. Many of these rulers, notably in 15th-century Italy, employed engineers to come up with new systems for both defence and attack. Undoubtedly the most famous such engineer was Leonardo da Vinci (1452–1519), military engineer to the Duke of Milan. Among the thousands of pages of da Vinci's notes are to be found an astonishing number of prescient plans for modern-day inventions: tanks, submarines, helicopters, and a whole range of firearms.

The Renaissance was ushered in by one of the most influential inventions in history: the development of the movable type printing press by the printer Johannes Gutenberg (*c.* 1397–1468) of Mainz, Germany. The Gutenberg Bible, the first book to be printed using this process, appeared in 1454. The English printer William Caxton set up a press in England in 1476. As well as disseminating religious knowledge over a far wider scale, the invention enabled the speeding up of the transfer of technological advances from one country to another.

Steam power and the Industrial Revolution

The ever-increasing use of metals made mining the focus of much effort in the 15th century, one of the biggest problems being that of adequately draining the mines. Pumping water fast enough and in sufficient volume was also a problem facing those wanting to create more agricultural land from poorly drained areas, and to supply towns with their needs. By the mid-17th century, a number of patents had been granted to water pumps which used a new, remarkably versatile source of power: steam.

English inventor and engineer Thomas Savery (*c.* 1650–1715) demonstrated an engine for 'raising of water and occasioning motion to all sorts of mill works, by the impellant force of fire', to the Royal Society of London in 1699. A far more efficient machine, the atmospheric engine, was invented *c.* 1712 by English blacksmith Thomas Newcomen (1663–1729). It was a beam engine that could pump water from deep underground and came into use to maintain the mines of Staffordshire, Cornwall, and Newcastle in workable condition.

Iron was first smelted over 3,000 years ago, but for most of that time it was produced in the form of malleable or wrought iron. Cast iron appeared some time in the 15th century with the introduction of the blast furnace using charcoal as a fuel. Iron making was revolutionized when the English iron manufacturer Abraham Darby (1677–1717) bought a foundry at Coalbrookdale, Shropshire where he successfully used coke instead of charcoal in the furnace and perfected techniques for casting in sand. The Coalbrookdale works went on to be hugely influential, casting cylinders for the first steam engines, making the first iron rails and, most famously, demonstrating the structural possibilities of iron by building the world's first iron bridge across the River Severn in 1779.

Newcomen steam engines grew in popularity during the 18th century, and it was while repairing a model one used on the physics course at Glasgow University in 1763 that the young Scottish mechanical engineer James Watt (1736–1819) saw how major improvements could be made, improvements that led to Watt's condenser engine completely replacing the earlier model by 1800.

Through its predominance as a manufacturing market, Britain was able to reap rich rewards but competition from overseas urged on the hunt for greater efficiency of production. Technologists were at the forefront of this effort, with English inventor John Kay's (1704–1780) invention of the flying shuttle for the weaving of textiles. Spinning was improved by the English weaver James Hargreaves' (1720–1778) spinning jenny. English industrial pioneer Richard Arkwright's (1732–1792) spinning machine was the centrepiece of the first cotton factory of 1771, and semi-automated mass production as a technique was born. The Industrial Revolution, which is arguably still under way, had begun.

The end of the 18th century saw English engineer Richard Trevithick (1771–1833) experimenting with the use of steam to provide motive power for boats, road vehicles, and locomotives on iron rails. Dogged by bad luck, his ideas did not achieve the success or acclaim they deserved, and the world had to wait longer than it should have for the full exploitation of steam in such applications. French engineers had more success initially, with Nicholas Joseph Cugnot (1725–1804) developing a three-wheeled steam-powered tractor around 1770, capable of 6 kph/3.5 mph. The Montgolfier brothers Joseph-Michael (1740–1810) and Jacques Etiènne (1745–1799) were responsible for the first sustained human flight, aboard a hot-air balloon 1763.

By the time Trevithick left England for Peru, his high-pressure engine had shown that steam was amply capable of providing a mobile power source. He returned to find his place as the greatest engineer in the new technology usurped by others. Most famous of these was English engineer George Stephenson (1781–1848), whose steam locomotives were responsible for the setting up of the first practical public railway ever built, in 1825, between Stockton and Darlington.

Improved communications

The transport revolution began in the 17th century with the construction by the Frenchman Pierre-Paul Riquet (1604–1680) of the Canal du Midi in France, opened in 1681. Riquet was working as a collector of the salt tax in the Languedoc region when he became aware of the need for good transport in the area and began canal promotion in the 1660s. In the UK, 3,220 km/2,000 mi of waterway were constructed between 1760 and 1830. In France, improved roads were built by civil engineer Pierre Trésaguet (1716–1796). Britain's appalling roads were improved by Scottish civil engineer Thomas Telford (1757–1834) whose works included the suspension bridge over the Menai Staits, and the scientific approach to road construction devised by another Scottish civil engineer John McAdam (1756–1836) radically improved this aspect of Britain's infrastructure.

Steam power also found its way into ocean-going vessels. US artist, engineer, and inventor Robert Fulton (1765–1815) returned from Europe, where he had seen what steam engines were capable of, to set up the first regular steamship service, between New York and Albany, in 1807. Whether steam-powered ships were capable of a longer journey, in particular across the Atlantic Ocean, was a major debating point when French-born British inventor and engineer Isambard Kingdom Brunel's (1806–1859) *Great Western* succeeded in sailing to New York without refuelling, in 1838. Brunel went on to design and launch the first iron ship (and the first to use a screw propellor, rather than paddle wheels) the *Great Britain,* and the colossal *Great Eastern,* whose steadiness and manoeuvrability were put to use in a key event in another field of technology altogether: the laying of the Atlantic telegraph cable in 1865.

Such very long-distance communication was made possible by advances in the understanding of electricity, and the translation of this into devices that could transmit and receive messages at the speed of light. The first successful system was brought out by English physicist Charles Wheatstone (1802–1875), in which electrical signals deflected magnetized needles indicating letters of the alphabet. The development of a communication system using a code of dots and dashes to represent the letters by US artist Samuel Morse (1791–1872) proved so successful it was still in use at the end of the 20th century.

The problem of making cheap steel was solved almost simultaneously by ironmaster William Kelly (1811–1888) in the United States and English engineer and inventor Henry Bessemer (1813–1898) in Britain. The process involved blowing air through molten pig iron to drive out the carbon. The Crimea War led to English engineer and inventor William Armstrong (1847–1908) developing the wrought-iron, breech-loading gun with rifled barrel.

The wars affecting the United States also assisted the development of weapons on that side of the Atlantic. The war against Mexico, which broke out in 1846, accelerated the revolution in small arms manufacture initiated by US inventor Samuel Colt (1814–1862), who had produced the first revolver in 1836. The Civil War of 1861–65 prompted US inventor Richard Gatling (1818–1903) to develop the rapid-fire gun that bears his name.

The internal combustion engine

It was around the middle of the 19th century that attention began to shift from steam to gas and other combustible materials as a means of providing motive power. As early as 1833, an engine that ran on an inflammable mixture of gas and air had been described, and a number of the fundamental principles of the fuel-powered engine had been described by the time Belgian-born French engineer and inventor Jean Lenoir (1822–1900) began building engines using the system which operated smoothly in 1860. Together, German engineers Nikolaus Otto (1832–1891) and Eugen Langen (1833–1895) managed by 1877 to solve the basic problems facing the development of the four-stroke internal combustion engine. This work led to the development of the modern motor car, and of powered flight. German engineer Gottlieb Daimler (1834–1900) was to join the pair as an engineer, leaving in 1883 to develop lighter, more efficient high-speed engines capable of driving cycles and boats, as well as automobiles.

German engineer Rudolph Diesel (1858–1913) experimented with internal combustion engines during the 1890s; by using the heat developed by compression of the fuel-air mixture, rather than a spark from an ignition system, to ignite the mixture, Diesel succeeded in producing an engine that could use cheaper fuels than the Otto cycle engines. However, the high pressures produced in the engines required the use of very heavy-gauge metal, with consequent weight–power ratio problems. Later advances in metallurgy enabled this disadvantage to be significantly reduced, with the result that the Diesel engine is still used in a wide range of vehicles today.

Work in the USA, as well as in Germany, succeeded in bringing the power of the internal combustion engine to bear on the problem of powered flight. Early work on the flow of air over gliders by German engineer Otto Lilienthal (1848–1896) and others established a body of knowledge needed to supplement the work of earlier enthusiasts such as English inventor George Cayley (1773–1857), who had defined the basic aerodynamic forces acting on a wing as early as 1799.

Internal combustion engines, coupled to balloons to form airships, were in use by the turn of the century. The first flight of a heavier-than-air machine powered by a light, efficient internal combustion engine was constructed by US aeronautical engineers the Wright brothers, Orville (1871–1948) and Wilbur (1867–1912), on 17 December 1903. As well as finding a suitable engine for such a machine, the Wrights had succeeded in solving the problem of controlling the aeroplane in all three axes.

Electricity as a source of power

Despite the success of the internal combustion engine in powering a wide range of machines, the use of water and steam as power sources remained important in another major field of technology: the generation of electricity. Water had been used to drive turbines of increasing efficiency for many years when English mechanical engineer Charles Parsons (1854–1931) adapted the basic design of a water turbine to enable a jet of steam to impart its kinetic energy to a series of turbine blades which then rotate. By combining this rotation with the ability of a dynamo to convert rotary motion into electric power, the electric generator was born. The first ever turbine-powered generating station was set up in 1888, using four Parsons turbines each developing 75 kW/100 hp. Direct use of the mechanical power developed by steam was made in Parson's 44 tonne/44.7 ton Turbinia, whose turbine engine developed 1.5 MW/2,000 hp, enabling it to travel at 60 kph/37 mph in 1897.

Work by the inventors and electrical engineers Joseph Swan (1828–1914) in England and Thomas Edison (1847–1931) in the United States finally resulted in the creation of a long-lasting light source powered by electricity – the filament lamp – around 1880.

While Germany and the United States were quick to use electrical power to bring about a revolution in their industrial processes, the availability of cheap labour and concentration on waning industries based on traditional raw material inhibited the adoption of electrical power in Britain. Concentration on telegraphic technology and the generation of electric illumination did have an indirect advantage, however. The invention of the two-electrode electric valve, the diode, by English electrical engineer John Fleming (1849–1945) provided a new outlet for the vacuum bulb technology. Such developments as radio communication, radar, television, and the computer all benefited from this.

Radio communications

Communication over long distances without the use of cables – 'wireless' communication – had been a practical possibility from the day when the electromagnetic wave physicist Scottish physicist James Clerk Maxwell's (1831–1879) theory combining electrical and magnetic phenomena had been investigated by German physicist Heinrich Hertz (1857–1894) in 1888. Both transmitters and detectors of these radio waves were developed until Italian physicist Guglielmo Marconi (1874–1937) succeeded in transmitting messages over a few yards using electromagnetic waves in 1895. By 1901, he had succeeded in sending signals right across the Atlantic.

More sophisticated communication was made possible by US inventor Lee De Forest (1873–1961) and Canadian physicist Reginald Fessenden (1866–1932), and their invention of the triode amplifier and amplitude modulation respectively. These advances enabled speech and sound to be transmitted over very long distances, and gave birth to modern communications.

Developments of the war years

World War I (1914–1918) saw the use of technology on an unprecedented scale. Although many of the advances then simply led to the deaths of hundreds of thousands of troops, many later found major applications in peacetime. An excellent example of this is provided by the development of the nitrogen fixation process to an industrial scale by German chemist Karl Bosch (1874–1940). This enabled the Germans to manufacture explosives such as TNT without relying on foreign imports of nitrogen-bearing materials, capable of being blockaded by the allies. In peacetime, the process allowed the cheap manufacture of fertilizers, equally vital to the survival of a country.

The war also had a profound effect on the aircraft industry. Starting the war as chiefly reconnaissance vehicles, the aircraft became directly involved in the fighting by the end, and mass production of tens of thousands became necessary. Governments spent money on research, accelerating advances in aerodynamics and power systems enormously. Civil aviation, begun by the Germans before the war, benefited, initially using modified military aircraft.

As with the automobile, engineers started to look at new power sources for the aircraft. The use of gas turbines was put forward in 1926, a suggestion turned into reality by English engineer and inventor Frank Whittle (1907–1996) in 1930. By combining a gas turbine with a centrifugal compressor, he created the jet engine.

The inter-war years saw considerable advances in rocket technology: an area of engineering that was to enable humans to leave the planet of their origin. The Chinese had used solid-fuelled rockets in battles as early as 1232; their direct ancestors are still to be seen strapped to the central booster of the Space Shuttle. It was Russian theoretician Konstantin Tsiolkovskii (1857–1935) who pointed out that liquid propellants had distinct advantages of power and controllability over solid fuels. The American astronautics pioneer US physicist Robert Goddard (1882–1949) succeeded in launching the first liquid-fuelled rocket in 1926. Just 35 years later, Soviet engineers used a liquid-filled booster to send the first human being into Earth orbit. Eight years after that a human being set foot on another celestial body for the first time.

Entertainment

By the 1920s, a number of devices born in research laboratories had become established as massively popular forms of entertainment. The work of US inventor and businessman George Eastman (1854–1932), US electrical engineer and inventor Thomas Edison (1847–1931) and others brought photography and sound-and-motion 'movie' pictures to millions. A working system of television was devised by Scottish inventor John Logie Baird (1888–1946) and shown in 1925, while the modern electronic system later adopted as standard for television was demonstrated by Russian-born US electronics engineer Vladimir Zworykin (1889–1982) in 1929. The British Broadcasting Corporation's forebear, the British Broadcasting Company, was formed in 1922, transmitting radio programmes to the public on a national scale. In 1936, experimental television broadcasts were made by the BBC from Alexandra Palace near London.

Atomic power

The growth in the use of electrical power put greater emphasis on ways of generating it cheaply. The United States in particular built many large storage dams, producing electricity by hydroelectric turbine technology. By 1920 some 40% of electricity in the USA was generated by this means. But developments in particle physics during these years were beginning to show that the fundamental constituents of matter would be capable of providing another, far more concentrated, form of energy: atomic power. By 1939 and the outbreak of World War II, a number of physicists had begun to appreciate the possibilities offered by 'chain reactions' involving the fission of unstable, chemical elements such as uranium.

The first electronic computers

The war itself again proved to be a sharp stimulus for the refinement of old ideas and the development of new ones. Radar, devised by Scottish physicist and engineer Robert Watson-Watt (1892–1973) in 1935, was developed into a national defence system against aircraft that were growing ever faster and more deadly. More sophisticated weapons and ever more complex message-encoding systems resulted in the development of early electronic computers. These were needed to rapidly sift through data and perform arithmetical operations upon it, and also to carry out numerical integrations which were particularly useful in the precise calculation of the trajectories for artillery shells. The pioneering work on mechanical computing machines by English mathematician Charles Babbage (1792–1871) in the early 1800s was transformed by the introduction of electronic devices by US electrical engineer Vannevar Bush (1890–1974) and others during the war years.

The inter-war ideas of jet and rocket propulsion were used in the development of fighter aircraft and missiles such as the 'V' (*vergeltung*) 1 and 2, powered by ram-jet and liquid fuel respectively. Most devastating of all was the use of the chain reaction of atomic energy in an uncontrolled explosive device against the Japanese in 1945. Although the use of the atomic bomb finally ended World War II, the world still lives under the threat of their use, in still more deadly form, to this day. The use of the first atomic bombs has tended to overshadow another event in the development of atomic power that took place during the war. This was the setting up of the first controlled atomic chain reaction in 1942 by a team of scientists at the University of Chicago, which paved the way for the peaceful use of atomic power to generate electricity. The first nuclear electricity was generated by the Experimental Breeder Reactor in Idaho, USA 1951. The world's first commercial scale nuclear power station entered service at Calder Hall, UK 1956.

The age of the microchip

While the demand for yet more electric power grew among industrial nations mass-producing cars, ships and aircraft, ways of reducing the complexity and power consumption of electronic devices such as computers, radios, and televisions were being sought. Most crucial of these was the invention of the transistor, by US physicists John Bardeen (1908–1991), William Shockley (1910–1989), and Walter Brattain (1902–1987) at Bell Laboratories in the USA, in 1948. These tiny semiconductor-based devices could achieve the rectification and amplification of the thermionic valves of the pre-war years at a fraction of the power consumption. The use of such devices in still smaller form allowed the miniaturization of electronic devices.

In 1958 the first integrated circuit, which contained the components of a complete circuit on a single piece of silicon, was built by US electrical engineer Jack Kilby (1923–) in the USA. The first microprocessor, a complete computer on a chip was designed by US computer engineer Ted Hoff (1937–) in 1971, containing 2,250 components. By 1990 memory chips capable of holding 4 million bits of information were being mass-produced in Japan. In 1993 the US chip manufacturer Intel introduced the Pentium microprocessor chip which was about five times more powerful than earlier processors and had 3.1 million transistors on a silicon square 15 mm/0.6 in across.

Computing power packed into smaller volumes has been the driving force of many areas of technology over the last 40 years. It made possible the era of human space flight, where keeping the mass of all components to a minimum is vital. Telecommunication satellites, such as *Telstar,* (which in 1962 transmitted the first live television pictures across the Atlantic), weather satellites and planetary probes were all made possible as a result of the semiconductor breakthrough. New materials capable of withstanding the rigours of space were also produced, many of which found use back on Earth.

Computer-aided design and manufacture have enabled new ideas in fields from architecture to aircraft manufacture to be tried out, tested, and produced far more quickly and cheaply. The influence of the computer is felt in everyday life, from the diagnosis of disease by the CAT scanner invented by English scientist Godfrey Hounsfield (1919–) to the production of bank statements. The first home computers were introduced in the early 1980s. In the 1990s, tens of millions of home computers are linked to the worldwide Internet. Home computer users have also powered the market for CD-ROM, which are able to hold vast amounts of information.

The first optical fibre cable, capable of carrying digital signals, was installed in California 1977. In 1988 the International Services Digital Network (ISDN), an international system for sending signals in digital format along optical fibres and coaxial cable, was launched in Japan. The first transoceanic optical fibre cable, capable of carrying 40,000 simultaneous telephone conversations, was laid between Europe and the USA in 1989. In 1992 videophones, made possible by advances in image compression and the development of ISDN, were introduced in the UK. The Japanese began broadcasting high-definition television in 1989. All digital high-definition television was demonstrated in the USA in 1992.

The laser was another significant development invented 1960 by US physicist Theodore Maiman (1927–), and often used in robotic welders. Holograms, or three-dimensional pictures, became practicable after the development of laser technology.

Power sources for the future

Even if, between them, the robot and the computer free us from manual labour, power will still be needed in vast quantities to process the raw material from which goods are produced. There is still considerable interest in finding ways of generating cheap power. Using nuclear fission has proved only a partial answer to the question of what will replace the burning of hydrocarbons such as coal to generate electricity. Public concern about both its inherent safety and the toxic waste produced has cast a shadow over the long-term future of fission-generated electricity.

Engineers and physicists in Europe, the United States, the former Soviet Union, and Japan are currently studying the generation of power by nuclear fusion. Using hydrogen and its isotopes derived from seawater, they hope to be able to mimic the reactions that have kept the Sun burning for thousands of millions of years, although the engineering difficulties presented by trying to keep a plasma stable at a temperature of 100 million degrees are immense. In 1994 the Tokamak Fusion Test Reactor at Princeton University produced 9 megawatts of power. It needed 33 megawatts to drive it and power production lasted for only 0.4 seconds. Nevertheless, the hot plasma of deuterium and tritium was so well behaved that researchers believe output can be boosted.

Others are turning to less exotic sources of energy, such as the wind, solar energy, and tides, to find better ways of exploiting them and generating cheap, clean power. The world's largest photovoltaic power station was plugged into the power grid at Davis, California, USA, in 1993. A wind farm in California's San Bernados Mountains, uses over 4,000 wind turbines to supply electricity to the Coachella Valley in south California.

The end of the 20th century

As the 20th century came to an end, the greatest advances were being made in information technology. Fast communication began with the facsimile or fax, transmitting documents by telephone or telex networks, and continued with electronic mail (e-mail). The most exciting development was the establishment of the Internet, a worldwide information service available to all computer users. It had its origins in a system developed by the US Defense Department in 1969 to link its various research departments. The non-military network was opened up in 1984, and within ten years there were an estimated 20 million users. Originally the network offered 1.5 Mbit/s, but this had risen to 1000 Mbit/s by 1996. Research continues on ever faster processors, with such innovatory ideas as making use of the changes in quantum states at the subatomic level. Traditional civil engineering also had its triumph with the completion of a scheme first proposed in 1802, for a tunnel to link England and France. The Channel Tunnel, opened in 1994, has two tunnels for trains and a service tunnel. It is 50 km/30 mi long, and a new Anglo French border was established 40 m/130 ft below the sea.

MATHEMATICS

MOST ANCIENT CIVILIZATIONS had the means to make accurate measurements, to record them in writing, and to use them in calculations involving elementary addition and subtraction. Apparently for many of them that was sufficient. It seems, for example, that the ancient Egyptians relied on simple addition and subtraction even for calculations of area and volume, although their number system was founded upon base 10, as ours is today. They certainly never thought of mathematics as a subject of potential interest or study for its own sake.

Not so the Babylonians. Contemporaries of the Egyptians, they nevertheless had a more practical form of numerical notation and were genuinely interested in improving their mathematical knowledge. (Perversely, however, their system used base 10 up to 59, after which 60 became a new base; one result of this is the way we now measure time and angles.) By about 1700 BC the Babylonians not only had the four elementary algorithms – the rules for addition, subtraction, multiplication, and division – but also had made some progress in geometry. They knew what we now call Pythagoras' theorem, and had formulated further theorems concerning chords in circles. This even led to a rudimentary understanding of algebraic functions.

The ancient Greeks

Until the very end of their own civilization, the ancient Greeks had little use for algebra other than within a study of logic. After all, to them learning was as interdisciplinary as possible. Even the philosopher Thales of Miletus (lived *c.* 585 BC), regarded as the first named mathematician, considered himself a philosopher in a school of philosophers; mathematics was peripheral. The Greeks' attitude of scientific curiosity was, however, to result in some notable advances in mathematics, especially in the endeavour to understand why and how algorithms worked, theorems were consistent, and calculations could be relied on. It led in particular to the notion of mathematical proof, in an elementary but no less factual way. Mathematician and philosopher Pythagoras (lived *c.* 530 BC), having proved the theorem now called after him, imbued mathematics with a kind of religious mystique on the basis of which he became a rather unsuccessful social reformer. Others became fascinated by solving problems using a ruler and compass, in which an outline of the concept of an irrational number (such as π) inevitably appeared. Further investigations of curves followed, and resulted in the first suggestions of what we now call integration. Such geometrical studies were often applied to astronomy. A corpus of various kinds of mathematical knowledge was beginning to accumulate.

The person who recorded much of it was the mathematician Euclid (*c.* 330–260 BC). His work *The Elements* is intended as much as a history of mathematics as a compendium of knowledge, and was massive therefore in both scope and production. It contained many philosophical elements (as we would now define them) and astronomical hypotheses, but the exposition of the mathematical work was masterly, and became the style of presentation emulated virtually to this day. Euclid's geometry, especially, became the standard for millennia: mathematicians still distinguish between Euclidean and non-Euclidean geometry. He even included discussion and ideas on spherical geometry. Unfortunately, some of *The Elements* was lost, including the work on conic sections.

Conic sections seem to have been a source of fascination to many ancient Greek mathematicians. Archimedes (*c.* 287–212 BC), one of the most practical men of all time, used the principle of conic sections in an investigation into how to solve problems of an algebraic nature. A little later, Apollonius of Perga (lived *c.* 230 BC), wrote definitively of the subject, adducing a considerable number of associated theorems and including relevant proofs. The significance of part of this extra material was established only at the end of the 19th century.

The Romans and their successors

After about 150 BC, the study of astronomy dominated the scientific world. Consequently, for a while, little mathematical progress was made except in the context of the cosmological theories of the time. (There was accordingly some significant research into spherical geometry and spherical trigonometry.) It was then too that Roman civilization briefly flourished and began to recede – again with little effect on the status of mathematics. Surprisingly, however, after about 400 years, the Alexandrian Diophantus (lived *c.* AD 270–280) devised something of extreme originality: the algebraic variable, in which a symbol stands for an unknown quantity. Equations involving such indeterminates – Diophantus included one indeterminate per equation, needing thus only one symbol – are now commonly called Diophantine equations.

Alexandria thus became the centre for mathematical thought at the time. Very shortly afterwards, the mathematician, astronomer, and geographer Pappus (lived *c.* 320) deemed it time again for a compilation of all known mathematical knowledge. In *The Collection* he revised, edited, and expanded the works of all the classic writers and added many of his own proofs and theorems, including some well-known problems that he left unsolved. It is this work more than any other that ensured the survival of the mathematics of the Greeks until the Renaissance about a thousand years later.

In the meantime the initiative was taken by the Arabs, whose main sphere of influence was, significantly, farther east. They were thus in contact with Persian and Indian scientific schools, and accustomed to translating learned texts. Both Greek and Babylonian precepts were assimilated and practised – the best known proponent was Al-Khwarizmi (lived *c.* 840), whose work was historically important to later mathematicians in Europe. The Arabs devised accurate trigonometrical tables (primarily for astronomical research) and continued the development of spherical trigonometry; they also made advances in descriptive geometry.

It was through his learning in the Arab markets of Algeria that the medieval merchant from Pisa Leonardo Fibonacci (or Leonardo Pisano, *c.* 1180–1250), brought much of contemporary mathematics back to Europe. It included – only then – the use of the 'Arabic' numerals 1 to 9 and the 'zephirum' (0), the innovation of partial numbers or fractions, and many other features of both geometry and algebra. From that time, hundreds of translators throughout Europe (especially in Spain) worked on Latin versions of Arab works and transcriptions. Only when Europe had regained all the knowledge and, so to speak, updated itself could genuine development take place. The effort took nearly 400 years before any truly outstanding advances were made – but may be said to be directly responsible for the overall updating and advance in science that then came about, known as the Renaissance.

The 16th century

One of the first instances of genuine progress in mathematics was the means of solving cubic equations, although acrimonious recriminations over priority surrounded its initial publication. One particularly charismatic contender – Italian mathematician and physicist Niccolo Fontana (*c.* 1499–1557), usually known as Tartaglia – besides being a military physicist, was also an inspirational figure in the propagation of mathematics. The means of solving quartic equations was discovered soon afterwards.

Within another 20 years, the French mathematician François Vieta (1540–1603) was improving on the systematization of algebra in symbolic terms and expounding on mathematical (as opposed to astronomical) applications of trigonometry. It was he, if anyone, who initiated the study of number theory as an independent branch of mathematics. At the time of Vieta's death, Henry Briggs (1561–1630) in England was already professor of geometry; a decade later he combined with Scottish mathematician John Napier (1550–1617), the deviser of 'Napier's bones', to produce the first logarithm tables using the number 10 as its base, a means of calculation commonly used until the late 1960s but now outmoded by the computer and pocket calculator. Simultaneously, the German astronomer Johannes Kepler was publishing one of the first works to consider infinitesimals, a concept that would lead later to the formulation of the differential calculus.

The 17th century

It was in France that the scope of mathematics was then widened by a group of great mathematicians. Most of them met at the scientific discussions run by the director of the convent of Place Royale in Paris, Fr Marin Mersenne (1588–1648). To these discussions sometimes came the philosopher-mathematician René Descartes (1596–1675), the lawyer and magistrate Pierre de Fermat (1601–1675), the physicist and mathematician Blaise Pascal (1623–1662), and the architect and mathematician Gérard Desargues (1591–1661). Descartes was probably the foremost of these in terms of mathematical innovation, although it is thought that Fermat – for whom mathematics was an absorbing but part-time hobby – had a profound influence upon him. His greatest contribution to science was in virtually founding the discipline of analytical (coordinate) geometry, in which geometrical figures can be described by algebraic expressions. He applied the tenets of geometry to algebra, and was the first to do so, although the converse was not uncommon. Unfortunately, Descartes so much enjoyed the reputation his mathematical discoveries afforded him that he began to envy anyone who then also achieved any kind of mathematical distinction. He therefore regarded Desargues – who published a well-received work on conics – not only as competition but actually as retrogressive. When Pascal then publicly championed Desargues (whom Descartes had openly ridiculed), putting forward an equally accepted form of geometry now known as projective geometry, matters became more than merely unfriendly.

In the meantime, Fermat took no sides, studied both types of geometry, and was in contact with several other European mathematicians. In particular, he used Descartes' geometry to derive an evaluation of the slope of a tangent, finding a method by which to compute the derivative and thus being considered by many the actual formulator of the differential calculus. Part of his study was of tangents as limits of secants. With Pascal he investigated probability theory, and in number theory he independently devised many theorems, one of them now famous as Fermat's Last Theorem.

It is now known that at about this time in Japan, a mathematician called Seki Kowa (*c.* 1642–1708) was independently discovering many of the mathematical innovations also being formulated in the West. Even more remarkably, he managed to change the social order of his time in order to popularize the subject.

Three years after Pascal died, a religious recluse haunted by self-doubt, English physicist and mathematician Isaac Newton (1642–1727), was obliged by the spread of the plague to his university college in Cambridge to return home to Woolsthorpe in Lincolnshire and there spend the next year and a half in scientific contemplation. One of his first discoveries was what is now called the binomial theorem, which led Newton to an investigation of infinite series, which in turn led to a study of integration and the notion that it might be achieved as the opposite of differentiation. He arrived at this conclusion in 1666, but did not publish it. More than seven years later, in Germany, philosopher, and mathematician Gottfried Leibniz (1646–1716) – who had possibly read the works of Pascal – arrived at exactly the same conclusion, and did publish it. He received considerable acclaim in Europe, much to Newton's annoyance, and a priority argument was very quickly in process. Naively, Leibniz submitted his claim for priority to a committee on which Newton was sitting, so the outcome was a surprise to no one else, but it was in fact Leibniz's notation system that was eventually universally adopted. It was not until 1687 that Newton's studies on calculus were published within his massive *Principia Mathematica*, which also included much of his investigations into physics and optics. Leibniz went on to try to develop a mathematical notation symbolizing logic, but although he made good initial progress it met with little general interest, and despite his energy and status he died a somewhat lonely and forlorn figure.

Another who died in even worse straits was an acquaintance of both Newton and Leibniz: French mathematician Abraham de Moivre (1667–1754), a Huguenot persecuted for his religious background to the extent that he could find no professional position despite being a first-class and innovative mathematician. He met his end broken by poverty and drink – but not before he had formulated game theory, reconstituted probability theory, and set the business of life insurance on a firm statistical basis.

Leibniz's work on calculus was greatly admired in Europe, and particularly by the great Swiss mathematician family domiciled in Basle: the Bernoullis. The eldest of three brothers, Jakob (or Jacques, 1654–1705), actually corresponded with Leibniz; the youngest, Johan (or Jean, 1667–1748), was recommended by the Dutch physicist Christiaan Huygens to a professorship at Groningen. Both brothers were fascinated by investigating possible applications of the new calculus. Unhappily, their study of special curves (particularly cycloids) using polar coordinates proceeded independently along identical lines and resulted in considerable animosity between them. When Jakob died, however, Johan succeeded him at Basle, where he educated his son Daniel – also a brilliant mathematician – whose great friends were Swiss mathematicians Leonhard Euler (1707–1783) and Gabriel Cramer (1704–1752).

The 18th century

Euler may have been the most prolific mathematical author ever. He had amazing energy, a virtually photographic memory and a gift for mental calculation that stood him in good stead late in life when he became totally blind. Not since Descartes had anyone contributed so innovatively to mathematical analysis – Euler's *Introduction* (1748) is considered practically to define in textbook fashion the modern understanding of analytical methodology, including especially the concept of a function. Other works introduced the calculus of variations and the now familiar symbols π, e, and i, and systematized differential geometry. He also popularized the use of polar coordinates, and explained the use of graphs to represent elementary functions.

It was his friend Swiss natural philosopher and mathematician Daniel Bernoulli (1700–1782) who had originally managed to secure a position for him in St Petersburg. When, in 1766, Euler returned there from a post at the Prussian Academy, his place in Berlin was taken by the Italian-born French mathematician Joseph Lagrange (1736–1813) whose ideas ran almost parallel with Euler's. In many ways Lagrange was equally as formative in the popularizing of mathematical analysis, for although he might not have been as energetic or outrightly creative as Euler, he was far more concerned with exactitude and axiomatic rigour, and combined with this a strong desire to generalize. The publication of his studies of number theory and algebra were thus models of precise presentation, and his mathematical research into mechanics began a process of creative thought that has not ceased since. One immediate result of the latter was to inspire his friend and fellow-Frenchman the mathematician and theoretical physicist Jean le Rond d'Alembert (1717–1783) to great achievements in dynamics and celestial mechanics. It was d'Alembert who first devised the theory of partial differential equations.

Towards the end of Lagrange's life, when he was already ailing, he became professor of mathematics at the institution which for the next 50 years at least was to exercise considerable influence over the progress of mathematics; the newly-established École Polytechnique in Paris. Two of his contemporaries there were French astronomer Pierre Laplace (1749–1827) and French mathematician Gaspard Monge (1746–1818). Laplace became famous for his astronomical calculations, Monge for his textbook on geometry; both were acquaintances of Napoleon Bonaparte – as was Joseph Fourier (1768–1830), the physicist who demonstrated that a function could be expanded in sines and cosines through a series now known as the Fourier series.

It was one of Gaspard Monge's pupils – French military engineer Jean-Victor Poncelet (1788–1867) – who first popularized the notion of continuity and outlined contemporary thinking on the principle of duality. And it was one of Laplace's colleagues (whom he disliked), French mathematician Adrien Legendre (1752–1833), who took over where Lagrange left off, and researched into elliptic functions for more than 40 years, eventually deriving the law of quadratic reciprocity and, in number theory, proving that π is irrational.

The 19th century

Legendre's investigations into elliptic integrals were outdated almost as soon as they were published by the work of the Norwegian mathematician Niels Abel (1802–1829) and the mathematician and mathematical physicist German Karl Jacobi (1804–1851). Jacobi went on to make important discoveries in the theory of determinants: he was a great interdisciplinarian. The tragically short-lived Abel has probably had the longer-lasting influence, in that he devised the functions now named after him. He was also unlucky in that his proof, that in general roots cannot be expressed in radicals, was discovered simultaneously and independently by the equally tragic French mathematician Evariste Galois (1811–1832), who only just had time before his violent death to initiate the theory of groups. Further progress in function theory was made by the French mathematician Augustin Cauchy (1789–1857), a prolific mathematical writer who in his works pioneered many modern mathematical methods, developing in particular the use of limits and continuity. He also originated the theory of complex variables, based at least partly on the work of Swiss mathematician Jean Argand (1767–1822), who had succeeded in representing complex numbers by means of a graph.

By this time, however, the centre of mathematics in Europe was undoubtedly Göttingen, where the great German mathematician Karl Gauss (1777–1855) had long presided. Sometimes compared with Archimedes and Newton, Gauss was indisputably not only a mathematical genius who made a multitude of far-reaching discoveries – particularly in geometry and statistical probability – but was also an exceptionally inspirational teacher who inculcated in his pupils the need for meticulous attention to proofs. Late in his tenure at Göttingen, three of his pupils/colleagues were German mathematicians Lejeune Dirichlet (1805–1859), Bernhard Riemann (1826–1866), and Julius Dedekind (1831–1916). There could not have been a more influential quartet in the history of mathematics: the work of all four provides the basis for a major part of modern mathematical knowledge.

Gauss himself was most interested in geometry. Swiss mathematician Jakob Steiner (1796–1863) in Germany was trying to remove geometry from the 'taint' of analysis as propounded by the French, but Gauss went further and decided to investigate geometry outside the scope of that described by Euclid. It was a momentous decision – made almost simultaneously and quite independently by Russian mathematician Nikolai Lobachevsky (1792–1856) and Hungarian mathematician János Bolyai (1802–1860). Between them they thus derived non-Euclidean geometry. The ramifications of this were widespread and fast-moving. In Ireland, mathematician William Hamilton (1805–1865) suggested the concept of *n*-dimensional space; in Germany, mathematician Hermann Grassmann (1809–1877) not only defined it but went on to use a form of calculus based on it. But it was Gauss's own pupil, Riemann, who really became the arch-apostle of the subject. He invented elliptical hyperbolic geometries, introduced 'Riemann surfaces' and redefined conformal mapping (transformations) explaining his innovations with such enthusiasm and accuracy that the modern understanding of time and space now owes much to his work.

Meanwhile Dirichlet – who succeeded Gauss when the great man died and himself became an influential teacher – and Dedekind concentrated more on number theory. Dirichlet slanted his teaching of mathematics towards applications in physics, whereas Dedekind was determined to arrive at a philosophical interpretation of the concept of numbers. Such an interpretation was thought likely to be of use in the contemporary search for a mathematical basis for logic. English mathematician George Boole (1815–1864) had already attempted to create a form of algebra intended to represent logic that, although not entirely successful, was stimulating to others.

As the study of geometry expanded rapidly, the importance of algebra also increased accordingly. Riemann was influential; German mathematician Karl Weierstrass (1815–1897) provided important redefinitions in function theory; but in algebraic terms development was next most instigated by the English mathematician Arthur Cayley (1821–1895) who discovered the theory of algebraic invariants even as he carried out research into *n*-dimensional geometry. The principles of topology were being established one by one even though the branch itself was not yet complete. Norwegian mathematician Sophus Lie (1842–1899) made important contributions to geometry and to algebra – and indirectly to topology – with the concept of continuous groups and contact transformations, and Cayley went on to invent the theory of matrices. French geometrician Gaston Darboux (1842–1917) revised popular thinking about surfaces. German mathematician Felix Klein (1849–1925) – an influential figure in his time – unified all the geometries within his *Erlangen Programme* (1872). But it is German mathematician Felix Hausdorff (1868–1942) who is actually credited with the formulation of topology.

Dedekind finally achieved his goal and axiomatized the concept of numbers – only for his axioms to be (albeit apologetically and acknowledgedly) 'stolen' from him by Italian mathematician Giuseppe Peano (1858–1932). The axioms, however, may have inspired – among others – Hausdorff to conceive the idea of point sets in topology, and Danish-born German mathematician Georg Cantor (1843–1918) to define set theory (the basis on which most mathematics is taught in schools today) and transfinite numbers, and certainly caused a revival of interest in number theory generally. German mathematician Immanuel Fuchs (1833–1902) reformulated much of function theory while attempting to refine Riemann's method for solving differential equations. His pupil, French mathematician Henri Poincaré (1854–1912) – similarly fascinated by Riemann's work – made many conjectures that were later useful in the investigation of topology and of space and time, but less successfully spent years researching into what are now called integral equations, only to discover after they were finally axiomatized by Swedish mathematician Ivar Fredholm (1866–1927) that he had done all the work without perceiving the answer.

The 20th century

It has been claimed that more professional mathematicians lived in the 20th century than had lived in all of recorded time. Certainly the enterprise of mathematics has grown remarkably, so much so that two figures active around the turn of the 20th century, Poincaré and the German mathematician, philosopher, and physicist David Hilbert (1862–1943), are claimed to be among the last mathematicians who had a broad and deep knowledge of the whole of the field.

A genuine polymath and an enthusiastic teacher, Hilbert expanded virtually all branches of mathematics, especially in the interpretation of geometric structures implied by infinite-dimensional space. Hilbert's plenary address to one of the very early International Congress of Mathematicians conferences (held in Paris in 1900) consisted of presenting 23 problems which he believed would occupy the world's mathematicians for the coming century. His predictions proved very generative of future work, and to solve a 'Hilbert problem', as they became known, was a guarantee of professional accolades of the highest order. The types of problems he proposed ranged from the quite specific to the sort 'develop a general theory of ...'. During the 20th century mathematics became gradually either more theoretical or more practical. Theoretically interest swung towards finding features in common between disparate mathematical structures. French mathematician Henri Lebesgue (1875–1941) devised a concept of measure that contributed greatly to the theory of abstract spaces. Andrei Kolmogorov (1903–1987) and others not only related this to probability theory but thereby to problems of statistical mechanics and the clarification of the ergodic theorems provided by US mathematician George Birkhoff (1884–1944) in 1932.

The 20th century was the century of generalization and abstraction, and of the study of structure by means of its systematization and classification. In an important sense, the 20th century was the century of algebra, not only as an ever-growing subject matter but also as the very language of much of pure mathematical proof (in much the same way that geometry was both subject and language for proof for the ancient Greeks). It is also a story of the development of very general and abstract theories, rather than the solution of individual problems, and the provision of very abstract settings for such theoretical accounts.

Such theories often developed as hybrids of earlier areas of study as the methods, concepts and techniques of abstract algebra were successfully deployed against previously resistant problems in a range of different mathematical domains. Systematic work of this sort was done in a variety of rapidly developing fields: algebraic geometry, algebraic number theory, algebraic topology, and even algebraic algebra (category theory).

Modern mathematics (as it became known) has concerned itself with such abstract mathematical structures, drawing on algebraic concepts that have become almost hegemonic, including groups, rings, and fields. As an example of some of these trends, one of the key problems in group theory (a branch of abstract algebra) became the classification of all finite simple groups. In 1980 mathematicians completed the classification, a problem which had taken over a 100 mathematicians more than 35 years, covering over 14,000 pages in mathematical journals.

In consequence of this broadening of scope and scale, an increasing amount of mathematical work has moved away from being that of an individual mathematician to teams of collaborators in the same or different institutions. One notable exception to this came with the announcement in October 1994 by Andrew Wiles (1953–), an English mathematician working at Princeton University, New Jersey, USA, that he had solved Fermat's Last Theorem, an unproven central result in number theory, working in complete secrecy for seven years. The theorem's proof had eluded mathematicians for over 300 years, and Wiles's proof drew heavily on a wide range of mathematical results and ideas, particularly from algebraic number theory. Starting in the 1930s and subsequently, the Bourbaki group of mathematicians, (comprising some of the most renowned European mathematicians of the period, such as the French algebraic geometer André Weil (1906–1998) commenced publishing a very influential set of books, entitled *Elements of Mathematics.* They saw it as the encyclopedia of mathematics, starting with set theory and abstract algebra, rather than geometry as Euclid had with his *Elements* some 2,300 years earlier. A whole generation of pure mathematicians in a number of countries gained their mathematical training using these works. They helped to shape and orientate mathematics towards the pure, the abstract, and the detached from the problems and phenomena of the material world, which had served since time immemorial as one key source of mathematical challenge.

Yet, as in previous generations, pure mathematical work, developed from queries and curiosity from within mathematics itself, found unexpected applications. In algebraic topology, work in knot theory was used to work on problems arising from the physical folding of the DNA molecule in space. And in the 1950s French mathematician René Thom (1923–) categorized abstract geometric surfaces called manifolds using the notion of cobordism (for which he won a Fields medal, the mathematical equivalent of a Nobel prize). It was during the 1970s that Thom and English mathematician Christopher Zeeman (1925–) produced applications of this work in what became known as catastrophe theory. In particular, they looked at notions of stability, continuity and discontinuity in physical phenomena and human systems, emerging from an area called non-linear dynamics.

Chaos theory

A modern computer-based relative to this work is chaos theory. The central discovery, actually made in 1961 by US meteorologist Edward Lorenz (1917–), is that random behaviour can arise in systems whose mathematical description contains no hint whatever of randomness. The geometry of chaos can be explored using theoretical techniques of topology, but the most vivid images are obtained using computer graphics. The geometric structures of chaos are called fractals; they have the same detailed form on all scales of magnification, a phenomenon called self-similarity.

Polish-born French mathematician Benoit Mandelbrot (1924–) produced the first fractal images in 1962, using a computer that repeated the same mathematical pattern over and over again. He wrote of investigating 'the geometry of nature', in some ways harking back to Galileo's claim over 300 years earlier that 'the book of the universe is written in the language of geometry.' In 1975, US mathematician Mitchell Feigenbaum (1945–) discovered a new universal constant (approximately 4.669201609103) that is important in chaos theory. Order and chaos, traditionally seen as opposites, are now viewed as two aspects of the same basic process, that is the evolution of a system over time.

The study of statistics and probability was also taken up with new enthusiasm for more practical applications. English mathematician and biometrician Karl Pearson (1857–1936) refined Gauss's ideas to derive the notion of standard deviation. Danish mathematician Agner Erlang (1878–1929) used probability theory in a highly practical way to aid the efficiency of the circuitry of his capital's telephone system. US mathematician Alonzo Church (1903–1995) defined a 'calculable function' and by so doing clarified the nature of algorithms. Following this, another US mathematician George Dantzig (1914–) was able to set up complex linear programmes for computers. Such progress is being maintained, sometimes now as a result of using the machines themselves to devise further advances.

Computer scientists have devised symbolic computation systems which manipulate algebraic expressions in the same way that a human mathematician would do, only faster and more accurately. The result might be called 'computer-assisted mathematics'. A good example is the proof in 1972 of the four-colour theorem by US mathematicians Kenneth Appel and Wolfgang Haken. In 1852 English mathematician Francis Guthrie conjectured that no more than four colours need be used in order to ensure that no two adjacent colours on a map share the same colour. Mathematicians quickly proved that five colours would suffice, but had no success whatsoever in reducing the number to four. A direct attack by computer would not be possible, for how could a computer consider all possible maps? But Appel and Haken came up with a list of around 2,000 particular maps, and showed that if each had a rather complicated property, then the conjecture must be true. The computer then checked this property, case by case.

The boundaries around the edge of mathematics are not completely clear, and computer incursions are only one such instance. The 20th century saw a resurgence of work in philosophical issues, in particular a renewal of the quest to find a relationship between mathematics and logic. German logical philosopher and mathematician Gottlob Frege (1848–1925) had devised a system of symbolic logic. However, his pride turned to ashes when English philosopher and mathematician Bertrand Russell (1872–1970) pointed out to him an internal, and fundamental, inconsistency. Russell, with his teacher and friend English mathematician and philosopher Alfred North Whitehead (1861–1947), attended lectures on this topic given by given by Peano; together Russell and Whitehead then published a large work on the foundations of mathematics, entitled *Principia Mathematica.* It had an immediate impact, and remained influential. And several different philosophical theories were dealt a heavy blow by the theorem formulated in 1930 by Austrian-born US philosopher and mathematician Kurt Gödel (1906–1978). This stated that the overall consistency (completeness) of a mathematics that contained the arithmetic of whole numbers cannot itself be proved mathematically. Thus, mathematical concepts and techniques could not be deployed to prove the consistency of mathematics itself. In some sense, paradox is a fundamental aspect of mathematics.

PHYSICS

PHYSICS IS A branch of science in which the theoretical and the practical are firmly intertwined. It has been so since ancient times, as physicists have striven to interpret observation or experiment in order to arrive at the fundamental laws that govern the behaviour of the universe. Physicists aim to explain the manifestations of matter and energy that characterize all things and processes, both living and inanimate, extending from the grandest of galaxies down to the most intimate recesses of the atom.

The history of physics has not been a straight and easy road to enlightenment. The exploration of new directions sometimes leads to dead ends. New ways of looking at things may result in the overthrow of a previously accepted system. Not Aristotle's system, nor Newton's, nor even Einstein's was 'true'; rather statements, or 'laws', in physics satisfy contemporary requirements or – in the existing state of knowledge – contemporary possibilities. The question that physicists ask is not so much 'Is it true?' as 'Does it work?'

Physics has many strands – such as mechanics, heat, light, sound, electricity, and magnetism – and, although they are often pursued separately, they are also all ultimately interdependent. To pursue the history of physics, therefore, it is necessary to follow several separate chains of discovery and then to find the links between them. The story is of frustration and missed opportunities as well as of genius and perseverance. But however complex it may appear, all physicists seek or have sought to play a part in the evolution of an ultimate explanation of all the effects that occur throughout the universe. That goal may be unattainable but the thrust towards it has kept physics as alive and vital today as it was when it originated in ancient times.

Force and motion

The development of an understanding of the nature of force and motion was a triumph for physics, one which marked the evolution of the scientific method. As in most other branches of physics, this development began in ancient Greece.

The earliest discovery in physics, apart from observations of effects like magnetism, was the relation between musical notes and the lengths of vibrating strings. The religious philosopher Pythagoras (*c.* 582–*c.* 497 BC) found that harmonious sounds were given by strings whose lengths were in simple numerical ratios, such as 2:1, 3:2 and 4:3. From this discovery the belief grew that all explanations could be found in terms of numbers. This was developed by mathematician and philosopher Plato (*c.* 427–*c.* 347 BC) into a conviction that the cause underlying any effect could be expressed in mathematical form. The motion of the heavenly bodies, Plato reasoned, must consist of circles, since these were the most perfect geometric forms.

Reason also led the philosopher Democritus (*c.* 470–*c.* 380 BC) to propose that everything consisted of minute indivisible particles called atoms. The properties of matter depend on the characteristics of the atoms of which it is composed, and the atoms combine in ways that are determined by unchanging fundamental laws of nature.

A third view of the nature of matter was given by the Greek polymath Aristotle (384–322 BC), who endeavoured to interpret the world as he observed it, without recourse to abstractions such as atoms and mathematics. Aristotle reasoned that matter consisted of four elements – earth, water, air and fire – with a fifth element, the ether, making up the heavens. Motion occurred when an object sought its rightful place in the order of elements, rocks falling through air and water to the earth, air rising through water as bubbles and fire through air as smoke.

There was value in all these approaches and physics has absorbed them all to some degree. Plato was essentially correct; only his geometry was wrong, the planets following elliptical, not circular, orbits. Atoms do exist as Democritus foretold and they do explain the properties of matter. Aristotle's emphasis on observation (though not his reasoning) was to be a feature of physics and many other sciences, notably biology, of which he may be considered the founder.

These ideas were, however, mainly deductions based solely on reason. Few of them were given the test of experiment to prove that they were right. Then came the achievements of mathematician and physicist Archimedes (*c.* 287–212 BC), who discovered the law of the lever and the principle of flotation by measuring the effects that occur and deduced general laws from his results. He was then able to apply his laws, building pulley systems and testing the purity of the gold in King Hieron's crown by a method involving immersion.

Archimedes thus gave physics the scientific method. All subsequent principal advances made by physicists were to take the form of mathematical interpretations of observations and experiments. Archimedes developed the method in founding the science of statics – how forces interact to produce equilibrium. But an understanding of motion lay a long way off. In the centuries following the collapse of Greek civilization in around AD 100, physics marked time. The Arabs kept the Greek achievements alive, but they made few advances in physics, while in Europe the scientific spirit was overshadowed by the 'Dark Ages'. Then in about 1200, the spirit of enquiry was rekindled in Europe by the import of Greek knowledge from the Arabs. Unfortunately, progress was hindered somewhat by the fact that Aristotle's ideas, particularly his views on motion, prevailed. Aristotle had assumed that a heavy object falls faster than a light object simply because it is heavier. He also argued that a stone continues to move when thrown because the air displaced by the stone closes behind it and pushes the stone. This explanation derived from Aristotle's conviction that nature abhors a vacuum (which is why he placed a fifth element in the heavens).

Aristotle's ideas on falling bodies were probably first disproved by the Flemish scientist Simon Stevinus (1548–1620), who is believed to have dropped unequal weights from a height and found that they reached the ground together. At about the same time Italian physicist and astronomer Galileo (1564–1642) measured the speeds of 'falling' bodies by rolling spheres down an inclined plane and discovered the laws that govern the motion of bodies under gravity. This work was brought to a brilliant climax by English physicist and mathematician Isaac Newton (1642–1727), who in his three laws of motion achieved an understanding of force and motion, relating them to mass and recognizing the existence of inertia and momentum. Newton thus explained why a stone continues to move when thrown; and he showed the law of falling bodies to be a special case of his more general laws. Newton went on to derive, from existing knowledge of the motion and dimensions of the Earth–Moon system, a universal law of gravitation which provided a mathematical statement for the laws of planetary motion discovered empirically by the German astronomer Johann Kepler (1571–1630).

Newton's laws of motion and gravitation, which were published in 1687, were fundamental laws which sought to explain all observed effects of force and motion. This triumph of the scientific method heralded the Age of Reason – not the Greek kind of reasoning, but a belief that all could be explained by the deduction of fundamental laws upheld by observation or experiment. It was to result in an explosion of scientific discovery in physics that has continued to the present day. In the field of force and motion, important advances were made with the discovery of the law governing the pendulum and the principle of conservation of momentum by Dutch physicist and astronomer Christiaan Huygens (1629–1695) and the determination of the gravitational constant by English natural philosopher Henry Cavendish (1731–1810).

The behaviour of matter

Physics is basically concerned with matter and energy, and investigation into the behaviour of matter also originated in ancient Greece with Archimedes' work concerning flotation. As with force and motion, Simon Stevinus made the first post-Greek advance with the discovery that the pressure of a liquid depends on its depth and area. This achievement was developed by French mathematician and physicist Blaise Pascal (1623–1662), who found that pressure is transmitted throughout a liquid in a closed vessel, acting perpendicularly to the surface at any point. Pascal's principle is the basis of hydraulics. Pascal also investigated the mercury barometer invented in 1643 by the Italian physicist and mathematician Evangelista Torricelli (1608–1647) and showed that air pressure supports the mercury column and that there is a vacuum above it, thus disproving Aristotle's contention that a vacuum cannot exist. The immense pressure that the atmosphere can exert was subsequently demonstrated in several sensational experiments by German physicist Otto von Guericke (1602–1686).

Solid materials were also investigated. The fundamental law of elasticity was discovered by English physicist Robert Hooke (1635–1703) in 1678 when he found that the stress (force) exerted is proportional to the strain (elongation) produced. English physicist and physician Thomas Young (1773–1829) later showed that a given material has a constant, known as Young's modulus, that defines the strain produced by a particular stress.

The effects that occur with fluids (liquids or gases) in motion were then explored. Swiss philosopher and mathematician Daniel Bernoulli (1700–1782) established hydrodynamics with his discovery that the pressure of a fluid depends on its velocity. Bernoulli's principle explains how lift occurs and led eventually to the invention of heavier-than-air flying machines. It also looked forward to ideas of the conservation of energy and the kinetic theory of gases. Other important advances in our understanding of fluid flow were later made by Irish physicist George Stokes (1819–1903), who discovered the law that relates motion to viscosity, and Austrian physicist Ernst Mach (1838–1916) and German physicist Ludwig Prandtl (1875–1953), who investigated the flow of fluids over surfaces and made discoveries vital to aerodynamics.

The effects of light

The Greeks were aware that light rays travel in straight lines, but they believed that the rays originate in the eyes and travel to the object that is seen. Euclid (*c.* 330–260 BC), Hero (lived AD 60) and Ptolemy (lived 2nd century AD) were of this opinion although, recognizing that optics is essentially a matter of geometry, they discovered the law of reflection and investigated refraction.

Optics made an immense stride forward with the work of Arabian scientist Alhazen (*c.* 965–1038), who was probably the greatest scientist of the Middle Ages. Alhazen recognized that light rays are emitted by a luminous source and are then reflected by objects into the eyes. He studied images formed by curved mirrors and lenses and formulated the geometrical optics involved. Alhazen's discoveries took centuries to filter into Europe, where they were not surpassed until the 17th century. The refracting telescope was then invented in Holland in 1608 and quickly improved by Galileo and Kepler, and in 1621 Dutch physicist Willebrord Snell (1580–1626) discovered the laws that govern refraction.

The discovery of the spectrum

The next major steps forward were taken by Newton, who not only invented the reflecting telescope in 1668, but a couple of years earlier found that white light is split into a spectrum of colours by a prism. Newton published his work on optics in 1704, provoking great controversy with his statement that light consists of a stream of particles. Huygens had put forward the view that light consists of a wave motion, an opinion reinforced by the discovery of diffraction by Italian physicist Francesco Grimaldi (1618–1663). Such was Newton's reputation, however, that the particulate theory held sway for the following century. In 1801 Young discovered the principle of interference, which could be explained only by assuming that light consisted of waves. This was confirmed in 1821, when French physicist Augustin Fresnel (1788–1827) showed from studies of polarized light, which had been discovered by another French physicist Étienne Malus (1775–1812) in 1808, that light is made up of a transverse wave motion, not longitudinal as had previously been thought.

Newton's discovery of the spectrum remained little more than a curiosity until 1814, when German physicist and optician Joseph von Fraunhofer (1787–1826) discovered that the Sun's spectrum is crossed by the dark lines now known as Fraunhofer lines. Fraunhofer was unable to explain the lines, but he did go on to invent the diffraction grating for the production of high-quality spectra and the spectroscope to study them. An explanation of the lines was provided by the German physicist Gustav Kirchhoff (1824–1887), who in 1859 showed that they are caused by elements present in the Sun's atmosphere. With German chemist Robert Bunsen (1811–1899), Kirchhoff discovered that elements have unique spectra by which they can be identified, and several new elements were found in this way. In 1885, Swiss mathematics teacher Johann Balmer (1825–1898) derived a mathematical relationship governing the frequencies of the lines in the spectrum of hydrogen. This later proved to be a crucial piece of evidence for revolutionary theories of the structure of the atom.

Meanwhile, several scientists investigated the phenomenon of colour, notably Young, German physicist and physiologist Hermann von Helmholtz (1821–1894) and Scottish physicist James Clerk Maxwell (1831–1879). Their research led to the establishment of the three-colour theory of light, which showed that the eye responds to varying amounts of red, green and blue in light and mixes them to give particular colours. This led directly to colour photography and other methods of colour reproduction used today.

The speed of light

The velocity of light was first measured accurately in 1862 by French physicist Jean Foucault (1819–1868), who obtained a value within 1% of the correct value. This led to a famous experiment performed by German-born US physicist Albert Michelson (1852–1931) and US physicist and chemist Edward Morley (1838–1923) in which the velocity of light was measured in two directions at right angles. Their purpose was to test the theory that a medium called the ***ether*** existed to carry light waves. If it did exist, then the two values obtained would be different. The Michelson–Morley experiment, performed in 1881 and then again in 1887, yielded a negative result both times (and on every occasion since), thus proving that the ether does not exist.

More important, the Michelson–Morley experiment showed that the velocity of light is constant regardless of the motion of the observer. From this result, and from the postulate that all motion is relative, German-born US theoretical physicist Albert Einstein (1879–1955) derived the special theory of relativity in 1905. The principal conclusion of special relativity is that in a system moving relative to the observer, length, mass and time vary with the velocity. The effects become noticeable only at velocities approaching light; at slower velocities, Newton's laws hold good. Special relativity was crucial to the formulation of new ideas of atomic structure and it also led to the idea that mass and energy are equivalent, an idea used later to explain the great power of nuclear reactions. In 1915 Einstein published his general theory of relativity, in which he showed that gravity distorts space. This explained an anomaly in the motion of Mercury, which does not quite obey Newton's laws, and it was dramatically confirmed in 1919 when a solar eclipse revealed that the Sun's gravity was bending light rays coming from stars.

Electricity and magnetism

The phenomena of electricity and magnetism are believed to have been first studied by the ancient Greek philosopher Thales (624–546 BC), who was considered by the Greeks to be the founder of their science. Thales found that a piece of amber picks up light objects when rubbed, the action of rubbing thus producing a charge of static electricity. The words 'electron' and 'electricity' came from this discovery, *elektron* being the Greek word for amber. Thales also studied the similar effect on each other of pieces of lodestone, a magnetic mineral found in the region of Magnesia. It is fitting that the study of electricity and magnetism originated together, for the later discovery that they are linked was one of the most important ever made in physics.

No further progress was made, however, for nearly 2,000 years. The strange behaviour of amber remained no more than a curiosity, though magnets were used to make compasses. From this, French scientist and scholar Petrus Peregrinus (lived 13th century) discovered the existence of north and south poles in magnets and realized that they attract or repel each other. English physician and physicist William Gilbert (1544–1603) first explained the Earth's magnetism and also investigated electricity, finding other substances besides amber that produce attraction when rubbed.

Then French chemist Charles Du Fay (1698–1739) discovered that substances charged by rubbing may repel as well as attract in a similar way to magnetic poles, and US scientist Benjamin Franklin (1706–1790) proposed that positive and negative charges are produced by the excess or deficiency of electricity. French physicist Charles Coulomb (1763–1806) measured the forces produced between magnetic poles and between electric charges and found that they both the same inverse square law.

A major step forward was taken in 1800, when Italian physicist Alessandro Volta (1745–1827) invented the battery. A source of current electricity was now available and in 1820 Danish physicist Hans Oersted (1777–1851) found that an electric current produces a magnetic field. This discovery of electromagnetism was immediately taken up by English physicist Michael Faraday (1791–1867), who realized that magnetic lines of force must surround a current. This concept led him to discover the principle of the electric motor in 1821 and electromagnetic induction in 1831, the phenomenon in which a changing magnetic field produces a current. This was independently discovered by US physicist Joseph Henry (1797–1878) at the same time.

Meanwhile, important theoretical developments were taking place in the study of electricity. In 1827, French physicist André Ampère (1775–1836) discovered the laws relating magnetic force to electric current and also properly distinguished current from tension, or EMF. In the same year, German physicist Georg Ohm (1789–1854) published his famous law relating current, EMF and resistance. Kirchhoff later extended Ohm's law to networks, and he also unified static and current electricity by showing that electrostatic potential is identical to EMF.

In the 1830s, German mathematician, physicist, and philosopher Carl Gauss (1777–1855) and German physicist Wilhelm Weber (1804–1891) defined a proper system of units for magnetism; later they did the same for electricity. In 1845 Faraday found that materials are paramagnetic or diamagnetic, and Irish physicist Lord Kelvin (1824–1907) developed Faraday's work into a full theory of magnetism. An explanation of the cause of magnetism was finally achieved in 1905 by French physicist Paul Langevin (1872–1946), who ascribed it to electron motion.

Electricity and magnetism were finally brought together in a brilliant theoretical synthesis by James Clerk Maxwell. From 1855 to 1873 Maxwell developed the theory of electromagnetism to show that electric and magnetic fields are propagated in a wave motion and that light consists of such an electromagnetic radiation. Maxwell predicted that other similar electromagnetic radiations must exist and, as a result, German physicist Heinrich Hertz (1857–1894) produced radio waves in 1888. X-rays and gamma rays were discovered accidentally soon after.

The nature of heat and energy

The first step towards measurement – and therefore an understanding – of heat was taken by Galileo, who constructed the first crude thermometer in 1593. Gradually these instruments improved and in 1714 Polish-born Dutch physicist Daniel Fahrenheit (1686–1736) invented the mercury thermometer and devised the Fahrenheit scale of temperature. This was replaced in physics by the Celsius or Centigrade scale proposed by Swedish astronomer, mathematician, and physicist Anders Celsius (1701–1744) in 1742.

At this time, heat was considered to be a fluid called caloric that flowed into or out of objects as they got hotter or colder, and even after 1798 when US-born physicist Count Rumford (Benjamin Thompson; 1753–1814) showed the idea to be false by his observation of the boring of cannon, it persisted. Earlier Scottish physicist and chemist Joseph Black (1728–1799) had correctly defined the quantity of heat in a body and the latent heat and specific heat of materials, and his values had been successfully applied to the improvement of steam engines. In 1824, French physicist Sadi Carnot (1796–1832), also a believer in the caloric theory, found that the amount of work that can be produced by an engine is related only to the temperature at which it operates.

Carnot's theorem, though not invalidated by the caloric theory, suggested that, since heat gives rise to work, it was likely that heat was a form of motion, not a fluid. The idea also grew that energy may be changed from one form to another (that is from heat to motion) without a change in the total amount of energy involved. The interconvertibility of energy and the principle of the conservation of energy were established in the 1840s by several physicists. German physicist Julius Mayer (1814–1878) first formulated the principle in general terms and obtained a theoretical value for the amount of work that may be obtained by the conversion of heat (the mechanical equivalent of heat). Helmholtz gave the principle a firmer scientific basis and English physicist James Joule (1818–1889) made an accurate experimental determination of the mechanical equivalent. German theoretical physicist Rudolf Clausius (1822–1888) and Kelvin developed the theory governing heat and work, thus founding the science of thermodynamics. This enabled Kelvin to propose the absolute scale of temperature that now bears his name.

The equivalence of heat and motion led to the kinetic theory of gases, which was developed by Scottish physicist John Waterston (1811–1883), Clausius, Maxwell, and Austrian theoretical physicist Ludwig Boltzmann (1844–1906) between 1845 and 1868. It gave a theoretical description of all effects of heat in terms of the motion of molecules.

During the 19th century it also came to be understood that heat may be transmitted by a form of radiation. Pioneering theoretical work on how bodies exchange heat had been carried out by Swiss physicist Pierre Prévost (1751–1839) in 1791, and the Sun's heat radiation had been discovered to consist of infrared rays by German-born British astronomer William Herschel (1738–1822) in 1800. In 1862 Kirchhoff derived the concept of the perfect black body – one that absorbs and emits radiation at all frequencies. In 1879 Austrian physicist Josef Stefan (1835–1893) discovered the law relating the amount of energy radiated by a black body to its temperature, but physicists were unable to relate the frequency distribution of the radiation to the temperature. This increases as the temperature is raised, causing an object to glow red, yellow and then white as it gets hotter. English physicist Lord Rayleigh (1842–1919) and German physicist Wilhelm Wien (1864–1928) derived incomplete theories of this effect, and then in 1900 another German physicist Max Planck (1858–1947) showed that it could be explained only if radiation consisted of indivisible units, called quanta, whose energy was proportional to their frequency.

Planck's quantum theory revolutionized physics. It showed that heat radiation and other electromagnetic radiations including light must consist of indivisible particles of energy and not of waves as had previously been thought. In 1905 Einstein found a ready explanation of the photoelectric effect using quantum theory, and the theory was experimentally confirmed by German-born US physicist James Franck (1882–1964) in the early 1920s.

Low temperature physics

Another advance in the study of heat that took place in the same period was the production of low temperatures. In 1852 Joule and Kelvin found the effect named after them is used to produce refrigeration by adiabatic expansion of a gas, and Scottish physicist and chemist James Dewar (1842–1923) developed this effect into a practical method of liquefying gases from 1877 onwards. Dutch physicist Heike Kamerlingh-Onnes (1853–1926) first produced temperatures within a degree of absolute zero and in 1911 he discovered superconductivity. A theoretical explanation of superconductivity had to await the work of US physicists John Bardeen (1908–1991), Leon Cooper (1930–), and John Schrieffer (1931–). Their ideas, the 'BCS theory', explained superconductivity as the result of electrons coupling in pairs, called Cooper pairs, that do not undergo scattering by collision with atoms in a conductor. In 1986 IBM researchers in Zurich, Georg Bednorz (1950–) and Alex Müller (1927–), produced superconductivity in metallic ceramics at relatively high temperatures, around 35K. The theoretical explanation of high-temperature superconductivity was still being developed in the early 1990s.

In 1995 scientists at the University of Colorado discovered a new form of matter when they used lasers and magnetic fields to cool rubidium atoms to within 20 billionths of a degree of absolute zero. For a brief time, the atoms lost their individuality and behaved as if they were part of a single giant atom. This type of material is called a Bose–Einstein condensate.

Sound

Sound is the one branch of physics that was well established by the Greeks, especially by Pythagoras. They surmised, correctly, that sound does not travel through a vacuum, a contention proved experimentally by Guericke in 1650. Measurements of the velocity of sound in air were made by French physicist and philosopher Pierre Gassendi (1592–1655) and in other materials by German physicist August Kundt (1839–1894). Another German physicist, Ernst Chladni (1756–1827), studied how the vibration of surfaces produces sound waves, and in 1845 Austrian physicist Christian Doppler (1803–1853) discovered the effect relating the frequency (pitch) of sound to the relative motion of the source and observer. The Doppler effect is also produced by light and other wave motions and has proved to be particularly valuable in astronomy.

The structure of the atom

The existence of atoms was proved theoretically by chemists during the 19th century, but the first experimental demonstration of their existence and the first estimate of their dimensions was made by French physicist Jean Perrin (1870–1942) in 1909.

The principal direction taken in physics in this century has been to determine the inner structure of the atom. It began with the discovery of the electron in 1897 by English physicist J J Thomson (1856–1940), who showed that cathode rays consist of streams of minute indivisible electric particles. The charge and mass of the electron were then found by Irish mathematical physicist John Townsend (1868–1937) and US physicist Robert Millikan (1868–1953).

Meanwhile, another important discovery had been made with the detection of radioactivity by French physicist Antoine Becquerel (1852–1908) in 1896. Three kinds of radioactivity were found; these were named alpha, beta, and gamma by New Zealand-born British physicist Ernest Rutherford (1871–1937). Becquerel recognized in 1900 that beta particles are electrons. In 1903 Rutherford explained that radioactivity is caused by the breakdown of atoms. In 1908 he identified alpha particles as helium nuclei, and in association with German physicist Hans Geiger (1882–1945) produced the nuclear model of the atom in 1911, proposing that it consists of electrons orbiting a nucleus. Then in 1914 Rutherford identified the proton and in 1919 he produced the first artificial atomic disintegration by bombarding nitrogen with alpha particles.

Rutherford's pioneering elucidation of the basic structure of the atom was aided by developments in the use of X-rays, which had been discovered in 1895 by German physicist Wilhelm Röntgen (1845–1923). In 1912 German physicist Max von Laue (1879–1960) produced diffraction in X-rays by passing them through crystals, showing X-rays to be electromagnetic waves, and English physicist Lawrence Bragg (1890–1971) developed this method to determine the arrangement of atoms in crystals. His work influenced English physicist Henry Moseley (1887–1915), who in 1914 found by studying X-ray spectra that each element has a particular atomic number, equal to the number of protons in the nucleus and to the number of electrons orbiting it.

In 1913, Danish physicist Niels Bohr (1885–1962) achieved a brilliant synthesis of Rutherford's nuclear model of the atom and Planck's quantum theory. He showed that the electrons must move in orbits at particular energy levels around the nucleus. As an atom emits or absorbs radiation, it moves from one orbit to another and produces or gains a certain number of quanta of energy. In so doing the quanta give rise to particular frequencies of radiation, producing certain lines in the spectrum of the radiation. Bohr's theory was able to explain the spectral lines of hydrogen and their relationship, found earlier by Balmer.

Wave-particle theory

These discoveries, made so quickly, seemed to achieve an astonishingly complete picture of the atom, but more was to come. In 1923, French physicist Louis de Broglie (1892–1987) described how electrons could behave as if they made up waves around the nucleus. This discovery was developed into a theoretical system of wave mechanics by Austrian physicist Erwin Schrödinger (1887–1961) in 1926 and experimentally confirmed in the following year. It showed that electrons exist both as particles and waves. Furthermore it reconciled Planck's quantum theory with classical physics by indicating that electromagnetic quanta or photons, which were named and detected experimentally in X-rays by US physicist Arthur Compton (1892–1962) in 1923, could behave as waves as well as particles. A prominent figure in the study of atomic structure was German physicist Werner Heisenberg (1901–1976), who showed in 1927 that the position and momentum of the electron in the atom cannot be known precisely, but only found with a degree of probability or uncertainty. His uncertainty principle follows from wave-particle duality and it negates cause and effect, an uncomfortable idea in a science that strives to reach laws of universal application.

The next step was to investigate the nucleus. A series of discoveries of nuclear particles accompanying the proton were made, starting in 1932 with the discovery of the positron by US physicist Carl Anderson (1905–1991) and the neutron by English physicist James Chadwick (1891–1974).

This work was aided by the development of particle accelerators, beginning with the voltage multiplier built by English physicist John Cockcroft (1897–1967) and Irish physicist Ernest Walton (1903–1995), which achieved the first artificial nuclear transformation in 1932. It led to the discovery of nuclear fission by German radiochemist Otto Hahn (1879–1968) in 1939 and the production of nuclear power by Italian-born US physicist Enrico Fermi (1901–1954) in 1942.

Fusion power, the release of energy by combining lighter nuclei into heavier ones, was achieved in the first hydrogen bomb, exploded in 1952. The struggle to harness fusion energy for peaceful purposes still continues.

The development of accelerators has continued until it has produced the largest scientific instrument ever built. The Large Electron-Positron (LEP) collider at CERN, the European Laboratory for Particle Physics near Geneva, Switzerland, is a ring 27 km/16 mi in circumference. Bunches of particles are accelerated to close to the speed of light, and then circulate, perhaps for hours, in storage rings, before being allowed to collide head-on with another stream of particles circulating in the opposite direction. Early in the 21st century the LEP's tunnel will also house the Large Hadron Collider (LHC) which will accelerate protons and other particles through a trillion volts.

Particle physics

Much of modern physics has been concerned with the behaviour of elementary particles. The first major theory in this area was quantum electrodynamics (QED), developed by US physicists Richard Feynman (1918–1988) and Julian Schwinger (1918–1994), and by Japanese physicist Sin-Itiro Tomonaga (1906–1979). This theory describes the interaction of charged subatomic particles in electric and magnetic fields. It combines quantum theory and relativity and considers charged particles to interact by the exchange of photons. QED is remarkable for the accuracy of its predictions – for example, it has been used to calculate the value of some physical quantities to an accuracy of ten decimal places, a feat equivalent to calculating the distance between New York and Los Angeles to within the thickness of a hair.

By 1960 the existence of around 200 elementary particles had been established, some of which did not behave as theory predicted. They did not decay into other particles as quickly as theory predicted, for example. To explain these anomalies, US theoretical physicist Murray Gell-Mann (1929–) developed a classification for elementary particles, called the eightfold way. This scheme predicted the existence of previously undetected particles. The omega-minus particle found in 1964 confirmed the theory. In the same year Gell-Mann suggested that some elementary particles were made up of smaller particles called quarks which could have fractional electric charges. This idea explained the eightfold classification and now forms the basis of the standard model of elementary particles and their interactions.

The details of the standard model have been confirmed by experiment. In 1991 experiments at CERN, the European particle physics laboratory at Geneva, confirmed the existence of three generations of elementary particles, each with two quarks and leptons (light particles) as predicted by the standard model. In 1995 researchers at Fermilab discovered the top quark, the final piece of evidence in support of the standard model.

Quantum chromodynamics (QCD) is the mathematical theory, similar in many ways to quantum electrodynamics, which describes the interactions of quarks by the exchange of particles called gluons. The mathematics involved is very complex and although a number of successful predictions have been made, as yet the theory does not compare in accuracy with QED.

The success of the mathematical methods of QED and QCD encouraged others to use these methods to unify the theory of the fundamental forces. Pakistani theoretical physicist Abdus Salam (1926–1996) and US physicists Steven Weinberg (1933–) and Sheldon Glashow (1932–) demonstrated that at high energies the electromagnetic and weak nuclear force could be regarded as aspects of a single combined force, the electroweak force. This was confirmed in 1983 by the discovery of new particles predicted by the theory.

In the 1980s a mathematical theory called string theory was developed, in which the fundamental objects of the universe were not pointlike particles but extremely small stringlike objects. These objects exist in a universe of ten dimensions, although for reasons which are not yet understood, only three space dimensions and one time dimension are discernible. There are many unresolved difficulties, but some physicists think that string theory, or some variant of it, could develop into a 'theory of everything' that explains space-time, together with the elementary particles and their interactions, within one comprehensive framework.

The convergence of particle physics and cosmology

Physical theory, even in its incomplete state, can give an account of the history of the universe from a fraction of a second after its birth. The Big Bang took place about 15 billion years ago. Cosmologists work out what happened at each succeeding instant by using the knowledge gained in particle accelerators, where the energies involved in particle collisions momentarily equal those prevailing at various stages of the Big Bang.

At the birth of the universe space, time, matter, and energy appeared together: the first split second was a chaos of radiation and particles existing fleetingly, in a universe swelling from the size of an atom. Though the universe had a limited volume, there was nothing outside it – space curved back on itself in a way that only the equations of relativity can describe. The cosmic fireball expanded, thinned, and cooled.

Present-day theories can be applied only back to a time 10^{-43} second after the beginning: before this, time and space are ripped apart by the incredible density and temperature (10^{32}K) of the universe. There was only one 'superforce' acting, and we have no theory to deal with it yet. After this time, gravity appeared as a separate force.

At 10^{-35} second the strong force separated out. At 10^{-12} second, electromagnetism split off from the weak force. The universe was a soup of exotic particles, including single quarks, as well as high-energy gamma photons.

When the universe was a millionth of a second old, its temperature had fallen to 10 trillion K, and quarks began to combine, forming protons, neutrons, and mesons. As the temperature fell still further these were able to combine and stay together, forming various types of light nuclei. This process ceased after three minutes when the expanding plasma had thinned and cooled too much. It was a few hundred thousand years later that fell into orbits around the nuclei to form atoms.

Physics in the 21st century

Beyond the orbit of the Moon, the Next Generation Space Telescope (NGST) will probe the early universe, and its results are bound to pose new problems for physics. Satellite systems using laser beams 5 million km/3 million mi
long may be set in space to search for gravity waves – ripples in space-time from violent events in the universe, such as the explosion of stars or the merging of black holes.

With such equipment physicists will address the many problems still unsolved at the turn of the century. For example, are matter and antimatter true 'mirror-images' of one another? Experimental results suggesting that they behave slightly differently in certain interactions were reported in 1999. Such a difference could explain why antimatter had almost completely disappeared from the universe within fractions of a second of the Big Bang, despite the fact that matter and antimatter should have been produced in equal amounts initially.

But physicists are sure that the matter we can directly observe is only a fraction of all the matter that exists. Some of the unknown 'dark matter' reveals its presence by its gravitational effects on the movements of galaxies. Some of it may be just cold, dark ordinary matter, mostly hydrogen gas. But physicists conjecture that it may consist of an undiscovered form of matter, WIMPs, or weakly interacting massive particles. These would interact with each other and with ordinary matter by gravitation alone. The search is on for such matter, both in the laboratory and in space.

Evidence was reported in 1998 that the expansion of the universe is accelerating. According to conventional physics the gravitational pull of the galaxies on each other should slow down the expansion. But if the new results are right, then there is a repulsive force between masses that increases with distance, and which is driving the galaxies apart ever faster.

Even without such phenomena to complicate things, the standard model of elementary particle interactions has many difficulties. For example, many quantities, such as the masses of the particles and the strengths of the interactions, are unexplained. The search is on for the Higgs boson, a particle whose existence was suggested by the British theoretical physicist Peter Higgs (1929–), as well as physicists in Belgium. If the Higgs boson is found it will explain how the masses of particles arise. The fact that elementary particles fall into three generations is another mystery that physicists hope to solve soon.

Biographies

(Abbe to Leavitt)

Abbe, Ernst (1840–1905) German physicist who, working with German optician Carl Zeiss (1816–1888), greatly improved the design and quality of optical instruments, particularly the compound microscope. He indirectly had a great influence in various physical sciences, particularly in biology where the improved resolving power of his instruments permitted researchers to observe microorganisms and internal cellular structures for the first time.

Abbe was born in Eisenach, Thuringia, Germany, on 23 January 1840, the son of a spinning-mill worker. On a scholarship provided by his father's employers he attended high school (graduating in 1857) and then went on to study physics at the University of Jena. He gained his doctorate from Göttingen University in 1861 and three years later became a lecturer in mathematics, physics, and astronomy at Jena, being appointed professor in 1870. In 1866 he began his association with Carl Zeiss, an instrument manufacturer who supplied optical instruments to the university and repaired them. Abbe was appointed director of the Astronomical and Meteorological Observatory at Jena in 1878. Two years earlier he had become a partner in Zeiss' firm, and in 1881 Abbe invited Otto Schott (1851–1935), who had studied the chemistry of glasses and manufactured them, to go to Jena, and the famous company of Schott and Sons was founded in 1884. On the death of Zeiss in 1888 Abbe became the sole owner of the Zeiss works. He established the Carl Zeiss Foundation in 1891, and in 1896 he formalized the association between the Zeiss works and Jena University by making the company a cooperative, with the profits being shared between the workers and the university. He died in Jena on 14 January 1905.

The success of the Jena enterprise arose largely from the right combination of talents: Zeiss as the manufacturer; Abbe as the physicist/theoretician, who performed the mathematical calculations for designing new lenses; and Schott the chemist, who formulated and made the special glasses needed by Abbe's designs. Abbe worked out why, contrary to expectation, a microscope's definition decreases with a reduction in the aperture of the objective; he found that the loss in resolving power is a diffraction effect. He calculated how to overcome spherical aberration in lenses – by a combination of geometry and the correct types of glass. He also explained the phenomenon of coma – first recognized in 1830 by Joseph Jackson Lister (1786–1869), the father of the English surgeon – in which even a corrected lens displays aberration when the object is slightly off the instrument's axis. It was overcome by applying Abbe's 'sine condition', producing the so-called aplanatic lens. Finally Abbe calculated how to correct chromatic aberration, using Schott's special glasses and, later, fluorite to make microscope objective lenses, culminating in the apochromatic lens system of 1886.

In 1872 he developed the Abbe substage condenser for illuminating objects under high-power magnification. Among his other inventions were a crystal refractometer and, developed with French physicist Armand Fizeau, an optical dilatometer for measuring the thermal expansion of solids.

Abel, Neils Henrik (1802–1829) Norwegian mathematician who, in a very brief career, became the first to demonstrate that an algebraic solution of the general equation of the fifth degree is impossible.

Abel was born in Finnöy, a small island near Stavanger, on 5 August 1802. He was educated by his father, a Lutheran minister, until the age of twelve, when he was enrolled in the cathedral school at Christiania (Oslo). There his flair for mathematics received little encouragement until he came under the tutelage of Bernt Holmboe in 1817. Holmboe put Abel in touch with Swiss mathematician Leonhard Euler's calculus texts and introduced him to the work ofFrench mathematicians Joseph Lagrange and Pierre Laplace. Abel's imagination was fired by algebraic equations theory and by the time that he left school in 1821, to enter the University of Oslo, he had become familiar with most of the body of mathematical literature then known. In particular he had, during his last year at school, began to work on the baffling problem of the quintic equation, or general equation of the fifth degree, unsolved since it had been taken up by Italian mathematicians early in the 16th century.

Because of his father's death in 1820, Abel arrived at the university virtually penniless. Fortunately, his talent was apparent, and he was given free rooms and financial support by the university. Since the university offered no courses in advanced mathematics, most of Abel's research was done on his own initiative. In 1823 he published his first paper. It was an unimportant discussion of functional equations, but another paper published in that year heralded the arrival of a highly original mind in the world of mathematics, although it went unregarded at the time. In it Abel provided the first solution in the history of mathematics of an integral equation. All the while he remained obsessed by the problem of the quintic equation. During his last year at school he had sent the Danish mathematician Ferdinand Deger his 'solution' to the problem, only to receive from Deger the advice to abandon that 'sterile' question and turn his mind to elliptic transcendentals (elliptic integrals). Deger was kind enough, even so, to ask Abel for examples of his solution and this request proved to be fruitful. For when Abel began to construct examples he discovered that it was no solution at all. He therefore wrote a paper demonstrating that a radical expression to represent a solution to fifth- or higher-degree equations was impossible. After three centuries a niggling question had been resolved. Yet when Abel sent his demonstration to the German mathematician Karl Gauss he received no reply. Nor was anyone else much interested and Abel was forced to publish the paper himself.

In 1825, taking advantage of a government grant to enable scholars to study foreign languages abroad, Abel went to Berlin. There he met Leopold Crelle, a privy councillor and engineer much taken with problems in mathematics. Together they brought out the first issue of *Crelle's Journal,* which was to become the leading 19th-century German organ of mathematics. (The first issue consisted almost entirely of seven papers by Abel.) A year later Abel moved on to Paris, where he wrote his famous paper 'Mémoire sur une propriété générale d'une classe très-étendue de fonctions transcendantes'. It dealt with the sum of the integrals of a given algebraic function and presented the theorem that any such sum can be expressed as a fixed number of these integrals with integration arguments that are algebraic functions of the original arguments. Abel sent the manuscript to the French Academy of Sciences and was deeply disappointed when the referees – who alleged that the manuscript was illegible! – did not publish it. He returned to Berlin a disheartened man. Low in funds and unable to get a post at the university there, he accepted Crelle's offer to edit the journal. In 1827 he published the longest paper of his career, the 'Recherche sur les fonctions elliptiques'. He also suffered his first attack of the tuberculosis that was to kill him. At the end of the year he returned to Norway, where he lived in gradually deteriorating health until his death, at Froland, on 6 April 1829.

Abel, in addition to this work on quintic equations, transformed the theory of elliptic integrals by introducing elliptic functions, and this generalization of trigonometric functions became one of the favourite topics of 19th-century mathematics. It led eventually to the theory of complex multiplication, with its important implications for algebraic number theory. He also provided the first stringent proof of the binomial theorem. A number of useful concepts in modern mathematics, notably the Abelian group and the Abelian function, bear his name. Yet it was only after his death that his achievement was publicly acknowledged. In 1830 the French Academy awarded him the Grand Prix, which he shared with Karl ⟡Jacobi, the German mathematician who had (independently) made important discoveries about elliptic functions. And 11 years later the academy finally came round to publishing the 'illegible' manuscript of 1826.

Abetti, Giorgio (1882–1982) Italian astrophysicist best known for his studies of the Sun.

Abetti was born in Padua on 5 October 1882. He studied at the universities of Padua and Rome and earned a PhD in the physical sciences. In 1921 he was appointed professor at the University of Florence, where he remained until his retirement in 1957. From 1921 until 1952 he was director of the Arcetri Observatory in Florence. During his tenure at Florence, Abetti travelled to Cairo in 1948 to serve as a visiting professor at the university there, and he toured the USA in 1950.

Abetti's research contributions quickly earned him a prominent and respected position among Italian scientists. He was awarded the Silver Medal of the Italian Geographical Society in 1915, the Reale Prize of the Academy of Lincei in 1926, and the Janssen Gold Medal of the Ministry of Public Instruction in 1937. He was a member of the Socio Nazionale, the Academy of Lincei in Rome, and the Royal Society of Edinburgh and the Royal Astronomical Society in the UK.

Abetti's research was in the field of astrophysics, with particular emphasis on the Sun. He participated in numerous expeditions to observe eclipses of the Sun, and led one such expedition to Siberia to observe the total solar eclipse of 19 June 1936. He was well known for his influential popular text on the Sun, and he wrote a handbook of astrophysics, published in 1936, and a popular history of astronomy, which appeared in 1963.

Adams, John Couch (1819–1892) English astronomer who was particularly skilled mathematically. His ability to deal adeptly with complex calculations helped him to discover, independently of French astronomer Urbain ⟡Leverrier, the planet Neptune.

Adams was born in Landeast, Cornwall, on 5 June 1819. His mathematical talents and interest in astronomy were apparent from an early age, and in 1839 he won a scholarship to Cambridge University. He graduated with top honours in 1843 and took up a fellowship at St John's College. When this lapsed in 1853 he was given a life fellowship at Pembroke College. St Andrew's University in Aberdeen appointed Adams to the chair of mathematics in 1858, but he returned to Cambridge a year later to become Lowdean Professor of Astronomy and Geometry, a post he held until his death.

The initial public acclaim for the discovery of Neptune went to Leverrier, but Adams nevertheless received many honours. He was awarded the Royal Society's Copley Medal (its highest honour) in 1848 and was elected Fellow of the Royal Society a year later. He was made a member of the Royal Astronomical Society and served it twice as president. He was awarded its Gold Medal for his later research into lunar theory. He succeeded James Challis as director of the Cambridge Observatory, but – always a modest man – he declined the offer of a knighthood from Queen Victoria, and also turned down the position of Astronomer Royal, pleading old age. He died in Cambridge on 21 January 1892.

The planet Uranus had been discovered by William Herschel in 1781. Its path was carefully studied during the first orbit after discovery, and it soon became clear that early predictions of the motions of Uranus were incorrect. On the basis of Isaac Newton's gravitational theory, certain aberrations in the orbit were accounted for as the result of perturbations caused by Jupiter and Saturn, but these were insufficient to explain the magnitude of Uranus' deviation from its predicted orbital path. This suggested that either the gravitational theory was incorrect or that an as-yet-undetected planet lay beyond the orbit of Uranus. The mathematical calculations necessary to solve the mystery were taken up independently by Adams and Leverrier. Adams had become interested in this problem while still an undergraduate, but it was not until he had completed his studies in 1843 that he had the time to focus his full attention on it. By 1845 he had determined the position and certain characteristics of this hypothetical planet. He attempted to convey the information to the new Astronomer Royal, George Airy, but the significance of Adams' findings was not fully appreciated. A search for the new planet was not instigated for nearly a year and was carried out by James Challis at Cambridge.

Meanwhile, in France, Leverrier had followed the same lines of thought as Adams and had sent his figure to Johann Galle at the Berlin Observatory. It was Leverrier's good fortune that Galle had just received a new and improved map of the sector of the sky in which the planet could be located. As a result, Galle was able to find the planet, which was later named Neptune, within a few hours of beginning his search on 25 September 1846. It later transpired that Challis had observed the new planet on a number of occasions, but had failed to recognize that it was new because of his inferior maps. The discovery of Neptune was credited to Leverrier, although not without much nationalistic acrimony on both sides of the Channel.

Adams' later work included research into lunar theory and terrestrial magnetism, as well as observations of the Leonid meteor shower. Later, he improved the findings of French astronomer Pierre ◊Laplace. This resulted in a reduction of 50% in the then current value for the secular acceleration of the Moon's mean motion.

Adams' contributions to observational astronomy, as well as his improvements to the accuracy of many mathematical constants, made him deeply respected – not only for the value of his work, but also for his modest attitude towards his achievements.

Adams, Walter Sydney (1876–1956) US astronomer who was particularly interested in stellar motion and luminosity. He developed spectroscopy as a valuable tool in the study of stars and planets.

Adams was born on 20 December 1876 at Antioch, Syria, where his parents were serving as missionaries. His early education was provided by his parents, who taught him much about ancient history and classical languages. In 1885 his parents returned to the USA for the sake of their children's education. When Adams entered Dartmouth College, Massachusetts, he had to choose between his love of classics and mathematical sciences. He graduated in 1898 and went to the University of Chicago for his postgraduate studies.

He studied celestial mechanics, publishing a paper during his first year on the polar compression of Jupiter. The next year was spent under George ◊Hale at the Yerkes Observatory, where Adams made a number of studies including one on the measurement of radical velocity. He went to Munich, Germany, the following year. Hale then invited him to return to Yerkes, which he did, and Adams spent the next three years working on stellar spectroscopy.

In 1904 he assisted Hale in the establishment of the Mount Wilson Observatory above Pasadena in California. Mount Wilson gradually became a renowned research centre. Adams served as deputy director, under Hale, from 1913 to 1923, when he took over as director. Adams was a member of many scientific organizations in both the USA and Europe. He was honoured for his achievements by being elected president of the American Astronomical Society in 1931. He died on 11 May 1956.

Adams' early work on radial velocities had used the 100-cm/40-in refractor at Yerkes, at the time the largest in the world. At Mount Wilson he was able to use larger and more sophisticated equipment. The first area that Adams investigated at Mount Wilson was the spectra obtained from sunspots as compared to those obtained from the rest of the solar disc and from laboratory sources. He found that the temperature, pressure, and density of a source affects the relative intensities of its spectral lines. This and other information enabled him to demonstrate that sunspots have a lower temperature than the rest of the solar disc. Adams also studied solar rotation by means of Doppler displacements.

In 1914 Adams turned to the spectroscopy of other stars. He found that luminosity and the relative intensities of particular spectral lines could distinguish giant stars from dwarf stars. Spectra could also be used to study the physical properties, motions, and distances of stars. This use of the intensity of spectral lines to determine the distance of stars has been termed spectroscopic parallax.

Adams was involved in a long-term project with other astronomers to determine the absolute magnitudes of stars; they found the value for 6,000 stars. A second long-term collaborative project was the determination of the radial velocities of more than 7,000 stars. This work led to an improved understanding of the behaviour and evolution of stars.

In 1915 Adams made a spectroscopic study of the small companion star of Sirius B. He identified it as a white dwarf containing about 80% of the mass of the Sun in a volume approximately the same as that of the Earth and thus having a density more than 40,000 times that of water. Adams demonstrated that the companion star was hotter than our Sun and not cold, as everyone had assumed. In 1920 Arthur Eddington in the UK suggested that if Sirius B were indeed so dense it would produce a powerful gravitational field and show a red shift (as predicted by Albert Einstein's general theory of relativity). In 1925 Adams reported a displacement of 21 km/13 mi per second, thus corroborating Einstein's theory.

During the 1920s and 1930s Adams studied the atmospheres of Mars and Venus, reporting in 1932 the presence of carbon dioxide in the atmosphere of Venus and, in 1934, the occurrence of oxygen in concentrations of less than 0.1% on Mars. He was involved in many other research projects, and he also made an important contribution in his capacity as director of the Mount Wilson Observatory. He was responsible for the design and installation of the 254-cm/100-in and 500-cm/200-in telescopes at Mount Wilson and Palomar. Adams was a fine scholar and administrator. Astronomy matured as a science during his active research years, a development to which he was an important contributor.

Addison, Thomas (1793–1860) English physician and endocrinologist who was the first to correlate a collection of symptoms with pathological changes in an endocrine gland. He described a metabolic disorder caused by a deficiency in the secretion of hormones from the adrenal glands (caused, in turn, by atrophy of the adrenal cortex), a condition now called Addison's disease. He is also known for his discovery of what is now called pernicious (or Addison's) anaemia.

Addison was born in April 1793 in Longbenton, Northumberland, and studied medicine at Edinburgh University, graduating in 1815. He then moved to London, where he was appointed a surgeon at the Lock Hospital. He also studied dermatology under Thomas Bateman (1778–1821) during his first years in London. In about 1820 Addison entered Guy's Hospital as a student, despite being a fully qualified physician, and remained there in various positions for the rest of his life, becoming assistant physician in 1824, lecturer in Materia Medica in 1827, and a full physician in 1837. While at Guy's Hospital, Addison collaborated with Richard ◊Bright, who also made important contributions to medicine. Addison's mental health deteriorated and he committed suicide on 29 June 1860 in Brighton, Sussex.

Addison gave a preliminary account of the condition now known as Addison's disease in 1849 in a paper entitled 'On anaemia: disease of the suprarenal capsules', which he read to the South London Medical Society. The paper went unnoticed, despite which Addison extended his original account in *On the Constitutional and Local Effects of Disease of the Suprarenal Capsules* (1855), in which he gave a full description of Addison's disease (characterized by abnormal darkening of the skin, progressive anaemia, weakness, intestinal disturbances, and weight loss) and differentiated it from pernicious anaemia (characterized by anaemia, intestinal disturbances, weakness, and tingling and numbness in the extremities). He also pointed out that Addison's disease is caused by atrophy of the suprarenal capsules (later called the adrenal glands). (Pernicious anaemia is caused by inadequate absorption of vitamin B_{12}.)

Addison also described xanthoma, (flat, soft spots that appear on the skin, usually on the eyelids) and wrote about other skin diseases, tuberculosis, pneumonia, and the anatomy of the lung. In collaboration with John Morgan (1797–1847), he wrote *An Essay on the Operation of Poisonous Agents Upon the Living Body* (1829), the first work on this subject to be published in English. In 1839 the first volume of *Elements of the Practice of Medicine*, written by Addison and Richard Bright was published. In this volume (which was, in fact, written almost entirely by Addison – Bright was to have been the principal contributor to the second volume, which was never published) Addison gave the first full description of appendicitis.

Adler, Alfred (1870–1937) Austrian psychiatrist who broke away from the theories of Sigmund ◊Freud, setting up the Individual Psychology Movement. He placed 'inferiority feeling' at the centre of his theory of neuroses.

Adler was born in Vienna on 7 February 1870, the son of a corn merchant. He obtained his MD from the University of Vienna in 1895 and worked for two years as a physician at Vienna General Hospital. His interests soon turned towards mental disorders, and by 1902 he had made contact with Freud. He played a major part in the development of the psychoanalytical movement, and was president of the Vienna Psychoanalytical Society. But by 1907 he had shifted his theory away from Freud's emphasis on infantile sexuality towards power as the origin of neuroses; in 1911 Adler, and a number of others, left the Freudian circle and founded the Individual Psychology Movement. By the late 1920s Adler was making many trips to the USA, where he proved to be a popular lecturer; in 1927 he became a visiting professor at Columbia University. In 1935 he decided to make the USA his permanent home and he became professor of psychiatry at the Long Island College of Medicine, New York. He died from a heart attack on 28 May 1937 in Aberdeen, Scotland, during a lecture tour.

The essence of Adler's theories differed from Freud's in that he thought that power not sex was the important factor in neurotic disorders. He popularized the term 'inferiority feeling' – later changed to 'inferiority complex' – and felt that much neurotic behaviour is a result of feelings of inadequacy or inferiority caused by, for instance, being the youngest in a family or being a child who is trying to compete in an adult world. In an attempt to overcome these inferiority feelings the patient overcompensates, often at the expense of normal social behaviour or, as Adler put it, 'social interest'. Adler's belief led on to his idea that a person can realize this ambition alone, which has consequences on the way in which a psychiatrist helps a patient if help is needed. His impact was less forceful than those of Carl Jung or Freud, and even though his psychology made good sense it lacked adequate definition and rigour of method. Adler summarized his theories in *Practice and Theory of Individual Psychology* (1927).

Adler's more practical work included the setting up of a system of child-guidance services in the schools in Vienna, which lasted until 1934, when they were closed by the Austrian Fascist government.

Adrian, Edgar Douglas (1889–1977) 1st Baron Adrian of Cambridge.

English physiologist known for his experimental research in electrophysiology and, in particular, nerve impulses. He was one of the first scientists to study the variations in electrical potential of nerve impulses amplified by thermionic valve amplifiers, and was also one of the first to study the electrical activity of the brain. He shared the 1932 Nobel Prize for Physiology or Medicine with Charles Sherrington for his work on the function of nerve cells.

Adrian was born in London on 30 November 1889, the son of a lawyer. After attending Westminster School, he won a scholarship to Trinity College, Cambridge, where he studied natural sciences. During World War I he went to St Bartholomew's Hospital in London to study medicine, graduating in 1915 and then working on nerve injuries and shell shock (battle fatigue) at Queen's Square and later at the Connaught Military Hospital in Aldershot. Adrian was much in demand as a lecturer from 1919 – when he returned to Cambridge University – especially on such subjects as sleep, dreams, hysteria, and multiple sclerosis. From 1937 until 1951 he was professor of physiology at Cambridge. He was awarded the Order of Merit in 1942 and was president of the Royal Society 1950–55. In that same year he was raised to the peerage and took the barony of Cambridge. He was master of Trinity College, Cambridge, from 1951 until his retirement in 1965. He died in London on 4 August 1977.

In 1912 nothing was known about electrochemical transmission within the nervous system. Adrian's most ambitious work in the pre-war years was to attempt to prove that the intensity of a nerve impulse at any point in a normal nerve is independent of the stimulus or of any change in intensity that might have occurred elsewhere.

The mechanism of muscular control was a subject of clinical interest in wartime, and Adrian studied and wrote on the electrical excitation of normal and denervated muscle. He showed that, with normal muscle, the time factor in excitation – known as the chronaxie – is very short and that it increases by a factor of 100 in denervated muscle after the nerve endings have degenerated. With L R Yealland of Queen's Square he worked on the application of a method of treatment based on suggestion, re-education, and discipline in cases of psychosomatic disorders.

Between 1925 and 1933 Adrian successfully recorded trains of nerve impulses travelling in single sensory- or motor-nerve fibres. This work was a turning point in the history of physiology. He began to use valve amplifiers and found that in a single nerve fibre the electrical impulse does not change with the nature or strength of the stimulus. He also discovered that some sense organs, such as those concerned with touch, rapidly adapt to a steady stimulus whereas others, such as muscle spindles, adapt slowly or not at all. His work at this time included the recording of optic-nerve impulses in the conger eel, investigations of the action of light on frogs' eyes, and researching the problem of pain and the responses of animals to speech.

Between 1933 and 1946 he worked on the ways in which the nervous system generates rhythmic electrical activity. He was one of the first scientists to use extensively the recently devised electroencephalograph (EEG) – a system of recording brain waves – to study the electrical activity of the brain. This system has since proved an invaluable diagnostic aid – for example, in the diagnosis of epilepsy and the location of cerebral lesions. The last years of his research life, from 1937 to 1959, were spent studying the sense of smell.

In 1932 Adrian published *The Mechanism of Nervous Action* and in 1947 *The Physical Background of Perception,* based on the Waynflete Lectures he had given in Oxford the preceding year.

Agassiz, (Jean) Louis Rodolphe (1807–1873) Swiss palaeontologist who developed the idea of the ice age.

The son of a Protestant pastor, Agassiz was born in Motier. He received his medical and scientific training at Zürich, Heidelberg, Munich, and Paris, where he fell under the spell of Georges ◊Cuvier and embraced his pioneering application of the techniques of comparative anatomy to palaeontology. His momentous *Researches on Fossil Fish* (1833–44) used Cuvierian comparative anatomy to describe and classify over 1,700 species; ichthyology was his preferred specialty. Travelling in 1836 in his native Alps, he developed the novel idea that glaciers, far from being static, were in a constant state of almost imperceptible motion. Finding rocks that had been shifted or abraded, presumably by glaciers, he inferred that in earlier times much of northern Europe had been covered with ice sheets. Agassiz's *Studies on Glaciers* (1840) developed the original concept of the ice age, which he viewed as a cause of extinction, demarcating past flora and fauna from those of the present. Agassiz's geological principles thus entailed a mode of catastrophism, fundamentally opposed to the extreme uniformitarianism promoted by Scottish geologist Charles ◊Lyell.

Agassiz rose to become professor at Harvard in the USA in 1847, where he worked for the rest of his career, founding the Museum of Comparative Zoology and building a huge collection (a quite obsessional collector, his accumulating habits ruined his purse and wrecked his marriage). His last major project was *Contributions to the Natural History of the United States* (1857–62), an exhaustive study of the natural environment of the USA.

Like his mentor Cuvier, Agassiz was always anti-evolutionist. Adducing religious, philosophical, and palaeontological arguments, he proved one of the staunchest adversaries of Darwin's theory of descent by natural selection.

Agnesi, Maria Gaetana (1718–1799) Italian mathematician who is noted for her work in differential calculus. Agnesi's solution to a curve, in English mistakenly called the 'witch of Agnesi', is still to be found in textbooks.

Agnesi Italian mathematician Maria Agnesi. *Mary Evans Picture Library*

Agnesi was born on 16 May 1718 in Milan (then in the Habsburg Empire). She was the first of 21 children in a wealthy, literate family. Her father, Pietro Agnesi, is thought to have been a professor of mathematics and Maria was a child prodigy who spoke French at the age of five and by nine had mastered Latin, Greek, German, Spanish, and Hebrew, resulting in her being known as the 'oracle of the seven tongues'. At the age of nine she also delivered, and later published, an hour-long discourse in Latin on women's rights to higher education to a learned gathering. In her teenage years Agnesi ran the household, mastered mathematics, taught her younger brothers, and was hostess at her father's gatherings of distinguished academics.

In 1738 Agnesi published a collection of essays on natural science and philosophy called *Propositiones Philosophicae.* Many of these essays were based on the discussions she had heard in her home and in them she often made mention of her passionate belief in the education of women.

Her most important work, *Istituzioni Analitiche/Analytical Institutions*, began – it is believed – as a textbook for her brothers; however, it grew into a book that caused a sensation in academic circles when its first volume was published in 1748. In it, Agnesi systematically brought together and analysed the works of various mathematicians writing in different languages. The book was divided into four parts – the first dealt with the analysis of finite quantities and the problems of maxima, minima, tangents, and inflection points; the second discussed infinitely small quantities; the third was about integral calculus; and the final part was about the inverse method of tangents and differential equations.

The so-called 'witch of Agnesi', a minor part of the book but the one for which Agnesi became best known, is the name given to a bell-shaped curve whose equation she calculated:

$$y = \frac{a^3}{x^2 + a^2}$$

The curve was originally called a *versiera* from the Latin *vertere* 'to turn', but this became corrupted to the Italian *avversiera* and translated as 'wife of the devil'. Agnesi was herself known in English as the 'witch of Agnesi' for many years.

Agnesi's book was very well received in mathematical circles. The Académie Française admired her work but did not admit her. However, she was elected to the Bologna Academy of Sciences and, in 1750, Pope Benedict XIV had her appointed professor of higher mathematics at the University of Bologna. It is a matter of debate whether she took up these posts but it is thought possible that she worked there until her father's death to please him and that all her mathematical work had this impetus. Certainly when he died she gave up mathematics altogether.

Agnesi had always been a shy and retiring individual and it is thought that she would rather have been in a convent than attending her father's soirées. She was very religious and after her father's death she converted the house into a hospital, devoting herself to the sick and homeless, especially women. She was asked by the Archbishop to take charge of the Pio Instituto Trivulzio for the ill and infirm when it was opened in 1771, and worked there until her death at the age of 81.

Agricola, Georgius (1494–1555) Latinized form of Georg Bauer. German mineralogist who pioneered mining technology.

Born in Glauchau in Saxony, Agricola trained in medicine, first in Leipzig and later in Italy. He served for many years as town physician in Joachimstal. Involvement with the medicinal use of minerals sparked his curiosity about the products of the Earth, and he soon developed an interest in the local mining ventures. Agricola quickly made himself an authority on mining, metal extraction, smelting, assaying, and related chemical processes. His *The Nature of Fossils* (1546) advances one of the first comprehensive classifications of minerals. Familiar with previous writers on mining like Bermannus, Agricola went on to explore the origins of rocks, mountains, and volcanoes. His most renowned work, *On Metals/De Re Metallica* (1556) is an indispensable survey, lavishly illustrated with woodcuts, of the smelting and chemical technology of the time and of the state of the mining industry. Drawing intelligently upon the *Pirotechnia* of the Italian Vannucio Biringuccio, Agricola's work became a standard text, being translated into Italian and German.

Aiken, Howard Hathaway (1900–1973) US computer and data-processing pioneer who invented the Harvard Mark I and Harvard Mark II computers, the prototypes of modern digital computers.

Aiken was born in Hoboken, New Jersey, on 9 March 1900. He studied engineering at the University of Wisconsin and graduated in 1923. He then took a job with the Madison Gas and Electric Company, where he remained until 1927, when he went to Chicago to work for the Westinghouse Electric Manufacturing Company. In 1931 he left Westinghouse to take up a research post in the department of physics at the University of Chicago.

The rest of the decade he spent in research, both at Chicago and at Harvard. He received his PhD from Harvard in 1939 and was appointed an instructor in physics and communication engineering. He quickly rose to become a full professor in applied mathematics and remained at Harvard until 1961, when he was appointed professor of information technology at the University of Miami. He died at St Louis, Missouri, on 14 March 1973.

When Aiken began his research into computer technology in the 1930s the subject was still in its infancy. Simple, manual calculating machines had been in use since the mid-17th century, but they were too elementary and too slow to meet the military and industrial requirements of the 20th century. It was, indeed, the US navy that started Aiken on the career for which he became world famous. His early research at Harvard was sponsored by the Navy Board of Ordnance and in 1939 he and three other engineers from the International Business Machines (IBM) Corporation were placed under contract by the navy to develop a machine capable of performing both the four basic operations of addition, subtraction, multiplication, and division and also referring to stored, tabulated results.

Aiken played the central role in the development for the navy of the first program-controlled calculator in the world – the Automatic Sequence Controlled Calculator, or Harvard Mark I, which was completed in 1944. It was principally a mechanical device, although it had a few electronic features; it was 15 m/49 ft long, 2.5 m/8 ft high and weighed more than 30 tonnes. Addition took 0.3 seconds, multiplication 4 seconds. It was able to manipulate numbers with up to 23 decimal places and to store 72 of them. Information was fed into the machine by tape or punched cards and output was produced in a similar form. The calculator's chief functions were to produce mathematical tables and to assist the ballistics and gunnery divisions of the military.

On the completion of the Mark I, Aiken was posted to the Naval Proving Ground at Dahlgren, Virginia, to begin working on improving his invention. There the Mark II was completed in 1947. It was a fully electronic machine containing 13,000 electronic relays. It was also much faster than its predecessor, requiring only 0.2 seconds for addition and 0.7 seconds for multiplication. Moreover, it could store 100 10-digit figures and their signs.

For his great achievements, sufficient to earn him the name of the pioneer of modern computers, Aiken was given the rank of commander in the US Navy Research Department.

Airy, George Biddell (1801–1892) English astronomer who, as Astronomer Royal for 46 years, was responsible for greatly simplifying the systematization of astronomical observations and for expanding and improving the Royal Observatory at Greenwich.

Airy was born in Alnwick, Northumberland, on 27 July 1801. His father, a collector of taxes and excise duties, was periodically transferred from one part of the country to another, with the result that his son was educated in a number of places. From 1814 to 1819 he attended Colchester Grammar School, where he was noted for his incredible memory (on one occasion he recited from memory 2,394 lines of Latin verse). In 1819 he became a student at Trinity College, Cambridge, and three years later took a scholarship there. He graduated in 1823 at the top of his class in mathematics. The following year he was elected a fellow of Trinity College and became an assistant tutor in mathematics. The physics of light and optics began to interest him and he was the first to describe the defect of vision – later termed astigmatism – from which he also suffered. In 1826 he became professor of mathematics at Cambridge and in the same year, having become interested in astronomy, published *Mathematical Tracts on Physical Astronomy,* which became a standard work. He was elected professor of astronomy and director of the Cambridge Observatory in 1828, and was then appointed Astronomer Royal in 1835, a post that he held until 1881. During this period he sat on many commissions and supervised the cataloguing of geographical boundaries. He was awarded the Copley and Royal medals by the Royal Society and was its president from 1827 to 1873. He was five times president of the Royal Astronomical Society, twice receiving its Gold Medal, and received various honorary degrees. He died in Greenwich, London, on 2 January 1892.

While Airy was director of the Cambridge Observatory, it flourished under his control; he introduced a much improved system of meridian observations and set the example of reducing them in scale before publishing them. As Astronomer Royal, Airy had the Royal Observatory at Greenwich re-equipped and many innovations were made. He supervised the gigantic task of reducing in scale all the planetary and lunar observations made at Greenwich between 1750 and 1830. In 1847 he had erected the alt-azimuth (an instrument he devised to calculate altitude and azimuth) for observing the Moon in every part of the sky. Airy also introduced new departments to the Observatory; in 1838 he created one for magnetic and meteorological data and, in 1840, a system of regular two-hourly observations was begun. Other innovations included photographic registration in 1848, transits timed by electricity in 1854, spectroscopic observations from 1868, and a daily round of sunspot observation using the Kew heliograph in 1873.

Airy's skills in mathematics were called upon in the exact mapping of geographical boundaries: he was responsible for establishing the border between Canada and the USA and later of the Oregon and Maine boundaries. He also established exact determinations of the longitudes of Valencia, Cambridge, Edinburgh, Brussels, and Paris. Airy's scientific expertise was also utilized on during the launch of the *SS Great Eastern,* the laying of the transatlantic telegraph cable, and the construction of the chimes of the clock in the tower at Westminster ('Big Ben'). During 1854 he supervised several experiments in Harton Colliery, South Shields, to measure the change in the force of gravity with distance below the Earth's surface.

Throughout all his additional duties Airy never allowed his work with the Royal Observatory to suffer and it was due to his enthusiasm and hard work that the Greenwich Observatory grew in importance both nationally and internationally.

Aitken, Robert Grant (1864–1951) US astronomer whose primary contribution to astronomy was the discovery and observation of thousands of double stars.

Aitken was born in Jackson, California, on 31 December 1864. He took his degree at Williams

College, where he forsook his earlier plans to enter the ministry for his interest in astronomy. He taught at Livermore College from 1888 until 1891, when he was made professor of mathematics at the University of the Pacific. From 1895 onwards he worked at the Lick Observatory on Mount Hamilton, first as assistant astronomer and ultimately as director of the Observatory from 1930 until his retirement in 1935.

Aitken's work on binary systems brought him widespread recognition and many honours. He was a member of numerous professional bodies, often holding positions of responsibility within them. These included the chairmanship (from 1929 to 1932) of the astronomy section of the National Academy of Sciences. He died in Berkeley, California, on 29 October 1951.

At first, Aitken's research at Lick was in many fields, but his interest soon focused on double stars. He began a mammoth survey of double stars in 1899 and this was not finished until 1915. During the early years of the project he was assisted by W J Hussey, and between them they discovered nearly 4,500 new binary systems. Their primary tool was the 91-cm/36-in refractor. Aitken then began a thorough statistical examination of this vast amount of information, which he published first in 1918 and then revised in 1935. His work lay not merely in the discovery of new binary stars, but also in determining their motions and orbits.

Aitken's other famous contribution was his revision of S W Burnham's catalogue of double stars, first published in 1906. This was completed in 1927. Aitken was also interested in, and contributed to, the popularization of astronomy, especially after his retirement.

Alder, Kurt (1902–1958) German organic chemist who, with Otto ◊Diels, developed diene synthesis, a fundamental process that has become known as the Diels–Alder reaction. It is used in organic chemistry to synthesize cyclic (ring) compounds, including many that can be made into plastics and others – which normally occur only in small quantities in plants and other natural sources – that are the starting materials for various drugs and dyes. This outstanding achievement was recognized by the award of the 1950 Nobel Prize for Chemistry jointly to Alder and Diels.

Alder was born on 10 July 1902 in the industrial town of Königshütte (Krolewska Huta) in Upper Silesia, which was then part of Germany. He was the son of a schoolteacher and began his education in his home town. When the region became part of Poland at the end of World War I, the Alder family moved to Berlin. There Kurt Alder finished his schooling and went on to study chemistry, first at the University of Berlin and later at Kiel, where he worked under Otto Diels. Alder became a chemistry reader at Kiel in 1930 and a professor in 1934, but two years later he began a four-year period in industry as research director of I G Farben at Leverkusen on the northern outskirts of Cologne. He returned to academic life in 1940 as professor and director of the Chemical Institute at the University of Cologne, where he remained for the rest of his life. He died in Cologne on 20 June 1958.

The first report of the diene synthesis, stemming from work in Diels' laboratory at Kiel, was made in 1928. The Diels–Alder reaction involves the adding of an organic compound that has two double bonds separated by a single bond (called a conjugated diene) to a compound with only one, activated double bond (termed a dienophile). A common example of a conjugated diene is butadiene (but-1,2:3,4-diene) and of a dienophile is maleic anhydride (*cis*-butenedioic anhydride). These two substances react readily to form the bicyclic compound tetrahydrophthalic anhydride (cyclohexene-1:2-dicarboxylic anhydride) – one of the reactions originally reported by Diels and Alder in 1928.

Azo-diesters, general formula $RCO_2.N:N.CO_2R$, also act as dienophiles in the reaction, as can other unsaturated acids and their esters. With a cyclodiene, the synthesis yields a bridged cyclic compound. One or two reactions of this type had been reported in the early 1900s, but Diels and Alder were the first to recognize its widespread and general nature. They also demonstrated the ease with which it takes place and the high yield of the product – two vitally important factors for successful organic synthesis. In association with Diels and later with his own students, Alder continued to study the general conditions of the diene synthesis and the overall scope of the method for synthetic purposes. In his Nobel prize address, Alder listed more than a dozen different dienes of widely differing structure that participate in the reaction. He also showed that the reaction is equally general with respect to dienophiles, provided that their double bonds are activated by a nearby group such as carboxyl, carbonyl, cyano, nitro, or ester. Many of the compounds studied were prepared for the first time in Alder's laboratory.

Alder was a particularly able stereochemist, and he showed that the diene addition takes place at double bonds with a *cis* configuration – that is, where the two groups substituting the double bond are both on the same side of it, as opposed to the *trans* isomer, with the groups on opposite sides.

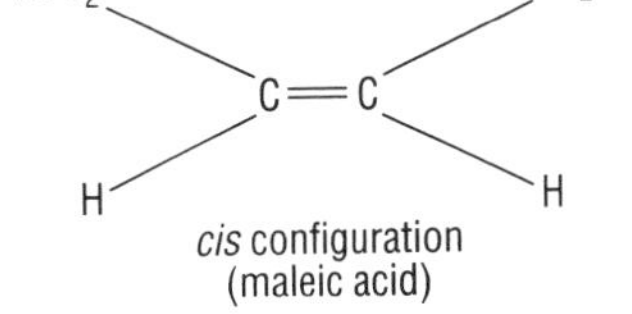

HCO_2 CO_2H
C=C
H H

cis configuration
(maleic acid)

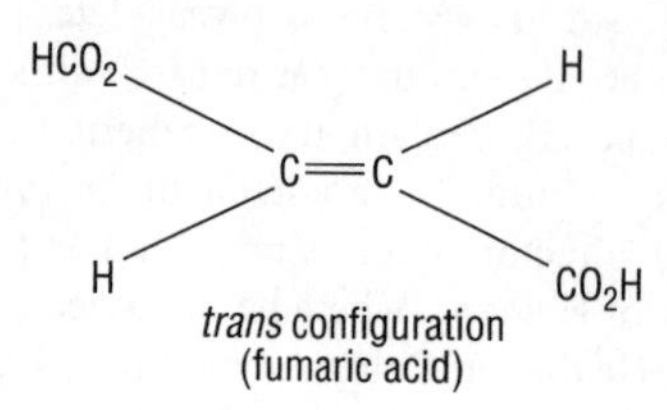

trans configuration
(fumaric acid)

That is why maleic acid (*cis*-butenedicarboxylic acid) reacts whereas its isomer fumaric acid (*trans*-butenedicarboxylic acid) does not. The stereospecific nature of the Diels–Alder reaction has thus become useful in structural studies for the detection of conjugated double bonds.

The bridged-ring, or bicyclic, compounds formed by using cyclic dienes (such as cyclopentadiene) are closely related to many naturally occuring organic compounds such as camphor and pinene, which belong to the group known as terpenes. The diene synthesis stimulated and made easier the understanding of this important group of natural products by providing a means of synthesizing them. Indeed, the ease with which the reaction takes place suggests that it may be the natural biosynthetic pathway. It has been found to be relevant in connection with quinone (vitamin K) – whose synthetic analogues are used to stimulate blood clotting – and anthraquinone type dyes now used universally. Many other commercial products have been made possible by Alder's work, including drugs, insecticides, lubricating oils, synthetic rubber, and plastics. He made a great contribution to synthetic organic chemistry at a time when it was effecting a great transition in industry and science.

Aleksandrov, Pavel Sergeevich (1896–1982) Russian mathematician who was a leading expert in the field of topology and one of the founders of the theory of compact and bicompact spaces.

He was born in Bogorodsk (now Noginok), near Moscow, on 7 May 1896. He studied mathematics at Moscow University, graduating in 1917, and was appointed a lecturer there in 1921. In 1929 he was appointed professor of mathematics. From 1932 to 1964 he was president of the Moscow Mathematical Society. He received five Orders of Lenin and was awarded the State Prize in 1942.

Although he began his career by studying set theory and the theory of functions, Aleksandrov worked principally in the development of topology. He introduced many of the basic concepts of this relatively new branch of mathematics, notably the notion that an arbitrarily general topological space can be approximated to an arbitrary degree of accuracy by simple geometric figures such as polyhedrons. Of great importance, too, were his investigations into that branch of topology known as homology, which examines the relationships between the ways in which spatial structures are dissected. He formulated the theory of essential mappings and the homological theory of dimensionality, which led to a number of basic laws of duality relating to the topological properties of an additional part of space.

Aleksandrov was always greatly interested in the dissemination of mathematical knowledge and in broad collaboration in seeking it. Much of his topological work was done within a group of colleagues and students whom he gathered round him. Being one of the few Soviet scientists given freedom to travel abroad, he, by his numerous visits to European universities, did much to carry new ideas back and forth between the East and the West. His passion for international cooperation led him to supervise the publication of an English–Russian dictionary of mathematical terminology in 1962.

Alexander, Hattie Elizabeth (1901–1968) US bacteriologist and paediatrician who became the leading authority on the treatment of bacterial meningitis.

Hattie Alexander was born in Baltimore, Maryland, on 5 April 1901, the second in a family of eight children. She was educated at the Western High School for Girls where she shone at sports and athletics. She won a partial scholarship to Goucher College but was known by her lecturers as an underachiever. However, she graduated (as a 'C' student) in 1923 with a bachelor's degree. None of her tutors could have suspected that she would become a world famous bacteriologist and paediatrician.

Alexander's ambition was to become a doctor and after graduation she worked for three years as a bacteriologist in the Maryland Public Health Services to save money for medical school, also studying physics at night. She achieved her aim in 1926 when she was admitted to Johns Hopkins University receiving her MD in 1930 with an outstanding academic record. She then took up an internship in paediatrics at the Harriet Lane Home in Baltimore where her interest in meningitis began. Her next internship was in the Babies Hospital of the Columbia-Presbyterian Medical Center in New York, where she was given an appointment in the department of paediatrics. She remained at the department for the rest of her life, becoming full professor in 1958. She died of cancer on 24 June 1968.

Alexander was a challenging teacher who rejected the traditional lecturing approach, expecting her students to question everything and present convincing evidence to support their hypotheses. Her pursuit of excellence in all she did was an inspiration to her students.

Treatment for influenza meningitis at this time was ineffective and the disease was fatal in 100% of cases.

Alexander worked with Michael Heidelberger, an immunochemist, to combine their experiences and discover a cure. They injected bacilli from the spinal fluid of infected children into healthy rabbits and from the antibody produced by the rabbits they developed an antiserum that could completely cure the disease. In two years fatalities had dropped from 100% to 20%. Continuing her work in antibiotic therapy, Alexander managed to reduce the fatality rate still further to 10%. For this achievement she was awarded the $500 E Mead Johnson Award for Research in Pediatrics in 1942.

From this early triumph, Alexander moved on to study bacterial genetics. With her assistant, Grace Leidy, she studied changes in the genetic characteristics of DNA arising from bacterial resistance to antibiotics. Their experiments were partly responsible for the acceptance of the study of DNA in medical and scientific communities.

Alexander set up a team in 1953 to discover whether genes in viruses are biochemically similar to those of bacteria. The results led her to believe that the same principles present in the genetic traits of microorganisms could be applied to the control of genetic traits in human cells. She suggested that this control could be applied to the growth responsible for tumours.

Alexander was admired by students and colleagues for having the courage of her convictions and for her scientific objectivity. She was awarded numerous honours, wrote 65 research papers, and contributed to a number of paediatric textbooks. She was awarded the Stevens Triennial Prize by the trustees of Columbia University in 1954 for an outstanding essay; was, in 1961, the first woman to win the Oscar B Hunter Memorial Award of the American Therapeutic Society; and in 1964 became the first woman elected to president of the American Pediatric Society.

Alfvén, Hannes Olof Gösta (1908–1995) Swedish astrophysicist who made fundamental contributions to plasma physics, particularly in the field of magnetohydrodynamics (MHD) – the study of plasmas in magnetic fields. For his pioneering work in this area he shared the 1970 Nobel Prize for Physics with the French physicist Louis Néel (1904–).

Alfvén was born in Norrköping on 30 May 1908 and was educated at the University of Uppsala, from which he gained his PhD in 1934. In 1940 he joined the Royal Institute of Technology, Stockholm, becoming professor of electronics in 1945 then professor of plasma physics in 1963, this latter chair having been specially created for him. In 1967, however, after disagreements with the Swedish government, he obtained a professorship at the University of California, San Diego. Later he divided his time between the University of California and the Royal Institute. He died on 2 April 1995.

Alfvén made his most important contributions in the late 1930s and early 1940s. Investigating the interactions of electrical and magnetic fields with plasmas (highly ionized gases containing both free positive ions and free electrons) in an attempt to explain sunspots, he formulated the frozen-in-flux theorem, according to which a plasma is – under certain conditions – bound to the magnetic lines of flux passing through it; later he used this theorem to explain the origin of cosmic rays. In 1939 he went on to propose a theory to explain aurorae and magnetic storms, a theory that greatly influenced later ideas about the Earth's magnetosphere. He also devised the guiding centre approximation, a widely used technique that enables the complex spiral movements of a charged particle in a magnetic field to be calculated relatively easily. Three years later, in 1942, he postulated that a form of electromagnetic wave would propagate through plasma; other scientists later observed this phenomenon in plasmas and in liquid metals. Also in 1942 Alfvén developed a theory of the origin of the planets in the Solar System. In this theory (sometimes called the Alfvén theory) he hypothesized that planets were formed from the material captured by the Sun from an interstellar cloud of gas and dust. As the atoms were drawn towards the Sun they became ionized and influenced by the Sun's magnetic field. The ions then condensed into small particles which, in turn, coalesced to form the planets, this process having occurred in the plane of the solar equator. This theory did not adequately explain the formation of the inner planets but it was important in suggesting the role of MHD in the genesis of the Solar System.

Although Alfvén studied MHD mainly in the context of astrophysics, his work has been fundamental to plasma physics and is applicable to the use of plasmas in experimental nuclear fusion reactors.

Alhazen, Abu Alī al-Hassan ibn al-Haytham (c. 965–1038) Arabian scientist who made significant advances in the theory and practice of optics. He was probably the greatest scientist of the Middle Ages and his work remained unsurpassed for nearly 600 years until the time of Johannes ◊Kepler.

Alhazen was born in Basra (Al Basra, now in Iraq). He made many contributions to optics, one of which was to contest the Greek view of Hero and Ptolemy (who flourished in the 2nd century AD) that vision involves rays that emerge from the eye and are reflected by objects viewed. Alhazen postulated that light rays originate in a flame or in the Sun, strike objects, and are reflected by them into the eye. He studied lenses and mirrors, working out that the curvature of a lens accounts for its ability to focus light. He measured the refraction of light by lenses and its reflection by mirrors, and formulated the geometric optics of image formation by spherical and parabolic mirrors. He used a pin-hole as a 'lens' to construct a primitive camera obscura. Alhazen also tried to account for the occurrence of rainbows, appreciating that they are formed in the atmosphere, which he estimated extended for about 15 km/9 mi above the ground. He wrote many scientific works, the chief of them being *Opticae thesaurus* which was published in 1572 from a 13th-century Latin translation of his original Arabic version.

Alhazen spent part of his life in Egypt, where he fell foul of the tyrannical (and mad) Caliph al-Hakim. In a foolhardy attempt to impress the caliph, Alhazen claimed he could devise a method of controlling the flooding of the River Nile. To escape the inevitable wrath of the caliph for nonfulfilment of the promise, Alhazen pretended to be mad himself and had to maintain the charade for many years until 1021, when al-Hakim died. Alhazen died in Cairo in 1038.

Alpher, Ralph Asher (1921–) US scientist who carried out the first quantitative work on nucleosynthesis and was the first to predict the existence of primordial background radiation.

Alpher, the youngest of four children and the son of a building contractor, was born in Washington, DC, in 1921. His initial interest in science was stimulated by his English teacher, Matilde Eiker, who was also an amateur astronomer, and by his chemistry teacher, Sarah Branch. Due to economic circumstances and the advent of World War II, Alpher was forced to continue his education as a night-school student, receiving his BSc from George Washington University in 1943 and his PhD in 1948. His PhD research topic was nucleosynthesis in a Big Bang universe, which was carried out under the supervision of George ◊Gamow. During World War II, Alpher worked at the Naval Ordnance Laboratory and after the war he joined the Applied Physics Laboratory of Johns Hopkins University. Here he took part in a varied research programme that, besides cosmology, included cosmic-ray physics and guided-missile aerodynamics. In 1955 Alpher took up a post at the Central Electric Research Laboratory, where besides his professional duties he continued his vocational involvement in cosmological research. Since 1986 he has been research professor of physics at Union College, Schenectady, New York.

Having graduated from George Washington University, Alpher worked with George Gamow and Robert Herman (1914–) on a series of papers that sought to explain physical aspects of the Big Bang theory of the universe. In 1948 Alpher and Gamow published the results of their work on nucleosynthesis in the early universe. They included the name of Hans ◊Bethe as a co-author of this paper, and their new theory became popularly known as the alpha–beta–gamma theory, appropriate for a theory on the beginning of the universe. Also in 1948, Alpher, together with Robert Herman, predicted the existence of the pervasive relic cosmic black-body radiation. They postulated that this radiation must exist, having originated in the early stages of the Big Bang with which the universe is thought to have begun. This primordial radiation was finally detected by Arno A ◊Penzias and Robert W ◊Wilson in 1965 and was found to have a temperature of 3K (–270 °C/–454 °F). Alpher and Herman had originally theorized that the radiation would have a temperature of approximately 5K (–268 °C/–450 °F), which was remarkably close to the actual value observed.

The existence of this low-temperature radiation that permeates the entire universe is now regarded as one of the major pieces of evidence for the validity of the Big Bang model of the universe; thus Alpher's early cosmological work has had a profound impact towards our understanding of the nature of the universe.

al-Sufi (903–986) Persian astronomer whose importance lies in his compilation of a catalogue of 1,018 stars with their approximate positions, magnitudes, and colours.

Little is known about the life of al-Sufi, but it has been established that he was a nobleman whose love of his country's folklore and mythology and interest in mathematics led him to the study of astronomy.

Alter, David (1807–1881) US inventor and physicist whose most important contribution to science was in the field of spectroscopy.

Alter was born in Westmoreland County, Pennsylvania, on 3 December 1807. He had little early schooling but in 1828 he entered the Reformed Medical College, New York City, to study medicine, in which he graduated in 1831. Thereafter he spent the rest of his life experimenting and making inventions, working alone and using home-made apparatus. He died in Freeport, Pennsylvania, on 18 September 1881.

Alter made his most important contribution to physics in 1854, when he put forward the idea that each element has a characteristic spectrum, and that spectroscopic analysis of a substance can therefore be used to identify the elements present. He also investigated the Fraunhofer lines in the solar spectrum. Although the significance of this work was not recognized in the USA at that time, his idea was experimentally verified in about 1860 in Germany by Robert ◊Bunsen and Gustav ◊Kirchhoff and today spectroscopic analysis is extensively used in chemistry for identifying the component elements of substances and in astronomy for determining the compositions of stars.

Alter devoted most of his life, however, to making inventions, which included a successful electric clock, a model for an electric locomotive (which was not put into production), a new process for purifying bromine, an electric telegraph that spelt out words with a pointer, and a method of extracting oil from coal (which was not put into commercial practice because of the discovery of oil in Pennsylvania).

Alvarez, Luis Walter (1911–1988) US physicist who won the 1968 Nobel Prize for Physics for developing the liquid-hydrogen bubble chamber and detecting new resonant states in particle physics. Discoveries made with the hydrogen bubble chamber were instrumental in the prediction of quarks. Alvarez also made many other breakthroughs in fundamental physics, accelerators, and radar, and was well known for his studies of the pyramids and suggestion that a meteor impact led to the extinction of the dinosaurs.

Alvarez was born on 13 June 1911 in San Francisco. He went to the University of Chicago to study chemistry but changed to physics and stayed there to complete a PhD. His first major discovery, with Arthur Compton, was the discovery of the 'east–west' effect in cosmic rays, proving them to be positive. He then moved to Ernest Lawrence's Radiation Laboratory at the University of California, Berkeley, where he spent the rest of his career. There he discovered that the capture of electrons by the nucleus of an atom is a beta-decay process, and that helium-3 is stable but hydrogen-3 (tritium) is not. Alvarez also made important contributions to the study of the spin dependence of nuclear forces and, working with Felix Bloch, measured the magnetic moment of the neutron.

During the war he moved to the Massachusetts Institute of Technology where he developed the VIXEN radar system for the airborne detection of submarines, phased-array radars, and the ground-controlled approach (GCA) radar that enabled aircraft to land in conditions of poor visibility. Alvarez received the US government's most prestigious aviation award, the Collier Trophy, for these achievements. Alvarez later worked on the atomic bomb project – with Enrico Fermi at Chicago and in the explosives division at Los Alamos – and participated in the Hiroshima mission.

After the war he returned to Berkeley and built the first practical linear accelerator (a 32-MeV proton linac) and invented the tandem electrostatic accelerator. He also devised, but never built, the microtron for accelerating electrons. During the Korean War, Alvarez and Lawrence became convinced that the US needed to produce its own plutonium and built another accelerator to 'breed' plutonium. This machine was later used for nuclear physics.

In 1953 Alvarez changed direction once again when he met Donald ◊Glaser, inventor of the bubble chamber detector for particle physics (and winner of the 1960 Nobel prize). Glaser had been using a small, 2.5-cm/1-in glass bulb full of diethyl ether. Alvarez decided to build a massive 183-cm/72-in chamber containing liquid hydrogen. His next idea was to automate the analysis of the particle tracks captured in the chamber. He also developed automatic scanning and measuring equipment whose output could be stored on punched cards and then analysed using computers. Alvarez and co-workers used the bubble chamber to discover a large number of new short-lived particles ('resonances') including the K (the first meson resonance) and the Ω (omega) meson. These experimental findings were crucial in the development of the 'eightfold way' model of elementary particles, and subsequently the theory of quarks, by Murray ◊Gell-Mann. The techniques developed by Alvarez became standard in high-energy laboratories all over the world.

There is no democracy in physics. We can't say that some second-rate guy has as much right to opinion as Fermi.

LUIS ALVAREZ *IN* D S GREENBERG, THE POLITICS OF PURE SCIENCE *1967*

In later life, Alvarez moved away from conventional physics, using cosmic rays to search for hidden chambers in the Egyptian pyramids and his knowledge of shock waves to study the Kennedy assassination. Best-known of these researches was the discovery he made with his son Walter, a geologist, of unexpectedly high concentrations of an isotope of iridium in the thin layer of clay separating Cretaceous and Tertiary rocks (the K–T boundary). Alvarez postulated that the iridium must have come from a giant meteorite impact some 65 million years ago, and that the resulting dust in the atmosphere must have so changed the climate that the dinosaurs, who lived at that time, must have become extinct. The first half of the hypothesis is now widely accepted. Alvarez died on 1 September 1988.

Alzheimer, Alois (1864–1915) German neuropathologist who was the first to describe the degenerative illness affecting the nerve cells of the frontal and temporal lobes of the cerebrum of the brain that is now known as Alzheimer's disease.

Born in Markbreit, Bavaria in 1864, Alois Alzheimer excelled in science at school and went on to study at the medical schools at Tübingen and Würzburg from where he graduated with an MD in 1887 at the age of 23. He was appointed clinical assistant at the state asylum in Frankfurt-am-Main where he became interested in working on the cortex of the human brain. Here he collaborated with Franz Nissl in researching the pathology of the nervous system and they published *Histologic and Histopathologic Studies of the Cerebral Cortex* in six volumes between 1904 and 1918.

In 1904 Alzheimer joined the Munich Psychiatric Clinic of Emil Kraepelin (1856–1926) and it was at a meeting in 1907 that he presented the case of a 51-year old woman with symptoms of depression, hallucinations, and memory loss that rapidly degenerated into severe dementia. At the postmortem on her death, which was apparently caused by this dementia at the age of 55, Alzheimer noted the presence of two abnormalities in the brain. The first was the presence of neuritic plaques similar to those found in elderly people, and now known to be composed of degenerating nerve terminals, reactive glial cells, and fibrous material called amyloid. The second abnormality he found was the neurofibrillary tangle, a fibrous structure within the nerve cells that showed up with the use of a silver stain. It was Emil Kraepelin who suggested naming the condition after Alzheimer.

As well as his research on dementia, Alzheimer made many advances in the field of histology and made important contributions to the study of epilepsy, brain tumours, Huntington's chorea, and alcoholic delirium. He was also a well-loved teacher.

Alzheimer was appointed professor of psychiatry and director of the Psychiatric and Neurologic Institute at the University of Breslau (now Wroław, Poland). He continued his research there for the next three years but it is thought that he never recovered from the complications of a severe cold that he caught on the journey from Munich and he died in 1915 at the age of 51 of rheumatic heart disease and cardiac failure. He is buried in the Jewish cemetery in Frankfurt-am-Main, next to his wife.

Ambartsumian, Viktor Amazasp (Amazaspovich) (1908–1996) Armenian astronomer whose chief contribution was to the theory of stellar origins.

Ambartsumian was born in Tbilisi, Georgia, on 18 September 1908. He studied at the University of Leningrad and Pulkovo Observatory, and then went on to found, within the university, the Soviet Union's first department of astrophysics. In 1946, he was appointed director of the Byurakan Astrophysical Observatory in Armenia, a position he was to keep until 1988, when he became its honorary director. Ambartsumian was president of the Armenian Academy of Sciences from 1947 and its honorary president from 1993 until his death on 12 August 1996.

Ambartsumian proposed the manner in which enormous catastrophes might take place within stars and galaxies during their evolution. The radio source in the constellation Cygnus had been associated with what appeared to be a closely connected pair of galaxies, and it was generally supposed that a galactic collision was taking place. If this were the case, such phenomena might account for many extra-galactic radio sources. Ambartsumian, however, presented convincing evidence in 1955 of the errors of this theory. He suggested instead that vast explosions occur within the cores of galaxies, analogous to supernovae, but on a galactic scale.

Ampère, André-Marie (1775–1836) French physicist, mathematician, chemist, and philosopher who founded the science of electromagnetics (which he named electrodynamics) and gave his name to the unit of electric current.

Ampère French physicist, mathematician, chemist, and philosopher André-Marie Ampère who proposed the theory and basic laws of electromagnetism. Ampère was a remarkable mathematician; by the age of 13 he had already written a paper on conic sections. The basic SI unit of electrical current was named after him. *Mary Evans Picture Library*

Ampère was born in Polémieux, near Lyon, on 22 January 1775. The son of a wealthy merchant, he was tutored privately and was, to a great extent, self-taught. His genius was evident from an early age, particularly in mathematics, which he taught himself and had mastered to an extremely high level by the age of about 12. The later part of his youth, however, was severely disrupted by the French Revolution. In 1793 Lyons was captured by the republican army and his father – who was both wealthy and a city official – was guillotined. Ampère taught mathematics at a school in Lyon from 1796 to 1801, during which period he married (in 1799); in the following year his wife gave birth to a son, Jean-Jacques-Antoine, who later became an eminent historian and philologist. In 1802 Ampère was appointed professor of physics and chemistry at the Ecole Centrale in Bourg, then, later in the same year, professor of mathematics at the Lycée in Lyon. Two years later his wife died, a blow from which Ampère never really recovered – indeed, the epitaph he chose for his gravestone was *Tandem felix* ('Happy at last'). In 1805 he was appointed an assistant lecturer in mathematical analysis at the Ecole Polytechnique in Paris where, four years later, he was promoted to professor of mathematics. Meanwhile his talent had been recognized by Napoleon, who in 1808 appointed him inspector general of the newly formed university system, a post he retained until his death. In addition to his professorship and inspector-generalship, Ampère taught philosophy at the University of Paris in 1819, became assistant professor of astronomy in 1820 and was appointed to the chair in experimental physics at the Collège de France in 1824 – an indication of the breadth of his talents. He died of pneumonia on 10 June 1836 while on an inspection tour of Marseille.

Ampère's first publication was an early contribution to probability theory – *Considérations sur la théorie mathématique de jeu/Considerations on the Mathematical Theory of Games* (1802), in which he discussed the inevitability of a player losing a gambling game of chance against an opponent with vastly greater financial resources. It was on the strength of this paper that he was appointed to the professorship at Lyon and later to a post at the Ecole Polytechnique.

In the period between his arrival in Paris in 1805 and his famous work on electromagnetism in the 1820s, Ampère studied a wide range of subjects, including psychology, philosophy, physics, and chemistry. His work in chemistry was both original and topical but in almost every case public recognition went to another scientist; for example, his studies on the elemental nature of chlorine and iodine were credited to the English chemist Humphry Davy. Ampère also suggested a method of classifying elements based on a comprehensive assessment of their chemical properties, anticipating to some extent the development of the periodic table later in that century. And in 1814 he independently arrived at what is now known as Avogadro's hypothesis of the molecular constitution of gases. He also analysed Boyle's law in terms of the isothermal volume and pressure of gases.

Despite these considerable and varied achievements, Ampère's fame today rests almost entirely on his even greater work on electromagnetism, a discipline that he, more than any other single scientist, was responsible for establishing. His work in this field was stimulated by the finding of the Danish physicist Hans Christian ◊Oersted that an electric current can deflect a compass needle – that is, that a wire carrying a current has a magnetic field associated with it. On 11 September 1820 Ampère witnessed a demonstration of this phenomenon given by Dominique Arago at the Academy of Sciences and, like many other scientists, was prompted to hectic activity. Within a week of the demonstration he had presented the first of a series of papers in which he expounded the theory and basic laws of electromagnetism (which he called electrodynamics to differentiate it from the study of stationary electric forces, which he called electrostatics). He showed that two parallel wires carrying current in the same direction attract each other, whereas when the currents are in opposite directions, mutual repulsion results. He also predicted and demonstrated that a helical 'coil' of wire (which he called a solenoid) behaves like a bar magnet while it is carrying an electric current.

In addition, Ampère reasoned that the deflection of a compass needle caused by an electric current could be used to construct a device to measure the strength of the current, an idea that eventually led to the development of the galvanometer. He also realized the difference between the rate of passage of an electric current and the driving force behind it; this has been commemorated in naming the unit of electric current the ampere (a usage introduced by Lord Kelvin in 1883). Furthermore, he tried to develop a theory to explain electromagnetism, proposing that magnetism is merely electricity in motion. Prompted by Augustus ◊Fresnel (one of the originators of the wave theory of light), Ampère suggested that molecules are surrounded by a perpetual electric current – a concept that may be regarded as a precursor of the electron-shell model.

The culmination of Ampère's studies came in 1827, when he published his famous *Mémoire sur la théorie mathématique des phénomènes électrodynamiques uniquement déduite de l'expérience/Notes on the Mathematical Theory of Electrodynamic Phenomena Deduced Solely from Experiment,* in which he enunciated precise mathematical formulations of electromagnetism, notably Ampère's law – an equation that relates the magnetic force produced by two parallel current-carrying conductors to the product of their currents and the distance between the conductors. Today Ampère's law is usually stated in the form of calculus: the line integral of the magnetic field around an arbitrarily chosen path is proportional to the net electric current enclosed by the path.

Ampère produced little worthy of note after the publication of his *Mémoire* but his work had a great impact and stimulated much further research into electromagnetism.

Amsler-Laffon, Jakob (1823–1912) Swiss mathematical physicist who designed and manufactured precision instruments for use in engineering.

Amsler-Laffon was born Jakob Amsler at Stalden bei Brugg on 16 November 1823 and educated locally until the age of 19, when he went to Jena to study theology. Theology absorbed his interest for only one year, however, and in 1844 he went to the university at Königsberg (now Kaliningrad in Russia) to study mathematics and physics. He received his doctorate in 1848, after which he worked briefly at the observatory in Geneva. The next eight years were spent in teaching, first at the University of Zürich, 1849–1851, and then at the Gymnasium in Schaffhausen. But in the middle of this period Amsler abandoned his interest in pure science and became involved in the design of scientific instruments. He established a factory for the manufacture of his designs at Schaffhausen in 1854. In that year he also married Elise Laffon, the daughter of a drugs manufacturer, and added her surname to his own. He devoted himself to his factory for the rest of his life, until his death at Schaffhausen on 3 January 1912.

Amsler-Laffon's best idea was his first, the design for an improved tool to measure the areas inside curves. This was his polar planimeter. Earlier models of tools to measure the surface of spheres had been based on Cartesian coordinates, but they were bulky and expensive. Amsler-Laffon's design, based on a polar coordinate system, was not only more delicate and more flexible than its predecessors, but was also much cheaper to manufacture. It could be used in the determination of Fourier coefficients and was thus particularly valuable use to shipbuilders and railway engineers. By the time he died, his factory had produced more than 50,000 polar planimeters.

Anaximander the Elder (*c.* 611–547 BC) Greek philosopher who formulated some basic natural philosophical views, often in opposition to his teacher, ◊Thales of Miletus. Though his writings are lost, he is credited by later writers with having developed major new ideas at the dawn of natural philosophy. He was the first Greek to handle a gnomon or a sundial with a vertical needle (it had been evolved earlier in the Middle East). Use of the gnomon enabled him, it seems, to ascertain the lengths and angles of shadows, and thereby to measure the length of the years (and thereby) of the seasons, by settling the times of the equinoxes and the solstices.

Anaximander viewed the universe as boundless and composed of an elementary substance of unspecified attributes. Recognizing that the Earth's surface was curved, he maintained it to be cylindrical, with its axis running east to west. With this cylindrical shape, the Earth's height was one third of its breadth, and it floated in space, poised motionless. Anaximander explained the starry heavens by presuming the Earth was encircled by bands of condensed air with vents in it; at these points the fire that was assumed to be trapped within them became visible. He regarded the relative distances of those stars to be an important scientific question. He taught that perpetual rotation in the universe created cosmic order by sorting heavier from lighter materials and packing them in concentric layers. He may have been the first Greek to map the whole known world.

Anaximander has been credited with naturalistic views of the origins of life and of humankind. Apparently he taught that life originated with primitive forms in the seas, that in time adjusted themselves to living on dry land. Human beings too had arisen from lower animals. Overall, he seems to have shared the early Greek philosophical urge to explain all creation within a tiny number of general laws.

Anderson, Carl David (1905–1991) US physicist who did pioneering work in particle physics, notably discovering the positron – the first antimatter particle to be found – and the muon (or mu-meson). He received many honours for his work, including the 1936 Nobel Prize for Physics, which he shared with Victor Hess.

Anderson was born in New York City on 3 September 1905, the son of Swedish immigrants. He was educated at the California Institute of Technology, from which he gained a BSc in physics and engineering in 1927 and a PhD in 1930. Thereafter he remained at the institute for the rest of his career, as a research fellow from 1930 to 1933, assistant professor of physics from 1933 to 1939, and professor of physics from 1939 until his retirement in 1976. After retiring he was made an emeritus professor of the California Institute of Technology.

Anderson's first research – performed for his doctoral thesis – was a study of the distribution of photoelectrons emitted from various gases as a result of irradiation with X-rays. Then, as a member of Robert ◊Millikan's research team, he began in 1930 to study gamma rays and cosmic rays, extending the work originally published by Skobelzyn, who had photographed tracks of cosmic rays made in a cloud chamber. Anderson devised a special type of cloud chamber that was divided by a lead plate in order to slow down the particles sufficiently for their paths to be accurately determined. Using this modified chamber he measured the energies of cosmic and gamma rays (by measuring the curvature of their paths) in strong magnetic fields (up to about 2.4 teslas). In 1932, in the course of this investigation, Anderson reported that he had found positively charged particles that occurred as abundantly as did negatively charged particles, and that in many cases several negative and positive particles were simultaneously projected from the same centre. Anderson initially thought that the positive particle was a proton but, after determining that its mass was similar to that of an electron, concluded that it was a positive electron; he then suggested the name positron for this antimatter particle. Working with Neddermeyer, Anderson also showed that positrons can be produced by the irradiation of various materials with gamma rays. In 1932 and 1933 other established scientists – notably Patrick Blackett, James Chadwick, and the Joliot-Curies – independently confirmed the existence of the positron and, later, elucidated some of its properties.

In 1936 Anderson contributed to the discovery of another fundamental particle, the muon. While studying tracks in a cloud chamber, he noticed an unusual track that seemed to have been made by a particle intermediate in mass between an electron and a proton. Initially it was thought that this new particle was the one whose existence had previously been predicted by Japanese physicist Hideki ◊Yukawa (his hypothetical particle was postulated to hold the nucleus together and to carry the strong nuclear force). Anderson named the particle he had discovered the mesotron, which later became shortened to meson. Further studies of the meson, however, showed that it did not readily interact with the nucleus and therefore could not be the particle predicted by Yukawa. In 1947 English physicist Cecil ◊Powell discovered another, more active type of meson that proved to be Yukawa's predicted particle. Anderson's particle – the role of which is still unclear – is now called the muon (or mu-meson) to distinguish it from Powell's particle, which is called the pion (or pi-meson).

Anderson, Philip Warren (1923–) US physicist who shared the 1977 Nobel Prize for Physics with Nevill Mott and John Van Vleck for his theoretical work on the behaviour of electrons in magnetic, non-crystalline solids.

Anderson was born in Indianapolis on 13 December 1923 and was educated at Harvard University, from which he gained his BS in 1943, MS in 1947 and PhD in 1949; he did his doctoral thesis under Van Vleck. Anderson's studies were interrupted by military service during part of World War II: from 1943 to 1945 he worked at the Harvard Naval Research Laboratories in Washington, DC, becoming a chief petty officer in the US navy. After obtaining his doctorate in 1949, he joined the Bell Laboratories in New Jersey, becoming its consulting director of physics research in 1976. In addition to this appointment he was appointed Joseph Henry Professor of Physics at Princeton University in 1975. Anderson also visited several foreign universities: he was a Fulbright lecturer at Tokyo University 1952–53; an overseas fellow at Churchill College, Cambridge, 1961–62; and visiting professor of theoretical physics at Cambridge 1967–75. He was a fellow of Jesus College, Cambridge, 1969–75, and was made an honorary fellow in 1978.

Anderson has made many varied contributions, although most of his work has been in solid-state physics. While studying under Van Vleck at Harvard, he investigated the pressure-broadening of spectral lines in spectroscopy and developed a method of deducing details of molecular interactions from the shapes of spectral peaks. In the late 1950s he devised a theory to explain superexchange – the coupling of the spins of two magnetic ions in an antiferromagnetic material through their interaction with a non magnetic anion situated between them – and then went on to apply the Bardeen–Cooper–Schrieffer (BCS) theory (see John ◊Bardeen) to explain the effects of impurities on the properties of superconductors. In the early 1960s he investigated the interatomic effects that influence the magnetic properties of metals and alloys, devising a theoretical model (now called the Anderson model) to describe the effect of the presence of an impurity atom in a metal. He also developed a method of describing the movements of impurities within crystalline substances; this method is now known as Anderson localization.

In addition, Anderson has studied the relationship between superconductivity, superfluidity and laser action – all of which involve coherent waves of matter or energy – and predicted the existence of resistance in superconductors. Of more immediate widespread practical application, however, is his work on the semiconducting properties of inexpensive, disordered glassy solids; his studies of these materials indicate that they could be used instead of the expensive crystalline semiconductors now used in many electronic devices, such as computer memories, electronic switches, and solar-energy converters.

Andrews, Thomas (1813–1885) Irish physical chemist, best known for postulating the idea of critical temperature and pressure from his experimental work on the liquefaction of gases, which demonstrated the continuity of the liquid and gaseous states. He also studied heats of chemical combination and was the first to establish the composition of ozone, proving it to be an allotrope of oxygen.

Andrews was born in Belfast on 19 December 1813, the son of a linen merchant. He attended five universities (acting on the advice of a physician friend of his father), beginning at the age of 15 at the University of Glasgow; after only a year there he published two scientific papers. Then in 1830 he went to Paris where, like his contemporary Louis Pasteur, he studied under the French organic chemist Jean Baptiste Dumas. He also became acquainted with several famous scientists of that time, including Joseph Gay-Lussac and Henri Becquerel. He returned to Ireland, but to Trinity College, Dublin, and then went from there via Belfast to Edinburgh. In 1835 he graduated from Edinburgh as a qualified doctor and surgeon, with a thesis on the circulation and properties of blood.

Even while following his medical studies, Andrews continued to experiment in chemistry, although he declined professorships in that subject at both the Richmond School of Medicine and at the Park Street School of Medicine, Dublin, preferring to devote his time to the private medical practice he had established in Belfast. He did, however, lecture on chemistry for a few hours each week at the Royal Belfast Academical Institution. It was during this time that he began work on a study of the heats of chemical combination, and in 1844 his paper on the thermal changes that accompany the neutralization of acids by bases won him the Medal of the Royal Society (of which he became a fellow five years later). By this time he was one of the leading scientific figures in the British Isles. In 1845 he was appointed vice president designate of the projected Queen's College, Belfast, in order that he might contribute to its foundation and philosophy, and in 1849 he became its professor of chemistry. He held both posts until 1879, when ill-health forced his retirement. He died in Belfast six years later, on 26 November 1885.

Andrews' research was concentrated into three main channels: the heat of chemical combination, ozone, and changes in physical state. He was only one of many mid-19th century investigators of thermochemistry: his Russian contemporary Germain ◊Hess was carrying out similar experiments at St Petersburg. Andrews' chief contribution was the direct determination of heats of neutralization and of formation of halides (chlorides, bromides, and iodides), but the law of constant heat summation was finally worked out by, and is now named after, Hess.

Before Andrews began to study ozone, it was postulated that the gas was either a compound of oxygen or that it was an oxide of hydrogen that contained a larger proportion of oxygen than does water. Andrews proved conclusively that ozone is an allotrope of oxygen, that from whatever source it is 'one and the same body, having identical properties and the same constitution, and is not a compound body but oxygen in an altered or allotropic condition'. Ozone is triatomic, with molecules represented by the formula O_3.

Many other scientists had tried to explain the relationship between gases and liquids, but none had really come to grips with the fundamentals. It is for his meticulous experimental work in this area that Andrews is best remembered. He constructed elaborate equipment in which he initially investigated the liquefaction of carbon dioxide, exploring the state of the substance (gas or liquid) over a wide range of temperatures and pressures. By 1869 he had concluded that if carbon dioxide is maintained at any temperature above 30.9 °C/87.6 °F, it cannot be condensed into a liquid by any pressure no matter how great. This discovery of a critical temperature (or critical point) soon enabled other workers – such as Raoul Pictet (1846–1929) in Geneva and Louis ◊Cailletet in France, both of whom independently in 1877 liquefied oxygen – to liquefy gases that had previously been thought to be 'non-condensible', the so-called permanent gases. Hydrogen, nitrogen, and air were also liquefied by applying pressure to the gases once they had been cooled to below their critical temperatures. Andrews also worked out sets of pressure–volume isotherms at temperatures above and below the critical temperature, and brought a sense of order to what had previously been a chaotic branch of physical chemistry.

Ångström, Anders Jonas (1814–1874) Swedish physicist and astronomer, one of the early pioneers in the development of spectroscopy.

Ångström was born in Lögdö on 13 August 1814, the son of a chaplain. He was educated at the University of Uppsala, which awarded him his doctorate in physics in 1839; he began to lecture there in the same year. In 1843 he was appointed an observer at the Uppsala Observatory. In 1858 he was elected to the chair of physics at the university, a post he held until his death, in Uppsala, on 21 June 1874.

Ångström's first important work was an investigation into the conduction of heat and his first important result was to devise a method of measuring thermal conductivity, which demonstrated it to be proportional to electrical conductivity.

Then, in 1853, he published his most substantial and influential work, *Optical Investigations,* which contains his principle of spectrum analysis. Ångström had studied electric arcs and discovered that they yield two spectra, one superimposed on the other. The first was emitted from the metal of the electrode itself, the second from the gas through which the spark passed. By applying Leonhard ◊Euler's theory of resonances Ångström was then able to demonstrate that a hot gas emits light at the same frequency as it absorbs it when it is cooled.

Ångström's early work provided the foundation for the spectrum analysis to which he devoted the rest of his career. He was chiefly interested in the Sun's spectrum, although in 1867 he investigated the spectrum of the aurora borealis, the first person to do so. In 1862 he announced his inference (in fact, it amounted to a discovery) that hydrogen was present in the Sun. In 1868 he published *Researches on the Solar System*, in which he presented measurements of the wavelengths of more than a hundred Fraunhofer lines (dark lines in the Sun's spectrum, which had been discovered by Joseph von ◊Fraunhofer some 50 years earlier). The lines were measured to six significant figures in units of 10^{-8} cm/3.9 × 10^{-9} in. The unit of measure for wavelength of light, called the angstrom (Å, equal to 10^{-10} m/3.3 × 10^{-10} ft), was officially adopted in 1907, though it has now been largely replaced by the nanometre (10^{-9} m/3.3 × 10^{-9} ft).

Another of Ångström's important contributions was his map of the normal solar spectrum, published in 1869, which remained a standard reference tool for 20 years.

Anning, Mary (1799–1847) English palaeontologist who found the first ichthyosaur skeleton, the first plesiosaur, and a pteradactyl. Her contributions to the new science of palaeontology were immense but were largely unrecognized because of her social position and her gender.

Anning was born in 1799 in Lyme Regis on the south coast of England, the daughter of a cabinet-maker who collected fossils in his spare time. She and her brother Joseph were the only surviving children of a family of ten. From her father she inherited a passion for fossil hunting along the cliffs of Lyme Regis, which were, and still are, rich in fossils from the Jurassic era. After her father died in 1810, the family was destitute and they eked out a living selling fossils. They opened a fossil shop in Lyme Regis, which proved to be a considerable tourist attraction. The tongue twister 'she sells sea shells on the sea shore' is thought to refer to Mary or her mother. The family's fossils were enthusiastically collected by museums and scientists and also bought for the large private collections of European nobles.

In the early 1820s the Annings attracted the interest of a professional fossil collector, Thomas Birch. He felt that a family who had found so many fine things should not be facing such financial difficulties and he sold his personal collection of fossils to support them. He attributed many discoveries to Mary and her family but it has been difficult for historians to trace many of her fossils, since they ended up in museums and personal collections without credit being given to their original discoverers.

Mary did gain brief recognition for discovering, with her brother, the first *Ichthyosaurus* specimen known to the London scientific community. Her brother discovered the head of a marine reptile on the shore between Lyme and Charmouth and a year later Mary (aged 12) carefully unearthed an entire ichthyosaur 5 m/17 ft long. She found several other fine ichthyosaur skeletons and also, most importantly, in 1828 the first plesiosaur. This find was endorsed by Georges ◊Cuvier, the French anatomist, conferring a previously elusive legitimacy on the Annings as fossilists.

At a time when upper-class literate gentlemen scholars from London received all the credit for geological discoveries, it was hardly surprising that the unpublished achievements of a self-taught young woman from a deprived background went unnoticed. Although they eventually won the respect of contemporary scientists, lack of proper documentation means that the achievements of the Anning family have been lost to history. Mary Anning accumulated skills, experience, and knowledge and according to the diary of Lady Harriet Silvester who visited her in 1824, '... by reading and application she has arrived to that greater degree of knowledge as to be in the habit of writing and talking with professors and other clever men on the subject, and they acknowledge that she understands more of the science than anyone else in the kingdom'.

In 1938 Mary received an annuity from the British Association for the Advancement of Science. The Geological Society of London collected a stipend for her and, in 1846, she was named the first honorary member of the new Dorset County Museum, a year before her death from breast cancer. The Geological Society (which did not admit women until 1904) published her obituary in their journal.

Antoniadi, Eugène Marie (1870–1944) Turkish-born French astronomer who had a particular interest in the planet Mars and later became an expert also on the scientific achievements of ancient civilizations.

Antoniadi was born in Istanbul (then Constantinople) in 1870. He became interested in astronomy as a young man and in 1893 he went to Juvisy-sur-Orge in France, where he worked at the Observatory with Nicolas Flammarion. He later moved to Meudon where he continued his research at the observatory there. He became a French citizen in 1928, and was appointed director of the Mars section of the British Astronomical Association. He died in Meudon on 10 February 1944.

Antoniadi began to make astronomical observations in 1888 at home and while visiting the Greek islands. He was interested in the nearby celestial system – the planets of our Solar System – but was frustrated by the primitive instruments available to him. When he moved to Juvisy he was able to use the 42-cm/16.5-in telescope there with which, in 1893, he and Flammarion observed faint spots on the surface of Saturn. This observation stimulated a vigorous debate with the US astronomer Edward ◊Barnard, who claimed the spots to be illusory. Antoniadi and Flammarion were vindicated when in 1902 Barnard discovered one of these spots himself.

Antoniadi's chief interest was, however, the planet Mars. When he was at Meudon Observatory, he took advantage of a favourable opposition of Mars to observe it, using the 84-cm/33-in telescope. He detected an apparent spot on the planet's surface, but soon realized that it was due merely to an optical effect caused by the diffraction of light by the Earth's atmosphere. His scepticism was not easy to announce, because there was at that time great interest in Giovanni ◊Schiaparelli's suggestion, seized upon by astronomers such as Percival ◊Lowell, that there was an intricate pattern of canals on the surface of Mars suggestive of advanced technology. Antoniadi eventually proposed that these 'canals' were also an optical illusion, produced by the eye's linking of many tiny surface details into an apparently meaningful pattern. In 1924 he was able, however, to confirm Schiaparelli's value for the rotational period of Mars.

Antoniadi's later work included research into the angle of the axis of rotation of Venus and the behaviour and properties of the planet Mercury. He published a book on the planet – *La Planète Mercure* (1934) – and then turned to a study of the history of astronomy and, in particular, to the work of the ancient Greek and Egyptian astronomers.

Apgar, Virginia (1909–1974) US physician and anaesthetist best known for developing the Apgar newborn scoring system, which increased child survival rates worldwide.

Virginia Apgar was born in Westfield New Jersey on 7 June 1909. Her father had a basement laboratory in the house in which he built a telescope and experimented with electricity and radio waves. At a young age Agpar set her heart on a career in medicine. After high school she entered Mount Holyoke College, emerging with a bachelor's degree in 1929. She then enrolled at the College of Physicians and Surgeons at Columbia University and in spite of financial hardship, gained a medical degree in 1933. Coming fourth in her graduation class, she earned an internship at Columbia but soon realized that she was unlikely to find a job in that male-dominated profession, so, in 1935, she took up a two-year residency in anaesthesia. She was taken on as director of the anaesthesia division at Columbia in 1938 and was the first woman to become a full professor at Harvard in 1949.

In the next ten years Apgar began to focus her research on anaesthesia used during childbirth. Understanding that the first moments of a baby's life could be crucial to its survival and recognizing the need to identify those babies at risk, she created a test to score a baby's heart rate, respiration, muscle tone, colour, and reflexes, to be used one minute after birth. The test was known as the Apgar newborn scoring system and it is still used worldwide.

Apgar made further advances in the perinatal field, researching the effects on newborns of anaesthesia given to women during childbirth. Using a catheter in the umbilical artery, she discovered that the anaesthetic cyclopropane had a noticably negative effect on the condition of the babies. The publication of her findings led many doctors to immediately discontinue the use of cyclopropane.

In 1959 Apgar left Columbia University to take a master of public health degree at Johns Hopkins University. She was then hired as head of the division of congenital birth defects at the charity March of Dimes organization, becoming head of the research programme in 1969. In 1973 she became a very successful fundraiser and vice president for medical affairs at the charity.

Apgar received numerous awards and four honorary degrees for her achievements in medicine and teaching, and was named Woman of the Year in 1973 by the *Ladies' Home Journal.* As well as being a pioneer in her field, she was known as an inspiring teacher. She never married and even at the age of 60 was still relentless in her pursuit of knowledge, studying genetics at Johns Hopkins. She died in New York City at the age of 65 on 7 August 1974.

Apollonius of Perga (*c.* 245–*c.*190 BC) Greek mathematician whose treatise on conic sections represents the final flowering of Greek mathematics.

Apollonius was born early in the reign of Ptolemy Euergetes, king of Egypt, in the Greek town of Perga in southern Asia Minor (now part of Turkey). Little is known of his life. It is thought that he may have studied at the school established by ◊Euclid at Alexandria, especially since much of his work was built on Euclidean foundations.

Apollonius' fame rests on his eight-volume treatise, *The Conics,* seven volumes of which are extant. The first four books consisted of an introduction and a statement of the state of mathematics provided by his predecessors. In the last four volumes Apollonius put forth his own important work on conic sections, the foundation of much of the geometry still used today in astronomy and ballistic science.

Apollonius described how a cone could be cut so as to produce circles, ellipses, parabolas, and hyperbolas;

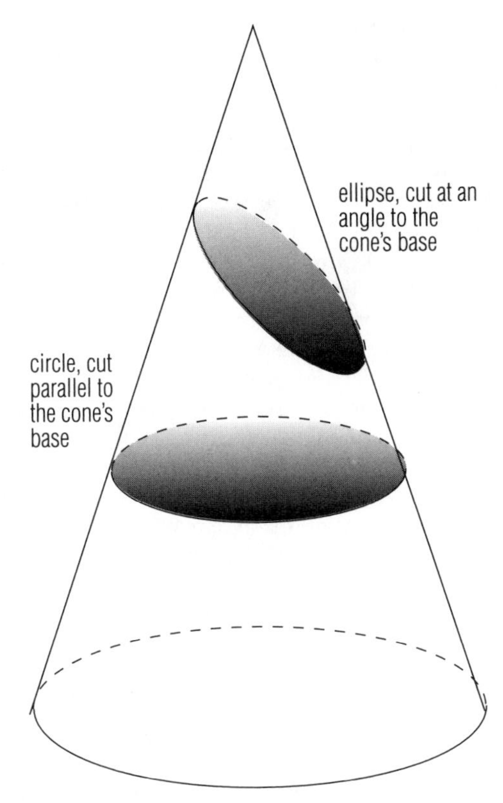

Apollonius of Perga Conic sections were studied avidly by ancient Greek mathematicians. Archimedes in particular wrote several contemplative articles on the subject, but it was Apollonius whose work became the definitive description.

the last three terms were coined by him. He investigated the properties of each and showed that they were all interrelated because, as he stated, 'any conic section is the locus of a point that moves so that the ratio of its distance, *f*, from a fixed point (the focus) to its distance, *d*, from a straight line (the directrix) is a constant'. Whether this constant, *e*, is greater than, equal to, or less than one determines which of the three types of curve the function represents. For a hyperbola $e > 1$; for a parabola $e = 1$; and for an ellipse $e < 1$. At the time, Apollonius' discoveries lay in the realm of pure mathematics; it was only later that their immensely valuable application became apparent, when it was discovered that conic sections form the paths, or loci, followed by planets and projectiles in space.

Other than *The Conics*, only one treatise of Apollonius survives; it is entitled *Cutting off a Ratio*. It was found written in Arabic and was translated into Latin in 1706, but is of little mathematical significance.

Apollonius' brilliant concept of geometry was a milestone in the understanding of mechanics, navigation, and astronomy. Above all, his work on epicircles and ellipses played a major part in ◊Ptolemy's working out of the cosmology that would dominate western astronomy from the 2nd century AD to the 16th century.

Appert, Nicolas (*c.* 1750–1841) French confectioner and inventor who originated the modern process of thermal sterilization of food in sealed containers; he has become known as the pioneer of canning.

Appert was born in Chalons-sur-Marne, just east of Paris. He was self-educated and, as the son of an innkeeper, at an early age learned methods of brewing and pickling. He served his apprenticeship as a chef and confectioner at the Palais Royal Hotel in Chalons and was later employed by the Duke and Duchess of Deux-Ponts. By 1780 he had settled in Paris where he became widely known as a confectioner. The feeding of Napoleon's armies and the navy was becoming a problem and new methods of preventing food decay were needed urgently. In 1795, the French Directory offered a prize for a practical method of preserving food, which encouraged Appert to begin a 14-year period of experimentation. In 1804, with the financial backing of De la Reyniere, he opened the world's first canning factory – the House of Appert – in Massy, south of Paris. By 1809 he had succeeded in preserving certain foods in glass bottles that had been immersed in boiling water. After his methods had been favourably tested and approved by the navy and the Consulting Bureau of Arts and Manufacturers, he was awarded the prize of 12,000 francs on 30 January 1810. One of the conditions of the award was that he should write and publish a detailed description of the processes he used. In 1811 he published *The Art of Preserving All Kinds of Animal and Vegetable Substances for Several Years*. The following year he was presented with a gold medal by the Society for the Encouragement of National Industry, and ten years later was given the title 'Benefactor of Humanity'. The fall of Napoleon, however, marked the end of his financial success, and his factories were destroyed by enemy action. He died in poverty in Massy on 3 June 1841. The cannery he founded continued to operate until 1933.

At the time Appert began his investigations into the preservation of perishable foods, chemistry was in its infancy and bacteriology was unknown. The only reference Appert had to similar work was to that of Italian physiologist Lazzaro ◊Spallanzani in 1765, on the preservation of food by heat sterilization. Appert's final successful results were produced, therefore, by trial and error – with a little insight. He based his methods on the heating of food to temperatures above 100 °C/212 °F, using an autoclave (which he perfected) and then sealing the food container to prevent putrefaction. Initially, he used glass jars and bottles stoppered with corks and reinforced with wire and sealing wax, but in 1822 he changed to cylindrical tin-plated steel cans. He experimented with about 70 foods until he achieved his objective.

In addition to his work on food preservation, Appert was also responsible for the invention of the bouillon cube and he devised a method of extracting gelatin from bones without using acid; he also popularized the use of cylindrical containers for preserved foods. Appert's work was the foundation for the development of the modern canning industry, although he himself could give no scientific explanation for the effectiveness of his methods. It was not until about 1860 that the biological causes of food decay became known as a result of the research begun by Louis ◊Pasteur.

Appleton, Edward Victor (1892–1965) English physicist famous for his discovery of the Appleton layer of the ionosphere which reflects radio waves and is therefore important in communications. He received many honours for his work, including a knighthood in 1941 and the Nobel Prize for Physics in 1947.

Appleton was born on 6 September 1892 in Bradford, Yorkshire. He attended Barkerend Elementary School 1899–1903, then won a scholarship to Hanson Secondary School. A gifted student, he also won a scholarship to St John's College, Cambridge, in 1910 and graduated with first class honours in 1913. After a short period of postgraduate research with William Henry Bragg, Appleton became a signals officer in the Royal Engineers with the outbreak of World War I in 1914. This aroused his interest in radio and he began to investigate radio propagation when he returned to Cambridge after the war. In 1919 he was elected a fellow of St John's College, and in the following year was appointed an assistant demonstrator at the Cavendish Laboratory. In 1924, when only 32 years old, he was appointed Wheatstone Professor of Physics at King's College, London, a post he held until 1936, when he was made Jacksonian Professor of Natural Philosophy at Cambridge. In 1939 he became secretary of the Department of Scientific and Industrial Research, in which position he gained a reputation as an adviser on government scientific policy. During World War II, he was involved in the development of radar and of the atomic bomb. In 1949 he was made principal and vice chancellor of Edinburgh University, a position he held until his death (in Edinburgh) on 21 April 1965. While at Edinburgh, he founded the *Journal of Atmospheric Research,* which became known as 'Appleton's journal'; he remained its editor-in-chief for the rest of his life.

Appleton began his research into radio when he returned to Cambridge after World War I. Initially he investigated (with Balthazar van der Pol, Jr) thermionic vacuum tubes – on which he wrote a monograph in 1932 – then in the early 1920s he turned his attention to studying the fading of radio signals, a phenomenon he had encountered while a signals officer during World War I.

The first transatlantic radio transmission had been made by Guglielmo Marconi in 1901, and to explain why this was possible (that is, why the radio waves 'bent' around the Earth and did not merely go straight out into space) Oliver ◊Heaviside and Arthur Kennelly postulated the existence of an atmospheric layer of charged particles (now called the Kennelly–Heaviside layer or E layer) that reflected the radio waves. Working with New Zealand graduate student Miles Barnell – and using the recently set up BBC radio transmitters – Appleton proved the existence of the Kennelly–Heaviside layer. By periodically varying the frequency of the BBC transmitter at Bournemouth and measuring the intensity of the received transmission 100 km/62 mi away, Appleton and Barnell found that there was a regular 'fading in' and 'fading out' of the signals at night but that this effect diminished considerably at dawn as the Kennelly–Heaviside layer broke up. They also noticed, however, that radio waves continued to be reflected by the atmosphere during the day but by a higher-level ionized layer. By 1926 this layer, which Appleton measured at about 250 km/155 mi above the Earth's surface (the first distance measurement made by means of radio) had became generally known as the Appleton layer (it is now also known as the F layer).

Appleton continued his studies of the ionosphere (as the charged layers of the atmosphere above the stratosphere are called), showing how they are affected by the position of the Sun and by changes in the sunspot cycle. He also calculated their reflection coefficients, electron densities, and their diurnal and seasonal variations. Furthermore, he showed that the Appleton layer is strongly affected by the Earth's magnetic field and that although further above the Earth, it has a greater density and temperature than does the Kennelly–Heaviside layer.

Appleton's research into the atmosphere was of fundamental importance to the development of radio communications, and his experimental methods were later used by the British physicist Robert ◊Watson-Watt – with whom Appleton had collaborated on several projects – in his development of radar.

Arago, (Dominique) François (1786–1853) French scientist who made contributions to the development of many areas of physics and astronomy, the breadth of his work compensating for the absence of a single product of truly outstanding quality. He was closely involved with André ◊Ampère in the development of electromagnetism and with Augustin ◊Fresnel in the establishment of the wave theory of light. Arago's political commitment demanded much time during his latter years, but he maintained a continuous flow of scientific investigations until almost the end of his life.

Arago was born in Estagel on 26 February 1786. He studied at the Ecole Polytechnique in Paris and was then appointed to the Bureau of Longitudes. He travelled to the south of France and Spain with Jean Biot (1774–1862) in 1806, where they intended to measure an arc of the terrestrial meridian. Biot returned to France in 1807, but Arago continued his work amidst a deteriorating political situation. His return to France in 1809 was somewhat enlivened by a shipwreck and his subsequent near-escape from being sold into slavery in Algiers.

To get to know, to discover, to publish – this is the destiny of a scientist.

FRANÇOIS ARAGO *ATTRIBUTED REMARK*

In the same year, Arago was elected to the membership of the French Academy of Sciences and became professor of analytical geometry at the Ecole Polytechnique, a post he held until 1830. He became a Fellow of the Royal Society of London in 1818, which awarded him the Copley Medal in 1825 for his work on electromagnetism.

The year 1830 was one of several changes for Arago. He resigned his post at the Ecole Polytechnique and succeeded Jean Fourier as permanent secretary to the Academy of Sciences. He also became director of the Paris Observatory and deputy for Pyrénées Orientales, a commitment he retained until 1852.

Arago's political affiliation was with the extreme left, and the political turbulence of 1848 saw him elected to a ministerial position in the provisional government. It was under his administration that slavery was abolished in the French colonies. He resigned his post of astronomer in 1852 upon the coronation of Emperor Napoleon III, refusing to take an oath of allegiance to the emperor. Arago's reputation protected him but he died soon afterwards in Paris on 2 October 1853.

Arago's one area of sustained scientific effort was the study of the nature of light. The controversy over the question of whether light behaves as a stream of particles or as a wave motion was one of the most hotly debated of the time. Arago initially sided with Biot in the particulate camp, but later took the other view with Baron Humboldt, Augustin, Fresnel, and others. In 1811, he invented the polariscope, with which he was able to measure the degree of polarization of light rays. From 1815, Arago worked with Fresnel on polarization and was able to elucidate the fundamental laws governing it. Fresnel's mathematical expertise complemented Arago's experimental ability in establishing the wave theory of light, though, because of difficulties in explaining its transmission through the aether, Arago could not accept Fresnel's assertion that light moves in transverse waves.

In 1838, Arago published a method for determining the speed of light using a rotating mirror. He was interested to find out how the speed of light is affected by travelling through a medium such as water that is dense in comparison with air. The experiment was a crucial one in determining whether light is a wave motion or not, but sadly difficulties with his laboratory equipment, the 1848 revolution, and finally the loss of his eyesight in 1850 prevented Arago from completing the experiment. Léon ◊Foucault was able to do this for him, and he obtained results which confirmed the wave theory before Arago's death.

In 1820, Arago announced to the Academy of Sciences some observations on the effect of an electric current on a magnet that had been obtained by Danish physicist Hans Oersted. Arago himself turned to the study of electromagnetism, inspiring Ampère to do the same and producing many interesting results himself. Arago found in 1820 that an electric current produces temporary magnetization in iron, a discovery crucial to the later development of electromagnets and electric relays and loudspeakers. Then, in 1824, he discovered that a rotating nonmagnetic metal disc – for example, of copper – deflects a magnetic needle placed above it. This was a demonstration of electromagnetic induction, explained by Michael ◊Faraday in 1831.

Arago also investigated the compressibility, density, diffraction, and dispersion of gases; the speed of sound, which he found to be 331.2 m/1,087 ft per second; lightning, of which he found four different types; and heat. His studies in astronomy included investigations of the solar corona and chromosphere, measurements of the diameters of the planets, and a theory that light interference is responsible for the 'twinkling' of stars.

Arago's energy and enthusiasm, while directing him into a multiplicity of endeavours, acted as a catalyst in the achievement of several fundamental advances in the study of light and electromagnetism.

Archimedes (*c.* 287–212 BC) Greek mathematician and physicist, generally considered to be the greatest in the ancient world. The details of his personal life and many of the stories surrounding his achievements are of dubious authenticity. A biography of his life written by Heracleides (a friend of Archimedes) has been lost, so modern historians of science have to rely on the mathematical treatises that Archimedes is known to have published and on accounts of his life by Greeks who lived after his time. The reliability of at least parts

Archimedes Greek mathematician and physicist Archimedes, who founded the sciences of statistics and hydrostatics. His greatest achievement in mathematics was to devise formulae for the areas and volumes of plane and solid figures, such as parabolas, spheres, and cylinders. *Mary Evans Picture Library*

of these accounts must be questioned, particularly those dealing with Archimedes' military exploits against the Romans during their siege of his home town Syracuse shortly before his death.

Archimedes is believed to have been born in 287 BC in Syracuse, Sicily, then a Greek colony. This date is based on the claim of a 12th-century historian that Archimedes survived to the age of 75, for the date of his death in 212 BC is not questioned. His father was Phidias, an astronomer, and the family was a noble one possibly related to that of King Hieron II of Syracuse.

Alexandria was a great centre of learning for mathematicians, and Archimedes travelled there to study under Conon (lived *c.* 250 BC) and other mathematicians who had in turn studied under Euclid. Unlike most other disciples, Archimedes did not remain in Alexandria but returned to his home where he devoted the rest of his life to the serious study of mathematics and physics and, by way of recreation, to the design of a variety of mechanical devices that brought him great fame.

He chose only to publish the results of his scientific studies since only these, in his eyes, were worthy of serious consideration. His ability to invent and construct machinery was exploited during the Romans' siege of Syracuse 215–12 BC. The siege is reported to have lasted such a long time because Archimedes was able to hold the Roman fleet at bay with a series of weapons that allegedly set fire to the ships or even caused them to capsize.

When the Roman sack of the city eventually came, Archimedes comported himself as the truly disinterested philosopher that he was. A Roman soldier came across him kneeling over a mathematical problem that he was examining, scratched out in the sand at the marketplace. Despite orders that Archimedes be taken alive and treated well, the impatient soldier – unable to remove him from his study – killed him on the spot.

Archimedes had decreed that his gravestone be inscribed with a cylinder enclosing a sphere together with the formula for the ratio of their volumes, this being the discovery that he regarded as his greatest achievement. His wish was evidently granted for Cicero found this gravestone in 75 BC, possibly confirming Plutarch's view that Archimedes himself thought highly only of his theoretical endeavours and disdained his practical inventions.

In physics, Archimedes is best known for his establishment of the sciences of statics and hydrostatics. In the field of statics, he is credited with working out the rigorous mathematical proofs behind the law of the lever. The lever had been used by other scientists, but it was Archimedes who demonstrated mathematically that the ratio of the effort applied to the load raised is equal to the inverse ratio of the distances of the effort and load from the pivot or fulcrum of the lever. Archimedes is credited with having claimed that if he had a sufficiently distant place to stand, he could use a lever to move the world.

This claim is said to have given rise to a challenge from King Hieron to Archimedes to show how he could move a truly heavy object with ease, even if he couldn't move the world. In answer to this, Archimedes developed a system of compound pulleys. According to Plutarch's *Life of Marcellus* (who sacked Syracuse), Archimedes used this to move with ease a ship that had been lifted with great effort by many men out of the harbour onto dry land. The ship was laden with passengers, crew, and freight, but Archimedes – sitting at a distance from the ship – was reportedly able to pull it over the land as though it were gliding through water.

The best known result of Archimedes' work on

Eureka! I have found it!

ARCHIMEDES REMARK, QUOTED IN VITRUVIUS POLLIO DE ARCHITECTURA IX

Archimedes The Archimedes screw, a spiral screw turned inside a cylinder, was once commonly used to lift water from canals. The screw is still used to lift water in the Nile delta in Egypt, and is often used to shift grain in mills and powders in factories.

hydrostatics is the so-called Archimedes principle, which states that a body immersed in water will displace a volume of fluid that weighs as much as the body would weigh in air. Archimedes is said to have discovered this famous principle when causing water to overflow from a bath. He was so overjoyed at the idea that he ran naked through the town crying 'Eureka!' which, roughly translated, means 'I've got it!'. His interest in the problem is said to have been stimulated by the problem of determining whether King Hieron's new crown was pure gold or not. The king had ordered that this be checked without damaging the crown in any way. Archimedes realized that if the gold had been mixed with silver (which is less dense than gold), the crown would have a greater volume and therefore displace more water than an equal weight of pure gold. The story goes that the crown was in fact found to be impure, and that the unfortunate goldsmith was executed.

> *Give me but one firm place on which to stand, and I will move the earth.*
>
> ARCHIMEDES *ON THE LEVER, QUOTED IN* PAPPUS ALEXANDER

Among Archimedes' inventions was a design for a model planetarium able to show the movement of the Sun, Moon, planets, and possibly constellations across the sky. According to Cicero, this was captured as booty by Marcellus after the sack of Syracuse. The Archimedes screw, an auger used to raise water for irrigation, is also credited to Archimedes, who is supposed to have invented it during his days in Egypt, though he may simply have borrowed the idea from others in Egypt.

Archimedes wrote many mathematical treatises, some of which still exist in altered forms in Arabic. Among the areas he investigated were the value for π (pi). Archimedes' approximation was more accurate than any previous estimate – he stated that the value for π lay between $\frac{223}{71}$ and $\frac{220}{70}$, the average of these two values being less than three parts in ten thousand different from the modern approximation for π. He also examined the expression of very large numbers, using a special notation to estimate the number of grains of sand in the universe. The result – 10^{63} – may be far from accurate, but it showed that large numbers could be considered and handled effectively. Archimedes also evolved methods to solve cubic equations and to determine square roots by approximation, as well as formulas for the determination of the surface areas and volumes of curved surfaces and solids. In the latter area, Archimedes' work anticipated the development of integral calculus, which did not come for another 2,000 years. In his mathematical proofs Archimedes employed the methods of exhaustion and *reductio ad absurdum,* which were perhaps originally developed by Eudoxus and were also used by Euclid.

Surprisingly, in view of his reputation, Archimedes' work was not widely known in antiquity and little advance was made on it. It was preserved by Byzantium and Islam, from where it spread to Europe from the 12th century. Archimedes then had a profound effect on the history of science, for his method of finding mathematical proof to substantiate experiment and observation became the method of modern science introduced by Simon Stevinus, Johannes Kepler, Galileo, and Evangelista Torricelli, among others in the late 16th and early 17th centuries.

Argand, Jean Robert (1768–1822) Swiss mathematician who invented a method of geometrically representing complex numbers and their operations.

Argand was born in Geneva on 18 July 1768. Almost nothing is known of his life except that he was working in Paris as a bookseller and living there with his wife, son, and daughter in 1806, the year in which he published his method. He appears to have been entirely self-taught as a mathematician and to have had no contact with any other mathematicians of any standing. By dint of his diligent pursuit of what was for him simply a hobby he hit upon a happy idea at just the right time in the history of mathematics. He devised his method in 1806 and thereafter did nothing again mathematically important. He died in obscurity at Paris on 13 August 1822.

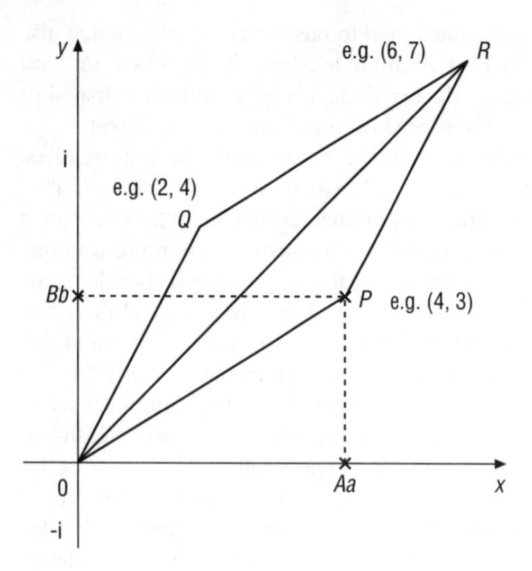

Argand The Argand diagram is a graphic representation of complex numbers and their addition. A complex number of the form $a + bi$, in which a and b are real numbers and i is the square root of −1, is represented by the point (a, bi). Complex numbers are added by means of a parallelogram construction.

The idea of giving geometric representation to complex numbers had already been worked out by the Norwegian mathematician Caspar Wessel (1745–1818) and by the German Karl ⟩Gauss when, in 1806, an anonymous book, *Essai sur une manière de représenter les quantités imaginaires dans les constructions géométriques,* was published in a small, privately printed edition. But neither Wessel nor Gauss had published their ideas, so that Argand, the anonymous author, deserves the credit for the system that the book propounded, sometimes wrongly attributed to Gauss but properly known as the Argand diagram. Yet Argand's name might never have come to light but for a curious set of circumstances. Before publishing his book he had outlined his ideas to French mathematician Adrien-Marie Legendre. Some years later Legendre mentioned the system in a letter to the brother of J R Français, a lecturer at the Imperial College of Artillery in Paris. Français discussed the new system in a paper published in the journal *Annales de mathématiques* in 1813 and, acting from curiosity and kindness, asked the anonymous inventor of it to come forward and make himself known. Argand did so and a paper of his was published in a later issue of the *Annales* for that year.

In his book, Argand adopted Descartes' practice of calling all multiples of $\sqrt{-1}$ 'imaginary'. That pure imaginary numbers might be represented by a line perpendicular to the axis of real numbers had been suggested in the 17th century by English mathematician John Wallis. Argand went further, to demonstrate that real and imaginary parts of a complex number could be represented as rectangular coordinates. It had for some time been usual to picture real numbers – negative and positive – as corresponding to points on a straight line. Of the models studied by Argand and discussed in his book, one used weights, progressively removed from a beam balance, to represent the generation of negative numbers by repeated subtractions. Another simply subtracted portions from a sum of money. Argand argued that these models showed that distance could be considered separately from direction in constructing a geometrical representation, such distance being 'absolute'. Furthermore, whether a negative quantity were considered as real or imaginary depended on the sort of quantity being measured. Argand was thus able to use these distinctions – between direction and absolute distance, and between real and imaginary negative quantities – to construct his diagram.

The diagram is a graphic representation of complex numbers of the form $a + b$i, in which a and b are integers and i is $\sqrt{-1}$. One axis represents the pure imaginary numbers (those belonging to the bi category) and the other the real numbers (those belonging to the a category); it is thus possible to plot a complex number as a set of coordinates in the field defined by the two axes.

Argand was an amateur and had a somewhat patchy knowledge of mathematics. It is clear, for instance, from his discussion of the central problem, whether all rational functions, $f(a + bi)$, could be reduced to the form $A + B$i, where i, b, A, and B are real, that he understood the work of Lagrange, Euler, and d'Alembert; but he also revealed that he did not know that Euler had so reduced $\sqrt{-1}^{\sqrt{-1}}$, since he cited that expression as one that could not be reduced. It is also true that, had Argand never lived, his idea would have come to light through the work of Gauss at about the same time. It would nevertheless be a grave injustice to deny him his honourable, if minor, place in the history of mathematics.

Argelander, Friedrich Wilhelm August (1799–1875) Prussian astronomer whose approach to the subject was one of great resourcefulness and thoroughness. His most enduring contribution was the publication of the *Bonner Durchmusterung/Bonn Survey* (1859–1862) of more than 300,000 stars in the northern hemisphere. Its value is such that it was reprinted as recently as 1950.

Argelander was born on 22 March 1799 in Memel, East Prussia (now Klaipeda in Lithuania). He studied in Elbing and Königsberg. His initial plan had been to study economics and politics, but lectures by Friedrich ⟩Bessel soon fired his interest in astronomy. He worked under Bessel and in 1922 was awarded a PhD for a thesis in which he described work he had done as part of Bessel's systematic evaluation of bright stars in part of the northern hemisphere. In the same year Argelander earned the title of lecturer.

In 1823 he went to Åbo (Turku) in Finland. He worked as an astronomical observer there, under difficult conditions, for four years until the observatory was destroyed by a fire that swept the town in 1827. He became professor of astronomy at the University of Helsinki in 1828, and from 1832 until 1836 he was director of the observatory there.

During the upheavals that followed the Napoleonic wars, the young princes of the Prussian kingdom had lived for a few years with the Argelander family. In 1836, when Argelander went to the University of Bonn as professor of astronomy, the grateful crown prince (who later became King Friedrich Wilhelm IV) promised Argelander a magnificent new observatory, which was eventually constructed under Argelander's supervision. Meanwhile, he continued his own studies under primitive conditions.

Argelander's work was of such impeccable standard that he earned an impressive international reputation. He was elected to virtually every prominent European scientific academy and was a member of scientific organizations in the USA. He was an active member of the Astronomische Gesellschaft, serving as chair of its governing body 1864–67. He died in Bonn on 17 February 1875.

Argelander's early astronomical studies were a continuation of Bessel's work on the mapping of stellar positions, so that his systematic approach was established from the beginning. After his move to Åbo, Argelander began to concentrate on the proper motion of stars – that is, the movement of stars measured in seconds of arc per year, relative to one another. Argelander considered this movement by analogy to a ship moving among a fleet: the farther vessels appear to be almost stationary relative to those nearby, although they are all moving. He studied the proper motion of more than 500 stars and was able to publish the most accurate catalogue of the day on the subject.

His next major area of study was a continuation of a preliminary investigation done by William ◊Herschel in 1783 on the movement of the Sun through the cosmos. Herschel's study had involved the proper motion of only seven stars. Argelander realized that observations of many more stars would be necessary before a firm conclusion about the direction of movement of the Sun, if indeed there were any, could be made. Only a large quantity of data would enable the dual effects of the movement of 'fixed' stars and the movement of the Sun to be distinguished. Argelander's conclusions, based on nearly 400 stars, confirmed Herschel's results. He found that the Sun is indeed moving towards the constellation of Hercules.

In 1843 Argelander published his *Uranometrica Nova,* based on studies made exclusively with the naked eye, because the Bonn Observatory was still under construction. The most important innovation in this study was the introduction of the 'estimation by steps' method for determining stellar magnitudes. It relied exclusively on the sensitivity of the trained eye in comparing the brightness of neighbouring stars.

In 1850 the system was elaborated by N G Pogson, who found that each step along the scale meant a change in brightness 2.5 fold. Bright stars have low numbers, for example 1 or 2, and dim stars have high numbers, such as 9. Stars of magnitude 7 and above are not visible to the naked eye. Extremely bright bodies have magnitudes with negative values, such as the full Moon with a value of –11 or the Sun with a value of –26.7.

Argelander's next project was an extension of Bessel's study of stars in the northern sky. At first he neglected stars up to a certain magnitude, which limited the usefulness of his data since it made it inadequate for statistical analysis. This was remedied during the late 1850s, with the help of E Schönfeld and A Kruger, and resulted in the publication of the *Bonner Durchmusterung.* This catalogued the position and brightness of nearly 324,000 stars, and although it was the last major catalogue to be produced without the aid of photography, it represents the cornerstone of later astronomical work. Argelander also initiated a mammoth project that required the cooperation of many observatories and was aimed at improving the accuracy of positional data recorded in the survey.

Argelander's work was characterized by its grand scope and admirable thoroughness. His contributions were fundamental to many later astronomical studies.

Aristarchos (*c.* 320–*c.* 250 BC) Greek mathematician and astronomer.

Aristarchos was born on the island of Samos in about 320 BC. He was born before Archimedes, although Aristarchos and Archimedes certainly knew of each other. Little is known of Aristarchos' life, but it is thought most likely that he studied in Alexandria under Strato of Lampsacos (*c.* 340–270 BC), before the latter succeeded Theophrastus, as head of the Athenian Lyceum (originally founded by Aristotle) in 287 BC. Aristarchos died in Alexandria *c.* 250 BC.

The only work by Aristarchos that still exists is *On the Magnitude and Distances of the Sun and Moon.* This document describes the first attempt, by means of simple trigonometry, to measure these sizes and distances. The measurements were not very accurate and they were improved upon a century later by ◊Hipparchus. It

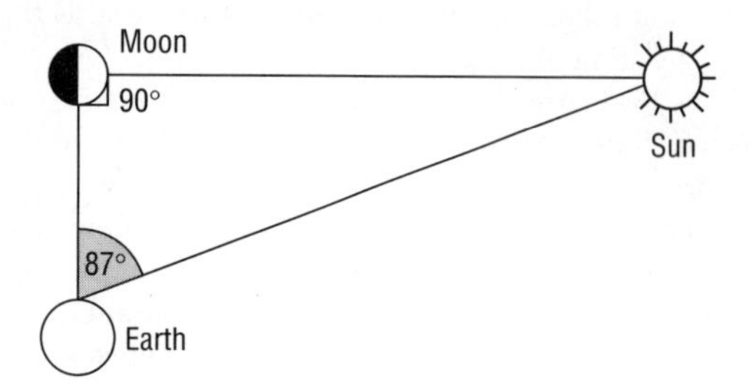

Aristarchos Aristarchos calculated the distance between the Earth and the Sun (in terms of the Earth–Moon distance) by measuring the angle to the Sun when the Moon was exactly half full (and its angle to the Sun was therefore 90°).

was, nevertheless, probably as a consequence of the figures Aristarchos obtained – however inaccurate – that he first began to conceive of his revolutionary cosmological model. In the text of his essay Aristarchos described a right-angled triangle with the Earth, the half-illuminated Moon, and the Sun at the corners; the Moon was positioned at the right-angle, and the hypotenuse ran between the Sun and the Earth. Aristarchos reasoned that since the angle at the Earth corner could be measured, the angle at the Sun could be deduced.

Based on six astronomical hypotheses Aristarchos obtained 18 propositions that described, among other things, the Sun–Earth and Moon–Earth distances in terms of Earth diameters. The circumference (and therefore the diameter) of the Earth had already been calculated with considerable accuracy; in addition, his basic mathematics was sound, but because Aristarchos was unable to make accurate measurements he arrived at incorrect results. Not only did he have difficulty in knowing exactly when the Moon was half-illuminated (when it forms a right-angle with the Sun and Earth), he also underestimated the angle formed by the Sun, the Earth, and the Moon at the Earth corner.

This miscalculation led to the erroneous conclusion that the distance between the Sun and the Earth was only 18–20 times the distance between the Moon and the Earth; in fact, the correct multiple is 397 times. Aristarchos also grossly miscalculated the diameter of the Sun as being only seven times that of the Earth. Even so, it still struck him as strange that a larger body should orbit a smaller one, as was assumed by the geocentric cosmological model.

◊Heraklides of Pontus had earlier proposed the idea, which had been accepted by many, that Mercury and Venus orbit the Sun. Aristarchos carried the argument further and suggested that the Earth also orbits the Sun. Aristarchos' model, the first heliocentric one to be proposed, described the Sun and the fixed stars as stationary in the cosmos, and the planets – including the Earth – as travelling in circular orbits around the Sun. He further stated that the apparent daily rotation of the sphere of stars is due to the Earth's rotation on its axis as it travels along its orbit. He anticipated the most powerful argument against his theory by stating that the reason no stellar parallax (change in position of the stars) was observed from one extreme of the orbit to the other is that even the diameter of the Earth's orbit is insignificant in relation to the vast dimensions of the universe.

Aristarchos' model is recorded in letters, in a book by Archimedes, and in the writings of Plutarch. The Polish astronomer ◊Copernicus, in the 16th century, certainly knew of Aristarchos' heliocentric model, but he deliberately suppressed reference to it, perhaps so as not to compromise his claims to originality.

Aristarchos is said to have carried out other astronomical research, including an observation of the summer solstice in 281 BC (according to Ptolemy). He may also have designed the skaphe – an improved sundial consisting of a hollow hemisphere with a vertical needle protruding from the base, which cast shadows to indicate the time.

We know today that Aristarchos was substantially correct in his views. But his theory made little impact on his contemporaries, since the powerful philosophical, religious, and astronomical ideas of the time were all based on a geocentric view of the universe. Even Aristarchos' initial measurements of the distances and sizes of the Earth, Moon, and Sun were probably based on a geocentric model. The new model, however, demanded that the universe be considered to have dimensions that exceeded the imagination of and were unacceptable to Aristarchos' fellow cosmologists. He was accused of impiety, but perhaps even more damning was the inability of his model to account for a number of astronomical anomalies and the unequal duration of the seasons. It was never suspected that the introduction of elliptical rather than circular orbits would have gone a long way to resolving these difficulties.

Virtually alone among the Greek astronomers, Aristarchos proposed a cosmological model based not on mathematical harmony, but on observed physical 'reality'. This achievement is something of a paradox since he is generally credited with being a mathematician rather than a descriptive astronomer.

Aristotle (384–322 BC) Greek polymath, one of the most imaginative and systematic thinkers in history, whose writings embraced virtually every aspect of contemporary thought, including cosmology.

Aristotle was born in Stagirus, a port on the

Chalcidic peninsula of Macedonia, in 384 BC. His father, Nichomachus, was court physician to Amyntas III (sometimes called Amyntas II), king of Macedonia, and it seems probable that he introduced Aristotle to the body of medical and biological knowledge at an early age. Nichomachus died in Aristotle's youth and Aristotle was placed in the care of a guardian, who sent him to Athens in 367 BC to study at Plato's Academy. Plato's death *c.* 347 BC coincided with a wave of anti-Macedonian fervour in Athens, a combination of events that induced Aristotle to leave the city and go on an extensive tour of Asia Minor, where for the first time he engaged in a serious study of natural history.

In 342 BC King Philip II invited Aristotle to the Macedonian court to become tutor to the crown prince, the future Alexander the Great. Shortly after Alexander came to the throne in 336 BC, Aristotle returned to Athens, where he established his own school, the Lyceum (known also as the Peripatetic School from Aristotle's habit of lecturing while walking in the garden) in 335 BC. At the Lyceum, Aristotle established a zoo (stocked with animals captured during Alexander's Asian campaigns) and a library. The latter formed the basis of the great library established in Alexandria by the Ptolemies. The death of Alexander in 323 BC and another upsurge of anti-Macedonian sentiment in Athens suddenly made Aristotle's position uncomfortable. Largely because of his association with Antipater, the Macedonian regent and general, Aristotle was politically suspect. He was charged with impiety and, rather than suffer the fate of Socrates, he withdrew to Chalcis (now Khalkis), north of Athens, where he died in 322 BC.

Aristotle's writings, which have come down to us only in later, edited versions of his notes, lectures, and publications, cover philosophy, logic, politics, physics, biology, and cosmology. Among his many scientific and philosophical treatises are the *Organon,* a collection of treatises on logic; the *Physica,* on natural science; the *Historia animalium,* a classification of animals; and *De incessu animalium,* on the progression of animals. His major writings on cosmology are brought together in the four-volume text *De caelo/Of the Heavens.* Aristotle rejected the notion of infinity and the notion of a vacuum. A vacuum he held to be impossible because an object moving in it would meet no resistance and would therefore attain infinite velocity. Space could not be infinite, because in Aristotle's view, adopted from the work of ◊Eudoxus of Cnidus and Callippus (*c.* 370–*c.* 300 BC), the universe consisted of a series of concentric spheres that rotated around the centrally placed, stationary Earth. If the outermost sphere were an infinite distance from the Earth, it would be unable to complete its rotation within a finite period of time, in particular within the 24-hour period in which the stars – fixed, as Aristotle believed, to the sphere – rotated around the Earth.

> *What we have to learn to do,*
> *we learn by doing.*
> ***Aristotle*** NICOMACHAEAN ETHICS BK II

Aristotle's cosmos – geocentric and broadly speaking mechanical – differed only in details from the model proposed by Eudoxus and Callippus. Callippus posited 33 spheres; Aristotle added 22 new spheres, then amalgamated some of them to reach a total of 49. This clumsy model, which was unable to account even for eclipses, was partly replaced by the Ptolemaic system based on epicycles (see ◊Ptolemy). In Aristotle's system the outermost sphere contained the fixed stars. Then followed the spheres of Saturn, Jupiter, Mars, the Sun, Venus, Mercury, and, closest to the Earth, the Moon. Each of these had several spheres in order to account for all their movements. The outermost sphere was controlled by divine influence and indirectly it determined the movement of all the inner spheres. The original motive power of the universe was thus removed from the centre, where the Pythagoreans had placed it.

According to Aristotle's laws of motion, bodies moved upwards or downwards in straight lines. Of ◊Empedocles' four natural elements in the universe, earth and water fell, air and fire rose. To explain the motion of the heavenly spheres, therefore, Aristotle introduced a fifth element, ether, whose natural movement was circular. Aristotle thus posited that the laws of motions governing the celestial bodies above the Moon were different from the laws that governed bodies beneath the Moon.

Aristotle's work in astronomy also included proving that the Earth was spherical. He observed that the Earth cast a circular shadow on the Moon during an eclipse and he pointed out that as one travelled north or south, the stars changed their positions. Since it was not necessary to travel very far to observe this effect, it was clear that the Earth was a sphere, and a rather small one at that. As a result Aristotle was able to make a tolerably fair estimate of the Earth's diameter, overestimating it by only 50%.

Arkwright, Richard (1732–1792) English inventor and manufacturing pioneer. His spinning machinery, which replaced and increased the speed and efficiency of handspinning, transformed the textile industry.

Born in Preston, Lancashire, on 23 December 1732, Arkwright was the youngest of seven surviving children of poor parents. He had little formal education but was taught to read by an uncle and educated himself until quite late in life.

Arkwright first worked as a barber–wigmaker using his own process for preparing and dyeing wigs. When the wig trade declined he turned his attention to engineering, and especially to the textile industry which was then being industrialized following John Kay's invention of the flying shuttle in 1733, the introduction of a carding machine in 1760, and of James Hargreaves' spinning jenny in 1764. The jenny partly did away with old-fashioned handwheel spinning but only for spinning the weft. The roving process was still performed by hand, since the jenny did not have sufficient strength for the warp, the longitudinal threads.

In 1767, Arkwright, probably unaware of Hargreaves' work, began to develop a machine that could spin by rollers. With this, Arkwright applied a new principle rather than simply mechanizing handwheel methods. His spinning frame consisted of four pairs of rollers acting by tooth and pinion. The top roller was covered with leather, to enable it to take hold of the cotton material. The lower was fluted longitudinally to let the cotton pass through it. One pair of rollers revolved more quickly than its corresponding pair, and so the rove was drawn to the requisite fineness for twisting. This was accomplished by spindles, or flyers, placed in front of each set of rollers.

Arkwright's spinning frame was patented in 1769. He set up a small factory at Nottingham, followed two years later by another at Cromford, Derbyshire. His partners there included Jedediah Strutt and Samuel Need, owners of a patent for the manufacture of ribbed stockings. At Cromford, the machines were run by water power, and the spinning frame therefore became known as the 'water frame'; it was later renamed 'throstle' with the advent of steam power in 1785.

The stockings woven on the spinning frame, from yarn of hard, firm texture and smooth consistency, were superior to those woven from handspun cotton. Other manufacturers who had declined to use the yarn were later to regret their decision. In 1773 Arkwright produced the first cloth made entirely from cotton, when he used thread as warp for the manufacture of calico: previously, the warp was of linen, and only the weft was of cotton.

The new material attracted enthusiastic demand, especially after 1774, when a special act of Parliament was passed exempting Arkwright's goods from the double duty imposed on cottons by an act of 1736. (The 1736 act had been designed to protect woollen manufacturers against the calico cottons of India, where the British East India Company was active in trade).

By 1775, Arkwright's factories, sited countrywide, were employing a sequence of machines to carry out all cotton manufacturing operations, from carding, drawing, and roving to spinning. The prosperity of the cotton industry was thus assured, and Arkwright's inventions were later adapted, with equal success, for the woollen and worsted trades. With his comprehensive machinery, Arkwright was an improver and adapter rather than an inventor. To a great extent he incorporated the ideas of others, including fundamental and original ideas, which was why the comprehensive patent he took out in 1775 was later rescinded. This did not detract from Arkwright's achievement, however, and in 1786 his work was acknowledged by King George III with a knighthood. Further recognition followed in 1787, when Arkwright was made high sheriff of Derbyshire.

Arkwright was the archetypal product of a new industrial age that enabled men of ingenuity and determination to rise to eminence from a state of poverty. Apart from his technical achievements, he was also notable for his organization of large-scale factory production and the division of labour between machines and work force. By 1782 Arkwright employed 5,000 workers. He died, rich and respected, at Cromford on 3 August 1792.

Armstrong, Edwin Howard (1890–1954) US electronics engineer who worked almost exclusively in radio, developing the superheterodyne receiver and frequency modulation (FM) radio transmission.

Armstrong was born in New York City on 18 December 1890, the son of a publisher. As a child he had an intense interest in machines and mechanisms and by the age of 14, stimulated by reports of the work of the Italian electrical engineer Guglielmo ◊Marconi, he was constructing wireless (radio) circuits for transmitters and receivers. After graduating from high school he attended the engineering department of Columbia University, New York, gaining a degree in electrical engineering in 1913. He became an instructor at Columbia, and during World War I he worked in the laboratories of the US Signal Corps in Paris. In 1918, after the end of the war, he returned to Columbia University as an assistant to the physicist Michael Pupin (1858–1935), whom he eventually succeeded as professor of electrical engineering in 1934. During World War II he again carried out research for the military. His inventions earned him a great deal of money – he became a millionaire – but also involved him in expensive, protracted litigation and promotional costs. His health declined and he became depressed; at the end of January 1954 he committed suicide by jumping from the window of his New York apartment.

In 1912, while he was still a student at Columbia, Armstrong designed and built a regenerative, or feedback, radio receiving circuit that made use of the amplifying ability of the triode thermionic valve, called the Audion by its inventor Lee ◊De Forest in 1906. The same circuit acted as an oscillator (transmitter) at high amplifications. He patented it, and for 14 years fought a series of legal battles about priority with De Forest. Armstrong finally lost the struggle in the Supreme Court, although the legal interpretation was not accepted by most of the scientists and engineers of the time, who continued to uphold Armstrong's claim and awarded him the Franklin Medal for his invention.

The superheterodyne receiving circuit was developed by Armstrong during World War I in an attempt to make a receiver that could detect the presence of aircraft by means of the electromagnetic radio waves given off by the sparking of the ignition systems of their engines. Again he patented the idea, and sold patent rights and licences to major manufacturers, including RCA in the USA. During the 1920s, earnings from these sources made him a rich man.

At that time, radio broadcasting used amplitude modulation (AM), in which the audio signal varies the amplitude of a transmitted carrier wave. Electrical disturbances in the atmosphere caused by storms or electrical machinery also modulate the carrier wave, resulting in static interference at the receiver. In 1933 Armstrong, working with Pupin, developed and patented a method of radio broadcasting in which the transmitted signal is made to modulate the frequency of the carrier wave over a wide waveband. This method, called frequency modulation (FM), is unaffected by static and is capable of high-fidelity sound reproduction; it remains the basis of quality radio, television, microwave, and satellite transmissions, although the high frequencies employed are generally limited to line-of-sight distances. FM broadcasting did not gain ground until after World War II, and then only after Armstrong had used much of his fortune in promoting it and in another round of legal actions about patent rights. Frustration and disenchantment with the attendant commercial, legal, and financial problems probably led to his suicide.

Armstrong, William George (1810–1900) English engineer and inventor who revolutionized the manufacture of big guns in the mid-18th century. He was also responsible for pioneering developments in hydraulic equipment.

Armstrong was born in Newcastle upon Tyne on 26 November 1810. He attended private schools in Newcastle and Whickham, and later completed his formal education at a grammar school in Bishop Auckland. Following this, he studied law in London. In 1833 he returned to Newcastle and was engaged in private practice as a solicitor. In his spare time he carried out numerous scientific experiments.

In 1839 he constructed an overshot waterwheel and soon afterwards he designed a hydraulic crane, which contained the germ of the ideas underlying all the hydraulic machinery for which he subsequently became famous. Abandoning his law practice in 1847, he founded an engineering works at Elswick to specialize in building hydraulic cranes. In 1850 he invented the hydraulic pressure accumulator and in 1854 he designed submarine mines for use in the Crimean War. A year later Armstrong designed a three-pounder gun with a barrel made of wrought iron wrapped round an inner steel tube. In 1859 he established the Elswick Ordnance Company for making the Armstrong gun for the British army. In the same year, he was appointed chief engineer of rifled ordnance at Woolwich, and received a knighthood. A government decision to revert to the production and use of muzzle-loading guns led to his resignation from Woolwich. He improved the design of his original gun and, by 1880, had completed the design of a 150-mm/6-in breech-loading gun with a wire-wound cylinder. This design was adopted by the British government and was the prototype of all subsequent artillery.

In 1882 Armstrong established a new shipbuilding yard at Elswick for the construction of warships. Five years later he was raised to the peerage and created Baron Armstrong of Craigside. He also entered into partnership with Joseph ◊Whitworth to form the famous Armstrong–Whitworth company at Openshaw, Manchester.

Although Armstrong initially chose law as his career, he soon switched to engineering. Even before he had abandoned his law practice, he had constructed an overshot waterwheel, having become engrossed with waterwheels as a source of power when, on a fishing holiday in Dentdale, he saw an overshot water wheel and observed that only about a twentieth of the energy available was being utilized. His original invention of 1839 was improved upon by his production of a rotary water motor.

His invention of a hydroelectric machine, which generated electricity by steam escaping through nozzles from an insulated boiler, originated in a report he read of a colliery engineman who noticed that he had received a sharp electric shock on exposing one hand to a jet of steam issuing from a boiler with which his other hand was in direct contact.

Armstrong's first hydraulic crane depended simply on the pressure of water acting directly on a piston in a cylinder. The resulting movement of the piston produced a corresponding movement through suitable gears. The first example was erected on the quay at Newcastle in 1846 – the pressure being obtained from the ordinary water mains of the town. The merits and advantages of this device soon became widely appreciated.

In 1850 a hydraulic installation was required for a new ferry station at New Holland on the Humber estuary. The absence of a water mains of any kind, coupled with the prohibitive cost of a special reservoir because of the character of the soil, made it necessary for Armstrong to invent a fresh piece of apparatus. This became known as the hydraulic accumulator. It consisted of a large cylinder containing a piston that could be loaded to any desired pressure – the water being pumped in below it by a steam engine or other prime mover. With various modifications, this device made possible the installation of hydraulic power in almost any situation. In particular it could be used on board ship, and its application to the manipulation of heavy naval guns made it among the most important of Armstrong's inventions.

The Elswick works had originally been founded for the manufacture of this hydraulic machinery, but it was not long before it became the birthplace of a revolution in gunmaking. Modern artillery dates from 1855 when Armstrong's first gun made its appearance. This weapon embodied all the essential features that distinguish the ordnance of today from the cannon of the Middle Ages.

There had been little change in heavy guns for 500 years. Their barrels were cast in bronze and they were loaded from the muzzle. Armstrong used the advances of the 19th century in metallurgy and chemistry, together with his own inventive genius. His gun was built up of rings of metal shrunk upon an inner steel barrel. It was loaded at the breech and it was rifled, and it threw not a round ball but an elongated shell. The British army adopted the Armstrong gun in 1859. The UK had thus originated an armament superior to that possessed by any other nation at that time. But in 1863 defects in the breech mechanism caused Armstrong to spend 17 years of his life improving the original design, which included the use of steel wire; Armstrong perceived that to coil many turns of wire round an inner barrel was a logical extension of the large hooped method that had been used in the gun designed by him in 1855.

Arp, Halton Christian (1927–) US astronomer known for his work on the identification of galaxies.

Arp was born on 21 March 1927 in New York City. He was educated at Harvard University, where he obtained his BA in 1949. Four years later he gained a PhD at the California Institute of Technology and became a Carnegie Fellow at the Mount Wilson and Palomar Observatory in 1953. He was a research associate at the University of Indiana 1955–57, and for the next eight years was assistant astronomer at the Mount Wilson and Palomar Observatory, of the Carnegie Institute in Washington, and at the California Institute of Technology. From 1965–69 he was astronomer at those institutions. In 1969 he became astronomer at the Hale Observatory in California and in 1960 a visiting professor of the National Science Foundation. In the 1990s he moved to Germany, researching at the Max Planck Institute of Astrophysics at Garching, near Munich. He is a member of several astronomical associations and was chair of the Los Angeles Chapter of the Federation of American Scientists and of Sigma XI, in 1965. He also received several awards for his achievements in the field of astronomy.

In 1956, while at Indiana University, Arp established the ratio between the absolute magnitude of novae at maximum brightness and the speed of decline of magnitude. Since then he has published several papers and in 1965 wrote the *Atlas of Peculiar Galaxies.*

During his research on globular clusters, globular cluster variable stars, novae, Cepheid variables, extragalactic nebulae, and so on, Arp has attempted to relate the listings of galaxies to radio sources and has compiled a catalogue of radio sources from the galaxies shown in the Palomar sky atlas; the optical identification of these sources can now be done fairly accurately. The most interesting of the radio sources found so far are quasars, which are characterized by a strong emission in the ultraviolet part of the spectrum. Arp is working on the question of whether the high red shifts in the spectrum of quasars are due to their great age and remote distance or, as he suspects, to their being ejected at high velocities from explosions within much nearer galaxies.

Arp has also carried out the first photometric work on the Magellanic Clouds – the nearest extragalactic system. In his research on pulsating novae, particularly those in the Andromeda nebula, Arp has demonstrated that there is a close relationship between the maximum magnitude and the luminosity of novae so that it is now possible to obtain absolute luminosities for novae fairly easily using light curves (graphs relating apparent magnitude to time).

Arp has attempted also to obtain better data on RV Tauri stars – variables that are much brighter than other cluster stars, many of which lie well outside the clusters. In doing this research he investigated the whole series of variable stars with periods of more than one day, in a number of globular clusters, and related their magnitudes with those of cluster-type variables in the same cluster. Arp's results are referred to as the zero-point of the cluster-type variables.

In the continuing search for explanations for and identification of phenomena in the universe, Arp has done a great deal to aid the classification of information as well as to provide a basis from which other astronomers may work to increase knowledge of these and other, as yet unexplained phenomena.

Arrhenius, Svante August (1859–1927) Swedish physical chemist who first explained that in an electrolyte (a solution of a chemical dissolved in water) the dissolved substance is dissociated into electrically charged ions. The electrolyte conducts electricity because the ions migrate through the solution. Although later modified, Arrhenius' theory of conductivity in solutions has stood the test of time. It was a major contribution to physical chemistry, ultimately acknowledged by the award to Arrhenius of the 1903 Nobel Prize for Chemistry.

Arrhenius was born in Uppsala on 19 February 1859, the son of a surveyor and estate manager who was also a supervisor at the local university. He was a brilliant student, entering Uppsala University at the age of 17 to study chemistry, physics, and mathematics. After graduating in 1878, he stayed to write his doctoral thesis, but became dissatisfied with the teaching at Uppsala and went to Stockholm to study solutions and electrolytes under Erik Edlund (1819–1888).

In 1884 he submitted his thesis, which contained the basis of the dissociation theory of electrolytes (although he did not at that time use the term dissociation), together with many other novel theories that aroused only suspicion and doubt in his superiors. The largely theoretical document was not welcomed by the academics, who were devoted experimentalists; Arrhenius later boasted that he had never performed an exact experiment in his life and preferred to take a general view of relationships from the results of many approximate experiments. The thesis (written in French) was awarded only a fourth class, the lowest possible pass, but Arrhenius sent copies of it to several eminent chemists, including Friedrich Wilhelm Ostwald (at Riga), Rudolf Clausius (Bonn), Lothar Meyer (Tübingen), and Jacobus ◊van't Hoff (Amsterdam). Ostwald's offer to Arrhenius of an academic appointment moderated the scepticism of the Uppsala authorities, who finally offered him a position and, later, a travelling fellowship. This enabled him to spend some time with other scientists working in the same field, such as Friedrich Kohlrausch, Hermann Nernstat Würzburg, Ludwig Boltzmann at Graz, and Jacobus van't Hoff, whose solution theories paralleled that of Arrhenius.

In 1891 Arrhenius was offered a professorship at Giessen in Germany as successor to Justus von Liebig, but he declined in preference for an appointment at the Royal Institute of Technology in Stockholm. Four years later he became professor of physics at a time when his work was attracting the attention of scientists throughout Europe, if not in his native Sweden – his election to the Swedish Academy of Sciences had to wait another six years until 1901 (the same year in which van't Hoff was awarded the first Nobel Prize for Chemistry). He again declined a German professorship, in Berlin, in 1905 and took instead the specially created post of director of the Nobel Institute of Physical Chemistry (Stockholm), where he remained until shortly before his death on 2 October 1927.

The Arrhenius theory of electrolytes (1887) is concerned with the formation, number, and speed of ions in solution. The key to the theory is the behaviour of the dissolved substance (solute) and the liquid (solvent), both of which are capable of dissociating into ions. It postulates that there is an equilibrium between undissociated solute molecules and its ions, whose movement or migration can conduct an electric current through the solution. Its chief points may be summarized as follows:

(a) An electrolytic solution contains free ions (that is, dissociation takes place even if no current is passed through the solution).

(b) Conduction of an electric current through such a solution depends on the number and speed of migration of the ions present.

(c) In a weak electrolyte, the degree of ionization (dissociation) increases with increasing dilution.

(d) In a weak electrolyte at infinite dilution, ionization is complete.

(e) In a strong electrolyte, ionization is always incomplete because the ions impede each other's migration; this interference is less in dilute solutions of strong electrolytes.

Apart from (e), regarding strong electrolytes (a difficulty not resolved until the work of Peter ◊Debye and Erich ◊Hückel in the 1920s), Arrhenius' theory is still largely accepted.

In another notable achievement, Arrhenius adapted van't Hoff's work on the colligative properties of non-electrolyte solutions. He found that solutions of salts, acids, and bases – electrolytes – possess greater osmotic pressures, higher vapour pressures, and lower freezing points than van't Hoff's calculated values but explained the discrepancies in terms of ionic dissociation by taking into account the number of solute ions (as opposed to molecules) present. In 1889 Arrhenius suggested that a molecule will take part in a chemical reaction on collision only if it has a higher than average energy – that is, if it is activated. As a result, the rate of a chemical reaction is proportional to the number of activated molecules (not to the total number of molecules, or concentration) and can be related to the activation energy.

After 1905 Arrhenius widened his research activities. For example, he applied the laws of theoretical chemistry to physiological problems (particularly immunology); once again initial criticism was replaced by universal acceptance. With N Ekholm he published papers on cosmic physics concerning the northern lights (aurora borealis), the transport of living matter ('spores') through space from one planet to another, and the climatic changes of the Earth over geological time – pointing out the 'greenhouse effect' brought about by carbon dioxide in the atmosphere. Arrhenius became more and more respected by the world of science and was much sought after for meetings, lectures, and discussions throughout the world. In his latter years he had to rise at 4 am in order to maintain his scientific activities, and this consistent hard work probably contributed to his death at the age of 68.

Artin, Emil (1898–1962) Austrian mathematician who made important contributions to the development of class field theory and the theory of hypercomplex numbers.

He was born in Vienna on 3 March 1898, but his secondary education was at Reichenburg in Bohemia (now part of Czechoslovakia). His undergraduate study at the University of Vienna was interrupted when he was called up for military service in World War I; in 1919 he went to the University of Leipzig, where he received his PhD in 1921. From 1923 to 1937 he lectured at the University of Hamburg in mathematics, mechanics, and the theory of relativity. He and his family emigrated to the USA in 1937. There Artin lectured at the University of Indiana 1938–1946 and Princeton 1946–1958. In 1958 he returned to Hamburg, where he died on 20 December 1962.

Artin's early work was concentrated on the analytical and arithmetical theory of quadratic number fields and in the 1920s he made a number of major advances in this field. In his doctoral thesis of 1921 he formulated the analogue of the Riemann hypothesis about the zeros of the classical zeta function, studying the quadratic extension of the field of rational functions of one variable over finite constant fields, by applying the arithmetical and analytical theory of quadratic numbers over the field of natural numbers. Then in 1923, in the most important discovery of his career, he derived a functional equation for his new-type L-series. The proof of this he published in 1927, thereby providing, by the use of the theory of formal real fields, the solution to David ◊Hilbert's problem of definite functions. The proof produced the general law of reciprocity – Artin's phrase – which included all previously known laws of reciprocity going back to Karl Gauss and which became the fundamental theorem in class field theory. Between the statement of his theory in 1923 and its publication in 1927, Artin made two other important theoretical advances. His theory of braids, given in 1925, was a major contribution to the study of nodes in three-dimensional space. A year later, in collaboration with Schrier, he succeeded in treating real algebra in an abstract manner, defining a field as real–closed if it itself was real but none of its algebraic extensions were. He was then able to demonstrate that a real–closed field could be ordered in an exact manner and that, in it, typical laws of algebra were valid.

Although the fires of his genius burned less brightly after 1930, Artin continued to work at a high level. In 1944 his discovery of rings with minimum conditions for right ideals – now known as Artin rings – was a fertile addition to the theory of associative ring algebras, and in 1961 he published his *Class Field Theory*, a rounded summation of his life's work as one of the leading creators of modern algebra.

Aston, Francis William (1877–1945) English chemist and physicist who developed the mass spectrograph, which he used to study atomic masses and to establish the existence of isotopes. For his unique contribution to analytic chemistry and the study of atomic theory he was awarded the 1922 Nobel Prize for Chemistry.

Aston was born on 1 September 1877 at Harbourne, Birmingham, the son of a merchant. He went to school at Malvern College, where he excelled at mathematics and science, and then to Mason College (which later became the University of Birmingham) to study chemistry. There from 1898 to 1900 he studied optical rotation with P F Frankland (1858–1946). Aston then left academic life for three years to work with a firm of brewers, although during his spare time he continued to experiment with discharge tubes. He returned to Mason College in 1903 to study gaseous discharges, before moving in 1909 to the Cavendish Laboratory, Cambridge, where J J ◊Thomson was also investigating positive rays from discharge tubes. Thomson and Aston examined the effects of electric and magnetic fields on positive rays, showing that the rays were deflected – one of the basic principles of the mass spectrograph.

Aston's researches were interrupted by World War I, during which he worked at the Royal Aircraft Establishment, Farnborough, on the treatment of aeroplane fabrics using dopes (lacquers). He escaped injury in 1914 after crashing in an experimental aircraft. After the war he returned to Cambridge and improved his earlier equipment, and the mass spectrograph was born. He went on to refine the instrument and apply it to a study of atomic masses and isotopes. Aston continued to live and work in Cambridge, where he died on 20 November 1945.

Between 1910 and 1913, Thomson and Aston showed that the amount of deflection of positive rays in electric and magnetic fields depends on their mass. The deflected rays were made to reveal their positions by aiming them at a photographic plate. Thomson was seeking evidence for an earlier theory of William ▷Crookes that the nonintegral atomic mass of neon (20.2) was caused by the presence of two very similar but different atoms. Each was expected to have a whole number atomic mass, but their mixture would result in an average nonintegral value. The Cambridge scientists found that the two paths of deflected positive rays from a neon discharge tube were consistent with atomic masses of 20 and 22. Aston attempted to fractionate neon and in 1913 made a partial separation by repeatedly diffusing the gas through porous pipeclay.

Aston's first improvement to the apparatus, made in 1919, caused the positive-ray deflections by both electric and magnetic fields to be in the same plane. The image produced on the photographic plate became known as a mass spectrum, and the instrument itself as a mass spectrograph.

Its principle is relatively simple. A beam of positive ions is produced by an electric discharge tube (in which a high voltage is passed between electrodes in a glass tube containing rarefied gas), which has holes in its cathode to let the accelerated ions pass through. The beam passes between a pair of electrically charged plates, whose electric field deflects the moving ions according to their charge-to-mass ratio, e/m (where e is the charge on the ion – usually 1 or 2 – and m is its mass). Lighter ions are deflected most, whereas those of largest mass are deflected least. The now separate ion streams then pass through a magnetic field arranged at right angles to the electric field, which deflects them still further in the same plane. The streams strike a photographic plate, where they expose a series of lines that constitute the mass spectrum. The position of a line depends on the ion's mass, and its intensity depends on the relative abundance of the ion in the original positive-ion beam.

The work with neon established that two spectral lines are produced on the plate, one about nine times darker (on a positive print) than the other. Calculation showed that these correspond to two types of ions of atomic masses 20 and 22. There are nine times as many of the former as of the latter, giving a weighted average atomic mass of about 20.2 (the value originally reported in 1898 by William Ramsay and Morris Travers, the discoverers of neon). Aston stated that there must be two kinds of neon atoms which differ in mass but not in chemical properties – that is, that naturally occurring neon gas consists of two isotopes.

Over the next few years Aston examined the isotopic composition of more than 50 elements. Most were found to have isotopes – tin has ten with atomic masses that are whole numbers (integers). In 1920, using the first mass spectrograph, he determined the mass of a hydrogen atom and found it to be 1% greater than a whole number (1.01). (Twelve years later in the USA Harold Urey discovered deuterium, an isotope of hydrogen with mass 2.) With an improved spectrograph, accurate to 1 part in 10,000, Aston confirmed that some other isotopes also show small deviations from the whole-number rule. The slight discrepancy is the packing fraction. (For example, the particles that make up four atoms of hydrogen are the same as those in one atom of helium, but in helium they are 'packed' and have 1% less mass. This mass defect is now known to be the source of the thermonuclear energy released during the fusion of hydrogen to form helium.)

Aston's interests also included astronomy, particularly observations of the Sun and its eclipses. His knowledge of photography made him a valuable member of the expeditions that studied eclipses in Sumatra in 1925, Canada in 1932, and Japan in 1936. But Aston will be remembered for his development of the mass spectrograph, which became an essential tool in the study of nuclear physics and later found application in the determination of the structures of organic compounds.

Atkinson, Robert D'escourt (1898–1982) Welsh astronomer and inventor.

Atkinson was born on 11 April 1898 at Rhayader. He was educated at Oxford University, where he obtained a BA in 1922, and then at the University of Göttingen in Germany, where he gained a PhD in 1928. He was a demonstrator in physics at the Clarendon Laboratory at Oxford 1922–26 and an assistant at the Technical University in Berlin 1928–29. He became assistant professor at Rutgers University in New Jersey 1922–34, then associate professor 1964–73, and later adjunct professor of astronomy there.

Atkinson was a member of the Harvard University/Massachusetts Institute of Technology eclipse expedition to the Soviet Union in 1936, and during World War II he was with the mine-design department of the British Admiralty. From 1944–46 he served with Ballistic Research Laboratory in Maryland.

Between 1952–55 Atkinson designed the astronomical clock at York Minster in England, and he designed a standard time sundial at Indiana University in 1977. He was a member of the British National Committee for Astronomy 1960–62 and belonged to the American Physical Society, the American Astronomical Society, the Royal Astronomical Society, the British Astronomical Association, and the Royal Institute of Navigation. He was awarded the Royal Commission Award to Inventors in 1948 and the Eddington Medal of the Royal Astronomical Society in 1960. In 1977 the International Astronomical Union named a minor planet (1,827 Atkinson) in his honour.

Atkinson's research was in the field of atomic synthesis, stellar energy, and positional astronomy. He was also deeply involved in instrument design.

Many scientists had been concerned with the problem of discovering how the Sun has maintained a reasonably steady yet high rate of radiation for at least 3 million years. The problem essentially was that there was no known physical or chemical process that could generate radiation from the materials that make up the Sun at so great a rate over so long a period of time, nor was there enough energy released by the contraction of the Sun under its own gravitation. Astronomers therefore began to look elsewhere for a process that could explain the mystery. In 1924 the English astronomer Arthur ◊Eddington, whose field of study was the internal make-up of stars, was computing what conditions must be like beneath the surface of stars in order that the basic laws of physics be obeyed. Eddington suggested that the only possibility was a process whereby atoms were broken down inside the central core of a star, converting matter into energy. He was supported in this view by Atkinson, who in 1932 was working at the Royal Greenwich Observatory. In that year there were new results from the physicists Ernest Rutherford, John Cockcroft, and Ernest Walton, who had just succeeded in splitting the central core, or nucleus, of an atom. Atkinson was able to work out a theoretical model of the way in which matter could be annihilated. Not only did he determine the amount of energy released from atomic reactions within stars, but he was also able to suggest the kinds of reactions necessary to produce the vast quantities of radiation required.

With the enormous amount of research into nuclear physics carried out since World War II, astronomers now have a much greater insight into how stars evolve and how long their evolution takes. There are still many questions to be answered, but it is now accepted that energy is generated in the Sun by a process of nuclear fusion. The central core of an atom fuses with the nuclei of other atoms, forming a new and heavier atom and at the same time releasing a vast amount of energy. Calculations have shown that this energy is more than enough to keep a star like the Sun radiating for billions of years.

Robert Atkinson's contributions were fundamental to our basic understanding of how stars like the Sun work and how they evolve.

Attenborough, David Frederick (1926–) English naturalist, film-maker, and author who is best known for his wildlife films, which have brought natural history to a wide audience.

Attenborough was born on 8 May 1926. He was educated at Wyggeston Grammar School in Leicester and Clare College, Cambridge, where he read zoology. From 1947–49 he did two years' military service in the Royal Navy, after which he became an editorial assistant in an educational publishers. In 1952 he joined the BBC Television Service as a trainee producer, then in 1954 he went on his first expedition, to West Africa. During the next ten years he made annual trips to film and study wildlife and human cultures in remote parts of the world; these expeditions were recorded in the *Zoo Quest* series of television programmes and books. From 1965–68 Attenborough was controller of BBC2, then from 1969–72 he was director of television programmes for the BBC and a member of its board of management. Despite the large amount of administrative work involved in these posts, he still managed to undertake several filming expeditions. His next major achievement was the television series *Life on Earth,* which was first shown in 1979. In this huge project, which took three years to complete, Attenborough attempted to outline the development of life on Earth – from its very beginnings to the present day – using plants and animals found today to illustrate the process of evolution. The series and its associated book (which also first appeared in 1979) met with great popular and critical acclaim and set new standards for the presentation of natural history to nonspecialists.

He later made two complementary, equally successful television series: *The Living Planet* (1983), which dealt with ecology and the environment, and *The Trials of Life* (1990), which described life cycles. *The Private Life of Plants* (1995) and *The Life of Birds* (1998) described the diversity and adaptive strategies of plants and birds, respectively. He was knighted in 1985.

Audubon, John James Laforest (1785–1851) French-born US ornithologist who painted intricately detailed studies of birds and animals. He was also an ardent conservationist.

Audubon was born in Les Cayes, Santo Domingo (now Haiti), on 26 April 1785. He was the illegitimate son of a French sea captain and a Creole woman who died soon after his birth. His father, who was also a planter, sent him back to France to his home near Nantes, where Audubon and his half-sister were taken in by the captain's wife, who had no children of her own. The couple legally adopted him in 1794. The young Audubon acquired a deep interest for natural history, painting, and music. He was educated locally and in Paris, where he had six months' tuition at the studio of the well-known painter Jacques Louis David. By 1803 young men in France were being conscripted for Napoleon's army, but the 18-year-old Audubon avoided the draft by making a timely emigration to the

USA to take up the running of his father's properties near Philadelphia. In 1808 he married and opened a store in Louisville, Kentucky, but he was a casual and poor businessman, his time being so passionately absorbed by nature; he was even imprisoned for debt in 1819 and declared a bankrupt.

Despite his financial problems, Audubon maintained a successful marriage and travelled throughout the USA collecting and painting the wildlife around him, while his wife worked as a teacher and governess to help to support them. He also painted portraits and even street signs, and gave lessons in drawing and French. By 1825 he had compiled his beautiful set of bird paintings, but US publishers were not interested. The following year Audubon set sail for the UK, where the Havells engraved his plates which he published by subscription. His talent was lauded and he attracted much publicity by appearing in English society wearing the outlandish clothes so suitable for his travels in the wilds of the USA. Even so, he was elected a Fellow of the Royal Society in 1830. After 13 years in the UK he went back to the USA. He died in New York on 27 January 1851.

Before Audubon, most painters of birds used stylized techniques; stuffed birds were often used as subjects. Audubon painted from life and his compositions were startling, his detail minute. *The Birds of America* was published in the UK in 87 parts between 1827 and 1838. On his return to the USA in 1839 he published a bound edition of the plates with additions. He illustrated *Viviparous Quadrupeds of North America* (1845–1848), compiling the text 1846–54 with his sons and John Bachman. Audubon was one of the earliest naturalists to pioneer conservation, and the various Audubon societies of today are named in his honour.

Auer, Carl (1858–1929) Austrian chemist and engineer; see ◊von Welsbach, Freiherr.

Avery, Oswald Theodore (1877–1955) Canadian-born US bacteriologist whose work on transformation in bacteria established that DNA (deoxyribonucleic acid) is responsible for the transmission of heritable characteristics. He also did pioneering research in immunology – again working with bacteria – proving that carbohydrates play an important part in immunity. Avery's achievements gained him many honours, including election to the National Academy of Science in the USA and to the Royal Society of London.

Avery was born on 21 October 1877 in Halifax, Nova Scotia, the son of a clergyman, but spent most of his life in New York City, where he was taken by his father in 1887. After qualifying in medicine in 1904 at Columbia University, Avery spent a brief period as a clinical physician but soon moved to the Hoagland Laboratory in Brooklyn in order to research and lecture in bacteriology and immunology. In 1913 he transferred to the Rockefeller Institute Hospital in New York, where he remained until he retired in 1948. Avery died in Nashville, Tennessee, having moved there on his retirement.

Avery's work on transformation – a process by which heritable characteristics of one species are incorporated into another species – is generally considered to be his most important contribution and was stimulated by the research of F Griffith, who in 1928 published the results of his studies on *Diplococcus pneumoniae,* a species of bacteria that causes pneumonia in mice. Griffith found that mice contracted pneumonia and died when they were injected with a mixture of an encapsulated strain of dead *D. pneumoniae* (living encapsulated bacteria are lethal to mice) and living unencapsulated bacteria (which have no protective outer capsule to resist antibodies and therefore do not cause pneumonia), despite the fact that, separately, each of the mixture's components is harmless. From the corpses he then isolated virulent, living encapsulated bacteria. These findings led Griffith to postulate that a transforming principle from the dead encapsulated bacteria had caused capsule development in the living, unencapsulated bacteria, thereby making them virulent. Moreover, when these living bacteria reproduced, the offspring were encapsulated, suggesting that the transforming principle had become incorporated into their genetic constitution.

Initially, Avery dismissed Griffith's findings but when they were supported by later studies, Avery and his colleagues Colin MacLeod and Maclyn McCarthy began investigating the nature of the transforming principle. They started experimenting in the early 1940s, working on *D. pneumoniae.* They obtained a pure sample of the virulent, living, encapsulated bacteria which were then killed by heat treatment. The bacteria's protein and polysaccharide (which makes up the capsule and is also found within the cells) were then removed and the remaining portion was added to living, unencapsulated bacteria. It was found that the progeny of these bacteria had capsules, so the active transforming principle still remained and was neither a protein nor a polysaccharide. Finally Avery extracted and purified the transforming principle and used various chemical, physical, and biological techniques to identify it. His analysis proved conclusively that DNA was the transforming principle responsible for the development of polysaccharide capsules in the unencapsulated bacteria.

Avery's discovery (which he published in 1944) was extremely important because for the first time it had been proved that DNA controls the development of a cellular feature – in this case the polysaccharide capsule – and implicated DNA as the basic genetic material of the cell. Other researchers later confirmed that DNA controls the development of cellular features in different organisms, and also established that it is the fundamental molecule involved in heredity. Moreover, Avery's work stimulated interest in DNA, eventually leading to the determination of its structure and method of replication by Francis ⟡Crick and James ⟡Watson in the early 1950s.

Avery's early work also involved pneumococci (bacteria that cause pneumonia), but was in the field of immunology. He demonstrated that pneumococci could be classified according to their immunological response to specific antibodies and that this immunological specificity is due to the particular polysaccharides that constitute the capsule of each bacterial type. This research established that polysaccharides play an important part in immunity and led to the development of sensitive diagnostic tests to identify the various types of pneumococcus bacteria.

Avicenna (980–1037) Arabic Abu 'Ali Al-Husain Ibn Abdallah Ibn Sina. Arab philosopher, physician and scientist. He was one of the main interpreters of Aristotle to the Arab world. He was physician to several sultans and a vizier in Hamadan, Persia. He was the author of over 250 books on philosophy, religion, and science.

Born at Afshana, near Bokhara in Central Asia (now in Uzbekistan), Avicenna was conventionally educated in the Koran and classical Arabic texts. A child prodigy, he then turned to astronomy and mathematics. He also learned Greek and mastered all the contemporary medical knowledge. At the age of 16 he began to practise medicine. When he was 18 he was appointed court physician to Prince Nuh ibn Mansur, gaining access to the Samanid court library, which he took full advantage of. When the Samanids were overthrown, he travelled for a time taking odd jobs – he was a jurist a Koranj, an administrator at Rayy, and physician to Prince Shams al-Dawlah at Hamadan. At Rayy he began his five-volume medical text book, *al Qānūn/ The Canon* in which he set down all previous medical knowledge and also his own ideas. This work was beautifully presented and overshadowed the work of al-Rāzī, 'Alī Ibn al-'Abbās, and even the Greek physician ⟡Galen. Indeed he was honoured in his time by the name 'Jālīnūs-al-Islām (Galen in Islam). *Al Qānūn* was well received by physicians and was taken as the authoritative work on medicine and many physicians took it to be complete and sufficient to their needs. Historians argue that this attitude, through no fault of Avicenna's, meant that Arabic and even European medicine remained static for several centuries. *Al Qānūn* became a standard text in Europe and the Middle East. It was translated into Latin by Gerard de Cremona between 1473 and 1486, and was the textbook at the universities of Montpellier and Louvain until as late as 1650.

Avicenna synthesized a number of currents of thought in an original way – the Koran and its theological ramifications, including thoedicy, cosmogony, anthropology, and eschatology; science, which included Greek astronomy, the circular movement of the heavenly spheres, the hierarchy of the cosmos, and the four elements; and philosophy, which relied on a melange of Aristotelianism with elements of Platonism and aspects of the Persian tradition. His major philosophical work *Al-Shifā'/The Cure (of ignorance)* or *Book of Healing,* much of which was based on the ideas of Aristotle and other Greek philosophers, was a four-part encyclopedia comprising logic, physics, mathematics, music, astronomy, and metaphysics.

Avicenna attempted to integrate science and religion into one great vision that would explain the formation of the universe, as well as more earthly problems including the organization of states. He was often involved in the politics of the warring states in which he lived and several times had to escape capture and imprisonment for his views. In 1022, Prince al-Dawla died and Avicenna left Hamadan taking refuge in the court of Prince 'Ala ad-Dawla, where he continued his prolific writings. He died of a mysterious illness while on a campaign with the prince in 1037. This was apparently a colic but it is thought that he may have been poisoned by one of his servants.

Avicenna's legacy is a bibliography of nearly 270 titles covering a wide range of topics. Among many other things he studied and wrote on the elements, symbology, minerals, optics and vision, time and motion, and anticipated Newton's law. His books also include an autobiography which was finished by al-Jūzjānī, one of his disciples.

Avogadro, Amedeo (1776–1856) Conte de Quaregna. Italian scientist who shares with his contemporary Claud Berthollet the honour of being one of the founders of physical chemistry. Although he was a professor of physics, he acknowledged no boundary between physics and chemistry and based most of his findings on a mathematical approach. Principally remembered for the hypothesis subsequently known as Avogadro's law (which states that, at a given temperature, equal volumes of all gases contain the same number of molecules), he gained no recognition for his achievement during his lifetime. He lived in what was a scientific backwater, with the result that his writings received scant examination or regard from the leading authorities of his day.

Avogadro was born in Turin on 9 June 1776. He

began his career in 1796 by obtaining a doctorate in law and for the next three years practised as a lawyer. In 1800 he began to take private lessons in mathematics and physics, made impressive progress, and decided to make the natural sciences his vocation. He was appointed as a demonstrator at the Academy of Turin in 1806 and professor of natural philosophy at the College of Vercelli in 1809, and when in 1820 the first professorship in mathematical physics in Italy was established at Turin, Avogadro was chosen for the post. Because of the political turmoil at that time the position was subsequently abolished, but calmer times permitted its re-establishment in 1832 and two years later Avogadro again held the appointment. He remained at Turin until his retirement in 1850. When he died there on 9 July 1856 his European contemporaries still regarded him as an incorrigibly self-deluding provincial professor of physics.

In 1809 the French chemist Joseph ◊Gay-Lussac had discovered that all gases, when subjected to an equal rise in temperature, expand by the same amount. Avogadro therefore deduced (and announced in 1811) that at a given temperature all gases must contain the same number of particles per unit volume. He also made it clear that the gas particles need not be individual atoms but might consist of molecules, the term he introduced to describe combinations of atoms. No previous scientists had made this fundamental distinction between the atoms of a substance and its molecules.

Using his hypothesis Avogadro provided the theoretical explanation of Gay-Lussac's law of combining volumes. It had already been observed that the electrolysis of water (to form hydrogen and oxygen) produces twice as much hydrogen (by volume) as oxygen. He reasoned that each molecule of water must contain hydrogen and oxygen atoms in the proportion of 2 to 1. Also, because the oxygen gas collected weighs eight times as much as the hydrogen, oxygen atoms must be 16 times as heavy as hydrogen atoms. It also follows from Avogadro's hypothesis that a molar volume of any substance (that is, the volume whose mass is one gram molecular weight) contains the same number of molecules. This quantity, now known as Avogadro's number or constant, is equal to 6.022045×10^{23}.

Leading chemists of the day paid little attention to Avogadro's hypothesis, with the result that the confusion between atoms and molecules and between atomic masses and molecular masses continued for nearly 50 years. In 1858, only two years after Avogadro's death, his fellow Italian Stanislao ◊Cannizzaro showed how the application of Avogadro's hypothesis could solve many of the major problems in chemistry. At the Karlsruhe Chemical Congress of 1860 Avogadro's paper of 1811 was read again to a much wider and more receptive audience of distinguished scientists. One of the most impressed was the young German chemist Lothar ◊Meyer. He found this final establishment of order in place of conflicting theories one of the great stimuli that eventually led him in 1870 to produce his most detailed exposition of the periodic law. A year later his namesake Viktor ◊Meyer used Avogadro's law as his principal yardstick in theoretically explaining the nature of vapour density.

It is interesting to analyse why such a fundamental and potentially useful work as Avogadro's lay fallow for nearly half a century. Various factors contributed to the delay. To begin with, Avogadro did not support his hypothesis with an impressive display of experimental results. He never acquired, nor did he deserve, a reputation for accurate experimental work; his contemporaries did not therefore regard him as a brilliant theoretician, merely as a careless experimenter. Also Avogadro extended his hypothesis to solid elements – and lacking experimental evidence he relied on analogy. So that whereas he was correct in considering molecules of oxygen and hydrogen to be diatomic, he had little justification for making a similar assumption about carbon and sulphur. His speculative treatment of metals in the vapour state (in his second paper of 1814) did little to advance his cause, revealing an excess of theorizing at the cost of attention to detail.

Furthermore, Avogadro's idea of a diatomic molecule was at odds with the dominant dualistic outlook of Jöns ◊Berzelius. According to the principles of electrochemistry, two atoms of the same element would have similar electric charges and therefore repel rather than attract each other (to form a molecule). During the 50 years after Avogadro's original hypothesis most activity was being devoted to organic chemistry, whose analysis and classification was based chiefly on masses, not volumes. And even when Avogadro's work was translated and published, it tended to appear in obscure journals, perhaps as a result of his modesty and his geographical isolation from the mainstream of the chemistry of his time.

Further Reading

Copley, George Novello, 'The Law of Avogadro-Gerhardt and 'Van der Waals's Forces', *Sch Sci Rev*, 1939, 21, p 869.

Meldrum, Andrew Norman, *Avogadro and Dalton: The Standing in Chemistry of Their Hypotheses*, Aberdeen University Studies, University of Aberdeen, 1904, no 10.

Morselli, Mario, *Amedeo Avogadro: A Scientific Biography*, D Reidel Publishing Co, 1984.

Ayrton, Hertha (Phoebe Sarah) Marks (1854–1923) English physicist and inventor who worked in electricity and wrote the standard textbook *The Electric Arc*. In 1904 she was the first woman to read a paper at the Royal Society and she received the Hughes Medal for her research into the electric arc and sand ripples.

Phoebe Sarah Marks was born in Portsea, near Plymouth, England on 28 April 1854, the third of eight children and eldest daughter of Levi and Alice Marks. Her father was a clockmaker and jeweller who had fled from anti-Semitic persecution in Poland. He died in 1861 leaving the family in debt.

Ayrton was sent to live with her aunt and uncle, Marion and Alphonse Harzog, in London in 1863 and was educated at their school with her cousins, who taught her mathematics and philosophy and introduced her to local intellectual circles. As a teenager, Sarah took the name Hertha after the Teutonic earth goddess and became an agnostic. She supported herself from the age of 16 as a tutor and by embroidery work, sending money to her mother in Portsea, but she always aspired to a university education.

In 1873 Ayrton met Barbara Bodichon, a suffragist and one of the founders of Girton College, Cambridge, who became a firm friend and benefactor and through whom she met George Eliot, William Morris, and other influential figures. Supported by Bodichon, she passed the Cambridge University Examination for Women in 1874 and completed the Cambridge tripos in 1881, having shown an aptitude for experimentation and the design of scientific instruments. Back in London, she again supported herself by teaching and in 1884 she patented an instrument for dividing a line into any number of equal parts. Following her developing interest in electricity she enrolled, again with the help of Bodichon, in Finsbury Technical College, London, where she studied physics and electrical engineering with Professor William ◊Ayrton, a pioneer of higher technical education and supporter of women's rights.

A relationship developed and in 1885 Hertha Marks married Ayrton, a widower with a daughter. She pursued her interests and lectured to women on practical electricity. However, ill health, the increased domestic and social responsibilities of marriage, and the birth of her own daughter, Barbara, left her no time for her professional work for a few years.

Barbara Bodichon died in 1891, leaving Ayrton deeply grieved. However the sum she was left by Bodichon allowed her to support her mother and employ a housekeeper and by 1893 Ayrton was back at work.

Inspired by her husband's experiments Ayrton began her own investigations into the electric arc. She discovered the cause of the anomalous behaviour of electric arcs, which seemed to defy Ohm's law, publishing articles in the *Electrician* and presenting papers to the Association for the Advancement of Science, which were later collected in her seminal work on the subject – *The Electric Arc* (1902). The Institute of Electrical Engineers awarded her a prize of £10 in 1899 for her paper on the hissing of the electric arc, allowing her to read the paper herself and electing her their first woman member. In the same year she demonstrated the arc at a Royal Society meeting, and presided over the science section at the International Congress of Women. In 1900 she read a paper on the arc in Paris at the International Electrical Congress. She continued investigating the practical uses of the electric arc for searchlights, cinema projectors, and studio lights and took out eight patents 1913–18. She helped to design the anti-aircraft searchlights used in both world wars.

Around the turn of the century Ayrton's husband's health was failing and she spent more and more time caring for him. It was when he was convalescing by the sea in 1901 that she took the opportunity to begin analysing the sand ripples caused by the oscillatory movement of the water. This led to further study of hydrodynamics and in 1915 she developed the Ayrton fan, a device for removing poisonous gases from the trenches, of which over 100,000 were supplied to the troops. The fan was later used to improve the ventilation of mines, sewers, and warships.

Ayrton was active in the suffrage movement and took part in the 1910 demonstrations. She cared for the hunger strikers and militant feminist activists. She felt a special affinity with her friend Marie ◊Curie, a woman in a similar position to herself, and gave her and her daughters an anonymous refuge in 1912 to recover from illness and stress. After World War I she joined the Labour Party and was a founder member of the National Union of Scientific Workers. She was also involved with the International Federation of University Women. Ayrton is remembered as a charming and vivacious woman with considerable artistic talent. Her husband died in 1908 and she herself died of septicaemia on 26 August 1923, while on holiday at North Lancing, Sussex.

Further Reading

Mason, Joan, 'Hertha Ayrton (1854–1923) and the admission of women to the Royal Society of London', *Notes and Records of the Royal Society of London*, 1991, v 45, pp 201–220.

Sharp, Evelyn, *Hertha Ayrton, 1854–1923: A Memoir*, E Arnold, 1926.

Ayrton, William Edward (1847–1908) English physicist and electrical engineer who invented many of the prototypes of modern electrical measuring instruments. He also created in 1873 the world's first laboratory for teaching applied electricity in Tokyo.

Ayrton was born in London on 14 September 1847, the son of a barrister. He was educated at University College School 1859–64, when he entered University College, London. In 1867 he obtained an honours degree in mathematics and joined the Indian Telegraph Service. He was sent to Glasgow to study electricity under William Thomson (later Lord Kelvin) and, after practical study at the works of the Telegraph and Maintenance Company, he went to Bombay in 1868 as an assistant superintendent.

In 1872 he returned to England and was placed in charge of the Great Western telegraph factory under Thomson and Fleeming Jenkin. A year later he accepted the chair of physics and telegraphy at the new Imperial Engineering College in Tokyo – founded by the Japanese government and at that time the world's largest technical university. There he created the first laboratory in the world for teaching applied electricity.

In 1878 Ayrton returned to England and in 1879 he became professor at the City and Guilds of the London Institute for the advancement of technical education. His first class consisted of one man and a boy. From 1881–84 he acted as professor of applied physics at the new Finsbury Technical College and became the first professor of physics and electrical engineering at the new Central Technical College – now the City and Guilds College, South Kensington. He held this post until he died in London on 8 November 1908.

In 1881 Ayrton and John Perry invented the surface-contact system for electric railways, which, together with Fleeming Jenkin, they applied to 'telpherage' (a system of overhead transport) and a line based on this system was installed in 1882 at Glynde in Sussex. In that year they also brought out the first electric tricycle.

There followed a whole series of new, portable electrical measuring instruments including the ammeter (so-named by its inventors), an electric power meter, various forms of improved voltmeters, and an instrument used for measuring self and mutual induction. In this, great use was made of an ingeniously devised flat spiral spring that yields a relatively large rotation for a small axial elongation. These instruments served as the prototypes for the many electrical measuring instruments that came into use in countries all over the world as electrical power became generally employed for domestic and industrial purposes. Ayrton's instruments gave electrical engineers the means of measuring almost every electrical quantity they had to deal with, and his electric meter was the only one to be awarded prizes at the Paris Exhibition in 1899.

Besides his contribution to the advancement of the practical aspect of electrical engineering, Ayrton was also a great teacher of the subject. His system of teaching was adopted and extended throughout the profession. He published many scientific papers and a book entitled *Practical Electricity.*

His wife Hertha also achieved renown for her researches into the electric arc and she had the distinction of becoming the first woman member of the Institute of Electrical Engineers.

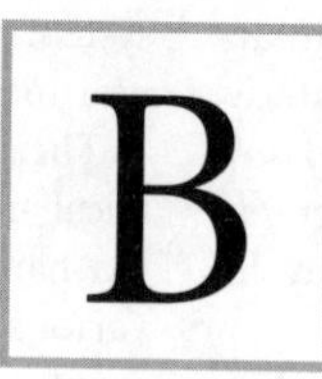

Baade, Walter (1893–1960) German-born US astronomer who is known for his discovery of stellar populations and whose research proved that the observable universe is larger than had originally been believed.

Baade was born in Shröttinghausen on 24 March 1893, the son of a schoolteacher. He studied at Münster and at Göttingen universities, obtaining a PhD from the latter in 1919. For the next eleven years he worked at Hamburg University in the Bergedorf Observatory. In 1931 he emigrated to the USA and joined the staff of the Mount Wilson Observatory, Pasadena. He left in 1948 and went to the nearby Mount Palomar Observatory where he worked until 1958, when he returned to Germany. The following year he became Gauss Professor at Göttingen University. He died there on 25 June 1960.

Baade's contributions to astronomy were numerous. In 1920 he discovered the most distant known asteroid, Hidalgo, whose orbit goes out as far as that of Saturn. In 1948 he found the innermost asteroid, Icarus, whose orbit comes within 18 million miles of the Sun, even closer than Mercury. At Mount Wilson Baade worked with Fritz ◊Zwicky and Edwin ◊Hubble on supernovae and galactic distances.

During the wartime blackout of Los Angeles, in 1943, Baade made his most important discovery. He made use of the enforced darkness to study the Andromeda galaxy with a 2.5-m/8.2-ft reflecting telescope. Until that time, Hubble had managed to view only the bright blue giant stars in the spiral arms of the galaxy, and a bright haze in its centre. Baade was able to observe, for the first time, some of the stars in the inner regions of the galaxy and he found that the most luminous stars towards the centre are not blue–white but reddish. He proposed that there exist two groups of stars with differing structures and origins. The bluish stars on the edge of the galaxy, called Population I stars, were distinguished from the reddish ones in the inner regions, Population II. Population I stars are young and formed from the dusty material of the spiral arms – hydrogen, helium, and heavier elements; Population II stars are old and were created near the nucleus and contain fewer heavy elements.

After the war, the 5-m/16.4-ft reflecting telescope was introduced and Baade continued his research using this instrument. He found that both stellar populations contain Cepheid variable stars (of which there are more than 300 in Andromeda). He also discovered that the period–luminosity curve established for Cepheid variables by Harlow ◊Shapley and Henrietta ◊Leavitt applied only to Population II Cepheids. This discovery meant that two types of Cepheids existed.

Baade's findings had even greater implications. In the 1920s the distances of the outer galaxies had been calculated by Hubble using light curves for Population I Cepheids. Baade decided that these calculations were incorrect and redrew a period–luminosity curve for Population I Cepheids, revealing that they are much brighter than had been previously thought. His discovery showed that the dim bluish-white Cepheids seen in the spiral arms of Andromeda were much farther away than believed and that Andromeda was not 800,000, but more than 2 million, light years distant. Also, since Hubble had used his incorrect distance of Andromeda to gauge the size of the universe, the new findings meant that the universe and all the extragalactic distances in it were at least double the size previously calculated. The increased distance of the Andromeda galaxy and others meant that to appear so bright they must be larger than had been thought. Astronomers realized that our own galaxy is smaller than Andromeda and not the pre-eminent galaxy they had imagined.

The enlarged scale of the universe stimulated the construction of far-reaching radio telescopes. Astronomers had known that a strong radio source existed in the sky but it could not be located with the 5-m/16.4-ft telescope. With a radio telescope, however, Baade discovered the source to be a distorted galaxy colliding with another galaxy in the constellation Cygnus. The interstellar dust created by the collision and the resulting radio waves could be detected clearly even though they were 260 million light years away.

Baade contributed a great deal to our knowledge of the universe. His main interests were extragalactic nebulae as stellar systems but he also studied variable stars in our own galaxy, in globular clusters, and in the Andromeda nebula, and by doing so stimulated interest in the theory of stellar interiors as the basis of theoretical interpretations of stellar evolution.

Babbage, Charles (1792–1871) English mathematician, one of the greatest pioneers of mechanical computation.

Babbage English mathematician Charles Babbage, the pioneer of machine computing. Unfortunately, lack of funding lead to many of his projects remaining unfinished. He wrote many books on mathematical and statistical subjects, and tried unsuccessfully to devise a system of forecasting the winners of horse races, losing heavily in the process. *Mary Evans Picture Library*

Babbage was born in Totnes, Devon, on 26 December 1792. His father was a banker who left him a large inheritance, and throughout his life Babbage was financially secure. He developed an interest in mathematics as a young boy and in 1810 went to Cambridge University to study mathematics. There he became a close friend of the future astronomer, John ◊Herschel, and, convincing himself that Herschel would be placed top in the honours examinations, chose not to be placed second and took only a pass degree, in 1814. While at Cambridge Babbage, Herschel, and other undergraduates founded the Analytical Society, and it was in the society's rooms one evening, that Babbage is recorded as saying (while looking over some error-filled logarithm tables) 'I am thinking that all these mathematical tables might be calculated by machinery'.

The year after he left Cambridge, Babbage wrote three papers on 'The calculus of functions' for publication by the Royal Society and in the following year, 1816, he was elected a fellow of the society. Some time later, while making a tour of France, he examined the famous French logarithms which had been recently calculated and which were the most accurate tables then known. They had, however, required the combined efforts of nearly a hundred clerks and mathematicians. Mathematical tables of all kinds, including logarithmic tables, were of great use for astronomical, commercial, and especially, navigational purposes. On his return to London, therefore, Babbage set about developing his ideas for a cheaper and more accurate method of producing tables by mechanical computation and automatic printing.

By 1822 he had ready a small calculating machine able to compute squares and the values of quadratic functions. It worked on the method of differences, an example of which involves subtracting one square from the preceding one to obtain a first difference, and then the first difference from the next above it to obtain a second difference, which is always two.

By working backwards, adding the second difference to the first difference, and then adding the higher of the squares used in the subtraction that had produced the first difference, the next square is always obtained. Using second-order differences of this kind Babbage's first machine could produce figures to six places of decimals.

With the backing of the Royal Society, Babbage was able to persuade the government to support his work to devise a much larger difference engine to calculate navigational and other tables. He was elected to the Lucasian chair of mathematics at Cambridge in 1826, a post he held until 1835, but he continued to live in London and did not perform the usual professorial duties of lecturing and teaching. He devoted himself entirely to his machine, which he proposed to make work to sixth-order differences and 20 places of decimals. The construction of the large difference engine was a laborious and lengthy operation. New tools had to be designed to previously unknown tolerances, many of them being made for Babbage by the great pioneer of precision engineering, Joseph ◊Whitworth. The project cost the government £20,000, but it was eventually abandoned, partly because of problems of friction (the arch-enemy of the engineer), partly because of personality clashes, but chiefly because before it was completed Babbage had hit upon a better idea.

The whole of the developments and operations of analysis are now capable of being executed by machinery. ... As soon as an Analytical Engine exists, it will necessarily guide the future course of science.

Charles Babbage Passages from the Life of a Philosopher *1864*

That idea was the analytical engine. The difference engine could perform only one function, once it was set up. The analytical engine was intended to perform many functions; it was to store numbers and be capable of working to a programme. In order to achieve this, Babbage borrowed the idea of using punched cards from the French engineer Joseph ◊Jacquard, who had invented such cards in 1801 to programme carpetmaking looms to weave a pattern. Babbage's machine, begun in 1833, was to have a mill to carry out arithmetical operations, a memory unit to store 1,000 numbers of 50 digits, and the programme cards, linked together to direct the machine. The cards were of three kinds: those to supply the store with numbers, those to transfer numbers from mill to store or store to mill, and those to direct the four basic arithmetical operations. In order to unite the programmes for his cards Babbage devised a new mathematical notation.

Babbage's machine had not been completed by the time of his death, on 18 October 1871. And although his son H P Babbage carried on the enterprise 1880–1910, the fact is that its complexity was beyond the engineering expertise of the day and its very conception beyond the grasp of society to spend the amount of time and money required to build it. Nevertheless, it was Babbage's machine – although, being decimal not binary and requiring the use of wheels, not strictly digital in the modern sense – which Howard ◊Aiken used more than 70 years after Babbage's death as the basis for his development of the Harvard Mark I Calculator.

Babbage was a true representative of the Victorian age, the age of the steam engine and the industrial revolution. He wished to harness science to the practical improvement of society. Hence, dissatisfied by the practical usefulness of the Royal Society, he had a hand in founding the Astronomical Society (1820), the British Association for the Advancement of Science (1831), and the Statistical Society of London (1834). The same passion for improvement led him to investigate the operation of the Post Office and, on finding that most of its costs derived from the handling of letters not their transport, to recommend (what Rowland Hill introduced as the penny post) that it should simplify its procedure by introducing a single rate. But his greatest idea, the mechanical computation of tables, remained ahead of the practical possibilities of his time.

Further Reading

Buxton, Harry Wilmot, *Memoir of the Life and Labours of the Late Charles Babbage Esq, FRS,* The Charles Babbage Institute Reprint Series for the History of Computing, MIT Press /Tomash, 1988, v 13.

Collier, Bruce, *Charles Babbage and the Engines of Perfection,* Oxford Portraits in Science series, Oxford University Press, 1998.

Dubbey, John Michael, *The Mathematical Work of Charles Babbage,* Cambridge University Press, 1978.

Hyman, Anthony, *Charles Babbage: Pioneer of the Computer,* Oxford University Press Paperback series, Oxford University Press, 1984.

Moseley, Maboth, *Irascible Genius: A Life of Charles Babbage, Inventor,* A M Kelley, 1965.

Swade, Doron, *Charles Babbage and his Calculating Engines,* Science Museum, 1991.

Babcock, George Herman (1832–1893) US co-inventor of the first polychromatic printing press, but chiefly remembered for the Babcock–Wilcox steam boiler, devised with his partner, Stephen ◊Wilcox.

Born in Unadilla Forks, near Otego, New York, on 1 June 1832, Babcock inherited his engineering and mechanical expertise from both sides of his family. When he was 12, the Babcock family moved to Westerly, Rhode Island. Babcock first went to work with his father, Asher, in daguerrotype and job printing for newpapers. It was during this period up to 1854, that Babcock and his father invented the polychromatic printing press. The father-and-son team also invented a job printing press that is still manufactured today.

Moving to Brooklyn, New York, in 1860, Babcock held many posts first in the offices of a patents solicitor then with the Mystic Iron Works and afterwards as chief draughtsman at the Hope Iron Works, Providence, Rhode Island.

At Providence, Babcock and Stephen Wilcox met, and became involved in steam engineering. Together they tackled the contemporary problems of steam boilers. Since their introduction in the 18th century, fire-tube boilers had been bedevilled by structural problems. Explosions – often fatal – were a constant danger when large quantities of water and steam were contained in boilers not built to withstand the strains imposed upon them. In the 19th century the growing use of steam power and of engines operating at ever higher pressures, increased the incidence of such disasters.

The water-tube boiler was developed as a safer replacement and it was to the improvement of this early model that Babcock and Wilcox applied themselves. Their design for a sectionally headed boiler was one of the earliest with automatic cutoff, being based on a safety water tube patented by Wilcox in 1856. This was the first engine to have front and rear water spaces connected by slanting tubes, the steam space being situated above. The Babcock–Wilcox boiler patented in 1867 employed cast-iron steam generating tubes placed in vertical rows over a grate. Steel or wrought iron tubes connected them with headers leading to the separating drum where the water was separated from the steam and recirculated.

The new boiler, considerably more powerful than its predecessors, was able to withstand very high pressures and also ensured a high standard of protection against explosions. It was first manufactured at Providence and then in New York, where the firm of Babcock and Wilcox was incorporated in 1881. The boilers were also built in New Jersey, at Elizabethport, and at a specially designed plant at Bayonne.

Patent restrictions on the design of steam boilers made the Babcock–Wilcox model expensive, and it fell behind the competition when the patents expired and the cheap engines invaded the market. The high quality of the boiler, however, ensured its continuation and the design was still being used 70 years after its inception. The firm of Babcock and Wilcox still exists (as the Babcock and Wilson Power Generation Group), and today manufactures modern high-quality steam boilers.

Babcock actively promoted his company's product by lecturing at technical institutes and colleges, an activity for which his impressive presence and personality made him especially suitable. In 1887, Babcock was president of the American Society of Mechanical Engineers, and was also seen frequently at meetings of technical societies. A sincerely religious man, a member of the Seventh Day Baptists, Babcock took great personal interest in the welfare of his employees and in his public duties: these included the presidencies of the board of trustees of Alfred University and, at his home of Plainfield, New Jersey, of the board of education and the public library.

Babcock was alert and energetic to the end of his life. He died on 16 December 1893.

Babcock, Harold Delos (1882–1968) US astronomer and physicist whose most important contributions were to spectroscopy and the study of solar magnetism.

Babcock was born on 24 January 1882 in Edgerton, Wisconsin. Much of his schooling was done at home, but in 1901 he enrolled at the College of Electrical Engineering of the University of California in Berkeley, and he was awarded a BA in 1907. By that time he had already taken up employment at the National Bureau of Standards. In 1908 Babcock taught a course in physics at the University of California. A year later George Hale invited him to work at the Mount Wilson Observatory.

The only breaks in Babcock's service at Mount Wilson 1909–48 came during the two world wars. During World War I Babcock worked for the Research Information Service of the National Research Council; during World War II he served as a consultant on several programmes, including the Manhattan Project. In 1933 he was elected to the National Academy of Sciences. Babcock retired formally in 1948, but he remained active in the supervision of the ruling engine for the 508-cm/200-in Hale telescope. He died on 8 April 1968.

As a child, Babcock had been interested in electricity, radio, and photography. In his first job, at the National Bureau of Standards, he investigated the problems concerning electrical resistance. His early astronomical work was in stellar photography, as part of an international research programme on the structure of the Galaxy being coordinated by Jacobus ◊Kapteyn. Babcock also collaborated with Walter ◊Adams in his spectroscopic studies.

The Sun was always a subject of Babcock's particular interest. He made an investigation of the Zeeman effect (whereby a magnetic field causes a substance's spectral lines to be split) in chromium and vanadium – important elements in the solar spectrum. Babcock's next major concern was the establishment of a standard spectrum for iron. This study was part of a large programme to determine standards for astronomical spectra.

During the 1920s much of Babcock's research dealt with the production of a revised table of wavelengths for the solar spectrum. In 1928 he published a list that included 22,000 spectral lines, and this list was extended in 1947 and again in 1948.

Babcock was also the director of a project aimed at devising an engine that could reliably and accurately rule gratings for the new telescope at Mount Wilson. The superior Babcock gratings that were eventually produced were installed in all of the observatory's spectrographs, including that of the 508-cm/200-in telescope.

There had been considerable interest among astronomers in the measurement of the Sun's magnetic field, although no reliable information had been obtained by 1938, when Babcock turned to the problem. He had little success until 1948 when, in collaboration with his son H W Babcock, the solar magnetic field was measured. They used an instrument of their own design which was dubbed the 'solar magnometer' and which exploited the Zeeman effect to produce a continuously changing record of the Sun's local magnetic fields. These surface fields are only weak, but they could be observed satisfactorily with the new instrument. They also studied the Sun's general magnetic field and the relationship between sunspots and local magnetic fields. Babcock was a skilled observational astronomer with a flair for the more practical side of his subject. His contributions to solar spectroscopy and magnetism were original and thorough.

Bacon, Francis (1561–1626) Baron Verulam, Viscount St Albans. English politician and philosopher of science.

Son of Sir Nicholas Bacon, courtier and Keeper of the Great Seal, Francis Bacon first attended Trinity College, Cambridge, and was trained in the law with a view to following the same path as his father. Politically unsuccessful under Elizabeth I, he finally gained a succession of offices under James I, eventually becoming Lord High Chancellor in 1618. Presently convicted of taking bribes, he was banished from court and office in 1621. Thereafter he was able to devote himself wholeheartedly to his scientific interests.

In works like the *Advancement of Learning* (1605) and the *Novum Organum* (1620), Bacon advanced challenging views of scientific method. He fiercely criticized Aristotle and the deductive mode supposedly followed ever since under the mental tyranny of scholasticism. He instead promoted 'induction', laying emphasis on the exhaustive collection of empirical data and its rigorous processing via a logical mill until general causes and conclusions emerge. Bacon deprecated dogmatic *a priori* reasoning and the wayward play of fancy: true science must be built on the solid foundation of fact. Though Bacon is barely remembered for his own scientific investigations, his ideas proved extraordinarily influential. His insistence upon the primacy of sense and experience underpinned the commitment of a later generation of English scientists such as Robert Boyle and Isaac Newton, to rejection of continental rationalism, and also suggested the Royal Society's 'experience first' motto of *nullius in verba* (on the authority of no person). His opinion that scientific knowledge should be applied to utilitarian purposes ('for the glory of God and the relief of man's estate') appealed to 17th-century Puritans and to the reforming minds of the Enlightenment. Until late in the 19th century, Bacon and his faith in facts were regularly held aloft as the banner of English science.

Further Reading

Davy, Martin, *The Life of Francis Bacon,* Kessinger Publishing Company, 1996.
Wormaid, B H, *Francis Bacon: History, Politics and Science, 1561–1626,* Cambridge University Press, 1993.
Zagorin, Perez, *Francis Bacon,* Princeton University Press, 1998.

Bacon, Roger (c. 1220–c. 1292) English philosopher and scientist who was among the first medieval scholars to realize and promote the value of experiment in reaching valid conclusions.

It is not certain when or where Bacon was born. One tradition says that he was born in Bisley in Gloucestershire, while another says he came from Ilchester in Somerset. His date of birth can only be guessed from his writing, in 1267, that he had learnt the alphabet forty years before. Bacon was educated in philosophy and mathematics at Oxford, and then lectured at the faculty of arts in Paris from about 1241. After about 1247, he returned to Oxford where he was inspired by the English scholar Robert Grosseteste (*c.* 1168–1253) to cultivate the 'new' branches of learning – languages, mathematics, optics, alchemy, and astronomy. In about 1257, he entered the Franciscan order and, while a friar, appealed to Pope Clement IV to allow sciences a higher status in the curriculum of university studies. At the pope's request, Bacon in 1266–67 wrote three volumes of an encyclopedia of all the known sciences: *Opus majus, Opus minus,* and *Opus tertium.* In about 1268, he published fragments of another encyclopedia – *Communia naturalium* and *Communia mathematica,* and in 1272, *Compendium philosophiae.* Following the death of Pope Clement IV in 1268, Bacon lost papal protection and was imprisoned by the Franciscans at some time 1277–79, possibly for criticisms of the order's educational practices. Bacon languished in prison for as long as 15 years. His last work, on theology, was published in 1292. He is said to have died in the same year and was buried in the Franciscan church in Oxford.

Bacon in his writings states that he spent huge sums of money in acquiring 'secret' books, in constructing instruments and tables, training assistants, and in getting to know knowledgeable people. The experiments he performed and the knowledge he acquired earned him the title of *doctor mirabilis.* However little, if any, of his work was original. His main contribution was in optics, in which Robert Grosseteste acquainted Bacon with the work of the 11th-century Arabian scientist ◊Alhazen. Bacon advocated the use of lenses as magnifying glasses to aid weak sight, and speculated that lenses might be used to make an instrument of great magnifying power – an indication of the telescope.

Bacon also described some of the properties of gunpowder in a volume written sometime before 1249, but there is no indication that he had any idea of its possible use as a propellant, only that it would explode and so could be used in warfare. He described early diving apparatus, in a passage where he referred to 'instruments whereby men can walk on sea or river beds without danger to themselves' in 1240.

Bacon was an ardent supporter of calendar reform, suggesting the changes necessary to improve the calendar that were carried out by Pope Gregory XIII in 1582. He also promoted the use of latitude and longitude in mapmaking and, by estimating the distance of India from Spain, may have inspired Columbus' voyage to America, which was a search for a shorter westward route to India.

Bacon is sometimes referred to as one of the 'forerunners' of science, but investigation does not show his scientific contribution to be very great. His importance in the history of science is due to his insistence on the use of experiment to prove an argument, a method that was not to flower until some three centuries later. Bacon also referred to the laws of reflection and refraction as being 'natural laws' – that is, of universal application – a concept vital to the development of science.

Further Reading

Hackett, Jeremiah M, *Roger Bacon and the Sciences: Commemorative Essays,* Studien und Texte zur Geistesgeschichte des Mittelalters series, Brill, 1997, v 57.

Redgrove, H Stanley, *Roger Bacon: Christian Mystic and Alchemist,* Holmes Publishing Group, 1994.

Westacott, E, *Roger Bacon in Life and in Legend,* Banton Press, 1993.

Baekeland, Leo Hendrik (1863–1944) Belgian-born US industrial chemist famous for his invention of Bakelite, the first commercially successful thermosetting plastic resin.

Baekeland was born in Ghent on 14 November 1863. He was a brilliant pupil at school and at the age of only 16 won a scholarship to the University of Ghent, from which he graduated in 1882. Two years later, still aged only 21, he was awarded his doctorate after studying electrochemistry at Charlottenburg Polytechnic. In 1887 he became professor of physics and chemistry at the University of Bruges, and in 1888 returned to Ghent as assistant professor of chemistry. The next year he got married and then went on a tour of the USA for his honeymoon, with financial help from a travelling scholarship. He decided to settle in the USA and took a job as a photographic chemist, setting up as a consultant in his own laboratory in New York City in 1891.

He returned briefly to Europe in 1900 to study at the Technische Hochschule at Charlottenburg. His development of Bakelite came after he had gone back to the USA, and was announced in 1909, the year in which he founded the General Bakelite Corporation, later to become part of the Union Carbide and Carbon Company. He continued his chemical researches and in 1924 was elected president of the American Chemical Society. He died in Beacon, New York, on 23 February 1944.

Baekeland's first chemical invention was a type of photographic printing paper, which he called Velox, that could be developed under artificial light. He began manufacturing it in 1893 at Yonkers, New York, and in 1899 George Eastman's Kodak Corporation bought the invention and the manufacturing company (the Nepera Chemical Company) from Baekeland for $1 million.

German chemist Johann ◊Baeyer had discovered the resin formed by the condensation reaction between formaldehyde (methanal) and phenol in 1871, and in the early 1900s Baekeland began to investigate it as a possible substitute for shellac. He could find no solvent for the resin, but discovered that it can be produced in a hard, machinable form that can also be moulded by casting under heat and pressure. On initial heating the material melts (becomes plastic) and then sets extremely hard and will not melt on further heating. Bakelite is a good insulator, and soon found use in the manufacture of electrical fittings such as plugs and switches. The General Bakelite Corporation merged with two other companies in 1922 and seven years later was incorporated into the Union Carbide group.

Baer, Karl Ernest Ritter von (1792–1876) Estonian embryologist famous for his discovery of the mammalian ovum, who made a significant contribution to the systematic study of the development of animals.

Baer was born on 29 February 1792 on his father's estate in Piep. The size of the family – he was one of ten children – forced his parents to send him to live with his paternal uncle and aunt, although his father was a wealthy landholder and district official. On his return home at the age of seven, he was privately tutored until 1807, when he attended a school for members of the nobility for three years.

He then went to the University of Dorpat, the local university, where he was taught by Karl Burdach (1776–1847), the professor of physiology there, who had a great influence on his life. He graduated with a medical degree in 1814 and went to Vienna in Austria for a year. During the following year he spent some time in Germany studying comparative anatomy at the University of Würzburg, where he met Ignaz Döllinger (1770–1841), who was professor of anatomy there. It was Döllinger who first introduced Baer to embryology. In 1817, at the invitation of his old teacher Burdach, Baer joined him at the University of Königsberg (now Kaliningrad in Russia), where he taught zoology, anatomy, and anthropology. Two years later he was appointed assistant professor of zoology. In 1820 he married Auguste Medem, and later had six children. He became restless at Königsberg, and in 1834 moved to St Petersburg in Russia, where he took up the appointment of librarian of the foreign division at the Academy of Sciences. In 1837 he led the first of many expeditions into Novaya Zemlya, in Arctic Russia, where he was the first naturalist to collect plant and animal specimens. He later led expeditions to Lapland and to the Caspian Sea. In 1846 he became the professor of comparative anatomy and physiology at the Medico-Chirurgical Academy in St Petersburg. He retired from the academy in 1862 but continued working for them until 1867 as an honorary member. He died on 28 November 1876 at Dorpat in Estonia.

At Würzburg, Döllinger had suggested that Baer study the blastoderm of chick embryos removed from the yolk, but the cost of a sufficient number of eggs for observation and someone to look after the incubator was too high and he left the investigation to his more affluent friend Christian Pander (1794–1865). Baer carried on Pander's research and applied it to all vertebrates. In 1817 Pander had described the formation of three layers in the vertebrate embryo – the ectoderm, endoderm, and mesoderm. Baer developed a theory regarding these germ layers in which he conceived that the goal of early development is the formation of these three layers, out of which all later organs are formed. At the same time he proposed the 'law of corresponding stages', which contradicted the popular belief that vertebrate embryos develop in stages similar to adults of other species. Instead, he suggested that the younger the embryos of various species are, the stronger is the resemblance between them. He demonstrated this fact by deliberately leaving off the labels of embryo species and saying: 'I am quite unable to say to what class they belong. They may be lizards, or small birds, or very young mammalia, so complete is the similarity in the mode of formation of the head and trunk in these animals. The extremities are still absent, but even if they existed, in the earliest stage of the development we should learn nothing, because all arise from the same fundamental form.' From this demonstration he formed his concept of epigenesis, that an embryo develops from simple to complex, from a homogeneous to a heterogeneous stage.

In 1827, Baer published the news of his most significant discovery – the mammalian ovum. William ⇨Harvey before him had tried to find it, dissecting a deer, but had searched for it in the uterus. Baer found the egg inside the Graafian follicle in the ovary of a bitch belonging to Burdach, which had been offered for the experiment. Baer's publication stated that 'every animal that springs from the coition of male and female is developed from an ovum, and none from a simple formative liquid'.

In his observations of the embryo, Baer discovered the extraembryonic membranes – the chorion, amnion, and allantois – and described their functions. He also identified for the first time the notochord, a gelatinous, cylindrical cord that passes along the body of the embryo of vertebrates (and in the embryos or adults of other chordates). In the lower vertebrates it forms the entire back skeleton, whereas in the higher ones the backbone and skull are developed around it. He revealed the neural folds, and suggested that they were the beginnings of the nervous system, and described the five primary brain vesicles.

On his expeditions, Baer made a significant geological discovery concerning the forces that cause a particular formation on riverbanks in Russia. His study of fishes, made at the same time and described in his *Development of Fishes* (1835), stimulated the development of scientific and economic interest in fisheries in Russia.

Baer collected skull specimens for his lectures on physical anthropology, the measurements of which he recorded. In 1859, the same year that Charles ⇨Darwin published *On the Origin of Species,* Baer published independently a work that suggested that human skulls might have originated from one type.

Baer's publication *De ovi mammalium et hominis genesi/On the Mammalian Egg and the Origin of Man* (1827) and his two volume *Über Entwicklunggeschichte der Thiere/On the Development of Animals* (1828 and 1837) paved the way for modern embryology and gave a basis for new and scientific interpretation of embryology and biology.

Baeyer, Johann Friedrich Wilhelm Adolf von (1835–1917) German organic chemist famous for developing methods of synthesis, the best known of which is his synthesis of the dye indigo. His major contribution to the science was acknowledged by the award of the 1905 Nobel Prize for Chemistry.

Baeyer was born in Berlin on 31 October 1835, the son of the Prussian general Johann Jacob von Baeyer, who later became head of the Berlin Geodetic Institute. He began his university career at Heidelberg in 1853 where he studied chemistry under Robert Bunsen and Friedrich Kekulé. He gained his PhD at Berlin in 1858 after working in the laboratory of August Hofmann, and two years later took an appointment as a teacher at a technical school in Berlin. He became professor of chemistry at the University of Strasbourg in 1872 and three years later was appointed to succeed Justus von Liebig as professor of chemistry at Munich, where he stayed for the rest of his career. He died at Starnberg, near Munich, on 20 August 1917.

Baeyer began his researches in the early 1860s with studies of uric acid, which led in 1863 to the discovery of barbituric acid, later to become the parent substance of a major class of sedative drugs. In 1865 he turned his attention to dyes. His student Karl Graebe (1841–1927) synthesized alizarin in 1868 (at the same time as William ⇨Perkin), and in 1871 Baeyer discovered phenolphthalein and fluorescein. He also found the resinous condensation product of phenol and formaldehyde (methanal), which Leo ⇨Baekeland later developed into the thermosetting plastic Bakelite.

In 1883 Baeyer determined the structure of indigo by reducing it to indole using powdered zinc. He had already, in 1880, devised a method for its synthesis, which was more lengthy than the commercial method later used when synthetic indigo began to be manufactured in 1890. In 1888 he carried out the first synthesis of a terpene.

His work with ring compounds and the highly unstable polyacetylenes led him to consider the effects of carbon–carbon bond angles on the stability of organic compounds. He concluded that the more a bond is deformed away from the ideal tetrahedral angle, the more unstable it is – known as Baeyer's strain theory. It explains why rings with five or six atoms are much more common, and stable, than those with fewer or more atoms in the ring. He also noticed that the aromatic character of the six-carbon benzene and its analogues is lost on reduction and saturation of the carbon atoms.

Baily, Francis (1774–1844) English astronomer who is best known for his discovery of the phenomenon called 'Baily's beads'.

Baily was born in Newbury, Berkshire, on 28 April 1774. He began a seven-year apprenticeship in 1788 with a firm of merchant bankers in London, but as soon as his apprenticeship ended he set out to explore unsettled parts of North America. On his return to England in 1798 he became a very successful stockbroker. Astronomy, however, took up an increasingly important part of his life. He was a founder, in 1820, and the first vice president of the Astronomical Society of London (later the Royal Astronomical Society); he was elected a Fellow of the Royal Society in 1821. He finally gave up his job as a stockbroker in 1825 and became a full-time astronomer. He was a member of numerous scientific bodies and received several distinguished awards for his contributions to astronomy, among them two gold medals from the Royal Astronomical Society. He died in London on 30 August 1844.

Baily began to publish his astronomical observations in 1811. He was the author of an accurate revised star catalogue in which he plotted the positions of nearly 3,000 stars. These positions were used for the determination of latitude, and for this work Baily was awarded the Astronomical Society's gold medal in 1827.

In 1836, on 15 May, Baily observed a total eclipse of the Sun from Scotland. He noticed that immediately before the Sun completely disappeared behind the Moon (and also just as it began to emerge from behind the Moon) light from the Sun appeared as a discontinuous line of brilliant spots. These 'spots' have been named 'Baily's beads' and are caused by sunlight showing through between the mountains on the Moon's horizon as it moves across the Sun's disc. Baily travelled to Italy in 1842 and was again able to see his 'beads' during a solar eclipse.

Baily did other research, including a redetermination of the mean density of the Earth using the methods of Henry ◊Cavendish. He also measured the Earth's elliptical shape. He earned his second gold medal from the Astronomical Society for these studies.

Baily's sighting was not the first of the 'bead' phenomenon, but his description of it and of the rest of the 1836 eclipse was so exciting that it sparked greatly renewed interest in eclipses, which persists to this day.

Bainbridge, Kenneth Tompkins (1904–1996) US physicist best known for his work on the development of the mass spectrometer.

Bainbridge was born in Cooperstown, New York, on 27 July 1904. He was educated at the Massachusetts Institute of Technology and at Princeton University, where he gained his MA in 1927 and a PhD in 1929. After working at the Bartol Research Foundation and at the Cavendish Laboratory in Cambridge, England, he became assistant professor of physics at Harvard University in 1934, associate professor in 1938, and professor from 1946. From 1975 he was professor emeritus. Bainbridge held a number of important concurrent appointments. He worked in the radiation laboratory at MIT 1940–43 and then in the Los Alamos National Laboratory until 1945, when he directed the first atomic bomb test. He was awarded the Presidential Certificate of Merit for his work on radar in 1948. He died on 14 July 1996.

The mass spectrometer or mass spectrograph was invented by the English chemist Francis ◊Aston in 1919. In it, a beam of ions is deflected by electric and magnetic fields so that ions of the same mass are brought to a focus at the same point, enabling the masses of the ions in the beam to be determined. It proved the existence of isotopes. Aston's mass spectrometer was a velocity-focusing machine, meaning that it focused beams of varying velocity but not varying direction. A direction-focusing machine, which focused beams of uniform velocity but varying direction, was developed by Arthur Dempster (1886–1950). Bainbridge's achievement in mass spectroscopy was to develop a double-focusing machine in 1936. It used successive electric and magnetic fields arranged in such a way that ion beams which are non-uniform in both direction and velocity can be brought to a focus. By the mid-1950s, instruments based on this principle were able to separate ions which differ in mass by only one part in 60,000.

Bainbridge was responsible for many of the innovations in the modern mass spectrometer, one of the most useful of analytical tools in physics, chemistry, geology, meteorology, biology, and medicine.

Baird, John Logie (1888–1946) Scottish inventor who was the first person to televise an image, using mechanical (non-electronic) scanning. He also gave the first demonstration of colour television.

Baird Scottish inventor John Logie Baird who pioneered the development of television. He provided the first experimental television programme to the BBC on 30 September 1929. *Mary Evans Picture Library*

Baird was born in Helensburgh, Dunbartonshire, on 13 August 1888. He was educated at Larchfield Academy and later took an engineering course at the Royal Technical College, Glasgow. He then studied at Glasgow University, but World War I interrupted his final year there. Rejected as physically unfit for military service, Baird became a superintendent engineer with the Clyde Valley Electrical Power Company. In 1918 he gave up engineering because of ill-health and set himself up in business, marketing successfully such diverse products as patent socks, confections, and soap in Glasgow, London, and the West Indies. Persistent ill-health, leading to a complete physical and nervous breakdown in 1923, forced him to retire to Horsham in Sussex.

Many inventors had patented their ideas about television – the electrical transmission of images in motion simultaneously with accompanying sound – but only a few, including Baird, pursued a practical study of the problem based on the use of mechanical scanners. In 1907 Boris Rosing had proposed that, in a television system that used mirror-scanning in the camera, a cathode-ray tube with a fluorescent screen should be fitted into the receiver. In 1911 A A Campbell Swinton had suggested that magnetically deflected cathode-ray tubes should be used both in the camera and the receiver.

On his retirement, Baird concentrated on solving the problems of television. Having little money, his first apparatus was crude and makeshift, set up on a washstand in his attic room. A tea-chest formed the base of his motor, a biscuit tin housed the projection lamp, and cheap cycle-lamp lenses were incorporated into the design. The whole contraption was held together by darning needles, pieces of string, and scrap wood. Yet within a year he had succeeded in transmitting a flickering image of the outline of a Maltese cross over a distance of a few metres.

Baird took his makeshift apparatus to London where, in one of two attic rooms in Soho, he proceeded to improve it. In 1925 he achieved the transmission of an image of a recognizable human face and the following year, on 26 January, he gave the world's first demonstration of true television before an audience of about 50 scientists at the Royal Institution, London. Baird used a mechanical scanner that temporarily changed an image into a sequence of electronic signals that could then be reconstructed on a screen as a pattern of half-tones. The neon discharge lamp Baird used offered a simple means for the electrical modulation of light at the receiver. His first pictures were formed of only 30 lines repeated approximately ten times a second. The results were crude but it was the start of television as a practical technology.

By 1927, Baird had transmitted television over 700 km/435 mi of telephone line between London and Glasgow and soon after made the first television broadcast using radio, between London and the SS *Berengaria* , halfway across the Atlantic Ocean. He also made the first transatlantic television broadcast between the UK and the USA, when signals transmitted from the Baird station in Coulson, Kent, were picked up by a receiver in Hartsdale, New York.

By 1928 Baird had succeeded in demonstrating colour television. The simplest way to reproduce a colour image is to produce the red, blue, and green primary images separately and then to superimpose them so that the eye merges the three images into one full-colour picture. Baird used three projection tubes arranged so that each threw a picture on to the same screen. By using only one amplifier chain and one cathode-ray tube that sequentially amplified the red, blue, and green signals, he overcame the problem of overregistration of the three images and matched the three channels. He used two rotating discs – each with segments of red-, green-, and blue-light filters, rotating synchronously before the camera tube and the receiver tube. Each primary-coloured filter remained over the tube face for the period of one field. Although partly successful, Baird's method had two major drawbacks. One was that the picture being transmitted consisted mainly of tones of one hue, say green, then the other two fields (red and blue) showed as black and each green field was succeeded by black ones, resulting in excessive flickering. The other was that the system required three times the bandwidth available.

Baird's black-and-white system was used by the BBC in an experimental television service in 1929. At first, the sound and vision were transmitted alternately, but by 1930 it was possible to broadcast them simultaneously. In 1936, when the public television service was started, his system was threatened by one promoted by Marconi–EMI. The following year the Baird system was dropped in favour of the Marconi electronic system, which gave a better definition.

Despite his bitter disappointment, Baird continued his experimental work in colour television. By 1939 he had demonstrated colour television using a cathode-ray tube which he had adapted as the most successful method for producing a well-defined and brilliant image. Baird's inventive and engineering abilities were widely recognized. In 1937, he became the first British subject to receive the gold medal of the International Faculty of Science. The same year, he was elected a Fellow of the Royal Institute of Edinburgh, where a plaque was erected to commemorate his demonstration of true television in 1926. Baird also became an Honorary Fellow of the Royal Society of Edinburgh, a Fellow of the Physical Society, and an Associate of the Royal Technical College.

He continued his research on stereoscopic and large screen television until his death, at Bexhill-on-Sea, Sussex, on 14 June 1946.

Further Reading

Exwood, Maurice, *John Logie Baird: 50 Years of Television*, IERE (Institution of Electric and Radio Engineers), History of Technology Monograph series, Institution of Electric and Radio Engineers, 1976.

Hallett, Michael, *John Logie Baird and Television*, Priory Press, 1978.

Hutchinson, Geoff, *Baird: The Story of John Logie Baird 1886–1946*, G Hutchinson, 1985.

McArthur, Tom and Waddell, Peter, *Vision Warrior: The Hidden Achievements of John Logie Baird*, Scottish Falcon Books series, 1990.

McArthur, Tom and Waddell, Peter, *The Secret Life of John Logie Baird*, Hutchinson, 1986.

Rowland, John, *The Television Man: The Story of John Logie Baird*, Roy Publishers, 1967.

Baker, Alan (1939–) English mathematician whose chief work has been devoted to the study of transcendental numbers.

Baker was born in London on 19 August 1939 and studied mathematics at the University of London, where he received his BSc in 1961. He then did graduate work at Cambridge and was awarded a PhD in 1964. He remained at Trinity College, Cambridge, for the next ten years, as a research fellow 1964–68 and as director of studies in mathematics 1968–74. In 1974 he became professor of pure mathematics in the university. He was elected a Fellow of the Royal Society in 1973. Visiting professorships have taken him to many parts of the USA and Europe. In 1978 he was appointed as the first Turán lecturer of the János Bolyai Mathematical Society in Hungary and in 1980 he was elected a Foreign Fellow of the Indian National Science Academy.

Since his early research days, Baker has been chiefly interested in transcendental numbers (numbers that cannot be expressed as roots or as the solution of an algebraic equation with rational coefficients). In 1966 he extended Joseph ◊Liouville's original proof of the existence of transcendental numbers by means of continued fractions, by obtaining a result on linear forms in the logarithms of algebraic numbers. This solution opened the way to the resolution of a wide range of diophantine problems and in 1967 Baker used his results to provide the first useful theorems concerning the theory of these problems. He obtained explicit upper bounds to Thue's equation, $F(x, y) = m$, where F denotes a binary irreducible form, and also to Mordell's equation, $y^2 = x^3 + k$.

In 1969 he achieved the same result for the hyperelliptic equation, $y^2 = f(x)$. For this work he was awarded the Fields Medal at the International Congress of Mathematicians at Nice in 1970.

Apart from individual papers, Baker's most important publication is *Transcendental Number Theory* (1975). Baker's work has greatly enriched the many branches of mathematics influenced by the development of transcendental-number theory. It has led to an important new series of results on exponential diophantine equations: his theory provides bounds of 10^{500} or more and it has been shown that these are sufficient in simple cases to calculate the complete list of solutions. The theory has also been used to solve some classical problems of Karl Gauss, to assist in the approximation of algebraic numbers by rationals (an investigation begun by Liouville in 1844), and to inspire new lines of research in elliptic and Abelian functions. Baker is thus at the very forefront of contemporary work on number theory.

Baker, Benjamin (1840–1907) English engineer most famous for his design of the Forth Bridge in Scotland.

Baker was born on 31 March 1840 at Keyford, Frome, Somerset. He attended the grammar school at Cheltenham until he was 16 when he was apprenticed at Neath Abbey ironworks. In 1860 he left this firm and became assistant to William Wilson, who designed

Victoria Station in London. Two years later he joined the staff of Sir John Fowler, (becoming his partner in 1875) and became particularly involved with the construction of the Metropolitan and District lines of the London Underground. This project demanded considerable ingenuity to overcome the many hazards caused by difficult soils, underground water, and the ruins of Roman and other civilizations. Baker incorporated an ingenious energy-conservation measure in the construction of the Central Line: he dipped the line between stations to reduce the need both for braking to a halt and for the increase in power required to accelerate away.

During the 1870s there was much interest in extending the Scottish East Coast Railway from Edinburgh to Dundee. This required the building of bridges across the Forth and Tay. The original Tay Bridge, which had been built by Sir Thomas Bouch of wrought iron lattice girders, collapsed under an express train during a storm on the night of 28 December 1879. This structure had been a considerable accomplishment, and indeed Bouch had built many successful bridges to similar designs. However it had been built without sufficient allowance for the force of the wind when the train was on the centre span. In addition, certain elements in its design were unsuitable, and inadequate supervision had blighted its construction. When the Tay disaster struck Bouch had already begun work on a bridge across the Forth.

It was only with great reluctance that the government authorities allowed an attempt to build a bridge across the Forth, but this time Baker was in charge. Baker's design had several features that made a repetition of the disaster unlikely. Mild steel, which was considerably stronger than the same weight of wrought iron, had become available through the new Siemens open-hearth process. This meant that the structure would be relatively light. Better estimates of the wind force that structures were required to withstand in storms were also available. Nevertheless, a rail bridge was a very difficult undertaking since the Forth was 60 m/200 ft deep. A site allowing the use of the little island of Inchgarvie as a foundation for the central pier was chosen; the two main spans were then 521 m/1,710 ft long each. A cantilever structure, which supports the bridge platform by projecting girders, was used since a suspension bridge, in which the platform hangs from catenaries, was not considered sufficiently stable in high winds for a railway. Indeed Bouch's original design had been for a stiffened suspension bridge, but this was abandoned after the Tay Bridge disaster.

The bridge was opened on 4 March 1890 by Edward, Prince of Wales. It has been in service ever since. For this achievement Baker was knighted by Queen Victoria.

Although the Forth Bridge made Baker famous, he has many other projects, both in the UK and abroad, to his credit. He worked on the Hudson River tunnel and at the docks at Hull, and played a prominent part in engineering development work in Egypt – he was involved with the Aswan Dam, which realized the dream of desert irrigation. A little while before he began work on the Forth Bridge, he designed the large wrought iron vessel in which Cleopatra's Needle, the obelisk on the Thames Embankment, was brought to England. The Needle was lost at sea, but when found later it was safely preserved within the hull of Baker's ship.

In his later years. Baker built up a large practice and was held in great esteem as a successful engineer who respected the theory of engineering. Although he had only a little formal education in this, he expressed it with great practical ability and artistry. He was made a Fellow of the Royal Society in 1880. He died, a bachelor, on 19 May 1907.

Balmer, Johann Jakob (1825–1898) Swiss mathematics teacher who devised mathematical formulae that give the frequencies of atomic spectral lines.

Balmer was born in Lausen, near Basle, on 1 May 1825, the eldest son of a farmer who was also a member of the local administration of Basle. He went to school in Liestal, Basle, then in 1844 entered the Technische Hochschule in Karlsruhe, where he studied mathematics, geology, and architecture. He subsequently spent a short time at the University of Berlin, then in 1846 returned to his old school in Basle to teach technical drawing while working towards his PhD, which he was awarded in 1849 by the University of Basle. In the following year he began teaching mathematics and Latin at a girls' school in Basle, a post he held for 40 years until his retirement in 1890. From 1865–90 he was also a part-time lecturer at Basle University, teaching projective geometry, which was the subject of a textbook he published in 1887. Balmer died in Basle on 12 March 1898.

Balmer was a mathematician by education and vocation and was not trained in physics. Nevertheless, he became interested in spectroscopy and – encouraged by J E Hagenbach-Bischoff, a professor at Basle University – began investigating the apparently random distribution of spectral lines. Spectroscopy was a relatively new discipline in the last quarter of the 19th century and, although it was known that excited molecules of a substance emit discrete, characteristic spectral lines, many scientists had failed to find a mathematical relationship between these lines. But in 1885 Balmer – then aged 60 – published an equation that described the four visible spectral lines of hydrogen (all that were then known) and also predicted the existence of a fifth line at the limit of the visible spectrum, which was soon detected and measured. He further predicted the existence of other hydrogen spectral lines beyond the visible spectrum. These lines were later found and named after their discoverers: the Lyman series (in the ultraviolet part of the spectrum), the Paschen series (in the infrared part of the spectrum), the Brackett series (infrared), and the Pfund series (infrared). The five lines in the visible part of the hydrogen spectrum are now known as the Balmer series, and are described by the formula:

$$\nu = R\,(1/2^2 - 1/n^2)$$

Where ν is the wave number, R is the Rydberg constant, and $n = 3, 4, 5$ ...

Balmer arrived at his formula (which has been modified only slightly in the light of later findings) solely from empirical evidence and offered no explanation as to why it gave the correct results. But it was later of crucial importance to Danish physicist Niels ◊Bohr, who used it to support his model of atomic structure. Moreover, not until atomic theory had developed still further was a full explanation of Balmer's equation possible.

Balmer's last paper, which was published in 1897, contained equations to describe the spectral lines of helium and lithium.

Banks, Joseph (1743–1820) English naturalist who, although making relatively few direct contributions to scientific knowledge himself, did much to promote science, both in the UK and internationally.

Banks was born on 13 February 1743 in London, the son of William Banks of Revesby Abbey in Lincolnshire. Born into a wealthy family, Banks was educated at Harrow and Eton public schools and then at Oxford University. At that time the university curriculum was biased towards the classics, but Banks was more interested in botany so he employed Israel Lyons (1739–1775), a botanist from Cambridge University, as a personal tutor in the subject. After graduating in 1763, Banks moved to London in order to meet other scientists. Meanwhile his father had died in 1761, leaving Banks a large fortune, which he inherited when he came of age in 1764. In 1776 he made his first voyage, to Labrador and Newfoundland, as naturalist on a fishery-protection ship. He collected many plant specimens during the trip and, on his return to England, was elected to the Royal Society of London.

In 1768 preparations were being made for an expedition to the southern hemisphere to observe the transit of Venus in 1769. Banks obtained the position of naturalist on the voyage and accompanied by several artists and an assistant botanist, Daniel Solander (1736–1782), set sail in the *Endeavour* – commanded by Captain James Cook – in 1768; Banks paid for his assistants and all the equipment he needed out of his own pocket, at a cost of about £10,000. After the astronomical observations had been completed (the transit was observed from Tahiti), the expedition proceeded on its second objective, to search for the large southern continent that was then thought to exist. During this part of the voyage the expedition explored the coasts of New Zealand and Australia. Banks' plant-collecting activities at the first landing place in Australia (near present-day Sydney) gave rise to the name of the area – Botany Bay. He also studied the Australian fauna, discovering that almost all of the mammals are Marsupials. The expedition returned to England in 1771 and Banks brought back a vast number of plant specimens, more than 800 of which were previously unknown. (Banks kept a journal of the expedition, part of which was published, although not until long after his death, but he did not write an account of his scientific findings on the voyage.) On his return, Banks found himself a celebrity and was summoned to Windsor Castle to give a personal account of his travels to King George III; this visit was the start of a lifelong friendship with the king, which helped Banks to establish many influential contacts.

In 1772 Banks went on his last expedition, to Iceland, where he studied geysers. In 1778 he was elected president of the Royal Society (perhaps because of his influence in high places), an office he held until his death 42 years later. As president, Banks re-established good relations between the Royal Society and the king, who had previously quarrelled with the society over the issue of the best shape for the ends of lightning conductors. He also brought several wealthy patrons into the society and helped to develop its international reputation.

As a result of the friendship between Banks and the king, the Royal Botanic Gardens at Kew – of which Banks was the honorary director – became a focus of botanical research. Banks sent plant collectors to many countries in an attempt to establish at Kew as many different species as possible. He also conceived of Kew as a major centre for the practical use of plants, to which end he initiated several important projects, including the introduction of the tea plant into India from its native China, and the transport of the breadfruit tree from Tahiti to the West Indies. This latter project, however, was initially unsuccessful because of the famous mutiny on the *Bounty,* which was carrying the breadfruit trees. At the king's request, Banks also played an active part in importing merino sheep into the UK from Spain; after initial difficulties, the breed was later successfully introduced into Australia.

Banks' voyage to Australia on the *Endeavour* stimulated a lifelong interest in the country's affairs, and he was instrumental in establishing the first colony at Botany Bay in 1788. Thereafter he greatly assisted the growth of the colony and was in regular correspondence with its various governors.

Banks was a generous patron who gave financial assistance to several talented young scientists, notably Robert ⇨Brown, who later became an eminent botanist although he is better known today as the discoverer of Brownian motion. Banks also made his large home in Soho Square, London, a renowned meeting place for scientists and prominent figures from other fields. In addition, his international prestige did much to promote the exchange of ideas among scientists in many countries. He also obtained safe passages for many scientists during the American War of Independence and during the Napoleonic wars, and petitioned on behalf of scientists who had been captured.

Banks received many honours during his life, including a baronetcy in 1781 and membership of the Privy Council in 1797. When he died on 19 June 1820, in Isleworth, near London, he left an extensive natural history library and a collection of plants regarded as one of the most important in existence, both of which are now housed in the British Museum.

Further Reading

Carter, Harold Burnell, *Sir Joseph Banks (1743–1820): A Guide to Biographical and Bibliographical Sources,* St Paul's Bibliographies (in association with British Museum, Natural History), 1987.

Gascoigne, John, *Science in the Service of Empire: Joseph Banks, the British State and the Uses of Science in the Age of Revolution,* Cambridge University Press, 1998.

O'Brian, Patrick, *Joseph Banks: A Life,* University of Chicago Press, 1997.

Banneker, Benjamin (1731–1806) US mathematician, astronomer, and surveyor who is chiefly known for his almanacs published in the 1790s.

The son of a freed slave, Benjamin Banneker was born on 9 November 1731 near the Patapsco River in Baltimore County. He received no formal education apart from attending a nearby Quaker school for several seasons, where he showed a great interest and ability in mathematics. Throughout most of his life he worked on his parents' tobacco farm, managing it after the death of his father in 1759, and taught himself mathematics and astronomy. He retired from farming in 1790 to devote all his time to his studies and remained at the farm until his death on 9 October 1806.

Banneker became known in 1753, at the age of 21 when, having studied only a pocket watch, he constructed a striking clock. This was the first clock of its kind in America and operated for more than 40 years. He was better known, however, for his almanacs which were published 1792–97. Having borrowed instruments and texts on astronomy from his neighbour George Ellicott in 1789, he taught himself the subject and learned to calculate an ephemeris (a numerical description of the orbits of celestial bodies) and to make projections for lunar and solar eclipses. He compiled an ephemeris for 1791, which he incorporated into an almanac but which was not published; the following year, however, his ephemeris was published as *Benjamin Banneker's Pennsylvania, Delaware, Maryland and Virginia Almanack and Ephemeris, for the year of our Lord, 1792.* Banneker sent a manuscript copy to the secretary of state, Thomas Jefferson, accompanied by a 12-page letter defending the mental capacities of black people and urging the abolition of slavery; this, with its acknowledgement by Jefferson, was reprinted in the almanac for 1793 and formed part of a long correspondence between the two men. He continued to complete the ephemerides for almanacs each year until 1804, although the later ones were never published. He also wrote a dissertation on bees and did a study of locust plague cycles.

Banneker US mathematician, astronomer and surveyor Benjamin Banneker. *Moorland-Spingarn Research Centre*

As well as working on his almanacs, in 1791 Banneker was appointed scientific assistant to Major Andrew Ellicott, George Ellicott's cousin, to survey the Federal Territory for the establishment of the new capital, Washington. Working with Major Pierre Charles L'Enfant they defined the boundaries, streets, and the major buildings of the new city, and it is said that when L'Enfant left the USA, taking his plans with him,

Banneker reproduced them from memory in two days. In 1970, Banneker Circle, adjoining L'Enfant Plaza, in Washington, DC was named after him.

Further Reading

Allen, Will W, *Banneker: The Afro-American Astronomer,* Ayer Co, 1992.

Bedini, Silvio A, *The Life of Benjamin Banneker: The First African American Man of Science,* Maryland Historical Society, 1998.

Conley, Kevin, *Benjamin Banneker,* Grolier, 1990.

Conley, Kevin, *Benjamin Banneker,* Black Americans of Achievement series, Chelsea House Publishers, 1989.

Litwin, Laura Baskes, *Benjamin Banneker: Astronomer and Mathematician,* African-American Biographies series, Enslow Publishers, 1999.

Banting, Frederick Grant (1891–1941) Canadian physiologist who discovered insulin, the hormone responsible for the regulation of the sugar content of the blood (an insufficiency of which results in the disease diabetes mellitus). For this achievement he was awarded the Nobel Prize for Physiology or Medicine in 1923, which he shared with the Scottish physiologist John Macleod (1876–1935).

Banting was born in Alliston, Ontario, on 14 November 1891, the son of a farmer. He went to the University of Toronto in 1910 to study for the ministry, but changed to medicine and obtained his medical degree in 1916. He served overseas as an officer in the Canadian Medical Corps during World War I, and was awarded the Military Cross for gallantry in 1918. After the war he held an appointment at the University of Western Ontario, but in 1921 returned to the University of Toronto to carry out research into diabetes. In 1930 the Banting and Best Department of Medical Research was opened at the University of Toronto, of which Banting became director. He was knighted in 1934. While serving as a major in the Canadian Army Medical Corps in 1941 he was killed in an air crash at Gander, Newfoundland.

At the University of Western Ontario, Banting became interested in diabetes, a disease (often fatal at that time) characterized by a high level of blood sugar (glucose) and the appearance of glucose in the urine. In 1889 Mehring and Minkowski had shown that the pancreas was somehow involved in diabetes because the removal of the pancreas from a dog resulted in its death from the disease within a few weeks. Other workers had investigated the effect of tying off the pancreatic duct in rabbits, which resulted in atrophy of the pancreas apart from small patches of cells – the islets of Langerhans. The rabbits did not, however, develop the diuretic condition of sugar in the urine.

It had therefore been suggested that a hormone, called insulin (from the Latin for 'island'), might be concerned in glucose metabolism and that its source might be the islets of Langerhans. But efforts to isolate the hormone from the pancreas failed because the digestive enzymes produced by the pancreas broke down the insulin when the gland was processed. In 1921 Banting went to discuss his ideas on the matter with John Macleod, the chair of the physiology department (and an expert on the metabolism of carbohydrates) at the University of Toronto. Macleod was unenthusiastic but agreed to find Banting a place for research in his laboratories.

Banting reasoned that if the pancreas were destroyed but the islets of Langerhans retained, the absence of digestive enzymes would allow them to isolate insulin. With Charles Best (1899–1978), one of his undergraduate students, he experimented on dogs. They put several of the animals into two groups; each dog in one group had the pancreatic duct tied, and those in the other group were depancreatized. After several weeks, they removed the degenerated pancreases from the dogs of the first group, extracted the glands with saline and injected the extract into the dogs of the second group, which by then had diabetes and were in poor condition. They took regular blood samples from the diabetic dogs and found that the sugar content dropped steadily as the condition of the dogs improved.

These results encouraged Banting. He obtained foetal pancreas material from an abattoir, thinking that it might contain more islet tissue. With his assistants he set about extracting an active product, but purification proved to be very difficult. Eventually reasonably pure insulin was produced and commercial production of the hormone started. By January 1922 it was ready for use in the Toronto General Hospital. The first patient was a 14-year-old diabetic boy who showed rapid improvement after treatment. They also discovered that the dose of insulin could be reduced by regulating the amount of carbohydrate in the patient's diet.

When Banting and Macleod were awarded the Nobel prize for this work Banting, feeling strongly that Best had made a valuable contribution, divided his share of the money with him. Banting's discovery of insulin and his attempts to purify the crude material led eventually to the commercial production of insulin, which has saved the lives of many diabetics.

Further Reading

Best, Charles Herbert, 'Frederick Grant Banting', 1891–1941, *Obituary Notices of Fellows of the Royal Society,* 1942–44, v 4, pp 21–26.

Bliss, Michael, *Banting: A Biography,* University of Toronto Press, 1992.

Levine, Israel E, *The Discoverer of Insulin: Dr Frederick G Banting,* Copp Clark Pub Co, 1959.

Stevenson, Lloyd G, *Sir Frederick Banting,* W Heinemann Medical Books, 1947.

Webb, Michael (ed), *Frederick Banting, Discoverer of Insulin*, Scientists and Inventors Series, Copp Clark Pitman, 1991.

Bardeen, John (1908–1991) US physicist whose work on semiconductors, together with William Shockley (1910–1989) and Walter Brattain (1902–1987), led to the first transistor, an achievement for which all three men shared the 1956 Nobel Prize for Physics. Bardeen also gained a second Nobel prize in 1972, which he then shared with John Robert Schrieffer (1931–) and Leon ◊Cooper for a complete theoretical explanation of superconductivity.

Bardeen, who is the only person thus far to be awarded two Nobel prizes for physics, was born in Madison, Wisconsin on 23 May 1908. He began his career with a BS in electrical engineering at the University of Wisconsin awarded in 1928. For the next two years, he was a graduate assistant working on mathematical problems of antennas and on applied geophysics, but it was at this stage that he first became acquainted with quantum mechanics. He moved to Gulf Research in Pittsburgh, where he worked on the mathematical modelling of magnetic and gravitational oil-prospecting surveys, but the lure of pure scientific research became increasingly strong, and in 1933 he gave up his industrial career to enrol for graduate work with Eugene Wigner (1902–1995) at Princeton, where he was introduced to the fast-developing field of solid-state physics. His early studies of work functions, cohesive energy, and electrical conductivity in metals were carried out at Princeton, Harvard, and the University of Minnesota. From 1941–45, he returned to applied physics at the Naval Ordnance Laboratory in Washington, DC, where he studied ship demagnetization and the magnetic detection of submarines.

In 1945 Bardeen joined the Bell Telephone Laboratory, and his work on semiconductors there led to the first transistor, an achievement rewarded with his first Nobel prize in conjunction with Walter Brattain and William Shockley. He moved to the University of Illinois in 1951, where in 1957, along with Bob Schrieffer and Leon Cooper, he developed the microscopic theory of superconductivity (the BCS theory) that was to gain him a second Nobel prize in 1972. After 1975, he was an emeritus professor at Illinois, concentrating on theories for liquid helium-3, which have analogies with the BCS theory.

The electrical properties of semiconductors gradually became understood in the late 1930s with the realization of the role of low concentrations of impurities in controlling the number of mobile charge carriers. Current rectification at metal–semiconductor junctions had long been known, but the natural next step was to produce amplification analagous to that achieved in triode and pentode valves. A group led by William Shockley began a programme to control the number of charge carriers at semiconductor surfaces by varying the electric field. John Bardeen interpreted the rather small observed effects in terms of surface trapping of carriers, but he and Walter Brattain successfully demonstrated amplification by putting two metal contacts 0.05 mm/0.002 in apart on a germanium surface. Large variations of the power output through one contact were observed in response to tiny changes in the current through the other. This so-called point contact transistor was the forerunner of the many complex devices now available through silicon-chip technology.

Ever since 1911, when Heike ◊Kamerlingh Onnes first observed zero electrical resistance in some metals below a critical temperature, physicists had sought a microscopic interpretation of this phenomenon of superconductivity. The methods that proved successful in explaining the electrical properties of normal metals were unable to predict the effect. At very low temperatures, metals were still expected to have a finite resistance due to the scattering of mobile electrons by impurities. Bardeen, Cooper, and Schrieffer overcame this problem by showing that electrons pair up through an attractive interaction, and that zero resistivity occurs when there is not enough thermal energy to break the pair apart.

Normally electrons repel one another through the Coulomb interaction, but a net attraction may be possible when the electrons are imbedded in a crystal. The ion cores in the lattice respond to the presence of a nearby electron, and the motion may result in another electron being attracted to the ion. The net effect is an attraction between two electrons through the response of the ions in the solid. The pairs condense out in a kind of phase transition below a temperature T_c (typically of the order of a few kelvin), and the involvement of the ions of the lattice in the interaction is confirmed by the dependence of T_c on the isotopic mass in the metal – that is, T_c varies with the variation of atomic vibration frequencies in the solid. The requirement of a finite energy to break up a so-called Cooper pair leads to a small energy gap below which excitations are not possible.

The BCS theory is amazingly complete, and explains all known properties associated with superconductivity. Although applications of superconductivity to magnets and motors were possible without the BCS theory, the theory is important for strategies to make T_c as high as possible – if T_c could be raised above liquid-nitrogen temperature, the economics of superconductivity would be transformed. In addition, the theory was an essential prerequisite for the prediction of Josephson tunnelling with its important applications in magnetometers, computers, and determination of the fundamental constants of physics.

Both of Bardeen's achievements have important consequences in the field of computers. The invention of the transistor led directly to the development of the integrated circuit and then the microchip, which has made computers both more powerful and more practical. Superconductivity enables the basic arithmetic and logic operations of computers to be carried out at much greater speeds and may lead to the development of artificial intelligence.

Barkla, Charles Glover (1877–1944) English physicist who made important contributions to our knowledge of X-rays, particularly the phenomenon of X-ray scattering. He received many honours for his work on ionizing radiation, including the 1917 Nobel Prize for Physics.

Barkla was born on 7 June 1877 in Widnes, Lancashire. After studying at the Liverpool Institute, he went to University College, Liverpool, in 1895, where he studied under Oliver Lodge. He obtained his BSc in 1898 and his MSc in the following year. In the autumn of 1899 he went to Trinity College, Cambridge, where he researched under J J Thomson at the Cavendish Laboratory, but transferred to King's College, Cambridge, in 1901 in order to sing in the choir. Nevertheless, he refused the offer of a choral scholarship, which would have enabled him to remain at Cambridge, and in 1902 returned to University College, Liverpool, as Oliver Lodge Fellow. In 1904 he was awarded a DSc and was regularly promoted, being appointed a special lecturer in 1907. He was made professor of physics at King's College, London, in 1909 and remained there until 1913, when he became professor of natural philosophy at Edinburgh University, a position he held until his death on 23 October 1944.

Barkla's first major piece of research, which was carried out while he was a student at the Cavendish Laboratory, involved measuring the speed at which electromagnetic waves travel along wires of different thickness and composition. During his third year at Cambridge, he began investigating secondary radiation – the effect whereby a substance subjected to X-rays re-emits X-rays – a subject that he spent most of his subsequent career studying. He published his first paper on this phenomenon in 1903. In this paper he announced his finding that for gases of elements with a low atomic mass the secondary scattered radiation is of the same average wavelength as that of the primary X-ray beam to which the gas is subjected. He also found that the extent of such scattering is proportional to the atomic mass of the gas concerned. The more massive an atom, the more charged particles it contains and it is these charged particles that are responsible for the X-ray scattering. Thus Barkla's work was one of the first indications of the importance of the amount of charge in an atom (rather than merely its atomic mass) in determining an element's position in the periodic table – a significant early step in the evolution of the concept of atomic number.

In 1904 Barkla found that, unlike the low-atomic-mass elements, the heavy elements produced secondary radiation of a longer wavelength than that of the primary X-ray beam. He also showed that X-rays can be partially polarized, thus proving that they are a form of transverse electromagnetic radiation, like visible light.

In 1907 Barkla began his most important research into X-rays, working at Liverpool with C A Sadler. They found that secondary radiation is homogeneous and that the radiation from the heavier elements is of two characteristic types. They also showed that these characteristic (that is, of a specific wavelength) radiations are emitted only after a heavy element is exposed to X-radiation 'harder' (that is, of shorter wavelength, and therefore more penetrating) than its own characteristic emissions. This finding was the first indication that X-ray emissions are monochromatic.

Barkla named the two types of characteristic emissions the K-series (for the more penetrating emissions) and the L-series (for the less penetrating emissions). He later predicted that other series of emissions with different penetrances might exist, and an M-series – radiation with even lower penetrance than that of the L-series – was subsequently discovered. After about 1916, however, Barkla devoted his research to investigating a J-series of extremely penetrating radiations, the existence of which he had suggested. But the results of these studies could not be confirmed by other workers and the existence of the J-series was not and still is not, part of accepted theory. Nevertheless Barkla continued to adhere to his theory of the J-phenomenon, as a result of which he became increasingly isolated from the rest of the scientific community during his later years.

Barnard, Christiaan Neethling (1922–) South African cardiothoracic surgeon who performed the first human heart transplant, on 3 December 1967 at Groote Schuur Hospital in Cape Town.

Barnard was born on 8 November 1922 in Beaufort West, South Africa. He attended the local high school and then in 1940 went to the University of Cape Town, where he received his medical degree in 1946. After working in private practice 1948–51, he became senior resident medical officer at the City Hospital in Cape Town for two years. In 1953 Barnard was appointed registrar at Groote Schuur Hospital, later to be the place of his most important work. In 1956 he was awarded a scholarship to the University of Minnesota, Minneapolis, USA; he returned to South Africa two years later, taking with him a heart–lung machine. He became director of surgical research at Groote Schuur and the University of Cape Town, and from 1968–83 was head of cardiothoracic surgery. Since 1984 he has been professor emeritus at the University of Cape Town and, since 1985, has been senior consultant and scientist in residence at the Oklahoma Heart Center, Oklahoma City.

Barnard's early research involved experiments with heart transplants in dogs. His success convinced him that similar operations could be performed on human patients. In December 1967 Denise Duvall, a 25-year-old woman, was critically injured in a road accident in Cape Town, and after it was established that her brain was irreparably damaged, permission was obtained for her heart to be donated for transplant purposes. The recipient was a man in his fifties, Louis Washkansky, whose heart – in Barnard's words – was 'shattered and ruined'.

X-ray motion pictures (angiograms) were taken of Washkansky's heart by injecting radio-opaque dye into each side of it using catheters inserted into the veins and arteries. These films were taken to prepare the surgical team for the operation on Washkansky. Once the donor heart had been removed, Barnard cut away part of it so that it would fit what remained of the recipient's heart. Two holes were made in the donor heart, one through which the venae cavae could enter and one for the pulmonary veins. The edges of these holes were stitched onto the waiting part of Washkansky's heart. The difference in size of the hearts did not matter because the openings in the donor heart could be enlarged to match the recipient's heart.

Surgically this first transplant was a success, but Washkansky died 18 days after the operation from double pneumonia – probably contracted as a result of the immunosuppressive drugs administered to prevent his body rejecting the new heart. Barnard continued to perform heart transplants, improving his methods all the time. Unfortunately the number of operations performed by him decreased because of the worsening arthritis in his hands.

Open-heart surgery was first introduced in South Africa by Barnard, and he further developed cardiothoracic surgery with new designs for artificial heart valves. His other achievements have included the discovery that intestinal artresia – a congenital deformity in the form of a hole in the small intestine – is the result of an insufficient supply of blood to the foetus during pregnancy. It was a fatal defect before Barnard developed the corrective surgery.

Barnard's techniques for heart-transplant surgery have been adopted and developed by many surgeons, and as the methods improve they can give a new lease of life to those suffering from fatal heart conditions.

Further Reading

Barnard, Christiaan, *The Second Life: Memoirs*, Hodder & Stoughton, 1993.

Cooper, David (ed), *Chris Barnard: By Those Who Know Him*, Vlaeberg Pub, 1992.

Hawthorne, Peter, *The Transplanted Heart: The Incredible Story of the Epic Heart Transplant Operations*, Hugh Keartland, 1968.

Leipold, L Edmond, *Dr Christiaan N Barnard: The Man with the Golden Hands*, Men of Achievement series, Denison, 1971.

Malan, Marais, *Heart Transplant: The Story of Barnard and the 'Ultimate in Cardiac Surgery'*, Voortrekkerpers, 1968.

Barnard, Edward Emerson (1857–1923) US observational astronomer whose keen vision and painstaking thoroughness made him an almost legendary figure.

Barnard was born in Nashville, Tennessee, on 16 December 1857. His family was poor and by the time he was nine years old Barnard had begun to work as an assistant in a photographic studio. The techniques he learned were to be invaluable in his later career. Barnard's fascination with astronomy led him to take a job in the observatory at Vanderbilt University. He took some courses but spent most of his time using the telescopes.

In 1877 Barnard went to California in order to work at the Lick Observatory when it opened in 1888. He was awarded a DSc from Vanderbilt in 1893, although he had never formally graduated. In 1895 he took up the chair of practical astronomy at the University of Chicago and became astronomer at the Yerkes Observatory. He participated in the expedition to Sumatra to observe the solar eclipse of 1901.

Barnard's many discoveries brought him worldwide respect and many honours. He received awards from the most prestigious scientific organizations, and was elected to their membership. He died in Williams Bay, Wisconsin, on 6 February 1923.

Barnard's early astronomical studies were made with a 12.7-cm/5-in telescope which he purchased in 1878. It was for the discovery of comets that Barnard first began to establish a reputation. He discovered his first comet on 5 May 1881 and by 1892 he had found 16. He also investigated the surface features of Jupiter, the *Gegenschein* (a faint patch of light visible only at certain times of the year and whose nature is still not certain), nebulae, and other celestial bodies.

His most dramatic discovery came on 9 September 1892 when, by blocking out the glow of the parent planet, Barnard discovered the fifth satellite of Jupiter and the first to be found since the four Galilean satellites. This was the last satellite to be discovered without the aid of photography. The fifth moon orbits inside all the others, which now number more than 20.

Barnard's later discoveries included the realization that the apparent voids in the Milky Way are in fact dark nebulae of dust and gas and the sighting in 1916 of the so-called 'Barnard's Runaway Star', which has a proper motion of 10 seconds of arc per year (faster than any star known until 1968).

Although Barnard's lack of mathematical flair prevented him from making profound contributions to theoretical advances in astronomy, he was one of the most eminent observational astronomers of his time.

Barr, Murray Llewellyn (1908–1995) Canadian anatomist and geneticist known for his research into defects of the human reproductive system, and particularly chromosomal defects.

Barr was born in Belmont, Ontario, on 20 June 1908, the son of a farmer (who was originally from Ireland). He attended the University of Western Ontario, where he gained his BA in 1930, his MD in 1933, and his MSc in 1938. Apart from serving as a medical officer with the Royal Canadian Air Force during World War II, he spent his entire career at the University of Western Ontario, where he became head of the department of anatomy in the Health Sciences Centre. He died in May 1995.

In 1949, working with Ewart Bertram, Barr noticed that the nuclei of nerve cells in females have a mass of chromatin (the nucleoprotein of chromosomes, which stains strongly with basic dyes) whereas those in males do not. He also found that this sex difference occurs in the cells of most mammals.

Improvements in cell-culture methods made the closer examination of human chromosomes possible; for example, in 1954 it was discovered that the chromosome number in human beings is 46 and not 48, as had previously been thought. From Barr's investigations, the sex chromatin (called the Barr body) is now known to be one of the two X chromosomes in the cells of females; it is more condensed than the other chromosomes and is genetically inactive. The other X chromosome in females is attenuated and genetically active in resting cells.

Before the discovery of sex chromatin, the nature of the sex-chromosome complex (XX female, or XY male) could be detected at cell level only by direct

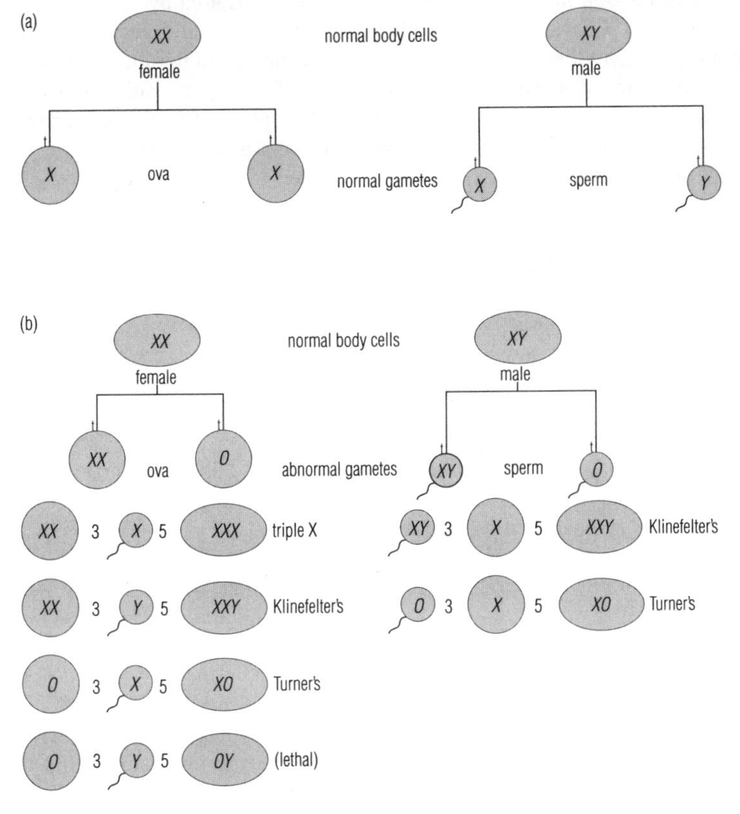

Barr (a) Meiosis with the formation of normal gametes; (b) chromosomal defects arise from fertilization of abnormal gametes by normal gametes.

examination of chromosomes in dividing cells. The new comparative method using stained sex chromatin offered a much-needed investigative and diagnostic procedure for patients with developmental anomalies of the reproductive system. Abnormalities of the sex chromosome complex often result in disorders such as Turner's syndrome (usually in females) and Klinefelter's syndrome. The former occurs when the gamete lacks a second sex chromosome so that the complex is XO. Klinefelter's syndrome, occurring in males only, results usually from a chromosomal complex of XXY.

Barr and his colleagues also devised a buccal smear test by rubbing the lining of the patient's mouth (the buccal cavity) and examining the cells obtained for chromosomal defects. This test is now used extensively to screen patients, including newborn babies, and has proved useful in the differential diagnosis of several kinds of hermaphroditism. Barr's research has thus been invaluable in simplifying diagnostic tests for chromosomal defects.

Barrow, Isaac (1630–1677) English mathematician, physicist, classicist, and Anglican divine, one of the intellectual luminaries of the Caroline period.

Barrow was born in London in October 1630. His father, a linen-draper to Charles I, sent him to Charterhouse as a day boy, but there he achieved little beyond gaining a reputation as a bully, and he was removed to Felstead School. In 1643 he was entered for Peterhouse, Cambridge, where an uncle was a fellow, but by the time that he went up to university in 1645 his uncle had moved to Trinity College and it was there that Barrow entered as a pensioner. He received his BA in 1648 and a year later was elected a fellow of Cambridge. In 1655 his former tutor, Dr Dupont, retired from the regius professorship of Greek; he wished Barrow, his former pupil, to succeed him. But the appointment was not offered to Barrow. His reputation then was more for mathematics than classics, and he was very young. Even so, it may be true that he was barred from the chair by Cromwell's intervention. Certainly Barrow had never concealed his royalist opinions and he was out of sympathy with the prevailing republican air of Cambridge. He decided to leave England for a tour of the continent and to help finance his trip sold his library.

He remained abroad for five years, returning to England only with the restoration of Charles II in 1660. Immediately he took Anglican orders and was elected to the regius professorship previously denied him. He was also appointed professor of geometry at Gresham College, London, and in 1663 became first Lucasian Professor of Mathematics at Cambridge. In 1669 he resigned the Cambridge chair in favour of Isaac Newton and a year later was made a DD by royal mandate. During the 1660s his lectures on mathematics at Gresham College formed the basis of his mathematical reputation; thereafter his energies were devoted more to theology and preaching. He was made master of Trinity in 1675 and died two years later, on 4 May 1677.

In his time Barrow was considered second only to Newton as a mathematician. Now he is best remembered as one of the greatest Caroline divines, whose sermons and treatises, especially the splendid *Treatise on the Pope's Supremacy* (1680), have gained a permanent place in ecclesiastical literature. Certainly he was an admirable teacher of mathematics, and although it is not true that Newton was his pupil, Newton attended his lectures and later formed a fruitful friendship with him. His mathematical importance is slight, the *Lectiones mathematicae,* delivered at Gresham between 1663 and 1666 and published in 1669, being marred by his insistence that algebra be separated from geometry and his desire to relegate algebra to a subsidiary branch of logic. His geometry lectures were read by few and had little influence.

More important were his lectures on optics. Most of his work in this field was immediately eclipsed by Newton's, but there is no doubt that Newton was greatly inspired by Barrow's work in the field, and to Barrow is due the credit for two original contributions: the method of finding the point of refraction at a plane interface, and his point construction of the diacaustic of a spherical interface. Barrow was a man of great powers of concentration and original thought. If he failed to reach the highest class of mathematics, the reason may well be that he spread his intellectual interests so broadly.

Bartlett, Neil (1932–) English chemist who achieved fame by preparing the first compound of one of the rare gases, previously thought to be totally inert and incapable of reacting with anything.

Bartlett was born in Newcastle upon Tyne on 15 September 1932. He attended the University of Durham, gaining his PhD in 1957. A year later he took an appointment at the University of British Columbia, Canada, and in 1966 became professor of chemistry at Princeton in the USA. From 1969 he held a similar position at the University of California, Berkeley, becoming professor emeritus from 1993.

In Canada in the early 1960s Bartlett was working with the fluorides of the platinum metals. He prepared platinum hexafluoride, PtF_6, and found that it is extremely reactive. It reacts with oxygen, for example, to form the ionic compound $O_2^+ PtF_6^-$. In 1962 he reacted platinum hexafluoride with xenon, the heaviest of the stable rare gases, and obtained xenon platinofluoride (xenon fluoroplatinate, $XePtF_6$), the first chemical compound of a rare gas. Other compounds of xenon followed, including xenon fluoride (XeF_4) and xenon oxyfluoride ($XeOF_4$). Other chemists soon made compounds of krypton and radon. It is for this reason that this and other modern books use the term 'rare gases' to describe the helium group of elements, not the former terms 'inert gases' or 'noble gases', for inert or noble they no longer are, due to the pioneer work of Bartlett.

Barton, Derek Harold Richard (1918–1998) English organic chemist whose chief work concerned the stereochemistry of natural compounds. He showed that their biological activity often depends on the shapes of their molecules and the positions and orientations of key functional groups. For this achievement he shared the 1969 Nobel Prize for Chemistry with the Norwegian Odd Hassel (1897–1981).

Barton was born in Gravesend on 8 September 1918. He was educated at Tonbridge School and graduated from Imperial College, London, in 1940, gaining a PhD in organic chemistry two years later. He held various professorships: at Birkbeck College, London 1953–55, Glasgow 1955–70, and Imperial College, London, 1970–78. In 1978 he became emeritus professor of organic chemistry at the University of London, the same year that he was appointed director of the Institute for the Chemistry of Natural Substances at Gif-sur-Yvette in France. He was knighted in 1972.

While lecturing in the USA at Harvard 1949–50, Barton studied the different rates of reaction of certain steroids and their triterpenoid isomers (substances with the same composition but differing in the way their atoms are joined and arranged in space). He deduced that the difference in the spatial orientation of their functional groups accounts for their behaviour, and so developed a new field in organic chemistry that became known as conformational analysis. Barton realized that in a complex system where the conformation is fixed, the reactivity of a given group depends on whether it is attached to the main molecule in an axial or an equatorial position. He discovered important correlations between the chemical reactivity and conformation of various groups in steroids and terpenes (which are structurally very similar).

Barton went on to examine many natural products, including phenols. For example, in 1956 he challenged the generally accepted structure of the substance known as Plummerer's ketone, showing how it could be formed by the oxidative coupling of two phenolic residues.

He realized the biosynthetic importance of this reaction, concluding that the structures of many phenols and alkaloids could be explained and predicted. He devised new ways of preparing oxyradicals and studied various natural products that contain the dienone group, predicting that if a hydrogen atom of the same molecule is spatially orientated near to a generated oxyradical, intermolecular elimination of the hydrogen atom is preferred.

Barton also studied photochemical routes and unravelled the complex transformations that take place during photolysis. In 1959 (at Cambridge, Massachusetts) he devised a simple synthesis of the naturally occurring hormone aldosterone. He also worked on the antibiotics tetracycline and penicillin. He thus contributed greatly to the study of natural products and their formation, and enabled a rational interpretation to be made of much stereochemical information.

Bascom, Florence (1862–1945) US geologist who was an expert in crystallography, mineralogy, and petrography, and the founder of the geology department at Bryn Mawr College, Pennsylvania, that was responsible for the training of the foremost women geologists of the early 20th century.

Born in Williamstown, Massachusetts, in 1862, Florence was the youngest of the six children of suffragist and schoolteacher Emma Curtiss Bascom and William Bascom, professor of philosophy at Williams College. Her father, a supporter of suffrage and the education of women, later became president of the University of Wisconsin to which women were admitted in 1875. Florence enrolled there in 1877 and with other women was allowed limited access to the facilities but was denied access to classrooms filled with men. In spite of this she gained a BA in 1882, a BSc in 1884, and an MS in 1887. When Johns Hopkins graduate school opened to women in 1889, Bascom was allowed to enrol to study petrology on condition that she sat behind a screen to avoid disrupting the male students. With the support of her advisor George Huntington Williams and her father she managed to become the second woman to gain a PhD in geology (the first being Mary Holmes at the University of Michigan in 1888) in 1893.

Bascom's interest in geology had been sparked off by a driving tour she took with her father and his friend Edward Orton, a geology professor at Ohio State. It was an exciting time for geologists with new areas opening up all the time. Bascom was also inspired by her teachers at Wisconsin, Roland Irving and Charles van Hise, and G H Williams at Johns Hopkins, experts in the new fields of metamorphism and crystallography. Bascom's PhD thesis was a petrographical study of rocks that had previously been thought to be sediments but which she proved to be metamorphosed lava flows.

While studying for her PhD, Bascom became a popular teacher, passing on her enthusiasm and rigour to her students. She taught at the Hampton Institute for

Negroes and American Indians and at Rockford College, becoming instructor and associate professor at Ohio State University in geology and petrography 1892–95. Moving to Bryn Mawr College, where geology was considered subordinate to the other sciences, she spent two years teaching in a store room while building up a considerable collection of fossils, rocks, and minerals. While at Bryn Mawr she took great pride in passing on her knowledge and training a generation of women who would become successful – including crystallographer Mary Porter, the petrologists Eleanora Knopf and Anna Jonas Stose, palaeontologist Julia Gardner and glacial geomorphologist Ida Ogilvie. At Bryn Mawr she rose rapidly becoming reader (1898), associate professor (1903), professor (1906), and finally professor emeritus from 1928 till her death in 1945 in Northampton, Massachusetts.

Bascom became, in 1896, the first woman to work as a geologist on the US Geological Survey, spending her summers mapping formations in Pennsylvania, Maryland, and New Jersey and spending her winters analysing slides. Her results were published in Geographical Society bulletins. In 1924 she became the first woman to be elected a fellow of the Geographical Society of America and went on, in 1930, to become the first woman vice president. She was associate editor of the *American Geologist* 1896–1905 and achieved a four-star place in the first edition of *American Men and Women of Science* (1906), showing how highly regarded she was in her field.

Bascom was the author of over 40 research papers. She was an expert on the crystalline rocks of the Appalachian Piedmont and she published her research on Piedmont geomorphology and the provenance of surficial deposits. Geologists in the Piedmont area still value her contributions and she is still a powerful model for women struggling for status in the field of geology today.

Bassi, Laura (1711–1778) Italian physicist and the first woman to become a professor of physics at any university. She was also one of the first scholars in Italy to teach Newtonian natural philosophy.

Bassi was born in Bologna into a wealthy family. Her father was a lawyer and she was educated at home by the family doctor. Bassi was an outstanding student of mathematics, philosophy, anatomy, natural history, Greek, Latin, French, and logic. She was awarded a doctorate in philosophy from the University of Bologna at the age of 21 by Cardinal Lambertini after she had held a public philosophical disputation against five scholars. She then studied mechanics, hydraulics, natural history, and anatomy at the university. She was later made professor of anatomy and lectured to large classes of students.

Bassi married Dr Guiseppe Veratti in 1738 and had 12 children. While bringing up the children she continued lecturing in experimental physics, studied mathematics, and gave private lessons from home.

Bassi was best known for her teaching but she did publish two dissertations in Latin in the *Commentarius* of the Bologna Institute: 'De problemate quodam mechanico' and 'De problemate quodam hydrometrico'. In spite of her chair in anatomy, the university senate tried to curtail her lectures and restrict her to lecturing at public functions. She succeeded, however, in petitioning for a higher salary and wider responsibilities.

Bassi was very religious and devoted a great deal of her time to helping the poor. The senate of Bologna coined a medal in her honour and she was given the chair of experimental physics at the university in 1776 at the age of 65.

Bates, H(enry) W(alter) (1825–1892) English naturalist and explorer whose discovery of a type of mimicry (called Batesian mimicry) lent substantial support to Charles Darwin's theory of natural selection.

Bates was born on 8 February 1825 in Leicester, the son of a clothing manufacturer. He received little formal education, leaving school when he was 13 years old to work in his father's stocking factory, but he was interested in natural history and devoted much of his spare time to private study. In 1844 he met Alfred Russel ◊Wallace and aroused in him an interest in entomology, and they planned a joint venture to the Amazon region of South America to study and collect its flora and fauna, aiming to pay their expenses by selling the specimens they collected. In 1848 they arrived in Brazil at Pará (also called Belém) near the mouth of the River Amazon, and for the next two years they worked together, thereafter separately. Wallace returned to England in 1852 but Bates remained in South America until 1859, during which time he explored much of the River Amazon. After returning to England, he spent several years organizing the specimens he had collected and writing about his observations, discoveries, and explorations. In 1864 he was appointed assistant secretary to the Royal Geographical Society in London, a post he held until his death in London, on 16 February 1892.

During his Amazon exploration, Bates collected a vast number of specimens, including more than 14,000 species of insects, more than half of which were previously unknown. He travelled continually up and down the Amazon waterways, usually spending only a few days at each stopping-place to collect specimens. These had to be prepared and preserved – a difficult task in the hot, humid conditions of the Amazon rainforest –

Bates English naturalist and explorer H W Bates who studied the phenomenon of mimicry in animal coloration. An amateur botanist and entomologist, Bates worked for several years in his father's stocking factory before making his career in natural history. *Mary Evans Picture Library*

before being sent to his agent in England for sale. He also managed to find time to write to Charles Darwin, who used Bates's findings in the development of his theory of natural selection.

After returning to England, Bates presented a paper to the Linnaean Society in 1861 entitled 'Contributions to an Insect Fauna of the Amazon valley', in which he outlined his observations of mimicry in insects. He had discovered that several different species of butterflies have almost identical patterns of colours on their wings, and that some are distasteful to bird predators whereas others are not. Further, he suggested that the latter types, influenced by natural selection, mimic the distasteful species and thus increase their chances of survival. This form of mimicry is now called Batesian mimicry. (Subsequently Fritz Müller (1821–1897), a German-born Brazilian zoologist, discovered that, in some cases, all the species of similarly coloured butterflies are distasteful to predators, a phenomenon known as Müllerian mimicry.)

Bates's paper was well received and he was asked to write an account of his experiences in South America. The result was *The Naturalist on the River Amazon* (1863), a two-volume work (with an introduction by Charles Darwin) in which Bates described both his explorations and his scientific findings. The book quickly sold out and a second edition was published. But in this second edition, which has been reprinted many times, much of the scientific material was omitted, which has tended to diminish Bates's reputation as a scientist.

Bateson, William (1861–1926) English geneticist who was one of the founders of the science of genetics (a term he introduced), and a leading proponent of Mendelian views after the rediscovery in 1900 of Gregor Mendel's work on heredity. Bateson also made important contributions to embryology.

Bateson was born on 8 August 1861 in Whitby, Yorkshire. He was educated at Rugby School and St John's College, Cambridge, from which he graduated in natural sciences in 1883. He then travelled to the USA, remaining there for two years doing embryological research. During this period he met W K Brooks of Johns Hopkins University, who interested Bateson in evolution, which he spent the rest of his life studying. On his return to the UK, Bateson spent several years investigating the fauna of salt lakes and undertaking other research into evolution and heredity. In 1908 he became the first professor of genetics at Cambridge University, but left this post in 1910 to be the director of the newly established John Innes Horticultural Institution at Merton, Surrey, where he remained until his death on 8 February 1926. In addition to his directorship, Bateson was Fullerian Professor of Physiology at the Royal Institution 1912–14 and a trustee of the British Museum from 1922.

Bateson's first important research was his embryological work in the USA. Studying the small, wormlike marine creature *Balanoglossus,* he discovered that although its larval stage is similar to that of the echinoderms, it also possesses a dorsal nerve cord and the beginnings of a notochord. Thus he demonstrated that *Balanoglossus* is a primitive chordate, which was the first indication that chordates had evolved from echinoderms – a theory now widely accepted.

His interest having turned to evolution while in the USA, Bateson spent the years immediately following his return to the UK investigating the fauna of the salt lakes of Europe, central Asia, and northern Egypt. The result of these studies was his book *Material for the Study for Variation* (1894), in which he put forward his theory of discontinuity to explain the long process of evolution. According to this theory, species do not develop in a predictable sequence of very gradual changes but instead evolve in a series of discontinuous 'jumps'. This theory was unacceptable to the traditional, biometrical evolutionists who maintained that there were no breaks in nature's pattern, and so Bateson began a series of breeding experiments to find corroborative evidence for his theory. When Mendel's work on heredity was rediscovered in 1900, Bateson translated Mendel's paper into English (as *Experiments on Hybrid Plants)*, and found that Mendel's work provided him with the supportive evidence he was seeking for his discontinuity theory. Bateson also assumed the task of publicizing and defending the highly controversial discoveries of Mendel. The long debate finally culminated in 1904 when, as president of the zoological section of the British Association, Bateson succeeded in vindicating Mendel's findings at a meeting in Cambridge.

Thereafter Bateson continued with his breeding experiments, the results of which he described in his *Mendel's Principles of Heredity* (1908). In this book he showed that certain traits are consistently inherited together, an apparent contradiction to Mendel's findings; this phenomenon is now known to result from genes being situated close together on the same chromosome – a phenomenon called linkage. Towards the end of his life Bateson proposed his own vibratory theory of inheritance, based on the physical laws of force and motion, but this theory has met with little acceptance from other scientists.

Bayliss, William Maddock (1860–1924) English physiologist who discovered the digestive hormone secretin, the first hormone to be found, and investigated the peristaltic movements of the intestine. He received many honours in recognition of his work, including the Royal Medal 1911 and Copley Medal of the Royal Society 1919, and in 1922 he was knighted.

Bayliss was born in Wolverhampton, Staffordshire, on 2 May 1860. He began studying medicine at University College, London, from 1881, but turned from medicine towards physiological research and entered Wadham College, Oxford, in 1885. After graduation he returned to University College where he began his main research. In 1893 he married the sister of Ernest ◊Starling, the man with whom he worked during his major discoveries. In 1903 he became a Fellow of the Royal Society, and in 1912 a professorship of general physiology was created for him at University College. He was a longstanding member of the Physiological Society, first as its secretary and then as its treasurer. He died in Hampstead, London, on 27 August 1924.

Bayliss discovered the hormone secretin in 1902 when working with Starling. He made an extract of a piece of the inner lining (mucosa) of the duodenum, which had already had hydrochloric acid introduced to it. When the extract was injected into the bloodstream the pancreas was stimulated to secrete digestive juices. Bayliss tried injecting hydrochloric acid intravenously, but no pancreatic secretion occurred. He then severed the nerves serving a loop of duodenum so that it was isolated from the pancreas except via the blood supply. Acid was introduced into the duodenum, and the pancreas produced secretions. Bayliss thus concluded that as hydrochloric acid (from the stomach's digestive juices) passes into the duodenum during the normal digestive process, the duodenal mucosa release a chemical (the hormone secretin) into the bloodstream which, in turn, makes the pancreas secrete its juices. The role of hormones in physiology is now commonly accepted, but at the time Bayliss' discovery was a breakthrough. He then went on to study the activation of enzymes, particularly the pancreatic enzyme trypsin.

Bayliss and Starling also worked on the nerve supply to the intestines, and on pressures within the venous and arterial systems. Bayliss did independent research into vasomotor reflexes. His method of treating patients suffering from surgical shock with saline to replace blood loss was widely used during World War I on injured troops. In 1915 he published *Principles of General Physiology,* which rapidly became a standard work.

Beadle, Tatum, and Lederberg **George Wells Beadle (1903–1989), Edward Lawrie Tatum (1909–1975), and Joshua Lederberg (1925–)** US scientists who shared the 1958 Nobel Prize for Physiology or Medicine for their pioneering work in the field of biochemical genetics.

Beadle was born on 22 October 1903 in Wahoo, Nebraska, and was educated at the University of Nebraska, graduating in 1926. After obtaining his doctorate in genetics at Cornell University, New York, in 1931, he went to the laboratory of Thomas Hunt Morgan at the California Institute of Technology, where he researched into the genetics of the fruit fly (*Drosophila melanogaster)*. In 1935 he went to Paris, where he continued his work on *Drosophila* at the Institut de Biologie Physico-Chimique, collaborating with Boris Ephrussi. Beadle returned to the USA in 1936 and taught genetics for a year at Harvard University. From 1937–46 he was professor of biology at Stanford University, California, and it was during this period that he collaborated with Tatum on the work that was to gain them the Nobel prize. Beadle was appointed professor and chair of the division of biology at the California Institute of Technology in 1946 and remained there until 1961, when he became chancellor of the University of Chicago. In 1968 he retired from the university in order to direct the American Medical Association's Institute for Biomedical Research.

Tatum was born on 14 December 1909 in Boulder, Colorado, and was educated at the University of Wisconsin (where his father was head of the pharmacology department), from which he graduated in 1931 and gained his doctorate in 1934. From 1937–41, after

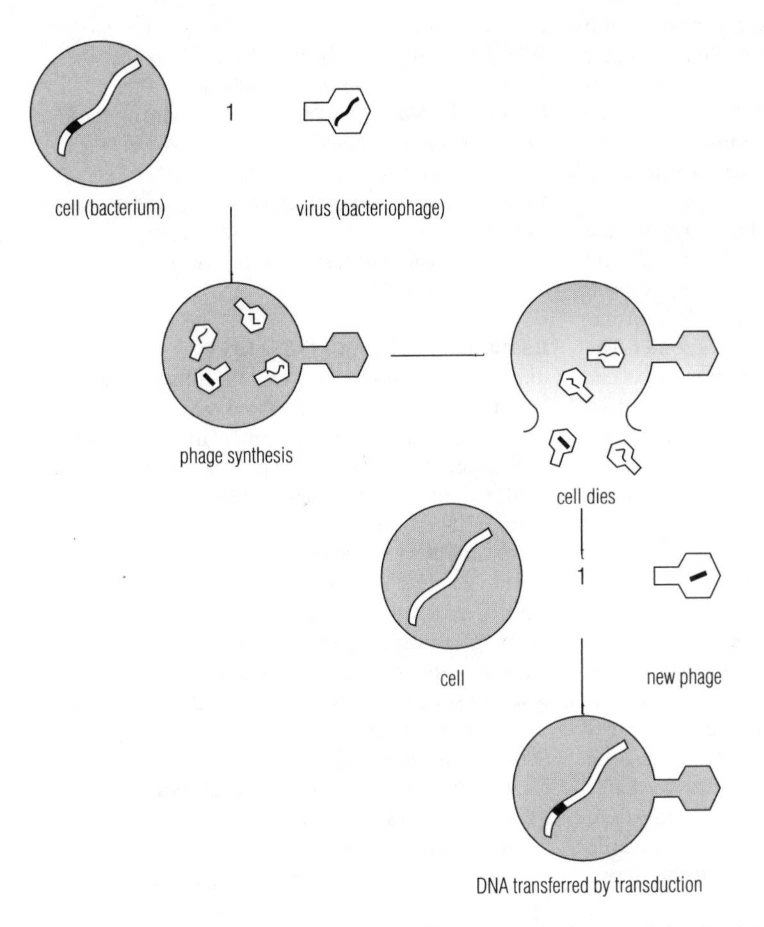

Beadle, Tatum, and Lederberg Beadle showed how a bacteriophage can bring about the transfer of DNA by transduction.

a period of postdoctoral study in the Netherlands, he was a research assistant at Stanford University, where he worked with Beadle. Tatum joined the faculty of Yale University in 1945 – as associate professor of botany until 1946, then as professor of microbiology 1946–48. While at Yale he worked with Lederberg. Tatum returned to Stanford University in 1948 as professor of biology, becoming professor of biochemistry there in 1956. In 1957 he went to the Rockefeller Institute for Medical Research (now Rockefeller University), New York City. Tatum died on 5 November 1975 in New York City.

Lederberg was born on 23 May 1925 in Montclair, New Jersey. He studied at Columbia University, graduating in 1944, then did postgraduate work at Yale University, gaining his doctorate in 1947. It was while he was at Yale that Lederberg worked with Tatum. In 1947 Lederberg joined the University of Wisconsin – initially as assistant professor of genetics, then as professor of medicine and genetics, and finally as chair of the genetics department. In 1959 he moved to Stanford University, where he was professor of genetics and biology and chair of the genetics department and, from 1962, director of the Kennedy Laboratories of Molecular Medicine.

The research that eventually led to the award of a Nobel prize had its origins in Beadle's early work on *Drosophila.* While at Morgan's laboratory, Beadle realized that genes influence heredity by chemical means. Later, working with Ephrussi in Paris, he showed that the eye colour of *Drosophila* is a result of a series of chemical reactions under genetic control. Because of the complexity of the relationship between genes and metabolic processes in *Drosophila* – itself a relatively simple organism – when Beadle returned to the USA and continued his research at Stanford, he used the red bread mould *Neurospora crassa,* which is even simpler than *Drosophila. Neurospora* can be cultured in a medium containing sugar and biotin as the only organic components, plus a few inorganic salts. Working with Tatum, Beadle subjected colonies of *Neurospora* to X-rays and studied the changes in the nutritional requirements of – and therefore in the enzymes formed by – the mutant *Neurospora* produced by the irradiation. By repeating the experiment with various mutant strains and culture mediums, Beadle and Tatum deduced that the formation of each individual enzyme is controlled by a single, specific gene. This one-gene–one-enzyme concept found wide applications in biology and virtually created the science of biochemical genetics.

After Tatum moved to Yale in 1945, he applied the mutation-inducing technique developed by Beadle and himself to bacteria. Using *Escherichia coli*, Tatum, working with Lederberg, showed that genetic information can be passed from one bacterium to another. The discovery that a form of sexual reproduction can occur in bacteria meant that these organisms could be used for research much more extensively than previously thought possible and, in fact, bacteria are now as important as *Drosophila* and *Neurospora* in genetic research.

Tatum's collaboration with Lederberg ended in 1947, when Lederberg went to the University of Wisconsin and continued to research into bacterial genetics. In 1952 he published a paper in which he revealed that bacteriophages (viruses that attack bacteria) can transfer genetic material from one bacterium to another, a phenomenon Lederberg termed transduction. This discovery further increased the usefulness of bacteria in genetic research. Later Lederberg diversified into other fields of investigation, devising means of identifying and classifying organic compounds by mathematical graph theory and developing statistical methods of studying human biology.

Beaufort, Francis (1774–1857) British hydrographer famous for his meteorological developments.

Born in Ireland, the son of a cleric of Huguenot origin, Beaufort followed his father in developing an early love of geography and topography. He joined the East India Company in 1789; a year later he enlisted in the Royal Navy, and spent the next 20 years in active service.

In 1806 Beaufort drew up the wind scale named after him. This ranged from 0 for dead calm up to storm force 13. He specified the amount of sail that a full-rigged ship should carry under the various wind conditions. It was taken up in the early 1830s by Robert FitzRoy (the captain of the *Beagle* on which Darwin sailed round the world) and officially adopted by the Admiralty in 1838. Modifications were made to the scale when sail gave way to steam.

Beaufort did major surveying work especially around the Turkish coast in 1812. In 1829 he became hydrographer to the Royal Navy, promoting voyages of discovery such as that of English botanist Joseph ◊Hooker on the *Erebus*.

Further Reading

Friendly, Alfred, *Beaufort of the Admiralty: The Life of Francis Beaufort, 1774–1857*, Hutchinson, 1977.

Garbett, L G, 'Admiral Sir Francis Beaufort and the Beaufort scales of wind and weather', *Quart J Roy Meteorol Soc*, 1926, v 52, pp 161–168.

Kinsman, Blair, *An Exploration of the Origin and Persistence of the Beaufort Wind Force Scale*, Technical Report/ Chesapeake Bay Institute, Reference series, Chesapeake Bay Institute, 1968.

Beaumont, William (1785–1853) US surgeon who did important early work on the physiology of the human stomach by taking advantage of a bizarre surgical case that he treated early in his career. He established that digestion is a chemical process, and his work encouraged other researchers to study the physiology of digestion.

Beaumont was born in Lebanon, Connecticut, on 21 November 1785. He was the son of a farmer, and worked as a schoolteacher in Champlain, New York, before going to study medicine at St Albans, Vermont. He was granted a licence to practise, becoming an assistant surgeon in the army in 1812. He resigned his commission after three years and practised in Plattsburgh, resuming his army career in 1820 when he was sent to the frontier post of Fort Mackinac, Michigan. In 1834 he was transferred to St Louis, Missouri, and he worked there for the rest of his life, although he left the army in 1839. During this time he became professor of surgery at St Louis University. He died in St Louis on 25 April 1853.

On 6 June 1822 a young French Canadian trapper named Alexis St Martin was accidentally shot in the left side of his back at close range, causing severe injury where the shot passed right through his abdomen. Beaumont was at Fort Mackinac, and treated the trapper swiftly and skilfully. The young man survived although he retained a permanent traumatic fistula, or hole, between his stomach and the outside of his abdomen. Beaumont looked after St Martin for two years, and in 1825 began a series of experiments and observations on the behaviour of the human stomach under various circumstances. Through the fistula he was able to extract and analyse gastric juice and stomach contents at various stages of digestion, observe changes in secretions, and note the muscular movements of the stomach. But he so hounded the unfortunate St Martin that he left (and eventually outlived Beaumont by nearly 30 years). Beaumont published his findings in *Experiments and Observations on the Gastric Juice* (1833).

Beaumont's work predated by many years the use of endoscopic examinations of the stomach and was the first well-documented and accurate observation of the digestive processes of a living human being.

Further Reading

Burns, Virginia, *William Beaumont: Frontier Doctor,* Enterprise Press, 1989.

Cohen, I Bernard (ed), *The Career of William Beaumont and the Reception of His Career,* Three Centuries of Science in America series, Arno Press, 1980.

Horsman, Reginald, *Frontier Doctor: William Beaumont, America's First Great Medical Scientist,* Missouri Biography Series, University of Missouri Press, 1995.

Osler, William, *William Beaumont: A Backwood Physiologist,* Notes from the Editors series, Gryphon Editions, 1980.

Becquerel, Antoine Henri (1852–1908) French physicist who discovered radioactivity in 1896, an achievement for which he shared the 1903 Nobel Prize for Physics with Pierre and Marie ◊Curie. The Curies did not participate in Becquerel's discovery but investigated radioactivity and gave the phenomenon its name.

Becquerel was born in Paris on 15 December 1852 and educated at the Ecole Polytechnique and Ecole des Ponts et Chaussées, where he received a training in engineering. In 1875, he began private scientific research, investigating the behaviour of polarized light in magnetic fields and in crystals, linking the degree of rotation to refractive index. Both Becquerel's grandfather and father were respected physicists with positions at the Museum of Natural History and other institutions. On their deaths, in 1878 and 1891 respectively, Becquerel succeeded to their posts. He became a member of the Academy of Sciences in 1889 and a professor at the museum in 1892 and at the Ecole Polytechnique in 1895.

Becquerel then began the work for which he is remembered, not necessarily because of his position but because of the discovery of X-rays made by Wilhelm ◊Röntgen early in 1896. This prompted Becquerel to investigate fluorescent crystals for the emission of X-rays, and in so doing he accidentally discovered radioactivity in uranium salts in the same year. Pierre and Marie Curie then searched for other radioactive materials, which led them to the discovery of polonium and radium in 1898.

Becquerel subsequently investigated the radioactivity of radium, and showed in 1900 that it consists of a stream of electrons. In the same year, Becquerel also obtained evidence that radioactivity causes the transformation of one element into another. Following his award of the 1903 Nobel Prize for Physics jointly with the Curies, Becquerel became vice president (1906) and president (1908) of the Academy of Sciences. He died soon after on 25 August 1908 in Brittany.

Becquerel French physicist Henri Becquerel who won the Nobel Prize for Physics in 1903, together with Pierre and Marie Curie, for the discovery of radioactivity. Among other subjects, he also carried out research into the influence of the Earth's magnetic field on its atmosphere. *Mary Evans Picture Library*

Becquerel's discovery of radioactivity was prompted by the mathematician Henri ◊Poincaré, who told Becquerel that X-rays were emitted from a fluorescent spot on the glass cathode-ray tube used by Röntgen. This immediately suggested to Becquerel that X-rays might be produced naturally by fluorescent crystals, with which he was familiar through his father's interest in fluorescence. He therefore placed some crystals of potassium uranyl sulphate on a photographic plate wrapped in paper, and put it in sunlight to make the crystals fluoresce. When he developed the plate, Becquerel found it to be fogged, showing that a radiation resembling X-rays had penetrated the paper and exposed the plate. Becquerel then tried to repeat the experiment to make further investigations, but the weather was cloudy and the uranium crystals would not fluoresce as there was no sunlight. He put a wrapped plate and the crystals into a drawer and waited. The weather did not improve and Becquerel impatiently decided to develop the plate. To his astonishment, the plate had been strongly exposed to radiation. Clearly it was not connected with fluorescence, but was emitted naturally by the crystals all the time.

Becquerel studied the radiation and found that it behaved like X-rays in penetrating matter and ionizing air. He showed that it was due to the presence of uranium in the crystals, and subsequently found that a disc of pure uranium metal is highly radioactive. This led Pierre and Marie Curie to isolate the radioactive elements polonium and radium. Becquerel later subjected the radiation from radium to magnetic fields and was able to prove by the amount of deflection that it must consist of the electrons that had been discovered by J J ◊Thomson in England in 1897. Becquerel also discovered that radioactivity could be removed from a radioactive material by chemical action, but that the material subsequently regained its radioactivity.

Becquerel's discovery of radioactivity and its investigation by himself and the Curies caused a revolution in physics. It marked the beginning of nuclear physics by showing that atoms, and then nuclei within atoms, are made up of smaller particles. Furthermore, the spontaneous regeneration of radioactivity observed by Becquerel was evidence that one element can be transformed into another with the production of energy. A full explanation of radioactivity was achieved by Ernest ◊Rutherford, leading eventually to nuclear fission and the commercial production of nuclear energy.

Behring, Emil Adolph von (1854–1917) German physician and immunologist who won the first Nobel Prize for Physiology or Medicine for his work on serum therapy against diphtheria and tetanus.

Behring was born on 15 March 1854 in Hansdorf, Prussia (now in Poland), where his father was a schoolteacher. He was educated at the Friedrich Wilhelm Institute of Military Medicine in Berlin, where he received a free medical training in return for subsequent service in the army medical corps. While still in training, Behring became interested in the possibility of preventing disease by hygienic and disinfectant methods; these concerns received further promotion whilst serving with a cavalry regiment in Posen (Poznań), now in Poland. During the next few years he combined clinical military duties with further research and training, especially in the use of disinfectants.

In 1889, after completing his army service, he became the assistant of the bacteriologist Robert ◊Koch at his Institute for Hygiene in Berlin. There Behring was joined by a young Japanese worker Shibasaburo ◊Kitasato and they discovered that healthy guinea pigs injected with serum prepared from the blood of animals infected with diphtheria remained healthy even when subsequently injected directly with diphtheria bacillus. Behring recognized that there was a substance in the transferred serum that rendered the poisonous toxins harmless, and this substance he termed antitoxin. From this Behring and Kitasato developed safe clinical preparations of their antitoxin serum and defined effective dosages for this new serum therapy. In its first year of use in Berlin children's hospitals in 1891, mortality from diphtheria dropped by more than two-thirds. A modification of the technique, by the French scientist Emil Roux (1853–1933) to produce antitoxin serum from horses rather than from guinea pigs improved large-scale production, and serum therapy for diphtheria and also tetanus became available worldwide. Behring adopted Roux's procedure and was supported in the production of serum by a German chemical firm.

In 1893 he was appointed professor of hygiene in Berlin but the follwing year, in the wake of his deteriorating relationship with Koch, he moved to the University of Halle and in the year after that to the University of Marburg, in an attempt to establish his own research institute. In 1901 he was awarded the first-ever Nobel Prize for Physiology or Medicine, and was elevated to the Prussian nobility. Throughout this period his research continued into developing methods of standardizing serum antitoxins and he began to investigate the problem of tuberculosis. He recognized that the bovine and human forms of the disease were caused by the same microorganism, thus identifying the danger to humans of drinking contaminated milk. He also introduced early vaccination techniques against diphtheria and tuberculosis, and during World War I his tetanus vaccines saved the lives of millions of German soldiers, for which Behring was, most unusually for a civilian, awarded an Iron Cross. He died in Marburg on 31 March 1917, honoured and remembered as a great benefactor to children whose lives were saved by his antitoxins and vaccines against diphtheria and tetanus.

Further Reading

De Kruif, Paul; Harry G Grover (ed), *Microbe Hunters*, Harcourt, Brace, and Co, 1937.

Satter, Heinrich, *Emil von Behring*, Inter Nationes, 1967.

Beijerinck, Martinus Willem (1851–1931) Dutch botanist who in 1898 published his finding that an agent smaller than bacteria could cause diseases, an agent that he called a virus (the Latin word for poison).

Beijerinck was born in Amsterdam on 16 March 1851. His earliest scientific interest was botany, but he graduated in 1872 with a diploma in chemical engineering from the Delft Polytechnic School, where one of his friends was Jacobus van't Hoff (who later won the first Nobel Prize for Chemistry in 1901). After graduating, Beijerinck taught botany to provide himself with a living while he studied for his doctorate, which he gained in 1877. He then became interested in bacteriology and took a job as a bacteriologist with an industrial company. In order to learn more about the subject, he travelled extensively throughout Europe. In 1895 he returned to the Delft Polytechnic

School, where he taught and carried out research for the rest of his career. He died on 1 January 1931 in Gorssel in the Netherlands.

In the early 1880s Beijerinck began studying the disease that stunts the growth of tobacco plants and mottles their leaves in a mosaic pattern (now called the tobacco mosaic virus disease). He tried to find a causative bacterium but was unsuccessful. This research stimulated his interest in bacteriology, however, and led to his taking a job as an industrial bacteriologist. While working in this capacity he discovered one of the types of nitrogen-fixing bacteria that live in the nodules on the roots of leguminous plants.

After he returned to academic life in 1895, Beijerinck resumed his study of the tobacco mosaic disease and again tried to isolate a causative agent. He pressed out the juice of infected tobacco leaves and found that the juice alone was able to infect healthy plants, but he could not detect a bacterial pathogen in the juice nor could he culture a microorganism from it. Furthermore, he found that the juice remained infectious even after he had passed it through a filter that removed even the smallest bacteria. He was also certain that the causative agent was not a toxin because he could infect a healthy plant and from that plant infect another healthy plant, continuing this process indefinitely – therefore the infective agent had to be capable of reproduction.

Louis Pasteur had earlier postulated the existence of pathogens too small to be visible under the microscope; and Dimitri Ivanovski, a Russian bacteriologist, had in 1892 observed that tobacco mosaic disease could be transmitted by a filtered juice but he thought that there was a flaw in his filter and still believed the disease to be bacterial. It was Beijerinck who first published his findings and stated that the tobacco mosaic disease is caused by a non-bacterial pathogen. He believed that the filtered juice of the infected plants was itself alive, and he called the causative agent a filterable virus. Thus Beijerinck was the first to recognize the existence of a class of pathogens now known to cause a wide range of diseases in animals and plants, as well as in human beings. He was, however, mistaken in his belief that the virus was a liquid; in 1935 the US biochemist Wendell Stanley (1904–1971) demonstrated that viruses are particulate.

Further Reading

Bos, Pieter, et al (eds), *Beijerinck and the Delft School of Microbiology*, Delft University Press, 1995.

Iterson, G L van; den Dooren de Jong, L E; and Kluyver, A J, *Martinus Willem Beijerinck: His Life and Work*, Science Tech, 1983.

Békésy, Georg von (1899–1972) Hungarian-born US scientist who resolved the longstanding controversy on how the inner ear functions. For his discovery concerning the mechanism of stimulation within the cochlea, he received the 1961 Nobel Prize for Physiology or Medicine (the first physicist to do so).

Békésy was born in Budapest on 3 June 1899, where his father was a member of the diplomatic service. He went to the University of Bern in 1916, graduated in 1920, and then enrolled at the University of Budapest where he took his PhD in physics in 1923. For the next 23 years he worked in the laboratories of the Hungarian Telephone System. During this time he was also employed at the central laboratories of Siemens and Halske AC in Berlin, and from 1932–39 he was a lecturer at the University of Budapest. He was appointed special professor there 1939–40 and a full professor 1940–46. In 1946 he emigrated to Sweden, disturbed by the Soviet occupation of Hungary, and worked at the Karolinska Institute in Stockholm. He held the title of research professor there, although he did not actually take up the post because, in 1947, he emigrated to the USA to become a research lecturer at the Psycho-Acoustic Laboratory at Harvard. From 1949–66 he served as a senior research fellow in psychophysics. He then went to the University of Hawaii where he took up the appointment of professor of sensory sciences, and remained there until his death on 13 June 1972.

While working as a telecommunications engineer for the Hungarian Telephone System, Békésy decided that to determine what frequency range a new cable should be able to carry, he would investigate how the human ear actually receives sound. He researched the functioning of the eardrum by gluing two mirrors to it and beaming light and sound into the ear. In this way he was able to observe reflections of the movements of the membrane when it was activated by sound waves.

In another series of experiments, Békésy observed how the auditory ossicles in the middle ear – the hammer, anvil, and stirrup – pick up the vibrations transmitted to them by the eardrum, and how they relay these messages to the cochlea in the inner ear. It had long been known that nerves in the cochlea pick up sound signals and transmit them along the auditory nerve to the brain for interpretation. It was also known that the cochlea consists of a spiral-shaped channel through which runs a thin partition known as the basilar membrane, containing groups of fine fibres. German physiologist Hermann von ◊Helmholtz had postulated a theory, which was widely accepted at that time, that each fibre had a natural period of vibration and responded only to sounds that vibrated at that period. He claimed that each group of fibres stimulated different nerve

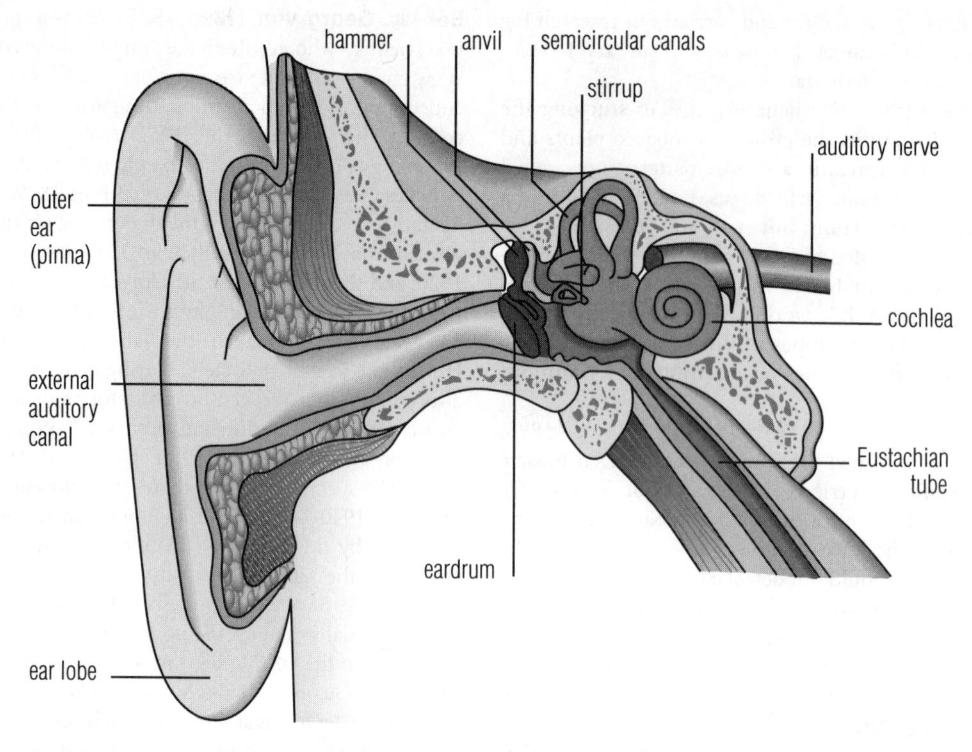

Békésy Békésy's study of human hearing resulted in an analysis of the mechanism of the cochlea and the role of the basilar membrane within it.

endings, which thus enabled the brain to distinguish specific frequencies.

Békésy's painstaking and original work disproved the Helmholtz theory. He constructed models of the cochlea and also worked with cadavers whose auditory mechanisms he stimulated electrically. Békésy had to design new instruments and develop new techniques in order to experiment on the delicate cochlea. He devised extremely fine drills and probes, and the scissors he used had blades of only a few hundredths of a millimetre in length. He reached the cochlea by grinding a small opening in the skull, and revealing part of the basilar membrane. By substituting a saline solution containing fine aluminium particles for the fluid in the cochlea and by using stroboscopic illumination, he was able to observe and measure for the first time a phenomenon he called the 'travelling wave'.

Békésy found that the innermost ossicle, the stirrup, acts as a lid on an opening in the cochlea, called the oval window (fenestra ovalis). As sound vibrations cause the stirrup to move it exerts pressure on the fluid within the cochlea, and these vibrations are transmitted to the basilar membrane in the form of travelling waves. He established that as the waves pass along the basilar membrane the entire membrane vibrates. Each wave causes maximum vibration at different sections of the membrane according to its frequency. High-frequency waves produced by high-pitched sounds reach their peak on that part of the basilar membrane nearest the stirrup. Low-frequency waves, from low sounds, attain a maximum amplitude farther along the membrane.

Later, Békésy extended his interests to visual and tactile sensations, which he measured and recorded. He developed an audiometer that determines whether deafness is caused by damage to the brain or to the ear, so that the appropriate treatment could be determined at an earlier stage.

Békésy also devised a means by which he enabled the skin to 'hear'. His apparatus consisted of a greatly enlarged version of a cochlea – a long tube filled with fluid and with a membrane running along its length. If the forearm of a person were pressed against the membrane, the skin of the arm felt the high and low sounds sent through the tube at distinctly different positions along the arm. Békésy's experiments pointed the way to methods that could enable the totally deaf to hear through tactile sensations, and improved modern interpretations of deafness.

Bell, Alexander Graham (1847–1922) Scottish-born US scientist who invented the telephone. He

Bell Scottish-born US scientist and educator Alexander Graham Bell, who invented the telephone. Bell had many other interests and was a prolific inventor of recording equipment and a dedicated teacher of the deaf. *Mary Evans Picture Library*

became the first person to transmit the human voice by electrical means and the telephone system he initiated is now used worldwide.

Bell was born in Edinburgh on 3 March 1847. Until the age of 11 he was taught at home by his mother. After graduating from the Edinburgh Royal High School at the age of 15, he went to London and lived with his grandfather, who was widely known as a speech tutor. From him Bell gained a knowledge of the mechanisms of speech and sound. In 1867, after a year of teaching and studying at Bath, he became an assistant to his father, who had originated the phonetic 'visible speech' system for teaching the deaf.

After the death of his second brother in 1870 the family moved to Brantford, Ontario, for health reasons. In April 1871, Bell started giving instruction in 'visible speech' to teachers of the deaf in and around Boston, Massachusetts, and by 1873 he was professor of vocal physiology at Boston University, a post he held for four years.

Bell's interest in speech and sound began at a very early age. As a boy he had constructed an automaton – using rubber, cotton, and a bellows – simulating the human organs of speech. He had even experimented with the throat of his pet Skye terrier, attempting to turn cooperative growls into words.

In 1874 he was granted patents on a multiple or harmonic telegraph for sending two or more messages simultaneously over the same wire. Elisha Grey had already developed the telegraph and, at this time, several individuals had become aware of the possibility of transmitting the human voice by electrical means. Bell, with Thomas A Watson as his assistant, had devised a transmitter for all the various voice frequencies by placing a magnetized reed in the centre of a circular diaphragm that was set to vibrate at the human voice. The vibrating reed induced varying electrical oscillations in an associated electromagnet. The same device, at the distant end of the circuit, functioned as a receiver.

The following year, Bell and Watson improved on this, and on 10 March 1876 Bell became the first person ever to transmit speech from one point to another by electrical means. The first message travelled only the distance from one room to the next, but two months later the first 'long-distance' voice message travelled 13 km/8 mi from Paris to Brantford, Ontario; before the end of the year, this distance had been increased to 229 km/143 mi. The patent for the telephone invented by Bell was granted in 1876. From then on, the telephone system spread rapidly across the country.

Mr Watson, come here; I want you.

FIRST COMPLETE SENTENCE SPOKEN OVER THE TELEPHONE **ALEXANDER GRAHAM BELL** MARCH 1876

With the financial independence he had gained through the success of his telephone system Bell was able to set up a laboratory at Baddeck Day, on the Bras d'Or lakes of Cape Breton Island, Nova Scotia. Among the first fruits of his work here was the photophone, which used the photoresistive properties of selenium crystals to apply the telephone principle to transmitting words in a beam of light. Bell thus achieved the first wire-less transmission of speech and from these principles evolved the photoelectric cell and other developments. Had coherent light (lasers) or optical fibres been available to him, Bell might have been able to develop the photophone into something useful. As it was, he seemed to recognize its potential instinctively for in in 1921, in an age of intercontinental radio, he declared that the photophone was his greatest invention. The spectrophone, for carrying out spectrum analysis by means of sound, followed.

Bell established the Volta Laboratory in Washington and patented the gramophone and wax recording cylinder which were commercially successful improvements on Thomas Edison's first phonograph and cylinders of metal foil. The laboratory also experimented with flat disc records, electroplating records, and impressing permanent magnetic fields on records – the embryonic tape recorder. Bell's entire share of $200,000 from the sale of some of his patents served to guarantee the perpetuation of the Volta Bureau, which he had formed for all conceivable kinds of research into deafness.

In 1881, Bell developed two telephonic devices for locating metallic masses (usually bullets) in the human body. One, an induction balance method, was first tried out on President Garfield, who was assassinated in 1881, while the other, a probe, was widely used until the advent of X-rays. At the Baddeck Bay laboratory, he built hydrofoil speed boats and sea-water converting units. He made kites capable of carrying people, employing a tetrahedral design for strength and lift.

Some idea of the true breadth of Bell's inventiveness can be gained by his lesser-known developments. These included an air-cooling system, a special strain of sheep (which he claimed produced twin or triplet lambs in more than half the births), the forerunner of the iron lung, and a sorting machine for punch-coded census cards.

In 1882 Bell became a citizen of the USA and moved to Washington, DC. In 1888, he became a founder member of the National Geographic Society.

The commercial success of his many inventions made Bell a wealthy man. He was also the recipient of many awards. In 1876, he received the Gold Medal awarded by the judges at the Centennial Exhibition in Philadephia where his prototype telephone was first shown. In 1880 he received the Volta Prize in France and was made an officer of the French Legion of Honour. From 1898–1903 he was president of the National Geographic Society, and in 1898 he was appointed regent of the Smithsonian Institution.

In 1917 a massive granite and bronze memorial was erected in Bell's honour at Brantford, Ontario. After his death on 2 August 1922 at his winter home at Baddeck Bay, a museum was built by the Canadian government at Baddeck as a permanent reminder of his achievements.

Further Reading

Grosvenor, Edwin S, *Alexander Graham Bell: The Life and Times of the Man Who Invented the Telephone,* Abradale Books, 1997.

Mackay, James A, *Alexander Graham Bell: A Life,* Wiley, 1998.

Matthews, Tom, *The World and All That's in It: A Photobiography of Alexander Graham Bell,* National Geographic Society, 1999.

Pasachoff, Naomi and Gingerich, Owen (ed), *Alexander Graham Bell: Making Connections,* Oxford Portraits in Science series, Oxford University Press, 1998.

Petrie, Roy, *Alexander Graham Bell,* Iowa State University Press, 1998.

Bell, Charles (1774–1842) Scottish anatomist and surgeon who carried out pioneering research on the human nervous system. He gave his name to Bell's palsy, an extracranial paralysis of the facial nerve (the VIIth cranial nerve) – not the same as the long thoracic nerve of Bell, which he also named and which supplies a muscle in the chest wall.

Bell was born in Edinburgh in November 1774. His brother, John Bell, was a renowned surgeon who taught him anatomy. After qualifying in 1799 Bell became a surgeon at the Edinburgh Royal Infirmary. He went to London in 1804 as a lecturer, and in 1812 was appointed surgeon at the Middlesex Hospital, London. He became professor of anatomy and surgery at the Royal College of Surgeons in 1824 and four years later was invited to become the first principal of the medical school at University College, London. He was knighted in 1831. In 1836 he became professor of surgery at the University of Edinburgh. He died at Hallow, Worcestershire, on 28 April 1842.

Bell carried out meticulous dissections and made the important discovery that nerves are composite structures, each with separate fibres for sensory and motor functions. His findings first appeared in a short essay 'Idea of a new anatomy of the brain' (1811); his main written work was *The Nervous System of the Human Body* (1830). The chief significance of Bell's discovery was the impetus it gave to other researchers in neurology.

Further Reading

Bell, Charles, *Letters of Sir Charles Bell Selected from His Correspondence with His Brother George Joseph Bell,* J Murray, 1870.

Cohen, John, *Sir Charles Bell (1774–1842): A Memoir,* TeamPrint, 1975.

Geldzahler, Henry, *Charles Bell: The Complete Works, 1970–1990,* Abrams, 1991.

Gordon-Taylor, Gordon and Walls, Eldred Wright, *Sir Charles Bell, His Life and Times,* E & S Livingstone, 1958.

Ollerenshaw, Robert, *Charles Bell: Anatomist, Surgeon, Artist, and Philosopher (1774–1842),* British Medical Association, 1951.

Bell, Patrick (1799–1869) Scottish clergyman who invented one of the first successful reaping machines. It is probable that later commercially successful machines owed much to his pioneering work.

Bell was born in April 1799 on a farm of which his father was a tenant in the parish of Auchterhouse, a few miles northwest of Dundee. He studied for the ministry at St Andrew's University, and it was there in 1827 that he turned his attention to the construction of a machine that he (like many other inventors) thought would considerably reduce the labour of the grain harvest. Bell died in 1869 in the parish of Carmylie, Arbroath, of which he had been ordained minister in 1843.

Harvesting equipment has progressed most since the beginning of the 19th century when Bell made his important contribution. Since then, methods have advanced from the use of the primitive sickle to the self-propelled combine harvester. It is said that the grain-harvesters of Egypt in 3000 BC could have been used by most farm workers 4,800 years later without any additional training being needed.

It was the invention of the reaper that opened the way to the complete mechanization of the grain harvest, and many attempts were made from the end of the 18th century to cut corn by machine.

The way most of these machines performed has been lost to history and it was not until Bell's machine appeared that the record becomes clearer. He started trials in deep secrecy inside a barn on a crop that had been planted by hand, stalk by stalk. In 1828, he and his brother carried out night-time trials, which were a success, leading them to exhibit the machine the following year. In the years to 1832 at least 20 machines were produced, 10 of them cutting 130 ha/320 acres in the UK and 10 going for export. Six reapers were exhibited at the Great Fair in New York in 1851. That year also saw the Great Exhibition in London, at which reapers by US inventors Obed ◊Hussey and Cyrus ◊McCormick were exhibited, both of them showing similarities to the Bell machines. This was the turning point for mechanical harvesting; mechanization gradually invaded farms and the sickle and scythe were virtually ousted by the 20th century.

Bell's reaper was pushed from behind by a pair of horses and the standing cereals were brought on to the reciprocating cutter bar by horizontally revolving rods similar to the reels seen on modern combine harvesters. The cut cereal fell on to an inclined rotating canvas cylinder and was sheaved and stooked by hand. One of Bell's machines was used on his brother's farm for many years until, in 1868, it was bought by the museum of the Patents Office where it was afterwards kept. He did not take out a patent and the design was improved upon and reintroduced as the 'Beverly Reaper' in 1857.

In recognition of his services to agriculture, Bell was presented with £1,000 and a commemorative plate by the Highland Society, the money being raised mainly from the farmers of Scotland. He also received the honorary degree of LLD from the University of St Andrews.

Although Bell did not achieve the fame of others such as McCormick, his work was of fundamental value and importance in the successful advancement of agricultural methods to the now well-established mechanized farming used today.

Bell Burnell, (Susan) Jocelyn (1943–) Northern Irish astronomer who discovered pulsating radio stars (pulsars) in the 1960's. – an important astronomical discovery of the 1960s.

Jocelyn Bell was born in Belfast on 15 July 1943. The Armagh Observatory, in which her father was architect, was sited near her home and the staff there were particularly helpful and offered encouragement when they learned of her early interest in astronomy. From 1956–61 she attended the Mount School in York. She then went to the University of Glasgow, receiving her BSc degree in 1965. In the summer of 1965 she began to work for her PhD under the supervision of Antony ◊Hewish at the University of Cambridge. It was during the course of this work that the discovery of pulsars was made. Having completed her doctorate at Cambridge, she went on to work in gamma-ray astronomy at the University of Southampton and from 1974–82 she worked at the Mullard Space Science Laboratory in X-ray astronomy. In 1982 she was appointed a senior research fellow at the Royal Observatory, Edinburgh, where she worked on infrared and optical astronomy. She was head of the James Clark Maxwell Telescope section, responsible for the British end of the telescope project based in Hawaii. In 1991 she was appointed professor of physics and departmental chair at the Open University, Milton Keynes, In England. A winner of the Royal Astronomical Society's prestigious Herschel Medal in 1989, she has made significant contributions in the fields of X-ray and gamma-ray astronomy.

She spent her first two years in Cambridge building a radio telescope that was specially designed to track quasars – her PhD research topic. The telescope that she and her team built had the ability to record rapid variations in signals. It was also nearly 2 ha/4.5 acres in area, equivalent to a dish with a diameter of 150 m/500 ft, making it an extremely sensitive instrument. The sky survey began when the telescope was finally completed in 1967 and Bell was given the task of analysing the signals received. One day, while scanning the charts of recorded signals, she noticed a rather unusual radio source that had occurred during the night and been picked up in a part of the sky that was opposite in direction to the Sun. This was curious because strong variations in the signals from quasars are caused by solar wind and are usually weak during the night. At first she thought that the signal might be due to a local interference, but after a month of further observations it became clear that the position of the peculiar signals remained fixed with respect to the stars, indicating that it was neither terrestrial nor solar in origin. A more detailed examination of the signal showed that it was, in fact, composed of a rapid set of pulses that occurred precisely every 1.337 seconds. The pulsed signal was as regular as the most regular clock on Earth.

One attempted explanation of this curious phenomenon was that it represented an interstellar beacon sent out by extraterrestrial life on another star and so initially it was nicknamed, LGM, for Little Green Men. Within a few months of noticing this signal, however, Bell located three other similar sources. They too pulsed at an extremely regular rate but their periods varied over a few fractions of a second and they all originated from widely spaced locations in our Galaxy. Thus it seemed that a more likely explanation of the signals was that they were being emitted by a special kind of star – a pulsar.

Since the astonishing discovery was announced, other observatories have searched the heavens for new pulsars and some 500 are now known to exist in our Galaxy (although a million or so may exist), their periods ranging from thousandths of a second to four seconds. It is thought that neutron stars are responsible for the signal. These are tiny stars, only about 7 km/10 mi in diameter, but they are incredibly massive. The whole star and its associated magnetic field are spinning at a rapid rate and the rotation produces the pulsed signal.

Beltrami, Eugenio (1835–1899) Italian mathematician whose work ranged over almost the whole field of pure and applied mathematics, but whose fame derives chiefly from his investigations into theories of surfaces and space of constant curvature, and his position as the modern pioneer of non-Euclidean geometry.

Beltrami was born in Cremona on 16 November 1835. From 1853–56 he studied mathematics at the University of Pavia. After graduating he was engaged as secretary to a railway engineer, but in his spare time he continued his research in mathematics and in 1862 he published his first paper, an analysis of the differential geometry of curves. In the same year he was appointed to the chair of complementary algebra and analytical geometry at the University of Bologna. The rest of his life was spent in the academic world: until 1864 at Bologna, as professor of geodesy at Pisa 1864–66, at Bologna again 1866–73, as professor of rational mechanics at the new University of Rome 1873–76, and at Pavia 1876–91 as professor of mathematical physics. In 1891 he returned to Rome, where he continued to teach until his death on 4 June 1899. A year before he died he was appointed president of the Accademia dei Lincei and was made a member of the Italian senate.

Beltrami's career may be divided into two parts. After 1872 he devoted himself to topics in applied mathematics, but in the earlier period he worked chiefly in pure mathematics, on problems in the differential geometry of curves and surfaces. Work had been done in this field earlier, notably by Saccheria a century before and by Russian mathematician Nikolai ◊Lobachevsky in the first half of the 19th century. But their work had had little influence, since their contemporaries were unimpressed by the possibilities of non-Euclidean geometry. The publication of Beltrami's paper 'Saggio di interpretazione della geometria non-euclidia' (1868) is therefore a landmark in the history of mathematics. It advanced a theory of hyperbolic space that laid the analytical base for the development of non-Euclidean geometry.

In an earlier paper of 1865, Beltrami had shown that on surfaces of constant curvature (and only on them) the formula $Ds^2 = Edu^2 + 2Fdudv + Gdv^2$ can be written such that the geodesics are represented by linear expressions in u and v. For positive curvature R^{-2} the formula would be $Ds^2 = R^2(v^2 + a^2)du^2 - 2uvdudv + (u^2 + a^2)dv^2 \times (u^2 + v^2 + a^2)^{-2}$, the geodesics behaving like the great circles of a sphere. In 1868 he went further and showed that by changing R to iR and a to ia, a new formula for ds^2 was obtained, one which defined the surfaces of constant curvature $-R^{-2}$ and which presented a new type of geometry for the geodesics of constant curvature inside the region $u^2 + v^2 < a^2$.

These demonstrations were not greatly different from what Lobachevsky had shown 40 years earlier, but what Beltrami did was to present the theories in terms that were acceptable within the existing Euclidean framework of the subject. He demonstrated that the concepts and formulae of Lobachevsky's geometry are realized for geodesics on surfaces of constant negative curvature. He showed also that there are rotation surfaces of this kind – and to these he gave the name 'pseudospherical surfaces'. He also demonstrated the usefulness of employing differential parameters in surface theory, thereby beginning the use of invariant methods in differential geometry.

After 1872 Beltrami switched his attention to questions of applied mathematics, especially problems in elasticity and electromagnetism. His paper 'Richerche sulle cinematice dei fluidi' (1872) was an important development in the field of elasticity. But his lasting fame rests on his signal achievement in overcoming the prevailing mid-19th century suspicions of non-Euclidean geometry and, by bringing it into the mainstream of mathematical thought, opening wide-ranging fields of new inquiry.

Benz, Karl (1844–1929) German engineer who designed and built the first commercially successful motor car.

Benz was born in Karlsruhe on 26 November 1844, the son of an engine driver, Johann Georg Benz, who died when Karl was two. He was educated at the gymnasium and polytechnic in his home town and began his career as an ordinary worker in a local machine shop when he was 21. Benz appears to have consciously shaped his career, moving from mechanical work on steam engines to design work with a more general engineering company in Mannheim, and then to a larger firm of engineers and ironfounders. In 1871 he returned to Mannheim, married Berta Ringer the following year, and opened a small engineering works in partnership with August Ritter.

After severe financial difficulties, Benz, now on his own, produced a two-stroke engine of his own design in 1878, and in 1883, after more financial difficulties, attracted enough support to found a new firm, Benz and Co. In 1885, he produced what is generally recognized to have been the first vehicle successfully propelled by an internal-combustion engine.

The motor car he produced stemmed from over 150 years of experimental work by many engineers working in many different fields. Benz was the first person to bring together the many threads to produce and exploit a commercially viable road vehicle. The evolution of the modern motor car really began when Joseph ◊Cugnot built his steam-driven gun carriage in 1769. Many other steam-driven vehicles followed, including that of Richard ◊Trevithick in 1802. Whether the Belgian Etienne ◊Lenoir or the Austrian Siegfried Marcus was the first to produce a carriage driven by an internal-combustion engine is a matter for dispute. Lenoir's machine of 1862 was driven by one of his gas engines and is said to have moved at 5 kph/3 mph. Marcus's machine is said to have run in 1868 and was certainly exhibited at the Vienna Exhibition in 1873. Neither, however, came to anything; Lenoir's had to carry about its own supply of town gas and was immensely heavy, while the Marcus engine had no clutch and was difficult to start.

In 1878 Benz produced his first two-stroke, 0.75 kW/1hp engine in the factory he had founded at Mannheim in 1872. The commercial success of this enabled him to found a new company, Benz and Co, and to experiment with the construction of motor vehicles as well as engines. As it happened, the timing was critical because Nikolaus ◊Otto and Alphonse Beau de Rochas (1815–1893) had each independently developed four-stroke engines. As a result of litigation between the two inventors, the Otto cycle became available to Benz and he was thus able to build a suitable power unit for his three-wheeled Tri-car.

In the spring of 1885, he produced what is generally regarded as the world's first vehicle successfully propelled by an internal-combustion engine. It used an Otto cycle four-stroke engine which gave about 0.56 kW/0.75 hp at 250 rpm and achieved a speed of up to 5 kph/3 mph during a journey of 91 m/299 ft on private ground adjoining the Mannheim workshop. Benz firmly believed that this vehicle would be a completely new system and not simply a carriage with a motor replacing the horse. The engine had a massive flywheel and was mounted horizontally in the rear, using electric ignition by coil and battery. The cooling system consisted simply of a cylinder jacket in which the water boiled away, being topped up as necessary. It had a carburettor of Benz's own design, which vapourized the fuel over a hot spot.

By the autumn, the prototype Tri-car was covering 1 km/0.6 mi at 12 kph/7.5 mph and had become a familiar sight on the streets of the town. The production model Tri-car appeared in 1886–87 and had a 1 kW/1.5 hp single-cylinder engine. The following year a Tri-car with an occasional extra seat and a 1.5 kW/2 hp twin cylinder engine appeared. Although the three-seat version was available for the Munich exhibition of 1888, Benz decided to exhibit a two-seater and won a gold medal for it.

Like most innovators, Benz had to contend with apathy and official hostility. There was also little demand for his Tri-cars at that time because of the public's general rejection of such 'monsters' and the severe restrictions placed on their use on public roads. British law, for instance, framed mainly with road-running steam engines in mind, required that all such vehicles have three drivers, be preceded by a man carrying a red flag, and move not faster than 6.4 kph/4 mph. This kind of control did nothing to promote sales. However, there was sufficient financial interest shown in France to enable Benz to improve his vehicles still further.

Benz laid down his first four-wheeled prototype in 1891 and by 1895, he was building a range of four-wheeled vehicles that were light, strong, inexpensive, and simple to operate. These automobiles had engines of 1–5.5 kW/1.5–6 hp and ran at speeds of about 24 kph/15 mph. Between 1897–1900 improved models appeared in increasing numbers – in particular, the Benz Velo 'Comfortable', of which over 4,000 were sold. At £135 each, they found a steady sale.

Benz and Co was now a thriving concern and, in 1899, it was turned into a limited company. Although

Benz retired from the board at the time of the transformation, the company he founded grew to become world famous for its production of high-performance cars. Under the gifted Hans Nibel as chief designer, the firm produced very successful racing cars and luxury limousines. In 1926, the company merged with the other famous German firm of Daimler – a firm that had developed at the same time as Benz and Co, along very similar lines and from similar beginnings – to form the world-famous Daimler–Benz company.

Benz died on 4 April 1929, at Ladenburg.

Further Reading

Diesel, Eugen; Goldbeck, Gustav; and Schildberger, Friedrich, *From Engines to Autos: Five Pioneers in Engine Development and their Contributions to the Automotive Industry*, Henry Regnery, 1960.

Nixon, St John C, *The Invention of the Automobile (Karl Benz and Gottlieb Daimler)*, Country Life Ltd, 1936.

Berg, Paul (1926–) US molecular biologist who shared (with Walter Gilbert and Frederick Sanger) the 1980 Nobel Prize for Chemistry for his work in genetic engineering, particularly for developing DNA recombinant techniques that enable genes from simple organisms to be inserted into the genetic material of other simple organisms. He is also well known for advocating restrictions on genetic engineering research because of the unpredictable, even dangerous, consequences that might result from uncontrolled DNA recombinant experiments.

Berg was born on 30 June 1926 in New York City. He was educated at Pennsylvania State University, from which he graduated in 1948, and at the Western Reserve University, from which he obtained his doctorate in 1952. From 1952–54 he was an American Cancer Society Research Fellow at the Institute of Cytophysiology in Copenhagen and at the School of Medicine at Washington University, St Louis. Between 1955–74 he held several positions at Washington University: from 1955–59 he was an assistant then associate professor in the microbiology department of the School of Medicine; from 1959–69 he was professor of microbiology; and from 1969–74 he was chair of the microbiology department. In 1970 he was also appointed Willson Professor of Biochemistry at the Medical Center of Stanford University, California.

Berg's early work concerned the mechanisms involved in intracellular protein synthesis. In 1956 he identified an RNA molecule (later known as a transfer RNA) that is specific to the amino acid methionine. He then began his Nobel prizewinning work in which he perfected a method for making bacteria accept genes from other bacteria. This genetic-engineering technique for DNA recombination can be extremely useful for creating strains of bacteria to manufacture specific substances, such as interferon. But there are also considerable dangers in the controlled use of these methods – a new, highly virulent pathogenic microorganism might accidentally be created, for example. Berg became aware of this danger and campaigned for strict controls on certain types of genetic-engineering experiments. As a result, an international conference was held in California, followed by the publication in 1976 of guidelines to restrict genetic engineering research.

Berg has also studied how viral and cellular genes interact to regulate growth and reproduction, and has investigated the mechanisms of gene expression in higher organisms.

Bergius, Friedrich Karl Rudolf (1884–1949) German industrial chemist famous for developing a process for the catalytic hydrogenation of coal to convert it into useful hydrocarbons such as petrol and lubricating oil. For this achievement he shared the 1931 Nobel Prize for Chemistry with Karl Bosch.

Bergius was born in Goldschmieden, near Breslau, Silesia (now Wrocław in Poland), on 11 October 1884, the son of the owner of a chemical factory. He studied chemistry at the universities of Breslau and Leipzig, gaining his doctorate in 1907. He did postdoctoral research with Herman ◊Nernst at Berlin and then at Karlsruhe Technische Hochschule with Fritz ◊Haber, who introduced him to high-pressure reactions. From 1909 he was professor of chemistry at the Technische Hochschule in Hannover. He then founded a private research laboratory in Hannover and in 1914 went to work for Goldschmidt AG in Essen, where he remained until the end of World War II. He lived for a while in Austria, then went to Spain, before finally settling in Argentina in 1948, where he held an appointment as a technical adviser to the government. He died in Buenos Aires on 30 March 1949.

In 1912 Bergius worked out a pilot scheme for using high pressure, high temperature, and a catalyst to hydrogenate coal dust or heavy oil to produce paraffins (alkanes) such as petrol and kerosene. The commercial process went into production in the mid-1920s, and became important to Germany during World War II as an alternative source of supply of petrol and aviation fuel. The process yielded nearly 1 tonne of petrol from 4.5 tonnes of coal. He also discovered a method of producing sugar and alcohol from simple substances made by breaking down the complex molecules in wood; the rights to the process were purchased by the German government in 1936. He continued this work in Argentina, and found a way of making fermentable sugars and thus cattle food from wood.

Bernard, Claude (1813–1878) French physiologist whose research and teaching were vitally important in founding experimental physiology as a separate discipline, distinct from anatomy, in the middle part of the 19th century.

Bernard was born on 12 July 1813 in St Julien, in the Beaujolais region of France. The son of a wine grower, Bernard originally wanted to be a playwright. On the advice of a theatre critic he started to study medicine, qualified in 1839, and became a research assistant to the physiologist François Magendie at the College de France in Paris. He graduated MD in 1843 but never practised medicine, preferring to develop his career in experimental physiology. He experienced great difficulties in obtaining suitable positions at the beginning of his career but in 1854 a chair of general physiology was created for him at the Faculty of Sciences in Paris, and the following year he succeeded Magendie to become professor of medicine at the Collège de France. He received numerous honours during his lifetime including the Légion d'Honneur and, after his death on 10 February 1878, was given a state funeral.

Bernard performed a series of important experiments on the physiology of digestion, showing that pancreatic secretions were important in fat metabolism, revealing the importance of the digestive activities of the small intestine, and investigating the mechanisms of nervous control of gastric secretion. He also discovered glycogen from experiments on the perfused liver; revealed the function of nerves that control the dilation or contraction of blood vessels, and investigated the physiology of fetal tissues and the nutritive role of the placenta. He made major investigations into the role of drugs such as curare and opium alkaloids and their effects on the nervous system. His most important contribution to physiological theory was the concept of the 'milieu intérieur' – that life requires a consistent internal environment that is maintained by physiological mechanisms. Bernard was an ardent teacher of the new experimental physiology and young physiologists from around the world went to Paris to train in his laboratory. His major didactic work was *Introduction to the Study of Experimental Medicine* (1865), which provided a comprehensive treatise on the role of experimental research as the basis for medicine.

Bernays, Paul (1888–1977) English-born Swiss mathematician who was chiefly interested in the connections between logic and mathematics, especially in the field of set theory.

Bernays was born in London on 17 October 1888, but grew up in Berlin, where he attended the Köllnisches Gymnasium 1895–1907. As an undergraduate he studied mathematics, philosophy, and theoretical physics at the universities of Berlin and Göttingen. In 1912 he presented his postdoctoral thesis at Zürich, where he continued to do research until 1917. In that year he was invited to become David ◊Hilbert's assistant at Göttingen. In 1919 he received his *venia legendi*, or right to lecture, for the university at Göttingen and he remained there as a lecturer without tenure until 1933. In that year he became one of the early Jewish victims of the Nazi regime in Germany, when his right to lecture was withdrawn. Hilbert employed him privately for six months, but Bernays then decided to take advantage of his father's adopted Swiss nationality and move to Switzerland. For several years he had to make do with short-term teaching appointments at the Technical High School in Zürich, before that institute granted him a *venia legendi* in 1939. He became an extraordinary professor there in 1945 and joined the editorial board of the philosophical journal *Dialectica*. In the 1950s and 1960s he was several times a visiting professor at the University of Pennsylvania and the Princeton Institute for Advanced Study.

Bernays's early interests ranged over a wide area of problems in mathematics. His doctoral thesis of 1912 was on the analytic number theory of binary quadratic forms and his postdoctoral thesis, later in the same year, dealt with function theory, in particular Picard's theorem (see Charles Emile ◊Picard). He then became interested in 'axiomatic thoughts', and it was after hearing Bernays lecture on this subject at Zürich in the autumn of 1917 that Hilbert invited him to Göttingen as his assistant to work on the foundations of arithmetic. There he wrote his thesis on the axiomatics of the propositional calculus in Bertrand Russell's *Principia Mathematica;* this was published in abridged form in 1926.

Bernays's most enduring work was in the field of mathematical logic and set theory. He first presented his principles of axiomatization in a talk to the Mathematical Society at Göttingen in 1931, but hesitated to publish his opinions because he was troubled by the thought that axiomatization was an artificial activity. His fullest treatment of the subject was given in lectures at the Princeton Institute for Advanced Study 1935–1936.

Bernays made a significant contribution to the theory of sets and classes. In his treatment of the subject, classes are not given the status of real mathematical objects. This represents a fundamental divergence from the theory of John ◊Von Neumann. Bernays modified the Neumann system of axioms to remain closer to the original ◊Zermelo structure. He also used some of the set-theoretic concepts of Friedrich Schröder's logic and some of the concepts of the *Principia Mathematica* that have become familiar to logicians. In Bernays's theory there are two kinds of individuals, 'sets' and 'classes': a 'set' is a multitude forming a real mathematical object, whereas a 'class' is a predicate to be regarded only with respect to its extension.

For all his own doubts about the validity of axiomatization, Bernays's arrangement of sets and classes is now widely believed to be the most useful, and by his study of the work of such mathematicians as Von Neumann, Hilbert, and Abraham ◊Fraenkel, he made a major contribution to the modern development of logic.

Bernoulli, Daniel (1700–1782) Swiss natural philosopher and mathematician, whose most important work was in the field of hydrodynamics and whose chief contribution to mathematics was in the field of differential equations.

Bernoulli was born in Grönigen on 9 February 1700, the son of the mathematician Jean Bernoulli and nephew of the mathematician Jacques ◊Bernoulli. As a young boy he first studied philosophy and logic, obtaining his baccalaureate by the age of 15 and his master's degree by the age of 16. He had also been given some mathematical training by his father and uncle, but in 1717 he went to Switzerland to study medicine. He received his doctorate for a thesis on the action of the lungs in 1721. His interest in mathematics seems then to have quickened. In 1724 he published his *Exercitationes Mathematicae* and in 1725 he was appointed to the chair of mathematics at the St Petersburg Academy. Bernoulli left Russia in 1732 and in the following year became professor of anatomy and botany at the University of Basle. He remained there, although after 1750 as professor of natural philosophy, until his retirement in 1777. He died in Basle on 17 March 1782.

Bernoulli was a scientific polymath. During his career he won ten prizes from the French Academy, for papers on subjects that included marine technology, oceanology, astronomy, and magnetism. In physics, Bernoulli made an outstanding contribution in *Hydrodynamica*, which is both a theoretical and practical study of equilibrium, pressure, and velocity in fluids. Bernoulli showed, in the principle given his name, that the pressure of a fluid depends on its velocity, the pressure decreasing as the velocity increases. This effect is a consequence of the conservation of energy and Bernoulli's principle is an early formulation of the idea of conservation of energy. *Hydrodynamica* also contains the first attempt at a thorough mathematical explanation of the behaviour of gases by assuming they are composed of tiny particles, producing an equation of state that enabled Bernoulli to relate atmospheric pressure to altitude, for example. This was the first step towards the kinetic theory of gases achieved a century later.

In mathematics, his interests probably stemmed, in part at least, from his close friendship with Leonhard Euler (1717–1783) and French mathematician Jean d'Alembert, who introduced him to problems associated with vibrating strings, in which connection Bernoulli did much of his work on partial differential equations. Perhaps his single most striking achievement in pure mathematics was to solve the differential equation of J F Riccati (1676–1754). Other achievements were to demonstrate how differential calculus could be used in problems of probability and to do some pioneering work in trigonometrical series and the computation of trigonometrical functions. He also showed the shape of the curve known as the lemniscate $(x^2 + y^2)^2 = a^2(x^2 - y^2)$, where a is constant and x and y are variables.

Bernoulli was a competent mathematician and a first-rate physicist, and his use of mathematics in the investigation of problems in physics makes him one of the founders of mathematical physics.

Bernoulli, Jacques (Jakob) (1654–1705) and Jean (Johann) (1667–1748) Swiss mathematicians, each of whom did important work in the early development of calculus.

The brothers were born in Basle – Jacques on 27 December 1654 and Jean on 7 August 1667. Although Jacques was originally trained in theology and expected to pursue a career in the church, he made himself familiar with higher mathematics – especially the work of Descartes, John Wallis, and Isaac Barrow – and on a trip to England in 1676 met Robert Boyle and other leading scientists. He then decided to devote himself to science, became particularly interested in comets (which he explained by an erroneous theory in 1681), and in 1682 began to lecture in mechanics and natural philosophy at the University of Basle. During the next few years he came to know the work of Gottfried ◊Leibniz and to begin a correspondence with him. In 1687 he was made professor of mathematics at Basle and he held the chair until his death, at Basle, on 10 August 1705.

Jean Bernoulli originally studied medicine, but he was instructed in mathematics by his elder brother and, before he received his doctorate for a thesis on muscular movement, he had already spent some time in Paris (1691) giving private tuition in mathematics. In 1694, the year in which his doctorate was awarded, he was appointed professor of mathematics at the University of Gröningen, where he remained until 1705, when he succeeded his brother in the chair at Basle. In 1730 he was awarded a prize by the French Academy of Sciences for a paper that sought (unsuccessfully, as d'Alembert was able to show) to reconcile Descartes' vortices with Kepler's third law. His son Daniel ◊Bernoulli became one of the first mathematical physicists. Jean Bernoulli died in Basle on 1 January 1748.

Both Jacques and Jean wrote papers on a wide variety of mathematical and physical subjects, but their chief importance in mathematical history rests on their work on calculus and on probability theory. It is often difficult to separate their work, even though they never published together. Thus, for example, Jean's solution to the problem of the catenary, published in 1691, had been all but given in Jean's analysis of the problem in 1690. Jean's *Leçons de calcul differentiel et integral* was written for his French pupil G de l'Hôpital (1661–1704) in 1691 and he was outraged when Hôpital later published the substance of Jean's teaching in his *Analyse des infiniments petits* without acknowledging his debt to his teacher. Jacques' most important papers were those on transcendental curves (1696) and isoperimetry (1700, 1701) – it is here that the first principles of the calculus of variations are to be found. It is probable that these papers owed something to collaboration with Jean. His other great achievement was his treatise on probability *Ars Conjectandi,* which contained both Bernoulli numbers and the Bernoulli theorem and which was not published until 1713, eight years after his death.

Together the two brothers advanced knowledge of calculus not simply by their own work but also by giving spirited public support to Leibniz in his famous quarrel with Newton and thereby helping to establish the ascendancy of the Leibnizian calculus on the continent.

Bernstein, Jeremy (1929–) US mathematical physicist who is well known for his popularizing books on various topics of pure and applied science for the lay reader.

Bernstein was born in Rochester, New York, on 31 December 1929. He received his BA from Harvard in 1951 and his PhD from the same university in 1955. From 1957–60 he was attached to the Institute for Advanced Study at Harvard. In 1962 he was appointed an associate professor in physics at New York University. In 1967 he became professor of physics at the Stevens Institute of Technology at Hoboken, New Jersey. He is currently an adjunct professor at Rockefeller University, New York, and is on the board of trustees at the Aspen Center for Physics, Colorado.

Bernstein's most important advanced work has been in the field of elementary particles and their currents. In particular he has sought to give a mathematical analysis and description to the behaviour of elementary particles. But although he is a competent mathematician who has worked at the very frontier of elementary-particle theory, his greatest gift is the ability to write lucidly about difficult subjects for the nonspecialist. He was a staff writer for *The New Yorker* for over thirty years – his best-known article for that magazine being 'The Analytical Engine: Computers Past, Present and Future', a witty guide to the history and theory of computers. His many books include a general survey of the historical progress of scientific knowledge, *Ascent* (1965), a biography of Albert Einstein (1973), and *An Introduction to Cosmology* (1997).

It is unusual for a scientist to both conduct research at the highest level of specialist theory and write clearly and entertainingly for the public, and Bernstein's singular achievement in this regard was suitably recognized when he was awarded the Westinghouse prize for scientific writing in 1964.

Berthelot, Pierre Eugène Marcelin (1827–1907) French chemist best known for his work on organic synthesis and in thermochemistry.

Berthelot was born in Paris on 27 October 1827, the son of a doctor. At first he studied medicine at the Collège de France, graduating in 1851; he then took up the study of chemistry. He worked at the Collège as assistant to his former tutor Antoine Balard, under Jean Baptiste Dumas and Henri Regnault, gaining his doctorate in 1854 for a thesis on the synthesis of natural fats, which extended the work of Michel ◊Chevreul. From 1859–65 he was professor of organic chemistry at the Ecole Supérieur de Pharmacie and in 1865 he returned to the Collège de France to take up a similar appointment, which he retained until his death. In 1870–71, during the siege of Paris in the Franco-Prussian War, he was consulted about the defence of the capital and became president of the Scientific Defence Committee, supervising the manufacture of guns and explosives. Thereafter he took an increasing part in politics. He became inspector of higher education in 1876, president of the Committee on Explosives in 1878, a senator in 1881, and minister for public instruction in 1886; he was foreign minister 1895–96. In 1889 he succeeded Louis Pasteur as secretary of the French Academy of Sciences. He died in Paris on 18 March 1907, on the same day as his wife.

All of Berthelot's early research concerned organic synthesis. He first studied alcohols, showing in 1854 that glycerol is a triatomic alcohol; he combined it with fatty (aliphatic) acids to make fats, including fats that do not occur naturally. This work provided increasing justification for the view that organic chemistry deals with all the compounds of carbon (including Berthelot's synthetic fats) and not just compounds formed and found in nature. He continued his research by investigating sugars, which he identified as being both alcohols and aldehydes. Using crude but effective methods he also synthesized many simple organic compounds, including methane, methyl alcohol (methanol), formic acid (methanoic acid), ethyl alcohol (ethanol), acetylene (ethyne), and benzene; he also made naphthalene and anthracene. His work during the 1850s was summed up in his book *Chimie organique fondée sur la synthèse* (1860).

Berthelot began his studies of thermochemistry in 1864. Parallelling the work of Germain ◊Hess he measured the heat changes during chemical reactions, inventing the bomb calorimeter to do so and to study the speeds of explosive reactions. He introduced the terms 'exothermic', to describe a reaction that evolves heat, and 'endothermic' for a reaction that absorbs heat. In 1878 he published *Mécanique chimique* followed by *Thermochimie* (1897), which put the science of thermochemistry on a firm footing.

In 1883 Berthelot established an experimental farm at Meudon, southwest of Paris. He discovered that some plants can absorb atmospheric nitrogen, investigated the action of nitrifying bacteria, and began to determine the details of the nitrogen cycle. But Berthelot was not a theorist; he was at his best carrying out practical work in the laboratory, and even led the opposition to the theory of atoms and molecules championed by Stanislao ◊Cannizzaro in the early 1860s.

Berthollet, Claude Louis (1748–1822) French chemist with a wide range of interests, the most significant of which concerned chemical reactions and the composition of the products of such reactions. He proposed that reactivity depends on the masses of the reactants (similar to the modern law of mass action) but that the composition of the product or products can vary, depending on the proportions of the reacting substances (contrary to the law of definite proportions). He was a champion of his contemporary Antoine ◊Lavoisier – although not of his political views – and had a desire to put science at the service of humanity's practical needs.

Berthollet was born of French parents on 9 December 1748 in the then-Italian region of Savoy. In 1768 he qualified as a physician at the University of Turin, moving to Paris four years later to study chemistry under Pierre Macquer (1718–1784) while continuing his medical studies, receiving his French qualification in 1778. While private physician to Mme de Montesson in the household of the Duc d'Orléans he carried out research in the laboratory at the Palais Royale.

After the death of Macquer in 1784 Berthollet was appointed inspector of dyeworks and director of the Gobelins tapestry factory. In 1787 he collaborated with Lavoisier on the publication of *Méthode de nomenclature chimique,* which incorporated the principles of the 'new chemistry' of Lavoisier. He taught chemistry to Napoleon and went with him to Egypt in 1798. There he observed the high concentration of sodium carbonate (soda) by Lake Natron on the edge of the desert. He reasoned that, under the prevailing physical conditions, sodium chloride in the upper layer of soil had reacted with calcium carbonate from nearby limestone hills – the beginning of his theory that chemical affinities are affected by physical conditions, in this case the heat and high concentration of calcium carbonate. He became a senator in 1804 but ten years later voted against Napoleon, and after the Reformation he became a count. Berthollet died on 6 November 1822 at Arcueil, near Paris.

Berthollet's proposal that chemical compounds do not have a constant composition brought him into conflict with Louis ◊Proust, who in 1799 put forward his law of definite proportions (which states that 'all pure samples of the same chemical compound contain the same elements combined together in the same proportions by mass'). It turned out that Berthollet's severe (but nonacrimonious) criticisms of Proust were based on imprecise distinctions between compounds, solutions, and mixtures, as well as on the inaccurate analyses of impure compounds. For example, he suggested that lead and oxygen could combine in almost any proportion, but it is now known that he was making and analysing mixtures of various lead oxides. Although the controversy between Berthollet and Proust ended mainly in Proust's favour, Berthollet's views were not entirely wrong, although at that time they were based on false evidence. Non-stoichiometric compounds (also called Berthollide compounds), with a variable composition, have been studied since 1930. For example, it is now believed that lattice deficiencies in iron can account for ferrous sulphide – iron (II) sulphide, FeS – with compositions that vary between $FeS_{1.00}$ and $FeS_{1.14}$ (or $Fe_{1.00}S$ and $Fe_{0.88}S$).

Berthollet also found himself disagreeing with Lavoisier, and would not concur with the theory that all acids contain oxygen. In this he was correct, but shared with Karl Scheele the false assumption that chlorine was not an element but consists of oxygenated hydrochloric acid. He did, however, introduce the use of chlorine as a bleaching agent (which led him to devise a volumetric analytical method for estimating the chlorine content of a bleaching solution by titration against a standard solution of indigo). Berthollet also investigated chlorates, suggesting that the oxidizing properties of potassium chlorate could make it a replacement for potassium nitrate in gunpowder and so produce a more powerful explosive. A public demonstration of this idea in 1778 resulted in disaster and the deaths of some of the onlookers.

In other wideranging studies Berthollet devised a method of smelting iron and making steel; he correctly determined the compositions of ammonia, prussic (hydrocyanic) acid, and sulphuretted hydrogen (hydrogen sulphide); and he made various discoveries in organic chemistry. On balance, Berthollet was right more often than he was wrong – and even when he was wrong, his arguments with Proust and Lavoisier stimulated chemical thought, ultimately to the benefit of the science.

Berthoud, Ferdinand (1727–1807) Swiss clockmaker and a maker of scientific instruments. He improved the work of English horologist John ⟩Harrison, devoting 30 years' work to the perfection of the marine chronometer, giving it practically its modern form.

Son of the architect and judiciary Jean Berthoud, Ferdinand Berthoud was born near Couvet. He was apprenticed to his brother, Jean-Henri, at the age of 14. In 1745 he went to Paris and in 1764 was appointed Horologer de la Marine. He made over 70 chronometers using a wide variety of mechanisms, and wrote ten volumes on the subject, many of them of considerable importance in his field.

In the early 18th century, clockmakers devoted much effort to the construction of chronometers that could be used at sea. This was because if navigators did not know the time at the zero meridian it was impossible for them to plot their precise position. Since the different astronomical methods of measuring longitude gave inadequate results, the solution seemed to lie in finding ways of making very accurate clocks whose workings would not be disturbed by the motion of the ship.

John Harrison was undoubtedly the first maker of a timepiece that could be used satisfactorily at sea. In 1735 he completed his first marine chronometer; this was tested at sea and gave results that were good enough for further financial help to be given to its maker but not good enough for the reward offered under an act passed by the British Parliament in 1714.

Harrison's chronometer number 4 was completed in 1759 and was tested at sea. While this met the requirements of the act, the responsible authorities in England, the Board of Longitude, still demanded numerous trials. Tested over long distances, his chronometer consistently showed variations, involving errors in the calculation of longitudes, that were less than the maximum of half a degree stipulated by the Board of Longitude. It was not until 1772, when Harrison was 80, and after long years of argument that he received the whole award.

Harrison had already obtained remarkable results when Berthoud began his investigations, which were conducted mainly from 1760 to 1768. He was a skilled clockmaker by the time he began to devote himself to the problem of marine chronometers and his work was guided by wide experience as well as his outstanding practical ability.

The range of parts constructed by Berthoud is large but the complete clocks made for his experiments numbered only seven or eight. His clocks number 1 and 2 were made in 1760 and 1763 respectively. Though number 2 was noticeably smaller than the first, these two instruments were still rather bulky. Berthoud put two circular balance wheels oscillating in opposite directions into these designs, which connected to one another by means of a toothed wheel. Their axes rested on roller bearings. A bimetallic grid iron, already widely used by clockmakers since its invention by Harrison, compensated for variations in the length of the spiral.

The escapement was made more complicated in clock number 2, in which Berthoud introduced an equalizing remontoir to compensate for errors caused by variations in the driving force of the escapements. A spring barrel fitted with a fuse was initially chosen as the source of motive power, a system still used in modern marine chronometers. However, Berthoud temporarily abandoned this system soon afterwards. In his third watch, also made in 1763, Berthoud used a bimetallic strip for thermal compensation and a single balance wheel. Retaining the single balance wheel suspended by a wire and guided by rollers, he then returned to the gridiron method of compensation for the construction of his subsequent clocks. Numbers 6 and 8 (1767) of these ensured his success.

In this series the motive power was produced by weights placed on a vertical metal plate which descended the length of three brass columns. An escapement with ruby cylinders was used, all parts being relatively compact. The mechanism was above the gridiron, the two occupying the top part of the clock, which was enclosed in a long glass cylinder. The three brass columns guiding the descent of the weight took up almost the whole height of this cylinder. A horizontal dial covered the glass box.

Clocks 6 and 8 were tested at sea in 1768 and 1769 and showed variations in their working of the order of from 5 to 20 seconds a day.

We known of only five chronometers made by John Harrison, who did not reach his goal until his last years. Berthoud, on the other hand, was able to continue working for another 30 years after his initial success. Abandoning the roller suspension of the balance wheel, he adopted a pivot suspension. He came back to compensation by bimetallic strips, eliminating the cumbersome gridiron and used a balance wheel compensated by four small weights. Two of these weights were fixed and the two others were carried by a bimetallic strip which moved towards or away from the arbor according to variations in temperature. Berthoud also eventually returned to the spring drive. His nephew, Pierre-Louis Berthoud (1754–1813) succeeded him in his work as a marine clock maker and began to industrialize the construction of chronometers as an industry in France.

Berzelius, Jöns Jakob (1779–1848) Swedish chemist, one of the founders of the science in its modern form. He contributed to atomic theory, devised chemical symbols, determined atomic masses, and discovered or had a hand in the discovery of several new elements. He became renowned as a teacher and gained a reputation as a world authority; such was his influence that other scientists were wary of contradicting him. Indeed the obstinacy with which he clung to his own theories, especially in later life, may well have retarded progress in some areas despite his many magnificent achievements, particularly in theoretical chemistry.

Berzelius was born on 20 August 1779 at Vaversunda, Ostergotland, of an ancient Swedish family that had long associations with the church. His father, a teacher at the gymnasium in nearby Linköping, died when Berzelius was only four years old and his mother married a pastor, Anders Ekmarck. Berzelius and his sister were brought up with the five Ekmarck children and educated by their stepfather and by private tutors. When in 1788 his mother also died, the nine-year-old Berzelius moved again to live with a maternal uncle. He attended his father's old school but quarrelled with his cousins and six years later left the family to become a tutor on a nearby farm, where he developed a strong interest in collecting and classifying flowers and insects. He had been destined for the priesthood, but decided in 1796 to study natural sciences and medicine at Uppsala University.

Berzelius Engraving of Swedish chemist Jöns Jakob Berzelius who accurately determined atomic weights, introduced modern chemical symbols, and discovered and isolated several chemical elements. He studied many organic substances, and coined the term 'protein'. *Mary Evans Picture Library*

Berzelius had to interrupt his studies to earn some money. His stepbrother aroused in him an interest in chemistry, but he received little encouragement at the university and began to experiment on his own. In the summer of 1800 he was introduced to Hedin, the chief physician at the Medivi mineral springs, and began his first scientific work – an analysis of the mineral

	Alchemists	Dalton	Berzelius
Phosphorus	[symbol]	[symbol]	P
Sulphur	🜍	⊕	S
Iron (ferrum)	♂	⦶	Fe
Hydrogen	unknown	⊙	H
Oxygen	unknown	○	O
Water	▽	⊙○	H^{2}O later H_2O

Berzelius Berzelius's system of chemical notation replaced the unwieldy pictorial symbols devised by the alchemists and by John Dalton.

content of the water. Hedin also secured for Berzelius the unpaid post of assistant to the professor of medicine and pharmacy at the College of Medicine in Stockholm, and when the professor died in 1807 Berzelius was given the post. The college became an independent medical school, the Karolinska Institute, in 1810, the same year that Berzelius was appointed president of the Swedish Academy of Sciences. On his wedding day in 1835 he was made a baron by the king of Sweden, in recognition of his status as the most influential chemist of the era. Berzelius died on 7 August 1848.

As an experimenter Berzelius was meticulous. Papers he published between 1810 and 1816 describe the preparation, purification, and analysis of about 2,000 chemical compounds. In the course of this work he improved many existing methods and developed new techniques. Quantitative analysis on such a broad scale established beyond doubt Dalton's atomic theory and Proust's law of definite proportions. It also laid the foundation of Berzelius' determination of the atomic masses of the 40 elements known at that time – a prodigious task in which he was aided by the work of several contemporaries, such as Eilhard ◊Mitscherlich (isomorphism), Pierre ◊Dulong and Alexis ◊Petit (specific heats), and Joseph ◊Gay-Lussac (combining volumes); Mitscherlich and Dulong were his former students. In common with other scientists of the time, Berzelius rejected Avogadro's hypothesis, which led him to confuse some atomic masses and molecular masses. Nevertheless most of the atomic masses in the table he published in 1828 closely correspond to the modern accepted values.

His dealing with such a variety of elements and compounds led Berzelius to simplify chemical symbols. Alchemists had used an elaborate pictorial presentation, and John Dalton had devised a system based on circular symbols that could be combined to represent compounds. Berzelius discarded both of these and introduced the notation still used today, in which letters (sometimes derived from Latin names) represent the elements and, combined with numbers if necessary, constitute the chemical formulae of compounds.

The invention of the voltaic cell at the beginning of the 19th century opened up a new field of research – electrochemistry. Berzelius' work in this area, although by no means as comprehensive as that of Humphry ◊Davy, lent support to his dualistic theory, which stated that compounds consist of electrically and chemically opposed parts. He considered the parts to be stable groups of atoms, which he called 'radicals'. Although Berzelius' extension of the theory was later proved to be incorrect, it did contain the basis of the modern theory of ionic compounds.

Berzelius also made his mark in the discovery of new elements. With the German chemist Wilhelm Hisinger (1766-1852) he found cerium in 1803. In 1815 he believed that he had isolated a second new element from a mineral specimen and named it thorium. Subsequent experiments revealed that the 'element' could be broken down into yttrium and phosphorus. His disappointment was softened by his discovery of selenium in 1818, silicon in 1824, and in 1829 a fourth new element, extracted from its ore by reduction with potassium, which he called thorium.

Several chemical terms in use today were coined by Berzelius. He noted that some reactions appeared to work faster in the presence of another substance which itself did not appear to change, and postulated that such a substance contained a 'catalytic force'. Platinum, for example, was well endowed with such a force because it was capable of speeding up reactions between gases. Although he appreciated the nature of catalysis, Berzelius was unable to give any real explanation of the mechanism.

In the early 19th century it became apparent that elements could be grouped by similar chemical properties. Chlorine, bromine, and iodine formed such a grouping. Each of these elements could be found as salts in sea water, consequently Berzelius invented the name 'halogens' (salt formers) to collectively describe the family. The other two halogens were isolated later – fluorine by French chemist Henri Moissan in 1886 and astatine by Emilio ◊Segrè (1852–1907) in 1940.

A different branch of chemistry, concerned with substances derived from living things, was capturing the interest of scientists of the day. Berzelius referred to this new sphere as 'organic chemistry' and expounded the belief that organic compounds arose from the operation of a 'vital force' in the living cell; synthesis was therefore impossible. Then in 1828 Friedrich ◊Wöhler (previously a student of Berzelius) prepared the organic compound urea from the inorganic salt ammonium cyanate. With great reluctance Berzelius eventually abandoned his vital force theory.

The word 'isomerism' was also introduced by Berzelius, to describe substances that have the same chemical composition but different physical properties. He encountered the phenomenon when working with the salts of racemic and tartaric acids. Later Mitscherlich showed that, of the two, only tartarates rotate the plane of polarized light (are optically active); in 1848 Louis Pasteur resolved the racemates, which are an equimolecular mixture of two optically active tartarates (and have given their name to all such combinations, which are now known as racemic mixtures).

Throughout his career Berzelius recognized the importance to chemical progress of disseminating information. His *Textbook of Chemistry* (1803) was received with acclaim and was soon accepted as the definitive work for students of the time. In addition to publishing numerous research papers of his own, he collated the work of other chemists and acted as editor of an annual review of chemistry which was published 1821–49.

Further Reading

Jorpes, Johan Erik, *Jacob Berzelius: His Life and Work,* University of California Press, 1970.
Melhado, Evan Marc, *Jacob Berzelius: The Emergence of His Chemical System,* Lychros-Bibliotek, Almqvist & Wiksell International, 1980, v 34.
Melhado, Evan Marc and Frangsmyr, Tore (eds), *Enlightenment Science in the Romantic Era: The Chemistry of Berzelius and Its Cultural Setting,* Cambridge University Press, 1992.

Bessel, Friedrich Wilhelm (1784–1846) German astronomer who in 1838 first observed stellar parallax, and who set new standards of accuracy for positional astronomy. His measurement of the positions of about 50,000 stars enabled the first accurate calculation of interstellar distances to be made. Bessel was the first person to measure the distance of a star other than the Sun. From the parallax observation of 61 Cygni he calculated the star to be about six light years distant, thus setting a new lower limit for the scale of the universe.

The son of a government employee, Bessel was born on 22 July 1784 in Minden. He began work at the age of 15 as an apprentice in an exporting company. During this period in his life, an unhappy one, he dreamed of escape and decided to travel. With this end in view, he studied languages, geography, and the principles of navigation: this led to an interest in mathematics and, eventually, astronomy. In 1804 he wrote a paper on Halley's comet in which he calculated the comet's orbit from observations made over a period of about a year. He sent the paper to Heinrich ⌽Olbers who was so impressed that he arranged for its publication and obtained a post for Bessel as an assistant at Lilienthal Observatory. There Bessel worked under the early lunar observer Johann Schröter (1745–1816). After only four years the Prussian government commissioned Bessel to construct the first large German observatory at Königsberg, where in 1810 he was appointed professor of astronomy. Bessel's whole life was devoted first to the completion of the observatory in 1813 and then its direction until his death in 1846.

Bessel's work laid the foundations of a more accurate calculation of the scale of the universe and the sizes of stars, galaxies, and clusters of galaxies than any previous method had done. In addition, he made a fundamental contribution to positional astronomy (the exact measurement of the position of celestial bodies), to celestial mechanics (the movements of stars), and to geodesy (the study of the Earth's size and shape). Bessel enlarged the resources of pure mathematics by his introduction and investigation of what are now known as Bessel functions, which he first used in 1817 to determine the motions of three bodies moving under mutual gravitation. Seven years later he developed Bessel functions more fully for the study of planetary perturbations. Bessel played a great part in the final establishment of a scale for the universe in terms of the Solar System and terrestrial distances. These naturally depended upon accurate measurement of the distances of the nearest stars from the Earth.

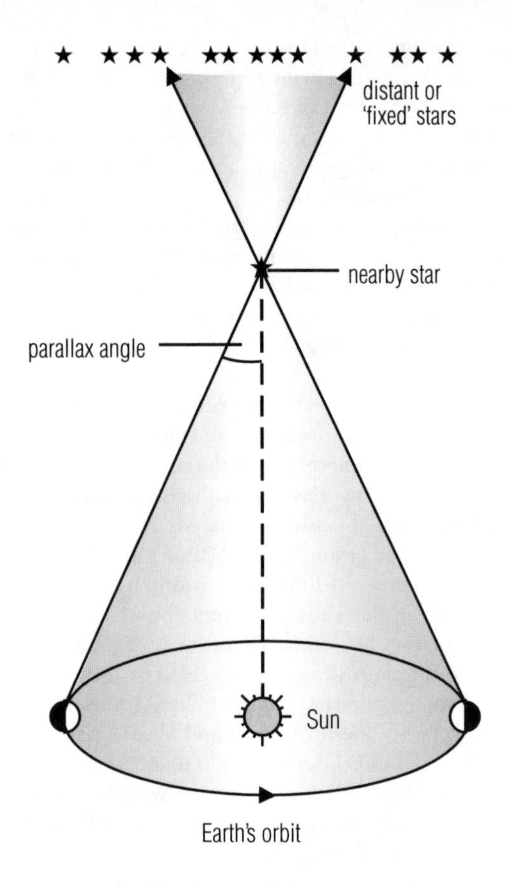

Bessel The parallax of a star, the apparent change of its position during the year, can be used to find the star's distance from the Earth. The star appears to change its position because it is viewed at a different angle in July and January. By measuring the angle of parallax, and knowing the diameter of the Earth's orbit, simple geometry can be used to calculate the distance to the star. The technique was developed by the German astronomer Friedrich Bessel.

Bessel's contributions to geodesy included a correction in 1826 to the seconds pendulum, the length of which is precisely calculated so that it requires exactly one second for a swing. Between 1831–32 he directed geodetical measurements of meridian arcs in East Prussia, and in 1841 he deduced a value of $\frac{1}{299}$ for the ellipticity of the Earth, or the amount of elliptical distortion by which the Earth's shape departs from a perfect sphere. He was the first to make effective use of the heliometer, an instrument designed for measuring the apparent diameter of the Sun.

Bessel also introduced corrected observations for the so-called personal equation, a statistical bias in measurement that is characteristic of the observer himself and must be eliminated before results can be considered reliable. He concerned himself greatly with accuracy, to the point of making a systematic study of the causes of instrumental errors. His own corrected observations were far more accurate than previous ones and his methods pointed the way to a great advancement in the study of the stars.

Bessel's later achievements were possible only because he first established the real framework of the scale of the universe through his accurate measurement of the positions and motions of the nearest stars, making corrections for errors caused by imperfections in his telescope and by disturbances in the atmosphere. Having established exact positions for about 50,000 stars, he was ready to observe exceedingly small but highly significant motions among them. Choosing 61 Cygni, a star barely visible to the naked eye, Bessel showed that the star apparently moved in an ellipse every year. He explained that this back-and-forth motion, called parallax, could only be caused by the motion of the Earth around the Sun. His calculation indicated a distance from Earth to 61 Cygni of 10.3 light years. When Olbers received these conclusions, on his 80th birthday, he thanked Bessel who, he said, 'put our ideas about the universe on a sound basis'. Bessel was honoured for this achievement by, among others, the Royal Astronomical Society.

One of Bessel's major discoveries was that two bright stars Sirius and Procyon execute minute motions that could be explained only by the assumption that they had invisible companions to disturb their motions. The existence of such bodies, now called Sirius B and Procyon B, was confirmed with more powerful telescopes after Bessel's death. He also contributed to the discovery of the planet Neptune. He published a paper in 1840 in which he called attention to small irregularities in the orbit of Uranus which he had observed and which were caused, he suggested, by an unknown planet beyond. Bessel's minor publications numbered more than 350 and his major ones included a multivolume series *Astronomische Beobachtungen auf der K Sternwarte zu Königsberg* (1815–1844).

Realizing early in his life where his potential lay, Bessel succeeded in achieving it in a profession entirely different from the one in which he had started out. Olbers said that the greatest service he himself had rendered astronomy was that he recognized and furthered Bessel's genius.

Bessemer, Henry (1813–1898) English engineer and inventor of the Bessemer process for the manufacture of steel, first publicly announced at a meeting of the British Association in 1856. His process was the first cheap, large-scale method of making steel from pig iron. Bessemer had many other inventions to his credit but none is comparable to that with which his name is linked and which earned him a fortune.

Bessemer English engineer and inventor Henry Bessemer, who invented the Bessemer process for steel manufacture. He developed the technique in order to produce guns for the Crimean War, but demand for steel increased hugely with the development of the railways, and Bessemer became a millionaire. *Mary Evans Picture Library*

With the French Revolution, the French Huguenot Bessemer family moved to England and their son Henry was born in Charlton, near Hitchin in Hertfordshire, on 19 January 1813. He inherited much skill and enterprise from his father who was associated with the firm founded by William Caslon, one of the pioneers in the development of movable type for printing. His earliest years were spent in his father's workshop where he found every chance to develop his inclinations as an inventor. At the age of 17 he went to London where he put to use his knowledge of easily fusible metals and casting in the production of artwork. His work was noticed and he was invited to exhibit at the Royal Academy. In about 1838 Bessemer invented a typesetting machine and a little later he perfected a process for making imitation Utrecht velvet. In this his combined mechanical skill and artistic capacity proved useful, for he not only had to design all the machinery but he also had to engrave the embossing rolls.

In about 1840, encouraged by his great friend the printer Thomas de la Rue, Bessemer turned his attention to the manufacture of bronze powder and gold paint. At that time Germany had the monopoly in this industry, having learned its secrets from China and Japan. Bessemer's product was at least equal to that of Germany and one-eighth of the price. Between 1849–53 he became interested in the process of sugar refining and obtained no fewer than 13 patents for machinery for this. He also invented a new method of making lead pencils. But it was the bronze-powder and gold-paint process, the secrets of which were kept in the Bessemer family for 40 years, that provided the capital needed to set up the small ironworks in St Pancras, London, where his experiments led to the invention of the Bessemer steelmaking process. This was his greatest achievement and at the time was of enormous industrial importance.

With the Crimean War of the early 1850s Bessemer turned his energies to the problems of high gas pressures in guns which were probably caused by early attempts at rifling. When he offered his services to the British military commanders they showed no interest, so Bessemer turned to the UK's ally, France. The French military, which had used rifled weapons intermittently for years, expressed interest, and Napoleon III encouraged Bessemer to experiment further. With weapons that fire from rifled barrels, the projectile has to fit tightly or it is not set spinning by the helical grooves in the barrel. The tight fit leads to high pressures, and many early weapons exploded, killing the gun crew. Bessemer set out to find a form of iron strong enough to resist these pressures.

At that time steel was expensive and cast iron, although very hard, was extremely brittle. The carbon in cast iron could be laboriously removed to form practically pure wrought iron that was not brittle but ductile. Steel, with a carbon content between that of cast iron and wrought iron, was both hard and tough. But, to make steel, cast iron had to be converted into wrought iron and carbon had then to be added – a laborious and time-consuming process. Bessemer considered the contemporary method of converting cast iron to wrought iron. A carefully measured quantity of iron ore was added to cast iron and then heated to the molten stage, when the oxygen in the iron ore combined with the carbon in the cast iron to form carbon monoxide (which was burned off), leaving pure iron.

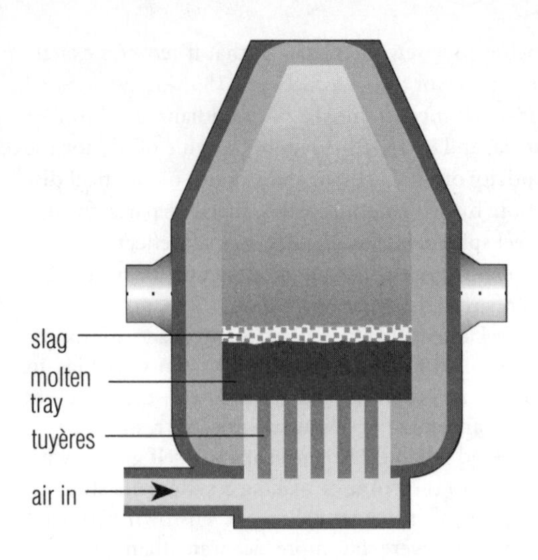

Bessemer In a Bessemer converter, a blast of high-pressure air oxidizes impurities in molten iron and converts it to steel.

Bessemer considered adding the oxygen directly as a blast of air, although it seemed likely that the cold air would cool and therefore solidify the molten iron. When he tried the process, however, he found that the reverse was true: the blast of air burned off the carbon and other impurities and the heat generated served to keep the iron molten – indeed the temperature was raised. By stopping the process at the right time he found that he could produce steel without the intermediate wrought-iron stage, and that its cost was reduced dramatically.

Ironmakers were enthusiastic when in 1856 Bessemer announced his discovery, and vast sums of money were invested in equipment for the new process. But instead of becoming a hero, Bessemer was derided, because the steel that was produced was of a very poor grade. In his original experiments Bessemer had used phosphorus-free ore, while the ironmakers had used ore containing phosphorous. Bessemer assured the ironmakers that good steel could be made if the appropriate ore was used, but they would not listen. In 1860 Bessemer erected his own steel works in Sheffield, importing phosphorous-free iron ore from Sweden. The high-grade steel he made was sold for a fraction of the current price. Ironmasters, feeling the sting of competition, applied for licences and the royalties made Bessemer very rich. He retired at the comparatively early age of 56 but continued inventing. One of his later schemes was a stabilized saloon for an ocean-going ship, the theory being that seasick passengers would not feel the motion of the vessel. The scheme was not a success and Bessemer had his fingers severely burnt, losing around £34,000 in the venture.

The recognition Bessemer received for his steelmaking process was however richly deserved for his method of making cheap steel benefited the whole world. He was not allowed to accept the Legion of Honour offered to him by the French Emperor, but he received many other honours and became a Fellow of the Royal Society in 1879. He was knighted in the same year for services to the Inland Revenue 40 years before when, before he was 20, he played an important part in the construction of the first method for stamping deeds, a method that was adopted by the British government. He died in Denmark Hill, London, on 15 March 1898. No fewer than six towns in the USA were named after him.

Further Reading

Barraclough, Kenneth, *Steelmaking Before Bessemer,* Ashgate Publishing Company, 1984, 2 vols.
Bessemer, Henry, *Sir Henry Bessemer, FRS: An Autobiography,* Ashgate Publishing Company, 1989.
Bodsworth, C (ed), *Sir Henry Bessemer: Father of the Steel Industry,* Institute of Materials,Whitakers, 1998.

Bethe, Hans Albrecht (1906–) German-born US physicist and astronomer, famous for his work on the production of energy within stars for which he was awarded the 1967 Nobel Prize for Physics.

Bethe was born in Strasbourg (now in France) on 2 July 1906, the son of a university professor. He was educated at the universities of Frankfurt and Munich, and gained a PhD from the latter in 1928. From 1928–29 he was an instructor in physics at the University of Frankfurt, and then at Stuttgart. He went on to lecture at the universities of Munich and Tübingen 1930–33. With the rise to power of Adolf Hitler in Germany, Bethe moved to the UK in 1933 and spent a year at the University of Manchester. He was a fellow of Bristol University 1934–35, when he was appointed assistant professor at Cornell University, Ithaca, USA. He became a professor there in the summer of 1937 and later became a naturalized US citizen.

From 1943–46 Bethe was chief of the Theoretical Physics Division of the Los Alamos Science Laboratory in New Mexico and he has been a consultant to Los Alamos since 1947. He has been a leading voice in emphasizing the social responsibility of the scientist and after World War II he served as part of the US delegation in Geneva during long negotiations with the Soviet Union on the control of nuclear weapons. He holds honorary doctorates from a large number of universities all over the world. He was awarded the Morrison Prize of the New York Academy of Sciences in 1938 and 1940, the American Medal of Merit in 1946, the Draper Medal of the National Academy of Sciences in 1948, the Planck Medal in 1955 and 1961, and the Fermi Award for his part in the development and use of atomic energy. He was made a foreign member of the Royal Society in 1957 and won the National Medal of Science in 1976. He is a Fellow of the American Physical Society and served as its president in 1954.

When Bethe went to the UK in the 1930s, he worked out how high-energy particles emit radiation when they are deflected by an electromagnetic field. This work was important to cosmic-ray studies. In 1938 he made his most important contribution to science when he worked out the details of how nuclear mechanisms power stars. Carl von ◊Weizsäcker was independently reaching the same conclusions in Germany. These nuclear mechanisms were to answer the questions that had concerned Hermann von ◊Helmholtz and William Thomson (Lord ◊Kelvin) 75 years earlier.

Bethe's mechanism began with the combining of a hydrogen nucleus (a proton) with a carbon nucleus. This initiates a series of reactions, at the end of which the carbon nucleus is regenerated and four hydrogen nuclei are converted into a helium nucleus. Hydrogen acts as the fuel of the star and helium is the 'ash'; carbon serves as a catalyst. As stars like the Sun are mostly made up of hydrogen, there is ample fuel to last for thousands of millions of years. The amount of helium present indicates that the Sun has already existed for billions of years.

Bethe later proposed a second scheme that involves the direct combination of hydrogen nuclei to form helium in a series of steps that proceed at lower temperatures. When hydrogen is converted into helium, either directly or by means of the carbon, nearly 1% of the mass of the hydrogen is converted into energy. Even a small amount of mass produces a great deal of energy, and the loss of mass in the Sun is enough to account for its vast and seemingly eternal radiation of energy.

With the discovery of neutron stars, Bethe turned again to astrophysical research in 1970. These stars are held by gravity at such a high density that protons fuse with electrons to produce the neutrons that constitute nearly all matter.

Although Bethe was primarily concerned with the rapidly developing subject of atomic and nuclear processes, he has also investigated the calculation of electron densities in crystals, using classical mathematical methods, and the order–disorder states in alloys. He has also concerned himself with the operational conditions in nuclear reactors and the detection of underground explosions by means of seismographic records. Apart from his contributions to atomic theory, therefore, his concern with the implications of its practical application have earned him worldwide respect within the field.

Further Reading

Bernstein, Jeremy, *Prophet of Energy: Hans Bethe*, Dutton, 1981.
Bethe, Hans Albrecht, *Selected Works of Hans A Bethe*, World Scientific Publishing Co, 1997.
Marshak, Robert Eugene and Blaker, J Warren (eds), *Perspectives in Modern Physics: Essays in Honor of Hans A Bethe on the Occasion of his 60th Birthday*, Interscience Publishers, 1966.
Molinari, A and Ricci, R A (eds), *From Nuclei to Stars: A Meeting in Nuclear Physics and Astrophysics Exploring the Path Opened by H A Bethe*, North-Holland, 1986.

Betti, Enrico (1823–1892) Italian mathematician who was the first to provide a thorough exposition and development of the theory of equations formulated by French mathematician Evariste ⬨Galois.

Betti was born near the Tuscan town of Pistoia on 21 October 1823. His father died when he was very young and he received his early education from his mother. After obtaining his BA in physical and mathematical sciences at the University of Pisa, he taught for a time at a secondary school at Pistoia before being appointed to a professorship at Pisa in 1856; he held the professorship for the rest of his life. Betti was also much interested in politics. He had fought against Austria at the battles of Curtatone and Montanara during the first wars of Italian independence and in 1862 he became a member of the new independent Italian parliament. He entered the government as undersecretary of state for education in 1874 and served in the senate after 1884. He died at Pisa on 11 August 1892.

Betti's early, and most important, work was on algebra and the theory of equations. In papers published in 1852 and 1855 he gave proofs of most of Galois's major theorems. In so doing he became the first mathematician to resolve integral functions of a complex variable into their primary factors. He also developed the theory of elliptical functions, demonstrating (in a paper of 1861) the theory of elliptical functions that is derived from constructing transcendental entire functions in relation to their zeros by means of infinite products. In thus providing formal demonstration to Galois's statements and in drawing out some of their implications, Betti greatly advanced the transition from classical to abstract algebra.

In 1863 a change occurred in Betti's mathematical interests. In that year the German mathematician Bernhard ⬨Riemann went to Pisa. He became a close friend of Betti and directed Betti's mind to mathematical physics, especially to problems of potential theory and elasticity. Betti studied George ⬨Green's methods of attempting to integrate ⬨Laplace's equation (the foundation of the theory of potentials) and applied them to the study of elasticity and heat. The result was the paper of 1878 in which he gave the law of reciprocity in elasticity theory that became known as Betti's theorem. Along the way, conducting research into 'analysis situs' in hyperspace in 1871, he also did valuable work on numbers characterizing the connection of a variety, these later becoming known as 'Betti numbers'.

Betti also played a principal part in the expansion of mathematics teaching in Italian schools, in particular lending his enthusiastic advocacy to the restoration of Euclid to a central place in the secondary school curriculum. His lectures at the University of Pisa also inspired a generation of Italian mathematicians, the most famous of them being Vito ⬨Volterra.

Bhabha, Homi Jehangir (1909–1966) Indian theoretical physicist who made several important explanations of the behaviour of subatomic particles. He was also responsible for the development of research and teaching of advanced physics in India, and for the establishment and direction of the nuclear power programme in India. He commanded wide respect in the international scientific community both for his contributions in the scientific sphere and also for his formidable skills as an administrator. He held many positions of responsibility at home and abroad, particularly in organizations concerned with the development of peaceful uses of atomic energy.

Bhabha was born in Bombay on 30 October 1909. He attended a number of schools in Bombay before entering Gonville and Caius College, Cambridge, in 1927 to study mechanical engineering. There his tutor in mathematics was Paul ⬨Dirac, who originated the relativistic electron theory that led to the prediction of antiparticles. Bhabha became fascinated by mathematics and theoretical physics and so, after earning his first-class honours degree in 1930, he began to do research at the Cavendish Laboratories in Cambridge. He was also then able to tour Europe, meeting scientists of such importance as Wolfgang Pauli and Enrico Fermi and also visiting Niels Bohr's laboratory in Denmark. He was awarded his PhD in 1935, and remained at Cambridge until 1939 when he returned to India for a holiday.

He was still in India when World War II broke out. As he was unable to return to Cambridge, a readership at the Bangalore Institute of Science was created for him and he was put in charge of a department investigating cosmic rays. The renowned Indian physicist C V ⬨Raman was the director of the institute, and he had a profound influence on Bhabha. Bhabha determined to remain in India and advance the development of science and technology there.

In 1944 he proposed the establishment of a centre for the training of scientists in advanced physics and for the cultivation of research in that field as a stimulus for industrial development. The Tata Institute of Fundamental Research was established at Bombay in 1945 with Bhabha as director, a position he held until his death.

Bhabha later took on many responsibilities in the establishment of the Indian atomic energy programme and also became a figure of great importance in the international scientific community, serving as president of the United Nations Conference on the Peaceful Uses of Atomic Energy, first held in Geneva in 1955, and as president of the International Union of Pure and Applied Physics 1960–63. He died on 24 January 1966 in a plane crash on Mont Blanc.

Bhabha studied at the Cavendish Laboratories at a time of great advances in the understanding of the structure and properties of matter. He made major contributions to the early development of quantum electrodynamics, a part of high-energy physics. His first paper concerned the absorption of high-energy gamma rays in matter. A primary gamma ray dissipates its energy in the formation of electron showers. In 1935, Bhabha became the first person to determine the cross section (and thus the probability) of electrons scattering positrons. This phenomenon is now known as Bhabha scattering.

Bhabha also studied cosmic rays and in 1937 suggested that the highly penetrating particles detected at and below ground level could not be electrons. In 1946, they were in fact found to be mu-mesons. Bhabha also put forward a theory proposing the existence of vector mesons, which were later identified in nuclear interactions. In 1938, he suggested a classic method of confirming the time dilation effect of the special theory of relativity by measuring the lifetimes of cosmic-ray particles striking the atmosphere at very high speeds. Their lifetimes were found to be prolonged by exactly the amount predicted by relativity.

Bhabha was a skilled theoretician and a master administrator. His love for his native land stimulated him to the cultivation of educational and research facilities of the highest standard, and furthermore to the awakening of his government's awareness of the potential importance of atomic energy. He did everything in his power to ensure that India would become self-sufficient in all stages of the nuclear cycle so as not to be dependent on other nations. He spoke for all developing nations when he said in 1964 at the Third United Nations Conference on the Peaceful Uses of Atomic Energy that 'no power is as expensive as no power'.

Further Reading

Anderson, Robert S, *Building Scientific Institutions in India: Saha and Bhabha,* Occasional Paper Series, Centre for Developing Area Studies, 1975, v 11.

Cockroft, John and Menon, M G K, *Homi Jehangir Bhabha, 1909–1966,* The Royal Institution of Great Britain, 1967.

Kulakarni, Raghunata Purushottama, *Homi Bhabha: Father of Nuclear Science in India,* Popular Prakashan, 1969.

Swarup, Govind, *Homi Bhabha Commemorative Issue,* Current Science series, Current Science, 1991, v 60.

Bichat, (Marie François) Xavier (1771–1802) French physician of the Revolutionary era.

Born in Thoirette, Jura, a doctor's son, Bichat also studied medicine, first in Lyon, but his education was soon interrupted by military service in the Revolution. In 1793, at the height of the Terror, Bichat settled in Paris. From 1797 he taught medicine, working from 1801 at the Hotel Dieu, Paris's huge general hospital for the poor.

His greatest contribution to medicine and physiology stemmed from his insight that the diverse organs of the body contained particular tissues or (his favourite word) 'membranes'. He described 21 such membranes, including muscle, connective, and nerve tissue. Bichat maintained that in the case of a diseased organ, it was generally not the whole organ but only certain tissues that were affected. While establishing the centrality of the study of tissues (histology), Bichat distrusted the microscope and made little use of it; his analysis of tissues consequently did not include any understanding of their cellular structure.

Executing his researches with enormous fervour during the last years of his short life – he performed over 600 post mortems – Bichat may be seen as a link between the morbid anatomy of Giovanni ◊Morgagni and the later cell pathology of German pathologist Rudolf Virchow (1821-1902). His lasting importance lay in the simplification he brought to anatomy and physiology, by showing that the complex structures of organs were to be understood in terms of their elementary tissues.

Bickford, William (1774–1834) English leather merchant who invented the miner's safety fuse. He made a major contribution to safety and productivity in mines and quarries, and even after electric ignition was introduced in 1952 the majority of charges were set off using fuses not very different from the one patented by him in 1831.

Bickford was born in Devon, and having tried unsuccessfully to carry on a currier's business (dressing and colouring tanned leather) in Truro, he set up as a leather merchant in Tuckingmill near Camborne in Cornwall. He was deeply distressed by the high casualty rate and terrible injuries suffered by local tin miners, and set out to discover a safe means of igniting charges. His first attempt failed, but his second resulted in a reliable fuse.

Gunpowder had been used for blasting since the early 1600s. The powder was put into a brass ball or 'pulta', the outside being covered with cotton soaked in saltpetre and dipped in molten pitch and sulphur. The powder was fired from outside through a small hole drilled through the brass case. Once the pitch was lit, the pulta was pushed into a crack in the rock.

The first borehole blasts were carried out by Casper Weindl in a mine near Ober-Biberstollen, about 120km/75 mi north of Budapest. In England, gunpowder was probably first used for blasting at the Ecton copper mine near the Derbyshire–Staffordshire border. The shot holes were filled with powder and then 'stemmed' or, as a contemporary account puts it: blocked 'by stones and rubbish ramm'd in (except a little place that is left for a Train) the powder by the help of that train being fir'd'. The method had spread to quarries by the early 1670s.

By the end of the 19th century an extremely popular type of fuse made of goose quills was being used. The quills were cut so that they could be inserted one into the other and then filled with powder. Such fuses could be ignited directly, that is without any delaying element such as the sulphur mannikin. However, the quills were often broken or pushed apart during stemming so that an irregular or even a damp column of powder was left to act as a very uncertain fuse. Quill fuses frequently apparently failed and then rekindled so that the miner, who went to inspect the apparently extinct fuse, was injured in the blast.

Bickford's safety fuse provided a dependable means for conveying flame to the charge so that the danger of such hang fires was virtually eliminated. Its timing (the time required for a given length to burn) was more accurate and consistent than that of its predecessors, and it had much better resistance to water and to general abuse. The burning section was protected by the stemming in each shot hole so that several holes could be fired at a time without the fusing of the last being destroyed by the blast from the first.

This and other techniques which the new fuse allowed increased not only safety, but also productivity, quickly making it popular among both miners and management. Many different types were subsequently made for volley firing, for use in flammable atmospheres, and for other applications. Bickford went into partnership with Thomas Davey, a working miner and a Methodist class leader, to construct the machinery for fuse production.

The first method used a funnel that trickled black powder into the centre of 12 yarns as they were spun. The process was discontinuous, producing 20 m/65 ft lengths of semifuse which were then 'countered' by twisting on a second set of yarns in the opposite direction to make a second 'rod' that would not unwind. The fuse was then covered with a layer of tar and resin. Later the process became continuous so that the length produced was limited only by interruptions in the process. This method, with refinements, is still in use today.

Bickford fell seriously ill in 1832 and took no further part in the exploitation of his invention. After his death in 1834 a fusemaking factory was built at Tuckingmill and continued in production until July 1961. Factories were later set up in Lancashire and also in the USA, France, Saxony, Austria–Hungary, and Australia, the organization becoming one of the major units of Explosive Trades Limited, later to become Imperial Chemical Industries (ICI).

Bigelow, Erastus Brigham (1814–1879) US inventor–industrialist who devised, among many such machines, the first power loom for weaving ingrain carpets and a loom for manufacturing Wilton and Brussels carpets.

Bigelow's beginnings at West Boylston, Massachusetts, where he was born on 2 April 1814, were so impoverished that he was forced to go out to work at the age of ten. Bigelow's father, a small farmer, also worked as a chairmaker and wheelwright. Until 1834, Bigelow did any work he could find – on neighbouring farms, playing the violin in the church orchestra or at local country dances, as a store clerk, and as a stenography teacher – nevertheless he hankered after a formal education. He had hopes of entering Harvard University as a medical student and afterwards, of a professorial or literary career.

Though these hopes never materialized, Bigelow's natural talent for mechanics and mathematics could not be suppressed. At the age of eight he mastered arithmetic without the aid of a teacher. Bigelow produced his first loom for weaving coach lace to trim stagecoach upholstery in 1837, at the age of 23. This was followed by more looms for the manufacture of ginghams, silk brocatelle, counterpanes, and other figured fabrics.

In about 1839 Bigelow met Alexander Wright, a Scots mechanic settled in the USA who had recently set up a small carpetweaving business with three looms and 20 workers. Wright had heard of Bigelow's mechanized coach-lace loom and encouraged Bigelow to look into the possibility of inventing a power loom for ingrain-type carpets. Wright discovered that, despite 40 years of experiment in Europe, no satisfactory machine with this capability had yet been devised.

Like his predecessors, Bigelow faced many complex problems: how to set needles for the mechanical interweaving of up to three piles at a time, how to provide accurate timing for the takeup beam of the fabric, how to create a firm and even selvage, how to control the smooth surface with repeating patterns of uniform length so that they matched when seamed, and how to control the timing of several shuttle boxes.

The resultant loom had a long gestation and most of the cost was borne by the Clinton Company that Bigelow and his brother Horatio had set up in 1843 to manufacture ginghams. In exchange, the company required exclusive rights to the machine, which, after several modifications, was at last ready for use in 1846. Bigelow's power loom proved capable of producing up to 16.5 sq m/20 sq yd a day of two-ply goods and 11.7 sq m/14 sq yd of three-ply. It was a great improvement on his earlier machines, which could produce only the simplest of patterns, allowing patterns with large flowers and sweeping foliage, asymmetrically arranged, to be produced. There was some popular criticism about the 'unnaturalness of walking on flowers' and geometric patterns were suggested instead. However, patterns with large floriated scrolls of the type woven by Bigelow's machines were ideal for incorporating the serrations around the leaf and flower-edges which 'hid' the uneven joins of the interchange of coloured threads. Bigelow's power loom for weaving Brussels and Wilton carpets was developed 1845–51 and, together with his other machines, brought the Clinton Company great riches. The town of Clinton, Massachusetts, which had started humbly as a factory village, grew up around the Bigelow plant. Later, other Bigelow mills were established at Lowell, Massachusetts, and Humphreysville, Connecticut. Bigelow's looms transformed the carpet industry in the USA, which until then had been outpaced by the UK, where weaving skills were greater and labour less costly. In fact, in the mid-19th century, the USA imported more British carpeting – an average of 550,000 sq m/660,000 sq yd a year – than was produced by all US factories put together. Ironically, when imported into the UK, Bigelow's looms gave the British an advantage over their French rivals because of their costcutting and because carpets could now be made of virtually any colour, to virtually any pattern.

Bigelow, who died in Boston on 6 December 1879, helped found the Massachusetts Institute of Technology in 1861. Six volumes of his English patents, with the original drawings, are preserved by the Massachusetts Historical Society.

Binet, Alfred (1857–1911) French psychologist who is best known for his pioneering work on the development of mental testing, particularly the testing of intelligence.

Binet was born on 8 July 1857 in Nice. He went to Paris in 1871 to study law, but became interested in the work of the neurologist Jean-Martin ◊Charcot on hypnosis and abandoned law in 1878 to study first neurology and later psychology at the Salpêtrière Hospital in Paris. He remained there until 1891, when he went to work in the physiological psychology laboratory at the Sorbonne in Paris. He became director of the laboratory in 1895 and held this post until his death on 18 October 1911.

Binet was principally interested in applying experimental techniques to the measurement of intellectual abilities and, after joining the Sorbonne, began to devise various tests in an attempt to gain an objective measure of mental ability. After experimenting with various combinations of different tests, Binet, with Théodore Simon, published in 1905 the Binet–Simon intelligence test. This test, which was designed to measure intellectual ability in children, required the subject to perform such tasks as naming objects, copying designs, and rearranging disordered patterns. The subject was then given a mental-age score according to how well he or she had performed in the tests compared with previously established norms for various age groups. This was one of the first attempts at objectively measuring intelligence and, because of its usefulness, quickly became adopted in France and other countries. The original Binet–Simon test subsequently underwent many revisions, the most notable of which was the scoring of intelligence as an intelligence quotient (IQ) – calculated as the ratio of mental age to chronological age, multiplied by 100 – introduced in 1916 by Lewis Terman (1877–1956), a US psychologist working at Stanford University, in the adaptation known as the Stanford–Binet test. (The Stanford–Binet test was later revised several times and, in its latest form, is still used today.)

Binet wrote several books on mental processes and reasoning ability, notably *L'Etude expérimentale de l'intelligence/Experimental Study of Intelligence* (1903), a study of the mental abilities of his two daughters, and, with Simon, *Les Enfants anormaux/Mentally Defective Children* (1907). In addition he devised several tests that involved interpreting a subject's response to various visual stimuli (such as inkblots and pictures), the forerunners of some types of modern personality tests. Binet also studied and wrote about hypnosis and hysteria.

Birkhoff, George David (1884–1944) US mathematician who made fundamental contributions to the study of dynamics and formulated the 'weak form' of the ergodic theorem.

Birkhoff was born in Overisel, Michigan, on 21 March 1884. He studied at the Lewis Institute (now the Illinois Institute of Technology) 1896–1902, when he entered the University of Chicago. He then went to Harvard, where he received his BA in mathematics in 1905. In 1907 he received a PhD for his thesis on boundary problems from the University of Chicago. He taught at the University of Michigan and Princeton before being appointed an assistant professor at Harvard in 1912. He was a full professor at Harvard from 1919 until his death. Birkhoff was president of the American Mathematical Society in 1925 and of the American Association for the Advancement of Science in 1937. He was awarded the Bocher Prize in 1923 for his work on dynamics and the AAAS Prize in 1926 for his investigation of differential equations. He died at Cambridge, Massachusetts, on 12 November 1944.

Birkhoff's early work was on integral equations and boundary problems, and his investigations led him into the field of differential and difference equations. He developed a system of differential equations that is still inspiring research, while his work on difference equations was notable for the prominence which he gave to the use of matrix algebra.

Birkhoff's high reputation derives chiefly, however, from his investigation of the theory of dynamical systems such as the Solar System. After grounding himself thoroughly in Jules ◊Poincaré's celestial mechanics, he began to examine the motion of bodies in the light of his work on asymptotic expansions and boundary value problems of linear differential equations. In 1913 – one of those exhilarating moments in mathematics – he proved Poincaré's last geometric theorem on the three-body problem, a problem with which Poincaré had grappled with unsuccessfully. Birkhoff's formulation ran as follows: 'Let us suppose that a continuous one-to-one transformation T takes the ring R, formed by concentric circles Ca and Cb of radii a and b ($a > 0$), into itself in such a way as to advance the point of Ca in a positive sense, and the point of Cb in a negative sense, and at the same time preserve areas. Then there are at least two invariant points.'

With John ◊Von Neumann, Birkhoff was chiefly responsible for establishing, in the 1930s, the modern science of ergodics. He arrived, indeed, at the statement of his 'positive ergodic theorem', or what is known as the 'weak form' of ergodic theory, just before Von Neumann published his 'strong form' of it. By using the Lebesque measure theory Birkhoff transformed the Maxwell–Boltzmann hypothesis of the kinetic theory of gases, which was undermined by the number of exceptions found to it, into a vigorous principle.

Birkhoff also made a number of valuable contributions to related problems in other fields to which he was led by his ergodic investigations. One such was his paper of 1938 'Electricity as a Fluid' which, although consistent with Einstein's special relativity, found no need of the general curvilinear coordinates of the general theory of relativity. Throughout his life Birkhoff continued to argue that Einstein's general relativity was an unhelpful theory and his 1938 paper did much to provoke thought and research on the subject.

Few 20th-century mathematicians achieved more than Birkhoff and he was, in addition, the most important teacher of his generation. Many of the USA's leading mathematicians did their doctoral or postdoctoral research under his direction. His standing, generally acknowledged, as the most illustrious US mathematician of the early 20th century is deserved.

Bjerknes, Vilhelm Firman Koren (1862–1951) Norwegian scientist who created modern meteorology.

Bjerknes came from a talented family. His father was professor of mathematics at the Christiania University (now Oslo University) and a highly influential geophysicist who clearly shaped his son's studies. Bjerknes held chairs at Stockholm and Leipzig before founding the Bergen Geophysical Institute in 1917. By developing hydrodynamic models of the oceans and the atmosphere, Bjerknes made momentous contributions that transformed meteorology into an accepted science. Not least, he showed how weather prediction could be put on a statistical basis, dependent on the use of mathematical models.

During World War I, Bjerknes instituted a network of weather stations throughout Norway; coordination of the findings from such stations led him and his coworkers to develop the highly influential theory of polar fronts, on the basis of the discovery that the atmosphere is made up of discrete air masses displaying dissimilar features. Bjerknes coined the word 'front' to delineate the boundaries between such air masses. Among much else, the 'Bergen frontal theory' explained the generation of cyclones over the Atlantic, at the junction of warm and cold air wedges. Bjerknes's work gave modern meteorology its theoretical tools and methods of investigation.

Black, Joseph (1728–1799) Scottish physicist and chemist whose most important contribution to physics was his work on thermodynamics, notably on latent heats and specific heats. He is classed with Henry Cavendish and Antoine Lavoisier as one of the pioneers of modern chemistry. He is remembered for his discovery of carbon dioxide, which he called 'fixed air'.

Black was born in Bordeaux, France, on 16 April 1728. His father was from Belfast, but of Scottish descent, and was working in Bordeaux in the wine trade. Black went to Belfast to be educated, and then on to Glasgow University to study natural sciences and medicine. He moved to Edinburgh in 1751 to finish his studies, gaining his doctor's degree in 1754. Two years later he took over from William Cullen, his chemistry teacher at Glasgow, and was also offered the chair in anatomy. He soon changed this position for that of professor of medicine, and also practised as a physician. In 1766 he again followed Cullen as professor of chemistry at Edinburgh University. Black died on 10 November 1799.

In Black's doctorate he described investigations in 'causticization' and indicated the existence of a gas distinct from common air, which he detected using a balance. He was therefore the founder of quantitative pneumatic chemistry and preceded Antoine Lavoisier in his experiments. In a more detailed account of his work, published in 1756, Black described how carbonates (which he called mild alkalis) become more alkaline (are causticized) when they lose carbon dioxide, whereas the taking up of carbon dioxide reconverts caustic alkalis into mild alkalis. Black identified 'fixed air' (carbon dioxide) but did not pursue this work. He also discovered that it behaves like an acid; is produced by fermentation, respiration and the combustion of carbon; and had guessed that it is present in the atmosphere. He also discovered the bicarbonates (hydrogen carbonates).

Until about 1760 Black devoted his research to chemistry but thereafter most of his work was in physics. He noticed that when ice melts it absorbs heat from its surroundings without itself undergoing a change in temperature, from which he argued that the heat must have combined with the ice particles and become latent. In 1761 he experimentally verified this hypothesis, thereby establishing the concept of latent heat (in the example of melting ice, the latent heat of fusion). In the following year he determined the latent heat of formation of steam (the latent heat of vaporization). Also in 1762 he described his work to a literary society in Glasgow, but did not publish his findings. In addition, he observed that equal masses of different substances require different quantities of heat to change their temperatures by the same amount, an observation that established the concept of specific heats (relative heat capacities).

Black, Max (1909–) Russian-born US philosopher and mathematician, one of whose concerns has been to investigate the question 'what is mathematics?'.

Black was born in Baku, Azerbaijan, on 24 February 1909, and he received his higher education in England where he studied philosophy, gaining his BA at Cambridge University in 1930 and his PhD from London University in 1939. From 1936–40 he was a lecturer at the University of London Institute of Education. He went to the USA in 1940 to take up a post in the department of philosophy at the University of Illinois. He became a naturalized US citizen in 1948. He moved from the University of Illinois to Cornell in 1946 and was Susan Lin-Sage Professor of Philosophy and Humane Letters there 1954–77, when he retired. In 1970 he was vice president of the International Institute of Philosophy.

Black's analysis led him to describe mathematics as the study of all structures whose form may be expressed in symbols. Within that broad spectrum there are three main schools of mathematics: the logical, the formalist, and the intuitional. The logical considers that all mathematical concepts, such as numbers or differential coefficients, are capable of purely logical definition, so that mathematics becomes a branch of logic. The formalist, rejecting the notion that all mathematics can be expressed as logical concepts, looks upon mathematics as the science of the structure of objects and concerns itself with the structural properties of symbols, independent of their meaning. The formalist approach has been especially fruitful in its application to geometry. The third school, the intuitional, by laying less emphasis on symbols and more on thought, considers mathematics to be grounded on the basic intuition of the possibility of constructing an infinite series of numbers. This approach has had most influence in the theory of sets of points.

Black has thus done little work in mathematics itself, but his writings, such as *The Nature of Mathematics* (1950) and *Problems of Analysis* (1954), have been a major contribution to the philosophy of mathematics.

Blackett, Patrick Maynard Stuart (1897–1974) Baron Blackett. English physicist who made the first photograph of an atomic transmutation and developed the cloud chamber into a practical instrument for studying nuclear reactions. This achievement gained him the 1948 Nobel Prize for Physics. Blackett also discovered the phenomenon of pair production of positrons and electrons in cosmic rays.

Blackett was born in Croydon, Surrey, on 18 November 1897. He did not initially go to university, but joined the Royal Navy in 1912 as a naval cadet. During World War I, he took part in the battles of the Falkland Islands and Jutland, and designed a revolutionary new gunsight. Attracted to science, Blackett resigned from the navy after the war ended and embarked on a science course at Cambridge University, obtaining a BA degree in 1921. Continuing at the university, Blackett started research with cloud chambers under Ernest ◊Rutherford in the Cavendish Laboratory. In 1924, Blackett succeeded in obtaining the first photographs of an atomic transmutation, which was of nitrogen into an oxygen isotope. He continued to develop the cloud chamber and in 1932, with assistance from Guiseppe Occhialini, he designed a cloud chamber in which photographs of cosmic rays were taken automatically. Early in 1933, the device confirmed the existence of the positron (positive electron) proposed by Carl ◊Anderson.

Blackett became professor of physics at Birkbeck College, London, in 1933 and continued his cosmic-ray studies, demonstrating in 1935 the formation of showers of positive and negative electrons from gamma rays in approximately equal numbers. In 1937, he succeeded Lawrence ◊Bragg at the University of Manchester, continuing with his cosmic-ray studies. During World War II, Blackett initiated the main principles of operational research, and afterwards returned to university life. At Manchester, his research team produced many important discoveries in studies involving cosmic radiation. Particles with a lifespan of 10^{-10} seconds were discovered. They became known as strange particles and included the negative cascade hyperon.

As a consequence of his research into cosmic rays, Blackett became interested in the history of the Earth's magnetic field and turned to the study of rock magnetism. He was appointed head of the physics department at Imperial College, London, in 1953, where he built up a research team specializing in rock magnetism. On Blackett's advice, the government formed a ministry of technology in 1974, and he became president of the Royal Society in 1965 and a life peer in 1969. Blackett died on 13 July 1974.

In 1919, Rutherford explained the transmutation of elements from experiments in which he bombarded nitrogen with alpha particles. The oxygen atoms and protons produced were detected by scintillations in a screen of zinc sulphide. Blackett used a cloud chamber to photograph the tracks formed by these particles, taking more than 20,000 pictures and recording some 400,000 tracks. Of these, eight showed that a nuclear reaction had taken place, confirming Rutherford's explanation of transmutation. To reduce the huge number of observations, Blackett and Occhialini invented a cloud chamber that was automatically triggered by the arrival of a cosmic ray likely to cause a nuclear reaction in the chamber. Two geiger counters were attached to the chamber, the passage of the ray creating a current in the counters that operated the chamber. Pair production (the formation of showers of positrons and electrons from gamma rays), which Blackett discovered, was the first evidence that matter may be created from energy as Albert ◊Einstein had predicted in his special theory of relativity. The two particles subsequently rejoin and annihilate each other to form a gamma ray, thus converting mass back into energy.

Blackwell, Elizabeth (1821–1910) English-born US physician who became, in 1849, the first woman to receive a medical degree in the USA and, in 1869, the first woman in the UK to be recognized as a qualified physician.

Blackwell was born in Bristol, England, into a family of nine children. Her parents were Liberal dissenters who emigrated to the USA in 1832. After the death of her father in 1839, Elizabeth and her mother ran a private school to support the family. Elizabeth's spare time was spent reading medical books. She decided that she wanted to study medicine properly, partly because she was unhappy with contemporary inequalities between men and women in education, but also because she wanted to put a barrier between herself and marriage and could see that the taking up of a 'male' career would provide just such a barrier. She relished the idea of the struggle she would have in pursuit of her ambition.

Blackwell moved to Philadelphia in 1847 where she studied anatomy privately. She applied to several medical schools and was turned down by them all until she was accepted by Geneva College, New York, 'by accident'. The professors put her application to the students who treated it as a joke and accepted her. In fact, it seems that she was well received by the (all-male) student body, it was the professors and qualified doctors who saw her as a threat. However, she succeeded and was awarded her MD in front of 20,000 people in 1849.

In her search for further qualifications Blackwell moved to Europe and as a midwifery student in Paris contracted purulent ophthalmia from a baby, which left her blind in one eye. This meant that she was unable to fulfil her dream of becoming a surgeon. Following this personal tragedy she worked with James ◊Paget, one of the founders of modern pathology, at St Bartholomew's Hospital, London.

In 1853 Blackwell moved back to the USA and set up a dispensary in a tenement district of New York before gaining a state charter in 1857 to found the New York Infirmary for Indigent Women and Children. She was joined in this venture by her younger sister Emily, also a qualified doctor, and Marie Elizabeth Zakrzwska, a Polish-German doctor who had emigrated to the USA because of the adverse conditions for women in medicine in Germany. The three of them ran the hospital, sharing nursing, housework, medical, and surgical work. Zakrzwska looked after the administration and devised a system for patients' records.

In 1868, Elizabeth and Emily opened a women's medical college which was attached to the hospital. Elizabeth was professor of hygiene and Emily was dean and professor of obstetrics and women's diseases. Elizabeth also travelled during this time giving lectures on hygiene and publishing them. The college was to run until 1899 when women were accepted as medical students at Cornell University.

Because Elizabeth had previously practised in England she was the first woman to appear on the first British Medical Register of 1859. She moved back to London in 1869 leaving Emily to run the New York infirmary and college until 1899. She founded the National Health Society in London and helped to found the London School of Medicine for Women where she was professor of gynaecology 1874–1907.

Blackwell opposed vaccination and animal experimentation and in London she met Charles Kingsley, a Christian Socialist concerned with improving the conditions of working-class life, and she converted to his cause. She never married but in 1854 she adopted Kitty Barry, a seven-year-old orphan, who became her constant companion.

Further Reading

Baker, Rachel, *The First Woman Doctor: The Story of Elizabeth Blackwell, M D,* G G Harrap, 1946.

Blackwell, Elizabeth, *Pioneer Work in Opening the Medical Profession to Women,* Longmans, Green, and Co, 1895.

Chambers, Peggy, *A Doctor's Alone: A Biography of Elizabeth Blackwell, the First Woman Doctor, 1821–1910,* Bodley Head, 1956.

Wilson, Dorothy Clarke, *Lone Women: The Story of Elizabeth Blackwell, the First Woman Doctor,* Hodder & Stoughton, 1970.

Blakemore, Colin (1944–) English physiologist who has made advanced studies of how the brain works, especially in connection with memory and the senses.

Blakemore was born in Stratford-upon-Avon on 1 June 1944 and educated at King Henry VIII School in Coventry. In 1962 he won a scholarship to Corpus Christi College, Cambridge, to study natural sciences and, in particular, medicine; he graduated in 1965. Blakemore then declined a scholarship to St Thomas's Hospital, London, and went instead to study physiological optics at the Neurosensory Laboratory at the University of California in Berkeley; he obtained his PhD in 1968. He was a demonstrator at Cambridge University 1967–72, when he was appointed lecturer in physiology, a position he held until 1979. He then went to work at Oxford University as Waynflete Professor of Physiology. He received the Royal Society's Michael Faraday Award in 1989.

In his experimental work, Blakemore has shown that cells in the visual cortex of the brain of a new-born kitten are able to detect visual outlines. But if the kitten is kept at a critical period in an environment with only, say, vertical lines, it will later prove to have in the cortex only cells that can 'recognize' these patterns and not others. Blakemore suggests that it is possible that the inherited DNA of genes already contains the capacity to synthesize RNA, the protein that is involved in the storage of any new remembrance.

Blakemore is known for his ability to explain complex science to the lay individual. This talent was demonstrated, for example, by his award-winning educational film *The Visual Cortex of the Cat,* in his Reith lectures of 1976, and in his publication *Mechanics of the Mind* (1977). In this book he explains the mechanics of sensation, sleep, memory, and thought, and discusses the philosophical questions of human consciousness, the evolution of thinking about body and mind, the relationship between art and perception, and the origin and function of language. He argues that an individual's system of knowledge, expertise, and ethical standards has evolved gradually and that the resulting 'collective mind' is a functional extension of all the human brains that have contributed to it.

Bloch, Konrad Emil (1912–) German-born US biochemist whose best-known work has been concerned with the biochemistry and metabolism of fats (lipids), particularly reactions involving cholesterol.

His research into the biosynthesis of cholesterol has produced a better understanding of this complex substance, whose presence in the human body is of supreme importance but whose excess is thought to be dangerous. For his work on the mechanism of cholesterol and fatty-acid metabolism he shared the 1964 Nobel Prize for Physiology or Medicine with Feodor ◊Lynen.

Bloch was born on 21 January 1912 at Niesse, Germany, and educated at the Munich Technische Hochschule, from which he graduated as a chemical engineer in 1934. In 1936 he emigrated to the USA where two years later he received his doctorate in biochemistry from Columbia University. Bloch became a US citizen in 1944 while serving on the faculty of the Columbian University College of Physicians and Surgeons. He joined the staff of the University of Chicago in 1946 and became professor of biochemistry there in 1952. He moved to a similar position at Harvard two years later and in 1956 became a member of the National Academy of Sciences.

Cholesterol, the most abundant sterol in animal tissue, was discovered in 1812 and is now one of the best-known (and commonest) steroids in the human body. It occurs either free or as esters of fatty acids, being found in practically all tissues but being most abundant in the brain, nervous tissue, and adrenal glands and, to a lesser extent, in the liver, kidneys, and skin. Its name arose because of its occurrence in gallstones (from the Greek for 'bile solid'). Cholesterol has the molecular formula $C_{27}H_{46}O$ and, like most steroids, has a structure based on phenanthrene. It is important to the body because it is a component of all cell membranes and is the metabolic precursor of many compounds with various physiological functions, including vitamin D, cortisone, and the male and female sex hormones.

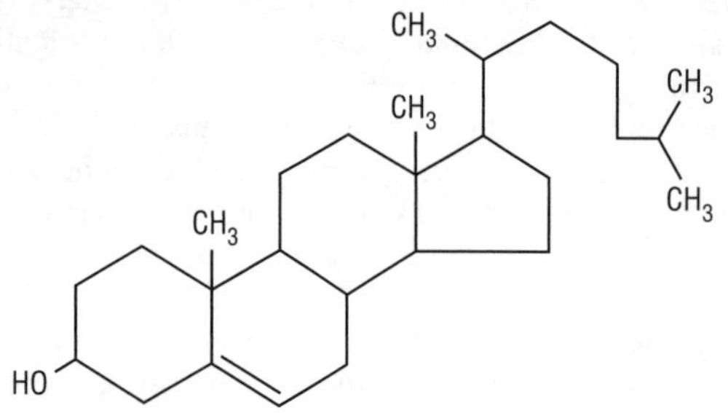

Bloch Structural formula of cholesterol.

In the early 1940s Bloch took the first steps that were to lead to today's extensive knowledge of cholesterol. Working first with David Rittenberger and later with Henry Little, he demonstrated that carbon atoms of carbon-labelled acetate (ethanoate) fed to rats was incorporated into cholesterol in the animals' livers. Using acetic acid (ethanoic acid) labelled with deuterium he showed for the first time that it is this acid, a compound having only two carbon atoms, that is the major precursor of cholesterol. This discovery was the first of a long series that elucidated the biological synthesis of the steroid. Later, using acetic acid labelled at one carbon atom with radioactive carbon-14, Bloch demonstrated which of the 27 carbon atoms in cholesterol is derived from each of the two carbon atoms of acetic acid.

The overall conversion of acetic acid to cholesterol requires 36 distinct chemical transformations, which occur in various tissues but principally in the liver. The route begins with the conversion of three molecules of acetic acid to form a five-carbon compound and carbon dioxide. Then six of the five-carbon compounds combine to form the long-chain unsaturated 30-carbon compound squalene, which, after cyclization, forms langesterol; the langesterol is finally converted to cholesterol.

Bloch's work undoubtedly paved the way to the successful tracing of the numerous metabolic changes that take place in the biosynthesis of cholesterol. It has important applications to medicine, because it is now thought that high levels of cholesterol in the bloodstream can cause it to be deposited on the inner walls of arteries (atherosclerosis), where it narrows the vessels and increases the chances of blood clotting.

Bode, Johann Elert (1749–1826) German mathematician and astronomer who contributed greatly to the popularization of astronomy.

Bode was born in Hamburg on 19 January 1747 into a well-educated family. He taught himself astronomy and was skilled mathematically. He was publishing astronomical treatises while still in his teens, one of which remained in print for nearly a century.

In 1772 Bode joined the Berlin Academy as a mathematician, overseeing the publication of the academy's yearbook and ensuring the accuracy of its mathematical content; he worked on all the yearbooks from 1776 to 1829. He was appointed director of the Astronomical Observatory in 1786. He supervised the renovation of the observatory, but he was unable to bring the standard of work there up to that of many other observatories because of the relatively simple equipment at his disposal. In 1784 he was appointed Royal Astronomer and elected to the Berlin Academy. He retired as director of the observatory in 1825 and died in Berlin on 23 November 1826.

Bode's early work at the Berlin Academy concerned the improvement of the accuracy of the mathematical content of the yearbook: the low standard had been depressing sales, upon which the academy's finances largely relied. The yearbook's popularity soon increased. In addition to astronomical tables it included information about observations and scientific developments elsewhere in the world.

Bode also worked on the compilation of two atlases, the *Vorstellung der Gestirne* and the *Uranographia,* which was a massive work describing the positions of more than 17,000 stars and including for the first time some of the celestial bodies discovered by William Herschel. It was Bode who named Herschel's new planet 'Uranus'.

Bode is best known for the law named after him, even though he did not first state it, but merely popularized work already done by german mathematician Johann Titius (1729–1796). The law, also known as the Titius–Bode rule, is a mathematical formula that approximately described the distances of all then known planets from the Sun. The series had no basis in theory, but it was accepted as an important finding at the time because of the near-mystical reverence attached to numbers and geometric progressions in descriptions of the universe. The discovery of the planet Neptune by Urbain Leverrier and John Couch Adams in 1846 disrupted the series and it lost its value.

Bode's main contribution to astronomy lay in the spreading of information about the subject to people from a wide range of backgrounds.

Boerhaave, Hermann (1668–1738) Dutch physician and chemist of tremendous learning who dominated and greatly influenced various branches of science in Europe.

Boerhaave was born in Verhout, near Leiden, on 31 December 1668, the son of a minister. He intended entering the church, and in 1684 went to the University of Leiden to study theology. While there he also studied philosophy, botany, languages, chemistry, and medicine. He qualified in natural philosophy in 1687 and gained his PhD two years later. Medicine and chemistry became his predominant interests and he entered the University of Haderwijk, from which he graduated in medicine in 1693. He went back to Leiden in 1701 as a physician, and also began teaching. In 1709 he took the chair of medicine and botany at Leiden, and was also made a professor of physic in 1714, as well as professor of chemistry in 1718. He was elected to the French Academy in 1728, and became a Fellow of the Royal Society in 1730. He died in Leiden on 23 September 1738.

For a man of such immense academic distinction and knowledge, Boerhaave made few original discoveries, although he did describe the structure and function of the sweat glands, and was the first to realize that smallpox is spread by contact. He was, however, an excellent tutor and re-established the technique of clinical teaching, taking his students to the bedsides of his patients. During his time at Leiden it became a famous centre of medical knowledge, attracting students from throughout Europe.

Boerhaave's writings remained authoritative works for nearly a century. In 1708 he published a physiology textbook *Institutiones Medicae* (a classification of diseases with their causes and treatment), followed by the *Book of Aphorisms* in the next year. In 1710 an *Index Plantarum* was published, followed by *Historia Plantarum* – a collection of his botanical lectures compiled by his ex-students. In 1724 his students published *Institutiones et Experimenta Chemiae,* a breakdown of Boerhaave's lectures on chemistry. He produced the official version of the lectures in 1732 called *Elementia Chemiae,* which presented a clear and precise approach to the chemistry of the day and remains his most famous work.

Further Reading

Boerhaave, Hermann and Lindeboom, Gerrit Arie, *Boerhaave's Correspondence,* Analecta Boerhaaviana series, Brill, 1962, v 3.

Boerhaave, Hermann; Kegel-Brinkgreve, E; and Luyendija-Elshout, A M (eds), *Orations,* Publications of the Sir Thomas Browne Institute, Brill Academic Publishers, 1983.

Johnson, Samuel and Schoneveld, Cornelius W, *The Life of Dr Boerhaave,* Academic Press, 1994.

Lindeboom, Gerrit Arie, *Herman Boerhaave: The Man and His Work,* Methuen, 1968.

Underwood, Edgar Ashworth, *Boerhaave's Men at Leyden and After,* Edinburgh University Press, Edinburgh.

Bohr, Mottelson, and Rainwater, Aage Niels Bohr (1922–), Ben Roy Mottelson (1926–), and (Leo) James Rainwater (1917–1986) Danish and US physicists who shared the 1975 Nobel Prize for Physics for their work on the structure of the atomic nucleus.

Aage Bohr is the son of Niels Bohr and was born in Copenhagen on 19 June 1922 (the year that his father won the Nobel Prize for Physics). He was educated at the University of Copenhagen, and then worked for the Department of Science and Industrial Research in London 1943–45. From 1946 he worked at his father's Institute of Theoretical Physics (now the Niels Bohr Institute) in Copenhagen and from 1963–70 was its director. Since 1956 he has been professor of physics at the University of Copenhagen and from 1975–81 he was director of Nordita (Nordic Institute for Theoretical Atomic Physics). In addition to the Nobel Prize for Physics awarded in 1975, Bohr won the Atoms for Peace Award in 1969 and the Rutherford Medal in 1972. He has published *Rotational States of Atomic Nuclei* (1954) and *Nuclear Structure* (1969 and 1975) jointly with Ben Mottelson.

Ben Mottelson was born in Chicago, Illinois, on 9 July 1926. He was educated at Purdue University, where he gained a PhD in 1950. From 1950–53, he was at the Institute of Theoretical Physics in Copenhagen, and he then held a position at CERN (European Centre for Nuclear Research) in a theoretical study group formed in Copenhagen. He has been professor at Nordita in Copenhagen since 1957. Mottelson became a Danish citizen in 1971, and has published several books and many scientific papers jointly with Bohr.

James Rainwater was born on 9 December 1917 in Council, Idaho. He gained a physics degree at the California Institute of Technology in 1939 and then read for his advanced degree at Columbia University. He remained there, rising to become professor of physics in 1952, a position he held until his death in 1986. During the period 1942–46, Rainwater worked for the Office of Scientific Research and Development (OSRD) and then the Manhattan Project, and he was director of the Nevis Cyclotron Laboratory 1951–53 and again 1956–61.

Aage Bohr, Ben Mottelson, and James Rainwater shared the 1975 Nobel Prize for Physics for their 'discovery of the connection between the collective motion and the particle motion in atomic nuclei and the development of the theory of the structure of the atomic nucleus, based on this connection'. The three men had been working in loose collaboration for nearly 25 years. Niels Bohr had proposed that the particles in the nucleus of an atom are arranged like molecules in a drop of liquid. Exceptions to this rule were found and this led to the belief that the nuclear particles are arranged in concentric shells. Further studies showed that perhaps these shells were not spherically symmetrical. This suggestion was both unexpected and unattractive. In 1950, Rainwater wrote a paper in which he observed that most of the nuclear particles form an inner nucleus, while the other particles form an outer nucleus. Each set of particles is in constant motion at very high velocity and the shape of each set affects the other set. He postulated that if some of the outer particles moved in similar orbits, this would create unequal centrifugal forces of enormous power, which could be strong enough to permanently deform an ideally symmetrical nucleus. At that time Aage Bohr was working with Rainwater as a visiting professor at the University of Columbia and was impressed by this theory. He began to work out a more detailed explanation of Rainwater's work after he returned to Denmark, and during the next three years he published results that he and his associate Ben Mottelson had obtained experimentally. These results proved the theoretical work.

The work of the three men achieved a deep understanding of the atomic nucleus and paved the way for nuclear fusion.

Bohr Danish physicist Niels Bohr, one of the major figures of atomic research. Successfully combining Rutherford's classical model of the atom with Planck's quantum theory, he developed a model of the atom that revolutionized atomic theory.
AEA Technology

Bohr, Niels Henrik David (1885–1962) Danish physicist who established the structure of the atom. For this achievement he was awarded the 1922 Nobel Prize for Physics. Bohr made another very important contribution to atomic physics by explaining the process of nuclear fission.

Bohr was born in Copenhagen on 7 October 1885. His father, Christian Bohr, was professor of physiology at the University of Copenhagen and his younger brother Harald became an eminent mathematician. Niels Bohr was a less brilliant student than his brother but a careful and thorough investigator. His first research project, completed in 1906, resulted in a precise determination of the surface tension of water and gained him the gold medal of the Academy of Sciences. In 1911, he was awarded his doctorate for a theory accounting for the behaviour of electrons in metals.

In the same year, Bohr went to Cambridge, England, to study under J J Thomson, who showed little interest in Bohr's electron theory so, in 1912, Bohr moved to Manchester to work with Ernest ◊Rutherford, who was making important investigations into the structure of the atom. Bohr developed models of the atom in which electrons are disposed in rings around the nucleus, a first step towards an explanation of atomic structure.

Bohr returned to Copenhagen as a lecturer at the university in 1912, and in 1913 developed his theory of atomic structure by applying the quantum theory to the observations of radiation emitted by atoms. He then went back to Manchester to take up a lectureship offered by Rutherford, enabling him to continue his investigations in ideal conditions. However, the authorities in Denmark enticed him back with a professorship and then built the Institute of Theoretical Physics in Copenhagen for him. He became director of the institute in 1920, holding this position until his death. The institute rapidly became a centre for theoretical physicists from throughout the world, and such figures as Wolfgang ◊Pauli and Werner ◊Heisenberg developed Bohr's work there, resulting in the theories of quantum and wave mechanics that more fully explain the behaviour of electrons within atoms.

The year 1922 marked not only the award of the Nobel Prize for Physics but also a triumphant vindication of Bohr's atomic theory, which he used to predict the existence of a hitherto-unknown element. The element was discovered at the institute and given the name hafnium.

In the 1930s, interest in physics turned towards nuclear reactions and in 1939 Bohr proposed his liquid-droplet model for the nucleus that was able to explain why a heavy nucleus could undergo fission following the capture of a neutron. Working from experimental results, Bohr was able to show that only the isotope uranium-235 would undergo fission with slow neutrons.

Our task is not to penetrate into the essence of things, the meaning of which we don't know anyway, but rather to develop concepts which allow us to talk in a productive way about phenomena in nature.
Neils Bohr *Letter to H P E Hansen 20 July 1935*

When Denmark was occupied by the Germans in 1940 early in World War II, Bohr took an active part in the resistance movement. In 1943, he escaped to Sweden with his family in a fishing boat – not without danger – and then went to the UK and on to the USA. He became involved in the development of the atomic bomb, helping to solve the physical problems involved, but later becoming a passionate advocate for the control of nuclear weapons. Among his efforts to persuade politicians to adopt rational and peaceful solutions was a famous open letter addressed to the United Nations in 1950 pleading for an 'open world' of free exchange of people and ideas.

In 1952 Bohr was instrumental in creating the European Centre for Nuclear Research (CERN), now at Geneva, Switzerland. He died in Copenhagen on 18 November 1962. In addition to his scientific papers, Bohr published three volumes of essays: *Atomic Theory and the Description of Nature* (1934), *Atomic Physics and Human Knowledge* (1958), and *Essays 1958–1962 on Atomic Physics and Human Knowledge* (1963).

Bohr's first great inspiration came from working with Rutherford, who had proposed a nuclear theory of atomic structure from his work on the scattering of alpha rays in 1911. It was not, however, understood how electrons could continually orbit the nucleus without radiating energy, as classical physics demanded. Ten years earlier Max ◊Planck had proposed that radiation is emitted or absorbed by atoms in discrete units, or quanta, of energy. Bohr applied this quantum theory to the nuclear atom to explain why elements emit radiation at precise frequencies that give set patterns of spectral lines. He postulated that an atom may exist in only a certain number of stable states, each with a certain amount of energy; the emission or absorption of energy may occur only with a transition from one stable state to another. Electrons normally orbit the nucleus without emitting or absorbing energy. When a transition occurs, an electron moves to a lower or higher orbit depending on whether it emits or absorbs energy. In so doing, a set number of quanta of energy are emitted or absorbed at a particular frequency. Bohr developed these ideas to show that the nuclei of atoms are surrounded by shells of electrons, each assigned particular sets of quantum numbers according to their orbits. Bohr's theory was used to determine the frequencies of spectral lines produced by elements and succeeded brilliantly. It also enabled him to explain the groups of the periodic table in terms of elements with similar electron structures, which led to the prediction and discovery of hafnium.

How wonderful that we have met with a paradox. Now we have hope of making progress.
Neils Bohr *Quoted in A Paiss,* Niels Bohr's Times *1991*

In developing a model for the nucleus, Bohr conceived of the nuclear particles being pulled together by short-range forces rather as the molecules in a drop of liquid are attracted to one another. The extra energy produced by the absorption of a neutron may cause the nuclear particles to separate into two groups of approximately the same size, thus breaking the nucleus into two smaller nuclei – as happens in nuclear fission. The model was vindicated when Bohr correctly predicted the differing behaviour of nuclei of uranium-235 and uranium-238 from the fact that the number of neutrons in each nucleus is odd and even respectively.

Niels Bohr gained not only a love of science from his father but also a philosophical insight into the nature of knowledge that enabled him to question accepted theories and seek new explanations. By reconciling Rutherford's nuclear model of the atom with Planck's quantum theory, he was able to produce a valid model for the atom completely at odds with classical physics. However, this did not prevent him from using a classical model to explain the structure and behaviour of the nucleus. Our present knowledge of the atom and the nucleus thus rests on the fundamental discoveries made by Bohr's restless and ingenious mind.

Further Reading

Blaedel, Niels, *Harmony and Unity: The Life of Niels Bohr,* Science Tech Publishers – Springer-Verlag, 1988.

Faye, Jan and Folse, Henry J Jr (eds), *Niels Bohr and Contemporary Philosophy,* Boston Studies in the Philosophy of Science, Kluwer Academic Publishers, 1994, v 153.

Pais, Abraham, *Niels Bohr's Times: In Physics, Philosophy, and Polity,* Clarendon Press, 1991.
Rozental, S, *Niels Bohr: His Life and Work as Seen by His Friends and Colleagues,* Elsevier Science, 1985.
Spangenburg, Ray and Moser, Diane K, *Niels Bohr: Gentle Genius of Denmark,* Makers of Modern Science series, Facts on File, 1995.
Whitaker, Andrew, *Einstein, Bohr, and the Quantum Dilemma,* Cambridge University Press, 1996.

Bok, Bart (1906–1992) Dutch astrophysicist best known for his discovery of the small, circular dark spots in nebulae (now known as Bok's globules).

Bok was born in the Netherlands in 1906 and educated at the University of Leiden 1924–26 and the University of Groningen 1927–29. He went to the USA in 1929, having gained a Robert Wheeler Wilson Fellowship in Astronomy at Harvard. He gained his PhD in 1932 and remained at Harvard as assistant professor 1933–39, associate professor 1939–46, and Robert Wheeler Wilson Professor of Astronomy 1947–57. In 1957 he went to Australia as professor and head of the department of astronomy at the Australian National University until 1966, being also director of the Mount Stromlo Observatory near Canberra at the same time. Bok's next appointment was as professor of astronomy at the University of Arizona 1966–74. He became professor emeritus on his retirement.

Bok served as president of a commission in the International Astronomical Union and president of the American Astronomical Society 1972–74. He was a member of a number of learned societies and received several medals. He published *The Distribution of the Stars in Space* with Priscilla Bok in 1937 and *The Milky Way* with F W Wright in 1944.

Photographs had shown that the Milky Way was dotted with dark patches or nebulae. Bok also discovered small, circular dark spots, which were best observed against a bright background. Measurements of their dimensions and opacity suggested that their masses were similar to that of the Sun. Bok suggested that the globules were clouds of gas in the process of condensation and that stars might be in the early stages of formation there. His work thus broadened our understanding of the nature of stellar birthplaces.

Boksenberg, Alexander (1936–) English astronomer and physicist who devised a new kind of light-detecting system that can be attached to telescopes and so vastly improve their optical powers. His image photon-counting system (IPCS) has revolutionized observational astronomy, enabling Boksenberg and others to study distant quasars, which may help towards a deeper understanding of the early phases and nature of the universe.

Boksenberg was born on 18 March 1936, the elder of two sons. He attended the Stationers' Company's School in London and was encouraged by his parents, who owned a shop, to study for a place at university. He gained a BSc in physics from London University and in 1957 began research at University College, London, into the physics of atomic collisions, for which he was awarded his PhD in 1961. Boksenberg then joined a research group at University College that was studying the ultraviolet spectra of stars using rocket- and satellite-borne instruments. It was during this period that he became interested in applying his knowledge of physics to astronomy. He also saw the need to improve the instrumentation being carried aboard space vehicles and began to specialize in image-detecting systems. He became a lecturer in physics in 1965, and in 1968 his innovative work on detectors led to his involvement in the British design team producing the instrumentation for the Anglo-Australian 4-m/13-ft telescope then being constructed at Sydney Springs in New South Wales, Australia. He devised a fundamentally new approach to optical detection in astronomy with his image photon-counting system.

In 1969 Boksenberg set up his own research group at University College to work on two main topics: optical astronomy, mainly using IPCSs he built for the Anglo-Australian telescope and for the 5-m/16.4-ft Hale telescope at Mount Palomar, California; and ultraviolet astronomy, using instrumentation he designed for use on high-altitude balloon-borne platforms and on satellites, particularly the *International Ultraviolet Explorer (IUE)* satellite observatory. In 1975 he was promoted to reader in physics, a year later he was awarded the first senior fellowship of the United Kingdom Science Research Council, and in 1978 became professor of physics and a Fellow of the Royal Society. From 1981–1993 he was director of the Royal Greenwich Observatory, and from 1993–96 was director of the Royal Observatories. Since 1996 he has been a research professor at the University of Cambridge, and a research fellow at both the University of Cambridge and of London. In 1982 he was awarded an honorary doctorate by the Paris Observatory.

During his time at University College, Boksenberg's work in ultraviolet astronomy using balloons yielded the first results for the important parameters of electron density and metal abundance in the local interstellar medium. For the European observatory satellite *TD-1A,* launched in 1972, he conceived a simple means of using the sky-scanning action of the satellite to produce a passive spectrum-scanning of the main ultraviolet sky-survey instrument, which greatly increased the scientific value of the project. He also designed and worked on a new ultraviolet television detector system for the *International Ultraviolet Explorer* satellite launched in 1978, and led the pioneering work on the complex Sun-baffle systems for *TD-IA* and *IUE* to enable the telescopes to observe faint astronomical objects in full orbital sunlight. Both satellites were widely used internationally and contributed greatly to the advancement of astronomy. Boksenberg's own main contribution was to a study of galactic haloes and the nuclei of active galaxies and quasars using the *IUE.*

Boksenberg's development of the IPCS sprang from his considerations of the workings of the human eye, which operates not only as an optical device but also relies on the correcting effect of the processing and memory functions of the retina and the brain. Rather than recording light with a photographic emulsion, he saw the potential of literally detecting the locations of the individual photons of light collected by a telescope from the faint astronomical object being studied, and building up the required image in a computer memory. By using an image intensifier coupled to a television camera he detected and amplified photons by a factor of 10^7. He then treated the signals in a special electronic processor, which passed the photon locations to a computer with a large digital memory, both to store the accumulating image and to present the incoming results as an instantaneous picture – a great advantage to an astronomer who would otherwise have to wait for a photographic image to form and be processed. The picture on the screen appears as an accumulating series of dots, each dot representing a photon that is counted, analysed, and stored by the computer.

By 1973 Boksenberg's IPCS was ready to be tested and in the autumn of that year he went to Mount Palomar and attached it to the spectrograph at the Coudé focus of the largest telescope in the world at that Time – the 5-m/16.4-ft Hale telescope. He used it in collaboration with Wallace Sargent (1935–) of the California Institute of Technology to observe absorption lines in the spectra of quasars. Within 30 minutes they saw spectra that would normally take three nights or more of exposure time to reveal. Sargent, an established scientist and astronomer, immediately recognized the enormous validity and potential of Boksenberg's photon detector, and on this basis his technique was generally accepted and applied to all modern telescopes, including the 2.4-m/7.8-ft Space Telescope.

Boksenberg, being interested in the overall nature of the universe, continued to collaborate with Sargent, mainly using the 5-m/16.4-ft Hale telescope and the IPCS in the study of the most distant quasars. The radiation from these has taken billions of years to reach us and so by studying their light, emitted way back in time, Boksenberg and his colleague use quasars to elucidate the early nature of the universe. Quasars also provide clues to the story of galactic evolution because they seem to be intimately connected with the central cores of galaxies.

Since 1973 Boksenberg and Sargent have been particularly interested in studying the absorption lines in the spectra of quasars, which they discovered are not a manifestation of the quasar itself but a reflection of the state of the universe – galaxies and intergalactic gas – that exists between the quasar and the Earth. They can thus provide direct information on the nature and evolution of the universe.

As director of the Royal Greenwich Observatory, Boksenberg was responsible for building a major new British observatory on Las Palmas in the Canary Islands. He also took part in plans to build the next generation of telescope. With the development of idealized light-detecting systems, such as the IPCS, using electronic devices to receive and record photons, optical astronomy has experienced a major resurgence and there is a recognized need to increase telescope apertures beyond the few metres currently available at each of the world's major observatories. The most favourable design of future ground-based optical telescopes, in terms of minimal cost and most efficient observational powers, seems to be the 'multi-mirror' candidate – a telescope composed of separate mirrors that combine to produce a light-collecting area that can be equivalent to a single mirror of about 20 m/66 ft in diameter.

Boltzmann, Ludwig (1844–1906) Austrian theoretical physicist who contributed to the development of the kinetic theory of gases, electromagnetism, and thermodynamics. His work in these fields led him to consider phenomena in terms of probability theory and atomic events, which led to the establishment of the branch of physics now known as statistical mechanics.

Boltzmann was born in Vienna on 20 February 1844 and educated at Linz and Vienna. He studied at the University of Vienna under Josef ◊Stefan and Josef Loschmidt (1821–1895), and received his PhD in 1866 from that institution. In 1867 Boltzmann became an assistant at the Physikalisches Institut in Vienna, after which he held a series of professorial posts. He was professor of theoretical physics at the University of Graz 1869–73, of mathematics at the University of Vienna 1873–76, and of experimental physics at Graz 1876–79. He then became director of the Physikalisches Institut, moving in 1899 to Munich to become professor of theoretical physics at the university there. In 1894 he went back to Vienna to succeed Stefan as professor of theoretical physics.

During these years there was a heated and sometimes unpleasant scientific debate between the 'atomists', championed by Boltzmann, and the 'energeticists', one of whose spokesmen was Wilhelm ◊Ostwald. Ernst ◊Mach, one of the 'energeticists', tried to reconcile the two schools of thought in an attempt to tone down the debate.

Boltzmann moved to the University of Leipzig in 1900, where he served as professor of theoretical physics. He suffered from depression and made an unsuccessful suicide attempt. He returned to his previous post at Vienna in 1902, with the added honour of the chair of natural philosophy from which Mach had just retired because of ill health. In 1904 Boltzmann travelled to the USA, lecturing at the World's Fair in St Louis and visiting Stanford and Berkeley, two prestigious campuses of the University of California. He still felt depressed at the hostility he sensed from some of his scientific colleagues, despite the fact that the revolutionary discoveries being made at the time about the structure of atoms and the properties of matter would ultimately support his theories. He committed suicide at Duino, near Trieste, on 5 September 1906.

Mechanics, dynamics, and electromagnetism were all being developed when Boltzmann was a student. He made contributions to all of these areas during the course of his career. One of his most important sources of stimulation was the work of James Clerk ◊Maxwell, whose investigations in the field of electromagnetism were not well known by European scientists. Boltzmann contributed greatly to the dissemination of Maxwell's work, particularly through a book on the subject published in 1891.

Boltzmann wrote his first paper on the kinetic theory of gases in 1859 while he was only a student. In 1868 he published a paper on thermal equilibrium in gases, citing and extending Maxwell's work on this subject. He was concerned with the distribution of energy among colliding gas molecules. He derived an exponential formula to describe the distribution of molecules which relates the mean total energy of a molecule to its temperature. It includes the constant k, now known as the Boltzmann constant, which is equal to 1.38066×10^{-23} joules per Kelvin. This constant has become a fundamental part of virtually every mathematical formulation of a statistical nature in both classical and quantum physics.

Another important area of Boltzmann's investigation was thermodynamics. The second law of thermodynamics had been formulated in 1850 by Rudolph ◊Clausius and William Thomson (Lord ◊Kelvin). Boltzmann sought to find a mathematical description for the demonstration of the tendency of a gas to reach a state of equilibrium as the most probable state. In 1877 he published his famous equation $S = k\log W$ (which was later engraved on his tombstone) describing the relation between entropy and probability.

Boltzmann also determined a theoretical derivation for Josef Stefan's experimentally derived law of black-body radiation. Stefan had shown that the energy radiated by a black body is proportional to the fourth power of its absolute temperature, and this is now often known as the Stefan–Boltzmann law. Boltzmann's investigation of the phenomenon was based on the second law of thermodynamics and Maxwell's electromagnetic theory.

Boltzmann's most enduring contribution arose from his pioneering work in the field of statistical mechanics. He had begun his study of the equipartition of energy, which resulted in the Maxwell–Boltzmann distribution law, early in his career. He demonstrated that the average amount of energy required for atomic motion in all directions is equal, and so formulated an equation for the distribution of atoms due to collision. This led to the foundation of statistical mechanics. This discipline holds that macroscopic properties of matter such as conductivity and viscosity can be understood, and are determined by, the cumulative properties of the constituent atoms. Boltzmann held that the second law of thermodynamics should be considered from this viewpoint. Boltzmann's work on the relationship between probability and entropy and on the inevitable and irreversible processes that must therefore occur, influenced Willard ◊Gibbs, Max ◊Planck, and others.

Boltzmann was a theoretician with great intuitive powers and vision. It is a tragedy that he did not live to see his work vindicated and the remarkable advances achieved by the 'atomists' of the 20th century.

Further Reading

Blackmore, John T, *Ludwig Boltzmann: His Later Life and Philosophy,* Boston Studies in the Philosophy of Science series, Kluwer Academic Publishers, 1995, 2 vols (v 168 and 174 in the series).

Broda, Engelbert and Gay, Larry, *Ludwig Boltzmann: Man, Physicist, Philosopher,* Ox Bow Press, 1983.

Cercignani, Carlo, *Ludwig Boltzmann: The Man Who Trusted Atoms,* Oxford University Press, 1998.

Bolyai, (Farkas) Wolfgang (1775–1856) and János (1802–1860) Hungarian mathematicians, father and son, the younger of whom was one of the founders of non-Euclidean geometry.

Wolfgang Bolyai was born in Nagyszeben, in Hungary (now Sibiu, Romania), on 9 February 1775. He studied mathematics at the University of Göttingen, where he fell into friendship with Karl ◊Gauss. For the rest of his life the elder Bolyai was a professional mathematician, teaching at the

Evangelical Reformed College at Nagyszeben and then at the college at Marosvásárhely (now Tîrgu Mures, Romania) until his retirement in 1853. He died at Marosvásárhely on 20 November 1856.

János Bolyai was born in Koloszvár, in Hungary (now Cluj, Romania), on 15 December 1802. He was first taught by his father and at an early age showed a marked talent both for mathematics and for playing the violin. At the age of 13, by which time he had mastered calculus and analytical mechanics, he entered the college at Marosvásárhely where his father was teaching. He remained there for five years, concentrating on mathematics, but also becoming an adept swordsman. Then, in 1818, against the wishes of his father (who wanted him to study under Gauss at Göttingen), he entered the Royal College of Engineers at Vienna. He graduated in 1822 and joined the army engineering corps, rising eventually to the rank of lieutenant, second class. Increasingly he fell victim to attacks of fever until, in 1833, he was retired from the army with a small pension. He returned to his father's house at Marosvásárhely and lived there, a semi-invalid, until his death on 27 January 1860.

The mathematical lives of the father and son were closely entwined. It was because of his son's growing interest in higher mathematics that Wolfgang was inspired to write the book on which his posthumous fame rests, the *Tentatem Juventum/Attempt to Introduce Studious Youth into the Elements of Pure Mathematics.* Completed in 1829, but not published until 1832, it was a brilliantly suggestive survey of mathematics, although (and to Wolfgang's chagrin) overlooked by his contemporaries. Of greater importance for his son's future was Wolfgang's obsession with the hoary problem of finding a proof for Euclid's fifth postulate – that there is only one line through a point outside another line that is parallel to it, or, in lay terms, that parallel lines do not meet.

In 1804 Wolfgang thought that he had found a proof of the axiom. He sent it to Gauss, who pointed out a flaw in the argument and returned it. Undaunted Wolfgang continued his quest, and János caught his enthusiasm. By about 1820, however, János had become convinced that a proof was impossible; he began instead to construct a geometry that did not depend upon Euclid's axiom. Over the next three years he developed a theory of absolute space in which several lines pass through the point *P* without intersecting the line *L*. He developed his formula relating the angle of parallelism of two lines with a term characterizing the line. In his new theory Euclidean space was simply a limiting case of the new space, and János introduced his formula to express what later became known as the space constant.

János described his new geometry in a paper of 1823 called 'The Absolute True Science of Space'. His father, unable to grasp its revolutionary meaning, rejected it, but sent it to Gauss. To the surprise of both father and son, Gauss replied that he had been thinking along the same lines for more than 25 years. Gauss had published nothing on the matter, however, and János' paper was printed as an appendix to his father's *Tentatem* in 1832.

János's paper was a thorough and consistent exposition of the foundations of non-Euclidean geometry. He was therefore cast down when its publication received little attention and when he discovered that Nikolai ◊Lobachevsky had published his account of a very similar geometry (also ignored) in 1829.

Deeply disappointed at their lack of recognition, the Bolyais retired into semi-seclusion. In 1837 the failure of their joint paper to win the Jablonov Society's Prize plunged them deeper into dejection. Thereafter Wolfgang did no serious mathematical research, and although János dabbled in problems connected with the relationship between pure trigonometry and spherical trigonometry, both died in relative obscurity. It was not until 30 years later that the work of Eugenio ◊Beltrami and Felix ◊Klein at last put into proper perspective János Bolyai's place as one of the pioneers of modern mathematics.

Bolzano, Bernardus Placidus Johann Nepomuk (1781–1848) Czech philosopher and mathematician who made a number of contributions to the development of several branches of mathematics.

Bolzano was born in Prague on 5 October 1781. He went to the University of Prague, where he studied philosophy, mathematics, and physics until 1800, when he entered the theology department. He was ordained in 1804. He did not abandon his mathematical interests, however, and in 1804 he was recommended for the chair of mathematics at the university. In 1805 he was appointed as the first professor to the new chair of philosophy at Prague. For the next 14 years he lectured mainly on ethical and social questions, although also on the links between mathematics and philosophy. He was much admired by the students not only for his intellectual abilities but also for his forthright expression of liberal and Czech nationalist views. He became dean of the philosophy faculty in 1818. But by then his opinions were bringing him into disfavour with the Austro-Hungarian authorities and in 1819, despite the backing of the Catholic hierarchy, he was suspended from his professorship and forbidden to publish. A five-year struggle ensued, ending only in 1824 when Bolzano, resolute in his refusal to sign an imperial order of 'recantation', resigned his chair. He retired to a small village in southern Bohemia. He returned to Prague in 1842 and died there on 18 December 1848.

Owing to the opposition of the imperial authorities, most of Bolzano's work remained in manuscript during his lifetime. It was not until the publication of the manuscripts in 1962 that the range and importance of his research was fully appreciated. Early in his career he worked on the theory of parallels, based on Euclid's fifth postulate. He found several faults in Euclid's method, but until the development of topology nearly a century later, these difficulties could not be resolved. Bolzano also formulated a proof of the binomial theorem and, in one of his few works published in his lifetime (1817), attempted to lay down a rigorous foundation of analysis. One of the most interesting parts of the book was his definition of continuous functions.

During the 1830s Bolzano concentrated on the study of real numbers. He also formulated a theory of real functions and introduced the non-differentiable 'Bolzano function'. He was also able to prove the existence and define the properties of infinite sets, work later of much use to Julius ◊Dedekind when he came to produce his definition of infinity in the 1880s.

Bolzano fell short of making any really fundamental breakthroughs in mathematics, but he was one of the most accomplished and wide-ranging of 19th-century mathematicians.

Bond, George Phillips (1825–1865) US astronomer whose best work was on the development of astronomical photography as an important research tool. His research was carried out exclusively at the observatory founded by his father, William Cranch ◊Bond.

Bond was born in Dorchester, Massachusetts, on 20 May 1825. He obtained a BA at Harvard and immediately began to work at the Harvard College Observatory. He served first as assistant astronomer to his father and then, upon the latter's death, as director of the observatory. In 1865, Bond became the first citizen of the USA to be awarded the Gold Medal of the Royal Astronomical Society of London, for his beautifully produced text on Donati's comet. He died in Cambridge, Massachusetts, on 2 February 1865.

The discovery of Hyperion (Saturn's eighth satellite) and the Crêpe Ring around Saturn were Bond's first major findings with his father. Since it was possible to see stars through the Crêpe Ring (a dim ring inside the two bright rings), Bond concluded that the rings were liquid, not solid, as most astronomers believed at the time.

During the late 1840s the Bonds worked on developing photographic techniques for astronomy. Since taking pictures even in full daylight often required long exposures in those days, photography at night was an arduous process indeed. Poor-quality daguerrotypes of the Moon had been taken in the early 1840s, but by 1850 the Bonds were able to take pictures of impressive quality. Improved techniques enabled Bond and Fred ◊Whipple to take a picture of Vega, the first star to be photographed, in 1850. In 1857 Bond also became the first person to photograph a double star, Mizar, with the aid of wet collodion plates. Bond suggested that a star's magnitude could be quantitatively determined by measuring the size of the image it made. A bright star would affect a greater area of silver grains.

Bond also made numerous studies of comets. He discovered 11 new comets and made calculations on the factors affecting their orbits. He is, however, best remembered for his work on photography and, indeed, is often credited as the pioneer of techniques of astronomical photography.

Bond, William Cranch (1789–1859) US astronomer who, with his son, George Phillips ◊Bond, established the Harvard College Observatory as a centre of astronomical research.

Bond was born into a poor family in Falmouth, Maine, on 9 September 1789. He was needed in the family business and he received little formal education. He worked in the shop as a watchmaker and displayed remarkable manual dexterity and mechanical ingenuity. The solar eclipse of 1806 was the stimulus that introduced him to the study of astronomy, which became an ever more absorbing hobby. Bond was one of the independent observers who discovered the comet of 1811. He was commissioned by Harvard College to investigate the equipment at observatories in England during a trip he made there in 1815.

In the absence of an observatory in the vicinity of his home, Bond had converted one of the rooms in his house. It became the best private observatory of his day and in 1839 Harvard invited him to move it into their premises (although he was not offered any stipend for doing so). Bond thus became the first director of the Harvard College Observatory, a post he held until his death. He was awarded the formal title of observer, and given an honorary MA in 1842.

Public interest in astronomy was aroused by the comet of 1843 and the observatory received sufficient funding to equip itself with a 38-cm/15-in refracting telescope. Bond continued to make observations and to design equipment until his death in Cambridge, Massachusetts, on 29 January 1859.

In addition to his early observations of comets and other celestial bodies, Bond established an international reputation for his work on chronometers. He worked not only on their design, but also on fixing the rate of the mechanisms of chronometers used, for example, in navigation.

From the observatory he also studied the Solar System, sunspots, and the nebulae in the constellations of Orion and Andromeda. It is difficult to distinguish the contributions made by Bond from those of his son during the later years of the career of Bond senior. They collaborated on the development of photographic techniques for the purposes of astronomy, succeeding in obtaining superior photographs of the Moon (exhibited in London in 1851 to enthusiastic audiences) and took the first photographs of stars.

The two Bonds discovered Hyperion (the eighth satellite of Saturn) in 1848 and the Crêpe Ring around Saturn in 1850. This is the faint ring inside the two bright rings of Saturn. Their observation that stars could be seen through the Crêpe Ring led to their conclusion that the rings of Saturn are not solid. The Crêpe Ring had probably been observed already by other astronomers, including H Kater in 1825 and Frederich von Struve in 1826. William Lassell, who missed being credited for the discovery of Hyperion by finding it a few days after the Bonds, also observed the Crêpe Ring shortly after it was described by them.

William Cranch Bond's chief contributions to astronomy were his careful observations and innovative designs, and his founding of an eminent research centre.

Bondi, Hermann (1919–) Austrian-born British scientist who was trained as a mathematician, but who went on to make important contributions to many disciplines both as a research scientist and as an enthusiastic administrator. He is best known in astronomy for his development, with Thomas ◊Gold and Fred ◊Hoyle, of the steady-state theory concerning the origin of the universe.

Bondi was born on 1 November 1919 in Vienna. At home he developed an early interest in mathematics. He taught himself the rudiments of calculus and theoretical physics and, after briefly meeting Arthur Eddington who was visiting Vienna, he decided to go to Cambridge University to study mathematics. He was recognized as a mathematician of considerable talent and was awarded an exhibition in his first year. He earned a BA in 1940, despite being caught up in the general order of May 1940 that required 'enemy aliens' resident in the UK to be interned for security reasons. While in internment Bondi met Thomas Gold; they were to become close friends and scientific associates.

Bondi returned to Cambridge in 1941 to become a research student. He began to do naval radar work for the British Admiralty in 1942 and through this work he met Fred Hoyle. Gold soon joined them and, inspired by Hoyle, who was already an established astrophysicist, the three discussed cosmology and related subjects in their spare time. This collaboration continued after the war.

Trinity College elected Bondi to a fellowship in 1943 on the basis of his first astrophysical research. He acquired British citizenship in 1947 and became an assistant lecturer 1945–48 and university lecturer 1943–54. He was visiting professor to Cornell University in 1951 and he went to Harvard in 1953. After a tour of US observatories, he returned to England to take up the chair of applied mathematics at King's College, London. He was elected Fellow of the Royal Society in 1959.

He has held advisory posts in the Ministry of Defence, in particular on the National Space Committee, the European Space Research Organization and the Natural Environment Research Council. From 1983 he has been master of Churchill College, Cambridge. Bondi is the author of a number of books on cosmology and allied subjects. His scientific contributions have been recognized by his election to prominent scientific organizations and his public service by honours such as a knighthood in 1973.

Bondi is perhaps best known for his proposal, together with Gold and Hoyle, of the steady-state theory. This is a cosmological model that explains the expansion of the universe not as a consequence of a singularity (as proposed by the Big Bang model), but as a feature of the universe as it has always been and always will be. The model requires that matter be continually created – albeit at a rate of only 1 gram per cubic decimetre per 10^{36} years – in order to keep the density of matter in the universe constant. The steady-state model was felt not only to resolve an apparent discrepancy between the age of the universe and of our Earth, but also to be a simpler theory than the Big Bang model. The steady-state theory was not in any way contradicted by facts then available.

The model created something of a sensation and stimulated much debate for, while its ideas were revolutionary, it was fully compatible with existing knowledge. The orthodox model of the day placed the origin of all elements in an early superhot stage of the universe. Hoyle stated the now universally accepted theory of the origin of the elements in observed types of stars. This was the greatest triumph of the steady-state theory, but evidence against the theory in general soon began to accumulate. In 1955 came evidence that the universe had once been denser than it is today. In 1965 came the identification of a universal 'background' radiation that was readily accounted for as a remainder of an early hot state of the universe. A further severe difficulty for the steady-state theory is that the universe seems to contain more helium than the theory predicts. Thus, today few scientists regard the steady-state theory as a serious competitor to the Big Bang theory of the origin of the universe.

Bondi's other contributions have been to the study of stellar structure, relativity, and gravitational waves. He demonstrated that gravitational waves are compatible with and are indeed a necessary consequence of the general theory of relativity. He was also able to describe the likely characteristics and physical properties of gravitational waves.

Since the 1960s, Bondi has been primarily concerned with more administrative duties. He organized the rebuilding of King's College and the establishment of the Anglo-Australian telescope. He was also an adviser on the Thames Barrage project. Bondi is a versatile and talented scientist, whose contributions are notable for their originality, insight, and scope.

Further Reading

Bondi, Hermann, *Science, Churchill, and Me: The Autobiography of Hermann Bondi, Master of Churchill College, Cambridge,* Pergamon Press, 1990.

Boole, George (1815–1864) English mathematician who, by being the first to employ symbolic language and notation for purely logical processes, founded the modern science of mathematical logic.

He was born in Lincoln on 2 November 1815. He received little formal education, although for a time he attended a national school in Lincoln and also a small commercial school. His interest in mathematics appears to have been kindled by his father, a cobbler with a keen amateur interest in mathematics and the making of optical instruments. Boole also taught himself Greek, Latin, French, German, and Italian. At the age of 16 he became a teacher at a school in Lincoln; he subsequently taught at Waddington; then, at the age of 20, he opened his own school. All the while his spare time was devoted to studying mathematics, especially Newton's *Principia* and Lagrange's *Mécanique analytique.* He was soon contributing papers to scientific journals and in 1844 he was awarded a Royal Society medal. In 1849, despite his lack of a university education, he was appointed professor of mathematics at the newly founded Queen's College in Cork, Ireland. He held the chair until his death on 8 December 1864.

Boole's first essay into the field of mathematical logic began in 1844 in a paper for the *Philosophical Transactions* of the Royal Society. In it Boole discussed ways in which algebra and calculus could be combined, and the discussion led him to the discovery that the algebra he had devised could be applied to logic. In a pamphlet of 1847 he announced, against all previous accepted divisions of human knowledge, that logic was more closely allied to mathematics than to philosophy. He argued not only that there was a close analogy between algebraic symbols and those that represented logical forms but also that symbols of quantity could be separated from symbols of operation. These were the leading ideas which received their fuller treatment in Boole's greatest work, *An Investigation of the Laws of Thought on which are Founded the Mathematical Theories of Logic and Probabilities* (1854).

Boole English mathematician George Boole, who devised the system of Boolean algebra, which is used today in designing computers. The son of a shopkeeper, Boole was largely self-taught. He was appointed to the chair of mathematics at Queen's College, Cork in 1849. *Mary Evans Picture Library*

It is not quite true to say that Boole's book reduced logic to a branch of mathematics; but it did mark the birth of the algebra of logic, later known as Boolean algebra. The basic process of Boole's system is continuous dichotomy. His algebra is essentially two-valued. By sub-dividing objects into separate classes, each with a given property, it enables different classes to be treated according to the presence or absence of the same property. Hence it involves just two numbers, 0 and 1. This simple framework has had far-reaching practical effects. Applying it to the concept of 'on' and 'off' eventually produced the modern system of telephone switching, and it was only a step beyond this to the application of the binary system of addition and subtraction in producing the modern computer.

Later mathematicians modified Boole's algebra. Friedrich ◊Frege, in particular, improved the scope of mathematical logic by introducing new symbols, whereas Boole had restricted himself to those symbols already in use. But Boole was the true founder of mathematical logic; it was on the foundations that he laid that Bertrand ◊Russell and Alfred ◊Whitehead attempted to build a rigidly logical structure of mathematics.

Further Reading

Barry, Patrick D, *George Boole: A Miscellany,* Cork University Press, 1969.

Boole, George; De Morgan, Augustus; and Smith, G C, *The Boole–De Morgan Correspondence 1842–1864*, Oxford Logic Guides series, Clarendon Press, 1982.
MacHale, Desmond, *George Boole: His Life and Work*, Profiles of Genius series, Boole Press, 1985, v 2.

Booth, Hubert Cecil (1871–1955) English mechanical and civil engineer best known for his invention of the vacuum cleaner in 1901.

Booth was born on 4 July 1871 in Gloucester, where he was educated at the college school and, later, the county school. He studied engineering at the City and Guilds Institute 1889–92.

In 1901, the same year in which he formed his own engineering consultancy, Booth conceived the principle of his vacuum cleaner after witnessing a somewhat self-defeating operation at St Pancras Station, London – the cleaning of a Midland Railway train carriage by means of compressed air, which simply blew a great cloud of dust around, allowing it to settle elsewhere. Booth realized that a machine to suck in the dust and trap it so that it could be disposed of afterwards would circumvent this unhygienic problem. Booth demonstrated his idea to his companions at St Pancras by placing his handkerchief over his mouth and sucking in his breath. A ring of dust particles appeared where his mouth had been. There was also, of course, a use for Booth's invention in the cleaning of houses, which was then done with handbrushes or with simple sweepers that pushed dust into a box, or with feather dusters that just moved dirt from one place to another. Booth's cleaner replaced these methods of cleaning carpets, upholstery, and surfaces with the air suction pump principle. In his machine, one end of the tube was connected to the pump, while the other, with nozzle attached, was pushed over the surface being cleaned. The cleaner incorporated an air filter to cleanse the air passing through and also served to collect the dust.

Because of the large size and high price of early vacuum cleaners and the fact that few houses had mains electricity, Booth initially offered cleaning services rather than machine sales. The large vacuum cleaner, powered by petrol or electric engine and mounted on a four-wheeled horse carriage was parked in the street outside a house while large cleaning tubes were passed in through the windows. The machine was such a novelty that society hostesses held special parties at which guests watched operatives cleaning carpets or furniture. Transparent tubes were provided so that the dust could be seen departing down them. Booth's machines received a popular boost when they were used to clean the blue pile carpets laid in Westminster Abbey for the coronation of King Edward VII in 1902.

Smaller, more compact indoors vacuum cleaners followed, but until the first electrically powered model appeared in 1905, two people were required to operate them – one to work the pump by bellows or a plunger, the other to handle the cleaning tube. Booth formed his British Vacuum Cleaner Company in 1903, running it in parallel with his equally successful engineering consultancy. His work in the latter area had begun with a post as a draughtsman with the marine-engine company Maudslay, Sons and Field. There, Booth worked on the design of engines for two Royal Navy battleships. At this time Booth's keen insight into technological problems and his original flair were noticed by W B Bassett, one of Maudslay's directors. Bassett was interested in great wheels, and in 1894 he chose Booth to work first on the wheel at Earl's Court, London and afterwards on similar structures in Blackpool and Vienna, and on the 61 m/200 ft diameter great wheel in Paris.

Booth's design principles were, in essence the same as those governing the design of modern long-span suspension bridges. Later in 1902, Booth directed the erection of the Connel Ferry Bridge over Loch Etive, Scotland.

From 1903 until his retirement in 1952, three years before his death in Croydon, Surrey, on 14 January 1955, Booth remained chairman of the British Vacuum Cleaner Company. He took special interest in the industrial potential of his famous invention and personally pioneered the development of cleaning installations in large industrial establishments. Both here and in the domestic context – even though he regarded the home vacuum cleaner as something of a toy – Booth was most gratified by the fact that his machines allowed higher standards of hygiene and counteracted dust-related diseases.

Borel, Emile Félix-Edouard-Justin (1871–1956) French mathematician whose lasting reputation derives from his rationalization of the theory of functions of real variables.

Borel was born in Saint-Affrique on 7 January 1871. At an early age he showed such a strong aptitude for mathematics that he was sent away from his native village to a lycée at Montauban. In 1890 he entered the Ecole Polytechnique in Paris, where he so distinguished himself that on his graduation in 1893 he was appointed to the faculty of mathematics at the University of Lille. In 1894 he received his DSc from the Ecole Normale Supérieure. For the next few years Borel proved himself to be a prolific writer of highly valuable papers and in 1909 the Sorbonne created a chair in function theory especially for him. A year later he took on the additional duty of becoming deputy director, in charge of science, at the Ecole Normale Supérieure.

Borel's professional career was interrupted by World War I, during which he took part in scientific and technical missions on the front. The war also marked a turning-point in his life. When it was over he took less interest in pure mathematics and more in applied science. He also became involved in politics. From 1924–36 he was a radical-socialist member of the national chamber of deputies, serving as minister of the navy in 1925. He was also in these years one of the moving spirits behind the establishment of the National Centre for Scientific Research (he received the institute's first gold medal in 1955). He was, too, one of the founding members of the Henri Poincaré Institute, serving as its first director from 1928 until his death. In 1936 Borel left active politics and four years later he retired from his chair at the Sorbonne. World War II drew him back into public life: in 1940 he was taken briefly into custody by the occupying German forces and on his release he joined the Resistance movement. In 1945 he was awarded the Resistance Medal. Thereafter Borel lived in retirement until his death, in Paris, on 3 February 1956.

Borel's first papers appeared in 1890 and it was in the 1890s that he did his most important work – on probability, the infinitesimal calculus, divergent series, and, most influential of all, the theory of measure. In 1896 he created a minor sensation by providing a proof of Picard's theorem (see Charles ◊Picard), an achievement that had eluded a host of mathematicians for nearly 20 years. In the 1920s he wrote on the subject of game theory, before John ◊Von Neumann (generally credited with being the founder of the subject) first wrote on it in 1928. But he will be remembered, above all, for his theory of integral functions and his analysis of measure theory and divergent series. It is this work that established him, alongside Henri ◊Lebesgue, as one of the founders of the theory of functions of real variables.

Born, Max (1882–1970) German-born British physicist who pioneered quantum mechanics, the mathematical explanation of the behaviour of an electron in an atom. His version of quantum mechanics had a short reign as the most popular explanation and was soon displaced by the equivalent and more convenient wave mechanics of Erwin ◊Schrödinger, which is in use today. The importance of Born's work was eventually recognized, however, with the award of the 1954 Nobel Prize for Physics, which he shared with German physicist Walther Bothe (1891–1957). (Bothe's share of the award was for his work on the detection of cosmic rays and was not connected with Born's achievements in quantum mechanics.) Born also made important advances in the understanding of the dynamics of crystal lattices.

Born was born in Breslau, Germany (now Wrocław, Poland) on 11 December 1882 into a wealthy Jewish family. His father was professor of anatomy at the University of Breslau, which Born later attended as well as the University of Göttingen, gaining his doctorate in physics and astronomy at Göttingen in 1907.

After periods of military service and various academic positions in Göttingen and Berlin, Born was appointed professor of physics at Frankfurt am Main in 1919. During this period, he became an expert on the physics of crystals. Following a meeting with Fritz ◊Haber, who developed the Haber process for the synthesis of ammonia, Born became aware that little was known about the calculation of chemical energies. Haber encouraged Born to apply his work on the lattice energies of crystals (the energy given out when gaseous ions are brought together to form a solid crystal lattice) to the formation of alkali metal chlorides. Born was able to determine the energies involved in lattice formation, from which the properties of crystals may be derived, and thus laid one of the foundations of solid-state physics.

In 1921, Born moved from Frankfurt to the more prestigious university at Göttingen, again as professor. Most of his research was then concerned with quantum mechanics, in particular with the electronic structure of atoms. He made Göttingen a leading centre for theoretical physics and together with his students and collaborators – notably Werner ◊Heisenberg – he devised a system called matrix mechanics that accounted mathematically for the position of the electron in the atom. He also devised a technique, called the Born approximation method, for computing the behaviour of subatomic particles that is of great use in high-energy physics.

In 1933, with the rise to power of Adolf Hitler, Born left Germany for Cambridge, England, and in 1936 he became professor of natural philosophy at Edinburgh University. He became a British citizen in 1939, and remained at Edinburgh until 1953, when he retired to Germany. The following year, Born received his Nobel prize which, somewhat ironically, was for his discovery published in 1926 that the wave function of an electron is linked to the probability that the electron is to be found at any point. Born died at Göttingen on 5 January 1970.

I am now convinced that theoretical physics is actual philosophy.

MAX BORN AUTOBIOGRAPHY

Born was inspired by Niels ◊Bohr to seek a mathematical explanation for Bohr's discovery that the quantum theory applies to the behaviour of electrons in atoms. In 1924 Born coined the term 'quantum mechanics' and the following year built on conceptual work by Werner Heisenberg to relate the position and momentum of the electron in a system called matrix mechanics. Wolfgang ◊Pauli immediately used the system to calculate the hydrogen spectrum and the results were correct. But about a year after this discovery, in 1926, Erwin Schrödinger expressed the same theory in terms of wave mechanics, which was not only more acceptable mathematically but enabled physicists to visualize the position of the electron as a wave motion around the nucleus, unlike Born's purely mathematical treatment. Born, however, was able to use wave mechanics to make a statistical interpretation of the quantum theory and expressed the probability that an electron will be found at a particular point as the square of the value of the amplitude of the wave at that point. This was the discovery for which he was belatedly awarded a Nobel prize.

Further Reading

Beller, Mara, 'Born's probabilistic interpretation', *Boston Studies in History and Philosophy of Science,* 1990, v 21, p 563.

Born, Max, *My Life: Recollections of a Nobel Laureate,* Taylor and Francis, 1978.

Bosch German chemist Karl Bosch, who developed the process for industrial synthesis of ammonia. Bosch shared the Nobel Prize for Chemistry in 1931 with Friedrich Bergius for his work on high-pressure synthesis techniques. *Mary Evans Picture Library*

Bosch, Karl (1874–1940) German chemist who developed the industrial synthesis of ammonia, leading to the cheap production of agricultural fertilizers and of explosives.

Bosch was born on 27 August 1874 in Cologne, the eldest son of an engineer. He showed an early aptitude for the sciences and, following a year gaining practical work experience as a metal-worker, he went to the University of Leipzig, where he studied chemistry under the noted organic chemist Johannes Wislicenus; he took his doctorate in 1898.

The year after graduating Bosch took a job with the major German chemicals company Badische Anilin und Sodafabrik (BASF) and by 1902 was working on methods of 'fixing' the nitrogen present in the Earth's atmosphere, a subject for which he was to become famous. At that time, the only major sources of nitrogen compounds essential for the production of fertilizers and explosives were in the natural deposits of nitrates in Chile, thousands of kilometres from industrial Europe. Industrial production of ammonia using the nitrogen in the air would end this dependency on foreign sources.

In 1908 Bosch learned of the work of his fellow countryman Fritz ◊Haber, who had also been considering the problem. Haber had studied ways of combining nitrogen with hydrogen to form ammonia under the influence of high pressure and metal catalysts. Bosch's employers seized on the work, with its promise of endless supplies of nitrogen compounds, and made him responsible for making Haber's process commercially viable.

He set up a team of chemists and engineers to study the processes Haber had succeeded in producing under laboratory conditions, and to scale them up. For example, the carbon-steel vessels used by Haber to combine the nitrogen and hydrogen were attacked by the hydrogen, causing failure under the high temperature and pressure conditions needed for the reaction to take place. Alloy steel replacements were brought in by Bosch and his team. To produce the required volumes of hydrogen, Bosch also introduced the water-gas shift reaction, where carbon monoxide is combined with steam to produce carbon dioxide and hydrogen. Different catalysts were also investigated.

By the time World War I broke out in 1914, Bosch had completed what was then the largest ever feat of chemical engineering, and the BASF ammonia plant at Oppau was producing 36,000 tonnes of the material in sulphate form using the Haber–Bosch process. Following the heavy demands for its product from the military, the plant was expanded, and another much larger factory was set up in Leuna. At the end of the war, Bosch acted as technical advisor to the German delegation at the armistice and peace conferences.

Following World War I, BASF continued to profit from the work of Bosch, remaining leaders in the technology during the 1920s. Despite further involvement in scientific work on the synthesis of methyl alcohol and of petrol from coal tar, Bosch himself became increasingly involved in adminstration. He was chair of the vast industrial conglomerate IG Farbeninindustrie AG after its formation from the merger of BASF with other major German industrial concerns in 1925. By 1935, he was chair of its supervisory board. Illness prevented his close involvement in the group in later years (which were darkened by the rise of the Nazis).

Bosch was also recognized for his work by the scientific community, which bestowed numerous honorary degrees on him including, in 1931, the Nobel Prize for Chemistry, jointly with his compatriot Friedrich ◊Bergius for their work on high-pressure synthesis reactions.

An essentially withdrawn character, Bosch shunned public appearances, and published little. He died in Heidelberg on 26 April 1940.

Boulton, Matthew (1728–1809) English manufacturer whose financial support and enthusiasm were of importance in the promotion of James ◊Watt's steam engine.

Boulton was born on 3 September in Birmingham, and it was near there in 1762 that he built his Soho factory. He produced small metal articles such as buckles, buttons, gilt and silver wares, Sheffield plate, and the like, having succeeded to his father's business of silver stamping three years earlier. The Soho factory was original in combining workshops of different trades with a warehouse and merchanting lines. Boulton wished to obtain not only the best workers but the finest artistry for his products, and to this end sent out agents to procure him the best examples of artwork not only in metal but also in pottery and other materials.

The growth of the factory led to an increased need for a motive power other than water, which was poorly supplied at Soho. This resulted in a meeting with James Watt, who had just developed a steam engine which Boulton was convinced would prove the answer to his wants. In 1769, when Watt's partner became bankrupt, Boulton took his place. Six years later the two men became partners in a steam-engine business, obtaining a 25-year extension of the patent. The inventiveness of Watt and the commercial enterprise of Boulton secured the steam engine's future, but not before Boulton had brought himself to the verge of bankruptcy through his support. Helped by the engineer William ◊Murdock, they established the steam engine by erecting pumps in machines to drain the Cornish tin mines. Boulton foresaw a great industrial demand for steam power and urged Watt to develop the double-action rotative engine patented in 1782 and the Watt engine of 1788 to drive the lapping machines in his factory. The testing period of the steam engine was a long one and Boulton was more than 60 years old before it began to make a profitable return.

In 1786 Boulton applied steam power to coining machines, obtaining a patent in 1790. So successful was the process that as well as his home market Boulton supplied coins to foreign governments and to the East India Company. In 1797 he was commissioned to reform the copper currency of the realm. One result of this highly successful venture was to make the counterfeiting of coins much more difficult. Boulton also supplied machines for the Royal Mint, near Tower Hill, London, and these continued in efficient operation until 1882.

In 1785 Boulton was made a Fellow of the Royal Society, but submitted no papers to it subsequently. A friendly and generous man, he was acquainted with all the leading scientists of his day. He died in Birmingham on 17 August 1809.

Further Reading

Delieb, Eric and Roberts, Michael, *The Great Silver Manufactory: Matthew Boulton and the Birmingham Silversmiths,* Studio Vista, 1971.

Keir, James, *Memoir of Matthew Boulton,* City of Birmingham School of Printing, College of Arts and Crafts, 1947.

Smiles, Samuel, *Lives of Boulton and Watt: Principally from the Original Soho Mss: Comprising also, A History of the Invention and Introduction of the Steam-Engine,* Routledge–Thoemmes Press, 1996.

Bourdon, Eugène (1808–1884) French instrumentmaker who developed the first compact and reliable high-pressure gauge.

Born in Paris on 8 April 1808, Bourdon was set to follow his father and become a merchant, but more practical inclinations led him to set up, at the age of 24, his own instrument and machine shop. By 1835 he had established himself at Faubourg du Temple in Paris, where he was to work for the next 37 years.

Bourdon was born into the age of steam, with the opening of George Stephenson's Liverpool to Manchester Railway taking place in 1830. Bourdon's work reflected this: in the same year that his new instrument works opened in 1832, he presented a model steam engine (complete with glass cylinders) to the Société d'Encouragement pour l'Industrie National. He was to make more than 200 small steam engines at his works, chiefly for demonstration purposes.

Following Richard ◊Trevithick's work with high-pressure steam engines around the turn of the

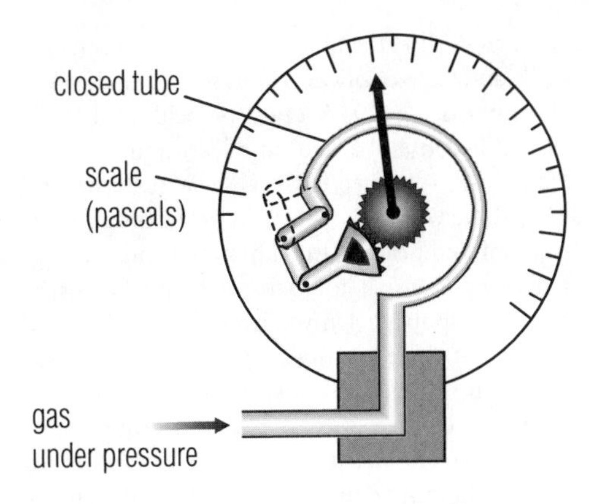

Bourdon The most common form of Bourdon gauge is the C-shaped tube. However, in high-pressure gauges spiral tubes are used; the spiral rotates as pressure increases and the tip screws forwards.

previous century (James ◊Watt's earlier engines worked at atmospheric pressure), the problem of measuring high pressures in such engines became apparent. Watt had used a fairly imprecise device for showing the variation of pressure in his relatively low-pressure systems, but the major difficulty lay at the higher pressures. As one atmosphere pressure is equivalent to 760 mm/30 in of mercury in a U-tube manometer, there was clearly a problem of providing a compact yet accurate pressure gauge.

Bourdon's solution to the problem was simple and ingenious, and is described in the patent for his 'metallic manometer' of 18 June 1849. An ordinary bicycle pump can be used to demonstrate the principle. By pumping air into the rubber connection dangling at the end of the pump, the connector becomes horizontal. This is because the pressure in the tube exceeds that outside (which is usually atmospheric), and the initially oval cross section of the tube becomes circular, causing the tube to straighten.

In the Bourdon gauge, a metal tube, sealed at one end, is subjected at the other to the pressure through a link to a quadrant rack, geared by a small pinion fixed to the back of a pointer. This moves across the calibrated scale. By using tubes of suitable thickness, a wide range of pressures can be measured using a variety of gauges, from fractions of an atmosphere to as much as 8,000 atmospheres, in both liquids and gases. It remains the most widely used gauge for measuring so wide a range of fluid pressures.

Bourdon was, apparently, led to his invention by noticing the change that occurred in a lead cooling coil under internal pressure. Just two years after its invention, the Bourdon gauge had become so renowned for its practicality and versatility that it won for its inventor the Legion of Honour.

By 1872, at the age of 64, Bourdon decided to pass over the management of his business to his sons. Nevertheless, he continued to experiment with instruments during the following 14 years, until his death on 29 September 1884 from a fall while testing a new form of anemometer.

Boveri, Theodor Heinrich (1862–1915) German biologist who performed valuable early work on chromosomes. Most of his life's work was devoted to investigating those processes by which an individual develops from the reproductive material of its parents.

Boveri was born in Bamberg on 12 October 1862, the second of four sons of a physician. He was educated at Bamberg 1868–75 and then for the next six years at the Realgymnasium in Nürnberg. He went to Munich University to study history and philosophy, but soon decided to change to natural sciences, and gained his doctorate in 1885. He then began a five-year fellowship at the Zoological Institute in Munich doing research into cytology under Richard Hertwig (1850–1937), and became a lecturer in zoology and comparative anatomy in 1887. He was appointed professor of zoology and comparative anatomy as well as director of the Zoological–Zootomical Institute in Würzburg in 1893. Four years later he married Marcella O'Grady, a US biologist, and they had one child – Margret – who became a writer and journalist. Apart from visits to the zoological station at Naples, Boveri spent the rest of his working life at Würzburg. He declined the position offered to him in 1913 of director of the Kaiser Wilhelm Institute for Biology in Berlin because of his failing health. He died in Würzburg on 15 October 1915, after successive physical breakdowns and recurrent depression.

Boveri's first piece of research was for his thesis in 1885, which had been on the structure of nerve fibres, but soon afterwards his interest was turned towards cell biology by Hertwig, when he worked as his assistant in Munich.

In the early 1880s the Belgian cytologist Edouard van Beneden (1846–1910) was investigating eggs of the nematode worm *Ascaris megalocephala*, which inhabits the intestine of horses. He discovered that the chromosomes of its offspring are derived in equal numbers from the nuclei of the ovum and sperm, that is equally from the two parents. It was known that the fusion of the nuclei of the ovum (egg) and the sperm is the essential feature of fertilization, and that this fusion eventually leads to the creation of all the nuclei of the body. In 1884 it was concluded that the cell nucleus holds the fundamental elements of heredity. It was also shown that the chromosomes split into daughter cells and that the chromosome number is constant for each species.

Inspired by the experiments of Beneden, Boveri did some of his own and described some aspects of the development of eggs and the formation of polar bodies (minute bodies which are produced during the division of an unfertilized ovum). He then demonstrated that the nuclei of *Ascaris* ova contain finger-shaped lobes that are chromosomes (each egg contains only two to four chromosomes), and also that chromosomes are separated into the daughter cells during the division that follows fertilization by a central piece in the sperm cell, which he termed the centrosome.

In 1889 Boveri experimented on sea urchins' eggs, fertilizing nucleated and non-nucleated fragments, and found that both types could develop normally. He also discovered that those occasional non-fertilized fragments containing only a nucleus were also able to develop normally. He showed that at fertilization the ovum and the sperm incorporate the same number of chromosomes each in the creation of the new individual. He was eventually able to demonstrate that cytoplasm plays an important part in development and went on to show that it is not a specific number, but a specific assortment of chromosomes that is responsible for normal development, indicating that individual chromosomes possess different qualities.

Boveri's powers of observation as a microscopist were remarkable, but so were his theories. In 1914, he theorized that tumours may become malignant as the result of abnormal chromosome numbers, and was the first to view the tumour as a cell problem. He also tried to explain, on the basis of an irregular chromosome distribution, a condition in bees in which male and female characteristics are mosaically distributed in different parts of the body. He discovered segmental excretory organs in the lancelet *Amphioxus* (an organism believed to be close to the type from which the vertebrates evolved).

Boveri enriched biological science with some fundamental discoveries and fruitful new conceptions. His theory of chromosomal individuality provided the working basis of nearly all cytological interpretations of genetic phenomena and still holds true today.

Bowden, Frank Philip (1903–1968) Australian physicist and chemist who worked mainly in the UK. He began his research career in electrochemistry, but early on became interested in surface studies and went on to make major contributions to the study of friction, lubricants, and surface erosion as well as other related subjects.

Bowden was born in Hobart, Tasmania, on 2 May 1903. He received his BSc at the University of Tasmania in 1924. He continued his studies there and then, in 1927, went to Cambridge University, where he began research in the department of colloid science. In 1935 Bowden became director of studies in natural sciences at Gonville and Caius College. He was then made a lecturer at Cambridge in 1937, a post he held until 1946.

In 1939 Bowden embarked on a lecture tour of the USA. He decided to return to the UK via Australia, and he was still there when World War II broke out in 1945. He was appointed head of the lubricants and bearings section for the Council for Scientific and Industrial Research at Melbourne University (which conducted research of great value to the war effort). When the war ended Bowden resigned his post in order to return to Cambridge to set up a similar research group there, and from 1946–65 he was director of the laboratory for physics and chemistry of solids. A chair in surface physics was created for him in 1966. Bowden died in Cambridge on 3 September 1968.

Bowden's work in Tasmania and his early work at Cambridge was in the field of electrochemistry; specifically he was interested in the establishment of the hydrogen overpotential and in the effects of impurities on this process. He also worked on electrode kinetics. Although his electrochemical investigations continued until 1939, he began to publish papers on friction as early as 1931.

His early work in this topic demonstrated that the area of contact between two solid surfaces depends not only on details of their surface features but also on the magnitude of the load applied to them and on the hardness of the two surfaces. Since even a clean smooth metallic surface is irregular on an atomic scale, the area of contact where strong intermetallic forces might form is small. Therefore, sliding produces friction over only a small fraction of the total area of the surface, but can produce exceedingly high temperatures and even induce melting in hot spots. Another implication of this work was that a lubricant might be decomposed by heat at exactly the place where it is most needed.

Bowden's interest in skiing led him to realize that the thin layer of water between a ski or an ice-skate and the snow or ice is produced not by pressure due to the weight on them but to friction-induced heat caused by irregularities in the sliding surfaces. A variety of industries saw applications for Bowden's research in more serious areas, and his war research covered a broad range of areas of military significance. Machine and tool lubricants, flame-throwing fuels, the accurate measurement of shell velocities for gun calibration, and the casting of aircraft bearings all came within their scope. Several ways of creating hot spots in a variety of materials were investigated, which had applications in the prevention of explosions due to frictional heating in munitions factories.

These research topics continued to be investigated by the research group that Bowden built up in

Cambridge after the war. They took advantage of new techniques such as scanning electron microscopy, electron diffraction, and high-speed photography in their investigations. The development of Teflon-coated metal appliances and devices was greatly accelerated by results obtained by Bowden and his team.

Bowen, Ira Sprague (1898–1973) US astrophysicist who is best known for his study of the spectra of planetary nebulae. He showed that strong green lines in such spectra are due to ionized oxygen and nitrogen under extreme conditions not found on Earth.

Bowen was born in New York State in 1898 and graduated from Oberlin College in 1919. He was an assistant in the physics department at the University of Chicago 1919–21, when he joined the California Institute of Technology. He gained his doctorate in 1926 and subsequently held the posts of instructor 1921–26, assistant professor 1926–28, associate professor 1928–30, and professor 1931–45. From 1946–64 he was director of the Mount Wilson and Palomar observatories. Bowen was a member of a number of learned societies and was awarded several medals. He was elected to the National Academy of Sciences in 1936.

In the 1860s William ◊Huggins had noticed strong green lines in the spectra of planetary nebulae. These were attributed either to complex atomic spectra or to an element previously unknown and given the name 'nebulium'.

With a greater understanding of the way in which spectral lines occur, astronomers began to doubt if nebulium really existed. They suspected that the spectral lines might be produced by a gas of extremely low density.

A spectral line is produced when an electron in an atom transfers itself from one energy level to another. Spectral analysis can determine the energy levels between which the electrons are moving, since strong lines are produced where it takes place easily ('permitted' transitions) and weak lines where it takes place with difficulty ('forbidden' transitions).

Bowen suggested that the strong green lines in the spectra of planetary nebulae might be caused not by permitted transitions in the hypothetical nebulium, but by forbidden transitions in known elements under conditions not produced in the laboratory. He calculated that the wavelengths of three of the spectral lines of nebulium were the same as those that would be produced from transitions within the lowest energy levels of doubly ionized oxygen, O(III). He compared his calculated wavelengths of forbidden transitions with those observed in the spectra of the nebulae and found that the strongest lines were produced by forbidden transitions of singly and doubly ionized oxygen, O(II) and O(III), and singly ionized nitrogen, N(II).

In 1938 Bowen constructed an ingenious piece of apparatus known as the image slicer for use with the slit spectrograph. His unmasking of nebulium led to the identification of other puzzling spectral lines, particularly those associated with the corona of the Sun, previously attributed to another hypothetical element, 'coronium'. Research into the chemical composition and physical properties of the Sun and other celestial bodies was stimulated by Bowen's work.

Bowen, Norman Levi (1887–1956) Canadian geologist whose work helped found modern petrology.

Born in Kingston, Ontario, Bowen was educated at the local Queen's University. Developing a love of laboratory geology, for his graduate studies he moved to the recently founded Geophysical Laboratory in Washington, DC, publishing from 1912 a steady stream of findings on the experimental melting and crystallization behaviour of silicates and similar mineral substances. In the course of more than 40 years of high-quality research centred on crystallization experiments, Bowen made distinguished contributions to modern understanding of the chemical and petrographical properties of igneous rocks.

Working alongside O F Tuttle and J F Schairer, Bowen in particular demonstrated the physicochemical principles governing the formation of magmas by partial melting and the fractional crystallization of magmas. Thanks to such research, igneous petrology ceased to be the essentially descriptive science created in the late 19th century by Rosenbusch and Zirkel and assumed a new experimental vigour and a sounder basis in physics and chemistry. Bowen particularly highlighted the importance of study of the evolution of magma, setting out his views in *The Evolution of Igneous Rocks* (1928), and being known as the head of the 'magmatist school' of Canadian geology.

Boyle, Robert (1627–1691) Irish natural philosopher and one of the founders of modern chemistry. He is best remembered for the law named after him, which states that, at a constant temperature, the volume of a given mass of gas is inversely proportional to the pressure upon it. He was instrumental in the founding of the Royal Society and a pioneer in the use of experiment and the scientific method.

Boyle was born on 25 January 1627 in Lismore Castle, County Waterford, the fourteenth child and seventh son of the Earl of Cork. He learned to speak French and Latin as a child and was sent to Eton College at the early age of eight. In 1641 he visited Italy, returning to England in 1644. He joined a group known as the Invisible College, whose aim was to cultivate the 'new philosophy' and which met at Gresham

Boyle Irish chemist and natural philosopher Robert Boyle, who formulated Boyle's Law, which states that the pressure and volume of a gas are inversely proportional. In 1680 he declined the offer of President of the Royal Society as he was unwilling to take the necessary oaths. *Mary Evans Picture Library*

College, London, and in Oxford, where Boyle went to live in 1654. The Invisible College became, under a charter granted by Charles II in 1663, the Royal Society of London for Improving Natural Knowledge, and Boyle was a member of its first council. (He was elected president of the Royal Society in 1680, but declined the office.) He moved to London in 1668 where he lived with his sister for the rest of his life. He died there on 30 December 1691.

Boyle's most active research was carried out while he lived in Oxford. By careful experiments he established the law that now bears his name. He determined the density of air and pointed out that bodies alter in weight according to the varying buoyancy of the atmosphere. He compared the lower strata of the air to a number of sponges or small springs that are compressed by the weight of the layers of air above them. In 1660 these findings were published in a book *The Spring of Air,* which also gave us the word 'elastic' in its present meaning(Boyle's law is not stated clearly until the revised edition was published in 1662).

A year later Boyle published *The Sceptical Chymist,* in which he criticized previous researchers for thinking that salt, sulphur, and mercury were the 'true principles of things'. He advanced towards the view that matter is ultimately composed of 'corpuscles' of various sorts and sizes, capable of arranging themselves into groups, and that each group constitutes a chemical substance. He successfully distinguished between mixtures and compounds and showed that a compound can have very different qualities from those of its constituents.

Also in about 1660 Boyle studied the chemistry of combustion, with the assistance of his pupil Robert ♢Hooke. They proved, using an air pump, that neither charcoal nor sulphur burns when strongly heated in a vessel exhausted of air, although each inflames as soon as air is re-admitted. Boyle then found that a mixture of either substance with saltpetre (potassium nitrate) catches fire even when heated in a vacuum and concluded that combustion must depend on something common to both air and saltpetre. Further experiments involved burning a range of combustible substances in a bell jar of air enclosed over water. But it was left to Joseph ♢Priestley in 1774 to discover the component of air that vigorously supports combustion, which three years later Antoine ♢Lavoisier named 'oxygen'.

The term 'analysis' was coined by Boyle and many of the reactions still used in qualitative work were known to him. He also introduced certain plant extracts, notably litmus, for the indication of acids and bases. In 1667 he was the first to study the phenomenon of bioluminescence, when he showed that fungi and bacteria require air (oxygen) for luminescence, becoming dark in a vacuum and luminescing again when air is re-admitted. In this he drew a comparison between a glowing coal and phosphorescent wood, although oxygen was still not known and combustion not properly understood. Boyle also seems to have been the first to construct a small portable box-type camera obscura in about 1665. It could be extended or shortened like a telescope to focus an image on a piece of paper stretched across the back of the box opposite the lens.

In 1665 Boyle published the first account in England of the use of a hydrometer for measuring the density of liquids. The instrument he described is essentially the same as those in use today. He can also be credited with the invention of the first match. In 1680 he found that by coating coarse paper with phosphorus, fire was produced when a sulphur-tipped splint was drawn through a fold in the treated paper. Boyle experimented in physiology, although 'the tenderness of his nature' prevented him from performing actual dissections. He also carried out experiments in the hope of changing one metal into another, and was instrumental in obtaining in 1689 the repeal of the statute of Henry IV against multiplying gold and silver.

Besides being a busy natural philosopher, Boyle was interested in theology and in 1665 would have received the provostship of Eton had he taken orders. He learned Hebrew, Greek, and Syriac in order to further his studies of the scriptures, and spent large sums on biblical translations.

Boyle accomplished much important work in physics, with Boyle's law, the role of air in propagating sound, the expansive force of freezing water, the refractive powers of crystals, the density of liquids, electricity, colour, hydrostatics and so on. But his greatest fondness was researching in chemistry, and he was the main agent in changing the outlook from an alchemical to a chemical one. He was the first to work towards removing the mystique and making chemistry into a pure science. He questioned the basis of the chemical theory of his day and taught that the proper object of chemistry was to determine the compositions of substances. His great merit as a scientific investigator was that he carried out the principles of Francis ⇨Bacon, although he did not consider himself to be a follower of him or of any other teacher. After his death, his natural history collections were passed as a bequest to the Royal Society.

Further Reading

Boyle, Robert and Birch, Thomas, *Robert Boyle: The Works*, Olms, 1965–66, 6 vols.

Hall, Marie Boas, *Robert Boyle and Seventeenth-Century Chemistry*, Kraus Reprint – Cambridge University Press, 1968.

Hunter, Michael Cyril William (ed), *Robert Boyle Reconsidered*, Cambridge University Press, 1994.

Jacob, James Randall, *Robert Boyle and the English Revolution: A Study in Social and Intellectual Change*, Studies in the History of Science series, B Franklin, 1977, v 3.

More, Louis Trenchard, *The Life and Works of the Honourable Robert Boyle*, 1944.

Sargent, Rose-Mary, *The Diffident Naturalist: Robert Boyle and the Philosophy of Experiment*, Science and Its Conceptual Foundations series, University of Chicago Press, 1995.

Boys, Charles Vernon (1855–1944) English inventor and physicist who is probably best known for designing a very sensitive apparatus (based on Henry ⇨Cavendish's earlier experiment) to determine Newton's gravitational constant and the mean density of the Earth. He received many honours for his work, including a knighthood in 1935.

Boys, the son of a clergyman, was born in Wing, Rutland, on 15 March 1855. He was educated at Marlborough College, then in 1873 went to the Royal School of Mines, London, where he was taught physics by R Guthrie (1833–1886) and chemistry by Edward Frankland (1825–1899). Although he received little formal instruction in mathematics, Boys taught himself calculus and designed an integrating machine that mechanically drew graphs of antiderivatives. He graduated in mining and metallurgy in 1876 and then, after a brief period working in a colliery, became an assistant to Guthrie, his former teacher, at the Royal College of Science in London. He was appointed assistant professor of physics at the Royal College in 1889, but, opportunities for further promotion being rare, he resigned in 1897 to work for the Metropolitan Gas Board as a referee. In 1920 he became one of the three gas referees that served all of the UK. He held this post until his formal retirement in 1939, although he continued to serve as an adviser to the Gas Board until 1943. In addition to his official appointments, from 1893 Boys pursued a lucrative business but he had sufficient spare time to develop his scientific interests and to write several popular science books on subjects as diverse as soap bubbles, weeds, and natural logarithms. He was also president of the Royal Society 1916–17. He died in Andover, Hampshire, on 30 March 1944.

Boys is known mainly as an inventor and designer of scientific instruments, in which area his most important contribution was probably his improved, extremely sensitive torsion balance, which he used to elaborate on Henry Cavendish's experiment to determine the gravitational constant and the mean density of the Earth. The novel feature of Boys's apparatus was his use of an extremely fine quartz fibre – ingeniously made by firing an arrow to the end of which was attached molten quartz – in the torsion balance. The properties of these fibres are such that Boy's balance was more sensitive, smaller, and quicker to react than was Cavendish's original apparatus (which used copper wire). Testing his improved design in the basement of the Clarendon Laboratories in Oxford in 1895, Boys determined Newton's gravitational constant and calculated the mean density of the Earth as 5.527 times the density of water.

Boys's other work included the invention in 1890 of a 'radiomicrometer' to detect infrared radiation. This extremely sensitive instrument, consisting of a thermocouple and a galvanometer, was intended to measure the heat radiated by the planets in the Solar System. Using this device he calculated that Jupiter's surface temperature did not exceed 100 °C/212 °F; it is now thought to be below –130 °C/–202 °F. He also developed a special camera to record fast-moving objects such as bullets and lightning flashes; designed a scientific toy using soap bubbles (called the Rainbow Cup, this device was later manufactured); and devised a calorimeter for determining the calorific value of coal gas (this subsequently became the standard instrument used to measure the calorific value of fuel gas in the UK). In addition, he performed a series of experiments on soap bubbles, which increased knowledge of surface tension and of the properties of thin films.

Bradley, James (1693–1762) English astronomer of great perception and practical skill. He was the third

Astronomer Royal and the discoverer of nutation and the aberration of light, both essential steps towards modern research into positional astronomy.

Bradley was born in Sherborne, Dorset, in March 1693. He entered Balliol College, Oxford, in 1711 and studied theology. He gained a BA in 1714, but he had by this time developed a fascination for astronomy through contact with his uncle, J Pound, who was an amateur astronomer and a friend of Edmond Halley. Bradley pursued his interest in astronomy after graduating and was made a Fellow of the Royal Society in 1718. In 1719 he became a vicar in Bridstow, and soon after was appointed chaplain to the Bishop of Hertford. He resigned his position in 1721 to become Savilian Professor of Astronomy at Oxford.

From then on Bradley devoted his whole career to astronomy. He lectured at the university until shortly before his death and pursued an active research programme. His most brilliant work was done during the 1720s, but he published material of exceptionally high standard throughout his life. In 1742, upon Halley's death, Bradley was appointed Astronomer Royal. In that position, he sought to modernize and re-equip the observatory at Greenwich, and he embarked upon an extensive programme of stellar observation. He was awarded the Copley Medal of the Royal Society in 1748 and served on the society's council 1752–62. He was a member of scientific academies in several European countries.

Bradley was unusually reluctant to publish his results until he had confirmed his ideas over periods of observation that sometimes exceeded 20 years. His catalogue of more than 60,000 observations made during the last years of his career was eventually published in two volumes in 1798 and 1805. He died in Chalford, Gloucestershire, on 13 July 1762.

Bradley's earliest astronomical observations were concerned with the determination of stellar parallax. Such measurement was the goal of many astronomers of his day because it would confirm ◊Copernicus' hypothesis that the Earth moved around the Sun. Copernicus himself (echoing Aristarchus 1,800 years before him) had stated that this parallax could not be detected because even the distance from one end of the Earth's orbit to the other was negligible compared with the enormous distance of the stars themselves. Nevertheless, Bradley and his contemporaries sought to observe parallactic displacement of the nearer stars compared to those at greater distances.

Bradley worked from 1725–26 with S Molyneux at the latter's private observatory in Kew. They chose to observe Gamma Draconis, and found that within a few days there did seem to be a displacement of the star. However, the displacement was not only too large but was in a different direction from that which would have been expected from parallactic displacement. Bradley studied this displacement of Gamma Draconis and other stars for more than a year and observed that this was the general effect. It took him some time to realize that the displacement was simply a consequence of observing a stationary object from a moving one, namely the Earth. The telescope needed to be tilted slightly in order to compensate for the movement of the Earth on its orbit around the Sun.

Bradley called this effect the 'aberration' of light, and he measured its angle to be between 20 and 20.5 seconds (the modern value being 20.47 seconds). From the size of this angle Bradley was able to obtain an independent determination of the velocity of light (308,300 km/191,578 mi per second compared with the modern value of 299,792 km/186,281 mi per second), confirming Ole ◊Römer's work of 1769. A conclusion of more immediate significance, however, was that the Copernican concept of a moving Earth had been confirmed. Bradley had failed to find the proof through measuring parallactic displacement, but he had proved it by means of aberration.

This discovery allowed Bradley to produce more accurate tables of stellar positions, but he found that even when he considered the effect of aberration his observations on the distances of stars were still variable. He studied the distribution of these variations and deduced that they were caused by the oscillation of the Earth's axis, which in turn was caused by the gravitational interaction between the Moon and the Earth's equatorial bulge, so that the orbit of the Moon was sometimes above the ecliptic and sometimes below it. Bradley named this oscillation 'nutation', and he studied it during the entire period of the revolution of the nodes of the lunar orbit (18.6 years) 1727–48. At the end of this period the positions of the stars were the same as when he started.

Perhaps as a result of his knowledge of Römer's work on the determination of the speed of light using the satellites of Jupiter, Bradley then turned his attention to a study of this planet. He measured its diameter and studied eclipses of its satellites.

The fruits of Bradley's observations were more accurate than those of his predecessors because of his discovery of the effects of aberration and nutation. Bradley was a skilful astronomer with unusual talents in both the practical and theoretical aspects of the subject.

Bragg, William Henry (1862–1942) and (William) Lawrence (1890–1971) British physicists, father and son, who pioneered and perfected the technique of X-ray diffraction in the study of the structure of crystals. For this work they were jointly awarded the 1915 Nobel Prize for Physics.

William Bragg was born in Westward, Cumberland, on 2 July 1862. In 1881 he went to Cambridge University, obtaining a first-class degree in mathematics in 1885. He was immediately appointed professor of mathematics and physics at the University of Adelaide, South Australia. It is said that he wondered why he was offered this appointment since he knew little physics, although he put this to rights by reading important texts on the long sea journey to Australia. For many years he did no research but concentrated on lecturing, although he did apprentice himself to a firm of instrumentmakers and subsequently made all the apparatus he required for practical laboratory teaching. Then in 1904 Bragg became president of the physics section of the Australian Association for the Advancement of Science and his opening address – on radioactivity – acted as the stimulus to begin original research. He found that radium produces alpha particles with a variety of energy ranges, confirming Ernest ◊Rutherford's theory that the various disintegration products of the radioactive series should produce radioactivity of differing energy. Rutherford proposed Bragg for a fellowship of the Royal Society and in 1908 Bragg was appointed professor of physics at Leeds University, returning to the UK in 1909.

Bragg began to work on X-rays and, together with his son Lawrence, became convinced that X-rays behave as an electromagnetic wave motion. Using the skills at instrument making he had gained in Australia, Bragg constructed the first X-ray spectrometer in 1913. Both men used it to determine the structures of various crystals on the basis that X-rays passing through the crystals are diffracted by the regular array of atoms within the crystal.

During World War I, the Braggs' work on X-rays virtually ceased and William Bragg devoted himself mainly to submarine detection. Both father and son were, however, awarded the 1915 Nobel Prize for Physics for their discovery, and in the same year William Bragg became professor of physics at University College, London. After the war, he turned to the X-ray analysis of organic crystals. He was knighted in 1920, and became director of the Royal Institution in 1923 (instituting the famous series of Christmas lectures there) and then president of the Royal Society in 1935. He died in London on 10 March 1942.

Physicists use the wave theory on Mondays, Wednesdays and Fridays, and the particle theory on Tuesdays, Thursdays and Saturdays.

WILLIAM HENRY BRAGG ATTRIBUTED REMARK

Lawrence Bragg was born in Adelaide on 31 March 1890. He studied mathematics at Adelaide University and continued in this subject at Trinity College, Cambridge. Then in 1910 he switched to physics at his father's suggestion, becoming interested in the X-ray work of Max von ◊Laue, who claimed to have observed X-ray diffraction in crystals, and repeated many of von Laue's experiments. Lawrence Bragg was able to determine an equation, now known as Bragg's law, that enabled both him and his father to deduce the structure of crystals such as diamond, using the X-ray spectrometer built by his father. In recognition of this achievement, Lawrence Bragg shared the 1915 Nobel Prize for Physics with his father, becoming the youngest person (at the age of 25) ever to receive the prize.

Lawrence Bragg then went on to determine the structures of such inorganic substances as silicates. In 1919, he became professor of physics at the University of Manchester and from 1938–54 was professor of physics at Cambridge. He was knighted in 1941. Like his father before him, Bragg became director of the Royal Institution in 1954 and he also devoted much energy to the popularization of science. He retired in 1966 and died in Ipswich on 1 July 1971.

Until the Braggs applied X-rays to the study of crystal structure, crystallography had not been concerned with the internal arrangement of atoms but only with the shape and number of crystal surfaces. The Braggs' work immediately gave a method of determining the positions of atoms in the lattices making up the crystals because they were able to relate the distance between the crystal planes or layers of atoms d to the angle of incidence of the X-rays θ and their wavelength λ, obtaining the simple equation $n\lambda = 2d \sin\theta$, where n is an integer. From the diffraction patterns obtained by passing X-rays through crystals, the Braggs were able to determine the dimensions of the crystal planes and thus the structure of a crystal. Furthermore, it gave a method for the accurate determination of X-ray wavelengths.

The Braggs' pioneering discovery led to an understanding of the ways in which atoms combine with each other and also revolutionized mineralogy and later molecular biology, in which X-ray diffraction was crucial to the elucidation of the structure of DNA.

Brahe, Tycho (1546–1601) Danish astronomer, sometimes known by his first name only, who is most noted for his remarkably accurate measurements of the positions of stars and the movements of the planets.

Tycho was born of aristocratic parents in Knudstrup in 1546. He was brought up by his paternal uncle, from whom he learnt Latin, and in early life he studied law and philosophy. A political career was planned for him,

Brahe Engraving of Danish astronomer Tycho Brahe, the greatest astronomical observer in the era before the optical telescope. Brahe made accurate measurements of the positions of the celestial bodies. He carried out much of his work from an observatory he built on Hveen, an island which was presented to him by the king of Denmark. *Mary Evans Picture Library*

but in 1560 Tycho observed a solar eclipse and was so fascinated by what he saw that he spent the rest of his life studying mathematics and astronomy.

Being of a noble family, Tycho did not need a university degree to establish himself in a profession, but he attended the University of Copenhagen and studied ethics, music, natural sciences, philosophy, and mathematics. From the beginning of his astronomical career he made a series of significant observations. Having seen the eclipse, he obtained a copy of Stadius' *Ephemerides,* which was based on the Copernican system. Observing a close approach of Jupiter and Saturn in 1563, Tycho noticed that it occurred a month earlier than predicted. He set about the preparation of his own tables. In 1564 he began observing with a radius, or cross-staff consisting of an arm along which could slide the centre of a crosspiece of half its length. Both arms were graduated and there was a fixed sight at the end of the larger arm that was held near the eye. To measure the angular distance between two objects, Tycho set the shorter arm at any gradation of the longer arm and moved a sight along the shorter arm until he saw the two objects through it and a sight at the centre of the transversal arm. The required angle was then obtained from the gradations and a table of tangents.

When his uncle died in 1565, Tycho travelled and studied in Germany – at Wittenburg and Rostock, where he graduated from the university in 1566. While he was at Rostock, it is said that he lost the greater part of his nose in a duel with another nobleman over a point of mathematics, and thereafter wore a false nose made of silver. After making a number of observations in Rostock, he moved to Basle before entering the intellectual life of Augsburg in 1569. Having returned home because of his father's ill-health, Tycho noticed one night in November 1572 a star in the constellation of Cassiopeia that was shining more brightly than all the others and which had not been there before. With a special sextant of his own making, Tycho observed the star until March 1574, when it ceased to be visible. His records of its variations in colour and magnitude identify it as a supernova.

In 1576 King Frederick II offered Tycho the island of Hven for the construction of an observatory. This was the first of its kind in history. Tycho's reputation grew and scholars from throughout Europe visited him.

Having observed a great comet in 1577, Tycho refuted Aristotle's theory of comets. He concluded that certain celestial bodies were supralunar, having no parallax and remaining stationary like fixed stars. Many other scientists had abandoned the Aristotelian theory in favour of the belief that something new could be created in the heavens and not necessarily out of the substances of the Earth. Tycho claimed that Aristotle's 'proof' had been based on meditation, not mathematical observation or demonstration. Tycho's main objective became to determine the comet's distance from the Earth. He was also concerned with its physical appearance – colour, magnitude, and the direction of the tail. He came to the conclusion that the comet's orbit must be elongated, a controversial suggestion indeed since it meant that the comet must have passed through the various planetary spheres, and it could not do that unless the planetary spheres did not exist. This possibility went against Tycho's most cherished beliefs. He could not abandon the ideas of his Greek predecessors, although he was the last great astronomer to reject the heliocentric theory of Copernicus. He tried to compromise, suggesting that, with the exception of the Earth, all the planets revolved around the Sun.

He prepared tables of the motion of the Sun and determined the length of a year to within less than a second, making calendar reform inevitable. In 1582 ten days were dropped, the Julian year being longer than the true year. To prevent further accumulations, the Gregorian calendar was adopted thereafter.

Tycho lost his patronage on the death of the king and he left for Germany in 1597. He settled in Prague at the invitation of the emperor and found a new assistant, Johannes ◊Kepler. Kepler loyally accepted and propounded Tycho's tables and data and continued his work with what were to be results of great importance. Many of Tycho's great contributions to science live on and he is remembered in particular for the improvements he made to almost every important astronomical measurement.

Further Reading

Chapman, Allan, *Astronomical Instruments and Their Users: Tycho Brahe to William Lassell,* Ashgate Publishing Company, 1996.

Gade, John Allyne, *The Life and Times of Tycho Brahe,* Princeton University Press for the American-Scandinavian Foundation, 1947.

Rosen, Edward, *Three Imperial Mathematicians: Kepler Trapped Between Tycho Brahe and Ursus,* New Horizon series, Abaris, 1986.

Taton, Rene and Wilson, Curtis, *Planetary Astronomy from the Renaissance to the Rise of Astrophysics. Part A: Tycho Brahe to Newton,* Cambridge University Press, 1989.

Thoren, Victor E, *The Lord of Uraniborg: A Biography of Tycho Brahe,* Cambridge University Press, 1990.

Bramah, Joseph (1748–1814) English engineer. He took out patents for 18 inventions, of which the hydraulic press was probably to be the most significant. Nevertheless, his training of a whole generation of engineers in the craft of precision engineering at the dawn of the Industrial Revolution was probably an even greater legacy to his country than any of his individual inventions.

Bramah was born Joe Brammer on 13 April 1748, in Stainborough, near Barnsley, Yorkshire. He began his employment by working on his father's farm but at the age of 16 he was made lame in an accident. He became apprenticed to a carpenter and cabinet maker, and on completing his apprenticeship made his way to London to set up his own business. A stream of inventions followed, one of the most useful being the flushing water closet he produced in 1778. When the patent was taken out he changed the spelling of his name to the more fashionable sounding Bramah. In 1784 he patented his most celebrated invention, the Bramah lock. Designed to foil thieves, a specimen of the lock was exhibited in a shop window in 1784 with a 200-guinea reward for anyone who could succeed in picking it. Bramah kept the money for the rest of his lifetime, 67 years passing before the lock was finally opened by a mechanic after 51 hours' work. Such an effective lock could be produced only by high precision engineering and machine tools of the finest quality.

To assist him with his work Bramah took into his employ a young blacksmith named Henry ◊Maudslay (later to become the friend and partner of Marc Isambard Brunel), whose mechanical skill was at least equal to Bramah's. In 1795 Bramah produced the hydraulic press. This device makes use of Pascal's law: pressure exerted upon the smaller of two pistons results in a greater force on the larger one, both cylinders being connected and filled with liquid. For an important part of his press – the seal that ensured water-tightness between the plunger and the cylinder in which it worked – he was particularly indebted to Maudslay who, then only 19, produced the leather U-seal, which expands under the fluid pressure.

The possibilities water offered as a means of propulsion were always in the forefront of Bramah's mind. In 1785 he suggested the locomotion of ships by means of screws; in 1790 and 1793 he constructed the hydraulic transmission of power. Among Bramah's other inventions were a machine for numbering bank notes, a beer pump, and machines for making paper and for the manufacture of aerated waters. He also produced a machine that made nibs for pens.

Bramah died in London on 9 December,1814, and was buried in Paddington churchyard. His press laid the foundation for a whole technology, applications of which include the car-jack, presses for baling waste paper and metal, and the hydraulic braking system for cars and other vehicles which ensures that the brakes on all the wheels operate simultaneously and evenly. Massive girders could be jacked into place, providing a powerful tool for such bridge-builders as Robert Stephenson, who used hydraulics to position the massive tubular spans of the bridges over the River Conway and the Menai Straits. Extrusion and forging presses still employ the principles for which Bramah laid the working foundations.

Branly, Edouard Eugène Désiré (1844–1940) French physicist and inventor who preceded Guglielmo ◊Marconi in performing experiments resulting in the invention of wireless telegraphy and radio.

Branly was born in Amiens on 23 October 1844, the son of a teacher. He entered the Ecole Normale Supérieure in Paris in 1865, from which he gained his licence in physical and mathematical sciences in 1867 and his agrégé in physics and natural science in 1868. He then joined the staff of the Lycée de Bourges but almost immediately afterwards returned to Paris to take charge of the physics laboratory at the Sorbonne, becoming its adjunct director in 1870. He then submitted his doctoral thesis to the Sorbonne, gaining a medical qualification in 1872 and his doctorship in 1873. In 1876 he was appointed professor of physics at the Ecole Supérieure de Sciences of the Catholic Institute in Paris through the influence of Abbot Hulst, who was then working at the Catholic Institute, and Henri-Etienne ◊Saint-Claire Deville, an eminent

French chemist. From 1897–1916 Branly then worked as a medical professor of electrotherapy. It was not until 1898 that his work was recognized, with the award of the Houllevignes Prize by the French Academy of Sciences. Thereafter he received many other honours and, after his death in Paris on 24 March 1940, a national funeral was held for him in Nôtre Dame Cathedral.

Branly researched in various subjects, notably electricity, electrostatics (on which he wrote his doctoral thesis), magnetism, and electrical dynamics. His first task at the Catholic Institute, however, was to set up (with limited financial resources) a physics laboratory. (It was not until Frère Coty made a large donation in 1932 that the physics laboratory was adequately equipped, and all of Branly's work was carried out in very poor laboratory conditions.) His most important work – on wireless telegraphy – was performed in 1899, when he demonstrated the coherer, an invention of his that enabled radio waves from a distant transmitter to be detected. Once he had established the principle, however, he did not develop it further and the practical points were later taken up by Marconi, who was to share the 1909 Nobel Prize for Physics with Karl Braun for the invention of wireless telegraphy.

Braun, Emma Lucy (1889–1971) US botanist, an early pioneer in recognizing the importance of plant ecology and conservation.

She was born in Cincinnati on 19 April 1889 and with her elder sister Annette (1884–1978) was encouraged by her parents in both informal nature study and formal academic work. She graduated from the University of Cincinnati, gaining a master's degree in geology in 1912, and was awarded a PhD in botany in 1914, three years after her sister had also achieved the degree. She remained in academic positions at the university, becoming professor of plant ecology in 1946, until her early retirement in 1948. She lived with her sister, an entomologist, and continued research work until the end of her life, the two setting up a home laboratory and an experimental garden. She died in her Cincinatti home on 5 March 1971.

Although Braun did produce laboratory-based work, her major advances were made from field studies. Her work in ecology concentrated on the vegetation of a selected variety of habitats in Ohio and Kentucky. An early taxonomic study provided a detailed catalogue of the flora of the Cincinnati region, which she then compared with that of the same region a century earlier. This approach became very influential for analysing regional changes in flora over a period of time. Her field studies of plant distribution combined with her interests in geology also led her to consider several innovative theories in the evolution of forest communities and their survival during periods of glaciation. Her laborious studies of the forests of the area over almost 30 years were distilled in a classic work *Deciduous Forests of Eastern North America* (1950).

Braun also contributed to the growing conservation movement, stressing the importance of preserving natural habitats. She established a local Wild Flower Preservation Society and wrote and campaigned to save natural areas and to create nature reserves. She became president of the Ohio Academy of Science and the Ecological Society of America, the first woman to achieve both positions, and was honoured by the Botanical Society of America.

Braun, Karl Ferdinand (1850–1918) German physicist who is best known for his improvements to Guglielmo ◊Marconi's system of wireless telegraphy, for which he shared (with Marconi) the 1909 Nobel Prize for Physics. He also made other important contributions to science, including the discovery of crystal rectifiers and the invention of the oscilloscope.

Braun was born in Fulda on 6 June 1850. He was educated at the universities of Marburg and Berlin, gaining his doctorate from the latter in 1872. His first job was assistant to Quinke at Würzburg University, after which he successively held positions at the universities of Leipzig, Marburg, Karlsruhe, Tübingen, and Strasbourg; while at Strasbourg 1880–83 he founded an Institute of Physics. In 1883 he was elected to the chair in physics at the Karlsruhe Technische Hochschule, a post he held until 1885, when he became professor of physics at Tübingen University. In 1895 he returned to Strasbourg University as professor of physics and director of the Institute of Physics. In 1917 he went to the USA to testify in litigation about radio patents, but when the USA entered World War I he was detained as an alien in New York City and died there shortly afterwards, on 20 April 1918.

Braun began to study radio transmission in the late 1890s in an attempt to increase the transmitter range to more than 15 km/9 mi (the maximum range then possible). He thought that the range could be increased by increasing the transmitter's power but, after working on Hertz oscillators, found that lengthening the spark gap to increase the power output was effective only up to a certain limit, beyond which power output actually decreased. He therefore devised a sparkless antenna (aerial) circuit in which the power from the transmitter was magnetically coupled (using electromagnetic induction) to the antenna circuit, instead of the antenna being connected directly in the power circuit. This invention, which Braun patented in 1899, greatly improved radio transmission, and the principle of magnetic coupling has since been applied to all similar transmission systems, including radar and television. Later Braun developed directional antennas.

Braun also discovered crystal rectifiers. In 1874 he published a paper describing his research on mineral metal sulphides, some of which, he found, conduct electricity in one direction only. His discovery did not find an immediate application, but crystal rectifiers ('cat's whiskers') were used in the crystal radio receivers of the early 20th century – until they were superseded by more efficient valve circuits.

Braun's other principal contribution to science was his invention in 1895 of the oscilloscope, which he used to study high-frequency alternating currents. In his oscilloscope – which was basically an adaptation of the cathode-ray tube – Braun used an alternating voltage applied to deflection plates to deflect an electron beam within a cathode-ray tube. The Braun tube (as his invention was initially called) became a valuable laboratory instrument and was the forerunner of more sophisticated oscilloscopes and of the modern television and radar display tubes.

Bredig, Georg (1868–1944) German physical chemist who contributed to a wide range of subjects within his discipline but is probably best known for his work on colloids and catalysts.

Bredig was born on 1 October 1868 in Glogau, Lower Silesia (now Glogow, Poland). After qualifying he went to work as an assistant in the laboratory of the German chemist Wilhelm ⇨Ostwald in Leipzig, and it was there that he did much of his significant work. He held a series of academic appointments in physical chemistry: at Heidelberg, Germany, 1901–10; Zürich, Switzerland, in 1910; and from 1911 at the Karlsruhe Hochschule, Germany. He went to the USA in 1940, and died in New York City on 24 April 1944.

While working with Ostwald, Bredig collaborated with him on the accumulation of experimental data with which to validate Ostwald's dilution law (which states that, for a binary electrolyte, the equilibrium constant K_c of a chemical reaction has the same value at all dilutions). The equilibrium constant depends not on the dilution, but on the chemical nature of the particular acid or base. Ostwald confirmed the law for 250 acids, and Bredig provided the comparable data for 50 bases.

The variation in the relative atomic mass of lead from various sources, the transition metals (on which he worked with Jacobus van't Hoff), and catalytic action formed other areas of his research. Bredig also supervised overseas chemistry students, such as the British chemist Nevil Sidgwick.

In the field of catalysis Bredig's particular study was the catalytic action of colloidal platinum and the 'poisoning' of catalysts by impurities. His most important contribution was a method of preparing colloidal solutions (lyophobic sols) using an electric arc, which he devised in 1898.

There are two ways of producing particles of colloidal size: larger particles can be broken down (dispersed) or smaller particles can be made to aggregate. Bredig's arc method is a dispersion technique. An electric arc is struck between metal electrodes immersed in a suitable electrolyte – for example, platinum, gold, or silver electrodes in distilled water containing an alkali. The colloidal particles are thought to be produced mainly by rapid condensation of the vapour of the arc, and they may be in the form of the metal or its oxide. A later extension of the method developed by Theodor ⇨Svedberg in the early 1900s uses an alternating current and produces sols of greater purity.

Brenner, Sydney (1927–) South African-born British molecular biologist noted for his work in the field of genetics.

Brenner was born on 13 January 1927 at Germiston, near Johannesburg, the son of an emigrant from Lithuania. He was educated there and studied at the University of the Witwatersrand, where he gained his MSc in 1947 and his MB and BCh in 1951. He then went to the UK and studied for a PhD at Oxford, which he received in 1954. In that year he also worked in the Virus Laboratory of the University of California in Berkeley, and from 1955–57 was a lecturer in physiology at the Witwatersrand University. From 1957 he researched in the Molecular Biology Laboratory of the Medical Research Council, Cambridge, and in 1980 was appointed its director. In 1996 he became head of the Molecular Sciences Institute, La Jolla, California.

Brenner's first research was on the molecular genetics of very simple organisms. Since then he has spent seven years on one of the most elaborate efforts in anatomy ever attempted, investigating the nervous system of nematode worms and comparing the nervous systems of different mutant forms of the animal. The nematode that lives in the soil and which feeds on or in roots can be as little as 0.5 mm/0.02 in in length. Brenner's experiments have included cutting a soil nematode into 20,000 extremely thin slices and, one at a time, projecting a long succession of electron micrographs onto a screen. The animal's nerves are traced in each picture by an electronic pen which automatically feeds information to a computer for storage. Brenner's reason for gathering and processing all this information is to compare the 'wiring' of the nervous system of normal nematodes with that of mutant ones which show peculiarities in behaviour. About one hundred genes are involved in constructing the nervous system of a nematode and most of the mutations that occur affect the overall design of a section of the nervous system. These genes are therefore of an organizational type and regulate the routing of the nervous system during the growth of the animal. The nematode is a simple animal although its make-up is extremely complicated. The amount of effort that has been put into this study indicates how much biologists still have to find out about the exact organization of living tissues.

Progress in science depends on new techniques, new discoveries, and new ideas, probably in that order.
SYDNEY BRENNER NATURE *1980*

Brenner is also interested in tumour biology and in the use of genetic engineering for purifying proteins, cloning genes, and synthesizing amino acids. His experiments have given and continue to give a great impetus to molecular biology.

Brewster, David (1781–1868) Scottish physicist who investigated the polarization of light, discovering the law named after him for which he was awarded the Rumford Medal by the Royal Society in 1819. He also helped to popularize science in his writings, and is perhaps best known as the inventor of the kaleidoscope.

Brewster was born in Jedburgh on 11 December 1781. His education was extended by reading his father's university notes on physics, his services as secretary to the local minister Thomas Somerville, and his friendship with James Veitch, an amateur astronomer. Brewster entered the University of Edinburgh in 1794 but never took his degree. He continued his studies, this time in divinity, and was awarded an honorary MA in 1800. He was later licensed to preach but was not ordained, and turned to publishing and teaching as means of making a living.

One of Brewster's major concerns was increasing the public awareness of the importance of science. He edited a number of scientific periodicals and wrote many books and articles on science, including entries in the *Encyclopaedia Britannica.* He was also instrumental in the foundation of several academic organizations including the Edinburgh School of Arts in 1821, the Royal Scottish Society for Arts in 1821, and the British Association for the Advancement of Science in 1831.

His achievements in optics were recognized by his election to many prestigious international scientific societies, including the Royal Society in 1815 and the French Institute. Brewster was knighted in 1832 and in 1859 made principal and later vice chancellor of Edinburgh University. He died on 10 February 1868 in Allerby in Scotland.

With James Veitch, Brewster built many optical devices such as microscopes and sundials, developing an expertise that resulted in the invention of the kaleidoscope in 1816. In trying to improve lenses for telescopes, he became interested in optics and particularly in the polarization of reflected and refracted light. In 1813 he demonstrated, by studying the polarization of light passing through a succession of glass plates, that the index of refraction of a particular medium determines the tangent of the angle of polarization for light that transverses it. Brewster then sought an expression for the polarization of light by reflection and found, in 1815, that the polarization of a beam of reflected light is greatest when the reflected and refracted rays are at right angles to each other. This is known as Brewster's law, and it may be stated in the form that the tangent of the angle of polarization is numerically equal to the refractive index of the reflecting medium when polarization is maximum.

Brewster then worked on the polarization of light reflected by metals, and established the new field of optical mineralogy. By 1819 he had classified most crystals and minerals on the basis of their optical properties. During the 1820s he studied colour in the optical spectrum, finding that it could be divided into red, yellow, and blue regions, and worked on absorption spectroscopy of natural substances, greatly extending the number of dark lines identified in spectra.

Brewster was an advocate of the corpuscular theory of light (that it consists of a stream of corpuscles, or particles) on the basis of his experimental results. Although forced to admit that the wave theory did provide excellent explanations for certain observed phenomena, he clung to his views as he was philosophically unable to accept the wave theory because it assumed the existence of a hypothetical 'ether'.

Bridgman, Percy Williams (1882–1961) US physicist famous for his work on the behaviour of materials at high temperature and pressure, for which he won the 1946 Nobel Prize for Physics.

Bridgman was born on 21 April 1882 in Cambridge, Massachusetts. His father was a journalist and a social and political writer. He went to Harvard in 1900 and was awarded a PhD in 1908. He then began research work in the Jefferson Research Laboratory and spent his entire research life at Harvard University, starting as assistant professor in 1913. He became professor in 1919, Hollis Professor of Mathematics and Philosophy in 1927, and Higgins Professor in 1950. He retired to become professor emeritus in 1954. Suffering from an incurable disease, Bridgman killed himself at his home in Randolph, New Hampshire, on 20 August 1961.

At Harvard, it became clear that Bridgman was an excellent experimentalist and was skilled at handling machine tools and at manipulating glass. His experimental work on static high pressure began in 1908, and was at first limited to pressures of around 6,500 atmospheres. He gradually extended it to more than 100,000 atmospheres and eventually to about 400,000 atmospheres. Because this field of research had not been explored before, Bridgman had to invent much of his own equipment. His most important invention was a special type of seal in which the pressure in the gasket always exceeds that in the pressurized fluid. The result is that the closure is self-sealing – without this, his work at high pressure would not have been possible. He was later able to use the new steels and alloys of metals with heat-resistant compounds.

Bridgman's work involved the measurement of the compressibilities of pressurized liquids and solids, and measurements of physical properties of solids such as electrical resistance. His discoveries included that of new high-pressure forms of ice. With the increase in range of possible pressures, new and unexpected phenomena appeared. He discovered that the electrons in caesium rearrange at a certain transition pressure. He pioneered the work to synthesize diamonds, which was eventually achieved in 1955. His technique was used to synthesize many more minerals and a new school of geology developed, based on experimental work at high pressure and temperature. Because the pressures and temperatures that Bridgman achieved simulated those deep below the ground, his discoveries gave an insight on the geophysical processes that take place within the Earth. His book *Physics of High Pressure* (1931) still remains a basic work.

Bridgman was an individualist. He published 260 papers, only two of which were with a co-author. He devoted himself to his scientific research, refusing to attend faculty meetings or to serve on committees. He disliked lecturing and did it badly. In 1914, however, during a course of lectures on advanced electrodynamics, Bridgman realized that many ambiguities and obscurities exist in the definition of scientific ideas and moved into the philosophy of science, publishing *The Logic of Modern Physics* in 1927.

Briggs, Henry (1561–1630) English mathematician, one of the founders of calculation by logarithms.

Briggs was born in Warley Wood, in Halifax, Yorkshire, in February 1561. He attended a nearby grammar school and in 1577 went to St John's College, Cambridge. He was made a scholar in 1579 and received his BA in 1581. He was elected a fellow of the college in 1588 and was appointed lecturer and examiner in 1592. It was a tribute to his rare abilities that, in an age when the universities paid scant attention to mathematics, he was appointed professor of geometry at the newly established Gresham College, London, in 1596. He held the chair until 1620, resigning it because a year earlier he had accepted the invitation of Henry Savile (1549–1622) to succeed him as professor of astronomy at Oxford. He was elected a fellow of Merton College, where he died on 26 January 1630.

In 1616 Briggs wrote a letter to James Usher, later archibishop of Armagh, informing him that he was wholly absorbed in 'the noble invention of logarithms'. Two years earlier John ◊Napier had published his discovery of logarithms and as soon as Briggs learned of it he formed an earnest desire to meet the great man, for which purpose he travelled to Edinburgh in 1616. On this and subsequent visits the two men worked together to improve Napier's original logarithms which, having in modern notation $\log N = 10^7 \log_e(10^7/N)$, were in need of simplification. It seems most probable that the idea of having a table of logarithms with ten for their base was originally conceived by Briggs. And although both men published separate descriptions of the advantages of allowing the logarithm of unity to be zero and of using the base 10, the first such logarithmic tables were published by Briggs in 1617. They were published under the title *Logarithmorum chilias prima,* and were followed in 1624 by the *Arithmetica Logarithmica,* in which the tables were given to 14 significant figures. In fact, the logarithms of Briggs (and of Napier) were logarithms of sines, a reflection of both men's interest in astronomy and navigation, fields in which accurate and lengthy calculations using that trigonometrical function were everyday matters. It is owing to the way that sines were then considered that the large factor 10^9 was important and that Briggs's tables of 1624 took the form $10^9\log_{10}N$.

For that reason Briggs's logarithms were one thousand million times 'larger' than those in modern tables. Despite that, however, and despite the fact that many mathematicians (including Kepler) subsequently calculated their own tables, Briggs's tables remain the basis of those used to this day.

Bright, Richard (1789–1858) English physician who was the first to describe the kidney disease known as Bright's disease, which is actually a rather vague term – now largely obsolete – sometimes used to denote any of several different kidney disorders that share a number of the same symptoms.

Bright was born on 28 September 1789 in Bristol and was privately educated in Exeter and Edinburgh. In 1809 he began studying medicine at Edinburgh University but interrupted his studies to travel to Iceland. On returning to the UK he resumed his medical training at Guy's Hospital and St Thomas's Hospital in London, receiving his medical degree from

Edinburgh University in 1813. After spending several years touring Europe – during which time he worked in several European hospitals – Bright was appointed assistant physician at Guy's Hospital in 1820. Four years later he became a full physician at Guy's, a post he held for the rest of his life. In 1837 he was also appointed Physician Extraordinary to Queen Victoria. Bright died in London on 16 December 1858.

Bright's principal interest lay in disorders of the kidneys, in which area he initiated the use of biochemical studies by working with chemists to demonstrate that urea is retained in the body in kidney failure. He also correlated symptoms in patients with the pathological changes he later found in postmortem examinations of these same people. Using these methods he found that albuminuria (the presence of the protein albumin in the urine) and oedema (accumulation of fluid in the body) are associated with pathological changes in the kidneys – a condition that came to be called Bright's disease. Later, however, it was discovered that several different kidney disorders produce these symptoms (although the most common cause is glomerulonephritis – inflammation of the glomeruli) and the term Bright's disease is little used today. Bright first published his findings in *Reports of Medical Cases* (1827 and 1931) and subsequently in the first volume of *Guy's Hospital Reports* (1836), which he helped to establish.

In addition to his studies of kidney disorders, Bright investigated jaundice, nervous diseases, and abdominal tumours. He also collaborated with Thomas ◊Addison, a contemporary at Guy's Hospital, in writing *Elements of the Practice of Medicine* (1839); in fact Addison wrote most of this volume, and the second volume – to which Bright was to have been the principal contributor – was never published.

Further Reading

Berry, Diana and Mackenzie, Campbell, *Richard Bright, 1789–1858: Physician in an Age of Revolution and Reform,* Eponymist in Medicine series, Royal Society of Medicine Press Ltd, 1992.

Kark, R M, *Physician Extraordinary: Dr Richard Bright (1789–1858),* Horn of the Moon Enterprises, 1986.

Rooney, Patrick J; Szebenyi, Béla; and Bálint, Géza P, Richard Bright's 'Travels from Vienna through Lower Hungary: a glimpse of medicine and health care in the early 19th century', *Can Bull Med Hist,* 1993, v 10, pp 87–96.

Brindley, James (1716–1772) English engineer who, in spite of the most formidable handicaps, became a pioneer of canal building.

Brindley was born near Buxton, Derbyshire, in 1716. Apprenticed to a millwright at the age of 17, he appears to have been completely devoid of promise even at this humble pursuit until the erection of a paper mill with certain novel features brought out in him a remarkable mechanical sense. As a result he was put in charge of his master's shop.

On his master's death Brindley set up his own business at Leek, Staffordshire, where he was soon running a thriving business repairing old machinery and installing new machines. Wedgwoods, then only a small pottery firm, employed him to construct flint mills. He completed the machinery for a silk mill at Congleton in Cheshire and had some limited success in improving the machinery then used to draw water from mines, although this problem was not effectively solved until the advent of the steam engine.

In 1759 Brindley was engaged by the Duke of Bridgewater to construct a canal to transport coal from the duke's mines at Worsley to the textile manufacturing centre of Manchester. He pursuaded the duke to change his plan of using locks and rivers and approve instead a revolutionary scheme, which included a subterranean channel extending from the barge basin at the head of the canal into the river, and an aqueduct – 12 m/40 ft high – carrying the canal over the River Irwell. Brindley's mechanical skill enabled him to construct impervious banks by 'puddling' clay, and the canal simultaneously acted as a mine drain, a feeder for the main canal at the summit level, and a barge-carrying canal.

Brindley's achievement, remarkable enough for any man, is made still more noteworthy by his almost complete lack of any formal schooling. He was virtually illiterate, barely able to write his name. He made no calculations or drawings of the tasks he set himself but worked out everything in his head. He was, in a most extreme form, a natural engineer.

The success of the Worsley scheme established Brindley as the leading canal builder in England. His next commission was to construct the Bridgewater Canal linking Manchester and Liverpool, after which came others – the most important being the Grand Union Canal, which connected Manchester with the Potteries in the Midlands (completely transforming the area – the population of the Potteries trebled between 1760 and 1785); the Oxford Canal; the old Birmingham and the Chesterfield canals; the Staffordshire and Worcestershire canals; and the Coventry Canal. All these were designed and with only one exception executed by Brindley. In all he constructed 584 km/360 mi of canals.

It is said of Brindley that when faced with an apparently intractable problem he would go home and think it over in his bed. This must not lead one to assume, however, that he took a phlegmatic attitude to his work. He died from his excessive and arduous labours on 27 September 1772, at Turnhurst, Staffordshire.

Further Reading

Bode, Harold, *James Brindley: An Illustrated Life of James Brindley, 1716–1772,* Shire Publications, 1980.
Evans, Kathleen, *James Brindley, Canal Engineer: A New Perspective, with Particular Reference to His Family Background,* Churnet Valley Books, 1998.
Meynell, Laurence, *James Brindley: The Pioneer of Canals,* W Laurie, 1956.
Smiles, Samuel, *Lives of the Engineers Vermuyden – Muddelton – Perry – James Brindley: Early Engineering,* Murray, 1904.

Brinell, Johann August (1849–1925) Swedish metallurgist who developed what became known as the Brinell hardness test, a rapid nondestructive method of estimating metal hardness.

Brinell was born in Småland and attended technical school at Borås. On leaving school he worked as a mechanical designer. In 1875 he was appointed chief engineer at the ironworks at Lejofors, and it was there that he became interested in metallurgy. In 1882 he became chief engineer at the Fagersta ironworks. While there he studied the internal composition of steel during heating and cooling, and devised the hardness test, which was put on trial at the Paris Exhibition of 1990. The test is based on the impression left by a small hardened steel ball after it is pushed into a metal with a given force. With minor innovations the test remains in use today. Brinell also carried out investigations into the abrasion resistance of selected materials. He died in Stockholm in 1925.

Based on the idea that a material's response to a load placed on one small point is related to its ability to deform permanently, the hardness test is performed by pressing a hardened steel ball (Brinell test) or a steel or diamond cone (Rockwell test) into the surface of the test piece. The hardness is inversely proportional to the depth of penetration of the ball or cone.

Brongniart, Alexandre (1770–1847) French geologist and palaeontologist. He was first to classify the Tertiary formations. Brongniart was responsible for gathering together a body of knowledge that was essential to establishing geology as a science in the first two decades of the 19th century. His work was of crucial importance to the elucidation of the main outlines of the history of the Earth and life on Earth.

Born in Paris on 5 February 1770, Alexandre Brongniart was the son of the eminent Parisian architect Alexandre-Theodore Brongniart (1739–1813) and Anne-Louise Degremont. The young Alexandre studied at the Ecole des Mines and then at the Ecole de Medécin. He then worked as assistant to his uncle, a professor of chemistry, before serving in the forces medical department in the Pyrenees. In 1794 he returned to Paris and became a mining engineer and in 1797 he became professor of natural history at the Ecole Centrale des Quatre-Nations. He became chief mining engineer in 1818 and in 1822 succeeded Rene ◊Haüy to the chair of mineralogy at the Natural History Museum in Paris. He was an approachable and generous teacher, allowing his students access to his collections on Sundays, and also entertaining distinguished scientists at his evening salons.

Brongniart was director of the Sèvres porcelain factory from 1800 till his death in 1847. As a young man he had learnt ceramic techniques in England and published a work on improving the art of enamelling. Towards the end of his life, he again turned his attention to ceramics, publishing his last major work, *Traité des arts ceramiques* in two volumes in 1844. He had also been responsible for the revival of the Sèvres factory by introducing rich new colours. He died in Paris on 7 October 1847.

Early in his career Brongniart divided reptiles into four groups – Chelonia, Ophida, Batrachia, and Sauria – according to careful comparative study of their anatomy. He published this work in his *Essai d'une classification naturelle des reptiles* (1800). This system was essentially the same as the one in use today. It was in collaboration with the French zoologist Georges ◊Cuvier that he produced perhaps his most significant piece of work on the geology of the Parisian region. They published their findings jointly in *Essai sur la géographie mińeralogique des environs de Paris* (1811), being among the first geologists to identify strata within rock formations by their fossil content. They identified a constant order of fossil sequences over the whole Paris region and established the principle whereby the changing fossil record can be related to the relative age of the rock strata. They interpreted the alternation of freshwater and seawater strata around Paris as an indication of catastrophic processes in the area. Brongniart published this important conclusion in a separate memoir 'Sur les terrains qui paraissent avoir ete formes sous l'eau douce' (1810). He also described freshwater deposits far from Paris. Brongniart's method of using fossils as a tool for stratigraphy was soon adopted by geologists in many other areas.

Brongniart had travelled widely in Europe and was thus able to apply his new knowledge to his earlier experiences. He identified a similarity in the age of seemingly different geological structures such as London clay and Paris chalk by applying his rule that stressed the importance of using of fossil evidence over lithology (rock type). He described his methodology in another memoir 'Sur les caracteres zoologiques des formations' (1821).

In 1822 Brongniart elucidated the zoological and geological relations of trilobites in his *Histoire naturelle des crustaces fossiles,* which was the first published full length study of trilobites. In this work, which was later to be important in the understanding of Palaeozoic stratigraphy, he classified a wide variety of trilobites from all over Europe and as far as America. Brongniart's last major geological work, *Tableau des terrains qui composent l'ecorce du globe* (1829), attempted to tackle problems that continue to puzzle geologists to the present day. It was the culmination of his life's work and was a valuable list of the fossils that characterize the various types of rock formation. His work laid the foundations of geological research and was the basis for many of the discoveries of the first half of the 19th century.

Bronowski, Jacob (1908–1974) Polish-born British scientist, journalist, and writer, originally trained as a mathematician, who won international recognition as one of the finest popularizers of scientific knowledge in the 20th century.

Bronowski was born in Poland on 18 January 1908, fled with his family to Germany when, in World War I, Russia occupied Poland, and moved in 1920 to England and became a naturalized British citizen. He studied mathematics at Jesus College, Cambridge, where he also edited a literary magazine and published some unremarkable verse, graduating as senior wrangler (with the highest marks). He was awarded his PhD in 1933 and a year later was appointed senior lecturer at University College, Hull. He remained there until after the outbreak of World War II, when in 1942 he joined Reginald Stradling's Military Research Unit at the Home Security Office. His principal job was to forecast the economic effects of bombing. After the war he conducted statistical research at the Ministry of Works until 1950, when he was appointed director of the National Coal Board's research establishment. From 1959–63 he served as director general of process development for the board. During these years as a government official, he was continually extending the range of his intellectual pursuits. In particular he devoted himself to studying the development of Western science and thought. In 1953 he was visiting professor of history at the Massachusetts Institute of Technology. His last appointment was as senior fellow at the Salk Institute for Biological Studies in California, a post he took up in 1964. He died in San Diego on 22 August 1974.

Of his appointment to the Salk Institute Bronowski said that he was a 'mathematician trained in physics, who was taken into the life sciences in middle age by a series of lucky chances'. In fact he will probably be least remembered as a biologist. His true métier was for explaining to a large public the broad canvas of European intellectual history. His first published work of note was *The Poet's Defence* (1939) and at one time it might have been thought that he would come to specialize in literary subjects. After World War II he wrote several plays for radio, most memorably two in 1948, *Journey to Japan* and *The Face of Violence,* the latter of which won the Italia Prize in 1951 as the best radio play in Europe. He never lost his interest in literature (*William Blake and the Age of Revolution* appeared as late as 1965), but from the early 1950s his main interest turned towards broader intellectual and scientific themes.

The Common Sense of Science (1951) represented Bronowski's first attempt to bring the mysteries of science within the ken of nonscientific readers and was notable for the manner in which it displayed the history and workings of science around three central notions – cause, chance, and order. In the early days of the Cold War he used the pages of the *New York Times* to discuss, in accessible language, both the technology of nuclear science and the moral questions raised by the development of nuclear weapons. An extension of his newspaper articles was the book *Science and Human Values* (1958).

Bronowski was deeply concerned about the general effects on society of the widening division between the arts and the sciences, a phenomenon given great publicity by the famous Leavis–Snow controversy, and he was at pains to do what he could to narrow the divide, while at the same time bringing the specialist conclusions of scholars in both the sciences and the humanities to a wide public. The result was two of his finest popularizing works: *The Western Intellectual Tradition* (1960), an illuminating survey of the growth of political, philosophical, and scientific knowledge from the Renaissance to the 19th century written with Bruce Mazlish, and the brilliant 13-part BBC television documentary, *The Ascent of Man,* issued as a book in 1973.

When he was president of the British Library Association 1957–58, Bronowski said, in his inaugural address, that for public libraries to serve in the general expansion of a society's culture, writers must make the language of science comprehensible to the nonspecialist. It is his great distinction that he practised, with great wit and erudition, what he preached.

Brönsted, Johannes Nicolaus (1879–1947) Danish physical chemist whose work in solution chemistry, particularly electrolytes, resulted in a new theory of acids and bases.

Brönsted was born on 22 February 1879 in Varde, Jutland, the son of a civil engineer. He was educated at local schools before going to study chemical engineering at the Technical Institute of the University of Copenhagen in 1897. He graduated two years later and then turned to chemistry, in which he qualified in 1902. After a short time in industry, he was appointed an assistant in the university's chemical laboratory in 1905, becoming professor of physical and inorganic chemistry in 1908. In his later years he turned to politics, being elected to the Danish parliament in 1947. He died on 17 December in that year, before he could take his seat.

Brönsted's early work was wide ranging, particularly in the fields of electrochemistry, the measurement of hydrogen-ion concentrations, amphoteric electrolytes, and the behaviour of indicators. He discovered a method of eliminating potentials in the measurement of hydrogen-ion concentrations, and devised a simple equation that connects the activity and osmotic coefficients of an electrolyte, and another that relates activity coefficients to reaction velocities. From the absorption spectra of chromic – chromium(III) – salts he concluded that strong electrolytes are completely dissociated, and that the changes of molecular conductivity and freezing point that accompany changes in concentration are caused by the electrical forces between ions in solution. He related the stages of ionization of polybasic acids to their molecular structure, and the specific heat capacities of steam and carbon dioxide to their band spectra. In 1912 he published work with Herman ◊Nernst on the specific heat capacities of steam and carbon dioxide at high temperatures. Two years later he laid the foundations of the theory of the infrared spectra of polyatomic molecules by introducing the so-called valency force-field. Brönsted also applied the newly developed quantum theory of specific heat capacities to gases, and published papers about the factors that determine the pH and fertility of soils.

In 1887 Svante ◊Arrhenius had proposed a theory of acidity that explained its nature on an atomic level. He defined an acid as a compound that could generate hydrogen ions in aqueous solution, and an alkali as a compound that could generate hydroxyl ions. A strong acid is completely ionized (dissociated) and produces many hydrogen ions, whereas a weak acid is only partly dissociated and produces few hydrogen ions. Conductivity measurements confirm the theory, as long as the solutions are not too concentrated.

In 1923 Brönsted published (simultaneously with Thomas Lowry in the UK) a new theory of acidity that has certain important advantages over that of Arrhenius. Brönsted defined an acid as a proton donor and a base as a proton acceptor. The definition applies to all solvents, not just water. It also explains the different behaviour of pure acids and acids in solution. Pure dry liquid sulphuric acid or acetic (ethanoic) acid does not change the colour of indicators nor react with carbonates or metals. But as soon as water is added, all of these reactions occur.

In Brönsted's scheme, every acid is related to a conjugate base, and every base to a conjugate acid. When hydrogen chloride dissolves in water, for example, a reaction takes place and an equilibrium is established:

$$\underset{\text{Acid 1}}{HCl} + \underset{\text{Base 2}}{H_2O} \leftrightarrow \underset{\text{Acid 2}}{H_3O^+} + \underset{\text{Base 1}}{Cl^-}$$

HCl is an acid for the forward reaction, but the hydroxonium ion (H_3O^+) is an acid in the reverse reaction; it is the conjugate acid (acid 2) of water (base 2). Similarly, the chloride ion (Cl^-, base 1) accepts protons in the reverse reaction to form its conjugate acid (HCl, acid 1). In this theory, acids are not confined to neutral species or positive ions. For example, the negatively charged hydrogen sulphate ion can behave as an acid:

$$HSO_4^-(aq) + H_2O(l) \leftrightarrow H_3^+O + SO_4^{2-}(aq)$$

It donates a proton to form the hydroxonium ion.

Brouwer, Luitzen Egbertus Jan (1881–1966)

Dutch mathematician who founded the school of mathematical thought known as intuitionism.

Brouwer was born in Overschie on 27 February 1881. He studied mathematics at the University of Amsterdam, and on receiving his BA was appointed an external lecturer to the university in 1902. He remained at that post until 1912, when he was appointed professor of mathematics, a chair that he held until his retirement in 1951. His singular contribution to mathematics earned him numerous honours, most notably the Knighthood of the Order of the Dutch Lion in 1932. He died in Blaricum on 2 December 1966.

Brouwer's first important paper, a discussion of continuous motion in four-dimensional space, was published by the Dutch Royal Academy of Science in 1904, but the greatest early influence on him was Gerritt Mannoury's work on topology and the foundations of mathematics. This led him to consider the quarrel between Henri Poincaré and Bertrand Russell on the logical foundations of mathematics, and his doctoral dissertation of 1907 came down on the side of Poincaré against Russell and David ◊Hilbert. He took the position that, although formal logic was helpful to describe regularities in systems, it was incapable of providing the foundation of mathematics.

For the rest of his career Brouwer's chief concern remained the debate over the logical, or other, foundations of mathematics. His inaugural address as professor of mathematics at Amsterdam in 1912 opened new ground in this debate, which had begun with the work of Georg ◊Cantor in the early 1880s. In particular Brouwer addressed himself to problems associated with the law of the excluded middle, one of the cardinal laws of logic. He consistently took issue with mathematical proofs (so-called proofs, as he saw them) that were based on the law. In 1918 he published his set theory, which was independent of the law, explaining the notion of a set by the introduction of the idea of a free-choice sequence.

Having rejected the principle of the excluded middle as a useful mathematical concept, Brouwer went on to establish the school of intuitional mathematics. Put simply, it is based on the premise that the only legitimate mathematical structures are those that can be introduced by a coherent system of construction, not those that depend upon the mere postulating of their existence. So, for example, the intuitionist principle denies that it makes sense to talk of an actual infinite totality of natural numbers; that infinite totality is something that requires to be constructed.

Brouwer's work did not create an overnight sensation. But when, in the late 1920s, Kurt ◊Gödel broke down Hilbert's foundation theory, it gained great pertinence. The result of Gödel's work was the theory of recursive functions, and in that field of mathematics Brouwer's work was of such fundamental significance that his intuitional theories and analysis have continued to be at the very centre of research into the foundations of mathematics.

Brown, Ernest William (1866–1938) English mathematician with a particular interest in celestial mechanics and lunar theory.

Brown was born in Hull on 29 November 1866. He was awarded a scholarship to Christ's College, Cambridge, where he was introduced to problems in lunar theory by George Darwin (1845–1912). Brown gained a BA in 1887, and from 1889–95 he held a fellowship at Christ's, although in 1891 he went to the USA to teach mathematics at Haverford College in Pennsylvania. He was professor of mathematics at Haverford 1893–1907, and then at Yale University until his retirement as professor emeritus in 1932.

Brown's work on the motions of the Moon earned him an international reputation. He received numerous honorary degrees and awards from scientific organizations such as the National Academy of Sciences and the Royal Astronomical Society. He was an active participant within the professional societies of which he was a member. Brown died in New Haven, Connecticut, on 22 September 1938.

The effect of gravity on the motions of the planets and smaller members of the Solar System was the major research interest of Brown's career. His work first focused on lunar motion, and he produced extremely accurate tables of the Moon's movements. Unable to account for the variation in the Moon's mean longitude, he proposed that the observed fluctuations arose as a consequence of a variable rate in the rotation of the Earth.

Brown was also interested in the asteroid belt. It had been proposed that asteroids might at one time have been part of one planet. Some astronomers attempted to compute the possible orbit of such a parent planet on the basis of the distribution of the asteroids, but Brown was highly critical of this approach.

One of his concerns during the later years of his career was the calculation of the gravitational effect exerted by the planet Pluto on the orbits of its nearest neighbours, Uranus and Neptune.

Brown's work on gravity did much to increase our understanding of the relationship of members of the Solar System.

Brown, Robert (1773–1858) Scottish botanist whose discovery of the movement of suspended particles has proved fundamental in the study of physics.

Brown was born in Montrose on 21 December 1773, the son of an Episcopalian priest. He studied medicine

Brown Scottish botanist Robert Brown, pictured when he was Keeper of Botany in the British Museum, London. Brown discovered the phenomenon known as Brownian motion whilst observing the movement of pollen grains in water. He was the first to observe the presence of the nucleus in living cells. *Mary Evans Picture Library*

at Edinburgh University but did not obtain his degree. He subsequently held the position of assistant surgeon in a Scottish infantry regiment, but soon revealed that his true interest lay in botany. In the late 1790s he was introduced to the well-known English botanist Joseph ◊Banks, who allowed him the free use of his library and collections. Shortly afterwards Brown resigned from the army in order to accept the post of naturalist on an expedition under Captain Matthew Flinders, on the *Investigator,* to survey the coast of the recently discovered Australian continent. He voyaged from 1801 to 1805 and on his return to England published, in 1810, the first part of his studies on the flora he had discovered on his Antipodean journey. The poor sales of the book discouraged him and he left the rest unpublished. In the same year, he was appointed librarian to Joseph Banks, a post he held until Banks's death in 1820. Banks bequeathed to Brown the full use of the library and its collections for life. In 1827, in compliance with the stipulations of Banks's will, he agreed to the transfer of the books and specimens to the British Museum and was appointed curator of the botanical collections there. He died in London on 10 June 1858.

In 1791 Brown submitted his first paper to the Natural History Society. It was a highly detailed classification of the plants he had collected in Scotland, with accompanying notes and observations. This list was to win him many introductions in the scientific world of his day. It was not until 1828, however, that he made one of his greatest contributions to science, published in the Edinburgh New Philosophical Journal. The paper was entitled 'A Brief Account of Microscopical Observations Made in the Months of June, July, and August 1827 on the Particles Contained in the Pollen of Plants, and on the general existence of active molecules in organic and inorganic bodies' and it was in this paper that Brown set out his observations on the 'Brownian movement', or 'motion', that perpetuates his name. The concept arose from his observation that very fine pollen grains of the plant *Clarkia pulchella* move about in a continuously agitated manner when suspended in water. This phenomenon is true for any small solid particles suspended in a liquid or gas and can be viewed in a bright light through a microscope. Brown was able to establish that the constant movement was not purely biological in origin because inorganic materials such as carbon and various metals are equally subject to it, although he could not find the cause of the movement. During his lifetime there was no shortage of theories to explain his discovery, but it was not until the 20th century that the question was answered.

Brown also published papers on Asclepiadaceae (1809) and on Proteaceae (1810), and wrote on the propagatory process of the gulf-weed and on the anatomy of fossilized plants. He also described the organs and mode of reproduction in orchids. In 1831, while investigating the fertilization of both Orchidaceae and Asclepiadacea, he discovered that a small body, fundamental in the creation of plant tissues, occurs regularly in plant cells – he called it a 'nucleus', a name which is still used. Another significant revelation Brown made was the identification of the difference between gymnosperms and angiosperms.

Brown's various papers on his findings and opinions in every division of botanical science made him the outstanding authority on plant physiology of his day, and he did much to improve the system of plant classification by describing new genera and families. His observation of Brownian movement was important in showing how molecular motion forms the basis of kinetic theory.

Brown, Robert Hanbury (1916–) English radio astronomer who was involved with the early development of radio astronomy techniques and who has since participated in designing a radio interferometer that permits considerably greater resolution in the results provided by radio telescopes.

Brown was born on 31 August 1916 in Aruvankadu, India. After studying engineering at Brighton Polytechnic College in Sussex, he was awarded an external degree by the University of London. He went on to do some postgraduate research at the City and Guilds College before joining a radar research team under the

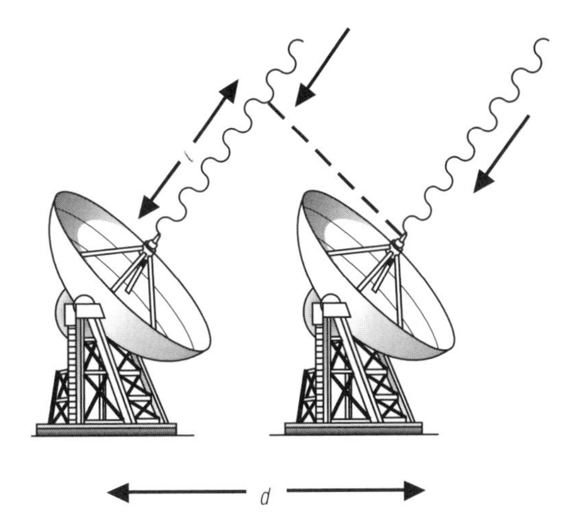

Brown Brown developed inferferometry in radioastronomy, which uses two (or more) radio telescopes receiving waves from the same radio source whose path length differ by *l*, and which are out of phase, resulting in interference when they are combined. Usually one of the telescopes is mounted on rails so that it can be moved to vary the baseline *d*.

auspices of the Air Ministry in 1936. During World War II, Brown took an active part in the radar research programme. A member of the British Air Commission, he also worked at the Naval Research Laboratory, Washington, DC. At the conclusion of the war, he briefly became a private radar consultant, but in 1949 he joined the staff at the Jodrell Bank Observatory in Cheshire, and began to carry out research into radio astronomy. In 1960 he was made a Fellow of the Royal Society and appointed professor of radio astronomy at the Victoria University, Manchester, where he remained until 1962. In 1964 he took up the chair of astronomy at the University of Sydney, becoming emeritus professor of physics in 1981.

Radio waves of cosmic origin were first detected accidentally by Karl ◊Jansky in 1931 while he was investigating a problem in communications for the Bell Telephone Company. Eighteen years later at Jodrell Bank Observatory, then under the direction of Bernard Lovell, Brown joined a team actively engaged in using radio methods for the investigation of the origin of meteor showers. He became one of the first astronomers to construct a radio map of the sky. Such a map could be compiled using data collected at night or during the day (unlike optical astronomy, which requires clear night-time conditions for observation purposes), and revealed features quite different from those found using optical telescopes.

In addition to the examination of radio emission from structures within the Solar System and our own Galaxy, Brown investigated possible emissions from extragalactic sources. In 1949, with C Hazard, he detected radio waves emanating from M31, the Andromeda nebula, at a distance of 2.2 million light years. But radio telescopes of the time lacked sufficient resolution to pinpoint a radio source accurately enough to identify that source through an optical telescope. It took three more years for Brown and his colleagues to devise the radio interferometer, which greatly improved resolution. Using the device, Brown measured the size of Cassiopeia A and Cygnus A – both very strong radio sources. Walter ◊Baade and Rudolph ◊Minkowski were then able to relate the more accurate radio locations given to their own optical observations, and as a result Cygnus A became the first radio source traced to a definite optical identification – even though it had a magnitude (brightness) of only 17.9.

In 1956 Brown devised a further refinement to radio astronomy, in the form of the technique of intensity interferometry. Since then he has used the stellar interferometer at Narrabi Observatory (in Australia) to study the sizes of hotter stars.

The early work carried out by Brown at Jodrell Bank contributed to the development of the 76-m/249-ft radio telescope, for a long time the largest steerable radio telescope in the world. Other types of radio telescopes have been developed since then, such as the 300-m/985-ft dish at Arecibo, in Puerto Rico; the 5-km/3-mi radio telescope at Cambridge (consisting of eight 13-m/42.6-ft dishes); and the VLA (Very Large Array) in the USA.

Brunel, Isambard Kingdom (1806–1859) English engineer and inventor. The only son of Marc Isambard ◊Brunel, he pursued a similar career, marked by hugely ambitious projects unparalleled in engineering history, becoming ever more ambitious as the years progressed.

Born in Portsmouth in 1806, he was sent to France at the age of 14 to the College of Caen in Normandy and later to the Henri Quatre school in Paris. Brunel was appointed resident engineer on his father's Thames Tunnel enterprise when only 19. This promising start was abruptly ended when he was seriously injured by a sudden flood of water into the tunnel. While recuperating from this accident at Bristol, he entered a design competition, submitting four designs for a suspension bridge over the River Avon. The judge, Thomas ◊Telford, rejected all of them in favour of his own. After many battles and a second contest, one of Brunel's designs was accepted. Work on the bridge began in 1833, but owing to lack of funds it was not completed until after its designer's death.

In 1833 Brunel was appointed to carry out improvements on the Bristol docks to enable heavily loaded merchant ships to berth more easily. It was while working on this project that Brunel's interest in the potential of railways was fired. The famous Rainhill trials, at which George Stephenson's *Rocket* had triumphed, had been held four years previously.

Brunel completed a survey for constructing a railway from London to Bristol, which was to be known (with a grandiloquence typical of Brunel) as the Great Western Railway. This characteristic love of the outsize was again evident in Brunel's decision to adopt a broad gauge of 2.1 m/7 ft for his locomotives, a choice that had the advantage of offering greater stability at high speeds. This size was in contrast to Stephenson's 'standard' of 1.44 m/4 ft 8 in.

In all, Brunel was responsible for building more than 2,600 km/1,600 mi of the permanent railway of the west of England, the Midlands, and South Wales. He also constructed two railway lines in Italy and acted as advisor on the construction of the Victoria line in Australia and on the East Bengal railway in India.

The bridges that Brunel designed for his railways are also worthy of note. Maidenhead railway bridge had the flattest archheads of any bridge in the world when it was opened, and Brunel's use of a compressed-air caisson to install the pier foundations for the bridge helped considerably to win acceptance of the

Brunel English engineer and inventor Isambard Kingdom Brunel next to the anchor chain of his last and largest steamship, the *Great Eastern*. Work on the *Great Eastern* fell into financial difficulty when it was realised that its launch on the Thames would be more expensive than expected because it was built broadside on to the river, and that its consumption of coal was twice that calculated. *Mary Evans/Institution of Civil Engineers*

compressed-air technique in underwater and underground constructions. However, many of Brunel's viaducts were workaday timber structures that used cheap, readily available materials and were designed so that renewal of members was quick and simple. They were only replaced when timber for repair rose to an uneconomic price. Of all the railway bridges Brunel produced, the last and the greatest was to be the Royal Albert, crossing the River Tamar at Saltash. It has two spans of 139 m/455 ft and a central pier built on the rock, 24 m/80 ft above the high-water mark. The bridge was opened in 1859, the year of Brunel's death.

No sooner, it was said, had Brunel provided a new land link between Bristol and London through his Great Western Railway than he decided to extend the link to New York. The means of achieving his aim came in the shape of ships: the *Great Western,* launched in 1837, the *Great Britain* of 1843, and the *Great Eastern* of 1858. Each was the largest steamer in the world at the time of its launch.

Brunel's predilection for large vessels was not simply the outcome of his love for the outsize – there was sound engineering reasoning behind his designs. At the time, there was a strong body of scientific opinion that held that ships could never cross the Atlantic under steam alone because they could not carry enough coal for the journey. And building a larger ship was no remedy, it was argued, because doubling the size of the vessel doubled the drag forces it had to overcome, thus doubling the power needed and therefore the amount of coal required. Brunel was probably the only one of his time who could see the fallacy of this argument. This lay in the fact that the coal-carrying capacity of a ship is roughly proportional to the cube of the vessel's leading dimension, while the water resistance increases only as the square of this dimension. The voyage of the *Great Western* proved Brunel right – when it docked after its first transatlantic voyage it had 200 tonnes of coal left in its bunkers.

The *Great Western,* with 2,340 tonnes displacement, was a timber vessel driven by paddles. Its crossing of the Atlantic in the unprecedented time of 15 days brought, after initial wariness, the most enthusiastic acclaim and established a regular steamship between the UK and the USA.

Brunel's next ship, the *Great Britain* of 3,676 tonnes displacement, represented a great advance in the design of the steamship. It had an iron hull and was the first ship to cross the Atlantic powered by a propellor. The value of its revolutionary hull was made clear on its first voyage, when it was beached in Dundrum Bay on the Southern Coast of Ireland. It remained there for the best part of a year without suffering serious structural damage. As a passenger ship, the *Great Britain* was an unprecedented success, remaining in service for 30 years, sailing to San Francisco, journeying regularly to Australia, and even serving as a troopship.

Then, in what was thought to be the final chapter of its life, the ship was badly damaged off Cape Horn in 1866, managing to struggle to the Falkland Islands only to be condemned. It lay, a hulk that refused to rot, in Sparrow cove until it was salvaged by the *Great Britain* project, set up in 1968. Through the efforts of the enthusiasts, who towed it to Montevideo and from there to Bristol, the *Great Britain* entered the dock where it was made on 19 July 1970, exactly 127 years from the day it was floated out.

On 31 January 1858 Brunel witnessed the spectacular sideways launching of his last ship, the *Great Eastern,* which was to remain the largest ship in service until the end of the 19th century. Well over ten times the tonnage of his first ship, it was 211 m/692 ft in length, had a displacement of 32,513 tonnes, and was the first ship to be built with a double iron hull. It was driven by both paddles and a screw propellor.

Initially the *Great Eastern* was beset with problems. There was constant engine trouble and the day after the ship set out on a new commissioning trial, Brunel – who was too ill to be on board – was struck down with paralysis. Unable to delegate responsibility, the work and worry of his many other enterprises had finally broken his health. He died at his home in London on 15 September 1859, having heard that an explosion aboard *Great Eastern* had apparently brought this most ambitious enterprise to nothing. Despite the damage this was not the end of his great ship. It was used successfully as a troop ship, and its greatest moment came in 1866 when, under the supervision of the great physicist Lord Kelvin, it was used to lay down the first successful transatlantic cable.

Brunel was elected to the Royal Society in 1830, at the age of 24, and was made a member of most of the leading scientific societies in the UK and abroad. These honours, however, seem scant reward for such a giant in an age that bred such engineering giants. He was the last, and the greatest, of them all.

Further Reading

Jenkins, David and Jenkins, Hugh, *Isambard Kingdom Brunel: Engineer Extraordinary,* Priory Press, 1977.

Rolt, L T C, *Isambard Kingdom Brunel,* Pelican Biographies series, Penguin, 1970.

Tames, Richard, *Isambard Kingdom Brunel, 1806–1859: An Illustrated Life,* Lifelines series, Shire Publications, 1972, v 1.

Vaughan, Adrian, *Isambard Kingdom Brunel: Engineering Knight-Errant,* John Murray, 1991.

Brunel, Marc Isambard (1769–1849) French-born British inventor and engineer who is best remembered for his success in overcoming the age-old problem of tunnelling in strata beneath water. His most notable achievement was the construction of the Thames Tunnel.

Born in Hacqueville in Normandy, Brunel served six years in the French navy, but left France in 1793, then at the peak of Revolutionary fervour, because of his royalist sympathies. His new home was the USA where he practised as an architect and civil engineer in New York, finally becoming the city's chief engineer. Here, besides surveying, canal engineering and constructing buildings, he advised on the improvement of the defences of the channel between Staten Island and Long Island, built an arsenal, and designed a cannon foundry.

While he was in the USA, Brunel developed an improved method for manufacturing ships' pulley blocks: at the time a 74-gun ship used 1,400 blocks, all of which had to be made by hand. Acting on the advice of a friend, Brunel sailed for England in 1799 and took the drawings of his invention to Henry ⌽Maudslay in London. Maudslay, a pioneer of machine tools and a former pupil of Joseph Bramah, was impressed by Brunel's designs with the result that the 43 machines operated by ten men produced blocks superior in quality and consistency to those that previously 100 men had made by hand. Historically it was a portent of what was soon to become a universal trend: specialist machine tools, each performing one of a series of operations and taking over almost all the many labours formerly performed by hand.

Brunel was a tireless and prolific inventor. Among the gadgets he devised were machines for sawing and bending timber, bootmaking, knitting stockings, printing, copying drawings, and manufacturing nails.

In 1814 the first of a series of disasters overtook Brunel when fire badly damaged his sawmill at Battersea. The heavy loss this occasioned him brought to the surface the financial incompetence with which his partners were conducting his enterprises. Brunel soon found himself heavily in debt, a condition that worsened when in the following year the final defeat of Napoleon at Waterloo brought peace to Europe, the government accordingly ceasing to pay for a workshop producing army boots. In 1821 Brunel was imprisoned for several months for debt, being released only when friends eventually obtained from the government a grant of £5,000 for his discharge on the condition that he should remain in England.

In 1825 Brunel started work on his last and longest commission: the construction of a tunnel under the Thames in London to link Rotherhithe with Wapping. Brunel took out a patent on a revolutionary tunnelling shield in 1818, a move that hastened his subsequent appointment as engineer of the Thames Tunnel Company. Although the project ended in success it underwent many reverses, took 18 years to complete, and shattered Brunel's health long before its completion.

At the time of the tunnel's construction, nothing so ambitious had been attempted before. The shield Brunel had designed covered the area to be excavated and consisted of 12 separate frames comprising altogether 36 cells in each of which a workman could be engaged at the workface independently of the others. The whole device obtained its propulsion from screw power which drove it forwards in 114 mm/4.5 in steps (the width of a brick) as the work progressed.

All too often, however, excavation had to stop. Five times water burst through the thin layer of earth beneath which the diggers worked, forcing a halt. Fortunately the shield always held, preventing total catastrophe. The long delays caused by these reverses put the scheme's finances under a severe strain. At one point, the whole operation came to a halt through lack of funds, the tunnel was bricked up, and for seven years no more work was done. When the operation was resumed, a much larger shield was introduced to cover the 120 mm/400 ft of the tunnel already constructed. The excavations took place only 4 m/14 ft under the riverbed at its lowest point.

The Thames tunnel was opened in 1843, and was the first public subaqueous tunnel ever built. It had a horseshoe cross section; its total length was 406 m/1,506 ft and its width 11 m/37 ft by height 7 m/23 ft. Brunel, elected to the Royal Society in 1814, was knighted in reward for his labours in 1814. The public flocked to go through the long-awaited tunnel and in three and a half months more than a million people had passed through it. The first trains used the tunnel in 1865. It is now part of the London Underground system.

Buchner, Eduard (1860–1917) German organic chemist who discovered non-cellular alcoholic fermentation of sugar – that is, that the active agent in the reaction is an enzyme contained in yeast, and not the yeast cells themselves. For this achievement he was awarded the 1907 Nobel Prize for Chemistry.

Buchner was born in Munich on 20 May 1860, of an old Bavarian family of scholars. His father was a professor of forensic medicine and obstetrics, as well as being editor of a medical journal. When Buchner graduated from the Realgymnasium he served in the field artillery before going to study chemistry at the Munich Technische Hochschule. His studies were again interrupted – for financial reasons this time – and he spent four years working in the canneries of Munich and Mombach. In 1884, with the assistance of his elder brother Hans, a bacteriologist, he resumed his academic training in the organic section of the chemical laboratory of the Bavarian Academy of Sciences in Munich, where he worked under Johann von Baeyer.

While he was studying chemistry, Buchner also worked in the Institute for Plant Physiology under the Swiss botanist Karl von Nägeli (1817–1891). He obtained his doctorate in 1888 and was appointed teaching assistant to Baeyer at the Privatdozent. In 1893 he succeeded Theodor Curtius (1857–1928) as head of the section for analytical chemistry at the University of Kiel and became associate professor there in 1895. Later professorships included appointments at Tübingen (1896), Berlin (1898), Breslau (1909), and Würzburg (1911). In 1914 he served in the German army as a captain in the ammunition supply unit and was promoted to major in 1916. He was recalled to Würzburg to teach for a short time but returned to the front in Romania on 11 August 1917. He was killed by a grenade four days later at Foçsani.

It was while Buchner was working at the Institute for Plant Physiology that he first became interested in the problems of alcoholic fermentation, and in his first paper (1886) he came to the conclusion that Louis ⌽Pasteur was wrong in his contention that the absence of oxygen was a necessary prerequisite for fermentation. In 1858 Moritz Traube had proposed that all fermentations were caused by what he termed 'ferments' – definite chemical substances that he thought were related to proteins and produced by living cells; in 1878 Willy Kühne (1837–1900) called these substances 'enzymes'. Many researchers, including Pasteur, had tried to liberate the fermentation enzyme from yeast.

In 1893 Buchner and his elder brother found that the cells of microorganisms were disrupted when they were ground with sand. After yeast had been treated in this way, it was possible to use a hydraulic press to squeeze out a yellow viscous liquid, free from cells. The Buchners were using the liquid for pharmaceutical studies (not for experiments on fermentation) and wished to add a preservative to it. As the juice was being used in experiments on animals, antiseptics could not be used and so Buchner added a thick sugar syrup to stop any bacterial action. He fully expected the sugar to act as a preservative, as it usually does, but to his surprise it had the opposite effect and carbon dioxide was produced. Thus the sugar had fermented, producing carbon dioxide and alcohol, in the same way as if whole yeast cells had been present. He named the enzyme concerned zymase.

Invertase, another enzyme of yeast, has been known since 1860 but zymase is different in that it is less stable to heat and catalyses a more complex reaction. Buchner was fortunate that he chose the correct type of yeast. It was soon realized that the conversion of sugar into alcohol by means of yeast juice is a series of stepwise reactions, and that zymase is really a mixture of several enzymes. It was to be 40 years before the process was fully understood, through the work of Arthur ⌽Harden, Otto ⌽Meyerhof, and others.

Buchner's other main research concerned aliphatic diazo compounds. Between 1885 and 1905 he published 48 papers that dealt with the synthesis of nitrogenous compounds, especially pyrazole. He also synthesized cycloheptane compounds.

Buckland, William (1784–1856) English geologist and palaeontologist, a pioneer of British geology.

Born in Axminster, the son of a clergyman, Buckland attended Oxford University and took holy orders. However, from an early age he had shown unbounded enthusiasm for the youthful science of geology. Appointed reader in mineralogy in 1813, he became reader in geology in 1818, while his ecclesiastical career culminated in his elevation in 1845 to the deanery of Westminster.

Always an enthusiastic fieldworker and a renowned lecturer, his geological investigations blossomed in three distinct, though related, areas. First, he made major contributions to the descriptive and historical stratigraphy of the British Isles, inferring from the vertical succession of the strata a stage-by-stage temporal development of the globe's crust. In this, he built on the pioneering stratigraphical work of William ◊Smith and the palaeontology of Georges ◊Cuvier. Second, he became a celebrated palaeontologist, using the techniques of Cuvierian comparative anatomy. In this respect, his greatest achievement lay in his reconstruction of *Megalosaurus*, and, in his book *Relics of the Deluge* (1823), in exploring the geological history of Kirkdale Cavern, a hyena cave den in Yorkshire. Third, he long sought to discover evidence for catastrophic transformations of the Earth's surface in the geologically recent past, as indicated by features of relief, fossil bones, erratic boulders, and gravel displacement. In his *Geology Vindicated* (1819), he confidently attributed such items to the Biblical Flood – an assertion he felt forced, however, to withdraw some 15 years later in the light of fresh evidence, adduced by Charles ◊Lyell and others, of the power of gradual and regular geomorphological processes. Buckland's interest in such evidence of violent change nevertheless helped him to become a leading and early British exponent of Louis ◊Agassiz's glacial theory.

Buffon, Georges-Louis Leclerc, Comte de (1707–1788) French naturalist who compiled the vast encyclopedic work *Histoire naturelle générale et particulière.*

Buffon was born in Montbard on 7 September 1707 and was educated at the Jesuits' College in Dijon. He graduated in law in 1726 and took the opportunity to study mathematics and astronomy. He travelled a great deal, and spent some time in England where science was undergoing a renaissance. Buffon set himself the task of translating the works of Newton and Hales into French. In 1732 Buffon's mother died and left him a handsome legacy, so the young man was financially stable enough to devote himself entirely to his scientific interests. Buffon was elected associate of the French Academy of Sciences in 1739 and took up the appointment of keeper of the Botanical Gardens (Jardin du Roi), a post that stimulated his interest in natural history. He was a prolific writer, and he turned his skills towards compiling what would eventually be a 44-volume work on natural history encompassing both the plant and the animal kingdoms. He became a member of the French Academy in 1753 and was made a count in 1771 and a Fellow of the Royal Society in 1739. He died in Paris on 16 April 1788 after a long and painful illness.

Le génie n'est qu'une grande aptitude à la patience. *Genius is only a great aptitude for patience.*

GEORGES-LOUIS BUFFON *ATTRIBUTED REMARK*

Buffon's encyclopedia was the first work to cover the whole of natural history and it was extremely popular. He wrote in a clear and interesting style – which he regarded as more important than originality – and did not personally originate a great deal of the material. He was aided by several eminent naturalists of the time, and organized the sometimes confusing wealth of material into a coherent form.

Although Buffon's work was inclined to generalizations, he proposed some innovatory and stimulating theories. He suggested that a cosmic catastrophe initiated the Earth's beginnings, and that its existence was far older than the 6,000 years suggested by the book of Genesis. He observed that some animals retain parts that are vestigial and no longer useful, suggesting that they have evolved rather than having been spontaneously generated. Theories such as these could have caused a furore at a time when it was strongly believed that the creation of the world and humanity had occurred as defined in the Bible, and even though Buffon wrote with political care he did upset the authorities and had to recant. It was not until the theories of Charles ◊Lyell and Charles ◊Darwin that people began to take such ideas seriously.

Buffon's encyclopedia did, however, arouse a great interest in natural science, which carried through to the early part of the 19th century.

Bullard, Edward Crisp (1907–1980) English geophysicist who, with US geologist Maurice ◊Ewing, is generally considered to have founded the discipline of marine geophysics. He received many honours for his work, including a knighthood in 1953 and the Vetlesen Medal and Prize (the earth sciences' equivalent of a Nobel prize) in 1968.

Bullard was born in Norwich on 21 September 1907 into a family of brewers. He was educated at Repton School and Clare College, Cambridge, from which he graduated in physics in 1929. In 1931, after two years' research at the Cavendish Laboratory, Cambridge, he became a demonstrator in Cambridge University's Department of Geodesy and Geophysics. During World War II he researched into methods of demagnetizing ships and of sweeping acoustic and magnetic mines. He then became the British Admiralty's assistant director of naval operational research under Patrick Blackett, and later served on several committees concerned with combating German V1 and V2 flying bombs. He continued to advise the Ministry of Defence for several years after the war. At the end of his wartime military service he returned to Cambridge as a reader in geophysics and soon became head of geophysics but, disillusioned by the university's tardiness in promoting growth in this subject area, went to Canada in 1948 to become professor of geophysics at the University of Toronto. He returned to England two years later, however, to take up the directorship of the National Physical Laboratory at Teddington. In 1957 he returned to Cambridge as head of geodesy and geophysics, a position he held until his official retirement in 1974. On retiring he moved to La Jolla, California, where he continued to teach at the University of California (of which he had been a professor since 1963). He also advised the US government on nuclear waste disposal. In addition to his academic and advisory posts, he played an active part in his family's brewing business and was a director of IBM (UK) for several years. Bullard died on 3 April 1980 in La Jolla.

Bullard's earliest work was to devise a technique (involving timing the swings of an invariant pendulum) to measure minute gravitational variations in the East African Rift Valley. Also before World War II he investigated the rate of efflux of the Earth's interior heat through the land surface and – influenced by the work of Ewing, a professor at Columbia University – he pioneered the application of the seismic method to study the sea floor. After the war, while at Toronto University, Bullard developed his 'dynamo' theory of geomagnetism, according to which the Earth's magnetic field results from convective movements of molten material within the Earth's core. Also in this period – working at the Scripps Institute of Oceanography, California, in his summer vacations – he devised apparatus for measuring the flow of heat through the deep sea floor.

After returning to Cambridge he played a large part in developing the potassium–argon method of rock-dating. He also studied continental drift – before the theory became generally accepted. Using a computer to analyse the shapes of the continents, he found that they fitted together reasonably well, especially if other factors such as sedimentation and deformation were taken into account. Later, when independent evidence for these factors had been found, Bullard's findings lent considerable support to the continental drift theory.

Bunsen, Robert Wilhelm (1811–1899) German chemist who pioneered the use of the spectroscope to analyse chemical compounds. Using the technique, he discovered two new elements, rubidium and caesium. He also devised several pieces of laboratory apparatus, although he probably played only a minor part (if any) in the invention of the Bunsen burner.

Bunsen was born on 31 March 1811 at Göttingen, the son of a librarian and linguistics professor at the local university. He studied chemistry there and at Paris, Berlin, and Vienna, gaining his PhD in 1830. He was appointed professor at the Polytechnic Institute of Kassel in 1836, and subsequently held chairs at Marburg (1838) and Breslau (1851) before becoming professor of experimental chemistry at Heidelberg in 1852. He remained there until he retired. Bunsen never married, and ten years after retiring he died, on 16 August 1899.

Bunsen's first significant work, begun in 1837, was on cacodyl compounds, unpleasant and dangerous organic compounds of arsenic; a laboratory explosion cost Bunsen the sight of one eye and he nearly died of arsenic poisoning. He did, however, stimulate later researches into organometallic compounds by his student, the British chemist Edward Frankland (1825–99). In 1841 he devised the Bunsen cell, a 1.9-volt carbon–zinc primary cell that he used to produce an extremely bright electric arc light. Then, in 1844, he invented a grease-spot photometer to measure brightness (by comparing a light source of known brightness with that being investigated). His contribution to the improvement of laboratory instruments and techniques gave rise also to the Bunsen ice calorimeter, which he developed in 1870 to measure the heat capacities of substances that were available in only small quantities.

Bunsen's first work in inorganic chemistry made use of his primary cell. Using electrolysis, he was the first to

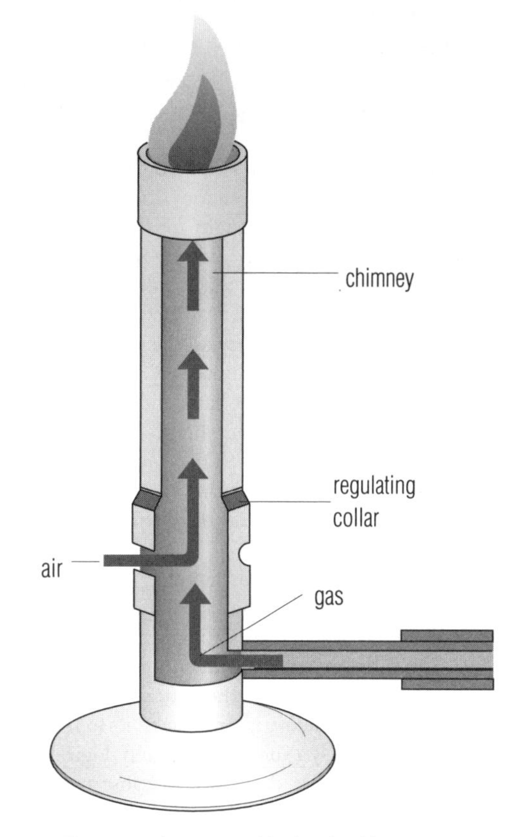

Bunsen The Bunsen burner, used for heating laboratory equipment and chemicals. The flame can reach temperatures of 1,500°C/2,732°F and is at its hottest when the collar is open.

isolate metallic magnesium, and demonstrate the intense light produced when the metal is burned in air. But his major contribution was the analysis of the spectra produced when metal salts (particularly chlorides) are heated to incandescence in a flame, a technique first advocated by the US physicist David ◊Alter. Working with Gustav ◊Kirchhoff in about 1860, Bunsen observed 'new' lines in the spectra of minerals, which represented the elements rubidium (which has a prominent red line) and caesium (blue line). Other workers using the same technique soon discovered several other new elements.

The Bunsen burner, probably used to heat the materials for spectroscopic analysis, seems to have been designed by Peter Desdega, Bunsen's technician. Gas (originally coal gas, but any inflammable gas can be used) is released from a jet at the base of a chimney. A hole or holes at the base of the chimney are encircled by a movable collar, which also has holes. Rotation of the collar controls the amount of air admitted at the base of the chimney; the air–gas mixture burns at the top. With the air holes closed, the gas burns with a luminous, sooty flame. With the air holes open, the air–gas mixture burns with a hot, nonluminous flame (and makes a characteristic roaring sound).

Burali-Forte, Cesare (1861–1931) Italian mathematician who is famous for the paradox named after him and for his work on the linear transformations of vectors.

Burali-Forte was born in Arezzo on 13 August 1861. He received his BA in mathematics from the University of Pisa in 1884 and then taught for three years at the Technical School in Sicily before being appointed extraordinary professor at the Academia Militare di Artiglieria e Genio in Turin. At Turin he lectured on analytical projective geometry. In the years 1894–96 he served as assistant to Giuseppe ◊Peano at the University of Turin, and he later did much to make known Peano's work on mathematical logic, especially by his expanded edition (with his own interpolations) of Peano's *Logica Mathematica.* He remained at the Academia Militare until his death, in Turin, on 21 January 1931.

Burali-Forte published his famous paradox in 1897: 'To every class of ordinal numbers there corresponds an ordinal number which is greater than any element of the class'. This discovery dealt a sudden, and severe, blow to the developing science of mathematical logic – that is, to the notion that mathematics (or at least its foundations) could be adequately expressed in purely logical terms. What Burali-Forte had done was to expose a contradiction in Georg Cantor's theory of infinite ordinal numbers, and in 1902 Bertrand ◊Russell demonstrated that this contradiction was of a fundamental logical character and could not be overcome by minor changes in the theory. It was thus Burali-Forte who brought to the fore the threat which such paradoxes posed to the foundations of mathematical logic.

Burali-Forte's chief interest, and major accomplishments, however, lay in the field of vector analysis. Much of this work was done in collaboration with Roberto Marcolongo. In 1904 they published a series of papers on the unification of vector notation, including in this work a comprehensive analysis of all the notations that had been proposed for a minimal system. Five years later they produced their own proposals for a unified system of vector notation. Having thus laid the groundwork, they began in 1909 to study the linear transformation of vectors. Of great importance was Burali-Forte's simplification of the foundations of vector analysis by the introduction of the notion of the derivative of a vector with respect to a point, which led to new applications of the theory of vector analysis and, in particular, to more efficient treatment of such operators as the Lorenz transformations.

In 1912–13 Burali-Forte published more volumes on linear transformations and demonstrated their application to such things as the theory of mechanics of continuous bodies, hydrodynamics, optics, and some problems of mechanics. His great ambition was to produce an encyclopedia of vector analysis and its applications, but he did not live to complete this work. His last contribution, a paper on differential projective geometry, was finished in 1930, shortly before his death.

Burbidge, Geoffrey (1925–) and (Eleanor) Margaret (born Peachey) (1919–) British husband-and-wife team of astrophysicists distinguished for their work, chiefly in the USA, on nucleosynthesis – the creation of elements in space – and on quasars and galaxies.

Geoffrey Burbidge studied physics at Bristol University (graduating in 1946), then combined lecturing and research for his PhD at University College, London, before going to the USA as Agassiz Fellow at Harvard University. From 1952 to 1953 he was a research fellow at the University of Chicago; he then returned to England as a research fellow at the Cavendish Laboratories, Cambridge. In 1955 he went back to the USA as Carnegie Fellow at the Mount Wilson and Palomar observatories, and in 1957 he joined the department of astronomy of the University of Chicago as assistant professor. Appointments followed at the University of California, San Diego, as associate professor 1962–63 and professor of physics 1963–84 and from 1988. He is at present director of the Kitt Peak National Observatory, Arizona.

Margaret Burbidge studied at University College, London, and gained her PhD for research at the

University of London Observatory, where in 1948 she became assistant director and then acting director 1950–51. Travelling to the USA, she held a fellowship from the International Astronomical Union at the Yerkes Observatory, University of Chicago. From 1955 to 1957, she was a research fellow at the California Institute of Technology. Her next appointment was at the Yerkes Observatory again, firstly as Shirley Farr Fellow and later, in 1959, as associate professor. Three years later she moved to the University of California, San Diego, as research astronomer before becoming professor of astronomy from 1964, being granted leave of absence to be director of the Royal Greenwich Observatory 1972–73.

Together the Burbidges published *Quasi-Stellar Objects* in 1967. Their work in nucleosynthesis followed the discovery of the spectral lines of the unstable element technetium in red giant stars by Paul Merrill (1889–1961) in 1952. Because technetium is too unstable to have existed for as long as the stars themselves, the discovery provided the first evidence for the actual creation of elements. The paper published by the Burbidges, William Fowler, and Fred Hoyle in 1957 began with the premise that at first stars consisted mainly of hydrogen and that most of the stars now visible are in the process of producing helium from hydrogen and releasing energy as starlight. They then suggested that as stars age some of their helium is 'burned' to form other elements, such as carbon and oxygen. The carbon and oxygen may trap hydrogen nuclei (protons) to form more complex nuclei, or may trap helium nuclei (alpha particles) to produce magnesium, silicon, sulphur, argon, and calcium.

The Burbidges and their colleagues distinguished five additional processes; one is the e-process, in which elements such as iron, nickel, chromium, and cobalt are formed at a high temperature. Up to this point, the 'iron peak', the buildup results in energy being released. Beyond the iron peak more energy is required to create heavier elements. In a supernova, a massive star exploding, this energy is available. Prior to the supernova, the star has medium-weight elements. It becomes unstable and nuclei trap neutrons so rapidly (the r-process) that newly formed nuclei do not have time to shed electrons. There is subsequent explosion and heavier elements such as selenium, bromine, krypton, tellurium, iodine, xenon, osmium, iridium, platinum, gold, uranium, and a number of unfamiliar elements are formed. From theoretical considerations, the collaborators calculated the proportions of the different heavy elements that would be most likely to be formed in a supernova. Observations indicate that the distribution of heavy elements could be explained by their production in supernovae. They believe a slow process (s-process) in red giants also builds up heavy elements.

The Burbidges also researched into quasars, quasi-stellar objects originally detected by their strong radio emissions and believed to be travelling away from the Earth at immense speed. Quasars give off ultraviolet radiation, and the Doppler effect, caused by their receding at great speed, results in the spectrum's being shifted towards the red (red shift), so that only faint points of blue light from the quasar reach Earth. Determining the spectra of suspected quasars is a laborious task. Margaret Burbidge measured the red shifts of a number of objects found by means of the screening process devised by Martin ◊Ryle and Allan Sandage (1926–), and in the process she detected objects without radio radiation but with large red shifts; they are now placed in the general group of quasars.

In 1963, the Burbidges and Sandage reviewed the evidence for intense activity in the nuclei of radio galaxies, quasi-stellar objects, and Seyfert galaxies (those with a small bright nucleus, fainter arms, and broad spectral emission lines). In 1970, using evidence gained from observations, Geoffrey Burbidge and Arthur Wolfe calculated that the stars emitting light in elliptical galaxies could not account for more than 25% of the mass. They produced arguments to indicate that black holes, from which light cannot escape, are the most likely source of the missing mass.

Another important paper, published by the Burbidges jointly with Solomon and Stritmatter, described the discovery that four quasars (listed in the third Cambridge catalogue) lie within a few arc minutes of bright galaxies. They suggested that the quasars and galaxies were linked in some way. Evidence found since then – examples of quasars located on opposite sides of a galaxy – tends to support this belief, which is accepted by several notable astronomers.

Burkitt, Denis Parsons (1911–1993) Northern Irish surgeon best known for his description of the childhood tumour named after him, Burkitt's lymphoma. He is also known for stressing the importance of roughage in the diet.

Burkitt was born on 28 February 1911 at Enniskillen, near Lough Erne. He was educated at the local school and later at schools in Anglesey and Cheltenham. When he was 18 years old he entered Dublin University to study engineering, but later turned his interests to medicine. He worked as a surgeon in the armed forces and became a Fellow of the Royal College of Surgeons in 1938 in Edinburgh. In 1946 he was accepted into the Colonial Service and eventually became senior consultant to the Ministry of Health in Kampala, Uganda, in 1961.

In 1957 Burkitt examined a child in Kampala. This was his first case of the lymphoma that typically affects the face and jaw, presenting several swellings.

Subsequent observations convinced him that these diversely distributed lumps were of a single tumour type. This has now been histologically confirmed by the presence of 'starry sky' cells within the tumours. Burkitt undertook a 15,000-km/9,300-mi safari with two other doctors, Edward Williams and Clifford Nelson, to discover whether there is a geographical correlation with the incidence of Burkitt's disease. Their research eventually showed that the lymphoma is commonest in areas of certain temperature and rainfall, in fact, where malaria is endemic. It is also associated with the presence of antibodies to the Epstein–Barr virus. Burkitt's work in the comparatively new field of geographical pathology was acknowledged in 1972 when he was elected a Fellow of the Royal Society.

He also became well known for his theories that a high-fibre diet prevents many of the common ailments of the Western world, such as appendicitis, diverticular disease, and carcinoma of the bowel, all of which are rarely encountered among the African peoples, for example. Burkitt pioneered the popular trend of high-roughage diets.

Burnet, (Frank) Macfarlane (1899–1985)
Australian immunologist whose research into viruses inspired his theory that antibodies could be produced artificially in the body in order to develop a specific type of immunity, which led to the concept of acquired immunological tolerance, particularly important in tissue-transplant surgery. For this work he was awarded the 1960 Nobel Prize for Physiology or Medicine, which he shared with Peter ◊Medawar.

Burnet was born on 3 February 1899 at Traralgon, Victoria. After graduating in biology from Geelong College, Victoria, he obtained his medical degree at Melbourne University in 1923 and for the following year was resident pathologist at Melbourne Hospital. He studied at the Lister Institute, London, 1926–27 and in that year gained his PhD from the University of London. He then returned to Australia as assistant director of the Walter and Eliza Hall Institute for Medical Research, Melbourne, and held this position until 1944, when he became the institute's director. Burnet was knighted in 1951. He was appointed emeritus professor of Melbourne University in 1965 and made an Honorary Fellow of the Royal College of Surgeons in 1969.

Early in his career Burnet did extensive research on viruses. He was the first to investigate the multiplication mechanism of bacteriophages (viruses that attack bacteria) and devised a method for identifying bacteria by the bacteriophages that attack them. This work was of immense importance, particularly 20 years later, when bacteriophages were first used as research tools in genetics and molecular biology.

In 1932 Burnet developed a technique for growing and isolating viruses in chick embryos, a technique that was to be used as a standard laboratory procedure for more than 20 years. Burnet's work on the chick embryo increased interest in the specific character of an embryo by which it seemed to be unable to resist virus infection or to produce any antibodies against viruses. Early attempts were made in Burnet's laboratory to use the chick embryo to show that tolerance could be produced artificially.

The idea of man as a dominant animal of the earth whose whole behaviour tends to be dominated by his own desire for dominance gripped me. It seemed to explain almost everything.
Macfarlane Burnet Dominant Manual *1970*

As a result of his virus research, Burnet became interested in immunology and in 1949 he predicted that an individual's ability to produce a particular antibody to a particular antigen was not innate, but was something that developed during the individual's life. In 1951 Medawar carried out the experiments that confirmed this theory.

Burnet's second major contribution to immunology was made in 1957 – his highly controversial 'clonal selection' theory of antibody formation, which explains why a particular antigen stimulates the production of its own specific antibody. According to Burnet, there is a region in the genes of the cells that produce antibodies that is continually mutating, such that each mutation leads to a new variant of antibody being produced. Normally the cells that produce a particular variant are few, but if the antibody they produce suddenly finds a target, then they multiply rapidly to meet this demand, and the other, useless, variants die out.

In recent years there has been a large amount of research on the immune response and on the many ways of imitating nature's way of reacting without inoculating the embryo. In particular, work has been carried out on the production of tolerance by drugs such as 6-mercaptopurine, in the hope that it will prove effective in surgical organ transplants. Burnet's investigations have stimulated research into the way viruses cause infection. His own research helped to eradicate diseases such as myxomatosis and isolate organisms such as *Rickettsia burneti,* which causes Q fever in sheep and cattle.

Burnet's publications include *Viruses and Man* (1953) and *The Clonal Selection Theory of Acquired Immunity* (1959).

Burnside, William (1852–1927) English applied mathematician and mathematical physicist whose interest turned in his later years to a profound absorption in pure mathematics. In this field he was particularly prominent for his research in group theory and the theory of probability.

Burnside was born in London on 2 July 1852, and was orphaned when only a young child. He proved so gifted in mathematics at Christ's Hospital that in 1871 he won a scholarship to St John's College, Cambridge. He transferred to Pembroke College after two years and graduated with high honours in 1875, winning the first Smith's Prize and being given the post of fellow and lecturer at the College. Apart from his normal lecturing duties, Burnside began to give advanced courses on hydrodynamics to other groups outside the college.

In 1885 he left Cambridge to take up the chair of mathematics at the Royal Naval College at Greenwich where, in addition to his teaching routine, he often accepted the responsibility of being an examiner for the universities and the civil service. It was at this stage that his interests began to move towards group theory. Even as he formalized the institution of instruction at three different levels at the Royal Naval College, he was putting together his thoughts and his papers for a book that finally appeared in 1897. A revised form of that work is now regarded as a classic. A member of the Royal Society from 1893, Burnside served on the society's council 1901–03 and was awarded its Royal Medal. After World War I he began to reduce his scholarly output and retired from the Royal Naval College in 1919. He continued to write on mathematical subjects, and a nearly completed manuscript on probability theory was found after his death and published posthumously. He died in West Wickham, Kent, on 21 August 1927.

Although it was well outside the scope of Burnside's undergraduate courses, his first publication (produced while he was at Cambridge) was in the field of elliptic functions. His study of elliptic functions led him, over the years, to study the functions of real variables and the theory of functions in general. One of his most influential papers, written in 1892, was a development of some work by Henri Poincaré on automorphic functions.

Burnside's lectures at the Royal Naval College were on a surprisingly wide range of subjects. He varied the content and the standard of his lectures according to whether he was addressing the junior (ballistics), senior (dynamics and mechanics), or the advanced (hydrodynamics and kinetics) section of trainees.

His research interests in the meantime included differential geometry and the kinetic theory of gases. During the early 1890s, references to group theory began to enter his work, and by the middle of the decade his study of automorphic functions had brought him fully into the field. He became particularly concerned with the theory of the discontinuous group of finite order. In 1897 he published the first book on group theory to appear in English. The papers that he and other mathematicians wrote over the ensuing years produced such major advances in the subject that a revised edition was soon necessary. It was issued in 1911, and is today considered to be a standard work.

Work on the theory of probability began only in 1918, by which time Burnside had virtually ended his career. Nevertheless he was sufficiently interested in the subject to draft a manuscript on it, which – although incomplete at the time of his death – was of so high a standard that it was published.

Bush, Vannevar (1890–1974) US electrical engineer and scientist who developed several mechanical and mechanical–electrical analogue computers that were highly effective in the solution of differential equations. During World War II he was scientific advisor to President Roosevelt and was instrumental in the initiation of the atomic bomb project. Later he greatly influenced the development of postwar science and engineering in the USA, being instrumental in the setting-up and running of the Office of Scientific Research and Development and its successor, the Research and Development Board.

Bush was born in Everett, Massachusetts on 11 March 1890. His father had first been a sailor but later became a cleric. During his childhood Vannevar developed an interest in practical things and constructed a radio receiver when they were almost unknown. He attended a local high school and then went on to Tufts College. He was forced to supplement the little money he had to support himself by washing dishes and giving private coaching in mathematics. In 1913 Bush received BSc and MSc degrees, the latter being for a thesis on a machine for plotting the profile of the land. He patented this device, which combined a gear mechanism mounted between two bicycle wheels which recorded the vertical distance the machine had risen while a paper was moved forward proportional to the distance travelled; a pen recorder drew the profile. After graduation he obtained a job as an engineer with the General Electric Company, but was made redundant shortly afterwards. This misfortune inspired him to read for a doctorate in engineering at the Massachusetts Institute of Technology (MIT). He proved himself so able at his task that he gained his degree in 1916 after only one year of study. From 1932 he held senior positions at MIT, including that of dean of the engineering school. He died of a stroke on 28 June 1974.

During World War I he worked on a magnetic device for detecting submarines. In 1919 he returned to MIT and was appointed associate professor of power transmission. In his research on the distribution of electricity, many problems arose which contained differential equations. Such equations, containing differential coefficients, can be difficult and time-consuming to solve in an explicit way, even if a solution is possible – an algebraic solution is of little use to a practising engineer who needs numbers with which to work. In about 1925 Bush began to construct what he called the product integraph. It contained a linkage to form the product of two algebraic functions and represent it mechanically. He also devised a watt–hour meter, a direct ancestor of the electricity meter now found in nearly all homes, which integrated the product of electric current and voltage. By similarly using current and voltage to represent equations, his 'calculus' machine was able to evaluate integrals involved in the solution of differential equations.

The product integraph, while suitable for solving the problems Bush had encountered, was limited to the solution of first-order differential equations. A device capable of the solution of second-order equations would be much more useful to electrical engineers, and to scientists generally. There was, however, a complication in coupling two watt–hour meters together. Bush found the solution by coupling one of the meters to a mechanical device known as a Kelvin integrator, after the physicist Lord Kelvin.

In 1931 Bush began work on an almost totally mechanical, and very much more ambitious machine known as the differential analyser. This machine had six integrators, three input tables, and an output table, which showed the graphical solution of an equation. The Bush analyser was the model for developments in the mechanical world, and many similar machines were built. One, a large differential analyser at Manchester University, was completed in 1935 and is now an exhibit in the Science Museum, London.

Following the success of the differential analyser, Bush built several specific-purpose analysers for solving problems related to electrical networks and for the evaluation of particular types of integrals. Then he built another large analyser, this time with many of the operations electrified, and a tape input that reduced the time taken for setting up a problem from days to minutes.

Bush's machines, and others that had been built in the USA, were used during World War II for military purposes, particularly the calculation of artillery range tables. The same military applications spawned the new machine, the digital electronic computer, which was to be the downfall of mechanical analogue devices.

One of Bush's other inventions was a cipher-breaking machine that played a large part in breaking Japanese codes. One of his less practical devices was a bird perch that dropped the birds the garden owner did not want to encourage. There were also many more patented and unpatented inventions to Bush's name.

Further Reading

Burke, Colin, *Information and Secrecy: Vannevar Bush, Ultra, and the Other Memex,* Scarecrow Press, 1993.

Genuth, Joel, 'Microwave radar, the atomic bomb and the background to US research priorities in World War II', *Sci Technol Hum Val,* 1988, v 13, pp 276–289.

Goldberg, Stanley, 'Inventing a climate of opinion: Vannevar Bush and the decision to build the bomb', *Isis,* 1992, v 83, pp 429–452.

Owens, Larry, 'Vannevar Bush and the differential analyzer: The text and context of an early computer', *Technol Cult,* 1986, v 27, pp 63–95.

Zachary, G Pascal, *The Endless Frontier: Vannevar Bush, Engineer of the American Century,* The Free Press, 1997.

Cailletet, Louis Paul (1832–1913) French physicist and inventor who is remembered chiefly for his work on the liquefaction of the 'permanent' gases: he was the first to liquefy oxygen, hydrogen, nitrogen, and air, for example.

Cailletet was born in Chatillon-sur-Seine on 21 September 1832, the son of an ironworks owner. He was educated at the college in Chatillon, the Lycée Henry IV in Paris, and the Ecole des Mines (also in Paris), after which he returned to Chatillon to manage his father's ironworks. Shortly afterwards he began his metallurgical studies, and later extended his research to the problems of liquefying gases. He died in Paris on 5 January 1913.

Investigating the causes of accidents that occurred during the tempering of incompletely forged iron, Cailletet found that many were due to the highly unstable state of the iron while it was hot and had gases dissolved in it. Also in the field of metallurgy he analysed the gases from blast furnaces. Other scientists had drawn off the gases under conditions that resulted in gradual cooling, which enabled the dissociated components in the gases to recombine. Cailletet, using a new technique by which the gases were cooled suddenly as soon as they had been collected, showed that the gases comprised of a large proportion of finely divided carbon particles, carbon monoxide, oxygen, hydrogen, and a small proportion of carbon dioxide – a composition different from that obtained by the old sampling method. As a result of these and other metallurgical studies, Cailletet developed a unified concept of the role of heat in changes of state of metals, and confirmed the views with which Antoine ◊Lavoisier had introduced his *Traité élémentaire de chimie* in 1789. In 1883 Cailletet was awarded the Priz Lacaze of the French Academy of Sciences for his work in metallurgy.

Cailletet's best-known work, however, was on the liquefaction of gases. Until his studies (which were completed by 1878) several gases – including oxygen, hydrogen, nitrogen, and air – were considered to be 'permanent' because nobody had succeeded in liquefying them, despite numerous attempts involving the use of what were then considered to be extremely high pressures. Cailletet realized that the failure of these attempts was due to the fact that the gases had not been cooled below their critical temperatures (the concept of critical temperature – the temperature above which liquefaction of a gas is impossible – was put forward by the Irish chemist Thomas ◊Andrews by 1869). In order to obtain the necessary amount of cooling, Cailletet employed the Joule–Thomson effect (the decrease in temperature that results when a gas expands freely), compressing a gas, cooling it, then allowing it to expand to cool it still further. Using this method he liquefied oxygen in 1877 and several other 'permanent' gases in 1878, including hydrogen, nitrogen, and air. (Raoul Pictet (1846–1929), working independently, also liquefied oxygen in 1877.)

Cailletet's other achievements included the installation of a 300-m/985-ft-high manometer on the Eiffel Tower; an investigation of air resistance on falling bodies; a study of a liquid oxygen breathing apparatus for high-altitude ascents; and the construction of numerous devices, including automatic cameras, an altimeter, and air-sample collectors for sounding-balloon studies of the upper atmosphere. He was elected president of the Aéro Club de France for these accomplishments.

Cairns, Hugh John Forster (1922–) English virologist known for his research into cancer.

Cairns was born in Oxford on 21 November 1922, the son of Professor Sir Hugh Cairns, a physician and fellow of Balliol College, Oxford. He attended Edinburgh Academy from 1933 to 1940, when he went to Balliol and gained his medical degree in 1943. His first appointment was in 1945 as surgical resident at the Radcliffe Infirmary, Oxford, and the next five years were spent in various appointments in London, Newcastle, and Oxford. From 1950–51 Cairns was a virologist at the Hall Institute in Melbourne, Australia, and he then went to the Viruses Research Institute, Entebbe, Uganda. In 1963 he became director of the Cold Spring Harbor Laboratory of Quantitative Biology in New York, a position he held until 1968. He then took professorships at the State University of New York and with the American Cancer Society. From 1973–80 he was in charge of the Mill Hill Laboratories of the Imperial Cancer Research Fund, London. He then moved to the USA to become professor of microbiology 1980–91 at the Harvard School of Public Health, Boston.

One of Cairns's first pieces of research in the early 1950s was into penicillin-resistant staphylococci, and their incidence in relation to the length of a patient's stay in hospital. He found that the rapid rise in their incidence was caused by continuous cross-infection with a few strains of the bacteria rather than repeated instances of fresh mutations. These findings are now generally accepted, although they were not at the time.

Cairns has also done much research on the influenza virus and in 1952 discovered that the virus is not released from the infected cell in a burst– as is a bacteriophage – but in a slow trickle. This evidence has since been found also to be true for the polio virus. In the following year he showed that the influenza virus particle is completed as it is released through the cell surface (also unlike a phage). This discovery has since been confirmed by electron microscopy and isotope-incorporation techniques. In 1959 Cairns succeeded in carrying out genetic mapping of an animal virus for the first time. In 1960 he showed that the DNA of the vaccinia virus is replicated in the cytoplasm of the cell (rather than in its nucleus) and that each infecting virus particle creates a separate DNA 'factory'.

Cairns's investigations into DNA have also led him to look at the way that DNA replicates itself and to compare the rates of replication of DNA in mammals with those in the bacterium *Escherichia coli.* He has found that mammalian DNA is replicated more slowly than that of *E. coli,* but is replicated simultaneously at many points of replication.

His later work studied the link between DNA and cancer, some forms of which may be caused by the alkylation of bases in the DNA. He showed that bacteria are able to inhibit the alkylation mechanism in their own cells, and later demonstrated this ability in mammalian cells. A similar mechanism probably prevents a high incidence of DNA mutations in human beings despite the presence of alkylating agents in the environment. Cairns made many important advances in the study of cancer and its relation to society. He has also spent much time in fundraising for cancer research.

Further Reading

Frankel, G J, *Hugh Cairns: First Nuffield Professor of Surgery, University of Oxford,* Oxford University Press, 1991.

Cajori, Florian (1859–1930) Swiss-born US historian of mathematics. His books dealt with the history of both elementary and advanced mathematics, as well as the teaching of mathematics (including its importance in education). Cajori was also the author of biographies of eminent mathematicians.

Cajori was born on 28 February 1859 in St Aignan, near Thusius. At the age of 16 he emigrated to the USA, where he took up studies at the University of Wisconsin. He was awarded his bachelor's degree in 1883, and in 1885 was offered the position of assistant professor of mathematics at Tulane University in New Orleans. There he continued his own studies in parallel with his teaching activities, earning his master's degree in 1886, and his PhD in 1894. In 1889 Cajori moved to Colorado College at Colorado Springs in order to take up the position of professor of physics, but returned to Tulane University in 1898 to become professor of mathematics, a position he held until 1918. He served also as dean of the department of engineering 1903–18. In 1918 Cajori moved to the University of California at Berkeley to take up the chair of history of mathematics, a post he retained until his death. He died at Berkeley on 14 August 1930.

Cajori's influence on the modern perception of the development of mathematics was profound, and his works are frequently quoted to this day. His reputation is founded mainly on his many books on the history of mathematics, although a number of his works – notably his edited version of Newton's *Principia mathematica* (published posthumously) – have been subject to some criticism for their interpretation of historical material. His two-volume *History of Mathematical Notations* (1928–29) is, however, still very much a standard reference text. He also compiled *A History of Physics* (1899).

Calkins, Mary Whiton (1863–1930) US psychologist who created a method of memorization called the right associated method. She was the founder of the psychology department at Wellesley College, Boston, Massachusetts, and the first female president of the American Psychological Association in 1905 and of the American Philosophical Association in 1918.

Born on 30 March 1863 in Hartford, Connecticut, Mary Whiton Calkins was the eldest of the five children of Wolcott and Charlotte Grosvenor Calkins. Her father was a Presbyterian minister and her mother a Puritan. Most of her childhood was spent in Buffalo, New York, and the family moved to Newton, Massachusetts, in 1880 where her father had been offered a congregational pastorate. When Mary graduated from high school in Newton, her graduation essay was entitled 'The apology Plato should have written: a vindication of the character Xantippi'.

In 1882, she entered Smith College but the illness and death of her younger sister forced her to stay at home the following year. She tutored her siblings and studied Greek and returned to Smith graduating with majors in both classics and philosophy.

After Calkin's graduation the whole family took a year's trip to Europe, during which time she tutored her younger brothers. She also met Abby Leach at this time who was an instructor at Vassar and who encouraged her to become a teacher. Immediately on her return to the USA, Wellesley College asked her to fill in temporarily for their Greek teacher. She worked there for the next 42 years, teaching various subjects including Greek, philosophy, and the new experimental psychology that had broken away from philosophy. Calkins had no training in psychology but was given the new position in view of her previous success at Wellesley and on condition that she study it and philosophy for a year.

In 1890, places for women to study psychology at graduate level were rare but she eventually managed to get a place at Clark University, Worcester, Massachusetts, where special arrangements were made for her to study psychology under Edmund C Sanford. It took a petition from her father, a letter from the president of Wellesley College, and pressure from professors William James and Josiah Royce to persuade the principal of Harvard to allow her to attend philosophy lectures there but she still was not entitled to register as a student. Calkins passed the exams easily to achieve her PhD but was denied the honour because she was a woman.

With the help of Sanford, Calkins set up the first psychology laboratory in a women's college at Wellesley in 1891, the year she began teaching psychology there. She subsequently did more graduate work in psychology at Harvard with Hugo Munsterberg, while continuing teaching. Munsterberg petitioned for her to be admitted as a PhD candidate but she was again refused and in 1902 declined the offer of a doctorate from Radcliffe College, Cambridge, Massachusetts, on the grounds that it had not existed when she was studying at Harvard. In 1927 a group of Harvard alumni petitioned the university to grant her a degree but they were again denied.

She became associate professor at Wellesley in 1895 and full professor from 1898 until her retirement in 1929. She performed many experiments in her laboratory at Wellesley and is best known for her invention of the paired-association technique for studying memory, her original research on dreams, and the development of a system of self-psychology on which she published her first article in 1900 and which she developed over the next three decades. She presented and defended theories that moved away from behaviourism towards a more philosophical system of psychology.

She was a prolific writer, producing over a hundred papers on a wide range of philosophical and psychological subjects and four books including *The Persistent Problems of Philosophy* (1907) and *A First Book in Psychology* (1909), which was widely used in colleges and universities in the USA. She became the first woman president of the Psychological Association in 1905 and also of the American Philosophical Association in 1918 and was awarded honorary membership of the British Psychological Association in 1928. She was awarded two honorary degrees – a doctor of letters from Columbia in 1909 and a doctor of laws from Smith in 1910.

Calkin was diagnosed with cancer in 1926 and lived out her last years in great pain. She retired from Wellesley in 1929 and died at her home in Newton, Massachusetts on 17 February 1930.

Callendar, Hugh Longbourne (1863–1930)
English physicist and engineer who carried out fundamental investigations into the behaviour of steam. One of the results was the compilation of reliable steam tables that enabled engineers to design advanced steam machinery.

Callendar was born on 18 April 1863 in Hatherop, Gloucestershire. He was educated at Marlborough where he was joint editor of the school magazine. At Marlborough he spent his spare time reading science although he was a classics scholar. He obtained first-class honours degrees in classics (1884) and mathematics (1885) at Cambridge, then studied physics under J J Thomson. After becoming a fellow of Trinity College, Cambridge, in 1886, with a substantial research grant, he started to study medicine, devised his own shorthand system, then entered Lincoln's Inn to study law in 1889.

From 1888–93 he was professor of physics at the Royal Holloway College, Egham. Thomson persuaded him to take up science in earnest, and in 1893 he accepted a professorship in physics at McGill College, Montréal. After six months there, he returned to marry Victoria Mary Stewart, whom he had met at Cambridge. They returned to Montréal together and in 1898 the family, which now included three children, left Montréal when Callendar was offered the chair of physics at University College, London.

In 1901 much larger chemistry and physics departments were built for the Royal College of Science (now part of Imperial College) in South Kensington and Callendar was appointed the new professor of physics, a chair he held for 29 years. In 1905 he went to Spain to observe a total eclipse of the Sun for the Royal Society, using his own design of shielded coronal thermopile.

He moved his family to Ealing in 1906 and was elected president of the Royal Physical Society for the 1910–1911 term, having been treasurer for ten years. In 1912 he was president of Section A of the British Association.

During World War I, Callendar was a consultant to the Board of Inventions, which received more than 100,000 'war-winning' ideas. In 1902 he published his great treatise. 'The properties of steam and thermodynamic theory of turbines', and was invited by Sir Charles Parsons to become a consultant.

In addition to his main research on steam, he served as director of engine research for the Air Ministry from

1924. Callendar died of pneumonia on 21 January 1930 in London; his wife survived him by 28 years.

Unlike most gases, steam used to power boilers, engines, turbines, and other equipment does not follow easily interpreted laws. Its behaviour must therefore be predicted by the use of tables or graphs. Steam tables had been produced in different countries from experimental results but the researches did not take into account the laws of thermodynamics, so the tables were unreliable. Callendar was the first to compile accurate tables, and these accelerated the development of turbines and other equipment dramatically.

While he was at Cambridge Callendar's main research was on the platinum resistance thermometer with which he obtained an accuracy of 0.1 °C/14 °F in 1,000 °C/1,832 °F – about a hundred times better than previous results. It was not until 40 years later, in 1928, that the method was adopted as an international standard.

This work led to recording temperatures on a moving chart, a principle now fundamental to any branch of science or industry that requires a continuous record of temperatures.

Callendar's research topics were varied, most of them connected with thermodynamics. He carried out experiments on the flow of steam through nozzles, producing much information of great value to steam turbine designers. With the collaboration of Howard Turner Barnes in Montréal, he developed his method of continuous electric calorimetry used to measure the specific heat of liquids. He also worked on anti-knock additives for fuels.

Callendar was not only an experimental physicist but also a talented engineer and mechanic (he converted the Stanley steam car to run on compressed air). He gave engineers fundamental tools with which they advanced the state of engineering, leading to the highly efficient plant in operation in the late 20th century.

Calne, Roy Yorke (1930–) English surgeon who developed the technique of organ transplants in human patients, and pioneered kidney transplant surgery in the UK.

Calne was born on 30 December 1930 and was educated at Lancing College and later at Guy's Hospital Medical School, London, where he qualified with distinction in 1953. He held a junior post at Guy's Hospital for one year before serving with the Royal Army Medical Corps 1954–56. After military service Calne spent two years at the Nuffield Orthopaedic Centre, Oxford, as a senior house surgeon, and then became surgical registrar at the Royal Free Hospital, London, until 1960. From 1960–61 he went to the Peter Bent Brigham Hospital, Harvard Medical School, and on his return to the UK was appointed lecturer in surgery at St Mary's Hospital, London. In 1962 he became senior lecturer at the Westminster Hospital, where he remained until accepting the appointment of professor of surgery at Cambridge University in 1965. He was elected a Fellow of the Royal Society in 1973, and was knighted in 1986.

The idea of removing a diseased organ and grafting on a healthy one is ancient. This concept was not finally realized, however, until the middle of the present century; Peter ◊Medawar demonstrated in 1957 how the rejection of tissue grafts could be prevented, and Joseph Murray and his team working in Boston, USA, successfully transplanted a kidney from one identical twin to the other, whose kidneys were afflicted with an incurable disease. Calne further developed the technique of kidney transplants, decreasing the possibility of rejection.

He has also carried out liver transplants. The liver is a complicated organ with many vital functions, and liver transplants presented serious technical problems. These have now been overcome, although liver transplants are still not carried out to the same extent as kidney transplants. One reason is that the patient cannot be kept in reasonable health while waiting for a suitable transplant – unlike a kidney patient who can be treated by dialysis (using a kidney machine).

Once the surgical techniques for tissue transplants were perfected, research centred around the development of specific immunosuppressive drugs to prevent the donor organs being rejected by the recipient. Calne has persevered with his operations, despite the ethical arguments surrounding this type of surgery, and has encouraged many developments in transplant techniques.

Calvin, Melvin (1911–1997) US chemist who worked out the biosynthetic pathways involved in photosynthesis, the process by which green plants use the energy of sunlight to convert water and carbon dioxide into carbohydrates and oxygen. For this achievement he was awarded the 1961 Nobel Prize for Chemistry.

Calvin was born of Russian immigrant parents on 8 April 1911 in St Paul, Minnesota. He graduated from Michigan College of Mining and Technology in 1931 and was awarded his PhD by the University of Minnesota in 1935. For the next two years he carried out research at the University of Manchester, England, and then returned to the USA as an instructor at the University of California. He remained there, becoming assistant professor in 1941, associate professor in 1945, professor in 1947, and finally university professor of chemistry in 1971. He died on 8 January 1997.

Calvin began work on photosynthesis in 1949, using radioactive carbon-14 as a tracer to investigate the conversion of carbon dioxide into starch. It was already

Calvin US chemist Melvin Calvin who pioneered the use of carbon-14 labelled carbon dioxide in studying the process of photosynthesis. Calvin was awarded the Nobel Prize for Chemistry in 1961. *Mary Evans Picture Library*

known that there were two interdependent processes: the light reaction, in which a plant 'captured' energy from sunlight, and the dark reaction (which proceeds in the absence of light), during which carbon dioxide and water combine to form carbohydrates such as sugar and starch. Calvin studied the latter reaction in a single-celled green alga called *Chlorella*. He showed that there is in fact a cycle of reactions (now called the Calvin cycle) in which the key step is the enzyme-catalysed carboxylation of the phosphate ester of a pentose (5-carbon) sugar, ribulose diphosphate (RuDP), to form the 3-carbon phosphoglyceric acid (PGA). This acid is then reduced to the 3-carbon glyceraldehyde phosphate (GALP), with the formation also of triose phosphate (TP) and hexose and its phosphate. The reduction and phosphorylation of PGA involves a reducing agent, NADPH, and the energy-rich compound adenosine triphosphate, ATP, which are derived from the photochemical light reaction. Finally, another enzyme catalyses the generation of ribulose monophosphate (RuMP), which a second ATP-induced phosphorylation reconverts to ribulose diphosphate (RuDP). The sequence of reactions is also called the reductive pentose phosphate cycle.

Cameron, Alastair Graham Walter (1925–) Canadian-born US astrophysicist responsible for theories regarding the formation of the unstable element technetium within the core of red giant stars and of the disappearance of Earth's original atmosphere.

Born in Winnipeg, Cameron gained a BSc from the University of Manitoba in 1947 and a PhD from the University of Saskatchewan in 1952. While working as a research officer for Atomic Energy Canada, he emigrated to the USA in 1959 (becoming naturalized in 1963). He then successively became senior research fellow at the California Institute of Technology, Pasadena; senior scientist of the Goddard Institute for Space Studies in New York; and professor of space physics at Yeshiva University, New York City. In 1973 he became professor of astronomy at Harvard University.

Following the discovery of Paul Merrill (1889–1961) in 1952 of the spectral lines denoting the presence in red giants of technetium (Tc) – an element too unstable to have existed for as long as the giants themselves (thus indicating the actual creation and flow of technetium in the stellar core) – it was Cameron's suggestion that Tc^{97} (mean lifetime 2.6×10^{6} years) might result from the decay of a nucleus of molybdenum, Mo^{97}, a usually stable nuclide that becomes unstable when it absorbs an X-ray photon at high temperatures.

Cameron also suggested that the Earth's original atmosphere was blown off into space by the early solar 'gale' – as opposed to the present weak solar 'breeze' – with its associated magnetic fields.

Campbell, William Wallace (1862–1938) US astronomer and mathematician, now particularly remembered for his research into the radial velocities of stars.

Born into a farming family in Hancock, Ohio, Campbell taught for a short while after completing his schooling; he then decided to continue his education at the University of Michigan in 1882. Although he enrolled to study engineering, he became keenly interested in astronomy and studied avidly under J M Schaeberle, who was responsible for the Michigan University Observatory. Campbell received his degree in 1886 and became professor of mathematics at the University of Colorado. He returned to the University of Michigan in 1888 to take up the post of instructor in astronomy, then moved again in 1891, this time to the newly established Lick Observatory, California. He served first as a staff astronomer 1891–1901 and then as director of the observatory 1901–30.

During Campbell's tenure at the Lick Observatory he was responsible for much of the spectroscopic work undertaken and was an active participant in and organizer of seven eclipse expeditions to many parts of the world. His administrative talents were also exercised during the period 1923–30, when he served as president of the University of California. He retired from both posts in 1930, and in the following year was elected president of the National Academy of Sciences. His most significant achievement in this office was the establishment of the influential Scientific Advisory Committee, which serves to improve links between the National Academy and the US government.

Failing health and the fear of complete loss of his faculties led him to commit suicide on 14 June 1938 in Berkeley, California.

Campbell's talent for observation was apparent from early in his career: one of his earliest interests in astronomy was the computation of the orbits of comets. His spectroscopic observations of Nova Auriga in 1892 enabled him to describe the changes in its spectral pattern with time. He also made spectroscopic studies of other celestial bodies and was active in the design of the Mills spectrograph, which was available for use from 1896.

It was in 1896 that Campbell initiated his lengthiest project, the compilation of a vast amount of data on radial velocities. He was aware that this would be of interest not merely for its own sake, but also for the determination of the motion of the Sun relative to other stars. He did not, however, anticipate that the programme would also lead to the discovery of many binary systems nor that the data would later be used in the study of galactic rotation nor that the programme itself would encourage the improvement of several techniques. Campbell published a catalogue of nearly 3,000 radial velocities in 1928.

The project has also led to the establishment of an observatory in Chile, which contributed data for the radial velocities programme 1910–29.

Campbell was an uncompromising scientist, even when it meant attracting criticism. He went against the popular opinion of the time in reporting his observations on the absence of sufficient oxygen or water vapour in the Martian atmosphere to support life as found on Earth. Other astronomers, who had been less careful in the design of their observations and interpretation of their results disagreed but his findings were supported by later work – most spectacularly by *Viking 1*'s Mars landing in 1976.

Another important result obtained by Campbell was a confirmation of the work done in 1919 by Arthur ◊Eddington on the deflection of light during an eclipse, which supported the general theory of relativity. The positive result that Campbell obtained in 1922 was arrived at only after two previous attempts (in 1914 and in 1918) which had been frustrated by poor weather conditions and by the use of inadequate equipment.

Campbell's contributions to astronomy spanned several fields, but were perhaps most notable in spectroscopy.

Cannizzaro, Stanislao (1826–1910) Italian chemist who, through his revival of Avogadro's hypothesis, laid the foundations of modern atomic theory. He is also remembered for an organic reaction named after him, the decomposition of aromatic aldehydes into a mixture of the corresponding acid and alcohol.

Cannizzaro was born on 13 July 1826 in Palermo. He studied chemistry at the universities of Palermo, Naples, and Pisa, where in 1845 he became assistant to Raffaele Piria (1815–1865), who worked on salicin (preparing salicylic acid) and glucosides. In 1848 Cannizzaro joined the artillery to fight in the Sicilian Revolution, was condemned to death, but in 1849 escaped to Marseille and went on to Paris. There he worked with Michel Chevreul and F Cloêz (1817–1883). In 1851 he synthesized cyanamide by treating an ether solution of cyanogen chloride with ammonia, and in the same year became professor of physics and chemistry at the Technical Institute of Alessandria, Piedmont. It was there that he discovered the Cannizzaro reaction. He was appointed professor of chemistry at Genoa University in 1855, followed by professorships at Palermo 1861–71 and Rome. He became a senator in 1871 and eventually vice president, pursuing his interest in scientific education. He died in Rome on 10 May 1910.

Cannizzaro's reaction involves the treatment of an aromatic aldehyde with an alcoholic solution of potassium hydroxide. The aldehyde undergoes simultaneous oxidation and reduction to form an alcohol and a carboxylic acid. It is an example of a dismutation or disproportionation reaction, and finds many uses in synthetic organic chemistry. Cannizzaro also investigated the natural plant product santonin, used as a vermifuge, which he showed was related to naphthalene.

His greatest contribution to chemistry was made in 1858 when he revived Avogadro's hypothesis and insisted on a proper distinction between atomic and molecular masses. The pamphlet he published was distributed at the Chemical Congress at Karlsrühe in 1860. Cannizzaro pointed out that once the molecular mass of a (volatile) compound had been determined from a measurement of its vapour density, it was necessary only to estimate, within limits, the atomic mass of one of its elemental components. Then by investigating a sufficient number of compounds of that element, the chances were that at least one of them would contain only one atom of the element concerned, so that its equivalent mass (atomic mass divided by valency) would correlate with its atomic mass. Despite objections by a group of French chemists led by Henri ◊Saint-Claire Deville (who studied abnormal vapour densities of substances such as ammonium chloride and phosphorus pentachloride, and were reluctant to account for these in terms of thermal dissociation), Cannizzaro's proposal was soon widely accepted.

Cannizzaro's contribution to atomic theory paved the way for later work on the periodic law and on an understanding of valency. The Royal Society recognized its significance with the award in 1891 of its Copley Medal.

Further Reading

Bradley, John, *Before and After Cannizzaro: Philosophical Commentary on the Development of the Atomic and Molecular Theories,* Whitaker, 1991.

Rocke, Alan J, *Chemical Atomism in the Nineteenth Century: From Dalton to Cannizzaro,* Ohio State University Press, 1984.

Cannon, Annie Jump (1863–1941) US astronomer renowned for her work in stellar spectral classification, with particular reference to variable stars.

She was born in Dover, Delaware, and attended local schools, showing aptitude for scientific study; she gained her bachelor's degree at Wellesley College in 1884. After a protracted period spent at her home in Dover, Cannon returned to Wellesley College at about the age of 30 to take postgraduate courses. A protégée of Edward ◊Pickering, the director of the Harvard Observatory, Cannon became a special student in astronomy at Radcliffe College in 1895 and was made an assistant at the Harvard College Observatory in 1896 – a post she held until 1911. From then until 1938 Cannon was curator of astronomical photographs at the Harvard Observatory. In 1938 she was appointed William Cranch Bond Astronomer and Curator. She retired in 1940, but continued in active research.

Cannon was the first woman to receive an honorary DSc from Oxford University, and she received several other honorary degrees from other universities in the USA and in Europe.

Cannon's return to academic life in 1894 was to research in physics rather than astronomy and into the uses of X-rays, recently discovered by Wilhelm Röntgen. A year later, at Harvard, her interests had inclined towards stellar spectroscopy in the field of astronomy.

One of Cannon's particular interests was the phenomenon of variable stars. Hipparchus in the 2nd century BC had established the concept of a continuous sequence of stellar magnitudes based on the assumption that a star's brightness was constant with time. The observation of variable stars, whose brightness sometimes changed quite dramatically, upset this scheme. (A star's brightness could change for any of several reasons; for example, a bright star might be orbited by a dim luminous star, which obscures it at regular intervals: an eclipsing variable.) Cannon studied photographs to record details of variable stars, and discovered 300 new variable stars. She also kept a detailed index-card record of all her information, which has served as an invaluable tool for many succeeding research astronomers.

Edward Pickering and Williamina ◊Fleming had in 1890 established a system for classifying stellar spectra. It allowed each spectrum to be allocated to one of a series of categories labelled alphabetically A to Q; the groups are related to the stars' temperatures and their compositions. In 1901 Cannon reformed this system: she subdivided the letter categories into ten subclasses, based on details in the spectra. With time the system became further modified, some letters being dropped, others rearranged. The sequence on which Cannon eventually settled ran O, B, A, F, G, K, M, R, N, and S. Stars in the O, B, A group are white or bluish; those in the F, G group are yellow; those in the K group are orange; and those in the M, R, N, S group are red. Our Sun, for instance is yellow and its spectrum places it in the G group.

In 1901 Cannon published a catalogue of the spectra of more than 1,000 stars, using her new classification system. She went on to classify the spectra of over 300,000 stars. Most of this work was published in a ten-volume set completed in 1924. It described almost all stars with magnitudes greater than nine. Her later work included classification of the spectra of fainter stars.

The ten-volume catalogue of stellar spectra stands as her greatest contribution to astronomy. It enabled Cannon to demonstrate that the spectra of virtually all stars can be classified easily into a few categories that follow a continuous sequence. Cannon's work was characterized by great thoroughness and accuracy. Her interest lay primarily in the description of the stars as they were observed; her legacy to astronomy was a vast body of accurate and carefully compiled information.

Further Reading

Greenstein, George, 'The Ladies of Observatory Hill: Annie Jump Cannon and Cecilia Payne-Gaposchkin', *Amer Sch,* 1993, v 62, pp 437–446.

Kidwell, Peggy Aldrich, 'Three women of American astronomy', *Amer Scient,* 1990, v 78, pp 244–251.

Cantor, Georg Ferdinand Ludwig Philip (1843–1918) Danish-born German mathematician and philosopher who is now chiefly remembered for his development of the theory of sets, for which he was obliged to devise a system of mathematics in which it was possible to consider infinite numbers or even transfinite ones.

Cantor was born on 3 March 1843 to Danish parents living in St Petersburg, Russia. The family moved to Germany when Cantor was 11, and he was educated at schools in Wiesbaden and Darmstadt, where he showed exceptional talent in mathematics. He then attended the universities of Zürich and Berlin – obtaining his doctorate in 1867 – before moving to Halle University to take up a position as member of staff in 1869. He remained at Halle for the rest of his life, as extraordinary professor from 1872, and as professor of mathematics from 1879. He founded the Association of German Mathematicians, was its first president 1890–93, and was also responsible for the first International Mathematical Congress in Zürich in 1897. Although he received a few honorary degrees and other awards, he did not gain great recognition during his lifetime. Indeed, controversy over some of his work may have contributed to the deep depression and mental illness he suffered towards the end of his life, particularly after 1884. He died in the psychiatric clinic of Halle University on 6 January 1918.

Cantor's early work was on series and real numbers, a popular field in Germany at the time. In a study on the Fourier series – a well-known series that enables functions to be represented by trigonometric series – he extended the results he obtained and developed a theory of irrational numbers. It was in this connection that he exchanged correspondence with Richard ⇨Dedekind, who later became famous for his definition of irrational numbers as classes of fractions. With Dedekind's support, Cantor investigated sets of the points of convergence of the Fourier series, and derived the theory of sets that is the basis of modern mathematical analysis (now more commonly called set theory). His work, fundamental to subsequent mathematics and mathematical logic, contains many definitions and theorems that are now referred to in textbooks on topology.

For the theory of sets, Cantor had to arrive at a definition of infinity, and had therefore to consider the transfinite; for this consideration he used the ancient term 'continuum'. He showed that within the infinite there are countable sets and there are sets having the power of a continuum, and proved that for every set there is another set of a higher power – a realization that was of great importance to the continued development of general set theory. Cantor's definitions were necessarily crude: he was breaking new ground. He left refininements to his successors.

Some of Cantor's other ideas and studies were distinctly odd, particularly in the realm of physics. He considered metaphysics and astrology to be a science, for example – a science into which mathematics, and especially set theory, was capable of being integrated. As probably the last Platonist among serious mathematicians, he also insisted that the atoms of the universe were countable.

Cantor's was in its way a unique contribution to the science of mathematics; he opened up a complete new area of research that at the same time was fundamental to basic mathematics.

Carathéodory, Constantin (1873–1950) German mathematician who made significant advances to the calculus of variations and to function theory.

Carathéodory was born in Berlin on 13 September 1873; his parents were of Greek extraction. He showed an aptitude for mathematics from an early age and attended the Belgian Military Academy 1891–95. He then worked in Egypt for the British Engineering Corps on the building of the Asyut Dam. Carathéodory returned to Germany in 1900, first attending the University of Berlin and then in 1902 moving to Göttingen University, where Felix Klein, David Hilbert, and Hermann Minkowski had built up an excellent mathematics department.

Carathéodory was awarded his PhD in 1904 and qualified as a lecturer a year later. He taught at the University of Bonn for four years and then in 1909 was appointed to a professorship at the University of Hanover. In 1910 he transferred to a similar position at the University of Breslau (now Wrocław, Poland). In 1913 he moved back to Göttingen, and five years later he returned to the University of Berlin.

The Greek government invited Carathéodory to supervise the establishment of a new university and he went to Smyrna in 1920, but his efforts were destroyed by fire two years later. He took a post at the University of Athens, where he taught until 1924 before becoming professor of mathematics at the University of Munich. He remained there for the rest of his life, apart from one year, 1936–37, in the USA as visiting professor at the University of Wisconsin. He died in Munich on 2 February 1950.

Carathéodory's work covered several areas of mathematics, including the calculus of variations, function theory, theory of measure, and applied mathematics. His first major contribution to the calculus of variations was his proposal of a theory of discontinuous curves. From his work on field theory he established links with partial differential calculus, and in 1937 he published a book on the application to geometrical optics of the results of his investigations into the calculus of variations.

One of Carathéodory's most significant achievements – also the subject of a book (1932) – was a simplification of the proof of one of the central theorems of conformal representation. It formed part of his work on function theory, which extended earlier findings of Picard and Schwarz. In measure theory he developed research begun in the 1890s by Emile Borel and his student Henri Lebesgue, work which he summarized in 1918 in a text on real functions.

Carathéodory's interest also extended beyond pure mathematics into the applications of the subject, particularly to mechanics, thermodynamics, and relativity theory. A mathematician of diverse talents, he can thus be seen to have enlarged the understanding of several disciplines.

Cardozo, William Warrick (1905–1962) US physician and paediatrician who is remembered for his pioneering investigations into sickle-cell disease.

William Cardozo was born on 6 April 1905. After attending public schools in Washington, DC, and the Hampton Institute, he went to Ohio State University where he received his AB in 1929 and MD in 1933. He was an intern at City Hospital, Cleveland, and a resident in paediatrics at Provident Hospital, Chicago, spending a year in each, and then he had a General Education Board fellowship in paediatrics 1935–37 at the Children's Memorial Hospital and Provident Hospital. In 1937 he started private practice in Washington, DC, and was appointed part-time instructor in paediatrics at Howard University College of Medicine and Freedmen's Hospital, later being promoted to clinical assistant professor and then clinical associate professor. In 1942 he was certified by the American Board of Paediatrics and in 1948 became a Fellow of the American Academy of Paediatrics. Cardozo was also a school medical inspector for the District of Columbia Board of Health for 24 years. He died from a heart attack on 11 August 1962; he was married with one daughter.

Cardozo is known for his pioneering research into sickle-cell disease (sickle-cell anaemia). His investigations were published in 'Immunologic studies in sickle-cell anaemia' in *Archives of Internal Medicine* in 1937, in which he concluded that the disease was inherited following Mendelian law and almost always occurred in black people or people of African descent; not all persons with sickle cells were necessarily anaemic and not all patients died of the disease. He also contributed articles to the *Journal of Paediatrics* on Hodgkin's Disease (1938) and *The Growth and Development of Black Infants* (1950).

Carnegie, Andrew (1835–1919) Scottish-born US industrialist whose willingness to adopt new methods of steelmaking was instrumental in advancing both the techniques and commercial potential of the iron and steel industry.

Carnegie was born in Dunfermline into a poor weavers's family, which, when he was 13 years old, emigrated to the USA. He was largely self-educated and began work as a telegraph messenger. After having held a variety of jobs and having saved some capital, he bought shares in a railroad company and land containing oil in Pennysylvania. These investments laid the foundation for his eventual huge fortune. After the American Civil War he became an iron manufacturer and built great iron and steel works in Pittsburgh. Apart from his fame as a steelmaker he is best known for his philanthropic activities, giving much of his fortune to the provision of libraries, a craft school, music centre, and many other public amenities in the USA, UK, and Europe. Carnegie scorned the word 'philanthropist', calling himself, in his later years, 'a distributor of wealth for the improvement of mankind'.

The US iron industry received a great impetus from the Civil War. Until this time the country had no great steel industry, but the sudden demand for war materials, railway supplies, and the like brought fortunes to the previously struggling ironmasters of Pittsburgh. Carnegie was 30 years old when the war ended and he had not yet begun his work in this field. It was not until 1873 that he concentrated on steel, having made a small fortune in oil and taken several trips to Europe selling railroad securities. His operations in bond selling, oil dealing and bridge-building were so successful that conservative Pittsburgh businessmen regarded him with a mixture of doubt and jealousy. Carnegie's European tours, however, had results of great consequence. He came into close touch with British steel makers – then the world's leaders – he became closely acquainted with the Bessemer process and formed a friendship with Henry ◊Bessemer, which was maintained until the latter's death.

Bessemer patented his process for the manufacture of steel in 1856, (based on the idea of blowing air through the molten steel to oxidize impurities), which the Carnegie Company adopted with great success. Then in 1867, William Siemens (1823–1883) invented the open-hearth process. Always adventurous, and with tremendous foresight, Carnegie scrapped most of the equipment used in the old processes and invested heavily in the new one. Pittsburgh is situated conveniently near to abundant supplies of coal, iron ore, and limestone and has become the leading iron- and steel-producing centre in the world.

Carnegie's success was the result of optimism, enthusiasm, and courage. He was not a gambler; he detested the speculative side of Wall Street. He did, however, make one gamble of titanic proportions – and won. He wagered everything he possessed on the industrial future of the USA and its economic potential. He was probably the most daring man in US industry; his insistence on having the most up-to-date machines, his readiness to discard costly equipment as soon as something better appeared has become a tradition in the steel trade.

Further Reading

Carnegie, Andrew, *The Autobiography of Andrew Carnegie,* Northeastern University Press, 1986.

Hendrick, B J, *Life of Andrew Carnegie,* Transaction Publishers, 1989.
Mackay, James A, *Andrew Carnegie: His Life and Times,* Jacaranda Wiley, 1998.
Meltzer, Milton, *The Many Lives of Andrew Carnegie,* Meltzer Biographies series, Franklin Watts, 1997.
Simon, Charnan, *Andrew Carnegie: A Library in Every Town,* Community Builders series, Children's Press, 1997.

Carnot, (Nicolas Léonard) Sadi (1796–1832) French physicist who founded the science of thermodynamics. He was the first to show the quantitative relationship between work and heat.

Sadi Carnot was born in Paris on 1 June 1796 into a distinguished family (his father Lazare Carnot and other relatives held important government positions). He was educated at the Ecole Polytechnique in Paris 1812–14 and then at the Ecole Genie in Metz until 1816. He became an army engineer, at first inspecting and reporting on fortifications and in 1819 transferring to the office of the general staff in Paris. Carnot had many interests, carrying out a wide range of study and research in industrial development, tax reform, mathematics, and the fine arts. He was particularly interested in the problems of the steam engine and, in 1824, he published his classic work *Reflections on the Motive Power of Heat,* which was well received. He was then forced to return to active service in the army in 1827, but was able to resign a year later in order to concentrate on the problems of engine design and to study the nature of heat. In 1831, he began to study the physical properties of gases, in particular the relationship between temperature and pressure. Sadi Carnot died suddenly of cholera on 24 August 1832.

In accordance with the custom of the time, Carnot's personal effects, including his notes, were burned after his death. Only a single manuscript and a few notes survived, and Carnot's work was virtually forgotten. However it was rediscovered by Lord ◊Kelvin, who confirmed Carnot's conclusions in his *Account of Carnot's Theorem* in 1849. He and Rudolf ◊Clausius then derived the second law of thermodynamics from Carnot's work.

In *Reflections,* Carnot reviewed the industrial, political, and economic importance of the steam engine. The engine invented by James Watt, although the best available, had an efficiency of only 6%, the remaining 94% of the heat energy being wasted. Carnot set out to answer two questions:

1. Is there a definite limit to the work a steam engine can produce and hence a limit to the degree of improvement of the steam engine?

2. Is there something better than steam for producing the work?

Engineers had worked on problems like these before, but Carnot's approach was new for he sought a theory, based on known principles, that could be applied to all types of heat engines. Carnot's theorem showed that the maximum amount of work that an engine can produce depends only on the temperature difference that occurs in the engine. (In a steam engine, the hottest part is the steam and the coldest part the cooling water.) It is independent of whether the temperature drops rapidly or slowly or in a number of stages, and it is also independent of the nature of the gas used in the engine. The maximum possible fraction of the heat energy that is capable of being converted into work is represented by Carnot's equation

$$\frac{(T_1 - T_2)}{T_2}$$

where T_1 is the absolute temperature of the hottest part and T_2 the temperature of the coldest part. Carnot's equation put the design of the steam engine on a scientific basis. Using experimental data and his own conclusions, he recommended that steam should be used over a large temperature interval and without losses due to conduction or friction.

In formulating his theorem, Carnot considered the case of an ideal heat engine following a reversible sequence known as the Carnot cycle. This cycle consists of the isothermal expansion and adiabatic expansion of a quantity of gas, producing work and consuming heat, followed by isothermal compression and adiabatic compression, consuming work and producing heat to restore the gas to its original state of pressure, volume, and temperature. Carnot's law states that no engine is more efficient than a reversible engine working between the same temperatures. The Carnot cycle differs from that of any practical engine in that heat is consumed at a constant temperature and produced at another constant temperature, and that no work is done in overcoming friction at any stage and no heat is lost to the surroundings – so that the cycle is completely reversible.

At the time he wrote *Reflections,* Carnot was a believer in the caloric theory of heat, which held that heat is a form of fluid. But this misconception of the nature of heat does not invalidate his conclusions. Some notes that escaped destruction after Carnot's death indicate that he later arrived at the idea that heat is essentially work, or rather work that has changed its form. He had calculated a conversion constant for heat and work and showed he believed that the total quantity of work in the universe is constant. It indicated that he had thought out the foundations of the first law of thermodynamics, which states that energy can never disappear but can only be altered into other forms of energy. Carnot's notes, however, remained undiscovered until 1878.

Carnot's work led Lord Kelvin, in 1850, to confirm and extend it. Rudolf Clausius made some modifications to it and Carnot's theorem became the basis of the second law of thermodynamics, which states that heat cannot flow of its own accord from a colder to a hotter substance.

The application of the science of thermodynamics founded by Sadi Carnot has been of great value to the production of power and also to industrial processes. To give one example, it has been useful in forecasting the conditions under which chemical reactions will or will not take place and the amount of heat absorbed or given out.

Carothers, Wallace Hume (1896–1937) US organic chemist who did pioneering work on the development of commercial polymers, producing nylon (a polyamide) and neoprene (a polybutylene, one of the first synthetic rubbers).

Carothers was born on 27 April 1896 in Burlington, Iowa, the son of a teacher. He attended schools in Des Moines, graduating from the North High School in 1914. His further studies were in accountancy and clerical practice at his father's college in Des Moines, until he entered Tarkio College, Missouri, in 1915 and specialized in chemistry. He later gained higher degrees in organic chemistry from the University of Illinois and Harvard, where he was appointed in 1926. In 1928 he accepted the post as head of organic chemistry research at the Du Pont research laboratory in Wilmington, Delaware. For his fundamental work on polymers he was elected to the US National Academy of Sciences in 1931. Carothers suffered from periods of depression that became more prolonged and severe as he grew older. He was deeply affected by the death of his sister in 1936, the same year in which he married. A few months later, on 29 April 1937, he committed suicide in Philadelphia, Pennsylvania.

Carothers began his work on polymerization and the structures of high-molecular-mass substances while at Harvard. Then at Du Pont's he carried out studies on linear condensation polymers, which culminated in 1931 with the development of nylon and neoprene. Much of his research effort was directed at producing a polymer that could be drawn out into a fibre. His first successful experiments involved polyesters formed from trimethylene glycol (propan-1, 3-diol) and octadecane dicarboxylic acid (octadecan-1, 18-dioic acid). But for finer fibres with enough strength (emulating silk) he turned to polyamides. Early attempts, made by heating amino-caproic acid (hexan-6-amino-1-oic acid), resulted in an unstable product containing ring compounds. The first polymer to be called nylon (strictly the trade-name Nylon 6,6) was made by heating hexamethylene diamine (hexan-1, 6-diamine) and adipic acid (hexan-1, 6-dioic acid). The product is a linear chain polymer that can be cold drawn after extrusion through spinnerets to orientate the molecules parallel to each other so that lateral hydrogen bonding takes place. The resultant nylon fibres are strong and have a characteristic lustre.

Carothers also worked on synthetic rubbers. His monomer was chlorobutadiene (but-2-chloro-1, 3-diene), which he first had to make by treating vinylacetylene (but-1-en-3-yne) with hydrogen chloride. Using a peroxide catalyst, the chloro compound polymerizes readily by a free-radical mechanism to form neoprene. This polymer, first produced commercially in 1932, is resistant to heat, light, and most solvents. In the years that followed, a whole range of useful polymers of the nylon and neoprene types were produced.

Carr, Emma Perry (1880–1972) US chemist, teacher and researcher, internationally renowned for her work in the field of spectroscopy.

Emma Carr was born in Holmesville, Ohio, on July 23 1880. After attending Coshocton High School, she

$$x\,H_2N{-}(CH_2)_6{-}NH_2 + x\,HOOC{-}(CH_2)_4{-}COOH \rightarrow [{-}NH{-}(CH_2)_6{-}NH{-}CO{-}(CH_2)_4{-}CO{-}]_x + 2x\,H_2O$$

hexamethylene diamine (hexan-1, 6-diamine)

adipic acid (hexan-1, 6-dioic acid)

nylon

Carothers The condensation polymerization of hexamethylene diamene and adipic acid to form nylon.

spent a year, 1898–99, at Ohio State University and then transferred to Mount Holyoke for a further two years' study; she continued as a chemistry assistant for three years and in 1905 finished her BS at the University of Chicago. Between 1905–08 Carr worked as an instructor at Mount Holyoke, after which she returned to Chicago and completed her PhD in physical chemistry in 1910. From 1910–13 she was associate professor of chemistry at Mount Holyoke after which she became professor of chemistry and head of the department, posts that she held until her retirement in 1946. Carr lived in South Hadley for 18 years until she moved in 1964 to the Presbyterian Home in Evanston, Illinois. She died there in 1972.

In 1913 Carr introduced a departmental research programme to train students through collaborative research projects combining both physical and organic chemistry. She achieved her reknown in spectroscopic research – under her leadership, she, her colleague Dorothy Hahn, and a team of students were among the first in the USA to synthesize and to analyse the structure of complex organic molecules using absorption spectroscopy. Her research into unsaturated hydrocarbons and far-ultraviolet vacuum spectroscopy led to grants from the National Research Council and the Rockefeller Foundation in the 1930s. She served as a consultant on the spectra for the *International Critical Tables,* and during the 1920s and 1930s she was three times a delegate to the International Union of Pure and Applied Chemistry. She was awarded four honorary degrees and was the first to receive the Garvan Medal, annually awarded to US women for achievement in chemistry.

Carrel, Alexis (1873–1944) French-born US surgeon whose work on joining blood vessels was key to the development of transplant surgery. He was awarded the Nobel Prize for Physiology or Medicine in 1912.

Alexis Carrel was born in Ste Foy-les-Lyon, France on 28 June 1873, and was educated at Lyon University, receiving a doctorate in medicine in 1900. In 1902 he began investigating techniques for anastomosis – the joining of blood vessels end to end to re-establish the circulation of the blood. He continued his work at the University of Chicago in 1904, moving to the Rockefeller Institute for Medical Research in New York in 1906.

Attempts to transplant organs or reattach severed limbs had failed in the past, partly because of the problems of successfully joining the blood vessels. Using animals, Carrel analysed all the possible complications and, with careful attention to detail in the use of instruments and sutures and the application of scrupulous asepsis, he was able to remove whole organs and either replace them in their original location, or attach them elsewhere in the body where they continued functioning. The studies he made were a major advance in vascular surgery and were used as a model for numerous new applications. His techniques were used in the transfusion of blood straight from donor to recipient (a procedure used before anticoagulants had been found to enable blood to be stored).

Carrel was awarded the Nobel prize for Physiology or Medicine in 1912 in recognition of this work. He and his colleagues also followed studies made by R G Harrison in 1907 on tissue culture. They made significant developments and demonstrated their tissue fragments in 1910.

During World War I, Carrel served in the French Army where he formulated, with the chemist, Henry Dakin, the Carrel–Dakin antiseptic, which successfully treated deep wounds by keeping them constantly irrigated. After the war he returned to the Rockefeller Institute and began work on methods of keeping tissues and organs alive outside the body. He kept heart tissue from a chick embryo alive for many years on artificial nutrients. He also worked with the aviator Charles Lindbergh on an artificial heart which could pump solutions through large organs.

Carrel's popular book *Man the Unknown* (1935) made known his views on the possible role of science in providing a method of organizing and improving society along authoritarian lines. During World War II, under the Vichy Government he founded and directed the Carrel Foundation for the Study of Human Problems which was in Paris. After the allied liberation of France he was charged with collaboration but died before his trial was arranged.

Carrel's contributions to the development of transplant surgery and other advances in medicine and biology make him a key figure in 20th-century science. He died in Paris on 5 November 1944.

Carrington, Richard Christopher (1826–1875) English astronomer who was the first to record the observation of a solar flare, and is now most remembered for his work on sunspots.

Carrington's family was in brewing, but neither the family business nor the church (for which he had been intended) attracted him. He realized very early that his interests lay in astronomy and scientific activities. He left Cambridge in 1849 and his first post was as observer at Durham, from where he made several reports to the *Monthly Notices of the Royal Astronomical Society* and to the *Astronomische Nachrichten* of Altona (mainly dealing with minor planets and comets), work that led to his election as a Fellow of the Royal Astronomical Society in 1851.

However, by 1852 he was impatient with the limited resources at Durham – he had in mind an ambitious programme of observation leading to a catalogue of circumpolar stars. In 1853 he set up his own house and observatory at Redhill, Surrey, with instruments made by W Simms. One of these was based on a larger Greenwich instrument; its telescope had a 12.7-cm/5-in aperture and a focal length of 1.68 m/5.5 ft.

By 1857 he had completed his *Catalogue of 3,735 Circumpolar Stars,* which was so highly regarded that it was printed by the Admiralty at public expense. The *Catalogue* won him the Gold Medal of the Royal Astronomical Society in 1859, and his election as a Fellow of the Royal Society shortly afterwards was fitting recognition of his qualities as an astronomer.

The death of his father in 1858 meant that Carrington had to take over the management of the Brentford Brewery, a substantial undertaking that entailed a reduction in his research activity. Nevertheless, in 1859 he recorded the first solar flare.

It is felt by some that he was disappointed by his failure to succeed James Challis (his mentor) as director of the Cambridge Observatory. At any rate his output declined in the 1860s and ill-health overtook him in 1865. He sold his business and his Redhill establishment and moved to Churt, near Farnham, Surrey, where he built another observatory containing some large telescopes.

Carrington is best known for his work on sunspots. He pursued the daylight project at Redhill for more than seven years (the original aim was for an 11-year period), in tandem with the work on the Redhill Catalogue. The sunspot cycle – an 11-year period between maxima of activity – had recently been discovered by Samuel ◊Schwabe and the connection with magnetic disturbances had been noted. A study of sunspot activity was thus highly topical and Carrington was keen to tidy up the mass of observations on the subject that had accumulated in the contemporary literature.

He required a simple yet accurate means of plotting sunspot positions and movements, and with much trial and error he arrived at a simple, elegant method. His system projected an image of the Sun of about 28 cm/11 in diameter using his 11.4-cm/4.5-in equatorial telescope. Crosswires at right angles were placed in the focus of the telescope inclined at 45° to the meridian; the exact angles were not important. The telescope was fixed and the Sun's image allowed to pass across the field; the times of contact of the Sun's limbs and spots were recorded. The method allowed the heliographic latitude and longitude of a sunspot to be determined without recourse to micrometers or clockwork mechanisms.

The principal results of this extended work were, first, to determine the position of the Sun's axis and, second, dramatically to show that the Sun's rotation is differential, that is, that it does not rotate as a solid body, but turns faster at the equator than at the poles. This conclusion was the result of observing the great systematic drift of the photosphere as seen in the drift of individual sunspots during the cycle. Carrington also derived a useful expression for the rotation of a spot in terms of heliographical latitude. An extensive account of all the observations was published, with the help of the Royal Society, in 1863. The complete cycle of work was, however, never accomplished by him.

An immensely practical and meticulous man, interested in international cooperation and the mutual contribution of ideas, Carrington in his work and his publications represents the true Victorian ideal of the investigative scientist.

Carson, Rachel Louise (1907–1964) US biologist, conservationist, and campaigner. Her writings on conservation and the dangers and hazards that many modern practices imposed on the environment inspired the creation of the modern environmental movement.

Carson was born in Springdale, Philadelphia, on 27 May 1907, and educated at the Pennsylvania College for Women, studying English to achieve her ambition for a literary career. A stimulating biology teacher diverted her towards the study of science, and she went to Johns Hopkins University, graduating in zoology in 1929. She received her master's degree in zoology in 1932 and was subsequently appointed to the department of zoology at the University of Maryland, spending her summers teaching and researching at the Woods Hole Marine Biological Laboratory in Massachusetts.

Family commitments to her widowed mother and orphaned nieces forced her to abandon her academic career and she worked for the US Bureau of Fisheries, writing in her spare time articles on marine life and fish, and producing her first book on the sea just before the Japanese attack on Pearl Harbor. During World War II she wrote fisheries information bulletins for the US Government and reorganized the publications department of what became known after the war as the US Fish and Wildlife Service. In 1949 she was appointed chief biologist and editor of the service. She also became occupied with fieldwork and wrote regular freelance articles on the natural world.

> *As cruel a weapon as the cave man's club, the chemical barrage has been hurled against the fabric of life.*
>
> ***Rachel Carson*** The Silent Spring *1962*

During this period she was also working on *The Sea Around Us,* which finally appeared in 1951 and was an immediate best-seller, being translated into several languages and winning several literary awards. Given a measure of financial independence by this success she resigned from her job in 1952 to become a professional writer. Her second book *The Edge of the Sea* (1955), an ecological exploration of the seashore, further established her reputation as a writer on biological subjects. Her most famous book *The Silent Spring* (1962) was a powerful denunciation of the effects of the chemical poisons, especially DDT, with which humans were destroying the earth, sea, and sky. Despite denunications from the influential agrochemical lobby, one immediate effect of Carson's book was a presidential advisory committee on the use of pesticides. By this time Carson was already seriously incapacitated by ill health and she died in Silver Spring, Maryland on 14 April 1964.

On a larger canvas, *The Silent Spring* alerted and inspired a new worldwide movement of environmental concern. Whilst writing about broad scientific issues of pollution and ecological exploitation, she also raised important issues about the reckless squandering of natural resources by an industrial world.

Further Reading

Burby, Liza N, *Rachel Carson,* Making Their Mark series, The Rosen Publishing Group, 1996.

Harlan, Judith, *Rachel Carson: Sounding the Alarm: A Biography of Rachel Carson,* People in Focus series, Silver Burdett Press, 1989.

Lear, Linda, *Rachel Carson: Witness for Nature,* Henry Holt & Company, 1997.

McKay, Mary A, *Rachel Carson,* Twayne's United States Authors series, Twayne Publishers, 1993.

Presnall, Judith, *Rachel Carson,* Lucent Books, 1995.

Sabin, Francene, *Rachel Carson: Friend of the Earth,* Troll Communications LLC, 1997.

Cartwright, Edmund (1743–1828) English clergyman who invented various kinds of textile machinery, the most significant of which was the power loom that helped initiate the Industrial Revolution.

Cartwright was born in Marnham, Nottinghamshire, and received his early education at Wakefield Grammar School. At the age of only 14 he went to University College, Oxford, and the regulations were changed to enable him to be awarded his BA earlier than usual (in 1764). In that same year he was elected a fellow of Magdalen College, gaining his MA two years later. He received the perpetual curacy of Brampton near Wakefield, and became rector of Goadby Marwood, Leicestershire, in 1779. He was prebendary of Lincoln from 1786 until his death in 1828.

Cartwright was an innovator, always curious and on the lookout for new ways of doing things. At Goadby Marwood he made agricultural experiments on his glebe land, and while on holiday at Matlock he visited the spinning mills of Richard ◊Arkwright at nearby Cromford. Arkwright had watched cotton weavers working in their homes. They used cotton thread from the side of the loom (the weft), but he noticed that they wove it in and out of Irish linen threads stretching lengthwise. When he asked the reason for this, he was told that they could not spin cotton thread that was strong and fine enough to use for the warp. This motivated Arkwright to invent the spinning frame and, watching it working, Cartwright remarked that Arkwright would have to set his wits to work to invent a weaving-mill. Soon after returning home Cartwright himself set about this task, devoting all his spare time and money to experiment.

Cartwright had never seen the working of the handloom and the first machine he made was an inadequate substitute for it. However, he patented it in 1785 and moved to Doncaster in the same year where his wife had inherited some property. There he continued to improve the simple water-driven machine, and visited Manchester to have it criticized by the local workers; he also tried to enlist the help of the local manufacturers. Disappointed in this hope and having taken out two more patents for further improvements in his loom, he set up a factory at Doncaster for weaving and spinning. His power loom now worked well and became the parent of all those in use today. It contained an ingenious mechanism that substituted for the hands and feet of the ordinary handloom weaver. There was a beam on which the required number of warp-ends was wound side by side, in perfect order. A device called a let-off motion held the warp-ends in place and let them go forwards only as required. The ends were threaded through eyes (loops) in sets of cords or wires called healds, and there was an apparatus that raised some sets of healds and lowered others, thus making a tunnel, called a shed, between the lower and upper warp ends.(The healds could be reversed so that the upper and lower layers of warp ends changed places.) The weft was carried to and fro through the shed by the shuttle. There was a device for pressing the weft up tightly against the already woven cloth, and another for keeping the cloth taut and rolling it up as fast as it was woven.

For centuries Yorkshire had been a principal seat of woollen manufacture, and in 1789 at Doncaster Cartwright invented a wool-combing machine that contributed greatly to cutting the cost of manufacture. Even in the earlier stages of its development, one machine did the work of 20 hand-combers. Petitions against its use poured into the House of Commons from the wool-combers – some 50,000 in number – and a committee was appointed to inquire into the matter; nothing came of the wool-combers' agitation.

Cartwright's Doncaster factory was enlarged when a steam engine was erected to power it, and in 1799 a Manchester firm contracted with Cartwright for the use of 400 of his power looms and built a mill where some of these were powered by steam. The Manchester mill was burned to the ground, probably by workers who feared to lose their jobs, and this catastrophe prevented other manufacturers from repeating the experiment. Cartwright's success at Doncaster was obstructed by opposition and by the costly character of his processes; in 1793, deeply in debt, he relinquished his works at Doncaster and gave up his property to his creditors. In 1807, however, 50 prominent Manchester firms petitioned the government to bestow a substantial recognition of the services rendered to the country by Cartwright's invention of the power loom. Cartwright too petitioned the House of Commons, which in 1809 voted him £10,000.

Carver, George Washington (*c.* 1860–1943) US agricultural chemist who revolutionized agriculture in the south of the country. He advocated the diversification of crops, crop rotation, and the cultivation of peanuts, from which he made over 300 products.

Carver was born about 1860 to slave parents near Diamond Grove in Missouri. He was educated at Minneapolis High School in Kansas and, having achieved an outstanding record, he received a scholarship to the Highland University, Kansas. He was, however, later rejected on account of his race. After working on the land and saving money he was accepted in 1887 by Simpson College, Iowa, and in 1891 entered Iowa Agricultural College from where he graduated in 1894 with a BS degree. After graduation, Carver was given an appointment in the faculty teaching agriculture and bacterial botany, and while pursuing research and conducting experiments into plant pathology, he obtained the MS in agriculture in 1896. In 1897 he transferred to the Tuskegee Institute, Alabama. He was made first director of agriculture and was also director of a research and experiment station. Carver remained at Tuskegee until his death from anaemia on 5 January 1943.

During his time in Iowa, Carver made important discoveries in the field of plant pathology; in 1897 he reported on new species of fungi that have since been named after him – *Taphrina carveri, Collectotrichum carveri,* and *Metasphaeria carveri.* At Tuskegee, Carver demonstrated the need for crop rotation and the use of leguminous plants, especially the peanut. Following his advice, farmers planted peanuts, which soon became the principal crop in the farming belt running from Montgomery to the Florida border. They were soon making more money from the peanut and its 325 by-products (including milk, cheese, face powder, printer's ink, shampoo, and dyes), which had been developed by Carver, than from tobacco and cotton. In 1921, following Carver's presentation to the Ways and Means Committee, the peanut was included in the Hawley–Smoot Tariff Bill to protect it from foreign competition.

Carver also discovered 118 products that could be made from the sweet potato and 75 products from the pecan nut. Carver's other work included developing a plastic material from soya beans that car manufacturer Henry Ford later used in part of his vehicle, and extracting dyes and paints from the clays of Alabama. He received three patents, for a cosmetic (1925), a paint and stain (1925), and a process for producing paints and stains (1927).

Carver was also an accomplished artist. He received many awards and honours for his outstanding work; he was elected a Fellow of the Royal Society of Arts, Manufactures, and Commerce of Great Britain (1916), he was awarded the Spingarn Medal (1923), the Theodore Roosevelt Medal 'for distinguished research in agricultural chemistry' (1939), and was chosen 'man of the year' by the International Federation of Architects, Engineers, Chemists, and Technicians (1941). He received honorary ScD degrees from Simpson College (1928) and from the University of Rochester (1941). Three months after his death he was posthumously awarded the New York City Teachers Union Medal, and 6 January 1946 was designated as George Washington Carver Day by a joint resolution of Congress.

Further Reading

Graham, Shirley, *George Washington Carver: Scientist,* Young Readers series, Africa World Press, 1996.

Kremer, Gary R, *George Washington Carver in His Own Words,* University of Missouri Press, 1990.

Neyland, James, *George Washington Carver: Scientist and Educator,* Black American series, Melrose Square Publishing Co, 1992.

Rogers, Teresa, *George Washington Carver: Nature's Trailblazer,* Earth Keepers series, Twenty-First Century Books, 1995.

Wellman, Sam, *George Washington Carver: Inventor and Naturalist,* Heroes of the Faith series, Barbour Publishing, 1998.

Cassegrain (*c.* 1650–1700) French inventor of the system of mirrors within many modern reflecting telescopes – a system by transference also sometimes used in large refraction telescopes.

Nothing is known for certain about the details of Cassegrain's life – not even his first name. Believed to have been a professor at the College of Chartres, he is

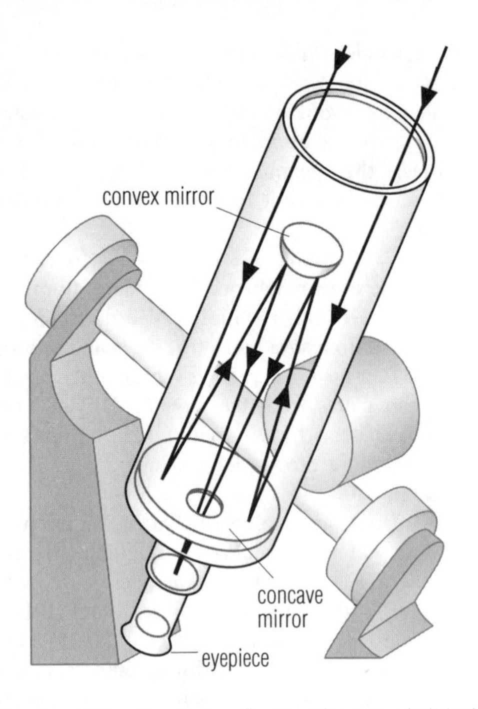

Cassegrain In a Cassegrain reflecting telescope, a hole in the centre of the concave main mirror allows light reflected by the convex secondary mirror to reach the eyepiece (or a camera).

variously credited with having been an astronomer, a physician, and a sculptor at the court of Louis XIV.

In the same year that he submitted a scientific paper concerning the megaphone to the Academy of Sciences in Paris, Cassegrain presented another paper in which he claimed to have improved on Newton's telescope design. Newton himself, however, suggested that the 'improvement' had been strongly influenced by the work of James Gregory (whose telescope had been described in *Optica Promota* in 1663).

Newton's own design employed a second, plane mirror to bring the reflected, magnified image out to the eyepiece through the side of the telescope. Cassegrain's telescope used an auxiliary convex mirror to reflect the image through a hole in the objective – that is, through the end of the telescope itself. One intention behind this innovation was further to increase the angular magnification.

An even more advantageous facet of this design was not realized until a century later, when Jesse Ramsden noted that it also partly cancelled out the spherical aberration, the blurring of the image caused by the use of two mirrors. The first practical reflecting telescope based on Cassegrain's design was 'Short's Dumpy' (focal length 609 mm/23.75 in, aperture 152 mm/5.93 in), built by J Short in the 18th century.

Cassini, Giovanni Domenico (Jean Dominique) (1625–1712) Italian-born French astronomer with a keen interest in geodesy.

Born in Perinaldo, Cassini studied in Vallebone and Genoa, displaying great talent in astronomy and mathematics. In 1644 he was invited to assist the Marquis Mavasia in his observatory at Panzano, near Bologna, and there he was introduced to the two prominent local astronomers, Giovanni Battistae Riccioli (1598–1671) and Francesco Maria ◊Grimaldi. Six years later, aged only 25, Cassini was made professor of astronomy at the University of Bologna; he remained there for 19 years.

In addition to his teaching duties at the university, Cassini was also called upon to serve a variety of civic and diplomatic duties. These included contributing to hydraulic projects, supervising construction work, and mediating in a dispute between Ferrara and Bologna over navigation rights on the River Reno. In 1669 Cassini departed for Paris at the invitation of King Louis XIV, who had nominated him as a member of the new French Academy of Sciences. Despite Pope Clement IV's insistence that his stay in France be only temporary, Cassini never returned to Italy.

The first task to confront Cassini upon his arrival in Paris was the construction of the Paris Observatory. He had been unable to convince the king or the architect (Claude Perrault) that certain aspects of the design were impractical, but he nevertheless took up the directorship of the observatory and assumed French citizenship. Cassini was extremely active in exploiting the work of astronomers in research outposts around the world, seeking to equip the observatory with the latest instruments, and making use of the skills of students of Galileo.

At the end of the century Cassini's health began to fail and his son Jacques took over an increasing share of his work. The elder Cassini lost his eyesight in 1710 and Jacques assumed the directorship of the observatory.

Cassini was renowned for his skills as an observational astronomer, which led him to many important discoveries. He was also extremely conservative in his approach to the more theoretical aspects of astronomy, and this conservatism led him frequently to propound the incorrect view. He refused to accept the Copernican cosmological model and rejected the concept of a finite speed of light (although its proof was demonstrated by Ole ◊Römer using Cassini's own data; it is likely that Cassini himself considered the possibility even prior to Römer's work). He also opposed a theory of universal gravitation and insisted (despite critical disagreement by Christiaan Huygens and Isaac Newton) that the Earth was flattened at the equator rather than at the poles. Despite these errors in judgement Cassini earned a well-deserved reputation as one of the finest astronomers of his day.

The best-known examples of Cassini's early work are a treatise on his observations of a comet made in

1652 and his design work for a meridian constructed at San Petrino in 1653. The meridian was used to make accurate observations of the Sun and enabled Cassini to publish improved tables in 1662.

During the period 1664–67 Cassini concentrated his efforts on determining the rotation periods of Mars, Jupiter, and Venus. In 1664 he found the rotation period of Jupiter to be nine hours from a study of Saturn. In 1675 he distinguished two zones within what was thought to be the single ring around Saturn. The dark central 'border' has since been named Cassini's division. Cassini correctly suggested that the rings were composed of myriads of tiny satellites, although it was not until the work of James Clerk ♢Maxwell in the 1850s that he was proved correct.

From 1671–79 Cassini made many observations of details on the lunar surface, which culminated in the production of a beautiful engraving of the Moon, presented to the French Academy in 1679. In 1672 Cassini took advantage of a good opposition of Mars to determine the distance between the Earth and that planet. He arranged for Jean Richer (1630–1696) to make measurements from his base in Cayenne, on the northeastern coast of South America, while Cassini made simultaneous measurements in Paris, which permitted them to make a triangulation of Mars with a baseline of nearly 10,000 km/6,200 mi. This derived a good approximation for the distance between the Earth and Mars, from which Cassini was able to deduce many other astronomical distances. These included the astronomical unit (AU), which Cassini found to be 138 million km/85.7 million mi, only 11.6 million km/7.2 million mi short.

Cassini's later work included a study (with N Fatio) on zodiacal light (1683), and a triangulation of the arcs of meridian aimed at resolving a controversy concerning the shape of the Earth.

Cassini's contributions to astronomy were original and plentiful, but his best work was of an observational rather than of a theoretical nature.

Cauchy, Augustin-Louis (1789–1857) French mathematician who did important work in astronomy and mechanics, but who is chiefly famous as the founder, with Karl ♢Gauss, of the modern subject of complex analysis.

Cauchy was born in Paris on 21 August 1789 and received his early education from his father, an accomplished classical scholar and a barrister in the *parlement* of Normandy. When he was little more than an infant he was taken by his family to live in the village of Arceuil, where his father went to escape the terror of the French Revolution 1793–1794. There he grew up with illustrious neighbours, the mathematician Pierre ♢Laplace and the chemist Claude-Louis ♢Berthollet, who together had established the famous Société d'Arceuil. The story runs that mathematician Joseph ♢Lagrange, who also met the young Cauchy, quickly recognized the boy's scientific flair, but advised his father to give him a firm literary education before showing him any mathematical texts. True or not, Cauchy's first real introduction to mathematics was delayed until he entered the Ecole Polytechnique, in Paris, in 1805. Two years later he entered the Ecole de Ponts et Chausses to study engineering, leaving in about 1809 to take employment first at the Ourcq Canal works, then at the Saint-Clou bridge, and finally, in 1810, at the Cherbourg harbour naval installations. In 1813 he returned to Paris, apparently for health reasons, and two years later he was appointed to the Ecole Polytechnique, where he was made a full professor in 1816.

In the same year the restoration of the Bourbons to the French throne after the Napoleonic interlude brought a great change in his fortunes. Gaspard Monge and Lazare Carnot, both of them republicans and Bonapartists, were expelled on political grounds from the Academy of Sciences and Cauchy was elected to fill one of the vacancies. In that year his paper on wave modulation won the Grand Prix of the Academy (renamed the Institut de France). That paper marked the real beginning of his fruitful years as a mathematician, years that gained the reward, some time before 1830, of his appointment to the chair of the faculty of science at the Collège de France.

In 1830 Charles X was overthrown by the July Revolution, and when Cauchy refused to take the new oath of allegiance he was forced to resign his chair. He went into exile at Fribourg, where he lived among a Jesuit community; they recommended him to the king of Sardinia and he was appointed to the chair of mathematical physics at the University of Turin. From 1833–38 he was tutor to Charles X's son at Prague. At last, in 1838, he returned to Paris to resume his professorship at the Ecole Polytechnique. From 1848–52 he was a professor at the Sorbonne. He died at Sceaux, outside Paris, on 23 May 1857.

In 1805 Cauchy provided a simple solution to the problem of Apollonius, namely to describe a circle touching three given circles; in 1811 he generalized Euler's theorem on polyhedra; and in 1816 he published his award-winning paper on wave modulation. His best work, however, was all done in the 1820s and was published in his three great treatises *Cours d'analyse de l'Ecole Polytechnique* (1821), *Résumé des leçons sur le calcul infinitésimal* (1823), and *Leçons sur les applications de calcul infinitésimal à la géométrie* (1826–28). Although he did other valuable research – in mechanics he substituted the concept of the continuity of geometrical displacements for the principle of the continuity of matter and in astronomy he described the motion of the asteroid Pallas – his vital contributions were contained in these three treatises.

Cauchy made the principles of calculus clearer by developing them with the aid of limits and continuity, and he was the first mathematician to provide a rigorous proof for the famous theorem of Brook Taylor (1685–1731). Taylor's theorem, enunciated in 1712, expands a function of x as an infinite series in powers of x. Cauchy's proof was of great usefulness, because the theorem is extremely helpful in finding the difference columns in books of tables. More generally, Cauchy's work in the 1820s provided a satisfactory basis for the calculus. Perhaps even more important, for future pure and applied mathematicians alike, was his monumental research into the fundamental theorems of complex functions. He provided the first comprehensive theory of complex numbers (still, at the beginning of the 19th century, not accepted by all mathematicians) in his *Cours d'analyse,* and in doing so made a vital contribution to the development of mathematical physics and, in particular, to aeronautics.

During his lifetime Cauchy published seven books and about 800 papers. He is credited with 16 fundamental concepts and theorems in mathematics and mathematical physics, more than for any other mathematician. For both his creative genius and his prolific output he is remembered as one of the greatest mathematicians in history.

Further Reading

Belhoste, B; Toomer, G J (ed), *Augustin-Louis Cauchy: A Biography,* Studies in the History of Mathematics and Physical Sciences series, Springer-Verlag, 1991, v 16.

Rosenfeld, B A, *History of Non-Euclidean Geometry,* 1987, v 12.

Smithies, Frank, *Cauchy and the Creation of Complex Function Theory,* Cambridge University Press, 1997.

Cavendish, Henry (1731–1810) English natural philosopher whose main interests lay in the fields of chemistry and physics. His chief experimental work concerned gases, although he also carried out fundamental experiments concerning electricity and gravitation. He made the first determination of the gravitational constant and thereby obtained the first values for the mass and density of the Earth. He is also usually credited with the discovery of hydrogen. Cavendish was one of the few scientists to approach Newton's standard in both mathematical and experimental skills and was a major figure in 18th-century science. He devoted his entire life to the acquisition of knowledge, but published only those results that satisfied him completely. Most of his work, especially his experiments with electricity, were unknown for a hundred years or more, so the immediate impact of his work was far less than it might have been.

Cavendish was born in Nice on 10 October 1731. He was of aristocratic descent, his paternal grandfather being the Duke of Devonshire and his maternal grandfather the Duke of Kent. Cavendish attended Dr Newcome's Academy in Hackney, London, and then went on to Peterhouse College, Cambridge, in 1749. He left in 1753 without a degree, which was not an unusual occurrence at that time, and spent the rest of his life in London. His father encouraged his scientific interests and introduced him to the Royal Society, of which he became a member in 1760. Despite his active participation in the scientific community, Cavendish was a recluse and shunned most social contact, making no attempt to use a fortune of the order of a million pounds bequeathed to him.

Cavendish published his first paper, which demonstrated the existence of hydrogen as a substance, in 1776. He received the Copley Medal of the Royal Society for this achievement. His subsequent papers were few and far between, and included, most notably, a theoretical study of electricity in 1771, the synthesis of water in 1784, and the determination of the gravitational constant in 1798. He died alone in London on 24 February 1810.

Little is known of Cavendish's work until the late 1760s, when he began experimenting with 'facticious airs' (gases that can be produced by the chemical treatment of solids or liquids). He studied 'fixed air' (carbon dioxide) produced by mixing acids and bases; 'inflammable air' (hydrogen) generated by the action of acids on metals; and the 'airs' produced during decay and fermentation. He measured the specific gravities of hydrogen and carbon dioxide, comparing them with that of 'common' (that is, atmospheric) air.

In 1783 Cavendish found that the composition of the atmosphere is the same in different locations and at different times. He also found that a small fraction of 'common air' seems to be inert – a hundred years later William ◊Ramsay was to show that this inert gas is mainly argon. A year later Cavendish demonstrated that water is produced when hydrogen burns in air, thus proving that water is a compound and not an element as had been suggested by early Greek scientists. By sending electric sparks through 'common air' he caused the nitrogen in it to combine with oxygen. When the gas produced was dissolved in water it produced nitric acid. He also showed that 'calcareous earth' dissolves in water containing carbon dioxide, to form what is now known as calcium bicarbonate (calcium hydrogen carbonate). He distinguished between the two oxides of arsenic, demonstrating that one contains more oxygen than does the other.

Cavendish's most important work in physics was on electricity and gravitation. His 1771 paper on the nature of electricity shows that he believed it to be an elastic fluid. He then worked on electricity for ten years, aiming to produce a sequel to Newton's *Principia* that would explain all electrical phenomena. But although this was his most concentrated research effort, Cavendish published nothing more about it. His fastidious attention to the details of his results and his thorough efforts to understand and unify all his observations frustrated this plan and he was not able to gain the overview that he sought. He tried unsuccessfully to uncover the relationship between force, velocity of current, and resistance, although he found that electric fields obey the inverse square law and was able to produce some valuable work on conductivity. Much of the work done by Michael Faraday, Charles Coulomb, and others during the next 50 years is foreseen in this early work by Cavendish, but none of his experiments were known until James Clerk Maxwell edited and published them in 1879.

During the latter part of the 1780s, Cavendish worked on the production of heat and determined the freezing points for many materials, including mercury. He relied on some of the early work he had done on latent heats. One of the practical outcomes from these experiments was the explanation for some anomalous readings obtained when using mercury thermometers at low temperatures.

The five papers that Cavendish published during the last 25 years of his life all had an astronomical theme. By far the most important of these appeared in 1798, when he announced his determination of Newton's gravitational constant, thereby deriving the density and mass of the Earth. Newton's law of gravitation contained two unknowns: the gravitational constant and the mass of the Earth. Determining one would give the other. In what has become known as the Cavendish experiment, the gravitational constant was found.

Cavendish used an apparatus that had been devised by John Michell (1724–1793). It consisted of a delicate suspended rod with two small spheres made of lead attached to each end. Two large stationary spheres were placed in a line at an angle to the rod. The gravitational attraction of the large spheres caused the small spheres to twist the rod towards them. The period of oscillation set up in the rod enabled Cavendish to determine the force of attraction between the large and small spheres, which led him to determine the gravitational constant for Newton's equation, and thus the density of the Earth (about 5.5 times that of water) and its mass (6×10^{24} kg/13×10^{24} lb). The sensitivity of this apparatus was extraordinary, for the gravitational force involved was 500 million times less than the weight of the spheres, and Cavendish's results were not bettered for more than a century.

Cavendish was a great scientist and was honoured by the naming of the Cavendish Laboratories at the University of Cambridge in his memory. His contributions to science are notable for their quality and diversity. Had he permitted all his results to be published, the rate of advancement of physical science would undoubtedly have been greatly accelerated. He stands today as one of the giants of modern science.

Cavendish, Margaret (1623–1673) Duchess of Newcastle. English natural philosopher, poet, essayist, and popularizer of science.

Margaret Lucas was born in St. John's, Colchester, Essex on 15 December 1623, the youngest of eight children. She was the daughter of a wealthy landowner, Sir Thomas Lucas, and received a minimal education at home. She had a keen interest in science but received no formal scientific education. On her father's death she inherited £10,000, but at the outbreak of the Civil War the family lost its estates. Two of her brothers died in battle and in 1641 Margaret became maid of honour to Queen Henrietta Maria at the court of Charles I. In 1643 she fled with the queen into exile in Paris. She hated court life, and her marriage in 1645 to William Cavendish, Duke of Newcastle, a noted patron of poets and playwrights, gave her a reason to leave it behind.

The Newcastles were forced to remain in impecunious exile because of their royalist sympathies, apart from 18 months when Margaret returned to England in a failed attempt to collect revenue from their estates. During all this time she was constantly writing plays, poems, and philosophy. William was mildly interested in mathematics and science, but it was his brother, Charles, who, during their long exile, sustained and helped Margaret to develop her consuming interest in science. In the society they kept in Paris and Antwerp, the ideas of Descartes, Gassendi, and Hobbes were often discussed and Margaret was introduced to mechanical philosophy. She became part of a scientific movement called atomism.

Margaret's first published works were *Philosophical Fancies and Poetical Fancies* (1653) and *Philosophical and Physical Opinions* (1655). At the Restoration of Charles II to the throne in 1660, the Newcastles returned from exile. On her return to England, Margaret found herself ridiculed as an eccentric for her ideas and dress. Nevertheless, she continued her writings, publishing *Orations of Divers Persons* in 1662, which contained the arguments of several women for freedom and equality but came to the conclusion that women's power lay in the romantic domination of men. In 1666, she published *Observations upon Natural Philosophy.* She was critical of the latest in science and also of the newly invented microscope, which she claimed distorted nature. She is probably best known for the biography of her husband, which she published in 1667 and which was ridiculed by Samuel Pepys at the time but later highly praised by Charles Lamb and Virginia Woolf. Her autobiography (1655) is one of the first written in English.

Margaret's later life was spent at Welbeck, Nottinghamshire, writing poetry and plays and following her interest in natural philosophy. In 1667 she visited the Royal Society to watch experiments performed by Boyle and Hooke. She was also interested in medicine and in treating herself and it is sometimes suggested that this was what led to her sudden death at Welbeck in 1673 at the age of 50.

Cayley, Arthur (1821–1895) English mathematician who was responsible for the formulation of the theory of algebraic invariants. A prolific writer of scholarly papers, he also developed the study of *n*-dimensional geometry, introducing the concept of the 'absolute', and devised the theory of matrices.

Cayley was born in Richmond, Surrey, on 16 August 1821, the son of a merchant and his wife who were visiting England from their home in St Petersburg, Russia. Cayley spent the first eight years of his life in Russia, and then attended a small private school in London, before moving to King's College School there. He entered Trinity College, Cambridge, as a 'pensioner' to study mathematics and became a scholar in 1840. He graduated with distinction in 1842. Awarded a fellowship at the college, he took up law at Lincoln's Inn in 1846 instead, prevented from remaining at Cambridge through his reluctance to take up religious orders – at that time a compulsory qualification. Cayley was called to the Bar in 1849 and worked as a barrister for many years before, in 1863, he was elected to the newly established Sadlerian Chair of Pure Mathematics at Cambridge. He occupied the post until he died in Cambridge on 26 January 1895.

Cayley published about 900 mathematical notes and papers on nearly every pure mathematical subject, as well as on theoretical dynamics and astronomy. Some 300 of these papers were published during his 14 years at the Bar, and for part of that time he worked in collaboration with James Joseph ◊Sylvester, another lawyer dividing his time between law and mathematics. Together they founded the algebraic theory of invariants (although in their later lives they drifted apart, until Cayley lectured at Johns Hopkins University, Baltimore, 1881–82 at Sylvester's invitation).

The beginnings of a theory of algebraic invariants may be traced first in the work of Joseph ◊Lagrange, who investigated binary quadratic form in 1773. Later, in 1801, Karl ◊Gauss studied binary ternary forms. A final impetus was provided by George ◊Boole, who, in a paper published in 1841, showed that all discriminants – special functions of the roots of an equation, expressible in terms of the coefficients – displayed the property of invariance. Two years later, Cayley himself published two papers on invariants; the first was on the theory of linear transformations. In the second paper he examined the idea of covariance, setting out to find 'all the derivatives of any number of functions which have the property of preserving their form unaltered after any linear transformations of the variables'. He was the first mathematician to state the problem of algebraic invariance in general terms, and his work immediately attracted a lot of interest from other mathematicians.

Over the next 35 years he wrote ten papers on what he called 'quantics' (which later mathematicians refer to as 'form') in which he gave a lively account of the theory as it was being developed. He used the term 'irreducible invariant' and defined it as an invariant that cannot be expressed rationally and integrally in terms of invariants of the same quantic(s) but of a degree lower in the coefficients than its own. At the same time he acknowledged that there are many circumstances in which irreducible invariants and covariants are limited. (His system was eventually simplified and generalized by David ◊Hilbert.)

Cayley developed a theory of metrical geometry that could be identified with the non-Euclidean geometry of such mathematicians as Nikolai Lobachevski, János Bolyai, and Bernhard Riemann. His geometry was the geometry of *n* dimensions. He introduced the concept of the 'absolute' into geometry, which links projective geometry with non-Euclidean geometry, and together with Felix ◊Klein, distinguished between 'hyperbolic' and 'elliptic' geometry – a distinction that was of great historical significance. When Cayley's 'absolute' was real, his distance function was that of hyperbolic geometry, and when 'absolute' was imaginary, the formulae reduced to Riemann's elliptic geometry.

Cayley also created a theory of matrices that did not need repeated reference to the equations from which their elements were taken, and established the principles for forming general algebraic functions of matrices. He went on to derive many important theorems of matrix theory. He claimed to have arrived at the theory of matrices via determinants, but he always made great use of geometrical analogies in his algebraic and analytical work.

He also laid down in general terms the elements of a study of 'hyperspace', and in 1860 devised a system of six homogeneous coordinates of a line. These are now more often known as Plücker's line coordinates because the same ideas were independently published – five years later – by Julius ◊Plücker (whose assistant was Cayley's former collaborator, Felix Klein).

Cayley wrote on almost every contemporary subject in mathematics, but completed only one full-length book. He clarified many of the theorems of algebraic geometry that had previously been only hinted at, and he was one of the first to realize how many different areas of mathematics were drawn together by the theory of groups. Awarded both the Royal Medal (1859) and the Copley Medal (1881) of the Royal Society, generally in demand for both his legal and his administrative skills, Cayley played a great part in bring mathematics in England back into the mainstream and in founding the modern British school of pure mathematics.

Cayley, George (1773–1857) English baronet who spent much of his life experimenting with flying machines, particularly kites and gliders. He eventually constructed a glider capable of carrying people, but never ventured into the realms of powered flight.

Cayley was born in Brompton, Yorkshire on 27 December, the son of wealthy parents. He received a good education and from an early age showed a keen observation and an enquiring mind. Throughout his life he could turn his attention to almost any problem with a degree of success. He is particularly associated with aeronautics and the teaching of engineering, and in later life he helped to found the Regent Street Polytechnic in London. Cayley died in Brompton on 15 December 1857.

Cayley first began experimenting with flight after patiently observing how birds use their wings. He realized that they have two functions: the first is a sort of sculling action by the wing tips, which provides thrust; the second is the actual lift, achieved by the shape of the wing, which we now refer to as an aerofoil. Air rushing faster over the curved surface of the upper wing creates low pressure and a sucking effect. As a result, the higher pressure on the undersurface of the wing gives lift.

His first attempt at a flying invention was a kite fitted with a long stick, a movable tail for some control, and a small weight at the front for balance. His idea was to create a design that would glide safely but with enough speed to give lift. Spurred on by the success of his first design, he wrote in his diary of how nice it was to see it in flight and 'it gave the idea that a larger instrument would be a better and safer conveyance down the Alps than even a sure-footed mule'.

In 1808, Cayley constructed a glider with a wing area of nearly 28 sq m/300 sq ft, and was probably the first person to achieve flight with a machine heavier than air. During the next 45 years he worked on many aspects of flight, including helicopters, streamlining, parachutes, and the idea of biplanes and triplanes. Eventually, in 1853, he built a triplane glider that carried his reluctant coachman 275 m/900 ft across a small valley – the first recorded flight by a person in an aircraft. Although delighted with the results he had attained, he realized that control of flight could not be mastered until a lightweight engine was developed to give the thrust and lift required.

The developments from Cayley's experiments are plain for everyone to see in the modern world, with the use of the aeroplane as a common means of transport. The first successful sustained flight was made by du Temple's clockwork model in 1857 (the year Cayley died) and the first actual crew-carrying powered flight was in 1874, but the plane did take off down a slope. It was another 16 years before a piloted plane managed a level-ground take-off, and this was Clément Ader's *Eole;* it hopped about 50 m/160 ft. True success came with the ◊Wright brothers and the key to their success was, as Cayley had predicted, a lightweight engine.

Celsius, Anders (1701–1744) Swedish astronomer, mathematician, and physicist, now mostly remembered for the Celsius scale of temperature.

Celsius was born on 27 November 1701 in Uppsala, where his father was professor of astronomy. In 1723 he became secretary of the Uppsala Scientific Society; by the age of 30 he was himself professor of astronomy there. It was at this time that he began to travel extensively in Europe, visiting astronomers and observatories in particular.

On his travels he observed the aurora borealis; he published some of the first scientific documents on the phenomenon in 1733. While in Paris he visited Pierre-Louis Maupertuis (1698–1759), who invited him to

Celsius Swedish astronomer, mathematician, and physicist Anders Celsius who introduced the Celsius scale of temperature. He first presented this system at the Swedish Academy of Sciences in 1742. *Mary Evans Picture Library*

join an expedition centred on Torneå in Lapland (now on the Finnish-Swedish border). It confirmed the theory propounded by Newton that the Earth is flattened at the poles. With knowledge and expertise gained in this way from the leading astronomers and scientists throughout Europe, Celsius returned to the University of Uppsala, where he built a new observatory – the first installation of its kind in Sweden. He died in Uppsala on 25 April 1744.

In 1742 Celsius presented a paper to the Swedish Academy of Sciences containing a proposal that all scientific measurements of temperature should be made on a fixed scale based on two invariable (generally speaking) and naturally occurring points. His scale defined 0 ° as the temperature at which water boils, and 100 ° as that at which water freezes. This scale, in an inverted form devised eight years later by his pupil Martin Strömer, has since been used in almost all scientific work. Generally known in most of Europe under the name of Celsius, in the UK the scale has also commonly been known as centigrade.

Celsius left several other important scientifc works, including a paper on accurately determining the shape and size of the Earth, some of the first attempts to gauge the magnitude of the stars in the constellation Aries, and a study of the falling water level of the Baltic Sea.

Cesaro, Ernesto (1859–1906) Italian mathematician whose interests were wide-ranging, but who is chiefly remembered for his important contributions to intrinsic geometry. His name is perpetuated in his description of Cesaro's curves, first defined in 1896.

Cesaro was born on 12 March 1859 in Naples, where he grew up and completed the first part of his education. At the age of 14 he joined his brother in Liège, Belgium, and entered the Ecole des Mines on a scholarship. After matriculation, he continued studying mathematics and published his first mathematical paper. On the death of his father in 1879, Cesaro returned to his family in Torre Annunziata for three years before going back to Liège on another scholarship. In 1883 he published a major mathematical paper, 'Sur diverses questions d'arithmétique', in the *Mémoires de l'Academie de Liège.* After some sort of disagreement with the educational authorities in Liège, however, he entered the University of Rome in 1884. There he wrote prolifically on a wide range of subjects. Two years later he became professor of mathematics at the Lycée Terenzio Mamiani, but left after one month to fill the vacant chair of higher algebra at the University of Palermo, where he remained until 1891. Finally he became professor of mathematical analysis at Naples, and held this post until his untimely death on 12 September 1906 as a result of injuries he received in attempting to rescue his son from rough seas near Torre Annunziata.

Cesaro's most important contribution to mathematics was his work on intrinsic geometry. He began his study of the subject while in Paris, and continued to develop it for the rest of his life. His earlier work is summed up in his monograph of 1896, the *Lezione di Geometrica Intrinsica* in which, commencing with Gaston ◊Darboux's method of a mobile coordinate trihedral (formed by the tangent, the principal normal, and the bi-normal at a variable point of a curve), Cesaro simplified the analytical expression and made it independent of extrinsic coordinate systems. He stressed the intrinsic qualities of the objects. In elaborating this method later, he pointed out further applications. In the *Lezione* Cesaro described the curves that now bear his name. He later included the curves devised by Koch (which are continuous but have no tangent at any point). The *Lezione* also deals with the theory of surfaces and multidimensional spaces in general. Much later on, Cesaro was able to emphasize the independence of his geometry from the axioms of parallels, and also established other foundations on which to base non-Euclidean geometry.

Cesaro's other work, particularly during his time at the University of Rome, covered topics ranging from elementary geometrical principles to the application of mathematical analysis; from the theory of numbers to symbolic algebra; and from the theory of probability to differential geometry. He also made notable interpretations of James Clerk Maxwell's work in theoretical physics.

Chadwick, James (1891–1974) English physicist who discovered the neutron in 1932. For this achievement he was awarded the 1935 Nobel Prize for Physics.

Chadwick was born in Bollington, Cheshire, on 20 October 1891. He began his scientific career at Manchester University, graduating in physics in 1911. Chadwick then continued at Manchester and, under Ernest ◊Rutherford, investigated the emission of gamma rays from radioactive materials. To gain further research experience, he went in 1913 to Berlin to work with Hans Geiger, the inventor of the geiger counter, where he discovered the continuous nature of the energy spectrum and investigated beta particles emitted by radioactive substances. Chadwick was then interned as an enemy alien on the outbreak of World War I, living and working in a stable for the duration of the war. He still managed to do original research, however, and investigated the ionization present during the oxidation of phosphorus and the photochemical reaction between chlorine and carbon monoxide.

At the end of the war, Rutherford invited Chadwick to Cambridge. During this period, he determined the atomic numbers of certain elements by the way in which alpha particles were scattered. He also established the equivalence of atomic number and atomic charge. With Rutherford, he produced artificial disintegration of some of the lighter elements by alpha-particle bombardment.

His most famous achievement, the discovery of the neutron, came in 1932 after its existence had been suspected by Rutherford as early as 1920. In experiments in which beryllium was bombarded by alpha particles, a usually energetic gamma radiation appeared to be emitted. It was more penetrating than gamma radiation from radioactive elements. Measurements of the energies involved and the conservation of energy and momentum suggested to Chadwick that a new kind of particle was being produced rather than radiation. The results pointed towards a neutral particle made up of a proton and an electron. Its mass should thus be slightly greater than that of the proton. Because the mass of the beryllium nucleus had not then been measured, Chadwick designed and carried out an experiment in which boron was bombarded with alpha particles. This produced neutrons, and from the mass of the boron nucleus and other elements and the energies involved, Chadwick determined the mass of the neutron to be 1.0067 atomic mass units, slightly greater than that of the proton.

In the same year, Chadwick became professor of physics at the University of Liverpool. He ordered the building of a cyclotron and, from 1939 onwards, used it to investigate the nuclear disintegration of the light elements. During World War II, he was closely involved with the atomic bomb, and much of the research and calculation for the British contribution to the Manhattan Project was carried out at Liverpool under his direction. From 1943, he led the British team with the project in the USA.

In 1945 Chadwick was knighted, and in the same year he returned to Liverpool to continue his own research and to develop a research school in nuclear physics. He returned to Cambridge as master of Gonville and Caius College in 1948, and stayed in this position until his retirement ten years later. He died on 24 July 1974.

The discovery of the neutron made by Chadwick led to a much deeper understanding of the nature of matter, explaining for example why isotopes of elements exist. It also inspired Enrico ◊Fermi and other physicists to investigate nuclear reactions produced by neutrons, leading to the discovery of nuclear fission.

Further Reading

Brown, Andrew, *The Neutron and the Bomb: A Biography of Sir James Chadwick*, Oxford University Press, 1997.

Gowing, Margaret, 'James Chadwick and the atomic bomb', *Notes Rec Roy Soc Lond*, 1993, v 47, pp 79–92.

Holt, J R, 'Reminiscences and discoveries: James Chadwick at Liverpool', *Notes Rec Roy Soc Lond*, 1994, v 48, pp 299–308.

Peierls, Rudolf, 'Reminiscences and discoveries: recollections of James Chadwick', *Notes Rec Roy Soc Lond*, 1994, v 48, pp 135–141.

Chain, Ernst Boris (1906–1979) German-born British biochemist who, in collaboration with Howard ◊Florey, first isolated and purified penicillin and demonstrated its therapeutic properties. Chain, Florey, and Alexander ◊Fleming shared the 1945 Nobel Prize for Physiology or Medicine – Chain and Florey for their joint work in isolating penicillin and demonstrating its clinical use against infection, and Fleming for his initial discovery of the *Penicillium notatum* mould. Chain also received many other honours for his work, including a knighthood in 1969.

Chain was born on 19 June 1906 in Berlin, the son of a chemist. He was educated at the Luisengymnasium, then at the Friedrich Wilhelm University in Berlin, from which he graduated in chemistry and physiology in 1930. After graduation he did research in the chemistry department of the Pathological Institute at the Charité Hospital in Berlin, but with the rise to power of Adolf Hitler in 1933, Chain emigrated to the UK. Initially he worked for a short time at University College, London, and then, on the recommendation of J B S ◊Haldane, he worked under Frederick Gowland ◊Hopkins at the Sir William Dunn School of Biochemistry at Cambridge University 1933–35. In that year Florey invited Chain to work with him at the Sir William Dunn School of Pathology at Oxford University as university demonstrator and lecturer in chemical pathology. In 1949 Chain was invited to be guest professor of biochemistry at the Istituto Superiore di Sanita in Rome; in the following year he accepted a permanent position as professor there and was also appointed scientific director of the International Research Centre for Chemical Microbiology. In 1961 he returned to the UK as professor of biochemistry at Imperial College, London, where he did much to ensure that the laboratories were equipped with modern facilities. On his retirement in 1973, Chain was made emeritus professor and senior research fellow of Imperial College. He died on 12 August 1979 in Ireland.

At Oxford University, Chain initially investigated the observation first made by Fleming in 1924 that tears, nasal secretion, and egg white destroyed bacteria. Chain showed that these substances contain an enzyme, lysozyme, which digests the outer cell wall of bacteria. In 1937, while preparing this discovery for publication, Chain found another observation of Fleming's, that the mould *Penicillium notatum* inhibits bacterial growth. In the following year, Chain, in collaboration with Florey, started research to try to isolate and identify the antibacterial factor in the mould. Chain first developed a method for determining the relative strength of a penicillin-containing broth by comparing its antibacterial effect (as shown on culture plates) with that of a standard penicillin solution, 1 cu cm/0.061 cu in of which is defined as containing one Oxford unit of penicillin. Then he developed a method of purifying penicillin without destroying its antibacterial effect. He found that the optimum time for extraction of the penicillin is when the mould is one week old; he also found that free penicillin is acidic and is therefore more soluble in certain organic solvents than it is in water. He then agitated the penicillin broth with acidified ether or amyl acetate, reduced the acidity of the solution until it was almost neutral, removed impurities, and evaporated the purified solution at a low temperature to give a stable form of the active substance. Chain and his co-workers found that 1 mg/0.015 grain of the active substance they had obtained contained between 40 and 50 Oxford units of penicillin and that, in a concentration of only one part per million, it was still able to destroy staphylococcus bacteria. Furthermore, they also showed that their purified penicillin was only minimally toxic and that its antibacterial effect was not diminished by the presence of blood or pus. With British biochemist E P Abraham (1913–1999), Chain then elucidated the chemical structure of crystalline penicillin, finding that there are four different types, each differing in their relative elemental constituents.

Chain also studied snake venoms and found that the neurotoxic effect of these venoms is caused by their destroying an essential intracellular respiratory coenzyme.

Challis, James (1803–1882) English astronomer renowned in his time for unconventional views concerning the fundamental laws of the universe, but now remembered more for an almost unbelievable lapse in scientific professionalism.

Challis was born in Braintree, Essex, on 12 December 1803. He attended a local school where he showed such promise that he won a place at a London school and later at Trinity College, Cambridge, which he entered in 1821. He graduated with top honours in 1825, and was a fellow of the college 1826–31 (and later from 1870–82). Challis was ordained in 1830 and served as rector at Papworth Everard, Cambridgeshire, 1830–52. He also succeeded George ◊Airy to the Plumian Professorship of Astronomy at Cambridge University in 1836 – a post he held until his death – and he served as director to the Cambridge Observatory 1836–61. A member of the Royal Astronomical Society, a Fellow of the Royal Society, and the author of several scientific publications, he died in Cambridge on 3 December 1882.

In 1844, John Couch ◊Adams – a young and enthusiastic astronomer and mathematician, and a recent graduate of Cambridge University – approached Challis to enlist his aid in obtaining data from Airy at the Greenwich Observatory regarding the known deviations in orbit of the planet Uranus. These were suspected of indicating the gravitational influence of a planet even farther out. With Challis's mediation, Adams received from Airy all the data the observatory possessed on Uranus for the period 1754–1830.

In September 1845 Adams supplied Challis and Airy with an estimated orbital path for the unknown planet and a prediction for its likely position on 1 October 1845. But Challis did not take the calculations seriously (saying later that he could not believe so youthful and inexperienced an astronomer as Adams would arrive at anything like a correct prediction), and Airy, through a series of mishaps, did not even see them until the following year.

By that time, in France, Urbain ◊Leverrier had performed calculations similar to those of Adams, and he was more successful than Adams in obtaining the cooperation of senior astronomers. Almost immediately after he sent his predictions to Berlin Observatory in September 1846, the new planet was discovered – by Johann Galle and Heinrich d'Arrest (1822–1875) – later to be called Neptune.

All Challis could do then was lamely to report that if he had indeed conducted a search at Adam's predicted position for 1 October 1845 he would have been within 2° of the planet's actual position and would almost certainly have spotted it.

Chamberlin, Thomas Chrowder (1843–1928) US geophysicist who asserted that the Earth was far older than was then believed.

A farmer's son, born in Illinois and brought up in Wisconsin, Chamberlin always claimed that his native terrain had shaped his geological thinking. Partly self-taught in science, Chamberlin joined the Wisconsin Geological Survey in 1873, and rose to become its chief geologist, publishing the *Geology of Wisconsin* (1877–83). He went on to work for the US Geological Survey before becoming professor of geology at Chicago in 1892, working there until his retirement in 1918.

Chamberlin's most important contribution to geological thinking lay in his bold attack on the British physicist Lord ◊Kelvin. Kelvin had postulated that the Earth was rather young (less than 100 million years), basing his views on the assumption, derived from the nebular hypothesis, that the Earth had steadily cooled from a molten mass. Chamberlin rebuked Kelvin for his dogmatic confidence in extrapolations from a single hypothesis, and stressed that geological reasoning must follow from a plurality of working hypotheses. He also believed geological evidence in any case suggested the Earth to be older than Kelvin had estimated. Chamberlin backed his refutation of the nebular hypothesis by developing (with the aid of the celestial physicist, F R Moulton (1872–1952)), the planetesimal hypothesis. This postulated a gradual origin, by accretion of particles, for the Earth and other planetary bodies – an origin for these bodies that was therefore cool and solid.

Chandrasekhar, Subrahmanyan (1910–1995) Indian-born US astrophysicist who was particularly concerned with the structure and evolution of stars. He is well-known for his studies of white dwarfs and the radiation of stellar energy.

Chandrasekhar was born on 19 October 1910 in Lahore, India (now in Pakistan). He went to Presidency College, University of Madras, from which he graduated with a BA in 1930. He continued his studies at Trinity College, Cambridge, gaining a PhD in 1933. There he studied under the physicist Paul Dirac. He left Trinity College in 1936 to take up a position on the staff of the University of Chicago, working in the Yerkes Laboratory. In 1938 he became assistant professor of astrophysics there and in 1952 was promoted to distinguished service professor. The following year he became a US citizen. He died on 21 August 1995.

Chandrasekhar's greatest contribution to astronomy was his explanation of the evolution of white dwarf stars, as laid out in his *Introduction to the Study of Stellar Structure* (1939). These stellar objects, which were first discovered in 1915 by Walter Sydney ◊Adams, are similar in size to the Earth. They have a very high density and are therefore very much more massive than the Earth. This enormous density is explained in terms of degeneracy – a consequence of the Pauli exclusion principle in which electrons become so tightly packed that their normal behaviour is suppressed; as stars evolve, they 'burn' their hydrogen, which is converted to helium and, eventually, heavier elements. During his work at Cambridge, Chandrasekhar suggested that when a star had burned nearly all its hydrogen, it would not be able to produce the pressure against its own gravitational field to sustain its size and would then contract. As its density increased during the contraction the star would build up sufficient internal energy to collapse its atomic structure into the degenerate state.

Not all stars, however, become white dwarfs. Chandrasekhar believed that – up to a certain point – the greater the mass of a star, the smaller would be the radius of the eventual white dwarf. But he also stated that beyond this point a large stellar mass would not be able to equalize the pressure involved and would explode. He calculated that stellar masses below 1.44 times that of the Sun would form stable white dwarfs, but those above this limit would not evolve into white dwarfs. This limit – known as the Chandrasekhar limit – was based on calculations involving the complete degeneracy of the stellar matter; the limit is now believed to be about 1.2 solar masses.

A certain modesty toward understanding nature is a precondition to the continued pursuit of science.

Subrahmanyan Chandrasekhar Interview 1984

Stars with masses above the Chandrasekhar limit are likely to become supernovae and rid themselves of their excess matter in a spectacular explosion. The remaining mass may form a white dwarf if the conditions of mass and pressure are suitable, but it is more likely to form a neutron star. Neutron stars were first identified by J Robert Oppenheimer and his co-workers in 1938. These stars are even more dense than white dwarfs, with an average radius of approximately 15 km/9 mi.

With the Polish astrophysicist Erich Schönberg, Chandrasekhar determined the Chandrasekhar–Schönberg limit of the mass of a star's helium core; if it is more than 10–15% of that of the entire star, the core rapidly contracts, often collapsing. Chandrasekhar also investigated the transfer of energy in stellar atmospheres by radiation and convection and the polarization of light emitted from particular stars.

Further Reading

Wald, Robert M (ed), *Black Holes and Relativistic Stars*, University of Chicago Press, 1998.

Wali, Kameshwar C, *Chandra: A Biography of S Chandrasekhar*, University of Chicago Press, 1991.

Wali, Kameshwar C (ed), *S Chandrasekhar: The Man Behind the Legend*, Imperial College Press, 1997.

Charcot, Jean-Martin (1825–1893) French neurologist whose studies of hysteria still excite controversy.

Born on 29 November 1825 in Paris, the son of a wheelwright, Charcot studied at the Paris Faculty of Medicine, graduating MD in 1853 with a doctoral thesis on chronic rheumatism and gout. In 1862 he became resident doctor at the Salpêtrière, where he built up a leading neurological department. In 1872 he was appointed professor of anatomical pathology at the Faculty of Medicine, ten years later moving to the chair for the study of nervous disorders at the

Salpêtrière, where the distinguished Joseph Babinski served as his director. Charcot died on 16 August 1893 in Niève, France.

Charcot was an ardent champion of the clinical anatomical method that systematically correlated the symptoms presented by the patient with the lesions discovered at autopsy. He was also committed to the view that all diseases (even apparently strange psychiatric conditions) were regular natural phenomena, whose laws could be discovered by medical science. Widespread observation of multiple cases (simple at a huge institution like the Salpêtrière), would thus crack the secrets of diseases. Over the course of a generation, Charcot published a series of memoirs that turned him into one of the world's pre-eminent neurologists. As well as portraying the neuropathy that became known as Charcot's disease, he produced classic descriptions of multiple or disseminated sclerosis; of amyotrophic lateral sclerosis; of cerebral haemorrhage; and of tabes dorsalis, a form of neurosyphilis. He studied Parkinson's disease and contributed to the investigation of poliomyelitis. His *Leçons sur les maladies du système nerveux faits à la Salpêtrière* (1872–73) laid his teachings on such subjects before a larger audience.

In his approach to brain function, Charcot vigorously supported the theory of cerebral localization, as developed by English neurologist Hughlings Jackson (1835–1911). He applied this theory to cases of Jacksonian epilepsy, aphasia, and Beard's neurasthenia. During the 1870s, he developed highly publicized work on hysteria. Far from being a psychogenic disorder or just a disease of women, Charcot regarded it as a general malady of neurological origin. Such views proved influential upon his pupil, Sigmund ◊Freud, not least because Charcot was also fascinated by the relations between hysteria and hypnotic phenomena. Critics widely accused Charcot of inadvertently 'training' the young women who were his main hysterical subjects. One of the founders of modern neurology, Charcot thus left the relations between neurology and psychiatry extremely obscure.

Further Reading

Charcot, Jean-Martin, *Charcot, the Clinician: The Tuesday Lessons,* Lippincott-Raven Publishers, 1987.

Goetz, Christopher G; Bonduelle, Michael; and Gelfand, Toby (eds), *Charcot: Constructing Neurology,* Oxford University Press, 1995.

Hunter, Dianne (ed), *The Makings of Dr Charcot's Hysteria Shows: Research Through Performance,* Studies in Theatre Arts series, Mellen University Press, 1998, v 4.

Chardonnet, (Louis-Marie) Hilaire Bernigaud (1839–1924) Comte de Chardonnet. French industrial chemist who invented rayon, the first type of artificial silk. He also worked on nitrocellulose (gun cotton).

Chardonnet was born into an aristocratic family on 1 May 1839 at Besançon, Doubs. He trained first as a civil engineer at the Ecole Polytechnique, Paris, and then went to work under Louis Pasteur, who was studying diseases in silkworms. This inspired Chardonnet to seek an artificial replacement for silk which he first patented in 1884. Five years later, at the Paris Exposition, he was awarded the Grand Prix for his invention. He opened his first factory, the Société de la Soie de Chardonnet, at Besançon in 1889, and in 1904 he built a second factory at Sarvar in Hungary. He died in Paris on 12 March 1924 at the age of 85.

Chardonnet began his experiments in 1878 but it was six years before he produced a satisfactory fibre. He prepared nitrocellulose (mainly cellulose tetranitrate) by treating a pulp made from mulberry leaves – the food plant of silkworms – with mixed nitric and sulphuric acids. The cellulose compound was dissolved in a mixture of ether and alcohol and the hot viscous solution forced through fine capillary tubes into cold water. The warm threads were stretched and dried in heated air.

The original nitrocellulose fibre was highly inflammable, and Chardonnet continued working to produce a fireproof version. By 1889 he had developed rayon, so-called because the brightness of the material was thought to resemble the emission of the Sun's rays. He later was able to make 35–40 denier threads (denier is the weight in grams of 9,000 m/29,500 ft of yarn; 9,000 m/29,500ft of 40 denier nylon weighs 40 g/1.4 oz) of tensile strength equivalent to that of natural silk.

Rayon was the first artificial fibre to come into common use. It was, admittedly, only modified cellulose but it pointed the way to the totally synthetic fibres developed about 50 years later by Wallace ◊Carothers and others. Today the term rayon is generally used for all types of fibres made from cellulose, although is most often applied to viscose yarns. The cellulose is usually derived from cotton or wood pulp.

As well as his development of artificial fibres, Chardonnet also spent some time working for the French government on the production of gun cotton, the original smokeless powder for cartridges and shells that exploits the material's high inflammability – the very feature that Chardonnet had to eliminate from his textile fibre. He also made minor contributions to studies of the absorption of ultraviolet light, telephony, and the behaviour of the eyes of birds.

Chargaff, Erwin (1905–) Austrian-born US biochemist noted for his work on nucleic acids and for the Chargaff rules that demonstrate the mathematical relationship between the nitrogenous bases of DNA (deoxyribonucleic acid).

Chargaff was born in Caernowitz, Bohemia (now in the Czech Republic) on 11 August 1905. He studied chemistry in Vienna, gaining a DPhil in 1928, and then at Yale University for two years. Chargaff returned to Europe, working in Berlin and then at the Pasteur Institute in Paris, but in 1935 the political situation in Europe forced him to emigrate to the USA, where he spent his whole career at Columbia University, New York, becoming full professor of biochemistry in 1952 and emeritus professor on his retirement in 1974.

Chargaff's initial work at Columbia was concerned with a range of biochemical subjects, including the coagulation of blood and the metabolism of fat in the body. Following the announcement in 1944 by Oswald ◊Avery that pure DNA is the factor causing the heritable transformation of bacteria, and Erwin ◊Schrödinger's suggestion in his book *What is Life?* (1944) that chromosomes carried a hereditary code, Chargaff concentrated his research on the DNA molecule. He hypothesized that there must be many more types of DNA molecules than had previously been believed and used the new techniques of paper chromatography and ultraviolet spectroscopy to study them. He made the finding that DNA was the same within a species but that there were great differences across species. This led him to believe that there must be as many types of DNA as there are species.

Chargaff found important contradictions in his research, and in 1950 he showed that the bases that formed DNA fell into complementary pairs. He also found that the number of purine bases (adenine and guanine) was always equal to the number of pyramidine bases (cytosine and thiamine) but that the numbers of adenine bases were equal to those of thiamine and that those of guanine were equal to those of cytosine. These 'Chargaff rules' were crucial to James ◊Watson and Francis ◊Crick's model building in the construction of the double helix.

In later life Chargaff has become disillusioned by science and critical of molecular biology, which in his opinion, actually impedes the flow of scientific explanation by claiming to be able to explain everything. His eloquent writings include *The Nucleic Acids* (1955, with J N Davidson), an autobiography *Heraclitean Fire: Sketches from a Life Before Nature* (1978), and *Serious Questions* (1986).

Further Reading

Chargaff, Erwin, *Heraclitean Fire: Sketches from A Life Before Nature*, Rockefeller University Press, 1978.

Chargaff, Erwin, *Serious Questions: An ABC of Skeptical Reflections*, Birkhauser, 1986.

Charles, Jacques Alexandre César (1746–1823)

French physicist and mathematician who is remembered for his work on the expansion of gases and his pioneering contribution to early ballooning.

Charles French physicist and balloonist Jacques Alexandre César Charles. His experiments on the expansion of gases with temperature lead to the formulation of Charles's Law, and then in 1783 to the first uncrewed ascent of a hydrogen balloon. *Mary Evans Picture Library*

Charles was born in Beaugency, Loiret, on 12 November 1746. He became interested in science while working as a clerk in the Ministry of Finance in Paris. Stimulated by Benjamin Franklin's experiments with lightning and electricity, he constructed a range of apparatus that he demonstrated at popular public lectures. He also experimented with gases. He was elected to the French Academy of Sciences in 1795 and later became professor of physics at the Paris Conservatoire des Arts et Métiers. He died in Paris on 7 April 1823.

The Montgolfier brothers made their first experiments with uncrewed hot-air balloons at Viadalon-les-Annonay in June 1783. On hearing about them, Charles tried filling a balloon with hydrogen, and with the brothers Nicolas and Anne-Jean Robert made the first successful (uncrewed) experiment in August 1783. In November of that year the Montgolfiers demonstrated their hot-air balloons in Paris, and on 1 December Charles and Nicolas Robert made the first human ascent in a hydrogen balloon. In later flights Charles ascended to an altitude of 3,000 m/9,846 ft. On a tide of public acclaim he was invited by King Louis XVI to move his laboratory to the Louvre – patronage that Charles was to regret ten years later during the French Revolution.

In about 1787 Charles experimented with hydrogen, oxygen, and nitrogen and demonstrated the constant expansion of these gases – that is, at constant pressure the volume of a gas is inversely proportional to its temperature. He found that a gas expands by 1/273 of its volume at 0 °C for each degree Celsius rise in temperature (implying that at –273 °C/–459.4 °F, now known as absolute zero, a gas has no volume). He did not publish his results, but communicated them to the French physical chemist Joseph ◊Gay-Lussac, who repeated the experiments and made more accurate measurements. Unknown to the two French scientists, John ◊Dalton in England was also about to embark on similar research. Dalton deduced the same gas law in 1802, but the first to publish (six months later) was Gay-Lussac. For this reason, the law became known in France as Gay-Lussac's law, but elsewhere it was, and still is, generally known as Charles's law. Incidentally, Gay-Lussac continued to emulate Charles by becoming a pioneer balloonist.

Charles devised or improved many scientific instruments. He invented a hydrometer and a reflecting goniometer, and improved the aerostat of Gabriel Fahrenheit and the heliostat of W J vans' Gravesande (1688–1742).

Charnley, John (1911–1982) English orthopaedic surgeon who appreciated the importance of applying engineering principles to the practice of orthopaedics. He is best known for his work on degenerative hip disease and total hip replacement, or arthroplasty. He also successfully pioneered arthrodeses (joint fusion) for the knee and hip. He was knighted in 1977 and awarded the Gold Medal of the British Medical Association in 1978.

Charnley was born on 29 August 1911 in Bury, Lancashire. He went to the local grammar school and then to Manchester University. His academic achievements in medicine were impressive – he was the only student to pass primary FRCS (the first stage in applying for fellowship of the Royal College of Surgeons) before graduating MB (in 1935), and he obtained his FRCS in 1936, only a year after qualifying. At the outbreak of World War II Charnley became a major in the Royal Army Medical Corps and spent some time in the Middle East. At Heliopolis he ran the army splint factory, turning out the Thomas splint for treating leg fractures among the soldiers. When the war ended he went back to Manchester as a lecturer, and in 1947 he became consultant orthopaedic surgeon at Manchester Royal Infirmary. He married in 1957 and had two children. In the mid-60s Charnley retired from the infirmary in order to devote his time to hip arthroplasty at the Centre of Hip Surgery at Wrightington Hospital, Lancashire, where he became director. He built the centre up to become the primary unit for hip replacement in the world, and surgeons from many countries visited Wrightington to observe the latest techniques. The Royal Society acknowledged his contributions to surgery in 1975 when he was made a Fellow. He died suddenly on 5 August 1982.

The replacement of the femoral head and acetabulum (socket) in the hip had been researched and tried by McKee and others to treat the painful condition of degenerative hip disease, but Charnley realized that the fundamental problem was one of lubrication of the artificial joint. He carried out research on the joints of animals and tried using the low-friction substance polytetrafluoroethylene (PTFE or Teflon), with great success at first. Teflon was eventually abandoned, but Charnley had learnt much – including the use of methyl methacrylate cement for holding the metal prosthesis or implant to the shaft of the femur. In 1962 the right high-density polythene was developed, and his results became increasingly successful.

For the treatment of rheumatoid arthritis Charnley devised a system for surgically fusing joint surfaces (arthrodesis) to immobilize the knee joint using an external compression device, which bears his name. A metal pin is passed through the lower femur and another through the upper tibia and these are clamped together externally to hold the bared joint surfaces together until the joint fuses, leaving it immobile but pain-free.

Throughout his career Charnley developed a series of highly practical and successful surgical instruments. In his fight against post-operative infection he used air 'tents' which allowed the surgeon and the wound to be kept in a sterile atmosphere throughout the operation.

Further Reading

Waugh, William, *John Charnley: The Man and the Hip*, Springer-Verlag, 1991.

Chase, Mary Agnes Meara (1869–1963) US botanist and suffragist who made outstanding contributions to the study of grasses, despite a lack of higher education and any formal qualifications. During the course of several research expeditions she collected many plants previously unknown to science, and her work provided much important information about naturally occurring cereals and other food crops. This knowledge could then be used by nutrition and agricultural scientists in developing disease-resistant and nutritionally enhanced strains.

Born Mary Agnes Meara on 20th April 1869 in Iroquois County, Illinois, she was the fifth of six children. After her father's death two years later, the family moved to Chicago, where she attended public school and worked at various jobs to help with household expenses. Aged 18, she married a newspaper editor named William Chase but was widowed the following year. She returned to employment as a proofreader and, through encouraging a nephew's botanical pursuits, developed an interest in the flora of her local area. In this she was assisted by the Reverend Ellsworth Hill, also a botanist, who guided her collecting and recording and employed her to draw specimens from his own collections. He also helped her to apply for more suitable positions, first as a meat inspector in the Chicago stockyards and in 1903 in Washington working for the US Department of Agriculture Bureau of Plant Industry and Exploration. In that position she worked with Albert Spear Hitchcock, the principal scientist in the division of agrostology (study of grasses), illustrating bureau publications and becoming first an assistant botanist and then a botanist in systematic agrostology. In 1936 she succeeded Hitchcock, with whom she had collaborated closely, and became the principal scientist for agrostology. She retired in 1939 and died on 24 September 1963 in Bethesda, Maryland.

Chase was particularly responsible for work in modernizing and extending the national collection of the grass herbarium that had been part of the US National Herbarium, although it was incorporated into the Smithsonian Institution in 1912. She travelled widely, collecting plants from several regions of North and South America, and also visiting European research institutes and herbaria during the 1920s. Several of her expeditions were self-financed, and it has been estimated that by the conclusion of her final collecting trip in 1940 she had collected more than 12,000 plants for the herbarium. She was also consulted by foreign officials and scientists for assistance in identifying grasses, and used these opportunities to acquire duplicates of 'type specimens' (from which the first descriptions of a new plant are made) from foreign collections. She donated her own extensive library to the Smithsonian Institution, wrote important monographs on grasses in the Western hemisphere, and was responsible for the authoritative *Manual of the Grasses of the United States* (1950). She also wrote popular accounts of her work, including the *First Book of Grasses* (1922). She was politically active in various reform movements, especially those for female suffrage, and on this account was jailed and forcibly fed during World War I.

Chevreul, Michel Eugène (1786–1889) French organic chemist who in a long lifetime devoted to scientific research studied a wide range of natural substances, including fats, sugars, and dyes.

Chevreul was born in Angers on 31 August 1786, the son of a surgeon. He went to Paris in 1803 when he was 17 years old to study chemistry at the Collège de France under Louis Vauquelin. He became an assistant to Antoine Fourcroy (1783–1791) in 1809 and a year later took up an appointment as an assistant at the Musée d'Histoire Naturelle. He was professor of physics at the Lycée Charlemagne 1813–30, after which he returned to the museum as professor of chemistry, succeeding his old tutor Vauquelin. In 1824 he was made a director of the dyeworks associated with the Gobelins Tapestry Factory, and in 1864 he became director for life of the Musée d'Histoire Naturelle. He died, aged 102, in Paris on 8 April 1889.

Chevreul's earliest research under Vauquelin was on indigo, a subject he was to return to later. He began his studies of fats in 1809 by first decomposing soaps (which at that time were made exclusively by the action of alkali on animal fats). By treating soaps with hydrochloric acid he obtained and identified various fatty acids, including stearic, palmitic, oleic, caproic, and valeric acids. He thus realized that the soapmaking process is the treatment of a glyceryl ester of fatty acids (that is, a fat) with an alkali to form fatty acid salts (that is, soap) and glycerol.

One of the most useful of the newly discovered acids was stearic acid, and in 1825 Chevreul and Joseph ◊Gay-Lussac patented a process for making candles from stearin (crude stearic acid), providing a cleaner and less odorous alternative to tallow candles. Chevreul determined the purity of fatty acids by measuring their melting points, and constancy of melting point soon became a criterion of purity throughout preparative and analytical organic chemistry. He also investigated natural waxy substances, such as

Soap-making (saponification):

glyceryl stearate	+	sodium hydroxide	→	sodium stearate	+	glycerol
(fat)		(alkali)		(soap)		

Chevreul's hydrolysis:

sodium stearate	+	hydrochloric acid	→	stearic acid	+	sodium chloride
(soap)		(acid)		(fatty acid)		(salt)

Chevreul By decomposing soaps Chevreul identified a number of fatty acids.

spermaceti, lanolin, and cholesterol (which did not yield fatty acids on treatment with hydrochloric acid).

During the many years he was working with fats Chevreul also studied other natural compounds. In 1815 he isolated grape sugar (glucose) from the urine of a patient suffering from diabetes mellitus. At the Gobelin dyeworks he discovered haematoxylin in the reddish-brown dye logwood and quercitrin in yellow oak; he also prepared the colourless reduced form of indigo. His interest in the creation of the illusion of continuous colour gradation by using massed small monochromatic dots (as in an embroidery or tapestry) later influenced Pointillist and Impressionist painters.

Further Reading

Chevreul, Michel-Eugène, *Principles of Harmony and Contrast of Colors and Their Applications to the Arts*, Schiffer Publishing Ltd, 1987.

Smeaton, William A, 'Michel Eugéne Chevreul (1786–1889): the doyen of French students', *Endeavour*, 1989, v 13, pp 89–92.

Cheyne and Stokes, John Cheyne (1777–1836) and William Stokes (1804–1878) Scottish and Irish physicians, respectively, who practised in Dublin and gave their name to Cheyne–Stokes breathing, or periodic respiration.

John Cheyne was born in Leith, Scotland, on 3 February 1777. He was educated at Edinburgh High School, and was formally apprenticed to his physician father at the age of 13. He qualified in 1795 and joined the British army as a surgeon. He returned four years later to take charge of an ordnance hospital at Leith and he began to take medicine seriously as a result of working for Charles ◊Bell. He visited Dublin in 1809, and decided to settle there. In 1811 he became a physician at Meath Hospital, where his practice flourished. He took the first professorial chair in medicine at the Royal College of Surgeons of Ireland in 1813, and was succeeded by Whitley Stokes, the father of William Stokes. Cheyne died in Newport Pagnell, Buckinghamshire, on 31 January 1836.

William Stokes studied clinical medicine at the Meath Hospital, then became a student in Edinburgh where he graduated in 1825. He went back to Dublin as physician to the Dublin General Dispensary, and later succeeded his father at the Meath Hospital. He died on 10 January 1878.

In 1818 Cheyne described the periodic respiration that occurs in and signifies patients with intracranial disease or cardiac disease. His paper described how the breathing ceases entirely for a quarter of a minute or more, then becomes perceptible and increases by degrees to quick, heaving breaths that gradually subside again. Stokes referred to Cheyne's paper in his famous book *The Diseases of the Heart and Aorta*, and thus their names became eponymous with the sign.

Stokes's name was also applied to Stokes–Adams attacks after his paper 'Observations on some cases of permanently slow pulse', which was published in 1846.

Child, Charles Manning (1869–1954) US zoologist who tried to elucidate one of the central problems of biology – that of organization within living organisms.

Child was born on 2 February 1869 in Ypsilanti, Michigan, where his grandfather was a physician, then three weeks later was taken home to Higganum, a small village in Connecticut where his father was a farmer. The last-born and only survivor of five sons, Child was taught by his mother until he was nine years old, when he went to Higganum District School. He then attended high school in Middleton, Connecticut, 1882–86, after which he studied zoology at the Wesleyan University in Middletown, from which he graduated in 1890 and obtained his MSc in 1892. While studying at university he continued to live with his parents so that he could run the farm for his father, who had previously been incapacitated by a cerebral haemorrhage. His parents died in 1892, and two years later Child went to Leipzig University to research for his doctorate, which he gained in the same year. After returning to the USA, he went to the newly established University of Chicago in 1895 and remained there for almost all of his academic career – as assistant 1895–1996, associate 1896–1998, instructor 1898–1905, assistant professor 1909–1916, and professor 1916–1934. After his retirement in 1934, Child was appointed professor emeritus. The only interruptions to his association with the University of Chicago were two sabbaticals – to Duke University as visiting professor in 1930, and to Tohoku University in Japan as visiting professor of the Rockefeller Foundation 1930–31. On retiring, Child moved to Palo Alto in California and became a guest at Stanford University. He remained there until his death on 19 December 1954.

Child's early work concerned the functioning of the nervous system in various invertebrates, but his interest soon turned to embryology, in which field he did some important research into cell lineage – tracing the fate of each cell in the early embryo. In 1900, however, he began a long series of experiments on regeneration in coelenterates and flatworms, a topic that occupied him for most of his career. Child believed that the regeneration of a piece of an organism into a normal whole resulted from the piece functioning like the missing parts. In 1910 he perceived that there is a gradation in the rate of physiological processes along the longitudinal axis of organisms, and in the following year he developed his gradient theory. According to this theory, each part of an organism dominates the region behind and is dominated by that in front. In general, the region of the highest rate of activity in eggs, embryos, and other reproductive regions becomes the apical end of the head of the larval form; in plants it becomes the growing tip of the shoot or of the primary root. Child also pointed out that regeneration is fundamentally the same as embryonic development, in that the dominant apical region is formed first then the other parts of the organism develop in relation to it. In 1915 Child demonstrated that the parts of an organism that have the highest metabolic rates are most susceptible to poisonous substances, but that these parts also have the greatest powers of recovery after damage.

Child's explanation of how the various cells and tissues in organisms are organized – by a gradation in the rate of physiological processes leading to relationships of dominance and subordination – may not be thought to be correct, but it was an important early contribution to the problem of functional organization within living organisms.

Chladni, Ernst Florens Friedrich (1756–1827) German physicist who studied sound and invented musical instruments, helping to establish the science of acoustics.

Chladni was born in Wittenberg, Saxony, on 30 November 1756. His father insisted that he study law at the University of Leipzig, from which he graduated in 1782. After his father's death in about 1785 Chladni changed to the study of science, concentrating on experiments in acoustics. He died in Breslau, Silesia (now Wrocław, Poland) on 3 April 1827, aged 70.

Chladni's interest in sound stemmed from his love of music. In 1786 he began studying sound waves and worked out mathematical formulae that describe their transmission. His best-known experiment made use of thin metal or glass plates covered with fine sand. When a plate was made to vibrate and produce sound (for example by striking it or stroking the edge of the plate with a violin blow), the sand collected along the nodal lines of vibration, creating patterns called Chladni's figures. In 1809 he demonstrated the technique to a group of scientists in Paris.

He also measured the velocity of sound in various gases by measuring the change in pitch of an organ pipe filled with a gas other than air (the pitch, or sound frequency, varies depending on the molecular composition of the gas). He invented various musical instruments, including ones he called the clavicylinder and the euphonium. The latter consisted of rods of glass and metal that were made to vibrate by being rubbed with a moistened finger; he demonstrated it at lectures throughout Europe.

In 1794 Chladni published a book about meteorites (which he collected) and postulated that they come from beyond the Earth as debris of an exploded planet. Nobody accepted his theory until 1803, when the French physicist Jean Biot (1774–1862) confirmed that meteorites do, in fact, fall from the sky.

Christoffel, Elwin Bruno (1829–1900) German mathematician who made a fundamental contribution to the differential geometry of surfaces, carried out some of the first investigations that later resulted in the theory of shock waves, and introduced what are now known as the Christoffel symbols into the theory of invariants.

Christoffel was born on 10 November 1829 in Montjoie (now Monschau), near Aachen. He studied at the University of Berlin, where he received his doctorate at the age of 27. Three years later he became a lecturer at the university before, in 1862, becoming a professor at the Polytechnicum in Zürich. After seven years there, he returned to Berlin to take the chair of mathematics at the Gewerbsakademie. In 1872 he became professor of mathematics at the newly founded University of Strasbourg, where he remained until his retirement in 1892. He died on 15 March 1900.

Christoffel's best-known paper annotated his investigation into the theory of invariants. Called 'Über die transformation der homogen differentialausdrücke zweiten grades' and published in 1869, the paper introduced the symbols that later became known as Christoffel symbols of the first and second order. The series of other symbols of more than three indices, including the four index symbols already introduced by Bernhard ⇨Riemann, are now known as the Riemann–Christoffel symbols. (The symbols of an order higher than four are obtained from those of a lower order by a process called co-variant differentiation.)

Christoffel is additionally remembered as the formulator of the theorem that also bears his name, and concerns the reduction of a quadrilateral form; the theorem was later incorporated by Gregorio ⇨Ricci-Curbastro and Tullio ⇨Levi-Civita in their tensor calculus.

Christoffel's contribution to the differential geometry of surfaces is contained in his *Allgemeine Theorie der geodätischen Dreiecke* (1868), in which he presented a trigonometry of triangles formed by geodesics on an arbitrary surface. He used the concept of reduced length of a geodesic arc, stating that when the linear element of the surface can be represented by $ds^2 = dr^2 + m^2dx^2$, where m is the reduced length of the arc r.

Inspired by Riemann – who was of a similar age – Christoffel's papers in 1867 and 1870 were on the conformal tracing of a simply connected area bounded by polygons on the area of a circle. In 1880 he showed algebraically that the number of linearly independent integrals of the first order on a Riemann surface is equal to the genus *p*. Later, in *Vollständige Theorie der Riemannschen* θ-*Function* (published posthumously), Christoffel gave an independent interpretation of Riemann's work on the subject.

In 1877, Christoffel published a paper on the propagation of plane waves in media with a surface discontinuity, and thus made an early contribution to shock-wave theory.

Church, Alonso (1903–1995) US mathematician who in 1936 published the first precise definition of a calculable function, and so contributed enormously to the systematic development of the theory of algorithms.

Church was born on 4 June 1903. Completing his education at Princeton University, and obtaining his PhD there in 1927, he joined the university staff and remained at Princeton for 40 years, finally occupying the chair of mathematics and philosophy. From 1967 he held a similar post at the University of California in Los Angeles. The author of many books on mathematical subjects, Church was a member of a number of academies and learned societies. He died on 11 August 1995, aged 92.

The concept of the algorithm, in the development of which Church played such a part, did not properly appear until the 20th century. Then, as the subject of independent study, the algorithm became one of the basic concepts in mathematics. The term denotes an exact procedure specifying a process of calculation that begins with an arbitrary initial datum and is directed towards a result that is fully determined by the initial datum. The algorithm process is one of sequential transformation of constructive entities: it proceeds in discrete steps, each of which consists of the replacement of a given constructive entity with another. (Familiar examples of algorithms are the rules for addition, subtraction, multiplication, and division in elementary mathematics.)

Luitzen ◊Brouwer and Hermann ◊Weyl did some tentative studies in the 1920s, and Alan ◊Turing later offered the first application of the algorithm concept in terms of a hypothetically perfect calculating machine. The solving of algorithmic problems involves the construction of an algorithm capable of solving a given set with respect to some other set, and if such an algorithm cannot be constructed, it signifies that the problem is unsolvable. Theorems establishing the unsolvability of such problems are among the most important in the theory of algorithms, and Church's theorem was the first of this kind. From Turing's thesis, Church proved that there were no algorithms for a class of quite elementary arithmetical questions. He also established the unsolvability of the solution problem for the set of all true propositions of the logic of prediction. Since Church's pioneering work, much further progress has been made: the Polish-born US mathematician Alfred Tarski (1902–1983), for example, obtained some important results. Today, the theory of algorithms is closely associated with cybernetics, and the concept is fundamental to programmed instruction in electronic computers.

Cierva, Juan de la (1895–1936) Spanish aeronautical engineer who invented the rotating-wing aircraft known as the autogyro.

Cierva was born in Murcia in southeastern Spain on 21 September 1895, the son of the Conservative politician, Juan de la Cierva y Penafiel (1864–1938). He was educated in Madrid at the engineering school called the Escuela Especial de Caminos, Canales y Puertos. During his six years there he also studied theoretical aerodynamics on his own, especially the work of Frederick ◊Lanchester. Soon after leaving school he followed his father into politics and was elected to the Cortes, the Spanish parliament, in 1919 and 1922. He showed little enthusiasm for politics, however, for his real interest was the designing of flying machines.

In 1919, he entered a competition to design a military aircraft for the Spanish government. His plan was for a three-engined biplane bomber with an aerofoil section of his own design. When it was tested in May 1919 engine failure caused it to stall in mid-air, and the plane crashed. This accident led Cierva to turn away from fixed-wing flying machines and to search for a machine with a rotating-wing mechanism that would be less vulnerable to engine failure. His first three designs, which all had blades fixed to the motor shaft, were unsuccessful. His fourth design introduced freely rotating wings. On 19 January 1923 the new gyroplane, to which Cierva gave the name Autogiro, was tested at Getafe, Spain, and it flew for 182 m/200 yd.

The autogyro consisted of one nose-mounted engine driving a conventional propeller, a fuselage, and a large, freely rotating rotor mounted horizontally above the fuselage. Allowing the blades of the motor to pivot on hinges, instead of being rigidly fixed to the shaft, largely solved the problem of uneven lift being generated by the advancing and retreating blades. In order to gain sufficient lift for the aurogyro to take off, rope was wound many times around the rotor shaft and then pulled by a gang of men to turn the rotor quickly.

After the initial success of 19 January three more test flights were quickly arranged and just two days later, on 21 January, the autogyro completed a 4-km/2.5-mi circuit in 3.5 minutes. The usefulness of the new machine to the police and maritime rescue services was immediately recognized and, after minor adjustments were made to eliminate teething troubles, full-scale production began in 1925 with the founding of the Cierva Autogyro Company in England. Cierva became technical director and on 18 September 1928 he flew one of the company's aircraft across the English Channel. He then flew one all the way to Spain; and in 1929 he demonstrated his new invention at the National Air Races held at Cleveland, Ohio. Cierva continued to exeriment and to test his own aircraft until he was killed in a crash at Croydon Aerodrome, just south of London, on 9 December 1936.

There is an essential difference between Cierva's machine and the modern helicopter. A helicopter has a powered rotor that rotates horizontally overhead, enabling the machine to rise vertically. As a result, it needs only a small space in which to take off and land. During flight, motion backwards and forwards is achieved by altering the inclination of the rotor blades. To stop the plane from rotating with the rotor, a small secondary rotor is is fitted to the tail; a helicopter can also hover. Cierva's autogyro, on the other hand, was designed in some respects like an ordinary aeroplane, with a propeller to pull it through the air. But instead of fixed wings on each side, it had a revolving rotor – a rotating wing – overhead to provide lift at slow forward speeds and allow it almost vertical descent, although it is not capable of vertical ascent nor the 360 ° manoeuvrability of the helicopter. It was this difference that constituted the basis of his invention and which made the autogyro the prototype – and provided some of the performance features – of the modern helicopter.

Clapp, Cornelia Mary (1849–1934) US biologist, who was one of the first women to earn a doctorate in science. Her research was instrumental in the development of marine biology and embryology.

Cornelia Mary Clapp was born in Montague, Massachusetts, on 17 March 1849, the eldest of a family of six. She was educated in private and public schools in Montague before entering Mount Holyoke Seminary, one of the first academies for women, in 1868. Graduating in 1871, she taught Latin at a boys school in Andalusia, Pennsylvania, for a year before being invited back to Mount Holyoke to teach. She taught mathematics for the first year, moving on to natural history (zoology) the following year. She also taught gymnastics for fifteen years in different schools. Clapp spent her spare time at the seminary with her colleague and former teacher Lydia Shattuck, painstakingly studying pond water, drop by drop under a microscope, in search of the amoeba. They found many other plants and animals as well as the amoeba.

In 1874 both Clapp and Shattuck were selected as summer students by the Anderson School on Penikese Island, Massachusetts, the first school of natural history by the sea and also one of the first places where women could obtain an advanced scientific education. The philosophy of the school and its founder, the Swiss naturalist Louis Agassiz, was that scientists should study nature itself not books. Clapp gave up her textbooks and began to base her teaching on the animals. She introduced the study of embryology by bringing a hen into the classroom to incubate a new egg every day. The 21 chicks that were hatched, ranging from one to 21 days old, were fine examples of the stages of their development and she put them in dishes and arranged them as a public exhibition. Clapp gained a reputation as an enthusiastic and inspiring teacher with a contagious vitality that she passed on to her students. She always used live subjects in her classes, on one occasion sending a line of fiddler crabs marching into the classroom.

Clapp joined a group of zoology professors on a walking tour of the southern USA in 1878, led by biologist David Starr Jordan. During the tour Clapp recalls having short hair and wearing short dresses, which enabled her to walk through water and cross rapids with the men. She travelled in Italy and Switzerland in 1879 with the same group.

From 1882 to 1883, Clapp worked on chick embryology with Professor William Sedgwick of MIT, and from 1883 to 1884 she worked on the earthworm with Professor Wilson of Williams College. In 1888 she began a long association with Woods Hole Marine Biology Laboratory in Massachusetts, to which she returned every summer for the rest of her life. She became a lecturer there and was on the board of trustees. While she was there she undertook a study of the toad fish – which became known as 'Dr Clapp's fish' amongst biologists.

Meanwhile, Clapp also achieved two PhDs – the first from Syracuse University in 1889 and the second (at the age of 47) from the University of Chicago in 1896. She then worked in Italy in 1901 at the Naples Station. Clapp was a member of the Society for American Zoologists, the Association of American Anatomists, and the American Association for the Advancement of Science. She was also among the only six women to appear in the first edition of *American Men of Science* (1906).

Finally retiring from Mount Holyoke in 1916, Clapp spent her winters in Florida and summers at Woods Hole. In 1923 the new laboratories at Mount Holyoke were dedicated to her. She was the last surviving member of the Penikese School when she died on 31 December 1934.

Clark, Wilfrid Edward Le Gros (1895–1971) English anatomist and surgeon who carried out important research that made a major contribution to the understanding of the structural anatomy of the brain.

Clark was born on 5 June 1895 in Hemel Hempstead, Hertfordshire. He went to Blundell's School in Tiverton and entered St Thomas's Hospital in 1912 on an entrance scholarship. He qualified in 1917 with a conjoint diploma, and joined the Royal Army Medical Corps without working his house appointments. He served in France until the end of World War I, after which he went back to St Thomas's as house surgeon to Sir Cuthbert Wallace. Clark became a Fellow of the Royal College of Surgeons in 1919. He took the post of principal medical officer in Sarawak, Borneo, to gain experience in practical surgery, and began research into the evolution of primitive primates. After successfully treating several local people for yaws, he became highly venerated and was tattooed on the shoulders with the insignia of the Sea Dyaks as a mark of their esteem. Clark returned to England in 1923 as an anatomy demonstrator at St Thomas's until he moved to St Bartholomew's Hospital in 1924 as a reader, then professor, of anatomy. He returned to St Thomas's as professor of anatomy in 1930, accepting the professorship of anatomy at Oxford in 1934, which he held until he retired in 1962. His work on primate evolution resulted in election to the Royal Society in 1935. During World War II his research was connected with the war effort despite his pacifist principles. After the war he created a new department of anatomy at Oxford which was finally opened in 1959. He was knighted in 1955, and was Arris and Gale Lecturer (1932), Hunterian Professor (1934 and 1945), and editor of the *Journal of Anatomy.* He died suddenly in Burton Bradstock, Dorset, on 28 June 1971 on a visit to a friend from his student days.

Clark had a profound influence on the teaching of anatomy. He moved away from the popular topographical approach, which encouraged students to learn repetitiously, and towards the importance of relating structure to function. His anatomy research was directed mainly towards the brain, and the relationship of the thalamus to the cerebral cortex. He also carried out further studies of the hypothalamus. His work on the sensory (largely visual) projections of the brain remains the basis of contemporary knowledge of this aspect of neuroanatomy.

His chief publications include: *Morphological Aspects of the Hypothalamus* (1938); *The Tissues of the Body* (1939); *History of the Primates* (1949); *Fossil Evidence of Human Evolution* (1955); and his autobiography *Chant of Pleasant Exploration* (1968).

Clausius, Rudolf Julius Emmanuel (1822–1888)
German theoretical physicist who is credited with being one of the founders of thermodynamics, and with originating its second law. His great skill lay not in experimental technique but in the interpretation and mathematical analysis of other scientists' results.

Clausius was born in Köslin in Pomerania (now Koszalin in Poland) on 2 January 1822. He obtained his schooling first at a small local school run by his father, and then at the Gymnasium in Stettin. He entered the University of Berlin in 1840, and obtained his PhD from the University of Halle in 1848. Clausius then taught at the Royal Artillery and Engineering School in Berlin, and in 1855 became professor of physics at the Zürich Polytechnic. He returned to Germany in 1867 to become professor of physics at the University of Würzburg, and then moved to Bonn in 1869 where he held the chair of physics until his death in Bonn on 24 August 1888.

In 1870 the Franco-Prussian war stimulated Clausius to organize a volunteer ambulance service run by his students. He was wounded during the course of these activities, and the injury caused him perpetual pain. This, combined with the death of his wife in 1875, probably served to reduce his productivity during his later years. However, his scientific achievements were rewarded with many honours, including the award of the Royal Society's Copley Medal in 1879. Clausius died in Bonn on 24 August 1888.

Sadi ◊Carnot, Benoit Clapeyron (1799–1864), and Lord ◊Kelvin had made contributions to the theory of heat and to changes of state. Clausius examined the caloric theory, eventually rejecting it in favour of the equivalence of heat and work. Drawing particularly on Carnot's and Kelvin's concept of the continuous degradation or dissipation of energy, Clausius formulated (in a paper published in 1850) the second law of thermodynamics and introduced the concept of entropy.

The word entropy derives from the Greek word for transformation, of which there are two types according to Clausius. These are the conversion of heat into work, and the transfer of heat from high to low temperature. Flow of heat from low to high temperature produces a negative transformation value, and is contrary to the normal behaviour of heat. Clausius deduced that transformation values can only be zero, which occurs only in a reversible process, or positive, which occurs in an irreversible process. Clausius therefore concluded that entropy must inevitably increase in the universe, a formulation of the second law of thermodynamics. This law can also be expressed by the statement that heat can never pass of its own accord from a colder to a hotter body.

Entropy is considered to be a measure of disorder, and of the extent to which energy can be converted into work. The greater the entropy, the less energy is available for work. Clausius was opposed in his views by a number of scientists, but James Clerk ◊Maxwell gave Clausius considerable support in scientific argument on the subject.

Clausius did other work on thermodynamics, for example by improving the mathematical treatment of Hermann ◊Helmholtz's law on the conservation of energy, which is the first law of thermodynamics; and by contributing to the formulation of the Clausius–Clapeyron equations that describe the relationship between pressure and temperature in working changes of state.

The second area to which Clausius made important contributions was the development of the kinetic theory of gases, which Maxwell and Ludwig ◊Boltzmann had done so much to establish. From 1857 onwards, Clausius examined the inner energy of a gas, determined the formula for the mean velocity and mean path-length of a gas molecule, provided support for Avogadro's work on the number of molecules in a particular volume of gas, demonstrated the diatomic nature of oxygen, and ascribed rotational and vibrational motion (in addition to translational motion) to gas molecules. Clausius also studied the relationship between thermodynamics and kinetic theory.

Clausius' third major research topic was the theory of electrolysis. In 1857, he became the first to propose that an electric current could induce the dissociation of materials, a concept which was eventually established by Svante ◊Arrhenius. Clausius also proposed that Ohm's law applies to electrolytes, describing the relationship between current density and the electric field.

Clausius was clearly a brilliant theoretician, but he showed a curious lack of interest in the developments that arose from his work. The results of Boltzmann, Maxwell, Willard Gibbs (1839–1903), and other scientists and the advancements in the field of thermodynamics, statistical mechanics, and kinetic theory seem to have gone completely unnoticed by Clausius in his later years.

One interesting consequence of the second law of thermodynamics is that as entropy increases in the universe, less and less energy will be available to do work. Eventually a state of maximum entropy will prevail and no more work can be done: the universe will be in a static state of constant temperature. This idea is called 'the heat death of the universe' and although it seems to follow logically from the second law of thermodynamics, this view of the future is by no means accepted by cosmologists.

Clifford, William Kingdon (1845–1879) English mathematician and scientific philosopher who developed the theory of biquaternions and proved a Riemann surface (see Bernhard ◊Riemann) to be topologically equivalent to a box with holes in it. His name is perpetuated in 'Clifford parallels' and 'Clifford surfaces'. In his philosophical studies he was much preoccupied with theories of evolution.

Clifford was born on 4 May 1845 in Exeter, Devon. Educated locally, he went to King's College, London, at the age of 15. Three years later he won a small scholarship and entered Trinity College, Cambridge, where his academic progress was phenomenal. In 1868 he was made a fellow of the college, and continued to live there until 1871, when he was appointed professor of applied mathematics at University College, London. He was elected a Fellow of the Royal Society in 1874, and became a prominent member of the Metaphysical Society. In 1876, however, he developed pulmonary tuberculosis and was obliged to live first in Algiers, and then in Spain, for the sake of his health. But his condition continued to deteriorate and, although he was able to make several trips to England for short periods, he went finally to Madeira in 1879, and died there on 3 March of that year.

Despite his connection with the city, Clifford was one of the first mathematicians to protest against the analytical methods of the 'Cambridge school'. Primarily a geometrician, regarding geometry as to all intents and purposes a branch of physics, Clifford had as his fundamental aim in his teaching the compelling of his students to think for themselves. He did much to revolutionize the teaching of elementary mathematics, and was responsible for introducing into England the geometrical and graphical methods of August ◊Möbius, Karl Culmann, and others.

It was through a generalization of the quaternions (themselves a generalization of complex numbers) formulated by William ◊Hamilton that Church derived his theory of biquaternions, associating them specifically with linear algebra. In this way representing motions in three-dimensional non-Euclidean space, and together with his suggestion in 1870 that matter itself was a kind of curvature of space, Church may be seen to have foreshadowed, in some respects, Einstein's general theory of relativity.

Clifford continued his studies in non-Euclidean geometry, with reference particularly to Riemann surfaces, for which he established some significant topological equivalences. He further investigated the consequences of adjusting the definitions of parallelism, and found that parallels not in the same place can exist only in a Riemann space – and he proved that they do exist. He showed how three parallels define a ruled second-order surface that has a number of interesting properties (which were subsequently examined by Bianchi and Felix Klein).

Clifford also achieved some renown as an agnostic philosopher.

Cockcroft, John Douglas (1897–1967) English physicist who, with Ernest ◊Walton, built the first particle accelerator and achieved the first artificial nuclear transformation in 1932. For this achievement, Cockcroft and Walton shared the award of the 1951 Nobel Prize for Physics.

Cockcroft was born in Todmorden, Yorkshire, on 27 May 1897. He was admitted to Manchester University in 1914 to study mathematics but left a year later to volunteer for war service. He returned to Manchester in 1918 to attend the College of Technology and study electrical engineering, having gained an interest in this subject during his army service as a signaller. He obtained an MSc Tech in 1922 and then went to Cambridge University, taking a BA in mathematics in 1924. Cockcroft then remained at Cambridge, working at the Cavendish Laboratory under Ernest Rutherford. There he collaborated with Pyotr Kapitza on the design of powerful electromagnets and then with Walton on the construction of a voltage multiplier to accelerate protons. With this instrument, the first artificial transformation – of lithium into helium – was achieved in 1932. Cockcroft and Walton then worked on the artificial disintegration of other elements such as boron.

During World War II, Cockcroft was closely involved with the development of radar and with the production of nuclear power, directing the construction of the first nuclear reactor in Canada. He returned to the UK in 1946 and was appointed the first director of the Atomic Energy Research Establishment at Harwell. Cockcroft was knighted in 1948, and in 1959 became master of Churchill College, Cambridge. He received the Atoms for Peace Award in 1961 and was made president of the Pugwash Conference, an international gathering of eminent scientists concerned with nuclear developments, shortly before he died at Cambridge on 18 September 1967, aged 70.

Cockcroft's background in electrical engineering was unusual for a nuclear physicist, but he was able to put it to good use when the need to increase the energy of particles used to bring about nuclear transformations became apparent in the late 1920s. Until then, only alpha particles emitted by radioactive substances were available. Cockcroft and Walton built a voltage multiplier to build up a charge of 710,000 volts and accelerate protons in a beam through a tube containing a high vacuum. They bombarded lithium in this way in 1932, and produced alpha particles (helium nuclei), thus artificially transforming lithium into helium. The production of the helium nuclei was confirmed by observing their tracks in a cloud chamber.

Cockcroft and Walton's voltage multiplier was the first of many particle accelerators. It was soon superseded by the cyclotron and led to accelerators of a wide range of energies. The development of the particle accelerator was essential to high-energy physics because it provided a means of studying subatomic particles that could not be produced in any other way, and it has also proved to be of great value in the production of radioactive isotopes.

Cockerell, Christopher Sydney (1910–1999) English engineer who invented the hovercraft in the 1950s. He also made a major contribution to aircraft radio navigation and communications.

Educated at Gresham's School, Holt, and Peterhouse, Cambridge, where he read engineering, Cockerell graduated in 1931 and spent two years in Bedford with the engineering firm of W H Allen and Sons. He then returned to Cambridge for two years of research into wireless, which was his hobby.

He joined the Marconi Wireless Telegraph Company in 1935, working on VHF transmitters and direction finders and, during World War II, on navigational and communication equipment for bombers and later on radar. During this period he filed 36 patents, the most interesting of which are frequency division (1935), the linearization of a transmitter by feedback (1937), pulse differentiation (1938), and various navigational systems.

In 1950 Cockerell left Marconi's and started up a boat-hire business on the Norfolk Broads, building boats and caravans. Trained as a development engineer, he set himself the task of trying to make a boat go faster. His first experiments were on the air lubrication of a hull and later on a 6.1-m/20-ft launch. A number of engineers had earlier suggested the use of air lubrication for the reduction of drag. Indeed in the 1870s the British engineer John Thorneycroft built test models to check the drag on a ship's hull with a concave bottom in which air could be contained between the hull and the water. Cockerell concluded after his first experiments with air lubrication that a major reduction in drag could be obtained only if the hull could be supported over the water by a really thick air cushion. This was because the fine structure of the upward

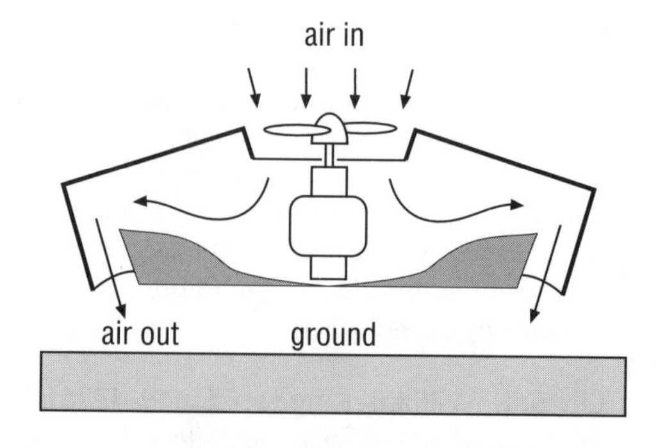

Cockerell Cockerell's original hovercraft design made use of a peripheral jet of compressed air to achieve lift.

pressure of the water, with the craft in motion over waves, varied so much that the pressure peaks broke through the layer of air. With a thin cushion, the best that could be attained was a 50% reduction in skin friction drag.

The first concept of a hovercraft was a sidewalled craft with hinged doors to contain the air which were capable of lifting over a wave. The next idea was a water curtain discharged inwards and downwards across the front and back of the craft. The third conception was to replace the water curtain by thin air jets across the bow and stern pointing towards the athwartships centreline of the craft. The next step for a faster amphibious craft, was a peripheral downward and inward-facing thin jet all round the periphery of the craft. Calculations showed that the power could be provided if the cushion pressure was kept down to reasonable limits. The power requirement was confirmed by some primitive experiments using tin cans as a simple way of constructing an annular jet. This concept was tried out in the experimental SR N1 in 1959.

The first true hovercraft was a 762-mm/2.5-ft balsawood model weighing 127.6 g/4.5oz, made in 1955 and powered by a model aircraft petrol engine. It could travel at 20.8 kph/13 mph over land or water. Cockerell's first hovercraft patent was filed on 12 December 1955 and in the following year he formed Hovercraft Ltd. He then began the thankless task of trying to interest manufacturers – shipbuilders said it was an aircraft and aircraft manufacturers said it was a ship.

Realizing that the hovercraft would have considerable military potential. Cockerell approached the Ministry of Supply, the Government's procurement agency for defence equipment. The air cushion vehicle was promptly classified 'secret' in 1956. This meant considerable delays but finally the secret leaked out, the project was declassified, and the National Research Development corporation decided to back the idea in 1958. Saunders Roe Ltd, a manufacturer with flying-boat background (and therefore both aircraft and marine experience) built the SR N1 that crossed the English Channel with the inventor on deck.

In the waiting period (1957) Cockerell had come to the conclusion that the hovercraft could not go fast in a seaway without diving into it. He came up with the idea of flexible skirts, which gave rise to much derision because nobody could believe that a piece of fabric could be made to support 100 tonnes with a mere 2.873 newtons per sq m/60 lb per sq ft, compared with, say, 206.843 newtons per sq m/30 lb per sq in for a motor-car tyre.

Manufacturers and operators in many parts of the world became interested and craft were made in the USA, Japan, Sweden, and France while in the UK other companies also began to manufacture. The SR N4 maintained commercial cross-Channel car ferry services in the severest weather conditions and the AP 1–88, the world's first diesel-powered hovercraft, went into service in early 1983.

Apart from passenger- and car-ferry applications, craft have been used for seismic surveys over shallow water or desert and in search and rescue operations. Military uses include carrying troops during amphibious assaults and as logistic support craft. Future military uses include mine counter-measure work, anti-submarine work, aircraft carrying, and missile launching.

Air cushion trains have a potential speeds of 480 kph/300 mph and the cost of track for these would be much less than that for conventional trains. However, because the air cushion requires considerable energy to maintain it, opinion has swung towards magnetic levitation (maglev) for advanced tracked transport.

Cockerell was appointed consultant (hovercraft) to the Ministry of Supply and has been director of and consultant to a number of firms working in the field. The winner of many awards and distinctions, Cockerell is a Fellow of the Royal Society and was knighted in 1969. In the 1970s and 1980s he interested himself in the generation of energy by wavepower and was chair of Wavepower Ltd 1974–82, thereafter acting as a consultant until 1988. In 1974 he became an Honorary Fellow of Peterhouse College, Cambridge, and in 1998 an Honorary Fellow of Downing College, Cambridge. He died on 1 June 1999.

Cockerill, Wilham (1759–1832) English engineer who is generally regarded as the founder of the European textile-machinery industry.

Cockerill was born in Lancashire, and throughout his childhood showed a remarkable talent for anything mechanical. Being brought up in a district greatly involved in the production of cloth, he naturally found an outlet for his particular genius constructing machinery for that use.

His working career began with the building of 'roving billies' and 'flying shuttles', but in 1794 he decided to seek his fortune elsewhere and chose Russia as the most likely place. He found employment in St Petersburg and, under the patronage of Catherine II, enjoyed a fair measure of success for as long as she ruled. On her death, however, her successor, the mentally unstable Paul I, imprisoned Cockerill for failing to complete a contract within the given time.

Eventually he escaped to Sweden and there tried to arouse an interest in a textile industry. The Swedes rejected his ideas, and so in 1799 he migrated to Belgium, where he established a flourishing business at Verviers as a manufacturer of textile machinery. In 1802 he was joined in partnership by a James Holden, but their association was short-lived and in 1807 Cockerill transferred his interests to Liège. There, together with his three sons William, Charles, and John – who also shared his enthusiasm for constructing machines – he built up a highly successful business making carding machines, spinning frames, and looms for the French woollen industry.

After retiring from the firm in 1814 he went to live at the home of his son Charles, at Aix-la-Chapelle (Aachen), and remained there until his death in 1832.

The invention of John Kay's flying shuttle in 1733 had helped to speed up the process of weaving in the English textile industry, but not until the latter half of the 1700s were other significant mechanical designs introduced to streamline the production further. Cockerill's childhood saw the coming of James Hargreaves's 'spinning jenny' in 1770, which enabled spinners to spin several threads at once; Richard Arkwright's water frame, which brought water power to the spinning machines; and the invention of a rotary carding machine to make the possibility of mass production a reality. By the time Cockerill left for Russia in 1794 he had mastered all the techniques of manufacturing the new machines.

Basing his designs on the English machines. Cockerill was able to build up a reputation for first-class workmanship and attention to detail. When, after his previous disappointments, he eventually founded his business in Liège he found a ready market in France. In fact, it was during the era of Napoleon that Cockerill achieved his greatest esteem and his efforts were to a large extent reponsible for breaking the monopoly England had previously held over the continental market.

Cohn, Ferdinand Julius (1828–1898) German botanist of distinction and one of the founders of bacteriology. He became the director of the world's first institute of plant physiology at Breslau University in 1868, and in 1872 published a treatise on bacteria.

Ferdinand Cohn was born in Breslau, Silesia (now Wrocław, Poland) on 24 January 1828 of poor parents. His father later became a successful businessman and was able to nurture Ferdinand's obvious talents. It is said that Ferdinand could read at the age of two and at three had grasped the basics of natural history. He started school at four and by the age of seven entered the Breslau Gymnasium. However, at 11 his pace was slowed by a hearing defect, which he managed to overcome by the time he entered the faculty of philosophy at Breslau University in 1842, aged 14. At that time he was unclear about the course he would take but, largely through the influence of professors Heinrich Goeppert and Christian Nees von Esenbeck, botany soon became his subject of choice. As a Jew, Cohn was barred from degree examinations at Breslau and his petitions to take the exams were denied, so in 1846 he moved to the University of Berlin, receiving a doctorate in botany in 1847.

In Berlin, Cohn met a number of inspirational teachers and was introduced to the study of microscopic organisms. He did not take an active part in the revolution in Berlin in 1848, although he was behind the revolutionaries. It is thought that his politics sometimes stood in the way of his career. In 1849 Cohn returned to the University of Breslau and was recognized as a *Privatdozent* the following year. He became extraordinary professor there in 1859 and ordinary professor in 1872. Cohn married Pauline Reichenbach in 1867. He died on 25 June 1898 in Breslau.

Cohn had argued for years for a botanical research institute to be set up – in fact it was the theme of his doctoral thesis – and he finally got his wish in 1868 when the world's first institute of plant physiology was opened at Breslau University. He was director of this institute and in 1870 he founded a journal, *Beitrage zur Biologie der Pflanzen,* devoted to the work carried out there.

In 1848 Cohn was asked by Professor Goeppert to devote himself to algae and make a contribution to a flora of Silesia for which he was collecting material. Cohn concentrated on the morphology and life history of microscopic algae and fungi and demonstrated that the protoplasm of plant and animal cells is similar. He was later inspired by the work of Louis Pasteur to study bacteria and his 1872 treatise, published in his journal, laid the foundations of bacteriology. He defined bacteria in his treatise, divided them into four groups and subdivided the groups into six genera. It was the first attempt ever made to classify bacteria and his system is still in use today. Cohn continued his work on bacteria until about 1885 when the study of bacteriology was becoming more medical. He was the first to demonstrate the 'fixity' of bacteriological species and to discover that some bacteria form spores that are resistant to harmful agents. He showed that some bacteria can be killed by boiling for 20 minutes and studied the effects of electric currents on bacteria.

In 1876 Cohn was contacted by Robert ◊Koch who had unravelled the complete life history of the anthrax bacillus. He arrived in Breslau shortly afterwards and demonstrated his work to Cohn and his colleagues for three days, convincing them that his results were true. Cohn published Koch's groundbreaking work in his journal.

In around 1885 Cohn returned to his work in plant physiology continuing Goeppert's classificatory work and publishing the first three volumes of the *Cryptogam-Flora of Silesia.* In addition, Cohn published popular lectures and a widely read volume *Die Pflanze* (1882) which contained history, biographical plans, and poetry, and served to popularize botany. In 1887, the University of Breslau gave him a new institute of plant physiology in the local botanical gardens. He was given an honorary doctorate by the University of Tubingen, and was a corresponding member of the Accademia dei Linnei in Rome, the Institut de France in Paris, and the Royal Society in London. He was the recipient of the Leewenhoek Gold Medal in 1885, and the Gold Medal of the Linnaean Society in 1895.

Colles, Abraham (1773–1843) Irish surgeon who observed and described the fracture of the wrist which bears his name.

Colles was born on 23 July 1773 in County Kilkenny, Ireland; he was educated at Kilkenny Preparatory School and Kilkenny College. In 1790 he entered Trinity College, Dublin, as a student of arts, but swiftly began his clinical training as a resident surgeon at Steeven's Hospital. He was granted his diploma of the Royal College of Surgeons in Ireland in 1795. He then went to Edinburgh and graduated at Edinburgh University in 1797. He returned to Dublin that year to set up in practice and began to teach anatomy and surgery. In 1799 he was appointed resident surgeon at Steeven's Hospital (where he eventually became governor) and was elected a member of the Royal College of Surgeons in Ireland. At the age of 29 he became president of the College. In 1804 the Surgeons' School at the College made him professor of anatomy and surgery (a position he held until 1836), and he gained his MA from the University of Dublin in 1832. He died at his home in Dublin on 16 November 1843.

In 1814 the paper on Colles's fracture was published describing the fracture of the distal (carpel) end of the radius bone in the forearm. This common fracture causes deformity and swelling of the wrist, but can be easily and successfully treated once diagnosed. It must be remembered that the diagnosis of fractures in those days was made on purely clinical grounds because X-rays had not then been developed. Thus his accurate description of the fracture was that much more impressive. In his original paper he advocated the use of tin splints to stabilize the wrist after closed reduction of the fracture. Nowadays the reduction is followed by the use of plaster of Paris casts but exactly the same principles apply.

Although Colles is best remembered for Colles's fracture, he was one of the greatest professors the Royal College of Surgeons in Ireland ever had. He was a brilliant anatomist and an excellent teacher who did much to make Dublin the leading medical centre it had become by the beginning of the 19th century.

Colles's other eponyms include Colles's fascia, Colles's space, the Colles ligament (of inguinal hernia), and Colles's law of the communication of (congenital) syphilis.

Colt, Samuel (1814–1862) US inventor of what was probably the most successful family of revolving pistols of his time, which revolutionized military tactics.

Colt US gunsmith Samuel Colt holding the revolver he invented. The Colt revolver was adopted for use by the US army in 1846, and was supplied in large numbers when the American Civil War broke out in 1861. *Mary Evans Picture Library*

Born on 19 July 1814 in Hartford, Connecticut, Colt had made up his mind as a boy to become an inventor but his early inventions (which included an abortive four-barrel rifle) were somewhat unreliable. One of his discoveries was that it was possible to fire gunpowder using an electric current. He applied this principle to an explosive mine, but after a disastrous public demonstration (which covered all the spectators with mud), he was sent off to Amherst Academy. Unfortunately, as a result of a fire caused by another of his experiments, he was asked to leave there as well. He then became apprenticed as a seaman.

During a journey to India in 1830, aboard the brig *Corlo,* Colt made some observations that were to change his life. In watching the helmsman and the wheel he operated, Colt noticed that whichever way the wheel turned, each of its spokes always lined up with a clutch that locked it into position. Colt conceived the idea that the mechanism of the helmsman's wheel could be applied to a firearm.

Grasping the opportunity presented by this, he left the sea at the age of 18 to raise money for a professionally built prototype. Ever resourceful, he travelled the USA as 'Dr Coult', giving demonstration of nitrous oxide (laughing gas) to willing audiences in a series of adventures worthy of Huckleberry Finn.

By 1835, Colt had perfected his revolver, which emulated the helmsman's wheel with its rotating breech that turned, locked, and unlocked by cocking the hammer. That year, Colt took out patents in the UK and France and, in 1836, in his native USA. On 5 March 1836 Colt set up a company in Paterson, New Jersey, the Patent Arms Manufacturing Company, where he produced the Colt Paterson, a five-shot revolver with folding trigger and a number of different rifles and shotguns based on the same revolving principle. These models were revolutionary in a market still dominated by the cumbersome 'pepperbox' type of revolver, with their revolving barrels. In the resulting conflict of ideas, and amid apathy among firearms buyers stemming from the price and unreliability of the early models, Colt's company failed in 1842.

Colt temporarily turned his attention to inventing electrically discharged submarine mines and running a telegraph business. Events soon restored him to his original course, however, when in 1846, the Mexican War broke out. Colt Paterson revolvers had come into the possession of the Texas Rangers, who persuaded the government to order Colt revolvers for use in the war. Colt had to be tracked down first, but was of course delighted with an order to make 1,000 pistols for the army. Having no factory of his own, Colt contracted the work out to Eli Whitney at Whitneyville, Connecticut, and before long the US army was receiving supplies of the new revolver. Captain Walker, who had succeeded in tracing Colt, contributed his own ideas to the .44 calibre revolver, which became known as the Colt Walker 'Dragoon'. The pioneers then starting to open up the US West adopted the Colt with alacrity, since it was superior to the one- or double-shot pistol against hostile native Americans and wildlife.

Colt used the profits from his early sales to buy land at Hartford, where in 1847 he rented premises and in 1854 built a large factory. Colt's Hartford company is still in existence; the factory burnt down in 1864, but was later rebuilt. Colt could now increase the types of firearms he produced, and by 1855 he had the largest private armoury in the world. One of Colt's new weapons was the .31 calibre Pocket Model first produced in 1848, and in 1851 he produced his famous percussion revolver, the .36 calibre Colt Navy. The Rootes .28 and the Army followed in 1860, and the Police Model in 1862.

By this time, the American Civil War had begun. Colt, whose pre-war sympathies were with the secessionist Confederate South, changed allegiance to the Unionist North once fighting started in 1860 and supplied thousands of guns to the US government.

While his patents (which he defended vigorously) lasted, Colt had a virtual monopoly of the firearms market. In time, however, his rivals found ways around the restrictions of the patent laws, and eventually the patents simply expired.

In the teeth of bitter opposition from established British arms manufacturers, Colt set up a factory in Pimlico, London, in 1852. He also decided to show his revolver at the Great Exhibition. Orders to equip the British armed forces, including those fighting in the Crimea, followed, with customers being supplied by a variety of models produced in both Hartford and London. The Pimlico factory closed down in 1856, but Colt retained an agency in London until 1913.

By the time of his death in Hartford, on 10 January 1862, Colt could leave behind a vast and efficient company, and a considerable personal fortune. He had succeeded in marketing a well-conceived product whose success stemmed from the mass production of interchangeable parts that needed little hand finishing in assembly. As a result, he made a major contribution to both the development of repeating firearms and to the process of mass production by machine tools.

Further Reading

Hosley, William, *Colt: The Making of an American Legend,* University of Massachusetts Press, 1996.

Rohan, Jack, *Yankee Arms Maker: The Incredible Career of Samuel Colt,* Harper and Brothers, 1935.

Compton, Arthur Holly (1892–1962) US physicist who is remembered for discovering the Compton effect, a phenomenon in which electromagnetic waves such as X-rays undergo an increase in wavelength after having been scattered by electrons. For this achievement he shared the 1927 Nobel Prize for Physics with the British physicist C T R Wilson. Compton was also a principal contributor to the development of the atomic bomb.

Compton was born in Wooster, Ohio, on 10 September 1892, the son of a philosophy professor at Wooster College who was also a Presbyterian minister. He was educated at his father's college, graduating in 1913 and going on to Princeton University, from which he gained his PhD three years later. He became a physics lecturer at the University of Minnesota, but left after a year to take up an appointment as an engineer in Pittsburgh for the Westinghouse Corporation. He travelled to the UK in 1919 and stayed for a year at Cambridge University, where he worked under Ernest

Rutherford. In 1920 he returned to the USA and went to Washington University, St Louis, as head of the physics department, moving to the University of Chicago in 1923 as professor of physics. Compton remained at Chicago for 22 years until, after the end of World War II in 1945, he returned to Washington University. He was chancellor of the university 1945–53 and then became professor of natural philosophy there, a position he held until 1961, when he became professor at large. He died in Berkeley, California, on 15 March 1962.

Compton began studying the scattering of X-rays by various elements in the early 1920s (using blocks of paraffin wax in which the carbon atoms' electrons deflected X-rays) and by 1922 he had noted the unexpected effects that the scattered X-rays show an increase in wavelength. He explained the Compton effect in 1923 by postulating that X-rays behave like particles and lose some of their energy in collisions with electrons, so increasing their wavelength and decreasing their frequency. He calculated that the change in wavelength, $\Delta\lambda$, is given by the equation $\Delta\lambda = (h/mc)(1 - \cos\theta)$, where h is Planck's constant, m is the rest mass, c is the velocity of light, and θ is the angle through which the incident radiation (for example, X-rays) is scattered.

The chief significance of Compton's discovery is its confirmation of the dual wave/particle nature of radiation (later extended by Louis ◊de Broglie in his hypothesis that matter can also have wave/particle duality). The behaviour of the X-ray, previously considered only as a wave, is explained best by considering that it acts as a corpuscle or particle – as a photon (Compton's term) of electromagnetic radiation. Quantum mechanics benefited greatly from this convincing interpretation. Further confirmation came from experiments using C T R Wilson's cloud chamber in which collisions between X-rays and electrons were photographed and analysed. They could be interpreted as elastic collisions between two particles and measurements of the particle tracks in the cloud chamber proved the correctness of the mathematical formulation of the Compton effect.

In the early 1930s several scientists were debating the nature of cosmic rays. Robert ◊Millikan had proposed that the rays are a form of electromagnetic radiation which, if this were true, would be unaffected by their passage from outer space through the Earth's magnetic field and should strike the Earth in undeflected straight lines. Other scientists, such as the German physicists Walther Bothe (1891–1957), postulated that the 'rays' are made up of streams of charged particles, which, therefore, should be deflected into curved paths as they traverse the Earth's magnetic field. Compton reasoned that the argument could be resolved by making cosmic-ray measurements at various latitudes, which he carried out during a long series of trips to various parts of the world to measure – or to organize dozens of other scientists to measure – comparative cosmic-ray intensities using ionization chambers. By 1938 he had collated the multitude of results and demonstrated a significant latitude effect by the Earth's field, proving that at least some component of cosmic rays consists of charged particles. He went on to confirm these findings by showing variations in cosmic-ray intensity with the rotation period of the Sun and with the time of day and time of year. One interpretation of these results suggests an extragalactic source for cosmic rays.

During World War II Chicago University was the prime location of the Manhattan Project, the effort to produce the first atomic bomb, and in 1942 Compton became one of its leaders (as director of the code-named Metallurgical Laboratory). He organized research into methods of isolating fissionable plutonium and worked with Enrico ◊Fermi in producing a self-sustaining nuclear chain reaction, which led ultimately to the construction of the bombs that, used against Japan, effectively ended World War II. Compton's book *Atomic Quest* (1956) summarized this part of his career.

Cook, James (1728–1779) English sea captain who made notable contributions to hydrography.

An agricultural labourer's son born in Marton, Yorkshire, on 27 October 1728. Cook obtained his early experience of the sea by sailing in a Whitby collier before joining the Royal Navy in 1755 as an ordinary seaman. Ambitious and clever, he quickly rose in the service. After much first-rate hydrographical surveying work, undertaken particularly around Newfoundland and the mouth of the St Lawrence River, Cook received in 1768 a commission, via the Royal Society, to take the *Endeavour* to the newly discovered Tahiti with various scientists, including the naturalist Joseph ◊Banks, to observe the transit of Venus. Observations of transits of the inner planets across the face of the Sun were highly valued amongst astronomers as a means of calculating the distance of the Sun from the Earth. On that voyage, Cook charted the east coast of Australia (exploring Botany Bay) and the entire coastline of New Zealand, clearly demonstrating that it consisted of two main islands. He brought back fascinating descriptions of the exotic civilization of the Tahitians. Cook formed the belief that the long-suspected 'unknown continent' of the southern hemisphere could not exist, or at least, could not be of immense dimensions. (He was under Admiralty orders to secure such a continent for the Crown, should it exist – Cook's voyages thus served both political and scientific purposes.) As a result of sailing at higher latitudes south than any previous captain, Cook's second expedition of 1772–75 further shrank the possible extent of a southern continent. In 1776, shortly after being made a fellow of the Royal Society, Cook embarked on his third expedition, which aimed to explore the possibility of a northern route between the Atlantic and the Pacific. The expedition reached a sad climax in his death at the hands of natives in Hawaii – ironically, because Cook was generally scrupulous in his treatment of indigenous peoples. He died in Kealakekua Bay on 14 February 1779.

Cook's own energies and talents, aided by improved sextants and John ◊Harrison's chronometer, ensured that a greater quantity of high-quality survey work and scientific research was accomplished on Cook's three expeditions than on any comparable expeditions. He set new standards of cartography and hydrography, and was the source of modern maps of the Pacific and its coasts.

Cook is also remembered as a sagacious captain, who took great care of his crews. He proved the value of fresh fruit and vegetables, cleanliness, and morale in preventing and treating scurvy, a major problem on long sea voyages at the time.

Further Reading

Gould, Rupert Thomas, *Captain Cook,* Duckworth, 1978.

Hough, Richard, *Captain James Cook: A Biography,* Replica Books, 1998.

Skelton, Raleigh Ashlin, *Captain James Cook After Two Hundred Years,* Published for the British Library by the British Museum Publications, 1976.

Stamp, Tom, *James Cook, Maritime Scientist,* Caedmon of Whitby Press, 1978.

Coolidge, Julian Lowell (1873–1954) US geometrician and a prolific author of mathematical textbooks in which he not only reported his results but also described the historical background together with contemporary developments.

Coolidge was born into a prominent family in Brookline, Massachusetts, on 28 September 1873. Completing his education, he studied at Harvard University where in 1895 he was awarded his bachelor's degree with top honours. Two years later he travelled to the UK and took a degree in natural sciences at Balliol College, Oxford. He then returned to the USA to teach mathematics at the Groton School, where one of his pupils was Franklin D Roosevelt. In 1900 he became an instructor in mathematics at Harvard, and was made a member of the faculty in 1902. In that same year he was given leave of absence in which to continue his own studies, and returned to Europe, studying in Paris, Griefswald, Turin, and Bonn. Having earned a PhD in 1904 at Bonn University, Coolidge went back to teaching at Harvard. Four years later he was made an assistant professor. His work was then disrupted by World War I, in which he served as a major in the army. After the war he was a liaison officer in France, where he organized and taught courses at the Sorbonne for the benefit of US soldiers still stationed in the country.

In 1918 Coolidge had been made a full professor of mathematics at Harvard, and on his return from France he remained there until his retirement in 1940. Even then, however, as his own war effort, he again took to giving courses to military personnel. Many honours were awarded to Coolidge in recognition of his contributions to mathematics. He died in Cambridge, Massachusetts, on 5 March 1954.

Coolidge's work lay in the field of geometry and the history of mathematics. His early research for his PhD thesis and his first book (1909) were on non-Euclidean geometry, and were strongly influenced by the work of Emilio Segrè and Study; in particular, Coolidge used Study's new method of approach to line geometry. He was especially interested in the use of geometry in the investigation of complex numbers.

Other areas of interest soon began to develop. Coolidge wrote his first paper on probability theory in 1909, in which he also examined certain problems in game theory. He produced several later studies on statistics, and all of this work was included in his 1925 book on probability – one of the first on the subject to be published in English, and as such it received considerable acclaim.

His first work on the algebraic theory of curves appeared in 1915. Stimulated by the topic, Coolidge elaborated on his initial investigation and eventually, in 1931, published a full-size book detailing his results. Work on two classical geometrical figures – the circle and the sphere – also led to the writing of a book published in 1916. Lectures given at the Sorbonne in 1919 on the geometry of the complex domain were expanded into book form and published in 1924. There were three further books: one in 1940 on geometrical methods, one in 1943 on conic sections, and the last a historical text in 1949. In addition to his books, Coolidge was also the author of many papers, the last of which was written in 1953.

Cooper, Leon Niels (1930–) US physicist who shared the 1972 Nobel prize with John ◊Bardeen and J Robert Schrieffer (1931–) for developing the 'BCS' theory of superconductivity. One of Cooper's key contributions to the theory, which explains why certain metals lose all electrical resistance at very low temperatures, was the suggestion that electrons in the metal form pairs. These are now known as Cooper pairs.

Cooper was born on 28 February 1930 in New York City where he attended the Bronx High School of Physics and Columbia University, obtaining a BA in 1951, an MA in 1953, and a PhD in 1954. At Columbia Cooper's speciality was quantum field theory – the interaction of particles and fields in subatomic systems. After a year at the Institute for Advanced Study in Princeton, Cooper moved to the University of Illinois in 1955 to work with John Bardeen, who was to share the Nobel prize in 1956 for his work on the transistor. Bardeen was now working on superconductivity. The electrical resistance of all metals decreases as they are cooled. This is because the thermal vibrations of the nuclei – which scatter the electrons carrying the current, and hence give rise to the resistance – decrease with temperature. (Resistance and conductivity are inversely related: that is, zero resistance is equivalent to infinite conductivity). But whereas the decrease in resistance is gradual in most metals, the resistance of superconductors suddenly disappears below a certain temperature. Experiments had shown that this temperature, the critical temperature, was inversely related to the mass of the nuclei, so Bardeen proposed that superconductivity depended on the interaction of electrons with these variations. Cooper showed that an electron moving through the lattice attracts positive ions, slightly deforming the lattice. This leads to a momentary concentration of positive charge that attracts a second electron. This is the Cooper pair. Although the electrons in the pair are only weakly bound to each other, BCS were able to show that they all formed a single quantum state with a single momentum. The scattering of individual electrons did not effect this momentum, and this lead to zero resistance. Cooper pairs could not be formed above a certain temperature, the critical temperature, and superconductivity broke down.

After a year at Ohio State University, in 1958 Cooper moved to Brown University, Rhode Island, and in 1968 published *The Meaning and Structure of Physics.* In his Nobel lecture in 1972, Cooper said that a theory is more than its applications, being 'an ordering of experience that both makes experience meaningful and is a pleasure to regard in its own right'. In 1978 Cooper also became director of the Centre for Neural Science at Brown and worked on developing a theory of the central nervous system. He has also worked on distributed memory and character recognition.

Copernicus, Nicolaus (Mikolaj Kopernigk) (1473–1543) Polish doctor and astronomer who, against the religion-reinforced tradition of many centuries, finally declared once and for all that the planet Earth is the centre neither of the universe, nor even of the Solar System. He was never a skilled observer and probably made fewer than one hundred observations in all, preferring to rely almost entirely on data accumulated by others. In his own time he was renowned probably more as a medical man and priest than as an astronomer.

Copernicus Nicolaus Copernicus, the founder of modern astronomy, who first proposed that the Earth orbits around the Sun. His theory challenged the accepted doctrine of the time, that the Earth was at the centre of the universe. *Mary Evans Picture Library*

Copernicus was born in Toruń in Ermland (under the Polish crown) on 19 February 1473, and at an early age attended St John's School there. After the death of his father in 1483, however, it was arranged – probably by his uncle (and patron) L Watzenrode – for him to study at the cathedral school in Wloclawek. From 1491–94 Copernicus studied mathematics and classics at the University of Krakow, under Brudzewski. Encouraged by Watzenrode to continue his studies, he travelled in 1496 to the University of Bologna, Italy, where he studied law and astronomy. In the latter subject he was taught by D M di Novara.

Despite his nephew's absence from Poland, Watzenrode (now himself a bishop) was able to arrange for Copernicus to be made canon of Frombork (Frauenburg), a post he retained for life. This job brought in sufficient income, combined with only light duties, to enable Copernicus to devote a great deal of time to astronomy.

When in 1500 he had completed his studies in Bologna, Copernicus moved on to Padua, where he continued his studies in law and Greek. He gained a doctorate in canon law in 1503 in Ferrara and then returned to Padua to study medicine. In 1506 he went home to Poland (where he remained for the rest of his life). From 1506 to 1512, when Watzenrode died, he not only worked for his uncle as his personal doctor and private secretary, but served in the cathedral chapter of Frombork.

After the death of Bishop Watzenrode, Copernicus was still not able to devote himself entirely to his ecclesiastical and astronomical interests. He also served on a number of diplomatic missions and as a financial advisor and administrator. Nevertheless, he wrote a brief outline of his new ideas in astronomy in about 1513; he privately circulated a more comprehensive version in 1530. Copernicus had the good sense to realize that he risked being branded a lunatic, or a heretic, and he was therefore reluctant to publish his theory more widely. G Joachim, known as Rhaeticus, encouraged him to present his work in book form, which he finally agreed to do. The book, *De revolutionibus orbium coelestrium,* prudently dedicated to Pope Paul III, was delayed in publication, so that Copernicus did not receive a copy of it until he was on his deathbed. Copernicus died at Frombork on 24 May 1543.

Copernicus began to make astronomical observations in March 1497, although it was not until about 1513 that he wrote the brief, anonymous text, entitled *Commentariolus,* in which he outlined the material he later discussed more fully. His main points were that the Ptolemaic system of a geocentric planetary model was complex, unwieldy and inaccurate. Copernicus proposed to replace ◊Ptolemy's ideas with a model in which the planets (*including* the Earth) orbited a centrally situated Sun. The Earth would describe one full orbit of the Sun in a year, while the Moon orbited the Earth. The Earth rotated daily about its axis (which was inclined at 23.5° to the plane of orbit), thus accounting for the apparent daily rotation of the sphere of the fixed stars.

Finally we shall place the Sun himself at the centre of the universe.

Nicolaus Copernicus DE REVOLUTIONIBUS ORBIUM COELESTIUM

This model was a distinct improvement on the Ptolemaic system for a number of reasons. It explained why the planets Mercury and Venus displayed only 'limited motion': their orbits were inside that of the Earth's. Similarly, it explained why the planets Mars, Jupiter, and Saturn displayed such curious patterns in their movements ('retrograde motion', loops, and kinks). These were all a consequence of their travelling in outer orbits at a slower pace than the Earth. The precession of the equinoxes, as discovered by ◊Hipparchus in the 2nd century BC, could be accounted for by the movement of the Earth on its axis.

Copernicus' theory was not, however, by any means perfect. He was unable to free himself entirely from the constraints of classical thinking and was able to imagine only circular planetary orbits. This forced him to retain the cumbersome system of epicycles, with the Earth revolving around a centre that revolved around another centre, which in turn orbited the Sun. It was the work of Johannes ◊Kepler, who introduced the concept of elliptical orbits, that rescued the Copernican model. Copernicus also held to the notion of spheres, in which the planets were supposed to travel. It was the Danish astronomer Tycho ◊Brahe who rid astronomy of that archaic concept.

In his greatest work, the *De revolutionibus,* Copernicus proposed that the atmosphere (at least part of it) rotated with the Earth about the planetary axis, so that the skies did not constantly flow westwards. He obtained fairly accurate estimates for the distances of the planets from the Sun in terms of the astronomical unit (the distance from Earth to the Sun). He echoed ◊Aristarchos in explaining the inability to observe stellar parallax from the extremes of the Earth's orbit around the Sun as being a consequence of the fact that the diameter of the orbit was insignificant in comparison with the distance from the Earth to the stars. (In fact, stellar parallax is detectable, but only with superior instruments; it was not observed until Friedrich ◊Bessel finally succeeded in 1838.)

A Osiander inserted a preface to *De revolutionibus* (without Copernicus' permission), stating that the theory was intended merely as an aid to the calculation of planetary positions not as a statement of reality. This served to compromise the value of the text in the eyes of many astronomers, but it also saved the book from instant condemnation by the Roman Catholic Church. *De revolutionibus* was not placed upon the index of forbidden books until 1616 (it was removed in 1835).

Copernicus relegated the Earth from being the centre of the universe to being merely a planet (the centre only of its own gravity and the orbit of its solitary Moon). This in itself forced a fundamental revision of the anthropocentric view of the universe and came as an enormous psychological shock to the whole of European culture. Copernicus' model could not be 'proved' right, because it contained several fundamental flaws, but it was the important first step to the more accurate picture built up by Brahe, Kepler, Galileo, and later astronomers. It is no exaggeration to view Copernicus' work as the cornerstone of the Scientific Revolution.

Further Reading

Gingerich, Owen, *The Eye of Heaven: Ptolemy, Copernicus, Kepler,* Masters of Modern Physics series, Springer-Verlag, 1992.

Rosen, Edward, *Copernicus and His Successors,* Hambledon Press, 1995.

Rosen, Edward, *Copernicus and the Scientific Revolution,* Anvil series, Krieger Publishing Company, 1984.

Speke, Shirley, *From Copernicus to Newton: A Study in the Development of the Concept of Science,* GC Book Publishers, 1995.

Cori, Carl Ferdinand (1896–1984) and Gerty Theresa (born Radnitz) (1896–1957) Czech-born US biochemists, husband and wife, who worked out the biosynthesis and degradation of glycogen, the carbohydrate stored in the liver and muscles. For this achievement they were jointly awarded the 1947 Nobel Prize for Physiology or Medicine, which they shared with the Argentine biochemist Bernardo Houssay (1887–1971).

Carl Cori was born on 5 December 1896 in Prague, then in Austria–Hungary. He was educated in Austria at the Trieste Gymnasium, and graduated in 1920 with a medical degree from the University of Prague. It was while he was a medical student that he met and married his classmate Gerty Radnitz. She was also born in Prague, on 15 August 1896, entered the medical school in 1914, and graduated in the same year as her husband. After having spent several years in war-torn Austria – Carl Cori was in the Austrian army during World War I – they emigrated to the USA in 1922 and became US citizens six years later.

From 1922–31 Carl Cori was a biochemist at the State Institute for Study of Malignant Diseases at Buffalo, New York. In 1931 he was appointed professor of biochemistry at Washington University School of Medicine in St Louis, Missouri. In the same year his wife became fellow and research associate in pharmacology and biochemistry there, a position she held until 1947. Gerty Cori died in St Louis on 26 October 1957. Carl Cori remained at St Louis until 1967, when he took up the appointment of biochemist at Massachusetts General Hospital, Harvard Medical School, Boston. He died on 20 october 1984, aged 87, in Cambridge, Massachusetts.

It was during the 1930s that the Coris began their researches on glycogen. Its basic structure was already known; it is a polysaccharide, a highly branched sugar molecule composed of several hundred glucose molecules linked by glycosidic bonds. Any excess food in an animal's diet is stored as glycogen or fat, and in times of shortage the animal makes use of these reserves. Glycogen is broken down in the muscles into lactic acid (2-hydroxypropanoic acid), as worked out by Otto ⇨Meyerhof about 20 years earlier, which when the muscles rest is reconverted to glycogen. The Coris set out to determine exactly how these changes take place.

It was tempting to assume that glycogen broke down into separate glucose molecules. But this hydrolysis would involve a loss of energy, which would have to be resupplied for the conversion back to glycogen. Gerty Cori found a new substance in muscle tissue, glucose-1-phosphate, now known as the Cori ester. Its formation from glycogen involves only a small amount of energy change, so that the balance between the two substances can easily be shifted in either direction. The second step in the reaction chain involves the conversion of glucose-1-phosphate into glucose-6-phosphate. Finally this second phosphate is changed to fructose-1, 6-diphosphate, which is eventually converted to lactic acid. The first set of reactions from glycogen to glucose-6-phosphate is now termed glycogenolysis; the second set, from glucose-6-phosphate to lactic acid, is referred to as glycolysis.

Coriolis, Gaspard Gustave de (1792–1843) French physicist who discovered the Coriolis force that governs the movements of winds in the atmosphere and currents in the ocean. Coriolis was also the first to derive formulae expressing kinetic energy and mechanical work.

Coriolis was born in Paris on 21 May 1792. From 1808, he studied there at the Ecole Polytechnique and the Ecole des Ponts et Chaussées, graduating in highway engineering. He returned to the Ecole Polytechnique in 1816 as a tutor, and then became assistant professor of analysis and mechanics. In 1829 Coriolis accepted a chair in mechanics at the Ecole Centrale des Arts et Manufactures and then, in 1836, he obtained the same position at the Ecole des Ponts et Chaussées. In 1838 he became director of studies at the Ecole Polytechnique. Often in poor health, he died in Paris on 19 September 1843.

From 1829, Coriolis was concerned that proper terms and definitions should be introduced into mechanics so that the principles governing the operation of machines could be clearly expressed and similar advances be made in technology as were being made in pure science. His teaching experience aided him greatly in this endeavour, and he succeeded in establishing the use of the word 'work' as a technical term in mechanics, defining it in terms of the displacement of force through a certain distance. He proposed a unit of work called the dynamode, which was equal to 1,000 kg m/ 7,200 lb ft, but it did not enter general use.

Coriolis then proceeded to give the name kinetic energy to the quantity he defined as $^1/_2mv^2$, making the mathematical expression of dynamics much easier.

Coriolis went on to investigate the movements of

moving parts in machines relative to the fixed parts and in 1835 made his most famous contribution to physics in a paper on the nature of relative motion in moving systems. Applying his ideas to the general case of rotating systems, Coriolis showed that an inertial force must be present to account for the relative motion of a body within a rotating frame of reference. This force is called the Coriolis force. It explains how the rotation of the Earth causes objects moving freely over the surface to follow a curved path relative to the surface, the Coriolis force turning them clockwise in the northern hemisphere and anticlockwise in the southern hemisphere. The Coriolis force therefore appears prominently in studies of the dynamics of the atmosphere and the oceans, and it is also an important factor in ballistics.

Coriolis was a scientist of acute perception who did much to clarify thought in the field of mechanics and dynamics. But for poor health, he might have realized his full potential and made an even greater contribution to science.

Corliss, George Henry (1817–1888) US engineer and inventor of many improvements to steam engines, particularly the Corliss valve for controlling the flow of steam to and through the cylinder(s).

Little is known about the early life of Corliss except that he was born on 2 June 1817 in Easton, New York. By the time he was in his late twenties he had moved to Providence, Rhode Island, where he became interested in steam engines, concerning himself particularly with attempts to improve their performance.

He took out the first of his many patents in 1849, for the Corliss valve. In 1856 he founded the Corliss Engine Company, which was to become one of the largest steam-engine manufacturers in the USA and, incidentally, helped to make Providence a major tool and machinery centre. His company designed and built the largest steam engine then in existence, to power all the exhibits in the Machinery Hall at the 1876 Philadelphia Centennial Exposition. With a weight of 630 tonnes/609 tons and a 9-m/29.5-ft flywheel, the engine was not surprisingly billed as the 'Eighth Wonder of the World'.

In the 1840s, the most advanced valves then in use suffered from several disadvantages inherent to their design. Most regulated steam flow by sliding a plate over the port into the piston chamber in such a way that while steam entered one end of the chamber it was exhausted from the other. Unequal wear on the sliding parts caused them to lose their steam-tight fit; they were also heavy to operate and, with the steam entering and exhausting through the same passage, they suffered considerable heat loss.

An essential feature of the Corliss valve was the separate inlet and exhaust port at each end of the cylinder (giving a total of four valve units per engine in twin cylinder engines). This saved heat loss and also cut down the 'dead' space at the ends of the chamber using older valves where the ports were located along the side of the chamber. Of equal significance to this was Corliss's realization that for maximum efficiency the valves must open and close as quickly as possible. This he achieved by a spring-loaded action.

Like any good invention the basic principle of the valve was very simple. On the first stroke, one inlet valve opened, allowing steam under pressure to enter that end of the cylinder and force the piston to move to where an exhaust valve was open. On the next stroke these two valves closed, and the other inlet and exhaust valves opened again, allowing steam to enter the cylinder and force the piston to move back in the other direction. Because of its compact size, the reduced wear on moving parts, and heat saved by having separate steam inlet and outlet paths, this valve greatly improved the efficiency of large steam engines. The success of Corliss's design is best shown by the fact that it continued in use for as long as large steam engines were manufactured, long after its inventor's death on 2 February 1888.

Cornforth, John Warcup (1917–) Australian organic chemist who shared the 1975 Nobel Prize for Chemistry with the Swiss biochemist Vladimir ◊Prelog for his work on the stereochemistry of biochemical compounds.

Cornforth was born on 7 September 1917 in Sydney. He began his academic training at the University of Sydney and then went to Oxford University, where he obtained his doctorate in 1941. For the next five years he worked with Robert ◊Robinson, who 30 years earlier had worked in Sydney and who was to receive the 1947 Nobel Prize for Chemistry for his work on plant alkaloids. At about this time Cornforth's hearing began to deteriorate and he was soon totally deaf. He worked for the British Medical Research Council 1946–62, when he became firector of the Milstead Laboratory of Chemical Enzymology, Shell Research Ltd. He remained there until 1975, when he accepted a professorship at the University of Sussex; he had previously been an associate professor at Warwick University 1965–71 and visiting professor at Sussex 1971–75. He was knighted in 1977.

In his researches, Cornforth studied enzymes, trying to determine specifically which group of hydrogen atoms in a biologically active compound is replaced by an enzyme to bring about a given effect. He painstakingly developed techniques to pinpoint a specific hydrogen component by using the element's three isotopes – normal hydrogen (^{1}H), deuterium (^{2}H), and tritium (^3H). Each isotope has a different speed of reaction and, by careful observation and many experiments, Cornforth was able to identify precisely which hydrogen atom was affected by enzyme action. He was able, for example, to establish the orientation of all the hydrogen atoms in the cholesterol molecule.

For him [the scientist], truth is so seldom the sudden light that shows new order and beauty; more often, truth is the uncharted rock that sinks his ship in the dark.

JOHN CORNFORTH NOBEL PRIZE ADDRESS 1975

Correns, Karl Franz Joseph Erich (1864–1933) German botanist and geneticist. He can be credited for rediscovering the work of Austrian biologist Gregor ◊Mendel and putting Mendelism back on the agenda in 1900, 16 years after Mendel's death.

Born in Munich on 19 September 1864, Karl was the only child of the painter Erich Correns. His mother was Swiss. Correns's education in Munich was cut short when he was orphaned and moved to Switzerland at the age of 17. Tuberculosis interrupted his schooling further and it was not until he was 21 that he entered Munich University, graduating in 1889. Moving to Tübingen University he began his research on the effect of foreign pollen in changing the visible characters of the endosperm (the nutritive tissue surrounding the plant embryo). Ten years later he was appointed assistant professor at Pfeffer's Institute in Leipzig, becoming full professor at Munster in 1909. From 1913 until his death on 14 February 1933 he was director of the Kaiser Wilhelm Institute for Biology in Berlin. His unpublished manuscripts, which had been stored at the institute, were destroyed in the bombing of Berlin in 1945.

While other geneticists were concerned with mutations and the practicality of plant breeding, one of his many projects in the 1890s was to concentrate on the effects of foreign pollen on the characteristics of the next generation of fruit and seeds – the xenia question. He crossed varieties of pea plants and maize, mistakenly concluding after three years that the xenia effect in maize (where the storage tissue of the seeds takes on different colours) was due either to an enzymatic influence of the embryo on the endosperm tissue or to a genuine hybridization. After four generations the colour changes in his pea crosses led him to assert correctly that only the cotyledons of the new embryo were involved, not the mother tissue. This gave him simpler ratios than the maize experiment and he was able to make an accurate explanation. It was only later that he read that Mendel had reached the same conclusions a generation earlier. Correns's work published in 1900 provided further proof of Mendel's experiments and brought Mendel into public view.

Correns's later research was devoted to discovering how widely Mendel's laws could be applied, and the extent of their validity. In 1902 he found a case of coupling in maize between self-sterility and the blue coloration of the aleurone layer (layer of protein in the seed), and he also produced a theory that allowed for the exchange of genes between homologous chromosomes, a theory later advanced by US geneticist Thomas Hunt ◊Morgan. In 1907 using *Bryonia,* Correns proved that sex is inherited in Mendelian fashion. In 1909 he demonstrated the first conclusive evidence, using variegated plants, for cytoplasmic (non-Mendelian) inheritance in which certain features of the offspring are determined by the cytoplasm in the egg cell. Correns was also the first to relate Mendelian segregation to meiosis and the first to obtain evidence of differential fertilization between gametes.

In terms of the subtle sophistication of his research techniques Correns was a good 20 years ahead of his fellow geneticists. He was a dedicated scientist, driven by a search for truth over quick results and he happily gave Mendel full credit for the work he had undertaken. In 1892 Correns had married Elizabeth Widmer, a niece of the botanist Karl Wilhem von Naegeli who had been one of his teachers. Mendel had sent a copy of one of his important papers to von Naegeli, who did not recognize its importance at the time.

Cort, Henry (1740–1800) English inventor who devised a method of producing high-quality iron using a reverberating furnace.

Cort was born in Lancaster, and at the age of 25 he left his home town to seek his fortune in London. He became an agent for the navy and was responsible for purchasing their guns. At this time (1765) all the best metal suitable for arms manufacture was imported from abroad, mainly from Russia, Sweden, and North America. Cort, realizing the potential, experimented to find a method of making this high-grade metal in England, and in 1775 he set up his own forge at Fontley, near Farnham.

By 1784 he was in a position to apply for a patent on his process of 'puddling and rolling', which allowed wrought-iron bars to be produced on a large scale and of a high quality. The process came at a particularly fortunate time for England's iron production, for with the advent of the Napoleonic Wars, pig-iron requirements rose from 40,000 tonnes/38,700 tons in 1780 to 400,000 tonnes/387,000 tons by 1820. But he was somewhat unwise in his choice of financial backer. He chose Samuel Jellicoe, a naval paymaster, who without Cort's knowledge obtained the money from public funds. When Jellicoe was found out he committed suicide and Cort, having handed over the rights to his patent as security for the capital, was left bankrupt. His patent was confiscated by the Admiralty and he was forced to watch the ironmasters of England grow wealthy on his hard work. He and his large family were reduced to living off a pension of £200 a year, which the state eventually granted him. Cort died in Hampstead, London, in 1800.

The skill of iron smelting has been known for centuries. Using a blast furnace fired by charcoal, the process remained unchanged until the early 1700s, when several people attempted to better the traditional ways. Three main factors limited production. The first was the choice of fuel. Using charcoal meant that the ironworks, of necessity, were sited near forests. The second was the source of power to work the bellows and the forge-hammer. A water wheel was a natural solution, but its efficiency depended upon the amount of water available and as a result production was often seasonal. The fluctuation of water supply also influenced the third factor, transport, which relied heavily on the rivers.

Coal answered the first problem, the steam engine was the solution for the power, and canals took over from natural waterways. Coal, however, although cheaper, still produced poor-quality iron and it was not until Abraham Darby used coal converted to coke at his Coalbrookdale works (and combined it with the right type of ore) that a pig iron was made that could be forged into wrought iron. Even so, the quality tended to be unreliable.

Cort experimented and found a solution in his 'puddling and rolling' process, which removed the impurities by stirring the pig iron on the bed of a reverberatory furnace. (This is a furnace with a low roof, so that the flames in passing to the chimney are reflected down on to the hearth, where the ore to be smelted can be heated without its coming into direct contact with the fuel.) The 'puddler' turned and stirred the mass until it was converted into a malleable iron by the decarburizing action of air circulating through the furnace. The iron was then run off, cooled, and rolled into bars with the aid of grooved rollers.

The significance of Cort's work was such that at last England did not have to rely on imported iron and could become self-sufficient. His method of manufacture combined previously separate actions into one process, producing high-class metal relatively cheaply and quickly and allowing iron production to increase to meet the growing needs of the industrial age.

Cotton, William (1786–1866) English inventor, financier, and philanthropist who greatly publicized the work of an earlier hydrographer and inventor, Joseph Huddart. He is probably best known, however, for inventing a knitting machine that paved the way for significant advances in the production of hosiery.

Cotton was born in Leyton, Essex, on 12 September 1786. On leaving school at the age of 15, he entered a counting-house as a clerk and all subsequent progress was as a result of self-education. In 1807 he was admitted as a partner in the firm of Huddart and Company in Limehouse, London. This business had been founded a few years earlier to promote the large-scale manufacture of an ingenious cordage-making machine designed by Joseph Huddart. The attention of Huddart was drawn to this field when, in the course of travelling for the East India Company, he saw a ship's cable snap. He thereupon worked on a method 'for the equal distribution of the strains upon the yarn'. When put on the market, his machine was to bring him an appreciable fortune.

Cotton did well in Huddart's firm and was eventually entrusted with its general management. He disposed of Huddart's machinery to the government in 1838. In the same year he wrote a memoir of Huddart with an account of his inventions, which was privately printed in 1855. In recognition of this publication he received a silver medal from the Institution of Civil Engineers. Cotton died in Leytonstone, Essex, on 1 December 1866.

In 1821 Cotton was elected a director of the Bank of England, a position he continued to hold until a few months before his death. He held the governership of the Bank of England 1843–46. A lasting memorial of his tenancy of this high office came about through his invention of the 'governor' and automatic weighing machine for sovereigns. It weighed these coins at a rate of 23 per minute, discharging the full and underweight specimens into separate compartments.

Although Cotton prospered in business, it is nevertheless as a philanthropist that he is chiefly remembered, notably for the building of schools, churches, and lodging houses in the East End of London. After his death he was commemorated by a painted window in St Paul's Cathedral, raised by public subscription.

It was not until 1864 that Cotton secured a patent for the knitting machine to be used in hosiery, the principal invention for which he is remembered. The machine he produced was remarkable for its adaptability: it had a straight-bar frame that automatically made fully fashioned stockings knitted flat and sewn up the back.

Subsequent improvements to Cotton's original machine have largely been directed at either increasing the capacity of the machine or producing fabric of even finer gauge. Increased capacity can be attained only by the use of larger and faster models, and finer gauge work requires finer needles and sinkers. A modern machine making fully fashioned silk or nylon hose may have as many as 40 needles to 2.5 cm/1 in and may produce 32 stockings at once. The reliability of such machines obviously depends on the accuracy with which they are made, and in this respect the hosiery industry owes much to improvements in engineering techniques. Towards the end of the 19th century the making of hosiery machines became an important industry in Germany and the USA, and in time the types used in these countries became also widely used in the UK.

Coulomb, Charles (1736–1806) French physicist who established the laws governing electric charge and magnetism. The unit of electric charge is named the coulomb in his honour.

Coulomb was born in Angoulême on 14 June 1736 and educated at the Ecole du Génie in Mézières, graduating in 1761 as a military engineer with the rank of first lieutenant. After three years in France, he was posted to Martinique to undertake construction work and remained there 1764–72. Postings followed within France to Bouchain, Cherbourg, Rochefort, and eventually to Paris in 1781.

In 1774, Coulomb had become a correspondent to the Paris Academy of Science. Three years later he shared the first prize in the academy's competition with a paper on magnetic compasses and, in 1781, gained a double first prize with his classic work on friction. In the same year, he was elected to the academy. During his years in Paris, his duties were those of an engineering consultant and he had time in hand for his physics research. The year 1789 marked the beginning of the French Revolution and Coulomb found it prudent to resign from the army in 1791. Between 1781–1806, he presented 25 papers (covering electricity and magnetism, torsion, and applications of the torsion balance) to the academy, which became the Insitut de France after the Revolution. He was also a contributor to several hundred committee reports to the academy on engineering and civil projects, machinery, and instruments. In 1801 he became president of the Institut de France; he died in Paris on 23 August 1806.

Coulomb took full advantage of his various postings to pursue a variety of studies, including structural mechanics, friction in machinery, and the elasticity of metal and silk fibres. In structural mechanics, he drew greatly on his experience as an engineer to investigate the strengths of materials and to determine the forces that affect materials in beams. In his study of friction, he extended knowledge of the effects of friction caused by factors such as lubrication and differences in materials and loads, producing a classic work that was not surpassed for 150 years. In both these fields, Coulomb greatly influenced and helped to develop engineering in the 19th century. He also carried out fundamental research in ergonomics, which led to an understanding not only of the ways in which people and animals can best do work but also, in influencing the subsequent research of Gaspard ◊Coriolis, to a proper understanding of the fundamental nature of work.

Coulomb's major contribution to science, however, was in the field of electrostatics and magnetism, in which he made use of a torsion balance he invented and which is described in a paper of 1777. He was able to show that torsion suspension can be used to measure extremely small forces, demonstrating that the force of torsion is proportional to the angle of twist produced in a thin stiff fibre. This paper also contained a design for a compass using the principle of torsion suspension, which was later adopted by the Paris Observatory.

Coulomb was very interested in the work of Joseph ◊Priestley on electrical repulsion, and in his paper of 1785 he discussed the adaptation of his torsion balance for electrical studies. He demonstrated that the force between two bodies of opposite charge is inversely proportional to the square of the distance between them. He also stated but did not demonstrate that the force of attraction or repulsion between two charged objects is proportional to the product of the charges on each. Coulomb's next paper, in 1787, produced proof of the inverse square law for both electricity and magnetism, and for both attractive and repulsive forces. A magnetic needle or charged pith ball was suspended from the torsion balance at a measured distance from a second similar needle or pith ball fixed independently, the torsion arm was deflected, and the period of the resulting oscillations was timed. The experiment was repeated for varying distances between the fixed and oscillating bodies under test. Coulomb showed that if certain assumptions hold, the forces are proportional to the inverse square of the period and the period will vary directly as the distance between the bodies. The assumptions were (1) that the electrical or magnetic forces behave as if concentrated at a point, and (2) the dimensions of the bodies must be small compared with the distance between them. The results are embodied in Coulomb's law, which states that the force between two electric charges is proportional to the product of the charges and inversely proportional to the square of the distance between them.

Coulomb went on to investigate the distribution of electric charge over a body and found that it is located only on the surface of a charged body and not in its interior.

Many of Coulomb's discoveries – including the torsion balance and the law that bears his name – had been made in essence by John Michell (1724–1793) and Henry ◊Cavendish in the UK. Michell died before completing his work and Cavendish neglected to publish his results, so Coulomb can truly be regarded as the principal scientist of his time. With his researches on electricity and magnetism, Coulomb brought this area of physics out of traditional natural philosophy and made it an exact science.

Coulson, Charles Alfred (1910–1974) English theoretical chemist whose major contribution to the science was his molecular orbital theory and the concept of partial valency. He also worked – and held professorships – in physics and mathematics and published hundreds of papers on topics as diverse as pure mathematics and the effects of radiation on bacteria.

Coulson was born in Dudley, Yorkshire, on 13 December 1910, one of twin sons. His family moved to Bristol in 1920 and he went to Clifton College, from which he won an open scholarship to Trinity College, Cambridge, in 1928. He graduated in mathematics three years later. He was awarded a research scholarship, and later an open scholarship, and worked with Lennard Jones, the first holder of the chair in theoretical chemistry. It was then that he was introduced to wave mechanics and quantum theory, to which he was to devote himself so successfully for much of his career. He remained at Cambridge until 1938, when he married and took up an appointment as senior lecturer in mathematics at the University of Dundee. In 1930 he had adopted deep Christian beliefs that were to lead him to become a pacifist; he was chairman of the charity Oxfam 1965–71.

In Dundee, Coulson taught and carried out research. His first book, *Waves* (1941), ran to seven editions and is still in worldwide use. In 1945 he moved to Oxford University to join the Physical Chemistry Laboratory as a 'theoretician', and soon also became a mathematics lecturer. Two years later he accepted the chair of theoretical physics at King's College, London. His second book, *Electricity*, was published in 1948 and was another immediate success.

In 1952, his third best seller, *Valence* was published. In the same year he returned to Oxford as a professor of mathematics. But he continued to teach chemistry and physics, beginning his famous summer schools in 1955. In 1963 he became curator of the Mathematics Institute (where he started the first university computing department), and was appointed its president in 1971. He further demonstrated his versatility by becoming Oxford's first professor of theoretical chemistry. He died in 1974 from an illness that had begun some years earlier.

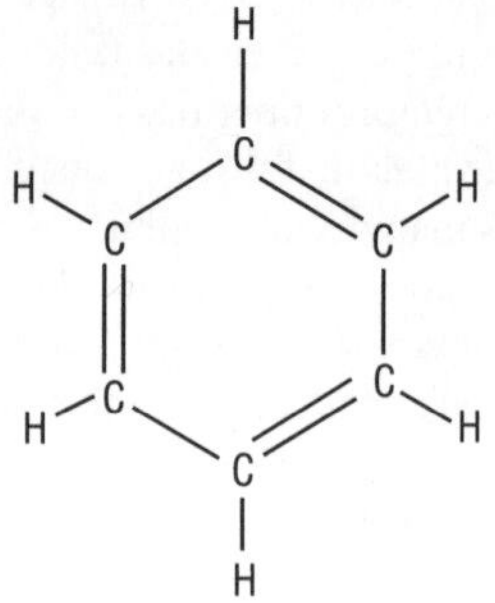

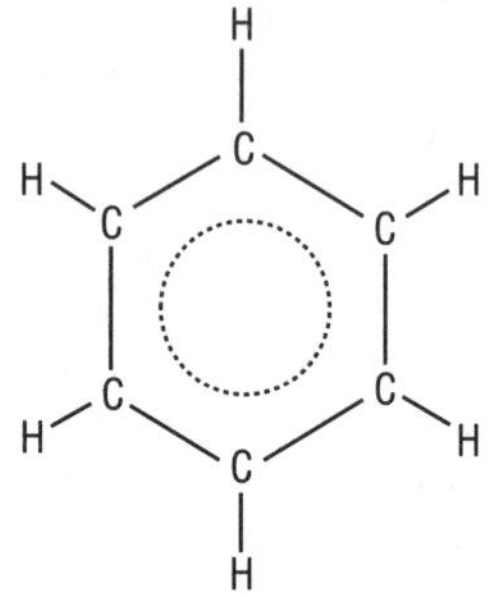

Coulson (a) 'Classical' (Kekulé) structure of benzene with alternate double and single carbon-carbon bonds that would not be equivalent. (b) Molecular orbital structure of benzene with six equivalent bonds; each carbon atom has three complete bonds to complete its valency of four.

Coulson's great contribution to chemistry was the way in which he influenced thinking about forces in molecules. Since J J Thomson's discovery of the electron and Ernest Rutherford's of the atomic nucleus, it had been obvious that chemical bonds must involve electrons, but nobody fully understood what form this involvement takes. The quantum theory of the time showed how electrons occupy energy levels in atoms, but it gave no real help with molecular forces. In the 1920s wave mechanics began to revolutionize the subject, but Erwin Schrödinger's famous wave equation becomes hopelessly complicated in all but the simplest cases. Coulson used powerful methods of approximation and computation to obtain some solutions of the wave equation for molecular systems.

The molecular orbital theory that Coulson developed is an extension of atomic quantum theory and deals with 'allowed' states of electrons in association with two or more atomic nuclei, treating a molecule as a whole. He was thus able to explain properly phenomena such as the structure of benzene and other conjugated systems, and invoked what he called partial valency to account for the bonding in such compounds as diborane – neither of which could be accounted for by the alternative valence-bond system proposed by Linus ◊Pauling and others.

Practical applications of Coulson's molecular orbital theory have included the prediction of new aromatic systems and accurate forecasting of bond lengths and angles.

Coulson also contributed significantly to the understanding of the solid state (particularly metals), such as the structure of graphite and its 'compounds'. He developed many mathematical techniques for solving chemical and physical problems while retaining the capacity to produce simple, intuitive models of the systems he was studying, and the ability to formulate these models in manipulatable mathematical terms.

Courant, Richard (1888–1972) German-born US mathematician who taught mathematics from a very early age, wrote several textbooks that are now standard reference works, and founded no fewer than three highly influential mathematical institutes.

Courant was born on 8 January 1888 in Lublintz, in Upper Silesia (now in Poland), the son of a businessman. He attended schools in Glatz, and then Breslau (Wrocław), where he showed exceptional talent and was soon teaching privately people who were several years above him at school. At the age of 16 Courant was earning so much money as a teacher that he remained in Breslau when his parents moved elsewhere; at school, however, he was told to give up his tutoring or be expelled. Instead, he left of his own accord, and began attending lectures unofficially at the university there. The following year, he entered the university on an official basis, but in 1907, on the advice of one of his friends, he moved on to the University of Göttingen, where he quickly distinguished himself. He became acquainted with David ◊Hilbert, and tutored Hilbert's son, becoming also Hilbert's assistant – a position that allowed him to be at the centre of one of the most thriving contemporary mathematical communities. In 1910 he was awarded his doctorate for an investigation into Dirichlet's principle. The following year he returned to Göttingen, married, and prepared the thesis to qualify him as a university teacher.

During World War I, Courant served as an infantryman (being promoted to lieutenant) until he was wounded in 1915. At about this time he interested the military authorities in a device for sending electromagnetic radiation through the Earth to carry messages; this was developed, and Courant spent the rest of the war involved with communications. He then returned to Göttingen for a few months before being appointed professor of mathematics at the University of Münster. However, he was soon recalled to the vacant chair at Göttingen (which only a few years earlier had been occupied by Felix Klein).

Courant spent more than ten years at Göttingen, gradually building up the mathematics department into an autonomous unit, and raising funds for new buildings. He found his position untenable, however, as a Jew in Hitler's Germany, and after a visit to the USA 1931–32, settled there in 1934, joining the teaching staff at New York University. Neither the university nor his position were similar to that which he had been used to in Germany, but from these small beginnings he again built the mathematical department into a renowned centre of research; the success he achieved was even greater than that in Göttingen, in that he was director of the Institute of Mathematical Sciences of New York University 1953–58, and in 1962 work was begun on a new institute, which was opened in 1965 as the Courant Institute of Mathematical Sciences. Courant retired in 1958 but still retained an interest in mathematics and mathematical education. He died in New Rochelle on 27 January 1972.

Much of what Courant achieved during his life was as a result of his great skill in administration and organization. At Göttingen, Felix Klein, although he had retired some years before, still enjoyed the privileged status of emeritus professor and wielded enormous influence in the structure of its mathematical teaching and organization. It was he who saw in Courant a worthy successor to himself as administrator. In his own characteristic way, Courant then set about developing the few rooms and the autonomous mathematical professors and other staff into a mathematical institute worthy of the reputation it already commanded throughout the world. By subtle changes (including that of the title on the stationery) he separated the mathematics department from the philosophical faculty. Then he persuaded the International Education Board to donate the money for a new building. It was unfortunate that after this was built and Courant appointed its first director in 1929 that Hitler came to power, because four years later Courant, being of Jewish descent, was suspended from duty. Nevertheless, once in the USA, Courant managed to do it all over again, this time more by paying close attention to the nurturing of graduates, and by publicizing the fact that his courses dealt with applied, as well as pure, mathematics.

Courant was also a very able writer on mathematical topics. By treating many of the subjects developed by his mentor David Hilbert, he prepared a book, *Methods of Mathematical Physics* (now universally known as the 'Courant–Hilbert'), which turned out to be just what was needed by physicists in their research on the quantum theory. The first volume was published while he was in Germany, the second after he had settled in the

USA. Its influence on the scientific community has been so great that it is still in use today. Another book, *Differential and Integral Calculus,* also in two volumes, has been used for 40 years as a university textbook, and is still in use in an updated form. Like the Courant–Hilbert it was soon translated into English. Courant was also sole or part author of a number of books on more specialized areas of mathematics. One general book still of great interest, however, is his *What is Mathematics?,* written in conjunction with Herbert Robbins and published in 1941.

Cousteau, Jacques-Yves (1910–1997) French diver, marine explorer, and documentary filmmaker who invented the aqualung and pioneered many of the diving techniques used today.

Cousteau was born on 11 June 1910 at Saint André in the Gironde. He became a gunner in the French navy shortly before the outbreak of World War II, and worked in Marseille for naval intelligence during the Nazi occupation. His interest in diving came in 1936 when he borrowed a pair of goggles and peered beneath the surface of the Mediterranean. He was instantly captivated by what he saw. Throughout the occupation he and his colleagues Frederick Dumas, Phillipe Taillez, and Emile Gagnan experimented with diving techniques in comparative peace.

Cousteau initially experimented with available naval equipment. Le Prieur had invented the compressed-air cylinder in 1933 that released a continuous flow of air through the face mask. This system restricted the diver to very short periods of time beneath the surface. Cousteau designed an oxygen rebreathing apparatus that would give him longer dives, but after two near-fatal accidents he abandoned his ideas. He also tested the Fernez equipment, which fed compressed air to the diver from the surface, and again was nearly killed in the process.

The turning point came in 1942, when Cousteau met Emile Gagnan, an expert on industrial gas equipment. Gagnan had designed an experimental demand valve for feeding gas to car engines. Together, Gagnan and Cousteau developed a self-contained compressed air 'lung', the aqualung. In June 1943 Cousteau made his first dive with it, achieving a depth of 18 m/60 ft.

In 1945, Cousteau founded the French navy's Undersea Research Group. Much of its early diving work involved locating and defusing the mines left behind by the Germans. He fitted out and commanded the Group's research ship, the *Elie Monnier,* on many oceanographic expeditions.

In 1951 he set out in the research ship *Calypso* on a four-year voyage of exploration beneath the oceans the world. He went on to make a series of television programmes about his many voyages aboard the ship, *The Undersea World of Jacques Cousteau* (1968–77), becoming a popular television personality as a result. He died on 25 June 1997.

The aqualung has changed very little since Cousteau's original design. A free-swimming, air-breathing diver can now descend to depths of over 60 m/200 ft and carry out work. Greater depths require extreme care and sophisticated gas mixtures to avoid danger.

Without Cousteau's contribution to the science of diving, the massive underwater tasks performed offshore for the oil industry and others would have been vastly more difficult. The foundations he laid for modern divers are being built upon constantly, allowing more difficult tasks to be carried out in ever more adverse conditions.

Further Reading

Cousteau, Jacques-Yves, *Undersea Discoveries of Jacques-Yves Cousteau,* Arrowood Press, 1989.

Markham, Lois, *Jacques-Yves Cousteau: Exploring the Wonders of the Deep,* Innovative Minds series, Raintree Steck-Vaughn Publishers, 1997.

Reef, Catherine, *Jacques Cousteau: Champion of the Sea,* Earth Keepers series, Twenty-First Century Books.

Cowling, Thomas George (1906–1990) English applied mathematician and physicist who contributed significantly to modern research into stellar energy, with special reference to the Sun.

Cowling was born on 17 June 1906 and educated at the Sir George Monoux School in Walthamstow, Essex. He attended Oxford University, where he studied mathematics, and spent many years teaching in various university posts in London, Swansea, Dundee, and Manchester before becoming professor of mathematics at University College, Bangor, in 1945. In 1948 he was appointed professor of mathematics at Leeds University, where he stayed until his retirement in 1970.

The author of several books on mathematics, Cowling became a Fellow of the Royal Society in 1947, and was made president of the Royal Astronomical Society for a two-year period in 1965.

Cowling's work was of important assistance in the discovery of the carbon–nitrogen cycle by Hans ◊Bethe in 1939. Bethe showed that the most significant source of energy in stars was the process by which four hydrogen atoms are converted into one helium atom, with carbon and nitrogen as intermediate products. This process was found to account satisfactorily for the rate of generation of the Sun's energy if its central temperature was 18.5 million degrees kelvin (18.5 million °C/ 33 million °F) – an estimate that corresponded well with the temperature of 19 million degrees kelvin (19 million °C/34 million °F) calculated following the theory of stellar structure previously proposed by Arthur ◊Eddington.

Less indirectly, Cowling was responsible for demonstrating the existence of a convective core in stars, suggesting thus that the Sun may behave like a giant dynamo whose rotation, internal circulation, and convection produce the immensely powerful electric currents and magnetic fields associated with sunspots. (The electric current required to produce the field of a large sunspot may be of the order of 10^{13} amps; magnetic fields similarly may reach several thousand gauss.) With the Swedish astronomer Hannes ◊Alfvén, Cowling showed that such currents and fields would be difficult to initiate in the solar atmosphere and are therefore likely to have existed since the Sun was first formed.

Cox, Allan Verne (1926–1987) US geophysicist whose palaeomagnetic research substantiated evidence of periodic reversals in the Earth's magnetic field, confirmed plate tectonic theories of continental drift and sea floor spreading, and led to the publication of a new geological time scale in 1982.

Born in Santa Ana, California in 1926, Cox was the son of a house painter. He served in the US Merchant Marine 1945–48 and, after graduating in chemistry from the University of California at Berkeley in 1951, joined the US Army for two years. He spent summers working in Alaska where he had developed an interest in geology and went back to do graduate work in 1954. His adviser at the time was US geophysicist John Verhoogen, who was interested in rock magnetism and was sympathetic to the radical notion of continental drift. Cox gained his PhD in 1959 and joined the US Geological Survey at Menlo Park, California, where he worked with US geologist Richard Doell. They wrote several papers together on rock magnetism in the early 1960s. Working with US geologist Brent Dalrymple they established a timescale that showed the irregular schedule of polarity changes in the Earth's past. Other geophysicists had noticed patterns of polarity change on either side of mid-ocean ridges, giving evidence of what was to become plate tectonics. Cox played a vital role in the discoveries that led to the revolution that transformed the earth sciences. His studies of palaeomagnetism and rock samples from around the world gave evidence of many reversals of the magnetic field in geological time. The geomagnetic timescale implied by this was published in *Nature* in 1963.

The magnetic field preserved in rocks holds much other information about the history of the Earth, apart from the simple reversal of magnetic direction, and Cox next turned his attention to this. In 1967 he moved to Stanford University, California, to become professor of geophysics, continuing to work on palaeomagnetism. He became a popular teacher, devising interesting research projects for undergraduates and attracting students to the field. He was made Green Professor of Geophysics in 1974 and became dean of the School of Earth Sciences in 1979.

Cox wrote over a hundred papers on plate tectonics and two popular books including *Plate Tectonics and Magnetic Reversals* (1973). He won several honours throughout his illustrious career including AGU's Fleming Medal (1969), the Vetlesen Medal of Columbia University (1971), and the Day Medal of the Geological Society of America (1975). He was elected to the National Academy of Sciences in 1974 and to the American Academy of Arts and Sciences and the American Philosophical Society. He was president of the American Geophysical Union 1978–80. He died in Stanford, California, in 1987.

Cramer, Gabriel (1704–1752) Swiss mathematician who is now chiefly remembered for Cramer's rule, Cramer's paradox, and for the concept of utility in mathematics. He was, however, an influential teacher, personally acquainted with some of the great mathematicians of his age, and a prolific editor of other people's writings.

Cramer was born on 31 July 1704 in Geneva. It was there that he was educated, and there too – at the age of 20 – that he shared the chair of mathematics at the Académie de la Rive with his friend Calandrini (he taught geometry and mechanics, and Calandrini taught algebra and astronomy). In 1727 Cramer travelled to Basle, where he met Leonhard Euler and his friend Daniel Bernoulli, and Daniel's father Jean Bernoulli – all famous mathematicians. During the next 18 months he visited London, Leiden, and Paris. (The next time he visited Paris, 20 years later, he took with him his pupil, the young Prince of Saxe-Gotha.) Cramer returned to Geneva, and in 1734 was appointed to the full chair of mathematics at de la Rive. In 1750 he was made professor of philosophy. In that year also, he published his major work, the *Introduction à l'analyse des lignes courbes algébriques*, in which 'Cramer's rule' provided a method for the solution of linear equations. The following year, however, he had an accident, after which he was advised to go to the south of France and rest. The journey there was itself too much for him; he died on the way, on 4 January 1752.

Cramer's work was acclaimed by his contemporaries and he received many honours. He was made a member of the Royal Society in London, and of the academies of Berlin, Lyon, and Montpellier.

Although Cramer made a number of original contributions to mathematics, the two he remains famous for are Cramer's rule and Cramer's paradox. His rule, published in 1750, was responsible for a revival in interest over the use of determinants. Determinants exist as part of a method to solve linear equations, and although the German mathematician Gottfried ◊Leibniz is most often credited with their discovery in 1693, there is some evidence that they were known to the Japanese mathematician ◊Seki Kowa some years before. But despite the fact that they were also referred to in the major work of Colin ◊Maclaurin in 1720, determinants had never received general attention until Cramer rediscovered them while working on the analysis of curves. The next year, Alexandre-Théophile Vandermonde developed Cramer's work; his results were in turn extended by Pierre Laplace and Joseph Lagrange; later, the foundations of modern determinant theory were laid by Augustin Cauchy, Karl Jacobi, and many others. Today, determinants are part of matrix theory, and are the means of classifying different systems of linear equations.

Cramer's paradox revolves around a theorem formulated by Colin Maclaurin. He stated that two different cubic curves intersect at nine points. Cramer pointed out that the definition of a cubic curve – a single curve – is that it is determined itself by nine points. But although he attempted an explanation, his was inadequate, and it was Leonhard Euler and others who derived a proper elucidation later.

Cramer's concept of utility now provides a connection between the theory of probability and mathematical economics. (That this was not the only interest Cramer showed in probability is revealed in his correspondence with Abraham de Moivre.)

As an editor of historic mathematical works, Cramer was indefatigable. In the last ten years of his life, he edited and published the collected works of Jean Bernoulli (whom he had met) and his brother Jacques Bernoulli (whom he had not), two volumes of correspondence between Jean Bernoulli and Gottfried Leibniz, and five volumes of the *Elementa* of Christian Wolf.

Crick, Francis Harry Compton (1916–) English biophysicist who shared the Nobel Prize for Physiology or Medicine with James Watson and Maurice ◊Wilkins for their work on determining the structure of DNA. Their discovery of the double helix is considered to be the most important biological advance of the twentieth century.

Crick was born in Nottingham on 8 June 1916 the older of two brothers. He went to Nottingham Grammar School and when the family moved to London in the 1920s, to Mill Hill School where he developed an interest in physics, chemistry, and mathematics. He then studied physics at University College, London, graduating with a BSc in 1937, and staying on for postgraduate work. However, his studies were interrupted by World War II and from 1940 he worked for the British Admiralty on the development of radar and magnetic mines for naval warfare. Staying on for two years after the war, he read and was inspired by Erwin ◊Schrödinger's *What is Life? The Physical Aspects of the Living Cell* which posited the idea of applying physics to the study of biology and proposed the investigation of genes at a molecular level. Excited by this, Crick decided in 1947 that he wanted to study biology rather than particle physics. With the financial help of his family and a studentship from the Medical Research Council he joined a Cambridge laboratory to study biology, organic chemistry, and X-ray diffraction techniques. By 1949 he was investigating the structure of proteins at the Cavendish Laboratory in Cambridge still driven by the idea of unravelling the mysteries of the genetic code.

Crick English molecular biologist Francis Crick who was awarded the Nobel Prize for Physiology or Medicine in 1962, together with Maurice Wilkins and James Watson, for his work on determining the structure of DNA. *Mary Evans Picture Library*

In 1951, a young US biologist, James Watson, joined the laboratory and he and Crick formed a close working relationship. They were both convinced that they

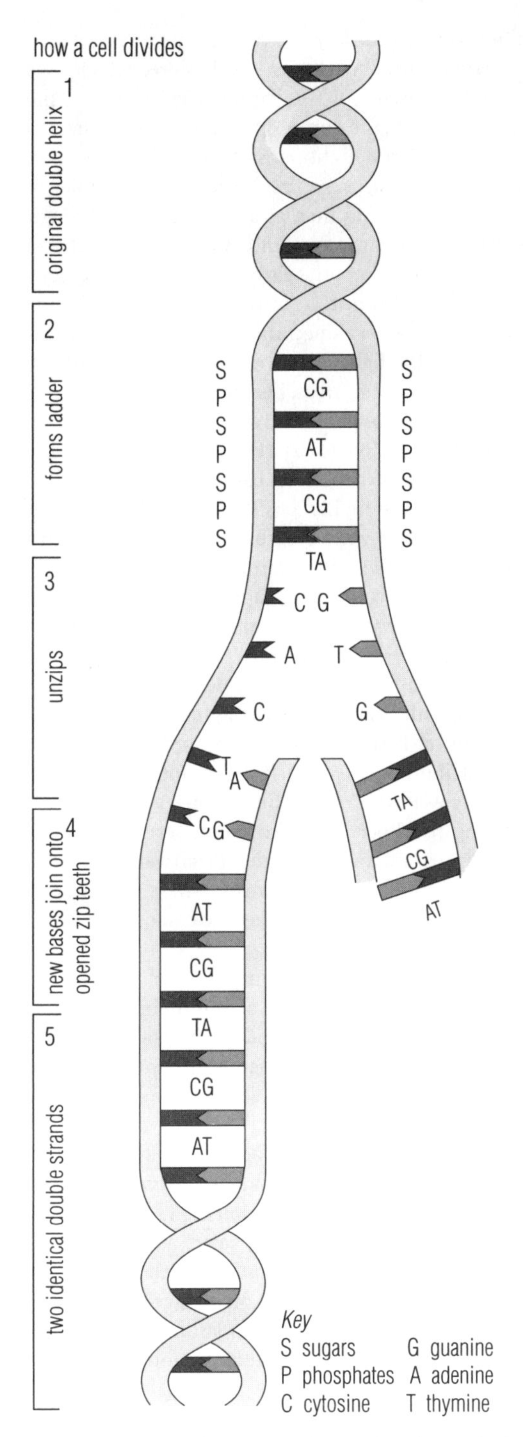

Crick DNA replicates by a process that involves the 'unzipping' of the parent double helix.

could reveal the way genes are passed on by determining the three-dimensional structure of DNA. In a race against other scientists pursuing the same goal, they made models based on research that had been done in various fields. By incorporating the results of Rosalind ◊Franklin's and Maurice Wilkins's X-ray diffraction studies, Edwin ◊Chargaff's discovery that nucleic acids contain only four different organic bases and Alexander ◊Todd's demonstration that nucleic acids contain sugar and phosphate groups, Crick and Watson postulated that DNA consists of a double helix consisting of two parallel chains of alternate sugar and phosphate groups linked by pairs of organic bases. They then built a series of accurate molecular models, eventually making one that incorporated all known features of DNA and which gave the same diffraction pattern as that found by Wilkins. They envisaged replication occurring by a parting of the two strands of the double helix, each organic base thus exposed, linking with a nucleotide (from the free nucleotides within a cell) bearing the complementary base. Thus two complete DNA molecules would eventually be formed by this step-by-step linking of nucleotides, with each of the new DNA molecules comprising one strand from the original DNA and one new strand. Their model also explained how genetic information could be coded – in a sequence of organic bases. Crick and Watson published their work on the proposed structure of DNA in 1953 and their hypothetical model has been confirmed by many other researchers and accepted as correct.

Crick continued to study genetic codes and viruses and also investigated protein synthesis. With Sydney ◊Brenner he demonstrated that each group of three adjacent bases (which he called codons) on a single DNA strand codes for one specific amino acid. He also helped to determine codons that code for each of the twenty main amino acids and he formulated the adaptor hypothesis later taken up by Paul ◊Berg and Robert Holley (1922–). He has been the recipient of numerous awards and honours including the Warren Triennial Prize Lecturer with Watson in 1959, the Lasker Foundation Award with Watson and Wilkins in 1960, the Prix Charles Leopold Meyer of the French Academy of Sciences in 1961, the Award of Merit of the Gairdner Foundation and the Research Corporation Award in 1962. Between 1960 and 1962 he was elected a fellow of the University, London, and of Churchill College, Cambridge, and Foreign Honorary Member of the American Academy of Arts and Sciences. In 1962 he also became director of Cambridge University's Molecular Biology Laboratory and a nonresident fellow of the Salk Institute in San Diego, California.

Crick married Ruth Dodd in 1940 and they had a son, but divorced in 1947. He married Odile Speed in 1949 and they had two daughters. He named their home in Cambridge 'The Golden Helix'. He gained his PhD from Gaius College, Cambridge in 1953 and remained there – apart from brief visiting lectureships in the US – until 1977 when he was appointed Kieckehefer Distinguished Professor at the Salk Institute. At Salk he developed an interest in studying the function of the brain. His book *Of Molecules and Men* (1966) describes the revolution in biochemistry, and *Life Itself: Its Origin and Nature* (1981) suggests that the seed of life on Earth could have come from another planet. His other books include *What Mad Pursuit* (1988) and *The Astonishing Hypothesis* (1994).

Further Reading

Carlisle, E Fred, 'Metaphoric references in science and literature: the examples of Watson and Crick and Roethke', *Centennial Review,* 1985, v 29, pp 281–301.

Crick, Francis, *What Mad Pursuit: A Personal View of Scientific Discovery,* Basic Books, 1988.

Newton, David, *James Watson and Francis Crick: Search for the Double Helix and Beyond,* Makers of Modern Science series, Facts on File, 1992.

Sherrow, Victoria, *Watson and Crick: Decoding the Secrets of DNA,* Partners II series, Blackbirch Press, 1995.

Crompton, Rookes Evelyn Bell (1845–1940)

English engineer who became famous as a pioneer of the dynamo, electric lighting, and road transport. During his long lifetime he also contributed to the development of various standards, both electrical and mechanical, including the British Association standards for small screw threads. In his later years, through experiment and design, he contributed to the design of the military tank.

Crompton was born near Thirsk, Yorkshire, on 31 May 1845. His early days were spent in nearby Ripon. When he was 11 he accompanied his parents to Gibraltar on HMS *Dragon,* which was commanded by his mother's cousin. He then sailed on to the Crimea as a cadet on the same ship, and visited his brother during the siege of Sevastopol. When he returned to school in the autumn of 1856 – still only 12 years old – he was one of the youngest decorated members of the Royal Navy, with the Crimea Medal and the Sevastopol Clasp. In 1858, after two years at Elstree, he entered Harrow School. During his holidays he built first a model steam-driven road locomotive, and then a full-sized one called *BlueBell.* On leaving school in 1864 he went to India as an army officer. During his service there he continued to develop his road vehicles and transferred out of his regiment so that he could concentrate his efforts on the development of steam locomotives for road haulage. The results were successful both in India and at home, but they were in many ways ahead of their time. The poor quality of roads, the cheapness of other forms of transport, and the developing railway system were to eclipse his pioneering work in this area.

On his return from service he became involved with the Stanton Iron Works in Derbyshire, which was owned by a branch of his family. He redesigned the works and brought some dynamos from France to power his electric lighting system. This innovation spread to other firms. At first he acted as a supplier of the lighting sets, and then as a manufacturer through his firm of Crompton and Company Electrical Engineers. These were the early days of electricity generation and Crompton found an area ready for exploitation. He developed and manufactured generating systems for lighting town halls, railway stations, and small residential areas. Direct current electricity of about 400 volts was generated and used with large storage batteries to allow the systems to operate smoothly within a range of demands. Later the supply of electricity became a highly profitable and fast-growing area, particularly with the coming of Joseph Swan's incandescent filament lamp, and Crompton's Kensington Court Electric Light Company Ltd was formed. He believed strongly in his direct-current system with batteries, and there was much competition between him and his younger contemporary Sebastian ◊Ferranti, with his rival alternating-current system. As time went on Ferranti's method proved the better for large distribution networks, but Crompton's machines fared very well within their limitations.

During the Boer War, Crompton served in South Africa as commandant of the Electrical Engineers' Royal Engineers Volunteer Corps, which was later to become the Royal Corps of Electrical and Mechanical Engineers (REME). At the beginning of the 20th century he returned to road transport, and contributed considerably to the principles of automobile engineering and the maintenance and design of roads that would withstand the new demands put upon them. He was a founder member of what is now the Royal Automobile Club (RAC).

His firm, which had manufactured all kinds of electrical machinery and instruments, became, after a merger in 1927, Crompton Parkinson Ltd. He continued as a director. In his later years he was internationally respected as an expert electrical engineer and a foremost authority on storage batteries. He used his influence to achieve standardization in industry generally, and was involved in the founding of the National Physical Laboratory and what is now the British Standards Institute. Heavy industry had the Whitworth Standard for screw threads, but there was no such standard for the small screws used in instruments. Crompton was a prime mover in the formation of the British Association committee that eventually developed the well-known BA Standards.

During World War I he was an adviser on the design and production of military tanks, having been appointed by Winston Churchill (who was then First Lord of the Admiralty). Crompton carried out much research of his own but the final vehicle embodied the work of several designers.

Many honours were awarded to him, including honorary membership and medals of the engineering Institutions. He was elected a Fellow of the Royal Society in 1933. Crompton died at Azerley Chase, Yorkshire, on 15 February 1940, aged 94 years.

Crompton, Samuel (1753–1827) English inventor whose machine for spinning fine cotton yarn – the spinning mule – revolutionized the industrial production of high-quality cotton textiles.

Crompton was born on 3 December 1753 on a small farm at Firwood near Bolton, Lancashire. His parents were poor and from an early age he was expected to work alongside them on their piece of land. When he was six years old his father died, and his mother was forced to rely heavily upon the cottage industry of spinning and weaving to support the family. Samuel had to help with this too, and he grew up familiar with all the associated problems. These experiences were to influence him for the rest of his life.

At the age of 21, with years of home spinning and weaving behind him, he resolved to try to design a better method for spinning the yarn than James Hargreaves's spinning jenny (patented in 1770). The jenny's yarn tended to break frequently and produced a coarse cloth that was commercially suitable only for the working-class market.

It took Crompton five years of hard work and all his money to develop a machine that span a yarn so fine and continuously that it revolutionized the cotton industry. The machine became known as a local wonder, and people flocked to Crompton's house to catch a glimpse of his spinning mule. Unfortunately, because he had used all his savings, he had no means of patenting the design and he lived in fear of someone stealing his idea. Eventually, he was persuaded to reveal his invention to a group of Bolton manufacturers, who promised to pay him a generous subscription in return. This never materialized other than the initial payment of £7 6s 6d (£67.32).

Aggrieved at his treatment by the Bolton manufacturers, he tried to set up his own company, but having 'sold' his new machine, he was soon outpaced by the other manufacturers (with better resources) and forced to sell up. By 1812 the boost given to the cotton industry by the introduction of Crompton's mule was at last recognized, and a sum of approximately £500 was raised by subscription. A national award was also sought, and this time £5,000 was offered, but Crompton was far from satisfied and returned home to Bolton from London, a broken man.

Once again he tried to enter business, this time as a partner in a cotton firm. Ultimately this too failed, and he was forced to turn to the generosity of friends for an annuity. He died on 26 June 1827 at Bolton, a saddened man.

Crompton's new invention, which contributed so much to the wealth of the textile industry, was called the mule because, just like the animal of that name, it was a hybrid. His machine used the best from the spinning jenny and from Richard Arkwright's water frame of 1768. The strong, even yarn it produced was so fine that it could be used to weave delicate fabrics such as muslin, which became particularly fashionable among the middle and upper classes, creating a new market for the British cotton trade. As a direct result of the better and faster method for spinning, which took the job out of the home and into the factories, coupled with the increase of raw cotton available from the USA, the cotton trade entered a golden age. The introduction of the power loom in the early 1800s further mechanized production and made the system of using home weavers practically obsolete. The Industrial Revolution had begun in the textile industry.

Cronin, James Watson (1931–) US physicist; see ⇨Fitch and Cronin.

Crookes, William (1832–1919) English physicist and chemist who, in a scientific career lasting more than 50 years, made many fundamental contributions to both sciences. He is best known in physics for his experiments with high-voltage discharge tubes and for inventing the Crookes radioscope or radiometer. A major achievement in chemistry was the discovery of the element thallium. Crookes never held a senior academic post, and carried out most of his researches in his own laboratory.

Crookes was born in London on 17 June 1832, the eldest of the 16 children of a tailor and businessman. Little is known of his early education but in 1848 he began a chemistry course at the new Royal College of Chemistry under August von Hofmann, who later made him a junior assistant, promoting him to senior assistant in 1851. Crookes's only academic posts were as superintendent of the Meteorological Department at the Radcliffe Observatory, Oxford 1854–55, and as lecturer in chemistry at Chester Training College 1855–56. He left Chester after only one year, dissatisfied at being unable to carry out original research and after having inherited enough money from his father to make him financially independent for life.

He returned to London and became secretary of the London Photographic Society and editor of its *Journal* in 1858. The following year he set up a chemical laboratory at his London home and founded as sole manager and editor the weekly *Chemical News,* which dealt with all aspects of theoretical and industrial chemistry and which he edited until 1906. Among his greatest achievements in chemistry was the discovery in 1861 of the element thallium using the newly developed spectroscope. He also applied this instrument to the study of physics. He was knighted in 1897 and made a Member of the Order of Merit in 1910. He died in London on 4 April 1919.

One of Crookes's earliest pieces of chemical research concerned the preparation and properties of potassium selenocyanate (potassium cyanoselenate (VI)), which he made using selenium isolated from 5 kg/11 lb of waste from a German sulphuric-acid works. Then in 1861, while examining selenium samples by means of a spectroscope – the method newly developed by Robert Bunsen and Gustav Kirchhoff – Crookes observed a transitory green line in the spectrum and attributed it to a new element, which he called thallium (from the Greek *thallos,* meaning a budding shoot). Over the next few years he determined the properties of thallium and its compounds; using thallium nitrate and a specially constructed sensitive balance he measured thallium's atomic mass as 203.715±0.0365 (modern value: 204.39).

This work involved weighing various samples of material in a vacuum, and he observed that when the delicate balance was counterpoised it occasionally made unexpected swings. He began to study the effects of light radiation on objects in a vacuum and in 1875 devised the radioscope (or radiometer). The instrument consists of a four-bladed paddlewheel mounted horizontally on a pinpoint bearing inside an evacuated glass globe. Each vane of the wheel is black on one side (making it a good absorber of heat) and silvered on the other side (making it a good reflector). When the radioscope is put in strong sunlight, the paddlewheel spins round. Although little more than a scientific toy, it defied attempts to explain how it works until James Clerk ◊Maxwell correctly showed that it is a demonstration of the kinetic theory of gases. The few air molecules in the imperfect vacuum in the radioscope bounce more strongly (with more momentum) off the heated, black sides of the vanes (than off the cooler, silvered sides), creating a greater reaction that 'pushes' the paddlewheel around. Crookes own observations were described in his paper 'Attraction and repulsion resulting from radiation' (1874).

During the 1870s Crookes's studies concerned the passage of an electric current through glass 'vacuum' tubes containing rarified gases; such discharge tubes became known as Crookes tubes. The ionized gas in a Crookes tube gives out light – as in a neon sign – and Crookes observed near the cathode a light-free gap in the discharge, now called the Crookes dark space. He named the ion stream 'molecular rays' and demonstrated how they are deflected in a magnetic field and how they can cast shadows, proving that they travel in straight lines. He made similar observations about cathode rays, but it was left to J J ◊Thomson to understand the true significance of such experiments and to discover the electron (in cathode rays) in 1897. Crookes also noted that wrapped and unexposed photographic plates left near his discharge tubes became fogged, but he did not follow up the observation, which was later the basis of Wilhelm ◊Röntgen's discovery of X-rays in 1895.

In the 1880s Crookes studied the phosphorescent spectra of rare-earth minerals, principally substances containing yttrium and samarium. He made the first references to what Frederick Soddy was to call isotopes. While experimenting with radium, he devised the spinthariscope. The instrument consists of a screen coated with zinc sulphide at the end of a tube fitted with a low-powered lens. When alpha particles emitted from a radioactive source (such as radium) hit the screen they produce a small flash of light.

Crookes's interests were very wide. Topics covered by his publications included chemical analysis; the manufacture of sugar from sugarbeet; dyeing and printing of textiles; oxidation of platinum, iridium and rhodium; use of carbolic acid (phenol) as an antiseptic in the treatment of diseases in cattle; the origin and formation of diamonds in South Africa; and the use of artificial fertilizers and their manufacture from atmospheric nitrogen. Crookes was never afraid of pursuing an idea counter to the trend of contemporary opinion. For several years, for example, he was very interested in spiritualism and published several papers that described experiments undertaken by a medium. To many of his scientific colleagues this was akin to heresy.

Croslin, Michael (1933–) US mechanical, electrical, and biomedical engineer who is best known for his invention of the Medtek 400, a computerized blood-pressure-measuring device.

Michael Croslin was born in Frederikstad, St Croix, in the US Virgin Islands. His parents abandoned him as a baby and he was brought up by a local family and given the name Miguel Britto. He ran away to the USA in 1945 at the age of 12, where he first worked in Georgia doing odd jobs. He also managed to get himself a place in a Jesuit school. He then moved to Wisconsin and was adopted by the Croslin family who gave him their name and enabled him to finish high school at the age of 14. Three years later he graduated from the University of Wisconsin with a BSc.

After his graduation in 1950, Croslin joined the US Air Force, serving in Korea before returning to the USA in 1955 to take another BSc degree at the University of New York in mechanical engineering in 1958, followed in 1963 by a master's degree in electrical engineering and a doctorate in biomedical engineering in 1968. Simultaneously with his PhD he gained a master's degree in business administration from Columbia University.

In 1978, Croslin founded a biomedical company, the Medtek Corporation, and produced the Medtek 400, a blood pressure measuring device. This differed from the traditional apparatus that measured the sound of pumping blood and could be unreliable, by measuring blood motion and giving a reading in seconds on a digitized screen.

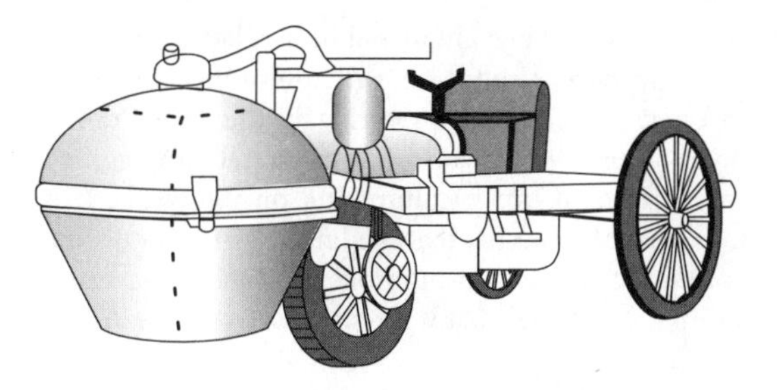

Cugnot Cugnot's steam carriage or tractor, invented 1769, was the first self-propelled road vehicle. It was capable of carrying four passengers at walking pace.

Cugnot, (Nicolas-) Joseph (1725–1804) French soldier and engineer who was the first to make a self-propelled road vehicle (in 1769). His three-wheeled machine was designed for towing guns: the front wheel was driven by a two-cylinder steam engine, and it could carry four people at a walking pace. As such it was the first automobile.

Cugnot was born in Void, Meuse, on 25 September 1725. As a young man he joined the French army and served for a time in Germany and Belgium, inventing a new kind of rifle used by French troops under Marshal du Saxe, who encouraged him to work on a steam-propelled gun-carriage. After serving in the Seven Years' War. Cugnot returned to Paris in 1763 as a military instructor. He also devoted his time to writing military treatises and exploring a number of inventions he had conceived during his campaigning, obtaining official help from the Duc de Choiseul, then minister of war. His major invention was a steam-propelled tractor for hauling artillery. He built two models, the first appearing in 1769 and the second in 1770. By that time he had also constructed a truck, now preserved in the Conservatoire des Arts et Métiers, Paris. The truck ran at a speed of 3–5 kph/2–3 mph before a large crowd of official spectators, and Cugnot was commissioned to make a larger one. The fall of the Duc de Choiseul led to the project being shelved and the truck was not tested further. Granted a pension in 1779 from the Ministry of War. Cugnot migrated to Brussels. The pension stopped at the outbreak of the French Revolution. In 1798 Napoleon asked the Institut de France to enquire into Cugnot's machine, but nothing came of this.

In 1698 the English inventor Thomas ◊Savery had invented a steam engine designed for pumping water out of mines. This was the first practical use of the steam engine, but the engine itself was inefficient and wasted a lot of steam. In 1690 the French physicist Denis ◊Papin designed a superior engine, employing high-pressure steam expansively without condensation. Cugnot improved on this, making an engine driven by the movements of two piston rods working from two cylinders (which, like the boiler, were made of copper); it carried no reserves of water or fuel. Although it proved the viability of steam-powered traction, the problems of water supply and pressure maintenance severely handicapped the vehicle.

Cugnot's vehicle was a huge, heavy tricycle and his model of 1769 was said to have run for 20 minutes while carrying four people, and to have recuperated sufficient steam pressure to move again after standing for 20 minutes. Cugnot was an artillery officer and the more-or-less steamtight pistons of his engine were made possible by the invention of a drill that accurately machined cannon bores. The first post-Cugnot steam carriage appears to have been that built in Amiens in 1790, although followers of Cugnot were soon on the road in other countries, notably in the UK. Steam buses were running in Paris about 1800. Oliver Evans of Philadelphia ran an amphibious steam dredger through the streets of that city in 1805. English exponents of the new form of propulsion became active, and by the 1830s the manufacture and use of steam road carriages had approached the status of an industry. James Watt's foreman William Murdock ran a model steam carriage on the roads of Cornwall in 1784, and Robert Fourness showed a working three-cylinder tractor in 1788. Richard Trevithick developed Murdock's ideas and at least one of his carriages with driving wheels 4.8 m/10 ft in diameter ran in London.

Between 27 February and 22 June 1831, steam coaches ran about 6,500 km/4,000 mi on the regular Gloucester–Cheltenham service, carrying some 3,000 passengers. Thus many passengers had been carried by steam carriage before the railways that were to cause their demise had accepted their first paying passenger. The decline did not mean that all effort in this field would be abandoned, and much attention was given to the steam tractor as a prime mover. Beginning in about 1868, the UK was the scene of a vogue for light, steam-powered personal carriages; if the popularity of these vehicles had not been hindered by legislation, it would certainly have resulted in the appearance of widespread enthusiasm for motoring in the 1860s rather than in the 1890s. Steam tractors, or traction engines, were also used in agriculture and on the roads. It is thus possible to argue that the line from Cugnot's first lumbering vehicle runs unbroken to the 20th-century steam automobiles that were made as late as 1926.

Curie, Marie (1867–1934), born Manya Sklodowska and Pierre (1859–1906) French scientists, husband and wife, who were early investigators of radioactivity. The Curies discovered the radioactive elements polonium and radium, for which achievement they shared the 1903 Nobel Prize for Physics with Henri ◊Becquerel. Marie Curie went on to study the chemistry and medical applications of radium, and was awarded the 1911 Nobel Prize for Chemistry in recognition of her work in isolating the pure metal.

Curie Polish scientist Marie Curie who, together with her husband, Pierre, discovered the nature of radioactivity, and isolated radium and poloium. She died of leukaemia in 1934 as a result of working with radioactive material for many years without any protection. Her laboratory notebooks are still radioactive today. *Mary Evans Picture Library*

Manya Sklodowska was born in Warsaw on 7 November 1867, at a time when Poland was under Russian domination after the unsuccessful revolt of 1863. Her parents were teachers and soon after she was born – their fifth child – they lost their teaching posts and had to take in boarders. Their young daughter worked long hours helping with the meals, but nevertheless won a medal for excellence at the local high school, where the examinations were held in Russian. No higher education was available so she took a job as a governess, sending part of her savings to Paris to help to pay for her elder sister's medical studies. Her sister qualified and married a fellow doctor in 1891 and Marie went to join them in Paris. She entered the Sorbonne and studied physics and mathematics, graduating top of her class. In 1894 she met the French chemist Pierre Curie and they were married the following year.

Pierre Curie was born in Paris on 15 May 1859, the son of a doctor. He was educated privately and at the Sorbonne, becoming an assistant there in 1878. He discovered the piezoelectric effect and, after being appointed head of the laboratory of the Ecole de Physique et Chimie, went on to study magnetism and formulate Curie's law (which states that magnetic susceptibility is inversely proportional to absolute temperature). In 1895 he discovered the Curie point, the critical temperature at which a paramagnetic substance become ferromagnetic. In the same year he married Manya Sklodowska.

From 1896 the Curies worked together on radioactivity, building on the results of Wilhelm ◊Röntgen (who had discovered X-rays) and Henri Becquerel (who had discovered that similar rays are emitted by uranium salts). Marie Curie discovered that thorium also emits radiation and found that the mineral pitchblende was even more radioactive than could be accounted for by any uranium and thorium content. The Curies then carried out an exhaustive search and in July 1898 announced the discovery of polonium, followed in December of that year with the discovery of radium. They eventually prepared 1 g/0.04 oz of pure radium chloride – from 8 tonnes of waste pitchblende from Austria. They also established that beta rays (now known to consist of electrons) are negatively charged particles.

On 19 April 1906, Pierre Curie was run down and killed by a horse-drawn carriage. Marie took over his post at the Sorbonne, becoming the first woman to teach there, and concentrated all her energies into research and caring for her daughters (one of whom, Irène, was to become a famous scientist and Nobel prizewinner; see Irène and Frédéric ◊Joliot-Curie). In 1910 with André Debierne (1874–1949), who in 1899 had discovered actinium in pitchblende, she isolated pure radium metal.

At the outbreak of World War I in 1914 Marie Curie helped to equip ambulances with X-ray equipment, which she drove to the front lines. The International Red Cross made her head of its Radiological Service. Assisted by Irène Curie and Martha Klein at the

Radium Institute she held courses for medical orderlies and doctors, teaching them how to use the new technique. By the late 1920s her health began to deteriorate: continued exposure to high-energy radiation had given her leukaemia. She entered a sanatorium at Haute Savoie and died there on 4 July 1934, a few months after her daughter and son-in-law, the Joliot-Curies, had announced the discovery of artifical radioactivity.

> *It would be impossible, it would go against the scientific spirit .... Physicists should always publish their researches completely. If our discovery has a commercial future that is a circumstance from which we should not profit. If radium is to be used in the treatment of disease, it is impossible for us to take advantage of that.*
>
> MARIE CURIE ON THE PATENTING OF RADIUM. DISCUSSION WITH HER HUSBAND, PIERRE, QUOTED IN EVE CURIE THE DISCOVERY OF RADIUM IN MARIE CURIE TRANSL V SHEEAN 1938

Throughout much of her life Marie Curie was poor and the painstaking radium extractions were carried out in primitive conditions. The Curies refused to patent any of their discoveries, wanting them freely to benefit everyone. The Nobel prize money and other financial rewards were used to finance further research. One of the outstanding applications of their work has been the use of radiation to treat cancer, one form of which cost Marie Curie her life.

Further Reading

Curie, Eve and Sheean, Vincent (ed), *Madame Curie: A Biography,* Doubleday, 1939.

Gribbin, John, *Curie in 90 Minutes,* Scientists in 90 Minutes series, Constable and Co, 1997.

Quinn, Susan, *Marie Curie,* Addison Wesley Longman, 1996.

Curl, Robert Floyd, Jnr (1933–) US chemist who with colleagues Richard ◊Smalley and Harold ◊Kroto discovered buckminsterfullerene, a form of carbon (carbon 60) in 1985. They shared the Nobel Prize for Chemistry in 1996 for their discovery.

Born in Alice, Texas on 23 August 1933, Robert Curl was the son of a methodist minister. The family lived in various southern Texan towns during his childhood. It was the gift of a chemistry set for Christmas when he was nine that sparked off his interest in chemistry. He received his BA from Rice Institute in 1954. Having read about Kenneth Pitzer's discovery of the barriers to internal rotation about single bonds, he decided to go to the University of California at Berkeley to work with him. After receiving his PhD from Berkeley in 1957 he moved on to Harvard and worked with US zoologist (1856–1939) E B Wilson in microscope spectroscopy for a year, after which he returned to Rice University as an assistant professor. He became full professor in 1967 and was chair of the chemistry department 1992–96.

Curl worked with Smalley and Kroto at Houston, Texas, on the discovery of buckminsterfullerene in 1985. They found a mass-spectrum signal for a molecule of exactly 60 carbon atoms (C60) in a perfect sphere in which 12 pentagons and 20 hexagons are arranged like the panels on a modern football hence the popular name buckyball. The structure also resembles architect Buckminster Fuller's geodesic dome, so Kroto came up with the name buckminsterfullerene, subsequently shortened to fullerene.

Since the discovery, other fullerenes have been identified with 28, 32, 50, 70, and 76 carbon atoms. New molecules have been made based on the buckyball enclosing a metal atom and buckytubes have been made consisting of cylinders of carbon atoms arranged in hexagons. In 1998 these were proved, in laboratories in Israel and the US to be 200 times tougher than any other known fibre. Possible uses for these new molecules are as lubricants, superconductors and as a starting point for new drugs. Research in C60 has continued at Sussex University, England, focusing on the implications of the discovery for several areas of fundamental chemistry including the way it has revolutionized perspectives on carbon based materials. The research is interdisciplinary involving Curl, Kroto, and Smalley among others, and covers the basic chemistry of fullerenes, fundamental studies of carbon and metal clusters, carbon microparticles and nanotubes, and the study of interstellar and circumstellar molecules and dust. Curl's own research involves the study of the molecules and reactions involved in combustion processes and will enable environmental emissions from vehicles, forest fires, chemical plants, and other sources to be monitored.

Curl is a member of the US Fraternities Phi Beta Kappa, Phi Lamda Upsilon, and Sigma Xi. He received the Clayton Prize of the Institute of Mechanical Engineers in 1957 with Pitzer, and with Smalley and Kroto he received the American Physical Society International Prize for New Materials in 1992.

Curtis, Heber Doust (1872–1942) US astronomer who became interested in astronomy only after he had begun a career in classics, but who went on to carry out important research into the nature of spiral nebulae.

Curtis was born in Muskegan, Michigan, on 27 June 1872. He attended a Detroit high school, where he displayed a flair for languages and an interest in the classics. He gained his BA at the University of Michigan in 1892 and his MA a year later. His first job was teaching Latin at his old high school, and then, aged only 22, he became professor of Latin at Napa College, California, where he shortly became aware of the availability of a refracting telescope and small observatory.

In 1897 Curtis abruptly changed the entire direction of his career and became professor of mathematics and astronomy at the University of the Pacific. He worked at the Lick Observatory, California, in 1898; in 1900 he became Vanderbilt Fellow at the Leander McCormick Observatory at the University of Virginia; and in 1902 he returned to become an assistant at the Lick Observatory. There he was promoted to assistant astronomer in 1904. In 1906, under the auspices of that observatory, he took charge of work being done at an observatory in Chile. Three years later, he returned to the Lick, where he worked as an astronomer until he retired from research work in 1920. His next appointment was as director of the Allegheny Observatory, a post he held until 1930, when he became director of the observatory at the University of Michigan. He died in Ann Arbor, Michigan, on 8 January 1942.

Curtis's first astronomical studies were of total solar eclipses in Thomaston, Georgia, in 1900, and in Solok, Sumatra, in 1901. But the main value of Curtis's early work at the Lick Observatory lay in his contributions to the programme for the measurement of stellar radial velocities, undertaken under the direction of William Campbell. He worked on this programme at Mount Hamilton 1902–06, and then in Chile 1906–09. For the following 11 years Curtis concentrated his efforts on the photography of spiral nebulae and on research into their nature.

Ever since Charles Messier had included 'nebulosities' in his catalogue of 1771, their precise composition had been the subject of dispute. There were two main schools of thought: that they were either giant star clusters far beyond our own Galaxy – as proposed by Richard Proctor (1837–1888) – or that they were merely clouds of debris. The scale of the universe itself was central to both points of view. Through his photography of spiral nebulae, Curtis began to appreciate the actual vastnesses of space and to incline towards Proctor's view of 'islands in the universe'.

He also noticed that on photographs of spiral nebulae viewed edge-on there was a dark line along the rim of each nebula. This suggested to Curtis a combination of the two theories: that spiral nebulae might indeed be complex galaxies like our own, and that such galaxies produced a cloud of debris that accumulated in the plane of the galaxy. If such a cloud of debris had also gathered outside our own Galaxy, this would explain the reported 'zone of avoidance' – spiral nebulae never appeared in the Milky Way (that is, in the plane of our own Galaxy). Spiral nebulae in that position, it now was evident, would simply be obscured by dust.

Following his appointment as director of the Allegheny Observatory in 1920, Curtis's research output declined, and his most important contribution to astronomy was in improving the modern understanding of the nature and position of spiral nebulae.

Cushing, Harvey Williams (1869–1939) US surgeon who pioneered several important neurosurgical techniques, made famous studies of the pituitary gland, and first described the chronic wasting disease now known as Cushing's syndrome (or disease).

Cushing was born on 8 April 1869 in Cleveland, Ohio, the fourth child in a family of physicians. He studied medicine at Yale and at the Harvard Medical School, graduating from the latter in 1895. He then spent about four years in practical training at the Massachusetts General Hospital, Boston, and Johns Hopkins Hospital, Baltimore, where he worked under William Halsted (1852–1922), a great innovator of surgical techniques. At about the turn of the century Cushing studied in Europe – under Emil Kocher (1841–1917) at Berne University and, briefly, under the famous neurophysiologist Charles Sherrington in England – after which he returned to the department of surgery at the Johns Hopkins University. From 1912–32 Cushing was professor of surgery at the Harvard Medical School and surgeon-in-chief at the Peter Bent Brigham Hospital in Boston. During this period he served in the Army Medical Corps in World War I and in 1918 was appointed senior consultant in neurological surgery to the American Expeditionary Force. In 1933 he became Sterling Professor of Neurology at Yale University, a post he held until his retirement in 1937. Cushing died on 7 October 1939 in New Haven, Connecticut. He bequeathed his large collection of books on the history of medicine and science to the Yale Medical Library.

Although Cushing is probably best known for his work on Cushing's syndrome, his major contribution was in the field of neurosurgery, which, until he introduced his pioneering techniques, was seldom successful. As a result of experimenting on the effect of artificially increasing intercranial pressure in animals, Cushing developed new methods of controlling blood pressure and bleeding during surgery. Moreover, his whole approach to medicine was characterized by painstaking carefulness: before operating he gave each of his patients an extremely thorough physical examination and took a detailed medical history. The operations themselves, which usually lasted for many hours, were performed with meticulous care and, over the years, were increasingly successful. In addition to developing neurosurgical techniques, Cushing wrote a description – still valid today – of the stages in the development of different types of intercranial tumours, classified such tumours, and published (in 1917) a definitive account of acoustic nerve tumours.

In 1908 Cushing began studying the pituitary gland and, after experimenting on animals, discovered a way of gaining access to this gland, which, being situated at the base of the brain and behind the nasal sinuses, is extremely difficult to approach surgically. As a result of this discovery it became possible to treat cases of blindness caused by tumours pressing on the optic nerve in the region of the pituitary gland. Cushing also investigated the effects of abnormal activity of the pituitary gland, establishing that hypopituitarism (undersecretion of pituitary hormones) in a growing person can cause a type of dwarfism and that hyperpituitarism (oversecretion of pituitary hormones) in adults can cause acromegaly (a form of gigantism characterized by excessive growth of the bones of the hands, feet, and face). As a result of his extensive studies of the pituitary gland, Cushing discovered the condition now called Cushing's syndrome, a rare chronic wasting disease with symptoms that include obesity of the face and trunk, combined with thin arms and legs; wasting of the muscles; atrophy of the skin, with the appearance of red lines on the skin; weakness; and accumulation of body fluids. Cushing attributed this disorder to a tumour of the basophilic cells of the anterior pituitary gland, but although this is one of the causes, the disorder is now known to be caused by any of several conditions that increase the secretion of glucocorticoids (particularly cortisol) by the adrenal glands, such as a tumour of the adrenal cortex itself.

Cushing was also interested in the history of medicine and in 1925 wrote a biography of William Osler (1849–1920) – one of the leading physicians of the time – which won him a Pulitzer Prize.

Further Reading

Aans (ed), *Bibliography of the Writings of Harvey Cushing*, American Association of Neurological Surgeons, 1993.

Aron, David C, 'The Path to the Soul: Harvey Cushing and surgery on the pituitary and its environs' in 1916, *Perspect Biol Med*, 1994, v 37, pp 551–565.

Arondess, Jeremiah A, 'Cushing and Osler: The evolution of a friendship', *Trans Stud Coll Physicians Phila*, 1985, v 7, pp 79–112.

Black, Peter M and Moore, M R (eds), *Harvey Cushing at the Brigham*, American Association of Neurological Surgeons, 1993.

Cuvier, Georges (1769–1832) French zoologist eminent for his role in the founding of modern palaeontology.

Cuvier was born at Montebéliard in the principality of Württemburg, on 23 August 1769. Cuvier, the son of a Swiss soldier, received his training in natural history at Stuttgart, before spending six years as a private tutor in Normandy. He came to Paris in 1795 as assistant to the professor of comparative anatomy at the Natural History Museum. In 1799, he was appointed professor of natural history at the Collège de France, and in 1802 professor at the Jardin des Plantes. Cuvier proved a key figure in trailblazing new classificationary approaches to natural history, including a total rejection of the old idea of the 'great chain of being'). But his supreme contribution lay in systematizing the laws of comparative anatomy and applying them to fossil vertebrates. Thanks to his work in this field, Cuvier was perhaps the most influential palaeontologist of the 19th century.

Cuvier's pursuits ranged widely through the animal kingdom. He divided invertebrates into three phyla, and conducted notable investigations into fish and molluscs. In two major texts, *Researches on the Fossil Bones of Quadrupeds* (1812) and the *Animal Kingdom* (1817), he reconstructed such extinct fossil quadrupeds as the mastodon and the paleotherium, applying the principles and practices of comparative anatomy. Undertaken in collaboration with Brongniart, his stratigraphical explorations of the tertiary rocks of the Paris Basin demonstrated that fossil flora and fauna were specific to particular strata (a parallel discovery to that of William ◊Smith in England). On the basis of these major conclusions in historical geology and vertebrate palaeontology, Cuvier judged that the history of the Earth had involved a chain of revolutions ('catastrophes') that had recurrently swept away entire living populations, their place being taken either by migration or by the creation of new species (in a manner that Cuvier tactfully chose never to specify). This theory, set out in his *Preliminary Discourse* (1812), expressly countered the evolutionary views of Jean Baptiste de ◊Lamarck and the palaeontologist Geoffrey Saint-Hilaire (1772–1844).

In later life, Cuvier was much concerned with scientific organization and education. He was councillor of state under Napoleon and later under Louis Philippe. In 1831 he was raised to the peerage of France, a rare honour for a Protestant. He died in Paris on 13 May 1832, aged 62.

Further Reading

Coleman, William, *Georges Cuvier, Zoologist: A Study in the History of Evolution Theory*, Harvard University Press, 1964.

Outram, Dorinda, *Georges Cuvier: Vocation, Science, and Authority in Post-Revolutionary France*, Manchester University Press, 1984.

Rudwick, M J S, *Georges Cuvier: Fossil Bones, and Geological Catastrophes: New Translations and Interpretations of the Primary Texts*, University of Chicago Press, 1997.

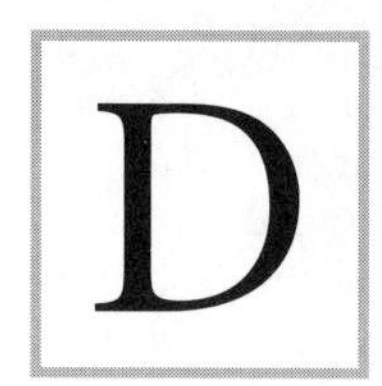

Daimler, Gottlieb Willhelm (1834–1900) German engineer who designed internal-combustion engines of relatively advanced performance for cars, developing the motor car possibly more than Carl Benz, (who had run a car at an earlier date).

Born at Schorndorf near Stuttgart on 17 March 1834, Daimler's technical education began in 1848 when he became a gunsmith's apprentice. Following a period at technical school in Stuttgart and factory experience in a Strasbourg engineering works, he completed his formal training as a mechanical engineer at the Stuttgart Polytechnic in 1859. He returned to Grafenstadt to do practical work for a while and then, sponsored by a leading Stuttgart benefactor, travelled to England where he worked for Joseph Whitworth. He then moved to France, where he may have seen Lenoir's newly developed gas engine.

Daimler spent the next ten years in heavy engineering. He joined Bruderhaus Maschinen-Fabrik in Reutlingen as manager in 1863, and there met Wilhelm ◊Maybach, with whom he was to be closely involved for the rest of his life.

Daimler's work on the internal-combusition engine began in earnest in 1872 when he teamed up with Nikolaus August ◊Otto, (later to become famous for the Otto cycle) and Peter Langen, at the Gasmotoren-Fabrik Deutz where Daimler was technical director. One of Daimler's first moves was to sign up Maybach as chief designer. Daimler was to work with Otto and Langen for the next ten years, studying gas engines (which resulted in Otto's historic patent of 1876) and perhaps also petrol engines.

Differences of opinion led to Daimler leaving the firm in 1881. After making a brief trip to Russia to study oil, he returned to Germany and bought a house in Cannstatt, a suburb of Stuttgart. It was in the summer house of this building that Daimler's first engines were built.

When he started work with Maybach, gas engines were being operated at 150–250 rpm. Daimler's first working petrol-fuelled unit, built in 1883, was an air-cooled, single-cylinder engine with a large cast-iron flywheel running at 900 rpm. With four times the number of power strokes per unit time, his engine had a very much greater output for a given size and weight. In itself, the use of petrol was not new: in 1870 Julius Hock in Vienna had built an engine working on Lenoir's principle. A piston drew in half-a-cylinder-full of mixture, which was fired when the piston was halfway down the cylinder. Without compression, the power produced was very low and the fuel consumption massive.

The genius of Daimler and Maybach lay in combining four of the elements essential to the modern car engine: the four-stroke Otto cycle, the vaporization of the fuel with a device similar to a carburettor, low weight, and high speeds. Lenoir had used electric ignition, but this proved unreliable; Daimler and Maybach used an igniter tube that was light, worked well, and operated independently of engine speed.

Daimler's second engine, which ran later the same year, was a 0.4-kW/0.5-hp vertical unit. It was fitted to a cycle in November 1885, (possibly even earlier) creating the world's first motor cycle. Daimler was apparently not impressed with the possibilities of motorized two-wheelers and went on to try his engine as the power source for a boat.

In 1889 Daimler produced two cars, and obtained a licence for Panhard and Levassor in Paris to sell them. The first was a light four-wheeler with a tubular frame and a vertical, single-cylinder, water-cooled engine in the rear. It also featured a novel four-speed gear transmission to the rear wheels, and engine-cooling water circulated through the frame, which acted as a radiator. The second car of 1889 had a belt drive and a vee-twin engine.

The two cars are important in that they show that Daimler had revised his earlier opinion that motor cars should be straight conversions of horsedrawn carriages. (Benz, on the other hand, had conceived his vehicle for motor-drive from the outset; nevertheless, Daimler's models were in many ways more advanced than the contemporary Benz models).

In 1886, Daimler approached Sarazin, a representative of Otto and Langen at the Deutz works, eager to increase sales of his engines overseas. Sarazin persuaded Panhard and Levassor to manufacture Daimler engines under licence, but died before they went into production. The firm succeeded in entering the motor industry with the Daimler licence, following the marriage of Levassor to Sarazin's widow in 1890.

The Daimler Motoren Gesellschaft was also founded in 1890, but Daimler and Maybach both retired the following year to concentrate on technical and commercial development work, only to rejoin in 1895.

A Daimler-powered car won the 1894 Paris to Rouen race, the first international motor contest, organized to promote the concept of motoring. Six years after this great success, on 6 March 1900, Daimler died from heart disease, and was buried in Cannstatt.

Dale, Henry Hallett (1875–1968) English physiologist, best known for his work on the chemical transmission of nerve impulses (particularly for isolating acetylcholine), for which he was awarded – jointly with Otto Loewi, a German pharmacologist – the 1936 Nobel Prize for Physiology or Medicine.

Dale was born on 9 June 1875 in London, the son of a businessman, and was educated at Tollington Park College, London, then at Leys School, Cambridge. He read natural sciences at Trinity College, Cambridge, graduating in 1898, then succeeded Ernest Rutherford in the Coutts–Trotter Studentship at Trinity College. In 1900 Dale began his clinical training at St Bartholomew's Hospital, London, gaining a bachelor of surgery degree in 1903 and a medical degree in 1907. While undergoing his clinical training, Dale continued his physiological studies in London 1902–04 under Ernest Starling and William Bayliss, first as a George Henry Lewes Student and later as a Sharpey Student in the department of physiology at University College, London. He also studied under Paul Ehrlich in Frankfurt for several months. In 1904 Dale accepted a post at the Wellcome Physiological Research Laboratories, becoming director there two years later. In 1914 he was appointed head of the department of biochemistry and pharmacology of the Medical Research Council, and from 1928 until his retirement in 1942 he was director of the National Institute for Medical Research. He died in Cambridge on 23 July 1968.

Dale received numerous British honours, including the Copley Medal of the Royal Society in 1937, a knighthood in 1943, the Order of Merit in 1944 and, at various times during his career, the presidencies of the Royal Society, the British Association, the Royal Society of Medicine, and the British Council. Furthermore, in 1959 the Society of Endocrinology struck the Dale Medal, an annual award, and in 1961 the Royal Society established the Henry Dale Professorship, bestowed by the Wellcome Trust, of which Dale had been the chairman 1938–60.

Dale's earliest research, performed while he worked at the Wellcome Physiological Research Laboratories, concerned the chemical composition and effects of ergot (a fungus that infects cereals and other grasses). In 1910, working with G Barger, he identified a substance in ergot extracts that produced dramatic effects, such as dilation of the arteries. Histamine, as this substance is now called, is found in all plant and animal cells and its release is associated with many of the symptoms of allergy.

In 1914 Dale isolated acetylcholine from biological material. Between 1921 and 1926 Otto Loewi and his co-workers showed that stimulation of the parasympathetic nerves in a perfused frog's heart (a heart that has an artificial passage of fluids through its blood vessels) resulted in the appearance of a substance that inhibited the action of a second heart that was receiving the perfused fluid from the first heart. This substance was later shown by Dale and Loewi to be acetylcholine, which is produced at the nerve endings of parasympathetic nerves. This finding provided the first definite proof that chemical substances are involved in the transmission of nerve impulses.

In addition to his research, Dale became concerned in his later years with the social effects of scientific developments. With Thorvald Madson of Copenhagen he was largely responsible for the adoption of an international scheme to standardize drugs and antitoxins. He was also concerned with preserving the apolitical nature of science and with the peaceful use of nuclear energy.

d'Alembert, Jean le Rond (1717–1783) French mathematician and theoretical physicist who was a great innovator in the field of applied mathematics,

d'Alembert Engraving of French mathematician and theoretical physicist Jean le Rond d'Alembert, who made major contributions to the field of mechanics. D'Alembert determined the general solution of the wave equation in 1747 by studying the properties of vibrating strings. His first name was taken from the church of St Jean le Rond in Paris, where he was abandoned as a baby. *Mary Evans Picture Library* discovering and inventing several theorems and principles – notably d'Alembert's principle – in dynamics and celestial mechanics. He devised the theory of partial differential equations and contributed many of the scientific articles that went into the first editions of Denis Diderot's (1713–1784) *Encyclopédie.*

D'Alembert was a foundling, discovered on the doorstep of a Paris church on 16 November 1717. Evidently the illegitimate son of a courtesan, d'Alembert nevertheless grew up well provided for, his accommodation and education financed by the chevalier Destouches (who is therefore generally supposed to have been his father). Following schooling under the Jansenists at Mazarin College, he studied law and was called to the Bar in 1738. However, he then spent a year engrossing himself in medical studies before deciding to devote the rest of his life to mathematics. This he did very successfully, distinguishing himself greatly and becoming personally acquainted with many famous scientists and literary figures of the time; in 1741 he was admitted as a member of the Academy of Sciences. He died in Paris on 29 October 1783.

D'Alembert's first mathematical work published was a paper on integral calculus (1739). The subject continued to fascinate him – in the next nine years he wrote two further papers, published in the *Mémoires* of the Academy of Berlin, that were fundamental to the development of calculus – and eventually led him to the discovery of the calculus of partial differences. Thereafter he applied his calculus to as many mathematical problems as he encountered.

Nevertheless, it is in the field of dynamics that d'Alembert remains best remembered. The principle that now bears his name was first published in 1743 in his *Traité de dynamique,* and was an extension of the third law of motion formulated by Isaac ◊Newton more than 50 years earlier: that for every force exerted on a static body there is an equal, opposite force from that body. D'Alembert maintained that the law was valid not merely for a static body, but also for mobile bodies. Within a year he had found a means of applying the principle to the theory of equilibrium and the motion of fluids; previously, such problems had always been solved by means of geometrical calculations. Within a further three years, and by then using also the theory of partial differential equations, he carried out important studies on the properties of sound, of air compression, and had also managed to relate his principle to an investigation of the motion of any body in a given figure.

It was natural for him, therefore, to turn his attention to astronomy. From the early 1750s, together with other mathematicians such as Leonhard Euler, Alexis-Claude Clairaut (1713–1765), Joseph Lagrange, and Pierre Laplace, he applied calculus to celestial mechanics. The problem they set themselves was to determine the motion of three mutually gravitating celestial bodies; in solving it they brought Newton's celestial mechanics to a high degree of sophistication, capable of explaining in detail all the peculiarities of celestial movements shown by contemporarily increasing accuracy of measurements. In particular, d'Alembert worked out in 1754 the theory needed to set Newton's discovery of the precession of the equinoxes on a sound mathematical basis. He determined the value of the precession and explained the phenomenon of the oscillation of the Earth's axis. At about the same time he also wrote an influential paper in which he gave accurate calculations of the perturbations in the orbits of the known planets.

It was also at that time that d'Alembert was persuaded by his friend Denis Diderot to contribute to his *Encyclopédie* – a work that was to contain a synthesis of all knowledge, particularly of new ideas and scientific discoveries. D'Alembert duly wrote on scientific topics, linking especially various branches of science. After a few years, however (when at least one volume had already appeared), the church in France denounced the project, and d'Alembert resigned his editorship. It may have been this, all the same, that spurred him into publishing no fewer than eight volumes of his mathematical investigations over the next 20 years.

Towards the end of his life, d'Alembert's friend Johann Lambert (1728–1777) announced that he had discovered a moon circling the planet Venus, and proposed that it should be named d'Alembert; he declined the honour very diplomatically – but whether because of (entirely justified) suspicions about the existence of such a satellite, or for other reasons, there is no means of knowing.

Dalton, John (1766–1844) English chemist, one of the founders of atomic theory. Some of his proposals have since proved to be incorrect, but his chief contribution was that he channelled the thinking of contemporary scientists along the correct lines, particularly in his method of using established facts to explain a new phenomenon.

Dalton was born in the village of Eaglesfield near Cockermouth, Cumbria, on or about 6 September 1766. He was the third of six children of a weaver, who was a devout Quaker and did not register the date of his son's birth. Dalton attended the village Quaker school and by the age of 12 was running it. He later became headmaster of a school in Kendal, before taking up a post in 1793 to teach mathematics and natural philosophy in Manchester. Dalton was largely self-taught, his Quaker beliefs excluding him from attending Oxford or Cambridge universities (at that time open only to members of the Church of England).

Dalton English chemist John Dalton who proposed the atomic theory of chemical reactions. Among his other scientific work he studied colour-blindness (he was colour blind himself), and kept a daily record of the weather. He was widely acclaimed for his work during his lifetime, and at his death tens of thousands of people filed past his coffin. *Mary Evans Picture Library*

Even before he moved to Manchester, a wealthy Quaker friend, the blind philosopher John Gough, had stimulated in Dalton an interest in meteorology and for 57 years (beginning in 1787) he kept a diary of observations about the weather. He gave lectures on this subject to the Manchester Literary and Philosophical Society, of which he became honorary secretary and later president. He determined that the density of water varies with temperature, reaching a maximum at 6.1 °C/42.5 °F; the modern value of this temperature is 4 °C). He also lectured about colour blindness, a condition he shared with his brother and which for a time was known as Daltonism. He resigned his lectureship in Manchester in 1799 in order to pursue his own researches, working as a private tutor to make a living. He did, however, remain as the society's secretary and was given accommodation in a house they bought for him. This house, still containing many of Dalton's records, was destroyed in a bombing raid in 1940 during World War II. He was awarded a government pension of £150 in 1833, which was doubled three years later. He died in Manchester on 27 July 1844.

From his interest in the weather, atmosphere, and gases in general, Dalton in 1803 proposed his law of partial pressures (which states that, in a mixture of gases, the total pressure is the sum of the pressures that each component would exert if it alone occupied the same volume). He also studied the variation of a gas's volume with temperature, concluding (independently of Joseph ⇨Gay-Lussac) that all gases have the same coefficient of thermal expansion. Gaseous diffusion and the solubility of gases in water were also the subjects of his experiments.

The work on the absorption of gases led Dalton to formulate his atomic theory – he considered that gases must be made up of particles that can somehow occupy spaces between the particles that make up water, and that in a mixture of gases the different particles must intermingle rather than separate into layers depending on their density. When presented in his book *New System of Chemical Philosophy* (1808), the idea that atoms of different elements have different weights was supported by a list of atomic weights (relative atomic masses) and his newly devised system of chemical symbols. Combinations of element symbols could be made to represent compounds.

Many of the atomic weights (confused with equivalent weights) were incorrect, for example oxygen's was 8 and carbon's 6, but a pattern had been established, introducing order to a science that was hitherto little more than a collection of facts.

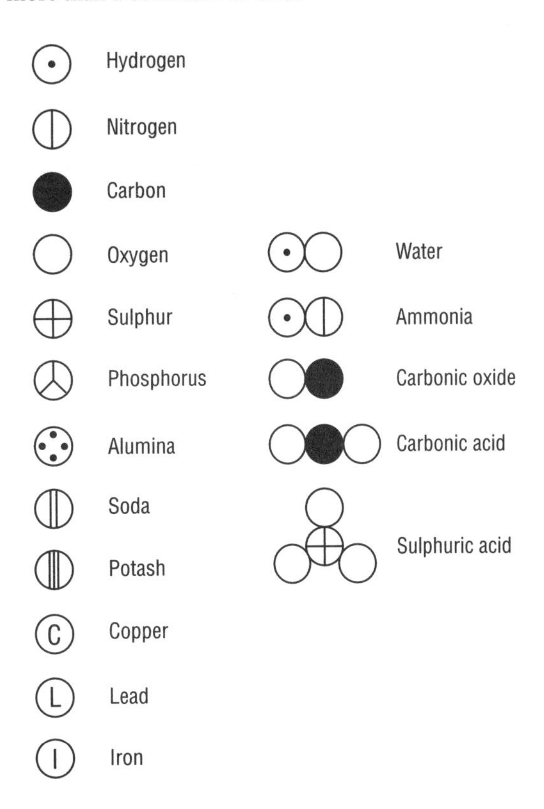

Dalton Dalton devised a set of pictorial symbols to represent individual elements; these could be combined to represent compounds of these elements.

Taking into account later work, Dalton's atomic theory may be summarized as follows:

(a) Matter cannot be subdivided indefinitely, because each element consists of indivisible particles called atoms.

(b) The atoms of the same element are alike in every respect, having the same weight (mass), volume, and chemical properties; atoms of different elements have different properties.

(c) In chemical combinations of different elements, atoms join together in simple definite numbers to form compound atoms (now called molecules).

His formula for water – one hydrogen atom combined with one oxygen atom – was wrong, although he was more fortunate with carbon monoxide ('carbonic oxide') and carbon dioxide ('carbonic acid'). But nevertheless he did bring a sort of order to the existing chaos, and provided a foundation for several generations of scientists. Several years later Jöns ◊Berzelius was to supersede Dalton's system with the chemical symbols and formulae still used today.

Throughout his life Dalton retained his Quaker habits and dress, and new acquaintances were often taken aback by his appearance. He continued to keep his diary, which eventually ran to 200,000 entries. He distrusted the results of other workers, preferring to rely on his own experiences. As he grew older he became almost a recluse, with few friends and deeply involved in his pursuit of knowledge. And although he shunned fame and glory, he became famous even outside the realms of science. When his coffin stood on public display in Manchester Town Hall, more than 40,000 people filed past to pay their respects.

Further Reading

Dickinson, Christine; Murray, Ian; and Carden, David (eds), *John Dalton's Colour Vision Legacy,* Taylor and Francis Ltd, 1997.

Greenaway, Frank, *John Dalton and the Atom,* Heinemann Books on the History of Science series, Heinemann, 1966.

Patterson, Elizabeth Chambers, *John Dalton and the Atomic Theory: The Biography of a Natural Philosopher,* The Science Study Series, Doubleday, 1970.

Smyth, A L, *John Dalton, 1766–1844: A Bibliography of Works by and About Him with an Annotated List of His Surviving Apparatus and Personal Effects,* Ashgate – Scholar Press, 1998.

Thackray, Arnold, *John Dalton: Critical Assessments of His Life and Science,* Harvard Monographs in the History of Science series, Harvard University Press, 1972.

Dana, James Dwight (1813–1895) US mineralogist, crystallographer, and geologist who is best known for his geosyncline theory of the origin of mountains. He was an active member of the American Association for the Advancement of Science and its president in 1854.

Dana was born in Utica, New York, on 12 February 1813. He was the son of a saddler and hardware merchant, and was educated at Yale where his interest in geology and mineralogy was nurtured by US chemist and geologist Benjamin Silliman.

Opportunities were limited at the time for scientists without independent means so on his graduation in 1833, Dana became a schoolmaster for midshipmen aboard the USS *Delaware.* While on board he saw his first volcano and wrote his first geological paper 'On the condition of Vesuvius in July 1834'. On his return to the USA, Silliman gave him an assistantship in his laboratory from where his career was truly launched when he became a scientific observer on the US exploratory expedition under US officer and explorer Charles Wilkes (1798–1877), which circled the globe 1838–42, conducting natural history surveys and charting the seas.

During this period Dana gained an experience and knowledge of natural history which was paralleled only by Charles ◊Darwin's. He was inspired by volcanoes and coral in the Pacific and evolved a theory of global tectonics that was one of the USA's first contributions to theoretical geology. His work supported Darwin's theories, and much of his research served to correct details of Darwin's work. In 1843 Dana refined Darwin's hypothesis of oceanic subsidence, by adding geomorphic evidence of variable island ages, and differential crustal responses, developing a theory of global history.

Dana's 400 publications include *A System of Mineralogy* (1837), *United States Exploring Expedition Under the Command of Charles Wilkes* (1846), *Corals and Coral Islands* (1872), *Manual of Geology* (1874), and *Hawaiian Volcanoes.* He retired from Yale in 1890, at the age of 77, but continued to produce books and scientific papers. He edited the *American Journal of Science* with Benjamin Silliman from 1846 until his death at New Haven on 14 April 1895.

Accepting the assumption that the Earth had been molten and had contracted as it cooled, Dana inferred that the common trend of linear island chains and mountain ranges from northwest and northeast indicated that cleavage lines originated during thermal contractions in the Archaean era. He thought these lines were a continued force in the evolution of the Earth's crust and concluded that subsidence increased from south to north. His thorough examination of the subject linked the smallest of the coral polyps in the volcanic islands of the Pacific with massive movements in the Earth's crust.

Realizing that volcanic activities were key to an understanding of the processes involved in the creation of the Earth, Dana visited many volcanic regions, and was the first of many geologists to study Mauna Loa and Kilauea in Hawaii. He gathered data by correspondence throughout his life and finally visited the islands again at the age of 74. His book *Characteristics of Volcanoes* (1890) successfully established the steps in the volcanic process, although it failed to determine their periodicity.

In response to Scottish geologist James ◊Hall's theory of mountains with (according to Dana) 'the mountains left out', Dana became an exponent of the geosyncline contraction hypothesis of mountain building in which contractive pressure buckled the continental margin and a downbuckle or geosyncline received thick sediment from the erosion of a complimentary unbuckle or geanticline, which eventually evolved into a continent. Dana considered North America to be a perfect example of the evolution of a continent which 'revealed God's plan of creation' better than any other.

Dana was one of the first US scientific specialists, and he was also part of the great creation debate. As a devout Christian he subscribed to the theory of catastrophe – in which entire species, with the exception of humans, were periodically destroyed. As a biologist and geologist he largely succumbed to Darwinism, although occasionally insisting on the possibility of divine intervention. He became progressively more accepting of social Darwinism in his later years. His theory of tectonics was influential, at least in the USA, until the appearance of the new tectonics in the 1960s.

Dancer, John Benjamin (1812–1887) English optician and instrumentmaker who applied his knowledge of physics to various inventions, particularly the development of microphotography.

Dancer was born in London on 8 October 1812. Both his father and grandfather were manufacturers of 'optical, philosophical, and nautical instruments'. In 1818 the family moved to Liverpool where his father, Joseph Dancer, was one of the founders of and a lecturer at the Liverpool Mechanical Institution. Dancer often assisted his father, and his interest in science grew. One instrument constructed by Josiah was a large solar microscope, which had a 30-cm/12-in condensing lens. Dancer used this equipment to view aquatic animals and he soon became an expert. On his father's death he took over control of the family business and the public lectures.

Dancer was a popular lecturer and he soon improved many of the standard laboratory practices of the period. He introduced unglazed porous jars in voltaic cells to separate the electrodes (previously the division was made of membranes from animal bladders and ox gullets). During the 19th century they were adopted as standard in Daniell and Leclanché cells, but unfortunately Dancer had not patented his invention.

He also devoted a lot of time to electrolysis during the 1830s; he became particularly expert in the electrodeposition of copper. He perfected a method of preparing a sheet of electrodeposited copper that retained features of a 'master' on which it was plated. Unfortunately for Dancer the invention was developed by Thomas Spencer into an early form of electrotype. He also improved on the Daniell cell by crimping or corrugating its copper plates. The power of the cell was considerably increased because of the greater electrode surface area.

Dancer was particularly interested in electrical circuits – in fact, in anything spectacular for inclusion in his lectures. Dancer used a Faraday voltameter with large platinum electrodes to prepare gases (hydrogen and oxygen) from water by electrolysis. By slight modification of the conditions he was able to prepare a colourless gas with a strong odour that caused coughing. The gas was not named initially, but was later (1839) identified as ozone by Christian Schonbein (professor of chemistry at Basle). The induction coil was a popular piece of equipment which was often claimed to have spurious medical benefits. Dancer incorporated the 'magnetic vibrator', similar to the spring make-and-break contact used in electric bells. He again did not patent his invention.

He began a series of experiments in 1839 based on Daguerre's and Fox Talbot's photography methods. By July 1840 Dancer had developed a method of taking photographs of microscopic objects, such as fleas, using silver plates. The photographic image was capable of magnification up to 20 times before clarity was lost. Dancer also gave 'magic lantern' shows, and these new plates were a considerable improvement over the painted slides then available. In the 1850s Scott Archer introduced a new process, the collodian method, and Dancer was not slow to realize its benefits in improving on his earlier techniques.

By 1856 Dancer had prepared many microphotographs and his work was exhibited throughout Europe. The clarity was excellent and by 1859 he was showing slides which carried whole pages of books 'in one-sixteen-hundredth part of a superficial inch'. Microdot photography was born. In all, the Dancer business produced for sale 512 different microphotographs, mounted on standard microscope slides. They included photographs of distinguished scientists of the era and portraits of the British royal family.

Dancer was an active member of the Manchester Literary and Philosophical Society, at whose meetings he presented many papers on a variety of scientific topics. He was a fellow of the Royal Astronomical Society and optician in Manchester to the Prince of Wales. He also constructed the apparatus with which James Joule determined the mechanical equivalent of heat.

In his later years he suffered from diabetes, which led to glaucoma of his eyes. Dancer died in November 1887, leaving behind a technique that has led to a method of storing huge quantities of data in a small space – microfilm and microfiche.

Daniell, John Frederic (1790–1845) English meteorologist, inventor, and chemist, famous for devising the Daniell cell, a primary cell that was the first reliable source of direct-current electricity. He received many honours for his work, including the Royal Society's Rumford and Copley medals, awarded in 1832 and 1837 respectively.

Daniell was born in London on 12 March 1790, the son of a lawyer. He received a private education, principally in the classics, after which he began working in a relative's sugar-refining factory. Early in his career he made the acquaintance of William Brande, professor of chemistry at the Royal Institution; the two men became lifelong friends, travelling together on several scientific expeditions and through their joint efforts reviving the journal of the Royal Institution. In 1831 Daniell was appointed the first professor of chemistry at the newly founded King's College, London, a post he held until his death (at a meeting of the Royal Society) on 13 March 1845.

Daniell made his best-known contribution to science – the cell named after him – in 1836, only nine years before his death. The Italian physicist Alessandro ◊Volta had, in 1797, invented a copper–zinc battery, which the English physicist William ◊Sturgeon had improved by using zinc and mercury amalgam; but both these simple, or voltaic, cells suffered the severe disadvantage of producing a current that diminished rapidly. Daniell found that this effect (polarization) was caused by hydrogen bubbles collecting on the copper electrode, thereby increasing the cell's internal resistance. In 1836 he proposed a new type of cell. The Daniell cell consists of a copper plate (the anode) immersed in a saturated solution of copper sulphate and separated by a porous barrier (Daniell used a natural membrane) from a zinc rod (the cathode) immersed in dilute sulphuric acid. The barrier prevents mechanical mixing of the solutions but allows the passage of charged ions. When the current flows, zinc combines with the sulphuric acid freeing positively charged hydrogen ions, which pass through the barrier to the copper plate. There they lose their charge and combine with copper sulphate, forming sulphuric acid and depositing copper on the plate. Negatively charged sulphate ions pass in the opposite direction and react with the zinc to form zinc sulphate, therefore no gases are evolved at the electrodes and a constant current is produced. Daniell's cell stimulated research into electricity – including his own investigations into electrolysis – and was later used commercially in electroplating and glyphography (a process giving a raised relief copy of an engraved plate for use in letterpress printing).

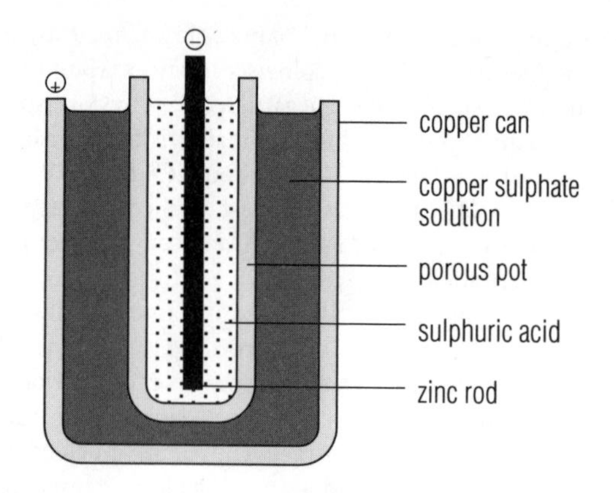

Daniell The Daniell cell was the first reliable battery, supplying a steady current for a long time. It quickly became the standard form of battery after 1836.

Daniell's other work included the development of improved processes for sugar manufacturing; investigations into gas generation by the distillation of resin dissolved in turpentine; and inventing a new type of dew-point hygrometer for measuring humidity (1820) and a pyrometer for measuring the temperatures of furnaces (1830). He also studied the behaviour of the Earth's atmosphere; gave an explanation of trade winds; researched into the meteorological effects of solar radiation and of the cooling of the Earth; suggested improvements for several meteorological instruments; and pointed out the importance of humidity in the management of greenhouses.

Dantzig, George Bernard (1914–) US mathematician who is an expert on (computer) linear programming and operations research. His work is now regarded as fundamental to many university courses in business studies, industrial engineering, and managerial sciences.

Born the son of a well-known mathematician on 8 November 1914 in Portland, Oregon, Dantzig completed his education at the University of Maryland, gaining his BA in 1936. He then attended the University of Michigan as a Horace Rackham Scholar for a year, earning his MA. During the latter part of World War II he was with the Statistical Control Headquarters of the USAF as chief of the Combat Analysis Branch, where he then, from 1946–52, became mathematical adviser. For the next eight years he was a research mathematician with the Rand Corporation at

Santa Monica, California, transferring in 1960 to the University of California at Berkeley as professor and chair of the Operations Research Center. Since 1966, among other consultative posts, his major position has been as CA Criley Professor of Operations Research and Computer Science at Stanford University, Palo Alto, California. The author of two influential books and many technical papers. Dantzig has received a number of awards and honorary degrees, including the National Academy of Sciences Award in Applied Mathematics and Numerical Analysis (in 1977).

A fundamental problem in economics involves the optimum allocation of scarce resources among competing activities – a problem that can be expressed in mathematical form. In 1947 Dantzig discovered that many such planning problems could be formulated as linear computer programs. He compounded this discovery in that at the same time he also devised an algorithm – known as the simplex method – that turned out to be remarkably efficient for the purpose. (His method is still the best way to resolve nearly all linear programmes of this type.) Moreover, Dantzig's discovery coincided with the development of the first successful computers, which meant that managers in industry were provided with a powerful and practical method for comparing a large number of interdependent alternative courses of action. Dantzig led the way in developing applications for the new linear programming approach. By 1972, a survey showed that a considerable proportion of all industrial organizations was using the simplex method of linear programming. The system has also had an impact on economics and statistics.

Subsequently, Dantzig has been involved in all the main areas of mathematical programming and other parts of operations research. He has worked on the development of techniques for dealing with large systems, and originated the 'decomposition principle' for solving large systems with block-diagonal structure. In 1971 it was possible by this method, and using an IBM 370/165 computer, to solve a linear programme with 282,468 variables and 50,215 equations in 2.5 hours.

The development of linear programming in the 1950s enabled the mathematical science of decision-making to be developed into the discipline now known as operations research or management science. Most universities now offer courses on operations research, nearly all emphasizing the importance of linear programming. Operations research is also important in the academic studies of many university departments of business sciences and industrial engineering.

Darboux, (Jean) Gaston (1842–1917) French geometrician who contributed immensely to the differential geometry of his time, and to the theory of surfaces. An innovator in much of his research, he was also an able teacher, a capable administrator, and an influential author in writing of his own studies. His name is perpetuated in the Darboux sums and the Darboux integrals.

Darboux was born on 13 August 1842 in Nîmes. Educated locally, at the age of 19 he sat the entrance examinations for both the Ecole Polytechnique and the Ecole Normale Supérieure in Paris. In both he came top; he chose to enter the Ecole Normale. During his studies there, he wrote his first paper (on orthogonal surfaces, in 1864). Two years later he extended this work, for which he received his doctorate from the Sorbonne. Thereafter, Darboux taught at the Lycée Louis le Grand 1866–72 before transferring to a similar post at the Ecole Normale. From 1873–78 he also held the chair of rational mechanics at the Sorbonne, as an assistant to Joseph Liouville (from whom he may have gained his interest in differential geometry). He then became assistant to Michel Chasles, the author of a standard reference work on geometry, and in 1880 succeeded him as professor of higher geometry. He held this post until he died, in Paris, on 25 February 1917. Elected a member of the Paris Academy of Sciences in 1884 (and for many years later its secretary), Darboux was made a Fellow of the Royal Society in 1902. He also won several other awards and honours.

Darboux's research concentrated on geometry – the major interest throughout his life. Nevertheless, only five years or so after his doctoral thesis on orthogonal surfaces (a subject he returned to time and time again), he published a paper on partial differential equations of the second order. What was particularly novel in his approach was that in order to further his examination he had devised a new method of integration. Within another five years – during which he also formulated the theory of a specific class of surface called a cyclide – he managed to complete a proof of the existence of integrals of continuous functions that had defeated Augustin-Louis Cauchy a generation before. Continuing his investigations, Darboux succeeded (in 1879) in defining the Riemann integral, in order to do which he derived the 'Darboux sums' and used the 'Darboux integrals'.

Between 1887–96 he published a collection of the lectures he had given at the Sorbonne, under the overall title *Lecons sur la théorie générale des surfaces et les applications géométriques du calcul infinitésimal.* The four volumes described all of his work to date, but dealt mainly with the application of analysis to curves and surfaces, and the study of minimal surfaces and geodesics. In a later work, *Lecons sur les systèmes orthogonaux et les coordonnées curvilignes* (1898), Darboux applied the theorem on algebraic integrals formulated by the Norwegian Niels Abel to orthogonal systems in *n* dimensions. Important among Darboux's other work were his papers on the theory of integrations, the theory of analytical functions, and his research into the problems involving the Jacobi polynomials.

Darby, Abraham (1677–1717) English ironmaster and engineer who devised a new way of smelting iron using coke rather than the much more expensive charcoal.

Darby was born in Wren's Nest near Dudley, Worcestershire, the son of a farmer. He served an engineering apprenticeship with a malt-mill maker in Birmingham, and on completing this in 1698 he set up in business on his own. In about 1704 he visited Holland and brought back with him some Dutch brass founders, establishing them at Bristol (at the Baptist Mill Brass Works) using capital from four associates who left to him the management of the business.

Believing that cast iron might be substituted for brass in some products, he tried with his Dutch workers to make iron castings in moulds of sand. At first the experiment failed but proved eventually successful when he adopted a suggestion made by a boy in his employment, named John Thomas, who consequently rose in his service and whose descendents were trusted agents of the Darby family.for about a hundred years. In April 1708 he took out a patent for a new way of casting iron pots and other ironware in sand only, without loam or clay, a process that cheapened utensils much used by poorer people and at that time largely imported from abroad. This decision led to protracted arguments with his associates, who finally refused to risk more money on the new venture. Darby dissolved his connection with them and, drawing out his share of the capital, he took a lease on an old furnace in Coalbrookdale, Shropshire, moving to Madely Court in 1709. There he prospered until his death on 8 March 1717.

In the 17th century the development of the iron industry in Britain was limited by two technical difficulties. Firstly, the growing demand for charcoal – then the only satisfactory fuel for blast furnaces – had forced up the price very considerably. The shortage of wood for conversion into charcoal had been caused by the heavy demands of shipbuilders and a rapid growth in the demand for iron-made objects (resulting in the denuding of the large areas of forest that had been a feature of the English landscape for thousands of years). The second difficulty was due to the attempts to improve efficiency by using bigger furnaces; these were frustrated by the fact that charcoal is too soft to support more than a relatively short column of heavy ore. Attempts were made to use coal in place of charcoal, but the presence in most coals of sulphur, which spoils the quality of the iron, resulted in only limited success; the claim put forward by Dud Dudley (1599–1684) that in 1619 he had successfully smelted iron by using coke is not now generally accepted.

Attention was therefore directed to coke – already used in the smelting of copper and lead – since the coking process eliminates the sulphur. Darby had considerable experience of smelting copper with coke at Bristol and had also used coke in malting during his apprenticeship days in Birmingham. (In malting, for different reasons, the presence of sulphur prevents the use of raw coal.) The old furnace that Darby converted to house the Bristol Iron Company was ideally situated on the River Severn, close to good local supplies of iron ore and good coking coal. Initially much of the iron was used for making pots and other hollow-ware. The quality of the molten iron made it possible for him to make thin castings that competed satisfactorily with the heavy brassware then in common use.

The advent of the Newcomen steam engine gave an important new market; some of the cylinders required for mine-pumping engines weighed as much as 6 tonnes with a length of 3 m/10 ft and a bore of 1.8 m/6 ft. By 1758 more than a hundred such cylinders had been cast. The steam engine, in turn, improved the manufacture of iron by giving a more powerful and reliable blast for the furnaces than water-power could supply.

The Coalbrookdale Works were much enlarged, their processes improved and increased and their operations extended under the second Abraham Darby. His son and successor, the third Abraham Darby, took over the management of the Coalbrookdale Works when he was about 18 and is memorable as having constructed the first iron bridge ever erected, the semicircular cast-iron arch across the River Severn near the village of Brosely at Coalbrookdale, opened for traffic in 1779 (and still standing today).

Dart, Raymond Arthur (1893–1988) Australian-born South African anatomist and palaeoanthropologist who made a significant contribution to the tracing of the early history of the human species.

Born in Brisbane on 4 February 1893, Dart studied medicine at the University of Sydney, where he qualified in 1917. He served in France in World War I before being appointed in 1922 to the chair of anatomy at the newly formed University of the Witwatersrand, Johannesberg, South Africa. Two years later he had a lucky find: one of his students brought him a fossil baboon skull, found in a lime quarry at Taung, Botswana. Excited, Dart took steps to ensure that similar finds would be preserved and despatched to him. Soon he received the skull of a previously unknown hominid, which Dart named *Australopithecus africanus* (southern ape of Africa). Dart saw *Australopithecus* as the 'missing link' between humans and the apes. This view received little support until Robert Broom found further hominid remains in the Transvaal in the 1930s. Physical anthropologists today believe that *Australopithecus africanus* lived about 1.2–2.5 million years ago, but it and other Australopithecines such as *A. afarensis, A. robustus,* and *A. ramidus* are not now considered to be the ancestors of modern humans but an evolutionary offshoot from a much earlier shared predecessor. Dart died in Johannesburg on 22 November 1988.

Further Reading

Tobias, Philip, *Dart, Taung and the Missing Link: An Essay on the Life and Work of Emeritus Professor Raymond Dart,* Witwatersrand University Press, 1984.

Washburn, S L, *Human Evolution after Raymond Dart,* Witwatersrand University Press, 1985.

Wheelhouse, Frances, *Raymond Arthur Dart: Pictorial Profile the Professor's Discovery of the Missing Link,* Transpareon Press, 1983.

Darwin, Charles Robert (1809–1882) English naturalist famous for his theory of evolution and natural selection as put forward in 1859 in his book *The Origin of Species.*

Darwin English naturalist Charles Darwin, author of one of the most influential scientific books ever published, *On the Origin of Species by Means of Natural Selection*, 1859. At a time when most people still believed in the literal truth of the Bible's account of creation, Darwin's idea that species had evolved gradually caused a storm of controversy. The first edition of the book sold out on the day of publication. *Mary Evans Picture Library*

Darwin was born in Shrewsbury on 12 February 1809. His father was a wealthy doctor, and his paternal grandfather was Erasmus Darwin, a well-known poet and physician; his mother's father was the pottery manufacturer Josiah Wedgwood. Darwin was educated locally from 1818, and when he left school in 1825 he attended Edinburgh University to study medicine. But he abhorred medicine and the science taught to him there disgusted him, and two years later his father sent him to Christ's College, Cambridge, to study theology – which he did not enjoy either. Natural history was his main interest, which was very much increased by his acquaintance with John Stevens Henslow, who was professor of botany at Cambridge. Henslow recommended to the Admiralty that Darwin should accompany HMS *Beagle* as a naturalist on its survey voyage of the coasts of Patagonia, Tierra del Fuego, Chile, Peru, and some Pacific islands. His father opposed this idea but, with the support of Wedgwood, Darwin sailed in the *Beagle* from Devonport on 27 December 1831 for a voyage of five years. On his return to England he found that some of his papers had been privately published during his absence and that he was regarded as one of the leading men of science. He published his findings on this epic voyage in the *Journal of Researches into the Geology and Natural History of the Various Countries Visited by HMS Beagle (1832–1836)* in 1839. In 1838 he was appointed secretary to the Geological Society, a position he retained until 1844. He married his cousin, Emma Wedgwood, in 1839, and the marriage produced ten children. He spent the rest of his life collating the findings made during the voyage and developing his theory for publication. He died on 19 April 1882 at Down, in Kent.

Before the voyage of the *Beagle* Darwin, like everyone else at that time, did not believe in the mutability of species. But in South America he saw fossil remains of giant sloths and other animals now extinct, and on the Galápagos Islands found a colony of finches that he could divide into at least 14 similar species, none of which existed on the mainland. It was obvious to him that one type must have evolved into many others, but how they did so eluded him. Two years after his return he read Malthus's *An Essay on the Principle of Population* (1798), which proposed that the human population is growing too fast for it to be adequately fed, and that something would have to happen to reduce it, such as war or natural disaster. This work inspired Darwin to see that the same principle could be applied to animal populations and he theorized that variations of a species that survive (while other members of the species do not) pass on the changed characteristic to their offspring. A new species is thereby developed which is fitter to survive in its environment than was the original species from which it evolved. Darwin did not make his ideas public at first, but put them into an essay in 1844 to which only his friend Joseph Hooker and a few others were privy.

In 1856 Darwin began writing fully about evolution and natural selection. Two years later he received a paper from a fellow naturalist, Alfred Russel ⇨Wallace, explaining exactly the same theory of evolution and natural selection. Unsure what to do, Darwin consulted his friends Charles Lyell and Hooker, who persuaded him to have the joint papers read in the absence of the authors before the Linnaean Society. The papers caused no stir, but Darwin was forced to speed up the completion of his work.

When I am obliged to give up observation and experiment I shall die.

CHARLES DARWIN QUOTED IN A MOORHEAD
DARWIN AND THE BEAGLE

The abstract of Darwin's findings was published in 1859, and was called *The Origin of Species by Means of Natural Selection or the Preservation of Favoured Races in the Struggle for Life.* It was very widely read, although many fellow scientists criticized it violently. Some considered that the book lacked a foundation of experimental evidence and was based purely on hypothesis; others were simply jealous. Many Christians were shocked by Darwin's work because it implied that the Biblical account of creation – if taken literally – is wrong, and that if evolution works automatically by natural selection then divine intervention plays no part in the lives of plants, animals, or humans.

When Darwin wrote the book, he avoided the issue of human evolution and merely remarked at the end that 'much light will be thrown on the origin of man and his history'. He did not seek the controversy he caused but his ideas soon caught the public imagination. After the publication in 1871 of *The Descent of Man and Selection in Relation to Sex,* in which he argued that people evolved just like other organisms, the popular press soon published articles about the 'missing link' between humans and apes. In fact what Darwin believed was that our ancestors, if alive today, would have been classified among the primates.

Darwin's name remains inseparably linked with the theory of evolution to this day. He never understood what actually caused newly formed advantageous characteristics to appear in animals and plants because he had no knowledge of heredity and mutations. The irony is that the key work on heredity by the Austrian monk Gregor ⇨Mendel was carried out during Darwin's own lifetime and published in 1865, but was neglected until 1900. Darwin's revolutionary publication, which is still widely read, marked a turning point in many of the sciences, including physical anthropology and palaeontology, and remains a source of strong controversy.

Further Reading

Bowler, Peter J, *Charles Darwin: The Man and His Influence,* Cambridge Science Biographies series, Cambridge University Press, 1996, v 1.

Bowler, Peter J, *Charles Darwin: The Man and His Influence,* Scientific Biographies series, Blackwell, 1990.

Howard, Jonathan, *Darwin,* Past Masters series, Oxford University Press, 1996.

White, Michael and Gribbin, John R, *Darwin: A Life in Science,* Simon and Schuster, 1995.

Darwin, Erasmus (1731–1802) English poet, naturalist, and physician. He was the grandfather of Charles ⇨Darwin and an inventor of mechanical machines.

Erasmus Darwin was born in Elston Hall, Nottinghamshire, on 12 December 1731, the fourth son in a family of four brothers and one sister. His father was a barrister and recorder at Lincoln. In 1741 Darwin was sent away to Chesterfield School, where, in his letters to his sister, he revealed a talent for writing. In 1750 he entered St John's College, Cambridge, with the Exeter scholarship and graduated in 1754 with a BA – the first of the junior optimes. From there he moved to Edinburgh to study medicine. Returning in 1758, he opened a medical practice in Nottingham but it failed to attract many patients, so he moved to Lichfield where he achieved much greater success.

Darwin was happily married to Mary Howard from 1757 until her death in 1770. He became so well known for his medical expertise that he was offered the position of personal physician by George III. He refused the offer, preferring to remain in Lichfield where he was a popular physician known for his poetry and his large botanical garden. He was a radical free thinker and met several influential people while he was in Lichfield including Jean-Jacques Rousseau, with whom he corresponded, and Samuel Johnson, with whom he did not get on.

Darwin and his friends Bolton, Watt, Wedgwood, and the Sewards held monthly discussion meetings at each others' houses. Darwin often elaborated on his theories and opinions in verse, which although it was considered bad poetry, was saved by the fact that it could always be seen to be informed by his considerable intellect. His long poem *The Botanic Garden* (1789–91) expounded the Linnaen system. *Zoonomia or the Laws of Organic Life* (1794–96) explained organic life in evolutionary terms similar to those set out later by Jean Baptiste de ⇨Lamarck. Darwin believed that reason was inferior to generation, but that evolution is brought about by a largely unconscious effort of will. Many of his ideas on evolution were taken from observation and anticipated later theories, although they were rejected by more sophisticated 19th-century evolutionists including his grandson. *Phytologia, or the Philosophy of Agriculture and Gardening* was published in 1799.

In 1781 Darwin married his second wife, a widow he had wooed with passionate poems while treating her children. They settled in Derby. Darwin was public spirited and charitable, and he was popular with his new family in spite of a reputation for being irascible and imperious. He set up a public dispensary in Lichfield and a philosophical society in Derby and in 1797, to help two illegitimate daughters who had opened a school at Ashbourne, he wrote *A Plan for the Conduct of Female Education in Boarding Schools,* which was full of useful advice. For many years he abstained from alcohol and was a vigorous campaigner for sobriety, persuading local gentry of the merits of water.

A larger-than-life, energetic man, Darwin also had time to be an inventor of mechanical contraptions. The carriage he invented brought him attention especially from R L Edgeworth who became a firm friend. It was fitted out for reading and writing and it was while he was out visiting patients that he wrote most of his poetry. The carriage was unfortunately involved in several accidents, in one of which he broke his kneecap causing a permanent disability.

Erasmus Darwin died suddenly of heart disease on 18 April 1802. He had three sons with his first wife. Tragically his first son died as a medical student from an infected wound and his second committed suicide in a fit of temporary insanity. His third son became the father of Charles Darwin. He had four sons and three daughters with his second wife. The eldest of the daughters became the mother of Francis ⟡Galton, the scientist and explorer, who had a monument erected to his grandfather at Lichfield Cathedral.

Further Reading

King-Hele, Desmond (ed), *The Letters of Erasmus Darwin,* Cambridge University Press, 1981.

King-Hele, Desmond, *Erasmus Darwin, 1731–1802: Master of Interdisciplinary Science,* Lichfield Science and Engineering Society, 1991.

King-Hele, Desmond, *Doctor of Evolution: The Life and Genius of Erasmus Darwin,* Faber & Faber, 1977.

McNeil, Maureen, *Under the Banner of Science: Erasmus Darwin and His Age,* Manchester University Press, 1987.

Priestland, Neal, *Erasmus Darwin: Philosopher, Scientist, Physician and Poet,* Ashbracken, 1990.

Davis, William Morris (1850–1934) US physical geographer who analysed landforms.

The son of a Philadelphia Quaker businessman, Davis studied science at Harvard University. After a spell as a meteorologist in Argentina, Davis served with the US North Pacific Survey before securing an appointment as a lecturer at Harvard in 1877. Becoming a professor, he taught at Harvard until 1912. During those 30 years, Davis became the most prominent US investigator of the physical environment. In three fields of science – meteorology, geology, and, most notably, geomorphology – he left enduring legacies. Above all, he proved an influential analyst of landforms. Building on experience gained in a classic study made in 1889 of the drainage system of the Pennsylvania and New Jersey region, he developed the organizing concept of the regular cycle of erosion, a theory that was to dominate geomorphology and physical geography for half a century. Davis proposed a standard stage-by-stage lifecycle for a river valley, marked by youth (steep-sided V-shaped valleys), maturity (flood-plain floors), and old age, as the river valley was imperceptibly worn down into the rolling landscape he termed a 'peneplain'. On occasion these developments, which Davis believed followed from the principles of Lyellian geology (see Charles ⟡Lyell), could be punctuated by upthrust, which would rejuvenate the river and initiate new cycles. The Davisian cycle presupposed an explicitly uniformitarian view of Earth history, in which the present was key to the past and piecemeal, and natural causes were paramount.

Davisson, Clinton Joseph (1881–1958) US physicist who made the first experimental observation of the wave nature of electrons. For this achievement, he was awarded the 1937 Nobel Prize for Physics with George ⟡Thomson, who made the same discovery independently of Davisson.

Davisson was born in Bloomington, Illinois, on 22 October 1881. He attended local schools before gaining a scholarship for his proficiency in mathematics and physics in 1902 to the University of Chicago, where he studied under Robert ⟡Millikan. In 1905 Davisson was appointed part-time instructor in physics at Princeton University while continuing his studies at Chicago, gaining a BS degree from Chicago in 1908, followed by a PhD in 1911 for a thesis on 'The thermal emission of positive ions from alkaline earth salts'. Davisson then spent six years as an instructor in the department of physics at the Carnegie Institute of Technology in Pittsburgh. He was able to spend the summer of 1913 in the Cavendish Laboratories at Cambridge, England, working under J J ⟡Thomson. Refused enlistment in the army in 1917, he accepted wartime employment in the engineering department of the Western Electric Company (later Bell Telephone) in New York. At the end of the war, although offered an assistant professorship at the Carnegie Institute, Davisson remained at Bell Telephone where he was able to work on research full-time. He stayed there until his retirement in 1946, when he became a visiting professor of physics at the University of Virginia, Charlottesville. He retired from this position in 1949 and died on 1 February 1958.

Davisson's work at Western Electric was concerned with the reflection of electrons from metal surfaces under electron bombardment. In April 1925, an accidental explosion caused a nickel target under investigation to become heavily oxidized. Davisson removed the coating of oxide by heating the nickel and resumed his work. He now found that the angle of reflection of electrons from the nickel surface had changed. Davisson and his assistant Lester Germer (1896–1971) suspected that the change was due to recrystallization of the nickel, the heating having converted many small crystals in the target surface into several large crystals.

A year later, Davisson attended a meeting of the British Association for the Advancement of Science where he received details of the theory proposed by Louis ◊de Broglie that electrons may behave as a wave motion. Davisson immediately believed that the effects he had observed were caused by the diffraction of electron waves in the planes of atoms in the nickel crystals, just as X-ray diffraction had earlier been observed in crystals by Max von ◊Laue and William and Lawrence ◊Bragg.

To resolve this question, Davisson and Germer used a single nickel crystal in their experiments. The atoms were in a cubic lattice with atoms at the apex of cubes, and the electrons were directed at the plane of atoms at 45 ° to the regular end plane. Electrons of a known velocity were directed at this plane and those emitted were detected by a Faraday chamber. In January 1927, results showed that at a certain velocity of incident electrons, diffraction occurred, producing outgoing beams that could be related to the interplanar distance. The wavelength of the beams was determined, and this was then used with the known velocity of the electrons to verify de Broglie's hypothesis. The first work gave results with an error of 1–2% but later systematic work produced results in complete agreement. Similar experiments at higher voltages using metal foil were carried out later the same year at Aberdeen University, Scotland, by George Thomson, the son of J J Thomson). The particle–wave duality of subatomic particles was established beyond doubt, and both men were awarded the 1937 Nobel Prize for Physics.

Davisson's experimental findings that electrons have a wave nature confirmed and established theoretical explanations of the structure of the atom in terms of electron waves. This understanding proved vital to investigations into the nature of chemical bonds, and to the production of high magnifications in the electron microscope.

Davy, Humphry (1778–1829) English chemist who is best known for his discovery of the elements sodium and potassium and for inventing a safety lamp for use in mines.

Davy was born on 17 December 1778 at Penzance, Cornwall, the son of well-to-do parents. He was educated in Penzance and, from 1793, in Truro, where he studied classics. But his father died a year later and, to help to support the family, the young Davy became apprenticed to a Penzance surgeon–apothecary, J Bingham Borlase. His interest in chemistry began in 1797 through reading Antoine Lavoisier's *Traité elémentaire,* and by 1799 he was working on the therapeutic uses of gases as an assistant at the Pneumatic Institute in Bristol.

Following Alessandro Volta's announcement in 1800 of the voltaic cell, Davy began his researches in electrochemistry. He moved to the Royal Institution in London in 1801, where he was influenced by Count Rumford and Henry Cavendish. He was knighted by the Prince Regent in 1812 and three days later married a wealthy widow named Jane Apreece. In 1813 he took on Michael ◊Faraday as a laboratory assistant, who accompanied him on a tour of Europe. When he returned in 1815 Davy designed his miner's safety lamp, which would burn safely even in an explosive mixture of air and fire damp (methane). He did not patent the lamp, a fact that was to lead to an acrimonious claim to priority by the steam locomotive engineer George Stephenson. Davy was created a baronet in 1818 and two years later he succeeded the botanist Joseph Banks as president of the Royal Society. He became seriously ill in 1827 and went abroad in 1828 to try to improve his health. He settled in Rome in early 1829 but suffered a heart attack and died in Geneva, Switzerland, on 29 May of that year.

While Davy was working in Bristol (1799) he prepared nitrous oxide (dinitrogen monoxide) by heating ammonium nitrate. He investigated the effects of breathing the gas, showing that it causes intoxication (although it was to be another 45 years before the gas was used as a dental anaesthetic). His early experiments on electrolysis of aqueous solutions (from 1800) led Davy to suggest its large-scale use in the alkali industry. He theorized that the mechanism of electrolysis could be explained in terms of species that have opposite electric charges, which could be arranged on a scale of relative affinities – the foundation of the modern electrochemical series. The climax of this work came in 1807 with the isolation of sodium and potassium metal by the electrolysis of their fused salts. Later, after consultation with Jöns Berzelius, he also isolated calcium, strontium, barium, and magnesium. His intensive study of the alkali metals provided proof of Antoine ◊Lavoisier's idea that all alkalis contain oxygen. In 1808 he first isolated boron, by heating borax with potassium.

Davy also initially supported Lavoisier's contention that oxygen is present in all acids. But in 1810, after doing quantitative analytical work with muriatic acid (hydrochloric acid) he disproved this hypothesis. He went on to show that its oxidation product, oxymuriatic acid (discovered in 1774 by Karl Scheele), is an element, which he named chlorine. He explained its bleaching action and later prepared two of its oxides and chlorides of sulphur and phosphorus. He also suggested that the element common to all acids is hydrogen not oxygen.

The eternal laws Preserve one glorious wise design; Order amidst confusion flows, And all the system is divine.

*FROM **HUMPHREY DAVY'S** NOTEBOOKS AT THE ROYAL INSTITUTION*

Davy was reluctant to accept the atomic theory of his contemporary John ◊Dalton, but in the face of mounting evidence finally concurred and attempted to apply the laws of definite and multiple proportions to various compounds. He determined the 'proportional weights' (relative atomic masses), of various elements, including chlorine at 33.9 (actual value 35.5), oxygen 15 (16), potassium 40.5 (39.1), and sulphur 30 (32).

The safety lamp of 1815 was designed after a series of laboratory experiments on explosive mixtures. Davy showed that a flame continues to burn safely in such a mixture if it is surrounded by a fine metal mesh to dissipate heat, if only a narrow air inlet is used, and if the air inside the lamp is diluted with an unreactive gas such as carbon dioxide.

Numerous other achievements can be attributed to this great scientist. Davy introduced a chemical approach to agriculture, the tanning industry, and mineralogy; he designed an arc lamp for illumination, an electrolytic process for the desalination of sea water, and a method of cathodic protection for the copper-clad ships of the day by connecting them to zinc plates. But his genius has been described as erratic. At his best he was a scientist of great perception, a prolific laboratory worker, and a brilliant lecturer. At other times he was unsystematic, readily distracted, and prone to hasty decisions. He was never trained as a chemist, and consequently his excellence in qualitative work was not always matched by quantitative skills. He sought and won many scientific honours, which he then jealously guarded, even going so far in 1824 as trying to oppose the election of his protégé Michael Faraday to the Royal Society.

Further Reading

Abbri, Ferdinando, 'Romanticism in science: science in Europe, 1790–1840' in: Poggi, Stefano and Bossi, Maurizio (eds), *Romanticism Versus Enlightenment: Sir Humphry Davy's Idea of Chemical Philosophy,* Kluwer Academic Publishers, 1994, pp 31–45.

Carrier, Elba O, *Humphry Davy and Chemical Discovery,* Immortals of Science series, 1967.

Golinski, Jan, 'Humphry Davy and the 'lever of experiment'' in: Le Grand, H E (ed), *Experimental Enquiries: Historical, Philosophical, and Social Studies of Experimentation in Science,* Kluwer Academic Publishers, 1990, pp 99–136.

Knight, David M, *Humphry Davy: Science and Power,* Cambridge Science Biographies series, Cambridge University Press, 1996.

Dawkins, (Clinton) Richard (1941–) British zoologist and evolutionary theorist.

Dawkins was born in Nairobi, Kenya, on 26 March 1941, and educated at Oundle School and Balliol College, Oxford, where he gained his doctorate under the ethologist Niko Tinbergen, who won the Nobel Prize for Physiology or Medicine in 1973. After two years as assistant professor of zoology at the University of California at Berkeley, Dawkins returned to Oxford as lecturer in animal behaviour, becoming reader in zoology in 1989 at the university, and a fellow of New College. He became Charles Simonyi Professor of Public Understanding of Science at Oxford in 1995.

His research has continued in ethology, the study of animal behaviour, and he has written and broadcast his ideas about the evolution of behavioural mechanisms widely. In 1976 he published *The Selfish Gene,* in which he argued that genes – not individuals, populations, or species – are the driving force of evolution and that animals such as humans are merely machines through which genes survive. He also suggested an analogous system of cultural transmission in human societies, and proposed the term 'mimeme', contracted to 'meme', as the unit of such a scheme. The book became an international popular success, and was translated into several languages. From it Dawkins's comments about religion attract particular attention as he considered the idea of God to be a meme with a high survival value.

His contentions were further developed in *The Extended Phenotype* (1982), primarily an academic work, and in *The Blind Watchmaker* (1986), *River Out of Eden* (1995), and *Climbing Mount Improbable* (1996), all of which achieved wide acclaim from literary as well as scientific critics. Following the success of his books Dawkins has presented scientific programmes on television, given the prestigious Royal Institution Christmas lectures for young people, and become a prominent public spokesperson for science in general and evolutionary ethology in particular.

De Beer, Gavin Rylands (1899–1972) English zoologist known for his important contributions to embryology and evolution, notably disproving the germ-layer theory and developing the concept of paedomorphism (the retention of juvenile characteristics of ancestors in mature adults). He received numerous honours for his work, including a knighthood in 1954.

De Beer was born on 1 November 1899 in London and, after military service in World War I, graduated from Oxford University. From 1926–38 he was Jenkinson Memorial Lecturer in Embryology at Oxford University, after which he served in World War II. In 1945 he became professor of embryology at University College, London, then in 1950 was appointed director of the Natural History Museum, a post he held until his retirement in 1960. De Beer died on 21 June 1972 in Alfriston, East Sussex.

De Beer's first major work was his *Introduction to Experimental Embryology* (1926), in which he observed that some vertebrate structures, such as certain cartilage and bone cells, are derived from the outer ectodermal layer of the embryo. This finally disproved the germ-layer theory, according to which cartilage and bone cells are formed from the mesoderm. Continuing his embryological investigations, De Beer described in *Embryos and Ancestors* (1940) his work showing that certain adult animals retain some of the juvenile characteristics of their ancestors, a phenomenon called paedomorphism. This finding refuted Ernst Haeckel's theory of phylogenetic recapitulation, according to which the embryonic development of an organism repeats the adult stages of the organism's evolutionary ancestors.

Turning his attention to evolution, De Beer then suggested that gaps in the fossil records of early ancestral forms are due to the impermanence of the soft tissues in these early ancestors. Also in the field of evolution, his studies of the fossil *Archaeopteryx*, the earliest known bird, led him to propose mosaic evolution – whereby evolutionary changes occur piecemeal – to explain the presence of both reptilian and avian features in *Archaeopteryx*.

De Beer also researched into the functions of the pituitary gland, and applied scientific methods to various historical problems, such as the origin of the Etruscans (which he traced using blood group data) and establishing the route taken by Hannibal in his march across the Alps (for which De Beer used pollen analysis, glaciology, and various other techniques).

de Broglie, Louis Victor (1892–1987) French physicist who first developed the principle that an electron or any other particle can be considered to behave as a wave as well as a particle. This wave–particle duality is a fundamental principle governing the structure of the atom, and for its discovery, de Broglie was awarded the 1929 Nobel Prize for Physics.

De Broglie was born in Dieppe on 15 August 1892, the second son of a noble French family and as such expected to have a distinguished military or diplomatic career. It was perhaps fortunate that his elder brother Maurice (1875–1960), who pioneered the study of X-ray spectra, had pursued his scientific interests to considerable success against the wishes of his family, his only compromise being a relatively short naval career. Nevertheless, in 1909 Louis entered the Sorbonne in Paris to read history. A year later, when he was 18, he began to study physics. It had been his intention to enter the diplomatic service, but he became so interested in scientific subjects – partly through the influence of Maurice, whom he helped in his extensive private laboratory at the family home – that he took a physics topic for his doctoral dissertation rather than that on French history which the Sorbonne first offered. This dissertation was submitted in 1924, but before that de Broglie began exploring the ideas of wave–particle duality. Following the award of his doctorate, de Broglie stayed on at the Sorbonne until 1928. He moved to the Henri Poincaré Institute in 1932 as professor of theoretical physics, retaining this position until 1962. From 1946, he was a senior adviser on the development of atomic energy in France. He died in Louveriennes on 19 March 1987.

In 1922, de Broglie was able to derive Planck's formula $E = h\nu$, where E is energy, h is Planck's constant, and ν is the frequency of the radiation, using the particle theory of light. This probably suggested the idea of wave–particle duality to him because it prompted the question of how a particle could have a frequency. Using this idea and ◊Einstein's mass–energy equation $E = mc^2$, he derived $E = mc^2 = h\nu$. Now, mc is the momentum of the particle and C/ν is the wavelength, λ, of the associated wave. Hence the momentum = h/λ. This relation between the momentum of the particle and the wavelength of the associated wave is fundamental to de Broglie's theory.

The extension of this idea from light particles (photons) to electrons and other particles was the next step. Niels Bohr in his model of the atom found that the angular momentum of an electron in an atom must be $nh/2\pi$, where n is a whole number. De Broglie showed that this expression for the angular momentum of the electron could be derived from his momentum–wavelength equation if an electron wave exactly makes up the circular orbit of the electron with a whole number of wavelengths and produces a standing wave, that is $n\lambda = 2\pi r$, where r is the radius of the orbit. Otherwise interference would take place and no standing wave would form. De Broglie's idea gave a further explanation of Bohr's model of the atom. Much of this work was described in his doctoral dissertation, although some of it was published in 1923.

If particles could be described as waves then they must satisfy a partial differential equation known as a wave equation. De Broglie developed such an equation in 1926, but found it in a form that did not offer useful information when it was solved. A more useful wave equation was developed Erwin ◊Schrödinger later in 1926.

The experimental evidence for de Broglie's theory was obtained by Clinton ◊Davisson and George ◊Thomson in 1927. These two scientists independently produced electron diffraction patterns, showing that particles can produce an effect that had until then been exclusive to electromagnetic waves such as light and X-rays. Such waves are known as matter waves.

De Broglie's discovery of wave–particle duality enabled physicists to view Einstein's conviction that matter and energy are interconvertible as being fundamental to the structure of matter. The study of matter waves led not only to a much deeper understanding of the nature of the atom but also to explanations of chemical bonds and the practical application of electron waves in electron microscopes.

Throughout his life, de Broglie was concerned with the philosophical issues of physics and he was the author of a number of books on this subject. He pondered whether the statistical results of physics are all that there is to be known or whether there is a completely determined reality which our experimental techniques are as yet inadequate to discern. During many of his years as a professional scientist, he inclined to the former view but his later writing suggests his belief in the latter.

Debye, Peter Joseph Willem (1884–1966) Dutch-born US physical chemist who in a long career made many important contributions to the science. He was awarded the 1936 Nobel Prize for Chemistry for his work on dipole moments and molecular structure.

Debye was born on 24 March 1884 in Maastricht. From 1900–05 he went to school at the Technische Hochschule over the border in Aachen, Germany, where he qualified as an electrical engineer. He then became assistant to Arnold Sommerfeld at the University of Munich, gaining his PhD in 1910. The next few years saw a remarkable progress from one distinguished post to another, starting in Zürich in 1910 where he succeeded Albert Einstein as professor of theoretical physics and culminating, via Utrecht in 1912 and Göttingen in 1914, in a return to Zürich in 1920 as professor of experimental physics and director of the Physics Institute. By 1927 Debye moved to Leipzig, where he took over from Friedrich Ostwald, and in 1934 he went to Berlin to supervise the building of the Kaiser Wilhelm Institute of Physics, which he renamed the Max Planck Institute.

In the late 1930s Debye found himself in difficulties with the Nazi authorities in Germany – mainly because of his Dutch nationality. He was lecturing at Cornell University in the USA in 1940 when Germany invaded the Netherlands, whereupon he accepted the post of professor and head of the chemistry department at Cornell. He became a US citizen in 1946 and formally retired in 1952, although he remained active until his death at the age of 82 on 2 November 1966 at Ithaca, New York.

Debye's first major contribution was a modification of Einstein's theory of specific heats to include compressibility and expansivity, leading to the expressions for specific heat capacities, $C_v = aT^3$, as T approaches absolute zero. The 'Debye extrapolation' incorporating these terms acknowledges the action of intermolecular forces. His studies of dielectric constants led to the explanation of their temperature dependence and of their importance in the interpretation of dipole moments as indicators of molecular structure. The unit of dielectric constant is now called the debye.

While he was at Göttingen in 1916, Debye followed on the work of Max von Laue and William and Lawrence Bragg on X-ray crystallography. He showed that the thermal motion of the atoms in a solid affects the X-ray interfaces and explained (using his specific heat theories) the temperature dependence of X-ray intensities. This work provided the basis for his observation with Paul Scherrer (1890–1969) that randomly orientated particles can produce X-ray diffraction patterns of a characteristic kind. Thus the need for comparatively large single crystals was avoided and powder X-ray diffraction analysis – now called the Debye–Scherrer method – became a new and versatile analytical tool.

At Zürich, Debye's work was dominated by his interest in electrolysis and the extension in 1923 of Svante ◊Arrhenius's theory of ionization developed by himself and Erich ◊Hückel. The Debye–Hückel theory compares the ordering of ions in solution to the situation in the crystalline state and postulates (a) that ionization is complete in a strong electrolyte and (b) that each ion is surrounded by a cluster of ions of opposite charge. The extent of this ordering is determined by the equilibrium between thermal motion and interionic forces. Also at Zürich, Debye used the quantum theory to derive a quantitative interpretation of the Compton effect (the small change in wavelength that occurs when an X-ray is scattered by collision with an electron). This laid the foundation for other researchers' work on electron diffraction.

In the 1930s Debye moved on to a study of the scattering of light by solutions. He showed that sound waves in a liquid can behave like a diffraction grating and developed techniques in turbidimetry that led to useful molecular-mass determinations for polymers. In the last phase of his researches Debye pursued his interest in polymers, investigating their behaviour in terms of viscosity, diffusion, and sedimentation, and he was involved in studies of synthetic rubber.

Debye was an excellent teacher as well as a brilliant experimentalist. But perhaps the outstanding feature of his long career was the very clear thinking that enabled him to persist with incomplete or inadequate theories until he had derived important generalizations.

Further Reading

Corson, Dale R; Salpeter, Edwin E; and Bauer, S H, 'Debye: an interview, *Science*', 1964, v 145, 554–559.

Debye, Peter J, *The Collected Papers of Peter J W Debye*, Ox Bow Press, 1988.

Dedekind, (Julius Wilhelm) Richard (1831–1916)

German mathematician, a great theoretician in some respects way ahead of his time, whose work on irrational numbers – in which he devised a system known as Dedekind's cuts – led to important and fundamental studies on the theory of numbers. A pupil and friend of outstanding mathematicians, he taught for many years and published some highly influential books.

Dedekind was born on 6 October 1831, the youngest son of a professional civil servant who worked at the Collegium Carolinum in Braunschweig (Brunswick). Educated locally, Dedekind showed aptitude in the sciences, particularly in physics. Nevertheless, at the age of 17, it was mathematics he went to study at the Collegium Carolinum for two years before entering the University of Göttingen, where he was taught by the ageing Karl ◊Gauss. Having received his PhD in 1852, he remained as an unpaid lecturer at the university for a few years before taking up the post of professor of mathematics at the Zürich Polytechnicum. In 1862 he returned to Brunswick and became professor at the Technische Hochschule – a position he held for the remainder of his life. He was director of the Hochschule 1872–75, and retired (as professor emeritus) in 1894. Dedekind became a member of several academies and received a large number of honorary degrees and other awards. He died in Brunswick on 12 February 1916.

At Göttingen, Dedekind developed a friendship with Bernhard ◊Riemann, who later became professor of mathematics there. He also met Lejeune ◊Dirichlet and, with Karl Gauss, the four of them formed a formidable mathematical quartet that profoundly influenced each other's ways of thinking. From Gauss, Dedekind learned about the method of least squares (in lectures he was able to recall vividly after 50 years); probably through Riemann's influence, Dedekind's thesis concerned the theory of integrals devised by Leonhard Euler; and from Dirichlet – himself to succeed Gauss and precede Riemann as professor of mathematics at Göttingen – Dedekind learned about the theory of numbers, potential theory, definite integrals, and partial differential equations. Both Riemann and Dedekind also gave lectures at the university, Riemann on Abelian and elliptic functions, Dedekind on the new beginnings of group theory (as advanced by Evariste Galois just before his death in 1832). Dedekind was no lecturer, however, and his seminars were very ill-attended.

He was nevertheless outstanding for his original contributions to mathematics. In 1858 he had succeeded in producing a purely arithmetic definition of continuity and an exact formulation of the concept of the irrational number. From this, and from his editing of Dirichlet's lectures in 1871 (to which he added a supplement establishing the theory of algebraic number fields), he derived the subject of the first of his three great publications, *Stetigkeit und irrationale Zahlen* (1872). In it he defined and explained the use of what are now called Dedekind's cuts – a device by which irrational numbers can be categorized as fractions – a completely original idea that has since passed into general use in the real number system.

Number theory continued to fascinate Dedekind. In his second great work, *Was sind und was sollen die Zahlen?* (1888), he elaborated on his attempt to derive a purely logical foundation for arithmetic, and devised a number of axioms that formally and exactly represented the logical concept of whole numbers. (The axioms were later wrongly attributed to the Italian mathematician of a generation ahead, Giuseppe Peano, by whose name they are still known.)

In his third great work, Dedekind returned again to one of his former interests in order to extend his previous research; he described the factorization of algebraic numbers using his new theory of the 'ideal' – the modern algebraic concept. Published in two sections, in 1879 and 1894, the work was fundamental in that Dedekind later further developed his theory of the ideal (determining the number of ideal classes in a field) and the subject was taken up by others.

Dedekind was also responsible for the publication of papers on a variety of other mathematically oriented subjects such as time-relationships and hydrodynamics; in 1897 and 1900 he introduced the concept of dual groups, which was eventually developed (well after his death) into the modern lattice theory.

de Duve, Christian René (1917–) English-born Belgian biochemist who discovered two organelles, the lysosome and the peroxisome. For this important contribution to cell biology he was awarded (jointly with Albert Claude and George Palade) the 1974 Nobel Prize for Physiology or Medicine.

De Duve was born on 2 October 1917 in Thames Ditton, Surrey, of Belgian parents who had taken refuge in England during the war. He was educated at the University of Louvain, Belgium, from which he graduated in medicine in 1941. He then held positions at the Nobel Institute, Stockholm, and Washington University before returning to Belgium in 1947. In 1951 he was appointed professor of biochemistry at the University of Louvain Medical School, a position he retained in 1962 when he became professor of biochemistry at the Rockefeller Institute (now Rockefeller University), New York City. He became an emeritus professor at Louvain in 1985 and at Rockefeller in 1988.

De Duve discovered lysosomes in the cytoplasm of animal cells in 1955, since when a similar organelle has been found in plant and fungal cells. As seen under the electron microscope, lysosomes in animal cells are usually spherical and are surrounded by a membrane. They are internally structureless but contain characteristic degradative enzymes, which can digest most known biopolymers, such as proteins, fats, and carbohydrates. After his discovery of the lysosome, de Duve found that the main role of lysosomes in the normal functioning of cells is intracellular digestion and he went on to describe the way in which this occurs. All cells require a supply of essential extracellular raw materials, some of which pass into the cell by diffusion. But with substances that are too large to diffuse into the cell, the cell membrane invaginates and surrounds the extracellular matter, forming a food vacuole; this process is called endocytosis. A lysosome then fuses with the food vacuole and releases its acid digestive enzymes into the vacuole (which at this stage is called a digestive vacuole or a secondary lysosome). The enzymes break down the material in the vacuole into molecules small enough to diffuse through the vacuole's wall into the cytoplasm. The undigested remnants contained within the residual body (as the vacuole is called at this stage) eventually pass out of the cell by exocytosis. In addition to intracellular digestion, lysosomes also play a part in the digestion and removal of dysfunctional organelles, a process known as autophagy.

Since the discovery of lysosomes and the elucidation of their role in intracellular processes, Gery Hers – a collaborator of de Duve – found in 1964 that malfunctioning of the lysosomes (which often involves an absence or insufficiency of one or more of the lysosomal enzymes) is associated with several diseases, some of which are hereditary.

As a result of his research on lysosomes, de Duve suspected the existence of another organelle, and in the 1960s he discovered the peroxisome. Almost identical to the lysosome in structure, the peroxisome is characterized by the enzymes it contains.

De Forest, Lee (1873–1961) US inventor of the triode valve, which contributed to the development of radio, radar, and television.

Born in Council Bluffs, Iowa, on 26 August 1873, De Forest was raised in Alabama, where his family moved in 1879. Even as a child, De Forest was keenly interested in machinery, building model trains and a blast furnace while still in his early teens. Despite pressure from his father (a Congregational minister who made great efforts to bring education to the local black community), he decided to follow his scientific inclinations, enrolling at the Sheffield Scientific School of Yale University in 1893. He was forced to supplement his scholarship and meagre allowance from his parents by taking menial jobs, but succeeded in gaining his doctorate in 1899. Called 'Reflection of Hertzian waves from the ends of parallel wires', his thesis was probably the first US dissertation to deal with radio.

De Forest's first appointment was with the Western Electric Company in Chicago. In the company's experimental laboratories, he devised ways of rapidly

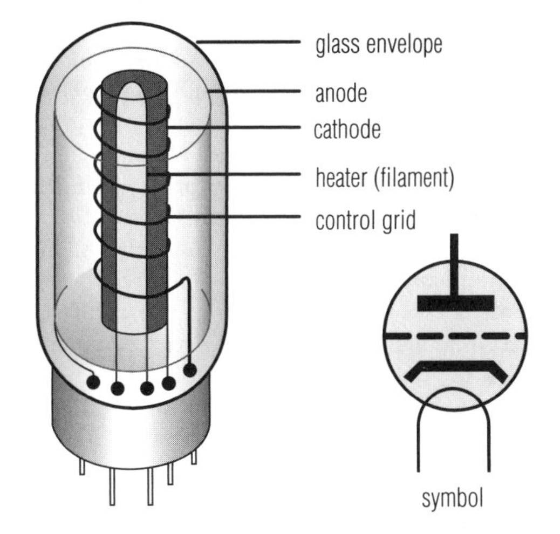

De Forest Lee De Forest developed the triode tube, a three-electrode vacuum tube in which a control grid allows the current flowing through the valve to be controlled by the voltage at the grid. The triode was used in amplifying circuits for radio and early computers, until largely superseded by transistors and other solid-state devices.

transmitting wireless signals, his system being used in 1904 in the first wireless news report (of the Russo-Japanese War).

Following this and other successes De Forest set up his own wireless telegraph company, but by 1906, following a number of serious misjudgements (he was twice defrauded by business partners), the firm went bankrupt. Then came his biggest breakthrough: the 'audion' detector, which he patented in 1907. This thermionic triode vacuum tube was based on the two-element (diode) valve device patented by John Ambrose ⇨Fleming in 1904. By adding a grid for a third electrode, De Forest had turned Fleming's valve into an amplifier as well as a rectifier. It made possible radios, radar, television, and even the earliest computers.

Even with such a major development behind it, De Forest's second company started to fold in 1909, again through internecine wrangling. He was indicted for attempting to use the US mail to defraud, by seeking to promote the 'worthless' audion tube, a charge of which he was later acquitted. He achieved much more favourable attention in 1910, however, by using his invention to broadcast the singing of Enrico Caruso at the Metropolitan Opera. In 1912 De Forest realized that by cascading the effect of a series of audion tubes, it would be possible to amplify high-frequency radio signals to a far higher degree than that achieved using single tubes. The use of this effect made possible long-range (consequently weak-signal) radio and telephonic communication. The same year saw his discovery of a way of using feedback to produce a means of transmission capable of sending both speech and music, and in 1916 he set up a radio station and was broadcasting news.

De Forest eventually sold his audion triode to American Telephone and Telegraph for $290,000, which used it to amplify long-distance communication. This company was to purchase many of De Forest's best inventions at very low prices, yet further proof of his lack of business acumen.

In 1923 De Forest demonstrated an early system of motion pictures. carrying a soundtrack called phonofilm. Unfortunately its poor quality, and lack of interest from filmmakers, led to its demise, despite the fact that the system that would later succeeded commercially was based on the same principles.

Even in the scientific world, De Forest was unlucky, being strongly recommended for the Nobel prize but failing to be awarded it. He ended his days quietly, with his third wife Marie, and died in Hollywood, California, on 30 June 1961.

De Havilland, Geoffrey (1882–1965) English pioneer of the early aircraft industry. He built his own aircraft six years after the Wright brothers' first flight, and then produced a line of mainly successful machines which culminated in the Comet, the first pure jet passenger airliner and the Comet 4, the first jetliner to fly the Atlantic.

De Havilland was born on 27 July 1882, near High Wycombe, where his father was a curate. As a schoolboy he was an enthusiastic builder of model engines. He graduated to the design and construction of racing steam cars and designed and built motor cycles, one of which became the basis of successful production machines, and in 1905 joined the Wolseley company in Birmingham.

In 1908 De Havilland was a draughtsman at the Motor Omnibus Company of Walthamstow, fascinated by news from abroad of flying. Now that flight had become a reality in the hands of the Wright brothers and Faman, De Havilland became convinced that he could design his own engine and aeroplane and teach himself to fly. He asked his friend and former marine engineering apprentice Frank Hearle to give up his job as a mechanic at another bus company and to join him in this project. Having obtained his legacy of £1,000 in advance from his grandfather, De Havilland rented a room and set about designing his first aero engine.

It was a flat-four water-cooled design giving 34 kW/45 hp at 1,500 rpm for a dry weight of 113 kg/250 lb. The Iris Car Company agreed to make a prototype for £220 and De Havilland rented the attic of a builder's workshop in Fulham, barely big enough for the span of the biplane wings he contemplated. The two men began construction of the aircraft, which was based on a tapering whitewood space-frame fuselage projecting almost equally in front and behind the 11-m/36-ft wings. It was to carry a 4.3-m/14-ft pair of elevators each side of the nose and a 3-m/10-ft plane at the tail, above which a rudder was stayed by a thin outrigger from the flat radiator forming the top wing centre-section. The front spar of the main planes differed noticeably in location from all previous biplanes, being more than 30 cm/1 ft back from the leading edge. Earlier biplanes, such as those of Wright, Farman, and Cody, had used the leading edge as a spar but this gave a high drag entry and a weak spar for the weight. De Havilland's design, using a location at a point of greater wing thickness in a better aerodynamic shape was to prove the ultimate standard.

By November 1909 work was nearly finished. De Havilland hired a lorry to transport the four wings, engine, fuselage, and other components to the Hampshire Downs where sheds had been bought at Seven Barrows from Moore-Brabazon (later Lord Brabazon).

De Havilland had never flown before (indeed, he had only ever seen one aircraft flying in the distance), and his first flight rather inevitably ended in the aircraft being wrecked. However, the salvaged aircraft Number One was the basis of the successful Number Two, even though only the engine had been left intact. In 1910 the Downs south of Newbury, Berkshire, saw De Havilland flying consistently for as long as 40 minutes around the countryside. He had done away with the twin propellers and mounted a single wooded screw direct on the crankshaft. The landing gear had been lightened and the main structure made lighter, simpler, and more robust.

To raise more money for further trials, De Havilland sold Number Two to the Army Balloon Factory (later called the Army Aircraft Factory) for £4,000, its two builders being taken on the staff. Although officialdom favoured balloons, De Havilland's aircraft was awarded an airworthiness certificate and officially designated FE1. (This stood for 'Faman Experimental No 1' because, like Faman's machine, it had a pusher propeller.) De Havilland produced FE2 with a Gnome rotary engine, then BE1 ('Blèriot Experimental' because it had a tractor propeller like Blèriot's machine) and a number of other designs.

As the war clouds were gathering in 1914, De Havilland joined the Aircraft Manufacturing Company, or Airco, as chief designer and produced DH1, a two-seater pusher, and DH2, a single-seat fighter with a single Lewis gun in the nose, which went into quantity production. Then came the DH4, a fast two-seat bomber that could hold its own against most fighters.

After World War I, De Havilland established his own company at the famous Stag Lane Works at Edgeware and during the interwar period produced a series of extremely successful light transport aircraft. The DH50, a civil version of the DH4, won a competition at Gothenburg, with Alan Cobham at the controls, but it was the Moth series, starting with the Cirrus Moth, that opened up aviation as no other aircraft had done before. To power these new machines De Havilland designed the 75 kW/100 hp Gipsy 1 engine, followed by the 97 kW/130 hp Gipsy Major, and established his own Engine Division. The Tiger Moth became the standard RAF trainer and De Havilland won the 1933 King's Cup air race in the three-seat Leopard Moth at 224 kph/140 mph.

Many records were established in DH aircraft, with the all-wood DH88 Comet racer monoplane, built in 1934 for the MacRobertson England–Australia Air Race, covering the 19,680 km/11,300 mi to Melbourne in 79 hours 59 minutes.

As a private venture, the De Havilland Company designed the famous all-wood Mosquito, which was at first rejected by the Air Ministry, losing six months of precious time. The highly versatile aircraft went into squadron service in September 1941 and a total of 7,781 were built. It was about 30 kph/20 mph faster than the Spitfire and so, like the DH4, could out-fly virtually anything in the air.

After World War II the De Havilland company put a range of jet-powered aircraft into production, many of which used the company's own engines. The world's first jet trainer was the De Havilland Vampire fighter and fighter–bomber, built under licence in Europe. The final development of this twin-boom jet fighter was the Sea Vixen.

The dangers facing pioneers in aeronautics were made tragically apparent to De Havilland on several occasions. On 27 September 1946, the experimental tailless DH108 Swallow broke up over the Thames estuary, killing his eldest son, Geoffrey. His second son, John, had been killed three years earlier in a mid-air collision in a Mosquito. A re-built Swallow subsequently captured the world air-speed record, averaging 968.37 kph/605.23 mph on a 100 km/62 mi closed circuit with John Derry at the controls. Derry died when a wing of his DH110 folded during a display at the Farnborough Air Show: 28 spectators were killed in the wreckage.

The world's first production jet airliner, the Comet, first flew on 27 July 1949, but after a triumphant entry into service in May 1952, a Comet crashed after take off from Rome in January 1954. A second crashed under the same circumstances in April, and caused the Comets to be withdrawn from service. After the most exhaustive investigation ever carried out on an aircraft, it was established that the pressurized cabin had ruptured due to the then unsuspected problems of low-cycle fatigue. The Comet eventually surmounted its problems, Comet 2s entering service with the Royal Air Force while the Comet 4 became the first jet airliner to operate transatlantic scheduled services, beating the Boeing 707 by a narrow margin.

Other civil projects, however, enjoyed notable success. He was knighted in 1944. De Havilland died in Stanmore, Middlesex, on 21 May 1965 having seen the company he had founded absorbed into the Hawker Siddeley conglomerate, which had also swallowed Armstrong Whitworth, Avro, Blackburn, Folland, Gloster, and Hawker.

Dehn, Max (1878–1952) German-born US mathematician who in 1907 provided one of the first systematic studies of what is now known as topology – the branch of mathematics dealing with geometric figures whose overall properties do not change despite a continuous process of deformation, by which a square is (topologically) equivalent to a circle, and a cube is (topologically) equivalent to a sphere.

Dehn was born on 13 November 1878 in Hamburg. He studied at Göttingen University under David

◊Hilbert and received his doctorate in 1900. He then became a teacher. At the outbreak of World War I he joined the army. After the war Dehn became professor of pure and applied mathematics at Frankfurt University, and remained there until 1935, when he fell victim to Adolf Hitler's anti-Semitism laws and he lost his position. Accordingly, he emigrated in 1940 to the USA, and occupied posts at the University of Idaho, the Illinois Institute of Technology, and St John's College, Annapolis. From 1945 Dehn worked at the Black Mountain College in North Carolina. He died there on 27 June 1952.

Influenced strongly by David Hilbert, Dehn's work was mainly concerned with a study of the geometric properties of polyhedra. His first major contribution was to demonstrate that whereas the postulate of Archimedes – that the sum of the angles of a triangle is not greater than two right angles – is not provable, a generalization of the related theorem proposed by Adrien Legendre – that the sum of the angles of any two triangles is identical – *is* provable.

In a famous address in 1900, David Hilbert presented 23 unsolved mathematical problems to the International Congress of Mathematicians. Dehn found a solution to one of them (concerning the existence of tetrahedra with equal bases and heights, but not equal in the sense of division and completeness).

In 1910 Dehn proved an important theorem on topological manifolds. The theorem came to be known as Dehn's lemma, but was later found not to apply in all circumstances. It nevertheless provided stimulation for considerable scientific discussion. Dehn continued to work on topological problems of transformation and isomorphism.

Dehn's later research concerned statistics and the algebraic structures derived from differently axiomatized projective planes. He also made a notable contribution with his published work on the history of mathematics.

De la Beche, born Beach, Henry Thomas (1796–1855) English geologist who secured the founding of the Geological Survey.

Born in London, the son of a military officer (Thomas Beach) whose wealth derived from Caribbean slave plantations, De la Beche originally trained for the army at the military school at Great Marlow, and served in the Napoleonic wars. Quitting the army and gentrifying his name, he turned himself into a gentleman amateur geologist, joining the Geological Society of London in 1817 and travelling extensively during the 1820s throughout the Britain and Europe. He was soon publishing widely in descriptive stratigraphy, above all on the Jurassic and Cretaceous rocks of the Devon and Dorset area. He also conducted important fieldwork on the Pembrokeshire coast and in Jamaica. He prided himself upon being a scrupulous fieldworker and a meticulous artist. Works like *Sections and Views Illustrative of Geological Phenomena* (1830) and *How to Observe* (1835) insisted upon the primacy of facts and sowed distrust of theories – views still being expounded in his final masterpiece, *The Geological Observer* (1851). Volumes like the *Manual of Geology* (1831) prove he was also an effective textbook author.

In the 1830s he evolved the plan of a government-sponsored geological study of Britain, region by region, on national and economic grounds. He was paid £500 out of state funds for a survey of Devon, which he personally undertook; and he quickly proved successful in persuading Whitehall to put these provisions on a more formal basis. In 1835, on the analogy of the Ordnance Survey, the Geological Survey was founded, with De la Beche as its first director. The survey flourished and expanded, and De la Beche's own career culminated in the establishment of a Mines Record Office and the opening, in 1851, under the aegis of the Geological Survey, of the Museum of Practical Geology and the School of Mines in Jermyn Street, London. For his services, De la Beche was knighted in 1842. He died in London on 13 April 1855.

de la Rue, Warren (1815–1889) British pioneer of celestial photography. Besides inventing the first photoheliographic telescope, he took the first photograph of a solar eclipse and used it to prove that the prominences observed during an eclipse are of solar rather than lunar origin.

De la Rue was born in Guernsey on 15 January 1815, the eldest son of Thomas de la Rue, a printer. Warren attended the Collège Sainte-Barbe in Paris before joining his father in the printing business. It was then that he first came into contact with science and technology. He was one of the first printers to adopt electrotyping and in 1851 he invented the first envelope-making machine. Initially he saw himself as an amateur chemist and, before making any achievements in the field of astronomy, he invented the silver chloride battery. He became a Fellow of the Royal Society, the Chemical Society and the Royal Astronomical Society. In later life, from 1868–83, he conducted a series of experiments on electric discharges through gases, but his results were inconclusive. He died in London on 19 April 1889.

De la Rue was introduced to astronomy by a friend, James ◊Nasmyth, who, like de la Rue, was a successful businessman besides being an inventor and telescope-maker. De la Rue began his research career in astronomy with the intention of producing more accurate and detailed pictures of the Moon and neighbouring heavenly bodies. His early observations of the Moon, the Sun, and Saturn were superbly drawn and their details were enhanced by de la Rue's innovative techniques in polishing and figuring the mirror of his own 33-cm/13-in reflecting telescope.

De la Rue's interest in new technologies led him to apply the art of photography that had been pioneered by Louis Daguerre to astronomy. He modified his telescope to incorporate a wet collodion plate. His first photographs were of the Moon, taken with exposures of 10–30 seconds. They were remarkably successful, considering that there was no drive fitted to de la Rue's telescope and that he had to guide the instrument by hand to hold the image steady on the photographic plate. The lack of a good drive made longer exposures impossible to achieve and so de la Rue postponed further research in celestial photography until he had built and equipped a new observatory.

At the new observatory de la Rue began a regular programme of astronomical photography, including a daily sequence of photographs of the Sun. He designed a photoheliographic telescope in connection with this project and took it to Spain in July 1860 to photograph a total eclipse of the Sun. This expedition was made in collaboration with Pietro ◊Secchi of the Collegio Romano, in order that photographs of prominences, only seen during the total phase of a solar eclipse, could be taken from two separate stations, 400 km/249 mi apart. The resulting photographic plates showed conclusively that the prominences were attached to the Sun and were not, as had been suggested, either effects of the Earth's atmosphere or the result of some unknown lunar phenomenon.

De la Rue's photoheliograph was subsequently set up at the Kew Observatory, sited about 8 km/5 mi from de la Rue's private observatory and home at Cranford, Middlesex. He used it to map the surface of the Sun and study the sunspot cycle. This work led to his being able to show that sunspots are in fact depressions in the Sun's atmosphere. De la Rue continued to use his reflecting telescope to take photographs of the Moon's surface and over a period of eight years his series of plates brought certain details to light that had never been noted before. This sequence turned out to be particularly relevant to a controversy initiated by Julius Schmidt, director of the Athens Observatory, who announced in 1866 that one of the lunar craters had disappeared.

De la Rue's particular talents were his understanding of technology and his innovative flare in designing instruments. As a result his observations were so accurate that they contributed to major advances in theoretical astronomy.

de Laval, Carl Gustaf Patrik (1845–1913) Swedish engineer who made a pioneering contribution to the development of high-speed steam turbines.

De Laval was born in Orsa on 9 May 1845, into a family that had emigrated to Sweden from France in the early 17th century. He was educated at the Stockholm Technical Institute and at Uppsala University. In the vigour and variety of his interests – and of his inventive talent – he has been likened to Thomas Edison. In 1878 he invented a high-speed centrifugal cream separator that incorporated a turbine, and the machine was successfully marketed and used in large dairies throughout the world. He also invented various other devices for the dairy industry, including a vacuum milking machine, perfected in 1913. De Laval's greatest achievement, however, lay in his further contribution to the development of the steam turbine, which he completed in 1890 after several years of experiments. In the absence of reliable data on the properties of steam, de Laval solved the problem of the high velocity by special features in the design of the wheel carrying the vanes of the turbine, and that of direction of the stream of particles by the form given to the nozzle through which the steam jet was produced. The turbine disc had a hyperbolic profile and was mounted on a flexible shaft, which ran well above the critical whirling speed. The machine had a convergent–divergent ('condi') exit nozzle. De Laval died in Stockholm on 2 February 1931.

De Laval's other interests ranged from electric lighting to electrometallurgy in aerodynamics. In the 1890s he employed more than 100 engineers in developing his devices and inventions, which are exactly described in the thousand or more diaries he kept.

The history of turbines goes back to ancient times, when people first began to use water and wind to perform useful tasks. The first device that could be classified as a steam turbine is generally attributed to ◊Hero of Alexandria is about the 1st century AD. Operating on the principle of reaction, this device achieved rotation through the action of steam issuing from curved tubes, or nozzles, in a manner similar to that of water in a rotary lawn sprinkler. Another steam-driven machine, described in about 1629, used a jet of steam impinging on blades projecting from a wheel, causing it to rotate. This 'motor', in contrast to Hero's reaction machine, operated on the impulse principle.

The first steam turbines having any commercial significance appear to have been those built in the USA by William Avery in 1831. His turbines consisted of two hollow arms attached at right angles to a hollow shaft, through which the steam could issue. The steam vents were openings at the trailing edge of the arms, so that rotation was achieved by the reactive force of the steam. Although about 50 of these turbines were made and used in sawmills, wood-working shops, and even on a locomotive, they were finally abandoned because of their difficult speed regulation and frequent need of repair. Among the prominent later inventors working in the steam-turbine field was Charles ◊Parsons. He soon recognized the advantages of employing a large number of stages in series, so that the release of energy from the expanding steam could take place in small steps. This principle opened the way for the development of the modern steam turbine. Parsons also developed the reaction-stage principle, in which pressure drop and energy release are equal, through both the stationary and moving blades.

In 1887 de Laval developed his small high-speed turbines with a single row of blades and a speed of 42,000 revolutions per minute. Although several of these were later employed for driving cream separators, he did not consider them practical for commercial application. De Laval turned to the development of reliable, single-stage, simple-impulse turbines. He is credited with being the first to employ a convergent–divergent type of nozzle in a steam turbine in order to realize the full potential energy of the expanding steam in a single-stage machine. During the period 1889–97 he built a large number of turbines (ranging in size from about five to several hundred horsepower), and it was de Laval who invented the special reduction gearing that allows a turbine rotating at high speed to drive a propellor or machine at comparatively slow speed, a principle having universal application in marine engineering.

Democritus (*c.* 460–*c.* 370 BC) Greek philosopher who is best known for formulating an atomic theory of matter and applying it to cosmology.

Little is known of Democritus' life. He was probably born in Abdera, Thrace, in about 460 BC. He became rich, and travelled far from his native Greece, particularly to Egypt and the East. In his long lifetime, Democritus is reputed to have written more than 70 works, although only fragments of them have survived. He died, aged 70, in about 370 BC.

Democritus believed that all matter – throughout the universe, solid or liquid, living or non-living – consists of an infinite number of tiny indivisible particles which he called atoms. According to the theory, atoms cannot be destroyed (an idea similar to the modern theory of the conservation of matter) and exist in a vacuum or 'void', which corresponds to the space between atoms. Atoms of a liquid, such as water, are smooth and round and so they easily 'roll' past each other and the liquid is formless and flows. A solid, on the other hand, has angular, jagged atoms that catch onto each other and hold them together as a solid of definite form. Atoms differ only in shape, position, and arrangement. Thus the differences in the physical properties of atoms account for the properties of various material substances. When atoms separate and rejoin, the properties of matter change, dissolving and crystallizing, dying and being reborn. Even new worlds can be created by the coming together in space of a sufficient number of similar atoms.

It is tempting, with the benefit of modern scientific knowledge, to overemphasize the apparent foresight of Democritus' theories. Remarkable although they undoubtedly were, they were a product of his mind and had no basis in experiment or scientific observation. Eventually they were superseded by other philosophies, and atomism was discarded as a theory until revived 800 years after Democritus by such scientists as ◊Galileo.

Further Reading

Cole, Thomas, *Democritus and the Sources of Greek Anthropology,* APA Philological Monographs series, Scholars Press, 1990.

McDonnell, John J, *The Concept of an Atom from Democritus to John Dalton,* Mellen University Press, 1992.

Said, Dibinga W, *The African Origins of Democritus's Atomic Thought,* The African Origins of Greek Philosophy series, Omenana, 1996,v 3.

de Moivre, Abraham (1667–1754) French mathematician who, despite being persecuted for his religious faith and subsequently leading a somewhat unstable life, pioneered the development of analytical trigonometry – for which he formulated his theorem regarding complex numbers – devised a means of research into the theory of probability, and was a friend of some of the greatest scientists of his age.

De Moivre was born on 26 May 1667 in Vitry-le-François, Champagne – a Huguenot (Protestant) in an increasingly intolerant Roman Catholic country. Although he first attended a local Catholic school, his next school was closed for being too evidently Protestant, and he then studied in Saumur, and finally (in 1684, at the age of 17) in Paris. With the revocation of the Edict of Nantes in the following year, however, he was imprisoned as a Protestant for 12 months; on his release he went immediately to England. In London he became a close friend of Isaac ◊Newton and Edmund ◊Halley. It was Halley who read de Moivre's first paper – on Newton's 'fluxions' (calculus) – to the Royal Society in 1695, and saw to his election to the Royal Society in 1697. (Forty years later he was elected a Fellow of the Berlin Academy of Sciences, and no less than fifty years later – in the year of his death – to the Paris Academy of Sciences.) In 1710 de Moivre was appointed to the grand commission through which the Royal Society tried to settle the dispute over the priority for the systematization of calculus between Gottfried ◊Leibniz and Newton. Although de Moivre was a distinguished mathematician, he spent his whole life in comparative poverty, eking out a precarious living by tutoring and acting as a consultant for gambling syndicates and insurance companies. In spite of the fact that he had powerful friends, he never obtained a permanent position; no matter how he begged his influential associates to help him secure a chair in mathematics, particularly at Cambridge, it was without success. Finally, at the age of 87, he gave in to lethargy and spent 20 hours of each day in bed. He died, nearly blind, on 27 November 1754 in London.

While de Moivre was studying in Saumur, he read mathematics almost secretly, and studied Christiaan Huygens's work on the mathematics of games of chance. It was not until he went to Paris that he received any thorough mathematical instruction and studied the later books of Euclid under the supervision of Jacques Ozanam (1813–1853). But his first view of Newton's *Principia mathematica* came even later, in London. He was so fascinated by it, it is said that he cut out the pages and read them as he walked along the street between tutoring one pupil and another.

He dedicated his own first book, *The Doctrine of Chances,* to Newton. (Subsequently, Newton – as he felt himself becoming more infirm with age – took to sending students to de Moivre.) A masterpiece, the work was published first in Latin in the *Philosophical Transactions* of the Royal Society, and then in expanded English versions in 1718, 1738, and 1758. Until 1711, the only texts on probability were the one by Huygens and another by Pierre de Montmort, published in Paris in 1708. When de Moivre first published, Montmort contested his priority and originality. Both men had made an approximation to the binomial probability distribution; now known as the normal or Gaussian distribution, it was the most important single discovery in the formulation of probability theory and was incorporated into statistical studies for the next 200 years. De Moivre was the first to derive an exact formulation of how 'chances' and stable frequency are related. He obtained from the binomial expansion of $(1 + 1)n$ what is now recognized as $n!$ (the approximation of Stirling's formula). With $n!$, de Moivre could sum the terms of the binomial from any point up to the central term. He seems to have been aware of the standard deviation parameter (σ) although he did not specify it. He also hinted at another approximation to the binomial distribution – which is now attributed to Siméon-Denis ◊Poisson, a century later – but in this case he seems not to have realized its potential in probability theory.

Perhaps with regard to his own constant state of penury, de Moivre also took a great interest in the analysis of mortality statistics, and laid the mathematical foundations of the theory of annuities, for which he devised formulae based on a postulated law of mortality and constant rates of interest on money. He worked out a treatment for joint annuities on several lives, and one for when both age and interest on capital have equal relevance. First published in 1725, his work became standard in textbooks of all subsequent commercial application. Again, however, he had to fight for copyright with Thomas ◊Simpson, who published a work on annuities in 1742, the year before de Moivre republished.

All his life, de Moivre published papers in other branches of mathematics; one of the subjects that particularly interested him was analytical trigonometry. In this field he discovered a trigonometric equation that is now named after him:

$$(\cos z + \mathrm{i} \sin z)^n = \cos nz + \mathrm{i} \sin nz$$

It was first stated in 1722, although it had been anticipated by related forms in 1707. It entails or suggests a great many valuable identities, and became one of the most useful steps in the early development of complex number theory.

Sadly, de Moivre died a disillusioned man; much of his work was valued only long after his death.

De Morgan, Augustus (1806–1871) British mathematician whose main field was the study of logic, an interest that led him into a bitter controversy with his contemporary William ◊Hamilton.

De Morgan was born on 27 June 1806 in Madura (now Madurai), India, the son of a colonel in the Indian Army. He entered Trinity College, Cambridge, in 1823 and graduated with a BA four years later. He disliked competitive scholarship and did not proceed to an MA degree nor did he become a candidate for a fellowship; neither would he comply with his parents' wishes that he should enter the church. De Morgan considered a career in medicine but decided instead to become a barrister, and entered Lincoln's Inn to study for the Bar. He changed his mind yet again and in 1828 applied for and obtained the position of the first professor of mathematics at the new University College, London, where he remained for 30 years. He died in London on 18 March 1871. After De Morgan's death, Lord Overstone bought his library of more than 3,000 books and presented them to the University of London.

De Morgan expended most of his energy on writing voluminous articles on mathematical, philosophical, and antiquarian matters. A major controversy arose from his tract on 'The structure of the syllogism', read to the Cambridge Philosophical Society in 1846 and subsequently incorporated into his book *Formal Logic,* published a year later. De Morgan had consulted William ◊Hamilton on the history of Aristotelian theory, and Hamilton accused him of appropriating his doctrine of 'quantification of the predicate' (and returned the copy of *Formal Logic* that De Morgan had presented to him).

In logic the expression 'every ... is' (or 'all ... is') is treated as a single syntactically unanalysable term, which in concatenation (linked) with two noun expressions forms a proposition. Traditionally, however, it has been held that 'every' (or 'all') modifies the way in which the subject should be construed. Logicians therefore suggested a similar modification of the predicate. This idea was not new – it had been suggested by Aristotle, only to be abandoned later. Some of his early commentators worked out that such modification could generate 16 different propositions, but that is as far as they went. De Morgan recognized these restrictions and succeeded in expanding syllogistic when he developed a logic of noun expressions in *Formal Logic* and in his *Syllabus of a Proposed System of Logic.* He also extended his syllogistic vocabulary using definitions, so giving rise to new kinds of inferences, both direct (involving one premise) and indirect (involving two premises). He was thus able to work out purely structural rules for transforming a premise or pair of premises into a valid conclusion.

Inferences that appeared to illustrate principles belonging to the logic of noun expressions – but which could not be accommodated in syllogistic – had been known before. But De Morgan initiated and developed a theory of relations. He devised a symbolism that could express such notions as the contradictory, the converse and the transitivity of a relation, as well as the union of two relations. Together with George ⇨Boole, De Morgan can be credited with stimulating the upsurge of interest in logic that took place in the mid-19th century.

Descartes French philosopher and mathematician René Descartes, who argued that mind and body are separate entities, and that this made possible human freedom and immortality. Trained at the Jesuit College in La Flèche, France, Descartes was a Catholic throughout his life, although much of his philosophy was subject to opposition by the ecclesiastical authorities. *Mary Evans Picture Library*

Descartes, René (1596–1650) French philosopher and mathematician whose work in attempting to reduce the physical sciences to purely mathematical principles – and particularly geometry – led to a fundamental revision of the whole of mathematical thought. So influential was his work that his astronomically erroneous description of the Solar System (in an endeavour not to offend the powerful Roman Catholic Church) held back astronomical research in continental Europe for decades.

Descartes was born in La Haye, Touraine, on 31 March 1596, the third son of Joachim Descartes, a councillor of the Parliament of Rennes in Brittany. When René was eight years old, he was sent to the Jesuit College at La Flèche, where he spent five years studying grammar and literature and then three years studying science, elementary philosophy, and theology; his favourite subject was mathematics. In 1612 he went to the University of Poitiers to study law and graduated four years later. Wanting to see the world, he joined the army of Prince Maurice of Nassau and used his mathematical ability in military engineering. A dream in November 1619 made him think that physics could be reduced to geometry and that all the sciences should be interconnected by mathematical links: he spent the next ten years applying this tenet to algebra.

Cogito, ergo sum. *I think, therefore I am.*

René Descartes Le Discours de la Méthode

Returning to France in 1622, Descartes sold his estate in Poitou in order to resume his travels, visiting scientists throughout France and western Europe. He finally settled in the Netherlands in 1629. Twenty years later he was invited to go to Sweden to instruct Queen Christina. On his arrival in Stockholm, he found that the somewhat whimsical queen intended to receive her instruction at 5 o'clock each morning. Unused to the cold of a Swedish winter, Descartes very shortly afterwards caught a severe chill and died on 11 February

1650. His remains were taken back to France and buried in the church of St Geneviève du Mont in Paris.

It was at the Jesuit College that mathematics became Descartes's favourite subject 'because of the certainty of its proofs and the logic of its reasoning'; he was surprised at how little had been built on such firm and logical foundations. Later, with the rector of Breda – the philosopher and mathematician Isaac Beeckman – he devised a way of approaching physics generally following mathematical principles.

Descartes's great work in mathematics – most of his publications concerned either philosophy or astronomy – was his *Geometry* (1637). Much of the book was revolutionary for its time but has now been long absorbed into standard textbooks of coordinate geometry. In it he provided a basis for analytical geometry – the geometry in which everything is reduced to numbers, so that a point is a set of numbers which are called its (Cartesian) coordinates, and a figure may be considered as an aggregate of points and described by formulae, equations, or inequalities. Today, analytical geometry has many practical applications, such as cartography and the construction of graphs.

In establishing analytical geometry, Descartes introduced constants and variables into conventional geometry in order to enable the properties of curves to be expressed as algebraic equations. He used algebra to resolve complicated problems in geometry, and he expressed algebraic results geometrically (in graphs). Although not the first to apply algebra to geometry, he was the first to apply geometry to algebra. He was also the first to classify curves systematically, separating 'geometric curves' (which can be precisely expressed as an equation) from 'mechanical curves' (which cannot). The geometric curves he then further subdivided into three groups of increasing complexity according to the degree of the equation – the simplest group contained the circle, the parabola, the hyperbola, and the ellipse. One of his more complex curves was the folium, $(x^2 + y^2)(y^2 + x(x + b)) = 4axy^2$, where x and y are variables and a is a constant; the result is a near loop.

Descartes's new geometry led to a concept of continuity, which in turn led to the theory of function and thence to the theory of limits.

It is not enough to have a good mind. The main thing is to use it well.

René Descartes LE DISCOURS DE LA MÉTHODE

In algebra he systematized the use of exponents (where the variable is itself a power; for example, ax), interpreted the idea of negative quantities, and enlarged on his 'rule of signs' for determining the number of negative and positive roots in (solutions to) an equation. He also resolved the long-standing problem of doubling the cube.

Descartes's work in mathematics was his greatest service to future science; his attempts to 'geometrize' nature and his contributions to pure mathematics were far more permanent in their significance than all his other scientific work.

Descartes set up a general theory of the universe, which was accepted in France for more than a hundred years. This was much longer than the theory deserved, but it was accepted because of Descartes's fame as a mathematician and philosopher. It had been Descartes's intention to write a work entitled *On the World,* founded on the Copernican system. But when he heard of the Roman Catholic Church's attack on ⇨Galileo, he gave up the idea. Some years later, he resolved his dilemma by proposing that the Earth did not move freely through space, but that it was carried round the Sun in a vortex of matter without changing its place in respect of surrounding particles. In this way, it could be said (through a slight stretch of the imagination) to be 'stationary'. In 1644 he published *Principia Philosophiae,* in which he assumed space to be full of matter that in the beginning had been set in motion by God, resulting in an immense number of vortices of particles of different sizes and shapes, which by friction had had their corners rubbed off. In this way, two kinds of matter were produced in each vortex: small spheres, which continued to move round the centre of motion, with a tendency to recede from it; and fine dust, which gradually settled at the centre and formed a star. Those particles that become channelled or twisted in passing through the vortex formed sunspots; these, declared Descartes, might eventually dissolve, or might form a comet, or might settle permanently in a part of the vortex that has a velocity equal to its own, and form a planet. In this way he was able to account for the origin of the Moon and other satellites. His theory was, of course, pure speculation, unsupported by any facts, and was an attempt to explain how the planets move round the Sun, able neither to move away nor to move closer. It does not explain any of the deviations of the planetary orbits, and only with difficulty accounts for the elliptical not circular form of the orbits.

Although his theory now seems rather illogical, it was significant in that the considerable following it had partly explains why Newton's theories and models were not generally accepted on the Continent until the middle of the 18th century.

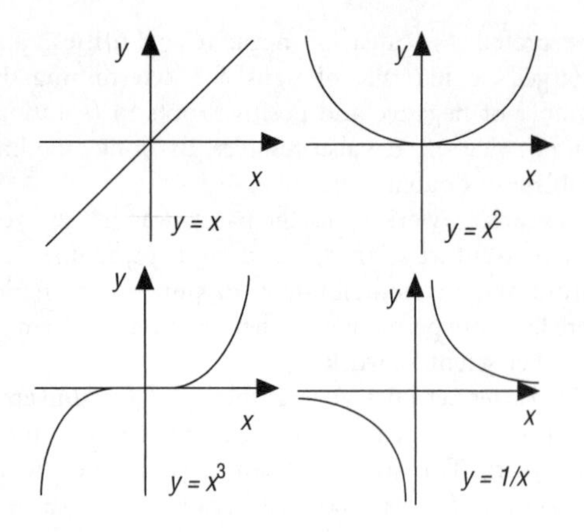

Descartes The use of graphs.

Further Reading

Cottingham, John, *Descartes,* Basil Blackwell, 1986.
Cottingham, John (ed), *The Cambridge Companion to Descartes,* Cambridge Companions to Philosophy series, Cambridge University Press, 1992.
Cottingham, John (ed), *Descartes,* Oxford Readings in Philosophy series, Oxford University Press, 1998.
Gaukroger, Stephen, *Descartes: An Intellectual Biography,* Clarendon Press, 1995.
Strathern, Paul, *Descartes in 90 Minutes,* I R Dee, 1996.
Voss, Stephen (ed), *Essays on the Philosophy and Science of René Descartes,* Oxford University Press, 1993.
Wilson, Margaret Dauler, *Descartes,* Arguments of the Philosophers series, Routledge, 1991.

de Sitter, Willem (1872–1934) Dutch mathematician, physicist, and astronomer whose wide knowledge and energetic application made him one of the most respected theoreticians of his time. He was particularly influential in English-speaking countries in bringing the relevance of the general theory of relativity to the attention of astronomers.

De Sitter was born in Sneek, in the Netherlands, on 6 May 1872. He attended the high school in Arnhem before going on to the University of Groningen. Although his primary interests were in mathematics and physics, he soon became interested in astronomy on learning of studies being conducted by Haga and Jacobus Kapteyn. De Sitter was invited to take his postgraduate courses at the Royal Observatory, Cape Town, South Africa, by David Gill and, since conditions for astronomical observations were excellent in South Africa, he sailed for the Cape in 1897.

Within two years de Sitter had collected sufficient data to enable him to return to the Netherlands to write his doctoral dissertation: he was awarded his PhD in 1901. From 1899–1908 he served as an assistant to Kapteyn, becoming professor of theoretical astronomy at the University of Leiden in 1908. In 1919 he took on the additional post of director of the observatory at Leiden, which was undergoing a programme of redevelopment and expansion. He died of pneumonia in Leiden on 19 November 1934.

De Sitter's early work in Cape Town consisted primarily of photometry and heliometry. Gill suggested that he study the moons of Jupiter, and his subsequent observations of the Jovian satellites were the first of many such investigations, the results of which he published during the course of his career. Taking advantage of data on Jupiter's moons dating back to 1688, in 1925 he produced a new mathematical theory on them. Four years later he published accurate tables describing the satellites. He also obtained an estimate for the mass of the parent planet itself.

Einstein's special theory of relativity appeared in 1905, but few astronomers recognized its importance to their field. In 1911 de Sitter wrote a brief paper in which he outlined how the motion of the constituent bodies of our Solar System might be expected to deviate from predictions based on Newtonian dynamics if relativity theory were valid. After the publication of Einstein's general theory of relativity in 1915 he expanded his ideas and presented a series of three papers to the Royal Astronomical Society on the matter. In the third paper, he introduced the 'de Sitter universe' (as distinct from the 'Einstein universe'). His model later formed an element in the theoretical basis for the steady-state hypothesis regarding the creation of the universe. De Sitter also noted that the solution Einstein presented for the Einstein field equation was not the only possible one. He therefore presented further models of a non-static universe: he described both an expanding universe and an oscillating universe (that is, one that alternately increases and decreases in diameter). One of several theoretical astronomers to develop this field, de Sitter thus contributed to the birth of modern cosmology.

There were two other areas of astronomical research to which he made important contributions. The first of these was the bringing up to date of many of the astronomical constants. He published his work on this subject in two papers, one in 1915 and the other in 1927. A third, incomplete, paper on the subject was published posthumously in 1938. Using geodetic and astronomical data in his analysis, de Sitter demonstrated that there is some variation in the rotation of the Earth, and presented suggestions for its mechanism. He suggested that tidal friction might affect the rotation of not only the Earth but also the Moon, but that some factors might affect the Earth alone.

Deslandres, Henri Alexandre (1853–1948) French physicist and astronomer, now remembered mostly for his work in spectroscopy and for his solar studies.

Deslandres was born in Paris on 24 July 1853. He studied at the Ecole Polytechnique in Paris, graduating in 1874 and then entering the army. He retired from the army in 1881 because of his strong interest in science, having attained the rank of captain. Deslandres worked at the Ecole Polytechnique and the Sorbonne 1881–89, when he began his astronomical career at the Paris Observatory. In 1897 he moved to the observatory at Meudon, where he became assistant director in 1906 and director in 1907. The Paris and Meudon observatories were combined in 1926 and in 1927 Deslandres became director of the Paris Observatory. He retired officially in 1929, but he continued to publish his research until 1947. He died in Paris on 15 January 1948.

Deslandres's early scientific work in spectroscopy led to the formulation of two simple empirical laws describing the banding patterns in molecular spectra; the laws were later found to be easily explained using quantum mechanics. When Deslandres joined the Paris Observatory his task was to organize spectroscopic research. He studied both planetary and stellar spectra, but he soon began to direct his attention to the study of the Sun.

In 1891 George Ellery ◊Hale and Deslandres independently made the same discovery; both turned to the development of a photographic device for studying the solar spectrum in more detail. Hale's spectrograph was ready a full year before Deslandres's version, but the latter's spectroheliograph was a more flexible device and Deslandres used it to full advantage over the ensuing years. Deslandres found his spectroheliograph particularly well adapted to studying the solar chromosphere.

Deslandres was a member of several expeditions to observe total solar eclipses. In 1902 he predicted that the Sun would be found to be a source of radio waves, and 40 years later this was confirmed. From 1908 onwards much of Deslandres's work on the Sun was done in collaboration with L d'Azabuja. After his official retirement Deslandres returned to spectroscopy, with a special interest in the Raman effect. Much of his later work was, however, outside the mainstream of contemporary research.

Desmarest, Nicolas (1725–1815) French naturalist who became a champion of volcanist geology.

A poor boy educated at the Oratorian school at Troyes, he moved to Paris, worked as a private tutor, and later occupied various minor government offices, spending some time as an industry inspector for the department of commerce and displaying expertise in technology. Gradually he staked out for himself a scientific career, publishing charts of the English Channel and developing skills in experimental philosophy. In the 1760s, he made extensive geological tours in Burgundy, Lorraine, Alsace, Franche-Comté, and Gascony, and became interested in the large basalt deposits of central France, discovered by Jean-Etienne Guettard (1715–1786) a decade before. He succeeded in tracing their origins to ancient volcanic activity in the Auvergne region. In 1768 he produced a detailed study of the geology and eruptive history of the volcanoes responsible. His work was important because it asserted categorically that prismatic basalts were igneous in origin – countering the widely held belief, advanced by his contemporary Abraham ◊Werner, that all rocks were sedimentary. Desmarest never however became a fully fledged plutonist, in his later career specifically refuting the views of James ◊Hutton. He emphasized the critical role of water in the shaping of the Earth's history. He did not believe that all rocks had an igneous origin, and was always cautious about making inferences from the former existence of volcanic activity to the role of central heat in the Earth's history. Indeed, Desmarest liked to see himself as a sound empiricist who eschewed speculative theories of the Earth.

Having written extensively for the *Encyclopédie*, in 1771 he was appointed a member of the Academy of Sciences.

Désormes, Charles Bernard (1777–1862) French physicist and chemist whose principal contribution was to determine the ratio of the specific heats of gases. He did this and almost all of his scientific work in collaboration with his son-in-law Nicolas Clément (1779–1841).

Désormes was born in Dijon, Côte d'Or, on 3 June 1777. He was a student at the Ecole Polytechnique in Paris from 1794, when it opened, and subsequently worked there as a demonstrator. Désormes met Clément at the Ecole Polytechnique, beginning a scientific collaboration that lasted until 1824. He left the Ecole in 1804 to establish an alum refinery at Berberie, Oise, with Clément and Joseph Montgolfier, who had earlier pioneered balloon flight.

Clément and Désormes published their most famous work on specific heat in 1819. This memoir led to the election of Désormes (but not of Clément) to the Academy of Sciences in Paris. Désormes's scientific productivity subsequently declined and in 1830 he turned his efforts entirely to politics. He was elected counsellor for Oise in 1830 and, after three consecutive attempts to enter Parliament in 1834, 1837, and 1842, finally was elected to the Constituent Assembly in 1848, in which he sat with the Republicans. He died in Verberie on 30 August 1862.

The period of Désormes's life in which he was actively involved in scientific study was comparatively short. It began in 1801, when he commenced his joint work with Clément. In this and the following year, they correctly determined the composition of carbon disulphide (CS_2) and carbon monoxide (CO). There was a certain amount of acrimonious debate with Claude Berthollet over the work on carbon disulphide, as he contended that it was in fact identical to hydrogen sulphide (H_2S). In 1806, Clément and Désormes made an important contribution in the field of industrial chemistry by the elucidation of all the chemical reactions that take place during the production of sulphuric acid by the lead chamber method, in particular clarifying the catalytic role of nitric oxide in the process. This improved the production of sulphuric acid, which was of great benefit to the chemical industry. In 1813 they made a study of iodine and its compounds.

From 1812 onwards, Clément and Désormes worked together on heat. Their first achievement in this field was to make an accurate estimate of the value for absolute zero. They also studied steam engines, showing that an engine would develop more power if the steam could undergo maximum expansion.

In 1819 Désormes and Clément published a classic paper describing the first determination of γ, the ratio of the specific heat capacity of a gas at constant pressure (C_p) to its specific heat capacity at constant volume (C_v). They did this by allowing a quantity of gas to expand adiabatically, thus lowering its temperature and pressure, and then to heat up to the original temperature without expanding, thus increasing its pressure. The ratio is found by a simple calculation involving the pressures that are produced. This method was not significantly improved until 1929.

Sadi ◊Carnot, who attended Clément's lectures, was probably inspired by the work of Clément and Désormes to develop his fundamental theorem on the motive power of heat.

De Vaucouleurs, Gérard Henri (1918–1995) French-born US astronomer who carried out important research into extragalactic nebulae.

Born in France on 25 April 1918, De Vaucouleurs was educated at the University of Paris and obtained his BSc in 1936. From 1945–50, he was a research fellow at the Institute of Astrophysics, National Centre for Scientific Research, before he went to Australia, to the Australian National University 1951–54, and, as observer, to the Yale–Columbia Southern Station in Australia 1954–57. He was awarded a DSc from the Australian National University in 1957. In 1957 De Vaucouleurs went to the USA to become astronomer at the Lowell Observatory in Arizona. A year later, he was appointed research associate at the Harvard College Observatory, and in 1960 he became associate professor. From 1965, he was professor of astronomy at the University of Texas, at Austin. He died on 7 October 1995.

The main object of De Vaucouleurs's research was to find a pattern in the location of nebulae – clusters of stars formerly thought to be randomly scattered across the sky as viewed from Earth. Yet as telescopes become more powerful, there is increasing evidence that in fact nebulae themselves tend to cluster together. Extragalactic nebulae, small and faint as they appear, are more numerous and seem distinctly grouped. The nebulae that are closer to us, of up to the 12th or 13th magnitude, form what are now called superclusters, and many occur in a definite band across the heavens.

In 1952, De Vaucouleurs, then at the Australian Commonwealth Observatory, Mount Stromlo, Canberra, began a thorough re-investigation of a local supercluster, using newer and more accurate data. He redetermined the magnitudes of many of the brighter southern nebulae and revised the magnitudes for most of the brighter catalogued nebulae following a modern photometric system. His aim was to obtain photometric consistency and completeness over the whole sky to approximately magnitude 12.5. In 1956, he published and discussed his material in great detail and it seemed to indicate that a 'local supergalaxy' exists, which includes our own Milky Way galaxy. He suggested a model in which the great Virgo cluster might be 'a dominant congregation not too far from its central region'. As evidence for its existence, De Vaucouleurs pointed out the similarity in position and extent of a broad maximum of cosmic radio noise, reported by other researchers both in the UK and the USA. (It may be that a local supercluster of bright nebulae including a galaxy may not be unique, since Harlow Shapley in 1934 reported among the fainter ones a distant double supergalaxy in Hercules.)

De Vaucouleurs's work in the southern hemisphere led him to suspect that there was another supergalaxy, from a great, elongated swarm of nebulae extending through Cetus, Dorado, Fornax, Eridanus, and Horologium. He estimated that its distance was only slightly greater than the Virgo cluster and he noted that the relative sizes and separations of this southern supergalaxy and the local supergalaxy appear to be comparable with Shapley's double supergalaxy in Hercules. Thus it seems that superclustering is a phenomenon that has some essential relevance to the structure of the universe.

de Vries, Hugo Marie (1848–1935) Dutch botanist and geneticist who is best known for his rediscovery (simultaneously with Karl Correns and Erich

Tschermak von Seysenegg) of Gregor Mendel's laws of heredity and for his studies of mutation.

De Vries was born on 16 February 1848 in Haarlem. He studied medicine at the universities of Heidelberg and Leyden, graduating from the latter in 1870. He then taught at the University of Amsterdam 1871–75, after which he went to work for the Prussian ministry of agriculture in Würzburg. In 1877 he taught at the universities of Halle and Amsterdam; he was appointed assistant professor of botany at Amsterdam in the following year and became a full professor there in 1881. De Vries remained at the University of Amsterdam – except for a visiting lectureship to the University of California in 1904 – until he retired in 1918. He died near Amsterdam on 21 May 1935.

De Vries began his work on genetics in 1886, when he noticed that some specimens of the evening primrose (*Oenothera lamarckiana*) differed markedly from others. He cultivated seeds of the various types in his experimental garden, and also undertook detailed research to discover the origin of the plants and the history of the species introduced into Europe. At the time it was believed that *O. lamarckiana* had been introduced from Texas and was known by Jean Lamarck under the name *O. grandiflora.* De Vries, however, discovered that the plant was unknown in the USA, from which he concluded that *O. lamarckiana* was a pure species. Continuing his investigations, he began a programme of plant-breeding experiments in 1892. Eight years later he formulated the same laws of heredity that – unknown to de Vries – Mendel had discovered 34 years previously; but while searching the scientific literature on the subject, de Vries came across Mendel's paper of 1866. De Vries's work went further than Mendel's had, however: he found that occasionally an entirely new variety of *Oenothera* appeared and that this variety reappeared in subsequent generations. De Vries first called these new varieties that appeared suddenly single variations but later called them mutations. He postulated that, in the course of evolution, a species produces mutants only during discrete, comparatively short periods (which he called mutation periods) in which the latent characters are formed. He also distinguished between mutants with useful characteristics (which he called progressive mutants) and those with useless or harmful traits (which he called retrogressive mutants), and proposed that only the progressive mutants contribute to the evolution of the species. Furthermore, he suggested that mutation was the means by which new species originated and that those mutations that were favourable for the survival of the individual persisted unchanged until other, more favourable mutations occurred. He considered the slight variations caused by environmental factors to be insignificant in the evolutionary process, mutations being the most important factor involved. De Vries's work on heredity and mutations – which he summarized in *Die Mutationstheorie* (1901–03; translated into English as *The Mutation Theory* 1910–11) was soon generally accepted and played an important role in helping to establish Darwin's theory of evolution.

Thomas Hunt ◊Morgan, in his work on the fruit fly *Drosophila,* also encountered unexpected new variations that were capable of breeding true, and he assumed that this was the result of a change in their genes; using de Vries's terminology, Morgan applied the term mutation to this spontaneous change in a gene. It has since been discovered that the mutations found by de Vries resulted from changes in the number of chromosomes in the new species, not from changes to the genes themselves. Today the term mutation refers to any change in the genetic material and includes many additional types of changes not known to de Vries.

De Vries's other major contribution concerned the physiology of plant cells. Experimenting on the effects of salt solutions of various concentrations on plant cells, he demonstrated in 1877 how plasmolysis of the cells can be used to establish a series of isotonic solutions.

Dewar, James (1842–1923) Scottish physicist and chemist with great experimental skills that enabled him to carry out pioneering work on cryogenics, the properties of matter at extreme low temperatures. Dewar is also remembered for his invention of the Dewar vacuum flask, which has been adapted for everyday use as the thermos flask.

Dewar was born in Kincardine-on-Forth on 20 September 1842. He went to local schools until he contracted rheumatic fever in 1852, which required a long convalescence at home. He was nevertheless able to enter the University of Edinburgh in 1859, where he studied physical science and later worked as a demonstrator. In 1867, Dewar presented models of various chemical structures at an exhibition of the Edinburgh Royal Society. This work interested Friedrich ◊Kekulé, who invited Dewar to Ghent in Belgium for the summer. Dewar then returned to Edinburgh, and continued as an assistant until 1873, holding a concurrent post as a lecturer at the Royal Veterinary College from 1869. The Jacksonian Professorship of Natural Experimental Philosophy and Chemistry at the University of Cambridge was offered to Dewar in 1875, followed two years later by the Fullerian Professorship of Chemistry at the Royal Institute in London. Dewar occupied both these chairs until his death. In 1877, he was also elected a Fellow of the Royal Society, which later awarded him the Rumford Medal for his research on the cryogenic properties of matter. Dewar was appointed to a government committee on explosives

1888–91, and in 1889 invented the important explosive cordite in collaboration with Frederick Abel (1827–1902). Dewar was knighted in 1904 and died in London on 27 March 1923.

Dewar's early work in Edinburgh concerned several branches of physics and chemistry. In 1867, he worked on the structures of various organic compounds and during his visit to Ghent may have suggested the structural formula for benzene that is credited to Kekulé. He also proposed formulas for pyridine and quinoline. In 1872, Dewar invented the vacuum flask as an insulating container in the course of an investigation of hydrogen-absorbed palladium. In collaboration with Peter Tait (1831–1901), he found that charcoal could be used as an absorbent to improve the strength of the vacuum. The vacuum prevented heat conduction or convection, so heat loss could occur only by radiation. The reflective silver layer on the walls of a vacuum flask are designed to minimize this.

When Dewar went to Cambridge in 1875, he began a long-term collaborative project on spectroscopy with George Liveing (1827–1924). They examined the correlation between spectral lines and bands and molecular states, and were particularly concerned with the absorption spectra of metals.

Dewar's move to the laboratories of the Royal Institution in London in 1877 marks the beginning of his main work on the liquefaction of gases and an examination of cryogenic properties. In that year, Louis ◊Cailletet and Raoul Pictet (1846–1929) in France had succeeded in liquefying oxygen and nitrogen, two gases that had resisted Faraday's attempts at liquefaction. Dewar first turned to the production of large quantities of liquid oxygen, examining its properties. He found in 1891 that both liquid oxygen and ozone are magnetic. His vacuum flasks were invaluable in this work because they enabled him to preserve the very cold liquids for longer than would otherwise have been possible.

Dewar developed the cooling technique used in liquefaction by applying the Joule–Thomson effect, whereby expansion of a compressed gas results in a lowering of temperature due to the energy used in overcoming attractive forces between the gas molecules. By subjecting an already chilled gas to this process, he became the first to produce liquid hydrogen in 1895, though only in small quantities; then by using bigger apparatus he made larger amounts in 1898. He proceeded to examine the refractive index of liquid hydrogen, and a year later succeeded in solidifying hydrogen at a temperature of –259 ° C/–434 °F.

Every gas had now been liquefied and solidified except one – helium. Dewar tried to liquefy helium, but was unsuccessful because his sample was contaminated with neon, which froze at the low temperature and blocked the apparatus. Heike ◊Kamerlingh Onnes managed to liquefy helium using Dewar's techniques in 1908, and Dewar was then able to achieve temperatures within a degree of absolute zero (–273 °C/–459 °F) by boiling helium at low pressure.

Dewar investigated properties such as chemical reactivity, strength, and phosphorescence at low temperature. Substances such as feathers, for example, were found to be phosphorescent at these temperatures. From 1892–95, he studied with John ◊Fleming (the inventor of the thermionic valve) the electric and magnetic properties of metals at low temperatures. They predicted that electrical resistance becomes negligible at extremely low temperatures, this phenomenon of superconductivity subsequently being detected by Kamerlingh Onnes in 1911.

During the World War I cryogenic research, being very expensive, had to be suspended. Dewar turned to the study of thin films, and later to the measurement of infrared radiation. He was a scientist with tremendous experimental flair and patient application.

Dicke, Robert Henry (1916–1997) US physicist who carried out considerable research into the rates of stellar and galactic evolution. Much of his work was innovatory and some remains controversial.

Born in St Louis, Missouri, on 6 May 1916, Dicke completed his education at Princeton University, graduating in 1938, before obtaining a PhD from the University of Rochester in 1941. That same year, he joined the staff of the Radiation Laboratory, Massachusetts Institute of Technology, where he remained until 1946. He then returned to Princeton, where he rose from being assistant professor to professor of physics, then to Cyrus Fogg Brachett Professor 1957–75, to chairman of the department of physics, and finally, 1975–84, to Albert Einstein Professor of Science. Among many prizes and honours he received was the Medal for Exceptional Science Achievement of the National Aeronautical and Space Administration in 1973.

In 1964 Dicke turned his attention to a version of the Big Bang theory known as the 'hot Big Bang'; he suggested that the present expansion of the universe had been preceded by a collapse in which high temperatures had been generated. Realizing that he should be able to test this hypothesis by detecting residual radiation in space at a wavelength of a few centimetres, Dicke and his colleagues started to build the equipment they needed. But before they were in a position to begin their measurements, Arno ◊Penzias and Robert ◊Wilson of Bell Telephone Laboratories announced they had detected an unexpected and relatively high level of radiation at a wavelength of 7 cm/2.8 in, with a temperature of about 3.5K (–270

°C/–453 °F). Dicke immediately proposed that this was cosmic black-body radiation from the hot Big Bang.

Dicke carried out experiments to verify the supposition of the general theory of relativity that a gravitational mass is equal to its inertial mass. He was able to establish the equality to an accuracy of one part in 10^{11}. In 1961 he put forward a theory (the Brans–Dicke theory) that the gravitational constant varied with time (by about 10^{-11} per year). Experiment has not supported this idea.

Dicksee, Cedric Bernard (1888–1981) English engineer who was a pioneer in developing the compression-ignition (diesel) engine into a suitable unit for road transport.

Dicksee received his technical education at the Northampton Engineering College 1906–10. He was chief designer at the Aster Engineering Company Ltd, Wembley, and in 1918 he joined the Austin Motor Company Ltd. The following year he went to Westinghouse Electric and Manufacturing Company in charge of design and development of engines for generating plants.

After seven years in the USA, Dicksee returned to the UK and joined the Association Equipment Company (AEC) Ltd – the manufacturing subsidiary of the London General Omnibus Company – in 1928 as engine designer, later becoming research engineer. When in 1929 the development of a compression-ignition engine suitable for road transport was started, this became his major activity and was the work for which he is best known.

During World War II Dicksee worked on combustion chamber design for the De Havilland series of jet engines. He returned to AEC at about the time the war ended and after leaving became a consultant to Waukesha for several years. He retired to Seaford, East Sussex, where he died.

When Dicksee began work, the London General Omnibus Company (LGOC) had two lorries at their depot fitted with massive three-cylinder Junkers oil engines. These were unsuitable for automotive use and within a year, he had his first engine running. It was a six-cylinder 8.1 litre/40 cu in unit with an aluminium crankcase and it entered service with the LGOC just before Christmas 1930.

After development with Ricardo and Company an improved version took to the road in September 1931. A smaller – 7.7 litre/35 cu in – engine followed in 1933 and soon became the standard for AEC chassis. A further development of this engine used combustion chambers of a toroidal shape and was also highly successful. This feature was subsequently adopted for larger engines in the company's range.

Dicksee's engines ran at speeds ranging from 1,800 to 2,400 (governed) rpm, which were higher than comparable engines of the day. With this performance, the way was opened for the adoption of compression-ignition engines instead of petrol engines for road transport. As a result, by the late 20th century there was scarcely a commercial vehicle of any size that did not have a high-speed compression-ignition engine as its power unit.

Dickson, Leonard Eugene (1874–1957) US mathematician who gave the first extensive exposition of the theory of fields. A prolific writer, he was also the author of a massive three-volume *History of the Theory of Numbers* (1919–23), now a standard work, in which he investigated abundant numbers, Diophantine equations, perfect numbers, and Fermat's last theorem.

Dickson was born in Independence, Iowa, on 22 January 1874. He received his BA at the age of 19 from the University of Texas, where he then taught for some months. Receiving his MA in 1894, he entered the University of Chicago where in 1896 he gained his doctorate in mathematics. In 1897 he became a postgraduate student at Leipzig and in Paris. The following year, Dickson returned to the USA and – for a year at a time – was appointed instructor in mathematics at the universities of California and Texas. In 1900 he was appointed assistant professor at the University of Chicago; promoted in 1907 to associate professor, he became professor in 1910. Apart from short periods as visiting professor at the University of California in 1914, 1918, and 1922, Dickson remained at the University of Chicago until his retirement in 1939. He died on 17 January 1957 in Harlingen, Texas.

For his prolific work in mathematics, Dickson received many awards and honours, and became a member of many influential societies. In 1913 he was elected to the National Academy of Sciences, and in 1916 became president of the American Mathematical Society, from whom he received the Cole Prize for his book *Algebren und ihre Zahlentheorie.*

Dickson's work in mathematics spanned many topics, including the theory of finite and infinite groups, the theory of numbers, algebras and their arithmetics (for this work he won a $1,000 prize), and the history of mathematics.

In his work on finite linear groups he generalized the results of Evariste Galois's studies, and those of Ernst Jordan and Jean-Pierre Serre, for groups over the field of n elements to apply to groups over an arbitrary finite field. He proved his modified version of the Chevalley theorem and published the first extensive exposition of the theory of finite fields.

During an investigation of the relationships between the theory of invariants and number theory, Dickson examined divisional algebra, particularly in the form systematized by Arthur Cayley, and he expanded the theorems of linear associative algebras formulated by Elie Cartan and Wedderburn.

Investigating the history of the theory of numbers, Dickson also studied the work of ◊Diophantus, who lived in Alexandria in the 3rd century AD. Diophantus assumed that every positive integer is the sum of four squares, a theory for which (in 1770) Waring attempted to derive an extension in the direction of higher powers. Using the results of Ivan Vinogradov, Dickson (in the 1930s) succeeded in proving Waring's theorem and made his contribution to Diophantine analysis, from which the complete criteria were obtained for the solution of:

$$ax^2 + by^2 + cz^2 + du^2 = 0$$

For any non-zero integer *a, b, c,* or *d.* In addition, Dickson gave the first complete proof that:

$$ax^2 + by^2 + cz^2 + du^2 + ev^2 = 0$$

Is always soluble in integers, if the non-zero integers *a, b, c, d,* and *e* are not all of a like sign.

Diels, Otto (1876–1954) German organic chemist who many fundamental discoveries, including (with Kurt ◊Alder) the diene synthesis or Diels–Alder reaction. He shared with Alder the 1950 Nobel Prize for Chemistry.

Diels was born in Hamburg on 23 January 1876 into a talented academic family. His father Hermann Diels (1848–1922) was professor of classical philology at Berlin University and his brothers Paul and Ludwig became professors of, respectively, philology and botany. During his school years Otto Diels became interested in chemistry and carried out a series of experiments with his brother Ludwig. In 1895 he went to his father's university to study chemistry under the great organic chemist Emil Fischer, obtaining his doctorate in 1899. He continued research as Fischer's assistant and became a lecturer in 1904. He moved to the Chemical Institute of the then Royal Friedrich Wilhelm University in 1914 and two years later was invited by the Christian Albrecht University in Kiel to serve as director of their Chemical Institute, where he remained for 32 years until he retired.

Diels married in 1909 and had three sons and two daughters; two of his sons were killed on the eastern front near the end of World War II. After the destruction of his institute by enemy action during the war, Diels planned to retire in 1945 but was persuaded to stay on to help with its rebuilding. He finally retired in 1948, and died in Kiel on 7 March 1954.

In 1906, while working in Berlin, Diels discovered carbon suboxide (tricarbon dioxide), which he prepared by dehydrating malonic acid (propandioic acid) with phosphorus pentoxide (phosphorus(V) oxide).

A year later he published the first edition of his textbook of organic chemistry, *Einführung in die organische Chemie,* which, because of its scope and clarity, became one of the most popular books of its kind.

Diels's investigation of the nature and structure of the biologically important compound cholesterol began as early as 1906. He isolated pure cholesterol from gallstones and converted it to 'Diels acid'. But it was not until 1927 that he successfully dehydrogenated cholesterol (by the drastic process of heating it with selenium at 300 °C/572 °F) to produce 'Diels hydrocarbon'. This substance was later shown to be 3-methyl-1, 2-cyclopentanophenanthrene ($C_{18}H_{16}$), an aromatic hydrocarbon closely related to the skeletal structure of all steroids, of which cholesterol is one. In 1935 he synthesized the $C_{18}H_{16}$ compound and showed it to be identical with Diels hydrocarbon. This work proved to be a turning point in the understanding of the chemistry of cholesterol and other steroids, although it was not until about 1955 that its structure was completely known.

Working with his assistant Kurt Alder, Diels spent much of the rest of his life developing the diene synthesis, which first achieved success in 1928 when they combined cyclopentadiene with maleic anhydride (*cis*-butenedioic anhydride) to form a complex derivative of phthalic anhydride. Generally, conjugated dienes (compounds with two double bonds separated by a single bond) react with dienophiles (compounds with one double bond activated by a neighbouring substituent such as a carbonyl or carboxyl group) to form a six-membered ring. For example, butadiene (but-1,3-diene) reacts with acrolein (prop-2-en-1-al) to give tetrahydrobenzaldehyde (cyclohex-4-en-1-carbaldehyde).

Applications of the Diels–Alder synthesis are numerous throughout organic chemistry and it is of great importance because many reactions of this type occur easily at low temperatures and give good yields.

Diels was considered always to be a reserved man, but one with a good sense of humour, and he was liked and respected by his students who enjoyed his well-planned lectures and experimental demonstrations.

Diesel, Rudolph Christian Karl (1853–1913) German mechanical engineer whose name will always be associated with the compression-ignition internal-combustion engines he invented.

Diesel was born in Paris on 18 March 1853, of parents who came from the Bavarian town of Augsburg. His father was a bookbinder. While receiving his early education in Paris, Diesel spent much of his spare time in the city's museums. He was particularly attracted to the Museum of Arts and Crafts, which had on permanent display Joseph Cugnot's steam-propelled gun carriage of 1769.

At the outbreak of the Franco-Prussian War in 1870, the Diesel family travelled to England and took up residence in London. Rudolph, however, was sent to Augsburg to continue his education with a relative who was a teacher at the local trade school. He progressed so well that he qualified for the Munich Polytechnic, where he proved to be a brilliant scholar. He shared part of his time there with Carl von Linde, the founder of modern refrigeration engineering, from whom he learnt the basic theory of heat engines.

Diesel became fascinated by engines and came to the conclusion that an engine four times as effective as a steam engine could be made by designing it to carry out the combustion within the cylinder while utilizing as large a temperature range as possible. Because the temperature obtained depends to a great extent on the pressure, a very high pressure must be used to compress the air before fuel injection to avoid premature explosion.

His ideas were published in a paper of 1893, one year after he had taken out his first patent in Berlin. In the early stages of his experiments he was lucky to escape being killed when a cylinder head blew off one of his prototype engines, but this did not deter him and he carried on to perfect a model that was capable of commercial exploitation. In 1899 he founded his own manufacturing company at Augsburg, which flourished despite Diesel having little or no business sense. He was, however, talented at putting over his ideas, and he gave a series of lectures in the USA in 1912, which were well attended. In 1913, at the height of his success, he vanished from the decks of a steamer crossing from Antwerp to England while on his way for consultations with the British Admiralty. His body was never found.

Like the petrol engine, the diesel engine is a form of internal-combustion engine but it differs in not having a carburettor to pre-mix air and fuel; nor does it require a spark for ignition. In the diesel engine, as the piston moves down, pure air is drawn into the cylinder and on the up stroke this air is compressed by a ratio of between 12:1 and 25:1 (much higher than that of the petrol engine, which is usually between 6:1 and 10:1). Compression, by increasing pressure, raises the temperature of the air to in excess of 540 °C/1,004 °F. When the piston nears the top of the compression stroke, an injector admits through a nozzle a fine spray of fuel, which mixes with the heated air and ignites spontaneously, the explosion moving the piston downwards. As the rate of inflow of air remains constant in a diesel engine, its power output is governed by the amount of fuel injected.

In his first engine, Diesel is thought to have used coal dust as a fuel, but he later discarded this along with several other types in favour of a form of refined mineral oil. Diesel engines will in fact run on many fuels but the most widely used today is distilled from petroleum closely related to kerosene.

Until the 1920s further development of the diesel engine took place mainly in Germany (principally by Karl Bosch), where it proved particularly useful during World War I for powering submarines. Designers such as Cummins of the USA and Gardner in the UK eventually adapted the invention for small boats, and later work by Cedric Dicksee and Harry Ricardo put it into a practical form for road use.

Diesel's death remains a mystery but his work lives on. A century after his birth more than half the world's tonnage of ships and a large proportion of the railways were diesel driven. The engine has been adapted for buses, tractors, and trucks as well as serving as an alternative to coal-burning boilers and fast-flowing water for driving alternators in the generation of electricity.

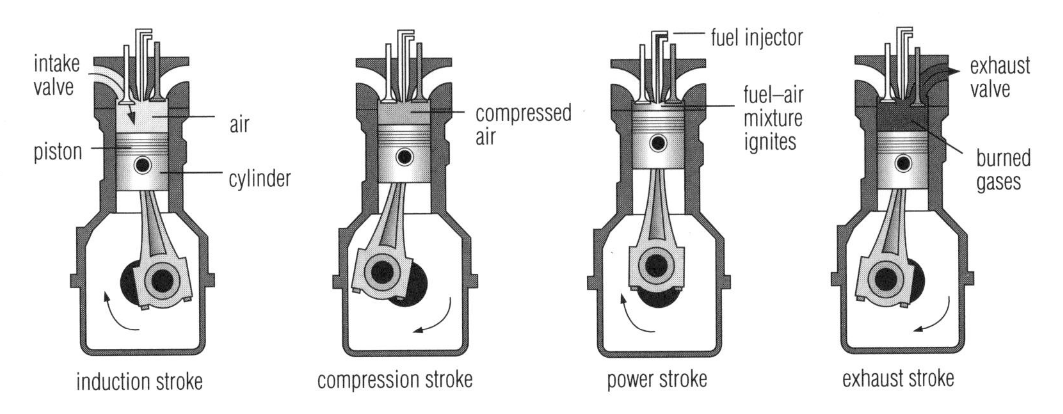

Diesel In a diesel engine, fuel is injected on the power stroke into hot compressed air at the top of the cylinder, where it ignites spontaneously. The four stages are exactly the same as those of the four-stroke or Otto cycle.

Further Reading
Grosser, Morton, *Diesel: The Man and the Engine*, Atheneum, 1978.
Nitske, W Robert and Wilson, Charles Morrow, *Rudolf Diesel: Pioneer of the Age of Power*, University of Oklahoma Press, 1965.
Thomas, Donald E, *Diesel Technology and Society in Industrial Germany*, University of Alabama Press, 1987.

Diophantus (lived c. AD 270–280) Greek mathematician who, in solving linear mathematical problems, developed an early form of algebra. Particularly innovative was his use of a symbol for an unknown quantity.

Very little is known about him or about his life. It is probable, however, that he was born in Alexandria (now in Egypt), a great Greek cultural centre, where he certainly lived. If the evidence of what is called 'Diophantus's riddle' is to be taken seriously, he lived at least to the age of 84 (which is the answer to the riddle).

To the Greeks, mathematics comprised two branches of study: arithmetic, the science of numbers, and geometry, the science of shapes. Towards the end of the Greek era – indeed, already under a declining Roman domination – Diophantus formulated his theories in a neglected field, the study of unknown quantities. His *Arithmetica*, according to its introduction, was compiled in 13 books (although the six that have come down to us are probably all that were completed); the work was translated by the Arabs, and it is through their word for his equations – 'the re-uniting of separate parts' – that we now call the system 'algebra'. Diophantus would merely have thought of it as abstract arithmetic.

In the solution of equations Diophantus was the first to devise a system of abbreviating the expression of his calculations by means of a symbol representing the unknown quantity. Because he invented only one symbol for the unknown, however, in equations requiring two or more variables his work can become extremely confusing in its repetition of that single symbol.

His main mathematical study was in the solution of what are now known as 'indeterminate' or 'Diophantine' equations – equations that do not contain enough facts to give a specific answer but enough to reduce the answer to a definite type. These equations have led to the formulation of a system for numbers, commonly called the theory of numbers, that is regarded as the purest branch of present-day mathematics. Using his method, the possibility of determining a type of answer rather than a specific one to a given problem has allowed modern mathematicians to approach the properties of various kinds of whole numbers (such as odds, evens, primes, and squares) with new insight. By then applying the use of infinite trains of numbers correlated through the Diophantine equation system, mathematicians have come to a new understanding of some of the basic rules that numbers follow.

Dirac, Paul Adrien Maurice (1902–1984) English theoretical physicist of great international standing. He played a pivotal role in the development of quantum electrodynamics, being responsible for the introduction of concepts such as electron spin and the magnetic monopole. Dirac's most important achievement was to predict the existence of antiparticles. For his contributions to theoretical physics, he shared the 1933 Nobel Prize for Physics with Erwin ◊Schrödinger.

Dirac was born in Bristol on 8 August 1902. He attended Bristol University 1918–21, taking a degree in engineering but also attending lectures on philosophy. He then spent two further years at Bristol studying mathematics, and in 1923 transferred to the University of Cambridge, where he was introduced to the work of Niels ◊Bohr and Ernest Rutherford on the structure of the atom. Dirac continued to study projective geometry, however, a tool he found most useful in his later research.

Dirac completed his PhD thesis in 1926. He then began to travel, visiting major centres for theoretical physics including the Institute of Theoretical Physics at Copenhagen and the University of Göttingen. There Dirac met and worked with the leading figures in the developing field of quantum mechanics, such as Niels Bohr, Wolfgang Pauli, Max Born, and Werner ◊Heisenberg, with whom Dirac had already built up a lengthy correspondence.

St John's College at Cambridge University elected Dirac a fellow in 1927, and in 1928 he formulated the relativistic theory of the electron from which he predicted the existence of the positron. The impact of the theory was profound and brought Dirac wide recognition. In 1932 he became Lucasian Professor of Mathematics at Cambridge, a post he held until 1969. In 1933 the positron was discovered and Dirac won the Nobel Prize for Physics. He was awarded the Royal Society's Royal Medal in 1939 and the Copley Medal in 1946. In 1969 he was made the first recipient of the Oppenheimer Prize. From 1971, Dirac was professor of physics at Florida State University. He died in Tallahassee, Florida, on 20 October 1984.

During the late 1920s, there was an intense burst of activity in the field of theoretical physics. In 1923 Louis ◊de Broglie proposed a model of the electron that ascribed to it wave-like properties. This was a major break from the Bohr model of the atom. Schrödinger extended de Broglie's work by considering the movement of a particle in an electromagnetic field, thereby establishing the subject now known as of wave mechanics. Heisenberg also worked on this subject and developed a non-relativistic model of the electron.

Dirac was perturbed by a number of features of Heisenberg's theory. He was dissatisfied with its non-relativistic form and wished to integrate it with special relativity. Consideration of these problems enabled Dirac in 1928 to formulate the relativistic theory of the electron. Apart from its mathematical elegance, an attribute Dirac constantly strove for in his work, the theory was also astonishingly fertile. The model was able to describe many quantitative aspects of the electron, including properties that were entirely new. These included the half-quantum spin and magnetic moment of the electron. Dirac's model was also able to account perfectly for certain anomalies in the hydrogen spectrum. Perhaps the most significant prediction to arise from Dirac's relativistic wave equation of the electron was the proposal of the positive electron, the positron.

Dirac found that the matrices describing the electron contained twice as many states as were expected. He proposed that positive energy states in the matrices described the electron, and that the negative energy states described a particle with a mass equal to that of an electron but with an opposite (positive) charge of equal strength, that is, an antiparticle of the electron. He predicted that an electron and its antiparticle, later called the positron, could be produced from a photon. This was soon confirmed by both Carl ◊Anderson and Patrick ◊Blackett, who independently discovered the positron in 1932 and 1933. This led to the prediction of the existence of other antiparticles, such as the antiproton. This was discovered by Emilio ◊Segrè and co-workers in 1955. Dirac's relativistic wave equation was also able to give a quantitative explanation for the Compton effect, and furthermore enabled Dirac to predict the existence of the magnetic monopole. This has proved to be more elusive than the positron and has not been irrefutably detected even to this day.

Dirac was eager to learn new mathematical methods, and he studied the mathematics of eigenvalues and eigenvectors that Schrödinger had found so useful. In examining the uses of these, he noticed that those particles with half-integral spins (such as the electron) that obeyed Pauli's exclusion principle also obeyed statistical rules different from the other particles. The latter class of particles obeyed statistical rules developed by Satyendranath Bose (1894–1974) and Albert Einstein. Dirac worked out the statistics for the other particles, only to discover that Enrico ◊Fermi had done very similar work already. Since the two scientists developed the theory independently, it is usually called Fermi–Dirac statistics to honour both. Fermi–Dirac statistics are of great value in nuclear and solid-state physics and are used, for example, to determine the distribution of electrons at different energy levels.

Dirac's later work was related principally to the field of mathematics. He worked on the large-number hypothesis. This hypothesis deals with pure, dimensionless numbers such as the ratio of the electrical and gravitational forces between an electron and a proton, which is 10^{39}. This ratio, perhaps coincidentally or perhaps not, also happens to be the age of the universe when it is expressed in terms of atomic units.

A theory with mathematical beauty is more likely to be correct than an ugly one that fits some experimental data. God is a mathematician of a very high order, and He used very advanced mathematics in constructing the universe.

Paul Dirac Scientific American *May 1963*

If there is some meaningful connection between these two values, there must be a connection between the age of the universe and either the electric force or the gravitational force. It is possible, therefore, that the gravitational force is not constant but is decreasing at a rate proportional to the rate of ageing of the universe.

Dirac was responsible for the formulation of equations that were essential to the development of quantum theory, and for the reconciliation of relativity with quantum theory resulting in the formulation of the relativistic wave theory for the electron. He was one of the giants of 20th-century physics.

Further Reading

Dirac, P A M; Dalitz, R H (ed), *The collected Works of P A M Dirac: 1924–1948,* Cambridge University Press, 1995.

Dutt, R and Ray, A K, *Dirac and Feynman: Pioneers in Quantum Mechanics,* Franklin Book Co, 1993.

Kragh, Helge, *Dirac: A Scientific Biography,* Cambridge University Press, 1990.

Kursunoglu, Behram N and Wigner, E P (eds), *Paul Adrien Maurice Dirac: Reminiscences About a Great Physicist,* Cambridge University Press, 1987.

Pais, Abraham (ed), *Paul Dirac: The Man and His Work,* Cambridge University Press, 1998.

Taylor, John (ed), *Tributes to Paul Dirac,* Hilger, 1987.

Dirichlet, (Peter Gustav) Lejeune (1805–1859)

German mathematician whose work in applying analytical techniques to mathematical theory resulted in the fundamental development of the theory of numbers. Also a physicist interested in dynamics, he knew many of the great scientists of his age and published a book of some historic importance.

Dirichlet was born in Düren on 13 February 1805, the son of the town postmaster. Precociously interested in mathematics (it is said that at the age of 12 he used his pocket money to buy mathematical books), in 1819 he was sent to the University of Cologne – where his teachers included the physicist Georg Ohm – and completed his final examination at the very early age of 16. He was then sent to France, the country of all the contemporarily great mathematicians. Arriving at the Collège de France, in Paris, he attended lectures for little more than a year before being appointed tutor to the children of the renowned General Fay; as such, he was treated as one of the family and met many of the most prominent figures in French intellectual life, particularly Joseph Fourier. In 1825 Dirichlet presented his first paper to the French Academy of Sciences.

General Fay's death, however, led Dirichlet to return to Germany, where he took a research post at the University of Breslau (now Wrocław). Later he moved to Berlin, teaching initially at the Military Academy, but was soon appointed professor also at the University of Berlin (still at the age of only 23). Dirichlet spent 27 years as a professor in Berlin, exerting a strong influence on German mathematics. He was an excellent teacher, despite being a modest and retiring man who shunned public appearances – unlike his lifelong friend Karl Jacobi. These two mathematicians stimulated and influenced each other, and when Jacobi's health forced him to move to Italy Dirichlet spent 18 months there with him, from 1843. Their presence caused a circle of leading German mathematicians to gather round them. In 1855, on the death of the great Karl Gauss, the University of Göttingen offered Dirichlet the prestigious vacant professorship. He accepted, but enjoyed the post for a mere three years, suffering a severe heart attack in the summer of 1858. He died the following spring, shortly after his wife had died, in Göttingen on 5 May 1859.

Dirichlet's first interest was number theory, and much of his work was on this topic. His first paper, written in France, concerned Diophantine equations of the form $x^5 + y^5 = kz^5$. Using the methods of this paper, Adrien Legendre succeeded only a few weeks later in appending a proof that Pierre de Fermat's famous equation $xn + yn = zn$ has no integral solution when $n = 5$.

Dirichlet was considerably influenced by Gauss, some of his early work being improvements on Gauss's proofs, but as his abilities developed, Dirichlet's intensive search for a general algebraic number theory substantially advanced this branch of mathematics with a number of very important papers. These included studies on quadratic forms, the number theory of irrational fields (including the integral complex numbers), and the theory of units. In 1837 Dirichlet presented his first paper on analytic number theory, giving a proof to the fundamental theorem that bears his name – any arithmetical series of integers $a \times n + b$, where a and b are relatively prime and $n = 0, 1, 2, 3 \ldots$, must include an infinite series of primes. Later papers included the analytical consideration of quadratic forms, studies of the theory of ideals, and the convergence of Dirichlet series, and introduced the deceptively simple *Schubfachprinzip* (the 'box principle') – a principle much used in the logic of modern number theory, which states: if in n boxes one distributes more than n objects, at least one box must contain more than one object.

In 1863 Dirichlet's *Vorlesungen über Zahlentheorie* was published posthumously by his friend and pupil Richard Dedekind. This summary of Dirichlet's work together with supplements by Dedekind is now considered one of the foundations on which the theory of ideals – the core of algebraic number theory – is based.

Alongside his theoretical work, Dirichlet also carried out a series of studies on analysis and applied mathematics. Important among these was an analysis of vibrating strings, in which he developed techniques now considered classic for the discernment of convergence. He also began to rewrite the vocabulary of mathematics. Whereas the mathematical concept of a function had previously been as an expression formulated in terms of mathematical symbols, Dirichlet introduced the modern concept of $y = f(x)$ as a correspondence that links each real x value with some unique y value denoted by $f(x)$.

Other papers included applications of Fourier series, a critique of Pierre Laplace's analysis of the stability of the Solar System, boundary values, and the first exact integration of the hydrodynamic equations.

Dirichlet's contributions to mathematics were both numerous and of different kinds; he made many important individual discoveries, but more important still was his method of approach – an essentially modern way of formulating or analysing mathematical problems, especially in number.

Dobzhansky, Theodosius (1900–1975) Russian-born US geneticist whose synthesis of Darwinian evolution and Mendelian genetics established evolutionary genetics as an independent discipline. He also wrote about human evolution and the philosophical aspects of evolution.

Dobzhansky was born on 25 January 1900 in Nemirov, Russia, the son of a mathematics teacher. His family moved to Kiev in 1910 and Dobzhansky first attended school there. In 1917 he went to Kiev University to study zoology and, after graduating in

1921, remained there to teach zoology until 1924, when he moved to Leningrad University as a teacher of genetics. Also in 1924 he married Natalia Sivertzev, whom he met while teaching at Kiev. In 1927 Dobzhansky went as a Rockefeller Fellow to Columbia University, New York City, where he worked with Thomas Hunt ◊Morgan, one of the pioneers of modern genetics. Morgan moved to the California Institute of Technology in 1928 and, impressed with Dobzhansky's ability, offered him a post teaching genetics there when his fellowship ended in 1929. Dobzhansky, who had become a US citizen in 1937, remained at the California Institute of Technology until 1940, when he returned to Columbia University as professor of zoology. He then worked at the Rockefeller Institute (later the Rockefeller University) from 1962 until his official retirement in 1971, after which he moved to the University of California at Davis, where he remained until his death on 18 December 1975.

Dobzhansky's most important contribution to genetics was probably his *Genetics and the Origin of Species* (1937), which was the first significant synthesis of Darwinian evolutionary theory and Mendelian genetics – areas in which there had been much progress since about 1920. This book was highly influential and established the discipline of evolutionary genetics.

Dobzhansky's other major contribution was his demonstration in the 1930s that genetic variability within populations is greater than was then generally thought. Until this work the consensus of opinion was that, in the wild state, most members of a species had the same 'wild-type' genotype and that each of the wild-type genes was homozygous in most individuals. Variant genes were usually deleterious mutants that rapidly vanished from the gene pool. Furthermore, when an advantageous mutation appeared, it gradually – over several generations – increased in frequency until it became the new, normal wild type. Working with wild populations of the fruit fly *Drosophila pseudoobscura*, Dobzhansky found that, in fact, there is a large amount of genetic variation within a population and that some genes regularly changed in frequency with the different seasons. Continuing this line of research, he also showed that many of the variant genes are recessives and so are not commonly expressed in the phenotype – a finding that disproved the original assumption of a high level of homozygosis in wild populations. Furthermore, he found that heterozygotes are more fertile and better able to survive than are homozygotes and, therefore, tend to be maintained at a high level in the population.

Dobzhansky also investigated speciation and, using the fly *D. paulistorum*, proved that there is a period when speciation is only partly complete and during which several races coexist. In addition, he wrote on human evolution – in which area his *Mankind Evolving* (1962) had great influence among anthropologists – and on the philosophical aspects of evolution in *The Biological Basis of Human Freedom* (1956) and *The Biology of Ultimate Concern* (1967).

Further Reading

Adams, Mark B (ed), *The Evolution of Theodosius Dobzhansky: Essays on His Life and Thought in Russia and America*, Princeton University Press, 1994.

Dobzhansky, Theodosius and Boesiger, Ernest; Wallace, Bruce (ed), *Human Culture: A Moment in Evolution*, Columbia University Press, 1983.

Levine, Louis (ed), *Genetics of Natural Populations: The Continuing Importance of Theodosius Dobzhansky*, Columbia University Press, 1995.

Dodgson, Charles Lutwidge (1832–1898) English writer and mathematician who, as Lewis Carroll, is famous as the author of *Alice's Adventures in Wonderland* and *Through the Looking Glass*, but was in fact also responsible in his publication of mathematical games and problems requiring the use of intelligent mental arithmetic for a general upsurge of interest in such pastimes. Several of his books of such puzzles suggest an awareness of the theory of sets – the basis on which most modern mathematical teaching is founded – that was being formulated by Dodgson's contemporary, Georg Cantor, but which did not become established until more than 20 years after Dodgson's death.

Dodgson was born in Daresbury, near Warrington, Cheshire, on 27 January 1832, the eldest son in a parish priest's family of 11 children. An acutely shy child with a pronounced stammer, he was educated at home until he reached the age of 12. He was then sent to Rugby School where, under the watchful eye of the Anglican prelate Archibald Tait (1811–1882), he displayed a natural talent for mathematics and an aptitude for divinity. He was awarded a place at Christ Church, Oxford, in 1850 and after taking courses in mathematics and classics, received his BA in 1854. The following year he was appointed lecturer in mathematics, and six years later he was ordained deacon in the Church of England (although he never in fact became a priest). Despite a great love of children – particularly girls – possibly also as a result of his shyness, Dodgson never married. Instead, he poured all his enthusiasm into writing and telling stories to the children of his friends. Under the pseudonym Lewis Carroll, he was eventually persuaded to publish two stories he had composed to amuse one little girl he especially favoured, Alice Liddell. *Alice's Adventures in Wonderland* (1865) and *Through the Looking Glass* (1872) became immensely popular, hailed as classics in the world of children's fiction. A life fellow of Christ Church, Dodgson gave up his lectureship in 1881 and concentrated on writing, both of mathematics and of children's fantasy. After the *Alice* books, however, he never again achieved such popularity. He died in Guildford, Surrey, on 14 January 1898.

Dodgson's enjoyment of mathematics and his affection for children are both reflected in the papers he wrote on the teaching of mathematics for the young. He was particularly interested in the use of number games, and made a compilation of a wide range of puzzles and brain teasers covering all aspects of the subject (including geometry, algebra, and graph work) that call for general intelligence to solve the problems rather than specialized knowledge. Number games were not a new idea (they probably originated with the ancient Greeks), but some time around the 15th century they had re-emerged; the Victorian society of the 19th century was ripe for them to regain considerable popularity. Dodgson saw their potential as teaching aids, and wrote about them as such, publishing several books including *Pillow Problems, The Game of Logic,* and *A Tangled Tale.* The chessboard featured in some of these games. With the publisher Edouard Lucas (1868–1938), Dodgson was responsible for the continued revival of such puzzles during the latter half of the 1800s.

Nevertheless, Dodgson also wrote a considerable number of serious and advanced papers on mathematical subjects (all of which Queen Victoria was apparently dumbfounded to receive, having ordered the author's complete works after reading *Alice in Wonderland).* He produced lengthy general syllabus textbooks, quite a few books on historical mathematics (particularly on Euclid and his geometry), and a number of specialized papers (such as his 'Condensation of determinants').

He also showed a keen interest in photography, and has been described as having an exceptional flair for it.

Further Reading

Abeles, Francine F, *The Mathematical Pamphlets of Charles Lutwidge Dodgson and Related Pieces, v 2: The Pamphlets,* Lewis Carroll Society of North America, 1995.

Dodgson, Charles Lutwidge; Wakeling, Edward (ed), *The Oxford Pamphlets, Leaflets and Circulars of Charles Lutwidge Dodgson,* The Pamphlets of Lewis Carroll series, University Press of Virginia, 1993, v 1.

Dollfus, Audouin Charles (1924–) French physicist and astronomer whose preferred method of research is to use polarization of light, for which method he is prepared to put up with some discomfort.

Dollfus was born on 12 November 1924; he studied at the Lycée Janson-de-Sailly and at the Faculty of Sciences in Paris, where he gained his doctorate in mathematical sciences. In 1946, he joined the Meudon Observatory (the astrophysical section of the Paris Observatory), becoming head of its Laboratory for Physics of the Solar System. Since 1965 he has been astronomer of the Paris Observatory (now emeritus).

Before the *Viking 1* probe landed on Mars in 1976, the mineral composition of the Martian deserts was a subject of considerable dispute. Dollfus checked the polarization of light by several hundreds of different terrestrial minerals to try to find one for which the light matched that polarized by the bright Martian desert areas. He found only one, and that was pulverized limonite (Fe_2O_3), which could be oxidized cosmic iron. (Another astronomer, Gerard Kuiper of the University of Chicago, however, did not agree with Dollfus's findings. In his work, iron oxides gave poor results and he obtained his closest match with brownish fine-grained igneous rocks.)

In pursuit of his detailed investigations into Mars, Dollfus made the first ascent in a stratospheric balloon in France.

By means of the polarization of light it is possible to detect an atmosphere round a planet or satellite. In 1950, at which time it was thought the planet Mercury, because of its small size, had probably lost its atmosphere through the escape of the molecules into space, Dollfus announced that he had detected a very faint atmosphere from polarization measurements carried out at the Pic-du-Midi Observatory in the French Pyrenees. This was also in contrast to theoretical expectations based on the kinetic theory of gases. Dollfus estimated that the atmospheric pressure at ground level was about 1 mm/0.039 in of mercury. (The nature of the gas making up this atmosphere is unknown, but it must be a dense, heavy gas. It is certain that the atmosphere on Mercury is not more than 1/300 that on Earth.)

Mercury shows faint shady markings, set against a dull whitish background, that were first observed by Giovanni Schiaparelli in 1889. Using the 60-cm/24-in refractor at the Pic-du-Midi Observatory, Dollfus, again in 1950, was able clearly to resolve spots about 300 km/186 mi apart.

Dollfus has also looked at the possibility of an atmosphere around the Moon. The rate of thermal dissipation into space of all but the heavier gases (which are cosmically very scarce) from the Moon is so high that an atmosphere cannot be expected. The most telling evidence is the complete absence of the twilight phenomena on the Moon. Any elongation of the points (or the cusps) of the Moon beyond 90 °, caused by scattered sunlight, should be detectable by polarization. But Bernard Lyot, and later Dollfus, proved that there was no detectable polarization.

In 1966 Dollfus discovered Janus, the innermost moon of Saturn, at a time when the rings – to which it is very close – were seen from Earth edgeways on (and practically invisible).

A practical astrophysicist, Dollfus has achieved remarkable results through patient and persistent research.

Domagk, Gerhard (1895–1964) German bacteriologist and a pioneer of chemotherapy who discovered the antibacterial effect of Prontosil, the first of the sulphonamide drugs. This important discovery led, in turn, to the development of a range of sulphonamide drugs effective against various bacterial diseases, such as pneumonia and puerperal fever, that previously had high mortality rates. For this achievement Domagk received many honours, including the 1939 Nobel Prize for Physiology or Medicine. At the time, however, Germans were forbidden by Adolf Hitler to accept such awards and Domagk did not receive his Nobel medal until 1947, by which time the prize money had reverted to the funds of the Nobel Foundation.

Domagk was born on 30 October 1895 in Lagow, Brandenburg (now in Poland), and studied medicine at Kiel University, graduating – after a period of military service during World War I – in 1921. In 1924 he became reader in pathology at the University of Griefswald then, in the following year, was appointed to a similar position at the University of Münster. In 1927 he accepted an invitation to direct research at the Laboratories for Experimental Pathology and Bacteriology of I G Farbenindustrie, Düsseldorf, a prominent German dye-making company. But he also remained on the staff of Münster University, which appointed him extraordinary professor of general pathology and pathological anatomy in 1928 and a professor in 1958. Domagk died on 24 April 1964 in Burgberg.

Following Paul Ehrlich's discovery of antiprotozoon chemotherapeutic agents, considerable advances had been made in combating protozoon infections but bacterial infections still remained a major cause of death. While working for I G Farbenindustrie, Domagk began systematically to test the new azo dyes in an attempt to find an effective antibacterial agent. In 1932 his industrial colleagues synthesized a new azo dye called Prontosil red, which Domagk found was effective against streptococcal infections in mice. In 1935 he published his discovery, but it received little favourable response. In the following year, however, the British Medical Research Council confirmed his findings, and shortly afterwards the Pasteur Institute in Paris found that the sulphonilamide portion of the Prontosil molecule is responsible for its antibacterial action. (This latter was an important finding because sulphonilamide is much cheaper to produce than is Prontosil.) Meanwhile Domagk had demonstrated the effectiveness of Prontosil in combating bacterial infections in humans. His daughter had accidentally infected herself while working on the clinical trials of Prontosil and, after the failure of conventional treatments, Domagk had cured her with Prontosil.

From about 1938 other sulphonamide drugs were produced that were effective against a number of hitherto serious bacterial diseases, but antibiotics were discovered shortly afterwards and they came to replace sulphonamides as the normal drugs used to treat bacterial infections. Nevertheless, sulphonamides and chemotherapy were – and still are – of great value, particularly in the treatment of antibiotic-resistant infections. In 1946 Domagk and his co-workers found two compounds (eventually produced under the names of Conteben and Tibione) that, although rather toxic, proved useful in treating tuberculosis caused by antibiotic-resistant bacteria. Subsequently Domagk attempted to find chemotherapeutic agents for treating cancer, but was unsuccessful.

Donati, Giovanni Battista (1826–1873) Italian astronomer whose principal astronomic interests were the study of comets and cosmic meteorology. He made important contributions to the early development of stellar spectroscopy and to the application of spectroscopic methods to the understanding of the nature of comets.

Donati was born in Pisa on 16 December 1826. He received his university training at the University of Pisa and began his career in astronomy at the observatory in Florence in 1852. He was first employed as an assistant, but in 1864 he succeeded Giovan Battista Amici (1786–1868) to the directorship of the observatory.

One of Donati's major responsibilities after assuming the directorship of the observatory was the supervision of the work at Arcetri, not far from Florence, where a new observatory was being set up. It was formally established a year before Donati's death. He was struck down by the plague and died in Florence on 20 September 1873.

Donati's active research career spanned little more than 20 years, but it was very productive. During the 1850s he was an enthusiastic comet-seeker, with six discoveries to his credit – the most dramatic of these was named after him. Donati's comet, which was first sighted on 2 June 1858, was notable for its great beauty. It had, in addition to its major 'tail', two narrow extra tails.

Donati then applied his talents and his time to the developing subject of stellar spectroscopy. He compared and contrasted the spectrum of the Sun with those of other stars, and then sought to use this technique to examine the properties and composition of comets. Donati found that when a comet was still distant from the Sun, its spectrum was identical to that of the Sun. When the comet approached the Sun it increased in magnitude (brightness) and its spectrum became completely different. Donati concluded that when the comet was still distant from the Sun, the light it emanated was simply a reflection of sunlight. As the comet approached the Sun the material in it became so heated that it emitted a light of its own, which reflected the comet's composition.

Shortly thereafter, William Muggins (1824–1910) reported that the tail of a comet contained carbon compounds. More definitive analyses of cometary make-up were written over the ensuing years, culminating with Whipple's report published in 1950.

Other areas of interest that engaged Donati's attention were atmospheric phenomena and events in higher zones, such as the aurora borealis. His most important research was, however, his pioneering efforts in the use of spectroscopy to elucidate the nature of comets.

Donkin, Bryan (1768–1855) English engineer who made several innovations in papermaking, printing, and food preservation.

Little is known of Donkin's childhood. He was born in Northumberland on 22 March 1768, and was later apprenticed to John Hall, a papermaker of Dartford, Kent. He went on to work for Hall and perfected a new type of papermaking machine that had been devised in 1798 by the Frenchman Nicolas Robert and later patented in England by Henry and Sealy Fourdrinier. This success led Donkin to investigate another recent invention from France, the preservation of food by bottling. He established his own company, even gaining royal approval by presenting samples to the Prince Regent in 1813. With this security behind him, he returned to the printing and paper trade, and invented the forerunner of the rotary press. By 1815 he had turned to civil engineering, and became a founder member of the Institution of Civil Engineers. When he died in London on 27 February 1855, he was remembered as much for his role in founding such official groups as for his engineering skills.

Donkin made his first practical Fourdrinier papermilling machine at Frogmore Hill, Hertfordshire, in 1803. He set up a factory in Bermondsey, South London, where he manufactured nearly 200 machines in all. Donkin's contribution to food preservation was to take Nicolas ◊Appert's bottling process and modify it to use metal cans instead of glass bottles. In printing, he tackled the problem of increasing the speed of presses. The original flat-bed press, with its back-and-forth movement, was too slow. Donkin arranged four (flat) formes of type around a spindle – a rudimentary rotary press. He introduced a composition of glue and treacle for the inking rollers, an innovation that was still widely used long after his press had been superseded.

Doppler, Christian Johann (1803–1853) Austrian physicist who discovered the Doppler effect, which relates the observed frequency of a wave to the relative motion of the source and the observer. The Doppler effect is readily observed in moving sound sources, producing a fall in pitch as the source passes the observer, but it is of most use in astronomy, where it is used to estimate the velocities and distances of distant bodies.

Doppler was born in Salzburg on 29 November 1803, the son of a stonemason. He showed early promise in mathematics, and attended the Polytechnic Institute in Vienna 1822–25. He then returned to Salzburg and continued his studies privately while tutoring in physics and mathematics. From 1829–33, Doppler went back to Vienna to work as a mathematical assistant and produced his first papers on mathematics and electricity. Despairing of ever obtaining an academic post, he decided in 1835 to emigrate to the USA. Then, on the point of departure, he was offered a professorship of mathematics at the State Secondary School in Prague and changed his mind. He subsequently obtained professorships in mathematics at the State Technical Academy in Prague in 1841, and at the Mining Academy in Schemnitz in 1847. Doppler returned to Vienna the following year and, in 1850, became director of the new Physical Institute and professor of experimental physics at the Royal Imperial University of Vienna. He died from a lung disease in Venice on 17 March 1853.

Doppler explained the effect that bears his name by pointing out that sound waves from a source moving towards an observer will reach the observer at a greater frequency than if the source is stationary, thus increasing the observed frequency and raising the pitch of the sound. Similarly, sound waves from a source moving away from the observer reach the observer more slowly, resulting in a decreased frequency and a lowering of pitch. In 1842, Doppler put forward this explanation and derived the observed frequency mathematically in Doppler's principle.

The first experimental test of Doppler's principle was made in 1845 at Utrecht in Holland. A locomotive was used to carry a group of trumpeters in an open carriage to and fro past some musicians able to sense the pitch of the notes being played. The variation of pitch produced by the motion of the trumpeters verified Doppler's equations.

Doppler correctly suggested that his principle would apply to any wave motion and cited light as an example as well as sound. He believed that all stars emit white light and that differences in colour are observed on Earth because the motion of stars affects the observed frequency of the light and hence its colour. This idea was not universally true as stars vary in their basic colour. However Armand ◊Fizeau pointed out in 1848 that shifts in the spectral lines of stars could be observed and ascribed to the Doppler effect and hence enable their motion to be determined. This idea was first applied in 1868 by William ◊Huggins, who found that Sirius is moving away from the Solar System by detecting a small red shift in its spectrum. With the linking of the velocity of a galaxy to its distance by Edwin ◊Hubble in 1929, it became possible to use the red shift to determine the distances of galaxies. Thus the principle that Doppler discovered to explain an everyday and inconsequential effect in sound turned out to be of truly cosmological importance.

Further Reading

Doppler, Christian Johann, *The Phenomenon of Doppler,* The Czech Technical University, Faculty of Nuclear Sciences and Physical Engineering, 1992.

Eden, A, *Search for Christian Doppler,* Blackwell, 1992.

Hearnshaw, J B, Doppler and Vogel: two notable adversaries in stellar astronomy, *Vistas Astron,* 1995, v 35, pp 157–177.

Draper, Henry (1837–1882) US amateur astronomer, noted for his work on stellar spectroscopy and commemorated by the Henry Draper Catalogue of stellar spectral types.

Draper was born in Virginia on 7 March 1837; his father, John William Draper, was a distinguished physician and chemist. He was educated at the University of the City of New York and entered the medical school there at the age of 17. By the time he was 20, he had completed the medical course, but since he had not reached the age required for graduation, he spent the following year travelling in Europe. During this period he visited, and was greatly influenced by, William ◊Parsons and his observatory in Parsonstown (now Birr) in Ireland. In 1860 Draper was appointed professor of natural science at the University of the City of New York, but the interests that he developed during his travels (telescope-making and photography) were to be woven into his professional career. He died unexpectedly, of double pleurisy, at his home in New York on 20 November 1882.

On returning from his travels in Europe, Draper began preparing his own glass mirror and by 1861 he had installed it in his new observatory on his father's estate at Hastings-on-Hudson, New York. Draper began his research career by making preliminary studies of the spectra of the more common elements and photographing the solar spectrum. By 1873 he had devised a spectrograph that was similar to William Huggins's visual spectroscope; it clarified the spectral lines by means of a slit and incorporated a reference spectra so that celestial elements could be identified more easily.

In 1874 Draper was asked to act as director of the photographic department of the US commission to observe the transit of Venus of that year. Draper's work was stimulated by the spectroscopic studies of Huggins and Lockyer in Europe and during the last years of his life he worked towards obtaining high-quality spectra of celestial objects. He studied the Moon, Mars, Jupiter, the comet 1881 III, and the Orion Nebula. He also succeeded in obtaining photographs of stars that were too faint to be seen with the same telescope by using exposure times of more than 140 minutes – exemplifying the advantages of photography in astronomy.

After Draper's sudden death, his widow established a fund to support further spectral studies. It was used by a team at Harvard College Observatory as part of a programme, begun in 1886, to establish a useful classification scheme for stars and a catalogue of spectra. The Harvard project was not completed until 1897 but the result was a comprehensive classification of stars according to their spectra, named the Henry Draper Catalogue.

Drew, Charles Richard (1904–1950) US surgeon chiefly remembered for his research into blood transfusion.

Born on 3 June 1904 in Washington, DC, Charles Drew was educated in the public schools of the city graduating with honours from Dunbar High School in 1922. After receiving a BA from Amherst College in 1926, he worked for two years as director of athletics and teacher of biology at Morgan State College. In 1928 he went on to McGill University Medical School and was awarded his MD and CM in 1933. Having completed his internship in Montréal General Hospital, in 1935 he went to Howard University Medical School as an instructor of pathology. In 1938 he was granted a research fellowship by the Rockefeller Foundation and spent two years at Columbia University, New York, and as a resident in surgery in the Presbyterian Hospital connected with the Medical School. He received a MedDSc degree from the University in 1940 – the first black person to receive this degree in the country. After working for the American Red Cross, Drew returned to Howard Medical School, where in 1942 he was made professor and head of the department of surgery and chief of surgery at the Freedmen's Hospital. In 1944 he was appointed chief of staff of the hospital. He remained there until his death following a motor accident, on 1 April 1950.

While at McGill University, Drew became interested in the problems of blood transfusion and it is his work in this field for which he is remembered. At the Presbyterian Hospital his research demonstrated that plasma had a longer life than whole blood and therefore could be better used for transfusion; he wrote a doctoral thesis 'Banked blood: a study in preservation', and was supervisor of the blood plasma division of the Blood Transfusion Association of New York City. In 1939 he established a blood bank and was in charge of collecting blood for the British army at the beginning of World War II. In 1941 he became director of the American Red Cross Blood Bank in New York City, which collected blood for the US armed forces. Drew resigned, however, when the Red Cross decided to segregate blood according to the race of the donor. Drew is also known for his teaching and training of surgeons and his publication of many papers in medical and scientific journals. For his work, Drew was awarded the Spingarn Medal (1944), honorary DSc degrees from Virginia State College (1945) and Amherst College (1947), and posthumously, the Distinguished Service Medal of the National Medical Association (1950). Several schools and medical centres have been named after him and a stamp was issued in his honour in 1981.

Further Reading

Crenshaw, Gwendolyn J (ed), *Charles Richard Drew: A Navigator on the River of Life,* AESOP Enterprises, 1991.

Jackson, Garnet N, *Charles Drew, Doctor,* Modern Curriculum, 1994.

Mahone-Lonesome, Robyn, *Charles Drew: Physician,* Black Americans of Achievement series, Chelsea House Publishers, 1990.

Shapiro, Miles, *Charles Drew: Founder of the Blood Bank,* Innovative Minds series, Raintree Steck-Vaughn Publishers, 1996.

Shoecraft, William D, *Dr Charles Drew,* Did You Know Publishings, 1994.

Dreyer, John Louis Emil (1852–1926) Danish-born Irish astronomer and author, best known for a biographical study of the work of the Danish scientist Tycho Brahe, and for his meticulous compilation of catalogues of nebulae and star clusters.

Dreyer was born in Copenhagen on 13 February 1852. Educated there, he displayed unusual talents in mathematics, physics, and history, although it was not until he was aged 14 that he read a book about Tycho Brahe and became keenly interested in astronomy – an interest that was encouraged by his friendship with Schjellerup, an astronomer at the Copenhagen University. Dreyer began his studies at the university in 1869, and by 1870 he had been given a key that allowed him free access to the instruments in the university observatory. In 1874 he was appointed assistant at William Parson's observatory in Parsonstown (now Birr) in Ireland. Four years later he took up a similar post at Dunsink Observatory at the University of Dublin, and four years later again, in 1882, he became director of the Armagh Observatory, where he remained until he retired in 1916. On his retirement he went to live in Oxford and, continuing his writing, made use of the facilities of the Bodleian Library. He died in Oxford on 14 September 1926.

Dreyer's earliest formal astronomical publication, published in 1872, was a description of the orbit of the first comet of 1870. After his move to Ireland in 1874, he became increasingly interested in making observations of nebulae and star clusters, a subject that occupied most of his time for the next 14 years. He was acutely aware of the element of error involved in astronomical observations of objects such as nebulae, and he published an important paper on the subject in 1876.

In 1877 Dreyer presented to the Royal Astronomical Society data on more than 1,000 new nebulae and corrections to the original catalogue on nebulae and star clusters compiled by John Herschel. He extended this work at Armagh, which led to the publication in 1886 of the Second Armagh Catalogue, with information on more than 3,000 stars. The Royal Astronomical Society then invited him to compile a comprehensive new catalogue of nebulae and star clusters to incorporate all the modern data and to supersede Herschel's old catalogue. This enormous task was completed in only two years, but the rapid accumulation of more information necessitated the publication of two supplementary indexes in 1895 and 1908. Together these three catalogues described more than 13,000 nebulae and star clusters and achieved international recognition as standard reference material.

The catalogue completed, Dreyer decided to write a biography of his hero, Tycho Brahe. It was published in 1890 and preceded a 15-volume series (1913–19) detailing all of Brahe's work. Dreyer's other writings included a history of astronomy (at that time the only authoritative and complete historical analysis), and an edition of the complete works of William Herschel.

On his retirement in 1916, Dreyer, a Fellow of the Royal Astronomical Society from 1875, was awarded the society's highest honour, the Gold Medal. A patient and skilled observational astronomer, Dreyer was also an excellent mathematician, a talented scholar, and an accomplished writer. He put all of these attainments to good use during his career, combining them to produce work of enduring quality.

Driesch, Hans Adolf Eduard (1867–1941) German embryologist and philosopher who is best known as one of the last advocates of vitalism, the theory that life is directed by a vital principle and cannot be explained solely in terms of chemical and physical processes. Nevertheless, he also made several important discoveries in embryology, although these have tended to be overlooked because of his mistaken belief in vitalism.

Driesch was born on 28 October 1867 in Bad Kreuznach, the son of a prosperous gold-merchant. He studied zoology, chemistry, and physics at the universities of Hamburg, Freiburg, Munich, and Jena, obtaining his doctorate from Jena – where he studied under Ernst Haeckel – in 1887. Coming from an affluent family, Driesch had no need of paid employment and he spent the next 22 years privately pursuing his embryological studies. After obtaining his doctorate, he travelled extensively in Europe and the Far East, spending nine years from 1891 working at the International Zoological Station in Naples. In 1899 he married Margarete Reifferscheidt; they later had two children, both of whom became musicians. Eventually Driesch settled in Heidelberg, and in 1909 he was appointed lecturer in philosophy at the university there, becoming professor of philosophy in 1911. Subsequently he was professor of philosophy at Cologne University 1920–21 and at Leipzig University 1921–35, when he was forced to retire by the Nazi regime. Driesch died in Leipzig on 16 April 1941.

In 1891 Driesch, experimenting with sea urchin eggs, discovered that when the two blastomeres of the two-cell stage of development are separated, each half is able to develop into a pluteus (a later larval stage) that is completely whole and normal, although of smaller than average size. Similarly he found that small, whole individuals can be obtained by separating the four cells of the four-cell stage of development. From these findings he concluded that the fate of a cell is not determined in the early developmental stages. Later, other workers discovered the same phenomenon in the early developmental stages of hydroids, most vertebrates, and certain insects. (It should be noted, however, that not all animal eggs behave in this way; for example, separation of the early embryonic cells of annelids, molluscs, and ascidians results in incomplete embryos.) Subsequently Driesch produced an oversized larva by fusing two normal embryos, and in 1896 he was the first to demonstrate embryonic induction when he displaced the skeleton-forming cells of sea urchin larvae and observed that they returned to their original positions. These findings provided a great impetus to embryological research but Driesch himself – unable to explain his results in mechanistic terms (principally because at that time very little was known about biochemistry)– came to believe that living activities, especially development, were controlled by an indefinable vital principle, which he called entelechy.

After his appointment as a lecturer in 1909, Driesch abandoned scientific research and devoted the rest of his life to philosophy.

Further Reading

Freyhofer, Horst, *Vitalism of Hans Driesch: The Success and Decline of a Scientific Theory,* Series 20: Philosophy, European University Studies, 1982, v 83.

Partenheimer, David, 'Henry Adam's scientific history and German scientists', *Engl Lang Notes,* 1990, v 27, no 3, pp 44–52.

Dubois, (Marie) Eugène François Thomas (1858–1940) Dutch palaeontologist who in 1891 discovered the remains of *Pithecanthropus erectus,* known as Java Man.

Dubois was born in Eijsden on 28 January 1858 and studied medicine and natural history at the University of Amsterdam. In 1886 he took the appointment of lecturer in anatomy there. A year later he joined the Dutch Army Medical Service and was posted to Java – then a Dutch possession – where he was commissioned by the Dutch government to search for fossils. The discoveries he made there brought him worldwide fame and he returned to Europe in 1895. Dubois took up a professorship in palaeontology, geology, and mineralogy at the University of Amsterdam in 1899. He retired in 1928 and died at Halen in Belgium on 16 December 1940.

The excavations of Pompeii and Herculaneum in 1748, and later, in 1870, of Troy were the beginnings of archaeology and encouraged attempts to piece together the history of human development. In 1857 the first Neanderthal skeleton was found and later Cro-Magnon skeletons were discovered in southwestern France. After Darwin published his *Origin of Species* in 1859, many people thought that the evolutionary principles he had outlined could equally apply to the origin of the human species.

The skeletons already found were undoubtedly those of an early species of human, but there was still a large gap in evolutionary terms between modern humans and the apes. Having first been involved in the comparative anatomy of vertebrates Dubois became fascinated with the problem of the 'missing link' in the evolutionary chain and was convinced that somewhere there existed its remains. He reasoned that such a missing link could have lived in an area where apes were still numerous, such as Africa or southeast Asia. His posting to Java gave Dubois the opportunity to investigate this theory.

The remains of extinct animals had been found in the deposits of volcanic ash on the banks of the Solo River in East Java and it was in the bone beds near the village of Trinil that he concentrated his search. In 1891 Dubois found teeth, a skullcap, and a femur. The skullcap was much larger than that of any living ape, and more primitive and apelike than that of

Neanderthal hominids; the bones were heavily ridged and the vault of the braincase extremely low, indicating that the size of the brain was far smaller than that of a modern human. The teeth were also intermediate between ape and human. The femur was definitely human, the ends of the bone and the straightness of the shaft suggesting that its owner had walked erect.

Dubois published a scientific description of the fragments he had found in 1894, and named them *Pithecanthropus erectus* ('erect ape-man') after the name given to the intermediate human by the German zoologist Ernst Haeckel.

The femur Dubois had found was discovered some distance away from the skullcap and could therefore have been from some other form of human. Many people were not convinced that Dubois had discovered the missing link, especially when further excavations at Trinil produced no more traces of *Pithecanthropus erectus.* In response to the controversy aroused by his findings, Dubois withdrew his discovery from the public until 1923. But between 1936 and 1939 the German archaeologist von Koenigswald was working in the Solo River valley farther upstream from Dubois's original discoveries, and found more skullcaps and a lower and an upper jaw. Later, a child's skull was discovered at Modjokerto on Java and is now believed to be from a young *Pithecanthropus erectus.*

Detailed measurements of casts of the skull of *Pithecanthropus erectus* indicated that the brain cavity had a volume of about 940 cu cm/57 cu in, whereas few apes have more than 600 cu cm/37 cu in and modern humans have about 1,500 cu cm/91.5 cu in. From these figures it was thought that *Pithecanthropus erectus* was almost exactly halfway between ape and human on the evolutionary scale, although it is now considered to be definitely more human and has been renamed *Homo erectus.* In contrast to this belief, in Dubois's later years he changed his ideas and stressed the ape-like similarities of his discovery rather than its hominid likenesses.

The history of human descent has been further pieced together as archaeological techniques have been refined and more discoveries have been made. But it cannot be doubted that one of the most important of these contributions has been that of Dubois.

Du Bois-Reymond, Emil Heinrich (1818–1896) German physiologist who showed the existence of electrical currents in nerves.

Du Bois-Reymond was born in Berlin on 7 November 1818. His father was a Swiss teacher who had settled in Berlin – the family was French-speaking. Talent ran in the family, his brother Paul (1831–1889) shining as a mathematician and making contributions to function theory. Du Bois-Reymond studied a wide range of subjects in Berlin for two years before finally choosing a medical career. Working under Johannes Müller, he graduated in 1843, soon plunging into research on animal electricity and especially on electric fishes. All through his career he was closely associated with the leading German investigators of human physiology: Theodor Schwann, Matthias Schleiden, Karl Ludwig and also the physicist Hermann Helmholtz.

In 1858 he succeeded Müller as professor of physiology, and was appointed head of the new Physiological Institute that opened in Berlin in 1877. Du Bois-Reymond's importance lay in his investigations of the physiology of muscles and nerves, and in his demonstrations of electricity in animals. He was adept in introducing improved techniques for measuring such neuroelectrical effects, first investigated by Luigi Galvani. By 1849 he had evolved a delicate multiplier for measuring nerve currents. Using his highly sensitive apparatus, he was able to detect an electric current in ordinary localized muscle tissues, notably contracting muscles. He observantly traced it to individual fibres, finding their interior was negative with regard to the surface. He showed the existence of electrical currents in nerves, correctly arguing that it would be possible to transmit nerve impulses chemically. Du Bois-Reymond died in Berlin on 26 December 1896.

Du Bois-Reymond's experimental methods proved the basis for almost all future work in electrophysiology. He held trenchant views about scientific metaphysics. He denounced the vitalistic doctrines that were in vogue amongst German scientists and denied that nature contained mystical life-forces independent of matter.

Dulong, Pierre Louis (1785–1838) French chemist, best known for his work with Alexis ◊Petit that resulted in Dulong and Petit's law, which states that, for any element, the product of its specific heat and atomic weight is a constant, a quantity they termed the atomic heat. In modern terms, the product of the specific heat capacity of an element (expressed in joules per gram per kelvin) and its relative atomic mass is about 25.

Dulong was born on or about 12 February 1785 in Rouen. He studied at the Ecole Polytechnique in Paris, training initially as a doctor. He married young and had to take a teaching post to keep his family and finance his research; one such (1811) was at the Ecole Normale and another (1813) was at the Ecole Vétérinaire at Alfort. He returned to Paris in 1820 to become professor of chemistry at the Faculté des Sciences, moving back to the Ecole Polytechnique, becoming its director of studies in 1830. He was elected to the physics section of the French Academy of Sciences in 1823. He died in Paris on or about 18 July 1838.

During Dulong's early work in chemistry he was an assistant to Claude Berthollet, and he studied the oxalates of calcium, strontium, and barium. In 1811 he discovered the explosive compound nitrogen trichloride, an accident with which cost him a finger and the sight in one eye. He resolved the contemporary dispute among chemists about the composition of phosphorus and phosphoric acids, identifying two new acids in the process. In 1815 he began working with Alexis Petit, and at first they applied their researches to the problem of measuring heat. They determined the absolute coefficient expansion of mercury, so improving the accuracy of mercury thermometers. They then explored the laws of cooling in a vacuum, work later extended by Josef ◊Stefan.

Then in 1818 Dulong and Petit began studying the specific heat capacity of elements, measuring this quantity for sulphur and 12 metals. When they multiplied each result by the element's atomic weight (relative atomic mass), they obtained values that were in close agreement with each other. They showed that an element's specific heat capacity is inversely proportional to its relative atomic mass (Dulong and Petit's law). Its chief application was in the estimation of the atomic masses of new elements.

After Petit died in 1820, Dulong continued to work on specific heat capacities, publishing his findings in 1829. He concluded that, under the same conditions of temperature and pressure, equal volumes of all gases evolve or absorb the same quantity of heat when they are suddenly expanded or compressed to the same fraction of their original volumes. He also deduced that the accompanying temperature changes are inversely proportional to the specific heat capacities of the gases at constant volume. He also collaborated with the French physicist Dominique ◊Arago on a study of the pressure of steam at high temperatures. In this rather hazardous research, working at pressures up to 27 atmospheres, their results were in agreement with Boyle's law. Dulong's last paper on the heats of chemical reaction was published in 1838, after his death.

Dumas, Jean Baptiste André (1800–1884) French chemist who made contributions to organic analysis and synthesis, and to the determination of atomic weights (relative atomic masses) through the measurement of vapour densities.

Dumas was born on or about 19 July 1800 in Alais (now Alès), Gard. He was educated at the local college and intended to enter the navy, but changed his mind after the overthrow of Napoleon I and became instead apprenticed to an apothecary, also in Alais. But he soon moved to Geneva, Switzerland, and in 1816 was working in the laboratory of a pharmacist there (who was investigating plant extracts). He studied also under the Swiss physicist Pierre ◊Prévost and the botanist Augustin Candolle (1778–1841), and in 1822 accepted an invitation from Alexander von Humboldt to go to Paris, where he took up an appointment at the Ecole Polytechnique and held the chair in chemistry at the Lyceum (later Athenaeum), succeeding André Ampère. The following year he became a lecturing assistant to the French chemist Louis Thénard (1777–1857) at the Ecole Polytechnique, whom he succeeded as professor of chemistry in 1835. Following the political upheavals of 1848 Dumas abandoned much of his scientific work for politics and public office. He served under Napoleon III as Minister of Agriculture and Commerce, Minister of Education, and Master of the Mint. After the deposition of Napoleon in 1871, Dumas left politics. He died in Cannes on 11 April 1884.

Dumas French organic chemist Jean Baptiste Dumas. Among his studies were the isolation of anthracene from coal tar, determining the formulae of menthol and camphor, and working out the theory of substitution from bleaching reactions. *Mary Evans Picture Library*

While he was an 18-year-old in Geneva, Dumas was involved in the study of the use of iodine to treat goitre (endemic in Switzerland at that time). Then under Prévost he – unsuccessfully – investigated the physiological effects of digitalis. They also studied blood and showed that urea is present in the blood of animals from which the kidneys have been removed, proving that one of the functions of the kidneys is to remove urea from the blood not to produce it.

In 1826 Dumas began working on atomic theory. He determined the molecular weights (molecular masses) of many substances by measuring their vapour densities and concluded that 'in all elastic fluids observed under the same conditions, the molecules are placed at equal distances' – that is, they are present in equal numbers.

His important work on the theory of substitution in organic compounds was inspired at a soirée at the Tuilleries when the candles gave off irritating fumes. The candle wax had been bleached with chlorine, some of which had been retained and during combustion was converted to hydrogen chloride. Dumas soon proved by experiments that organic substances treated with chlorine retain it in combination, and proposed that the chlorine had displaced hydrogen, atom for atom. He studied the action of chlorine on alcohol (ethanol) to produce chloral (trichloroethanal), which he decomposed with alkali to give chloroform (trichloromethane) and formic acid (methanoic acid). He also chlorinated acetic acid (ethanoic acid), to give trichloracetic acid (trichlorethanoic acid), thus proving his theory of substitution. He had shown that atoms of apparently opposite electrical charge had replaced each other, in opposition to the dualistic theory of organic chemistry proposed by Jöns ◊Berzelius, who in 1830 was at the height of his fame and influence.

Together with the Belgian chemist Jean ◊Stas, who was at that time his student, Dumas investigated the action of alkalis on alcohols and ethers, which led to a study of the acids produced by the oxidation of alcohols.

In 1833 Dumas worked out an absolute method for the estimation of the amount of nitrogen in an organic compound – which still forms the basis of modern methods of analysis. The nitrogen in a sample of known weight is eliminated in gaseous form and estimated by direct measurement. The sample is heated with cupric oxide (copper(II) oxide) and oxidized completely in a stream of carbon dioxide; the gaseous products of combustion are passed over a heated copper spiral and the nitrogen collected in a gas burette over concentrated potassium hydroxide solution.

Also with Stas, in 1849 he revised the atomic weight (relative atomic mass) of carbon to 12 (from Berzelius's value of 12.24). He went on to correct the atomic masses of 30 elements – half the total number known at that time – referring to the hydrogen value as 1. With Milne Edwards he investigated the way in which bees convert sugar to fat (wax). His last papers were on alcoholic fermentation (1872) and on the occlusion of oxygen in silver (1878).

Dunlop, John Boyd (1840–1921) Scottish veterinary surgeon who is usually credited with the invention of the pneumatic tyre (originally for bicycles).

Dunlop was born in Dreghorn, Ayrshire, on 5 February 1840. He studied veterinary medicine at Edinburgh University, before setting up practice in Ireland near Belfast in 1867. He devised a pneumatic tyre ten years later in an attempt to improve the comfort of his son's tricycle. It was so successful that in the following year (1888) he applied for the British patent and within three years he had founded his own company, with the encouragement of Harvey du Cros (whose sons were keen racing cyclists).

When Dunlop's business was already established, it was discovered that the tyre had previously been patented by another Scotsman, R W Thomson of Stonehaven, Kincardineshire. He had made a set of tyres as early as 1846 and fitted them to a horse-drawn carriage. They had been tested over 1,600 km/995 mi before they had needed replacing. Thomson's invention had gone practically unnoticed, whereas Dunlop's arrived at a crucial time in the development of transport. The bicycle was fast becoming recognized as more than a hobby, and the motor car was about to make its spectacular appearance on the roads. In 1896, after trading for only about five years, Dunlop sold both his patent and his business for £3 million, and Arthur Philip du Cros took over as managing director. He died in Dublin on 23 October 1921.

The key to Dunlop's invention was rubber. The rubber industry had become established in Europe in around 1830 with the development of vulcanization (the blending of india rubber with sulphur to produce a workable substance). In 1876 the English planter Henry Wickham travelled to the Amazon and collected seeds of wild rubber plants, which he brought back to Kew Gardens for propagation. The young plants were taken to Ceylon (Sri Lanka) and Malaya, where they formed the nucleus of the new rubber plantations.

Dunlop realized that rubber was the most suitable material for making tyres because it could stand up to wear and tear while retaining its resilience. His first simple design consisted of a rubber inner tube, covered by a jacket of linen tape with an outer tread also of rubber. The inner tube was inflated using a football pump and the tyre was attached by flaps in the jacket which were rubber-cemented to the wheel. Later he incorporated a wire through the edge of the tyre that secured it to the rim of the wheel. Whether or not Dunlop's invention was the first, he certainly pioneered the mass production of tyres and Fort Dunlop, the Dunlop Rubber Company's factory in Birmingham, remains today.

Du Toit, Alexander Logie (1878–1948) South African geologist who helped pave the way for theories of continental drift.

Born near Cape Town of a wealthy family of Huguenot descent, Du Toit studied at Cape Town, Glasgow, and the Royal College of Science, London.

He spent 17 highly fruitful years of his career from 1903 mapping for the Geological Commission of the Cape of Good Hope. At the height of his powers, and following a visit to South America, he developed a profound interest in Alfred ◊Wegener's theory of continental drift. In *A Geological Comparison of South America and South Africa* (1927), he systematically adumbrated the abundant similarities in the geologies of the two continents, suggesting that they had probably once been joined. Concepts of this kind were most thoroughly and famously stated in his significant *Our Wandering Continents* (1937), in which he maintained that the southern continents had in earlier times formed the supercontinent of Gondwanaland, which was distinct from the northern supercontinent of Laurasia. This notion, though initially deprecated, steadily grew in acceptance, and was to form one of the foundations for the synthesis of continental drift theory and plate tectonics that created the geological revolution of the 1960s. The most eminent geologist South Africa has produced, Du Toit was widely hailed as the world's finest field geologist.

Dyson, Frank Watson (1868–1939) English astronomer especially interested in stellar motion and time determination.

Dyson was born in Ashby-de-la-Zouch, Leicestershire, on 8 January 1868. He attended Bradford Grammar School and Trinity College, Cambridge, from which he graduated in 1889. He became a fellow of Trinity in 1891 and in 1894 he was made chief assistant at the Royal Greenwich Observatory. He left in 1906 to become Astronomer Royal for Scotland, but returned to Greenwich in 1910 to serve as Astronomer Royal for England. He retired in 1933, but remained active in research and writing. In addition to his many research publications, Dyson was the author of several general books on astronomy. He died off the coast of South Africa while on a sea voyage from Australia, on 25 May 1939.

Dyson's early research was concentrated on problems in gravity theory, but as soon as he started his work at the Greenwich Observatory he began a lengthy study of stellar proper motion in collaboration with William Thackeray. He was an active member of several expeditions to study total eclipses of the Sun, and in 1906 he published a book in which he discussed data he had obtained on these occasions on the spectrum of the solar chromosphere.

Dyson was one of a number of astronomers who confirmed the observations of Jacobus ◊Kapteyn on the proper motions of stars, which indicated that the stars in our Galaxy seemed to be moving in two great streams. These results were later realized to be the first evidence for the rotation of our Galaxy.

The measurement of time has always been an important function of the Greenwich Observatory, and Dyson was passionately interested in this aspect of his work. It was he who initiated the public broadcasting of time-signals by the British Broadcasting Corporation (BBC) over the radio, in the form of the familiar six-pip signal. (This was first broadcast in 1924 from Rugby.)

Another important research area for the Greenwich Observatory is the study of solar eclipses, and Dyson was active in the organization of expeditions to observe these. The most significant of these expeditions were the two he coordinated for the 1919 eclipse. They served as the occasion for Arthur ◊Eddington's famous confirmation of the gravitational deflection of light by the Sun, as predicted by Einstein's general theory of relativity.

Other areas to which Dyson made important contributions include the study of the Sun's corona and of stellar parallaxes. He was a talented astronomer and a skilled administrator.

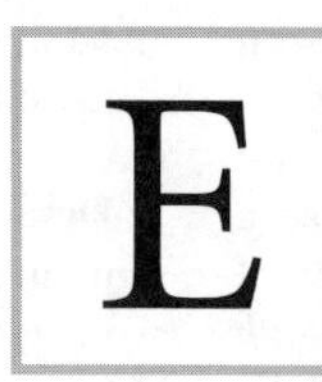

Eastman, George (1854–1932) US inventor, entrepreneur, and benefactor who founded the Kodak company. He started the 'press button' end of photography and brought it within range of virtually everybody's skill and pocket.

Born on 12 July 1854 in Waterville, New York, Eastman left school at 14 to earn a living and ease his family's financial hardships. He started as a messenger boy, studying accounting in the evenings. He saved $3,000, in 1879 patented a photographic emulsion coating machine, and the following year began mass production of dry plates.

Eastman had started experimenting with photographic emulsion in 1878. At that time, there were large numbers of photographic processes in use, the most popular of which was collodion-coated glass plates. These had to be prepared, sensitized, exposed, and developed in rapid succession, which meant that landscape photographers had to carry a complete darkroom about with them. Enlargement was not practicable so negatives had to be the size of the finished print – thus some cameras were the size of soap boxes.

Various methods to preserve the collodion were used but most were complicated and reduced the sensitivity of the plate. The first practical dry process introduced in 1855 was the collodion albumen plate, which needed six times the exposure of the wet plate. Gelatine emulsion, usually attributed to Maddox, appeared in 1871 and Burgess's gelatino-bromide prepared dry plates appeared in 1873. This ushered in a new era of photography in which the mobile darkroom was no longer required. Eastman began production of his own dry plates in 1880 in a rented loft of a building in Rochester, New York.

Stripping film, in which paper was used only to support the emulsion and was not destined to be part of the negative, was patented in 1855 by Scott Archer, but the technique was not used to any great extent until Eastman introduced it for the Eastman/Walker roller slide in 1886. Eastman's system consisted of a paper base, a layer of soluble gelatine, a layer of collodion, and a layer of sensitized gelatine emulsion.

After exposure, the roll was cut up into individual negatives, developed, and fixed. The emulsion side was then attached to glass plates and coated with glycerine. Hot water dissolved the gelatine so the paper could be stripped off. The image on the glass had to be transferred to a gelatine sheet for printing.

Eastman's roller slide fitted into existing cameras and he confidently expected that users of glass plates would soon change over to the new film. But this was not to be. He therefore decided to reach the general public with the Kodak camera, launched in 1888. This camera was loaded with a roll of the stripping film large enough for 100 exposures and sold, with shoulder strap and case, for $25. After use, the camera was sent to Rochester, where the film was developed by the complicated process and a new film loaded for $10.

Eastman followed this up in the next year with the first commercially available transparent nitrocellulose (celluloid) roll films. These had been produced since the mid 1850s, but until the advent of celluloid, none was successful. Unfortunately, on 2 May 1887 the Rev Hannibal Goodwin applied for a patent for a transparent roll film made of celluloid but the patent was not granted until September 1898. Meanwhile Eastman's company had captured the market with its own roll film. A long and complicated law suit between the company that had acquired Goodwin's interests (he had died in 1900), and Eastman's company was finally settled in March 1914 with the Eastman company paying $5 million to the owners of Goodwin's patent. Virtually the only change there has been since then is the replacement of the flammable nitrocellulose by nonflammable cellulose acetate. Eastman's roll-film replaced his stripping film in the Kodak camera and ushered in the era of press-button photography.

Eastman continued the popularization of photography by spooling his film so that the camera could be loaded in daylight, and did not have to be returned to the Rochester factory. A pocket-sized box camera was marketed for $1 in 1900. The 8 mm movie camera, (which appeared in 1932), colour film, panchromatic film, automatic exposure control, sound film, instant cameras– all these and many more innovations followed.

Eastman was the first to recognize the importance of the amateur market in photography, which until 1880 had been restricted by the need for darkroom work. Other manufacturers tried to produce rival cameras but due to Kodak's virtual roll-film monopoly, for two or three decades a hand camera, of whatever make, was simply known as a 'Kodak'.

Despite all these successes, Eastman still had severe private problems that led to his suicide on 14 March 1932.

Further Reading

Brayer, Elizabeth, *George Eastman: A Biography,* Johns Hopkins University Press, 1996.
Holmes, Burnham, *George Eastman,* Pioneers in Change series, Silver Burdett Press, 1992.
Joseph, Paul, *George Eastman,* Inventors series, ABDO Publishing Co, 1996.
Mitchell, Barbara, *Click!: A Story About George Eastman,* Creative Minds series, Lerner Publishing Group, 1986.

Eastwood, Alice (1859–1953) US botanist who provided critical specimens for professional botanists as well as advising travellers on methods of plant collecting and arousing popular support for saving native species.

Eastwood was born on 19 January 1859 in Toronto, daughter of Colin Skinner, steward of the Toronto Asylum for the Insane. Following the death of her mother, at the age of six she was taken into care by her uncle, William Eastwood. Alice Eastwood's passion for botany started at an early age; she learned the Latin names for plants from her physician uncle and gained gardening knowledge from a French priest at the convent where she lived for six years. After graduating from East Denver High School as valedictorian in 1879 she taught a variety of subjects at the school for the next ten years and studied plants in the Colorado Mountains. In 1890 she resigned from teaching and toured California and remote places in Colorado. In 1892 she received an assistantship at the California Academy of Sciences, San Francisco, and founded and ran the California Botanical Club there. In 1894, she became the curator of the herbarium at the academy and for a while acted as editor of the journal *Zoe;* she also carried out extensive field work in California. She remained at the academy until her retirement in 1949. She died from cancer in 1953.

Eastwood's early fieldwork led to the publication of *A Handbook of the Trees of California* (1905), and her time prior to the San Francisco earthquake of 1906 was spent enlarging the academy's botanical collection and segregating critical type specimens from the herbarium. Much, however, was destroyed in the earthquake and between 1906 and 1912, she devoted her time to rebuilding the collections, involving field trips to the coastal ranges and the Sierra Nevada and visits to botanical gardens around the world. Between 1912 and her retirement over 340,000 specimens were added to the herbarium. In 1932 she and her assistant John Thomas Howell founded and edited the journal *Leaflets of Western Botany;* she also assisted in editing the journal *Erythea.* Eastwood herself published over 300 items. She received awards from garden clubs and in 1950 was elected honorary president of the Seventh International Botanical Congress in Stockholm.

Further Reading

Ross, Michael E, *Flower Watching with Alice Eastwood,* Naturalist's Apprentice series, Lerner Publishing Group, 1997.

Eastwood, Eric (1910–1981) English electronics engineer who made major contributions to the development of radar for both military and civilian purposes.

Eastwood was born on 12 March 1910 and educated at Oldham High School, Lancashire. He went to Manchester University and studied physics under Lawrence Bragg, and then moved to Cambridge University to take his PhD. He taught physics for a while at the Collegiate School, Liverpool, before entering the Signals branch of the Royal Air Force during World War II, and becoming involved with the solution of technical problems concerning radar.

In 1945, Eastwood joined the Nelson Research Laboratory of the English Electric Company. In 1948 he transferred to the Marconi Research Laboratory and was its director for many years.

In 1962 he accepted the appointment of director of research for the English Electric Group of companies and, with the merger in 1968 of GEC and AEI, became director of research for the new company, General Electric, a position he held until his retirement in 1974.

Radar is a system using pulsed radio waves transmitted to and reflected back from a 'target' to measure its distance and direction. It was developed as a result of research carried out in the late 1800s and early 1900s. Heinrich Hertz first discovered radio waves in 1888 and found they could be focused into a beam. Over the next 20 years several other scientists and inventors experimented with their potential, the most successful being Guglielmo ◊Marconi, who in 1901 used these 'Hertzian waves' to send a message in Morse code across the Atlantic.

Further research on radio waves led to the first 'wireless' (radio) programmes of the early 1920s, but it took the probability of a war to rekindle interest in radio waves as a possible means of early detection of hostile aircraft. In 1935 the British Air Ministry asked Robert ◊Watson-Watt to investigate the concept of pulse radar. By the outbreak of World War II in 1939 there were radar stations along Britain's eastern coast, and these played a major part in the subsequent Battle of Britain.

Eastwood's part in the development of this system, and in the solving of its technical problems, left him ideally qualified to continue in this field of research. At the Marconi Research Laboratory he concentrated on the extension of the laboratory's interest in telecommunications, radar, and applied physics. He soon realized that the new radar systems, which the laboratory had been commissioned to develop, could also be used as powerful tools in research. With the aid of the Marconi experimental station at Bushy Hill, Essex, he applied radar methods to the study of various meteorological phenomena (such as the auroras) and carried out extensive investigations into the flight behaviour of birds and migration; his book *Radar Ornithology* was published in 1967.

Eckert, (John) Presper (1919–1995) US electronics engineer best known for his pioneering work on the design and construction of digital computers. He also held patents on numerous other electronic devices.

Eckert was born in Philadelphia, Pennsylvania on 9 April 1919. He attended the William Pema Charter School and went on to graduate in 1941 from the Moore School of Electrical Engineering at the University of Pennsylvania. He remained at the Moore School for five years as a research associate. During this time he worked on the design of radar ranging systems and then, to help with the complex calculations that these involve, turned to the design of electronic calculating devices. From 1942–46, with John William ◊Mauchly, he devised the Electronic Numerical Integrator and Calculator (ENIAC), one of the first modern computers.

ENIAC was first used in 1947, and proved to be successful although open to various improvements. Eckert had left Pennsylvania University the year before to become a partner in the Electric Control Company, and in 1947 began a three-year term as a vice president of the Eckert–Mauchly Computer Corporation. During this period design improvements were incorporated into new computer models.

In 1950, the company was incorporated in Remington Rand. Eckert became director of engineering within the Eckert–Mauchly division, becoming vice president in 1954. A year later the company came under the control of the Sperry Rand Corporation, and Eckert stayed on as vice president when the company became UNIVAC (Universal Automatic Computer Division) and later the Unisys Corporation. In 1989 he retired from Unisys but continued to act as a consultant for the company until his death on 3 June 1995.

Eckert's important contributions to computer design were recognized with awards and honorary degrees from various organizations and universities.

The need for improved accuracy and, particularly, speed in routine calculations became most apparent during World War II, when ballistic firing tables had to be rapidly recalculated to suit new weaponry and battle conditions. Eckert realized that normal calculators were ineffective and inefficient. With the assistance of Mauchly he used electronics to construct an integrator, and produced a flexible digital computer that could be used for calculating firing tables and much more.

The ENIAC, although only a prototype of present-day computers, incorporated many modern design features and could perform mathematical functions. It lacked a memory, but could store a limited amount of information. Its major drawbacks were its size (it weighed many tonnes and included thousands of resistors and valves) and its high running cost (it consumed 100 KW of electric power). Nevertheless it formed applications to various military, meteorological, and research problems.

ENIAC was superseded by BINAC, also designed in part by Eckert, which (by virtue of even more sophisticated design) was smaller and faster. In the early 1950s Eckert's group began to produce computers for the commercial market with the construction of the UNIVAC I. Its chief advance was the capacity to store programs.

Eddington, Arthur Stanley (1882–1944) English astronomer and writer who discovered the fundamental role of radiation pressure in the maintenance of stellar equilibrium, explained the method by which the energy of a star moves from its interior to its exterior, and finally showed that the luminosity of a star depends almost exclusively on its mass – a discovery that caused a complete revision of contemporary ideas on stellar evolution. He also demonstrated that a ray of light is deflected by gravity, thus confirming one application of Albert Einstein's general theory of relativity.

Eddington was born on 28 December 1882 in Kendal, Cumbria, although he spent his childhood in Weston-super-Mare, Somerset, and was educated there and at Owen's College, Manchester. In 1902 he won an entrance scholarship to Trinity College, Cambridge. Graduating three years later, he taught for a short time before being appointed chief assistant at the Royal Observatory, Greenwich. His seven-year stay there saw the beginning of his theoretical work.

In 1909 he was sent to Malta to determine the longitude of the geodetic station there and in 1912 he went to Brazil as the leader of an eclipse expedition. In 1913, Eddington returned to Cambridge to become Plumian Professor of Astronomy; shortly afterwards, he became director of the University Observatory. Eddington remained at Cambridge for the next 31 years.

During his lifetime, Eddington was considered to be one of the greatest astronomers of the age. In 1906, he was elected Fellow of the Royal Astronomical Society and he was its president 1921–23. In 1914, he was elected

Fellow of the Royal Society. He was knighted in 1938.

In the autumn of 1944 Eddington underwent major surgery, from which he never recovered. He died in Cambridge on 22 November of that year. After his death the Eddington Memorial Scholarship was established and the Eddington Medal was struck as an annual award.

Eddington published a large number of works. His first book, *Stellar Movements and the Structure of the Universe* (1914), is considered to be a model of scientific exposition. The final chapter of the book alone, entitled 'Dynamics of the Stellar System', marked the founding of an important branch of astronomical research. His book *The Internal Construction of the Stars* (1926) became one of the classics of astronomy. His report to the Physical Society in 1918, expanded into his *Mathematical Theory of Relativity,* was the work that first gave English-speaking people the chance to learn the mathematical details of Einstein's famous theory of gravitation. Eddington's ability as a writer not only served to introduce a whole generation to the science of astronomy; it also had a stimulating effect on other astronomers.

Science is one thing, wisdom is another. Science is an edged tool, with which men play like children, and cut their own fingers.

Arthur Eddington

Eddington involved himself in a great deal of practical work. He was the leader of the expedition to west Africa in 1919, where on 29 May on the island of Príncipe he observed the total eclipse of the Sun. The data he obtained served to verify one of the predictions contained in Einstein's general theory of relativity, that rays of light are affected by gravitation.

Eddington's first theoretical investigations were concerned with the systematic motion of the stars, but his great pioneering work in astrophysics began in 1916 when he started to study their composition. He established that in a star energy is transported by radiation, not by convection as had hitherto been thought. He also established that the mechanical pressure of the radiation was an important element in the maintenance of the star's mechanical equilibrium. Eddington showed that the equation of equilibrium must take into account three forces: gravitation, gas pressure, and radiative pressure. One of the major questions at that time was how the gas of which stars and the Sun are composed was prevented from contracting under the tremendous force of stellar gravity. Eddington decided that the expansive force of heat and radiation pressure countered the contractive force of gravity. He also concluded that, since the pressure of the stellar matter increased rapidly with depth, the radiative pressure must also increase, and the only way in which that could happen was through a rise in temperature. Eddington showed that the more massive a star, the greater the pressure in its interior, and so the greater the countering temperature and radiation pressure, and consequently the greater its luminosity. He had found that the luminosity of a great star depends almost exclusively on its mass and, in 1924, he announced his mass–luminosity law. This work was of outstanding importance and necessitated a complete revision of contemporary notions regarding stellar evolution.

From 1930, Eddington worked on relating the theory of relativity and quantum theory. He believed that he could calculate mathematically, without recourse to observation, all the values of those constants of nature that are pure numbers – for example, the ratio of the mass of the proton to that of the electron. In his posthumously published *Fundamental Theory,* he presented his calculations of many of the constants of nature, including the recession velocity constant of the external galaxies, the number of particles in the universe, the ratio of the gravitational force to the electrical force between a proton and an electron, the fine structure constant, and the velocity of light.

In his later years, Eddington dealt with the philosophy of science and discussed the question of what sort of knowledge it was that science conveys to people. He had himself contributed considerably to that knowledge.

Further Reading

Chandrasekhar, Subrahmanyan, *Eddington: The Most Distinguished Astrophysicist of His Time,* Cambridge University Press, 1983.

Kilminster, Clive William, *Eddington's Search for a Fundamental Theory: A Key to the Universe,* Cambridge University Press, 1994.

Edinger, Johanna Gabrielle Ottilie (Tilly) (1897–1967) German-born US palaeontologist internationally known for her research in the field of palaeoneurology.

Tilly Edinger was born in Frankfurt on 13 November 1897, the daughter of Ludwig Edinger, a leading neuroanatomist. Between 1916–18 she studied psychology, zoology, and geology at Heidelberg and Munich and then returned to Frankfurt to study for her doctorate on *Nothosaurus,* which she completed in 1921. Her interest in the biological interpretation of fossils was influenced by Friedrich Drevermann and after her doctorate she became his research assistant. In

1927 she became the curator of the vertebrate collection at the Natural History Museum of Senckenberg. With Hitler's rise to power, Edinger, of Jewish extraction, was forced to leave Germany. In 1939 she went to London where she worked as a translator; she then went to Cambridge, Massachusetts in 1940, having been offered work by Alfred S Romer, director of the Museum of Comparative Zoology at Harvard. She received fellowships from the Guggenheim Foundation 1943–44 and the American Association of University Women 1950–51; she also taught at Wellesley College 1944–45. In 1945 she became a US citizen. She died following a traffic accident on 26 May 1967.

Edinger was a leading figure in the field of 20th-century vertebrate palaeontology and laid the foundations for the study of palaeoneurology. In her two great works *Die fossilen Gehirne/Fossil Brains* (1929) and *The Evolution of the Horse Brain* (1948) she demonstrated that the evolution of the brain could be studied directly from fossil cranial casts. Her research shed new light on the evolution of the brain and showed that the progression of brain structure does not proceed at a constant rate in a given family but varies over time; also that the enlarged forebrain evolved several times independently among advanced groups of mammals and there was no single evolutionary scale. In 1962 she and three other authors published the two-volume *Bibliography of Fossil Vertebrates, Exclusive of North America, 1509–1927* and until her death she worked on *Paleoneurology 1804–1966.* An annotated bibliography, published posthumously in 1975. For her work, Edinger received honorary doctorates from Wellesley College in 1950, the University of Giessen in 1957, and a medical doctorate from the University of Frankfurt in 1964. After her death the Tilly Edinger Fund at the Museum of Comparative Zoology at Harvard University was established to support the writing of books on vertebrate palaeontology.

Edison US inventor Thomas Alva Edison, inventor of the electric light bulb, the phonograph, and the telephone transmitter. Edison preferred to work with practical trial and error rather than theoretical calculation. *Mary Evans Picture Library*

Edison, Thomas Alva (1847–1931) US electrical engineer and inventor. He took out more than 1,000 patents, the best known of which were the phonograph, the precursor of the gramophone, and the incandescent filament lamp.

Edison was born in the small town of Milan, Ohio, on 11 February 1847 and brought up in Michigan. Most of his tuition was provided by his mother – he received only three months' formal public elementary education. His lifelong interest in things technical was soon apparent – by the age of ten he had set up a laboratory in the basement of his father's house. By the age of 12, he was selling newspapers and candy on trains between Port Huron and Detroit, and three years later, in 1862, he had progressed to telegraph operator, a job he maintained throughout the Civil War and for a couple of years thereafter. During this period, in 1866 (at the age of 19), he took out a patent on an electric vote recorder, the first of a total of 1,069 patents.

Perceiving the need for rapid communications, made apparent by the recent war, he turned his inventive mind to problems in that field. His first success came with a tape machine called a 'ticker', which communicated stock exchange prices across the country. He sold the rights in this and other telegraph improvements to the Gold and Stock Telegraph Company for $30,000, using the money to equip an industrial research laboratory in Newark, New Jersey, which he opened in 1869.

From telegraphy, the transmission of coded signals across long distances, he then turned his attention to telephony, the transmission of the human voice over long distances. In 1876 he patented an electric transmitter system that proved to be less commercially successful than the telephone of Bell and Gray, patented a few months later. Typically, he was undeterred and not for the first time he applied his keenly inventive mind to improving someone else's idea. His improvements to their systems culminated in the invention of the carbon granule microphone, which so increased the volume of the signal that despite his deafness he could hear it.

With the money made from this invention he moved to Menlo Park, where he bought a house and equipped the laboratory that was to remain the centre for his research. In the following year, 1877, he invented the phonograph, a device in which the vibrations of the human voice were engraved by a needle on a revolving cylinder coated with tin foil. Thus began the era of recorded sound.

In the 1870s gas was the most advanced form of artificial lighting, the only successful rivals being various clumsy and expensive types of electrically powered arc lamp. While experimenting with the carbon microphone, Edison had toyed briefly with the idea of using a thin carbon filament as a light source in an incandescent electric lamp, an idea he returned to in 1879. His first major success came on 19 October of that year when, using carbonized sewing cotton mounted on an electrode in a vacuum (one millionth of an atmosphere), he obtained a source that remained aglow for 45 hours without overheating, a major problem with all other materials used. Even this success was not enough for him, and he and his assistants tried 6,000 other organic materials before finding a bamboo fibre that gave a bulb life of 1,000 hours. In 1883 he joined forces with Joseph Wilson ◊Swan, a chemist from Sunderland to form the Edison and Swan United Electrical Company Ltd.

Genius is 1% inspiration and 99% perspiration.

Thomas Edison Life *ch 24*

To produce a serious rival to gas illumination, a power source was required as well as a cheap and reliable lamp. The alternatives were generators or heavy and expensive batteries. At that time the best generators rarely converted more than 40% of the mechanical energy supplied into electrical energy. Edison made his first generator for the ill-fated Jeannette Arctic Expendition of 1879. It consisted of a drum armature of soft iron wire and a simple bipolar magnet, and was designed to operate one arc lamp and some incandescent lamps in series. A few months later he built a much more ambitious generator, the largest built to date; weighing 500 kg/1,103 lb, it had an efficiency of 82%. Edison's team were at the forefront of development in generator technology over the next decade, during which efficiency was raised above 90%. To complete his electrical system he designed cables to carry power into the home from small (by modern standards) generating stations, and also invented an electricity meter to record its use.

Edison became involved with the early development of the film industry in 1888. After persuading George Eastman to make a suitable celluloid film, he developed the high-speed camera and kinetograph, viewing the picture through a peephole. Although he had referred to the possibility of projecting the image, he omitted it from his patent – a rare error. He had dropped his interest in kinematography by 1893, but three years later resumed it when Thomas Armat developed a projector. They joined forces but Armat was commercially naive and the machine was advertised as Edison's latest triumph; the resulting split caused considerable patent litigation.

Edison's later years were spent in an unsuccessful attempt to develop a battery-powered car to rival the horseless carriages of Henry Ford. During World War I he produced many memoranda on military and naval matters for the Department of Operational Research.

When he died, aged 84, on 18 October 1931, Edison had come a long way from the ten-year-old boy with a laboratory in his father's basement to being probably the most prolific and practical inventive genius of his age. He was a man whose work has greatly influenced the world in which we live, particularly in the fields of communication and electrical power. On the day following his death, his obituary in the *New York Times* occupied four-and-a-half pages, an indication of the importance of Edison to the 20th-century world.

Further Reading

Dolan, Ellen M, *Thomas Alva Edison: Inventor,* Historical American Biographies series, Enslow Publishers, 1998.

Israel, Paul, *Edison: A Life of Invention,* Wiley-Liss, 1998.

Melosi, Martin V, *Thomas A Edison and the Modernization of America,* The Library of American Biography, HarperCollins, 1990.

Millard, A J, *Edison and the Business of Innovation,* Johns Hopkins Studies in the History of Technology, New Series, Johns Hopkins University Press, 1990, v 10.

Edlén, Bengt (1906–1993) Swedish astrophysicist whose main achievement lay in resolving the identification of certain lines in spectra of the solar corona that had misled scientists for the previous 70 years.

Edlén was born in Gusum in Ostergotland, south-eastern Sweden, on 2 November 1906. He was educated at Uppsala University and, in 1928, became an assistant in the physics department there. In 1936 he was appointed assistant professor of physics. In 1944 he moved to southern Sweden to become professor of physics at Lund University. He held this post until 1973, when he became emeritus professor.

During the eclipse of 1869, astronomers recorded the presence of a hitherto unknown series of spectral lines in the Sun's corona. Because they failed to identify the origin of these lines, they ascribed them to the presence of a new element, which they called 'coronium'. The origin of the lines was originally recorded as being located high above the Sun's surface, but similar lines were then discovered to originate nearer the Earth; these were accordingly attributed to another new element, which they called 'geocoronium'. Both the new elements were predicted to be much lighter than hydrogen, the lightest element known on Earth.

For 70 years all attempts to associate the coronal lines with known elements on Earth failed. Then, in the early 1940s, Edlén carried out a series of experiments and showed that, if iron atoms are deprived of some of their electrons, they can produce spectral lines similar to those produced by 'coronium'. He established that if half the normal number of 26 electrons of iron are removed, the effect produced is that of the green lines observed on the coronal line. Other lines were identified as iron atoms with different numbers of their electrons removed. Furthermore, it was found that similarly ionized atoms of nickel, calcium, and argon produced even more lines.

It was determined that such high stages of ionization would require temperatures of about 1,000,000 °C/ 1,800,000 °F and when, in the 1950s, it was verified that such high temperatures did exist in the solar corona, it became accepted that 'coronium' as a separate element did not exist and that the 'coronium' lines owed their existence to ordinary elements being subjected to extreme temperatures – temperatures so high that they caused the corona to expand continuously. It was also established that the lines formerly thought to be caused by the presence of 'geocoronium' are produced by atomic nitrogen emitting radiation in the Earth's upper atmosphere.

Edwards and Steptoe **Robert Geoffrey Edwards (1925–) and Patrick Christopher Steptoe (1913–1988)** British researchers – a physiologist and an obstetric surgeon, respectively – who devised a technique for fertilizing a human egg outside the body and transferring the fertilized embryo to the uterus of a woman. A child born following the use of this technique is popularly known as a 'test-tube baby'.

Robert Edwards was educated at the universities of Wales and Edinburgh and served in the British Army 1944–48. For the next three years he was at the University College of North Wales, Bangor, and from 1951–57 at the University of Edinburgh. That year Edwards went to the California Institute of Technology but returned to England the following year to the National Institute of Medical Research at Mill Hill, remaining there until 1962 when he took up an appointment at Glasgow University. A year later he moved again to the Department of Physiology at Cambridge. From 1985–89 he was professor of human reproduction at Cambridge, thereafter emeritus.

During his research in Edinburgh, Edwards successfully replanted mouse embryos into the uterus of a mouse and he wondered if the same process could be used to replant a human embryo into the uterus of a woman.

One common cause of infertility in women is disease or damage to the Fallopian tubes, which prevents eggs from being fertilized. Normally these tubes allow the mature egg, when released from the ovary, to travel to the uterus and if spermatozoa are present the egg will become fertilized on the way. Gregory Pincus (1903–1967), the US biologist who developed the contraceptive pill, showed that human eggs could mature outside the body, when they would be ready for fertilization. Edwards, by then working in Cambridge, was able to obtain human eggs from pieces of ovarian tissue removed during surgery. He found that the ripening process was very slow, the first division beginning only after 24 hours.

During the following year he studied the maturation of eggs of different species of mammals, and in 1965 attempted the fertilization of human eggs. He left mature eggs with spermatozoa overnight and found just one where a sperm had passed through the outer membrane, but it had failed to fertilize the egg. In 1967 Edwards read a paper by Steptoe describing the use of a new instrument, known as the laparoscope, to view the internal organs, which he saw had a possible application to his own research. At about this time, Bavister, a research student at Cambridge, who had been trying to fertilize hamster eggs, devised a successful culture solution. Edwards used some of this solution with the one he used for the culture of human eggs and achieved fertilization.

Patrick Steptoe was educated at King's College and St George's Hospital Medical School, London, qualifying in 1939. During World War II he served in the Royal Naval Volunteer Reserve and was a prisoner of war in Italy from 1941–43. He was appointed chief assistant obstetrician and gynaecologist at St George's Hospital, London, in 1947, and senior registrar at the Whittington Hospital, London, in 1949. From 1951–78 he was senior obstetrician and gynaecologist at Oldham General Hospital, and from 1969 director of the Centre for Human Reproduction.

The paper that interested Edwards described laparoscopy – Steptoe's method of exploring the interior of the abdomen without a major operation. Steptoe inserted the laparoscope through a small incision near the navel and by means of this telescope-like instrument, with its object lens inside the body and its eyepiece outside, he was able to examine the ovaries and other internal organs.

Early in 1968 Edwards and Steptoe met and arranged to collaborate. During the next few months they repeated experiments on the fertilization of human eggs. Steptoe treated volunteer patients with a fertility drug to stimulate maturation of the eggs in the ovary, while Edwards devised a simple piece of apparatus to be used with the laparoscope for collecting mature eggs from human ovaries. The mature eggs were removed and Edwards then prepared them for fertilization using spermatozoa provided by the prospective father. For a year they continued experiments of this kind until they were sure that the fertilized eggs were developing normally. The next step was to see if an eight-celled embryo would develop to the blastocyst state (the last stage of growth before it implants itself into the wall of the uterus); success was achieved.

In 1971, Edwards and Steptoe were ready to introduce an eight-celled embryo into the uterus of a volunteer patient who hoped to become pregnant, but this and similar attempts over a period of three years were unsuccessful. In 1975 an embryo did implant, but in the stump of a Fallopian tube where it could not develop properly and was a danger to the mother. It was removed, but it did demonstrate the basic technique to be sound. In 1977 it was decided to abandon the use of the fertility drug and remove the egg at precisely the right natural stage of maturity; an egg was fertilized and then reimplanted (a process called *in vitro* fertilization) in the mother two days later. The patient became pregnant and 12 weeks later the position of the baby was found to be satisfactory and its heartbeat could be heard. The last eight weeks of the pregnancy were kept under close medical supervision and a healthy girl – Louise Brown – was delivered by Caesarean section on 25 July 1978.

Edwards and Steptoe showed how one common cause of infertility may be overcome. In the UK, infertility due to nonfunctional Fallopian tubes affects several thousand women every year, of which only a half can be helped by conventional methods. *In vitro* fertilization has also been used to overcome the infertility in men that is due to a low sperm count. Edwards's research has further added to knowledge of the development of the human egg and young embryo, and Steptoe's laparoscope is a valuable instrument capable of wider application.

Eggen, Olin Jenck (1919–) US astronomer who has spent much of his working life in senior appointments all round the world. His work has included studies of high-velocity stars, red giants (using narrow- and broad-band photometry) and subluminous stars, and he has published some research on historical aspects of astronomy.

Born in Orfordville, Wisconsin, on 9 July 1919, Eggen graduated from Wisconsin University before becoming astronomer at the University of California in 1945. In 1956 he became the chief assistant at the Royal Greenwich Observatory in the UK. Maintaining his links with the USA, he was professor of astronomy at the California Institute of Technology 1960–63. He returned to Greenwich for a short time before serving as astronomer at the Mount Palomar Observatory 1965–66. In 1966 he went to Australia to take up the post of director of Mount Stromlo and Siding Spring Observatories, combining this with a professorship of astronomy at Australian National University's Institute of Advanced Study. Eggen remained at Mount Stromlo until 1977, when he moved to a position at the Observatory Interamericano de Cerro Tololo, Chile.

During the mid-1970s, Eggen completed a study – based on UBV photometry and every available apparent motion – of all red giants brighter than $V = 5^m0$. As a result he was able to classify these stars, categorizing them as very young discs, young discs, and old discs. A few remained unclassifiable (haloes).

He also systematically investigated the efficiency of the method of stellar parallax using visual binaries originally suggested by William Herschel in 1781, and reviewed the original correspondence of John Flamsteed and Edmond Halley.

Ehrenberg, Christian Gottfried (1795–1876) German naturalist who developed a scheme for the classification of the animal kingdom. He was the first scientist to study the fossils of microorganisms and is regarded as the founder of micropalaeontology.

Born in the small city of Delitzsch, near Leipzig, on 19 April 1795, Ehrenberg was the son of a municipal magistrate. His mother, Christiane Dorothea Becker, died when he was 13 years old. From the age of 14 Ehrenberg received a classical–philological education at a Protestant boarding school near Naumburg, passing his final examination in 1815. He then entered the University of Leipzig, studying theology to please his father, but later changing to study medicine and other sciences there and at Berlin where his interest in botanical and zoological studies was encouraged. He also developed an interest in microscopical techniques and was especially influenced by Karl Rudolphi and Heinrich Link.

Graduating with an MD in 1818, he was elected member of the Leopoldine German Academy of Researchers in Natural Sciences in 1818, and continued with the groundbreaking work on the sexual generation of moulds and mushrooms that he had started as an undergraduate. His findings were published in *Syzgites megalocarpu* (1819) and in the essay 'De mycetogenesis epistol' (1821).

The turning point in Ehrenberg's career came in 1820 when he and his friend Wilhelm Hemprich were sponsored by the University of Berlin and the Prussian Academy of Sciences to take part in a scientific expedition to Egypt, Libya, the Sudan, and the Red Sea. During this trip they collected and classified 34,000 animal and 46,000 plant specimens, including microorganisms. The expedition, led by Count Heinrich von Menu von Minutoli was poorly organized and badly funded, which meant that the travellers suffered deprivations and most of them, including Hemprich, became seriously ill. Ehrenberg returned to Trieste in 1825, the sole survivor of the original ten members. He was disappointed to find that many of his specimens had not survived the journey, labels and sketches were also missing and he felt unable to organize the collections on his own. He described the Middle East expedition and its animal and bird life in *Symbolae physicae* (1928) but was later reproached for not properly utilizing the collections.

In 1827 Ehrenberg became an assistant professor at the University of Berlin and later that year was elected a member of the Berlin Academy of Sciences. Encouraged and accompanied by Alexander von ◊Humboldt, he took part in a scientific expedition to Siberia in 1829, financed by Tsar Nicholas I. On this journey Ehrenberg gathered fish for the museums in St Petersburg, Paris, and Berlin and also made observations on living plankton. During the 1830s Ehrenberg did not receive the academic recognition he deserved and it was only in 1839, with the intercession of von Humboldt, that he achieved a full professorship at Berlin. However, he never took up a very active teaching role at the university.

Ehrenberg married twice. On his return from Siberia in 1931, he married Julie Rose, the niece of the Berlin mineralogist Gustave Rose, who had also been on the expedition. They had five children – one son and four daughters. His youngest daughter, Clara, helped him with his research. Julie died in 1848 and Ehrenberg was married again in 1852 to Karoline Friccius.

Ehrenberg's main interest was microscopical research and, following the Siberian expedition, he published several works, comparing material he had collected on the Middle Eastern expedition with material from Siberia, the Baltic, and the North Sea. He published works on the coral polyps of the Red Sea 1831–34, the medusae of the Red Sea 1834 and 1835, and in 1838 published a large work, *Die Infusionsthierchen als vollkommene Organismen,* on the subject of infusaria (infusion molecules) which included protozoa, bacteria, and algal forms such as diatoms. He described his methods of study, which included feeding infusaria with dyes to determine their anatomy, and also appended an atlas of 64 colour plates drawn by himself. From this research he erroneously regarded all infusaria as multicellular animals.

With the microscope, Ehrenberg discovered the single-celled fossils that made up certain geological strata. This work made him the founder of microgeology and micropalaeontology. His descriptions and classification of fossil protozoans was published as *Microgeology* (1854). His collection of samples and his manuscripts are available for study in the Museum fur Naturkunde in Berlin. Ehrenberg also participated in international oceanographic research projects, including James Ross's Antarctic expedition 1839–43 and an expedition to Novara, Italy, 1857–59.

Ehrenberg's achievements have stood the test of time in spite of the fact that he has attracted considerable criticism for not completing his work on the collections from the Middle East expedition, for not accepting Darwin's theory of evolution, and for not correcting his mistakes.

Ehrlich, Paul (1854–1915) German bacteriologist who founded chemotherapy – the use of a chemical substance to destroy disease organisms in the body. He was also one of the earliest workers on immunology, and through his studies on blood samples the discipline of haematology was recognized. In 1908, together with the Russian-born French bacteriologist Ilya Mechnikov, he was awarded the Nobel Prize for Physiology or Medicine for his work on serum therapy and immunity.

Ehrlich was born on 14 March 1854 in Silesia, which was then part of the Austro-Hungarian Empire, in a town called Strehlin (now Strzelin, in Poland). He studied in Breslau and Strasbourg, graduating in 1878 with a medical degree from the University of Leipzig. For the next six years he was a clinical assistant at the University of Berlin and then became head physician at the medical clinic in the Charité Hospital in Berlin; in 1884 he was promoted to professor there. Ehrlich spent two years in Egypt, 1886–88, to cure himself of tuberculosis. Successful, he returned to Berlin in 1889 where he set up a small private laboratory. The following year, he took up a professorial appointment at the University of Berlin. In 1891 he joined the Institute of Infectious Diseases, Berlin, as a researcher and five years later became director of the newly established Institute for the Investigation and Control of Sera, opened in Berlin by the German government, which had been impressed by his efforts. Ehrlich continued working in his laboratories until just before his death on 20 August 1915.

As a student, Ehrlich had shown an unusual fascination with chemistry. Encouraged by his teachers he worked on the use of aniline dyes in microscopic techniques and discovered a few dyes for selectively staining, and therefore simplifying the study of, bacteria. He made histological preparations and stained them with various combinations of dyes to observe the different effects of basic and acidic stains. While at the Charité Hospital Ehrlich was able to distinguish between a number of blood disorders by examining blood cells in his stained preparations. In this way he discovered 'mast cells' (connective tissue cells), and it was also while studying the staining of tubercle cells that he contracted a mild case of tuberculosis.

On his return from his curative stay in Egypt, Ehrlich teamed up with the German bacteriologist Emil von ◊Behring and the Japanese Shibasaburo ◊Kitasato to try to find a cure for diphtheria. Ehrlich had studied antigen–antibody reactions using toxic plant proteins on mice, gradually increasing the dose, and found that the mice developed specific antibodies in their blood. Litters bred from these immunized mice possessed a short-lived immunity, sustained by suckling from the immunized mothers. Behring and Ehrlich were able to produce antitoxins obtained from much larger mammals that had been immunized against the diphtheria organism; these antitoxins were concentrated and purified for use in clinical trials, and once Ehrlich had developed the correct dosage, in 1892, the antitoxin was ready for use. In 1894 it was tried on 220 children with diphtheria and achieved great success. Ehrlich then decided that antitoxins should be standardized, their potency described in terms of international units of antitoxin, and the distribution made in dried form in vacuum phials.

At the Institute for the Investigation and Control of Sera, the number of Ehrlich's staff allowed him to investigate his theory that chemical compounds could cure a disease and not merely alleviate the symptoms. This stage was the beginning of chemotherapy (Ehrlich's word). The search progressed for dyes that would stain only bacteria and not other cells, and from this research the team continued synthesizing and testing chemical substances that could seek out and destroy the bacteria without harming the human body. Ehrlich termed these compounds 'magic bullets'.

Ehrlich's first success developed from the use of trypan red to kill trypanosomes (parasitic protozoans that cause sleeping sickness) in infected mice. The results of the tests proved inconclusive, but Ehrlich decided that the active agents in trypan red were nitrogen compounds. Atoxyl – an arsenical organic compound, and therefore similar in chemical properties to its nitrogen analogues – had shown greater success in the treatment of sleeping sickness and Ehrlich believed that it should be possible to make more effective derivatives of the substance. The accepted formula for atoxyl was a benzene ring with one side chain; Ehrlich, however, believed it had two side chains. If the accepted formula was correct, any derivatives would be unstable; but if Ehrlich was correct, they would be stable. Ehrlich proved to be right. He and his staff prepared nearly a thousand derivatives of arsenic-containing compounds, testing each on animals. In 1907, they reached compound number 606 (dihydroxydiamino-arsenobenzene hydrochloride), which proved ineffective against trypanosomes and so was forgotten. But it was investigated again in 1909 and discovered to be effective, instead, against spirochaetes, the bacteria that cause syphilis. Ehrlich tried it on himself, without any harm, and in 1910 announced the discovery of the synthetic chemical, now called Salvarsan (arsphenamine), for treating syphilis.

Ehrlich devised scientific techniques for developing chemical cures that opened up new fields of research in 20th-century medicine, particularly in chemotherapy, haematology, and immunology. Innumerable lives have been saved and the economic and social effects of his work continue to be far-reaching.

Further Reading

Baumler, Ernst and Edwards, Grant, *Paul Ehrlich: Scientist for Life*, Hilmes and Meier Publishers, 1984.

Ehrlich, Paul, *The Population Bomb*, Buccaneer Books, 1997.

Ehrlich, Paul, *The Stork and the Plow: The Equity Answer to the Human Dilemma*, Yale University Press, 1997.

Ehrlich, Paul and Ehrlich, Anne, *Betrayal of Science and Reason: How Anti-Environmental Rhetoric Threatens Our Future*, Island Press, 1998.

Ehrlich, Paul and Ehrlich, Anne, *The End of Affluence*, Buccaneer Books, 1995.

Eiffel, Alexandre Gustav (1832–1923) French engineer now known chiefly for the 320-m/1,050-ft-high edifice he erected for the 1889 exhibition commemorating the 100th anniversary of the French Revolution.

Eiffel was born in Dijon in 1832. After attending the Ecole des Arts et Manufactures in Paris he won early fame by specializing in the design of large metal structures, notably the iron railway bridge over the Garonne at Bordeaux. Here he was one of the first to use compresssed air for underwater foundations, in 1867, he set up his own firm that constructed bridges, viaducts, harbour work,s and other large projects. His great arch bridges include the 159-m/530-ft span over the Douro River in Portugal, and the 165-m/550-ft span of the Garabit Viaduct in France. Other projects included the immense roof that covers the central station at Budapest, the Machinery Hall for the Paris Exhibition of 1867, the protective ironwork for the Statue of Liberty in New York Harbour, designed by Bartholdi, and the 84-m/277-ft dome for the observatory at Nice.

Eiffel began work on the famous wrought-iron tower for the Champs de Mars in 1886, using his experience of building high-level railway bridges. From a detailed set of plans, the 12,000 metal parts of the tower were all prefabricated and numbered for assembly. The majority of the 2.5 million rivets used were put in place before the structure was erected on the site. Work proceeded so smoothly that not one worker's life was lost through accidents on the scaffolding, and was completed (except for the lifts) in two and a quarter years.

The tower's cross-braced, latticed girder structure offers minimum wind resistance: the estimated movement of the structure with hurricane-force winds is only 22 cm/9 in. It is constructed from over 7,000 tonnes/6,900 tons (7,700 US tons) of wrought iron, resting upon four masonry piers. The piers are set in 2 m/7 ft of concrete on foundations carried down by the aid of caissons and compressed air to about 15 m/44 ft on the side next to the Seine and about 9 m/30 ft on the other side.

The tower has three well-marked stages. Below the first platform, which is placed at a height of 57 m/183 ft the four quadrilateral legs are linked by arches. At the second platform, 115 m/380 ft high, the legs join and the third platform is at a height of 276 m/905 ft. Above this platform are the lantern and the final terrace.

Originally, the tower was intended to be dismantled at the conclusion of the exhibition. Many writers and artists deplored its construction, with one describing it as 'a hideous hollow candlestick'. However the newly discovered possibilities of the tower as a radio transmitting station finally won the day and it was left standing. For some time it was by far the highest artificial structure in the world. It also showed what could be achieved by correct engineering design and paved the way for yet higher structures in the future.

With subsequent disaster for himself and his country. Eiffel participated in the Panama Canal enterprise, in the course of which he designed and partly constructed some huge locks. When the entire project collapsed in 1893. Eiffel was implicated in the scandal; he went to prison for two years and received a fine of 200,000 francs. In 1900 he took up meteorology and later, using wind tunnels, carried out extensive research in aerodynamics at the Eiffel Tower and afterwards at Auteuil where he constructed the first laboratory for the new science. He died in Paris in December, 1923.

Eigen, Manfred (1927–) German physical chemist who shared the 1967 Nobel Prize for Chemistry with Ronald Norrish and George Porter for his work on the study of fast reactions in liquids.

Eigen was born in Bochum, Ruhr, on 9 May 1927, the son of a musician. He was educated at Göttingen University and on his 18th birthday, one day after the formal ending of World War II, he was drafted to do military service with an anti-aircraft artillery unit. He later returned to Göttingen, gained his doctorate in 1951, and worked as a research assistant there for the next two years. In 1953 he moved to the Max Planck Institute of Physical Chemistry in Göttingen, becoming a research fellow in 1958, head of the department of biochemical kinetics in 1962, and eventually director of the institute 1964–70. From 1971 he was honorary professor at the Technical University of Göttingen.

> *A theory has only the alternative of being right or wrong. A model has a third possibility: it may be right, but irrelevant.*
>
> Manfred Eigen quoted in Jagdish Mehra (ed) The Physicist's Conception of Nature *1973*

Eigen studied fast-reaction kinetics by disturbing the equilibria in liquid systems using short changes of temperature, pressure, or electric field (whereas Norrish and Porter had used flashes of light to disturb equilibria in gaseous systems). He investigated particularly very fast biochemical reactions that take place in the body, trying to discover how rapidly reactions proceed among the working molecules of life and how a particular sequence of chemical units could come about by chance in the time available. With his colleague Ruthild Winkler he tried to relate chance and chemistry in processes that could have led to the origin of life on Earth. They questioned how molecules with the right kind of properties might form and what would be the simplest combination of molecules that could survive and evolve into the first primitive organisms. Eigen theorized that in the 'primeval soup' of the early Earth, cycles of chemical reactions would have occurred, one reproducing nucleic acids (which possess information but have a very limited chemical function) and one reproducing proteins (which ensured chemical function and reproduction of the information contained in the nucleic acids). He postulated that eventually a number of the nucleic acid cycles and proteins would have come to coexist and form a 'hypercycle'. By natural selection the best hypercycle would have eventually caused the first organism to evolve – a chance set of molecules coming together in a single drop – providing a possible theory of the chemical transition from nonlife to life.

Eijkman, Christiaan (1858–1930) Dutch physician and pioneer of the study of vitamins. He shared the 1929 Nobel Prize for Physiology or Medicine with

Frederick Gowland ◊Hopkins for his work on beriberi.

Christiaan Eijkman was born in Nijkerk in the Netherlands on 11 August 1858, the seventh child of a local headmaster. The family moved to Zaandam in 1859, where his father was head of a new school for advanced elementary education. Eijkman received his early education there. In 1875 he became a student at the Military Medical School of the University of Amsterdam, where he was trained as a medical officer for the Netherlands Indies Army. He passed all his examinations with honours.

Eijkman was assistant to professor of physiology T Place 1879–81, wrote his thesis, *On Polarisation of the Nerves,* and gained a doctor's degree with honours in 1883. He then began serving in the Dutch East Indies as an army medical officer but returned to the Netherlands in 1885 to recover from a serious bout of malaria. On his recovery, he worked in Robert Koch's bacteriological laboratory in Berlin where he met Pekelharing and Winkler who were about to depart on a Dutch government mission to study beriberi in Batavia (now Djakarta). Eijkman was taken on as an assistant. He remained in Batavia after the work of the commission was finished, as director of a new bacteriological laboratory 1888–96. He was also made director of the Javanese Medical School. In 1896 he returned to the University of Utrecht to take up the offer of the chair of hygiene and forensic medicine.

Eijkman had married Aaltje Wigeri van Edema in 1883, before departing to the Indies, but she died in 1886. In 1888, in Batavia, Eijkman married his second wife, Bertha van der Kemp and his son Pieter was born in 1890.

In Batavia, Eijkman made a discovery leading to an understanding of the nature of beriberi. He witnessed several severe outbreaks, in some of which the mortality rate was over 80%. The symptoms of beriberi are ascending paralysis, cardiac problems, and oedema and Eijkman and several of his colleagues had tried but failed to find the cause of it. It was by lucky coincidence that he stumbled on the truth – that the symptoms were caused by the lack of something rather than the presence of an infective organism. A disease broke out among the chickens at Eijkman's research laboratory which exhibited similar symptoms to human beriberi. The search for causative microorganisms led to a dead end and the disease disappeared for no apparent reason. Eijkman found that the disease had occurred during a five-month period when the chickens had been eating cooked, hulled, and polished rice from the hospital kitchens. A new chef had refused to continue this practice and the chickens had been put back onto unmilled rice.

Eijkman's discovery led to experiments on prisoners in the Far East, and the prevention and cure of human beriberi. However, Eijkman failed to unravel the true import of his discovery – that beriberi is a deficiency disease. He concluded that the bran that was removed in the hulled rice contained a protective agent which neutralized a natural toxin in the rice.

Eijkman's lectures were always informed by his considerable practical knowledge and his refusal to rely on dogma himself or to allow his students to. He did not solely confine himself to work at the university in Utrecht, however, engaging himself in local health and hygiene problems. As well as being involved in the problems of the water supply, school hygiene, and physical education, he was the founder of a society to fight tuberculosis. Eijkman held the John Scott Medal, Philadelphia; he was Foreign Associate of the Academy of Sciences in Washington and honorary fellow of the Royal Sanitary Institute in London. In 1907 he was appointed member of the Royal Academy of Sciences in the Netherlands. Following on Eijkman's work on beriberi, Hopkins identified the cause as a vitamin deficiency and he and Eijkman shared the 1929 Nobel Prize for Physiology or Medicine. However, it was not until after Eijkman's death in Utrecht on 5 November 1930 that Robert Williams identified the vitamin as thiamine (vitamin B1).

Eilenberg, Samuel (1913–1998) Polish-born US mathematician whose research in the field of algebraic topology led to considerable development in the theory of cohomology. He was also well known for his work in computer mathematics.

Eilenberg was born on 30 September 1913 in Warsaw, where he grew up and completed his education, gaining his master's degree in 1934 and his PhD in mathematics two years later at the University of Warsaw. He then emigrated to the USA where, in 1940, he joined the staff of the University of Michigan as an instructor; by 1946 he was associate professor of mathematics. In that year he was appointed professor of mathematics at the University of Indiana, where he remained for three years. After a series of visiting professorships – some in the USA, some in Europe and the Indian subcontinent – Eilenberg became professor of mathematics at Columbia University, New York, where he remained for the rest of his academic life. He died on 30 January 1998.

Eilenberg's main field of work was that of algebraic topology – a subject on which, with Norman Steenrod, he wrote a successful advanced textbook. Topology is the study of figures and shapes that retain their essential proportions even when twisted or stretched; in topology, therefore, a square is (topologically) equivalent to any closed plane figure – such as a circle – and a cube is even (topologically) equivalent to a sphere. Since Henri ◊Poincaré first developed the subject systematically in a series of papers written 1895–1905, topology theory has been elaborated at a rapid rate, and has considerably influenced other branches of mathematics. Eilenberg carried out valuable work in the area of topology which, although generally known as the algebraic topology, is sometimes called 'combinatorial' topology and is distinctive for the extensive use of algebraic techniques to solve topological problems. The basis on which algebraic topology is founded is homology theory – the study of closed curves, closed surfaces, and similar geometric arrangements in a given topological space. Much of Eilenberg's work was concerned with a modification of homology theory called cohomology theory – cohomology groups have properties similar to homology groups but have several important advantages. It is possible to define a 'product' of cohomology classes by means of which, together with the addition of cohomology classes, the direct sum of the cohomology classes of all dimensions becomes a ring (the cohomology ring). This is a richer structure than is available for homology groups, and allows finer results. Various other very complicated algebraic operations using cohomology classes can lead to results not provable in any other way – the Poincaré duality theorem, for example, is considerably easier to state precisely if cohomology groups are used.

Einstein US physicist Albert Einstein photographed in 1922, six years after the publication of his general theory of relativity. In 1919 Einstein observed a solar eclipse which confirmed his theory that light was bent by gravity. *Mary Evans Picture Library*

Einstein, Albert (1879–1955) German-born US theoretical physicist who revolutionized our understanding of matter, space, and time with his two theories of relativity. Einstein also established that light may have a particle nature and deduced the photoelectric law that governs the production of electricity from light-sensitive metals. For this achievement, he was awarded the 1921 Nobel Prize for Physics. Einstein also investigated Brownian motion and was able to explain it so that it not only confirmed the existence of atoms but could be used to determine their dimensions. He also proposed the equivalence of mass and energy, which enabled physicists to deepen their understanding of the nature of the atom and explained radioactivity and other nuclear processes. Einstein, with his extraordinary insight into the workings of nature, may be compared with Isaac Newton (whose achievements he extended greatly) as one of the greatest scientists ever to have lived.

Einstein was born in Ulm, Germany, on 14 March 1879. His father's business enterprises were not successful in that town and soon the family moved to Munich, where Einstein attended school. He was not regarded as a genius by his teachers; indeed there was some delay because of his poor mathematics before he could enter the Eidgenössosche Technische Hochschule in Zürich, Switzerland, when he was 17. As a student he was not outstanding and Hermann ◊Minkowski, who was one of his mathematics professors, found it difficult in later years to believe that the famous scientist was the same person he had taught as a student.

Einstein graduated in 1900 and after spending some time as a teacher, he was appointed a year later to a technical post in the Swiss Patent Office in Berne. Also in 1901 he became a Swiss citizen and then in 1903 he married his first wife, Mileva Marié. This marriage ended in divorce in 1919. During his years with the Patent Office, Einstein worked on theoretical physics in his spare time and evolved the ideas that were to revolutionize physics. In 1905 he published three classic papers on Brownian motion, the photoelectric effect, and special relativity.

Einstein did not, however, find immediate recognition. When he applied to the University of Berne for an academic position, his work was returned with a rude remark. But by 1909 his discoveries were known and understood by a few people, and he was offered a junior professorship at the University of Zürich. As his reputation spread, Einstein became full professor or Ordinariat, first in Prague in 1911 and then in Zürich in 1912, and he was then appointed director of the Institute of Physics at the Kaiser Wilhelm Institute in Berlin in 1914, where he was free from teaching duties.

The year 1915 saw the publication of Einstein's general theory of relativity, as a result of which Einstein predicted that light rays are bent by gravity. Confirmation of this prediction by the solar eclipse of 1919 made Einstein world famous. In the same year he married his second wife, his cousin Elsa, and then travelled widely to lecture on his discoveries. One of his many trips was to the California Institute of Technology during the winter of 1932. In 1933 Adolf Hitler came to power and Einstein, who was a Jew, did not return to Germany but accepted a position at the Princeton Institute for Advanced Study, where he spent the rest of his life. During these later years, he attempted to explain gravitational, electromagnetic, and nuclear forces by one unified field theory. Although he expended much time and effort in this pursuit, success was to elude him.

Imagination is more important than knowledge.

Albert Einstein On Science

In 1939, Einstein used his reputation to draw the attention of the US president to the possibility that Germany might be developing the atomic bomb. This prompted US efforts to produce the bomb, though Einstein did not take part in them. In 1940 Einstein became a citizen of the USA. In 1952, the state of Israel paid him the highest honour it could by offering him the presidency, which he did not accept because he felt that he did not have the personality for such an office. Einstein was a devoted scientist who disliked publicity and preferred to live quietly, but after World War II he was actively involved in the movement to abolish nuclear weapons. He died at Princeton on 18 April 1955.

Einstein's first major achievement concerned Brownian motion, the random movement of fine particles that can be seen through a microscope and was first observed in 1827 by Robert ◊Brown when studying a suspension of pollen grains in water. The motion of the particles increases when the temperature increases but decreases if larger particles are used. Einstein explained this phenomenon as being the effect of large numbers of molecules bombarding the particles. He was able to make predictions of the movement and size of the particles, which were later verified experimentally by the French physicist Jean ◊Perrin. Experiments based on this work was used to obtain an accurate value of the Avogadro number, which is the number of atoms in one mole of a substance, and the first accurate values of atomic size. Einstein's explanation of Brownian motion and its subsequent experimental confirmation was one of the most important pieces of evidence for the hypothesis that matter is composed of atoms.

Einstein's work on photoelectricity began with an explanation of the radiation law proposed in 1901 by Max ◊Planck. This is $E = h\nu$, where E is the energy, h is a number known as Planck's constant, and ν is the frequency of radiation. Planck had confined himself to black-body radiation, and Einstein suggested that packets of light energy are capable of behaving as particles called 'light quanta' (later called photons). Einstein used this hypothesis to explain the photoelectric effect, proposing that light particles striking the surface of certain metals cause electrons to be emitted. It had been found experimentally that electrons are not emitted by light of less than a certain frequency $\nu^{0;}$ that when electrons are emitted, their energy increases with an increase in the frequency of the light; and that an increase in light intensity produces more electrons but does not increase their energy. Einstein suggested that the kinetic energy of each electron, $^1/_2mv^2$, is equal to the difference in the incident light energy $h\nu$ and the light energy needed to overcome the threshold of emission $h\nu^0$. This can be written mathematically as:

$$^1/_2mv^2 = h\nu - h\nu^0$$

And this equation has become known as Einstein's photoelectric law. Its discovery earned Einstein the 1921 Nobel Prize for Physics.

Einstein's most revolutionary paper of 1905 contained the idea that was to make him famous, relativity. Up to this time, there had been a steady accumulation of knowledge that suggested that light and other electromagnetic radiation do not behave as predicted by classical physics. For example, no method had been found to determine the velocity of light in a single direction. All the known methods involved a reflection of light rays back along their original path. It had also proved impossible to measure the expected changes in the speed of light relative to the motion of the Earth. The Michelson–Morley experiment had demonstrated conclusively in 1881 and again in 1887 that the velocity of light is constant and does not vary with the motion of either the source or the observer. To account for this, Hendrik ◊Lorentz and George ◊FitzGerald independently suggested that all lengths contract in the direction of motion by a factor of $(1 - v^2/c^2)^1/_2$, where v is the velocity of the moving body and c is the speed of light.

The results of the Michelson–Morley experiment confirmed that no 'ether' can exist in the universe as a medium to carry light waves, as was required by classical physics. This did not worry Einstein, who viewed light as behaving like particles, and it enabled him to suggest that the lack of an ether removes any frame of reference against which absolute motion can be measured. All motion can only be measured as motion relative to the observer. This idea of relative motion is central to relativity, and is one of the two postulates of the special theory, which considers uniform relative motion. The other is that the velocity of light is constant and does not depend on the motion of the observer. From these two notions and little more than school algebra, Einstein derived that in a system in motion relative to an observer, length would be observed to decrease by the amount postulated by Lorentz and FitzGerald. Furthermore, he found that time would slow by this amount and that mass would increase. The magnitude of these effects is negligible at ordinary velocities and Newton's laws still held good. But at velocities approaching that of light, they become substantial. If a system were to move at the velocity of light, to an observer its length would be zero, time would be at a stop, and its mass would be infinite. Einstein therefore concluded that no system can move at a velocity equal to or greater than the velocity of light.

The unleashed power of the atom has changed everything save our modes of thinking and we thus drift toward unparalleled catastrophe.

Albert Einstein *Telegram sent to prominent Americans 24 May 1946*

Einstein's conclusions regarding time dilation and mass increase were later verified with observations of fast-moving subatomic particles and cosmic rays. Length contraction follows from these observations, and no velocity greater than light has ever been detected. Einstein went on to show in 1907 that mass is related to energy by the famous equation $E = mc^2$. This indicates the enormous amount of energy that is stored as mass, some of which is released in radioactivity and nuclear reactions – for example, in the Sun. One of its many implications concerns the atomic masses of the elements, which are not quite what would be expected by the proportions of their isotopes. These are slightly decreased by the mass equivalent of the binding energy that holds their molecules together. This decrease can be explained by, and calculated from, the famous Einstein formula.

Minkowski, Einstein's former teacher, saw that relativity totally revised accepted ideas of space and time and expressed Einstein's conclusions in a geometric form in 1908. He considered that everything exists in a four-dimensional space–time continuum made up of three dimensions of space and one of time. This interpretation of relativity expressed its conclusions very clearly and helped to make relativity acceptable to most physicists.

Science without religion is lame. Religion without science is blind.

Albert Einstein *Quoted in* A Pais *'Subtle is the Lord...': The Science and the Life of Albert Einstein 1982*

Einstein now sought to make the theory of relativity generally applicable by considering systems that are not in uniform motion but under acceleration. He introduced the notion that it is not possible to distinguish being in a uniform gravitational field from moving under constant acceleration without gravitation. Therefore, in a general view of relativity, gravitation must be taken into account. To extend the special theory, he investigated the effect of gravitation on light and in 1911 concluded that light rays would be bent in a gravitational field. He developed these ideas into his general theory of relativity, which was published in 1915. According to this theory, masses distort the structure of space–time.

Einstein was able to show that Newton's theory of gravitation is a close approximation of his more exact general theory of relativity. He was immediately successful in using the general theory to account for an anomaly in the orbit of the planet Mercury that could not be explained by Newtonian mechanics. Furthermore, the general theory made two predictions concerning light and gravitation. The first was that a red shift is produced if light passes through an intense gravitational field, and this was subsequently detected in astronomical observations in 1925. The second was a prediction that the apparent positions of stars would shift when they are seen near the Sun because the Sun's intense gravity would bend the light rays from the stars as they pass the Sun. Einstein was triumphantly vindicated when observations of a solar eclipse in 1919 showed apparent shifts of exactly the amount he had predicted.

Einstein later returned to the quantum theory. In 1909, he had expressed the need for a theory to reconcile both the particle and wave nature of light. In 1923, Louis ◊de Broglie used Einstein's mass–energy equation and Planck's quantum theory to achieve an expression describing the wave nature of a particle. Einstein's support for de Broglie inspired Erwin ◊Schrödinger to establish wave mechanics. The development of this system into one involving indeterminacy did not meet with Einstein's approval, however, because it was expressed in terms of probabilities and not definite values. Einstein could not accept that the fundamental structure of matter could rest on chance events, making the famous remark 'God does not play dice'. Nevertheless, the theory remains valid.

Albert Einstein towers above all other scientists of the 20th century. In changing our view of the nature of the universe, he has extended existing laws and discovered new ones, all of which have stood up to the test of experimental verification with ever-increasing precision. The development of science in the future is likely to continue to produce discoveries that accord with Einstein's ideas. In particular, it is possible that relativity will enable us to make fundamental advances in our understanding of the origin, structure, and future of the universe.

Further Reading

Bernstein, Jeremy, *Albert Einstein and the Frontiers of Physics,* Oxford Portraits in Science, Oxford University Press, 1996.
Brian, Denis, *Einstein: A Life,* Wiley, 1996.
Clark, Ronald William, *Einstein: The Life and Times,* Hodder and Stoughton, 1996.
Davies, P C W, *About Time: Einstein's Unfinished Revolution,* Simon & Schuster, 1995.
Fölsing, Albrecht, *Albert Einstein,* Penguin, 1998.
Phillips, Kenneth, *Relativity Explained,* Albion Scientific Series, Abbey Dene, 1991.
Schwartz, Joseph, *Einstein for Beginners,* Icon Books, 1992.
Steichel, John J, *Einstein's Miraculous Year: Five Papers That Changed the Face of Physics,* Princeton University Press, 1998.
White, Michael, *Einstein: A Life in Science,* Pocket Books, 1997.

Einthoven, Willem (1860–1927) Dutch physiologist and inventor of the electrocardiogram (ECG). He demonstrated that certain disorders of the heart alter its electrical activity in characteristic ways. He was awarded the 1924 Nobel Prize for Physiology or Medicine.

Willem Einthoven was born in Semarang, Java, on 21 May 1860. His father, who died in 1866, was the municipal physician there, and Willem's mother and her six children returned to the Netherlands, settling in Utrecht in 1870. Willem was educated there and became a medical student in 1879, gaining a PhD in medicine in 1885 with a thesis on stereoscopy through colour differentiation.

In 1886, Einthoven was appointed professor of physiology at Leiden at the age of 25. British physiologist Augustus Waller had recorded the electric currents generated by the heart. He published the curve for the action current of the heart in 1895 but announced that he could not calculate its true shape (using Lippmann's capillary electrometer, which was inaccurate for general use). Einthoven continued this work using a string galvanometer that he adapted from Deprez-d'Arsonval's mirror galvanometer and which made up for the inaccuracies of the capillary electrometer. A string galvanometer consists of a fine wire stretched between the poles of a magnet that uses the magnetic effect of electric currents to detect small changes in those currents. In 1896 while developing this machine and the photographic equipment required to register the movement of the wire he produced a curve that was later to be called an electrocardiogram and also registered the sounds of human and animal hearts.

In papers produced between 1903 and 1908, Einthoven mentioned the possibility of his machine being used in the diagnosis of heart disease, but his main concern was with the physical principles involved. In 1903 he defined the standard measures for use of the string galvanometer. His measurements were taken, not from the chest wall, but from two of the limbs of the subject used in three different combinations (right arm to left arm, right leg to left arm, and left leg to right arm). These are still the standard recordings. In 1906 Einthoven studied the electrocardiograms of heart patients in a nearby academic hospital by means of a cable 1.5 km/1 mi long. Through these 'telecardiograms' he gained knowledge of all forms of heart disease and could also register murmurs and other heart sounds with a second string galvanometer. With a string recorder and a string myograph he was able to prove the relationship between muscle contraction and the heart.

In 1924 Einthoven visited the USA on a lecture tour and when he returned to Leiden he was asked to register the currents of the cervical sympathetic nerve. He duly constructed a vacuum string galvanometer and succeeded in registering the action of the nerve in 1926. At the time of his death on 28 September 1927 he was still working on perfecting this technique, which was successfully achieved later with the invention of the cathode ray oscilloscope.

In 1886 Einthoven had married his cousin Frederique Jeanne Louise de Vogel and they had three daughters and one son. His son worked with him on his last physical experiment on the reception of radiotelegrams broadcast from a transmitter in Java. They synchronized a string of 0.1 micrometre diameter with the 40,000 vibrations of the transmitting wave and found the resonance point that allowed them to reproduce the telegrams on paper.

Electrocardiography is second only to the discovery of X-rays in physical medicine and Einthoven was awarded the 1924 Nobel prize for his invention. His last work, published posthumously was on the action current of the heart and appeared in *Bethe's Handbuch der normalen und pathologischen Physiologie.*

Eisenhart, Luther Pfahler (1876–1965) US theoretical geometrist whose early work was concerned with the properties of surfaces and their deformation; later he became interested in Riemann geometry from which he attempted to develop his own geometry theory. The author of several books detailing his results, he also wrote two books on historical topics.

Eisenhart was born on 13 January 1876 in York, Pennsylvania, the second son of the dentist who was also the founder of the Edison Electric Light and York Telephone Company. Educated locally, and to a high standard, he attended Gettysburg College (in southern Pennsylvania) 1892–96 where, for the last two years, he studied mathematics independently of his other work through guided reading. After a year's teaching at the college he went to Johns Hopkins University, Baltimore, in 1897, to carry out graduate studies, obtaining his PhD there three years later. He then began his life's work in mathematical research at Princeton University, retiring from there in 1945 after 45 years' successful study and teaching. Twenty years later it was there he died, on 28 October 1965.

One of Eisenhart's major achievements was to relate his theories regarding differential geometry to studies bordering on the topological. At the age of 25 he wrote one of the first characterizations of a sphere as defined in terms of differential geometry (the paper had the somewhat daunting title 'Surfaces whose first and second forms are respectively the second and first forms of another surface'). For the next 20 years he continued to develop his research, concentrating particularly on the subject of surface deformation. The theory of the deformation of surfaces was a part of the study of the properties of surfaces and systems of surfaces that was an especially popular area for geometrical research in continental Europe at that time, but Eisenhart was (apparently) the only person in the USA to devote his attention to it. It was he, nevertheless, who managed to formulate a unifying principle to the theory. The deformation of a surface involves the congruence of lines connecting a point and its image. Eisenhart's contribution was to realize that in all known cases, the intersections of these surfaces with the given surface and its image form a set of curves that have special properties. He wrote his account of the theory in 1923, in *Transformations of Surface.*

His work on surfaces led him then to study Riemann geometry – in which the properties of geometric space are considered locally rather than in one overall framework for the whole space. Eisenhart developed a geometry analogous to Riemann geometry, which he called non-Riemann geometry (although the term has since been used for several other forms of geometry), and wrote *Fields of Parallel Vectors in the Geometry of Paths* in 1922; it was followed by *Fields of Parallel Vectors in a Riemannian Geometry* (1925) and *Riemannian Geometry* (1926).

Interested in history, Eisenhart also wrote several other papers, including 'Lives of Princeton mathematicians' (1931), 'Plan for a university of discoverers' (1947), and 'The preface to historic Philadelphia' (1953).

Elion, Gertrude (1918–1999) US biochemist who developed several drugs and shared the Nobel Prize for Physiology or Medicine in 1988. In 1991 she was the first woman to be inducted in the National Inventors' Hall of Fame. Her name is on over 40 patents.

Gertrude Elion was born in New York City and was educated at Hunter College, graduating with a BA in chemistry in 1937. While studying at night school for her master's degree in chemistry she worked as a laboratory assistant, a food analyst, and a school teacher. Her ambition from a young age had been to find a cure for cancer.

World War II opened up laboratory jobs for women and after completing her master's degree in 1941, Elion worked as a biochemist in the Wellcome Research Laboratories in Tuckahoe, New York. It was here that she began her long association with George Herbert Hitchings, becoming his research associate in 1944. They were pioneers of pharmaceutical research, developing drugs to treat previously incurable diseases. She later became senior research chemist in the company and from 1967–83 was head of experimental therapy. Since retiring from Burroughs Wellcome in 1983 she has been emeritus scientist.

With Hitchings, Elion worked on the development of a drug that would stop the growth of bacteria or tumour cells by synthesizing compounds to inhibit DNA synthesis, and prevent the replication of unwanted cells. Elion started work on purines, which are important constituents of DNA, and the synthesis of nucleic acids. During this time she had to choose between continuing this work and a doctorate she had started at Brooklyn Polytechnic Institute. She chose to continue in her job and was rewarded later by several honorary doctorates for her research.

In the early 1950s, Elion and Hitchings's research into the chemistry of purines and pyramidines led to the development of drugs called purine antimetabolites, two of which were successful in treating leukaemia in rodents. One of these, 6MP, was tested on terminally ill children and found to produce complete remission, although this was sometimes temporary. This opened up whole new areas of research and 6MP is still used in combination with other drugs to treat children with acute leukaemia, with success in around 80% of cases.

From the study of 6MP, Elion developed a drug, azathioprene, to block the immune response that causes the rejection of transplanted kidneys. Azothioprene is also used to treat rheumatoid arthritis and systemic lupus. With her colleagues she also synthesized allopurinol, which inhibits the formation of uric acid and is used to treat gout and other diseases related to the overproduction of this substance.

In 1970 Elion and her team synthesized Acyclovir, which has been in use since 1981 and is very effective in the treatment of herpes in its various forms, including shingles (caused by varicella-zoster), and life-threatening herpes encephalitis. This preceded the successful development of AZT, the first AIDS drug, by researchers at Burroughs Wellcome who had been trained by Hitchings and Elion.

Since her retirement from Burroughs Wellcome in 1983, Elion, who lives in North Carolina, has taken up honorary lectureships, served on many advisory boards, and shared her experience with medical research students. She also works with the World Health Organization.

In 1988 Elion and Hitchings shared the Nobel Prize for Physiology or Medicine with Scottish pharmacologist James Black, and in 1991 she became the first woman to be inducted into the National Inventors' Hall of Fame. She was named in the Engineering and Science Hall of Fame and received the National Medal for Science. She won the Higuchi Memorial Award in 1995.

Elkington, George Richards (1801–1865) English inventor who pioneered the use of electroplating for finishing metal objects.

Elkington was born in Birmingham on 17 October 1801, and in 1818 he became an apprentice in the local small-arms factory; in due course he became its proprietor. With his cousin Henry Elkington he explored the alternatives to silver-plating from about 1832. The fire-gilding process of plating base metals, or more often silver, with a thick film of gold had been practised from early times. The article was first cleaned and then placed in a solution of mercurous nitrate and nitric acid, so that it acquired a thin coating of mercury. The surface was next rubbed with an amalgam of gold and mercury, in the form of a stiff paste held in a porous fabric bag, until a smooth coating of the pasty mixture had been applied. Finally, the article was heated on a charcoal fire to drive off the mercury, and the residual gold was burnished.

The process of plating base metals with silver and gold by electrodeposition was announced in a patent taken out by the Elkington cousins in 1840. This proposed the use of electrolytes prepared by dissolving silver and gold, or their oxides, in potassium or sodium cyanide solution. The articles to be coated were cleaned of grease and scale and immersed as cathodes in the solution, current being supplied through a bar of metallic zinc or other electropositive metal (anode); later silver or gold anodes were used, brass and German silver were suggested as the most suitable metals for plating.

A second patent, granted in 1842 to Henry Beaumont (an employee of the Elkingtons), covered some 430 additional salts of silver which it was thought might have application to electroplating. In the same year, John Stephen Woolrich obtained a patent for the use of a magnetoelectric machine that depended on Michael Faraday's discovery of electromagnetism in 1830, the plating solution being the soluble double sulphate of silver and potassium. Licences were first issued to the Elkingtons in 1843, but Thomas Prime of Birmingham appears to have been the first to use such plate commercially, by employing Woolrich's patented machine; Dr Percy, a famous metallurgist, claims to have conducted Faraday himself round Prime's works in 1845. Subsequently the Elkingtons took over Woolrich's patent, for which they paid him a royalty, and they were thereafter able to command a minimum royalty themselves of £150 from all who practised electroplating. In spite of this, however, electroplate rapidly supplanted Old Sheffield plate; the Sheffield directory of 1852 contained the last entry under this heading, with only a single representative remaining from the many who had practised the art five years previously.

Taking into partnership Joseph Mason, the founder of Mason College (subsequently Birmingham University), the Elkingtons established a large workshop in Newhall Street, Birmingham, which after a seven-year battle against the older methods of silver plating, at last won acceptance. The Elkingtons also successfully patented their ideas in France. George Elkington established large copper-smelting works in Pembrey, near Llanelli in South Wales, additionally providing houses for his workers and schools for their children, but the chief centre for his activities remained in Birmingham. He died at Pool Park, North Wales, on 22 September 1965.

Ellet, Charles (1810–1862) US civil engineer who designed the first wire-cable suspension bridge in the USA and became known as the 'American Brunel'.

Ellet was born at Penn's Manor, Pennsylvania, on 1 January 1810. Ellet's career began when he was appointed as a surveyor and assistant engineer on the Chesapeake and Ohio Canal in 1828, where he remained for three years. He then went to Europe and enrolled as a student at the Ecole Polytechnique in Paris, and continued to gather experience by studying the various engineering works taking place in France, Germany, and the UK.

He returned to the USA in 1832 and submitted to Congress a proposal for a 305-m/1,000-ft suspension bridge over the Potomac River at Washington, DC, but the plan was too advanced for its time and failed to receive government support. In 1842, over the Schuylkill River at Fairmount, Pennsylvania, he built his first wire-cable suspension bridge. Ellet introduced there a technique that was common in France, that of binding small wires together to make the cables; five of these latter supported the bridge at each side, the span being 109 m/358 ft.

Between 1846 and 1849 he designed and built for the Baltimore and Ohio Railway the world's first long-span wire-cable suspension bridge, crossing the Ohio River at Wheeling, West Virginia. The central span of 308 m/1,010 ft was then the longest ever built. However the bridge failed under wind forces in 1854 because of its overall aerodynamic instability. Ellet's towers remained standing, and the rest of the bridge was rebuilt by John Roebling (1806–1869), who later achieved fame with his own record-breaking activities in building long-span suspension bridges of wire cable of his own manufacture. (In 1956 the Wheeling Bridge was again under repair; Ellet's towers and anchorages and Roebling's cables and suspenders were retained, but the deck was entirely renewed.)

In 1847 Ellet received a contract to build a bridge over the Niagara River, only 3.2 km/2 mi below the falls. The result was a light suspension structure, and Ellet subsequently claimed to be the first person to cross the Niagara Gorge on the back of a horse (thanks to the bridge). This was, however, to prove to be another enterprise that turned sour for its promoter. A dispute over money led Ellet to resign in 1848, leaving the project uncompleted.

Following the outbreak of the American Civil War in 1861, Ellet produced a steam-powered ram that was used by the Union (Northern) forces with decisive effect against the Confederate army on the Mississippi River. In June 1862 Ellet personally led a fleet of nine of these rams in the Battle of Memphis. The Union side was victorious, but in the course of the fighting Ellet was fatally wounded. He died in Cairo, Illinois, on 21 June 1862.

Elsasser, Walter Maurice (1904–1991) German-born US geophysicist who pioneered analysis of the Earth's former magnetic fields.

Born in Mannheim, Germany on 20 March 1904, and educated at Göttingen, Elsasser left in 1933 following Hitler's rise to power and spent three years in Paris where he worked on the theory of atomic nuclei. After settling in 1936 in the USA and joining the staff of the California Institute of Technology, he specialized in geophysics. His magnetical researches in the 1940s yielded the dynamo model of the Earth's magnetic field. In this the field is explained in terms of the activity of electric currents flowing in the Earth's fluid metallic outer core. The theory premises that these currents are magnified through mechanical motions, rather as currents are sustained in power-station generators. It was Elsasser who pioneered analysis of the Earth's former magnetic fields, frozen in rocks. Taken up into the work of Drummond Hoyle Matthews, John Tuzo Wilson, and others, Elsasser's insights have subsequently proved crucial to the development of modern ideas of oceanic expansion and continental movement.

Elsasser served as professor of physics at the University of Pennsylvania from 1947; in 1962 he was made professor of geophysics at Princeton and between 1968 and 1974 held a research chair at the University of Maryland. A wide-ranging and speculative thinker, Elsasser also produced *The Physical Foundation of Biology* (1958) and *Atom and Organization* (1966).

Emeléus, Harry Julius (1903–1993) English chemist who made wide-ranging investigations in inorganic chemistry, studying particularly nonmetallic elements and their compounds.

Emeléus was born in London on 22 June 1903. He attended Hastings Grammar School and Imperial College, London. After graduation he went to Karlsruhe University where he met several of the German exponents of preparative inorganic chemistry. From 1929–31 he worked at Princeton University in the USA, and it was there that he met and married Mary Catherine Horton. He returned to Imperial College to continue his researches and in 1945 became professor of inorganic chemistry at Cambridge University, where he remained until he retired in 1970. He died on 2 December 1993.

Emeléus began his researches during his first period at Imperial College with a study of the phosphorescence of white phosphorus, showing that the glow was caused by the slow oxidation of phosphorus(III) oxide formed in a preliminary nonluminous oxidation. He also studied the inhibition of the glow by organic vapours. His work continued with spectrographic investigations of the phosphorescent flames of carbon disulphide, ether (ethoxyethane), arsenic, sulphur, and the phosphorescence of phosphorus(V) oxide illuminated with ultraviolet light. The results provided new information about the mechanisms of combustion reactions.

While at Princeton Emeléus worked on the photosensitization by ammonia of the polymerization of ethene (ethylene), the photochemical interaction of amines and ethene, and the photochemistry of the decomposition of amines and their reaction with carbon monoxide. This phase of his work, on chemical kinetics, helped to prepare him for the great career in chemistry that lay ahead.

On his return to Imperial College he began investigating the hydrides of silicon, especially the kinetics of the oxidation of mono-, di-, and trisilane. He also studied the isotopic composition of water from different sources. He developed a very accurate method of measuring densities and showed that naturally occurring water exhibits a small variation in deuterium content, and that distillation, freezing, and adsorption methods can all effect some degree of separation of the two isotopic forms. Continuing his work on silicon hydrides, he prepared tetrasilane, Si_4H_{10} (the silicon analogue of butane), by treating magnesium silicide with dilute hydrochloric acid, and went on to produce alkyl and aryl derivatives of the silanes. In 1938 Emeléus and John Anderson published *Modern Aspects of Inorganic Chemistry.*

When Emeléus moved to Cambridge in 1945 he started studying the halogen fluorides. He showed that the much sought after trifluoroidomethane, CIF_3 – the key to many synthetic processes – can be made by reacting iodine(V) fluoride with carbon(IV) iodide. He prepared polyhalides of potassium and demonstrated that bromine(III) fluoride could be used as a nonaqueous solvent in the study of acid/base reactions. In 1949 he prepared organometallic fluorides of mercury and went on to make various derivatives containing the methylsilyl group, CH_3SiH_2–. By 1959 he was working with the fluorides of vanadium, niobium, tantalum, and tungsten and much of his research in the 1960s concerned the fluoralkyl derivatives of metals. He summarized much of his work in *The Chemistry of Fluorine and its Compounds* (1969).

Emeléus received many honours but remained an extremely modest man, never claiming that the results of his work were of outstanding significance. However, his influence on inorganic chemistry was enormous.

Empedocles (*c.* 490–430 BC) Greek philosopher who was rather a man of many parts, being prominent in poetry, philosophy, politics, mysticism, and medicine.

Empedocles lived in Acragas (Agrigentum) in Sicily and seems to have been one of the earliest philosophers to embrace the view that terrestrial objects are made up of four elements or basic principles, that is, fire, air, water, and earth. He viewed these as united or divided by two forces, attraction and repulsion (or, more poetically, love and strife). Such views of the elemental construction of matter, later more fully developed by Aristotle, proved influential until the mechanical philosophy associated with the scientific revolution of the 17th century imposed a doctrine that viewed matter as corpuscular and governed by mathematically calculable mechanical forces.

Empedocles seems to have been influenced by Pythagoreanism and perhaps Orphism. Legend has it that Empedocles ended his life by jumping into the crater of Mount Etna, possibly in the belief that he would demonstrate his immortality. Perhaps for those reasons, he was a figure greatly admired by 19th-century Romantic poets.

Further Reading

Johnstone, Henry W, Jr, *Empedocles: The Fragments,* Greek Commentaries series, Bryn Mawr Commentaries, 1985.

O'Brien, Denis, *Empedocles' Cosmic Cycle: A Reconstruction from the Fragments and Secondary Sources.*

Wright, M R, *Empedocles: Extant Fragments,* Greek Texts series, Bristol Classical Press, 1995.

Encke, Johann Franz (1791–1865) German astronomer whose work on star charts during the 1840s contributed to the discovery of the planet Neptune in 1846. He also worked out the path of the comet that bears his name.

Born in Hamburg on 23 September 1791, Encke was the eighth child of a Lutheran preacher. As a child he was exceedingly proficient at mathematics and at the age of 20 he became a student at the University of Göttingen. His degree studies were interrupted by military service in the Wars of Liberation.

As a student Encke impressed the physicist Karl Gauss, who was instrumental in securing a post for him at a small astronomical observatory at Seeberg near Gotha. There the quality of his work was soon recognized and he rose in seniority from assistant to director. Encke then accepted the offer of a professorship at the Academy of Sciences in Berlin and the directorship of the Berlin Observatory in 1825. After 40 years in Berlin, Encke died there on 26 August 1865.

A fine mathematician who carried out continuous research on comets and the perturbations of the asteroids, Encke spent much of his time putting together the information with which to prepare new star charts. The compilation was from both old and new observations and many alternative sources. The charts, taking nearly 20 years to draw up, were complete in 1859 – but were soon improved upon by those prepared by Friedrich ◊Argelander. Nevertheless Encke's charts were of some value in that they pointed out the existence of Neptune and several minor planets.

Encke's most successful piece of work was on what subsequently became known as Encke's comet. This comet had been reported by Jean ◊Pons, but little was known of its behaviour. Encke showed that the comet had an elliptical orbit with a period of just less than four years.

One of his tasks at Berlin was to oversee the yearbook *Astronomisches Jahrbuch,* of which he was the editor for the period 1830–66. The books included large sections on minor planets, which increased their cost considerably. Encke was prepared to take the risk that high costs would diminish the market for the annual, since he considered that the data were worth publishing. He also used the books to publish many of his own mathematical determinations of orbits and perturbations, although he preferred to publish papers on planets of the Solar System in *Astronomische Nachrichten.*

During the later 1820s and the early 1830s Encke was responsible for the re-equipping and resiting of the Berlin Observatory. After raising the necessary financial support he installed a meridian circle, a large Fraunhofer refractor, and a heliometer. The observatory specialized in the observation of moveable stars.

Enders, John Franklin (1897–1985) US microbiologist who succeeded in culturing viruses in quantity outside the human body. Before this time, progress in research on viruses had been greatly hindered by the fact that viruses need living cells in which to grow. For his work on virus culture he was awarded the 1954 Nobel Prize for Physiology or Medicine, which he shared with Frederick Robbins (1916–) and Thomas Weller (1915–).

Enders was born on 10 February 1897 in West Hartford, Connecticut, the son of a banker. He was educated at Yale University but interrupted his studies to become a flying instructor during World War I and did not graduate until 1920. He then took up a business career but left it to study English at Harvard. Enders changed to study medicine, and finally obtained a doctorate in bacteriology in 1930. He remained at Harvard Medical School, progressing from instructor in 1930 to professor in 1962 and professor emeritus in 1968. From 1947–72 he was chief of research in the Division of Infectious Diseases at the Children's Hospital Medical Center, Boston. From 1972 he was chief of the Virus Research Unit at the hospital. He died in Waterford, Connecticut on 8 September 1985.

Viruses cannot be grown, as bacteria can, in nutrient substances, and so a method had been developed for growing them in a living chick embryo. Enders believed that he could improve on this method and reasoned that it was unnecessary to use a whole organism, but that living cells might be sufficient. In 1948 Enders and his colleagues Robbins and Weller prepared a medium of homogenized chick embryo and blood and attempted to grow a mumps virus in it. This experiment had been tried before, unsuccessfully, but at that time penicillin had not been available and it was penicillin that they added to suppress the growth of bacteria in the mixture. The experiment worked and they turned to the growth of other viruses.

The disease poliomyelitis attacked and debilitated many children at that time and they decided to investigate the virus responsible. Previously the polio virus could be grown only in the living nerve tissue of primates. But using their method, Enders managed to grow the virus successfully on tissue scraps obtained first from stillborn human embryos, and then on other tissue.

In the 1950s, the three produced a vaccine against the measles virus, which was improved and then produced commercially in 1963.

The virus culture technique developed by Enders and his co-workers enabled virus material to be produced in sufficient quantity for experimental work. The use of this technique meant that viruses could be more readily isolated and identified.

Erasistratus (lived *c.* 250 BC) Greek physician and anatomist regarded as the founder of physiology. He came close to discovering the true function of several important systems of the body that were not fully understood until nearly a thousand years later when physiologists had access to far more advanced methods of experimentation and dissection.

Erasistratus was born on the Aegean island of Ceos (now Khios). He learnt his skills in Athens and became court physician to Seleucus I, who governed western Asia. He then moved on to Alexandria where he taught and advanced some of the work of the Greek anatomist Herophilus. But the Egyptians, among whom he worked, were morally against the use of cadavers for dissection, so that after Erasistratus this type of anatomical research ceased until well into the 13th century.

Erasistratus dissected and examined the human brain, noting the convolutions of the outer surface, and observed that the organ is divided into larger and smaller portions (the cerebrum and cerebellum). He compared the human brain with those of other animals and made the correct hypothesis that the surface area/volume complexity is directly related to the intelligence of the animal. He traced the network of veins, arteries, and nerves and realized their topographical associations, but his conclusions ran too closely along the lines of popular opinion. He postulated that the nerves carry the 'animal' spirit, the arteries the 'vital' spirit, and the veins blood. He did, however, grasp a rudimentary principle of oxygen exchange, noting that air was taken from the lungs to the heart where it became vital spirit for distribution via the arteries (as vital spirit) to the brain and then via the nerves to the body as animal spirit. (If one reads vital spirit as haemoglobin, he was not far wrong.) He also put forward the idea of capillaries, explaining that the reason an artery bleeds when cut is because, as the vital spirit flows out, the blood rushes in from the veins through the capillaries to replace the vacuum created. He described the valves in the heart and condemned bloodletting as a form of treatment.

Erasistratus came near to discovering the principle of blood circulation (although he had it circulating in the wrong direction), but this mystery was not to be finally unravelled until Harvey's discoveries of the 17th century.

Eratosthenes (c. 276–194 BC) Greek scholar and polymath, many of whose writings have been lost, although it is known that they included papers on geography, mathematics, philosophy, chronology, and literature.

The son of Aglaos, Eratosthenes was born in Cyrene (now known as Shahhat, part of Libya). He underwent the equivalent of a university education in Athens before being invited, at the age of 30, by Ptolemy III Euergetes to become tutor to his son and to work in the library of the famous museum at Alexandria. On the death of Zenodotus in 240 BC he became the museum's chief librarian.

No single complete work of Eratosthenes, a premier scholar of his time, survives. The most important that remains is on geography – a word that he virtually coined as the title of his three-volume study of the Earth (as much as he knew of it) and its measurement. The work was concerned with the whole of the known world and divided the Earth into zones and surface features; parallels and meridian lines were used as a basis for establishing distances between places. It was accepted as the definitive work of its time (although it was criticized by Hipparchus for not making sufficient use of astronomical data). Eratosthenes greatly improved upon the inaccurate Ionian map.

The base line was a parallel running from Gibraltar through the middle of the Mediterranean and Rhodes to the Taurus Mountains in modern Turkey, onward to the Elburz range, the Hindu Kush, and to the Himalayas. At right angles to this line was a meridian passing through Heroe, Syene (now Aswan), Alexandria, Rhodes, and the mouth of the River Borysthenes (now Dniepr). The data available were mostly the notes and records of travellers and their estimates of days in transit, although some data about the height and angle of the Sun at Meroe, Alexandria, and Marseille had been collected.

Eratosthenes' measurements of the height of the Sun at Alexandria used the fact that Syene was on the Tropic of Cancer (where at midday on the summer solstice a vertical post casts no shadow, for the Sun is directly overhead). Alexandria and Syene were on the same meridian. Eratosthenes' measurements were made at midday, but with a thin pillar in the centre of a hemispherical bowl. He estimated that the shadow was 1/25 of the hemisphere and so was 1/50 of the whole circle. Since the rays of the Sun can be considered to be striking any point on the Earth's surface in parallel lines, by using alternate angles and incorporating the known distance between Alexandria and Syene of 5,000 stades, Eratosthenes was able to calculate that the total circumference of the Earth was just over 250,000 stades. There are several errors in his assumptions, but as the first attempt it was given much credit. The conversion of stades into modern units creates additional error, although a value of 46,500 km/28,900 mi is usually quoted. (The modern figure is usually accepted as 40,075 km/24,903 mi at the Equator.)

Eratosthenes also divided the Earth into five zones: two frigid zones around each pole with a radius of 25,200 stades on the meridian circle; two temperate zones between the polar zones and the tropics, with a radius of 21,000 stades; and a torrid zone comprising the two areas from the equator to each tropic, having a radius of 16,800 stades. The frigid zones he described as the 'Arctic' and 'Antarctic' circles of an observer on the major parallel of latitude (approximately 36 °N); these circles mark the limits of the circumpolar stars that never rise or set. Eratosthenes' model of the known world had a north–south length of 38,000 stades from the Cinnamon country to Thale, and an east–west length of 77,800 stades from eastern India to the Straits of Gibraltar.

Eratosthenes was familiar with the Earth and the apparent movement of the Sun, but he did little serious astronomical study. However, he also estimated the obliquity of the ecliptic as 11/83 of a circle, equivalent to 23 °51′.

In the mathematical area he was most successful in offering a solution to the famous Delian problem of doubling the cube. For his proof, Eratosthenes proposed an apparatus consisting of a framework of two parallel rulers with grooves along which could be slid three rectangular plates capable of moving independently of each other and able to overlap. In arithmetic he devised a technique called the Sieve for finding prime numbers.

Eratosthenes also spent a considerable period of his life establishing the dates of historical events. In his two major works *Chronography* and *Olympic Victors*, many of the dates he set for events have been accepted by later historians and have never been changed. (For example, the Fall of Troy was in 1184–1183 BC, and the First Olympiad took place in 777–776 BC.) He also wrote many books on literary criticism in a series entitled *On the Old Comedy*. Like much of his output it is referred to by contemporary and later scholars, but it did not survive the passage of time.

Erdös, Paul (1913–1996) Hungarian mathematical genius and probably the most prolific mathematician in history. He wrote over 1,000 publications with hundreds of co-authors. He had no home and no job and lived out of a suitcase.

Born in Budapest on 26 March 1913, both Paul Erdös's parents were mathematics teachers and he showed a prodigious mathematical talent from an early age. His two older sisters had died of scarlet fever a few days before he was born so his parents were very protective of him. His mother did everything for him and when his father was captured in a Russian offensive and sent to Siberia for six years she removed him from school and taught him herself. As a teenager he came to the attention of academics with his work on prime numbers and at 17 he had proved, in a more elegant proof than any mathematician before, Chebseyev's theorem that there is always at least one prime number between any number greater than one and its double. He entered the University of Budapest at the age of 17 and left with a PhD in 1934 at the age of 21. Sensing political trouble at home he moved to Manchester, England, on a four-year fellowship: the longest time he stayed in one place. He left England for America before the outbreak of war taking on several short-term jobs and returning to Hungary for short stints. Most members of his family remained in Hungary and were killed during the war. He was denied re-entry to America in the McCarthy era after a visit to a conference in Amsterdam and settled in Israel for several years in the 1950s until he was allowed a US visa in the 1960s.

For most of his life Erdös did not have a job. He never wanted material possessions and based his lifestyle on an old Greek saying that the wise man has nothing he cannot carry in his hands. He never had a bank account or credit card and lived out of a suitcase, travelling from one mathematical centre to another, and between one mathematical professor and another, relying on their hospitality and often turning up on his colleagues' doorsteps with his half-empty bag, saying 'My brain is open,' meaning that he wanted to stay for a while and collaborate with them. Remarkably most of them were delighted by these random visits and would provide him with meals and lodgings being amply rewarded by his presence and the chance to talk mathematics and share problems with him. Working at a furious pace he would exhaust his hosts and then move on. They would make sure he got the right flight to his next destination. At the last count Erdös had 458 published collaborators and another 4,500 mathematicians who had worked with him. From 1964 his mother, in her eighties by this time, began travelling with him. He was devastated by her death in 1971 and threw himself even more into his work, sometimes putting in a 20 hour day fuelled by coffee, caffeine tablets, and benzedrine. He was fond of quoting a remark of one of his friends – 'A mathematician is a machine for converting caffeine into theorems'.

Mathematician Ron Graham of Bell Labs, one of his collaborators, looked after his money, depositing all his cheques for lecture fees, prizes, grants, and so on, paying his bills and performing mundane tasks like making his dental appointments. He had a room set aside in his house for the job and was responsible for giving away most of Erdös's money. In 1984, Erdös won the $50,000 Wolf Prize for Mathematics and gave all of it, except $750, to relatives and to a scholarship fund in his mother's memory. He offered monetary rewards for the solutions to problems, ranging from $25 for easy ones to $10,000 for the impossible ones. The most he ever had to pay out was $1,000. On his death he left only $25,000.

Erdös's work was wide ranging, spanning many areas from probability to number theory and graph theory. He was best known for his work on Ramsey theory, which is concerned with the relationship between groups of points in a network – a network being anything from people at a party to stars in the sky. His philosophy of maths was that it is eternal because it has an infinity of problems. He invented the art of posing problems, the simpler the better, and delighted in posing them. Erdös never got involved in the competitive aspects of mathematics and was philosophical about a particular incident in the late 1940s in which a colleague cut him out.

Erdös envisaged his ideal death occurring just after a lecture in which an awkward member of the audience had a asked a tricky question to which he would reply 'I think I'll leave that to the next generation,' and fall down dead. In fact, he almost achieved his ideal – he had just given two lectures in Warsaw and was about to fly to Vilnius to give another when he died of a heart attack on 20 September 1996 at the age of 83. He had never married and left no immediate family.

Ericsson, John (1803–1889) Swedish-born US engineer and inventor who is best known for his work on naval vessels.

Ericsson was born on 31 July 1803 at Langban Shyttan in Varmland, Sweden, and between the ages of 13 and 17 served as a draughtsman in the Gotha Canal Works. He was then commissioned into the Swedish army, where he carried out map surveys. In 1826 he moved to London to seek sponsorship for a new type of heat engine he had invented (which used the expansion of superheated air as the driving force). This forerunner of the gas turbine was not successful, and Ericsson turned his attention to steam engines.

In 1829 he built the *Novelty,* a steam locomotive that competed unsuccessfully against George Stephenson's *Rocket* at the Rainhill Trials for adoption on the Liverpool and Manchester Railway. In 1839 Captain Stockton of the US navy placed an order for Ericsson to supply a small iron vessel fitted with steam engines and a screw propellor. The vessel was built and sailed to New York; Ericsson himself sailed out a few months later. He became a US citizen in 1848.

In 1851 he resumed his interest in the heat or 'caloric' engine. It was found to be too heavy for the ship he had built for it (immodestly called the *Ericsson),* making the vessel too slow. Only towards the end of his life did Ericsson construct small, efficient engines of this type.

A more successful line of development resulted from his work on the helical screw propellor, an interest he shared with his contemporary Isambard Kingdom ◊Brunel. Realizing that the paddle steamer was incapable of further development, Ericsson had built two small screw-driven ships in the UK in 1837 and 1839. In 1849 he built the *Princeton,* the first metal-hulled, screw-propelled warship and the first to have its engines below the waterline for added protection.

It was the outbreak of the American Civil War in 1861 that finally gave Ericsson the opportunity to demonstrate his skill as a naval engineer. His turreted iron-clad ship, the *Monitor,* was first offered to Napoleon III and only after he refused it did it go to Ericsson's adopted country. Equipped with a low freeboard and heavy guns, it was the first warship to have revolving gun turrets – a practice soon to be adopted universally. In the Battle of Hampton Roads in 1862, the *Monitor* defeated the Confederate ship *Morrimack.*

After the Civil War Ericsson continued to design warships, torpedoes, and a 35.6-cm/14-in naval gun, but he also devoted time to more peaceful pursuits. Among his inventions were an apparatus for extracting salt from seawater, fans for forced draught and ventilation, a shipboard depth-finder, a steam fire engine, and surface condensers for marine engines. Between the years 1870–85, he also explored the possiblity of using solar energy and gravitation and tidal forces as sources of power.

Not always successful in the realization of his ideas, Ericsson was a wilful and impetuous man, often far ahead of other engineers of the day. He died on 8 March 1889 in New York City, and in 1889 the VSS *Baltimore* took his body back to Sweden, in accordance with his last wishes.

Further Reading

Brophy, Ann and Gallin, Richard (ed), *John Ericcson and the Inventions of War,* History of the Civil War series, Silver Burdett Press, 1990.

de Kay, James T, *Monitor: The Story of the Revolutionary Ship and the Men Whose Invention Changed the Course of History,* Walker Publishing Co, 1997.

Erlang, Agner Krarup (1878–1929) Danish mathematician who, although he was extremely knowledgeable in many fields, might never have become famous if he had not become scientific adviser – and leader – of the Copenhagen Telephone Company's research laboratory. His application of the theory of probabilities to problems connected with telephone traffic made his name known all over the world, and the 'erlang' is now the unit of traffic flow. A meticulous mathematician, he published many influential papers and was also responsible for constructing a device to measure alternating electric current.

Erlang was born on 1 January 1878 in the village of Lonborg, near Tarm, in Jylland (Jutland), the son of a schoolmaster. Completing his education he studied mathematics and natural sciences at the University of Copenhagen, obtaining his MA in mathematics in 1901 (with astronomy, physics, and chemistry as secondary subjects). On leaving the university he worked as a teacher in various schools. During this time he won the award for solving the mathematical prize problem set by the University of Copenhagen, which, in that year, was concerned with Huygens's solution of infinitesimal problems. In 1908, at the age of 30, Erlang was appointed scientific collaborator and leader of the laboratory of the Copenhagen Telephone Company, where he remained for the rest of his life. He died suddenly on 3 February 1929, at the age of only 51. Single all his life, he devoted almost all his time to scientific study. He had a large library and collected mathematical, physical, and astronomical works in particular; his knowledge in these subjects was extensive, but he was also well versed in philosophy, history, and poetry. A modest man with an original mind, he was of an extremely kind and friendly disposition.

At the laboratory of the Copenhagen Telephone Company, Erlang came under the influence of Franz Johanssen, the managing director of the company and another mathematician who had himself, in the year before Erlang joined, published two short essays in which he dealt with problems of telephone traffic flow – such as congestion and waiting time – and in which he introduced probability calculations. Erlang took to the work immediately. Within a year he had published his first paper on the subject, in which he was able to arrive at an exact solution to another problem posed by Johanssen previously. And over the next few years, Erlang published a number of other papers on the theory of telephone traffic, which, because of their meticulous precision, became pioneer works.

It is rare for a telephone caller to get the engaged tone for any reason other than that the receiver at the call's destination is already in use. Yet it is possible through the links of the connection – the many coding selectors and digit selectors in use at any one time – for the call to find a selector already in use, which would automatically trip in the engaged tone. The fact that this is so rare is a result of the provision of switches in numbers based on the calculation that not more than one call in 500 at each selecting stage will fail because the equipment is engaged. To fulfil this standard, rules determining the amount of traffic to be carried per selector have been established by every telephone authority and company, following Erlang's formulae. Especially important are his formula for the probability of barred access in busy-signal systems – the so-called B-formula – and his formulae for the probability of delay and for the mean waiting time in waiting time systems. These formulae may be considered the most important within the theory of telephone traffic.

The erlang is the unit of telephone traffic flow; it is defined as 'the number of calls originated during a period, multiplied by the average holding time of a call, expressed in terms of the period: one erlang is therefore equivalent to the traffic flow in one circuit continuously occupied'.

Erlang also published studies of other mathematical problems. His work on logarithms and other numerical tables, in which he attempted to reduce the mean error to the lowest figure possible, resulted in the compilation of four- and five-figure tables that are now considered among the best available.

As the leader of the Laboratory of the Telephone Company, Erlang also investigated several assorted physio-technical problems that cropped up. In particular, he constructed a measuring bridge to meter alternating current (the so-called Erlang complex compensator) that was a considerable improvement on earlier apparatus of similar function. Of equal significance were his investigations into telephone transformers and telephone cable theory.

Esaki, Leo (1925–) Japanese physicist who shared the 1973 Nobel prize for his discovery of tunnelling in semiconductor diodes. These devices are now called Esaki diodes. Esaki spent most of his working life in the USA but returned to Japan in the 1990s as president of the University of Tsukuba.

Esaki was born on 12 March 1925 in Osaka. When he graduated from the University of Tokyo in 1947, he wanted to become a nuclear physicist but Japan lacked the particle accelerators needed to compete in this field and he moved to solid-state physics. Working for Sony in 1957, Esaki noticed while studying very small heavily-doped germanium diodes that sometimes, and unexpectedly, the resistance decreased as current increased. This was caused by 'tunnelling' – a quantum mechanical effect whereby electrons can travel (tunnel) through electrostatic potentials that they would be unable to overcome classically. These barriers have to be very thin for tunnelling to occur and Esaki was able to use this effect for switching and to build ultra-small and ultra-fast tunnel diodes. In 1959 Esaki received a PhD from Tokyo for this discovery. It also earned Esaki the 1973 Nobel prize, which he shared with Brian Josephson, who predicted supercurrents in superconducting diodes, and Ivar Giaever who discovered electron tunnelling in superconductors. In 1960 Esaki joined IBM's Thomas J Watson Research Center in Yorktown Heights, New York, where he became an IBM fellow, the company's highest research honour, in 1967. He continued to research the nonlinear transport and optical properties of semiconductors, in particular multilayer superlattice structures grown by molecular beam epitaxy techniques. In 1992 Esaki returned to Japan to become president of Tsukuba University. The return of the nation's only living physics Nobel laureate was front-page news in Japan.

Eskola, Pentti Eelis (1883–1964) Finnish geologist important in the field of petrology.

Born in Lellainen, the son of a farmer, on 8 December 1883, Eskola was educated as a chemist at the University of Helsinki, before specializing in petrology. Throughout his life he was fascinated by the study of metamorphic rocks, taking early interest in the Precambrian rocks of England. He died in Helsinki on 6 December 1964.

Fascinated by the mineral facies of rocks, Eskola was one of the first to apply physicochemical postulates on a far-reaching basis to the study of metamorphism, thereby laying the foundations of most subsequent studies in metamorphic petrology. Building largely on Scandinavian studies, Eskola was concerned to define the changing pressure and temperature conditions under which metamorphic rocks were formed. His approaches enabled comparison of rocks of widely differing compositions in respect of the pressure and temperature under which they had originated.

Euclid (lived *c.* 300 BC) Greek mathematician whose works, and the style in which they were presented, formed the basis for all mathematical thought and expression for the following 2,000 years (although they were not entirely without fault). He also wrote books on

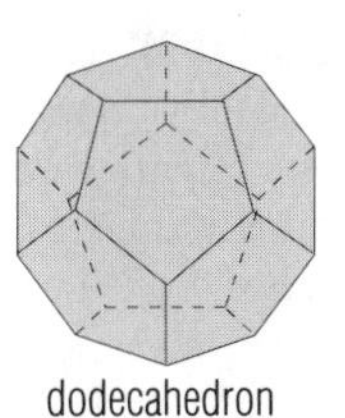

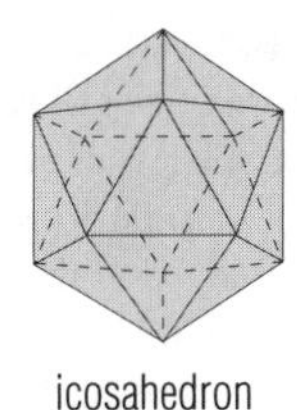

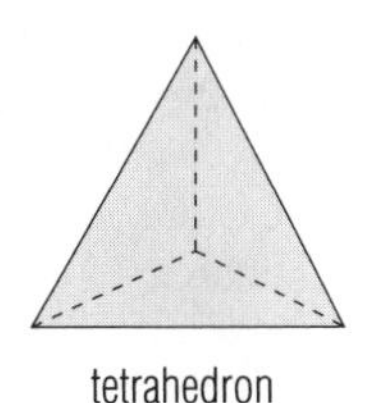

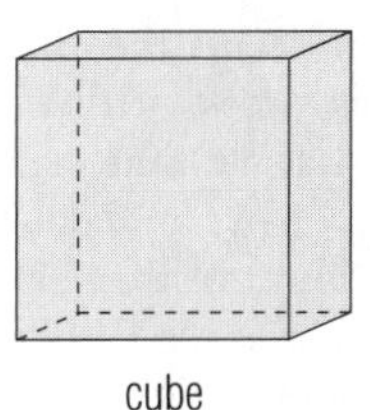

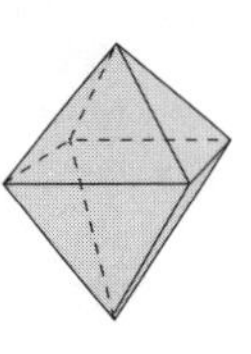

Euclid The five regular polyhedra or Platonic solids.

other scientific topics, but these have survived the passage of the centuries only fragmentarily or not at all.

Very little indeed is known about Euclid. No record is preserved of his date or place of birth, his education, or even his date or place of death. The influence of ◊Plato, who lived in the 4th century BC, is certainly detectable in his work – so Euclid must either have been contemporary or later. Some commentators have suggested that he attended Plato's Academy in Athens but, if so, it is likely to have been after Plato's death. In any case, it has been established that Euclid went to the recently founded city of Alexandria (now in Egypt) in around 300 BC and set up his own school of mathematics there. Fifty years later, however, Euclid's disciple ◊Apollonius of Perga was said to have been leading the school for some considerable time; it seems very possible, therefore, that Euclid died in around 270 BC.

Euclid's mathematical works survived in almost complete form because they were translated first into Arabic, then into Latin; from both of these they were then translated into other European languages. He employed two main styles of presentation: the synthetic (in which one proceeds from the known to the unknown via logical steps) and the analytical (in which one posits the unknown and works towards it from the known, again via logical steps). In his major work, *The Elements,* Euclid used the synthetic approach, which suited the subject matter so perfectly that the method became the standard procedure for scientific investigation and exposition for millennia afterwards. The strictly logical arrangement demanding the absolute minimum of assumption, and the omission of all superfluous material, is one of the great strengths of *The Elements,* in which Euclid incorporated and developed the work of previous mathematicians as well as including his own many innovations. The presentation was one of extreme clarity and he was rigorous, too, about the actual detail of the mathematical work, attempting to provide proofs for every one of the theorems.

The Elements is divided into 13 books. The first six deal with plane geometry (points, lines, triangles, squares, parallelograms, circles, and so on), and include hypotheses such as 'Pythagoras' theorem' which Euclid generalized. Books 7 to 9 are concerned with arithmetic and number theory. In Book 10 Euclid treats irrational numbers. And Books 11 to 13 discuss solid geometry, ending with the five 'Platonic solids' (the tetrahedron, octahedron, cube, icosahedron, and dodecahedron).

There is no royal road to geometry.

EUCLID TO PTOLEMY I, *QUOTED IN* PROCLUS COMMENTARY ON EUCLID, *PROLOGUE*

Euclid favoured the analytical mode of presentation in writing his other important mathematical work, the *Treasury of Analysis.* This comprised three parts, now known as *The Data, On Divisions of Figures,* and *Porisms.*

Euclid's geometry formed the basis for mathematical study during the next 2,000 years. It was not until the 19th century that a different form of geometry was even considered: 'accidentally' discovered by Saccheri in 1733, non-Euclidean geometry was not in any way defined until Nikolai Lobachevsky (in the 1820s), János Bolyai (in the 1830s), and Bernhard Riemann (in the 1850s) examined the subject. It is difficult to see, therefore, how Euclid's contribution to the science of mathematics could have been more fundamental than it was.

Further Reading

Artmann, B, *Euclid: The Creation of Mathematics,* Springer-Verlag, 1999.

Engelfriet, Peter M, *Euclid in China: The Genesis of the First Chinese Translation of Euclid's Elements, Books I–VI (Jihe Yuanben, Beijing, 1607) and Its Reception up to 1723,* Brill, 1998.

Hartshorne, Robin (ed), *Companion to Euclid,* Berkeley Mathematical Lecture Notes series, American Mathematical Society, 1997, v 9.

Kheirandish, E; Toomer, G J (ed), *The Arabic Version of Euclid's Optics,* Sources in the History of Mathematics and Physical Sciences series, Springer-Verlag, 1998, v 16.

Eudoxus of Cnidus (408–355 BC) Greek mathematician and astronomer who is said to have studied under Plato. Himself a great influence on contemporary scientific thought, Eudoxus was the author of several important works. Many of his theories have survived the test of centuries; work attributed to Eudoxus includes methods to calculate the area of a circle and to derive the volume of a pyramid or a cone. He also devised a system to demonstrate the motion of the known planets when viewed from the Earth.

Very little is known about Eudoxus' life, although it is recorded that he spent more than a year in Egypt, some of the time as the guest of the priests of Heliopolis. In a series of geographical books with the overall title of *A Tour of the Earth,* he later described the political, historical, and religious customs of the countries of the eastern Mediterranean area.

Primarily a mathematician, Eudoxus used his mathematical knowledge to construct a model of homocentric rotating spheres to explain the motion of planets as viewed from the Earth, which was at the centre of the system. The model was later extended by ◊Aristotle and Callippus (*c.* 370–*c.* 300 BC), and although superseded by the theory of epicycles, it was still widely accepted during the Middle Ages. Eudoxus was able to give close approximations for the synodic periods of the planets Saturn, Jupiter, Mars, Venus, and Mercury. The geometry of the model was impressive, but there are several weaknesses. Eudoxus assumed, for example, that each planet remains at a constant distance from the centre of its orbital circle, and in addition that each retrograde loop as seen from the Earth is identical with the last. Neither assumption complies with observation.

The model of planetary motion was published in a book called *On Rates.* Further astronomical observations were included in two other works, *The Mirror* and *Phenomena.* Subject later to considerable criticism by ◊Hipparchus, these books nevertheless established patterns where before none existed.

In mathematics Eudoxus' early success was in the removal of many of the limitations imposed by ◊Pythagoras on the theory of proportion. Eudoxus showed that the theory was applicable in many more circumstances. Subsequently he established a test for the equality of two ratios, and noted that it was possible to find a good approximate value for the area of a circle by the 'method of exhaustion': a polygon is drawn within the circle, the number of its sides is repeatedly doubled, and the area of each new polygon is found by using a simple formula. It was to Eudoxus that Archimedes attributed the discovery that the volume of a pyramid or a cone is equal to one-third of the area of the base times the perpendicular height.

Although none of Eudoxus' works has survived the passage of time in complete form, the mathematical skills he practised were important and influential both in his own age and for centuries afterwards.

Euler, Leonhard (or Leonard) (1707–1783) Swiss mathematician whose power of mental calculation was prodigious; so great was his capacity for concentrating on mathematical computation that even when he became totally blind towards the end of his life, he was able to continue his work without pause. With such ability and with true scientific curiosity, Euler – a brilliant teacher – expanded the scope of virtually all the known branches of mathematics, devising and formulating a considerable number of theorems and rules now named after him. He also enlarged mathematical notation.

Euler was born on 15 April 1707 in Basle, where he grew up and was educated. At the University of Basle he studied under Jean Bernoulli, obtaining his master's degree at the age of 16, in 1723. He then found it impossible – perhaps because he was so young – to gain a faculty position. Four years later, however, he was invited by his great friend and contemporary Daniel Bernoulli (son of Johann) to join him in Russia, at St Petersburg. Euler duly arrived there, spent three years at the Naval College, and then (in 1730) was appointed professor of physics at the Academy of Sciences. When Bernoulli returned home in 1733, Euler succeeded him as professor of mathematics. Shortly afterwards, through looking at the Sun during his astronomical studies, he lost the sight of his right eye. In 1741 he travelled to Berlin at the request of the emperor, Frederick the Great, and in 1744 became director of the Berlin Academy of Sciences. He remained there until 1766 when the Empress

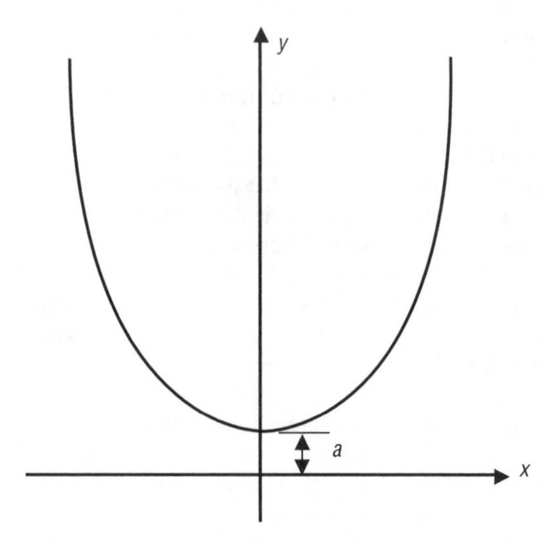

Euler A catenary is a transcendental curve that may be represented by the equation $y=(a/2)(e^{x/a} + e^{-x/a})$, where a is a constant and e is Euler's number (2.718...).

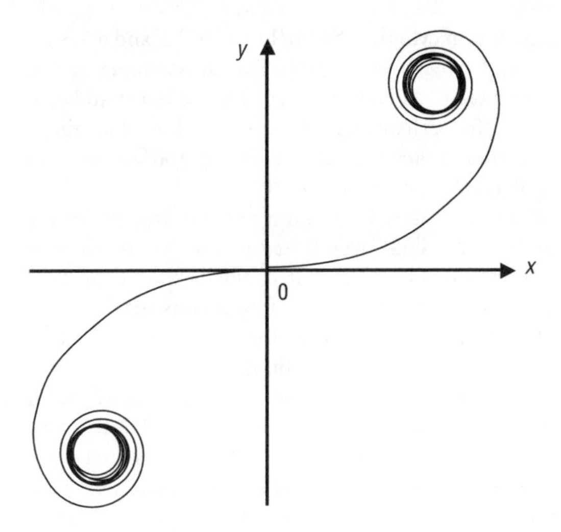

Euler Euler's spiral is sometimes alternatively called a clothoid.

Catherine the Great recalled him to St Petersburg to become director of the Academy of Sciences there. Soon after his return he lost the sight of his other eye, through cataracts, but retained his office, competently carrying out all his duties and responsibilities for another 15 years or more, until he died on 18 September 1783.

Euler contributed to all the classic areas of mathematics, although his most innovatory work was in the field of analysis, in which he considerably improved mathematical methodology and the rigour of presentation. He advanced the study of trigonometry, in particular developing spherical trigonometry. Following the analytical work of his friend Daniel Bernoulli, Euler demonstrated the significance of the coefficients of trigonometric expansions; Euler's number (e, as it is now called) has various useful theoretical properties and is also used in the summation of particular series.

Euler also studied algebraic series and demonstrated the importance of convergence. He applied algebraic methods instead of geometric ones (as used by Newton, Galileo, and Kepler previously), and improved differential and integral calculus, bringing them virtually to their modern forms. He used constant coefficients in the integration of linear differential equations and originated the calculus of partial differentials. And he became very interested in applying mathematical – and particularly analytical – principles to mechanics, especially celestial mechanics.

After his return to Russia in 1766 Euler carried out further research into the motion and positions of the Moon, and the gravitational relationships between the Moon, the Sun, and the Earth. His resulting work on tidal fluctuations took him into the realm of fluid mechanics, in which he successfully analysed motion in a perfectly compressible fluid. Further astronomical work brought him an award of £300 from the British government for his development of theorems useful in navigation (he had been a member of the Royal Society since 1746).

A prolific author of influential papers detailing his methods and his results, and a teacher of outstanding authority, Euler made a number of innovations both in mathematical concept – such as Euler's constant, Euler's equations, Euler's line, and Euler's variables – and in mathematical notation – he was responsible among other things, for the use of π, e in natural (Napierian) logarithms, i for imaginary numbers, and Σ for summation. Other interests of his included acoustics and optics.

Further Reading

Euler, Leonhard; Hunter, Henry (transl); Pyle, Andrew (introduction), *Letters of Euler to a German Princess: On Different Subjects in Physics and Philosophy*, Thoemmes Antiquarian Books, 1997, 2 vols.

Fraser, Craig G, 'The origins of Euler's variational calculus', *Arch Hist Exact Sci*, 1994, v 47, pp 103–141.

Gray, Jeremy, 'Leonhard Euler: 1707–1783', *Janus*, 1985, v 72, pp 171–192.

Hakfoort, Casper, *Optics in the Age of Euler: Conceptions of the Nature of Light, 1700–1795*, Cambridge University Press, 1995.

Evans, Alice Catherine (1881–1975) US microbiologist whose research into the bacterial contamination of milk led to the recognition of the danger of unpasteurized milk. As a result of her research the incidence of brucellosis was greatly reduced when the dairy industry accepted that all milk should be pasteurized.

Alice Evans was born in Neath, Pennsylvania, on 29 January 1881. Leaving Susquehanna Collegiate Institute in 1901 she taught in a school for four years, before going to Ithaca, New York, on a two-year nature-study course for rural teachers organized by Cornell University. During this time she became interested in science and continued her course, graduating with a BS degree in 1909; she was awarded an MS degree in 1910 by the University of Wisconsin, after which she took a research post at the dairy division of the US Department of Agriculture's Bureau of Animal Industry, studying the bacteriology of milk and cheese. In 1913 she transferred to the new laboratories in Washington, DC, where she researched the bacterial contamination of milk products. In 1918 Evans moved to the Hygienic Laboratories of the US Public Health Service as an assistant biologist to research into epidemic meningitis and influenza; she returned to the study of milk flora at the end of World War I. Evans retired in 1945 and moved to Arlington,

Virginia, in 1969 where she lived until her death in 1975 following a stroke.

Evans's renown as a microbiologist resulted from her discovery of a common origin for the disease brucellosis in humans and cattle, previously thought to be two separate diseases. Her research into the Bruce bacillus (affecting humans) and the Bang Bacillus (affecting cattle) showed that for practical purposes, these could be regarded as identical and either may cause the disease 'undulant fever'. Since humans contract the disease by handling infected animals or through drinking the milk, Evans warned of the dangers of drinking unpasteurized milk. Her results were published in the *Journal of Infectious Diseases* in 1918 and were at first ignored in US medical circles. They were, however, later confirmed by other scientists and by the late 1920s numerous cases of human brucellosis had been reported and the disease was recognised as a major threat, not only to those in close contact with animals but also through the food industry. By the 1930s the dairy industry accepted that all milk should be pasteurized.

Evans, Oliver (1755–1819) US engineer who developed high-pressure steam engines and various machines powered by them. He also pioneered production-line techniques in manufacturing.

Evans was born on 13 September in 1755 in Newport, Delaware, and at the age of 16 he was apprenticed to a wagon-maker. He read books on mathematics and mechanics, and became interested in steam engines. His first invention was a machine for cutting and mounting wire teeth in a leather backing to make devices for carding textile fibres before spinning; his machine produced 1,500 cards a minute. In 1780 he joined his two brothers at a flour mill at Wilmington, where he helped to build machinery that used water power to drive conveyors and elevators. As a result, he attained a degree of automation that allowed one person to operate the whole mill as a single production line. Other millers remained unimpressed by his invention, however.

Evans then moved to Philadelphia, where he spent more than ten years trying to develop a steam carriage. For some time he tried to couple one of James ◊Watt's steam engines to a wagon and as early as 1792 he was working on internally fired boilers for steam engines. But he had to abandon the enterprise because of lack of financial support, and eventually he turned his attention to the manufacture of stationary steam engines.

In 1786 and 1787 he successfully petitioned the legislatures of Pennsylvania and Maryland for exclusive patent rights to profit from his inventions, and by 1802 he had developed a high-pressure steam engine with a 15-cm/6-in cylinder, 95-cm/18-in stroke, and a 2.3-m/7.5-ft flywheel. It employed a 'grasshopper' beam mechanism and had an internally fired boiler and provision for exhausting spent steam into the air. It worked at a pressure of 3.5 kg/sq cm (50 psi) and ran at 30 revolutions per minute.

Two years later Evans built a steam dredger for use on the Schuylkill River. It had power-driven rollers as well as a paddle so that it could be moved on land under its own power. This amphibious machine, the *Orukter Amphibole,* was the forerunner of some 50 or so steam engines which he built in the next 15 years.

In 1806 Evans began to develop the Mars works for the manufacture of steam engines. In 1817 he completed his last work, a 17.4 kW/24 hp engine for a waterworks. On 11 April 1819 his principal workshop was destroyed in a fire started by a grudge-bearing apprentice. The consequent shock to Evans probably hastened his death in New York City a few days later on 15 April 1819.

Like many engineers Evans died with many bold dreams unrealized. In March 1815 he claimed, in the *National Intelligencer* (although many seriously question the claim), that he could have introduced steam carriages as early as 1773 and steam paddle boats by 1778. Less controversially, Evans published the *Young Millwright* and *The Miller's Guide* (1792) and, in 1805, a book called *The Abortion of the Young Steam Engineer's Guide.*

Ewing, (William) Maurice (1906–1974) US geologist, a major innovator in modern geology.

Born in Texas, on 12 May 1906, the son of a merchant, Ewing studied at the Rice Institute in Houston from 1923 and developed his geological interests by working for various oil companies. He moved to work at Lehigh University, Pennsylvania, in 1929 (first in physics, then in geology) and, in 1944, joined the Lamont–Doherty Geological Observatory, New York, the world's foremost geophysical research institution. From 1947 till his retirement, he was professor of geology at Columbia University, while also holding a position at the Woods Hole Oceanographic Institute. Ewing died in Lockney, Texas, on 4 May 1974.

Using marine 'sound-fixing' seismic techniques (SOFAR – 'sound fixing and ranging'), and pioneering deep-ocean photography and sampling, Ewing ascertained that the crust of the Earth under the ocean is much thinner (5–8 km/3–5 mi thick) than the continental shell (about 40 km/25 mi thick). Ewing also demonstrated that mid-ocean ridges were common to all oceans, and in 1957 further discovered the presence within them of deep central canyons. His studies of ocean sediment showed that its depth increases with distance from the mid-ocean ridge, which gave clear support for the sea-floor spreading hypothesis proposed by Harry ◊Hess in 1962. Mainly through his notions of oceanic rift valleys, Ewing provided several of the crucial jigsaw pieces for the plate tectonics revolution in geology in the 1960s.

Eyde, Samuel (1866–1940) Norwegian industrial chemist who helped to develop a commercial process for the manufacture of nitric acid, which made use of comparatively cheap hydroelectricity. He was a member of the Norwegian parliament, and in 1920 Norwegian minister in Poland.

Eyde was born in Arendal, Norway, on 29 October 1866, and received his higher education in Germany at the Charlottenburg High School in Berlin, where he gained a diploma in constructural engineering. He then worked as an engineer in various German cities, principally Hamburg. In partnership with the German engineer C O Gleim, he returned to Scandinavia to work on the construction of various railway and harbour installations. But increasingly his interest turned to the possibiltity of developing the electrochemical industry in his native Norway, where hydroelectric schemes were beginning to make cheap electrical energy available.

In 1901, while studying the problem of the fixation of nitrogen (the conversion of atmospheric nitrogen into chemically useful compounds), he came to know his compatriot Kristian Birkeland. The two men set up a small laboratory where they combined their efforts to discover the conditions necessary for the economic combination of nitrogen and oxygen (from air) in an electric arc to produce nitrogen oxides and, eventually, nitric oxide. Their method, known as the Birkeland–Eyde process, can be summarized by the following chemical equations:

$$N_2 + O_2 \rightarrow 2NO$$

(nitrogen and oxygen combine to give nitric oxide)

$$2NO + O_2 \rightarrow 2NO_2$$

(nitric oxide combines with oxygen to form nitrogen dioxide)

$$4NO_2 + O_2 + 2H_2O \rightarrow 4HNO_3$$

(oxidation of nitrogen dioxide in the presence of water produces nitric acid)

Their experiments resulted in the first commercial success in this area of research. The key to the process was an oscillating disc-shaped electric arc, produced by applying a powerful magnetic field to an arc formed between two metal electrodes by an alternating current. The electrodes were copper tubes cooled by water circulating inside them. In 1903 the Norwegische Elektrishe Aktiengeswllschaft constructed a small-scale plant at Ankerlökken, near Oslo. Two years later full-scale operation began at Notodden. Then interest in the work was shown by the company of Badische und Anilin Soda Fabrik, which, following on from German research on the thermodynamic equilibrium of nitrogen oxides, had employed O Schonherr to study the nitrogen–oxygen reaction in 1897. In 1904 he patented a method of producing a steadier arc than had been attained by Birkeland and Eyde.

By this time Eyde had obtained the hydroelectric rights on some waterfalls, and in 1904 he became administrative director of an electrochemical company, financed partly from a Swedish source that supported the Birkeland–Eyde process. In the following year he obtained extensive support from French financiers to found a hydroelectric company, which he directed with great skill until he retired from active participation in 1917.

Eyde died at Aasgaardstrand in Norway on 21 June 1940, shortly after writing his autobiography.

Eysenck, Hans Jurgen (1916–1997) German-born British psychologist, renowned for his controversial theories about a wide range of subjects, especially human intelligence.

Eysenck was born in Germany on 4 March 1916 and educated at various schools in Germany, France, and the UK. With the rise to power of Adolf Hitler in the 1930s he left Germany and went to the UK. He studied psychology at London University, graduating in 1938 and gaining his doctorate in 1940. During the rest of World War II he was senior research fellow and psychologist at the Mill Hill Emergency Hospital, London, and in 1946 he was appointed director of the psychology department at the Maudsley Hospital, Surrey. He went to the USA in 1949 as visiting professor at the University of Pennsylvania, and on his return to the UK in 1950 became reader in psychology at London University's Institute of Psychiatry. In 1954 he again went to the USA, as visiting professor at the University of California. In the following year he was appointed professor of psychology at London University's Institute of Psychiatry.

Eysenck investigated many areas of psychology, often producing highly controversial theories as a result of his studies. But it is his theory that intelligence is almost entirely inherited and can be only slightly modified by education that has aroused the greatest opposition. The concept of intelligence is difficult to define and even more difficult to measure – the commonly used intelligence tests, for example, are often criticized for being culturally biased, favouring well-educated white people and penalizing poorly educated black people. Eysenck attempted to devise a fairer, culture-free method for assessing intelligence. It involved neither problem-solving nor even conscious thought and therefore, Eysenck argued, it could not be criticized for being culturally biased. Basically his method involved subjecting a person to stroboscopic light flashes and buzzing sounds while simultaneously recording the electrical activity of the person's brain by means of an electroencephalograph. It has been known since the early 1970s that the pattern of brain waves is related to intelligence (as measured by conventional intelligence tests) but the correlation has been too approximate to be of practical use. Eysenck, however, claimed to be able to measure the brain waves accurately enough to give a very high correlation with conventional intelligence tests, which, if true, would mean that his method of measuring intelligence was as valid and useful as conventional tests. Using his method, Eysenck then did a cross-cultural comparison of intelligence and found that, on average, black people obtained significantly lower intelligence quotients than whites. Combining this finding with his theory that intelligence is predominantly inherited, he claimed that black people are inherently less intelligent than are whites. This claim met with great – sometimes violent – opposition and was widely criticized by educationalists and other psychologists, who stated that, because Eysenck had validated the results obtained by electroencephalography with those from conventional intelligence tests the two methods must contain the same biases and therefore Eysenck's method could not be considered culturally unbiased.

Eysenck also studied personality traits, anxiety and neurosis, the influence of violence shown on television on behaviour, and the psychology of smoking. In addition, he wrote many popular psychology books, in which he presented his contentious ideas in relatively simple, non-technical language.

Further Reading

Eysenck, Hans Jürgen, *Rebel With a Cause*, W H Allen, 1990.

Gibson, Hamilton Bertie, *Hans Eysenck: The Man and His Work*, Owen, 1981.

Modgil, Sohan and Modgil, Celia (eds), *Hans Eysenck: Consensus and Controversy*, Falmer International Master-Minds Challenged, Falmer Press, 1986, v 2.

Nyborg, Helmuth (ed), *The Scientific Study of Human Nature: Tribute to Hans J Eysenck at Eighty*, Pergamon, 1997.

Fabre, Jean Henri (1823–1915) French entomologist whose studies of insects, particularly their anatomy and behaviour, have become classics.

Fabre was born on 22 December 1823 in Saint-Léons in southern France, the son of a farmer who had left the land and set up a small business. Fabre's family was poor and for much of his early childhood he lived with his grandmother in the country, which fostered his interest in natural history. When he was seven years old he returned to Saint-Léons to attend the village school. Later, after passing through senior school, he won a scholarship to Avignon, from which he gained his certificate of education in 1842. While at school Fabre had to take part-time jobs to help to pay for his education and contribute to his family's income and so, on obtaining his certificate he immediately took a teaching post at the lycée in Carpentras, a small town in northeastern Avignon. After further studies at Montpellier, he gained his teaching licence (which enabled him to teach in higher schools) in mathematics and physics and in 1851 was appointed a physics teacher at a lycée in Ajaccio, Corsica. But soon afterwards he contracted a fever, which forced him to resign and return to the mainland. After he had recovered, he went to Paris to gain a degree, then returned to Avignon where, in 1852, he became professor of physics and chemistry at the lycée. He held this post for 20 years, eventually resigning because the authorities would not allow girls to attend his science classes.

Fabre decided to abandon his teaching career and moved to the village of Orange, where he embarked on a serious study of entomology. He was very poor and had virtually no equipment to help him in his studies, but by writing articles for scientific journals he eventually managed in 1878 to buy a small plot of waste land in Serignan, Provence. He built a wall around the plot, treating it as an open-air laboratory, and remained there for the rest of his life, pursuing his entomological studies and writing about his findings. Towards the end of his life he became world famous as an authority on entomology and, in 1910, many leading scientific figures visited him in Serignan for a celebration given in his honour. Fabre died in Serignan five years later, on 11 October 1915. After his death, the French National Museum of Natural History purchased Fabre's plot of land as a memorial to the man and his work.

Although he had written several previous articles, Fabre's first important paper was published in 1855. It was a detailed account of the behaviour of a type of wasp that paralyses its prey (mainly beetles and weevils), which it then carries to its nest to feed to its young. In 1857 he wrote another important paper describing the life cycle of the Meloidae (oil beetles) – hypermetamorphic beetles that begin life as larvae, then hatch into a second larval stage (called a triungulin) and climb onto particular types of flowers frequented by solitary bees of the genus *Anthophora.* When a bee visits the flower, the triungulin attaches itself to the bee and is carried to the bee's nest, where it passes through several other larval stages – feeding on honey – before finally developing into an adult beetle. The significance of this and other early work by Fabre was recognized by Charles Darwin, who quoted Fabre in *On the Origin of Species,* but Fabre himself did not accept the idea of evolution.

Fabre began his most famous work after he settled at Serignan where, in addition to writing numerous entomological papers, he embarked on his ten-volume *Souvenirs Entomologiques,* which took him 30 years to complete. Based almost entirely on observations Fabre had made in his small plot, this work is a model of meticulous attention to detail. It became a classic entomological work and also did much to revitalize interest in entomology.

Further Reading

Darrow, Clarence, *Insects and Men: Instinct and Reason – Reflections on the Observations and Discoveries of Henri Fabre, Naturalist,* Gordon Press Publishers, 1991.

Fabre, Jean H; Stawell (transl), *Fabre's Book of Insects,* Dover Publications, 1998.

Pasteur, Georges, 'Jean Henri Fabre', *Sci Amer,* 1994, v 271, no 1, pp 74–80.

Yavetz, Ido, 'Jean Henri Fabre and evolution: indifference or blind hatred?', *Hist Phil Life Sci,* 1988, v 10, pp 3–36.

Fabricius ab Aquapendente, Hieronymus (1537–1619) Latinized name of Girolamo Fabrizio. Italian anatomist and embryologist who gave the first accurate description of the semilunar valves in the veins and whose pioneering studies of embryonic development helped to establish embryology as an independent discipline.

Fabricius was born on 20 May 1537 in Aquapendente, near Orvieto in Italy, and studied first the humanities then medicine at the University of Padua, where he was taught anatomy and surgery by the eminent Italian anatomist Gabriele ◊Falloppio. After graduating in 1559, Fabricius worked privately as an anatomy teacher and surgeon until 1565, when he succeeded Fallopio as professor of surgery and anatomy at Padua University (the chair had been vacant since Fallopio's death in 1562). Fabricius remained at Padua University for the rest of his career, eventually retiring because of ill health in 1613. During his time at Padua, Fabricius built up an international reputation that attracted students from many countries, including William Harvey (who studied under him 1597–1602). He also helped to establish a permanent anatomical theatre at the university, a structure that still exists today. About 1596 Fabricius started to acquire an estate at Bugazzi where, after retiring, he remained until his death on 21 May 1619.

Fabricius's principal anatomical work was his accurate and detailed description of the valves in the veins. Although they had previously been observed and crudely drawn by other scientists, Fabricius publicly demonstrated them in 1579 in the veins of the limbs and in 1603 published the first accurate description – with detailed illustrations – of these valves in *De Venarum Ostiolis/On the Valves of the Veins.* He mistakenly believed, however, that the valves' function was to retard the flow of blood to enable the tissues to absorb nutriment.

Fabricius's most important completely original work was in embryology. In his treatise *De Formato Foetu/On the Formation of the Foetus* (1600) – the first work of its kind – he compared the late foetal stages of different animals and gave the first detailed description of the placenta. Continuing his embryological studies, Fabricius published *De Formatione Ovi et Pulli/On the Development of the Egg and the Chick* (1612), in which he gave a detailed, excellently illustrated account of the developmental stages of chick embryos. Again, however, Fabricius made some erroneous assumptions. He believed that the sperm did not enter the ovum, but stimulated the generative process from a distance. He also believed that both the yolk and the albumen nourished the embryo and that the embryo itself was produced from the spiral threads (chalaza) that maintain the position of the yolk. Nevertheless, his embryological studies were extremely influential and helped to establish embryology as an independent science.

Fabricius also investigated the mechanics of respiration, the action of muscles, and the anatomy of the larynx (about which he was the first to give a full description) and the eye (he was the first to describe correctly the location of the lens and the first to demonstrate that the pupil changes size).

Fabry, Charles (1867–1945) French physicist who specialized in optics. He is best remembered for his part in the invention and then in the investigations of the applications of the Fabry–Pérot interferometer and etalon.

Fabry was born in Marseille on 11 June 1867. He and his two brothers all became prominent mathematicians or physicists. Fabry attended the Ecole Polytechnique in Paris 1885–89, and despite becoming interested in astronomy along with his brothers, his studies under Macé de Lépinay took him into the field of optics. After completing his degree at the Ecole Polytechnique, Fabry went on to take a doctorate in physics at the University of Paris in 1892. He then did some teaching at lycées in Pau, Bordeaux, Marseille, Nevers, and Paris, before joining the Faculty of Science at the University of Marseille in 1894.

The next 25 years saw Fabry concentrating on research and teaching in Marseille, where he was elected to the chair of industrial physics in 1904. The Ministry of Inventions recalled him to Paris in 1914 in order to investigate interference phenomena in light and sound waves, and in 1921 Fabry returned to Paris on a more permanent basis as professor of physics at the Sorbonne. He was later also made professor of physics at the Ecole Polytechnique, and the first director of the Institute of Optics. In 1935 Fabry became a member of the International Committee on Weights and Measures at the Bureau de Longitudes, but he retired in 1937 and died in Paris on 11 December 1945.

Fabry's research was dedicated to devising methods for the accurate measurement of interference effects, and to the application of this technique to a broad range of scientific subjects. His interest in interference was apparent very early in his career, his doctoral thesis having been concerned with various theoretical aspects of interference fringes.

Alfred Pérot (1863–1925) and Fabry invented the interferometer that was named after them in 1896. It is based on multiple beam interference and consists of two flat and perfectly parallel plates of half-silvered glass or quartz. A light source produces rays that undergo a different number of reflections before being focused. When the rays are reunited on the focal plane of the instrument, they either interfere or cohere, producing dark and light bands respectively. When the light source is monochromatic – as, for example, with a laser or a mercury vapour lamp – the Fabry–Pérot interference fringes appear as sharp concentric rings. A non-monochromatic source produces a more complex pattern of concentric rings. If the two reflecting plates are fixed in position relative to one another, the device is called a Fabry–Pérot etalon. If the distance between the plates can be varied, the apparatus is a Fabry–Pérot interferometer.

These devices can be used to distinguish different wavelengths of light from a single source, and are accurate to a resolving power of one million. The Fabry–Pérot interferometer has a resolving power 10–100 times greater than a prism or a small diffraction grating spectroscope, and is more accurate that the device used by Albert ⟡Michelson and Edward ⟡Morley in their classic experiment intended to detect the motion of the Earth through the hypothetical ether.

Fabry and Pérot worked together for a decade on the design and uses of their invention. One of their achievements was the setting of a series of standard wavelengths. Then in 1906 Fabry began to collaborate with Henri Buisson. By 1912 study of the spectra of neon, krypton and helium enabled them to confirm a broadening of lines predicted by the kinetic theory of gases. In 1914, they used the interferometer to confirm the Doppler effect for light, not with stellar sources as had until then been necessary but in the laboratory.

Applications to astronomy were also explored, including the measurement of solar and stellar spectra, and photometry. An issue of practical importance in biology, medicine, astronomy, and, of course, physics was the identification of the material responsible for the filtering out of ultraviolet radiation from the Sun. In 1913 Fabry was able to demonstrate that ozone is plentiful in the upper atmosphere and is responsible for this.

Fahrenheit, Gabriel Daniel (1686–1736) Polish-born Dutch physicist who invented the first accurate thermometers and devised the Fahrenheit scale of temperature.

Fahrenheit was born in Danzig (now Gdansk), Poland, on 14 May 1686. He settled in Amsterdam in 1701 in order to learn a business, and became interested in the manufacture of scientific instruments. From about 1707 he spent his time wandering about Europe, meeting scientists and other instrumentmakers and gaining knowledge of his trade. In 1717 he set himself up as an instrumentmaker in Amsterdam, and remained in the Netherlands for the rest of his life.

Fahrenheit published his methods of making thermometers in the *Philosophical Transactions* in 1724 and was admitted to the Royal Society in the same year. He died in The Hague on 16 September 1736.

The first thermometers were constructed by ⟡Galileo at the end of the 16th century, but no standard of thermometry had been decided upon even a century later. Galileo's was a gas thermometer, in which the expansion and contraction of a bulb of air raised or lowered a column of water. Because it took no account of the effect of atmospheric pressure on the water level, it was very inaccurate. Guillaume Amontons (1663–1705) improved the gas thermometer in 1695 by using mercury instead of water and having the mercury rise and fall in a closed column, thus avoiding effects due to atmospheric pressure.

Fahrenheit's first thermometers contained a column of alcohol that expanded and contracted directly in the same way as modern thermometers. Fahrenheit came across this instrument in 1708 when he visited Ole Römer, who had devised a thermometer in 1701. Römer had developed a scale of temperature in which the upper fixed point, the boiling point of water, was 60° and the lower, the temperature of an ice–salt mixture, was 0°. Body temperature on this scale was 22.5° and the freezing point of water came to 7.5°.

Fahrenheit took up Römer's ideas and combined them with those of Amontons in 1714, when he substituted mercury for alcohol and constructed the first mercury-in-glass thermometers Fahrenheit found that mercury was more accurate because its rate of expansion, although less than that of alcohol, is more constant. Furthermore mercury could be used over a much wider temperature range than alcohol. In order to reflect the greater sensitivity of his thermometer, Fahrenheit expanded Römer's scale so that blood heat was 90° and the ice–salt mixture was 0°; on this scale freezing point was 30°. Fahrenheit later adjusted the scale to ignore body temperature as a fixed point so that the boiling point of water came to 212° and freezing point was 32°. This is the Fahrenheit scale that is still in use today.

Using his thermometer, Fahrenheit was able to determine the boiling points of liquids and found that they vary with atmospheric pressure. In producing a standard scale for the measurement of temperature as well as an accurate measuring instrument, Fahrenheit made a very substantial contribution to the advance of science.

Further Reading

Burnett, John, Fahrenheit's thermometers and letters, *Bull Sci Instr Soc*, 1988, v 16, pp 3–6.

Fahrenheit, Daniel Gabriel; Star, Pieter van der (ed), *Fahrenheit's Letters to Leibniz and Boerhaave*, Editions Rodopi, 1983.

Fairbairn, William (1789–1874) Scottish engineer who designed a riveting machine that revolutionized the making of boilers for steam engines. He also worked on many bridges, including the wrought-iron box-girder construction used first on the bridge across the Menai Straits in North Wales.

Fairbairn was born into a poor family on 19 February 1789, at Kelso. He received little early education, although he did learn to read at the local parish school. He started work when he was 14 years old when his family moved to a farm owned by the Percy Main colliery near Newcastle upon Tyne; Fairbairn became apprenticed to a millwright. He learned mathematics in his spare time and displayed his engineering ingenuity by constructing an orrery (a working model of the Solar System).

He finished his apprenticeship in 1811 having, in the meantime, become a friend of the engineers George and Robert Stephenson. He worked as a millwright at Bedlington, then took a series of jobs in London, Bath, Dublin, and Manchester. During this time he invented a sausage-making machine and a machine for making nails. In Manchester he worked on the construction of the Blackfriars Bridge and then set up as a manufacturer of cotton-mill machinery. In 1824 Fairbairn erected two watermills in Zürich, and later turned his attention to ship-building and, finally, bridge-building.

In 1862 he invented a self-acting planning machine for dealing with work up to 6 m/20 ft by 1.8 m/6 ft. He became an authority on mechanical and engineering problems and received many honours and awards, including a baronetcy in 1869. Fairbairn died on 18 August 1874, at Moor Park in Surrey and was buried at Prestwick, Northumberland.

From very humble beginnings, Fairbairn used his inventive skills and engineering ability to earn a fortune by the time he was 40 years old. He had acquired a sound reputation for producing machinery for the cotton mills and by that time employed about 300 workers. His reputation abroad had been enhanced when he solved the problem of an irregular water supply with his watermills in Switzerland.

In 1830 he was commissioned by the Forth of Clyde Company to build a light iron boat to run between Glasgow and Edinburgh. He then concentrated on shipbuilding, first in Manchester (where he built ships in sections) then from 1835 in Millwall on the River Thames, where his Millwall Iron Works employed some 2,000 people.

Fairbairn returned to Manchester, and in 1844 designed and built the first Lancashire shellboiler. It was constructed of rolled wrought-iron plates rivetted together by a machine of his own design. It was this expertise that led Robert Stephenson to consult Fairbairn over the building of the Menai Railway Bridge, which was constructed of wrought-iron plates. Built between 1846 and 1850, it was the longest railway bridge at the time with a continuous box girder (in which the trains ran) 461 m/1,511 ft long. Fairbairn's participation ended in 1849 after a misunderstanding about his position (he and Stephenson were both termed superintendent), and he published his own account of the construction.

Further Reading

Rinsley, Donald (ed), *Fairbairn and the Origins of Object Relations*, Guildford Publications, 1994.

Scharff, David E and Fairbairn Birtles, E (eds), *From Instinct to Self: Selected Papers of W R D Fairbairn, v 1: Clinical and Theoretical Papers*, Jason Aronson Publishers, 1995.

Fajans, Kasimir (1887–1975) Polish-born US chemist, best known for his work on radioactivity and isotopes and for formulating rules that help to explain valence and chemical bonding.

Fajans was born in Warsaw on 27 May 1887. He was educated at Leipzig, Heidelberg (where he gained his PhD in 1909), Zürich, and Manchester. From 1911–17 he worked at the Technische Hochschule at Karlsruhe, and between 1917 –35 he held appointments at the Munich Institute of Physical Chemistry, where he rose from assistant professor to be the director. In 1936 he emigrated to the USA and served as a professor at the University of Michigan, Ann Arbor. He became a US citizen in 1942. He died, aged 88, on 18 May 1975.

In 1913 Fajans formulated, simultaneously with but independently of Frederick ◊Soddy, the theory of isotopes – that is, elements that have the same atomic number but different atomic weights (relative atomic masses). He showed that uranium-X_1 (itself a decay product by alpha-ray emission of uranium-238) disintegrates by beta-ray emission into uranium-X_2, which he called 'brevium' on account of its short half-life; this latter isotope then undergoes further beta-decay to form uranium-234. An alpha particle is a helium nucleus ($^{4}_{2}He$), consisting of two protons and two neutrons; its loss from an element's nucleus results in a different element that is four mass units lighter and two less in atomic number. A beta particle is an electron so that its emission, on the other hand, results in an element of the same mass number but with an atomic number larger by 1. Thus the decay of uranium-238 by emitting first an alpha particle and then two beta particles should produce a new uranium isotope four mass units lighter: uranium-234. Fajans and Soddy were the first to explain this and other radioactive processes in terms of transitions between various isotopes.

Fajans's work in inorganic chemistry was equally important. He formulated two rules to account for the well-known diagonal similarities between elements in the periodic table in terms of the ease of formation of covalencies and electrovalencies (ionic valencies). The first rule states that covalencies are more likely to be formed as the number of electrons to be removed or donated increases, so that highly charged ions are rare or impossible. The removal (or donation) of a second electron must overcome the effect of charge due to the removal (or donation) of the first, and so on for each successive electron. As the number of electrons increases, the work required soon becomes impossibly great for chemical forces, and covalencies result instead.

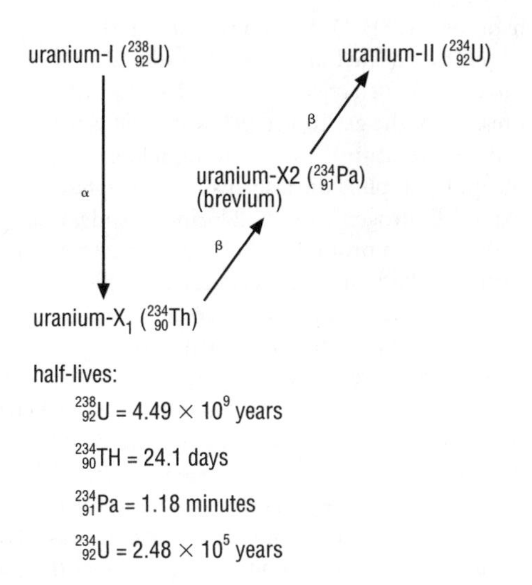

Fajans Fajans' scheme for the decay of uranium-I with modern symbols added. Uranium-X_1 and uranium-X_2 are actually isotopes of tjorium and protactinium.

Fajans's second rule states that electrovalencies are favoured by large cations and small anions. In a large atom, the outer electrons are farther from the attractive force of the positive nucleus and hence are more easily removed to form cations. In a small atom, on the other hand, an electron added to form an anion can approach more closely to the positive field of the nucleus and is therefore more strongly held than in a large anion.

The operation of the first rule in passing from Group I to Group II of the periodic table is approximately balanced by the effect of the second rule in passing from the first period to the second period. The most obvious examples of diagonal similarity are the following:

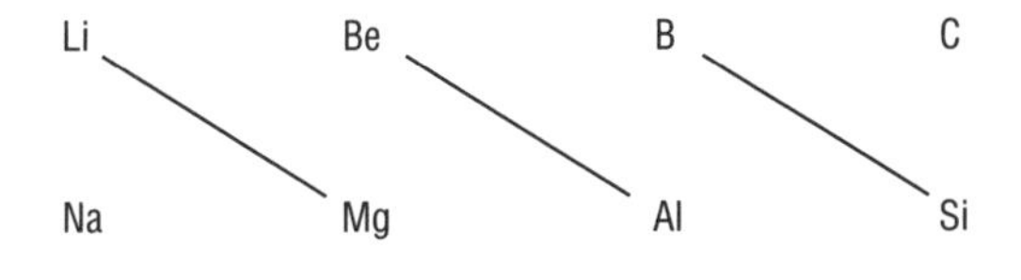

Lithium, for example, shows its similarity to magnesium in all those points that emphasize its difference from sodium.

In an extension of his work on radioactivity, Fajans estimated the ages of minerals from Norway by measuring the percentages of lead, the end-product of radioactive decay. In 1919 he did research on the energies of hydration of ions. Thermochemistry, theory of chemical forces, light absorption, and photochemistry are other areas in which he became an authority.

Falloppio, Gabriele (Gabriel Falloppius) (1523–1562) Italian anatomist who studied the female reproductive system, the brain, and the eye, and gave the first accurate description of the inner ear.

Born in Modena, Falloppio first intended to become a priest, but shifted to medicine, and was taught anatomy by Vesalius in Padua, becoming professor of anatomy at Pisa in 1548 and Padua in 1551. He died in Padua, on 9 October 1562.

He extended Vesalius's work and corrected its details. His discoveries included structures in the human ear and skull, explorations in the field of urology, and researches into the female genitalia. He coined the term 'vagina' and described the clitoris. He was the first to describe the tubes leading from the ovary to the uterus that were subsequently named after him. He failed, however, to grasp the function of the Fallopian tubes; it took a further two centuries before it was recognized that ova are formed in the ovary, passing down these tubes to the uterus. Falloppio also carried out investigations on the larynx, the eye, muscular action, and respiration. He was the teacher of ◊Fabricius ab Aquapendente.

Faraday, Michael (1791–1867) English physicist and chemist who is often regarded as the greatest experimental scientist of the 1800s. He made pioneering contributions to electricity, inventing the electric motor, electric generator and the transformer, and discovering electromagnetic induction and the laws of electrolysis. He also discovered benzene and was the first to observe that the plane of polarization of light is rotated in a magnetic field.

Faraday was born in Newington, Surrey, on 22 September 1791. His father was a poor blacksmith, who went to London to seek work in the year that Faraday was born. Faraday received only a rudimentary education as a child and although he was literate, he gained little knowledge of mathematics. At the age of 14, he became an apprentice to a bookbinder in London and began to read voraciously. The article on electricity in the *Encyclopedia Britannica* fascinated him in particular, for it presented the view that electricity is a kind of vibration, an idea that was to remain with Faraday. He also read the works of Antoine ◊Lavoisier and became interested in chemistry, and carried out what scientific experiments he could put together with his limited resources. He was aided by a manual dexterity gained from his trade, which also stood him in great stead in his later experimental work.

In 1810 Faraday was introduced to the City Philosophical Society and there received a basic grounding in science, attending lectures on most aspects of physics and chemistry and carrying out some experimental work. He also attended the Royal

Institution, where he was enthralled by the lectures and demonstrations given by Humphry ◊Davy. He made notes eagerly, assembling them at work into finely bound books. In 1812 Faraday came to the end of his apprenticeship and prepared to devote himself to his trade, not expecting to make a career in science. Almost immediately, however, there came an extraordinary stroke of luck. Davy was temporarily blinded by an explosion in a chemistry experiment and asked Faraday to help him until he regained his sight. When he recovered, Faraday sent Davy the bound notes of his lectures. Impressed by the young man, Davy marked him out as his next permanent assistant at the Royal Institution and Faraday took up this post in 1813.

This was remarkably good fortune for Faraday, because Davy was a man of wide-ranging interests and great scientific insight as well as a brilliant exponent of ideas. Furthermore, Davy undertook a tour of France and Italy 1813–15 to visit the leading scientists of the day, including the pioneer of current electricity Alessandro ◊Volta. Faraday accompanied Davy, gaining an immense amount of knowledge, and on his return to London threw himself wholeheartedly into scientific research.

Faraday remained at the Royal Institution and made most of his pioneering discoveries in chemistry and electricity there over the next 20 years. He became a great popularizer of science with his lectures at the Royal Institution, which he began in 1825 and continued until 1862. His fame grew rapidly, soon eclipsing even that of Davy, who became embittered as a result. But the strain of his restless pursuit of knowledge told and in 1839 Faraday suffered a breakdown. He never totally recovered but at the instigation of Lord ◊Kelvin, returned to research in 1845 and made his important discoveries of the effect of magnetism on light and developed his field theory. In the 1850s, Faraday's mind began to lose its sharp grip, possibly as a result of low-grade poisoning caused by his chemical researches, and he abandoned research and then finally lecturing. He resigned from the Royal Institution in 1862 and retired to an apartment provided for him at Hampton Court, Middlesex, by Queen Victoria. He died there on 25 August 1867.

Faraday was mainly interested in chemistry during his early years at the Royal Institution. He investigated the effects of including precious metals in steel in 1818, producing high-quality alloys that later stimulated the production of special high-grade steels. Faraday's first serious chemical discoveries were made in 1820, when he prepared the chlorides of carbon – C_2Cl_6 from ethane and C_2Cl_4 from ethylene (ethene) – substitution reactions that anticipated the work a few years later by Jean Baptiste ◊Dumas. In 1823 Faraday produced liquid chlorine by heating crystals of chlorine hydrate ($Cl_2.8H_2O$) in an inverted U-tube, one limb of which was heated and the other placed in a freezing mixture (liquefaction resulted because of the high pressure of the gas cooled below its relatively high critical temperature). He then liquefied other gases, including sulphur dioxide, hydrogen sulphide, nitrous oxide (dinitrogen oxide), chlorine dioxide, cyanogen, and hydrogen bromide. After the production of liquid carbon dioxide in 1825, Faraday used this coolant to liquefy such gases as ethylene (ethene), phosphine, silicon tetrafluoride, and boron trifluoride.

In the same year (1825) he made his greatest contribution to organic chemistry, the isolation of benzene from gas oils. He also worked out the empirical formula of naphthalene and prepared various sulphonic acids – later to have great importance in the industries devoted to dyestuffs and detergents. It was also at about this time that Faraday demonstrated the use of platinum as a catalyst and showed the importance in chemical reactions of surfaces and inhibitors – again foreshadowing a huge area of the modern chemical industry.

But Faraday's interest in science had been initiated by a fascination for electricity and he eventually combined the knowledge that he gained of this subject with chemistry to produce the basic laws of electrolysis in 1833. His researches, summed up in Faraday's laws of electrolysis, established the link between electricity and chemical affinity, one of the most fundamental concepts in science. It was Faraday who coined the terms anode, cathode, cation, anion, electrode, and electrolyte. He postulated that during the electrolysis of an aqueous electrolyte, positively charged cations move towards the negatively charged cathode and negatively charged anions migrate to the positively charged anode. At each electrode the ions are discharged according to the following rules:

(a) the quantity of a substance produced is proportional to the amount of electricity passed;

(b) the relative quantities of different substances produced by the same amount of electricity are proportional to their equivalent weights (that is, the relative atomic mass divided by the oxidation state or valency).

But his first major electrical discovery was made much earlier, in 1821, only a year after Hans ◊Oersted had discovered with a compass needle that a current of electricity flowing through a wire produces a magnetic field. Faraday was asked to investigate the phenomenon of electromagnetism by the editor of the *Philosophical Magazine*, who hoped that Faraday would elucidate the facts of the situation following the wild theories and opinions that Oersted's sensational discovery had aroused. Faraday conceived that circular lines of magnetic force are produced around the wire to explain the orientation of Oersted's compass needle, and therefore set about devising an apparatus that would demonstrate this by causing a magnet to revolve around an electric current. He succeeded in October 1821 with an elaborate device consisting of two vessels of mercury connected to a battery. Above the vessels and connected to each other were suspended a magnet and a wire; these were free to move and dipped just below the surface of the mercury. In the mercury were fixed a wire and a magnet respectively. When the current was switched on, it flowed through both the fixed and free wires, generating a magnetic field in them. This caused the free magnet to revolve around the fixed wire, and the free wire to revolve around the fixed magnet.

This was a brilliant demonstration of the conversion of electrical energy into motive force, for it showed that either the conductor or the magnet could be made to move. In this experiment, Faraday demonstrated the basic principles governing the electric motor and although practical motors subsequently developed had a very different form to Faraday's apparatus, he is nevertheless usually credited with the invention of the electric motor.

Faraday's conviction that an electric current gave rise to lines of magnetic force arose from his idea that electricity was a form of vibration and not a moving fluid. He believed that electricity was a state of varying strain in the molecules of the conductor, and this gave rise to a similar strain in the medium surrounding the conductor. It was reasonable to consider therefore that the transmitted strain might set up a similar strain in the molecules of another nearby conductor – that a magnetic field might bring about an electric current in the reverse of the electromagnetic effect discovered by Oersted.

Faraday hunted for this effect from 1824 onwards, expecting to find that a magnetic field would induce a steady electric current in a conductor. In 1824, François ◊Arago found that a rotating nonmagnetic disc, specifically of copper, caused the deflection of a magnetic needle placed above it. This was in fact a demonstration of electromagnetic induction, but nobody at that time could explain Arago's wheel (as it was called). Faraday eventually succeeded in producing induction in 1831. In August of that year, he wound two coils around an iron bar and connected one to a battery and the other to a galvanometer. Nothing happened when the current flowed through the first coil, but Faraday noticed that the galvanometer gave a kick whenever the current was switched on or off. Faraday found an immediate explanation with his lines of force. If the lines of force were cut – that is, if the magnetic field changed – then an electric current would be induced in a conductor placed within the magnetic field. The iron core in fact helped to concentrate the magnetic field, as Faraday later came to understand, and a current was induced in the second coil by the magnetic field momentarily set up as current entered or left the first coil. With this device, Faraday had discovered the transformer, a modern transformer being no different in essence even though the alternating current required had not then been discovered.

Faraday is thus also credited with the simultaneous discovery of electromagnetic induction, though the same discovery had been made in the same way by Joseph ◊Henry in 1830. However, busy teaching, Henry had not been able to publish his findings before Faraday did, although both men are now credited with the independent discovery of induction.

Faraday's insight enabled him to make another great discovery soon afterwards. He realized that the motion of the copper wheel relative to the magnet in Arago's experiment caused an electric current to flow in the disc, which in turn set up a magnetic field and deflected the magnet. He set about constructing a similar device in which the current produced could be led off, and in October 1831 built the first electric generator. This consisted of a copper disc that was rotated between the poles of a magnet; Faraday touched wires to the edge and centre of the disc and connected them to a galvanometer, which registered a steady current. This was the first electric generator, and generators employing coils and magnets in the same way as modern generators were developed by others over the next two years.

Faraday's next discoveries in electricity, apart from his major contribution to electrochemistry, were to show in 1832 that an electrostatic charge gives rise to the same effects as current electricity, thus proving that there is no basic difference between them. Then in 1837 he investigated electrostatic force and demonstrated that it consists of a field of curved lines of force, and that different substances take up different amounts of electric charge when subjected to an electric field. This led Faraday to conceive of specific inductive capacity. In 1838, he proposed a theory of electricity based on his discoveries that elaborated his idea of varying strain in molecules. In a good conductor, a rapid build-up and breakdown of strain took place, transferring energy quickly from one molecule to the next. This also accounted for the decomposition of compounds in electrolysis. At the same time, Faraday rejected the notion that electricity involved the movement of any kind of electrical fluid. In this, he was wrong (because the motion of electrons is involved) but in that this motion causes a rapid transfer of electrical energy through a conductor, Faraday's ideas were valid.

Faraday's theory was not taken seriously by many scientists, but his concept of the line of force was developed mathematically by Kelvin. In 1845, he suggested that Faraday investigate the action of electricity on polarized light. Faraday had in fact already carried out such experiments with no success, but this could have been because electrical forces were not strong. Faraday now used an electromagnet to give a strong magnetic field instead and found that it causes the plane of polarization to rotate, the angle of rotation being proportional to the strength of the magnetic field.

Several further discoveries resulted from this experiment. Faraday realized that the glass block used to transmit the beam of light must also transmit the magnetic field, and he noticed that the glass tended to set itself at right angles to the poles of the magnet rather than lining up with it as an iron bar would. Faraday showed that the differing responses of substances to a magnetic field depended on the distribution of the lines of force through them, and not on the induction of different poles. He called materials that are attracted to a magnetic field paramagnetic, and those that are repulsed diamagnetic. Faraday then went on to point out that the energy of a magnet is in the field around it and not in the magnet itself, and he extended this basic conception of field theory to electrical and gravitational systems.

Finally Faraday considered the nature of light and in 1846 arrived at a form of the electromagnetic theory of light that was later developed by James Clerk ◊Maxwell. In a brilliant demonstration of both his intuition and foresight, Faraday said 'The view which I am so bold to put forth considers radiation as a high species of vibration in the lines of force which are known to connect particles, and also masses of matter, together. It endeavours to dismiss the ether but not the vibrations.' It was a bold view, for no scientist until Albert ◊Einstein was to take such a daring step.

Michael Faraday was a scientific genius of a most extraordinary kind. Without any mathematical ability at all, he succeeded in making the basic discoveries on which virtually all our uses of electricity depend and also in conceiving the fundamental nature of magnetism and, to a degree, of electricity and light. He owed this extraordinary degree of insight to an amazing talent for producing valid pictorial interpretations of the workings of nature. Faraday himself was a modest man, content to serve science as best he could without undue reward, and he declined both a knighthood and the presidency of the Royal Society. Characteristically, he also refused to take part in the preparation of poison gas for use in the Crimean War. His many achievements are honoured in the use of his name in science, the farad being the SI unit of capacitance and the Faraday constant being the quantity of electricity required to liberate a standard amount of substance in electrolysis.

Further Reading

Bowers, Brian, *Michael Faraday and the Modern World,* EPA Press, 1991.
Cantor, Geoffrey and Gooding, David, *Michael Faraday,* Humanities Press International, 1996.
Cantor, Geoffrey N, *Faraday,* Macmillan, 1991.
Gooding, David, *Faraday Rediscovered,* Springer-Verlag, 1997.
Gribbin, John and Gribbin, Mary, *Faraday in 90 Minutes,* Scientists in 90 Minutes series, Constable and Co, 1997.
Thomas, John Meurig, *Michael Faraday, 1791–1867, and His Contemporaries,* National Portrait Gallery, 1991.
Thomas, John Meurig, *Michael Faraday and the Royal Institution: The Genius of Man and Place,* Hilger, 1991.

Feller, William (1906–1970) Croatian-born US mathematician largely responsible for making the theory of probability accessible to students of subjects outside the field of mathematics through his two-volume textbook on the subject. He was also interested in the theory of limits, but it was his work on probability theory that brought him widespread recognition.

Feller was born on 7 July 1906 in Zagreb, where he grew up, was educated, and attended the university, earning his BA in 1925. He then continued his studies at the University of Göttingen, Germany, where he was awarded a PhD in 1926. He moved to Kiel University, and in 1929 was put in charge of the laboratory of applied mathematics there. Four years later, Feller went to the University of Stockholm, where he served as a research associate until 1939, working as a consultant for economists, biologists, and others interested in probability theory.

At the outbreak of World War II in 1939, Feller emigrated to the USA. His first job was as executive editor of *Mathematical Reviews,* but he was soon appointed to the simultaneous post of associate professor at Brown University. In 1945 – by then a naturalized US citizen– he was made professor of mathematics at Cornell University, New York, and in 1950 Feller became Eugene Higgins Professor of Mathematics at Princeton University, a post he held until his death in 1970.

Feller had always been fascinated by the study of chance fluctuations. He came to the conclusion, early on, that the traditional emphasis placed on averages meant that insufficient attention was paid to random fluctuations, which could bear significant impact on processes under investigation. His serious study of probability theory began soon after his arrival at the University of Stockholm. Much of his effort focused on the nature and use of Markov processes (a mathematical description of random changes in a system that, for instance, can occur in either of two states; see Andrei ◊Markov). Feller used semigroups to develop a general theory of Markov processes, and was able to demonstrate the applicability of this tool to subjects in which probability theory had not usually previously been employed, for example in the study of genetics. A strong advocate of the renewal method, Feller's preoccupation with problems of methodology greatly influenced the style of his book. The work – on the introduction to and the applications of probability theory – was published in two volumes, the first in 1950 and the second in 1966, and received considerable acclaim. It may fairly be said to have been significant to the further development of study on the subject.

Feller was also keenly interested in limit theory, in which he first formulated the law of the iterated logarithm; he made a real contribution to the study of the central limit theorem.

Ferguson, Henry George (1884–1960) Northern Irish inventor and engineer who developed automatic draught control for farm tractors. This improved performance so dramatically that farmers worldwide could buy the small, inexpensive machines, thus precipitating a revolution in farming methods. Virtually every modern tractor incorporates some form of automatic draught control.

Ferguson was born near Growell, near Belfast. He left school at 14 to work on his father's farm. In 1902 he joined his brother in a car and cycle repair business in Belfast. He started to build his own aeroplane in 1908 and flew it on 31 December 1909, becoming one of the first Britons to do so.

He started his own motor business, imported tractors from the USA, and in 1917 he and William Sands designed their first plough. Another went into production in the USA, and on 12 February 1925 Ferguson patented the principle of draught contol. The Brown–Ferguson tractor, manufactured by David Brown, was launched in May 1936 and a redesigned version, built by Henry ◊Ford in the USA, was launched in June 1939.

In 1946, with British government backing, the famous TE20 Ferguson tractor, made by the Standard Motor Company in Coventry, was launched. In the USA, Ferguson and Ford fought a massive anti-trust suit, largely over a similar machine produced by Ford. Ferguson set up his own US plant, the first machine coming off the line on 11 October 1948.

After selling out to Massey–Harris in 1953, Ferguson entered into various negotiations and tried to interest motor manufacturers in a revolutionary design of car produced by Harry Ferguson Research. These attempts came to little, and Ferguson died at his home at Abbotswood, Stow-on-the-Wold, on 25 October 1960.

Tractive effort on smaller farms around the turn of the century was provided almost exclusively by draught animals. In addition to ploughs and harrows, there were horse-drawn binders and mowers and some barn machinery was driven by oil engines. Mechanized ploughing was carried out by traction engines, such as those produced by Fowler, one at each end of a field pulling a balance plough between them. This and other mechanized equipment was generally affordable only by large landowners.

In the 1880s two mobile steam engines for the farm appeared. They were small by the standards of the time but weighed 2 tonnes. Both were expensive, however, and farmers that could afford them preferred the steam-driven tackle. In Canada 10-tonne traction engines were being used. These heavy machines compacted the soil and produced a 'pan' that prevented proper drainage and inhibited root formation.

Lighter, petrol-engined tractors were being made by 1910, but they were simply mechanical horses used to pull trailed equipment. Driving them was dangerous because, if a plough hit an obstruction, they tended to rear up and could crush the driver. For the first Ferguson tractor, Ferguson designed a plough that coupled to the back of the tractor with a duplex hitch. This overcame the rearing problem, but there was little transfer of plough draught to the rear wheels. It was the principle of draught control that was the complete answer to the problem. The Ferguson System plough was built with two hitching points at about the normal level and a third hitch about 1 m/3 ft above the ground, connecting via a top link to a point about the same height on the tractor. When the plough hit an obstruction, the top link went into compression and tended to push the front of the tractor on to the ground instead of the reverse.

Ferguson incorporated a hydraulic system in the tractor for raising or lowering the plough out of or into work. This system was not new but using it to keep the plough at the correct working depth without the use of a depth wheel was a revolutionary step. The compression in the top link operated a sleeve valve in the hydraulic system, admitting varying amounts of oil to the pump. When the plough bit deep, it tended to tilt forwards, putting the top link in compression and opening the valve. This allowed more oil to be pumped into the hydraulic cylinder.

The cylinder raised the two bottom links, keeping the plough at an even depth. Thus the weight of the furrow slice was transferred to the tractor's back wheels, enabling it to exert greater effective traction. If the plough had a depth wheel, the draught was transferred to this wheel and not to the tractor's wheels. Ferguson's system meant that expensive, heavy machines were no longer necessary.

Ferguson, Margaret Clay (1863–1951) US botanist who made important contributions to the study of plant genetics, and as a teacher and administrator inspired generations of undergraduate students at Wellesley College.

She was born on 20 August 1863 in Orleans, New York State, the fourth of six children. She attended local schools and began teaching children herself when she was 14, whilst also studying at the local Wesleyan seminary. She attended Wellesley College 1888–91 as a special student studying chemistry and botany, and then returned to schoolteaching in Ohio. In 1893 the head of the department of botany at Wellesley, Professor Susan Hallowell, asked her to return as an instuctor in botany. In 1897 she decided to complete her formal education at Cornell University and gained her bachelor's degree in 1899 and her doctorate in 1901. She then returned to Wellesley, becoming head of the department of botany in 1902 and full professor in 1906, and remained there until 1938. She spent her retirement in New York State and Florida, and moved to California in 1946. She died in San Diego on 28 August 1951.

Ferguson proved to be a tireless teacher and administrator, emphasizing in particular the importance of practical experimental work. She supervised the building of new laboratory accommodation, and extensively developed the college herbarium and library facilities. Significantly she also designed and organized the erection of college greenhouses and associated laboratories in which students performed experiments in plant genetics, horticulture, and plant physiology; after her retirement these buildings were named after her. Ferguson's department became a major centre for botanical education and she has been credited with training more professional botanists at Wellesley than any other botanist. One of the most distinguished pupils to emerge from Wellesley was the Nobel prizewinner Barbara ◊McClintock.

Ferguson's doctoral research at Cornell had focused on the life history and reproductive physiology of a species of North American pine, which was published by the Washington Academy of Sciences. It was an innovative study of the functional morphology and cytology of a native pine and served as a model for such work for many years. Her research interests shifted, as did her teaching interests, towards plant genetics during the 1920s. She made a particular study of the genetics of *Petunia* emphasizing its use as a tool for studying the transfer of heredity from generation to generation. She extensively analysed the inheritance of features such as petal colour, flower pattern, and pollen colour, and built up a major database of genetic information. She continued to publish on the genetics of *Petunia* as research professor at Wellesley from 1930 until her formal retirement in 1932, after which her work was supported by the National Research Council for a further six years. She received many important honours and awards, including election as the first woman president of the Botanical Society of America (1929).

Fermat, Pierre de (1601–1675) French lawyer and magistrate for whom mathematics was an absorbing hobby. He contributed greatly to the development of number theory, analytical geometry, and calculus; carried out important research in probability theory and in optics; and was at the same time a competent classical scholar. Yet it is thanks only to his letters to various scientists and theoreticians that many of his accomplishments did not vanish into obscurity.

Fermat French mathematician Pierre de Fermat, who developed the theories of probability and of numbers. Fermat left virtually no written records, and his work can be studied only from his correspondence with the philosopher Descartes and other contemporaries. *Mary Evans Picture Library*

Born on 20 August 1601 in Beaumont de Lomagne, Fermat obtained a classical education locally. Between the ages of 20 and 30 he was in Bordeaux, possibly at the University of Toulouse. It was not, however, until he was 30 that he gained his bachelor's degree in civil law from the University of Orléans, set up a legal practice in Toulouse, and became commissioner of requests for the local parliament. In that parliament he was gradually promoted, gaining the high rank of king's counsellor in 1648, an office he retained until 1665. In 1652, however, he suffered a severe attack of the plague after which he devoted much of his time to mathematics, being particularly concerned with reconstructing some of the missing texts of the ancient Greeks such as

◊Euclid and ◊Apollonius of Perga. Curiously, he refused to publish any of his achievements, which were considerable despite the occasionally eccentric style in which they were presented. In increasing isolation, therefore, from the rest of the European mathematical community, Fermat lived to an old age. He died in Castres on 12 January 1675.

While Fermat was in Bordeaux, he became fascinated by the work of the mathematician François ◊Viète; it was from then that most of his mathematical achievements were attained. And it was through Viète's influence that Fermat came to regard number theory as a 'lingua franca' between geometry and arithmetic, and went on to make many significant discoveries in the field. Himself responsible for the development of number theory as an independent branch of mathematics, Fermat's work on the theory was later revived by Leonhard ◊Euler and continued to stimulate further research well into the 19th century. In 1657 Fermat published a series of problems as challenges to other mathematicians, in the form of theorems to be proved. All of them have since been proved – including 'Fermat's last theorem', the proof of which eluded mathematicians for more than 300 years. The theorem states that there is no solution in whole integers to the equation:

$$xn + yn = zn$$

Where n is greater than 2. In 1993, Andrew Wiles, an English mathematician at Princeton University, USA, announced a proof; this turned out to be premature, but he put forward a revised proof in 1994. Fermat's last theorem was finally laid to rest in June 1997 when Wiles collected the Wolfskehl prize (the legacy bequeathed in the 19th century for the problem's solution).

Fermat's technique in much of his work was 'reduction analysis', a reversible process in which a particular problem is 'reduced' until it can be seen to be part of a group of problems for which solutions are already known. Using this procedure, Fermat turned his attention to geometry. Unfortunately, analytical geometry was developed simultaneously both by Fermat (in letters written before 1636) and by the great René ◊Descartes (who published his *Géométrie* in 1637). There followed a protracted and bitter dispute over priority. The discipline permitted the use of equations to describe geometric figures, and Fermat demonstrated that a second-degree equation could be used to describe seven 'irreducible forms', each of which gave complete descriptions for different curves (such as parabolas and ellipses). He tried to extend this system into three dimensions to describe solids (in 1643), but was unsuccessful in the attempt beyond the establishment (in 1650) of the algebraic foundation for solid analytical geometry.

In 1636 he turned to the concept of 'infinitesimals' and applied it to equations of quadrature, the determination of the maxima and minima of curves, and the method of finding the tangent to a curve. All his work in these fields was superseded within 50 years through the development of calculus by Isaac ◊Newton and Gottfried ◊Leibniz; Newton did, however, acknowledge the importance of Fermat's work in the evolution of his own ideas.

Correspondence between Fermat and Blaise ◊Pascal resulted in the foundation of probability theory. Their joint conclusion was that if the probability of two independent events is respectively p and q, the probability of both occurring is pq.

In the field of optics, yet another disagreement with Descartes – this time on his law of refraction – led Fermat to investigate it mathematically. Ultimately obliged to confirm the law – but incidentally discovering the fact that light travels more slowly through denser mediums – Fermat also derived what is now known as 'Fermat's principle' – which states that light travels by the path of least duration – after making a study of the transmission of light through materials with different refractive indices.

Further Reading

Mahoney, Michael Sean, *The Mathematical Career of Pierre de Fermat, 1601–1665,* Princeton University Press, 1994.

Sakmar, Ismail A, *The Last Theorem of Pierre Fermat: A Study,* 1994.

Scharlau, W and Opolka, H, *From Fermat to Minkowski: Lectures on the Theory of Numbers and Its Historical Development,* Undergraduate Studies in Mathematics series, Springer-Verlag, 1984.

Singh, Simon, *Fermat's Last Theorem: The Story of a Riddle That Confounded the World's Greatest Minds for 358 Years,* Fourth Estate, 1998.

Fermi, Enrico (1901–1954) Italian-born US physicist best known for bringing about the first controlled chain reaction (in a nuclear reactor) and for his part in the development of the atomic bomb. He also carried out early research using slow neutrons to produce new radioactive elements, for which work he was awarded the 1938 Nobel Prize for Physics.

Fermi was born in Rome on 19 September 1901, the son of a government official. He was educated at the select Reale Scuola Normale Superior in Pisa (which he attended from 1918) and went on to the University of Pisa, receiving his PhD in 1929 for a thesis on X-rays. He then travelled to Göttingen University, where he worked under Max ◊Born, and to Leiden University, where he studied with P Ehrenfest. He became a mathematics lecturer at the University of Florence in 1924, and two years later he was appointed professor of

Fermi Italian-born US physicist Enrico Fermi who won the Nobel Prize for Physics in 1938 for his work on radioactivity. In 1942 Fermi built the first experimental controlled nuclear reactor, the forerunner of all nuclear power stations and nuclear weapons. He constructed it in the squash court of the University of Chicago. *Mary Evans Picture Library*

theoretical physics at Rome University. Fermi married a Jewish woman in 1928 and during the 1930s became alarmed by increasing antisemitism in Fascist Italy under Benito Mussolini. After the Nobel prize ceremony in Stockholm in 1938 Fermi did not return to Italy but went with his wife and two children to the USA, where he took up an appointment in New York at Columbia University. In 1941 he and his team moved to Chicago University where he began building a nuclear reactor, which first went 'critical' at the end of 1942. He became involved in the Manhattan Project to construct an atomic bomb, working mainly at Los Alamos, New Mexico. At the end of World War II in 1945 Fermi became a US citizen and returned to Chicago to continue his researches as professor of physics. He died there, of cancer, on 28 November 1954.

Fermi first gained fame soon after his Rome appointment with his publication *Introduzione alla fisica atomica* (1928), the first textbook on modern physics to be published in Italy. His experimental work on beta decay in radioactive materials provided further evidence for the existence of the neutrino (as predicted by Wolfgang ◊Pauli) and earned him an international reputation. The decay, which takes place in the unstable nuclei of radioactive elements, results from the conversion of a neutron into a proton, an electron (beta particle), and an antineutrino.

Following the work of Irène and Frédéric ◊Joliot-Curie, who discovered artificial radioactivity in 1934 using alpha-particle bombardment, Fermi began producing new radioactive isotopes by neutron bombardment. He found that a block of paraffin wax or a jacket of water round the neutron source produced slow, or 'thermal', neutrons that are more effective at producing such elements. This was the work that earned him the Nobel prize. He did, however, misinterpret the results of experiments involving neutron bombardment of uranium, and it was left to Lise ◊Meitner and Otto ◊Frisch in Sweden to explain nuclear fission in 1938.

If I could remember the names of all these particles I'd be a botanist.

Enrico Fermi quoted in R L Weber

More Random Walks in Science

In the USA Fermi continued the work on the fission of uranium (initiated by neutrons) by building the first nuclear reactor, then called an atomic pile because it had a moderator consisting of a pile of purified graphite blocks (to slow the neutrons) with holes drilled in them to take rods of enriched uranium. Other neutron-absorbing rods of cadmium, called control rods, could be lowered into or withdrawn from the pile to limit the number of slow neutrons available to initiate the fission of uranium. The reactor was built on the squash court of Chicago University, and on the afternoon of 2 December 1942 the control rods were withdrawn for the first time and the reactor began to work, using a self-sustaining nuclear chain reaction. Two years later the USA, through a team led by Arthur ◊Compton and Fermi, had constructed an atomic bomb, which used the same reaction but without control, resulting in a nuclear explosion.

Element number 100 – discovered in 1955, a year after Fermi died – was named fermium, and his name is also honoured in the fermi, a unit of length equal to 10^{-15} m.

Further Reading

Cooper, Dan, *Enrico Fermi and the Revolutions of Modern Physics,* Oxford Portraits in Science series, Oxford University Press, 1998.

Fermi, Laura, *Atoms in the Family: My Life With Enrico Fermi,* History of Modern Physics and Astronomy series, American Institute of Physics, 1987.

Gottfried, Ted, *Enrico Fermi,* Makers of Modern Science series, Facts On File, 1993.

Latil, Pierre de, *Enrico Fermi: The Man and His Theories,* A Profile in Science series, Souvenir Press, 1965.

Segrè, Emilio, *Enrico Fermi: Physicist,* University of Chicago Press, 1970.

Ferranti, Sebastian Ziani de (1864–1930) English electrical engineer who pioneered the high-voltage AC electricity generating and distribution system still used by most power networks. He also designed, constructed, and experimented with many other electrical and mechanical devices, including high-tension cables, circuit breakers, transformers, turbines, and spinning machines.

Ferranti was born in Liverpool on 9 April 1864. From his youth he was fascinated by machines and the principles by which they operate. After moving south he attended St Augustine's College in Ramsgate, Kent, and so impressed his teachers with his mechanical ideas that they set aside a room in which he could experiment. During this time he constructed an electrical generator. He left school in 1881 and took a job at the Siemens works in Charlton near London. He discovered that he could rotate and therefore mix the molten steel in a Siemens furnace by applying an electric current, and within a year he was supervising the installation of electric lighting systems for the company. A year after that – still only 18 years old – he was engineer to his own company, which designed and manufactured the Thompson–Ferranti alternator and installed lighting systems. The company was formed in partnership with Lord ◊Kelvin (Joseph Thompson) and a solicitor named Ince.

In 1886 Ferranti became engineer to the Grosvenor Gallery Company in London. The gallery had its own electricity generating system for lighting, and was also selling electricity to outside customers. Ferranti modified the system considerably to meet extra demand and, realizing the business potential, led the Grosvenor Company into the formation of a separate enterprise, the London Electric Supply Corporation Ltd, and suggested building a large generating station at Deptford. Extending an electricity supply to such a large area would, he argued, eventually become more economical and practicable than hundreds of small electrical enterprises serving limited areas. Most of the small systems used direct current of 200–400 volts together with storage batteries (accumulators), and were suitable only over short distances and when the demand for electricity fell within suitable limits. To achieve large-scale distribution, Ferranti proposed using alternating current (AC) at 10,000 volts, which was fed by mains to London, where step-down transformers reduced it to a voltage suitable for its purpose. The idea was revolutionary, because electricity at more than 2,000 volts was considered to be extremely dangerous. The cable for the mains was made to Ferranti's design, and produced in 6 m/20 ft lengths that were spliced together without the use of solder. The Deptford power station and its associated distribution network became the basic model for the future of electricity generation and supply.

In 1888 Ferranti married Gertrude Ince, the daughter of his solicitor partner. Three years later he left the London Electric Supply Corporation and concentrated on work as a consultant and on the development of his own company. This firm went on to design and build all kinds of electrical equipment, most of which was designed by Ferranti himself. He was also involved with heat engines of various kinds, turbines, cotton-spinning machines, and, during World War I, the design and manufacture of steel casings.

He became president of the Institution of Electrical Engineers in 1911 and a year later was awarded a DSc degree by the University of Manchester. In 1929 he was elected a fellow of the Royal Society. He enjoyed motoring and was very proud of his fast journey times. He died after an illness while on holiday in Zürich, Switzerland, on 13 January 1930.

Fessenden, Reginald Aubrey (1866–1932) Canadian physicist whose invention of radio-wave modulation paved the way for modern radio communication.

Born in East Bolton, Québec, on 6 October 1866, Fessenden studied at Trinity College School, Port Hope, Ontario, and Bishop's University at Lennoxville, Québec. It was only after he had moved to Bermuda, to take up the position of principal at the Whitney Institute there, that his interest in science really came alive. With little opportunity to follow up such interests in Bermuda, he left to go to New York, where he met Thomas ◊Edison. By his early twenties, Fessenden had become the chief chemist at Edison's laboratories at Orange, New Jersey.

In 1890 he left to join Edison's great rival, the Westinghouse Electric and Manufacturing Company, where he stayed for two years before returning to academic life, first at Purdue University, Lafayette, as professor of electrical engineering, and then at the Western University of Pennsylvania (now the University of Pittsburgh). It was there that Fessenden began major work on the problems of radio communication.

By 1900 he had overcome some major technical difficulties and succeeded in transmitting speech by radio, taking out the first patent for voice transmission the following year. Two key inventions of Fessenden's turned his 1900 experiment into the forerunner of modern radio communication. The first was that of modulation. In the early days of radio, experimenters such as Guglielmo ◊Marconi had sent messages using short bursts of signals to mimic Morse code. Fessenden realized that the amplitude of a continuous radio wave could be varied to mimic the variations of more complex wave patterns. By converting sound waves into variations of amplitude, it would therefore be possible to transmit voices and music. This is the principle of amplitude modulation.

Fessenden's other major invention was that of the heterodyne effect. In this, the received radio wave is combined with a wave of frequency slightly different to that of the carrier wave. The resulting intermediate frequency wave is easier to amplify before being demodulated to generate the original sound wave.

Two years after his initial broadcast, Fessenden organized the building of a 50-kHz alternator for radiotelephony by the General Electric Company. This was followed by his building a transmitting station at Brant Rock, Massachussetts. On Christmas Eve, 1906, the first amplitude-modulated radio message was broadcast; both words and music were transmitted. In the same year he established two-way radio communication between Brant Rock and Scotland.

By his death in Bermuda on 22 July 1932, Fessenden held 500 patents, a figure surpassed only by his former employer Edison. Amongst his patents are those for the sonic depth finder, the loop-antenna radio compass, and submarine signalling devices.

Feynman US physicist Richard Feynman, noted for the major theoretical advances he made in quantum electrodynamics. Feynman began working on the Manhattan Project while at Princeton University and then worked at Los Alamos 1943–46, on the development of the first atomic bomb. *Californian Institute of Technology*

Feynman, Richard Phillips (1918–1988) US physicist who shared the 1965 Nobel prize for his role in the development of the theory of quantum electrodynamics. He also made important contributions to the theory of quarks and superfluidity, was a noted teacher, and a self-styled 'curious character'. Feynman diagrams have become a standard way of representing particle interactions.

Feynman was born on 11 May 1918 in New York City. As a child he was fascinated by mathematics and electronics and became known as 'the boy who fixes radios by thinking'. He graduated from the Massachusetts Institute of Technology in 1939 and obtained a PhD from Princeton University in 1942. His supervisor was John Wheeler and his thesis, 'A Principle of Least Action in Quantum Mechanics', was typical of his first-principles approach to fundamental problems. During World War II Feynman worked at Los Alamos and was in charge of a group responsible for 'diffusion problems'. This involved large-scale computations (in the days before computers) to predict the behaviour of neutrons in atomic explosions. Working on the bomb, however, did not dampen his enthusiasm for practical jokes. Sadly, Feynman's first wife, Arlene, died of tuberculosis whilst he was working in Los Alamos.

After the war Feynman moved to Cornell University, where Hans ◊Bethe, who had also been at Los Alamos, was building up an impressive school of theoretical physicists. He continued developing his own approach to quantum electrodynamics (QED) before moving to California Institute of Technology (Caltech) in 1950. Feynman shared the Nobel prize with Julian Schwinger (1918–1984) and Sin-Itiro Tomonaga (1906–1979) for his work on QED. All three had independently developed methods for calculating the interaction between electrons, positrons (anti-electrons), and photons (light). The three approaches were fundamentally the same and QED remains the most accurate physical theory known. In Feynman's 'space–time' approach, different physical processes were represented as collections of diagrams showing how the particles moved from one space–time point to another. Feynman had rules for calculating the probability associated with each diagram, which he added to give the probability of the physical process itself.

Although Feynman only wrote 37 research papers in his career, a remarkably small number for such a prolific researcher, many consider that two discoveries he made at Caltech were also worthy of the Nobel prize. The first is the theory of superfluidity (frictionless flow) in liquid helium developed in the early 1950s; the second is Feynman's work on the weak interaction (with Murray ◊Gell-Mann) and the strong force, and his prediction that the proton and neutron are not elementary particles. Both particles are now known to be composed of quarks.

One does not, by knowing all the physical laws as we know them today, immediately obtain an understanding of anything much.

Richard Feynman The Character of Physical Law

A series of undergraduate lectures given by Feynman at Caltech, *The Feynman Lectures on Physics* (three volumes with R Leighton and R Sands; 1963), quickly became standard reference in physics. At the front of the lectures Feynman is shown indulging in one of his favourite pastimes, playing the bongo drum. Painting was another hobby. Feynman's fame increased further in 1986 when he was appointed to the commission to investigate the explosion on board the *Challenger* space shuttle in which seven astronauts died. In front of television cameras, he demonstrated how the failure of a simple rubber 'O-ring' seal, caused by the cold, was responsible for the disaster. His experiences on the commission feature heavily in his 'further adventures of a curious character' – *What Do You Care What Other People Think?* (1993). Feynman died in Los Angeles on 15 February 1988.

Further Reading

Feynman, Richard Phillips and Leighton, Ralph, *What Do You Care What Other People Think?: Further Adventures of a Curious Character,* Harper Collins, 1993.

Gleick, James, *Feynman,* Viking Penguin, 1999.

Gleick, James, *Genius: The Life and Science of Richard Feynman,* Pantheon Books, 1992.

Gleick, James, *Genius: Richard Feynman and Modern Physics,* Little Brown, 1992.

Goodstein, David L, *Feynman's Lost Lecture: The Motion of Planets Around the Sun,* Vintage, 1997.

Gribbon, John and Gribbon, Mary, *Richard Feynman: A Life in Science,* Penguin Books, 1998.

Hey, Anthony J (ed), *Feynman and Computation: Exploring the Limits of Computers,* Perseus Books, 1998.

Leighton, Ralph, *Tuva or Bust!: Richard Feynman's Last Journey,* Penguin, 1993.

Mehra, Jagdish, *The Beat of a Different Drum: The Life and Science of Richard Feynman,* Clarendon Press, 1996.

Fibonacci, Leonardo (c. 1180–c. 1250) Leonardo of Pisa. Italian mathematician whose writings were influential in introducing and popularizing the Indo-Arabic numeral system, and whose work in algebra, geometry, and theoretical mathematics was far in advance of the contemporary European standards.

Fibonacci was born in Pisa in about 1180, the son of a member of the government of the republic of Pisa. When Fibonacci was 12 years old, his father was made administrator of Pisa's trading colony in Algeria, and it was there – in a town now called Bougie – that he was taught the art of calculating, using the commercial North African medium of Indo-Arabic numerals. His teacher, who remains completely unknown, seems to have imparted to him not only an excellently practical and well-rounded fundamental grounding in mathematics, but also a true scientific curiosity.

Having achieved maturity, Fibonacci travelled extensively, both for business and for pleasure, spending time in Italy, Syria, Egypt, Greece, and elsewhere. Wherever he went he observed and analysed the arithmetical systems used in local commerce, studying through discussion and argument with the native scholars of the countries he visited. He returned to Pisa in about the year 1200 and began his mathematical writings. Little more is known of him, although in 1225 he won a mathematical tournament in the presence of the Holy Roman Emperor Frederick II at the court of Pisa. A marble tablet dated 1240 appears to refer to him as having been awarded an annual pension following his valuable accountancy services to the state. He is assumed to have died in Pisa in about 1250.

Two years after finally settling in Pisa, Fibonacci produced his most famous book, *Liber abaci/The Book of the Calculator.* In four parts, and revised by him a quarter of a century later (in 1228), it was a thorough treatise on algebraic methods and problems in which he strongly advocated the introduction of the Indo-Arabic numeral system, comprising the figures 1 to 9, and the innovation of the 'zephirum' – the figure 0 (zero). Dealing with operations in whole numbers systematically, he also proposed the idea of a bar (solidus) for fractions, and went on to develop rules for converting fraction factors into the sum of unit factors. (However, his expression of fractions followed the Arabic practice – on the *left* of the relevant integral.) At the end of the first part of the book, he presented tables for multiplication, prime numbers, and factoring numbers. In the second part, he demonstrated mathematical applications to commercial transactions. In part three he gave many examples of recreational mathematical problems of the type enjoyed today, leading up to a thesis on series from which, in turn, he derived what is now called the Fibonacci series. This is a sequence in which each term after the first two is the sum of the two terms immediately preceding it – 1, 1, 2, 3, 5, 8, 13, 21, ... for example – and which has been found to have many significant and interesting properties. And in the final part of the book Fibonacci, a student of Euclid, applied the algebraic method. The *Liber abaci* remained a standard text for the next two centuries.

In 1220 he published *Practica geometriae,* a book on geometry that was of fundamental significance to future studies of the subject, and that (to some commentators, at least) seems to be based on a work of Euclid now lost. In it, Fibonacci used algebraic methods to solve many arithmetical and geometrical problems. In *Flos/Flower*, published four years later, he considered indeterminate problems in a way that had not properly been carried out since the work of ◊Diophantus in the 2nd century AD, and again demonstrated Euclidean methodology combined with techniques of Chinese and Arabic origin (learned during his travels many years before) in solving determinate problems. In both *Liber quadratorum/The Book of Squares* and in a separate letter to the philosopher Theodorus, Fibonacci dealt with some problems set by John of Palermo (one of which was the one he solved in front of the emperor); his treatments show unusual mathematical skill and originality.

The complete works of Fibonacci were edited in the 19th century by B Boncompagni, and published in two volumes under the title *Scritti di Leonardo Pisano.*

Field, George Brooks (1929–) US theoretical astrophysicist whose main research has been into the nature and composition of intergalactic matter and the properties of residual radiation in space.

Field was born in Providence, Rhode Island, on 25 October 1929. Educated at the Massachusetts Institute of Technology, he graduated in 1951 and four years later gained a PhD in astronomy at Princeton University. In 1957 he was appointed assistant professor in astronomy at Princeton and he progressed to become associate professor. He was made professor of the University of California at Berkeley in 1965 and was chair of the department 1970–71. In 1972 he became professor of astronomy at Harvard University and from 1973 he was also director of the Center of Astrophysics at the Harvard College Observatory and the Smithsonian Astrophysical Observatory.

One of Field's major areas of research has been to investigate why a cluster of galaxies remains a cluster. It seems evident that such clusters ought to be rapidly dispersing unless they are stabilized in some way, presumably gravitationally by intergalactic matter that contributes from ten to thirty times more material than the galaxies themselves. Such matter has never been detected, although considerable research has been undertaken and is still in progress. From a consideration of the composition of galaxies in clusters, it would seem probable that the most likely substance of such intergalactic matter would be in the form of hydrogen, that around 27% by mass would be helium, and that a negligible fraction would be in the form of heavier elements. There is at present no means of detecting helium in intergalactic space – but intergalactic hydrogen does produce effects that are potentially detectable, and this is the work in which Field has been particularly involved.

Atomic hydrogen distributed intergalactically (in contrast to ionized hydrogen) would act both as an absorber and an emitter of radiation at a wavelength of 21 cm/8.3 in. Field first tried to find evidence of this absorption in 1958. He studied the spectrum of the radio source Cygnus A – the brightest extragalactic radio source in the sky – in the region of 21 cm/8.3 in, taking into account the known red shift associated with the expansion of the universe. The narrow range of wavelengths over which the intergalactic hydrogen would absorb is called the 'absorption trough'. A wavelength of 21 cm/8.3 in is remarkably long for an atom to absorb or emit, and from the point of view of the two energy levels involved, the hydrogen is immersed in a heat bath at an extremely high temperature – so that there are nearly as many atoms in the upper state as in the lower, and the absorption trough cannot be very exactly observed. Field's later results have given greater precision.

Field has also carried out research into the spectral lines in the spectra of stars. In the 1930s, A McKellar found absorption lines corresponding to interstellar cyanogen (CN) in the spectra of several stars. For the star Zeta Ophiuci, McKellar was able to obtain a measure of the relative number of molecules in the ground state and the first rotational state, which he defined in terms of an excitation temperature (the temperature at which the molecules would possess the observed degree of excitation if they were in equilibrium in a heat bath). At that time, however, excitation was assumed to occur only by collisions with other particles or by radiation with no thermal spectrum. But in 1966, microwave measurements of the cosmic background were made that suggested that this background has a black-body spectrum at all wavelengths. Field then wondered if the CN 'molecules' might after all be in a heat bath. He re-observed the spectrum of Zeta Ophiuci, and obtained an excitation temperature of 3.22 ± 0.15K, which corresponded well with the value of 2.3K (–270.8 °C/–454.9 °F) determined by McKellar. A number of other stars have since been analysed and all of them yield a temperature of about 3K (–270 °C/ –454 °F). There remains much more work on this subject to be done.

Fischer, Emil (1852–1919) German organic chemist who analysed and synthesized many biologically important compounds. He was awarded the 1902 Nobel Prize for Chemistry for his work on the synthesis of sugars and purine compounds.

Fischer was born in Euskirchen, near Bonn, on 9 October 1852, the son of a merchant. After leaving school he acceded to his father's wishes and joined the family business, but later left and in 1871 entered the

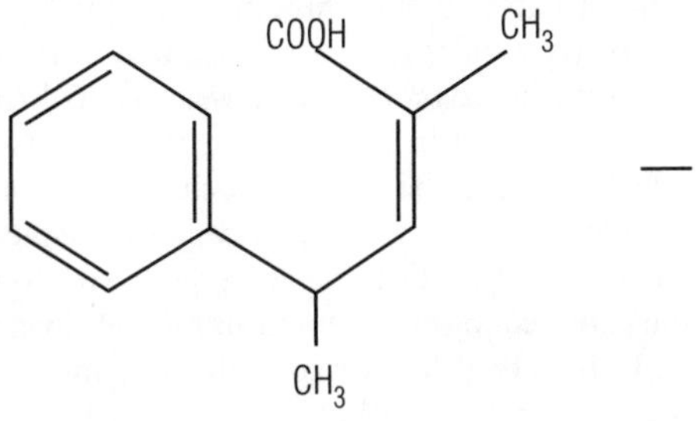

COOH CH_3

CH_3

derivative of 1-methylphenylhydrazone

Fischer Fischer's indole synthesis.

University of Bonn to study chemistry under Friedrich ◊Kekulé von Stradonitz. The following year he went to Strasbourg and graduated from the university there in 1874 with a doctoral thesis which was supervised by Johann von ◊Baeyer. He continued his studies at Munich where he became an unpaid lecturer in 1878 and a (paid) assistant professor in 1879. He then held professorships at Erlangen from 1882, Würzburg from 1885, and finally Berlin from 1892. Before his last move he married Agnes Gerlach. His wife died young but they had three sons, the eldest of whom, Hermann (see Hermann ◊Fischer), also became a distinguished organic chemist; two sons were killed in World War I. Fischer suffered a serious bout of mercury poisoning (during a brief incursion into inorganic chemistry) and the equally serious effects of phenylhydrazine poisoning. He contracted cancer and this fact, coupled with the death of his sons, led him to commit suicide on 15 July 1919. Fischer's early research, carried out with his cousin Otto Fischer, concerned the dye rosaniline and similar compounds, which they showed have a structure related to that of triphenylmethane. In 1875 Fischer discovered phenylhydrazine, but it was not until 1884 that he found it formed bright yellow crystalline derivatives with carbohydrates, a key reaction in the study of sugars. The derivatives are known as osazones, and Fischer obtained the same osazone from three different sugars – glucose, fructose, and mannose – demonstrating that all three have the same structure in the part of their molecules unaffected by phenylhydrazine. He went on to determine the structures of the 12 possible stereoisomers of glucose, the important group of sugars known collectively as hexoses. The naming of carbohydrates is a complicated process based on the chemical origin (D or L) and optical activity (+ or –) of the compound. Fischer based his nomenclature on dextrorotatory (+) glyceraldehyde and called this the D series. It was only the advent of X-ray analysis that confirmed that Fischer's arbitrarily assigned configurations are correct.

From about 1882 he began working on a group of compounds that included uric acid and caffeine. Fischer realized that they were all related to a hitherto unknown substance, which he called purine. Over the next few years he synthesized about 130 related compounds, one of which was the first synthetic nucleotide, a biologically important phosphoric ester of a compound made from a purine-type molecule and a carbohydrate. These studies led to the synthesis of powerful hypnotic drugs derived from barbituric acids, including in 1903 5,5-diethyl barbituric acid, which became widely used as a sedative.

In 1885 experiments with phenylhydrazine led to what is known as Fischer's indole synthesis, in which he heated a phenylhydrazone with an acid catalyst to produce a derivative of indole. Indole itself cannot, however, be obtained by this method.

Fischer's investigations into the chemistry of proteins began in 1899. He synthesized the amino acids ornithine (1,4-diaminopentanoic acid) in 1901, serine (1-hydroxy-2-aminobutanoic acid) in 1902, and the sulphur-containing cystine in 1908. He then combined amino acids to form polypeptides, the largest of which – composed of 18 amino-acid residues – had a molecular mass of 1,213. Later work included a study of tannins, which he carried out with the assistance of his son Hermann.

Fischer was involved with many aspects and branches of organic chemistry. He was a man of considerable insight, as exemplified by his description of the action of enzymes as a lock-and-key mechanism in which the enzyme model fits exactly onto the molecule with which it reacts. But he did not consider himself to be a theoretician; he believed in, and used, the synthetic methods of the practical organic chemist.

Fischer, Ernst Otto Fischer (1918–) German inorganic chemist who shared the 1973 Nobel Prize for Chemistry with English inorganic chemist Geoffrey Wilkinson for pioneering work (carried out independently) on the organometallic compounds of the transition metals.

Fischer was born in Munich on 10 November 1918, the son of Professor Karl T Fischer. He was educated at the Munich Technical University, from which he gained a diploma in chemistry in 1949 and a doctorate in 1952. He remained at Munich, becoming successively associate professor from 1957, professor from 1959, and finally, from 1964, professor and director of inorganic chemistry.

In about 1830 a Danish pharmacist described the compound $PtCl_2.C_2H_4$, which is now known to exist as a dimer with chloride bridges. This and the ion $[C_2H_4PtCl_3]^-$ were the first known organometallic derivatives of the transition metals. In 1951, both

Fischer and Wilkinson read an article in the journal *Nature* about a puzzling synthetic compound called ferrocene. Working independently, they came to the conclusion that each molecule of ferrocene consists of a single iron atom sandwiched between two five-sided carbon rings – an organometallic compound. A combination of chemical and physical studies, finally confirmed by X-ray analysis, revealed the compound's structure. In the ferrocene molecule, the two symmetrical five-membered rings are staggered with respect to each other, but in the corresponding ruthenium compound (called ruthenocene) they are eclipsed.

With this work came the general realization that transition metals can bond chemically to carbon, and other ring systems were then studied. The hydrocarbon cyclopentadiene behaves as a weak acid and with various bases forms salts containing the symmetrical cyclopentadienide ion $C_5H_5^-$. It also forms 'sandwich' compounds and, like other ring systems that behave in this way, has the 'aromatic' sextet of (six) electrons. All of the elements of the first transition series have now been incorporated into molecules of this kind and all except that of manganese have the ferrocene-type structure. Only ferrocene, however, is stable in air, the others being sensitive to oxidation in the order (of decreasing stability) nickel, cobalt, vanadium, chromium, and titanium. The cationic species behaves like a large monopositive ion; $(\pi\text{-}C_5H_5)Co^+$ is particularly stable.

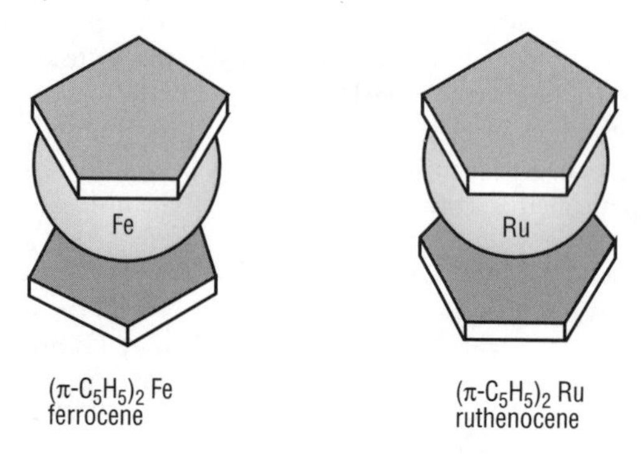

Fischer Structures of ferrocene and ruthenocene.

The recognition of the 'sandwich' concept initiated a vast amount of research, not only on cyclopentadienyl derivatives but on similar systems with four-, six-, seven-, and even eight-membered carbon rings. All of this work was stimulated by the revolutionary explanation by Fischer and Wilkinson of the previously unknown way in which metals and organic compounds can combine.

Fischer, Hans (1881–1945) German organic chemist who is best known for his determinations of the molecular structures of three important biological pigments: haemoglobin, chlorophyll, and bilirubin. For his work on haemoglobin he was awarded the 1930 Nobel Prize for Chemistry.

Fischer was born in Höchst-am-Main, near Frankfurt, on 27 July, 1881, the son of a chemical manufacturer. He studied chemistry at the University of Marburg, gaining his doctorate in 1904. He then went to the University of Munich to study medicine, qualifying as a doctor in 1908. He became a research assistant to his namesake Emil ◊Fischer at the University of Berlin, before taking up an appointment as professor of medical chemistry at the University of Innsbruck in 1915 as successor to Adolf Windaus (1876–1959). Three years later he held a similar post at Vienna, before taking over from Heinrich ◊Wieland at the Munich Technische Hochschule in 1921 as professor of organic chemistry. Towards the end of World War II in 1945, Fischer's laboratories were destroyed in an Allied bombing raid and in a fit of despair, like Emil Fischer before him, he committed suicide. He died in Munich on 31 March 1945.

In 1921 Fischer began investigating haemoglobin, the oxygen-carrying, red colouring matter in blood. He concentrated on haem, the iron-containing non-protein part of the molecule, and showed that it consists of four pyrrole rings (five-membered heterocyclic rings containing four carbon atoms and one nitrogen atom) surrounding a single iron atom. By 1929 he had elucidated the complete structure and synthesized haem.

He then turned his attention to chlorophyll, the green colouring matter in plants that Richard ◊Willstätter had isolated in 1910. He found that its structure is similar to that of haem, with a group of substituted porphins surrounding an atom of magnesium. This work occupied Fischer for much of the 1930s, after which he began to study the bile pigments, particularly bilirubin (the pigment responsible for the colour of the skin of patients suffering from jaundice). He showed that the bile acids are degraded porphins and by 1944 had achieved a complete synthesis of bilirubin. He also investigated the yellow plant pigment carotene, the precursor of vitamin A.

Fischer, Hermann Otto Laurenz (1888–1960) German organic chemist whose chief contribution to the science concerned the synthetic and structural chemistry of carbohydrates, glycerides, and inositols.

Fischer was born on 8 December 1888 in Würzburg, Bavaria, the eldest of the three sons of Emil Fischer, who at that time was professor of organic chemistry at the local university. The early death of his mother brought him and his two brothers into closer contact with their illustrious father. The two younger brothers went on to study medicine, while Hermann followed his father into organic chemistry.

Fischer began his undergraduate career at Cambridge University in 1907, but had to return to Germany the following year for military training. After a brief period in Berlin, he started his doctorial research work at Jena University under Ludwig Knorr. In 1912 he returned to the Chemical Institute of Berlin University to continue research with his father. Two years later the outbreak of World War I interrupted his research. Both his brothers were killed in the war and his father committed suicide in 1919, soon after Hermann returned to Berlin.

In 1922 Fischer married Ruth Seckels, and they had a daughter and two sons. With the rise of Adolf Hitler the Fischers left Berlin in 1932 and went to Basel in Switzerland. Then in 1937 he moved to the Banting Institute in Toronto, Canada, where he stayed until his final move to the USA in 1948, to the biochemistry department of the University of California at Berkeley. He became chair of the department and emeritus professor before retiring in 1956. Fischer died on 9 March 1960.

While working with Knorr at Jena, Fischer separated the keto and enol tautomers of acetyl-acetone (pentan-2,4-dione) by low-temperature crystallization. Then in Berlin his father (Emil Fischer), who was investigating tannins, assigned him the task of synthesizing some of the naturally occurring depsides (a depside is a condensation product formed from two hydroxy-aromatic acid molecules). He succeeded in producing various didepsides and diorsellinic acids.

Between 1920 and 1932 Fischer pursued two main lines of research. One was the study of quinic acid (tetrahydroxycyclohexanecarboxylic acid). By 1921 he had made various derivatives, but it was not until 1932 that he finally worked out its exact structure. The other research during this period dealt with the difficult chemistry of the trioses glyceraldehyde and dihydroxyacetone (2,3-dihydroxy-propanal and 1,3-dyhydroxy-propanone) and the related two-, three-, and four-carbon compounds. The crowning achievement of this work was the preparation by his assistant Erich Baer of DL-glyceraldehyde-3-phosphate.

Baer accompanied Fischer to Basel in 1932 and there they developed a practical method for the preparation of the enantiomorphous acetonated glyceraldehydes. They were able to make D-fructose and D-sorbose almost entirely free from their isomers D-psicose and D-tagatose, using the aldol reaction between unsubstituted D-glyceraldehyde and its ketonic isomer, dihydroxyacetone.

While at Basel Fischer continued to collaborate with Gerda Dangschat (who had remained at Berlin) on structural and configurational studies of shikimic acid (first isolated in 1885). By a series of degradation reactions they converted shikimic acid into 2-deoxy-D-arabino-hexonic acid, which finally located the position of the double bond in the former (proving it to be 3,4,5-trihydroxycyclohexene-1-carboxylic acid).

From 1937, at the Canadian Banting Institute, Fischer extended his work on glyceraldehydes to glycerides (esters of glycerol, or propan-1,2,3-triol). He prepared the first optically pure α-monoglycerides and α-glycerophosphoric acids (glycerol-1-phosphates) and demonstrated the action of lipase enzymes on these biologically important substances.

Despite the intervention of World War II and the distance between Berlin and Toronto, Fischer and Dangschat continued their work on the inositols (hexahydroxycyclohexanes). They succeeded in establishing the configuration of myoinositol and showed its relationship to D-glucose.

At Berkeley, Fischer carried on research into the inositols and other carbohydrates. He described the 12 years in California as the most pleasant in his life. His warm and friendly personality made him many friends there and among distinguished scientists throughout the world.

Fisher, Ronald Aylmer (1890–1962) English mathematical biologist whose work in the field of statistics resulted in the formulation of a methodology in which the analysis of results obtained using small samples produced interpretations that were objective and valid overall. His work revolutionized research methods in many areas, and found immediate and widespread use, particularly in genetics and agriculture.

Fisher was born in London on 17 February 1890. He attended Stanmore Park and Harrow schools before going to Gonville and Caius College, Cambridge, in 1909. He graduated in 1912, having specialized in mathematics and theoretical physics, and then spent an additional year at Cambridge studying statistical mechanics and researching into the theory of errors. The next six years were spent in various occupations – his poor eyesight made him unacceptable for military service during World War I – and he worked in an investment brokerage, as a teacher, and even as a farm labourer. At the end of the war, however, Fisher obtained a post at the Rothamstead Experimental Station, where he single-handedly ran a statistics department whose main job was to analyse a huge backlog of experimental data that had built up over more than 60 years. It was while at Rothamstead that he evolved many improvements to traditional statistical methods. His textbook on the subject, which appeared in 1925, was a landmark.

At Rothamstead, Fisher was able also to indulge in his second scientific passion, genetics: he bred poultry, mice, snails, and other creatures, and in his papers on the subject contributed to the contemporary understanding of genetic dominance. As a result, in 1933 he was appointed to the Galton Chair of Eugenics at University College, London. During World War II, however, his department was (ironically) evacuated to Rothamstead, and eventually disbanded. Fisher then became Balfour Professor of Genetics at the University of Cambridge in 1943. He was knighted nine years later. He officially retired from Cambridge in 1957, but stayed on until 1959, when a successor was found. Following a visit to the Mathematical Statistics Division of the Commonwealth Scientific and Industrial Research Organization in Adelaide, Fisher emigrated to Australia. He died in Adelaide on 29 July 1962.

Elected a fellow of the Royal Society in 1929, Fisher was awarded their Royal Medal in 1938, their Darwin Medal in 1948, and their Copley Medal in 1955.

Fisher's early work concerned the development of methods for the determination of the exact distributions of several statistical functions, such as the regression coefficient and the discriminant function. He improved the Helmut–Pearson χ^2 and Gosset's Z functions, modifying the latter to the now familiar t-test for significance. He evolved the rules for 'decision-making' that are now used almost automatically, and are based on the percentage deviation of the results of an experiment from the 'null hypothesis' (which assumes that events occur on an exclusively random basis). A deviation of 95–99% represented only a suggestive likelihood that the null hypothesis was incorrect; a deviation in excess of 99% indicated strongly that this was so.

Other statistical methods that Fisher originated include the analysis of variance, the analysis of covariance, multivariate analysis, contingency tables, and more. All his mathematical methods were developed for further application in fields such as genetics, evolution, and natural selection.

One of Fisher's first studies concerned the importance of dominant genes. A confirmed eugenicist, he looked on the study of human blood as an essential factor in his research. The department he established at the Galton Laboratories to investigate blood types made significant contributions to the final elucidation of the inheritance of Rhesus blood groups.

Fisher's methods have since been extended to virtually every academic field in which statistical analysis can be applied.

Fitch and Cronin, Val Logsdon Fitch (1923–) and James Watson Cronin (1931–) US physicists who shared the 1980 Nobel Prize for Physics for their work in particle physics.

Fitch was born in Merriman, Nebraska, on 10 March 1923. He was educated at McGill University, where he obtained his BEng in 1948. He gained a PhD from Columbia University in physics in 1954. From 1954 to 1960 Fitch rose from instructor to professor of physics at Princeton University, and in 1976 he became Cyrus Fogg Bracket Professor of Physics and head of department there. From 1984 he was James S McDonnel Distinguished University Professor. Fitch was a member of the US president's Scientific Advisory Committee 1970–73.

Cronin was born in Chicago, Illinois, on 29 September 1931. He was educated at the Southern Methodist University, obtaining a BS in 1951. He then gained an MS from the University of Chicago in 1953 and a PhD in physics in 1955. From 1955 to 1958, he was an assistant physicist at the Brookhaven National Laboratory. Cronin then went to Princeton University, becoming professor of physics in 1965. In 1971, he became professor of physics at the University of Chicago.

The discovery for which Fitch and Cronin received the 1980 Nobel prize for Physics was first published in 1964, and at that time was regarded as a bombshell in the field of particle physics. It is surprising that their work was not rewarded earlier.

Until 1964, it was not possible to distinguish unambiguously between matter and antimatter outside our own Galaxy and a few nearby galaxies. In 1964, Cronin, Fitch, and their colleagues had set up an experiment with the proton accelerator at the Brookhaven Laboratory in New York to study the properties of K^0 mesons. These are neutral, unstable particles with a mass approximately half that of a proton, and had been discovered earlier in the interactions of particles from outer space with the Earth's atmosphere. K^0 is a mixture of two 'basic states' that have a long and a short lifetime and are therefore called K^0_L and K^0_S, respectively. These two basic states can also mix together to form not K^0 but an antimatter particle (anti-K^0), and K^0 can oscillate from particle to antiparticle through either of its basic states. This is a unique phenomenon in the world of particle physics.

A rule called CP-conservation states how K^0_L and K^0_S should decay (C stands for conjugation and P for parity). Charge conjugation changes all particles to antiparticles. Parity concerns handedness and is like a mirror reflection – all positive points in a three-dimensional coordinate system (x, y, z) are changed to negative points $(-x, -y, -z)$ by a parity operation. The conservation rule means that these two operations when applied together do not alter an interaction, such as the decay of an unstable particle. What Cronin and Fitch found to their surprise was that a K^0_L meson can decay in such a way as to violate CP-conservation about 0.2% of the time. Their results were verified at other accelerators and many strange explanations were provided. The conclusion that had to be drawn was that decays of K^0_L mesons do violate CP-conservation and so are different from all other known particle interactions.

While it was relatively simple to confirm these results, it has proved much more difficult to explain them. The latest theories view particles as being built up from six basic different entities called quarks (evidence for all six of which had been discovered by 1995). Earlier theories based on four quarks did not explain CP-violation satisfactorily. CP-violation could explain why we exist at all. Scientists are puzzled why matter seems to dominate over antimatter, when all the theories suggest that matter and antimatter should have been formed in equal amounts. CP-violation could help to solve this problem, since if anti-particles decay faster than particles, they would totally disappear.

The work of Cronin and Fitch is recorded as a classic piece of research and their findings have had impact on an outstanding controversy about the symmetry of nature. They have shown for the first time that left–right asymmetry is not always preserved when some particles are changed in state from matter to antimatter.

FitzGerald, George Francis (1851–1901) Irish theoretical physicist who worked on the electromagnetic theory of light and radio waves. He is best known, however, for the theoretical phenomenon called the Lorentz–FitzGerald contraction which he and, independently, Hendrik ◊Lorentz proposed to account for the negative result of the famous Michelson–Morley experiment on the velocity of light.

FitzGerald was born in Dublin on 3 August 1851. He was a nephew of the physicist George Stoney, and received his initial education at home, tutored by the sister of the mathematician George ◊Boole. He entered Trinity College, Dublin, in 1867 (at the age of only 16) and graduated four years later. FitzGerald went on to study mathematical physics, and in 1877 received a fellowship of Trinity College. He became Erasmus Smith Professor of Natural and Experimental Philosophy there in 1881, an appointment he retained until he died, in Dublin, on 22 February 1901.

FitzGerald predicted that a rapidly oscillating (that is, alternating) electric current should result in the radiation of electromagnetic waves – a prediction proved correct in the late 1880s by Heinrich ◊Hertz's early experiments with radio, which FitzGerald brought to the attention of the scientific community in the UK.

At that time, one school of thought among physicists postulated that electromagnetic radiation – such as light – had to have a medium in which to travel. This hypothetical medium, termed the 'ether', was supposed to permeate all space and in 1887 the US physicists Albert ◊Michelson and Edward ◊Morley conducted an experiment intended to detect the motion of the Earth through the ether. They measured the velocity of light simultaneously in two directions at right angles to each other, but found no difference in the values – the velocity was unaffected by the Earth's motion through the ether. FitzGerald proposed in 1892 that the Michelson–Morley result – or lack of result – could be accounted for by assuming that a fast-moving object diminishes in length, and that light emitted by it does indeed have a different velocity but travels over a shorter path, and so seems to have a constant velocity no matter what the direction of motion. (Only an observer outside the moving system would be aware of the reduction in light velocity; within the system the contraction would also affect the measuring instruments and result in no change in the perceived velocity.) He worked out a simple mathematical relationship to show how velocity affects physical dimensions. The idea was independently arrived at and developed by the Dutch physicist Hendrik Lorentz in 1895, and it became known as the Lorentz–FitzGerald contraction. In 1905, four years after FitzGerald's death, the contraction hypothesis was incorporated and given a different interpretation in Albert ◊Einstein's general theory of relativity.

Fizeau, Armand Hippolyte Louis (1819–1896) French physicist who was the first to measure the velocity of light on the Earth's surface. He also found that light travels faster in air than in water, which confirmed the wave theory of light, and that the motion of a star affects the position of the lines in its spectrum.

Fizeau was born in Paris on 23 September 1819 into a wealthy family. He began to study medicine, his father being professor of pathology at the Paris Faculty of Medicine, but was forced to abandon his medical ambitions through ill health. Fizeau then turned to physics, studying optics at the College de France and taking a course with François ◊Arago at the Paris Observatory.

Fizeau's principal contributions to physics and astronomy cover a fairly short period of time and include the first detailed photographs of the Sun in 1845, the proposal that a moving light source such as a star undergoes a change in observed frequency that can be detected by a shift in its spectral lines in 1848, the first reasonably accurate determination of the speed of light in 1849, and the discovery that the velocity of light is greater in air than in water in 1850. Many of his discoveries were made in collaboration with Léon ◊Foucault.

In recognition of his work, Fizeau gained the first Triennial Prize of the Institut de France in 1856. He became a member of the Academy of Sciences in 1860, rising to become its president in 1878, and he was awarded the Rumford Medal of the Royal Society in 1866. Fizeau died in Venteuil on 18 September 1896.

Inspired by Arago, Fizeau began to research into the new science of photography in 1839, improving the daguerrotype process. The fruitful collaboration with Foucault commenced at this time, leading them to develop photography for astronomical observations by taking the first detailed pictures of the Sun's surface in 1845. The two men went on to study the interference of light rays and showed how it is related to the wavelength of the light. They also found, in 1847, that heat rays from the Sun undergo interference and that radiant heat therefore behaves as a wave motion.

These latter results greatly strengthened the view that light is a wave motion, an issue that was being hotly debated at the time following the experiments of Thomas ⌽Young and Augustin ⌽Fresnel that had established the wave theory earlier in the century. In 1838, Arago had proposed an experiment to settle the question by measuring the velocity of light in air and in water. The wave theory demanded that light travels faster in air and the particle theory required it to travel faster in water. Arago's proposal was to reflect light from a rotating mirror, the amount of deflection produced being related to the velocity of light.

In 1847, Fizeau and Foucault split up their partnership. Foucault persevered with Arago's suggestion but Fizeau decided to try a simpler method. In 1849, he sent a beam of light through the gaps in the teeth of a rapidly rotating cog wheel to a mirror 8 km/5 mi away. On returning, the beam was brought to the edge of the wheel, the speed being adjusted so that the light was obscured. This meant that light rays that had passed through the gaps were being blocked on their return by the adjacent teeth as they moved into the position of the gaps. The time taken for the teeth to move this distance was equal to the time taken for light travel 16 km/10 mi to the mirror and back. Fizeau's wheel had 720 teeth and rotated at a rate of 12.6 revolutions per second. By a simple calculation, he obtained the value of 315,000 km/195,741 mi a second for the speed of light. Although 5% too high, this was the first reasonably accurate estimate, previous values having been obtained by inaccurate astronomical methods.

In collaboration with Louis Bréguet (1804–1883), Fizeau now applied himself to Arago's proposal and set about measuring the speed of light in air and water. Using a rotating-mirror apparatus and dividing the light beam so that half passed through air and half through a tube of water, Fizeau showed in 1850 that light travels faster in air. Foucault reached the same conclusion in the same way at the same time, and the wave theory of light was finally confirmed.

Another important experiment conducted by Fizeau in 1851 was one in which he determined the amount of drift of light waves in a transparent medium which is in motion. According to a theory given by Fresnel, the velocity of drift of waves in a medium moving with velocity u is

$$\left(1 - \frac{1}{\eta^2}\right)u$$

Where η is the refractive index of the medium. Fizeau arranged two tubes side by side and water was forced at a considerable speed (as much as 7 m/23 ft per second) along one tube and back by the other, while a beam of light was split into two parts, which were sent through the tubes, one with the stream and the other against it. They were then brought together again and tested for interference produced by any difference in time traversed arising from the motion of the water. The result gave exactly the formula quoted above.

Earlier, in 1848, Fizeau made a fundamental contribution to astrophysics by suggesting that the Doppler effect would apply to the light received from stars, motion away from the Earth causing a red shift in the spectral lines and motion towards the Earth producing a blue shift. Fizeau may have been unaware of Christian ⌽Doppler's discovery of the effect in sound in 1842 and of his erroneous theory relating the colour of stars to their motion. Fizeau's discovery is now the basis of the principal method of determining the distances of galaxies and other distant bodies.

Fizeau was able to finance his experiments from his personal wealth, and this enabled him to develop experimental methods that were later refined by others, particularly Foucault and Albert ⌽Michelson, leading to accurate values for the velocity of light and the consequent development of relativity.

Flamsteed, John (1646–1719) English astronomer and writer who became the first Astronomer Royal based at Greenwich. His work on the stars, which formed the basis of modern star catalogues, was much admired by contemporary scientists, among whom were Isaac ⌽Newton and Edmond ⌽Halley. Like many professional scientists of his time, he was also a cleric.

Flamsteed was born on 19 August 1646 at Denby, near Derby, the son of a prosperous businessman. At the age of 16, ill-health forced him to leave Derby Free School and abandon, at least temporarily, his university ambitions. The next seven years he spent at home, educating himself in astronomy against the wishes of his father (who evidently saw him as his own successor in the family business). In 1670, however, he entered his name at Jesus College, Cambridge, and took an MA degree by letters-patent. He was ordained in the following year. Through the influence of Jonas Moore, a courtier of King Charles II, Flamsteed was made 'astronomical observator' at the newly established

Greenwich Observatory in 1675. In 1684 he was presented with the living of Burstow, Surrey, by Lord North. He died in Greenwich on 31 December 1719 and was succeeded in his post there by Edmond Halley.

Flamsteed began his astronomical studies at home by observing a solar eclipse on 12 September 1662, about which he corresponded with several other astronomers. When he decided to take his university degree in 1670, he sent his early studies to the Royal Society and they were published in *Philosophical Transactions.* This was enough to gain him general scientific recognition, but it was Jonas Moore who launched him in his career when he gave him a micrometer and promised him a good telescopic lens, thus enabling him to start serious practical work.

At the time Flamsteed's research began, 60 years had passed since ◊Galileo had made his discoveries, the star catalogue prepared by Tycho ◊Brahe more than a hundred years earlier was still the standard work, and Johannes ◊Kepler's laws of 1609 were only gradually being accepted. Flamsteed resolved to end the apparent stagnation in astronomical science. In 1672 he determined the solar parallax from observations of Mars when the planet was at its closest to the Sun, using the rotation of the Earth to establish a baseline. Four years later, and only two months after his appointment to Greenwich, he began observations that were to result in a 3,000-star British catalogue. His results improved on Brahe's work by a factor of 15, but at first they were only relative measurements, with no anchor in the celestial sphere. To assist him in his work, Jonas Moore further donated two chronometers and a 2.1-m/7-ft sextant, with which he made 20,000 observations 1676–89.

Settled in at Greenwich, Flamsteed became interested in the work on lunar theory published by Jeremiah Horrocks (1619–1641) some decades earlier. He brought Horrocks's constants up to date and revised his calculations no fewer than three times. The models in Flamsteed's third set of calculations were of fundamental importance in some of Newton's theoretical work. Following his revisions of lunar theory, Flamsteed also produced three different sets of tables describing the motion of the Sun. The first was issued before Flamsteed had any original observations on which he could base his parameters. His second gave a new determination of solar eccentricity, at almost the true value of 0.01675, and was published in his book *Doctrine of the Sphere* (1680). The third was printed in 1707 in Whiston's *Praelectiones astronomicae.* It included more detailed observations made possible by new equipment purchased after he inherited his father's estate in 1688.

With his improved facilities, including a mural arc, Flamsteed was able to make some very precise measurements: he determined the latitude of Greenwich, the slant of the ecliptic, and the position of the equinox. He also worked out an ingenious method of observing the absolute right ascension – a coordinate of the position of a heavenly body. His method, a great improvement on previous systems, removed all errors of parallax, refraction, and latitude. Having obtained the positions of 40 reference stars, he then went back and computed positions for the rest of the 3,000 stars in his catalogue.

Flamsteed also produced tables of atmospheric refraction, tidal tables, and supervised the compilation of the first table describing the inequality of the lunar elliptic following Kepler's second law.

A serious-minded man (possibly as a result of his constantly frail health), Flamsteed was never good at dealing with other people. Much of the last 20 years of his life was spent in controversy over the publication of his work. His results were urgently needed by Isaac Newton and Edmond Halley to test their theories, but Flamsteed was determined to withhold them until he was quite certain they were correct. A row with both Halley and Newton in 1704 eventually led to Flamsteed's work being unlawfully printed in 1712, but Flamsteed managed to secure and burn 300 copies of the printed production. Accordingly, the preparation of his great work, *Historia coelestis Britannica,* was completed by his assistants six years after his death, in 1725, and his *Atlas coelestis* was published even later, in 1729.

Further Reading

Flamsteed, John; Champan, Allan; and Johnson, Alison (eds), *Historia Coelestis Britannica,* National Maritime Museum, 1983.

Forbes, Eric G (compiler); Murdin, Lesley; Willmoth, Frances (ed), *The Correspondence of John Flamsteed, the First Astronomer Royal,* Institute of Physics Publishing, 1995–97, 2 vols.

Willmoth, Frances, *Flamsteed's Stars: New Perspectives on the First Astronomer Royal, 1646–1719,* Boydell & Brewer, 1997.

Fleming, (John) Ambrose (1849–1945) English electrical engineer whose many contributions to science included the invention of the thermionic diode valve, which proved to be a key electronic component in the early development of radio.

Fleming was born in Lancaster, Lancashire, on 29 November 1849, the son of a parson. When he was four years old his family moved to London, where his father continued his ministry in Kentish Town. Fleming was educated at University College School, and later at University College, London where he was awarded his BSc in 1870. His studies at the university were only part-time because for two years he worked as a clerk with a firm of stockbrokers. From 1872 to 1874 he

Fleming English electrical engineer and engineer Ambrose Fleming standing beside a wireless dynamo, which used the principals explained in Fleming's rules to generate electricity. *Mary Evans Picture Library*

studied at South Kensington, first under Edward Frankland and later under Frederick Guthrie.

For the next three years he taught science at Cheltenham College, then in 1877 he entered St John's College, Cambridge, having won an entrance exhibition (and saved £400 to pay the fees). At Cambridge he worked in the Cavendish Laboratory and studied electricity and advanced mathematics under James Clerk ◊Maxwell, author of the famous treatise on electricity and magnetism. In 1879 Fleming obtained his DSc and researched into electrical resistances. Three years later he was appointed to the newly created chair of mathematics and physics at University College, Nottingham, but resigned his post the following year to take up consulting work with the Edison Electric Light Company.

In 1883 Fleming was elected fellow of St. John's and in 1885 he was appointed professor at University College, London. Between 1889 and 1898 he published several important papers on the practical problems of the electrical and magnetic properties of materials at very low temperatures. During part of this time he made a careful study of the 'Edison effect' in carbon-filament lamps. In 1904 he produced experimental proof that the known rectifying property of a thermionic valve was still operative at radio frequencies, and this discovery led to the invention and production of what was first known as the 'Fleming valve'.

In 1905 Fleming described his electric-wave measurer or cymometer, and demonstrated it to the Royal Society. In 1874, he read the very first paper to the Physical Society on its foundation in that year; 65 years later in 1939, at the age of 90, he read his last paper to the same society. Fleming died on 18 April 1945 at his home in Sidmouth, Devon.

The value of Fleming's work was widely recognized. In 1892 he was elected a fellow of the Royal Society and received its Hughes Medal in 1910. In 1921 he was awarded the Albert Medal of the Royal Society of Arts and, in 1928, the Faraday Medal from the Instition of Electrical Engineers. He was knighted in 1929.

Although an avid experimenter, Fleming did not concern himself merely with the theoretical aspects of electrical science but took an active part in its practical application. As engineer and adviser to the Edison, Swan, and Ferranti electric lighting companies 1882–89, he was responsible for improvements in incandescent lamps, meters, and generators. For 26 years he was scientific consultant to the Marconi Wireless Telegraph Company and he designed many parts of their early radio apparatus – particularly those used by Guglielmo ◊Marconi in his pioneering transatlantic transmission in 1901.

In the early 1880s, Fleming had investigated the phenomenon known as the Edison effect – the escape of electrons or ions from a heated solid or liquid – but had abandoned the project as being of no practical value. In 1904 this early work on the 'one-way'

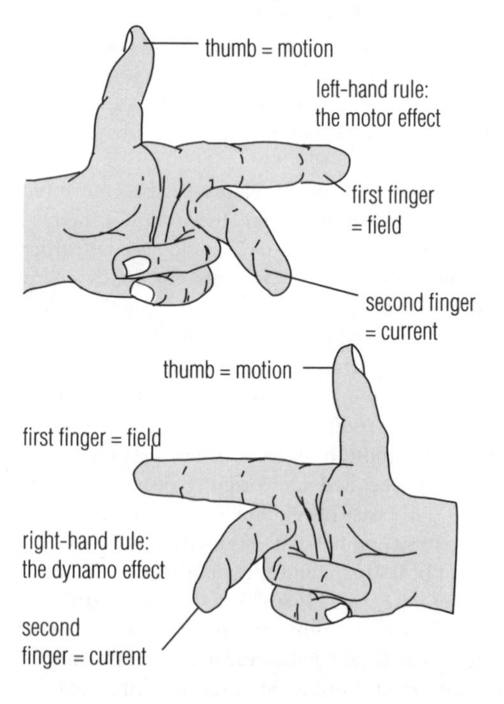

Fleming Fleming's rules give the direction of the magnetic field, motion, and current in electrical machines. The left hand is used for motors, and the right hand for generators and dynamos.

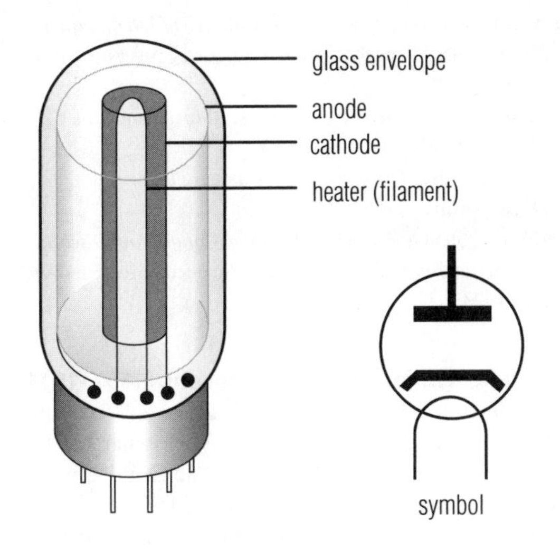

Fleming Fleming's work on electric lamps led him to invent the diode valve, a two-electrode vacuum tube in which a heated filament causes thermionic emission of electrons from the cathode. The valve was used as a rectifier and a detector in radio receivers until largely superceded by the semiconductor diode.

conductance of electricity in an incandescent lamp led to Fleming's most important practical achievement – the invention of the two-electrode thermionic rectifier, which became known as the Fleming valve or diode.

Fleming was searching for a more reliable detector of weak electric currents. He recalled his earlier experiments and made a new 'lamp' that had a metal cylinder surrounding the filament in a high vacuum. He found that it was very useful in detecting the very weak currents in radio receiving apparatus because it responded to currents alternating at very high frequencies. Previous instruments he had employed had been unable to do this because the electric 'waves' produced forces that tended to produce additional alternating currents.

He called his invention a valve because it was the electric equivalent of a check valve in a water-supply system, which allows water to pass in one direction only. In a like manner, the Fleming valve allowed electrical currents to pass in only one direction. It worked by allowing one of the electrodes – the cathode – to be kept hot so that electrons could evaporate from it into the vacuum. The other electrode – the anode – was left cool enough to prevent any appreciable evaporation of electrons from it. It was thus a device that permitted currents to flow essentially in one direction only when an alternating current was applied to it, and it revolutionized the early science of radio.

The Fleming valve has now been superseded by the transistor diode, an electronic device that utilizes the properties of single-crystal semiconductors. However, because he made possible a very significant advance in radio and television, Fleming's work will always remain an important milestone in electronic engineering. And despite the comparative relegation of the importance of his most famous practical contribution, his work in the theoretical and teaching aspects of electrical science remains undiminished.

Fleming, Alexander (1881–1955) Scottish bacteriologist who discovered penicillin, a substance produced by the mould *Penicillium notatum* and found to be effective in killing various pathogenic bacteria without harming the cells of the human body. Penicillin was the first antibiotic to be used in medicine. For this discovery, he shared the 1945 Nobel Prize for Physiology or Medicine with Howard ⇨Florey and Ernst ⇨Chain, who developed a method of producing penicillin in quantity.

Fleming was born on 6 August 1881 in Lochfield, Ayrshire, the son of a farmer. He was educated at Kilmarnock Academy and, after his father died in 1894, his poverty-stricken family sent him to London, where he first studied at the London Polytechnic Institute and then got a job as a clerk in a shipping office. While working there, and encouraged by his brother who was a doctor, he won a scholarship to study medicine at St Mary's Hospital Medical School, London, in 1902. He graduated four years later and remained at St Mary's in the bacteriology department for the rest of his career.

In his early years Fleming assisted the bacteriologist Almroth ⇨Wright, an association that continued when the two men were in the Royal Army Medical Corps and worked together in military hospitals during World War I. After the war, in 1918, Fleming returned to St Mary's as a lecturer, becoming director of the department of systematic bacteriology and assistant director of the inoculation department in 1920. He was appointed professor there and lecturer at the Royal College of Surgeons in 1928. He was knighted in 1944 and in 1946 became director of the Wright–Fleming Institute, where he continued to work until he retired in 1954. He died in London on 11 March 1955.

In 1928 Fleming made his major discovery quite by accident. He was working on the bacterium *Staphylococcus aureus* and had put aside some Petri dishes that contained the cultures. He later noticed that specks of green mould had appeared on the nutrient agar and that the bacterial colonies around the specks had disappeared. The effect on the bacteria was 'antibiosis' (against life). Fleming cultured the mould in nutrient broth and it formed a felt-like layer on the surface, which he filtered off. He tested the filtrate on a range of bacteria and found it killed some disease bacteria, but not all of them. He identified the mould as *Penicillium notatum,* a species related to that which grows on stale bread, and named the active substance it produced – the antibiotic element – 'penicillin'. Craddock, one of Fleming's assistants, grew some *Penicillium* in milk and ate the cheeselike product without any ill effects; no harm resulted, either, when mice and rabbits were injected with the material.

The purification and concentration of penicillin was, however, a chemical problem and Fleming was not a chemist. Two of his assistants made some progress but they left the matter unresolved until 1939, when Florey and Chain, in Oxford, isolated the substance and purified it. They published their results in 1940, and work began on the large-scale production of penicillin.

It was generally assumed that the original phenomenon that Fleming had observed was a common event, but Fleming was never able to produce the effect again. It has since been shown that a similar result is achieved only under very precise conditions, which are unlikely to be met during the routine inoculation and incubation of a bacterial plate.

Fleming also developed methods, which are still in use, of staining the spores and flagella of bacteria. He identified organisms that cause wound infections and showed how cross-infection by streptococci can occur among patients in hospital wards. He also studied the effects of different antiseptics on various kinds of bacteria and on living cells. His interest in chemotherapy led him to introduce Paul ◊Ehrlich's antibacterial agent Salvarsan into British medical practice.

In 1922 Fleming discovered the presence of the enzyme lysozyme in nasal mucus, tears, and saliva, where it catalyses the breakdown of carbohydrates surrounding bacteria and kills them. Fleming later showed it to be present in most body fluids and tissues. The enzyme thus helps to prevent infections, and has become a useful research tool for dissolving bacteria for chemical examination.

Penicillin, the first of the antibiotics, has been used with outstanding success in the treatment of many bacterial diseases, including pneumonia, scarlet fever, gonorrhea, diphtheria, and meningitis, and for infected wounds. Its discovery led to a scramble for further antibiotics in which streptomycin, chloromycetin, and the tetracyclines were discovered. Most antibiotics can now be made synthetically, and penicillin can be modified by chemical means for specific purposes.

Further Reading

Kaye, Judith, *The Life of Alexander Fleming,* Pioneers in Health and Medicine series, Twenty-First Century Books, 1995.

Macfarlane, Gwyn, *Alexander Fleming: The Man and the Myth,* Oxford Paperbacks series, Oxford University Press, 1985.

Malkin, John, *Sir Alexander Fleming: Man of Penicillin,* Alloway Publishing, 1981.

Maurois, André, *The Life of Sir Alexander Fleming, Discoverer of Penicillin,* Cape, 1959.

Otfinoski, Steven, *Alexander Fleming: Conquering Disease with Penicillin,* Makers of Modern Science series, Facts on File, 1993.

Fleming, Williamina Paton Stevens (1857–1911) Scottish-born US coauthor (with Edward ◊Pickering) of the first general catalogue classifying stellar spectra.

Fleming was born on 15 May 1857 in Dundee, where she was educated. She taught for a few years there before she married and emigrated to the USA in 1878. Shortly after her arrival in Boston her marriage broke up. She was then employed, in 1879, as an assistant to Edward Pickering, director of the Harvard College Observatory. Her work for him was as a 'computer' and copy editor, at which she was so successful that she was soon put in charge of twelve other 'computers'. In 1898 she was appointed curator of astronomical photographs. She died in Boston on 21 May 1911.

The project initiated by Pickering was simple in concept, but required meticulous dedication and patience. Photographs were taken of the spectra obtained using prisms placed in front of the objectives of telescopes. Although the use of the technique was restricted to stars about a certain magnitude, it yielded a wealth of information. In the course of her analysis of these spectra, Fleming discovered 59 nebulae, more than 300 variable stars, and 10 novae (which is even more impressive when it is recalled that at the time of her death in 1911 only 28 novae had been found).

The spectra of the stars observed in this manner could be classified into categories. Fleming designed the system adopted in the 1890 *Draper Catalogues,* in which 10,351 stellar spectra were listed in 17 categories ('A' to 'Q'). The majority of the spectra fell into one of six common categories; only 72 spectra accounted for the other 11 classes. This classification system represented a considerable advance in the study of stellar spectra, although it later was superseded by the work of Annie Jump ◊Cannon at the same observatory.

Fleming's special interest was in the detailed classification of the spectra of variable stars; she proposed a system in which their spectra were subdivided into 11 subclasses on the basis of further detailed spectral characteristics.

Florey, Howard Walter (1898–1968) Australian-born British bacteriologist who developed penicillin and made possible its commercial production. For this work he received the 1945 Nobel Prize for Physiology or Medicine, which he shared with Alexander ◊Fleming (who discovered penicillin) and Ernst ◊Chain.

Florey was born on 24 September 1898 in Adelaide. He was educated locally and read medicine at the University of Adelaide, qualifying in 1921 and winning a Rhodes scholarship to Oxford University to study physiology and pharmacology, where he worked under Charles ◊Sherrington. He spent a brief period at Cambridge in 1924 before going to the USA to study. In 1926 he returned to the UK as a researcher at the London Hospital, moving back to Cambridge in the following year as a lecturer in special pathology and later director of medical studies. Florey was appointed to the chair of pathology at Sheffield University in 1932 and became professor of pathology at Oxford in 1935 and head of the Sir William Dunn School of Pathology. He was knighted in 1944 and was president of the Royal Society 1960–65, the same year that he was made a life peer and member of the Order of Merit. He was elected provost of Queen's College, Oxford, in 1962 and died in Oxford on 21 February 1968.

At Oxford, Florey conducted investigations on antibacterial substances and during these he successfully purified lysozyme, the bacteriolytic enzyme discovered by Fleming. In 1939, continuing his research, he decided to concentrate on Fleming's unresolved problem of the purification of penicillin. Florey's co-worker, Ernst Chain, an accomplished biochemist, set about growing Fleming's strain of *Penicillium* and extracting the active material from the liquid culture medium. Chain and Florey extracted a yellow powder from the medium, but the process proved to be an extraordinarily difficult task and 18 months later they had collected only 100 mg/0.0035 oz of it. Florey and his team began a series of carefully controlled experiments on mice infected with standard doses of streptococci. The team found that a dilution of one in a million inhibited the growth of streptococci but was harmless to mice, growing tissue cells, and leucocytes. It was also discovered that the penicillin did not behave like an antiseptic or an enzyme, but blocked the normal process of cell division. The tests showed conclusively that penicillin could protect against infection but that the concentration of penicillin in the human body and the length of time of treatment were vital factors in the rate of success.

In 1940, during the early part of World War II, the German invasion of Britain seemed imminent; Florey and his colleagues smeared spores of the *Penicillium* culture on their coat linings so that, if necessary, any one of them could continue their research elsewhere. Further research was hindered by the great difficulty in producing enough penicillin for tests, and commercial firms were at that time too committed to vaccine production to participate. Florey's team persevered and improved their techniques, which resulted in a purer product suitable for preliminary trials on human beings. Only desperately ill patients with little hope of recovery were selected. The first patient was a police constable with a rampant infection of the face, head, and lungs. Within five days his improvement was miraculous, but he died one month later because it had been impossible to continue treatment long enough – the stock of penicillin was exhausted. The next five patients treated made complete recoveries.

The problem remained of producing penicillin in large quantities. Small-scale production continued in Florey's department, supplemented by minimal contributions from two commercial firms. In 1943 he went to Tunisia and Sicily and used penicillin successfully on war casualties. By 1945 studies had progressed far enough to show that antibacterial activity could take place using a dilution of one part in 50 million and, with the war over, large-scale commercial production of penicillin could begin.

Florey and his co-workers resumed their researches on other antibiotics. They discovered cephalosporin C, which later became the basis of some derivatives, such as cephalothin, which can be used as an alternative antibiotic to penicillin.

Florey was a great scientist with abundant energy, experimental skill, and a flair for choosing fruitful lines of research. He and his collaborators made penicillin available for therapeutic purposes, and were responsible for ushering in the era of antibiotic therapy.

Further Reading

Bickel, Lennard, *Howard Florey: The Man Who Made Penicillin*, Melbourne University Press, 1996.

Williams, Trevor I, *Howard Florey: Penicillin and After*, Oxford University Press, 1984.

Flory, Paul John (1910–1985) US polymer chemist who was awarded the 1974 Nobel Prize for Chemistry for his investigations of synthetic and natural macromolecules. With Wallace ◊Carothers he developed nylon, the first synthetic polyamide, and the synthetic rubber neoprene.

Flory was born on 19 June 1910 at Sterling, Illinois. He graduated from Manchester College, Indiana, in 1931 and gained his PhD from Ohio State University three years later. He then embarked on a career as an industrial research chemist, working at the Du Pont Experimental Station in Wilmington, Delaware (with Carothers) 1934–38 and at the Esso Laboratory Standard Oil Development Company in Elizabeth, New Jersey, 1940–43, before becoming director of fundamental research at the Goodyear Tire and Rubber

Company 1943–48. He had held a research associateship at Cincinnati for three years 1940–43, and in 1948 he again took up an academic post as professor of chemistry at Cornell University. He remained there until 1956 when he became executive director of research at the Mellon Institute, Pittsburgh. In 1961 he was made professor of chemistry at Stamford University, where he remained, eventually becoming emeritus professor. He died in Big Sur, California, on 9 September 1985.

Flory pioneered research into the constitution and properties of substances made up of giant molecules, such as rubbers, plastics, fibres, films, and proteins. He showed the importance of understanding the sizes and shapes of these flexible molecules in order to be able to relate their chemical structures to their physical properties. In addition to developing polymerization techniques, he discovered ways of analysing polymers. Many of these substances are able to increase the lengths of their component molecular chains and Flory found that one extending molecule can stop growing and pass on its growing ability to another molecule.

Working with Carothers, he prepared the polyamide Nylon 6,6 by heating a mixture of adipic acid (hexan-1,6-dioic acid) and hexamethylene diamine (hexan-1,6-diamine). They made Nylon 6 from caprolactam, showing it to be a polymer of the type $-[-CO(CH_2)_5NH-]n-$. Neoprene was made by polymerizing chloroprene (but-2-chloro-1,3-diene), and was soon followed by other synthetic rubbers made by polymerization and co-polymerization of various butenes. Flory's later researches looked for and found similarities between the elasticity of natural organic tissues – such as ligaments, muscles, and blood vessels – and synthetic and natural plastic materials.

Fontana, Niccolò Italian mathematician and physicist, nicknamed ⇨Tartaglia.

Forbes, Edward (1815–1854) British naturalist who made significant contributions to oceanography.

Born on the Isle of Man, on 12 February 1815, Forbes was a banker's son, who showed an early talent for natural history and for drawing. He studied medicine at Edinburgh, but soon grew absorbed in natural history, under the influence of Professor Robert Jameson. Taking a prominent role in the British Association in its early years, he became curator and later, in 1844, palaeontologist to the Geological Society of London and subsequently professor of natural history at Edinburgh; the last post in his tragically brief career was at the Royal School of Mines in London. He died in Wardic, near Edinburgh, on 18 November 1854.

Travelling widely in Europe and in the Near East, and in 1841 serving as naturalist on a naval expedition in the eastern Mediterranean, Forbes proved himself a tireless collector of fauna and flora, showing particular interest in the natural distribution of species. Molluscs interested him particularly; he was concerned with their taxonomy, and he also studied their migration habits and their different environments. A passionate palaeobotanist, Forbes divided British plants into five groups, and proposed that Britain had once been joined to the continent by a land-bridge. Plants had crossed over, he believed, in three distinct periods. Forbes also blazed trails in scientific oceanography. He discounted the contemporary conviction that marine life subsisted only close to the sea surface, spectacularly dredging a starfish from a depth of 400 m/1,300 ft in the Mediterranean. His *The Natural History of European Seas* (posthumously published in 1859) was a pioneering oceanographical text. It developed his favourite idea of 'centres of creation', that is, the notion that species had come into being at one particularly favoured location. Though not an evolutionist, Forbes's ideas could be commandeered for evolutionary purposes.

Ford, Henry (1863–1947) US automotive engineer and industrialist who, in the early 20th century, revolutionized the motorcar industry and manufacturing methods generally. His production of the Model-T popularized the car as a means of transport and made a considerable social and economic impact on society. His introduction of the assembly line (bringing components to the workers, rather than vice versa) gave impetus to the lagging Industrial Revolution.

Ford was born in Springwells, Michigan, on 30 July 1863. He attended rural schools and soon displayed a mechanical and inventive skill. Moving to Detroit at the age of 16, he obtained a job as a machinist's apprentice. During the next few years he worked for several different companies, and repaired watches and clocks in his spare time. After his apprenticeship Ford worked on the maintenance and repair of Westinghouse steam engines. In 1891 he was appointed chief engineer to the Edison Illuminating Company.

In about 1893 he constructed a one-cylinder petrol (gasoline) engine and went on to build his first car in 1896. Three years later he resigned from Edison's and joined the Detroit Automobile Company. He left there in March 1902 and, with some financial backing, formed the Ford Motor Company on 16 June 1903.

The first Ford car sold almost as soon as it was produced; further orders came in, and production rose rapidly. In 1906, because of a disagreement with his business associates (the Dodge brothers), Ford became the majority shareholder and president of the company, and by 1919 he and his immediate family held complete control of it.

Despite his success with the motorcar, Ford's non-industrial activities met with little success. An expedition to Europe he organized in December 1915 aimed at ending World War I, proved to be a fiasco. His attempt to run as a democrat for a Senate seat in Michigan and subsequent defeat left him bitter about alleged irregularities in his opponent's campaign.

However, other of his activities have brought him a great deal of credit as a benefactor. He created Greenfield Village as a monument to the simple rural world – a world that his automobiles had done so much to destroy. In it, he reconstructed the physical surroundings and the crafts of an earlier era. Near the village is the Henry Ford Museum containing his fine collection of antiques. Ford also restored the Wayside Inn of Longfellow's poem, and his important collection of early motion pictures were donated to the National Archives. He endowed the Ford Foundation, established in 1936, as a private, non-profitmaking corporation 'to receive and adminster funds for scientific, educational, and charitable purposes'. Ford finally gave up the presidency of his company in 1945. He died in Dearborn on 7 April 1947.

Motorcars were in their infancy when Henry Ford produced his first automobile in 1896 and decided to make his reputation in the field of racing cars. His determination to do this led him to leave the Detroit Automobile Company and to work on his own. In a memorable race at Grosse Point, Michigan, in October 1901, his victory brought him the publicity he sought. Barney Oldfield, also driving a Ford racer, added to Ford's reputation and in 1904 Ford himself drove his '999' to set a world record of 39.4 sec for 1 mi/1.6 km over the ice on Lake St Clair in January 1904.

The success of Ford's family cars was immediate. From the low-priced Model-N he went on to produce the Model-T, which first appeared in 1908. Over 19 years, 15 million were sold, and the car is regarded as having changed the pattern of life in the USA. It was one of the first cars to be made using assembly-line methods, and Henry Ford's name became a household word the world over.

The car itself was a sturdy black vehicle with a 4-cylinder 20 hp engine with magnetic ignition. A planetary transmission eliminated the gear-shift (and the danger of stripping the gears). 'Splash' lubrication was used and vanadium steel, of high tensile strength but easy to machine, was employed in many of the car's parts. Ford himself was responsible for the overall concept, and many of the basic ideas embodied in the construction of the Model-T were his own. In 1914 Ford became the first employer of mass labour to pay $5 a day minimum wage to all his employees who met certain basic requirements.

Dictatorial in his attitude, he later dismissed many key individuals who had helped to build the company's early success. He relinquished presidency of the company in 1909 to his son Edsel but strongly resisted changes in production despite an increasing loss of the market to up-and-coming competitors like the General Motors Corporation and Chrysler. Eventually, Ford acknowledged the inroads the newcomers were making on his Model-T. Characteristically, he set out to beat them with a new design, and in January 1928 he produced the Model-A.

The new car was the first to have safety glass in its windscreen as standard equipment. It was available in four colours and 17 body styles. Four-wheel brakes and hydraulic shock absorbers were incorporated in the car and it became a worthy successor to the Model-T. But Ford's previously undisputed leadership in the industry was not restored. Even the introduction of the V-8 engine – an engineering innovation at the time – did not halt the steady deterioration of Ford's share of the market.

When Edsel Ford died in 1943. Henry Ford resumed presidency of the company but in 1945 he surrendered it, for the last time, to his grandson and namesake, Henry Ford II.

Further Reading

Batchelor, Ray, *Henry Ford, Mass Production, Modernism, and Design*, Studies in Design and Material Culture series, Manchester University Press, 1994.

Bennett, Henry, *Ford: We Never Called Him Henry*, Tor Books, 1987.

Stidger, W L, *Henry Ford: The Man and His Motives*, Gordon Press Publishers, 1991.

Forsyth, Andrew Russell (1858–1942) Scottish mathematician whose facility with languages enabled him first of all to keep pace with, and even surpass, mathematical developments elsewhere in contemporary Europe, and then to translate such developments into English for the benefit of British mathematicians. Having done so – in an extremely important book – he was apparently unable then to maintain his precedence. Nevertheless, it was through his influence that in the UK the subject of the theory of functions dominated mathematical research for many years.

Forsyth was born in Glasgow on 18 June 1858, and obtained his initial education at Liverpool College. From there he won a scholarship to Trinity College, Cambridge, which he entered in 1877; lectures given there by Arthur ◊Cayley had a profound influence upon him and upon his general approach to mathematics. A dissertation by Forsyth published in the *Proceedings* of the Royal Society then led to the offer of a 'prize' fellowship at Trinity College, but no subsequent offer of a faculty position was forthcoming so in 1882 he took up the chair of mathematics at Liverpool College instead. Two years later, however, he returned to Trinity College as a lecturer, and remained there for the next 26 years. During that time he wrote several books and translated the works of others – his crucial *Theory of Functions* appeared in 1893 – and became considerably involved with the day-to-day administration of Cambridge University. In 1895 he was appointed to the Sadlerian Chair of Pure Mathematics.

In 1910, however, at the age of 52, Forsyth left Cambridge. Much of 1912 he spent lecturing in India, returning to England the following year to become chief professor of mathematics at Imperial College, London. Determined to renew his study of languages he retired early (in 1923), but within two years had reverted to his mathematical interests. After the publication of his last mathematical work in 1935, however, he again returned to linguistic studies. He died in London on 2 June 1942.

During his lifetime Forsyth received many honours and awards. A member of the Royal Society from 1886, he was presented with its Royal Medal in 1897.

Forsyth's early work was to systematize and develop the theory of double theta functions. He succeeded in demonstrating that such functions are related to the square roots of quintic and sextic polynomials in the same way that single theta functions are related to the square roots of cubic and quadratic polynomials. He also formulated a theorem that generalized a large number of identities between double theta functions; because this work was also carried out independently yet simultaneously by Henry Smith (1826–1883), the theorem is now called the Smith–Forsyth theorem.

Forsyth's *Theory of Functions* was intended as an advanced text that would introduce the main strands of continental mathematical study to British mathematicians, who were then tending to lag behind in terms of development and innovative creativity. In fact, more importantly, the book not only served to introduce the work of the European schools, but also brought together the work of all the various schools in a single volume – and as such was of considerable importance not merely in the UK but also in continental Europe, where it also achieved success in translation. In the UK, the book led to the introduction of concepts such as symbolic variant theory, Weierstrassian elliptic functions, and many more, and completely changed the nature of mathematical thinking. The developments that the book stimulated were rapid and, sometimes, fundamental, and sadly left Forsyth – who only five years previously had been publicly acknowledged as the most brilliant pure mathematician in the country – far behind. His skills belonged to older concepts. During his later years he wrote a number of books (some on ordinary, linear, and partial differential equations; one or two on Einstein's general theory of relativity) but he never again achieved the spectacular acclaim he had once enjoyed.

Fortin, Jean Nicholas (1750–1831) French instrumentmaker who made precision equipment for many of the most eminent French scientists of his time, although he is remembered today for the portable mercury barometer, which is named after him.

Little is known of Fortin's life. He was born in Mouchy-la-Ville, Ile de France, on 8 August 1750. He worked in Paris as a member of the Bureau de Longitudes, having been helped in his early career by the chemist Antoine ◊Lavoisier, for whom he made several scientific instruments. During his later years Fortin worked for the Paris Observatory, constructing instruments for astronomical studies and surveying. His only known publication is an abridgment of John ◊Flamsteed's *Atlas celeste,* which was published in 1776. Fortin died in Paris in 1831.

One of Fortin's major early achievements was his construction of a precision balance for Lavoisier; it consisted of a beam, 1 m/3.3 ft long, mounted on steel knife-edges and was able to measure weights as little as 70 mg/0.0025 oz. He made the first version of this balance in 1778, and another in 1799 for the Convention Committee on Weights and Measures; this latter version incorporated a comparator for standardizing weights. In the same year he adjusted the weight standard, the platinum kilogram, which was stored in the French National Archives.

Fortin is best known, however, for his barometers, although he did not make many of them. In 1800 he designed a portable mercury barometer that incorporated a mercury-filled leather bag, a glass cylinder in the cistern, and an ivory pointer for marking the mercury level. The mercury level could also be adjusted to the zero mark, and any barometer that possesses this feature is now known as a Fortin barometer. Fortin did not invent these features but he was the first to use them together in a sensitive portable barometer.

Fortin also made apparatus used by Joseph ◊Gay-Lussac in experiments on gas expansion; for Pierre ◊Dulong and François ◊Arago's investigation of the validity of the Boyle–Mariotte law; for Jean Biot (1774–1862) and Arago's expedition to Spain in 1806 to measure the arc of the terrestrial meridian; and numerous other instruments, including various clocks.

Foucault, (Jean Bernard) Léon (1819–1868) French physicist who invented the gyroscope, demonstrated the rotation of the Earth, and obtained the first accurate value for the velocity of light.

Foucault was born in Paris on 19 September 1819. He was educated at home because his health was poor, and went on to study medicine, hoping that the manual skills he developed in his youth would stand him in good stead as a surgeon. But Foucault soon abandoned medicine for science and supported himself from 1844 onwards at first by writing scientific textbooks and then popular articles on science for a newspaper. He carried out research into physics at his home until 1855, when he became a physicist at the Paris Observatory. He received the Copley Medal of the Royal Society in the same year, and was made a member of the Bureau des Longitudes in 1862 and the Académie des Sciences in 1865. He died of a brain disease in Paris on 11 February 1868.

Foucault's first scientific work was carried out in collaboration with Armand ◊Fizeau. Inspired by François ◊Arago, Foucault and Fizeau researched into the scientific uses of photography, taking the first detailed pictures of the Sun's surface in 1845. In 1847, they found that the radiant heat from the Sun undergoes interference and that it therefore behaves as a wave motion. Foucault parted from Fizeau in 1847, and his early work then propelled him in two directions.

Making the long exposures required in those early days of photography necessitated a clockwork device to turn the camera slowly so that it would follow the Sun. Foucault noticed that the pendulum in the mechanism behaved rather oddly and realized that it was attempting to maintain the same plane of vibration when rotated. Foucault developed this observation into a convincing demonstration of the Earth's rotation by showing that a pendulum maintains the same movement relative to the Earth's axis and the plane of vibration appears to rotate slowly as the Earth turns beneath it. Foucault first carried out this experiment at home in 1851, and then made a spectacular demonstration by suspending a pendulum from the dome of the Panthéon in Paris. From this, Foucault realized that a rotating body would behave in the same way as a pendulum and in 1852, he invented the gyroscope. Demonstrations of the motion of both the pendulum and gyroscope proved important to an understanding of the action of forces, particularly those involved in motion over the Earth's surface.

Foucault's other main research effort was to investigate the velocity of light. Both he and Fizeau took up Arago's suggestion that the comparative velocity of light in air and water should be found. If it travelled faster in water, then the particulate theory of light would be vindicated; if the velocity were greater in air, then the wave theory would be shown to be true. Arago

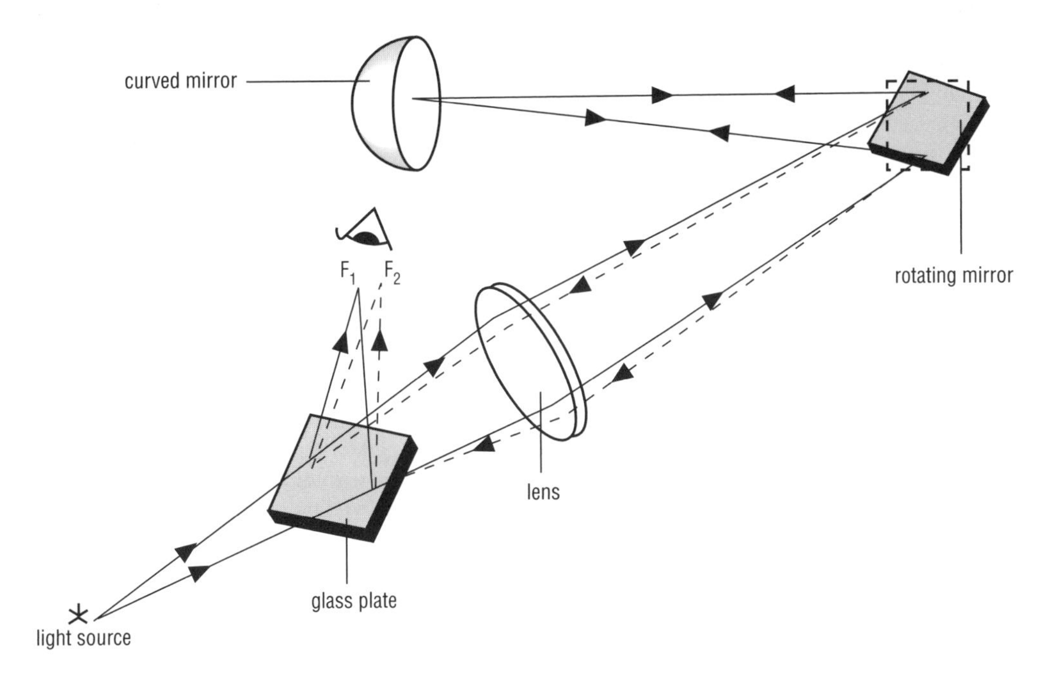

Foucault Foucault measured the velocity of light by directing a converging beam of light at a rotating plane mirror situated at the radius of curvature of a spherical mirror, which reflected the light back along the same path. The change in angle of the rotating mirror displaced the focus of the returning beam from F_1 to F_2, and from this displacement and the mirror's rotational speed the velocity of light was calculated.

had suggested a rotating-mirror method first developed by the English physicist Charles ◊Wheatstone for measuring the speed of electricity. It involved reflecting a beam of light from a rotating mirror to a stationary mirror and back again to the rotating mirror, the time taken by the light to travel this path causing a deflection of the image. The deflection would be greater if the light travelled through a medium that slowed its velocity. Fizeau abandoned this method after parting from Foucault and developed a similar method involving a rotating toothed wheel. With this, he first obtained in 1849 a fairly accurate numerical value for the speed of light.

Foucault persevered with the rotating-mirror method and in 1850 succeeded in showing that light travels faster in air than in water, just beating Fizeau to the same conclusion. He then refined the method and in 1862 used it to make the first accurate determination of the velocity of light. His value of 298,000 km/ 185,177 mi per second was within 1% of the correct value, Fizeau's previous estimate having been about 5% too high.

Foucault also interested himself in astronomy when he went to work at the Paris Observatory. He made several important contributions to practical astronomy, developing methods for silvering glass to improve telescope mirrors in 1857 and for accurate testing of mirrors and lenses in 1858. In 1860, he invented high-quality regulators for driving machinery at a constant speed and these were used in telescope motors and also in factory engines.

Foucault's outstanding ability as an experimental physicist brought great benefits to practical astronomy and also, in the invention of the gyroscope, led to an invaluable method of navigation. It is ironic that he missed the significance of an unusual observation of great importance. In 1848, Foucault noticed that a carbon arc absorbed light from sunlight, intensifying dark lines in the solar spectrum. This observation, repeated by Gustav ◊Kirchhoff in Germany in 1859, led immediately to the development of spectroscopy.

Fourier, (Jean Baptiste) Joseph (1768–1830)
French mathematical physicist whose particular interest was to try to describe the transfer of heat in purely mathematical terms. The formulation of equations in order to achieve this was a complex task that necessitated the development of new mathematical tools, and he was responsible in this way for the discovery of the Fourier series and the Fourier integral theorem, which have together led to the evolution of the modern process now known as harmonic analysis.

Fourier was born in Auxerre on 21 March 1768, the son of a tailor. Orphaned when very young, he obtained his education at the local military academy, and it was there that his interest in mathematics was first aroused. He then went on to a Benedictine school in St Bênoit-sur-Loire, but returned to Auxerre at the outbreak of the French Revolution and taught at his old school. He was arrested in 1794, only to be released a few months later after the execution of Robespierre. He next studied in Paris at the Ecole Normale for a short period, and in 1795 was made an assistant lecturer at the Ecole Polytechnique under Joseph ◊Lagrange and Gaspard ◊Monge.

In 1798 Fourier was selected to accompany Napoleon on his Egyptian campaign, and there conducted a variety of diplomatic affairs. Returning to France in 1801 he was appointed prefect of Isère in the south of the country. During this period he continued his mathematical studies on a part-time basis. Napoleon conferred the title of baron on Fourier in 1808, and later made him a count. Fourier was then made prefect of the *département* of Rhône, but resigned the post during Napoleon's Hundred Days in protest against the activities of the régime. Soon afterwards he obtained a post at the Bureau of Statistics and was able to devote all his energies to mathematics. He was elected to the French Academy of Sciences and made joint Secrétaire Perpétuel with Georges ◊Cuvier in 1822; he was also elected to the Académie Française and made a foreign member of the Royal Society. He died on 4 May 1830 as an indirect result of a disease he had contracted while serving in Egypt.

One of Fourier's most important contributions to both mathematics and physics was the use of linear partial differential equations in the study of physical phenomena as boundary-value problems. In order to comprehend and explain the conduction of heat under conditions of different temperature gradients, and in materials with different shapes and conductivities, Fourier developed what is now called the Fourier theorem. This enables the equation for the description of heat diffusion to be broken up into a series of simpler (trigonometric) equations, the sum of which equals the original. The Fourier series can be used to describe complex periodic (that is, repeating) functions, and so can be applied to many branches of mathematical physics. Light, sound, and other wavelike forms of energy can be studied using the Fourier theorem, and a developed version of this method is now called harmonic analysis. At the time, such was the creative brilliance of Fourier in using linear partial differential equations to this end, however, that for the following century or more nonlinear differential equations were hardly used at all in mathematical physics.

Fourier contributed to other areas of mathematics as well; for example, he laid the groundwork for the later development of dimensional analysis and linear programming. Fascinated since the age of 16 by the theory of equations, Fourier's work at the Bureau of Statistics also stimulated him to investigate the probability theory and the theory of errors.

Further Reading

Garber, Elizabeth, 'Reading mathematics, constructing physics: Fourier and his readers, 1822–1850' in: Kox, A J, Siegel and Daniel M (eds), *No Truth Except in the Details: Essays in Honor of Martin Klein,* Kluwer, 1995, pp 31–54.

Herivel, John, *Joseph Fourier: The Man and the Physicist,* Clarendon Press, 1975.

Transnational College of LEX Staff, *Who is Fourier?: A Mathematical Adventure,* Language Research Foundation, 1995.

Fourneyron, Benoit (1802–1867) French engineer who invented the first practical water turbine.

Fourneyron was born on 31 October 1802 in Saint-Etienne, Loir. He entered the New School of Mines at Saint-Etienne at the age of 15, having already acquired a good grounding in mathematical sciences from his father, who was a geometrician. Graduating from Saint-Etienne as the top pupil in his class, he went on to apply his skills to various projects including the development of the Le Creusot mines, oil exploration, and the building of a railway. He pioneered the manufacture of tin plate at Pont-sur-l'Ognon, Haute Saône, ending the British monopoly of the industry.

The fabrication of tin plate involved the use of a water wheel, the efficiency of which was very low. Fourneyman became consumed with the quest to produce a greatly improved water wheel, whose efficiency would surpass that of any model so far in existence. He succeeded in his ambition in 1827 and went on to produce, in 1855, an improved version of his original model. From there he built more than a hundred diverse models of hydraulic turbine, which were exported for use all over the world. He died in Paris on 8 July 1867.

The idea of using a stream of water to drive a wheel is very old and it is thought that the water wheel was invented in the 1st century BC. The first device that operated on the principle of reaction was a steam 'turbine' of ◊Hero of Alexandria in the 1st century AD.

Improvement in the design and efficiency of water wheels came slowly. By the early part of the 19th century, with the application of mathematics and a growing knowledge of hydraulics, the first reaction wheels of Leonhard ◊Euler and those of Claude Burdin were produced – but it was Fourneyron, one of Burdin's pupils, who first achieved success. His 1827 reaction turbine was 80% efficient and could develop about 4.5kW/6 hp.

Fourneyron's machine is generally recognized as opening the modern era of practical water turbines. It was essentially an outward-flow turbine. Water passed through fixed guide passages and hence into guide passages in the moveable outer wheel. When the water impinged on these wheel vanes, its direction was changed and it escaped round the periphery of the wheel. But the outward-flow turbine was essentially unstable because as water flowed through the fixed and moveable vanes it entered a region of successively increasing volume. The speed regulation of the turbine also presented difficulties.

Fourneyron patented an improved design that incorporated a three-turbine installation in 1832. However, his machine lost favour, being superseded in 1843 by the Jonval axial-flow machine.

Fourneyron's machines were still used in large commercial undertakings. All his earlier models were based on the free-flow efflux design, but he later realized the advantages of diffusing the outward flow and, in 1855, patented the outflow diffuser, the basis of which forms the modern-day inflow scroll case. Two turbines, each consisting of Fourneyron wheels keyed to one shaft, were used by the Niagara Falls Power Company in 1895. They were built into 49-m/160-ft wheel-pits dug into the supply channel at the top of the falls.

Fowler, William Alfred (1911–1995) US physicist and astronomer who, with Subrahmanyan ◊Chandrasekhar, won the 1983 Nobel Prize for Physics for his work on the nuclear reactions that play a role in the formation of chemical elements in the universe.

Fowler was born in Pittsburgh, Pennsylvania, on 9 August 1911. Obtaining his bachelor's degree in physics at Ohio State University in 1933, he went to the California Institute of Technology (Caltech), gained a PhD, and became a research fellow there in 1936. He remained at Caltech., rising from assistant professor to professor and, in 1970, instructor professor of physics, until his death in 1995. In 1989 he was awarded the Legion of Honour by President Mitterrand of France.

Fowler's work, in the main, concentrated on research into the abundance of helium in the universe. The helium abundance was first defined as the result of the 'hot Big Bang' theory proposed by Ralph ◊Alpher, Hans ◊Bethe, and George ◊Gamow in 1948, and corrected through the brilliant theoretical work of Chushiro ◊Hayashi in 1950. In addition to altering the time-scale proposed in the alpha–beta–gamma theory, Hayashi also showed that the abundance of neutrons at the heart of the Big Bang did not depend on the material density but on the temperature and the properties of the weak interreactions. Provided the density is great enough for the reaction between neutrons and protons to combine at a rate faster than the expansion rate, a fixed concentration of neutrons will be incorporated into helium nuclei, however great the material density is – producing a 'plateau' in the relationship between helium abundance and material density.

In 1967, Fowler – together with Fred ◊Hoyle and R Wagoner – made elaborate calculations of the percentage plateau abundance. His calculations took into account all the reactions that can occur between the light elements, and also considered the buildup of heavier elements; 144 different reactions were observed and the results analysed by computer. He and his collaborators claimed an accuracy of helium abundance to 1% and found that the percentage abundance of helium in this plateau is between 25% and 29%. Their calculations for the buildup of other elements such as deuterium and lithium agreed well with observations.

Fraenkel, Abraham Adolf (1891–1965) German-born Israeli mathematician who is chiefly remembered for his research and perception in set theory, and for his many textbooks.

Fraenkel was born on 11 February 1891 in Munich, where he grew up, was educated, and first attended university. He also studied at the universities of Marburg, Berlin, and Breslau (now Wrocław). In 1916 he became a lecturer at the University of Marburg, and in 1922 was appointed to the position of professor. Six years later, he taught for a year at the University of Kiel before going to Israel to teach at the Hebrew University of Jerusalem 1929–59. He showed a deep commitment to Zionism and Jewish culture, and throughout his life engaged in many Jewish social and educational activities. Fraenkel died in Jerusalem on 15 October 1965.

Fraenkel's early interest in the axioms (universally accepted facts) of mathematical theories led to his investigative studies of the axiomatic foundations of Hensel's p-adic numbers and of the theory of rings. He then became interested in the theory of sets (on which, in 1919, he wrote *Einleitung in die Mengenlehre,* a book that was well received and was reprinted several times).

Fraenkel became very involved with set theory as it had been formulated in 1908, in the axiomatic system put forward by Ernst ◊Zermelo. The axioms, however, included the hitherto unexplained notion of a 'definite property', and Fraenkel determined he should be the first to succeed amongst the several mathematicians attempting to overcome this difficulty.

In 1922 Fraenkel's solution proposed that Zermelo's definite property be replaced by the concept of a definition of function. He also omitted entirely Zermelo's axiom of subsets, which stated that if a property E is definite in a set m, there is a subset consisting of those elements x of m for which $E(x)$ is true. To replace this axiom, Fraenkel said instead that if m is a set and φ and ψ are functions, there are subsets m_E and m_E' consisting of those elements x of m for which $\varphi(x)$ is an element of $\psi(x)$ and $\varphi(x)$ is not an element of $\psi(x)$, respectively. By applying this axiom, Fraenkel demonstrated that Zermelo's axiomatics of choice (devised in 1904), can be treated independently by referring to an infinite set of objects that are not sets themselves. It turned out to be extremely complicated to prove this without referring to an external assumption. (It was not, in fact, successfully accomplished until 1963, when P Cohen proved it for a revised system combining the work of Zermelo, Fraenkel, and Thoralf ◊Skolem – calling it therefore the ZFS system.)

It was however Skolem's proposal for the explanation of Zermelo's definite property, published in 1923, that was ultimately accepted. His suggestion had the advantage over Fraenkel's in that it led more directly to a logical formulation of Zermelo's axioms (which, till then, existed only as intuitive statement).

Fraenkel nevertheless actively continued his development of the theory of sets, in which he showed considerable perception, evident in his papers and books. In 1953 he published *Abstract Set Theory,* and in 1958 *Foundations of Set Theory.* His research led him to posit an eighth axiom (to follow Zermelo's seventh), an axiom of replacement, which stated that if the domain of a single-valued function is a set, its counter-domain is also a set.

Later, John Von Neumann – the pioneer in computer mathematics – was to propose a ninth axiom, the axiom of foundation. It states that every non-empty set a contains a member b such that a and b have no members in common.

Fraenkel-Conrat, Heinz (1910–1999) German-born US biochemist who showed that the infectivity of bacteriophages (viruses that infect bacteria) is a property of their inner nucleic-acid component, not of their outer protein case.

Fraenkel-Conrat was born on 29 July 1910 in Breslau, Germany (now Wrocław, Poland), the son of a prominent gynaecologist. He studied medicine at the University of Breslau, graduated in 1933, and then – with the rise to power of Adolf Hitler – left Germany and went to the UK. He did postgraduate work at the University of Edinburgh and obtained his PhD in 1936 for a thesis on ergot alkaloids, after which he moved to the USA. He settled there and became a naturalized citizen in 1941. He went to the University of California in 1951, and became a professor there in 1955, becoming an emeritus professor in 1981.

In 1955 Fraenkel-Conrat developed a technique for separating the outer protein coat from the inner nucleic-acid core of bacteriophages without seriously damaging either portion. He also succeeded in reassembling the components and showed that these reformed bacteriophages are still capable of infecting bacteria. This work raised fundamental questions about the molecular basis of life. He then showed that the protein component of bacteriophages is inert and that the nucleic-acid component alone has the capacity to infect bacteria. Thus it seemed the fundamental properties of life resulted from the activity of nucleic acids.

Francis, James Bicheno (1815–1892) English-born US hydraulics engineer who played a crucial role in the industrial development of part of New England. He made significant contributions to the understanding of fluid flow and to the development the Francis-type water turbine for which he is remembered.

Francis was born on 18 May 1815 at Southleigh, Oxfordshire, the son of a railway superintendent and builder. After a short education at Radleigh Hall and Wantage Academy, he became assistant to his father on canal and harbour works. Two years later, he was employed by the Great Western Canal Company.

He travelled to the USA in search of greater opportunities, arriving in New York City in the spring of 1833. There he was employed by Major George Washington Whistler (1800–1849) on building the Stonington Railroad, Connecticut. A year later when Whistler became chief engineer to 'The Proprietors of the Locks and Canals on the Merrimack River', a corporation known simply as the 'Proprietors', Francis went with him to Lowell, Massachusetts.

In 1837 Whistler resigned and Francis succeeded him. On 12 July the same year Francis married Sarah Wilbur Brownell of Lowell. When the Proprietors decided (in 1845) to develop the river's water-powered facilities, Francis was made chief engineer and general manager. He travelled briefly to England in 1849 to study timber preservation methods and on his return turned his attention to developing water turbines. In 1855 his famous work, *The Lowell Hydraulic Experiments,* was published.

Francis wrote more than 200 papers for learned societies and was president of the American Society of Civil Engineers in 1880. He advised on a number of important dam projects and was a member of the Massachusetts state legislature, president of the Stonybrook Railroad for 20 years, and for 43 years a director of the Lowell Gas Light Company.

He retired from active business in 1885, and was succeeeded by one of his sons. Francis died on 18 September 1892 and was survived by his wife and six children.

The industrialization of New England resulted initially from water power rather than steam. The leading part Francis played in the exploitation of the Merrimack River was thus at the time more important than his work on turbines.

The Proprietors' corporation had been formed in 1792, originally to improve navigation. Realizing the potential, a Boston group purchased 160 ha/400 acres near the Pawtucket Falls, a site that soon developed into the town of Lowell. The company built a 290-m/950-ft dam on the river, which produced a 11-m/35-ft head and 29 km/18 mi of backwater, the pondage feeding 11 independent mills.

One of Francis's responsibilities was the measurement of the flows used by each of the manufacturing companies along the river to assess costs. He made numerous tests on sharp-crested weirs, and determined the numerical values in the Francis weir formula, the form of which was suggested by his colleague, Uriah Atherton Boyden (1804–1879). The second (1868) edition of Francis's work included his studies of measurements with weighted floats.

Francis's work on turbines started when the Proprietors acquired, in the late 1840s, an interest in the patent turbine designed by Samuel B Howd. This was a radial inflow (or 'centre-vent') machine, which was effective but inefficient. Significantly, however, Francis had built (in 1847) a model wheel similar to Howd's, and it, too, was somewhat inefficient. Two years later several inwardflow wheels of 170 kW/230 hp each were built from Francis's design. Tests showed peak efficiencies of nearly 80%.

The Francis wheels of the development days were an improvement on those of Howd, but only to a small degree do the so-called Francis turbines of today resemble Francis's original designs. At the outset they utilized purely radial flow runners and they had neither the familiar scroll case nor the draught tube of modern units. Later engineers developed the design into the forerunner of the modern mixed-flow unit.

The reason Francis's name continues to be associated with the design presumably stemmed initially from the widespread attention attracted by his book and then from the adoption of the designation by the German and Swiss firms that led in its scientific development later in the century.

Francis also devised a complete system of water supply for fire protection and had it working in the Lowell district for many years before anything similar was in operation anywhere else. He designed and built hydraulic lifts for the guard gates of the Pawtucket Canal and between 1875 and 1876 he reconstructed the Pawtucket Dam.

Francis was largely responsible for Lowell's rise to industrial importance. In retrospect, however, this is less notable than the experimental work he did in connection with the flow of fluids over weirs, and the establishment of the Francis formula. His work on the inward-flow turbine was significant and after his death the Canadian Niagara Power Company installed Francis turbines developing 7,643 kW/ 10,250 hp at the famous falls.

Franck, James (1882–1964) German-born US physicist who provided the experimental evidence for the quantum theory of Max ◊Planck and the quantum model of the atom developed by Niels ◊Bohr. For this achievement, Franck and his co-worker Gustav Hertz (1887–1975) were awarded the 1925 Nobel Prize for Physics.

Franck was born in Hamburg on 26 August 1882. When he left school, his father (a prosperous banker) sent him to Heidelberg University in 1901 to read law and economics as a preparation for his entry into the family firm, considering the status of scientists to be very lowly indeed. Fortunately, at Heidelberg Franck met Max ◊Born and a lifelong friendship began. Born, also from a wealthy Jewish family, had full parental approval for his career and this eventually convinced Franck's father to allow his son to follow a scientific career. At first it was to be geology, but this quickly turned to chemistry and then to physics when he went to Berlin University in 1902. It was there that Franck obtained his doctorate in 1906 for research into ionic mobility in gases.

Franck was awarded the Iron Cross during World War I, and from 1916 he worked at the Kaiser Wilhelm Institute of Physical Chemistry under Fritz ◊Haber on the study of gases, becoming head of the physics division there in 1918. Two years later, Franck became professor of experimental physics at the University of Göttingen, where Born had just taken the chair of theoretical physics. At Göttingen, Franck and Hertz undertook the work on the quantum theory that gained them the 1925 Nobel Prize for Physics. Franck remained there until 1933, when Adolf Hitler came to power. Although allowed to retain his position because of his distinguished war record, Franck was told to dismiss other Jewish members of his Institute in the university. Franck refused to do this and left Germany, going first to Denmark and then to the USA, where he became professor of physics at Johns Hopkins University in Baltimore in 1935. This was followed by a move to Chicago in 1938, where Franck was appointed professor of physical chemistry.

During World War II, Franck became a US citizen and carried out metallurgical work related to the production of the atomic bomb. He became aware of the devastating power of this weapon and, in a document that became known as the Franck Report, he and other scientists suggested that it should first be demonstrated to the Japanese on unpopulated territory. Franck retired from the University of Chicago in 1949. Numerous honours were accorded him in these late years by academics and universities in both the USA and Europe. The city of Göttingen, as part of its 1000th anniversary in 1953, made Franck and Born honorary citizens and Franck died there on 21 May 1964 while visiting friends.

In his major contribution to physics, Franck investigated the collisions of electrons with noble gas atoms and found that they are almost completely elastic and that no kinetic energy is lost. With Hertz, he extended this work to other atoms. This led to the discovery that there are inelastic collisions in which energy is transferred in definite amounts. For the mercury atom, electrons accept energy only in quanta of 4.9 electronvolts. For such collisions to be inelastic, the electrons need kinetic energy in excess of this figure. As the energy is accepted by the mercury atoms, they emit light at a spectral line of 2,537Å/2.5×10^{-7} m. This was the first experimental proof of Planck's quantum hypothesis that $E = h\nu$ where E is the change in energy, h is Planck's constant, and ν the frequency of light emitted. These experiments also tended to confirm the existence of the energy levels postulated by Bohr in his model of the atom. For this work, Franck and Hertz shared the 1925 Nobel Prize for Physics.

Franck also studied the formation, dissociation, vibration, and rotation of molecules. With Born he developed the potential-energy diagrams that are now common in textbooks of physical chemistry. From the extrapolation of data regarding the vibration of molecules obtained from spectra, he was able to calculate the dissociation energies of molecules. Edward Condon (1902–1974) interpreted this method in terms of wave mechanics, and it has since become known as the Franck–Condon principle.

During his later years at Göttingen and at Baltimore, Franck carried out experiments on the photodissociation of diatomic molecules in liquids and solids and this led to an interest in photosynthesis. Research in this field was dominated by organic chemists and biochemists and Franck found himself involved in much controversy. His research led him to believe in a two-stage mechanism within the same molecule for the photosynthetic process, when the established view was that two different molecules are involved.

Franklin, Benjamin (1706–1790) US scientist and statesman. He made an important contribution to physics by arriving at an understanding of the nature of electric charge as a presence or absence of electricity, introducing the terms 'positive' and 'negative' to describe charges. He also proved in a classic experiment that lightning is electrical in nature, and went on to invent the lightning conductor. Franklin also mapped the Gulf Stream, and made several useful inventions, including bifocal spectacles. In addition to being a scientist and inventor, Franklin is widely remembered as a politican. He played a leading role in the drafting of the Declaration of Independence and the US constitution.

Franklin was born in Boston, Massachusetts, of British settlers on 17 January 1706. He started life with little formal instruction and by the age of ten he was helping his father in the tallow and soap business. Soon, apprenticed to his brother, a printer, he was launched into that trade, leaving home shortly afterwards to try for himself in Philadelphia. There he set himself up as a printer and in 1724 was sent to London to prospect for presses and types. However, this turned out to be a ruse of the city governors to get rid of him – the reason is obscure. Nevertheless, Franklin, without funds or introductions, soon found himself work and put the next two years to good use in becoming a skilled printer.

Back in Philadelphia, Franklin's fortunes progressed. His own business prospered and he was soon active in journalism and publishing. He started the *Pennsylvania Gazette,* but is better remembered for *Poor Richard's Almanack,* a great collection of articles and advice on a huge range of topics, 'conveying instruction among the common people'. Published in 1732, it was a great success and brought Franklin a considerable income. Public affairs also proved to be his metier and gradually Franklin became enmeshed in all sorts of progressive undertakings. He was clerk of the Pennsylvania assembly as well as postmaster of Philadelphia, and founded the American Philosophical Society in 1743 and in 1749 a college that later became the University of Pennsylvania. He was elected to the Pennsylvania assembly in 1751 and as a politician became concerned with the government of the colony from Britain. These activities by no means affected his scientific investigations, however, and his major work on electricity was done in this period.

In 1757, Franklin travelled again to Britain, this time with proper credentials as the agent of the Pennsylvania assembly, and stayed there on and off until 1775, attending meetings of the Royal Society as well as campaigning for the independence of the American colonies as their leading spokesman in Britain. He was awarded the Copley Medal of the Royal Society in 1753 and elected to the society in 1756, the subscription of 25 guineas being waived in honour of his achievements.

Back in America, Franklin helped to draft the Declaration of Independence in 1776 and was one of its signatories. He then travelled to France to enlist help for the American cause in the Revolutionary War that followed, successfully organizing nearly all outside aid. He played a central part in the negotiation of the peace with the UK, signing a treaty that guaranteed independence in 1783. Franklin, though now well over 70, continued to play an active part in the affairs of the new nation. He became president of Pennsylvania, worked hard to abolish slavery, and in 1787 guided the constitutional convention to formulate and ratify the constitution. He died soon after in Philadelphia on 17 April 1790.

In 1746, his business booming, Franklin turned his thoughts to electricity and spent the next seven years in the execution of a remarkable series of experiments. Although he had little formal education, his voracious reading habits gave him the necessary background and his practical skills, together with an analytical yet intuitive approach, enabled Franklin to put the whole topic on a very sound basis. It was said that he found electricity a curiosity and left it a science.

By the time of Franklin's entry to the field, the notions of charged bodies, insulators, and conductors were established, though what was being 'charged' or 'conducted' was a matter of speculation. One of Franklin's earliest observations was of the ability of a pointed metal object to discharge an electrified conductor. A bodkin was used to discharge metal shot on a dry glass base. An earthed bodkin discharged the shot either by touching it or as much as 20 cm/8 in distant, but an insulated bodkin had no effect. A person on an insulated base could electrify a glass tube by rubbing, and 'communicate' the charge to another person similarly insulated. These experiments led Franklin to the fundamental conclusion that electricity is a single fluid that flows into or out of objects to produce electric charges. This naturally led to the introduction of the terms positive and negative, a positive charge being an excess of electricity and a negative charge a corresponding deficiency of electricity. But Franklin had no way of knowing in which direction the electric fluid moved, and he made an arbitrary choice of which bodies became positive and which negative. In fact, electric charge moves to his negatively charged bodies, which is why the electron is given a negative charge. However, Franklin made a fundamental discovery when he realized that the gain and loss of electricity must be balanced – the concept of conservation of charge. And his notion that the so-called electrical fire is fundamental to all matter, his 'one-fluid' theory, brings us right up to the 20th century.

The Leyden jar – an early capacitor invented in 1745 and consisting of a glass jar with a coating of metal foil on the outside and another on the inside – was an ideal proving ground for the clarification of these ideas. Franklin was able to show that its two coatings were oppositely charged, and that one had to be earthed while the other was being electrified. And finally he showed that the 'power is in the glass' – an appreciation of the importance of the dielectric later to be fully discussed by Michael ◊Faraday. This work led to the first plate condenser or capacitor, which contained glass sheets between lead plates – the Franklin panes. This device simplified electrical experiments by allowing the glass and lead components to be separated and increased in number, and gave rise to a 'battery' of condensers.

These fundamental ideas led to Franklin's most famous work of all – that on lightning – and his invention of the lightning conductor. Although not the first to wonder about thunder and lightning in connection with electricity, Franklin showed that the thunderclouds are indeed seats of electric charge whose behaviour is like charged bodies, and that pointed conductors can dissipate the charge or carry it to earth safely.

The sentry-box, or Philadelphia, experiment, which was devised by Franklin but first performed in France in 1752, caused a sensation. A man on an insulating base stood inside a sentry box, which had a pointed rod some 10 m/33 ft long fixed to its roof and insulated from it. The box itself was sited on a high building. Franklin suggested that a man could draw charge from a thundercloud by presenting an earthed wire to the rod, the man being protected by a wax handle. Sparks a few centimetres long were obtained. A later worker who did not fully observe the safety precautions laid down by Franklin was electrocuted. Later that year (1752), Franklin himself carried out his famous experiments with kites. By flying a kite in a thunderstorm, he was able to charge up a Leyden jar and produce sparks from the end of the wet string, which he held with a piece of insulating silk. The lightning conductor used everywhere today owes its origin to these experiments. Furthermore, some of Franklin's last work in this area demonstrated that while most thunderclouds have negative charges, a few are positive – something confirmed in modern times.

Of the rest of Franklin's work, staggering in its diversity, interest, and humanity, we can give only a summary here. His interest in atmospheric electricity led him to recognize the aurora borealis as an electrical phenomenon, postulating good conditions in the rarefied upper atmosphere for electrical discharges, and speculating on the existence of what we now call the ionosphere. An interest in meteorology followed naturally from this. Franklin explained tornadoes and waterspouts as being due to very rapid air circulation leading to low pressure in axial regions. He also pondered (long before the eruption of Krakatoa) on the effect of volcanic activity on weather.

Franklin's interest in heat and insulation led him to study clothing, air circulation, and the cooling effect of perspiration. He was very enthusiastic about health-improving activities, especially swimming, wrote on lead poisoning, gout, sleep, and was a late convert to inoculation. There was a fashion to apply electric-shock treatment in cases of paralysis but Franklin remained a sceptic and Franz Mesmer (1734–1815) and his followers were discredited. In a very modern comment, Franklin wondered about the possible psychological value of such treatment.

Franklin's Pennsylvanian fireplace – a stove with underfloor draught – was a great success in 1742. Franklin refused a patent and showed his magnanimity of nature when a London manufacturer made a lot of money out of a 'very similar' model. 'Not the first instance ...' he said quietly in his autobiography. It was recognition enough to be imitated.

Franklin was also influential in areas of physics other than electricity. Unfashionably for the time, he rejected the particle theory of light, being unable to account for the vast momentum that the particles should possess if they existed. This view inspired Thomas ◊Young to his fundamental work on the wave theory early in the following century, and the concept of light particles having momentum was used by Louis ◊de Broglie to establish the wave nature of the electron more than a century later still. But in considering heat, Franklin's views were less helpful to the course of science. Franklin suggested a famous experiment to demonstrate conductivity in which rods of different metals are heated at one end and wax rings placed at the same distance along the rods fall off at different times as heat spreads through the rods at different rates. This led Franklin to speculate that heat is a fluid like electricity, and aided development of the erroneous caloric theory of heat.

Always interested in the sea, Franklin produced the first chart of the Gulf Stream following observations made in 1770 that merchant ships crossed the Atlantic from east to west two weeks faster than did the mail ships – the former keeping to the side of the current, not fighting it. In 1775 he used a thermometer to aid navigation in the warm waters of the Gulf Stream and, as late as 1785, was devising ways of measuring the temperature at a depth of 30 m/98 ft. Finally Franklin also busied himself with such diverse topics as the first public library, bifocal lenses, population control, the rocking chair, and daylight-saving time.

Benjamin Franklin is arguably the most attractive and interesting figure in the history of science, and not only because of his extraordinary range of interests, his central role in the establishment of the USA, and his amazing willingness to risk his life to perform a crucial experiment – a unique achievement in science. By conceiving of the fundamental nature of electricity, he began the process by which a most detailed understanding of the structure of matter has been achieved.

Further Reading

Franklin, Benjamin; Seavey, Ormond (ed), *Autobiography and Other Writings*, Oxford University Press, 1998.

Friends of Franklin, Inc Staff, *Benjamin Franklin 1706–1790: A Chronology of the Eighteenth Century's*

Most Eminent Citizen, Kendall Hunt Publishing Company, 1996.
Jennings, Francis, *Benjamin Franklin, Politician: The Mask and The Man,* W W Norton, 1996.
Middlekauff, Robert, *Benjamin Franklin and His Enemies,* University of California Press, 1996.
Van Doren, Carl C, *Benjamin Franklin,* Viking Penguin, 1991.
Walters, Kerry S, *Benjamin Franklin and His Gods,* University of Illinois Press, 1998.

Franklin, Rosalind Elsie (1920–1958) English chemist and X-ray crystallographer who was the first to recognize the helical shape of DNA. Her work, without which the discovery of the structure of DNA would not have been possible, was built into James Watson and Francis ◊Crick's Nobel prizewinning description of DNA.

Rosalind Franklin was born in London on 25 July 1920, into a professional family. She was educated at St Paul's School, London, and won an exhibition scholarship in 1938 to study chemistry at Cambridge University. After graduating in 1941 she stayed on to carry out postgraduate study on gas-phase chromatography with Ronald ◊Norrish. From 1942 she studied the physical structure of coal for the British Coal Utilization Research Association, from where she moved to Paris in 1947 to research the graphitization of carbon at high temperatures at the Laboratoire Central des Services Chimique. She became a skilled X-ray crystallographer, and later applied these techniques to her study of DNA and viruses at King's College and then Birkbeck College, London.

In 1951 Franklin was appointed research associate to John Randall at King's College. Maurice ◊Wilkins and R Gosling at King's had obtained diffraction images of DNA that indicated a high degree of crystallinity and Randall gave Franklin the job of elucidating the structure of DNA.

Working with Gosling, Franklin used her chemical expertise to study the unwieldy DNA molecule and established that DNA exists in two forms – A and B – and that the sugar–phosphate backbone of DNA lies on the outside of the molecule. She also explained the basic helical structure of DNA, and produced X-ray crystallographic studies. However, she was not entirely convinced about the helical structure and was seeking further evidence in support of her theory.

Meanwhile, unknown to Randall, who had presented Franklin's data at a routine seminar, Franklin's work had found its way to her competitors Crick and Watson, at Cambridge University. They incorporated her work with that of others into their description of the double-helical structure of DNA, which was published in 1953 in the same issue of *Nature* that Franklin's X-ray crystallographic studies of DNA were published. Franklin was not bitter about their use of her research material but began writing a corroboration of the Crick–Watson model.

Finding Wilkins difficult to work with, Franklin left King's College and joined John Bernal's laboratory at Birkbeck College to work on the tobacco mosaic virus. She died of cancer on 16 April 1958 at the age of 37, four years before she could be awarded the Nobel Prize for Physiology or Medicine with Watson, Crick, and Wilkins in 1962. The Nobel prize cannot be awarded posthumously.

Fraunhofer, Joseph von (1787–1826) German physicist and optician who was the first person to investigate the dark lines in the spectra of the Sun and stars, which are named Fraunhofer lines in his honour. In so doing he developed the spectroscope, and he later became the first to use a diffraction grating to produce a spectrum from white light. These achievements were made and used by Fraunhofer mainly to improve his optical instruments, and his work laid the basis for subsequent German supremacy in the making of high-grade scientific and optical instruments.

Fraunhofer was born in Straubing on 6 March 1787. His education was limited and he started work in his father's glazing workshop at the age of ten. After his father's death in 1798, he was apprenticed to a Munich mirror-maker and glass-cutter, and in 1806 he entered the optical shop of the Munich Philosophical Instrument Company, which produced scientific instruments. There Fraunhofer developed an expertise in both the practice and theory of optics. Under Pierre Guinand, a master glassmaker, Fraunhofer acquired practical knowledge of glassmaking and combined this with his grasp of optical science to improve the fortunes of the company. This success brought him promotion within the firm and, by 1811 he had become a director.

Still highly active in business, Fraunhofer made his discovery of the dark lines in the Sun's spectrum in 1814, and invented the diffraction grating in 1821. In 1823, Fraunhofer accepted the post of director of the Physics Museum of the Bavarian Academy of Sciences in Munich. But he contracted tuberculosis two years later and died in Munich on 7 June 1826, at the early age of 39.

Although he had little formal education, Fraunhofer sought to understand optical theory and apply it to the practical work of constructing lens combinations of minimum aberration. At that time there was little high-quality crown and flint glass, and methods of determining the optical constants of glass were crude, limiting the size and quality of lenses that could be produced and also confining instrumentmakers to trial-and-error methods of construction. Fraunhofer approached lens-making according to optical theory and set out to determine the dispersion powers and refractive indices of different kinds of optical glass with greater accuracy. In collaboration with Guinand, he also sought to improve the quality of the glass used to make lenses.

In 1814, Fraunhofer began to use two bright-yellow bands in flame spectra to produce a source of monochromatic light to obtain more accurate optical values. His comparison of the effect of the light rendered by the flame with that of the Sun established the existence of profuse fine, dark lines across the solar spectrum. A small number of these had been seen by William Woolaston (1766–1828) in 1802, but Fraunhofer observed 574 lines between the red and violet ends of the spectrum.

In this way, Fraunhofer was able to make very accurate measurements of the dispersion and refractive properties of various kinds of glass, and in so doing he developed the spectroscope into an instrument for the scientific study of spectra.

In 1821, Fraunhofer examined the patterns that result from light diffracted through a single slit, and related the width of the slit to the angles of dispersion of the light. Extending from a large number of slits, he constructed a grating of 260 parallel wires and made the first study of spectra produced by diffraction gratings. The presence of the dark lines in the solar spectrum facilitated Frauenhofer's observation that a diffraction grating produces a greater dispersion of the spectra than a prism. He examined the correlation between the dispersion of the spectra and the spacing of the wires in the grating, establishing an inverse relationship between dispersion and the divisional distance separating the successive slits in the grating. His measurement of the dispersion enabled him to discover the wavelengths of light of specific colours and the dark lines.

Fraunhofer also constructed reflection gratings, enabling him to extend his studies to the effect of diffraction on oblique rays. By using the wave theory of light, he was able to derive a general form of the grating equation that is still in use today.

Fraunhofer never published his researches into glassmaking nor the methods that he developed for calculating and testing lenses, viewing them as trade secrets. This work enabled him to develop telescopes of unsurpassed quality, leading to important discoveries in astronomy. But Fraunhofer did publish his work on diffraction gratings, which was important in the development of spectroscopy, a vital tool in the elucidation of atomic structure much later. He also published his observations of the dark lines in the spectrum, although he could not provide an explanation of them. This was achieved by Gustav ◊Kirchhoff in 1859.

The dark lines crossing the Sun's spectrum indicate the presence in the Sun's atmosphere of certain elements in the vapour state. The vapours in the chromosphere are cooler than the lower photosphere of the Sun and they absorb their own characteristic wavelengths from the Sun's continuous spectrum according to Kirchhoff's law – a substance that emits light of a certain wavelength at a given temperature will also absorb light of the same wavelength at that temperature. Some of the wavelengths emitted are absorbed by the vapours in the chromosphere as the Sun's light passes through them, producing the dark lines. It therefore became possible to identify the elements in the Sun's atmosphere from a study of the Fraunhofer lines, and hydrogen and helium were both shown to be present – in fact, this was how helium was first discovered. The same method is also applied to other stars.

Fraunhofer's insistence on technical quality thus led to discoveries and developments in science on both a cosmological as well as an atomic scale.

Fredholm, Erik Ivar (1866–1927) Swedish mathematician and mathematical physicist who founded the modern theory of integral equations, and in his work provided the foundations upon which much of the extremely important research later carried out by David ◊Hilbert was based. Fredholm's name is perpetuated in several concepts and theorems.

Fredholm was born on 7 April 1866 in Stockholm, where he grew up and was educated. At the age of 19 he entered the Polytechnic Institute there, studying applied mathematics; his particular interest was the solution of problems of practical mechanics. After one year, however, he transferred to the University of Uppsala where in 1888 he received his bachelor's degree. He then returned to Stockholm. After ten years of further research and work, Fredholm finally obtained his PhD from Uppsala University – for a thesis on partial differential equations – and became a lecturer in mathematical physics at Stockholm University. Within the next five years he wrote his most important paper, on integral equations; for it, in 1903, he received the Wallmark Prize of the Swedish Academy of Sciences, and the Poncelet Prize of the Académie de France. In 1906 he was promoted to professor of rational mechanics and mathematical physics, a post he held until his death. He died in Stockholm on 17 August 1927.

Fredholm's success in deriving a theory of integral equations was in some respects a success in combining parts of the work of others with his own creative flair and a novel approach. He founded much of his theory on work carried out by US astronomer George Hill

(1838–1914) in 1877, who was investigating lunar motion. In his examination, Hill used linear equations involving determinants of an infinite number of rows and columns. In Fredholm's paper 'Sur une nouvelle méthode pour la résolution du problème de Dirichlet' (1900), he first developed the essential part of the theory of what is now known as Fredholm's integral equation; further, he went on to define and solve the Fredholm equation of the second type, involving a definite integral.

Such equations had been under scientific consideration for some years. Niels ◊Abel and Vito ◊Volterra had all put forward tentative or incomplete results: Henri ◊Poincaré had even arrived at Fredholm's solution but been unable to derive a proof of it, although in order to carry out his own work on partial differential equations 1895–96 he had been obliged to assume that it was correct.

Fredholm's novel approach in continuing his research led to his discovery, also in 1900, of the algebraic analogue of his own theory of integral equations. It was not until 1903, however, that he completed the solution, recognizing the analogous identity between the Fredholm integral equation and the linear-matrix vector equation $(I + F)U = V$, and showing that the analogy was complete.

Shortly afterwards, Fredholm's results were used by David Hilbert in Germany, who extended them in deriving his own theories – such as the theory of eigenvalues and the theory of spaces involving an infinite number of dimensions – that finally contributed fundamentally towards quantum theory.

Frege, Friedrich Ludwig Gottlob (1848–1925)
German logical philosopher and mathematician whose main purpose was to define once and for all an evolutionary connection between the fundamental rigour and mathematics in logic and the fundamental rigour and logic in mathematics. For this purpose he revised certain parts of mathematical notation in order to introduce total precision in logic, and he further revised philosophical vocabulary with the same intent.

His first books describing this work were successful – and to some extent gratifyingly revolutionary in effect. His later development of this in a two-volume work, however, was over-ambitious and was accounted a failure, unfortunately discrediting much of his earlier and his later work. Nevertheless, his final system of logic is now accepted as one of the greatest contributions in the field put forward in the century surrounding it.

Frege was born on 8 November 1848 in Wismar, Germany, and grew up and received his early education there. He then spent two years 1869–71 as a student in Jena before transferring to Göttingen University, where he studied physics, chemistry, mathematics, and philosophy, earning his PhD in mathematics in 1873. The next year he entered the faculty of philosophy there.

Frege's studies over the following years were crystallized in a book he published in 1879 on a new symbolic mathematical language he had devised, which he called *Begriffschrift.* In that year he returned to Jena to take up a teaching post, and remained there for the rest of his working life. In 1884, Frege published another important book on the foundations of mathematics, and again relied upon his *Begriffschrift.* He attempted to develop his ideas still further in another, ill-fated, two-volume text, the first volume of which appeared in 1893, and the second in 1903. These volumes, entitled *Grundgesetze der Arithmetik,* received a severe blow when, shortly before the second volume was due to be published, Frege was sent a letter that demonstrated to him that the entire mathematical system described in the books was in fact of no value; nobly, he included a postscript to that effect in the second volume. After this personal disaster, Frege continued to study mathematics but never with the same scope or depth. He retired in 1917, still writing further material, extending some of his previous studies in the period 1918–23. He died in Bad Keinen, Germany, on 26 July 1925.

At the beginning of what was to become his life's work, Frege was correctly convinced that in terms of absolute precision ordinary language is not sufficiently strict for the expression of mathematical concepts such as the definition of number, object, and function. Furthermore, he saw that the symbols already available to mathematicians were themselves not adequate for this purpose either, and so it would be necessary to create new ones – a vital step that mathematicians before him had resisted taking.

The resultant *Begriffschrift* (which translates literally as 'idea-script') was intended as a method for the analysis and representation of mathematical proofs. It has since been developed into modern mathematical symbolic logic, and Frege is generally – and only reasonably – credited as its originator. He introduced the symbols for assertions, implications, and their converse notions; he also introduced propositional logic and quantification theory, inventing symbols for 'and', 'or', 'if ..., then ...', and so on. Using his new 'language' he was able succinctly and unambiguously to express complex logical relations, and even – when Frege applied it to the theory of sequences – to define the ancestral relation. This represented a major development in mathematical induction, and was later further explored by mathematicians such as Bertrand ◊Russell and Alfred ◊Whitehead in the UK.

Frege incorporated improvements to the *Begriffschrift* into *Grundgesetze,* but was devastated to receive a letter from Bertrand Russell in 1902 – nine years after the appearance of the first volume – in which Russell asked Frege how his logical system coped with a particular logical paradox. To his chagrin, Frege's system was not able to resolve it – and since the system had been intended to be complete and contradiction-free, he was forced to acknowledge his system to be useless.

Although at the time Frege was largely discredited, his work today is seen as of considerable importance. His innovations have been useful in the development of symbolic logic, and even the problem posed by Russell was resolved by later logico-mathematicians.

Frege, nevertheless, in many ways simply stopped at that point. Despite the fact that he carried on working, and for quite a number of years, developments in early 20th-century mathematics – such as Hilbert's axiomatics – were apparently beyond Frege's scope. He was unable to accept these new ideas, even when David Hilbert himself tried to clarify the issue for him. Frege was, therefore, a mathematician with the most ambitious plans for the development of a rigorous foundation for mathematics in which, in his own eyes, he did not succeed in his own lifetime.

Fresnel, Augustin Jean (1788–1827) French physicist who established the transverse-wave theory of light in 1821.

Fresnel was born on 10 May 1788 in Broglie, Normandy. His parents provided his early education themselves and, at the age of 12, he entered the Ecole Centrale in Caen where he was introduced to science. In 1804 Fresnel entered the Ecole Polytechnique in Paris intending to become an engineer. Two years later he went to the Ecole des Ponts et Chaussées for a three-year technical course that included experience in practical engineering. Completing the course successfully, he became a civil engineer for the government.

Fresnel then worked on road projects in various parts of France, but became interested in optics. When Napoleon returned to France from Elba in 1815, Fresnel deserted his government post in protest. He was arrested and confined to his home in Normandy, but took advantage of this enforced leisure to develop his ideas on the wave nature of light into a comprehensive mathematical theory.

Napoleon's return proved to be short-lived and Fresnel was soon reinstated into government service. His scientific investigations were curtailed from then on and all his work was done in his spare time. But his achievements were great, rewarding Fresnel with unanimous election to the Académie des Sciences in 1823 and, just before he died, the Rumford Medal of the Royal Society. Fresnel had been continually plagued with ill-health, and eventually died of tuberculosis at Ville D'Avray, near Paris, on 14 July 1827 at the early age of 39.

Fresnel began his investigations into the nature of light convinced that the wave theory was correct because of its essential simplicity. He was not aware of earlier work by Christiaan ◊Huygens and Thomas ◊Young advocating a wave nature for light, and sought to establish his own ideas in a study of diffraction. Fresnel was soon able to demonstrate mathematically that the dimensions of light and dark bands produced by diffraction could be related to the wavelength of the light producing them if light were to consist of waves.

Huygens had believed that light consists of longitudinal waves like sound waves, and Fresnel endeavoured to refine his mathematical explanation on this basis. He ran into formidable difficulties but then became aware of new discoveries indicating that light may be polarized by reflection. Fresnel plunged into a mathematical explanation for this phenomenon and concluded in 1821 that polarization could occur only if light consists of transverse waves. Few agreed with him because of difficulties in conceiving of the nature of a medium to carry transverse waves. Fresnel was able to show, however, that his theory convincingly explained the phenomenon of double refraction and it rapidly gained acceptance.

Neither the particle theory nor the longitudinal wave theory could explain double refraction. Fresnel showed that light could be refracted through two different angles because one ray would consist of waves oscillating in one plane while the other ray consisted of waves oscillating in a plane perpendicular to the first one. The understanding of polarized light came to have important applications in organic chemistry (in optical isomerism) and was later carried beyond the visible spectrum to lay the groundwork for the theoretical uncovering of a wide range of radiation by James Clerk ◊Maxwell.

Fresnel is also remembered for the application of his new ideas on light to lenses for lighthouses. He produced a revolutionary design consisting of concentric rings of triangular cross section, varying the overall curvature to produce lenses that required no reflectors to produce a bright parallel beam.

Freud, Anna (1895–1982) Austrian-born English psychoanalyst best known for her work on the psychoanalysis of children. Her main contributions to psychoanalysis have been collected into eight volumes of her writings and her methods are still practised at the Anna Freud Centre in Hampstead, London.

Born in Vienna on 3 December 1895, the sixth child and youngest daughter of the founder of psychoanalysis, Sigmund ◊Freud, Anna left school early and taught at a primary school before becoming her father's amanuensis – secretary, companion, and student. She was her father's favourite daughter and became his confidante, working closely with him on the development of the theory of psychoanalysis and shielding him from intrusion. It was said that they were so close as to be almost telepathic. She became a member of the International Psychoanalytical Association in 1922 and began to practise as a psychoanalyst. In 1925 she became secretary of the Vienna Training Institute, headed by Helene Deutsch, and gave a series of lectures on childhood behaviour that led to the foundation of the *Kinderseminar.* Her method of analysis was of the 'continental' school, which believed that the methods used for adults needed to be modified for children. In that respect her analysis differed from that of Austrian child psychoanalyst Melanie Klein (1882–1960). Her book *The Ego and the Mechanisms of Defence* (1937) extends her father's work on anxiety as a function of ego.

Anna and her father escaped from the Nazis in 1938 to London where her father died in 1939. She spent the war years with her partner Dorothy Burlingham in pioneering work in child analysis and the exploration of normal child development. Together they produced three books: *Young Children in Wartime* (1942), *Infants without Families* (1943), and *War and Children* (1944). Anna Freud took for granted many of the Freudian formulations that have been rejected by feminist analysts, and wrote prolifically on penis envy and masturbation. She was critical of feminist terminology such as the 'social construction of gender', which she thought ignored the anatomical and psychic conflicts between the sexes. However, her work still has much to offer feminists, a fact that has been recognized over the years in contributions to *The Psychanalytic Study of the Child,* an annual journal she co-founded in 1945 and edited for several years.

The Hampstead Child Therapy Clinic was set up in 1952 and Anna remained its director until her death. She created a developmental profile of the child that did not give much importance to the pre-verbal stage and which stressed the importance of the psychological relationship between parent and child. Crucially, she never assumed a monolinear model of human development, but believed that each person follows a number of possible developmental lines. According to her theory the adult character is a register of progress along these lines – problems are the product of uneven progress along one or more of them. For some women, then, penis envy can have a critical effect whereas for others it will be minor; sibling rivalry can figure large in some and in others have very little bearing, and so on. Some lives are shaped by specific traumas and specific defences, inhibitions, and repressions are set up to counter their effects. Each stage has an important role to play in creating a rounded character and Freud argued against privileging one against another.

Anna Freud returned to Vienna for the first time in 1971 and received a standing ovation at the 27th International Psychoanalytical Congress. Among several honours she was awarded an honorary MD from Vienna in 1972. Her work continues at the Anna Freud Centre (formerly the Child Therapy Clinic) in Hampstead. She died in London on 9 October 1982.

Further Reading

Coles, Robert, *Anna Freud: The Dream of Psychoanalysis,* Radcliffe Biography series, Addison-Wesley, 1992.

Peltzmann, Barbara R, *Anna Freud: A Guide to Research,* Garland Reference Library of Social Science, Garland, 1990, v 610.

Peters, Uwe Henrik, *Anna Freud: A Life Dedicated to Children,* Schocken, 1985.

Sayers, Janet, *Mothers of Psychoanalysis: Helene Deutsch, Karen Horney, Anna Freud, Melanie Klein,* Norton, 1991.

Young-Bruehl, Elisabeth, *Anna Freud: A Biography,* Summit Books, 1988.

Freud, Sigmund (1856–1939) Austrian psychiatrist and the father of modern psychoanalysis. He was best known for his use of the free-association method in analysis, and for his ideas on the interpretation of dreams. His theories on child and adult sexuality shocked Europe and greatly influenced later psychology.

Freud Austrian psychiatrist Sigmund Freud, the founder of psychoanalysis, photographed in 1922. In 1938 Freud and his family emigrated to Hampstead, London to escape the Nazi occupation of Austria. The following year, at the age of 83, he died from cancer of the jaw and cheek, from which he had suffered for 16 years. *Mary Evans Picture Library*

Freud was born on 6 May 1856 at Freiberg, in Moravia (now Príbor in the Czech Republic), the son of a wool merchant. The family moved to Vienna when he was four, and at the age of 17 he entered Vienna University to read medicine. He graduated in 1881 and continued neurological research in the university laboratories under Ernest Brücke (1819–1892). In 1885 he went to Paris where he studied with Jean-Martin ⇨Charcot. The following year he returned to Vienna and set up private practice as a neurologist. He also married that year, and later had six children, one of whom was Anna Freud (1895–1982), who became a distinguished child psychoanalyst. Although derided by much of the medical profession, he gave psychoanalysis enough of an impetus to warrant the holding of the first psychoanalytic congress, in Salzburg in 1908. He was elected a member of the Royal Society in 1936. Two years later Nazi Germany invaded and occupied Austria and Freud had to leave. He moved to London and a year later, on 23 September 1939, he died of cancer of the jaw, from which he had suffered for 16 years.

While assisting the French neurologist Charcot, Freud became influenced by his use of hypnosis in trying to find an organic basis for hysteria. Charcot had been investigating areas of the brain responsible for certain nervous functions, and encouraged Freud's interest in the psychological aspects of neurology, in particular hysteria. Freud set up his practice in Vienna to study the psychological basis of nervous disorders. The Viennese physician Josef Breuer (1842–1925) had told Freud of an occasion when he had cured symptoms of hysteria by encouraging a patient to recollect, under hypnosis, the circumstances of the hysteria and to express the emotions that accompanied them. Following the methods used by his colleague, Freud treated his patients with hypnosis and formulated his ideas about the conscious and the unconscious mind. He believed that repressed thoughts were stored in the unconscious mind and affected a person without the source of the effect being known.

Freud used this method until about 1895 when he replaced hypnosis by the technique of free-association (the 'talking cure'), perhaps because he could not master the art of hypnosis. The free-association method allowed the patient to talk randomly and with little guidance. The patient relaxed to such an extent that thoughts came through that were previously hidden from the patient's conscious. The introduction of free-association marked the beginning of psychoanalysis in the sense used today. That is, that through a succession of periods of analysis, barriers that one puts up against knowledge of oneself are slowly broken down, with the help of the analyst.

Free-association led to the interpretation of dreams. Freud reasoned that dreams represented thoughts in the unconscious mind. He had noticed how often the train of thought in free-association included the recollection of a dream. By using free-association on the subjects of some of his own dreams he explained them as attempts to fulfil in fantasy some desire that he was repressing. The use of dream interpretation thus lay in revealing the contents of the unconscious, which are repressed when one is awake.

Freud drew a comparison between the symbolism of dreams and of mythology and religion, stating that religion was infantile (God as the father image) and neurotic (projection of repressed wishes); he claimed that it was unnecessary and retarded social maturity by perpetuating the projection of these desires.

Around 1905 Freud was collecting about him a following of young men such as Alfred ⇨Adler and Carl ⇨Jung, whom he influenced deeply. It was at this time also that the most controversial part of his work was publicized, connected with his ideas on sex as a cause of some neurotic disorders. Freud maintained that sexual gratification in childhood could carry through to adulthood resulting in psychological problems. An example he gave was that the first sexual impulses are felt during the suckling phase of an infant's life, and if these impulses become fixated, the child may mature into an adult dependent on the mother. In Freud's terms, the sexual nature of a child's relationship with its parents could result in an Oedipus complex involving dormant sexual feelings towards the mother and jealousy towards the father, feelings that could last through adulthood.

Freud saw the mind as operating on two levels: the primary level, characterized by symbolic thinking, which he called the id, and which was the source of all basic drives, and the secondary level, marked by logical thinking and characterized by a sense of reality, and critical faculties. The id suppressed by social mores becomes the superego, and eventually emerges as the ego.

Freud produced several publications detailing his theories. The best-known of these books are: *The Physical Mechanism of Hysterical Phenomena* (1893), *Studien über Hysterie/Studies in Hysteria* (1895), *Die Traumdeutung/The Interpretation of Dreams* (1900), and *Totem und Tabu/Totem and Taboo* (1913).

In his early years Freud also carried out pioneer work on cocaine and advocated its use as a local anaesthetic for mild pain. Only later, however, was it accepted and introduced into the medical practice.

Further Reading

Appignanesi, Richard, *Freud for Beginners,* Icon Books, 1992.

Badcock, C R (Christopher R), *Essential Freud,* Basil Blackwell, 1988.

DiCenso, James, *The Other Freud: Religion, Culture, and Psychoanalysis,* Routledge, 1999.

Ferris, Paul, *Dr Freud: A Life*, Pimlico, 1998.
Gay, Peter, *Reading Freud: Explorations and Entertainments*, Yale University Press, 1990.
Isbister, J N, *Freud: An Introduction to His Life and Work*, Polity – Blackwell, 1985.
Storr, Anthony, *Freud*, Past Masters series, Oxford University Press, 1996.
Wilson, Stephen, *Freud*, Pocket Biographies series, Sutton, 1997.

Freundlich, Herbert Max Finlay (1880–1941) German physical chemist, best known for his extensive work on the nature of colloids, particularly sols and gels.

Freundlich was born in Berlin-Charlottenburg on 28 January 1880, the son of a German father and a Scottish mother. Shortly after his birth the family moved to Biebrich, where his father had been made director of an iron foundry. At school Freundlich studied classics and showed a great aptitude for music, but by the time he left the school in 1898 he had opted for a career in science. In that same year he went to the University of Munich for a preliminary science course, then specialized in chemistry when he moved to the University of Leipzig. His thesis on the precipitation of colloidal solutions by electrolysis gained him a doctorate in 1903.

From 1903 to 1911 he assisted Wilhelm ⟩Ostwald at his Leipzig Institute for Analytical and Physical Chemistry. He then became professor of physical chemistry at the Technische Hochschule in Brunswick. At the outbreak of World War I in 1914 he was declared unfit for military service, so went to the Kaiser Wilhelm Institut in Berlin-Dahlem where he worked (under Fritz ⟩Haber) on the use of charcoal in gas masks. In 1919 he became the head of the institute's section devoted to colloid chemistry. He had married his first wife, Marie Mann, in 1908, but she died in 1917 during childbirth. Six years later he married Hella Gilbert.

Freundlich resigned his position when Adolf Hitler came to power in 1933 and went to the UK where, sponsored by Imperial Chemical Industries (ICI), he worked at University College, London. Later, in 1938, he went to the USA and took up the appointment of distinguished service professor of colloid chemistry at the University of Minnesota. He died in Minnesota on 30 March 1941.

While he was at Leipzig, Freundlich formulated what has become known as Freundlich's adsorption isotherm, which concerns the accumulation of molecules of a solution at a surface. It can be expressed as:

$$x/m = (kc)^{\frac{1}{n}}$$

Where x and m are the masses of material adsorbed and adsorbent, respectively, c is the equilibrium concentration of the solution, and k and n are constants.

Freundlich's other researches were devoted to all aspects of colloid science. He investigated colloid optics, the scattering of light by dispersed particles of various shapes. He studied the electrical properties of colloids, since electrostatic charges are largely responsible for holding colloidal dispersions in place.

He also carried out a major series of investigations on mechanical properties such as viscosity and elasticity, and studied the behaviour of certain systems under other types of mechanical forces. For example, he introduced the term 'thixotropy' to describe the behaviour of gels, the jelly-like colloids that show many strange properties intermediate between those of liquids and solids. One modern industrial application of this work has been the development of non-drip paints.

Freundlich was very much a chemist and was not particularly interested in formulating long mathematical equations to explain or justify his practical results. In his later years he was particularly interested in the commercial use of colloids in the rubber and paint industries.

Friedel, Charles (1832–1899) French organic chemist and mineralogist, best remembered for his part in the discovery of the Friedel–Crafts reaction. Throughout his career, he successfully combined his interests in chemistry and minerals.

Friedel was born on 12 March 1832 in Strasbourg. He was educated locally (where he studied under Louis ⟩Pasteur) and at the Sorbonne in Paris. He qualified in both chemistry and mineralogy and in 1856 was made curator of the collection of minerals at the Ecole des Mines. In 1871 he became an instructor at the Ecole Normale and from 1876 was professor of mineralogy at the Sorbonne. Following the death of Charles Adolphe ⟩Wurtz in 1884 he became also professor of organic chemistry and director of research, a post he held until he died in Montaubin on 20 April 1899.

Friedel's dual interests gave him a wide-ranging field for research, which he covered with great success. He made extensive inroads into the mysteries of the various alcohols, a subject that had received little attention until then. In 1871, working with R D da Silva, he synthesized glycerol (propantriol) from propylene (propene). Friedel met and became friends with US chemist James Mason Crafts (1839–1917), who arrived in Paris having spent a year doing postgraduate studies in Germany. Crafts returned to the USA to take up a professorship at Cornell University, but went back to Paris in 1874 and began carrying out research with Friedel.

In 1877 they discovered the Friedel–Crafts reaction, which uses aluminium chloride as a catalyst to facilitate the addition of an alkyl halide (halogenoalkane) to an aromatic compound. The reaction has proved to be extremely useful in organic synthesis and is now employed extensively in the industrial preparation of triphenylamine dyes.

From 1879 to 1887 Friedel concentrated on the attempted synthesis of minerals, including diamonds, using heat and pressure. He established the similarity in properties between carbon and silicon and therefore the similarity in structure of their long-chain polymeric compounds. This work paved the way for a better understanding of silica minerals and the use of silicates in industry.

Friedman, Aleksandr Aleksandrovich (1888–1925) Russian mathematician with a keen interest in applied mathematics and physics. He made fundamental contributions to the development of theories regarding the expansion of the universe.

Friedman was born in St Petersburg on 29 June 1888. An excellent scholar, both at school and at the University of St Petersburg, where he studied mathematics 1906–10, Friedman later served as a member of the mathematics faculty staff. In 1914 he was awarded his master's degree in pure and applied mathematics, and served with the Russian air force as a technical expert during World War I. Friedman also worked in a factory (which he later managed) that produced instruments for the aviation industry. From 1918 to 1920 he was professor of theoretical mechanics at Perm University, but he returned to St Petersburg in 1920 to conduct research at the Academy of Sciences. He died prematurely, after a life dogged by ill-health, on 16 September 1925.

Friedman's early research was in the fields of geomagnetism, hydromechanics, and, above all, theoretical meteorology. His work of the greatest relevance to astronomy was his independent and original approach to the solution of Albert ◊Einstein's field equation in the general theory of relativity. Einstein had produced a static solution, which indicated a closed universe. Friedman derived several solutions, all of which suggested that space and time were isotropic (uniform at all points and in every direction), but that the mean density and radius of the universe varied with time – indicating either an expanding or contracting universe. Einstein himself applauded this significant result.

Friese-Greene, William (1855–1921) English inventor and one of the pioneers of cinema photography.

Friese-Greene was born on 7 September 1855 in Bristol where he was educated at the Blue Coat School. He became interested in photography and in about 1875 opened a portrait studio in Bath. In the early 1880s he met a mechanic, JAR Rudge, who asked him to produce slides for a magic lantern (forerunner of the modern slide projector). This work awakened Friese-Greene's interest in moving pictures. In 1885 he opened a studio in London and met Mortimer Evans, an engineer. They decided to collaborate, and in 1889 Friese-Greene patented a camera that could take ten photographs per second on a roll of sensitized paper. Using his own apparatus, he was able to project a jerky picture of people and horse-drawn vehicles moving past Hyde Park Corner – probably the first time a film of an actual event had been projected on a screen. In 1890 he substituted celluloid film for the paper in the camera, and in the next few years he patented improved cameras and projectors. Friese-Green died in London on 5 May 1921.

The early story of moving pictures is obscure, complicated by claims and counterclaims. Certainly several inventors were working on similar lines at about the same time. In 1824, Peter Roget lectured to the Royal Society on the subject of persistance of vision, and projected onto a screen a series of still pictures at the rate of 24 per second, which gave the illusion of smooth and continuous movement. In the 1860s and 1870s there were various inventions for similarly projecting a series of stills, such as that of Heyl (in which transparencies were mounted on a glass disc and rotated). Faster camera shutter speeds and improved photographic emulsions enabled people to take sharper pictures of moving objects. Eadweard Muybridge (1830–1904) took a series of photographs of a racehorse by placing 24 cameras along a track, then projecting the pictures using an apparatus similar to Heyl's. Thomas ◊Edison designed a motion-picture machine that recorded pictures in a spiral on a cylinder, but it was unsatisfactory. In 1889 George ◊Eastman, founder of the Kodak company, produced roll film that solved part of Edison's problem, but the pictures could still be viewed only through a lens and not projected. Edison's invention was improved by others, and in France the brothers Auguste and Louis ◊Lumière developed a machine that functioned both as a camera and a projector. They arranged a show in London in 1896, the first time the public had been able to see moving pictures.

Although Friese-Greene's films were only short fragments and consisted of only ten pictures per second – inadequate to produce a convincing effect of movement – he had taken and projected 'moving' pictures before Edison, and his patent was judged by a US court to be the master patent. This brought him neither success nor financial gain, however. The same seems to have been true of his other inventions: a three-colour camera, moving pictures using a two-colour process, and machinery for rapid photographic processing and printing.

Frisch, Karl von (1886–1982) Austrian-born German ethologist who is best known for his studies of bees, particularly for his interpretation of their dances as a means of communicating the location of a food source to other bees. He was awarded many honours for his research into animal behaviour, including the 1973 Nobel Prize for Physiology or Medicine (shared with Konrad ◊Lorenz and Niko ◊Tinbergen).

Frisch was born on 20 November 1886 in Vienna. He was educated at the Schottengymnasium and the University of Vienna and then moved to Germany to the University of Munich, from which he gained his doctorate in 1910. In the same year he became a research assistant there, and then lecturer in zoology and comparative anatomy in 1912. From 1921 to 1923 he was professor and director of the Zoological Institution at the University of Rostock, then held the same post at the University of Breslau (now Wrocław in Poland)

He returned to the University of Munich in 1925 and established the Zoological Institution there. But the institution was destroyed during World War II, and so Frisch moved back to Austria, to the University of Graz, in 1946. He returned again to Munich in 1950, however, and remained there until his retirement in 1958. He died in Munich on 6 December 1982.

Frisch's most renowned work concerned communication among honey bees. By marking bees and following their movements in special observation hives, he found that foraging bees that had returned from a food source often perform a dance, either a relatively simple round dance or a more complex waggle dance. He also found that bees dance only when they have encountered a rich food source, and concluded that these dances are a means of communicating information to other bees about these sources. In the round dance, the returned forager moves round and round in a tight circle on the honeycomb. Other bees cluster round and touch the dancing bee with their antennae, thereby detecting the scent of the feeding place (bees' scent organs are located on their antennae). According to Frisch, the round dance tells the bees only that there is food within about 100 m/330 ft of the hive; it does not give the exact location. If the food is more than 100 m/330 ft away, the returned forager performs the waggle dance, in which the bee moves through a figure of eight, its abdomen waggling vigorously during the straight cross-piece of the figure of eight. Frisch claimed that the waggle dance conveys both the distance of the food from the hive and the direction of the food relative to the hive: the distance is proportional to the time taken to complete one circuit of the dance, and the direction of the food in relation to the Sun is indicated by the angle between the vertical (the dance is performed on a vertical honeycomb) and the crosspiece of the figure of eight.

Frisch published his findings in 1943, but doubt has since been thrown on his interpretation of the significance of the dances. Other researchers have found that bees emit various sounds, some of which seem to relate to the location of food. This does not imply that bees do not dance: they do, but their dances may not be the means by which they communicate.

In addition to his studies of bees' dances, Frisch also researched into the visual and chemical sense of bees and fish. In 1910 he began an investigation that demonstrated that fish can distinguish between differences in brightness and colour (the subject of his PhD thesis). And in 1919 he showed that bees can be trained to discriminate between various tastes and odours.

Frisch, Otto Robert (1904–1979) Austrian-born British physicist who first described the fission of uranium nuclei under neutron bombardment, coining the term 'fission' to describe the splitting of a nucleus.

Frisch was born in Vienna on 1 October 1904. He displayed surprising skills in mathematics at a very early age, but when he entered the University of Vienna in 1922 he studied physics, being awarded his PhD in 1926. The next year he began work in the optics division of the Physikalische Technische Reichsanstalt in Berlin on a new light standard to replace the candle power.

In 1930 Frisch moved to Hamburg to become an assistant to Otto ◊Stern, where they did research into molecular beams. Then in 1933 Frisch, being Jewish, was fired because of the racial laws the Nazis introduced following Adolf Hitler's rise to power. He moved to the department of physics at Birkbeck College, London, and a year later went to the Institute of Theoretical Physics in Copenhagen.

The German occupation of Denmark at the beginning of World War II forced Frisch to move once again and he was offered posts in the UK, first at Birmingham University and then in 1940 at Liverpool, where facilities such as a cyclotron were available for nuclear research. In 1943 he became a British citizen, which enabled him to become part of the delegation of British scientists sent to the Los Alamos nuclear research base in the USA. There he supervised the Dragon experiment, which led to the first atomic bomb. At the end of World War II, Frisch returned to the UK to become deputy chief scientific officer at the newly established Atomic Energy Research Establishment at Harwell, near Oxford. He remained there until 1947 when he became Jacksonian Professor of Natural Philosophy at Cambridge University. He retired in 1971 and died after an accident in Cambridge on 22 September 1979.

In 1938 Frisch was spending a holiday in Sweden with his aunt Lise ◊Meitner, herself a well-known physicist. She received a letter from two of her colleagues, Otto ◊Hahn and Fritz Strassmann (1902– 1980), which stated that the bombardment of uranium nuclei with neutrons had resulted in the production of three isotopes of barium, a much lighter element. Frisch and Meitner realized that this meant that the uranium nucleus had been cleaved. They calculated that the collision of a neutron with a uranium nucleus might cause it to vibrate so violently that it would become elongated. If the nucleus stretched too much, the mutual repulsion of the positive charges would overcome the surface tension of the nucleus and it would then break into two smaller pieces, liberating 200 MeV (2×10^8 electronvolts) of energy and secondary neutrons. Frisch coined the term 'nuclear fission' for this process, after the use of that term to describe cell division in biology.

During his wartime researches in the UK, Frisch worked on methods of separating the rare uranium-235 isotope that Niels ◊Bohr had calculated would undergo fission. He also calculated details such as the critical mass needed to produce a chain reaction and make an atomic bomb, and was instrumental in alerting the British government to the need to undertake nuclear research. Then at Los Alamos, Frisch designed the Dragon experiment, which involved dropping a slug of fissile material through a hole in a near-critical mass of uranium. This allowed the scientists to study many details of a near-critical mass that they would not otherwise have been able to investigate. At the first test explosion on 16 July 1945, Frisch conducted experiments from a distance of 40 km/25 mi.

When Frisch moved to Harwell after the war he did some mathematical work on chain reactions and supervised the building of the laboratories. At Cambridge he was mainly concerned with teaching and the development of automatic devices to evaluate information produced in bubble chambers, which are used to study particle collisions.

Frisch was a talented experimental physicist, who partly by virtue of his fortunate position – both geographical and temporal – was able to contribute to and observe some of the most dramatic developments in the history of science.

Frobenius, Georg Ferdinand (1849–1917) German mathematician who is now chiefly remembered for his formulation of the concept of the abstract group – a theory that proposed what is now generally considered to be the first abstract structure of 'new' mathematics. His research into the theory of groups and complex number systems was of fundamental significance, and he also made important contributions to the theory of elliptic functions, to the solution of differential equations, and to quaternions.

Frobenius was born on 26 October 1849 in Berlin, where he received his early education. His study of mathematics began when he attended Göttingen University from 1867; in only three years he gained a doctorate. In 1870 he returned to his former school in Berlin as a teacher, and stayed there for a year before moving to a school of higher standard and status in the same city. By this time he had already presented many papers on mathematics – including the publication of his method of finding an infinite series solution of a differential equation at a regular single point, and other papers on Niels ◊Abel's problem in the convergence of series and Pfaff's problem in differential equations – and had earned a fair reputation. The result was that he was appointed assistant professor at Berlin University in 1874, and in the following year became full professor at the Eidgenossische Polytechnikum in Zürich. Seventeen years later, in 1892, Frobenius returned to the University of Berlin in order to take up the post of professor of mathematics, where he remained for the rest of his working life. He died on 3 August 1917, in Charlottenberg.

It was in Berlin the first time, in a study of the work of Ernst ◊Kummer and Leopold ◊Kronecker, that Frobenius became interested in abstract algebra; his major contributions to the subject were published in 1879 (*Über Gruppen von vertauschbaren Elementen,* written in collaboration with Stickelberger) and in 1895 (*Über endliche Gruppen*). Later publications contained further development of group theory, the last of which (in 1906) – *Über die reellen Darstellungen der endlichen Gruppen* – was written with Issai Schur (1875–1941), together with whom Frobenius completed the theory of finite groups of linear substitutions of *n* variables.

By studying the different representations of groups and their elements, Frobenius provided a firm basis for the solving of general problems in the theory of finite groups. His methods were later continued by William ◊Burnside, and his results also proved useful to the development of quantum mechanics.

Froude, William (1810–1879) English engineer and hydrodynamiscist who first formulated reliable laws for the resistance that water offers to ships and for predicting their stability. He also invented the hydraulic dynameter for measuring the output of high-power engines. These achievements were fundamental to marine development.

Froude was educated at Buckfastleigh, Westminster School, and Oriel College, Oxford, where he obtained a first in mathematics and a third in classics in 1832. He remained at Oxford working on water resistance and the propulsion of ships and in 1838 became an assistant to Isambard Kingdom ◊Brunel on the building of the Bristol and Exeter Railway.

In 1839 he married Katherine Holdsworth of Widdicombe. He probably worked on the South Devon

Railway and was intimately connected with the ill-fated atmospheric railway. In 1846 he went to live with his father at Dartington Parsonage and began work in earnest on marine hydrodynamics. Brunel consulted him on the behaviour of the *Great Eastern* at sea and, on his recommendation, the ship was fitted with bilge keels.

When his father died in 1859 Froude moved to Paignton where he began his tank-testing experiments. In 1863 he started to build his own house, known as Chelston Cross, at Cockington, Torquay, and helped the local water authority with its supply problems. In 1867 he began his experiments with towed models. After grudging financial assistance from the Admiralty, he built another experimental tank near his home in 1871.

He described his hydraulic dynameter in a paper to the Institution of Mechanical Engineers in 1877, but did not live to see his machine work. It was built in 1878, the year in which Froude's wife died and in which he became seriously ill. He went on a voyage of recuperation to South Africa but caught dysentry and died on 6 May 1879 in Simonstown, where he was buried.

Froude was elected a fellow of the Royal Society in 1870 and his work was continued by his sons Richard and Robert. Robert built the towing tank for the Admiralty at Haslar, near Portsmouth, and Richard joined Hammersley Heenan to manufacture the dynamometer commercially.

Generally there are two modes in which vessels can travel: by displacement, in which they force their way through the water, and by planing, or skating on top of the water. In the displacement mode, the propulsive power is absorbed in making waves and in overcoming the friction of the hull against the water.

Froude's first successful experiments started in 1867 when he towed models in pairs, balancing one hull shape against the other. Initially he incorporated his findings into a single law, now known as Froude's law of comparison. This states that the entire resistance of similar-shaped models varies as the cube of their dimensions if their speeds are as the square root of their dimensions.

As Froude himself realized, this law becomes increasingly unreliable as the difference in size between the models increases. This is because the frictional resistance and the wavemaking resistance follow different laws. His law of comparisons is now only applied to the wavemaking component.

To estimate frictional resistance, Froude carried out tests in his tank at Torquay, where he towed submerged (to eliminate wavemaking resistance) planks with different surface roughness. He was able to establish a formula that would predict the frictional resistance of a hull with accuracy.

With these two analytical results, using only models and mathematics, Froude had found a reliable means of estimating the power required to drive a hull at a given speed. Model-testing had been tried before but was considered unreliable because previous workers had failed to appreciate that the two major components of ship resistance varied differently. Opponents of model testing maintained that the only way to gather the required information was to work on full-sized hulls. It was this anti-model lobby that was largely responsible for Froude's difficulties in persuading the Admiralty to part with £2,000 for building the Torquay tank.

Froude had done a large amount of theoretical work on the rolling stability of ships and when the Torquay tank was built, he was able to carry out model experiments, relating them to observations made on actual ships. His general deductions were challenged at the time but they were found to be correct and to this day are the standard exposition of the rolling and oscillation of ships.

Engine builders usually need to test their machinery before installation. The friction brake is useful only for lower-power applications, so Froude employed hydrodynamic principles to absorb 1,500 kW/2,000 hp at 90 rpm. His brake consisted of a rotor and stator, both of which were shaped in a series of semi-circular cups angled at 45 ° but of opposite pitch. The change of momentum when water passes from rotor to stator and back again creates a braking reaction. By measuring this braking reaction and the shaft speed, the power of the engine could be calculated.

Froude was a tireless experimenter who put one of the most powerful analytical tools into the hands of marine architects. He has been credited as the founder of ship hydrodynamics. If any one person can be said to have founded anything in engineering, this is probably accurate.

Fuchs, Immanuel Lazarus (1833–1902) German mathematician whose work on Bernhard ◊Riemann's method for the solution of differential equations led to a study of the theory of functions that was later crucial to Henri ◊Poincaré in his own important investigation of function theory. Fuchs's main scientific importance may be seen, therefore, as providing a sort of link between the 19th- and 20th-century ideas of mathematical development.

Fuchs was born on 5 May 1833 in Moschin (now in Poland). During his elementary education his mathematical talents were already apparent, and he went on to study at the University of Berlin, under Ernst ◊Kummer and Karl ◊Weierstrass. He gained his PhD in Berlin in 1858, and began teaching at local schools. In 1865 he became a lecturer at Berlin University, and only a year later was promoted to professor. In 1869 he became the professor of mathematics at the Artillery and Engineering School at Griefswald. Later, in 1874, he transferred to the University of Göttingen and then, in 1875, to that of Heidelberg to take up their chairs of mathematics. Returning to Berlin in 1882, he succeeded Weierstrass two years later as professor of mathematics there, and over the next 20 years held several administrative and academic posts of responsibility at the university. He died, in Berlin, on 26 April 1902.

Fuchs's interest in the theory of functions was first aroused during his student days in Berlin, but it was not until he took up his first professional appointment that he began to work seriously on his research. He produced a number of papers that were intended to develop Riemann's work on a method for solving differential equations. Fuchs's proposals were in contrast to those put forward earlier by Augustin ◊Cauchy, who used power series.

The first proof for solutions of linear differential equations of order *n* was developed from this study, as were the Fuchsian differential equations and the Fuchsian theory on solutions for singular points. His work in this field was of great importance to Poincaré's work on automorphic functions. Fuchs also carried out some research into number theory and geometry.

Fuller, Solomon Carter (1872–1953) US physician, neurologist, psychiatrist, and pathologist who is acknowledged as being the first black psychiatrist.

The son of state officials, Fuller was born in Monrovia, Liberia, on 11 August 1872. He went to the USA in 1889 and was educated at Livingstone College, Salisbury, North Carolina, receiving his BA in 1893. After studying medicine at Long Island College Hospital, Brooklyn, New York, and later at the Boston University School of Medicine, he was awarded his MD in 1897. Following a two-year internship at Westborough State Hospital, Massachusetts, he was appointed as a pathologist at the hospital and as a faculty member of the Boston University School of Medicine. He practised medicine in Boston and at his home in Framingham and taught pathology, neurology, and psychology at the university until he retired as professor emeritus in 1937. Fuller died in 1953; he was married to the sculptor Meta Warrick Fuller and had three sons.

Fuller is best known for his work in neuropathology and psychiatry. His postgraduate studies included training at the Carnegie Laboratory, New York, and, between 1904–05, at the University of Munich under Alois Alzheimer (1864–1915). He became well known for his work on degenerative brain diseases, including Alzheimer's disease, which he attributed to causes other than arteriosclerosis; this was further supported by medical researchers in 1953.

Fuller published widely in books and medical journals and in 1913 became the editor of *Westborough State Hospital Papers,* which specialized in mental diseases. He was a member of many medical and psychiatric societies. Fuller's work was acknowledged in the 1970s; in 1971 the Black Psychiatrists of America presented a portrait of Fuller to the American Psychiatric Association in Washington, DC, and in 1973 Boston University commemorated him in an all-day conference; the Solomon Carter Fuller Mental Health Center in Boston is named after him.

Fulton, Robert (1765–1815) US artist and engineer and inventor who built one of the first successful steamboats, propelled by two paddle-wheels.

Fulton was born on 14 November 1765 in Lancaster County, Pennsylvania. His artistic talent was evident at an early age, and at the age of 17 he moved to Philadelphia, where he became a successful portrait painter and miniaturist. Four years later, in 1786, he decided to go to London to study under the US artist Benjamin West. England changed Fulton's life. The country was involved in the Industrial Revolution and he was fascinated by all he saw and eventually, in 1793, he gave up art as a vocation in favour of engineering projects.

When Fulton was 14 he designed a small paddleboat and now he considered designing one with a steam engine to power it. In 1786 John Fitch (1743–1798) had demonstrated a steamboat in the USA, but it had not proved to be a success. The British government had placed a ban on the export of steam engines, but Fulton nevertheless contacted a British company about the possibility of purchasing an engine suitable for boat propulsion. Meanwhile he designed and patented a device for hauling canal boats over difficult country. He also patented machines for spinning flax (for linen), sawing marble, and twisting hemp (for rope), and he built a mechanical dredger for canal construction.

In 1796 Fulton went to France, where he experimented with fitting steam engines to ships, and by 1801 he had also carried out tests with the *Nautilus,* a submarine he had invented. But he failed to interest the French in his inventions, so by 1804 he tried the British government, again without success. In 1802 Fulton had met Robert Livingston (1746–1813), a former partner in another steamboat invention and then US minister to the French government. He and Fulton joined forces and in 1803 a steam engine was obtained from the British firm of Boulton and Watt; but it took three years to get permission to export it to the USA.

In New York, Fulton worked to install the new engine in a locally built vessel. Livingston favoured designing a propulsion system using a jet of water forced out at the stern under pressure, but Fulton settled on a paddlewheel on each side. In 1807, the paddlesteamer *Clermont,* with a 18-kW/24-hp engine fitted into its 30-m/100-ft hull, made its first successful voyage up the Hudson River at an average speed of 8 kph/5 mph. A large boatworks was built in New Jersey, and steamboats came into use along the Atlantic coast and later in the west. In 1815, Fulton began to build a steam-powered warship for the US navy.

Fulton died on 24 February 1815 in New York City.

Further Reading

Bowen, Andy R, *Head Full of Notions: A Story About Robert Fulton,* Lerner Publishing Group, 1996.

Sutcliffe, Alice, *Robert Fulton and the 'Clermont',* Reprint Services Corporation, 1993.

Funk, Casimir (1884–1967) Polish-born US biochemist, best known for his researches on dietary requirements, particularly of vitamins. He was the first to isolate nicotinic acid (also called niacin, one of the B vitamins) and his work did much to further studies of nutritional requirements.

Funk was born in Warsaw on 23 February 1884, the son of an eminent dermatologist. An able student, he was educated at the University of Bern, Switzerland, from which he gained his doctorate in organic chemistry in 1904, when he was still only 20 years old. He then worked at the Pasteur Institute in Paris, the Wiesbaden Municipal Hospital, the University of Berlin, and the Lister Institute in London, where he researched into the disease beriberi. In 1915 he emigrated to the USA, working in various industrial and university posts in New York; he became a naturalized US citizen in 1920.

In 1923, supported by the Rockefeller Foundation, he returned to Warsaw to take up the directorship of the biochemistry department of the State Institute of Hygiene but, because of the country's uncertain political situation, Funk left this post in 1927 and went to Paris. In the following year he worked in a consultatory capacity for a pharmaceuticals company and founded the Casa Biochemica, a privately funded institution concerned with research. With the German invasion of France in 1939, Funk abandoned this venture and returned to the USA, becoming a consultant with the United States Vitamin Corporation. In 1940 he became president of the Funk Foundation for Medical Research, a position he held until his death. He died in Albany, New York, on 20 November 1967.

Following the research of Frederick Gowland ◊Hopkins into what we now know to be deficiency diseases, such as beriberi and pellagra, Funk succeeded in 1911 at the Lister Institute in obtaining a concentrate that cured a pigeon disease similar to beriberi. He showed that the anti-beriberi concentrate was an amine, and in 1912 he discussed the causes of beriberi, scurvy, and rickets as deficiency diseases. He also suggested that the accessory food factors needed to prevent these disorders are all amines and so he named them vitamines (vital amines). It was later discovered that not all accessory food factors are amines, and Funk's term was shortened to 'vitamins'.

Funk continued to try to find the anti-beriberi factor for human beings and finally isolated nicotinic acid from rice polishings. But it did not cure beriberi, so he discarded it. (Later Otto ◊Warburg discovered that nicotinic acid prevents pellagra.) In 1934 Robert Williams (1886–1965) isolated about 10 g/0.3 oz of the anti-beriberi factor (now known as thiamin or vitamin B_1) from 1,000 kg/2,200 lb of rice polishings, and in 1936 Funk determined its molecular structure and developed a method of synthesizing it.

Funk also researched extensively into the function of hormones in animals, particularly that of male-sex hormones, and into the biochemical causes of cancer, diabetic disorders, and ulcerative conditions.

Gabor, Dennis (1900–1979) Hungarian-born British physicist and electrical engineer, famous for his invention of holography – three-dimensional photography using lasers – for which he received the 1971 Nobel Prize for Physics.

Gabor was born on 5 January 1900 in Budapest. He was educated at the Budapest Technical University and then at the Technishe Hochschule in Charlottenburg, Berlin. From 1924 to 1926 he was an assistant there, and for the next three years he held the position of research associate with the German Research Association for High-Voltage Plants. He was a research engineer for the firm of Siemens and Halske in Berlin from 1927 until he fled Nazi Germany for the UK in 1933. He then worked as a research engineer with the Thomson–Houston Company of Rugby 1934–38, and later became a British subject. In 1949, he joined Imperial College, London, as a reader in electronics. He was professor of applied electron physics 1958–67, when he became a senior research fellow. From 1976 until his death he was professor emeritus of applied electron physics at the University of London. He died in London on 8 February 1979.

Gabor first conceived the idea of holography in 1947 and developed the basic technique by using conventional filtered light sources. Because conventional light

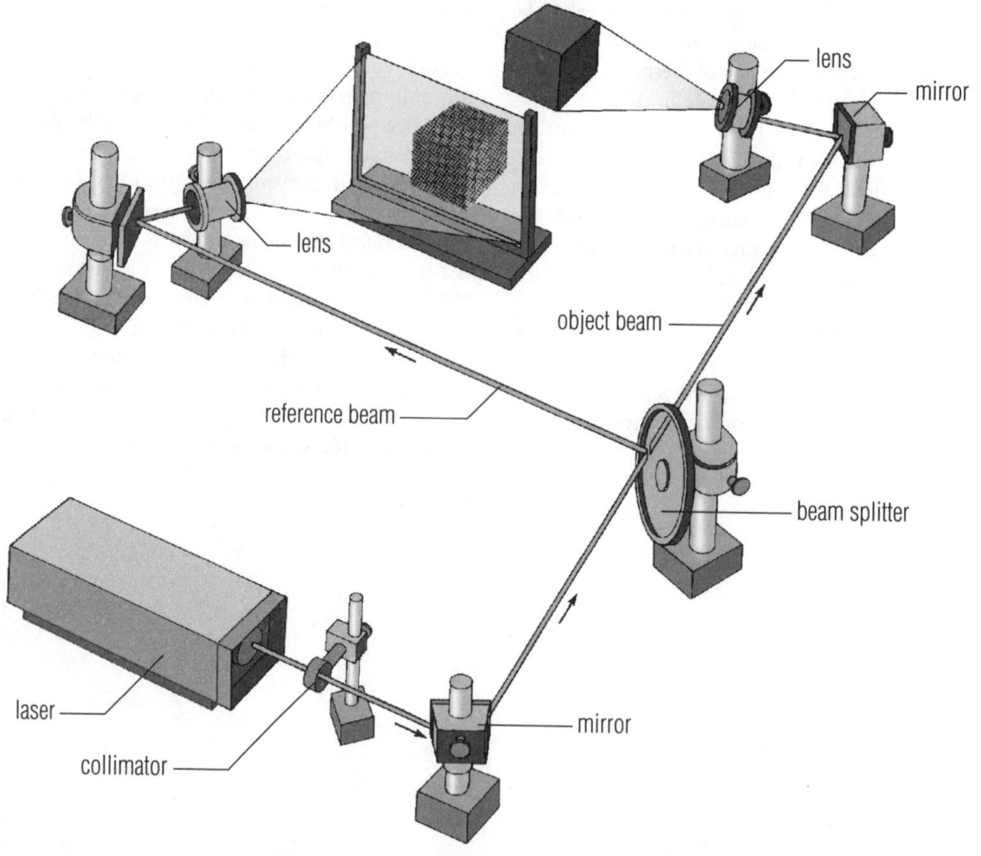

Gabor Recording a transmission hologram. Light from a laser is divided into two beams. One beam goes directly to the photographic plate. The other beam reflects off the object before hitting the photographic plate. The two beams combine to produce a pattern on the plate which contains information about the 3-D shape of the object. If the exposed and developed plate is illuminated by laser light, the pattern can be seen as a 3-D picture of the object.

sources provided too little light or light that was too diffuse, his idea did not become commercially feasible until the laser was demonstrated in 1960 and was shown to be capable of amplifying the intensity of light waves.

Holography is a means of creating a unique photographic image without the use of a lens. The photographic recording of the image is called a hologram. The hologram appears to be an unrecognizable pattern of shapes and whorls, but when it is illuminated by coherent light (as by a laser beam), the light is organized into a three-dimensional representation of the original object. Gabor coined the name from the Greek *holos* ('whole') and *gram* ('message'), because the image-forming mechanism that he conceived recorded all the optical information in a wavefront of light. In ordinary photography, the photographic image records the variations in light intensity reflected from an object, so that dark areas are produced where less light is reflected and light areas where more light is reflected. Holography records not only the intensity of light, but also its phase, or the degree to which the wavefronts making up the reflected light are in step with each other. The wavefronts of ordinary light waves are not in step – ordinary light is incoherent.

When Gabor began work on the holograph, he considered the possibility of improving the resolving power of the electron microscope, first by using the electron beam to make a hologram of the object and then by examining this hologram with a beam of coherent light. It is possible to obtain a degree of coherence by focusing light through a very small pinhole, but the resulting light intensity is then too low for it to be useful in holography. In 1960, the laser beam was developed. This has a high degree of coherence and also has high intensity. There are many kinds of laser beam, but two have special interest in holography, the continuous-wave (CW) and the pulsed laser. The CW laser emits a bright continuous beam of light of a single, nearly pure colour. The pulsed laser emits an extremely intense, short flash of light that lasts only about 10^{-8} second. Two US scientists, Emmett Leith and Juris Upatnieks have applied the CW laser to holography with great success, opening the way to many research applications. To achieve a three-dimensional image, the light streaming from the source must itself be photographed. If the waves of this light, with its many rapidly moving crests and troughs, are frozen for an instant and photographed, the wave pattern can then be reconstructed and will show the same three-dimensional character as the object from which the light is reflected.

Pulsed-laser holography is used in the study of chemical reactions, where optical properties of solutions often change. It is also used in wind-tunnel experiments, where it can be used to record refractive index changes in the air flow, created by pressure changes as the gas deflects around the aerodynamic object. This recording is done interferometrically (by observing interference fringes).

Apart from the invention of holography, Gabor's other work included research on high-speed oscilloscopes, communication theory, physical optics, and television. He took out more than a hundred patents for his invention and became renowned as an outstanding engineer and physicist of the 20th century.

Gajdusek, (Daniel) Carleton (1923–) US paediatrician and virologist who, with Baruch Blumberg (1925–), won the 1976 Nobel Prize for Physiology or Medicine for his work on identifying and describing 'slow virus' infections in humans based on his extensive studies of kuru, a disease of neural degeneration found in people in New Guinea.

Gajdusek was born on 9 September 1923 in Yonkers, New York, into a family of Central European extraction that put a high premium on education. He was educated at the universities of Rochester and Harvard and graduated in medicine in 1941, when he decided to specialize in paediatrics and completed internship and residencies in several children's hospitals. He also studied physical chemistry at the California Institute of Technology with the Nobel laureate Linus ◊Pauling, and won a research fellowship to work in the virology department at Harvard. Drafted in 1952, he served at the Walter Reed Army Institute for Research and continued virological research, also becoming interested in epidemiology, which he continued to study on his return to civilian life. He worked at the University of Tehran in Iran and with the future Nobel prizewinning immunologist Macfarlane ◊Burnet in Melbourne, Australia. From there he travelled to New Guinea and learned of the disease kuru, a fatal degeneration of the brain, restricted to a group of people living in a remote area. Gajdusek lived with these people, studying their language and culture and collecting post-mortem samples from kuru victims for laboratory analysis. In 1958 he returned to the USA to the National Institute of Neurological Disorders and Stroke in Bethesda, and in 1970 became head of the Central Nervous Systems Studies Laboratory, retiring in 1997. In April 1997 he was sentenced by a Maryland court to 18 months' imprisonment for sexually abusing a 15-year-old boy.

Gajdusek's researches into kuru continued for many years and he made frequent trips back to New Guinea. The affected people practised a form of ritual cannibalism, in which the women and children consumed the brains of the dead, including those who had died from the disease. (Education programmes in New Guinea have since eradicated the practices that caused kuru, and the disease had virtually disappeared by the late 1970s). Analyses of victims' brain tissues failed to reveal any signs of infective organisms, and Gajdusek and his collaborators injected extracts from such brains into the brains of chimpanzees. After about a year the animals began to display signs that characterized the human disease – shaking and trembling and loss of motor function, and the experiment was repeated, material from the brains of the ill animals being injected into further animals. In this way the illness was transmitted from animal to animal in a way that was similar to the course of the human disease.

The transmissability of the disease led Gajdusek to propose that kuru was caused by a 'slow virus' that exerts its effects after a very long incubation period and further work by his group showed that Creutzfeldt–Jakob disease (CJD) was caused by a similar agent. Since then, Stanley ◊Prusiner and others have shown that CJD, kuru and a number of similar conditions (known collectively as transmissible spongiform encephalopathies) are caused by non-living protein particles called prions.

Galen (*c.* AD 129–*c.* 199) Greek physician, anatomist, and physiologist whose ideas – as conveyed in his many books, more than 130 of which have survived – had a profound influence on medicine for some 1,400 years.

Galen was born in about AD 129 in Pergamum (now Bergama in Turkey), the son of an architect. The city of his birth contained an important shrine to Asclepius, the god of medicine, and Galen received his early medical training at the medical school attached to the shrine; he also studied philosophy at Pergamum. From about 148–157 he continued his studies at Smyrna (now Izmir) on the west coast of present-day Turkey, Corinth in Greece, and Alexandria in Egypt, after which he returned home to become chief physician to the gladiators at Pergamum. In 161 he went to Rome where he cured the eminent philosopher Eudemus, who introduced Galen to many influential people. Successful as a physician – and boastful of his successes – he soon became physician to Marcus Aurelius, then co-emperor with Lucius Verus. In about 166, however, Galen returned to Pergamum – according to Galen himself, because of the intolerable envy of his colleagues, although it is also probable that he left Rome to escape the plague brought back by Verus' army after a foreign war. About two years later Marcus Aurelius recalled Galen to Rome, and when Lucius Verus died of the plague in 169, Galen was appointed physician to Commodus, Marcus Aurelius' son. Little is known of Galen's life after this, although it is thought he remained in Rome until his death in about AD 199.

Galen was an expert dissector and his best work was in the area of descriptive anatomy. The dissection of human beings was then regarded as taboo and Galen made inferences about human anatomy from his many dissections of pigs, dogs, goats, and especially Barbary apes – although there is some evidence that he did dissect humans as well. He was a meticulous observer and his detailed descriptions of bones and muscles, many of which he was the first to identify, are particularly good; he also noted that many muscles are arranged in antagonistic pairs. He described the heart valves, distinguished seven pairs of cranial nerves, and noted the structural differences between arteries and veins. He also described the network of blood vessels under the brain, but although this network is present in many animals, it is not found in humans, and Galen erroneously assigned it an important role in his tripartite scheme of circulation in the human body. In addition he performed several vivisection experiments: he tied off the recurrent laryngeal nerve to show that the brain controls the larynx; he demonstrated the effects of severing the spinal cord at different points; and he tied off the ureters to show that urine passes from the kidneys down the ureters to the bladder. More importantly, he demonstrated that arteries carry blood, not air, thus disproving ◊Erasistratus' view, which had been taught for some 500 years.

In physiology, however, Galen was less successful. He thought that blood (the natural spirit) was formed in the liver and was then carried through the veins to nourish every part of the body, forming flesh in the body's periphery. Some of the blood, however, passed through the vena cava to the right-hand side of the heart, from where it then passed to the left-hand side through minute pores (the existence of which was pure conjecture on Galen's part) in the septum. In the left side the blood mixed with air brought from the lungs, and the resulting vital spirit was then carried around the body by the arteries. Some of this fluid went to the head, where, in the hypothetical network of blood vessels beneath the brain, the fluid was infused with animal spirit (which was thought to produce consciousness) and distributed to the muscles and senses by the nerves (which Galen thought were hollow). Thus Galen postulated a tripartite circulation system in which the liver produced the natural spirit, the heart the vital spirit, and the brain the animal spirit. Although this system is now known to be a complete misconception, it did provide a rational explanation of the facts known at the time and, because of the high esteem in which Galen was held until the Renaissance, persisted until William ◊Harvey proposed his theory of blood circulation in 1628. Another of Galen's mistaken beliefs that inhibited the progress of medical knowledge was his subscription to ◊Hippocrates' theory that human health relied upon a balance between the four humours (phlegm, black bile, yellow bile, and blood).

Galen was a prolific author who, in addition to scientific works, also wrote about philosophy and literature. He believed that Nature expressed a divine purpose, a belief that was reflected in many of his books, including his scientific texts. Belief in divine purpose became increasingly popular with the rise of Christianity (Galen himself was not a Christian), which fact, combined with the eclipse of science until the Renaissance, probably ensured the survival of many of Galen's works and accounted for the enormous influence of his ideas.

Further Reading

Brain, Peter, *Galen on Bloodletting: A Study of the Origins, Development, and Validity of his Opinions, with a Translation of the Three Works,* Cambridge University Press, 1986.

Durling, Richard J, *A Dictionary of Medical Terms in Galen,* Studies in Ancient Medicine series, Brill, 1993, v 5.

Galen, *On Examinations by Which the Best Physicians Are Recognized,* Corpus Medicorum Graecorum, Suplementum Orientale series, Akademie-Verlag, 1988, v 4.

Gilbreath, Allan, *Galen,* Ronin Enterprises, 1997.

Kudlien, Fridolf and Durling, Richard J (eds), *Galen's Method of Healing: Proceedings of the 1982 Galen Symposium,* Studies in Ancient Medicine series, Brill Academic Publishers, 1991, v 1.

Galileo, properly Galileo Galilei (1564–1643)

Italian physicist and astronomer whose work founded the modern scientific method of deducing laws to explain the results of observation and experiment. In physics, Galileo discovered the properties of the pendulum, invented the thermometer, and formulated the laws that govern the motion of falling bodies. In astronomy, Galileo was the first to use the telescope to make observations of the Moon, Sun, planets, and stars.

Galileo was born in Pisa on 15 February 1564. His full name was Galileo Galilei, his father being Vincenzio Galilei (*c.* 1520–1591), a musician and mathematician. Galileo received his early education from a private tutor at Pisa until 1575, when his family moved to Florence. He then studied at a monastery until 1581, when he returned to Pisa to study medicine at the university. Galileo was attracted to mathematics rather than medicine, however, and also began to take an interest in physics. In about 1583 he discovered that a pendulum always swings to and fro in the same period of time regardless of the amplitude (length) of the swing. He is said to have made this discovery by using his pulse to time the swing of a lamp in Pisa Cathedral.

Galileo remained at Pisa until 1585, when he left without taking a degree and returned home to Florence. There he studied the works of ◊Euclid and ◊Archimedes, and in 1586 extended Archimedes' work in hydrostatics by inventing an improved version of the hydrostatic balance for measuring specific gravity. At this time, Galileo's father was investigating the ratios of the tensions and lengths of vibrating strings that produce consonant intervals in music, and this work may well have demonstrated to Galileo that the validity of a mathematical formula could be tested by practical experiment.

In 1589 Galileo became professor of mathematics at the University of Pisa. He attacked the theories of ◊Aristotle that then prevailed in physics, allegedly demonstrating that unequal weights fall at the same speed by dropping two cannon balls of different weights from the Leaning Tower of Pisa to show that they hit the ground together. Galileo also published his first ideas on motion in *De motu/On Motion* (1590).

Galileo remained in Pisa until 1592, when he became professor of mathematics at Padua. He flourished at Padua, and refined his ideas on motion over the next 17 years. He deduced the law of falling bodies in 1604 and came to an understanding of the nature of acceleration. In 1609 he was diverted from his work in physics by

Galileo Italian physicist and astronomer Galileo Galilei was one of the founders of modern scientific method. In 1633 he was tried and imprisoned by the Roman inquisition for his support of the Copernican theory that the Earth revolves around the Sun. It was not until 1992 that the Vatican formally recognized that he was correct. *Mary Evans Picture Library*

reports of the telescope, which had been invented in Holland the year before. Galileo immediately constructed his own telescopes and set about observing the heavens, publishing his findings in *Sidereus nuncius/The Starry Messenger* (1610). The book was a sensation throughout Europe and brought Galileo immediate fame. It resulted in a lifetime appointment to the University of Padua, but Galileo rejected the post and later in 1610 became mathematician and philosopher to the Grand Duke of Tuscany, under whose patronage he continued his scientific work at Florence.

In 1612 Galileo returned to hydrostatics and published a study of the behaviour of bodies in water, in which he championed Archimedes against Aristotle. In the following year he performed the same service for Nicolaus ◊Copernicus, publicly espousing the heliocentric system. This viewpoint aroused the opposition of the church, which declared it to be heretical and in 1616 Galileo was instructed to abandon the Copernican theory. Galileo continued his studies in astronomy and mechanics without publicly supporting Copernicus although he was personally convinced. In 1624 Urban VIII became pope and Galileo obtained permission to present the arguments for the rival heliocentric and geocentric systems in an impartial way. The result, *Dialogue Concerning The Two Chief World Systems,* was published in 1632, but it was scarcely impartial. Galileo used the evidence of his telescope observations and experiments on motion to favour Copernicus, and immediately fell foul of the church again. The book was banned and Galileo was taken to Rome to face trial on a charge of heresy in April 1633. Forced to abjure his belief that the Earth moves around the Sun, Galileo is reputed to have muttered 'Eppur si muove' ('Yet it does move').

Galileo was sentenced to life imprisonment, but the sentence was commuted to house arrest and he was confined to his villa at Arcetri near Florence for the rest of his life. He continued to work on physics, and summed up his life's work in *Discourses and Mathematical Discoveries Concerning Two New Sciences.* The manuscript of this book was smuggled out of Italy and published in Holland in 1638. By this time Galileo was blind, but he still continued his scientific studies. In these last years, he designed a pendulum clock. Galileo died in Arcetri on 8 January 1642.

Galileo made several fundamental contributions to mechanics. He rejected the impetus theory that a force or push is required to sustain motion, and identified motion in a horizontal plane and rotation as being 'neutral' motions that are neither natural nor forced. Galileo realized that gravity not only causes a body to fall, but that it also determines the motion of rising bodies and furthermore that gravity extends to the centre of the Earth. He found that the distance travelled by a falling body is proportional to the square of the time of descent. Galileo then showed that the motion of a projectile is made up of two components: one component consists of uniform motion in a horizontal direction and the other component is vertical motion under acceleration or deceleration due to gravity. This explanation was used by Galileo to refute objections to Copernicus based on the argument that birds and clouds would not be carried along with a turning or moving Earth. He explained that a horizontal component of motion provided by the Earth always exists to keep such objects in position even though they are not attached to the ground.

Galileo came to an understanding of uniform velocity and uniform acceleration by measuring the time it takes for bodies to move various distances. In an era without clocks, it was difficult to measure time accurately and Galileo may have used his pulse or a water clock – but not, ironically, the pendulum – to do so. He had the brilliant idea of slowing vertical motion by measuring the movement of balls rolling down inclined planes, realizing that the vertical component of this motion is a uniform acceleration due to gravity while the horizontal component is a uniform velocity. Even so, this work was arduous and it took Galileo many years to arrive at the correct expression of the law of falling bodies. This was presented in the *Two New Sciences,* together with his derivation that the square of the period of a pendulum varies with its length (and is independent of the mass of the pendulum bob). Galileo also deduced by combining horizontal and vertical motion that the trajectory of a projectile is a parabola. He furthermore realized that the law of falling bodies is perfectly obeyed only in a vacuum, and that air resistance always causes a uniform terminal velocity to be reached.

The other new science of Galileo's masterwork was engineering, particularly the science of structures. His main contribution was to point out that the dimensions of a structure are important to its stability: a small structure will stand whereas a larger structure of the same relative dimensions may collapse. Using the laws of levers, Galileo went on to examine the strengths of the materials necessary to support structures.

Galileo's other achievements include the invention of the thermometer in 1593. This device consisted of a bulb of air that expanded or contracted as the temperature changed, causing the level of a column of water to rise or fall. Galileo's thermometer was very inaccurate because it neglected the effect of atmospheric pressure, but it is historically important as one of the first measuring instruments in science.

Galileo's astronomical discoveries were a tribute to both his scientific curiosity and his ability to devise new techniques with instruments. Within two years of first building a telescope he had compiled fairly accurate tables of the orbits of four of Jupiter's satellites and proposed that their frequent eclipses could serve as a means of determining longitude on land and at sea. His observations on sunspots are noteworthy for their accuracy and for the deductions he drew from them regarding the rotation of the Sun and the orbit of the Earth. He believed, however – following both Greek and medieval tradition – that orbits must be circular, not elliptical, in order to maintain the fabric of the cosmos in a state of perfection. This preconception prevented him from deriving a full formulation of the law of inertia, which he himself discovered although it is usually attributed to the contemporary French mathematician René ◊Descartes. Lacking the theory of Newtonian gravity, he hoped to explain the paths of the planets in terms of circular inertial orbits around the Sun.

... in my studies of astronomy and philosophy I hold this opinion about the universe, that the Sun remains fixed in the centre of the circle of heavenly bodies, without changing its place: and the Earth, turning upon itself moves round the Sun.
GALILEO LETTER TO CRISTINA DI LORENA 1615

Galileo was a pugnacious and sarcastic man, especially when confronted with those who could not or would not admit the validity of his observations and arguments. This characteristic symbolizes his position in the history of science, for with Galileo the idea that experiment and observation can be used to prove the validity of a proposed mathematical description of a phenomenon really became a working method in science. Although he built on the views and work of Archimedes and Copernicus, in his achievements Galileo can be considered to have founded not only classical mechanics and observational astronomy, but the method of modern science overall. In mechanics Isaac ◊Newton developed Galileo's work into a full understanding of inertia, which Galileo had not quite grasped, and arrived at the basic laws of motion underlying Galileo's discoveries.

Further Reading

Drake, Stillman, *Galileo: Pioneer Scientist,* University of Toronto Press, 1990.
Langford, Jerome J, *Galileo, Science, and the Church,* Ann Arbor Paperbacks series, University of Michigan Press, 1992.
Machamer, Peter K (ed), *The Cambridge Companion to Galileo,* Cambridge Companions to Philosophy series, Cambridge University Press, 1998.
Reston, James, *Galileo: A Life,* Cassell, 1994.
Sharratt, Michael, *Galileo: Decisive Innovator,* Cambridge University Press, 1996.

Galle, Johann Gottfried (1812–1910) German astronomer who was one of the first to observe Neptune and recognize it as a new planet.

Galle was born in Pabsthaus in Prussian Saxony (now in Germany) on 9 June 1812, the son of a turpentine-maker. He was educated at the University of Berlin, which he entered in 1830, and where he first met Johann ◊Encke. Graduating, he joined the staff of the Berlin Observatory, and in 1835 became assistant director under Encke. In 1851 he was appointed professor of astronomy and director of the Breslau Observatory (now in Wrocław, Poland), where he spent the rest of his working life until his retirement at the age of 83. Galle died at the age of 98, in Potsdam, Prussia, on 10 July 1910.

To gain his doctorate in 1845, Galle sent his dissertation to the director of the Paris Observatory, Urbain ◊Leverrier. By way of reply, Leverrier sent back his calculation of the position of a new planet, predicted mathematically from its apparent gravitational effect on Uranus, then the outermost known planet; Galle received Leverrier's final prediction in September 1846. Within one hour of beginning their search, Galle and his colleague Heinrich d'Arrest (1822–1875) had located Neptune, less than 1° away from the predicted position. In fact, they located it more from its absence as a star plotted on one of Encke's new star-charts, but the discovery was acclaimed nevertheless as a triumph for scientific theory as opposed to observation.

Galle was also the first to distinguish the Crepe Ring, a somewhat obscure inner ring around Saturn. First announced by Galle in 1838, its existence was forgotten until it was rediscovered independently by George ◊Bond at Harvard and by William Dawes (1799–1868) in England. It can be traced as a dusky band where it crosses the planet's globe. Saturn itself can be clearly seen through the Crepe Ring.

He also suggested a method of measuring the scale of the Solar System by observing the parallax of asteroids, first applying his method to the asteroid Flora in 1873. The method was eventually carried out with great success, but not until Galle had been dead for nearly two decades.

Galois, Evariste (1811–1832) French mathematician who, building on the work of Joseph ◊Lagrange, Karl ◊Gauss, Niels ◊Abel, and Augustin ◊Cauchy, greatly extended the understanding of the conditions in which an algebraic equation is solvable and, by his method of doing so, laid the foundations of modern group theory.

Galois was born in the village of Bourg-la-Reine, on the outskirts of Paris, on 25 October 1811. His family was of the highly respectable bourgeois class that came into its own with the restoration of the monarchy after the Napoleonic empire. His father was the headmaster of the local boarding school and mayor of the town; his mother came from a family of jurists, and it was she who gave Galois his early (chiefly classical) education. He first attended school at the age of 12, when he was sent to the Collège Louis-le-Grand in Paris as a fourth-form boarder in 1823. Another four years passed before Galois, who gained notoriety for resisting the strict discipline imposed at the school, had his mathematical imagination fired by the lectures of H J Vernier. He soon became familiar with the works of Legrange and Adrien ◊Legendre and by 1828 he was busy mastering the most recent work on the theory of equations, number theory, and elliptic functions. He quickly came to believe that he had solved the general fifth-degree equation. Like Abel before him, he discovered his error, and that discovery launched him on his search – ultimately successful and ultimately of momentous consequence for the future of mathematics – for a solution to the problem of the solubility of algebraic equations generally. By 1829 he had progressed far enough to interest Augustin Cauchy, who presented Galois' early results to the Academy of Science.

In a remarkably short period of time Galois, at the age of 17, had arrived very near the apex of existing mathematical thought. Then the first of the emotional disruptions that were to darken the few remaining years of his life occurred. His father, the victim of a political plot that unjustly discredited him, committed suicide in July 1829. A month later Galois, partly from his fiery impatience with the examiners' instructions, failed to gain entrance to the Ecole Polytechnique. He had therefore to be content with entering, in the autumn of 1829, the Ecole Normale Supérieure.

It was then that Galois learned that the ideas that were contained in the paper presented to the Academy of Science by Cauchy were not original. Shortly before, very similar ideas had been published by Abel in his last paper. Encouraged by Cauchy, Galois began to revise his paper in the light of Abel's findings. He presented the revised version to the academy in February 1830, with high hopes that it would gain him the Grand Prix. To his dismay the academy not only rejected his paper, but the examiner lost the manuscript. Galois was indignant at this ill-treatment, but four months later he succeeded in having a paper on number theory published in the prestigious *Bulletin des sciences mathématiques.* It was this paper that contained the highly original and diverting theory of 'Galois imaginaires'.

Thus, when the July Revolution that drove Charles X off the throne shocked French society, Galois had reason to be proud of his barely acknowledged mathematical genius and cause to feel estranged, both on his own and his father's account, from the stuffy officialdom that had cast its pall over France from 1815. He joined the revolutionary movement. In the next year he was twice arrested, in May 1831 for proposing a regicide toast at a republican banquet (Louis-Philippe having taken Charles's place as king) and again in July for taking part in a republican demonstration. For the second offence Galois was imprisoned for nine months. In prison he continued his mathematical research. But shortly after his release he became involved in a duel, perhaps from political reasons, perhaps from complications arising from a love affair. The event remains mysterious. Galois, it is evident, expected to be killed. On 29 May 1832, in a letter written in feverish haste, he outlined the principal results of his mathematical inquiries. He sent the letter to his friend Auguste Chevalier, almost certainly in the expectation that it would find its way to Karl Gauss and Karl ◊Jacobi in Germany. The duel took place on the following morning. Galois was severely wounded in the stomach and died in hospital on 31 May 1832.

So brief was Galois's life – he was only 20 when he died – and so dismissively were his ideas treated by the academy, that he was known to his contemporaries principally as a rather headstrong republican agitator. Cauchy, who alone sensed the importance of what Galois was doing, was out of France after 1830 and did not see any of the revisions or developments of Galois's first paper. Moreover, in his letter to Chevalier, Galois had time only to put down in concise form – without demonstrations – his most important conclusions. When he died, therefore, Galois's work amounted to fewer than a hundred pages, much of it fragmentary and nearly all of it unpublished.

The honour of rescuing Galois from obscurity belongs, in the first instance, to Joseph ◊Liouville, who in 1843 began to prepare his papers for publication and informed the academy that Galois had provided a convincing answer to the question whether first-degree equations were solvable by radicals. Finally, in 1846, both Galois's 1831 paper and a short notice on the solution of primitive equations by radicals were (thanks to Liouville) published in the *Bulletin.*

Galois's achievement, put tersely, was to arrive at a definitive solution to the problem of the solvability of algebraic equations, and in doing so to produce such a breakthrough in the understanding of fields of algebraic numbers and also of groups that he is deservedly considered as the chief founder of modern group theory. What has come to be known as the Galois theorem made immediately demonstrable the insolubility of higher-than-fourth-degree equations by radicals. The theorem also showed that if the highest power of x is a prime, and if all other values of x can be found by taking only two values of x and combining them using only addition, subtraction, multiplication, and division, then the equation can be solved by using formulae similar in principle to the formula used in solving quadratic equations.

Galois's work involved groups formed from the arrangements of the roots of equations and their subgroups, groups that he fitted into each other rather on the analogy of the Chinese box arrangement. This, his most far-reaching achievement for the subsequent development of group theory, is known as Galois theory. Along with other such terms – Galois groups, Galois fields, and the Galois theorem – it bears testimony to the lasting influence of his rejected genius.

Galton, Francis (1822–1911) English scientist, inventor, and explorer who made contributions to several disciplines, including anthropology, meteorology, geography, and statistics, but who is best known as the initiator of the study of eugenics (a term he coined).

Galton was born on 16 February 1822 near Sparkbrook, Birmingham, the youngest of nine children of a rich banker and a first cousin of Charles Darwin. Galton's exceptional intelligence was apparent at an early age – he could read before he was three years old and was studying Latin when he was four – but his achievements in higher education were unremarkable. Acceding to his father's wish, Galton studied medicine at Birmingham General Hospital and then King's College, London, but interrupted his medical studies to read mathematics at Trinity College, Cambridge. He returned to medicine, however, studying at St George's Hospital in London, but abandoned his studies on inheriting his father's fortune and spent the rest of his life pursuing his own interests. Shortly after the death of his father, Galton set out in 1850 to explore various uncharted areas of Africa. He collected much valuable information, and on his return to England wrote two books describing his explorations. These writings won him the Royal Geographical Society's annual gold medal in 1853, and three years later he was elected a fellow of the Royal Society.

Galton then turned his attention to meteorology. He designed several instruments to plot meteorological data, discovered and named anticyclones, and made the first serious attempt to chart the weather over large areas – described in his book *Meteorographica* (1863). He also helped to establish the Meteorological Office and the National Physical Laboratory.

Stimulated by Darwin's *On the Origin of Species* (1859), Galton then began his best-known work, concerning heredity and eugenics, which occupied him for much of the rest of his life. Studying mental abilities, he analysed the histories of famous families and found that the probability of eminent men having eminent relatives was high, from which he concluded that intellectual ability is inherited. He tended, however, to underestimate the role of the environment in mental development, since intelligent parents tend to provide their children with a mentally stimulating environment as well as with genes for high intelligence. Galton described his study of mental abilities in eminent families in *Hereditary Genius* (1869), in which he used the term 'genius' to define creative ability of an exceptionally high order, as shown by actual achievement. He formulated the theory that genius is an extreme degree of three traits: intellect, zeal, and power of working. He also formulated the regression law, which states that parents who deviate from the average type of the race in a positive or negative direction have children who, on average, also deviate in the same direction but to a lesser extent.

Continuing his research, Galton devised instruments to measure various mental abilities and used them to obtain data from some 9,000 subjects. The results were summarized in *Inquiries into Human Faculty and its Development* (1883), which established Galton as a pioneer of British scientific psychology. He was the first to use twins to try to assess the influence of environment on development, but more important was his quantitative and analytical approach to his investigations. In order to interpret his results, Galton devised new statistical methods of analysis, including correlational calculus, which has since become an invaluable tool in many disciplines, especially the social sciences.

As a result of his research into human abilities, Galton became convinced that the human species could be improved, both physically and mentally, by selective breeding, and in 1885 he gave the name 'eugenics' to the study of methods by which such improvements could be attained. He defined eugenics as 'the study of the agencies under social control which may improve or impair the racial qualities of future generations, physically or mentally.'

Galton also made a number of other contributions. He demonstrated the permanence and uniqueness of fingerprints and began to work out a system of fingerprint identification, now extensively used in police work. In 1879 he devised a word-association test that was later developed by Sigmund Freud and became a useful aid in psychoanalysis. Also, he invented a teletype printer and the ultrasonic dog whistle.

Galton was knighted in 1909, two years before his death on 17 January 1911 in Haslemere, Surrey. In his will he left a large bequest to endow a chair in eugenics at the University of London.

Galvani, Luigi (1737–1798) Italian anatomist whose discovery of 'animal electricity' stimulated the work of Alessandro ◊Volta and others to discover and develop current electricity.

Galvani was born in Bologna on 9 September 1737. He studied medicine at the University of Bologna, graduating in 1759 and gaining his doctorate three years later with a thesis on the development of bones. In that same year (1762) he married Lucia Galeazzi, a professor's daughter, and was appointed lecturer in anatomy. In 1772 he became president of the Bologna Academy of Science and in 1775 he took up the appointment of professor of anatomy and gynaecology at the university. The death of his wife in 1790 left him griefstricken. In 1797 he was required to swear allegiance to Napoleon as head of the new Cisalpine Republic, but Galvani refused and thereby lost his appointment at Bologna University. Saddened, he retired to the family home and died there on 4 December 1798.

Comparative anatomy formed the subject of Galvani's early research, during which he investigated such structures as the semicircular canals in the ear and the sinuses of the nose. He began studying electrophysiology in the late 1770s, using static electricity stored in a Leyden jar to stimulate muscular contractions in living and dead animals. The frog was already established as a convenient laboratory animal and in 1786 Galvani noticed that touching a frog with a metal instrument during a thunderstorm made the frog's muscles twitch. He later hung some dissected frogs from brass hooks on an iron railing to dry them, and noticed that their muscles contracted when the legs came into contact with the iron – even if there was no electrical storm. Galvani concluded that electricity was causing the contraction and postulated (incorrectly) that it came from the animal's muscle and nerve tissues. He summarized his findings in 1791 in a paper, 'De viribus electricitatis in motu musculari commentarius', that gained general acceptance except by Alessandro Volta, who by 1800 had constructed electric batteries consisting of plates (electrodes) of dissimilar metals in a salty electrolyte, thus proving that Galvani had been wrong and that the source of the electricity in his experiment had been the two different metals and the animal's body fluids. But for many years current electricity was called Galvanic electricity and Galvani's name is preserved in the word galvanometer, a current-measuring device developed in 1820 by André ◊Ampère, and in galvanization, the process of covering steel with a layer of zinc (originally by electroplating). Galvani's original paper was translated into English and published in 1953 as 'Commentary on the effect of electricity on muscular motion'.

Gamow, George (1904–1968) Russian-born US physicist who provided the first evidence for the Big Bang theory of the origin of the universe. He predicted that it would have produced a background of microwave radiation, which was later found to exist. Gamow was also closely involved with the early theoretical development of nuclear physics, and he made an important contribution to the understanding of protein synthesis.

Gamow was born in Odessa on 4 March 1904. He attended local schools and first became interested in astronomy when he received a telescope from his father as a gift for his 13th birthday. In 1922 he entered Novorossysky University, but he soon transferred to the University of Leningrad (now St Petersburg) where he studied optics and later cosmology, gaining his PhD in 1928. Gamow then went to the University of Göttingen, where his career as a nuclear physicist really began. His work impressed Niels ◊Bohr who invited Gamow to the Institute of Theoretical Physics in Copenhagen. There he continued his work on nuclear physics and he also studied nuclear reactions in stars. In 1929 Gamow worked with Ernest ◊Rutherford at the Cavendish Laboratory, Cambridge. His work there laid the theoretical foundations for the artificial transmutation of elements, which was achieved in 1932.

Gamow returned to Copenhagen in 1930. From 1931 to 1933 he served as master of research at the Academy of Science in Leningrad and then, after being denied permission by the Soviet government to attend a conference on nuclear physics in Rome in 1931, he used the Solvay Conference held in Brussels in 1933 as an opportunity to defect. He settled in the USA where he held the chair of physics at George Washington University in Washington 1934–56. He then held the post of professor of physics at the University of Colorado until his death. In 1948 Gamow was given top security clearance and worked on the hydrogen bomb project at Los Alamos, New Mexico.

Later in life, he gained considerable fame as an author of popular scientific books, being particularly well-known for the 'Mr Tompkins' series. He was awarded the Kalinga Prize by UNESCO in 1956 in recognition of the value of his popular scientific texts. Gamow died in Boulder, Colorado, on the 20 August 1968.

Gamow's work can be broadly divided into two main areas: his theoretical contributions to nuclear physics, and astronomy. Some of his studies fell outside both of these disciplines.

His first major scientific work was his theory of alpha decay, which he produced in 1928 while at the University of Göttingen. He was able to explain why it was that uranium nuclei could not be penetrated by alpha particles that have as much as double the energy of the alpha particles emitted by the nuclei. This, he proposed, was due to a 'potential barrier' that arose due to repulsive Coulomb and other forces. The model represented the first application of quantum mechanics to the study of nuclear structure, and was simultaneously and independently developed by Edward Condon (1902–1974) at Princeton University.

During his visits to Copenhagen in 1928 and 1930, Gamow continued his work on nuclear physics and in particular on the liquid drop model of nuclear structure. This theory held that the nucleus could be regarded as a collection of alpha particles that interacted via strong nuclear and Coulomb forces. His work under Rutherford in 1929 was concerned with the calculation of the energy that would be required to split a nucleus using artificially accelerated protons. This work led directly to the construction of a linear accelerator in 1932 by John ◊Cockcroft (1897–1967) and Ernest Watson (1903–), in which protons were used to disintegrate boron and lithium, resulting in the production of helium. This was the first experimentally produced transmutation that did not employ radioactive materials.

In collaboration with Edward ◊Teller, Gamow produced in 1936 the Gamow–Teller selection rule for beta decay. This has since been elaborated, but was one of the first formulations to describe beta decay.

During World War II, Gamow was intimately concerned with the Manhattan Project on the development of the atomic bomb, and later contributed significantly to research at Los Alamos that led to the production of the hydrogen bomb.

His astronomical studies were concerned mainly with the origin of the universe and the evolution of stars. He followed the model devised by Hans ◊Bethe on the mechanism by which heat and radiation are generated in the cores of stars (thermonuclear reactions), and postulated that a star heats up – rather than cools down – as its 'fuel' is consumed. In 1938 he related this theory to the Hertzsprung–Russell diagram. The following year, in collaboration with M Schoenberg, he investigated the role of neutrino emission in novae and supernovae. He then turned to the problem of energy production in red giant stars and in 1942 produced his 'shell' model to describe such stars.

Gamow's most famous contribution to astronomy began with his support in 1946 for the Big Bang theory of the origin of the universe proposed by Georges ◊Lemaître. With Ralph ◊Alpher he investigated the possibility that heavy elements could have been produced by a sequence of neutron-capture thermonuclear reactions. They published a famous paper in 1948, which became known as the Alpher–Bethe–Gamow (or alpha–beta–gamma) hypothesis, describing the 'hot Big Bang'. They suggested that the primordial state of matter – which they called ylem, the term Aristotle had given to the ultimate state of matter – consisted of neutrons and their decay products. This mixture of neutrons, protons, electrons, and radiation was almost unimaginably hot. As this matter expanded after the hot Big Bang, it cooled sufficiently for hydrogen nuclei to fuse to form helium nuclei (alpha particles) – which explained the abundance of helium in the universe (one atom in 12 is helium).

The model's inability to account for the presence – and evident creation – of elements heavier than helium was later vindicated by the work of Geoffrey and Margaret ◊Burbidge and their collaborators, who found some evidence of what they called nucleosynthesis, or nucleogenesis.

The hot Big Bang model indicated that there ought to be a universal radiation field as a remnant of the intense temperatures of the primordial Big Bang. In 1964 Arno ◊Penzias and Robert ◊Wilson detected isotropic microwave radiation. Research at Princeton University confirmed that this was 3K (–270°C/ –454°F) black-body radiation, as predicted by Gamow's model. Gamow had in fact postulated the temperature of this radiation as 25K (–248°C/ –414°F), but errors were found in his method that explained why his estimate was too high. The detection of this microwave radiation led most cosmologists to support the Big Bang model, in preference to the steady-state hypothesis proposed originally by Fred ◊Hoyle, Hermann ◊Bondi, and Thomas ◊Gold.

In the entirely different field of molecular biochemistry, Gamow contributed to the solution of the genetic code in which virtually all hereditable biological information is stored. The double-helix model for the structure of DNA had been published in 1953 by James ◊Watson and Francis ◊Crick. The double helix consists of a twisted double chain of four types of nucleotides, sugar residues, and phosphates. Gamow realized that if three nucleotides were used at a time, 64 different triplets could be constructed. These could easily code for the different amino acids of which all proteins are constructed. Gamow's theory was found to be correct, and in 1961 the genetic code was cracked and the meanings of the 64 triplets identified.

Gamow was a talented and wide-ranging scientist who, with his fundamental contribution to cosmology, was responsible for the first clear indication of the origin of the universe.

Gardner, Julia Anna (1882–1960) US geologist and palaeontologist internationally known for her work on stratigraphic palaeontology.

Gardner was born on 26 January 1882 in Chamberlain, South Dakota. She was educated at Drury Academy, North Adams, Massachusetts, and later at Bryn Mawr College, where she studied geology and palaeontology. She graduated with an AB in 1905 and, following a year's schoolteaching in Chamberlain, she received an AM in 1907. She was awarded a PhD from Johns Hopkins University in 1911 with her dissertation 'On certain families of Gastropoda from the Miocene and Pliocene of Virginia and North Carolina'. Between 1911 and 1917, Gardner worked as a research assistant in palaeontology in Johns Hopkins and also for the US Geological Survey (USGS). After joining the war effort in 1917, often serving near the front, she returned to the USA in 1920 and was employed by the USGS Coastal Plain Section, transferring in 1936 to the Palaeontology and Stratigraphy Section. Gardner remained with the USGS until her retirement in 1952, after which she was hired on a yearly contract basis until 1954. She died on 15 November 1960, following a stroke.

Gardner advanced quickly in the USGS; she became associate geologist in 1924 and geologist in 1928. Her work on the Cenozoic stratigraphic palaeontology of the coastal plain in Texas and the Rio Grande embayment in northeast Mexico led to the publication of *Correlation of the Cenozoic Formations of the Atlantic and Gulf Coastal Plain and the Caribbean Region* (1943) coauthored with C Wythe Cooke and Wendell Woodring; her work was important for petroleum geologists establishing standard stratigraphic sections for Tertiary rocks in the southern Caribbean. During World War II, Gardner joined the Military Geologic Unit where amongst other things, she helped to locate Japanese beaches from which incendiary bombs were being launched by identifying shells in the sand ballast of the balloons. On her retirement, Gardner was awarded the Interior Department's Distinguished Award. She was president of the Paleontological Society in 1952 and a vice president of the Geological Society of America in 1953.

Gassendi, Pierre (1592–1655) French philosopher, physicist, and priest, best known for his opposition to Aristotelian philosophy and his revival of the Epicurean system.

Born in Champtercier, near Digne in Provence, France on 22 January 1592, Gassendi was appointed to teach rhetoric in Digne at the age of 16, and, at the age of 19, to teach philosphy at Aix. Patrons at Aix recognized his studious nature and helped him to become doctor of theology at Aix where he studied Greek and Hebrew literature. He was appointed canon of the cathedral in about 1623 and provost in about 1625, which allowed him time to continue his studies.

Gassendi was opposed to Aristotelian philosophy. He advocated an atomic theory based on matter on the lines of the Epicurean system, which he attempted to integrate into a Christian philosophy. His inconsistency in this was criticized by both Christian and non-Christian scholars. He was in correspondence with Thomas Hobbes (1588–1679), Queen Christina of Sweden (1626–1689), Robert Fludd (1574–1637), and René ◊Descartes, and is known more for his exposition and criticism of the views of others than for his own original thought. His best-known book is his *Objections* to Descartes's *Meditations*, published in 1642.

Cardinal Richelieu (1585–1642) took an interest in Gassendi's work, and in 1645 he recommended that the king appoint Gassendi professor of mathematics at the College Royal of France in Paris. Gassendi accepted this position although he often travelled back to Provence because of ill-health. Most of his lectures in Paris were in astronomy, a field in which he had taken an amateur interest since 1618. He had kept an astronomical notebook which ran to 400 pages. He had recorded observations of a comet in 1618, the aurora borealis in 1621 (which he named), and a lunar eclipse in 1623. In 1631 he observed the transit of Mercury, predicted by Johannes ◊Kepler, through a Galilean telescope, projecting the Sun's image on to a piece of paper. He corresponded with Giovanni ◊Cassini, Galileo, Kepler, Johannes ◊Hevelius, and others and he was of the opinion that Copernicus' system was based on probabilities, and that Tycho ◊Brahe's system was the most acceptable. He wrote about Copernicus and in 1647 he published *Instituto astronomica.*

Gassendi was a devout Christian who never missed Mass. He was well-liked and his friends knew him for his humour and his adoption of Socratic methods of argument in debate. In 1655 he became seriously ill in Paris with intermittent fever and died quietly on 24 October, aged 65. He is buried at the chapel of a friend in St Nicolas des Champs in Paris and there is a marble bust of him there.

Gatling, Richard Jordan (1818–1903) US inventor who is best remembered for his invention of the Gatling gun, one of the earliest successful rapid-fire weapons.

Gatling was born in Hertford County, North Carolina, on 12 September 1818. The son of a well-to-do planter, he showed mechanical aptitude early in life and collaborated with his father on the invention of a machine for sowing cotton and for thinning out the young plants. He followed a varied career, his first job being in the county clerk's office. He then taught for a brief period before becoming a merchant.

In 1844, Gatling moved to St Louis, adapted his cotton-sowing machine for sowing rice and other grains and established his manufacturing centre there. Other factories were later set up in Ohio and Indiana.

Gatling's career took a new turn after an attack of smallpox. He entered Ohio Medical College,

Cincinnati, to study medicine, qualifying in 1850 but never practising. Instead, for a time he was concerned with railway enterprises and property, but he was still inventing – a hemp-breaking machine in 1850, a steam plough in 1857, and a marine steam ram in 1862.

Forms of rapid-fire weapons were in use as early as the 16th century, but not until about 1860, when breech loading became firmly established, did effective machine guns become possible. One of the first was the Reffye, which had 25 barrels round a common axis.

Gatling's gun, which he patented in 1862, consisted of ten parallel barrels arranged round a central shaft, each barrel firing in turn as it reached the firing position. An operating cam, driven through a hand-cranked worm gear, forced the locks forward and backward. On a single barrel, loading took place in the first position. The second, third, and fourth positions were used to force the cartridge into the breech, and the gun was fired into the fifth position. The sixth, seventh, eighth, and ninth positions were used for the extraction of the empty case, which was rejected at the tenth.

The gun was fed from a drum on top, which dropped the cartridges into the locks. Each barrel therefore fired as it approached the bottom of its circular journey. In 1862, the year of Gatling's patent specification, the gun was capable of firing 350 rounds per minute. There was no official interest in his invention, however, so he made further improvements to it, extending its range and increasing its firing rate. This still failed to attract official interest, but a few commanders during the Civil War purchased some with private funds. It was not until 1866, after the end of the war, that the Army Ordnance Department ordered 100. Meanwhile, they were being manufactured by the Colt Company and sold overseas to England, Russia, Austria, and South America, and were used in the war of 1870.

In that year, Gatling moved to Hartford, Connecticut, to supervise manufacture and by 1882 his gun could fire 1,200 rounds per minute. Gatling continued to develop his invention, producing greater rates of fire and driving it by different methods.

The Gatling was one of the earliest successful rapid-fire weapons but it cannot be regarded as the forerunner of the modern machine gun as it needed an external power source to drive the loading, firing, extraction, and ejection mechanism. The laurels for the first workable, truly automatic machine gun must go to Hiram Stevens ◊Maxim. The main contribution Gatling made to history was to alert the military mind to the possibilities of such weapons. Indeed, special tactics were developed in the Franco-German war of 1870 to counter Gatling-type weapons by massive artillery barrages. It thus failed to make any serious impact.

Gatling died, aged 84, in New York City on 26 February 1903.

Gauss, Karl (Carl) Friedrich (1777–1855) German mathematician, physicist, and astronomer whose innovations in mathematics proved him to be the equal of ◊Archimedes or Isaac ◊Newton.

Gauss was born in Braunschweig (Brunswick) Germany, on 30 April 1777, to a very poor, uneducated family. His father was a gardener, an assistant to a merchant, and the treasurer of an insurance fund. Gauss taught himself to count and read – he is said to have spotted a mistake in his father's arithmetic at the age of three. At elementary school, at the age of eight, he added the first 100 digits in his first lesson. Recognizing his precocious talent, his teacher persuaded his father that Gauss should be encouraged to train towards following a profession rather than learn a trade. At the age of 11, he went to high school and proved to be just as proficient in the classics as in mathematics. When he was 14 he was present at the court of the Duke of Brunswick in order to demonstrate his skill at computing; the duke was so impressed that he supported Gauss generously with a grant from then until his own death in 1806. In 1792, with the duke's aid, Gauss began to study at the Collegium Carolinum in Brunswick, and then he was taught at the University of Göttingen 1795–98. He was awarded a doctorate in 1799 from the University of Helmstedt, by which time he had already made nearly all his fundamental mathematical discoveries. In 1801 he decided to develop his interest in astronomy; by 1807 he pursued it so enthusiastically that not only was he professor of mathematics, he was also director of the Göttingen Observatory.

At about this time, he began to gain recognition from other parts of the world: he was offered a job in St Petersburg, was made a foreign member of the Royal Society in London, and was invited to join the Russian and French academies of science. Nevertheless, Gauss remained at Göttingen for the rest of his life, and died there on 23 February 1855.

Between the ages of 14 and 17, Gauss devised many of the theories and mathematical proofs that, because of his lack of experience in publication and his diffidence, had to be rediscovered in the following decades. The extent to which this was true was revealed only after Gauss's death. There are, nevertheless, various innovations that are ascribed directly to him during his three years at the Collegium Carolinum, including the principle of least squares (by which the equation curve best fitting a set of observations can be drawn). At this time he was particularly intrigued by number theory, especially on the frequency of primes. This subject became a life's work and he is known as its modern founder. In

1795, having completed some significant work on quadratic residues, Gauss began to study at the University of Göttingen, where he had access to the works of Pierre de ◊Fermat, Leonhard ◊Euler, Joseph ◊Lagrange, and Adrien ◊Legendre. He immediately seized the opportunity to write a book on the theory of numbers, which appeared in 1801 as *Disquisitiones arithmeticae*, generally regarded to be his greatest accomplishment. In it, he summarized all the work that had been carried out up to that time and formulated concepts and questions that are still relevant today.

He was still at university when he discovered, in 1796, that a regular 17-sided polygon could be inscribed in a circle, using ruler and compasses only. This represented the first discovery in Euclidean geometry that had been made in 2,000 years.

In 1799 Gauss proved a fundamental theorem of algebra – that every algabraic equation has a root of the form $a + bi$, where a and b are real numbers and i is the square root of –1. In his doctoral thesis, Gauss showed that numbers of the form $a + bi$ (called complex numbers) can be regarded as analogous to points on a plane.

The years 1800–10 were for Gauss the years in which he concentrated on astronomy. In mathematics he had had no collaborators, although he had inspired mathematicians such as Lejeune ◊Dirichlet and Bernhard ◊Riemann. In astronomy, however, he corresponded with many, and his friendship with Alexander von ◊Humboldt played an important part in the development of science in Germany. The discovery of the first asteroid, Ceres – and its subsequent 'loss' – by Guiseppe ◊Piazzi at the beginning of 1801 gave Gauss the chance to use his mathematical brilliance in another cause. He developed a quick method for calculating the asteroid's orbit from only three observations and published this work – a classic in astronomy – in 1809. The 1,001st asteroid to be discovered was named Gaussia in his honour. He also worked out the theories of perturbations that were eventually used by Urbain ◊Leverrier and John Couch ◊Adams in their independent calculations towards the discovery of Neptune. After 1817, he did no further work in theoretical astronomy, although he continued to work in positional astronomy for the rest of his life.

Gauss was also a pioneer in topology, and he worked besides on crystallography, optics, mechanics, and capillarity. At Göttingen, he devised the heliotrope, a type of heliostat that has applications in surveying. After 1831 he collaborated with Wilhelm ◊Weber on research into electricity and magnetism, and in 1833 they invented an electromagnetic telegraph. Gauss devised logical sets of units for magnetic phenomena and a unit of magnetic flux density is therefore called after him.

There is scarcely any physical, mathematical, or astronomical field in which Gauss did not work. He retained an active mind well into old age and at the age of 62, already an accomplished linguist. he taught himself Russian. The full value of his work has been realized only in the 20th century.

Further Reading

Buehler, W K, *Gauss: A Biographical Study*, Springer-Verlag, 1987.

Hall, Tord, *Carl Friedrich Gauss: A Biography*, MIT Press, 1970.

Merzbach, Uta C, *Carl Friedrich Gauss: A Bibliography*, Scholarly Resources, 1984.

Rassias, G M, *The Mathematical Heritage of C F Gauss*, World Scientific, 1991.

Gay-Lussac, Joseph Louis (1778–1850) French chemist who pioneered the quantitative study of gases and established the link between gaseous behaviour and chemical reactions. He also made important discoveries in inorganic chemistry, often parallelling (independently) the work of Humphry ◊Davy in England.

Gay-Lussac was born on 6 December 1778 at Saint-Léonard, Haute Vienne, the son of a judge (who was arrested in 1793 for his aristocratic sympathies). He entered the Ecole Polytechnique in Paris in 1797 and

Gay-Lussac Engraving of French chemist Joseph Louis Gay-Lussac, who was distinguished for his work on the properties of gases. Gay-Lussac developed several analytical methods, including an assay for silver. He was appointed chief assayer to the French Mint in 1829. *Mary Evans Picture Library*

graduated in 1800. His first interest was in engineering but in 1801 the great French chemist Claude ◊Berthollet invited Gay-Lussac to join him as an assistant at his house in Arcueil, the meeting place of many contemporary scientists, including Pierre ◊Laplace. From 1805 to 1806 he accompanied Alexander von ◊Humboldt on an expedition to measure terrestrial magnetism and in 1809, a year after his marriage to Geneviève Rojet, he became professor of chemistry at the Ecole Polytechnique and professor of physics at the Sorbonne. He held various government appointments, including that of superintendent of a gunpowder factory from 1818 and chief assayer to the Mint from 1829. In 1832 he was made professor of chemistry at the Musée National d'Histoire Naturelle. He was a member of the Chamber of Deputies for a short time in the 1830s, and was made a peer in 1839. He died in Paris on 9 May 1850.

Gay-Lussac began studying gases in collaboration with Louis Thénard. In 1802 he formulated the law of expansion of gases, which states that for a given rise in temperature all gases expand by the same fraction of their volume (Jacques ◊Charles had recognized equal expansion in 1787 but had not published his findings; nevertheless the relationship is now usually known as Charles's law). During balloon ascents in 1804, in which he established an altitude record that was to stand for 50 years, Gay-Lussac showed that the composition of air is constant up to heights in excess of 7,000 m/23,000 ft above sea level. With Humboldt he accurately determined the proportions of hydrogen and oxygen in water, showing the volume ratio to be 2:1; they also established the existence of explosive limits in mixtures of the two gases.

His greatest achievement concerning gases came in 1808 with the formulation of Gay-Lussac's law of combining volumes, which states that gases combine in simple proportions by volume and that the volumes of the products are related to the original volumes. His examples include:

$$\begin{array}{ccccc} HCl & + & NH_3 & \rightarrow & NH_4Cl \\ 1 & & 1 & & 1 \end{array}$$

and

$$\begin{array}{ccccc} 2CO & + & O_2 & \rightarrow & 2CO_2 \\ 2 & & 1 & & 2 \end{array}$$

Another fruitful area of Gay-Lussac's work with Thénard concerned the alkali metals, where they followed up the discovery of sodium and potassium by Davy by devising chemical means for producing the metals in quantity (heating fused alkalis with red-hot iron). They anticipated (by nine days) Davy's isolation of boron and independently investigated the properties of iodine, which had been discovered in 1811 by Bernard Courtois (1777–1838) but was named by Gay-Lussac in 1813. He first prepared iodine(I) chloride, ICl, and iodine(III) chloride, ICl_3. Studies of the hydrogen halides – hydrogen fluoride, hydrogen chloride, and hydrogen iodide – led to the extremely important conclusion that acids need not contain oxygen (as proposed by Antoine ◊Lavoisier) and to the naming of hydrochloric acid (previously known as muriatic acid). From his researches with Prussian blue Gay-Lussac prepared hydrogen cyanide and in 1815 discovered cyanogen, $(CN)_2$; he went on to develop the chemistry of the cyanides.

Another significant area of work was the development of volumetric analysis – which provided more evidence, if any were needed, of Gay-Lussac's quantitative approach to chemistry. In 1832 he introduced the method of estimating silver by titrating silver nitrate against sodium chloride. Later work included the analysis of vegetable and animal substances and a contribution to the manufacture of sulphuric acid, the Gay-Lussac tower, which recovers spent nitrogen oxides from the chamber process.

Geber Arabian alchemist also known as ◊Jabir ibn Hayyan.

Geiger, Hans Wilhelm (1882–1945) German physicist who invented the Geiger counter for detecting radioactivity.

Geiger was born in Neustadt, Rheinland-Pfalz, on 30 September 1882. He studied physics at the universities of Munich and Erlangen, obtaining his PhD from the latter institution in 1906 with a dissertation on gaseous ionization. In 1906 he became assistant to Arthur Schuster (1851–1934), professor of physics at the University of Manchester. Ernest ◊Rutherford succeeded Schuster in 1907 and Geiger stayed on as his assistant. They were later joined by Ernest Marsden, with whom Geiger collaborated.

In 1912 Geiger became head of the Radioactivity Laboratories at the Physikalische Technische Reichsanstalt in Berlin, after declining the offer of a post at the University of Tübingen. He established a successful research group, which was joined by such eminent scientists as Walther Bothe (1891–1957) and James ◊Chadwick. Geiger served in the German artillery during World War I, and then returned to the Reichsanstalt in Berlin.

In 1925 Geiger accepted a post as professor of physics at the University of Kiel. He moved in 1929 to the University of Tübingen, and in 1936 he became head of physics at the Technical University of Charlottenberg-Berlin as well as the editor of *Zeitschrift für Physik.* Illness reduced the pace of his

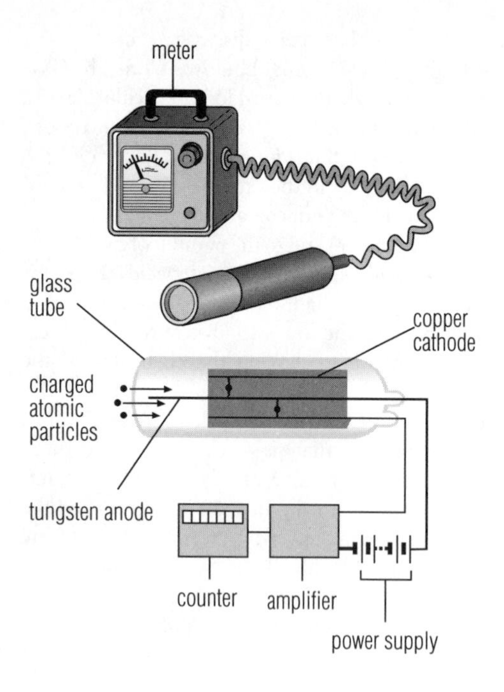

Geiger A Geiger–Müller counter detects and measures ionizing radiation (alpha, beta, and gamma particles) emitted by radioactive materials. Any incoming radiation creates ions (charged particles) within the counter, which are accelerated by the anode and cathode to create a measurable electric current.

scientific activities during most of World War II and then he lost his home and belongings during the Allied occupation of Germany. Geiger died a few months later on 24 September 1945 at Potsdam.

Geiger's early postdoctoral work in Manchester involved the application of his expertise in gaseous ionization to the field of radioactive decay. One direction which his work took was the development in 1908 of a device that counted alpha particles. The counter consisted of a metal tube containing a gas at low pressure and thin wire in the centre of the tube. The wire and tube were under a high voltage and each alpha particle arriving through a window at one end of the tube caused the gas to ionize, producing a momentary flow of current. The resulting electric signals counted the number of alpha particles entering the tube. With this counter, Geiger was able to show that approximately 3.4×10^{10} alpha particles are emitted per second by a gram of radium, and that each alpha particle has a double charge. Rutherford later demonstrated that the doubly charged alpha particle was in fact a helium nucleus.

In 1909, Marsden and Geiger studied the interactions of alpha particles from radium with metal reflectors. They found that most alpha particles passed straight through the metal, but that a few were deflected at wide angles to the beam. They showed that the amount of deflection depended on the metal that the reflector was made of and that it decreased with a lowering of atomic mass. They used metal reflectors of different thicknesses to demonstrate that the deflection was not caused merely by some surface effect. This work led directly to Rutherford's proposal in 1911 that the atom consists of a central, positively charged nucleus surrounded by electron shells. The observed deflection of the alpha particles was due to the positively charged particle interacting with the positively charged nuclei in the metal and being deflected by repulsive forces. Geiger and Marsden later published the mathematical relationship between the amount of alpha scattering and atomic mass.

Other subjects that Geiger studied under Rutherford included the relationship between the range of an alpha particle and its velocity in 1910; the various disintegration products of uranium in 1910 and 1911; and the relationship between the range of an alpha particle and the radioactive constant, which was determined with John Nuttall (1890–1958) in 1911.

In 1912, upon his return to Germany, Geiger began the first of many refinements to his radiation counter, enabling it to detect beta particles and other kinds of radiation as well as alpha particles. The modern form of the instrument was finally developed with Walther Müller in 1928, and it is now usually known as the Geiger–Müller counter. It can be operated so that radioactivity causes it to emit audible clicks, which can be recorded automatically. Geiger used this instrument to confirm the Compton effect in 1925 and to study cosmic radiation from 1931 onwards. Cosmic rays became a prolonged source of interest that occupied him for most of the rest of his career.

Geiger's contribution to science in discovering a simple and reliable way of detecting radiation not only enabled discoveries to be made in nuclear physics, but also afforded an instant method of checking radiation levels and finding radioactive minerals.

Gell-Mann, Murray (1929–) US theoretical physicist who was awarded the 1969 Nobel Prize for Physics for his work on the classification and interactions of subatomic particles. He is the originator of the quark hypothesis.

Gell-Mann was born in New York City on 15 September 1929. He showed strong scholastic aptitude at an early age and entered Yale University when he was only 15 years old. He received his bachelor's degree in 1948 and proceeded to the Massachusetts Institute of Technology, where he earned his PhD three years later. He spent the next year as a member of the Institute for Advanced Studies at Princeton University. In 1952 he joined the faculty of the University of Chicago as an instructor to work at the Institute for Nuclear Studies under Enrico ◊Fermi, becoming an assistant professor in 1953. In 1955, Gell-Mann moved to the California Institute of Technology at Pasadena as associate professor of theoretical physics, becoming professor a year later and, in 1966, Robert Andrews Millikan Professor of Theoretical Physics (emeritus from 1993).

From the early 1930s, the relatively simple concept of the atom as being composed of only electrons and protons began to give way to more complex models involving neutrons and then other particles. By the 1950s, the field was in a state of complete chaos following the discovery of mesons, which have masses intermediate between a proton and an electron, and then hyperons, which are even more massive than the proton. One of the most puzzling things was that the hyperons and some mesons had much longer lifetimes than was predicted by accepted theory at that time, although still only of the order of 10^{-9} seconds.

In an endeavour to bring order to the subject, Gell-Mann proposed in 1953 that the long-lived particles, and indeed other particles such as the neutron and the proton, should be given a new quantum number called the strangeness number. The strangeness number differed from particle to particle. It could be 0, –1, +1, –2, etc. He also proposed the law of conservation of strangeness, which states that the total strangeness must be conserved on both sides of an equation describing a strong or an electromagnetic interaction but *not* a weak interaction. This work was also done independently by Nishijima in Japan.

The law of conservation of strangeness formed an important theoretical basis for the subsequent theory of associated production, which was proposed by Gell-Mann in 1955. This model held that the strong forces that were responsible for the creation of strange particles could create them only in groups of two at a time (that is, in pairs). As the partners in a pair moved apart, the energy that would be required for them to decay through strong interactions would exceed the energy that had originally gone into their creation. The strange particles therefore survived long enough to decay through weak interactions instead. This model explained their unusually extended lifespan.

Gell-Mann used these rules to group mesons, nucleons (neutrons and protons), and hyperons, and was thereby able to form predictions in the same way that Dmitri ◊Mendeleyev had been able to make predictions about the chemical elements once he had constructed the periodic table. Gell-Mann's prediction of the existence of a particle, which he named xi-zero, to complete a doublet with xi-minus was soon rewarded with experimental verification.

The law of conservation of strangeness was also important in the formulation of SU(3) symmetry, a scheme for classifying strongly interacting particles that Gell-Mann proposed in 1961. This classification system itself formed part of the basis for yet another classification scheme entitled the eightfold way (named after the eight virtues of Buddhism). This model was devised by Israeli physicist Yuval Ne'emann (1925–) and Gell-Mann, and published in 1962.

The eightfold way was intended to incorporate all the new particles and the new quantum numbers. It postulated the existence of supermultiplets, or groups of eight particles that have the same spin value but different values for charge, isotopic spin, mass, and strangeness. The model also predicts the existence of supermultiplets of different sizes. The strongest support for the theory arose from the discovery in 1964 of a particle named omega-minus, which had been predicted by the eightfold way.

In 1964 Gell-Mann proposed a model for yet a further level of complexity within the atom. He proposed that particles such as the proton are not themselves fundamental but are composed of quarks. Quarks differ from all previously proposed subatomic particles because they have fractional charges of, for instance, $+^2/_3$ or $-^1/_3$. Quarks always occur in pairs or in trios and can never be detected singly, exchanging gluons that bind them together. Gluons are the quanta for interactions between quarks, a process that goes by the exotic name of quantum chromodynamics, and their behaviour is in some ways analogous to the exchange of protons between electrons in electromagnetic interactions (quantum electrodynamics). Six quarks have been predicted, and all, by 1994, had been indirectly detected. The names of the quarks are up, down, strange, charm, bottom, and top, and they have been ascribed properties of electric charge, strangeness, charm, bottomness, topness, baryon number, and so on.

Gell-Mann's work has been characterized by originality and bold synthesis. His models have been useful not only for their predictive value, but also for the work they have spurred others to do.

Germain, Sophie Marie (1776–1831) French mathematician who developed the modern theory of elasticity (the mathematical theory of the stress and strain that a material can sustain and still return to its original form) and made major contributions to numbers theory and acoustics.

Germain was born on 1 April 1776 in Paris, the second of three daughters of a prosperous silk merchant. She grew up in Paris at the time of the fall of the Bastille and the Reign of Terror and her home was a meeting place for liberal reformers. Germain was used to intellectual discussion from a young age.

Germain studied alone in her father's library from childhood. When she was 13 she was moved by the story of the death of ◊Archimedes at the hands of a Roman soldier and became fascinated by geometry. She was determined to become a mathematician and taught herself Greek and Latin and read the works of Isaac ◊Newton and Leonhard ◊Euler during the night. Out of concern for her health her parents doused the fire in her bedroom and took away her clothes and her light to try to prevent this but she still came downstairs and worked secretly by firelight wrapped in a quilt.

The Ecole Polytechnique opened in Paris in 1794 and although women were not admitted, she managed to obtain professors' lecture notes for several courses. Inspired by the work of Joseph ◊Lagrange on analysis and using the pseudonym M le Blanc she sent him a paper. Impressed by her originality he sought her out, and out of respect for her work, became her sponsor and mathematical counsellor, in spite of her gender and lack of formal education.

Germain then entered into a long correspondence with Adrien ◊Legendre and he included some of her discoveries in his writings. But her most productive and best known correspondence, in the guise of M. le Blanc, was with the German mathematician Karl ◊Gauss, who did not suspect her identity for several years. Between 1801–09, she wrote to him outlining her number theory proofs and he praised her highly, eulogizing her to his colleagues. When the French occupied his home town of Braunschweig in 1806, she feared for his safety and interceded on his behalf through a family friend who was a French commander. When Gauss discovered her true identity he was even more full of praise for her. Germain also worked on Pierre de ◊Fermat's last theorem at this time and proved that his equation does not hold for the case in which n is equal to 5. She wrote to Gauss explaining this but he failed to reply to her letter. She produced a theorem that has become known as Germain's theorem. Gauss later recommended her for an honorary doctorate from the University of Göttingen, but she died before it could awarded.

In 1809 Germain began work on the theory of the patterns formed by sand on vibrating plates. The German physicist Ernst ◊Chladni had demonstrated the phenomenon to Napoleon in 1808 and the emperor had been so impressed that he had offered a gold medal weighing 1 kg/2.2 lb to the first person to explain 'Chladni's figures'. Germain submitted the only entry in 1811 to the competition, but because of her lack of formal knowledge of analysis and the calculus of variation it was flawed. It was at her third attempt in 1815, again as the only entrant, that she finally won the award. To public disappointment, however, she did not appear at the presentation ceremony. She published her work privately in 1821 as *Recherches sur la theorie des surfaces élastiques.* She published memoirs on the theory of numbers and others on the theory of elasticity and her paper 'Considerations sur l'etat des sciences et des lettres aux differentes époques de leur culture' was a philosophical essay on the theme of unity of thought.

Germain was, probably quite rightly, of the opinion that she was not being taken seriously by the scientific community and when others from more privileged educational backgrounds took up the work she had begun on elasticity she was simply ignored, as was a subsequent paper she submitted to the Institut de France in 1825. When the Eiffel Tower was built in 1889 using in part her groundbreaking work on elasticity, she failed to receive a mention among the 72 contributors in the inscription.

Germain never married and was never offered a position in a university or fellowship of any society. Her father supported her financially throughout her life. She was diagnosed as having breast cancer in 1829 but continued working, completing papers on numbers theory and the curvature of surfaces before her death on 27 June 1831, in Paris.

Giacconi, Riccardo (1931–) Italian-born US physicist, the head of a team whose work has been fundamental in the development of X-ray astronomy.

Giacconi was born in Genoa on 6 October 1931. Educated in Italy, he obtained a doctorate in 1954 at the University of Milan, where he then took up the post of associate professor. Two years later he emigrated to the USA and became research associate at the University of Indiana at Bloomington, before taking up a similar position at Princeton in 1958. He then joined the company American Scientific and Engineering as senior scientist for a year until his naturalization papers came through in 1960. In 1963 he became chief of the Space Physics Division, and in 1973 he joined the Harvard–Smithsonian Center for Astrophysics, leading the construction and successful operation of the X-ray observatory HEAO-2. He was appointed the first director of the Space Telescope Science Institute, Baltimore, in 1981, and was director general of the European Southern Observatory, Germany, 1994–99.

In June 1962 a rocket was sent up by Giacconi and his group to observe secondary spectral emission from the Moon. But to the surprise of all concerned, it detected extremely strong X-rays from a source evidently located outside the Solar System. X-ray research has since become an important branch of astronomy, leading to the discovery of the existence of many previously unsuspected types of stellar and interstellar material, whose emissions lie in the X-ray band of the electromagnetic spectrum. Because much of the high-energy radiation is filtered out by the Earth's upper atmosphere, research has had to make considerable use of equipment carried aboard balloons or rockets; but the launching of *Uhuru*, in 1970, marked the beginning of a new era in that it was the first satellite devoted entirely to X-ray astronomy. Under the guidance of Giacconi, the team maintained a lead in the field. They devised and developed a telescope capable of producing X-ray images. Their years of experimental work in the laboratory and the more recent facility to use rocket-borne equipment employing solar power have resulted in instruments of increasing efficiency and angular resolution.

Giacconi has also worked a great deal with a Cherenkov detector, by means of which it is possible to observe the existence and velocity of high-speed particles, important in experimental nuclear physics and in the study of cosmic radiation.

Gibbs, (Josiah) Willard (1839–1903) US scientist who laid the foundation of modern chemical thermodynamics. He devised the phase rule and formulated Gibbs's adsorption isotherm.

Gibbs was born on 11 February 1839 in New Haven, Connecticut, into an academic family. His father was professor of sacred literature at the divinity school of Yale University, and Gibbs excelled at classics at school. He attended Yale in 1854, winning prizes for Latin and mathematics before graduating in 1858 at the age of only 19. During the next five years he continued his studies by specializing in engineering and in 1863 gained the first Yale PhD in this subject for a thesis on the design of gears. He then accepted teaching posts at Yale, first in Latin and then in natural philosophy. In 1866 he patented a railway braking system.

Also in 1866 Gibbs went abroad for three years to attend lectures (mainly in physics) in Paris, Berlin, and Heidelberg. In 1871, two years after his return to the USA, he was appointed professor of mathematical physics at Yale, a post he retained until his death despite offers from other academic institutions. He never married but lived with his sister and her family. He died in New Haven on 28 April 1903.

Gibbs did not publish his first papers until 1873, which were preliminaries to his 300-page series *On the Equilibrium of Heterogeneous Substances* (1876–78). In it he formulated the phase rule, which may be stated as:

$$f = n + 2 - r$$

where f is the number of degrees of freedom, n the number of chemical components, and r the number of phases – solid, liquid, or gas; degrees of freedom are quantities such as temperature and pressure that may be altered without changing the number of phases. Gibbs did not explore the chemical applications of the phase rule, later done by others who came to realize its importance. In the same work he also described his concept of free energy, which can be used as a measure of the feasibility of a given chemical reaction. It is defined in terms of the enthalpy, or heat content, and entropy, a measure of the disorder of a chemical system. From this Gibbs developed the notion of chemical potential, which is a measure of how the free energy of a particular phase depends on changes in composition (expressed mathematically as the differential coefficient of the free energy with respect to the number of moles of the chemical). The fourth fundamental contribution in this extensive work was a thermodynamic analysis that showed that changes in the concentration of a component of a solution in contact with a surface occur if there is an alteration in the surface tension – the Gibbs's adsorption isotherm.

All of these very technical discoveries now form part of the armoury of the physical chemist and thermodynamicist, together with their extension to electrochemistry and the subsequent developments of other scientists. But for many years Gibbs's work was unknown outside the USA, until it was translated into German by Wilhelm ◊Ostwald in 1891 and into French by Henri ◊Le Châtelier in 1899.

During his teaching studies Gibbs adapted the work of the mathematicians William Rowan ◊Hamilton and Hermann ◊Grassmann into a vector analysis that was both simple to use and easily applicable to physics, particularly electricity and magnetism. It was left to one of Gibbs's students, E B Wilson, to write a textbook on the subject, which was largely responsible for the popularization of vector analysis. Also during the 1880s Gibbs worked on the electromagnetic theory of light. From an entirely theoretical viewpoint and making very few assumptions he accounted correctly for most of the properties of light using only an electrical theory.

In his last major work, *Elementary Principles of Statistical Mechanics,* Gibbs turned his attention to heat and showed how many thermodynamic laws could be interpreted in terms of the results of the movements of enormous numbers of bodies such as molecules. His ensemble method equated the behaviour of a large number of systems at once to that of a single system over a period of time.

Giffard, Henri (1825–1882) French aeronautical engineer, famous for building the first steerable powered airship and making a successful flight in it.

Giffard, who studied engineering, was particularly interested in balloon flight, which had been pioneered in his native France. In the early 1850s he began to experiment with methods for steering balloons, which hitherto had been entirely at the mercy of the wind. In 1852 Giffard built his airship. It was a sausage-shaped gas bag 44 m/144 ft long and 12 m/52 ft in diameter, with a hydrogen capacity of 2,500 cu m/88,300 cu ft. The gondola was strung from a long pole or keel attached to the gasbag by ropes. It was powered by a small 2-kW/3-hp steam engine driving a three-bladed propeller – itself an innovation at the time. The airship was steered using a rudder, a canvas sail stretched over a bamboo frame and hinged to the keel. On 24 September 1852 Giffard took off from the Hippodrome in Paris and flew to Elancourt, near Trappes. His average speed was only about 5 kph/ 3 mph and he had problems with the steering.

Giffard gave up his experiments with airships but went on to other inventions, such as an injector to feed water into a steam engine boiler to prevent it running out of steam when not in motion. But his historic flight marked the real beginning of the human conquest of the air.

Gilbert, Walter (1932–) US molecular biologist who shared (with Paul ◊Berg and Frederick ◊Sanger) the 1980 Nobel Prize for Chemistry for his work in devising techniques for determining the sequence of bases in DNA. He also isolated the repressor substance postulated to exist by François ◊Jacob, André ◊Lwoff, and Jacques ◊Monod in their theory concerning the regulation of gene action.

Gilbert was born on 21 March 1932 in Boston, Massachusetts. He was educated at Harvard University, from which he graduated in physics in 1954, and at Cambridge University, England, from which he obtained his doctorate in mathematics in 1957. He then returned to Harvard, where he was a National Science Foundation Fellow in Physics 1957–58, and a lecturer and research fellow 1958–59. In 1960, however, influenced by James ◊Watson, Gilbert changed to biology, becoming professor of biophysics 1964–68, then professor of molecular biology 1969–72 at Harvard. He remained at Harvard, becoming, successively, American Cancer Society Professor of Molecular Biology 1972–81, senior associate in biochemistry and molecular biology 1982–84, professor of biology 1985–86, HH Timken Professor of Science 1986–87, chair of the department of cellular and developmental biology 1987–93, and Carl M Loeb University Professor from 1987.

Gilbert began his first major biological project in 1965, when he attempted to isolate and identify repressor substances involved in the regulation of gene activity. Working with Benno Muller-Hill, he devised a special experimental technique called equilibrium dialysis that enabled him to produce relatively large quantities of the repressor substance, which he then isolated and purified, and by late 1966 he had identified it as a large protein molecule.

After his work on the repressor substance, Gilbert then developed a method of determining the sequence of bases in DNA, which involved using an enzyme that breaks the DNA molecule at specific known points, thereby effectively excising predetermined fragments of the DNA. It was for developing this important investigative technique that he shared the 1980 Nobel Prize for Chemistry.

Gilbert, William (1544–1603) English physician and physicist who performed fundamental pioneering research into magnetism and also helped to establish the modern scientific method.

Gilbert was born in Colchester, Essex, on 24 May 1544. He was educated at St John's College, Cambridge, where he graduated in medicine in 1569 and later became a fellow. In about 1573 he settled in London, where he established a successful medical practice, and in 1599 he was elected president of the Royal College of Physicians. In the following year he was appointed personal physician to Queen Elizabeth I, and was later knighted for his services to the queen. On Elizabeth's death in March 1603 he was appointed physician to her successor, James I, but Gilbert held this position for only a short time because he died on 10 December of that year.

Although successful as a physician, Gilbert's most important contribution to science was his research into magnetism and electrical attraction, which he described in his book *De magnete, magneticisque corporibus, et de magno magnete tellure/Concerning Magnetism, Magnetic Bodies, and the Great Magnet Earth* (1600). This work gave a detailed account of years of investigations (on which he is thought to have spent some £5,000 of his own money), which were rigorously performed. The importance of this work was recognized by ◊Galileo, who also considered Gilbert to be the principal founder of the experimental method – before the work of Francis ◊Bacon and Isaac ◊Newton. Not only did Gilbert disprove many popular but erroneous beliefs about magnetism – that lodestone (naturally occurring magnetic iron oxide, the only magnetic material available until Gilbert discovered a method for making metal magnets) cured headaches and rejuvenated the body, for example – but he also discovered many important facts about magnetism, such as the laws of attraction and repulsion and magnetic dip. From his studies of the behaviour of small magnets, Gilbert concluded that the Earth itself acts as a giant bar magnet, with a magnetic pole (a term he introduced) near each geographical pole. He also showed that the strength of a lodestone can be increased by combining it with soft iron, that iron and steel can be magnetized by stroking them with lodestones, and that when an iron magnet is heated to red-heat, it loses its magnetism and also cannot be remagnetized while it remains hot. So extensive were Gilbert's investigations of magnetism that not until William ◊Sturgeon made the first electromagnet in 1825 and Michael ◊Faraday began his studies was substantial new knowledge added to the subject.

Gilbert also investigated static electricity. The ancient Greeks had discovered that amber (*elektron*), when rubbed with silk, can attract light objects. Gilbert demonstrated that other substances exhibit the same effect and that the magnitude of the effect is approximately proportional to the area being rubbed. He called these substances 'electrics' and clearly differentiated between magnetic attraction and electric attraction (as he called the ability of an electrostatically charged body to attract light objects). This work eventually led to the idea of electrical charge, but it was not until 1745 that it was discovered – by the French physicist Charles du Fay (1698–1739) – that there are two types of charge, positive and negative.

Gilbert also held remarkably modern views on astronomy. He was the first English scientist to accept Copernicus' idea that the Earth rotates on its axis and revolves around the Sun. He also believed that the stars are at different distances from the Earth and might be orbited by habitable planets, and that the planets were kept in their orbits by magnetic attraction.

Gilchrist, Percy Carlyle (1851–1935) English metallurgist who devised an inexpensive steel-making process.

Gilchrist was born in Lyme Regis, Dorset, on 27 September 1851, the son of a local barrister. He was educated at Felsted School, Essex, and later at the Royal School of Mines, where he acquired a sound knowledge of metallurgy. Between 1875 and 1877, together with his cousin Sidney Gilchrist ◊Thomas, he developed a method of producing low-phosphorus steel from high-phosphorus British ores (reducing the need – and cost – of using special imported ores). Thomas died soon afterwards but Gilchrist, who lived on for more than 50 years, became famous. He was vice president of the Iron and Steel Institute and in 1891 was elected a fellow of the Royal Society. He died on 16 December 1931.

Steel was a comparatively rare commodity until the 1850s, because its production was difficult and costly. Then in 1855 Henry ◊Bessemer invented his convector (in which air blown through molten pig iron oxidized impurities to produce a brittle, low-carbon steel). The addition of ferro-manganese gave a tougher product, but even so only low-phosphorus iron could be used, and most British ores were too high in phosphorus.

In 1870 Sidney Gilchrist Thomas was a junior clerk working in London for the Metropolitan Police but, like his cousin, he had an intense interest in natural science and metallurgy. After attending a series of lectures at the Birkbeck Institute, he joined Percy Gilchrist in trying to manfacture cheap steel from high-phosphorus ores. They used an old cupola to make a Bessemer-type convector and lined it with a paste made from crushed brick, sodium silicate, and water. The pig iron was added and melted before being subjected to prolonged 'blowing'. The oxygen in the blast of air oxidized carbon and other impurities, and the addition of lime at this stage caused the oxides to separate out as a slag on the surface of the molten metal. Continued blowing then brought about oxidation of the phosphorus, raising the temperature of the metal still further. When oxidation was complete (as judged by the colour of the flames coming from the convector), the cupola was tilted and the slag run off, leaving the molten steel to be poured into ingot moulds. The product became known at first as 'Thomas steel', and the age of cheap steel had arrived.

Gill, David (1843–1914) Scottish astronomer whose precision and patience using old instruments brought him renown before he achieved even greater fame for his pioneer work in the use of photography to catalogue stars.

Gill, whose father was a successful clockmaker, was born in Aberdeen on 12 June 1843. As the eldest surviving son, he was expected to take over the family business and was trained to do so by his father. After completing two years at Marischal College, where he was taught by the physicist and astronomer James Clerk ◊Maxwell, he went to Switzerland and studied clockmaking, becoming an expert in fine mechanisms, experienced in business methods, and fluent in French, all of which later stood him in good stead.

He then moved to Coventry, and later Clerkenwell in London, to continue his studies and then went back to Aberdeen, where he ran his father's business for ten years, gradually developing his interests in astronomy and astronomical techniques. In 1872 he became director of Lord Lindsay's private observatory at Dun Echt, 12 miles from Aberdeen. Seven years later he was appointed astronomer at the observatory at the Cape of Good Hope, South Africa. He was made Knight Commander of the Order of the Bath on 24 May 1900. He retired in 1906, for health reasons, and lived in London until he died of pneumonia on 24 January 1914.

On Gill's return to Aberdeen in 1862, he was given the use of a small telescope at King's College Observatory. Through business contacts he acquired a 30-cm/12-in reflector for the college with which he began to try to determine stellar parallaxes. This work was interrupted by Lord Lindsay's invitation to become director of his new observatory. Gill's job was to equip and supervise the building of this new observatory, which Lindsay was determined to make the best possible. In the course of this work, Gill met many important European astronomers and skilled instrument-makers, contacts that were to be very useful in his later position at the Cape. He also mastered the use of a heliometer, a rather old-fashioned instrument that he was to use a great deal and one that required great precision of hand and eye. The heliometer was first designed for measuring the variation of the Sun's diameter at different seasons, but it was also known to be one of the most accurate instruments for measuring angular distances between stars. Gill was perhaps the last great master of the heliometer and his measurements of the parallax of the Southern stars were unsurpassed for many years.

In 1872 Gill went on a six-year expedition to the island of Mauritius, with Lord Lindsay and others, in order to measure the distance of the Sun and other related constants particularly during the 1874 transit of Venus. The method Gill used was first proposed by Edmond ◊Halley, involving the amalgamation of the transit times of Venus across the face of the Sun as observed from various positions on the Earth, as widely spaced as possible. On Mauritius, through his skilful use of the heliometer, Gill observed the apparent near-approach of the asteroid Juno, and accurately deduced the value of the solar parallax. He used the same method of determining solar parallax on a private expedition. sponsored by the Royal Astronomical Society, on Ascension Island in 1877, when he measured solar parallax by considering the near approach of Mars. On both of these expeditions he used a 10-cm/4-in heliometer.

In 1879 he was appointed to the observatory at the Cape of Good Hope. This observatory had been built in 1820 in order to observe the southern hemisphere, but by 1879, when Gill took over, it was in poor condition. Most of the instruments were in need of repair and many of the results that had been obtained had not been published. Gill set to work at once. He was particularly unhappy about the Airy transit circle; although it was identical to the one at Greenwich, Gill designed an improved version that, when built in 1900, became the pattern for most transit circles built afterwards.

But it was the heliometer that Gill used most – at first a 10-cm/4-in one, then a 18-cm/7-in instrument installed in 1887. With the cooperation of many other astronomers, he made intensive investigations of several asteroids – Iris, Victoria, and Sappho – and in 1901 he made the first accurate determination of the solar parallax: 8.80‘. This figure was used universally in almanacs until it was superseded, in 1968, by a value of 8.794‘, a figure deduced by the analytical use of radar astronomy and using transmissions from one of the US Mariner space probes. With the heliometer he also measured the distances of 20 of the brighter and nearer southern stars.

The bright comet of 1882 was of great interest to astronomers. But it was on seeing a photograph of it that Gill realized it should be possible to chart and measure star positions by photography. At once he initiated a vast project, with the help of other observatories, to produce the *Cape Durchmusterung,* listing over 450,000 southern stars and the first major astronomical work to use photographic observation to catalogue stellar positions and brightness. Gill also served on the original council for the *International Astrographic Chart and Catalogue,* which was to give precise positions for all stars to the 11th magnitude. It was not completed until 1961, although all the photographs had been taken by 1900.

Gilman, Henry (1893–1986) US organic chemist best known for his work on organometallic compounds, particularly Grignard reagents.

Gilman was born on 9 May 1893 in Boston, Massachusetts. He grew up there and attended university in nearby Harvard, from which he graduated with top honours in 1915. After a year abroad, during which he visited academic centres in Zürich, Paris, and Oxford, he returned to Harvard to continue his postgraduate studies, and gained his PhD in chemistry in 1918.

He began his career as an instructor at Harvard, but in 1919 was offered a similar post in the chemistry department at the University of Illinois. Then an assistant professorship at the newly established faculty at Iowa State University prompted another move in the same year. He became professor of organic chemistry in 1923 and continued to teach at Iowa until 1947, when his vision deteriorated to such an extent that he became virtually blind.

Gilman's first paper on organometallic compounds, on Grignard reagents, was published in 1920. He went on to investigate the organic chemistry of 26 different metals, from aluminium, arsenic, and barium through to thallium, uranium, and zinc, and discovered several new types of compounds. He was the first to study organocuprates, now known as Gilman reagents, and his early work with organomagnesium compounds (Grignard reagents) led to experiments with organolithium compounds, later to play an important part in the preparation of polythene. Organogold compounds found applications in medicine.

The comprehensive study of methods of high-yield synthesis, quantitative and qualitative analysis, and uses of organometallic compounds represent only part – albeit the major one – of Gilman's research achievements. He also developed an international reputation for his work on organosilicon compounds, which again found many industrial applications. Other research interests included heterocyclic compounds, catalysis, resins, plastics, and insecticides.

Ginzburg, Vitalii Lazarevich (1916–) Russian scientist recognized as one of the leading astrophysicists of the 20th century.

Born in Moscow on 21 September 1916, Ginzburg completed his education at Moscow University, graduating in physics in 1938. He then studied as a postgraduate there and for a further two years at the Physics Institute of the Academy of Sciences. He was professor of physics at Gorky University 1945–68, and at the Moscow Technical Institute of Physics from 1968. A member of the Communist Party from 1944, Ginzburg won several Soviet prizes and honours, notably the Order of Lenin in 1966.

Ginzburg's first major success was his use of quantum theory in a study of the Cherenkov radiation effect (in which charged particles such as electrons travel through a medium at a speed greater than that of light in that medium) in 1940. This work was important for the development of nuclear physics and the study of cosmic radiation, and Ginzburg continued to theorize in this field. Seven years later, with Igor Tamm (1895–1971), one of the original interpreters of the Cherenkov effect), he formulated the theory of a molecule containing particles with varying degrees of motion, for which he devised the first relativistically invariant wave equation. With Lev ◊Landau, his former teacher, Ginzburg posited a phenomenological theory of conductivity. After 1950 he concentrated on problems in thermonuclear reactions. His work on cosmic rays – he was one of the first to believe that background radiation came from farther than within our own Galaxy – led him to a hypothesis about their origin and to the conclusion that they can be accelerated in a supernova.

Glaser, Donald Arthur (1926–) US physicist who invented the bubble chamber, an instrument much used in nuclear physics to study short-lived subatomic particles. For this development he was awarded the 1960 Nobel Prize for Physics.

Glaser was born on 21 September 1926 in Cleveland, Ohio, and educated there at the Case Institute of Technology. He graduated in 1946 and then went to the California Institute of Technology, where he gained his PhD three years later. He joined the physics department of the University of Michigan in 1949, and was professor of physics there from 1957. Two years later he took up a similar appointment at the University of California, Berkeley, and in 1964 became professor of physics and biology.

In the 1920s, the British physicist C T R ◊Wilson developed the cloud chamber for detecting and recording the presence of subatomic particles. It contained a saturated vapour that condensed as a series of liquid droplets along the ionized path left by a particle crossing the chamber. With the advances in particle accelerators in the early 1950s, nuclear physicists began to deal with fast high-energy particles, which could be missed by the cloud chamber. In 1952, while he was at the University of Michigan, Glaser built his first bubble chamber. It consisted of a vessel only a few centimetres across, containing superheated liquid ether under pressure. When the pressure was released suddenly, particles traversing the chamber left tracks consisting of streams of small bubbles formed when the ether boiled locally; the tracks were photographed using a high-speed camera. In later bubble chambers, liquid hydrogen was substituted for ether and chambers 2 m/6.6 ft across are now in use.

In the early 1960s, Glaser turned his attention from physics to molecular biology, as reflected in his change of appointment at Berkeley.

Glashow, Sheldon Lee (1932–) US particle physicist who proposed the existence of a fourth 'charmed' quark and later argued that quarks must be coloured. Insights gained from his theoretical studies enabled him to consider ways in which the weak nuclear force and the electromagnetic force could be unified as a single force (now called the electroweak force). He shared the 1979 Nobel Prize for Physics with Pakistani physicist Abdus ◊Salam and US physicist Steven ◊Weinberg for this work.

Sheldon Glashow was born in Manhattan on 5 December 1932, the son of immigrant Jewish parents from Tsarist Russia and the third of three brothers. After years of struggle his father became a plumber and was able to provide for the university education for his sons that he and his wife had missed. Sheldon's brothers became a dentist and a doctor but Sheldon wanted to be a scientist from an early age, and at the age of 15 he had a basement laboratory in which he carried out dangerous research on the synthesis of selenium halides. His parents always encouraged him in his scientific pursuits, only occasionally suggesting that he should take the safe route of becoming a doctor first.

Glashow was educated at the Bronx High School of Science where Weinberg was his classmate in a highly talented class. In 1950 he entered Cornell University (with Weinberg) obtaining a bachelor's degree in 1954 and moving to Harvard to gain an MA in 1955 and a PhD in 1958. His thesis at Harvard was called 'The vector meson in elementary particle decays', reflecting an early interest in electroweak synthesis.

Winning a postdoctoral fellowship, Glashow's intention was to work in Moscow but his visa never came, so he spent his fellowship 1958–60 at the Niels Bohr

Institute in Copenhagen where he discovered the SU(2) × U(I) structure of the electroweak theory and also later worked briefly, in 1964, with US physicist James D Bjorken (1934–) on a property that he called 'charm'. His research took him to the European Organization for Nuclear Research (CERN) in Geneva and then to the California Institute of Technology (Caltech) in Stanford where he spent several years working with Sidney Coleman on the eightfold way invented by Murray ◊Gell-Mann. He was a member of the faculty at Berkeley 1962–66, returning to Harvard as professor of physics where he has remained ever since except for periods spent at CERN (1968), Marseilles (1970), MIT (1974), and Texas (1982).

John Iliopoulos and Luciano Miani arrived at Harvard in 1969 as research fellows and joined Glashow in predicting the existence of 'charmed hadrons'. He later collaborated with Alvaro de Rujula and Howard Georgi in correctly predicting the presence of what he called 'charm' in neutrino physics or in e + e = annihilation, thus realizing that many strands of research were converging on one theory of physics.

Glashow shared the Nobel Prize for physics with Steven Weinberg and Abdus Salam in 1979 for what was cited as 'their contribution to the theory of the unified weak and electromagnetic interaction between elementary particles, including *inter alia* the prediction of the weak neutral current'. Working on Glashow's early model explaining the electromagnetic and weak nuclear forces, Weinberg and Salam developed a coherent theory that unified two of the four forces of physics: the electromagnetic interaction and the weak interaction, which they applied to leptons (electrons and neutrinos). Using Gell-Mann's theory that particles are made up of smaller particles called quarks, Glashow then postulated that a fourth quark was necessary and extended the Weinberg–Salam theory to other particles – baryons and mesons – by introducing the particle property 'charm'. Glashow predicted the existence of other new particles, which were found in the following years, and the quark theory was expanded to include a coloured quark. The theory of quantum chromodynamics is based on his approach.

Glashow has received several honorary degrees and many honours for his work including the Oppenheimer Memorial Medal in 1977 and the George Ledlie Award in 1978. He married Joan Alexander in 1972 and they have four children.

Glauber, Johann Rudolf (1604–1670) German chemist who lived at a time when chemistry was evolving as a science from the mysticism of alchemy. He prepared and sold what today would be called patent medicines and panaceas, one of which (Glauber's salt, sodium sulphate) still bears his name. He did make some genuine chemical experiments and preparations, although many of the results were misinterpreted (through ignorance) or simply ignored by his contemporaries.

Glauber was born in Karlstadt, Franconia, some time during 1604, the son of a barber. His parents died when he was young and he had little or no formal education, teaching himself and acquiring knowledge on his travels through Germany. He earned a living chiefly from selling medicines and chemicals. He lived for a while in Vienna, then in various places along the Rhine valley, before going to Amsterdam in 1648 and finally settling there in 1655 after another six-year sojourn in Germany. He designed and built a chemical laboratory in an Amsterdam house formerly occupied by an alchemist. He died in Amsterdam on or about 10 March 1670, possibly from the cumulative effects of poisonous substances with which he had worked and, no doubt, tried on himself as medicaments.

Some time around 1625 Glauber prepared hydrochloric acid by the action of concentrated sulphuric acid on common salt (sodium chloride). He found that the other product of the reaction, sodium sulphate, was a comparatively safe but efficient laxative, and could be sold as a cure-all for practically any complaint. He called it *sal mirabile* (miraculous salt), although it soon became known as Glauber's salt – and it is still the chief active ingredient of various patent medicines. A Stassfurt mineral, a double sulphate of sodium and calcium, is known as glauberite. He also prepared nitric acid by substituting saltpetre (potassium nitrate) for salt in the reaction with sulphuric acid. He made many metal chlorides and nitrates from the mineral acids. He discovered tartar emetic (antimony potassium tartrate), so-called because of its medicinal use in the days before antimony's toxicity was fully realized.

In his Amsterdam laboratory Glauber investigated and developed processes that could have industrial application, pursuing the principles he had outlined in his book *Furni novi philosophici* (1646). Many of his experiments were carried out in secret, so that the methods or their products could be sold exclusively. He produced organic liquids containing such solvents as acetone (dimethylketone) and benzene – although he did not identify them – by reacting and distilling natural substances such as wood, wine, vinegar, vegetable oils, and coal. One aim of his experiments was to improve the chemical techniques involved, and this aspect of Glauber's work was summarized in his book *Opera omnia chymica*, which was published in various versions 1651–61. He was conscious of the value to any country of its chemical and mineral raw materials, which he thought should be husbanded and exploited only for the good of the whole community. He outlined his views on a possible utopian future for Germany in *Teutschlands-Wohlfahrt/Germany's Prosperity.*

Goddard, Robert Hutchings (1882–1945) US physicist who pioneered modern research into rocketry. He developed the principle of combining liquid fuels in a rocket motor and the technique has been used subsequently in every practical space vehicle. He was the first to prove by actual test that a rocket will work in a vacuum and he was the first to fire a rocket faster than the speed of sound.

Goddard was born on 5 October 1882 in Worcester, Massachusetts. He was 17 when he began to speculate about conditions in the upper atmosphere and in interplanetary space. He considered the possibility of using rockets as a means of carrying research instruments.

In 1901, while studying for his BSc at Worcester Polytechnic Institute, he wrote a paper suggesting that the heat from radioactive materials could be used to expel substances at high velocities through a rocket motor to provide power for space travel. Two years later, after making his first practical experiments to determine the efficiency of rocket power, he proposed the use of high-energy propellants, such as liquid oxygen and hydrogen.

At Clark University in his home town, and at Mount Wilson, California, Goddard carried out experiments with naval signal rockets. He went on to design and build his own rocket motors. As an instructor in physics at Worcester Polytechnic, Goddard directed his experiments towards the development of rockets to explore the upper atmosphere. It was not until 1917 that he received financial aid from the Smithsonian Institution.

On the USA's entry into World War I, he turned his energies to investigating the military application of rockets, whereupon the US army provided funds. At Clark University, teaching physics in the early 1920s, Goddard switched his practical research from solid fuel to liquid propellants. On 1 November 1923 he fired the first liquid rocket in a test stand. Two years of development were rewarded with the historic first launching of a liquid-propelled rocket on 16 March 1926 at Auburn, Massachusetts.

During the next three years he improved engine performance and reduced the weight of his design. Instruments, and a camera to record them, were carried aloft for the first time on 17 July 1929. This significant flight attracted funds for a fuel-scale rocket-testing programme.

The greater altitudes and speeds his rocket attained dictated the need for a precise method of flight control and on 28 March 1935 Goddard successfully fired a rocket controlled by gyroscopes linked to vanes in the exhaust stream.

It was in 1937 that one of his rockets reached 3 km/1.8 mi, then the greatest altitude for a projectile. His work in New Mexico continued until World War II. In 1942, with the USA in the war, he joined the Naval Engineering Experimental Station at Annapolis, Maryland, continuing his research until his death on 10 August 1945.

Unlike his contemporaries at the dawn of the space age, Goddard pioneered the science of rocketry by practical experiment. He proved that liquid oxygen and a hydrocarbon fuel, such as petrol or kerosene, was the mixture that would provide the considerable thrust necessary. He built a motor in which to fire the fuels and used it in the first rocket of its kind, nearly five years before the first German experimental flights.

His rocket designs which flew in 1940, although smaller, were more advanced than the menacing V2 missile that was to fly two years later and was to be launched against European cities. Goddard's development of gyroscopically controlled steering vanes also predated German work by a number of years.

Goddard's other practical work led to the development of the centrifugal propellant pump. While in his early years he foresaw the use of the rocket for high-altitude research and wrote about space exploration, he was also aware of the military applications of his work. A rocket he developed for the US army was fired from a launching tube, but the war ended a few days after its successful demonstration and the need disappeared.

Through much of his career Goddard worked as a 'loner', many of his achievements being made single-handedly. As one of the few practitioners of space research, in a world that was still trying to grasp the significance of air travel, his work and writings were not taken seriously until 1929 and the flight of his first instrumented rocket. The finance that this achievement attracted (initially from the industrialist Daniel Guggenheim) enabled Goddard and his small team to undertake research during a period of 12 years. This was surpassed only (because it had practical military application) by the achievement of the army of German scientists under Dornberger, in their five years' work at Peenemunde to put the V2 strategic missile into service.

The full significance of the German work was not appreciated until the war's end, when Goddard was amazed to find such similarity with his designs. He died only months later and the military and space rocket programmes of both the USA and USSR were to be founded upon this German work.

It was not until 1960 that the US government recognized the value of Goddard's work. It admitted to frequent infringement of many of his 214 patents during the evolution of missile and satellite programmes. In that year his widow received $1,000,000 from federal funds.

In view of the USA's failure to recognize the significance of his work, at a time when its enemy was almost in a position to influence the course of the war with the rocket, it is ironic to record that Robert Goddard received more financial support for one project than any other single scientist up to World War II.

Further Reading

Coil, Suzanne M, *Robert Hutchings Goddard: Pioneer of Rocketry and Space Flight,* Makers of Modern Science Series, Facts on File, 1992.

Farley, Karin C, *Robert H Goddard,* Silver Burdett Press, 1992.

Lehman, Milton, *Robert H Goddard,* Da Capo Press, 1988.

Gödel, Kurt (1906–1978) Austrian-born US philosopher and mathematician who, in his philosophical endeavour to establish the science of mathematics as totally consistent and totally complete, proved that it could never be. This realization has been seen as fundamental to both mathematical and philosophical studies and concepts, and at the time was devastatingly revolutionary to other scientists working in the same endeavour.

Gödel was born on 28 April 1906 at Brünn, Moravia (now Brno in the Czech Republic), where his father was a businessman. Several childhood illnesses left him with a lifelong preoccupation with his health, but at school he worked hard and successfully: he is reputed never to have made a mistake in Latin grammar. At the age of 17 he went to the University of Vienna where he was at first unsure whether to read mathematics or physics, but settled for mathematics. At that time Vienna – and particularly its university – was the centre of activity in positivist philosophy and Gödel could not help but be influenced, and although he was apparently unimpressed at the meetings of the prestigious Viennese Circle of philosophers that he attended, his studies drifted from pure mathematics towards mathematical logic and the foundations of mathematics. In 1930 he was awarded his doctorate at the university, and in 1931 he published his most important paper, which was accepted by the university authorities as the thesis on which they licensed him to become an unsalaried lecturer there. Throughout the 1930s Gödel continued to work at the University of Vienna (except for a visit to Princeton in the USA 1933–34). Then, in 1938, when Austria became part of Germany, he was not appointed as a paid lecturer with his colleagues because it was thought, erroneously, that he was Jewish. Later the same year he married and travelled to the USA, where he spent a short time at the Institute for Advanced Study, Princeton, before returning to Austria. Finally, in 1939, he went back to the USA to settle at Princeton where, in 1953, at the age of 47, he was appointed professor. A quiet and unassuming man, Gödel was awarded many honours – although he also refused quite a number, particularly from Germany. He died in Princeton on 14 January 1978.

In 1930 Gödel was granted his doctorate for a dissertation in which he showed that a particular logical system (predicate calculus of the first order) was such that every valid formula could be proved within the system; in other words, the system was what mathematicians call complete. This research represented the way in which his studies were subsequently to take him. He then investigated a much larger logical system – that constructed by Bertrand ◊Russell and Alfred ◊Whitehead as the logical basis of mathematics, published as *Principia mathematica.* Accordingly the title of his licensiate paper, published in 1931, was 'On formally undecidable propositions of *Principia Mathematica* and related systems'. In it Gödel dashed the hopes of philosophers and mathematicians alike, and showed that there were systems upon which mathematics was based in which it was impossible to decide whether or not a valid statement or formula was true or false within the system. In more practical terms, it was not possible to show that such areas as arithmetic worked entirely within arithmetic; there was always the possibility of coming across something that could be either true or false.

Mathematics itself was unaffected by this bombshell. But the hopes for an absolutely perfect subject that could justify itself completely to even the most penetrating philosopher were utterly set back. Gödel himself felt that David ◊Hilbert's formalist school would save the day, and did not fully anticipate the effect his proof would have. Nevertheless, mathematical logic has subsequently made some progress – albeit mainly by attacking the subject from outside the logical system it embraces.

During his career Gödel made a number of other inspired contributions to the subject of mathematical logic, but none of the same importance as his earlier work. Because of his way of numbering statements in his famous proof, certain numbers were given the name Gödel numbers by fellow mathematicians. The other area of mathematics to which he contributed significantly was general relativity theory; this was probably due in part at least to his close friendship with Albert ◊Einstein at the Princeton Institute. Gödel solved some of Einstein's equations and constructed mathematical models of the universe. In one of these it was theoretically possible for a person to travel into his own past, but at the cost of such a large consumption of fuel as not to make the journey feasible. (According to the model, however, it might be possible for a person to send a message into his own past.)

Gödel kept detailed diaries that record numerous interests, including the laws of nature, the evolution of life, time travel, and ghosts and demonology, as well as his preoccupation with mathematics and philosophy.

Further Reading

Hofstadter, Douglas R and Gödel, Escher, *Bach: An Eternal Golden Braid,* Vintage Books, 1989.

Wang, Hao, *Reflections on Kurt Gödel,* MIT Press, 1990.

Goeppert-Mayer, Maria (1906–1972) German-born US physicist who shared the 1963 Nobel prize for discovering the shell model of nuclear structure.

Maria Goeppert was born on 28 June 1906 in Kattowitz, Upper Silesia (now Katowice in Poland). When Maria was four years old her father was appointed professor of paediatrics at the University of Göttingen and the family moved there. High inflation rates in Germany in the 1920s meant that the only school capable of preparing girls for the *abitur* exam was closed, so Maria's education had to be continued in private. She passed the exam in Hanover in 1924. Maria Goeppert entered Göttingen University as a mathematics student but was excited by quantum mechanics – Max ◊Born was on the staff in Göttingen – and quickly changed to physics. She obtained her doctorate in theoretical physics in 1930 and married Joseph Edward Mayer, a US physicist working in Göttingen with James ◊Franck. The couple moved to Johns Hopkins University in Baltimore, USA, where Maria worked on chemical physics and the colour of organic molecules, and to Columbia University in 1939, where she worked on isotope separation. In 1946 Maria was appointed a professor in the physics department at Chicago and given a post at the nearby Argonne National Laboratory. Such an appointment came as a relief as Maria had not been paid at Johns Hopkins and felt she had been kept on the sidelines at Columbia.

In 1945 Goeppert-Mayer had developed a 'little bang' theory of cosmic origin with Edward ◊Teller to explain element and isotope abundances in the universe and so became interested in the stability of nuclei. The 'liquid drop' model of the nucleus explained some properties but nuclei with certain numbers of neutrons (or protons) were known to be exceptionally stable. The stability of some of the light elements – helium (two neutrons and two protons), oxygen (eight each), calcium (20 each) – could be explained, but what about nuclei with 28, 50, 82 and 126 neutrons? The stability of nuclei with these so-called magic numbers of neutrons or protons had been demonstrated in experiments but could not be explained. Goeppert-Mayer and Hans ◊Jensen independently proposed a shell model in which the nucleons (protons and neutrons) moved in orbits (or shells) around the centre of the nucleus. This gave the neutrons orbital angular momentum, which then coupled to their spin (inbuilt angular momentum). A combination of the shell model and this spin–orbit coupling correctly predicted which nuclei were the most stable. Goeppert-Mayer and Jensen shared half of the 1963 Nobel prize – the other half went to Eugene Wigner (1902–1995) – and in 1955 they wrote a book *Elementary Theory of Nuclear Shell Structure.* In 1960 Goeppert-Mayer and her husband moved again, to take up professorships in physics and chemistry respectively at the University of California, La Jolla. She died in San Diego on 20 February 1972.

Gold, Thomas (1920–) Austrian-born US astronomer and physicist who has carried out research in several fields but remains most famous for his share in formulating, with Fred ◊Hoyle and Hermann ◊Bondi, the steady-state theory regarding the creation of the universe.

Gold was born in Vienna on 22 May 1920. He received his university training at Cambridge University, where he earned his bachelor's degree in 1942. Elected a fellow of Trinity College, Cambridge, in 1947, he lectured in physics at Cambridge until 1952. From then until 1956 he was chief assistant to Martin ◊Ryle, the discoverer of quasars, later to become Astronomer Royal. In 1956, Gold emigrated to the USA where, two years later, he became professor of astronomy at Harvard. At Cornell University from 1959, he was director of its Center for Radiophysics and Space Research until 1981 and professor of astronomy 1971–86 (emeritus from 1987). Gold has served as an adviser to NASA and is a member of the Royal Astronomical Society, the Royal Society, and the National Academy of Sciences.

The question of the conditions surrounding the beginning of the universe has fascinated astronomers for many centuries. The Big Bang theory, developed by George ◊Gamow and others, was parallelled in 1948 by the steady-state hypothesis put forward by Gold, Bondi, and Hoyle.

The steady-state theory assumes an expanding universe in which the density of matter remains constant. It postulates that as galaxies recede from one another, new matter is continually created (at an undetectably slow rate). The implications that follow are that galaxies are not all of the same age, and that the rate of recession is uniform.

Evidence began to accumulate in the 1950s, however, that the density of matter in the universe had been greater during an earlier epoch. In the 1960s, microwave background radiation at 3K (–270 °C/–454 °F) was detected, which was interpreted by most astronomers as being residual radiation from the primordial Big Bang. Accordingly, the steady-state hypothesis was abandoned by most cosmologists in favour of the Big Bang model.

Gold has also carried out research on a variety of processes within the Solar System, including studies on the rotation of Mercury and of the Earth, and on the Moon. In addition he has published some work on relativity theory.

Goldberg, Leo (1913–1987) US astrophysicist who carried out research, generally as one of a team, into the composition of stellar atmospheres and the dynamics of the loss of mass from cool stars.

Goldberg was born in Brooklyn, New York, on 26 January 1913. Completing his education, he gained his bachelor's degree in 1934 at Harvard University, where he immediately became assistant astronomer. Three years later he received his master's degree and moved to the University of Michigan, where he was special research fellow 1938–44. At the end of that time he was made a research associate both at Michigan University and at the McMath and Hulbert Observatory. Between 1945–60 he rose from assistant professor to professor of astronomy, and for almost all of that time he was also chair of the university's department of astronomy and director of the observatory. In 1960 he was appointed Higgins Professor of Astronomy at Harvard, a position he held until 1973. He was chair of the department of astronomy and director of Harvard College Observatory 1966–71. From 1971 to 1977 he was director – and in 1977 he became emeritus research scientist – of the Kitt Peak National Observatory in Arizona.

Goldberg's main subject of research was the Sun. As one of a team at the McMath and Hulbert Observatory, he contributed towards some spectacular films of solar flares, prominences, and other features of the Sun's surface. At Harvard, he and his colleagues designed an instrument that could function either as a spectrograph or as a spectroheliograph (a device to photograph the Sun using monochromatic light), that formed part of the equipment of Orbital Solar Observatory IV, launched in October 1967. He has also carried out research on the temperature variations and chemical composition of the Sun and of its atmosphere, in which he succeeded in detecting carbon monoxide (as predicted by Henry ◊Russell).

Goldring, Winifred (1888–1971) US palaeontologist who is known for her research on Devonian fossils and popularization of geology.

Goldring was born on 1 February 1888 and was raised near Albany, New York State. She was educated at the New York State Normal School from which she graduated in 1905 as valedictorian, and Wellesley College receiving her AB in 1909 and AM in 1912. She remained at Wellesley until 1914 as an instructor in petrology and geology and assisting in geography and field geology, as well as teaching geography at the Teachers' School of Science. Returning to Albany in 1914, Goldring took a temporary appointment in the Hall of Invertebrate Palaeontology at the New York State Museum. She remained at the museum and in 1920 was appointed associate palaeontologist. In 1939 Goldring was made state palaeontologist, and was the first woman to hold the position, which she kept until her retirement in 1954; she died in Albany on 30 January 1971 following a gastrointestinal illness.

Goldring became internationally known in the discipline of palaeobotany with her work on Devonian crinoids started in 1916 and published as *The Devonian Crinoids of the State of New York* (1923). During the late 1920s and 30s, as well as geologically mapping the Coxsackie and Berne quadrangles of New York, she developed and maintained the State Museum's public programme in palaeontology by popularizing geology through her stimulating museum exhibits and the publication of handbooks, notably the two-volume *Handbook of Paleontology for Beginners and Amateurs* (1929, 1931) and her *Guide to the Geology of John Boyd Thacher Park* (1933). Goldring's work led to recognition and honours. She was the first woman president of the Paleontological Society (1949) and was vice president of the Geological Society of America (1950); she was also awarded honorary degrees from Russell Sage College (1937) and Smith College (1957).

Goldschmidt, Victor Moritz (1888–1947) Swiss-born Norwegian chemist who has been called the founder of modern geochemistry.

Goldschmidt was born on 27 January 1888 in Zürich but moved to Norway with his family in 1900, where his father became professor of physical chemistry, at the University of Christiania (now Oslo). The family became Norwegian citizens in 1905. Goldschmidt completed his schooling in Christiania and continued at the university there, where he studied under the geologist Waldemar Brøgger and graduated in 1911. He was appointed professor and director of the Mineralogical Institute in 1914, a post he held until 1929, when he moved to Göttingen. The rise of anti-Semitism forced Goldschmidt to return to Norway in 1935, but soon World War II engulfed Europe and he had to flee again, first to Sweden and then to the UK, where he worked at Aberdeen and Rothamsted (on soil science). He returned to Norway after the end of the war in 1945, and died in Oslo on 20 March 1947.

Goldschmidt's doctoral thesis on contact metamorphism in rocks is recognized as a fundamental work in geochemistry. It set the scene for a huge programme of research on the elements, their origins, and their relationships, which was to occupy him for the next 30 years. He broke new ground when he applied the concepts of Willard ◊Gibbs's phase rule to the colossal chemical processes of geological time, which he considered to be interpretable in terms of the laws of chemical equilibrium. The evidence of geological change over millions of years represents a series of chemical processes on a scarcely imaginable scale, and even an imperceptibly slow reaction can yield megatonnes of product over the time scale involved.

Shortage of materials during World War I led Goldschmidt to speculate further on the distribution of elements in the Earth's crust. In the next few years he and his co-workers studied 200 compounds of 75 elements and produced the first tables of ionic and atomic radii. The new science of X-ray crystallography, developed by the William and Lawrence ◊Bragg after Max von ◊Laue's original discovery in 1912, could hardly have been more opportune. Goldschmidt was able to show that, given an electrical balance between positive and negative ions, the most important factor in crystal structure is ionic size. He suggested furthermore that complex natural minerals, such as hornblende $(OH)_2Ca_2Mg_5Si_8O_{22}$, can be explained by the balancing of charge by means of substitution based primarily on size. This led to the relationships between close-packing of identical spheres and the various interstitial sites available for the formation of crystal lattices. He also established the relation of hardness to interionic distances.

At Göttingen from1929, Goldschmidt pursued his general researches and extended them to include meteorites, pioneering spectrographic methods for the rapid determination of small amounts of elements. Exhaustive analysis of results from geochemistry, astrophysics, and nuclear physics led to his work on the cosmic abundance of the elements and the important links between isotopic stability and abundance. Studies of terrestrial abundance reveal about eight predominant elements. Recalculation of atom and volume percentages lead to the remarkable notion that the Earth's crust is composed largely of oxygen anions (90% of the volume), with silicon and the common metals filling up the rest of the space.

Goldschmidt was a brilliant scientist, with the rare ability to arrive at broad generalizations that draw together many apparently unconnected pieces of information. He also had a steely sense of humour. During his exile in the UK he carried a cyanide suicide capsule for the ultimate escape should the Germans successfully invade the UK. When a colleague asked for one he was told 'Cyanide is for chemists; you, being a professor of mechanical engineering, will have to use a rope.'

Further Reading

Kauffman, George B, Founder of geochemistry: Victor Moritz Goldschmidt, *Ind Chem*, 1988, v 9, no 1, pp 46–47.

Mason, Brian, *Victor Moritz Goldschmidt: Father of Modern Geochemistry*, Special Publication series, Geochemical Society, 1992, v 4.

Goldstein, Eugen (1850–1930) German physicist who investigated electrical discharges through gases at low pressures. He discovered canal rays and gave cathode rays their name.

Goldstein was born in Gleiwitz in Upper Silesia (now Gliwice in Poland) on 5 September 1850. He went to the University of Breslau in 1869 but after only one year moved to the University of Berlin to work with Hermann von ◊Helmholtz, first as his student and then as his assistant. Before obtaining his doctorate he became employed as a physicist at the Berlin Observatory in 1878. Following the award of his doctorate in 1881 for some of his important work on cathode rays, Goldstein continued at the observatory until 1890, spending the next six years at the Physikalische Technische Reichsanstalt. There then followed some time when he had a laboratory and assistants in his own home, until he was appointed head of the Astrophysical Section of Potsdam Observatory. Goldstein died in Berlin on 25 December 1930.

As Helmholtz's assistant, Goldstein worked on electrical discharges in gases. Goldstein coined the term 'cathode rays' to explain the glow first observed in an evacuated tube by Julius ◊Plücker in 1858. His first publication, in 1876, demonstrated that cathode rays could cast shadows and that the rays are emitted perpendicular to the cathode surface, enabling a concave cathode to bring them to a focus. There followed investigations by Goldstein and others that showed how cathode rays could be deflected by magnetic fields. This was the period when such phenomena were under investigation throughout Europe, and Goldstein made many contributions which aided the conclusion that cathode rays were fast-moving, negative particles or electrons. He himself believed that they were waves not unlike light.

In the course of his investigations, Goldstein studied many different arrangements of anode and cathode discharge circuits. In an experiment in 1886, he perforated the anode and observed glowing yellow streamers behind the anode emanating from the perforations. Since they were connected to the holes or canals, he termed them *Kanalstrahlen* or canal rays. These always moved away from the anode. Wilhem Wien (1864–1928), Goldstein's student, showed in

1898 that they could be deflected by electric and magnetic fields in such a way that they appeared to be positively charged. Earlier, Goldstein had been able to observe only slight deflections. Wien also calculated the charge-to-mass ratio of the rays and found this to be some 10,000 times greater than the corresponding ratio for cathode rays. Later, the rays were called positive rays by J J ◊Thomson, who also investigated their nature and led the way to their identification as ionized molecules or atoms of the gaseous contents of the discharge tube. Tubes with different contents formed different positive ions, and the different charge-to-mass ratios were detected by Thomson. Francis ◊Aston began investigations in this field and developed the mass spectrograph in 1919 to separate positive rays. From these beginnings has developed mass spectrometry, a method extremely useful in chemical analysis.

In 1905 Johannes ◊Stark, another student of Goldstein, used light from canal rays to demonstrate an optical Doppler shift in a terrestrial, rather than a stellar, source of light for the first time.

Goldstein later investigated the wavelengths of light emitted by metals and oxides when canal rays impinge on them, and observed that alkali metals show their characteristic bright spectral lines. He continued his investigation of the complex phenomena and gaseous discharge at the anode for many years, but his subsequent discoveries were interesting and curious rather than of great importance for physics. He compared his work with the natural phenomenon of the aurora borealis, which he believed was caused by cathode rays originating in the Sun. His last paper, published in 1928, gave a new discovery, the importance of which was almost completely missed. In a discharge tube containing nitrogen and hydrogen and possibly other gases, he observed that a trace of ammonia was present after the discharge. It was only much later that investigations along these lines were begun in earnest to see whether biologically important molecules, and hence life, could have originated in this way.

Golgi, Camillo (1843–1926) Italian cytologist and histologist who pioneered the study of the detailed structure of the nervous system, made possible by his development of a new staining technique. For his outstanding work in this field, Golgi shared the 1906 Nobel Prize for Physiology or Medicine with Santiago ◊Ramón y Cajal.

Golgi was born on 7 July 1843 in Corteno, near Brescia, the son of a physician. He studied medicine at the University of Pavia, graduating in 1865, then worked in a psychiatric clinic, before researching in histology. In 1872 he became principal physician at a hospital in Abbiategrasso near Milan, but managed to continue his histological research. In 1875 he was appointed lecturer in histology at the University of Pavia and professor the following year, a post he held until 1879, when he became professor of anatomy at the University of Siena. In the following year, however, he returned to Pavia University as professor of histology, subsequently becoming professor of general pathology, and remained there until his retirement in 1918. During his time at the university he took an active part in its administration, as dean of the faculty of medicine and later as president of the university. He also played an important role outside the university, becoming a senator in the Italian government in 1900. Golgi died on 21 January 1926 in Pavia.

Golgi Italian cytologist and histologist Camillo Golgi who, together with Santiago Ramón y Cajal, won the Nobel Prize for Physiology or Medicine in 1906 for his work on the nervous system. Golgi devised a technique for staining nervous tissue with silver that greatly advanced the study of cell histology. *Mary Evans Picture Library*

At the time Golgi was studying medicine there were no techniques suitable for the study of nerve cells, with the result that very little was known about their structure and function. In 1873, however, Golgi invented a new technique of staining cells with silver salts, a method he found to be particularly suitable for studying nerve cells because it allowed controlled staining of certain features in the cells, thereby enabling them to be studied in great detail. From 1873 onwards he published many articles on the results of his systematic studies – using his new technique – of the fine anatomy of the nervous system. He verified Wilhelm von Waldeyer's view that nerve cells are separated by synaptic gaps, and he also discovered a type of nerve cell (later called the Golgi cell) that connects many other cells by means of dendrites. In 1896, while studying the brain of a barn owl, he detected flattened cavities near the nuclear membrane; these structures – called Golgi bodies, the Golgi complex, or the Golgi apparatus – have since been found to be involved in intracellular secretion and transport.

From his examinations of different parts of the brain, Golgi put forward the theory that there are two types of nerve cells – sensory and motor – and that axons are concerned with the transmission of nerve impulses. He showed that there is a fine nerve network in the grey matter but could not say positively whether the filaments were joined or merely interwoven. Golgi made other contributions to the study of the nervous system, including the discovery of tension receptors in the tendons – now called the organs of Golgi.

Between 1885–93 Golgi investigated malaria, the causative agent of which – the protozoan *Plasmodium* – had been discovered by the French physician Charles ◊Laveran in 1880. (And by 1885 two Italian scientists had discovered the stages in the parasite's life cycle.)

Golgi's research verified a number of important facts about the disease and discovered that different species of *Plasmodium* are responsible for the two main types of intermittent fever – tertian fever, with attacks every three days, and quartan fever, with attacks every four days – and that the fever attacks coincide with the release into the bloodstream of a new generation of the parasites. He also put the results of his work to practical use by establishing a method of treatment that involved determining the type of malaria a person had contracted, and then giving the patient quinine (the only effective antimalarial drug then available) a few hours before the predicted onset of the fever.

Goodall, Jane (1934–) English zoologist best known for her studies of chimpanzees in the Gombe Stream Reserve in Tanzania. She is a leading authority on chimpanzees and a tireless campaigner on their behalf.

Goodall was born in London on 3 April 1934, one of two sisters; her father was a businessman and motor-racing enthusiast and her mother a novelist. She had a traditional middle-class upbringing and was educated at Uplands School in Bournemouth before taking secretarial courses and taking a job as a secretary at Oxford University at the age of 18. She had been fascinated by wildlife since she was a child and was always making notes and sketches of local birds and animals. She was very knowledgeable in zoology and ethology. Her dream was to travel to Africa to study the animals in their natural environment. In 1957, at the age of 23, after saving money by working in her spare time at a documentary film company, she achieved her trip to Kenya with a childhood friend. While she was there she met anthropologist Louis ◊Leakey and became his secretary at the Coryndon Museum in Nairobi where he was curator. Leakey took her to Olduvai Gorge, which was to become famous as the site of one of the most important anthropological finds this century. He also sent her to an island in Lake Victoria to study vervet monkeys.

Leakey believed that Goodall had the right temperament to, as he put it, 'forego the amenities of civilization for long periods of time without difficulty' and offered her a job studying 160 chimpanzees in the Gombe Stream Reserve on the eastern shores of Lake Tanganyika. She had no formal scientific education and his choice was generally disapproved of. But what she lacked in expertise she made up for in enthusiasm and she leapt at the chance. While Leakey looked for financial support for the venture, she went back to London and trained at the Royal Free Hospital and London Zoo while working on an animal documentary for British television. In 1960, she, her mother, and a cook set up camp in the Gombe Stream Reserve.

The following few years were spent patiently gaining the trust of a group of chimpanzees by appearing every day at the same time, some distance from their feeding place. After 2 years she was able to feed them on bananas and spent time in the trees with them, eating their foods and imitating them, while observing their behaviour. Through prolonged observation she has documented the social development of the individual in the community and the histories of several families. Goodall was awarded a PhD in ethology from Cambridge in 1965 and her doctoral thesis, 'Behaviour of the free ranging chimpanzee' was a detailed description of her first years in Gombe.

Her first book, *In the Shadow of Man* (1971), presents an idealized picture of a chimpanzee society – peaceful vegetarians, socializing with each other, making tools and fashioning grass to extract termites from termite mounds. She subsequently discovered that chimpanzees eat meat and use stones as weapons. She saw chimps stalking and killing other animals, such as antelopes and baby baboons, and she also recorded incidents of cannibalism. Her book, *Through a Window* (1990), reveals the darker side of chimpanzee nature.

In 1964 Jane Goodall married Baron Hugo van Lawick, a Dutch wildlife photographer who had visited her from the National Geographical Society to film her work, making a rare departure from Gombe for a honeymoon in Europe. Hugo brought their lifestyle to the living rooms of Britain and America with his television programmes featuring Jane, himself, the chimpanzees and their son, Hugo, who was born in 1967. Through her writing, television appearances and the challenges she laid down to academics and scientists about the nature of the relationship between chimps and humans and their societies, Goodall became quite a celebrity. In 1970 she and Hugo published a joint study of wild dogs, jackals, and hyenas, *Innocent Killers.* After her divorce, Goodall married Derek Bryceson in 1973, an administrator in Tanzania who was in charge of the wildlife parks, and who subsequently became a politician. He has since died.

From 1970 to 1975 Goodall was visiting professor in psychiatry at Stanford University, and since 1973 she has been honorary visiting professor at the University of Dar es Salaam in Tanzania. She has written numerous books, including books for children, and articles, both for academic journals and for the magazine *National Geographic.* She has been a tireless campaigner in the west and with African governments on behalf of chimpanzees in the wild and the preservation of their environment, and on behalf of chimpanzees in captivity. The Jane Goodall Institute for Wildlife Research was first set up in California by Goodall in cooperation with Princess Genevieve di Faustino in 1977 and eventually moved to Washington, DC, in 1998. It has centres around the world.

Among her numerous awards and honours, Goodall received the Gold Medal of Conservation from the San Diego Zoological Society in 1974, the J Paul Getty Wildlife Conservation Prize in 1984, The Schweitzer Medal of the Animal Welfare Institute in 1987 and the Unicef/Unesco Children's Book of the Year Award in 1989. *The Chimpanzee Family Book* for which she won the award was translated at her expense into Swahili and French and is distributed to children in Tanzania, Uganda, Burundi, and Congo who live in areas populated by chimpanzees. She also teaches and encourages children to appreciate conservation issues. She lectures around the world and works to support her own mission and the L S B Leakey Foundation.

Goodyear, Charles (1800–1860) US inventor who is generally credited with inventing the process for vulcanizing rubber.

Goodyear was born in New Haven, Connecticut, on 18 December 1800. When he came of age he entered his father's hardware business at Naugatuck, Connecticut, where he worked with enthusiasm, inventing various implements, including a steel pitchfork to replace the heavy iron type.

The firm became financially unstable, and by 1830 Goodyear realized that he would have to turn to something else. He chose to investigate india rubber and the problems associated with it, particularly how to make the rubber remain strong and pliable over a range of temperatures. He thought he had found the solution by mixing nitric acid with the rubber, and in 1836 secured a government order for a consignment of mail bags. But they would not stand up to high temperatures and were therefore useless. Goodyear was forced to start again.

In 1837 he bought out the rights of Nathaniel Hayward, who had had some success by mixing sulphur with raw rubber. After much patient experiment – and a deal of luck – Goodyear finally perfected the process he called vulcanization.

He obtained US patents in 1844, but both the UK and France refused his applications because of legal technicalities. His attempts to set up companies in both countries failed, and for a while he was imprisoned for debt in Paris. Eventually he returned to the USA, where many of his patents had been pirated by associates. Even his son, who had been working for him, decided to leave the ailing firm and Goodyear was forced to face his heavy debts alone. He died lonely and poverty-striken in New York City on 1 July 1860.

Various people throughout the western world tried to make rubber a more commercially viable material. Goodyear discovered the vulcanization process by accident. One day he was mixing rubber with sulphur and various other ingredients when he dropped some on top of a hot stove. The next morning the stove had cooled, the rubber had vulcanized, and he thought that all his problems had been solved.

The invention was highly significant at a time of industrial advancement. The new process made rubber a suitable material for such applications as belting and hoses, for which strength at high temperatures was the governing factor. It was also particularly valuable once the idea of rubber tyres was conceived, at first for bicycles and then for motorcars.

Goodyear's process was eventually superseded by more refined methods and by the development of synthetic rubbers. And although he failed to find wealth in his lifetime, his name still lives on and can be seen on motor tyres throughout the world.

Further Reading

Barker, P W, *Charles Goodyear – Connecticut Yankee and Rubber Pioneer: A Biography,* published privately, 1941.

Pierce, Bradford K, *Trials of an Inventor Life and Discoveries of Charles Goodyear,* Carlton and Porter, 1866.

Regli, Adolph C, *Rubber's Goodyear: The Story of a Man's Perseverance,* J Messner, 1941.

Gosset, William Sealey (1876–1937) English industrial research scientist, famous for his work on statistics.

Gosset was born in Canterbury on 13 June 1876 and educated at Winchester College, before going to New College, Oxford, to study mathematics and chemistry.

He received a first class in his mathematical moderations in 1897 and a first class on receiving his BA in natural sciences (chemistry) in 1899. On leaving Oxford he immediately joined the Guinness brewery firm in Dublin, the firm recently having adopted the policy of hiring a number of university-trained scientists to conduct research into the manufacture of its ale. He remained with the company in Dublin until 1935, when he was posted to London to direct the operations of the new Guinness brewery there. He was appointed the company's head brewer a few months before his death, at Beaconsfield, on 16 October 1937.

When Gosset arrived in Dublin he found that there was a mass of data concerning brewing – on the relationships between the raw materials, hops and barley, and the quality of the finished product and on the methods of production – which had been almost entirely left unanalysed. After a few years with the company he was able to persuade the owners that they would profit from more sophisticated mathematical analysis of a variety of processes from the production of barley to the fermentation of yeast. Accordingly, the company sent him to University College, London, in September 1906 to study for a year under the statistician Karl ◊Pearson. It was the experience of that year that turned Gosset into an outstanding statistical theorist, even though all the questions he asked were inspired directly by problems in the brewery trade.

Gosset's statistical techniques were simple: he relied on the mean, the standard deviation, and the correlation coefficient as his basic tools. Through all his work ran one theme, expressed in two formulae:

$$\sigma_x^2 - y = \sigma_x^2 + \sigma_y^2 - 2p\sigma_3\sigma_y$$

$$\sigma_x^2 + y = \sigma_x^2 + \sigma y + 2p\sigma_3\sigma_y$$

Gosset's amplification of these formulae opened the door to modern developments in the analysis of variance.

When Gosset began to examine the data at the Guinness brewery, it quickly became apparent to him that what was most needed was improved knowledge of the theory of errors. In 1904 he wrote a report for the company, 'Application of the law of error to the brewing industry'. His most famous work, published in 1908, was on the probable error of a mean. More than a century earlier Karl ◊Gauss had worked out a satisfactory method of estimating the mean value of a characteristic in a population on the basis of large samples; Gosset's problem was to do the same on the basis of very small samples, for use by industry when large sampling was too expensive or impracticable. For any large probability, that is one of 95% or more, Gosset was able to compute the error *e*, such that it is 95% probable that

$$(x - \mu) \leq e$$

Where x is the value of the sample, and μ is the mean. From this t-test was derived what came to be known as the Student's t-test of statistical hypotheses. The name of the test comes from the fact that Gosset published all his papers under the pseudonym 'Student'. The test consists of rejecting a hypothesis if, and only if, the probability (derived from t) of erroneous rejection is small.

Gosset hit on the statistic that has proved to be fundamental for the statistical analysis of the normal distribution, and he went further, to make the shrewd observation that the sampling distribution of such statistics is of basic importance in the drawing of inferences. In particular it opened the way to the analysis of variance, that branch of the subject so important to statisticians who came after him.

Gould, Stephen Jay (1941–) US palaeontologist and writer who teaches and researches in geology, evolutionary biology, and the history of science.

Gould was born in New York City on 10 September 1941 and graduated in 1963 from Antioch College in Ohio and then took a PhD at Columbia University in 1967. In that year he joined the staff at Harvard University and became professor of geology there in 1973 and Alexander Agassiz Professor of Zoology in 1981. He holds a concurrent post in the department of the history of science.

> *Nature's oddities are more than good theories. They are material for probing the limits of interesting theories about life's history and meaning.*
>
> STEPHEN JAY GOULD THE PANDA'S THUMB *1980*

Gould has written extensively on several aspects of evolutionary science, in both professional and popular books. His early research interests were focused on the evolutionary development and speciation of the land snail, and initiated his wider studies of animal form and function and the relationship between ontogeny and phylogeny. His book *Ontogeny and Phylogeny* (1977) provided a detailed scholarly analysis of his work on the developmental process of recapitulation. In *Wonderful Life* (1989), he drew attention to the diversity of the fossil finds in the Burgess Shale Site in Yoho National Park, Canada, which he interprets as evidence of parallel early evolutionary trends extinguished by chance rather than natural selection.

Collections of essays published in, for example, *Ever Since Darwin* (1977), *Hen's Teeth and Horses' Toes* (1983), and *Life's Grandeur* (1996) have all achieved critical acclaim and large readerships.

Graham, Thomas (1805–1869) Scottish physical chemist who pioneered the chemistry of colloids, but who is best known for his studies of the diffusion of gases, the principal law concerning which is named after him.

Graham was born on 21 December 1805 in Glasgow, the son of a successful local manufacturer. His father had hoped that his son would, after leaving school, enter the Presbyterian ministry but in 1819, when he was only 14 years old, Graham enrolled at Glasgow University to study science. He later transferred to Edinburgh University and graduated in 1824. He returned to Glasgow to teach at the Mechanics Institute, which had been founded a year or two earlier by George Birkbeck for teaching craftsmen the scientific principles of their trades. In 1830 Graham became professor of chemistry at Anderson's College, Glasgow. He left Scotland seven years later to take up a similar position at University College, London, where he remained until 1854. In 1841 he became the first president of the Chemical Society of London, itself the first national society devoted solely to the science of chemistry. In 1855 he was appointed master of the Royal Mint, a position once held by Isaac ◊Newton. He died in London on 16 September 1869.

Graham's early interest was the dissolution and diffusion of gases. In 1826 he discovered that very soluble gases do not obey Henry's law (which states that solubility is proportional to the pressure of the gas). He measured the rates at which gases diffused through a porous plug of plaster of Paris, through narrow glass tubes, and through small holes in a metal plate. By 1831 he had formulated Graham's law of diffusion, which states that the rate of diffusion of a gas is inversely proportional to the square root of its density.

In 1829 Graham turned his attention briefly to inorganic chemistry. He studied the glow of phosphorus and observed that it was extinguished by organic vapours and various gases. He went on to examine phosphorus compounds in general, particularly phosphine and salts of the various oxyacids. He distinguished ortho-, meta-, and pyrophosphates, which he prepared by fusing sodium carbonate with orthophosphoric acid. Graham had made the first detailed study of a polybasic acid.

In the 1850s, following his work on gases, Graham investigated the movement of molecules in solutions. He added crystals of a coloured chemical, such as cupric sulphate, to water and noted how long it took for the colour to spread throughout the solution. He observed that different chemicals took different times to disperse and that the dispersion rate increased with increasing temperature.

Then in 1861 he tried a technique similar to that which he had used for gases. He inserted a parchment barrier across a tank of water and added a coloured salt to the water on one side of it. He discovered that some of the coloured substance passed through the barrier. Repeating the experiment using glue or gelatin, he found that these substances did not pass through parchment. All the substances tested that could pass through also formed crystals, and Graham called this category crystalloids. Those that failed to cross the barrier did not form crystals and he called these colloids (from the Greek *kolla*, meaning glue). He distinguished between sols and gels (although he did not use these terms to describe them).

Using the same discovery, Graham developed a method of purifying colloids. The impure colloid was placed in a porous tube suspended in running water. The crystalloids (impurities) were washed away, leaving the purified colloid in the tube. He called the process dialysis, and it has since found a multitude of applications, from desalination equipment to artificial kidney machines.

Graham maintained his interest in gases, and in 1866 started a study of the occlusion of hydrogen by metals such as iron, platinum, and palladium. He observed that metal foils that freely absorb gas at low temperatures become permeable to hydrogen when heated.

Grandi, Guido (1671–1742) Italian mathematician famous for his work on the definition of curves – particularly curves that are symmetrically pleasing to the eye. It was he who devised the curves now known as the 'versiera', the 'rose', and the 'cliela' (after Cliela, Countess Borromeo), and his theory of curves also comprehended the means of finding the equations of curves of known form. He was mainly responsible, in addition, for introducing calculus into Italy.

At his birth, on 1 October 1671 in Cremona, his given names were Francesco Lodovico Grandi. At the age of 16 he entered the religious order of the Camaldolese and changed his first names to Guido. In 1694 he was appointed teacher of mathematics in the order's monastery in Florence, and it was there that he first became acquainted with the *Principia mathematica* of Isaac ◊Newton (published in England in 1687), which inspired him to devote much of the remainder of his life to the study of geometry. Nevertheless he became professor of philosophy at Pisa in 1700, and seven years later was given the post of honorary mathematician to the Grand Duke. In 1714 he became professor of mathematics at Pisa and established a considerable reputation as a teacher. The recipient of several awards and honours – a member of the Royal Society from 1709 – Grandi died in Pisa on 4 July 1742.

In his fascination for the study of curves, Grandi was influenced first by Newton; early in his career he determined the points of inflection in the conchoid curve. He examined also the studies published by Pierre de ◊Fermat, whose treatment he found somewhat limited, and by Christiaan ◊Huygens, who had revealed the most important properties of logistic curves in research based on the work of Evangelista ◊Torricelli. In 1701 Grandi devised a proof for Huygens's theorem.

However, Grandi's name will always be associated with the 'rose', the 'cliela', and the 'versiera' curves; his work on the first two was presented in a paper to the Royal Society in 1723 – ten years after he had corresponded with the great Gottfried ◊Leibniz on the subject. In 1728 he published his complete theory in *Fleores geometrica*, an attempt (among other things) to define geometrically the curves that make up the shapes of flowers, particularly multi-petalled roses.

What today is known as the 'rose' curve, Grandi called by its Greek name *rhodonea*. The polar equation for such a curve is:

$$r = a \sin k\theta$$

where k is an integer. Depending on whether k is an odd or an even number, there are k or $2k$ 'petals' on the rose; and depending on whether k is a rational or an irrational number, the number of petals is finite or infinite. The Cartesian coordinates for the curve are:

$$(x^2 + y^2)^3 = 4a^2x^2y^2$$

where a is a constant.

The 'cliela' curve Grandi described as the locus of P where P represents a point on a sphere of radius a where φ and θ are the longitude and co-latitude of P, with P moving such that $\theta = m\varphi$, where m is a constant.

Grandi defined the curve now called the 'versiera' (from the Latin *sinus versus)* by stating: given a circle with diameter AC, let BDM be a straight line perpendicular to AC at B and intersecting the circumference at D; let M be a point determined by the length of BM so that $AC: BM = AB: BD$; the locus of all such points, M, is the versiera. The curve is in some places more commonly known as the 'witch of Agnesi' as a result of a mistranslation and the attribution of its discovery to Maria Gaetana ◊Agnesi in a treatise of 1748.

Although the study of curves was Grandi's joy and passion, he did make other contributions to mathematics, notably in 1703 when his treatise on quadrature – using Leibniz's methods in preference to those of Francesco Cavalieri and Vincenzo Viviani – was responsible for the introduction of calculus into Italy.

Grandi also did some work in practical mechanics and his observations regarding hydraulics were utilized by the Italian government in such public works as the drainage of the Chiana valley and the Pontine marshes in central Italy.

Grassmann, Hermann Günther (1809–1877) German mathematician whose methods of presentation were so unclear that his innovations were never really appreciated during his lifetime despite their importance. He had greater success with his work in a completely different field: comparative linguistics.

Grassmann was born in Stettin, Pomerania (now Szczecin in Poland), on 15 April 1809. He studied first at home and then in local schools, before going to the University of Berlin, where he studied theology 1827–30. Instead of becoming a minister, however, he returned to Stettin and began to study mathematics and physics for an examination to enable him to teach at secondary schools. Poor marks in the examination meant that he could teach only younger students, from 1832. It was not until 1840 that he obtained the qualifications necessary to teach more advanced students. In 1842 he joined the staff of the Friedrich Wilhelm School in Stettin; five years later he became senior teacher there. Finally the Stettin high school appointed Grassmann to an important teaching post (which carried the title of professor) in 1852. Grassmann's numerous schoolbooks and his texts on more advanced mathematics and linguistics were not uniformly successful. The international mathematical community was slow to recognize the significance of Grassmann's work in mathematics, although he received much more immediate recognition for his work in linguistics, and in particular on Sanskrit. He was elected to the Göttingen Academy of Sciences in 1871, but the honorary doctorate he received in 1876 from the University of Tübingen was for his work in linguistics. He died in Stettin on 26 September 1877. After his death his work was popularized, partly through the efforts of his children, several of whom became prominent academics.

The examination paper that Grassmann sat in 1840 in order to become eligible to teach advanced students included a paper on tides. In his research at home while framing his answer to the paper, Grassmann discovered a new calculus that enabled him to apply parts of Pierre ◊Laplace's book on celestial mechanics to the paper on tides. The technique clearly had wider applications, so Grassmann decided to devote himself to exploring it further, and during the years 1840–44, concentrated on developing his method, which he called the theory of extension. It was one of the earliest mathematical attempts to investigate n-dimensional space, where n is greater than 3. He published the method in a book in 1844, but his vocabulary was so obscure that the book had virtually no impact at all in the short term.

Grassmann in the following year applied the method to reformulate Ampère's law (see André ◊Ampère), but again through poor exposition the work was ignored. (Thirty years later, Rudolf ◊Clausius independently found a similar improvement to Ampère's law, but gave credit to Grassmann for its earlier discovery.)

The next ten years saw Grassmann investigating the subject of algebraic curves, using his calculus of extension, but the papers he published on the subject once more made little impact. He was awarded a prize for a paper in topology by the Leipzig Academy of Sciences in 1847 – but the published version needed a note of clarification because of the essay's complex format.

The poor response of the mathematical community to Grassmann's work led him also to devote considerable energy to a completely different field of study. He learned and examined many ancient languages, such as Persian, Sanskrit, and Lithuanian. His books on this subject were considerably more successful, especially the glossary he published to the Hindu scriptures, the Rig-Veda, 1873–1875. From his investigations he derived a theory of speech.

Another area of his interest was the mixing of colours. His work in this field attracted some favourable comments from Hermann von ◊Helmholtz – the physiologist and expert on colour vision and colour blindness – whose work Grassmann had in fact criticized.

The book on extension theory that Grassmann published in 1844 was reprinted in a revised form in 1862, and again posthumously in 1878. Gradually it began to receive some recognition as a work of considerable value by German, French, and US mathematicians. A scientist on whom it had a particularly strong effect was Willard ◊Gibbs, who made use of it in his development of vector analysis.

Further Reading

Grassmann, Hermann Günther, *Hermann Günther Grassmann (1809–1877): Visionary Mathematician, Scientist and Neohumanist Scholar: Papers from a Sesquicentennial Conference,* Boston Studies in the Philosophy of Science series, Kluwer Academic Publishers, 1996.

Otte, Michael, 'The ideas of Hermann Grassmann in the context of the mathematical and philosophical tradition since Leibniz', *Hist Math,* 1989, v 16, pp 1–35.

Gray, Asa (1810–1888) US botanist who was the leading authority on botanical taxonomy in the USA in the 19th century and a pioneer of plant geography. He was also the chief US proponent of Charles Darwin's theory of natural selection.

Gray was born on 18 November 1810 in Sauquoit, New York, and studied medicine at Fairfield Medical School, Connecticut, teaching himself botany in his spare time. After graduating in 1831 he taught science at Bartlett's High School in Utica, New York, until 1834 when he became an assistant to John Torrey, a chemistry professor at the College of Physicians and Surgeons in New York. Also interested in botany, Torrey became a lifelong friend of Gray and the two men collaborated on several botanical projects. In 1835 Gray accepted the post of curator and librarian of the New York Lyceum of Natural History, which gave him more time to devote to botany and provided greater financial security than did his previous position. Gray was made botanist to the US Exploring Expedition in 1836, but he resigned in the following year because of delays in sailing. In 1838 he was appointed a professor at the newly established University of Michigan, but did not commence his duties because, in the same year, he travelled to Europe to acquire books for his library and to study specimens of US plants in European herbaria. On his return to the USA in 1839 he was occupied with writing and organizing the results of his studies. In 1842 Gray accepted the professorship of natural history at Harvard University, on the understanding that he could devote himself to botany. He held this position for 31 years, until he retired in 1873, during which time he made several journeys to Europe, meeting Charles Darwin in 1851, and donated his priceless collection of plants and books to Harvard University on condition that the university housed his collection in a special building; this led to the establishment of the botany department at Harvard University, and the botanical garden and herbarium were later named after him. Gray died on 30 January 1888 in Cambridge, Massachusetts.

Gray was a prolific author, writing more than 360 books, monographs, and papers, but is probably best known for his *Manual of the Botany of the Northern United States, from New England to Wisconsin and South to Ohio and Pennsylvania Inclusive* (1848), often called *Gray's Manual,* which was an extremely comprehensive study of North American flora and which also helped to establish systematic botany in the USA. This work went through several editions and is still a standard text in its subject area. Before the publication of *Gray's Manual,* however, Gray had written several other important works, including the two-volume *Flora of North America* (1838–43), which he coauthored with Torrey. This work was later expanded, under Gray's direction, and published in 1878 as the first volume of the *Synoptical Flora of North America.*

In addition to these contributions to botanical taxonomy, Gray also investigated the geographical distribution of plants, notably a comparison of the flora found in Japan, Europe, and northern and eastern North America. This knowledge of plant distribution proved useful to Darwin, who asked Gray to analyse his plant-distribution data in order to provide him with evidence for his theory of natural selection. Darwin also told Gray about his ideas on natural selection as early as 1857, and after the publication of *On the Origin of Species* (1859) Gray became a leading advocate of Darwin's theory. In his own work, Gray reached conclusions about variations that foreshadowed the work of Gregor ◊Mendel and Hugo ◊de Vries.

Further Reading

Dupree, A Hunter, *Asa Gray: American Botanist, Friend of Darwin*, Johns Hopkins University Press, 1988.
Fry, C George and Fry, Jon Paul, *Congregationalist and Evolution: Asa Gray and Louis Agassiz*, University Press of America, 1989.
Porter, Duncan M, 'On the Road to the *Origin* with Darwin, Hooker, and Gray *J Hist Biol*, 1993, v 26, pp 1–38.

Gray, Henry (1827–1861) British anatomist who compiled a book on his subject that, through its various editions and revisions, has remained the definitive work on anatomy for more than a hundred years.

Little is known of Gray's early life and education. His father was a private messenger to George IV and William IV. Gray became a 'perpetual student' at St George's Hospital, London, on 6 May 1845. In 1848 he was awarded the Triennial Prize of the Royal College of Surgeons. When Gray was 25 years old, the Royal Society acknowledged his contribution to anatomy by electing him a fellow, and the following year in 1853 he was awarded the Astley Cooper Prize of 300 guineas for a talk on the structure and function of the spleen. He was offered and accepted the post of demonstrator of anatomy at St George's Hospital. He became a fellow of the Royal College of Surgeons, and was curator of the St George's Museum. He was also lecturer in anatomy and in line for the post of assistant surgeon. At this point his brilliant and promising career was cut short. He tended a nephew suffering from smallpox, and contracted a particularly virulent strain of the disease. He died in London in 1861 at the age of only 34.

The first edition of what is now known as *Gray's Anatomy* was published in 1858. He had painstakingly and methodically learnt all his anatomy on the merits of his own dissections, and had the good fortune to persuade his friend H Vandyke Carter to do the drawings for him. Vandyke Carter was a demonstrator of anatomy as well as a draughtsman, and undoubtedly his beautiful illustrations were partly responsible for the success of the book. Gray prepared the second edition in 1860.

Gray's Anatomy was fundamentally different from other contemporary works of a similar nature because it was organized in terms of systems, rather than areas of the body. His layout and general philosophy remain in the current edition, although the book has been revised and updated nearly 50 times. Areas such as neuroanatomy have been greatly enlarged but the section that deals with, for example, the skeletal system is almost identical to Gray's original work. It remains a standard text for students and surgeons alike.

Green, George (1793–1841) English businessman and self-taught mathematician; through hard work and considerable creative and perceptive ability he made significant advances in both the physics and the mathematics of his time, although he was really achieving his proper status and recognition only at the time of his death. He is best remembered for his paper, published before he entered formal education, in which he introduced the term 'potential', now a central concept in electricity.

Green was baptized in Nottingham on 14 July 1793 (date of birth unknown). He had to leave school at an early age in order to assist at the family bakery, but his mathematical interests and abilities spurred him to continue his studies on his own, through reading. The death of his father and the ensuing sale of the family firm eventually made Green financially independent. He moved to Cambridge and in 1833 (at the age of 40) became a student at Caius College. Conducting his own studies in mathematics above and beyond those required by the curriculum, he graduated with honours in 1837. Further private researches led to the award to Green of a fellowship at Caius College – although he was not to be able to continue his work. Poor health led to his untimely death, in Sneinton, Nottinghamshire, on 31 March 1841.

Green's most famous paper was published in 1828. Only a few copies were printed, and these were circulated only privately and locally in Nottingham. It could easily have simply sunk out of sight without ever attracting the attention it deserved and eventually got. The essay dealt with a mathematical approach to electricity and magnetism. Its two outstanding features were the coining of the term 'potential', and the introduction of the Green theorem – which is still applied in the solution of partial differential equations – for instance, in the study of relativity. Green demonstrated the importance of 'potential function' (also known as the Green function) in both magnetism and electricity, and he showed how the Green theorem enabled volume integrals to be reduced to surface integrals.

Green went on to produce other important papers on fluids (1832, 1833), attraction (1833), waves in fluids (1837), sound (1837), and light (1837); his development of earlier work by Augustin ◊Cauchy on light reflection and refraction (1837) and on light propagation (1839) represented the best of his later work.

Green's work on electricity and magnetism might have been totally lost had it not been for William Thomson (later Lord ◊Kelvin), an undergraduate at the time of Green's death. He, after completing his studies, showed Green's essay of 1828 to a number of prominent physicists whom he knew personally. It stimulated great interest, and went on to influence scientists such as James Clerk ◊Maxwell and Kelvin himself, and was thus significant in the development of 19th-century theories of electromagnetism.

Greenstein, Jesse Leonard (1909–) US astronomer who has made important discoveries by combining his observational skills with current theoretical ideas and techniques. His early work involved the spectroscopic investigation of stellar atmospheres; later work included a study of the structure and composition of degenerate stars. He took part in the discovery of the interstellar magnetic field and in the discovery and interpretation of quasi-stellar radio sources – quasars.

Greenstein was born in New York City on 15 October 1909. He developed an interest in astronomy from an early age, although when he was young there were few popular astronomy textbooks and almost no professional astronomers in the USA. His grandfather, however, had an excellent library and from the age of eight the young Jesse began to use it; he was also encouraged by the gift of a small telescope. Greenstein admits that its location, overlooking New York City, was not ideal for observational purposes, but he nevertheless retained his childhood enthusiasm for astronomy throughout his life. He studied at Harvard and wrote a thesis on interstellar dust for his PhD, which he received in 1937. For the following two years he was a fellow of the National Research Council. In 1939 he accepted a post at the University of Chicago where, besides his teaching, he continued his research career at the Yerkes Observatory; he held this post for nine years, with a brief interlude during World War II, during which time he was involved in designing specialized optical instruments. In 1948 he joined the California Institute of Technology and also became a staff member of the Mount Wilson and Palomar observatories. Soon after joining the California Institute of Technology, he initiated its graduate school of astrophysics; he became chair of the faculty in 1965 and Lee A. Dubridge Professor of Astrophysics in 1971. He has been awarded several medals, including the Gold Medal of the Royal Astronomical Society, the Gold Medal of the Astronomical Society of the Pacific, and NASA's Distinguished Public Service Medal. After his retirement in 1980, Greenstein continued to write prolifically and carried on with his observational research work.

Greenstein began his career by spectroscopically studying stellar atmospheres to try to explain anomalies in the spectra of some stars. Using the spectrographs at the McDonald Observatory and later the Mount Wilson and Palomar observatories (now Hale), Greenstein developed a method of analysing the spectra of 'peculiar' stars by comparing them with the spectra of other average stars such as the Sun. His spectroscopic programme was closely linked with a parallel growth of new ideas in nuclear astrophysics that eventually led to the currently accepted theories of stellar evolution. In connection with the development of these theories, Greenstein independently suggested that the neutron-producing reaction in red giant stars was required for the production of heavy elements.

Greenstein rekindled his early postgraduate interest in the properties of interstellar dust by studying the nature of the interstellar medium with Leverett Davis. He developed the idea that space was pervaded by dominantly regular magnetic fields that aligned nonspherical, rapidly spinning, paramagnetic dust grains and so produced the interstellar polarization of light.

Greenstein was one of the leading figures in the explosive post-World War II development of radio astronomy in the USA. The rapid growth in this field resulted in the discovery of quasars in 1964, and Greenstein played a key role in the story of their discovery. He confirmed Maarten Schmidt's hypothesis that the emission lines of these peculiar bluish stellar objects could be explained by a shift in wavelength and he found this to be true of the spectrum of 3 C 48. He found that the lines were due to hydrogen provided that the velocity with which 3 C 48 was receding was just over 110,000 km/70,000 mi per second. In collaboration with Schmidt, Greenstein proposed a detailed physical model of the size, mass, temperature, luminosity, magnetic field, and high-energy particle content of quasars. The nature of quasars, the most luminous and enigmatic objects in the universe, still remains debatable, however.

During the 1970s Greenstein spent most of his time studying white dwarf stars. He collected quantitive information on their size, temperature, motion, and composition and by 1978 he had discovered some 500 of these degenerate stars. His research enabled him to pinpoint the problems of explaining the evolutionary sequence that links red giant stars with white dwarfs, thus initiating spectroscopic studies of such stars from space.

Greenstein is a leading figure in modern astronomy.

During the 1970s he guided both NASA and the National Academy of Sciences in their future policies and by chairing the board of directors of the Association of Universities for Research in Astronomy 1974–77 he has been influential in the research programmes of the Kitt Peak and Cerro Tololo observatories. He is the author of more than 380 technical papers and has edited several books, including *Stellar Atmospheres* (1960). During his career, Greenstein has changed fields of specialization many times, with the personal goal of applying new branches of physics to the study of the universe.

Grignard, (François Auguste) Victor (1871–1935) French organic chemist, best known for his work on organomagnesium compounds, or Grignard reagents. For this work he shared the 1912 Nobel Prize for Chemistry with Paul ◊Sabatier.

Grignard was born on 6 May 1871 in Cherbourg, the son of a sailmaker, and educated locally and at Cluny. His studies were interrupted for a year in 1892 while he did his military service (rising to the rank of corporal), after which he went to the University of Lyon and took a degree in mathematics. A former classmate then persuaded him to work in an organic chemistry laboratory under P A Barbier (1848–1922), where he quickly took to the new subject and began studying organomagnesium compounds, the subject of his PhD thesis in 1901. In 1905 he began lecturing in chemistry at Besançon and a year later he moved to Lyon. He was placed in charge of the organic chemistry courses at the University of Nancy in 1909, becoming a professor there in 1910. During World War I Grignard headed a department at the Sorbonne in Paris, concerned with the development of chemical warfare, going on a scientific trip to the USA in 1917. He succeeded Barbier as professor of chemistry at the University of Lyon in 1919, where he remained for the rest of his life. He died in Lyon on 13 December 1935.

Grignard's scientific career can be divided into three main periods. Up to 1914 he was concerned with the discovery, modifications, and applications of organomagnesium compounds (Grignard reagents). During the years 1914–18 he was engaged in research geared to the war effort, and afterwards he carried out a variety of researches in organic chemistry and concerned himself with chemistry education.

The general formula of a Grignard reagent is RMgX, where R is an alkyl radical, Mg is magnesium, and X is a halogen – usually chlorine, bromine, or iodine. It is used to add an alkyl group (–R) to various organic molecules. Grignard made his first reagents in about 1900 by treating magnesium shavings with an alkyl halide (halogenoalkane) in anhydrous ether (ethoxyethane). The discovery made a dramatic impact on synthetic organic chemistry; by 1905, 200 publications had appeared on the topic of Grignard reagents and within another three years the total had risen to 500.

Grignard reagents added to formaldehyde (methanal) produce a primary alcohol; with any other aldehyde they form secondary alcohols, and added to ketones give rise to tertiary alcohols. They will also add to a carboxylic acid to produce first a ketone and ultimately a tertiary alcohol.

The reagents have many industrial applications, such as the production of tetraethyl lead (from ethyl magnesium bromide and lead chloride) for use as an antiknock compound in petrol.

During World War I Grignard worked on the production of the poisonous gas phosgene (carbonyl chloride) and on methods for the rapid detection of the presence of mustard gas. After the war he started compiling his great *Traité de chimie organique,* a multivolume work that began publication in 1935. He continued research, investigating the organic compounds of aluminium and mercury; he also studied the terpenes.

Grimaldi, Francesco Maria (1618–1663) Italian physicist who discovered the diffraction of light and made various observations in physiology.

Grimaldi was born in Bologna on 2 April 1618 and at the age of 14 entered the order of the Society of Jesus, being educated at the Society's houses at Parma, Ferrara, and finally, in 1637, back at Bologna. As a Jesuit, his primary studies were in theology, which he finished in 1645. Three years later he became professor of mathematics at the Jesuit College in Bologna. There he acted also as assistant to the professor of astronomy, Giovanni Riccioli (1598–1671), whom he helped with astronomical observations. Grimaldi remained at the Jesuit College in Bologna, and died there on 28 December 1663.

Grimaldi observed muscle action and was the first to note that minute sounds are produced by muscles during contraction. But he is best known for his experiments with light. He let a beam of sunlight enter a darkened room through a small circular aperture and observed that, when the beam passed through a second aperture and onto a screen, the spot of light was slightly larger than the second aperture and had coloured fringes. Grimaldi concluded that the light rays had diverged slightly, becoming bent outwards or diffracted. On placing a narrow obstruction in the light beam, he noticed bright bands at each side of its shadow on the screen. He therefore added diffraction to reflection and refraction, the other known phenomena that 'bend' light rays. In modern terms, Grimaldi had observed a phenomenon that can be explained readily only if light is regarded as travelling in waves – contrary to the then accepted corpuscular theory of light. In 1665, two years after Grimaldi's death, a description of his experiments was published in the large work *Physicomathesis de lumine, coloribus, et iride/Physicomathematical Thesis of Light, Colours, and the Rainbow.*

Guericke, Otto von (1602–1686) German physicist and politician who invented the air pump and carried out the classic experiment of the Magdeburg hemispheres that demonstrated the pressure of the atmosphere. He also constructed the first machine for generating static electricity.

Guericke was born in Magdeburg on 20 November 1602 into a wealthy family, which afforded him wide educational opportunities. He attended several universities, going to Leipzig 1617–20, Helmstedt in 1620, Jena in 1621–22, and Leiden 1622–25. He studied law, science, and engineering, and on his return to Magdeburg was made an alderman of the city.

In 1631 Magdeburg sustained tremendous war damage, most of the population of 40,000 being butchered, and he moved to Brunswick and Erfurt where he worked as engineer to the Swedish government and later for that of Saxony as well. He was able to serve Magdeburg as envoy to various occupying powers, for as the fortunes of war swayed back and forth, French, Habsburg, and Swedish forces gained control. Guericke represented Madgeburg at various conferences that marked at least a lessening of hostilities in that part of Europe.

In 1646, Guericke became mayor of Magdeburg and his scientific achievements were made from this time onwards into the 1650s. He was ennobled and remained mayor until 1676. In 1681 he retired and went to live with his son in Hamburg, where he remained for the rest of his life, dying on 11 May 1686.

It is remarkable that, amidst all the political turmoil surrounding him, Guericke found time for deep philosophical speculation about the nature of space. At the time, ◊Aristotle's idea that 'nature abhors a vacuum' – that a vacuum could not exist – was being hotly debated. The phrase had been taken up by René ◊Descartes and had great currency, but in 1643 Evangelista ◊Torricelli had demonstrated that a vacuum could exist in the space above an enclosed column of mercury. His experiment could be explained by the atmosphere possessing weight, and Guericke decided to resolve matters by experiment. He set about constructing an air pump in an attempt to pump the air from a vessel and produce a vacuum. He postulated that if Descartes's ideas (the plenist assumption that space is matter) were correct, then an evacuated vessel should collapse. Guericke's first air pump was rather like a large bicycle pump and in 1647, after much trouble with seals, he succeeded in imploding a copper sphere from which he had pumped the air via an outlet at the bottom.

Although it seemed that Descartes was vindicated, Guericke still had his doubts. He built a stronger vessel and succeeded in evacuating it without causing it to collapse. He also demonstrated that the sound of a bell is muffled and that a candle is extinguished as the air is removed, and gradually the theory that a vacuum cannot exist was discarded.

Guericke then turned to demonstrating the immense force that the pressure of the atmosphere exerts on an evacuated vessel. One of his most famous experiments was in 1654. A cylinder containing a well-fitting piston was connected to a rope passing over a pulley. Fifty strong men were unable to move the piston more than halfway up the cylinder, and when the closed end was evacuated they could not prevent the piston from descending.

Guericke went on to make his classic air-pressure experiment, which was probably first performed at Magdeburg in 1657, in which two copper hemispheres with tight-fitting edges were placed together and the air pumped out. Two teams of horses were then harnessed to the hemispheres and proved unable to pull them apart. On admitting air to the hemispheres, they fell apart instantly.

Guericke's experiments and other observations on the elasticity of air led him to investigate the decrease of pressure with altitude, which Blaise ◊Pascal had demonstrated in his Puy de Dôme excursion in 1648. A natural consequence was the establishment of the link between atmospheric pressure and the weather, which Guericke pursued as far as making weather forecasts with a water barometer and proposing a string of meteorological stations contributing data to a forecasting system – all this in 1660.

Guericke also interested himself in cosmology and magnetism. While experimenting with a globe of sulphur constructed to simulate the magnetic properties of the Earth, he discovered that it produced static electricity when rubbed and went on to develop a primitive machine for the production of static electricity. Guericke also demonstrated the magnetization of iron by hammering in a north–south direction.

Another 'first' with which he is credited is the observation of coloured shadows; these he observed in the early morning when an object illuminated by a candle cast a blue shadow on a white surface.

Guericke's remarkable blend of political and scientific activities was to be echoed by such figures as Count ◊Rumford and Benjamin ◊Franklin in the next century – although in the latter's case, on an altogether grander scale. Guericke's discovery that the pressure of the atmosphere can be made to exert considerable force by the creation of a vacuum led directly to the development of steam engines in the 1700s.

Guettard, Jean-Etienne (1715–1786) French naturalist who pioneered geological mapping.

Born in Etampes, on 22 September 1715, Guettard rose to become keeper of the natural history collection of the Duc d'Orléans. He displayed a deep interest in many departments of botany and medicine, but he is remembered for his geological work. Extensive research in the field suggested that the rocks of the Auvergne district of central France were volcanic in nature, and Guettard boldly identified several peaks in the area as extinct volcanoes. Though, as a good Baconian empiricist, he later had doubts about this hypothesis, Guettard's studies proved a notable stimulus for later investigations, notably those of Nicolas ◊Desmarest. In further work, Guettard played an important part in stimulating the debate on the origin of basalt. Originally he had taken the view that columnar basalt was not volcanic in origin. Visits to Italy in the 1770s led him to modify his earlier views. Guettard was a humble man who consistently avoided wider speculations regarding the implications of his views for such issues as the age of the Earth.

Guettard was also an innovator in geological cartography. In 1746 he presented his first mineralogical map of France to the Academy of Sciences; and in 1766 he and Antoine ◊Lavoisier were commissioned to prepare a geological survey of France, though only a fraction was ever completed. He died in Paris on 6 January 1786.

Guillemin, Roger Charles Louis (1924–) French-born US endocrinologist. His research has included the isolation and identification of various hormones, for which he received the 1977 Nobel Prize for Physiology or Medicine, together with his co-worker Andrew Schally (1926–) and the US physicist Rosalyn ◊Yalow. Yalow developed radioimmunoassays of insulin, a technique that allows the analysis of minute amounts of substances such as hormones.

Guillemin was born in Dijon on 11 January 1924. In 1941 he obtained a BA from the University of Dijon, and the following year a BSc. He read medicine at the University of Lyon and gained his medical degree in 1949. During the following two years he was a resident intern at the University Hospital in Dijon, and then took up a professorial appointment at the Institute of Experimental Medicine and Surgery, Montréal, from which he gained his PhD in 1953. That year, Guillemin went to Baylor College of Medicine, Houston, Texas, and also became a naturalized US citizen. From 1960 to 1963 he was an associate director in the department of experimental endocrinology at the Collège de France, Paris. He returned to Baylor College and worked there until 1970 when he joined the Salk Institute in La Jolla, California as resident fellow until 1989 and as an adjunct research professor 1989–94. He also became Distinguished Science Professor of the Whittier Institute in La Jolla in 1989, and from 1995 was an adjunct professor of medicine at the University of California, San Diego.

While at the Baylor College, Guillemin attempted to prove the theory put forward by the British anatomist Geoffrey Harris that hormones secreted from the hypothalamus regulate those produced by the pituitary gland. Guillemin found that the brain controls the pituitary gland by means of hormones produced by central neurons (nerve cells) – the neurosecretory cells of the hypothalamus.

Guillemin was joined in his studies by the Lithuanian refugee Andrew Schally, who went to Baylor College in 1957. They tried to isolate the substance that regulates the secretion of the adrenocorticotrophic hormone (ACTH), which is produced by the pituitary gland, but they were defeated.

Schally left Baylor College in 1962 and went to New Orleans but both he and Guillemin turned their attention to the isolation of other hormones. In parallel investigations 1968–73 they isolated and synthesized three hypothalamic hormones that regulate the secretion of the anterior pituitary gland. The first of these was the tripeptide TRH (thyrotropin-releasing hormone), isolated in 1968. This hormone in turn induces the pituitary gland to secrete TSH (thyroid-stimulating hormone). After months of hard work they finally extracted 1 mg/0.000035 oz of the hormone, which they found to be a simple compound and easily synthesized. In 1971 Schally announced his discovery of LHRH (luteinizing-hormone-releasing hormone), which regulates the release in the pituitary gland of the luteinizing and follicle-stimulating hormones that control ovulation in women. Two years later, working at the Salk Institute, Guillemin discovered somatostatin, which inhibits the secretion of the growth hormone somatotropin, and of insulin and gastrin.

Guillemin continued his work at the Salk Institute, and in 1979 investigated the possible presence in the pituitary gland of opiate-like peptides. The peptide structure discovered is called ß-endorphin and represents the opiate-like effects of all pituitary extracts.

The discoveries of Guillemin and Schally are highly relevant to endocrinologists. Synthetic TRH is now used to treat conditions connected with deficiencies in the secretion of pituitary hormones; and somatostatin is under investigation for its possible use in treating diabetics, and for curing peptic ulcers.

Gurdon, John Bertrand (1933–) English molecular biologist who is probably best known for his work on nuclear transplantation.

Gurdon was born on 2 October 1933 in Dippenhall, Hampshire, and was educated at Eton College and then Christ Church College, Oxford, from which he graduated in zoology in 1956. From 1956 to 1960 he studied for his doctorate (in embryology) in the zoology department at Oxford University, then from 1960–62 he worked at the California Institute of Technology as a Gosney Research Fellow. He then returned to the UK, and from 1963 to 1964 was a departmental demonstrator in Oxford University's zoology department. Also in 1963 he was appointed a research fellow of Christ Church College. He became a lecturer in the zoology department in 1965, remaining there until 1972, when he was made a staff member of the Laboratory of Molecular Biology at Cambridge; seven years later he was appointed head of the cell biology division where he remained until 1983. Since 1983 he has been John Humphrey Plummer Professor of Cell Biology and since 1991 chair of the Wellcome Cancer Research Campaign Institue, both at Cambridge. In 1995 he was appointed master of Magdalene College, Cambridge.

Gurdon has worked mostly on the effects of transplanting nuclei and nuclear constituents, such as DNA, into enucleated eggs, mainly frogs' eggs. Using nuclei from somatic cells, he showed how the genetic activity changes in the recipient eggs; for example, after transplanting nuclei from differentiated somatic cells, Gurdon discovered that the somatic nuclei come to resemble the nuclei of fertilized eggs in both structure and metabolism. From this and other research, he concluded that the changes in gene activity induced by nuclear transplantation are indistinguishable from those that occur in normal early development. He also demonstrated how nuclear transplantation and microinjection techniques can be used to elucidate the intracellular movements of proteins, and has investigated the effects of known protein fractions on gene activity.

Gutenberg, Johann (*c.*1397–1468) German printer who invented moveable type, often regarded as one of the most significant technical developments of all time.

Gutenberg was born in Mainz, a small town on the Rhine. His exact birthdate is not known, although it is believed to have been between 1394 and 1399. He served an apprenticeship as a goldsmith, but his interest soon turned to printing. In partnership with his friend Andreas Dritzehn he set up a printing firm in Strasbourg some time in the late 1430s. Dritzehn died soon afterwards, and his brother sued for admission to the partnership in his place. Gutenberg was exposed to legal proceedings and from evidence given during the hearing it is thought that the two men may have already invented moveable type by 1438, although no printed work has survived to substantiate this assumption.

Gutenberg German inventor Johannes Gutenberg, who developed the technique of printing from movable metal type. This invention was used for the first printed Bible, the *Gutenberg Bible*, produced in 1456. *Mary Evans Picture Library*

Gutenberg returned to Mainz and persuaded the goldsmith and financier Johann Fust to lend him the money to set up a new press. With the security of financial backing he produced the so-called Gutenberg Bible, now regarded as the first major work to come from a printing press. His security was short-lived, however, when Fust brought a lawsuit to recover his loan. In November 1455 Gutenberg's printing offices were taken over in lieu of payment, and Fust installed his son-in-law Peter Schoeffer to operate the press.

Gutenberg may or may not have set up another press. Certainly an edition of Johann Balbus' Catholicon printed in Mainz in 1460 is often attributed to him, along with other lesser works. In 1462 Mainz was involved in a local feud, and in the upheaval Gutenberg was expelled from the city for five years before being reinstated, offered a pension, and given tax exemption. He died there on 3 February 1468.

The art of printing using moveable type is thought to have been originally invented nearly four centuries earlier in China, then a much more advanced civilization, but was unknown in the western world until the work of Gutenberg. Previously books had to be printed by the laborious method of carving out each page individually on a wood block, and this meant that only the very wealthy were able to afford them. Gutenberg made his type individually, so that each letter was interchangeable. After a page had been printed, the type could be disassembled and used again. He punched and engraved a steel character (letter shape) into a piece of copper to form a mould, which he then filled with molten type metal. The letters were in the Gothic style, nearest to the handwriting of the day, and of equal height.

Using paper made from cloth rags and vellum sheets he printed the famous '42-line' (42 lines to a column) Gutenberg Bible, thought to be the first major work to come from a press of this kind. There are 47 surviving copies of the book, of which 12 are printed on vellum. Many bear no printer's name or date, but from evidence found in two copies in the Paris Library it is concluded that they were on sale before August 1456. By 1500 more than 180 European towns had working presses of this kind, including William Caxton's in Westminster, London, set up in 1476.

Further Reading

Baxter, Clifford, *Was Gutenberg the True Inventor of Printing?,* Sandelswood, 1988.

Ing, Janet, *Johann Gutenberg and His Bible: A Historical Study,* Typophile Chap Books, Typophiles, 1998, v 58.

Kapr, Albert, *Johann Gutenberg: The Man and His Invention,* Scolar Press, 1996.

H

Haber, Fritz (1868–1934) German chemist who made contributions to physical chemistry and electrochemistry, but who is best remembered for the Haber process, a method of synthesizing ammonia by the direct catalytic combination of nitrogen and hydrogen. For this outstanding achievement he was awarded the 1918 Nobel Prize for Chemistry (presented in 1919).

Haber was born on 9 December 1868 in Bresslau, Silesia (now Wrocław, Poland), the son of a dye manufacturer. He was educated at the local school and at the universities of Berlin (under August von ◊Hofmann) and Heidelberg (where he was a student of Robert ◊Bunsen). He completed his undergraduate studies at the Technische Hochschule at Charlottenburg, where he carried out his first research in organic chemistry. After spending some time working for three chemical companies, he briefly entered his father's business, but after only a few months Haber returned to organic chemistry research at Jena.

He then obtained a junior post at Karlsruhe Technische Hochschule, becoming an unpaid lecturer there in 1896. He held the professorship of physical chemistry and electrochemistry at Karlsruhe from 1906 until 1911, by which time his reputation had grown to such an extent that he was made director of the newly established Kaiser Wilhelm Institute for Physical Chemistry at Berlin-Dahlem. During World War I he placed the institute's resources in the hands of the government and, being a staunch patriot, was bitterly disappointed by Germany's defeat in 1918. Then when Adolf Hitler rose to power in 1933 even Haber's genius and patriotism could not compensate for his Jewish ancestry and he was forced to seek exile in the UK, where he worked for a few months at the Cavendish Laboratory, Cambridge. Shortly afterwards, while on holiday in Switzerland, he died of a heart attack in Basel on 29 January 1934.

Haber German chemist Fritz Haber who discovered how to synthesize ammonia from atmospheric nitrogen. As head of the German chemical warfare service, Haber worked to develop poisonous gases and gas masks during World War I, but in 1933, following the rise of the Nazi party to power, he went into exile in Cambridge, England, rather than suffer the Nazi treatment of Jews. *Mary Evans Picture Library*

During the late 1890s Haber carried out important researches in thermodynamics and electrochemistry, studying particularly electrode potentials and the processes that occur at electrodes. He was an early, but largely unsuccessful, pioneer of fuel cells in which the electric current was produced by the atmospheric oxidation of carbon or carbon monoxide. He also worked on the electrodeposition of iron, which had applications in the making of plates for printing banknotes. His work during this period culminated with the publication of two influential books: *The Theoretical Bases of Technical Electrochemistry* (1898) and *The Thermodynamics of Technical Gas Reactions* (1905).

Haber used his extensive knowledge and experience of thermodynamics in his investigations of the synthesis of ammonia by the catalytic hydrogenation of atmospheric nitrogen. Although from 1904 he made extensive free energy calculations and performed many experiments, it was not until 1909 that the process was ready for industrial development under Karl ◊Bosch. Haber's early successful experiments used osmium or uranium catalysts at a temperature of 550 °C/1,022 °F (823K) and 150–200 atmospheres of pressure (2.2–2.9 $\times$ 10^3 psi/1.5–2.0 $\times$ 10^7 Pa). Later industrial processes used higher pressures and slightly lower temperatures, with finely divided iron as the catalyst. The nitrogen from air was compressed with producer gas (the source of hydrogen) and the ammonia formed was removed as a solution in water. In this version it is also known as the Haber–Bosch process.

At Karlsruhe Haber continued his work in electrochemistry and in 1909 constructed the first glass electrode (which measures hydrogen ion concentration – acidity, or pH – by monitoring the electrode potential across a piece of thin glass). One of his first tasks after the outbreak of World War I in 1914 was to devise a method of producing nitric acid for making high explosives, using ammonia from the Haber process (to make blockaded Germany independent of supplies of Chilean saltpetre – sodium nitrate). Later he became involved in the gas offensive and superintended the release of chlorine into the Allied trenches at Ypres. When the French later retaliated by using phosgene (carbonyl chloride, whose production was in the hands of Victor ◊Grignard), Haber was given control of the German Chemical Warfare Service, with the rank of captain, and organized the manufacture of lethal gases. He was also responsible for the training of personnel and developed an effective gas mask.

After the war Haber set himself the task of extracting gold from sea water to help to pay off the debt demanded by the Allies in reparation for the damage caused during the hostilities. Svante ◊Arrhenius had calculated that the sea contains 8,000 million tonnes of gold. The project got as far as the fitting out of a ship and the commencement of the extraction process, but the yields were too low and the project was abandoned in 1928.

Hadamard, Jacques (1865–1963) French mathematician whose contributions to the subject ranged over so wide a field that he gained recognition as one of the outstanding mathematicians of modern times.

Hadamard was born in Versailles on 8 December 1865 into a family well able to perceive and encourage his rare intellectual gifts. His father taught Latin at a lycée in Paris and his mother was an accomplished pianist and music teacher. Hadamard spent his undergraduate years at the Ecole Normale Supérieure in Paris, and after four years there gained his BA in mathematics in 1888. In 1890, while he was working on his doctoral thesis on function theory, he began to teach at the Lycée Buffon. He received his doctorate in 1892 and in the following year took up a teaching appointment at Bordeaux. In the next few years he was also an occasional lecturer at the Sorbonne.

In 1909 Hadamard was appointed professor of mathematics at the Collège de France in Paris, a chair that he held, together with subsequent appointments at the Ecole Polytechnique and the Ecole Centrale, until 1937. In order to escape from the disruptions of the German occupation of France during World War II, he went to the USA in 1941, then to London, where he took part in operations research for the Royal Air Force. In 1945 he returned to France, where he lived in retirement – cultivating his tastes as an amateur musician (Einstein once played in an orchestra he assembled) and as a collector of ferns and fungi. He died in Paris on 17 October 1963.

Rare are the mathematicians who have left a heritage as rich as Hadamard did; few are the fields of mathematics that have not been touched by his influence. His first major papers, those written in the mid-1890s and growing out of his doctoral thesis, dealt with the nature of analytic functions – that is, functions that can be developed as power series that converge. Hadamard created the theory of the detection and nature of singularities in the analytic continuation of a Taylor series. He began by giving a proof of the so-called Cauchy test for convergence of a power series ($\Sigma anxn$), and included the definition of an upper limit from first principles. By applying this method to Taylor series he arrived at results that are now standard textbook illustrations for the study of these series.

At about the same time he began to study the Riemann zeta function and in 1896 solved the old and famous problem relating to prime numbers of determining the number of primes less than a given number x. Hadamard was able to demonstrate that this number was asymptotically equal to $x/\log x$, which was the most important single result ever obtained in number theory.

Hadamard also became interested in the work of his friend, Vito ◊Volterra, on the 'functions of lines' – numerical functions that depend upon a curve or an ordinary function as their variable. In an ordinary function such as $y = f(x)$, the variable, x, is a simple number. In the kinds of functions on which Volterra was working – Hadamard named them 'functionals' – one might have, to take a simple example, $y = A(c)$, where A is the area of a closed curve c. By asking a bold question of these functions, or functionals, Hadamard created a new branch of mathematics. He asked whether it might be possible to extend the theory of ordinary functions to the case where the variable, or variables, would no longer be a number, or numbers. This required a redefinition, or at least a new generalization, of many concepts: continuity, derivative, and differential among them. In seeking to provide this new approach Hadamard gave birth to functional analysis, one of the most fertile branches of modern mathematics.

By extension from this work Hadamard came to investigate functions of a complex variable and, in so doing, to define a singularity as a point at which a function ceases to be regular. He was able to show, however, that the existence of a set of singular points may be compatible with the continuity of a function.

He named the region formed by such a set a 'lacunary space' and the study of such spaces has occupied mathematicians ever since.

Finally there is Hadamard's famous concept of 'the problem correctly posed'. Since it is often helpful, or necessary, to find an approximate solution (in physics, for example), a correctly posed problem, according to Hadamard, is one for which a solution exists that is unique for given data, but which also depends continuously on the data. This is the case when the solution can be expressed as a set of convergent power series. The idea has proved to be far more fruitful than Hadamard was able to envisage and has led mathematical analysts to consider different types of neighbourhood and continuity and so been fundamental to the development of the theory of function spaces, functional analysis.

Hadamard was a prolific mathematician. In his lifetime he published more than 300 papers on a great variety of topics. And few mathematicians influenced the growth of mathematics in so many different directions as he did.

Hadfield, Robert Abbott (1858–1940) British industrial chemist and metallurgist who invented stainless steel and developed various other ferrous alloys.

Hadfield was born in Sheffield, Yorkshire, on 29 November 1858; his father owned a small steel foundry. He received a good education and was trained locally as a chemist before taking up an appointment as an analyst. After a while he joined his father's firm (founded in 1872) and by the age of 24 was its manager. His father died in 1888 and Hadfield became chairman and managing director. He also continued his research into various steel alloys, which he had begun in the early 1880s, eventually publishing more than 150 scientific papers. His findings proved productive and commercially successful, and in recognition of his contribution to the British steel industry he was knighted in 1908, elected a fellow of the Royal Society a year later, and created a baronet in 1917. He died in London on 30 September 1940, leaving the well-known company of Hadfields in Sheffield as reminder of the family name.

The basic ingredient of steel is pig iron, the product of iron smelting in a blast furnace, which contains 2.5–4.5% carbon. In making ordinary mild steel, carbon (and other impurities) are oxidized to lower the carbon content. Hadfield carried out many experiments in which he mixed other metals to the steel. He found, for example, that a small amount of manganese gave a tough, wear-resistant steel suitable for such applications as railway track and grinding machinery. By adding nickel and chromium he produced corrosion-resistant stainless steels. Within 20 years silicon steels were available. Again the basic metal with the addition of 3–4% silicon produced an ideal metal for making transformer cones, for which high permeability is required. Today steels and other alloys can be tailor-made to suit almost any requirements.

Haeckel, Ernst Heinrich (1834–1919) German zoologist well known for his genealogical trees of living organisms and for his early support of Darwin's ideas on evolution.

Haeckel was born in Potsdam, Prussia (now in Germany), on 16 February 1834. His father was a lawyer in Merseburg, where Ernst Haeckel was educated. He studied medicine (although his main interest was botany) at the University of Würzburg and obtained his degree from the University of Berlin in 1857. There he was taught by Johannes Müller (1801–1858), who interested him greatly in zoology; he also studied under Rudolf Virchow (1821–1902). Haeckel travelled through Italy and then, after practising medicine for a year, became a lecturer at the University of Jena in 1861. The following year he was appointed extraordinary professor of comparative anatomy at the Zoological Institute in Jena. He founded the Phyletic Museum in Jena and the Ernst Haeckel Haus, which contains his archives as well as many personal mementos. In 1865 he took up the full professorship at Jena, a position he retained until he retired in 1909. He died in Jena on 8 August 1919.

In 1866 Haeckel met Charles Darwin and was completely convinced by his theory of evolution. He went further and, using Darwin's research, developed Haeckel's law of recapitulation, which concerns resemblances between the ontogeny (the development from fertilized egg to adult) of different animals. According to the law of recapitulation, during its ontogeny an individual goes through a series of stages similar to the evolutionary stages of its adult ancestors, which show characteristics of less highly evolved animals. In this way ontogeny recapitulates phylogeny (the developmental history of a species). An example of the evidence on which Haeckel founded this view is the series of gill pouches found both in birds and mammals at the embryo stage. These gill pouches are not present in adult mammals, although the slits are present in full-grown birds and fish from which the embryo forms were descended. Haeckel's theory was thought to make ontogeny relevant to evolution simply because he believed that it should be possible to find out, by studying the development of an individual, what its adult ancestors were like. The concept that Haeckel revived had already been refuted by Karl von ⇨Baer, who had shown that embryos resemble the embryos only and not the adults of other species, but

Haeckel's claims remained popular to the end of the 19th century.

In the same year that he met Darwin, Haeckel introduced a method of representing evolutionary history, or phylogeny, by means of tree-like diagrams. His method is still used today by animal systematists to show degrees of presumed relationship in the various groups and can be traced in present modified zoological classifications.

Haeckel also tried to apply Darwin's doctrine of evolution to philosophy and religion. He believed that just as the higher animals have evolved from the simpler forms of life, so the highest human faculties have evolved from the soul of animals. He denied the immortality of the soul, the freedom of the will, and the existence of a personal God. He now occupies no serious place in the history of philosophy, although he was widely read in his own day.

Haeckel held some ideas that are still accepted, one of them being his view that the origin of life lies in the chemical and physical factors of the environment, such as sunlight, oxygen, water, and methane. This theory has recently, as a result of laboratory experiments, been shown to be likely. He also believed that the simplest forms of life were developed by a form of crystallization. A further influence Haeckel had on science was the coining of the word 'ecology', to mean the study of living organisms in relation to one another and to the inanimate environment. As a field naturalist Haeckel was a man of extraordinary energy, and he gave much of interest to biology, even if the theory was tenuous and was typical of the extreme evolutionists of his era.

Further Reading

Haeckel, Ernst; Hartman, Richard (contributor), *Art Forms in Nature: The Prints of Ernst Haeckel,* Prestel Verlag, 1998.

Hoenigwald, Henry M and Wiener, Linda F (eds), *Biological Metaphor and Cladistic Classification: An Interdisciplinary Perspective,* University of Pennsylvania Press, 1987.

Mocek, Reinhard, 'Two faces of biologism: some reflections on a difficult period in the history of biology in Germany' in: Woodward, William R and Cohen, Robert S (eds), *World Views and Scientific Discipline Formation,* Kluwer, 1991, pp 279–291.

Rooy, Piet de, 'Of monkeys, blacks, and proles: Ernst Haeckel's theory of recapitulation' in: Breman, Jan (ed), *Imperial Monkey Business – Racial Supremacy: Social Darwinist Theory and Colonial Practice,* VU University Press, 1990, pp 7–34.

Hahn, Otto (1879–1968) German radiochemist who discovered nuclear fission. For this achievement he was awarded the 1944 Nobel Prize for Chemistry.

Hahn was born on 8 March 1879 in Frankfurt am Main and studied at the University of Marburg, gaining his doctorate at Marburg in 1901. From 1904 to 1905 he worked in London under William ◊Ramsay, who introduced Hahn to radiochemistry. In 1905 he moved to McGill University at Montréal, Canada, to further these studies with Ernest ◊Rutherford. The next year Hahn returned to Germany to work under Emil ◊Fischer at the University of Berlin. There he was joined in 1907 by Lise ◊Meitner, beginning a collaboration that was to last more than 30 years. In 1912 Hahn was made a member of the Kaiser-Wilhelm Institute for Chemistry in Berlin-Dahlem and in 1928 was appointed its director. He held this position until 1944, when the Max Planck Institute in Göttingen took over the functions of the Kaiser-Wilhelm Institute and Hahn became president.

Perturbed by the military uses to which nuclear fission had been put, Hahn was concerned in his later years to warn the world of the great dangers of nuclear weapons. In 1966 he was co-winner of the Enrico Fermi Award with Meitner and Fritz Strassmann (1902–). Hahn died in Göttingen on 28 July 1968.

Hahn's first important piece of research was his discovery of some of the intermediate radioactive isotopes formed in the radioactive breakdown of thorium. These were radiothorium (discovered in 1904) and mesothorium (1907). In 1918 Hahn and Meitner discovered the longest-lived isotope of a new element, which they called protactinium, and in 1921 they discovered nuclear isomers – radioisotopes with nuclei containing the same subatomic particles but differing in energy content and half-life.

In the 1930s, Enrico ◊Fermi and his collaborators showed that nearly every element in the periodic table may undergo a nuclear transformation when bombarded by neutrons. In many cases, radioactive isotopes of the elements are formed and slow neutrons were found to be particularly effective in producing many of these transformations. In 1934 Fermi found that uranium is among the elements in which neutron bombardment induces transformations. The identity of the products was uncertain, but it was suspected that artificial elements higher than uranium had been found.

Hahn, Meitner, and Strassmann investigated Fermi's work, believing the formation of radium to have occurred, and treated the bombarded uranium with barium. Radium is chemically very similar to barium and would accompany barium in any chemical reaction, but no radium could be found in the barium-treated fractions. In 1938 Meitner, being Jewish, fled from Germany to escape the Nazis, and Hahn and Strassmann carried on the search. Hahn began to wonder whether radioactive barium was being formed from the uranium when it was bombarded with neutrons, and that this was being carried down by the barium he had added. The atomic number of barium is so much lower than that of uranium, however, that the only way it could possibly have been formed was if the uranium atoms broke in half. The idea of this happening seemed so ridiculous that Hahn hesitated to publish it.

Meitner was now in exile in Copenhagen and with Otto ◊Frisch quickly confirmed Hahn's explanation, showing the reaction to be an entirely new type of nuclear process and suggesting the name nuclear fission for it. Meitner and Frisch communicated the significance of this discovery to Niels ◊Bohr, who was preparing to visit the USA. Arriving in January 1939, Bohr discussed these results with Albert ◊Einstein and others. The presence of barium meant that uranium had been split into two nearly equal fragments, a tremendous jump in transmutation over all previous reactions. Calculations showed that such a reaction should yield 10–100 times the energy of less violent nuclear reactions. This was quickly confirmed by experiment, and later the same year it was discovered that neutrons emitted in fission could cause a chain reaction to occur.

Fearing that Nazi Germany would put Hahn's discovery to use and build an atomic bomb, Einstein communicated with the US president and urged that a programme be set up to achieve a nuclear chain reaction. Fermi subsequently built the first atomic reactor in 1942 and the first atomic bomb was exploded in 1945, bringing World War II to a close. Hahn had in fact not worked on the production of nuclear energy for Germany, and the German effort to build a reactor and a bomb had not got anywhere.

Otto Hahn made a discovery of enormous importance for the human race for, as well as providing a new and immense source of power, it has given us enormous destructive ability, as he was well aware. He is commemorated in science in the naming of element 105 as hahnium.

Further Reading

Shea, William R (ed), *Otto Hahn and the Rise of Nuclear Physics,* The University of Ontario Series in Philosophy of Science, Reidel Pub Co, 1983 v 22.

Sime, Ruth Levin, 'The discovery of protactinium', *J Chem Educ,* 1986, v 63, pp 653–657.

Haldane, J(ohn) B(urdon) S(anderson) (1892–1964) English-born Indian physiologist famous also for his work in genetics.

Haldane was born in Oxford on 5 November 1892, the son of John Scott ◊Haldane, himself a well-known physiologist. From the early age of eight he was introduced to medicine and assisted his father. He went to Eton and was later educated at New College, Oxford. After gaining a degree in mathematics he did equally well in classics and philosophy. During World War I he served on the Western Front and in Mesopotamia, where he was wounded twice; he then returned to study physiology at New College in 1919. Two years later he moved to Cambridge to work under the English biochemist Frederick Gowland ◊Hopkins. In 1933 he took up the genetics chair at University College, London, and later the chair in biometry. Between 1927–36 he also held a part-time appointment at the John Innes Horticultural Institution at Merton, where he carried on the work of the previous director William ◊Bateson. Haldane was an outspoken Marxist during the 1930s and served for a time as chairman of the editorial board of the London *Daily Worker.* He worked for the Admiralty in World War II and left the Communist Party disappointed by the fame awarded to the Soviet biologist Trofim ◊Lysenko. He emigrated to India in 1957 in protest at the Anglo-French invasion of Suez and was appointed director of the Genetics and Biometry Laboratory in Orissa, which had excellent facilities. He became a naturalized Indian citizen in 1961, and died of cancer at Bhubaneshwar on 1 December 1964.

Haldane's interest in the field of genetics was first sparked by a lecture in 1901 on the discovery, a year earlier, of the work of Gregor ◊Mendel, outlining the basic laws of inheritance. Some years later, in 1910, he commenced his study of the laws of heredity as displayed by the 300 guinea pigs kept by his sister. He later published a paper on gene linkage in vertebrates and in 1922 propounded Haldane's law. In 1925 he researched the association of gene linkage and age, and published a paper on the mathematics of natural selection. He was convinced that natural selection and not mutation is the driving force behind evolution. In 1932 he estimated for the first time the rate of mutation of the human gene and worked out the effect of recurrent harmful mutations on a population. While he was at University College he continued his work on human genetics and in 1936 showed the genetic link between haemophilia and colour blindness.

> *Reality is not only more fantastic than we think, but also much more fantastic than we imagine.*
>
> *J B S Haldane Attributed remark*

While Haldane was still at school he helped his father with research on the physiology of breathing and aspects of respiration concerned with deep-sea diving and safety in mines. Haldane's interest in respiration led him, during World War I, to work with his father yet again, on the improvization of gas masks. After the war, at Oxford, he investigated how the presence of carbon dioxide in human blood is vital in the muscular regulation of breathing in varying conditions. In the course of their research Haldane and his colleague, Peter Davies, subjected themselves to eating large quantities of sodium bicarbonate and drank ammonium chloride to introduce hydrochloric acid into their bloodstream. They also experimented with fluctuating concentrations of sugar and phosphate in the bloodstream and in urine. During World War II, in 1942, Haldane and a friend spent two days in a submarine to test an air-purifying system. He also simulated conditions inside submarines and subjected himself to extremes of temperature and a concentration of carbon dioxide in the air.

In 1924, having been introduced to enzyme reactions by Hopkins, Haldane produced the first proof that they obey the laws of thermodynamics. In 1930 Haldane published *Enzymes,* which gave an overall picture of how enzymes work.

Haldane is most remembered as a geneticist and a proponent of the unity of the sciences. His papers, lectures, and broadcasts made him one of the world's best-known scientists.

Haldane, John Scott (1860–1936) Scottish physiologist well known for his investigations into the safety measures necessary in stressful working conditions such as coal-mining and deep-sea diving, and for his studies on the exchange of gases during respiration.

Haldane was born in Edinburgh on 3 May 1860 into the Cloan branch of the ancient Scottish Haldane family, originating in Gleneagles. He was educated at the Edinburgh Academy, and graduated in medicine from Edinburgh University in 1884. He then briefly attended the universities of Jena and Berlin, and was a demonstrator in physiology at the University of Dundee. In 1887 he became a demonstrator in physiology at Oxford University, and from 1907 until he resigned in 1913 was a reader there. He then became director of the Mining Research Laboratory, Doncaster (which moved to Birmingham in 1921), a position he held until 1928. Haldane lectured at Yale in 1916, at Glasgow University 1927–29, and at Dublin University in the following year. He was elected to the Royal Society in 1897 and received its Royal Medal in 1916. He died in Oxford on 15 March 1936.

Haldane was concerned about the artificial differences that then existed between theoretical and applied science and strove to combine the two. His first piece of research was on the composition of the air in houses and schools in Dundee, to determine the amount of carbon dioxide in inhaled air and respiratory volume. He applied the results of this work to an investigation of the health hazards to which coal miners were exposed underground, particularly the risk of suffocation. In a report in 1896 he highlighted the dangers of the presence of carbon monoxide in mines after a mine explosion. Haldane then researched the nature of the toxicity of carbon monoxide in a significant paper that demonstrated that this gas combines with the blood's haemoglobin in preference to oxygen, preventing the blood from transporting the latter, with lethal consequences. The full physiological implications of his findings were not fully appreciated in the medical world, however, for another 50 years.

Between 1892–1900 Haldane devised methods for studying respiration and the blood. In 1898 he developed in principle the Haldane gas analyser and, a few years later, he invented an apparatus able to measure gas content in blood using comparatively small samples of blood – the haemoglobinometer. Both methods are still used, although the blood gas apparatus has largely been superseded.

In 1905 Haldane published his idea that breathing is controlled by the effect of the concentration of carbon dioxide in arterial blood on the respiratory centre of the brain. He showed that in all but extreme conditions, the regulatory processes in breathing are less dependent on the oxygen content of inspired air than on its carbon dioxide content. Although this was his most significant work, its findings received little attention in the medical field until the late 1940s.

Haldane then attempted to provide a preventive solution to the causes of caisson disease (decompression sickness, commonly known as 'the bends'), suffered by divers and other workers who have to breathe compressed air at high pressures. In 1907 he announced a technique of decompression by stages, which allows a deep-sea diver to rise to the surface safely; it is still used today. In 1911 Haldane undertook an expedition to Pikes Peak, Colorado, with several US physiologists, to study the physiological effects of low barometric pressures.

Haldane's study of gas exchanges in the bloodstream demonstrated the extent to which the absorption by the body tissues of carbon dioxide and its exhalation by the lungs depends on the action of oxygen on haemoglobin. He also researched the response of the body's kidneys to water present in the blood, and the physiological function of perspiration.

The application of many of Haldane's findings contributed much to the safety and health of miners and divers. He also had a great interest in philosophical topics and wrote extensively on the relationship between science and philosophy. In the 25 years he was at Oxford he did more than anyone else to earn the Oxford school of physiology international acclaim. His concern was to show that there is a dynamic and constructive equilibrium between theoretical and applied science.

Hale, George Ellery (1868–1938) US astronomer who spent much of his life arranging funding for the construction of large telescopes, including the 1.5-m/60-in and 2.5-m/100-in reflectors at Mount Wilson, and the 5-m/200-in Mount Palomar telescope. Principally he was an astrophysicist, and he distinguished himself in the study of solar spectra and sunspots. He developed a number of important instruments for the study of solar and stellar spectra, including the spectroheliograph and the spectrohelioscope.

Hale was born in Chicago on 29 June 1868, the son of a man of considerable means. He began his education at the Oakland Public School and continued at Adam Academy, Chicago, before going on to the Massachusetts Institute of Technology to study physics, chemistry, and mathematics.

In 1892 Hale was appointed professor of astronomy at the University of Chicago, and at once set about the establishment of a new observatory. He persuaded C T Yerkes, a Chicago industrialist, to donate a large sum of money to fund the construction of the 1-m/40-in Yerkes refractor telescope. When this was completed in 1897, Hale moved in his astrophysical instruments from his own laboratory at Kenwood.

For a while in 1893 he worked with the physicists Hermann von ◊Helmholtz, Max ◊Planck, and August ◊Kundt at the University of Berlin. However, he eventually abandoned his plans to take a doctorate and never found the time afterwards to work towards this goal. He was, even at this time, a very distinguished man of science and later, in 1899, he was instrumental in the founding of the American Astronomical Society.

From 1904 Hale was director of the Mount Wilson Observatory. Eventually, however, overwork forced him to resign in 1923. Nevertheless, he persisted in pursuit of his goal of a 5-m/200-in telescope. Through his inspiration in the conception of the telescope, and his efforts in securing from the Rockefeller Foundation the vast amount of money necessary for its construction, the 5-m/200-in reflecting telescope on Mount Palomar came into being. He did not see the completion of his greatest dream, for he died in Pasadena on 21 February 1938; but ten years later, when it was completed, the telescope was dedicated to him.

Even when Hale was quite young his interest in astronomy was enormous. Through a friendship with the amateur astronomer Sherburne Burnham (1838–1921) he bought a second-hand 10-cm/4-in refracting telescope. Another friendship with Hough (1836–1909) allowed him to view the heavens, from time to time, through the old 47-cm/18.5-in refractor at the University of Chicago. Long before he became a student he constructed spectroscopes. When his father bought him a spectrometer he accurately measured the principal Fraunhofer lines in the solar spectrum. While he was a student he continued his observations and developed the idea of the spectroheliograph – a means of surveying the occurrence of a particular line in the Sun's spectrum. The instrument he constructed did not fit well onto his 10-cm/4-in telescope, so his father bought him a 30-cm/12-in refractor and also equipped a solar laboratory for him. From 1891 to 1895 he used this observatory extensively and made improvements to his instruments.

Hale's work on solar spectra was the stimulus for his construction of a number of specially designed telescopes, the most important of which was the Snow telescope (named after the benefactor, Helen Snow) eventually installed on Mount Wilson. Later Hale was also to seek benefactors to build a 150-cm/60-in reflector (completed in 1908) using the mirror blank his father bought, and later still he managed to persuade J D Hooker, a Los Angeles businessman, to finance the construction of the 250-cm/100-in reflector (completed in 1918).

Meanwhile he surveyed the solar chromosphere as well as the rest of the Sun. In 1905 he obtained photographs of the spectra of sunspots that suggested that sunspots were colder – rather than hotter, as had been suspected – than the surrounding solar surface. Three years later he detected a fine structure in the hydrogen lines of the sunspot spectra. By comparing the split spectral lines of sunspots with those lines produced in the laboratory by intense magnetic fields (exhibiting the Zeeman effect) he showed the presence of very strong magnetic fields associated with sunspots. This was the first discovery of a magnetic field outside the Earth. In 1919 he made another important discovery relating to these magnetic fields by showing that they reverse polarity twice every 22–23 years.

Hale had to resign as director of the Mount Wilson Observatory because of overwork, but he was not a man to rest, and shortly afterwards he adapted his spectroheliograph to allow an observer to view the spectra with his eye. This adaptation, which created the spectrohelioscope, involved much more than merely the replacement of photographic film by an eyepiece.

As is evident from his creation of numerous telescopes and observatories, Hale was a highly successful organizer. At one time he was elected to the governing body of Throop Polytechnic Institute in Pasadena, and through his influence this initially little-known institute with a few hundred students developed into the California Institute of Technology (Caltech), now famous throughout the world for its research and scholarship.

Further Reading

Osterbrock, Donald E, 'Founded in 1895 by George E Hale and James E Keeler: the *Astrophysical Journal* centennial', *Astrophys J*, 1995, v 438, pp 1–7.

Osterbrock, Donald E, *Pauper and Prince: Ritchey, Hale, and Big American Telescopes,* University of Arizona Press, 1993.
Wright, Helen, *Explorer of the Universe: A Biography of George Ellery Hale,* History of Modern Physics and Astronomy Series, American Institute of Physics, 1994, v 14.

Hales, Stephen (1677–1761) English clergyman who devoted his life to the careful investigation of scientific matters. He researched the physiology and growth of plants, in particular transpiration.

Hales was born on 17 September 1677 in Bekesbourne, Kent. He spent 13 years at Cambridge, 1696–1709, studying divinity at what is now Corpus Christi College. His long stay there allowed him to obtain a thorough grounding in all scientific disciplines, as well as to prepare for the priesthood. He was ordained in 1703 and in 1709 took up the post of curate at Teddington, where he remained for the rest of his life. He died there on 4 January 1761.

Hales's experiments on plants took place mainly between 1719 and 1725. In one experiment he cut off a vine at ground level in the spring, before the buds had burst. He attached a glass tube to the cut surface and showed that the sap rose up the tube to a height of 7.5 m/25 ft, indicating that the sap was under considerable pressure.

It is now known that this high 'root pressure' establishes a column of water between the roots and the buds in the spring. Once the leaves open, the column extends to the spongy tissues in the leaves, from which water evaporates. The transpirational stream is then set in motion, drawing water from the soil, through the plant and to the atmosphere via the leaves.

Hales also worked on vines in full leaf and was able to show that the pressure exerted by water in the transpiration stream is greatest when the plant is illuminated. He demonstrated this phenomenon by attaching his pressure-measuring apparatus to cut side-shoots (or roots) and taking measurements at various times of day. He further showed that water flows in a plant in one direction only, from root to stem to leaf. He used either a straight glass tube to measure the pressure of plant sap or a simple manometer filled with mercury. He calculated in this way the actual velocity of the sap and found that the rate of flow varies in different plants. In investigating the development of plants, Hales was able to show that most of the growth in plant size is made by the younger shoots. On any one section of shoot, he found that it is usually the part of the shoot between the nodes that increases in size the most.

Hales was particularly interested in processes that are common to both plants and animals, such as the nature of growth, and the role of water and air in the maintenance of life. His animal studies involved inserting probes into the veins and arteries of domestic animals and recording the height to which their blood rose up a glass tube connected to the probe, which gave a measure of blood pressure.

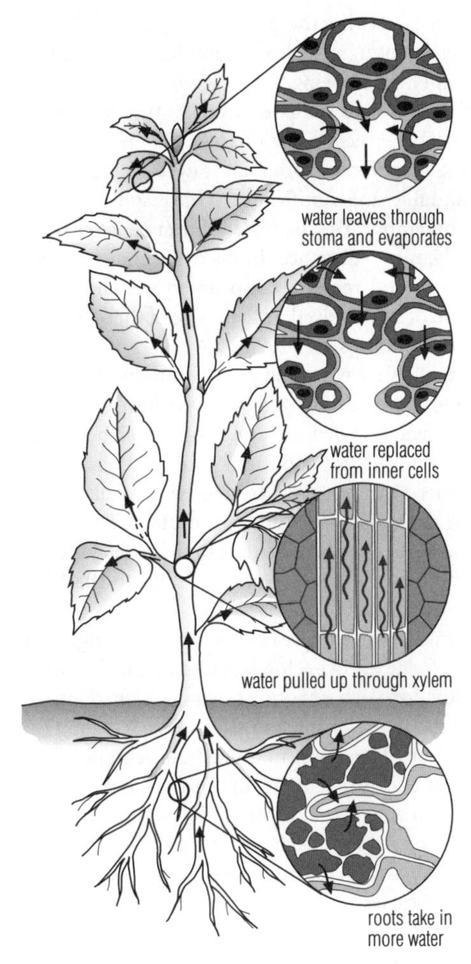

Hales The loss of water from a plant by evaporation is known as transpiration. Most of the water is lost through the surface openings, or stomata, on the leaves. The evaporation produces what is known as the transpiration stream, a tension that draws water up from the roots through the xylem, water-carrying vessels in the stem.

He went on to make wax casts of the left ventricles of the hearts of several animals to measure the blood capacities. He also estimated the rate of blood flow through the capillaries by injecting chemicals into various body organs, and watching their movement and the dispersion of the chemicals. He showed that the presence of foreign chemicals can change the rate of flow of blood through an organ and thereby demonstrated (although indirectly) that capillaries are capable of dilation or contraction. As a result of these experiments Hales became convinced of the fallacy of the popular belief that the pressure of the blood itself was powerful enough to effect muscular movement in animals and human beings.

Hales also worked on gases, from both a biological and physical chemical point of view. He made several important discoveries about air but was unable to understand fully the significance of his results because, like his contemporaries, he believed that all gases were a single chemical substance. In his experiments, Hales heated up several materials and noted the change in the volume of air involved in the process, thereby confirming the findings of previous workers that some solids liberate 'air' on heating whereas others absorb it. He devised an apparatus that allowed him to breath solely his own expelled air and found that he could continue to do this for about one minute. If, however, he introduced salt or tartar treated with an alkali, he could use the apparatus for $8^1/_2$ minutes. (The alkali mixture absorbed carbon dioxide, thereby delaying for several minutes the response of his lungs to the high carbon dioxide levels.)

Hales's work on air revealed to him the dangers of breathing 'spent' air in enclosed places, and he invented a ventilator for such situations, which he had introduced on naval, merchant, and slave ships; in hospitals; and in prisons. There was an immediate acknowledgement of the improvement in human health and survival following their introduction, although these favourable reactions foreshadowed an overemphasis on air quality as a factor in disease control in the 18th and 19th centuries.

Hales carried out many other investigations, such as examining stones taken from the bladder and kidney and suggesting possible chemical solvents for their non-surgical treatment. In the course of his investigations he also invented the surgical forceps.

Hales's research on plants together with his work on the nature and properties of air were published in his famous book *Vegetable Staticks* (1727). This book was subsequently republished in an enlarged form, containing an additional section on his findings on sap flow and blood and circulation in mammals (in 1733). The enlarged work was entitled *Statical Essays, Containing Haemastaticks, etc.* Hales's emphasis on the need for careful measurement of scientific phenomena may well have had as great an influence on his contemporaries as the results he obtained.

Hall, Asaph (1829–1907) US astronomer who discovered the two Martian satellites, Deimos and Phobos. He is also noted for his work on satellites of other planets, the rotation of Saturn, the mass of Mars, and double stars.

Hall was born in Goshen, Connecticut, on 15 October 1829, into a well-established and once-prosperous New England family. His father, a clock manufacturer, also called Asaph Hall, died in 1842 and to alleviate the family's financial difficulties the young Asaph became a carpenter's apprentice at the age of 16. However by 1854 he wanted to continue his education and so he enrolled at the Central College in McGrawville, New York. There he met his future wife, a determined suffragist, Chloe Angeline Stickney, and soon after their marriage in 1856, Hall, who by this time was determined to become an astronomer, took an extremely low-paid job at the Harvard College Observatory in Cambridge, Massachusetts. His status soon rose to that of an observer and being an expert computer of orbits he managed to supplement his meagre income by compiling almanacs and observing culminations of the Moon. In 1862 Hall accepted a post as assistant astronomer at the US Naval Observatory in Washington and within a year of his arrival he was given a professorship.

In 1877, Hall discovered that Mars had two satellites and this achievement won him the Lalande Prize, the Gold Medal of the Royal Astronomical Society in 1879, and the Arago Medal in 1893. He was elected to the National Academy of Sciences in 1875 and served as its home secretary for 12 years and its vice president for six years. In 1902 he served as president of the American Association for the Advancement of Science and he was associate editor of the *Astronomical Journal* 1897–1907. Following his retirement from the US Naval Observatory at the age of 62, he continued to work as an observer in a voluntary capacity until 1898, when he went to Harvard to take up a teaching post in celestial mechanics. After five years of teaching he retired to his rural home in Connecticut where he lived until his death on 22 November 1907.

Hall's first years at the US Naval Observatory were troubled by the unwholesome climate of the American Civil War. He spent this inital period, 1862–75, as an assistant observer of comets and asteroids. However, in 1875 Hall was given responsibility for the 66-cm/26-in telescope that had been built by Alvan Clark specifically for the US Naval Observatory. At the time, it was the largest refractor in the world. Hall's first discovery using this telescope was a white spot on the planet Saturn, which he used as a marker to ascertain Saturn's rotational period.

In the year 1877, Hall decided to use the superb image-forming qualities of the 66-cm/26-in refractor in a search for possible moons of Mars. It was a particularly good year for such a search because of the unusually close approach of Mars and Hall was guided by the parameters set by his theoretical calculations that indicated the possibility of a Martian satellite being close to the planet itself. Hall's first glimpse of a Martian moon, later named Deimos, occurred on

11 August 1877. By 17 August Hall had convinced himself that the object was definitely a satellite and on that same day he found the second satellite, Phobos. He disclosed his observations to Simon ◊Newcomb, the scientific head of the observatory, who erroneously believed that Hall had failed to recognize that the 'Martian stars' were satellites. Hence Newcomb took the undeserved credit for their discovery in the wide press coverage that followed.

During the following years Hall continued to work on the orbital elements of planetary satellites. In 1884 he showed that the position of the elliptical orbit of Saturn's moon, Hyperion, was retrograding by about 20 ° per year. In addition to his planetary satellite studies, Hall made numerous investigations of binary star orbits, stellar parallaxes, and the position of the stars in the Pleiades cluster. Owing to his discoveries and to the clarity and precision of his work on the satellites of Mars, Saturn, Uranus, and Neptune, which has been compared with that of Friedrich ◊Bessel, Hall has become generally known as 'the caretaker of the satellites'.

Hall, James (1761–1832) Scottish geologist, one of the founders of experimental geology.

The son and heir of Sir John Hall of Dunglass, Berwickshire, Hall spent much of the 1780s travelling in Europe. He undertook extensive geological observations in the Alps and in Italy and Sicily, studying Mount Etna. He was also won over to the new chemistry of Antoine ◊Lavoisier. Developing a warm friendship with James ◊Hutton, Hall set out to defend Hutton's plutonist geological theories (viewing that heat rather than water was the chief rock-building agent and shaper of the Earth's crust) by providing them with experimental proofs. By means of an innovative series of furnace investigations, he showed with fair success that Hutton had been correct to maintain that igneous rocks would generate crystalline structures if cooled very slowly (a possibility denied by supporters of Abraham ◊Werner's neptunist theory, which attributed stratification to water). Hall also demonstrated that there was a degree of interconvertibility between basaltine and granitic rocks; and that, even though subjected to immense heat, limestone would not decompose if sustained under suitable pressure. All these conclusions helped bolster Hutton's plutonian geology and, through being experimental findings, gave it a more modern appearance.

Hall, Philip (1904–1982) English mathematician who specialized in the study of group theory.

Born on 11 April 1904, in London, Hall attended Christ's Hospital until the age of 18, when he went to King's College, Cambridge. After receiving his degree, he became a fellow of the college in 1927. Six years later he became a university lecturer in mathematics, a post he held until 1951, when he became reader in algebra at the university. Hall then became Sadlerian Professor of Pure Mathematics in 1953 and remained as such until his retirement in 1967. As emeritus professor, his connections with the university did not cease upon his retirement, and in 1976 he was elected an honorary fellow of Jesus College.

In addition to his academic duties, Hall was an active participant in professional societies. The London Mathematical Society elected him to its presidency in 1955, and awarded him the De Morgan Medal and the Lamor Prize in 1965. He was elected a fellow of the Royal Society in 1942, and awarded its Sylvester Medal in 1961.

Hall's major contributions to mathematics came through his work on group theory. In 1928 he extended some work on group theory done by Ludwig ◊Sylow in 1872, which led him to a study of prime power groups. From this work he developed his 1933 theory of regular groups. In that paper he also presented material that forms part of the connection between group theory and the study of Lie groups (see Marius ◊Lie).

An investigation of the conditions under which finite groups are soluble led him in 1937 to postulate a general structure theory for finite soluble groups. In 1954 he published an examination of finitely generated soluble groups in which he demonstrated that they could be divided into two classes of unequal size.

Hall collaborated with G Higman in 1956 in a study of *p*-soluble groups; their work generated results that were important to later work on the theory of finite groups.

At the end of the 1950s Hall turned to the subject of simple groups, and later also examined non-strictly simple groups. His researches into the subject of group theory produced many valuable results that have contributed to the further development of this area.

Hall and Héroult Charles Martin Hall (1863–1914) and Paul Louis Toussaint Héroult (1863–1914) US chemist and French metallurgist, respectively. Hall developed a process for the commercial production of aluminium while his exact contemporary Héroult simultaneously but independently developed a similar process.

Hall was born on 6 December 1863 in Thompson, Ohio, and educated at Oberlin, where he was influenced by F F Jewett, a former pupil of the German chemist Friedrich ◊Wöhler who had succeeded in producing impure aluminium under laboratory conditions.

As a result, Hall became interested in the commercial manufacture of aluminium while still a student, perhaps attracted by the prospect of becoming wealthy and famous by bringing this versatile metal to the industrialized world. Just eight months after graduating in 1855, Hall had found that electrolysis was the best route to achieving his goal, had built his own batteries, and, in his home laboratory, had isolated the best compound for commercial production of aluminium: cryolite (sodium aluminium fluoride). By adding aluminium oxides to this compound in its molten state, and using small carbon anodes, a direct current gave a deposit of free molten aluminium on the carbon lining of the electrolytic cell. On 23 February 1886, the 22-year-old Hall was able to present his professor with buttons of aluminium he had himself prepared; these are still preserved by the Aluminium Company of America.

Despite this astonishing achievement, Hall had both financial and technical difficulties in achieving commercial success. A firm that took out a one-year option on the process found that the electrolytic cells failed after just a few days. Financial backing eventually came from the wealthy Mellon family, and the Pittsburgh Reduction Company (later to become the Aluminium Company of America) was formed to build commercially viable units, requiring no external heating during operation.

Meanwhile, in Europe, Héroult had arrived at the same process and was achieving similar success, the first British plant being based at Patricroft in 1890. Héroult was born in Thury-Harcourt, Normandy, in November 1863. He was still a student when he read Henri DeVille's account of his efforts to isolate aluminium by chemical reactions. Following a year spent at the Paris School of Mines, Héroult began to study the use of electrolysis to extract aluminum from compounds. Like Hall in the USA, Héroult used direct-current electrolysis to find the best combination of material for the production of aluminium, dissolving aluminium oxide in a variety of molten fluorides, and finding cryolite (sodium aluminium flouride) the most promising. Unlike Hall's apparatus, however, Héroult used one large, central graphite electrode in the graphite cell holding the molten material. As with Hall, Héroult succeeded in producing aluminum using his electrolytic apparatus at the age of 23. He patented the system in 1886. He also met with the same subsequent difficulty in commercializing the process. The system was taken up, however, by a joint German–Swiss venture at Neuhausen, and large-scale production got underway. Héroult also patented a method for the production of aluminium alloys in 1888.

By 1914 the Hall–Héroult process had brought the price of aluminium down from tens of thousands of dollars a kilogram to 40 cents in the space of 70 years.

As Hall had been told as a student, the process that he had succeeded in commercializing resulted in his becoming a multi-millionaire. On his death at Daytona Beach, Florida, on 27 December 1914, Hall recognized his debt to Oberlin by leaving several million dollars as a gift. Eight months earlier, on 10 April 1914, Héroult had died near Antibes, southern France.

Haller, Albrecht von (1708–1777) Swiss physiologist, the founder of neurology. By means of skilful investigations into the neuromuscular system he freed it from the remaining myths and superstitions of his day.

Haller was born in Bern on 16 October 1708. A sickly child, he spent most of his time indoors writing and studying. By the age of ten he had compiled a Greek dictionary and written on a number of scholarly topics. He was a medical student of the Dutch physician Hermann ◊Boerhaave at Leiden, graduating in 1727 and starting his own practice two years later. In 1736 he was appointed professor of medicine, anatomy, surgery, and botany at the new University of Göttingen, where he remained until he retired in 1753 to write. He died in Bern on 12 December 1777.

During the 18th century the idea prevailed that a mysterious force was associated with the nerves, which were believed to be liquid-filled tubes in the body. Haller rejected these notions, especially as they could not be observed experimentally. His own experiments concerning the relationship between nerves and muscles demonstrated the irritability of muscle. He found that if he applied a stimulus to a muscle, the muscle contracted, and that if he stimulated the nerve attached to the muscle there was a stronger contraction. He therefore deduced that it is the stimulation of a nerve that brings about muscular movement.

Haller carried out further experiments to show that tissues are incapable of feeling, but that nerves collect and conduct away impulses produced by a stimulus. Tracing the pathways of nerves, he was able to demonstrate that they always lead to the spinal cord or the brain, suggesting that these regions might be where awareness of sensation and the initiation of answering responses are located. Haller also experimented on the brains of animals and observed the reactions resulting from the damage or stimulation of various parts of the brain.

While carrying out his experiments, Haller discovered several processes of the human body, such as the role of bile in digesting fats. He also wrote a report on his study of embryonic development.

In 1747 Haller published *De respiratione experimenta anatomica/Experiments in the Anatomy of Respiration* and between 1757 and 1766 his eight-volume encyclopedia *Elementa physiologiae corporis humani/The Physiological Elements of the Human Body* was published.

Medicine was just one of Haller's many interests. Botany was another, and he was an avid collector of plants. He devised a botanical classification that equals that of Carolus ◊Linnaeus, and wrote a work on the Swiss flora.

Further Reading

Buess, Heinrich, 'Albrecht von Haller and his 'Elementa Physiologiae' as the beginning of pathological physiology', *Med Hist,* 1959, v 3, pp 123–131.

Sonntag, O (ed), *The Correspondence Between Albrecht von Haller and Horace-Benedict de Saussure,* Hogrefe and Huber Publishers, 1990.

Halley, Edmond (Edmund) (1656–1742) English mathematician, physicist, and astronomer who not only identified the comet later to be known by his name, but also compiled a star catalogue, detected stellar motion using historical records, and began a line of research that – after his death – resulted in a reasonably accurate calculation of the astronomical unit.

Halley was born in Haggerton, near London, on 8 November 1656. The son of a wealthy businessman, he attended St Paul's School (in London) and then Oxford University, where he wrote and published a book on the laws of Johannes ◊Kepler that drew him to the attention of the Astronomer Royal, John ◊Flamsteed. Flamsteed's interest secured for him, despite his leaving Oxford without a degree, the opportunity to begin his scientific career by spending two years on the island of St Helena, charting (none too successfully) the hitherto unmapped stars of the southern hemisphere. The result was the first catalogue of star positions compiled with the use of a telescope. On his return in 1678, Halley was elected to the Royal Society: he was 22 years old. For some years he then travelled widely in Europe, meeting scientists – particularly astronomers – of international renown, including Johannes ◊Hevelius and Giovanni ◊Cassini, before finally returning to England to settle down to research. He also became a firm friend of Isaac ◊Newton. It may have been through Newton's influence that Halley in 1696 took up the post of deputy controller of the Mint at Chester. Two years later he accepted command of a Royal Navy warship, and spent considerable time at sea. In 1702 and 1703 he made a couple of diplomatic missions to Vienna before, in the latter year, being appointed professor of geometry at Oxford. His study of comets followed immediately. He succeeded Flamsteed as Astronomer Royal in 1720 and held the post until his death, at Greenwich, on 14 January 1742.

Halley English mathematician, physicist, and astronomer Edmond Halley, who had a long career in which he discovered the proper motions of the stars, and the periodicity of comets, and made many other observations. He was appointed Astronomer Royal in 1720. *Mary Evans Picture Library*

In St Helena, Halley first observed and timed a transit of Mercury, realizing as he did so that if a sufficient number of astronomers in different locations round the world also timed their observations and then compared notes, it would be possible to derive the distance both of Mercury and of the Sun. Many years later he prepared extensive notes on procedures to be followed by astronomers observing the expected transit of Venus in 1761. No fewer than 62 observing stations noted the 1761 transit, and from their findings the distance of the Sun from the Earth was calculated to be 153 million km/95 million mi – remarkably accurate for its time (the modern value is 149.6 million km/92.9 million mi).

Astronomy was always Halley's major interest. In the 1680s and 1690s he prepared papers on the nature of trade winds, magnetism, monsoons, the tides, the relationship between height and pressure, evaporation and the salinity of inland waters, the rainbow, and a diving bell; for some of the time he was also helping Newton both practically and financially to formulate his great work, the *Principia;* but these activities were all incidental to the pleasure he took in observing the heavens.

One of Halley's first labours as professor at Oxford was to make a close study of the nature of comets. Twenty years earlier, the appearance of a comet visible to the naked eye had aroused great popular excitement; yet somehow Newton's *Principia* had ignored the subject. Now Halley, with Newton's assistance, compiled a record of as many comets as possible and charted their progress through the heavens. A major difficulty in determining the paths of comets arose from their being visible for only short periods, leaving by far the greater part of their journey explicable by any number of hypotheses. Some authorities of the time believed that comets travelled in a straight line; some in a parabola or a hyperbola; others suggested an ellipse. In the course of his investigations Halley became convinced that the cometary sightings reported in 1456, 1531, 1607 and, most recently, in 1682, all represented reappearances of the same comet. Halley therefore assumed that such a traveller through space must follow a very elongated orbit around the Sun, taking it at times farther away than the remotest of the planets. On the parabolic path that he calculated it should follow – and making due allowance for deviations from its 'proper' path through the attraction of Jupiter – Halley declared that this comet would appear again in December 1758. When it did, public acclaim for the astronomer (who by that time had been dead 16 years) was such that his name was irrevocably attached to it. In 1710 Halley began to examine the writings of ◊Ptolemy, the Alexandrian astronomer of the 2nd century AD. Throughout his life Halley was always keenly interested in classical astronomy (he made outstanding translations of the *Conics* of ◊Apollonius of Perga and the *Sphaerica* of Menelaus of Alexandria), and having catalogued stars himself, he paid special attention to Ptolemy's stellar catalogue. This was not Ptolemy's original; it was in fact 300 years older than it appeared to be, being a direct borrowing of the list compiled by ◊Hipparchus in the 3rd century BC. For all Ptolemy's shortcomings, however, Halley could not believe that he had been so negligent as to credit the stars with positions wildly at variance with the bearings they now occupied, 15 centuries later. The conclusion that Halley was forced to come to was that the stars had moved. Later Halley was able to detect such movements in the instances of three bright stars: Sirius, Arcturus, and Procyon. He was correct, too, in assuming that other stars farther away and consequently dimmer underwent changes of position too small for the naked eye to detect. More than a century had to pass before optical instruments achieved a sophistication sufficient to be able to detect such movement.

Further Reading

Cook, Alan H, *Edmond Halley: Charting the Heavens and the Seas,* Clarendon Press, 1998.

Halley, Edmond, *Correspondence and Papers of Edmond Halley,* Clarendon Press, 1932.

Mason, John W (ed), *Comet Haley: Investigations, Results, Interpretations,* Ellis Horwood Library of Space Science and Space Technology. Series in Astronomy, Ellis Horwood, 1990, 2 vols.

Ronan, Colin Alistair, *Edmond Halley: Genius in Eclipse,* Macdonald, 1969.

Hamilton, Alice (1869–1970) US physician and social reformer who pioneered the study of industrial toxicology.

Hamilton was born in New York State on 27 February 1869 into an affluent family and grew up on the family's estate in Indiana. She was educated at home in an intellectually stimulating family with a strong sense of social responsibility. The failure of her father's provisioning firm meant that she had to earn her living and she decided to study medicine and entered the University of Michigan, graduating in 1893. She then followed a course open to few men, let alone women; she studied bacteriology in Ann Arbor before travelling with her sister for further training in pathology and bacteriology at the universities of Leipzig and Munich, before returning to the Johns Hopkins Medical School for a final postgraduate year of study. She was then appointed professor of pathology at the Woman's Medical College of North Western University.

While working in Chicago she tried to combine medical practice and research with social commitment, the latter fostered by her involvement with Hull House, a radical settlement in which she first observed links between environment and disease. She was alerted to the dangers inherent in many industrial practices by an investigative journalist, and she decided to shift her work in this direction. In 1908 she was appointed to the Illinois Commission on Occupational Diseases, and two years later supervised a state-wide survey of industrial poisons, her own researches concentrating on lead, which was extensively used in many industrial processes. By laboriously examining hospital records, factories, and the workers themselves, she and her staff identified many hazardous procedures and consequent state legislature introduced safety measures in the workplace and medical examinations for workers at risk. The following year Hamilton was appointed special investigator for the US Bureau of Labor and rapidly became the leading authority on lead poisoning in particular and industrial diseases in general. Her reputation was such that she became the first woman professor at Harvard when she was appointed assistant professor of industrial medicine, a position she accepted in 1919, almost 30 years before Harvard accepted women as medical students. She continued with her fieldwork for the Department of Labor (as the Bureau had become) and her pioneering book *Industrial Poisons in the United States* (1925) established her reputation worldwide. She retired from Harvard in 1935 to live in Connecticut, but continued to be active in many fields. Her classic textbook *Industrial Toxicology* (1934) was extensively revised in 1949, and she published an autobiography *Exploring the Dangerous Trades* (1943). She died on 22 September 1970 at the age of 101, much honoured and admired.

Hamilton's medical work was strongly infused with social concern, and her political beliefs and activities were in turn influenced by many of the conditions she witnessed in the course of her researches. During and after World War I she attended International Congresses of Women and declared herself a pacifist. She was profoundly affected by the conditions of famine she witnessed in postwar Germany and a further visit to Nazi Germany in 1933 disturbed her deeply, but she continued to believe in pacifism until 1940, when she urged US participation in World War II. During the 1940s and 1950s she spoke out on controversial subjects such as contraception, civil liberties, and workers rights. Even in the 1960s she was still considered worthy of attention by the Federal Bureau of Investigation when she protested against US involvement in Vietnam.

Further Reading

Sicherman, Barbara, *Alice Hamilton: A Life in Letters*, Commonwealth Fund Publications, Harvard University Press, 1984.

Hamilton, William Rowan (1805–1865) Irish mathematician, widely regarded in his time as a 'new Newton', who created a new system of algebra based on quaternions.

Hamilton was born in Dublin, of Scottish parents, on the stroke of midnight, 3–4 August 1805. At about the age of three he was sent by his father, a solicitor, to Trim, where he was raised by his uncle (the local curate) and his aunt. He soon showed a remarkably precocious facility for languages, and made himself expert in Latin, Greek, Hebrew, French, Italian, and a number of oriental languages (in which he was encouraged by his father, who wished him to become a clerk with the East India Company). He never lost his interest in languages; throughout his life he wrote poetry and corresponded with Wordsworth, Coleridge, and Southey. But at the age of ten Hamilton came upon and read ◊Euclid, and from that moment mathematics became his chief love. By the age of 12 he was reading the work of Isaac ◊Newton and when he went to Trinity College, Dublin, to study classics and mathematics, he had mastered the work of the great French mathematicians, including Pierre ◊Laplace, and had begun to make experiments in physics and astronomy. In particular, he had begun to investigate the patterns produced by rays of light on reflection and refraction, known as caustics.

At Trinity College, Hamilton was far and away the outstanding scholar of his generation. He gained the highest grade of optime, not once, which would have been rare enough, but twice (in Greek and in mathematical physics), which was unprecedented. He also twice won the vice chancellor's gold medal for English verse. In only his second year as an undergraduate he presented a paper on caustics to the Royal Irish Academy in which he showed that light travels by the path of the least action and that its path can be treated as a function of the points through which it travels and expressed mathematically in one 'characteristic function' involving advanced calculus. The paper, which predicted the existence of conical refraction (later proved experimentally by H Lloyd) was published as 'The theory of systems of rays' by the academy in 1828.

This early work on caustics so impressed his contemporaries that in 1827, while he was still an undergraduate, Hamilton was appointed Andrews Professor of Astronomy at the university. With the chair went the title of Royal Astronomer of Ireland and the charge of the Dunsink observatory. In fact, Hamilton never gave his heart to astronomy; for the rest of his life he devoted himself to mathematical research.

Throughout much of his life Hamilton was closely involved in the work of the British Association for the Advancement of Science, and at a meeting of the association in Dublin in 1835 he was knighted by the Lord Lieutenant of Ireland. With the knighthood was awarded an annual pension of £200. Many international honours followed, and Hamilton had the distinction of being the first non-US citizen to be elected a fellow of the US National Academy of Sciences. In his latter years Hamilton became something of a recluse, working tirelessly at his home (unfinished meals were found among the piles of paper in his study when he died) and drinking too heavily. He died, of gout, at Dublin on 2 September 1865.

In addition to his work on caustics, Hamilton's most important research was in the field of complex numbers and in the branch of study that he created, the algebra of quaternions. In 1833 he began to seek an improved way of handling complex numbers of the form:

$$a + ib$$

Where $i = \sqrt{-1}$ and a and b are real numbers. He thought that, since a and ib were different types of quantity, it was wrong to connect them by a + sign; he also considered $\sqrt{-1}$ to be meaningless. He therefore devised new rules of addition, subtraction, multiplication, and division on the basis of considering $a + ib$ as a couple of real numbers represented by (a, b).

He also showed that the sum of two complex numbers could be represented by a parallelogram and that complex numbers could be used, in general, as a useful tool in plane geometry.

From couples Hamilton went on to investigate triples and this led him to his great work on quaternions. Hamilton first formally announced his definition of quaternions to the Irish Royal Academy in 1844. But it was not until he began to lecture on them in 1848 that his new algebra really began to take shape. The lectures were published in 1853. Hamilton's investigation of triples, with its obvious relevance to three-dimensional geometry, was a much more difficult task than his work on doubles. He found that many mathematical notions had to be sacrificed. Of these by far the most important was the commutative principle. Hamilton found out, during his research, that his quaternions had four components (hence the name), not three as he had expected. They took the form

$$a + bi + cj + dk$$

Where j and k were similar to i. The fundamental formula was:

$$I^2 = j^2 = k^2 = ijk = -1$$

So delighted was Hamilton at this discovery that he scratched the formula on the stonework of a bridge under which he was passing when it came to him. The result of this formula was to reveal that for quaternions the ordinary commutative principle of multiplication (that is $3 \times 4 = 4 \times 3$) did not work, for:

$$Ij = -ji.$$

This was perhaps Hamilton's most important contribution to mathematics, since it forced mathematicians to abandon their belief in the commutative principle as an axiom.

Hamilton hoped that quaternions would find applications in the solution of problems in physics in the way that vectors have proved to do. Indeed, it was Hamilton who coined the word 'vector' and it was he who made it possible to deal with lines in all possible positions and directions and freed them from dependence on Cartesian axes of reference. But although he wrote two large books on the subjects his hopes for quaternions were not fulfilled, and today their usefulness has been superseded, for most problems, by the development of vector and tensor analysis.

Hammick, Dalziel Llewellyn (1887–1966) English chemist whose major contributions were in the fields of theoretical and synthetic organic chemistry.

Hammick was born in West Norwood, London, the son of a businessman; when he was 12 years old his family moved to the country, which gave him a lifelong interest in rural pursuits. He was educated at Whitgift School and then won a scholarship to Magdalen College, Oxford; he graduated in 1908. From Oxford he went to work in the laboratory of Otto Dimroth in Munich, where in 1910 he gained a PhD for his researches on solubility. He returned to the UK and held teaching posts at Gresham's School in Holt, Norfolk, and at Winchester College. In 1921 he returned to Oxford as a fellow of Oriel College and a lecturer at Corpus Christi College. He remained at Oxford until his retirement; he died there on 17 October 1966.

Hammick's research activities began at Gresham's School, where he investigated the action of sulphur dioxide on various metal oxides. At Oxford he initially continued researching inorganic substances, particularly with regard to their solubilities. He investigated the dimorphism of potassium ethyl sulphate and the ternary system comprising water, sodium sulphite, and sodium hydroxide. He also studied sulphur and its compounds (such as carbon disulphide), and suggested structures for liquid sulphur and plastic sulphur (which later workers interpreted in terms of linear polymers).

By the mid-1920s, Hammick's attentions had begun to swing towards organic chemistry. In 1922 he showed that the sublimation of α-trioxymethylene results in the polymer polyoxymethylene; 40 years later this substance was to be used as a commercial polymer. In 1930 Hammick devised a rule to predict the order of substitution in benzene derivatives. Aromaticity and the role of electron attraction and donation of substituents already present in the benzene ring were not yet understood, but Hammick proposed the following useful rule for further substitution of the compound Ph-XY: 'If Y is in a higher group of the periodic table than X or is in the same group (Y is of a lower mass than X), then further substitution will take place *meta* to the group XY (giving 1,3 di-substitution). In all other cases, substitution goes *ortho* or *para* (giving 1,2-or 1,4-di-substitution); the nitroso group is an exception.'

Hammick spent much time investigating organic reaction rates. He studied, for example, the decarboxylation of quinaldinic acid (the alpha-, *ortho*-, or 2-carboxylic acid of quinoline) by aldehydes and ketones. The reaction, now known as the Hammick reaction, is used in the synthesis of larger molecules.

During World War II, when supplies of toluene for making explosives were short, Hammick directed his efforts at syntheses using benzene (obtained from the distillation of coal tar). Using alumina–silica catalysts, he achieved a 30% conversion of benzene to toluene.

Hancock, Thomas (1786–1865) English inventor who was influential in the development of the rubber industry.

Hancock, the son of a timber merchant and brother of Walter Hancock, one of the pioneers of steam road-carriages was born in Marlborough, Wiltshire on 8 May 1786. Both brothers were enterprising in their own particular fields. Thomas travelled to London in 1820 and took out a patent for applying caoutchouc, (now known as india rubber) to various articles of dress. Having secured the patent, he opened a factory in Goswell Road, London, and began manufacturing on a large scale, assisted at a later date by his brother.

It was a successful venture, and from 1820 to 1847 he took out a further 16 patents connected with working rubber. He also collaborated with Charles Macintosh (1766–1843) of Glasgow, famous for his waterproof cloth, over some aspects of manufacturing.

The Goswell Road factory was transferred to his nephew James Lyne Hancock in 1842, and Hancock himself applied his talents to reseach in his own private laboratory, which he had set up in Stoke Newington. Like Charles ◊Goodyear in the USA, he was interested in solving the problems of rubber's tackiness and inconsistency at different temperatures. After seeing some samples of Goodyear's products and reading an article about his work, Hancock adopted the heat process of vulcanization. In 1857 he published *Personal Narrative of the Origin and Progress of the Caoutchouc or India Rubber Manufacture in England.* He died in Stoke Newington, London, on 26 March 1865.

Joseph ◊Priestley, the discoverer of oxygen, introduced the term 'rubber' after using a small piece of the substance to erase a pencil mark on his notes. The prefix 'India' came from the place where it was first found, the West Indies. The first description of rubber in use came to Europe via Charles de la Condamine, who in 1735 was commissioned by the Academy of Sciences, Paris, to carry out some research in South America. He noted that the local people made items from a watertight material that they had previously prepared from a milky liquid spread out in the sun to cure.

One of Hancock's earliest and most important inventions was a 'masticator', a machine that kneaded the raw rubber to produce a solid block. He nicknamed this his 'pickle machine'.

He worked continuously on improvements to rubber, seeking the help and advice of Charles Macintosh. Like Goodyear in the USA, he found that the real problems lay in maintaining the strength and pliable nature of the substance under changing temperatures. Hancock finally mastered the technique after experimenting with sulphur additives and reading an article on Goodyear's work.

Handley Page, Frederick (1885–1962) English aeronautical engineer who founded a company that achieved a worldwide reputation for constructing successful military aircraft.

Handley Page was born in Cheltenham and as a boy determined to become an electrical engineer: by the age of 21 he was chief designer of an electrical company. In 1907 he presented a paper on the design of electrical equipment to the Institute of Electrical Engineers, and such was the interest created that he was offered a tempting post by the Westinghouse Company in the USA. But Handley Page was fascinated with the potential of the aeroplane and he became an early member of the (later Royal) Aeronautical Society. In 1908 he set up as an aeronautical engineer and a year later, with a capital of £10,000, he established the first private British company of this kind, using some sheds at Barking, Essex, as workshops and nearby waste ground as an aerodrome.

The outbreak of World War I in 1914 determined the line he was to follow. The need of the British Admiralty for an aircraft capable of carrying a heavy load of bombs led to his design of the first two-engined bombers. Igor ◊Sikorsky in Russia and Giovanni Caproni (1886–1957) in Italy had been thinking along similar lines, but Handley Page produced the first large aircraft. It first flew in December 1915 and before the war had ended had been enlarged into a four-engined bomber with a fully-laden weight of 13 tonnes. More than 250 of these aircraft were produced although the war ended before many saw active service.

After the war's end in 1918 Handley Page turned his attention to civil aircraft. For four years another company bearing his name used his machines to operate a service on routes from London to Paris and Brussels, later extended to Zürich. All that this experience proved commercially, however, was that civil aviation was not viable without government subsidies, and these were only to be had when Handley Page lost its identity by merging with Imperial Airways, the forerunner of BOAC. The manufacturing side of Handley Page's business continued, due largely to his development of the automatic slot safety device.

During the interwar years, in 1930, Handley Page produced the first 40-seat airliner. This was the Hercules, a four-engined plane that arose at about 145 kph/90 mph. It was extremely comfortable – often being compared with Pullman trains but the aircraft failed to attract overseas orders.

During World War II Handley Page produced another bomber, the Halifax, of which 7,000 were constructed. Work on the bomber continued after the end of the war in 1945, resulting in a four-engined jet of unusual design, the Victor, which made its first flight in 1952. For his services to the aircraft industry in peace and war, Handley Page was knighted in 1942. In 1960 he was awarded the Gold Medal of the Royal Aeronautical Society.

Hankel, Hermann (1839–1873) German mathematician and mathematical historian who, in a somewhat brief life, nevertheless made significant contributions to the study of complex and hypercomplex numbers and the theory of functions. A great innovator himself – his name is perpetuated in the Hankel functions – much of his work was also in developing that of others. In turn, his work later received considerable and useful development.

Hankel was born in Halle on 14 February 1839, but it was in Leipzig – where his father was professor of physics at the university – that he grew up and was educated. At the age of 21, Hankel went to Göttingen University, but transferred a year later to the University of Berlin, where he studied under Karl ◊Weierstrass and Leopold ◊Kronecker, receiving his doctorate in only 12 months for a thesis on a special class of symmetrical determinates. In 1867 he was appointed associate professor of mathematics at Leipzig. Taking up a similar post at Tübingen in 1869, he spent only four years there before he died, on 29 August 1873.

Hankel published many important works, the first of which was his *Theorie der complexen Zahlensysteme* (1867). The work dealt with the real, complex, and hypercomplex number systems, and demonstrated that no hypercomplex number system can satisfy all the laws of ordinary arithmetic. Hankel also revised Peacock's principle of the permanence of formal laws and developed the theory of complex numbers as well as the higher algebraic systems of August ◊Möbius, Hermann ◊Grassmann, and William Rowan ◊Hamilton. Presenting algebra as a deductive science, Hankel was one of the original few to recognize the importance of Grassmann's work. His researches into the foundations of arithmetic promoted the development of the theory of quaternions.

In *Untersuchungen über die unendlich oft oscillerenden und unstetigen Functionen* – another of Hankel's major works, concerning in particular the theory of functions – he presented a method for constructing functions with singularities at every rational point. The method was based on a principle of his own devising, regarding the condensation of singularities. He also explicitly stated that functions do not possess general properties; this work was an important advance towards modern integration theory.

In another work on the same topic Hankel provided an example of a continuous function that was non-differentiable at an infinite number of points.

The Hankel functions provide a solution to the Bessel differential equation, which had originally occurred in connection with the theory of planetary motions. Today the equation holds more relevance to the study of wave propagation, to problems in optics, to electromagnetic theory, to elasticity and fluid motions, and to potential and diffusion problems.

Hankel was also the first to suggest a method for assessing the magnitude, or 'measure', of absolutely discontinuous point sets (such as the set of only irrational numbers lying between 0 and 1). Subsequently developed by Georg ◊Cantor, Emile ◊Borel, Henri ◊Lebesgue, and (finally) Andrei Kolmogorov, the 'measure' theory of point sets has now been extensively applied to probability, cybernetics and electronics.

Apart from his purely scientific work, Hankel was also a noteworthy historian of mathematics, concentrating on the mathematics of the classical and medieval periods. One of his assertions was that it was the Brahmans – members of the the highest, priestly caste – of Hindustan who were the real inventors of algebra, in that they were the first to recognize the existence of negative numbers.

Harden, Arthur (1865–1940) English biochemist who investigated the mechanism of sugar fermentation and the role of enzymes in this process. For this work he shared the 1929 Nobel Prize for Chemistry with the German biochemist Hans von Euler-Chelpin (1873–1964).

Harden was born in Manchester on 12 October 1865, the third child of a local businessman. With his eight sisters, he was brought up in an austere Nonconformist family environment. At the age of seven he attended a private school in Manchester and then from 1876 to 1881 he was at Tettenhall College, Staffordshire. From there he moved to Owens College (now the University of Manchester) to study chemistry, graduating in 1885. In the following year he won a research scholarship to Erlangen University and was awarded his PhD in 1888 for his work on purification and properties of ß-nitrosonaphthylamine.

On his return to Owens College Harden became a lecturer and demonstrator, teaching chemistry and writing textbooks. In 1897 he became head of the chemistry section of the British Institute of Preventative Medicine, later called the Jenner Institute. In 1905 the chemistry and biology departments merged under his direction. In 1912 he was appointed professor of biochemistry at the University of London. His main researches were interrupted by World War I, when he turned his attention to the production of vitamins B and C as a contribution to the war effort. Harden was knighted in 1936 and died in Bourne End, Buckinghamshire, on 17 June 1940.

Harden began work on sugar fermentation in 1898, soon after joining the Jenner Institute. He used the bacterium *Bacillus coli* (now called *Escherichia coli*), hoping to detect metabolic differences that would help to distinguish between the various strains. He started studying the metabolism of yeasts in 1900, three years after Eduard ⌽Buchner had produced evidence for the existence of enzymes with his experiments on cell-free alcoholic fermentation of sugar. Harden showed that the occurrence of fermentation in the cell-free yeast extract even in the absence of sugar was caused by the action of the enzyme zymase on glycogen from the disrupted cells. The extract's loss of enzymic activity on standing he demonstrated to be caused by a second enzyme, a proteolytic (protein-splitting) substance that breaks down zymase.

Harden and his co-workers went on to show that zymase consists of at least two different substances: one heat-sensitive (probably a protein, the enzyme) and one heat-stable, which he called a co-ferment (now known as a coenzyme). He separated the two using Thomas ⌽Graham's dialysis method, proving that the coenzyme is not colloidal in nature. Hans von Euler-Chelpin later showed that it is diphosphopyridine nucleotide (DPN).

So far Harden had demonstrated that the fermentation process requires three ingredients: enzyme, coenzyme, and substrate. But with only these three the reaction soon slowed down. It was accelerated, however, by the unlikely addition of phosphates. Harden discovered the hexose sugar compound hexose diphosphate in the normal reaction mixture (he later found hexose monophosphate as well), thus proving that phosphorylation is an intermediate step in the fermentation process. This important finding stimulated great interest in intermediate metabolism, and later researchers were able to show the widespread importance of phosphate groups in many metabolic reactions.

After World War I Harden continued to investigate the details of fermentation. His approach to work always remained accurate, objective, and dispassionate. In fact his unwillingness to speculate combined with a sometimes narrow approach to a problem may have hampered his researches. In later life he devoted much of his energy to launching and promoting the *Biochemical Journal* and was its editor for many years.

Hardy, Alistair Clavering (1896–1985) English marine biologist who designed the Hardy plankton continuous recorder. His development of methods for ascertaining the numbers and types of minute sea organisms helped to unravel the intricate web of life that exists in the sea.

Hardy was born in Nottingham on 10 February 1896 and was educated at Oundle School and then at Exeter College, Oxford. He served in World War I as a lieutenant and later as a captain 1915–19. The following year he studied at the Stazione Zoologica in Naples and returned to the UK in 1921 to become assistant naturalist in the fisheries department of the Ministry of Agriculture and Fisheries. In 1924 he joined the *Discovery* expedition to the Antarctic as chief zoologist and on his return in 1928 he was appointed professor of zoology and oceanography at Hull University, where he founded the department of oceanography. In 1942 he was made professor of natural history at the University of Aberdeen and, two years later, became professor of zoology at Oxford. He held this post for 15 years, 1946–61, when he served as professor of field studies at Oxford. In recognition of his achievements in marine biology, he was knighted in 1957. He returned to Aberdeen in 1963 to take up a lectureship at the university.

Before Hardy's investigations, the German zoologist Johannes Müller had towed a conical net of fine-meshed cloth behind a ship and collected enough specimens from the sea to reveal an entirely new sphere for biological research. The really serious study of the sea began in the late 19th century with the voyage of HMS *Challenger*, the purpose of which was to investigate all kinds of sea life, and it returned to the UK with an enormous wealth of material. A German plankton expedition was led by Victor Hensen in 1899, who coined the word plankton to describe the minute sea creatures that, it has since been established, form the first link in the vital sea food-chain.

Hardy made his special study of plankton on the 1924 *Discovery* expedition. The aim of quantitative plankton studies is to estimate the numbers or weights of organisms beneath a unit area of sea surface or in a unit volume of water. Müller's original conical net of fine mesh is still the basic requirement in any instrument designed to collect specimens from the sea, but the drawback is that it can be used only from stationary vessels for collections at various depths, or from moving vessels whose speed must not exceed two knots. Faster speeds displace the small organisms by the turbulence created. Hardy developed a net that can be used behind faster-moving vessels and which increases enormously the area in which accurate recordings can be made.

The first of these nets was the Hardy plankton recorder, which, reduced to its simplest terms, is a high-speed net. It consists of a metal tube with a constricted opening, a fixed diving plane instead of a weight, and a stabilizing fin. The net itself is a disc of 60-mesh silk attached to a ring placed inside near the tail. It was designed for easy handling aboard herring-fishing vessels to be towed when the skipper was near the grounds chosen for the night's fishing. If the disc showed plenty of herring food, the chances of a good catch were high. The indicator is no longer used for this purpose, having been superseded by more modern echo-sounding equipment, but it served as the basis for Hardy's second invention, which is still used and through which it has been possible to map the sea life in the oceans of the world.

This improved instrument is known as the Hardy continuous plankton recorder, and was first developed at the oceanographic department that Hardy himself founded at the University of Hull in 1931. The instrument can be used by unskilled personnel aboard ship after being suitably prepared by scientists ashore and is later returned to the laboratory where scientific observations can be made under more ideal conditions.

While the ship tows it along at normal cruising speed the instrument continuously samples the plankton. As the instrument is towed, the water flowing through it drives a small propeller that, acting through a gearbox, slowly winds a long length of plankton silk mesh and draws it across the path of the incoming water. As the mesh, which is graduated in numbered divisions, collects the plankton, it is slowly wound round a spool. This roll of silk mesh is then met by another roll and both are wound together in sandwich form, with the plankton collected by the first roll safely trapped between the two. The whole is stored in a small tank filled with a solution that preserves the plankton for later detailed study in the laboratory.

The recorder can be used at a depth of 10 m/33 ft and at speeds ranging from 8–16 knots/15–30 kph. Regular surveys, initiated at Hull and developed in Edinburgh, now annually cover many thousands of kilometres in the Atlantic, North Sea, and Icelandic waters. The knowledge of plankton distribution, which is continuously being updated, is of major importance to the fishing industry because there is a vitally close relationship between the occurrence of plankton and the movements of plankton-feeding fish used for human consumption.

Hardy suggested that if just 25% of the pests that exist in the sea can be eliminated and fish can be allowed to have some 20% of the potential food supply instead of the 2% they have now, then any given area could support ten times the amount of fish it supports at present. Hardy's methods have therefore not only added to our overall knowledge of sea life but have also indicated ways in which it could be put to better use.

Hardy, Godfrey Harold (1877–1947) English mathematician who carried out research at a very advanced level in the fields of pure mathematics known as analysis and number theory. However, although there are few – if any – well-known results that can be directly linked with his name, his influence on early 20th-century mathematics was enormous. His book *A Course in Pure Mathematics* revolutionized the teaching of mathematics at senior school and university levels, and another book, *An Introduction to the Theory of Numbers,* written in conjunction with E Wright, was once described as one of the most often quoted in mathematical literature.

Hardy was born on 7 February 1877 at Cranleigh, Surrey, where his father was an art teacher at Cranleigh public school. Hardy attended the school and showed no little talent in mathematics and languages. Later he obtained a scholarship to Winchester School on the basis of his mathematical ability. There, considered too able in the subject to be taught as part of the usual class, he was tutored on his own. Completing his education, Hardy then went to Trinity College, Cambridge, where in 1896 he was awarded the Smith's Prize for mathematics and elected fellow of the college. His first results were published in 1900 and by 1906, when he became a permanent lecturer in mathematics, he had shown himself to be an accomplished research mathematician. Two years later he published his mighty work *A Course in Pure Mathematics,* in which his approach was based on emulation of Camille ◊Jordan's spiritedly modern *Cours d'analyse.* Hardy's book has ever since been the mathematical equivalent of a bestseller. By 1910, when he was elected a fellow of the Royal Society, his reputation was established worldwide. He was elected to the Savilian Chair of Geometry at Oxford in 1919, but returned in 1931 to Cambridge as Sadlerian Professor of Pure Mathematics. His fame as a mathematician was universal and recognized by many honours and awards. He spent two separate years as visiting professor at Princeton University and at the California Institute of Technology. At the age of 65 he retired from his position at Cambridge, and died on 1 December 1947.

Hardy's researches included such topics as the evaluation of difficult integrals and the treatment of awkward series of algebraic terms. Among his successes in number theory was his new proof of the prime number theorem. According to the theorem – orginally proved by the French mathematician Jacques ◊Hadamard – $\pi(x)$, the number of prime numbers not exceeding x, is said to approach $x/\log_e x$ when x approaches infinity. His investigations in this area led to the discovery of further important results in the theory of numbers. Other problems on which he worked, often with John Littlewood (1885–1977), were the ways in which numbers could be partitioned into simpler numbers. (For example, $6 = 5 + 1$, $4 + 2$, and $3 + 3$, and can thus be partitioned into two numbers in three ways. It can also be partitioned into three numbers in three ways, and four numbers in two ways.) Akin to these problems were the ways in which numbers could be 'decomposed' into squares and cubes, and so on. Hardy also worked on the conjecture put forward by Goldbach as to whether every even number is the sum of two prime numbers. (This is still an unsolved problem.) A more geometrical problem, yet one deep in number theory, was to examine the number of points with whole-number coordinates (or points at the vertices of an equal square lattice) inside or on the circumference of a given circle.

Apart from the books already cited, Hardy was the sole or joint author of several others, including *Inequalities* and *Divergent Series,* the latter of which was published after his death.

The purest of pure mathematicians, Hardy shunned all connections of mathematics with physics, technology, engineering – and especially with warfare. Even those branches of the subject that had peripheral applications were avoided. The only 'tarnishes' on his work were a note on a problem in genetics, which described what has become known as Hardy's law, and an interest in relativity theory.

Many leading mathematicians were influenced by him and considered him a great teacher. Besides his prolific collaboration with John Littlewood, he also had a fascinating mathematical association with the largely untutored Indian mathematical genius Srinavasa ◊Ramanujan.

Hargreaves, James (1720–1778) English weaver who invented the spinning jenny, a machine that enabled several threads to be spun at once.

Virtually nothing is known of Hargreaves's early life. He was probably born in about 1720 in Blackburn, Lancashire, and from 1740 to 1750 seems to have been a carpenter and hand-loom weaver at nearby Standhill. In about 1760 his skill led Robert Peel (the grandfather of the politician) to employ him to devise an improved carding machine. He is supposed to have invented the spinning jenny in about 1764 when he observed that a spinning-wheel, accidentally overturned by his daughter, continued to revolve, together with the spindle, in a vertical position. Seeing the machine in this unfamiliar up-ended state apparently made him think that if a number of spindles were placed upright and side by side, several threads might be spun at the same time.

At first the resulting jenny was used only by Hargreaves and his children to make weft for the family loom. But in order to supply the needs of a large family he sold some of his new machines. Spinners with the old-fashioned wheel became alarmed by the possibility of cheaper competition and in the spring of 1768 a mob from Blackburn and its neighbourhood gutted Hargreaves's house and destroyed his jenny and his loom. Hargreaves moved to Nottingham, where he formed a partnership with a Mr James, who built a small cotton mill in which the jenny was used.

Throughout his life Hargreaves was handicapped by having had neither formal education nor what has often proved to be an adequate compensation for it, a sound business sense. He was, however, at least aware of his deficiencies in the latter and with the aid of his partner took out a patent for the spinning jenny in 1770. And when he found out that many Lancashire manufacturers were using the machine, he brought actions for infringement of patent rights. They offered him £3,000 for permission to use it, but he stood out for £4,000. The case continued until his lawyer learned that Hargreaves had sold a number of jennies in Blackburn, and the lawyer withdrew his services.

Hargreaves continued his business partnership until James died on 22 April 1778 in Nottingham. Six years later there were 20,000 hand-jennies in use in England compared with 550 of the mechanized (spinning) mules. Hargreaves is said to have left property worth £7,000 and his widow received £4,000 for her share in the business. But after her death some of the children were extremely poor and Joseph Brotherton sought to raise a fund for them and found great difficulty in getting from the wealthy Lancashire manufactures sufficient money to save them from destitution.

The spinning jenny came, like so many successful inventions, at a time when the need for it was at a maximum. The flying shuttle, invented by John ◊Kay and first used in cotton weaving in about 1760, had doubled the productive output of the weaver, whereas that of the worker at the spinning-wheel had remained much the same. The spinning jenny at once multiplied eightfold the output of the spinner and, because of its simplicity, could be worked easily by children. It did not, however, entirely supersede the spinning-wheel in cotton manufacturing (and was itself overtaken by Samuel ◊Crompton's mule). But for woollen textiles the jenny could be used to make both the warp and the weft.

There is as about as much uncertainty surrounding the origins of the spinning jenny and its name as there is about its inventor. Some maintain that it was Thomas Highs, and not Hargreaves, who invented the jenny and that Richard ◊Arkwright (another contender in this confused history) stole part of Highs's idea in order to claim the invention was his own. But there is little conclusive evidence that it was Highs or Arkwright, rather than Hargreaves, who invented the jenny. As to the name of the machine, it is said that Highs had a daughter named Jane whereas Hargreaves definitely did not, and that Jane Highs gave her pet name – Jenny – to the machine. Were this so, it is difficult to believe that Hargreaves would have accepted without comment or protest the widespread use of a name that virtually branded his claim to be 'father' of the jenny as nothing less than imposture.

Harrison, John (1693–1776) English horologist and instrumentmaker who made the first chronometers that were accurate enough to allow the precise determination of longitude at sea, and so permit reliable (and safe) navigation over long distances.

Harrison was born in Foulky, Yorkshire, in March 1693, the son of a carpenter and mechanic. He learned his father's trades and became increasingly interested in and adept at making mechanisms and instruments. In 1726 he made a compensated clock pendulum, which remained the same length at any temperature by making use of the different coefficients of expansion of two different metals.

In 1714 the British government's Board of Longitude announced that it would award a series of prizes up to the princely sum of £20,000 to anyone who could make an instrument to determine longitude at sea to an accuracy of 30 minutes (half a degree). By 1728 Harrison had constructed his first marine chronometer. Encouraged by the horologist George Graham (1673–1751) he made various improvements, and in 1735 he submitted one of his instruments for the award. Two other chronometers, smaller and lighter than the first, were finished in 1739 and 1757 and also submitted before the board had completed trials on the first. Then in about 1760 Harrison perfected his number 4 marine chronometer, which was not much larger than a watch of the time and was finally subjected to the ultimate test at sea. On two voyages to the West Indies it kept accurate time to within 5 seconds over the duration of the voyage, equivalent to just over one minute of longitude; after five months on a sea voyage, it was less than one minute (of time) out of true.

A unique feature that contributed to the chronometer's accuracy – and subsequently incorporated into other chronometers – was a device that enabled it to be rewound without temporarily stopping the mechanism. In 1763 the government awarded Harrison the sum of £5,000, but he did not receive the remainder of the original 'prize' of £20,000 until ten years later, by which time the Admiralty had built its own chronometer to Harrison's design. Three years later Harrison died in London on 24 March 1776.

Further Reading

Andrewes, William (ed and introduction); Cooke, Alistair (contributor), *The Quest for Longitude: Collection of Historical Scientific Instruments Publications*, Harvard University Press, 1996.

Hobden, Heather and Hobden, Mervyn, *John Harrison and the Problem of Longitude*, Cosmic Elk, 1998.

Sobel, Dava, *Longitude: The True Story of a Lone Genius Who Solved the Greatest Scientific Problem of His Time*, Penguin, 1995

Harvey, Ethel Browne (1885–1965) US embryologist and cell biologist, particularly renowned for her discoveries of the mechanisms of cell division, using sea-urchin eggs as her experimental model.

Ethel Browne was born on 14 December 1885 in Baltimore, the youngest of five children in a family that was strongly supportive of women's education. Her father was a distinguished gynaecologist and two of her sisters became physicians. Browne was educated at Bryn Mawr School and the Women's College of Baltimore, before attending Columbia University from which she graduated in 1907 and achieved a PhD in 1913 for her work on germ cells in an aquatic insect. This marked the beginning of a lifetime's fascination with cellular mechanisms of inheritance and development. During World War I she worked as a laboratory assistant in biology and histology and in 1916 married Edmund Harvey, a professor of biology at Princeton. She subsequently worked part-time, and visited several marine laboratories, until 1928 when she became an instructor at New York University for three years. Most of her subsequent career was as an independent research worker attached to the biology department of Princeton, largely self-financed. She continued to work in marine laboartories, and was a frequent visitor to the Stazione Zoologica in Naples. Her work in cytology, the study of cells, was internationally recognised and she was awarded, amongst other honours, fellowships of the American Association for the Advancement of Science and the New York Academy of Sciences. She died in Falmouth, Massachusetts on 2 September 1965.

At a time when most research on cell development was directed towards understanding the function of the nucleus, which contained the chromosomes, Harvey's work concentrated on cell fertilization and the development of non-nuclear cell components in the cytoplasm. She undertook morphological studies and physiological experiments to examine the factors that affect the process of cell division and was able to stimulate division in fragments of sea urchin eggs that contained no nucleus. This was publicized by the popular press as creation without parental influences, but was an important scientific contribution to unravelling the connections between different cellular structures in controlling cell division and development.

Harvey, William (1578–1657) English physician who discovered that blood is circulated around the body by pulsations of the heart, a landmark in medical investigations. His work did much to pave the way for modern physiology.

Harvey was born in Folkestone, Kent, on 1 April 1578. He went to the King's School, Canterbury, and then attended Gonville and Caius College at Cambridge in 1593. He graduated with a BA in 1597 and extended his studies under ◊Fabricius ab Aquapendente at the university medical school in Padua, Italy, gaining his medical degree in 1602. He returned to London, built up a successful practice, and in 1609 he was appointed physician to St Bartholomew's Hospital, London, and served as a professor there 1615–43. In 1618 he became Physician Extraordinary to James I, and then Royal Physician, a position he retained until the death of Charles I in 1649. He was elected president of the College of Physicians in 1654 but was too old to accept, and he died three years later in Roehampton on 3 June 1657.

Harvey was deeply involved in medical research and his spare time was devoted to his consuming interest, the investigation of the movement of blood in the body. He had developed this interest while studying in Padua under Fabricius, who had discovered the valves in the veins but had not appreciated their significance. The old idea about blood movement, established by ◊Galen, was that food turned to blood in the liver, ebbed and flowed in vessels and, on reaching the heart, flowed through pores in the dividing wall (septum) from the right to the left side and was sent on its way by heart spasms. Andreas ◊Vesalius, who secretly dissected corpses, failed to find the pores in the heart's dividing wall, and concluded that Galen could never have dissected a human body.

Harvey was not at all convinced by Galen's explanation either. Examining the heart and blood vessels of about 128 mammals, Harvey found that the valve separating the auricle from the ventricle, on each side of the heart, is a one-way structure, as are the valves in the veins discovered by his tutor, Fabricius. For this reason he decided that the blood in the veins must flow only towards the heart. Harvey tied off an artery and found that it bulged with blood on the heart side; he then tied a vein and discovered that it swelled on the side away from the heart. He also calculated the amount of blood that left the heart at each beat. He worked out that in human beings it was about 60 cu cm/3.7 cu in per beat, which meant that the heart pumped out 260 litres/57 gal (68 US gal) of blood an hour. This amount would weigh more than 200 kg/440 lb – more than three times the weight of an average man. Clearly that was absurd, and therefore a much smaller quantity of the same blood must be circulating continuously around the body. Harvey demonstrated that no blood seeps through the septum and reasoned that it passes from the right side of the heart to the left through the lungs (pulmonary circulation).

The publication of these findings aroused the hostility Harvey had predicted, because to refute Galen was almost unthinkable in his time. His practice declined but he continued with his studies and, unlike many early scientists who made an outstanding discovery, he lived to see his work accepted.

The great classic Harvey published in 1628, *De motu cordis et sanguinis in animalibus/On the Motion of the Heart and Blood in Animals* (628), pointed the way for physicians who followed him. He also published *Exercitationes de generatione animalium/ Anatomical Exercitations Concerning the Generation of Living Creatures* (1651, translated 1653).

Harvey was one of the first to study the development of a chick in the egg. He also carried out many dissections to find out how mammalian embryos are formed, and many of the animals he dissected were the royal deer put at his disposal by Charles I. Harvey suspected that semen might be involved in the making of an embryo, but did not have the microscopic apparatus, later developed by Anton van ◊Leeuwenhoek, needed to study the tiny spermatozoa.

Harvey's discovery of the circulation of the blood marked the beginning of the end of medicine as taught by Galen, which had been accepted for 1,400 years. From then on, experimental physiology was to sweep away many erroneous ideas and replace them with personal observations made by experiment and careful measurement.

Further Reading

Frank, Robert Gregg, *Harvey and the Oxford Physiologists: Scientific Ideas and Social Interaction*, University of California Press, 1980.

French, Roger Kenneth, *William Harvey's Natural Philosophy*, Cambridge University Press, 1994.

Hamburger, Jean and Wright, Barbara, *The Diary of William Harvey: The Imaginary Journal of the Physician Who Revolutionized Medicine*, Rutgers University Press, 1992.

Hamburger, Jean; Wright, Barbara (transl), *Diary of William Harvey*, Rutgers University Press, 1992.

Harvey, William; Wear, Andrew (ed), *The Circulation of Blood and Other Writings*, Everyman, 1993.

Keynes, Geoffrey L; revised by Whitteridge, Gweneth and English, Christine, *A Bibliography of the Writings of Dr William Harvey, 1578–1657*, Norman Publishing, 1989.

McMullen, Emerson Thomas, *William Harvey and the Use of Purpose in the Scientific Revolution: Cosmos by Chance or Universe by Design?*, University Press of America, 1997.

Hausdorff, Felix (1868–1942) German mathematician and philosopher who is chiefly remembered for his development of the branch of mathematics known as topology, in which he formulated the theory of point sets. The author of an influential book on the subject, he wrote philosophical works as well, and contributed extensively to several other fields of mathematics.

Hausdorff was born in Breslau, Germany (now Wrocław, Poland), on 8 November 1868. He completed his secondary education in Leipzig and, after studying mathematics and astronomy in Berlin and Freiburg, returned to Leipzig to graduate there in 1891. For the next seven years he wrote and published a number of papers dealing with optics, mathematics, and astronomy; he also produced works on philosophical and literary themes. In 1902 he was appointed professor at Leipzig University, after which the subject of mathematics – and particularly set theory – dominated his output. Then appointed associate professor at Bonn University, he published his major work, *Grundzuge der Mengenlehre/Basic Features of Set Theory.* In 1913 Hausdorff took up the position of professor at Griefswald, but eight years later returned to Bonn for the remainder of his academic life; he retired at the statutory age of 67, in 1935. Even after his official retirement he continued to work on the theory of sets and in his studies of topology; his writing, however, was not published in Germany. Because of his Jewish faith, Hausdorff was listed in 1942 for internment. Rather than let that happen he committed suicide – with his wife and her sister – on 26 January of that year.

Topology is the study of figures and shapes that retain their essential proportions despite being 'squeezed' or 'stretched'. In this way, a square may be considered (topologically) equivalent to any closed plane figure and a cube may even be regarded as (topologically) equivalent to a sphere. In his *Grundzuge der Mengenlehre* Hausdorff formulated a theory of topological and metric spaces into which concepts previously advanced by other mathematicians fitted well. He proposed that such spaces be regarded as sets of points and sets of relations among the points, and introduced the principle of duality. This principle states that an equation between sets remains valid if the sets are replaced by their complements and the symbol for union is exchanged for that of intersection; and that an inequality (inequation) remains valid if each side is replaced by its complement and the symbol for union is exchanged for that of intersection, provided also the inclusion symbol is reversed.

Developing his point set theory, Hausdorff also created what are now called Hausdorff's neighbourhood axioms; there are four axioms:

1) to each point x there corresponds at least one neighbourhood U_x, each neighbourhood U_x containing the point x;

2) if U_x and V_x are two neighbourhoods of the same point x, their intersection contains a neighbourhood of x;

3) if U_x contains a point y, there is a neighbourhood U_y such that $U_y \subset U_x$;

4) if x is not equal to y, there are two neighbourhoods U_x and U_y such that the intersection of U_x and $U_y = 0$ (that is, such that U_x and U_y have no points in common).

From these axioms Hausdorff derived what are now called Hausdorff's topological spaces. A topological space is understood to be a set E of elements x and certain subsets S_x of E, which are known as neighbourhoods of x. These neighbourhoods satisfy the Hausdorff axioms. Later in the book – after dealing with the properties of general space – Hausdorff introduced further axioms involving metric and particular Euclidean spaces.

Hausdorff contributed extensively to several fields of mathematics including mathematical analysis, in which he proved some very important theorems on summation methods and properties of moments. He also investigated the symbolic exponential formula that he himself had derived. His main contribution to mathematics, however, was in the field of point set theory. Hausdorff's work in this area led to many results of primary importance, such as the investigation of general closure spaces, and what is now termed Hausdorff's maximal principle in general set theory.

Haüy, Rene-Just (1743–1822) French mineralogist, the founder of modern crystallography.

He was born on 28 February 1743 in St Jest-en-Chaussee, Oise, the son of an impoverished weaver, Haüy entered the priesthood before developing a passionate love of mineralogy and crystallography. He became professor of mineralogy in Paris in 1802. His two major works – the *Treatise of Mineralogy* (1801) and the *Treatise of Crystallography* (1822) – are widely acknowledged as foundational texts of modern crystallography. Haüy remained a priest, receiving protection from anticlerical attacks during the Revolutionary era thanks to Napoleon's patronage. He died in Paris on 3 June 1822.

The 17th and 18th centuries had generated many explanations as to why crystals had regular forms. Various sorts of shaping forces, sometimes chemical, sometimes physical, had been touted (though this might be seen as explaining the unknown in terms of the unknown). In 1670 Nicolaus ◊Steno had demonstrated the constancy of the angle between corresponding faces in the crystals of one substance, irrespective of crystal size or nature. In his vast crystal studies, Haüy was to build on the theories of Steno and his followers. His approach was principally morphological; he did not seek to explain the causes of the regularly varied forms of crystals, but rather sought to develop a geometrical classification of those forms – ideally, through the geometry of simple relationships between integers. Thus he demonstrated in 1784 that the faces of a calcite crystal could be regarded in terms of the standard stacking of cleavage rhombs, provided that the rhombs were assumed to be so small that the face looked smooth. Similar principles would lead to other crystal shapes being built up from suitable simple structural units. In broad terms, this remains the foundation of the modern view.

Haüy thus interpreted crystals as structured assemblages of secondary bodies (integrant molecules), which grouped themselves according to regular geometric laws. He proposed six primary forms: parallelepiped, rhombic dodecahedron, hexagonal dipyramid, right hexagonal prism, octahedron, and tetrahedron. Most of his research was devoted to establishing and explicating this typology of forms.

Hawking, Stephen William (1942–) English theoretical physicist and mathematician whose main field of research has been the nature of space–time and those anomalies in space–time known as singularities. His book *A Brief History of Time* (1988) gives a popular account of cosmology, and became an international bestseller.

Hawking was born in Oxford on 8 January 1942, and showed exceptional talent in mathematics and physics from an early age. At Oxford University he became especially interested in thermodynamics, relativity theory, and quantum mechanics – an interest that was encouraged by his attending a summer course at the Royal Observatory in 1961. When he completed his undergraduate course in 1962 (receiving a first-class honours degree in physics), he enrolled as a research student in general relativity at the Department of Applied Mathematics and Theoretical Physics at the University of Cambridge.

During his postgraduate programme Hawking was diagnosed as having ALS (amyotrophic lateral sclerosis), a rare and progressive neuromotor disease that handicaps motor and vocal functions. He was nevertheless able to continue his studies and to embark upon a distinguished and productive scientific career. He was elected fellow of the Royal Society in 1974, and became Lucasian Professor of Mathematics at Cambridge University in 1980. He was made a Companion of Honour in 1989.

From its earliest stages, Hawking's research has been concerned with the concept of singularities – breakdowns in the space–time continuum where the classic laws of physics no longer apply. The prime example of a singularity is a black hole, the final form of a collapsed star. During the later 1960s, Hawking – relying on a few assumptions about the properties of matter, and incidentally developing a mathematical theory of causality in curved space–time – proved that if Albert ◊Einstein's general theory of relativity is correct, then a singularity must also have occurred at the Big Bang, the beginning of the universe and the birth of space–time itself.

In 1970 Hawking's research turned to the examination of the properties of black holes. A black hole is a chasm in the fabric of space–time, and its boundary is called the event horizon. Hawking realized that the surface area of the event horizon around a black hole could only increase or remain constant with time – it could never decrease. This meant, for example, that when two black holes merged, the surface area of the new black hole would be larger than the sum of the surface areas of the two original black holes. He also noticed that there were certain parallels between the laws of thermodynamics and the properties of black holes. For instance, the second law of thermodynamics states that entropy must increase with time; the surface area of the event horizon can thus be seen as the entropy ('randomness') of the black hole.

Over the next four years, Hawking – with Carter, Israel and Robinson – provided mathematical proof for the hypothesis formulated by John Wheeler (1911–), known as the 'no hair theorem'. This stated that the only properties of matter that were conserved once it entered a black hole were its mass, its angular momentum, and its electric charge; it thus lost its shape, its 'experience', its baryon number, and its existence as matter or antimatter.

If we find why it is that we and the universe exist, it would be the ultimate triumph of human reason – for then we would know the mind of God.

STEPHEN HAWKING

Since 1974, Hawking has studied the behaviour of matter in the immediate vicinity of a black hole, from a theoretical basis in quantum mechanics. He found, to his initial surprise, that black holes – from which nothing was supposed to be able to escape – could emit thermal radiation. Several explanations for this phenomenon were proposed, including one involving the creation of 'virtual particles'. A virtual particle differs from a 'real' particle in that it cannot be seen by means of a particle detector, but can be observed through its indirect effects. 'Empty' space is thus full of virtual particles being fleetingly 'created' out of 'nothing', forming a particle and antiparticle pair that immediately destroy each other. (This is a violation of the principle of conservation of mass and energy, but is permitted – and predicted – by the 'uncertainty principle' of

Werner ◊Heisenberg.) Hawking proposed that when a particle pair is created near a black hole, one half of the pair might disappear into the black hole, leaving the other half, which might radiate away from the black hole (rather than be drawn into it). This would be seen by a distant observer as thermal radiation.

Hawking's present objective is to produce an overall synthesis of quantum mechanics and relativity theory, to yield a full quantum theory of gravity. Such a unified physical theory would incorporate all four basic types of interaction: strong nuclear, weak nuclear, electromagnetic, and gravitational.

Hawking's books include *The Large Scale Structure of Space–Time* (1973), with GFR Ellis; *Superspace and Supergravity* (1981); *The Very Early Universe* (1983); *General Relativity: An Einstein Centenary Survey,* with Werner Israel; and *300 Years of Gravity,* again with Israel. His popular works include *A Brief History of Time* (1988), which stayed in the *Sunday Times* UK bestsellers list for a record-breaking 237 weeks and is considered a landmark in popular scientific writing; *Black Holes and Baby Universes and Other Essays* (1993); and *The Nature of Space and Time* (1996), with Roger Penrose.

> *God not only plays dice, he throws them where they can't be seen.*
>
> ***Stephen Hawking*** A Brief History of Time *1988*

The properties of space–time, the beginning of the universe, and a unified theory of physics are all fundamental research areas of science. Hawking has made, and continues to make, major contributions to the modern understanding of them all.

Further Reading

Ferguson, Kitty, Hawking, Stephen W, *Stephen Hawking: Quest for a Theory of Everything: The Story of His Life and Work,* Bantam Press, 1992.

Filkin, David, *Stephen Hawking's Universe: The Cosmos Explained,* Basic Books, 1998.

Hawking, Stephen W, *A Brief History of Time,* Bantam Press, 1998.

Hawking, Stephen W, *Black Holes and Baby Universes, and Other Essays,* Bantam Press, 1997.

Hawking, Stephen W and Stone, Gene, *Stephen Hawking's A Brief History of Time: A Reader's Companion,* Bantam Press, 1992.

Hawking, Stephen W and Penrose, Roger, *The Nature of Space and Time,* Isaac Newton Institute Series of Lectures, Princeton University Press, 1996.

Strathern, Paul, *Hawking and Black Holes,* Big Idea series, Arrow, 1997.

White, Michael and Gribbin, John R, *Stephen Hawking: A Life in Science,* Penguin, 1998.

Wilkinson, David A, *God, the Big Bang and Stephen Hawking,* Monarch, 1993.

Haworth, (Walter) Norman (1883–1950) English organic chemist whose researches concentrated on carbohydrates, particularly sugars. He made significant advances in determining the structures of many of these compounds, for which he shared the 1937 Nobel Prize for Chemistry with the German organic chemist Paul ◊Karrer, becoming the first British organic chemist to be so honoured.

Haworth was born in Chorley, Lancashire, on 19 March 1883, the son of a factory manager. He left school at the age of 14 and joined his father's linoleum works, where he became interested in the dyestuffs used in floor coverings. In spite of parental disapproval, he continued his education with a private tutor and in 1903 realized his ambition to enter the chemistry department of the University of Manchester, where he studied under William Perkin, Jr (1860–1929). He gained his degree in 1906 and won a scholarship to Göttingen, where he did research with Otto Wallach (1847–1931). He was awarded his doctorate after only one year and returned to Manchester, gaining a DSc there in 1911 in the remarkably short time of only five years since his first degree.

Over the next few years Haworth held various university appointments – at Imperial College, London 1911–12, St Andrews 1912–20, and Durham 1920–25 – before accepting a professorship at Birmingham University, where he remained until he retired in 1948, a year after he received a knighthood. He died in Birmingham on his birthday in 1950.

Haworth's lifelong interest in carbohydrates began when he was at St Andrews. There he teamed up with T Purdie and J C Irvine, trying to characterize sugars by the methylation of their hydroxy groups. The work was interrupted by World War I, when the laboratory was turned over to the production of fine chemicals and drugs, which Haworth supervised. After the war he continued research, concentrating his team's efforts on the disaccharides. Again he used the technique of methylation followed by acid hydrolysis. Lactose, for example, forms an octamethyl derivative that on hydrolysis yields tetramethyl galactose and 2,3,6-trimethyl glucose, indicating the structural formula for lactose.

Later investigations suggested that the carbon atoms in hexoses usually have a ring structure and a great deal of Haworth's work at Birmingham was devoted to establishing the linkages in these rings (known as Haworth formulae). In 1926 he proposed that methyl glucose normally exists as what is now called a pyranose ring. The complex analytical procedures involved oxidation and methylation to give products that were further reacted to form simpler known substances that could be crystallized and identified. He also established that methyl glucose can have an alternative furanose structure. His book *The Constitution of the Sugars* (1929) became a standard work.

Using their knowledge of simple sugars, Haworth's research team was able to investigate the chain structures of polysaccharides, establishing the structures of cellulose, starch, and glycogen. Work on sugars led naturally to studies of vitamin C and by 1932 the whole research effort was directed to the determination of the structure and synthesis of this substance, which Haworth named ascorbic acid. The research resulted in an industrial process for synthesizing vitamin C for medical uses.

In the late 1930s Haworth studied the reactions of polysaccharides with enzymes and certain aspects of the chemistry of the hormone insulin. During World War II Haworth's laboratory became a primary producer of purified uranium for the war effort, which led to work on the preparation and properties of organic fluorine compounds. Many of the workers went to the newly opened experimental stations at Oak Ridge in the USA and Chalk River in Canada.

Hayashi, Chushiro (1920–) Japanese physicist whose research in 1950 exposed a fallacy in the 'hot Big Bang' theory proposed two years earlier by Ralph ◊Alpher, Hans ◊Bethe, and George ◊Gamow. Since that time, Hayashi has published many papers on the origin of the chemical elements in stellar evolution and on the composition of primordial matter in an expanding universe.

Hayashi was born in Kyoto on 25 July 1920. Completing his education at the University of Kyoto, from which he received a BSc in 1942, he became a research associate at the University of Tokyo for four years, before returning to Kyoto as research associate from 1946 to 1949. He then spent five years as assistant professor at Naniwa University, Osaka, until he once again returned to Kyoto University as a member of the physics faculty, becoming professor of physics in 1957. In 1995 he was awarded the Kyoto Prize for Basic Sciences.

It was George Gamow who, in 1946, proposed that the early dense stages of the universe were hot enough to enable thermonuclear reactions to occur. The first detailed calculation of the formation of helium in this way, in the 'hot Big Bang', was published two years later. The three collaborators hoped that their work might account not only for the observed abundance of helium in the universe but also for the distribution of other, heavier, elements. This theory assumed that matter was originally composed of neutrons, which, at very high temperatures and if the matter was of great enough density, combine with protons.

But Hayashi pointed out that at times in the Big Bang earlier than the first two seconds, the temperature would have been greater than 10^{10}K, which is above the threshold for the making of electron–positron pairs. The creation of one pair requires about 1 MeV of energy, and at 10^{10}K many protons have this much energy. The effect of this is radically to change the timescale of the alpha–beta–gamma theory. Neutrons can react with the thermally excited positrons through the so-called weak interactions to produce protons and antineutrinos, and the reaction time for this to happen (at temperatures above 10^{10}K) is less than about one second. Hayashi proposed that at such a temperature there was complete thermal equilibrium between all forms of matter and radiation. Below that temperature, the weak interactions cannot maintain the neutrons in statistical balance with the protons because the concentration of electron pairs is falling abruptly. This means that the ratio of neutrons to protons is 'frozen in' until a few hundred seconds have gone by and neutron decay becomes appreciable. The frozen-in ratio at a temperature below 10^{10}K is about 15% – and such a change in the ratio affects the helium abundances. The frozen-in abundance of neutrons does not depend on the material density but on the temperature and the properties of the weak interactions. So, provided the density is great enough for the combining reaction between neutrons and protons to occur faster than the expansion rate, a fixed concentration of neutrons will be incorporated into helium nuclei, however great the material density is, producing a 'plateau' in the relationship between helium abundance and material density.

Hayashi derived a percentage value for the abundance plateau, but more recent research by physicists such as Fred ◊Hoyle, William ◊Fowler, and R Wagoner suggests that his value is too high.

Heath, Thomas Little (1861–1940) English civil servant and historian of the ancient Greek mathematicians. For his considerable services to the Treasury Office he was knighted; nevertheless he received no little acclaim for his historical works – he became a fellow of the Royal Society in 1912, and of the Royal Academy in 1932 – and for a year he was president of the Mathematical Association.

Heath was born in 1861 in Lincolnshire. As a boy he attended the Caistor Grammar School and later studied at Clifton and Trinity colleges, Cambridge, where he obtained a first-class degree in classics – the first part in 1881 and the second in 1883. It was at this time that he won first place in an open competition for entry to the civil service, and was appointed a clerk in the Treasury in 1884. In 1885 he was made fellow of Trinity College, although it was not until 11 years later that he obtained his doctorate. In 1887 he became private secretary to the permanent secretary at the Treasury, and from 1891–94 held a similar post in relation to successive financial secretaries. He became principal clerk of the Treasury 1901–07, and in that year was himself appointed assistant secretary. In 1908 he published his first full-scale book, *Euclid's Elements*, a monumental work in three volumes. Five years later he was appointed joint permanent secretary to the Treasury and auditor of the civil list, a post he held until 1919, when he was made comptroller general and secretary to the Commissioners for the Reduction of the National Debt. He remained as such till 1926, when he retired from the civil service. Meanwhile, in 1921 he produced his two-volume *History of Greek Mathematics*, which has come to be regarded as the standard work on the subject in the English language. After leaving the civil service, Heath was appointed one of the University of Cambridge Commissioners, and was also a member of the Royal Commission on National Museums and Galleries 1927–29. He died on 16 March 1940 in Ashtead, Surrey.

No doubt Heath's interest in Greek mathematics, which made him one of the leading authorities on the subject, was the result of his university training in mathematics and the classics. He took a special interest in ◊Diophantus, whose *Arithmetica* he edited for an English-language edition. The work, *Diophantus of Alexandria: A Study in the History of Greek Algebra*, was published in 1885, and was more in the style of an essay.

Apollonius of Perga: A Treatise on Conic Sections was published in 1896. It is particularly interesting that in this work Heath used modern mathematical notation. It was followed, a year later, by an edition of the works of ◊Archimedes, in which Heath used the same treatment. (This was translated into German in 1914.) And Heath's version of Archimedes' *Method* appeared in 1912. In 1913, Heath produced *Aristarchus of Samos, the Ancient Copernicus*, a comprehensive account of ◊Aristarchos and his work on astronomy.

In the great *History of Greek Mathematics*, Heath dealt with his subjects mainly according to topics and not in chronological order as others had done before him. In the preface to the book, Heath justified such an approach by reference to the famous problem of doubling the cube, saying, 'If all recorded solutions are collected together, it is much easier to see the relations ... between them and to get a comprehensive view of the history of the problem.' Heath rewrote the *History* in 1931 and published it as a *Manual of Greek Mathematics*; in fact, however, it is really a mere condensed version of the original text.

Heathcoat, John (1783–1861) English inventor of lace-making machinery. His contribution was acknowledged by Marc Isambard ◊Brunel who said that Heathcoat (then aged 24) had devised 'the most complicated machine ever invented'.

Heathcoat was born in Duffield, near Derby, on 7 August 1783. He received an average education and, after completing his apprenticeship as a journeyman in the hosiery trade, he became a master mechanic at Hathorn in about 1803. He then set himself the task of constructing a machine that would do the work of the pillow, the multitude of pins, thread and bobbins, and the fingers of the hand lace-maker and supersede them in the production of lace – just as the stocking loom had replaced knitting needles in stocking-making.

Analysing the component threads of pillow lace, he classified them into longitudinal and diagonal. The former he placed on a beam as a warp. The remainder he reserved as weft with each thread worked separately, twisted around the warp thread to close the upper and lower sides of the mesh. He then devised the necessary mechanical features: bobbins to distribute the thread, the carriage and groove in which they must run, and their mode of twisting round the warp and travelling from side to side of the machine. The first square yard (0.8 sq m) of plain net from the machine was sold for £5. By the end of the 19th century its price had fallen to one shilling.

In 1805 Heathcoat settled in Loughborough – as a consequence his improved machine became known as the 'Old Loughborough'. Four years later he went into partnership with the aptly named Charles Lacy, a former point-net maker at Nottingham. Following this amalgamation the machine's capacity was so increased that by 1816 there were as many as 35 frames at work in the Loughborough factory. They also made a great deal of money from royalties paid by other companies for permission to use the machines.

On the night of 28 June 1816 an angry crowd of Luddites, fearful that the new machines would deprive them of their jobs, attacked Heathcoat's Loughborough factory and destroyed 35 frames, burning the lace that was on them. The company sued the county for damages and received £10,000 in compensation, on condition that the money was spent locally. Heathcoat refused to accept the condition; he had already received threats to his life and wanted to leave the district for good. He dissolved his partnership with Lacy and left for Tiverton in Devon, forfeiting his right to compensation.

At Tiverton events took a more favourable turn. With a former partner, John Boden, he bought a large water-powered mill on the River Exe. Heathcoat devised new frames that were wider and faster, and by using rotary power he lowered the cost of production.

He patented a rotary, self-narrowing stocking frame and put gimp and other ornamental threads into the bobbin by mechanical adjustment.

In 1821 the partnership with Boden was ended. Year by year Heathcoat took out patents for further inventions, continuing to make improvements in the textile trade until he retired in 1843. Also in 1832, with Henry Handley, he patented a steam plough to assist with agricultural improvements in Ireland. On 12 December of that year he was elected to represent Tiverton in the newly reformed Parliament and remained MP for the borough until 1859. He died in Tiverton on 18 January 1861.

Heaviside, Oliver (1850–1925) English physicist and electrical engineer who predicted the existence of the ionosphere, which was known for a time as the Kennelly–Heaviside Layer. Heaviside made significant discoveries concerning the passage of electrical waves through the atmosphere and along cables, and added the concepts of inductance, capacitance, and impedance to electrical science.

Heaviside was born in Camden Town, London, on 18 May 1850. His uncle was Charles ◊Wheatstone, who was a pioneer of the telegraph and may have stimulated in him an interest in electricity. Heaviside received only an elementary education, and was mainly self-taught. He took up employment with the Great Northern Telegraph Company at Newcastle upon Tyne in 1870, when he was 20. This employment lasted only four years, when he was forced to retire because of his deafness. Shortly after he is believed to have visited Denmark to study telegraph apparatus. On his return he was supported first by his parents and later in life he lived with his brother. Heaviside never obtained an academic position but received several honours, including fellowship of the Royal Society in 1891 and the first award of the Faraday Medal by the Institution of Electrical Engineers shortly before he died in Paignton, Devon, on 3 February 1925.

During his early years, Heaviside was absorbed by mathematics. He was familiar with the attempt by Peter Tait (1831–1901) to popularize the quaternions of William Rowan ◊Hamilton, but he rejected the scalar part of the quaternions in his notion of a vector. Heaviside's vectors were of the form:

$$\mathbf{v} = a\mathbf{i} + b\mathbf{j} + c\mathbf{k}$$

where $\mathbf{i}$, $\mathbf{j}$, and $\mathbf{k}$ are unit vectors along the Cartesian x-, y-, and z-axes respectively. When Heaviside became involved with work relating to James Clerk ◊Maxwell's famous theory of electricity, he wrote Maxwell's equations in vector form incorporating some discoveries of his own. Heaviside also used operator techniques in his expression of calculus and made much use of divergent series (those with terms whose sum does not approach a fixed amount). He made great use of these in his electrical calculations when many mathematicians were afraid to venture away from convergent series.

In his twenties and thirties, Heaviside became very interested in electrical science, first carrying out experiments and then developing advanced theoretical discussions. These were, in fact, so advanced that they were eventually dismissed by some mathematicians and scientists, and were classed as too abstract to be published in the electrical journals. He expressed his ideas in the three-volume *Electromagnetic Theory* (1893–1912). Much of this work extended Maxwell's discoveries but he made many valuable discoveries of his own. Heaviside visualized electrical ideas often in mechanical terms and thought of the inertia of machines as being similar to electrical inductance.

When Heaviside became involved with the passage of electricity along conductors, he modified Ohm's law (see Georg ◊Ohm) to include inductance and this, together with other electrical properties, resulted in his derivation of the equation of telegraphy. On considering the problem of signal distortion in a telegraph cable, he came to the conclusion that this could be substantially reduced by the addition of small inductance coils throughout its length, and this method has since been used to great effect.

In the third volume of his *Electromagnetic Theory*, Heaviside considered wireless telegraphy and, in drawing attention to the enormous power required to send useful signals long distances, he suggested that they may be guided by hugging the land and sea. He also suggested that part of the atmosphere may act as a good reflector of electrical waves. Arthur Kennelly (1861–1939) in the USA also made a similar suggestion and this layer, which was for some time known as the Kennelly–Heaviside layer, was subsequently shown to exist by Edward ◊Appleton. This part of the atmosphere, about 100 km/62 mi above the ground, is now known as the ionosphere.

Although from time to time Heaviside had his detractors, perhaps because he was unorthodox in his approach, he was later known and highly valued by the leading physicists and electrical engineers of his day.

Further Reading

Nahin, Paul J, *Oliver Heaviside: Sage in Solitude*, Institute of Electrical and Electronics Engineers, 1998.

Searle, G F C; Catt, Ivor (ed), *Oliver Heaviside, the Man*, CAM Publishing, 1987.

Heine, Heinrich Eduard (1812–1881) German mathematician and a prolific author of mathematical papers on advanced topics. He completed the formulation of the notion of uniform continuity and subsequently provided a proof of the classic theorem on uniform continuity of continuous functions. This theorem has since become known as Heine's theorem.

Heine was born on 16 March 1812 in Berlin, where his father was a banker. His first education was by tutors at home, and he then attended local high schools. Eventually he went to Göttingen University and studied under Karl ⟡Gauss. To complete his education he returned to Berlin and became a pupil of Lejeune ⟡Dirichlet, Jakob ⟡Steiner, and the more astronomically minded Johann ⟡Encke, receiving his PhD in 1842. He then spent a further couple of years at Königsberg, learning from Karl ⟡Jacobi and Franz Neumann (1798–1895). Finally, at the age of 32, Heine became an unpaid lecturer at Bonn University. Four years later, in 1848, he was appointed professor there, but in the same year he took up a similar post at Halle University. And it was at Halle he remained for the rest of his life; he died there on 21 October 1881.

During his lifetime he received few honours or awards. On the other hand, when offered the chair in mathematics at Göttingen University – perhaps the most prestigious appointment possible at the time – in 1875, he turned it down.

In all, Heine published more than 50 papers on mathematics. His specialty was spherical functions, Lamé functions, and Bessel functions. He published his most important work – *Handbuch der Kugelfunctionen* – in 1861; it was reissued in a second edition in 1881 and became the standard work on spherical functions for the next 50 years.

Although the name of his theory is sometimes linked with the name of Emile ⟡Borel (as the Heine–Borel theorem), it can be confidently argued that Borel's contribution was of lesser value. Borel formulated the covering property of uniform conformity, and proved it; Heine formulated the notion of uniform continuity – a notion that had escaped the attention of Augustin ⟡Cauchy, regarded as the innovator of continuity theory – and went on to prove the classic theorem of uniform continuity of continuous functions.

Heisenberg, Werner Karl (1901–1976) German physicist who founded quantum mechanics and the uncertainty principle. In recognition of these achievements, he was awarded the 1932 Nobel Prize for Physics.

Heisenberg was born on 5 December 1901 in Duisberg. He showed an early aptitude for mathematics and when, after leaving school, he came across a book on relativity by Hermann ⟡Weyl entitled *Space, Time, And Matter* (1918), he became interested in the mathematical arguments underlying physical concepts. This led Heisenberg to study theoretical physics under Arnold ⟡Sommerfeld at Munich University.

Heisenberg German physicist Werner Heisenberg, who developed the quantum theory and the principle of uncertainty. During World War II, Heisenberg worked for the Nazis on nuclear fission, but his team was many months behind the Allied atom-bomb project. After the war he worked on superconductivity. *Mary Evans Picture Library*

There he formed a lasting friendship with Wolfgang ⟡Pauli. He obtained his doctor's degree in 1923 and immediately became assistant to Max ⟡Born at Göttingen. From 1924 to 1926, Heisenberg worked with Niels ⟡Bohr in Copenhagen and then in 1927 he was offered the chair of theoretical physics at Leipzig at the age of only 26.

Heisenberg stayed at Leipzig until 1941 and then from 1942 to 1945 was director of the Max Planck Institute for Physics in Berlin. During World War II he was in charge of atomic research in Germany and it is possible that he may have been able to direct efforts away from military uses of nuclear power. In 1946 Heisenberg was made director of the Max Planck Institute for Physics in Göttingen, which in 1958 was moved to his home city of Munich. He held the post of director until 1970, and died in Munich on 1 February 1976.

In the early 1920s, there were burning questions relating to the quantum theory. Bohr had used the quantum concept of Max ⟡Planck to explain the spectral lines of atoms in terms of the energy differences of electron orbits at various quantum states, but there were weaknesses in the results. In 1922 Pauli began to throw doubt on them and Sommerfeld, although an optimist, could see some flaws in Bohr's work.

In 1923 Heisenberg went to Göttingen to work with Born and used a modification of the quantum rules to explain the anomalous Zeeman effect (see Pieter ◊Zeeman), in which single spectral lines split into groups of closely spaced lines in a strong magnetic field. Information about atomic structure can be deduced from the separation of these lines.

Heisenberg was concerned not to try to picture what happens inside the atom but to find a mathematical system that explained the properties of the atom – in this case, the position of the spectral lines given by hydrogen, the simplest atom. Born helped Heisenberg to develop his ideas, which he presented in 1925 as a system called matrix mechanics. By mathematical treatment of values within matrices or arrays, the frequencies of the lines in the hydrogen spectrum were obtained. This was the first precise mathematical description of the workings of the atom and with it Heisenberg is regarded as founding quantum mechanics, which seeks to explain atomic structure in mathematical terms.

The following year, however, Erwin ◊Schrödinger produced a system of wave mechanics that accounted mathematically for the discovery made in 1923 by Louis ◊de Broglie that electrons do not occupy orbits but exist as standing waves around the nucleus. Wave mechanics was a much more convenient system than the matrix mechanics of Heisenberg and Born and rapidly replaced it, though John ◊Von Neumann showed the two systems to be equivalent in 1944.

An expert is someone who knows some of the worst mistakes that can be made in his subject and how to avoid them.

Werner Heisenberg The Part and the Whole

Nevertheless, Heisenberg was able to predict from studies of the hydrogen spectrum that hydrogen exists in two allotropes – ortho-hydrogen and para-hydrogen – in which the two nuclei of the atoms in a hydrogen molecule spin in the same or opposite directions respectively. The allotropes were discovered in 1929.

In 1927 Heisenberg made the discovery for which he is best known – that of the uncertainty principle. This states that it is impossible to specify precisely both the position and the simultaneous momentum (mass multiplied by velocity) of a particle. There is always a degree of uncertainty in either, and as one is determined with greater precision, the other can only be found less exactly. Multiplying the degrees of uncertainty of the position and momentum yield a value approximately equal to Planck's constant. (This is a consequence of the wave–particle duality discovered by de Broglie.) Heisenberg's uncertainty principle negates cause and effect; it maintains that the result of an action can be expressed only in terms of the probability that a certain effect will occur. The idea was revolutionary and discomforted even Albert ◊Einstein, but it has remained valid.

Science clears the fields on which technology can build.

Werner Heisenberg Attributed remark

Another great discovery was made in 1927, when Heisenberg solved the mystery of ferromagnetism. In it he employed the Pauli exclusion principle, which states that no two electrons can have all four quantum numbers the same. Heisenberg used the principle to show that ferromagnetism is caused by electrostatic intereaction between the electrons. This was also a major insight that has stood the test of time.

Heisenberg was a brilliant scientist with a very incisive mind. He will be remembered with gratitude and respect for solving some of the great and complex problems in quantum mechanics that were so crucial at the beginning of this century for our understanding of nuclear physics.

Further Reading

Brennan, Richard P, *Heisenberg Probably Slept Here: The Lives, Times and Ideas of the Great Physicists of the 20th Century,* Wiley-Liss, 1997.

Cassidy, David C, *Uncertainty: The Life and Science of Werner Heisenberg,* W H Freeman, 1991.

Cassidy, David C and Baker, Martha (compilers), *Werner Heisenberg: A Bibliography of His Writings,* Berkeley Papers in the History of Science series, Office for History of Science and Technology, University of California, 1984, v 9.

Finkelstein, D, *Quantum Relativity: A Synthesis of the Ideas of Einstein and Heisenberg,* Texts and Monographs in Physics, Springer-Verlag, 1997.

Powers, Thomas, *Heisenberg's War: The Secret History of the German Bomb,* NAL Dutton, 1999.

Rose, Paul Lawrence, *Heisenberg and the Nazi Atomic Bomb Project,* University of California Press, 1998.

Helmholtz, Hermann Ludwig Ferdinand von (1821–1894) German physicist and physiologist who made a major contribution to physics with the first precise formulation of the principle of conservation of energy. He also made important advances in physiology, in which he first measured the speed of nerve impulses, invented the ophthalmoscope, and revealed the mechanism by which the ear senses tone and pitch. Helmholtz also made important discoveries in physical chemistry.

Helmholtz was born in Potsdam on 31 August 1821. His father was a teacher of philosophy and literature at the Potsdam Gymnasium and his mother was a descendant of William Penn, the founder of the state of Pennsylvania in the USA. Although a delicate child, Helmholtz thrived on scholarship; his father taught him Latin, Greek, Hebrew, French, Italian, and Arabic. He attended the local school and showed a talent for physics, but as his father could not afford a university education for him, Helmholtz embarked on the study of medicine, for which financial aid was available in return for a commitment to serve as an army doctor for eight years. In 1838 Helmholtz entered the Friedrich Wilhelm Institute in Berlin, where his time was spent not only on medical and physiological studies but also on chemistry and higher mathematics. He also became an expert pianist.

In 1842 Helmholtz received his MD and became a surgeon with his regiment at Potsdam, where he carried out experiments in a laboratory he set up in the barracks. His ability in science as opposed to medicine was soon recognized, and in 1848 he was released from his military duties. The following year, he took up the post of associate professor of physiology at Königsberg. In 1855 Helmholtz moved to Bonn to become professor of anatomy and physiology, and then in 1858 was appointed professor of physiology at the University of Heidelberg. In 1871 he took up the chair of physics at the University of Berlin, and in 1887 became director of the new Physico-Technical Institute of Berlin. His health then began to fail, and Helmholtz died in Berlin on 8 September 1894.

Throughout his life, Helmholtz's phenomenal intellectual capacity and grounding in philosophy were at the basis of his work. While training as a doctor, his supervisor was Johannes Müller (1801–1858), a distinguished physiologist who like Helmholtz was interested in the ideas promoted by philosophers. The abstract nature of the subject was obviously attractive to his superb brain. During his student days, Helmholtz rejected many of the ideas offered by the German philosopher Immanuel ◊Kant. Helmholtz's supervisor subscribed to Kant's Nature philosophy, which assumed that the organism as a whole is greater than the sum of its parts. Helmholtz's scientific work in many fields was guided by an effort to disprove these fundamentals. He was convinced, for example, that living things possess no innate vital force, and that their life processes are driven by the same forces and obey the same principles as nonliving systems.

Helmholtz's doctoral submission in 1842 was concerned with the relationship between the nerve fibres and nerve cells of invertebrates, which led to an investigation of the nature and origin of animal heat. In 1850 he measured the velocity of nerve impulses, stimulated by a statement by Müller that vitalism caused the impulses to be instantaneous. Helmholtz measured the impulse velocity by experiments with a frog and found it to be about a tenth of the speed of sound. He went on to investigate muscle action and found in 1848 that animal heat and muscle action are generated by chemical changes in the muscles.

This work led Helmholtz to the idea of the principle of conservation of energy, observing that the energy of life processes is derived entirely from oxidation of food. His deduction was remarkably similar to that made by Julius ◊Mayer a few years earlier, but Helmholtz was not aware of Mayer's work and when he did arrive at a formulation of the principle in 1847, he expressed it in a more effective way. Helmholtz also had the benefit of the precise experimental determination of the mechanical equivalent of heat published by James ◊Joule in 1847. He was able to derive a general equation that expressed the kinetic energy of a moving body as being equal to the product of the force and distance through which the force moves to bring about the energy change. This equation could be applied in many fields to show that energy is always conserved and it led to the first law of thermodynamics, which states that the total energy of a system and its surroundings remains constant even if it may be changed from one form of energy to another.

Helmholtz went on to develop thermodynamics in physical chemistry, and in 1882 he derived an expression that relates the total energy of a system to its free energy (which is the proportion that can be converted to forms other than heat) and to its temperature and entropy. It enables chemists to determine the direction of a chemical reaction. In this work, Helmholtz was anticipated independently by Willard ◊Gibbs and both scientists are usually given credit for it.

Helmholtz's work on nerve impulses led him also to make important discoveries in the physiology of vision and hearing. He invented the ophthalmoscope, which is used to examine the retina, in 1851 and the ophthalmometer, which measures the curvature of the eye, in 1855. He also revived the three-colour theory of vision first proposed in 1801 by Thomas ◊Young, who like Helmholtz was also a physician and physicist. He developed Young's ideas by showing that a single primary colour (red, green, or violet) must also affect retinal structures sensitive to the other primary colours. In this way Helmholtz successfully explained the colour of after-images and the effects of colour blindness. He also extended Young's work on accommodation. In acoustics, Helmholtz's contribution was to produce a comprehensive explanation of how the upper partials in sounds combine to give them a particular tone or timbre, and how resonance may cause this to happen – for example, in the mouth cavity to produce vowel sounds. He also formulated a theory of hearing, correctly suggesting that structures within the inner ear resonate at particular frequencies to enable both pitch and tone to be perceived.

Helmholtz also interested himself in hydrodynamics and in 1858 produced an important paper in which he established the mathematical principles that define motion in a vortex. Helmholtz will also be remembered for his electrical double layer theories. First proposed in 1879, they were initially applied to the situation of a solid immersed in a solution. He imagined that a layer of ions is held at a solid surface by an oppositely charged layer of ions in the solution. These ideas were later applied to solids suspended in liquids.

Helmholtz dominated German science during the mid-1800s, his wide-ranging and exact work bringing it to the forefront of world attention, a position Germany was to enjoy well into the following century. He served as a great inspiration to others, not least his many students. Foremost of these was Heinrich ◊Hertz, who as a direct result of Helmholtz's encouragement discovered radio waves. Helmholtz took classical mechanics to its limits in physics, paving the way for the radical departure from tradition that was soon to follow with the quantum theory and relativity. When this revolution did occur, however, it was ushered in mainly by German scientists applying the mathematical and experimental expertise upon which Helmholtz had insisted.

Further Reading

Cahan, David (ed), *Hermann von Helmholtz and the Foundations of Nineteenth-Century Science,* California Studies in the History of Science, University of California Press, 1993, v 12.

Helmholtz, Hermann von; Richard L Kremer (ed), *Letters of Hermann von Helmholtz to his Wife, 1842–1859,* Steiner, 1990.

Holmes, Frederic L, Olesko, Kathryn M, The values of precision in: Wise, M Norton (ed), *The Image of Precision: Helmholtz and the Graphical Method in Physiology,* Princeton University Press, 1995, pp 198–221.

Turner, R Steven, *In the Eye's Mind: Vision and the Helmholtz–Hering Controversy,* Princeton University Press, 1994.

Helmont, Jan Baptista van (1579–1644) Flemish chemist and physician.

Helmont was born into Flemish landed gentry in Brussels on 12 January 1579. He studied chemistry, medicine, and mysticism at the University of Louvain and received his MD in 1599. From 1600 to 1605 he travelled to Spain, Italy, France, and England to gain more medical knowledge and then carried out private research at his home at Vilvorde. Helmont took great interest in the controversy over the 'weapon salve' and the magnetic cure of wounds (applying ointment to the weapon rather than the wound). In 1621 he published his treatise *De magnetica vulnerum ... curatione* in which, although he did not reject the belief, he insisted that it was a natural phenomenon with no supernatural elements. He was arrested for his views and the church proceedings against him did not formally end until 1642. Helmont died in Vilvorde on 30 December 1644. He had married in 1605.

Although Helmont believed in alchemy, his work represents a transition to chemistry proper. He first emphasized the use of the balance in chemistry, giving him an insight into the indestructibility of matter and the fact that metals dissolved in acid were recoverable, not destroyed or transmuted. His experiments led to his belief that all matter was composed of water and air, which he demonstrated by growing a willow tree over five years in a measured quantity of earth; after this time, after adding only water, the tree had increased its weight by 164 pounds, while the earth it was in had lost little weight. It is claimed that Helmont was the first to use the term 'gas'; he applied chemical analysis to smoke, which he found to be different from air and water vapour and specific to the substance burned. He identified four gases: carbon dioxide, carbon monoxide, nitrous oxide, and methane. He was the first to take the melting point of ice and the boiling point of water as standards for temperature and was the first to employ the term 'saturation' to signify the combination of an acid and a base.

In medicine, Helmont was a follower of ◊Paracelsus and improved on his chemical medicines, notably mercury preparation. He discussed the effects of ether, demonstrated acid as the digestive agent in the stomach, made contributions to the understanding of asthma, and was among the founders of the modern ontological concept of disease. He rejected the humoral theory and its traditional therapy and used remedies that specifically considered the type of disease, the organ affected, and the causative agent. During his lifetime, Helmont published *Supplementum de spadanis fontibus* (1624), *Febrium doctrina inaudita* (1642), and *Opuscula medica inaudita* (1644). His works were collectively published posthumously by his son as *Ortus medicinae* (1648).

Further Reading

Coulter, Harris L, *Divided legacy, v 2: The Origins of Modern Medicine: J B Van Helmont to Claude Bernard,* North Atlantic Books, 1998.

Pagel, Walter, *Jean Baptista Van Helmont: Reformer of Science and Medicine,* Monographies on the History of Medicine series, Cambridge University Press, 1982.

Henry, Joseph (1797–1878) US physicist who carried out early experiments in electromagnetic induction.

Henry was born in Albany, New York, on 17 December 1797, the son of a labourer. He had little schooling, working his way through Albany Academy to study medicine and then, from 1825, engineering. He worked at the academy as a teacher, and in 1826 was made professor of mathematics and physics. He moved to New Jersey College (later Princeton) in 1832 as professor of natural philosophy, in which post he lectured in most of the sciences. He was the Smithsonian Institution's first director, from 1846, and first president of the National Academy of Sciences, from 1868, a position he held until his death, in Washington, on 13 May 1878.

Many of Henry's early experiments were with electromagnetism. By 1830 he had made powerful electromagnets by using many turns of fine insulated wire wound on iron cores. In that year he anticipated Michael ◊Faraday's discovery of electromagnetic induction (although Faraday published first), and two years later he discovered self-induction. He also built a practical electric motor and in 1835 developed the relay (later to be much used in electric telegraphy). In astronomy Henry studied sunspots and solar radiation, and his meteorological studies at the Smithsonian led to the founding of the US Weather Bureau. The unit of inductance was named the henry in 1893.

Further Reading

Moyer, Albert E, *Joseph Henry: The Rise of an American Scientist*, Smithsonian Institution Press, 1997.

Rothenberg, Marc (ed), *The Papers of Joseph Henry, v 6: The Princeton Years, January 1844–December 1846*, Smithsonian Institution Press, 1992.

Rothenberg, Marc (ed), *The Papers of Joseph Henry, v 7: The Smithsonian Years, January 1847–December 1849*, Smithsonian Institution Press, 1996.

Henry, William (1774–1836) English chemist and physician, best known for his study of the solubility of gases in liquids, which led to what is now known as Henry's law. He was also an early experimenter in electrochemistry, and he established that firedamp – the cause of many mining disasters – is methane.

Henry was born in Manchester on 12 December 1774. His father was a physician and industrial chemist who introduced the use of chlorine as a bleaching agent for textiles. A childhood accident affected Henry's growth and left him in poor health for the rest of his life. He went to a school run by a local clergyman and then to Manchester Academy, which was an offshoot of the famous Dissenting Academy at Warrington. (The Dissenting institutions offered a wide curriculum containing scientific, practical, and mathematical subjects, unlike the grammar schools of the time, which provided an education based almost entirely on Latin.)

On leaving the academy in 1790 Henry became secretary companion to Dr Thomas Percival, the founder of the influential Literary and Philosophical Society of Manchester. Five years later he went to Edinburgh University to study medicine, but left after a year to help his father. During this time he began his own chemical experiments, and published his first paper in 1797. He returned to Edinburgh in 1805 and graduated in medicine two years later, with a thesis on uric acid.

Henry suffered from a disorder that affected his hands and after unsuccessful surgery in 1824 he was unable to carry out chemical experiments; he returned to a study of medicine. But his continuing poor health led to severe depression and in Manchester on 2 September 1836 he committed suicide.

Henry's early chemical experiments concerned the composition of hydrogen chloride and hydrochloric acid. A series of chemistry lectures that he gave in Manchester in the winter of 1798–99 were later published as a successful textbook, *Elements of Experimental Chemistry*, which he expanded over the following 30 years and which by 1829 had run to its 11th edition. In 1803 he published the results of his research on the solubility of gases, outlining the basis of Henry's law, which in its modern form states that the mass of gas dissolved by a given volume of liquid is directly proportional to the pressure of the gas with which it is in equilibrium, provided the temperature is constant. In mathematical terms, $m = kp$, where m is the mass of gas, p the pressure exerted by the undissolved gas, and k a constant. There are many deviations from the law and it is closely obeyed only by very dilute solutions.

In 1805 Henry worked at determining the composition of ammonia. He confirmed that it contains hydrogen and nitrogen in the proportions suggested by Humphry ◊Davy and Claude ◊Berthollet, but showed that Davy was incorrect in his belief that it contains also oxygen, which for Davy was a necessary characteristic of all alkalis.

At the beginning of the 19th century coal gas was being introduced as a possible fuel for illumination. Henry worked for about 20 years on the analysis of inflammable mixtures of gases and attempted to find correlations between chemical composition and illuminative properties. During these investigations he confirmed John ◊Dalton's results of the composition of methane and ethane, and like Dalton showed that hydrogen and carbon combine in definite proportions to form a limited number of compounds. In this work Henry made use of the catalytic properties of platinum, newly discovered by Johann Döbereiner in 1821.

After 1824 Henry returned to medical investigations and studied contagious diseases. He believed that these were spread by chemicals that could be rendered harmless by heating; he used heat to disinfect clothing during an outbreak of Asiatic cholera in 1831. By the time of Henry's death, his son W C Henry was already an active chemist, publishing a work in 1836 on the poisoning of platinum catalysts.

Heraklides of Pontus (388–315 BC) Greek philosopher and astronomer who is remembered particularly for his teaching that the Earth turns on its axis, from west to east, once every 24 hours.

Born in Heraklea, near the Black Sea, Heraklides migrated to Athens and became a pupil of Speusippus, under the direction of ◊Plato, in the Academy. Nearly elected to succeed Speusippus as head of the Academy, Heraklides was at one stage left temporarily in charge of the whole establishment by Plato. He is said also to have attended the schools of the Pythagorean philosophers, and would thus have come into contact with his contemporary, ◊Aristotle. Although all his writings are lost, it is clear from those of his contemporaries that his astronomical theories were more advanced than those of other scientists of his age.

In proposing the doctrine of a rotating Earth, Heraklides contradicted the accepted model of the universe put forward by Aristotle. Aristotle had accepted that the Earth was fixed, and said that the stars and the planets in their respective spheres might be at rest. Heraklides thought it highly impossible that the immense spheres of the stars and planets rotated once every 24 hours. He also thought that the observed motions of Mercury and Venus suggested that they orbited the Sun rather than the Earth. He did not completely adopt the heliocentric view of the universe stated later by ◊Aristarchos (in around 280 BC). He proposed instead that the Sun moved in a circular orbit (in its sphere) and that Mercury and Venus moved on epicycles around the Sun as centre.

Apart from his astronomical work, Heraklides contributed to literary criticism and musicology. His less scientific interests included phenomena of the occult – trances, visions, prophesies, and portents – breakdowns in health, which he attempted to interpret in terms of the retribution of the gods, and reincarnation.

Herbig, George Howard (1920–) US astronomer who specializes in spectroscopic research into irregular variable stars, notably those of the T-Tauri group.

Born on 20 January 1920 in Wheeling, West Virginia, Herbig was educated at the University of California in Los Angeles, graduating in 1943. In 1944 he became a member of the staff at the Lick Observatory, California, beginning as an assistant. From junior astronomer he progressed to assistant astronomer, astronomer, and, from 1960 to 1963, assistant director; in 1967 he became professor of astronomy. In 1987 he moved to the University of Hawaii to become astronomer at its Institute of Astronomy.

Herbig's main area of research has been the nebular variables of which the prototype is T-Tauri. It is believed that the members of this group are in an early stage of stellar evolution. Most of them are red and fluctuate in light intensity; their associated nebulosities are also variable in brightness and structure, although the reason for this is not known.

With Bidelman and Preston, Herbig has worked on the spectra of these and other variable stars. In 1960 he drew attention to the fact that many of them have a predominance of lithium lines, similar to the abundance of lithium on Earth and in meteorites (although considerably more abundant than in the Sun). He concluded that both the planetary and T-Tauri abundance of lithium might represent the original level of this element in the Milky Way, but that the lithium in the Sun and other stars may have largely been lost through nuclear transformation. Herbig also showed that there seems to be a conservation of angular momemtum in such young, cool variable stars, and that T-Tauri variables move together in parallel paths within the obscuring cloud in which they were formed.

Herbig also worked on binary stars, which are in relative orbital motion because of their proximity and mutual gravitational attraction. He investigated the binary of shortest known orbital period, VV Puppis, which is an eclipsing binary of period 100 minutes.

Herbig has also investigated the spectra of atoms and molecules that originate in interstellar space. In 1904, Johannes Hartmann (1865–1936) found that the absolute lines of Ca(II) in the spectrum of Delta Orionis do not take part in the periodic oscillation of other lines. Since then other atoms and molecular combinations have been discovered to have originated in interstellar space, such as Na(I), Ca(I), K(I), Ti(II), CN, and CH. There are also a number of diffuse interstellar absorption lines that are as yet unidentified but which Herbig has succeeded in resolving into band lines.

Herbrand, Jacques (1908–1931) French mathematical prodigy who, in a life cut tragically short, still originated some innovatory concepts in the field of mathematical logic. Interested also in class-field theory, he remains best remembered for the Herbrand theorem.

Herbrand was born on 12 February 1908 in Paris, where he grew up and was educated, entering the Ecole Normale Supérieure at the age of 17 already precociously talented in mathematics. Three years later he published his first paper on mathematical logic for the Paris Academy of Sciences, and in the following year – 1929, when he was 21 – for his doctorate, he produced the paper in which he formulated Herbrand's theorem. His military service in the French army then occupied him for 12 months, at the end of which a Rockefeller scholarship enabled him to go to Berlin to study there, and then for a couple of months in Hamburg and Göttingen. He was killed in an Alpine accident on 27 July 1931.

For his doctor's degree, Herbrand introduced what is now considered to be his main contribution to mathematical logic, the theorem to which his name is attached. A fundamental development in quantification theory, Herbrand's theorem established a link between that theory and sentential logic, and has since found many applications in such fields as decision and reduction problems.

Herbrand was also fascinated by modern algebra and wrote a number of papers on class field theory, which deals with Abelian extensions of a given algebraic number field from properties of the field. He contributed greatly to the development of the theory that, until that time, had received scant attention since being devised by Leopold ◊Kronecker in the middle of the 19th century. After Herbrand's death, however, the theory was the subject of further development by later mathematicians.

Hermite, Charles (1822–1901) French mathematician who was a principal contributor to the development of the theory of algebraic forms, the arithmetical theory of quadratic forms, and the theories of elliptic and Abelian functions. Much of his work was highly innovative, especially his solution of the quintic equation through elliptic modular functions, and his proof of the transcendence of e. A man of generous disposition and a good teacher, Hermite in his work showed that the brilliance of the previous generation of mathematicians – Karl ◊Gauss, Augustin ◊Cauchy, Karl ◊Jacobi, and Lejeune ◊Dirichlet – was not to fade in the next.

Hermite was born, the sixth of the seven children of a minor businessman in Dieuze, Lorraine, on 24 December 1822. He grew up and was educated in Nancy, however. Completing his education, he went to Paris and first studied physics at the Henry IV Collège before he went to the Lycée Louis le Grand and became fascinated by mathematics. He was always hopeless at passing examinations – but his private reading at night of the works of Leonhard ◊Euler, Karl Gauss, and Joseph ◊Lagrange might well have distracted him from other preparation. From 1842 he studied at the Ecole Polytechnique for a year, but because he did not pass the examinations there either (and apparently also because he was congenitally lame), he was asked not to return. So it was not until 1848, at the age of 25, that he finally graduated, although by that time he already had some reputation as an innovative mathematician. He was appointed to a minor teaching post at the Collège de France. In 1856 two things happened: he was elected to the Paris Academy of Sciences, and he caught smallpox. Recovering from the latter, he had to wait until he was 47 years old before he finally attained the status of professor, first in 1869 at the Ecole Normale, then a year later at the Sorbonne. An honorary member of many societies, he received numerous awards and honours during his lifetime. He died in Paris on 14 January 1901.

Throughout his life Hermite exerted great scientific influence by corresponding with other prominent mathematicians. One of his first pieces of work was to generalize Niels ◊Abel's theorem on elliptic functions to the case of hyperelliptic ones. Communicating his discovery to Karl Jacobi in August 1844, he also discussed four other papers on number theory.

Between 1847 and 1851 he worked on the arithmetical theory of quadratic forms and the use of continuous variables. Then for ten years between 1854 and 1864 he worked on the theory of invariants.

Today he is remembered chiefly in connection with Hermitean forms (a complex generalization of quadratic forms) and with Hermitean polynomials – work carried out in 1873. In the same year, he showed that e, the base of natural logarithms, is transcendental. (Transcendental numbers are real or complex numbers that are not algebraic; we now know that most real numbers are transcendental.) It was by a slight adaptation of Hermite's proof for this that Ferdinand von ◊Lindemann in 1882 demonstrated the transcendence of π.

In 1872 and 1877 Hermite solved the Lamé differential equation, and in 1878 he solved the fifth-degree (quintic) equation of elliptic functions.

Hermite's scientific work was collected and edited by the later French mathematician Emile ◊Picard.

Hero of Alexandria (lived c. AD 60) Greek mathematician and engineer, the greatest experimentalist of antiquity. His numerous writings describe many of his inventions, formulae, and theories, some of them centuries ahead of their time. His famous aeolipile (a rudimentary steam turbine), which demonstrated the force of steam generated in a closed chamber and escaping through a small aperture, was not developed further for another 18 centuries.

A gifted mathematician, he adopted the division of the circle into 360°, orginally brought to the western world by ◊Hipparchus. He was also a famous teacher and at Alexandria founded a technical school with one section devoted entirely to research.

He regarded air as a very elastic substance that could be compressed and expanded, and successfully explained the phenomenon of suction and associated apparatus, such as the pipette and cupping glass; he also used a suction machine for pumping water. His writing on air compressibility and density and his assumption that air is composed of minute particles able to move relative to one another preceded Robert ◊Boyle by 1,500 years.

In mechanics he devised a system of gear wheels that could lift a 1,000 kg/2,200 lb weight by means of a mere 5 kg/11 lb. His work entitled *Mechanics,* the only existing copy of which is in Arabic, contains the parallelogram of velocity, the laws of levers, the mysteries of motion on an inclined plane and the effects of friction, and much data on gears and the positions of the centre of gravity of various objects. His construction of a variable ratio via a friction disc has been used to build a motor vehicle with a semi-automatic transmission.

Civil and construction engineers owe him a debt for his tables of dimensions used in building arches and in drilling tunnels and wells. Another of his famous books, *On the Dioptra,* describes a diversity of instruments, including a type of theodolite that employs a refined screw-cutting technique for use in boring tunnels; the book also describes a hodometer for measuring the distance travelled by a wheel.

Hero's book *Metrica* explains the measurement of plain and solid geometrical figures and discusses conic sections, the frustum of a cone, and the five regular, or Platonic, solids. It includes a formula for calculating the area of a triangle from the lengths of its sides, and a method of determining the square root of a non-square number.

Another of his books, *Pneumatics,* describes numerous mechanical devices operated by gas, water, steam, or atmospheric pressure, and siphons, pumps, fountains, and working automata in the likeness of animals or birds devised, it seems, just for amusement.

Hero was a genius in the techniques of measurement of all kinds and in founding the science of mechanics. After his death there was no further progress in physics for centuries. But for the Arabs, it is extremely likely that all record of his works would have vanished.

Héroult, Paul Louis Toussaint French metallurgist; see ◊Hall and Héroult.

Herschel, Caroline (1750–1848) German-born English astronomer who discovered 14 nebulae and 8 comets and worked on her brother William ◊Herschel's catalogue of star clusters and nebulae.

Caroline Herschel was born on 16 March 1750 into a working-class family of six children in Hannover, Germany. Her father worked as a gardener but was also a musician and eventually secured himself a position in the Prussian army band. He encouraged all of his children to study mathematics, French, and music. Herschel's growth was stunted at the age of ten by typhus and, assuming her to be unmarriageable because of this, her mother chose her to remain at home as a house servant. Her father took pity on her and secretly encouraged her to continue her education.

In 1772, Caroline's brother William visited Hannover and rescued her from life as a *hausfrau* in her parents' house, taking her to Bath, England, where he was an organist and choir master at the Octagon Chapel. Caroline trained as a singer and she became a successful soprano, performing solo in the *Messiah* and *Judas Maccabaeus.* William meanwhile spent all his spare time on astronomy and was obsessed with seeing further and further into space. With Caroline's help he produced and sold fine telescopes to finance his hobby, and eventually he managed to give up his job and devote all his time to astronomy. William trained Caroline in mathematics and she became his assistant, spending hours grinding and polishing the mirrors they used to collect light from distant objects.

The planet Uranus was discovered by William and Caroline on 13 March 1781 using a 6-m/20-ft telescope of their own manufacture. King George III, an enthusiastic patron of William's work, commissioned a telescope from them and William was sent as emissary to present it to Göttingen University. While he was away, Caroline independently discovered a comet, gaining her own standing in the scientific community. She was to go on to discover a total of eight comets.

In 1787 William was appointed Astronomer Royal and they moved to Old Windsor. Caroline received a salary of £50 a year from the king to work as William's assistant; the first time a woman in England had been recognized in a scientific position. From then on Caroline devoted herself to assisting William while also spending her spare time independently 'minding the heavens' and becoming an astronomer in her own right. She discovered 14 nebulae and catalogued them with the discoveries of her brother and English astronomer John ◊Flamsteed into a book published by the Royal Society – *Catalogue of 850 stars observed by Flamsteed but not included in the British Catalogue.* Also published by the Royal Society was *A General Index of Reference to every Observation of every Star in the above-mentioned British Catalogue.*

William married in 1788 and the family moved to Slough. He began spending less time in his observatory, leaving Caroline to continue her work alone. When William died in 1822, Caroline returned to Hannover to live with her younger brother Deitrich. She continued her meticulous work there and was very popular, especially with the Prussian aristocracy – it was an honour to be seen with her in public. In 1828 she was awarded a gold medal by the Royal Astronomical Society for *The Reduction and Arrangement in the Form of Catalogue, in Zones, of all the Star-clusters and Nebulae Observed by Sir William Herschel in His Sweeps,* a catalogue of 2,500 nebulae. She was made an honorary member of the Royal Astronomical Society and the Royal Irish Academy for her work and was also honoured by the King of Prussia with the Gold Medal of Science for her life's achievements.

William's son, John, also a prominent astronomer, was a lifelong friend and shortly before her death Herschel received a copy of his *Cape Observations,* which provided a southern counterpoint to William's and her own work. Wary of anything that might detract from her brother's reputation she observed, 'All that I am, all I know, I owe to my brother.' Caroline was 98 when she died on 9 January 1848, and was widely mourned by the scientific community.

Herschel, (Frederick) William (1738–1822) German-born English astronomer who contributed immensely to contemporary scientific knowledge. Through determined efforts to improve the quality of his telescopes, he was able to use the finest equipment of his time, which in turn permitted him to make many significant discoveries about the nature and distribution of stars and other bodies, both within the Solar System and beyond it. He was the most influential astronomer of his day.

Herschel German-born English astronomer William Herschel who discovered the planet Uranus in 1781 and infrared solar rays in 1801. He was a skilled telescope-maker and in 1789 he built the largest telescope in the world at that time in Slough, Berkshire, England. His work in cataloguing stars was continued by his son John Herschel. *Mary Evans Picture Library*

Herschel was born in Hanover, Germany, on 15 November 1738. At the age of 14 he joined the regimental band of the Hanoverian Guards as an oboist (as his father had before him), and four years later he visited England with the band. In 1757 he emigrated to the UK, going first to Leeds – where he earned his living copying music and teaching – and then, three years later, by commission from the Earl of Darlington, to Durham to become conductor of a military band there. From 1761 to 1765 Herschel worked as a teacher, organist, composer, and conductor in Doncaster. He then did similar work in Halifax for a year. In 1766 he was hired as organist at the Octagon Chapel in Bath, where he remained for the next 16 years; during this time he ran a tutorial service that helped to finance his growing interest in astronomy. In 1772 he went to Hanover to bring his sister Caroline Herschel back to England. She too became fascinated by astronomy and helped Herschel enormously both in the delicate task of preparing instruments and in making observations. The serious astronomical work began in 1773, with the building of telescopes and the grinding of mirrors. Herschel's first large reflector was set up behind his house in 1775.

Herschel's discovery of the planet Uranus in 1781 – the first planet to have been discovered in modern times – created a sensation: it signalled that Isaac ◊Newton's work had not covered everything there was to know about the universe. Herschel originally named the new planet 'Georgium Sidum' (George's Star) in honour of King George III, but the name Uranus – proposed by Johann ◊Bode of the Berlin Observatory – was ultimately accepted. Nevertheless, the king was flattered, and Herschel received a royal summons to bring his equipment to court for inspection. In the same year the Royal Society elected Herschel a fellow, and awarded him the Copley Medal. In 1782 he was appointed court astronomer, a post that carried with it a pension of 300 guineas per year. This enabled him to give up teaching and to move from Bath to Windsor, and then to Slough, although he continued to make telescopes for sale in order to supplement his income until 1788 (when he married a wealthy widow). In addition, the king provided grants for the construction of larger instruments; for instance, he provided £4,000 for a telescope with a focal length of more than 12 m/40 ft and a reflector that was 1 m/3 ft in diameter. (This telescope proved rather cumbersome and was not used after 1811, although for many years it remained the largest in the world.)

Herschel visited Paris in 1801, meeting Pierre ◊Laplace and Napoleon Bonaparte. Many honours were conferred upon him (he was knighted in 1816), and he was a member of several important scientific organizations. He became ill in 1808, but continued to make observations until 1819. In that year his only son, John ⇨Herschel, finished his studies at Cambridge and came to take over his father's work. Herschel died in Slough on 25 August 1822.

Herschel's first large telescope was a 1.8-m/6-ft Gregorian reflector that he built himself in 1774. In many ways it was better than other existing telescopes, and Herschel decided that its primary use would be to make a systematic survey of the whole sky. His first review was completed in 1779, when he immediately began a second survey with a 2.1-m/7-ft reflector. In this project he concentrated on noting the positions of double stars, that is, stars that appear to be very close together, perhaps only as a consequence of chance alignment with the observer. His first catalogue of double stars was published in 1782 (with 269 examples); a second catalogue (with 434 examples) appeared in 1785; and a third was issued in 1821, bringing the total number of double stars recorded to 848. ⇨Galileo and others had suggested that the motion of the nearer (and it was presumed therefore brighter) star relative to the more distant (thus fainter) star could be used to measure annual movements of the stars. Herschel's observations in 1793 demonstrated that a correlation between dimness and distance did not apply in all cases, and that in fact some double stars were so close together as to rotate round each other, held together by an attractive force. This was the first indication of gravity acting on bodies outside the Solar System.

Herschel's work on bodies within the Solar System included an accurate determination of the rotation period of Mars (in 1781): 24 h, 39 min, 21.67 sec – only two minutes longer than the period now accepted. By inventing an improved viewing apparatus for his telescope, particularly valuable when little light was available, in 1787 he discovered Titania and Oberon, two satellites of Uranus. Incorporating the 'Herschelian arrangement' into a massive telescope 12 m/39 ft long, Herschel then discovered two further satellites of Saturn (Enceladus and Mimas) on its first night of use. Saturn continued to be an object of great interest to him.

During the 1780s Herschel published a number of papers on the evolution of the universe from a hypothetical uniform initial state to one in which stars were clumped into galaxies (seen as nebulae). Herschel had become interested in nebulae in 1781, when he was given a list of a hundred of these indistinct celestial bodies compiled by Charles ⇨Messier. Herschel began looking for more nebulae in 1783, and the improved resolving power of his telescopes enabled him to see them as clusters of stars. His first catalogue of nebulae (citing no fewer than 2,500 examples) was published in 1802; an even longer catalogue (with 5,000 nebulae) appeared in 1820.

In 1800 Herschel examined the solar spectrum using prisms and temperature-measuring equipment. He found that there were temperature differences between the various regions of the spectrum, but that the hottest radiation was not within the visible range, but in the region now known as the infrared. This was the beginning of the science of stellar photometry. Using data published by Nevil ⇨Maskelyne on seven particularly bright stars, Herschel demonstrated that if the Sun's motion towards Argelander (a star in the constellation Hercules) was accepted, then the 'proper motion' of the seven stars was a reflection of the motion of the observer. This relegation of the Solar System from the centre of the universe was, in its way, analogous to the dethronement of the Earth from the centre of the Solar System by Copernicus. Herschel also measured the velocity of the Sun's motion.

An industrious and dedicated astronomer, and a very practical man, Herschel contributed enormously to the advance of scientific progress.

Further Reading

Hoskin, Michael A, *William Herschel and the Construction of the Heavens,* Oldbourne History of Science Library, 1963.

Moore, Patrick, *William Herschel, Astronomer and Musician of 19 New King Street, Bath,* PME Erwood in association with the William Herschel Society, 1981.

Sidgwick, J B, *William Herschel, Explorer of the Heavens,* Faber & Faber, 1953.

Herschel, John Frederick William (1792–1871)

English astronomer, the first to carry out a systematic survey of the stars in the southern hemisphere and to attempt to measure, rather than estimate, the brightness of stars. Besides this, he continued his father's studies of double stars, nebulae, and the Milky Way.

John Herschel was born in Slough on 7 March 1792, the only child of the astronomer William ⇨Herschel. Unsurprisingly, the young Herschel's career was strongly influenced by the fact that he was brought up by his father and his aunt Caroline Herschel, who were both devoted to astronomy. From the age of eight, after a short spell at Eton, Herschel was educated at a local private school. He then went to St John's College, Cambridge, to read mathematics. From 1816 to 1850 Herschel had no permanent paid post, but his scientific life was closely bound up with two royal societies. He became a fellow of the Royal Society in 1813, served as its secretary 1824–27, and won its Copley Medal in 1821 and 1847 and its Royal Medal in 1833, 1836, and 1840. He received the Lalande Prize of the French Academy in 1825 and the Gold Medal of the Astronomical Society in 1826 for the catalogue of double stars that he had compiled with James ⇨South. When Herschel died on 11 May 1871 he was mourned by the whole nation, not only as a great scientist, but also as a remarkable public figure.

After completing his studies and taking his MA in 1816, Herschel embarked on his scientific career. As one of his obituarists noted, he may well have taken up astronomy out of a sense of 'filial devotion'. His first paper, on the computation of lunar occultation, appeared in 1822, by which time he had moved on to systematically observing double stars. This work, a continuation of one of his father's projects, was carried out in collaboration with James South, who possessed two excellent refracting telescopes in London. William Herschel had demonstrated that the orbital motion of binary stars was due to their mutual attraction; John Herschel studied new and known double star systems, especially Gamma Virginis, in order to establish a method for determining their orbital elements.

Soon after his marriage to Margaret Brodie Stewart, in 1829, Herschel planned an expedition to the southern hemisphere. He chose to go to the Cape of Good Hope, partly because of its excellent astronomical tradition, built up as a result of Nicolas de ◊Lacaille's work in the 1850s, partly because, being located on the same meridian of longitude as eastern Europe, it made cooperative observations easier. Herschel arrived in Cape Town with his wife and the first three of their twelve children on 16 January 1834 and set himself the mammoth task of mapping the southern skies. By 1838 he had catalogued 1,707 nebulae and clusters, listed 2,102 pairs of binary stars, and carried out star counts in 3,000 sky areas. Besides simply mapping the stars, he made accurate micrometer measurements of the separation and position angle of many stellar pairs and produced detailed sketches of the Orion region, the Magellanic Clouds, and extragalactic and planetary nebulae. He also recorded the behaviour of Eta Carinae (whose nature is still not entirely understood) when it underwent a period of dramatic brightening in December 1837.

To ascertain the brightness of the stars he catalogued, Herschel invented a device called an astrometer. It enabled him to compare the light output of a star with an image of the full moon, whose brightness could be varied to match the star under observation with the main telescope. This attempt to ascertain absolute magnitude was a major step forward in stellar photometry. Besides his personal research work, Herschel also collaborated with Thomas Maclear, director of the Cape Observatory, in the observatory's routine geodetic and tidal work and in observations of Encke's and Halley's comets.

On his return to London on 15 May 1838, Herschel began to prepare the results of his African trip for publication. At the same time he pursued his interest in the art of photography. His interest in the subject led him to invent much of the vocabulary – positive, negative, and snapshot – that is associated with the craft today. His massive *Results of Astronomical Observations Made During the Years 1834–38 at the Cape of Good Hope* was published in 1847; two years later he produced *Outlines of Astronomy,* which became a standard textbook for the following decades. His *General Catalogue of Nebulae and Clusters,* now known as the *NGC,* still remains the standard reference catalogue for these objects. The last of his ambitious projects, *General Catalogue of 10,300 Multiple and Double Stars,* was published posthumously.

John Herschel was a celebrated scientist throughout his life and his name epitomized science to the public in much the same way as Albert ◊Einstein's did in the following century. After his death his reputation suffered a decline, and it rose again only when astronomers began to realize that he occupied the same commanding and innovative position for astronomers in the southern hemisphere as his father William did for astronomers in the north.

Further Reading

Buttman, Günther and Evans, David Stanley, *The Shadow of the Telescope: A Biography of John Herschel,* Lutterworth Press, 1974.

Crowe, Michael J; Dyck, David R; and Kevin, James J (eds), *A Calendar of the Correspondence of Sir John Herschel,* Cambridge University Press, 1998.

King-Hele, Desmond (ed), *John Herschel 1792–1871: A Bicentennial Commemoration,* Royal Society, 1992.

Schaaf, Larry John, *Out of the Shadows: Herschel, Talbot, and the Invention of Photography,* Yale University Press, 1992.

Hershey, Alfred Day (1908–1997) US biochemist who shared the Nobel Prize for Physiology or Medicine in 1969 with German-born US biologist Max Delbruck (1906–1981) and Italian-born US physician Salvador Luria (1912–1991) for his work using bacteriophages – viruses that infect bacteria – to demonstrate that DNA, not protein, is the genetic material. His experiments showed that viral DNA is sufficient to transform bacteria.

Alfred Hershey was born on 4 December 1908 in Owosso, Michigan. He studied at Michigan State College obtaining a BS in chemistry in 1930 and remained there to do his PhD thesis on the chemistry of the *Brucella* virus, receiving his doctorate in 1934. From 1934 until 1950 he was engaged in teaching and research at the department of bacteriology at the Washington University School of Medicine in St Louis, Missouri. In 1950 he became a staff member of the Carnegie Institute, New York, and he was appointed director of the Genetics Research Unit there in 1962. He married Harriet Davidson in 1945 and they had one son.

In Missouri Hershey worked under US bacteriologist J J Bronfenbrenner, one of the first bacteriologists in the USA to study bacteriophages, and with Delbruck and Luria, he became a founding member of the American Phage Society in the early 1940s. His discovery in 1946 of the existence of the exchange of genetic material in phages was the first laboratory demonstration of genetic recombination in viruses.

Hershey is best known for his work with Martha Chase at the Carnegie Institute. In 1952 they conducted the Hershey–Chase experiment that confirmed the hypothesis put forward by Oswald ◊Avery in 1944 that genes were made of DNA. Their experiment demonstrated that DNA is the genetic material by studying the T2 bacteriophage, a virus that infects the bacterium *Escherichia coli* (*E. coli*). In 1951 R Herriot had suggested that a bacteriophage acted like 'a little hypodermic needle full of transforming principles', which did not enter the cell itself but whose tail contacted the host and perhaps enzymatically cut a small hole through the outer membrane, allowing the nucleic acid of the virus head to flow into the cell.

Hershey and Chase confirmed Herriot's hypothesis by labelling the bacteriophage DNA with active phosphorus and the protein with radioactive sulphur. A sample of *E. coli* was infected with the radiolabelled bacteriophage for a short incubation and the two were then separated by centrifugation. The bacteria were found to be full of radioactive phosphorus but devoid of radioactive sulphur. They were also fully competent to produce progeny virus. Their conclusion was that 'a physical separation of the phage T2 into genetic and non-genetic parts is possible' and they published their findings in their 1952 paper 'Independent functions of viral proteins and nucleic acid in growth of bacteriophage'. This convinced the other members of the Phage Society that DNA was the genetic component of viruses, and was also influential on Francis Crick and James ◊Watson's work in Cambridge, England, on the structure of DNA. Hershey was awarded an honorary DSc by the University of Chicago in 1967 and an MD by Michigan State University in 1970. He received the 1958 Albert Lasker Award and the 1965 Kimber Genetics Award but it was not until 1969 that he was awarded the Nobel prize. Hershey died in Syosset, New York, on 22 May 1997.

Hertz, Heinrich Rudolf (1857–1894) German physicist who discovered radio waves. His name is commemorated in the use of the hertz as a unit of frequency, one hertz being equal to one complete vibration, or cycle, per second.

Hertz was born in Hamburg on 22 February 1857. As a child he was interested in practical things and equipped his own workshop. At the age of 15 he entered the Johanneum Gymnasium and, on leaving school three years later, went to Frankfurt to gain practical experience as the beginning of a career in engineering. Engineering qualifications were governed by a state examination, and he went to Dresden Polytechnic to work for this in 1876. During a year of compulsory military service 1876–77, Hertz decided to be a scientist rather than an engineer and on his return he entered Munich University. He began studies in mathematics, but soon switched to practical physics, which greatly interested him.

Hertz moved to Berlin in 1878 and there came into contact with Hermann von ◊Helmholtz, who immediately recognized his talents and encouraged him greatly. He gained his PhD in 1880 and remained at Berlin for a further three years to work as Helmholtz's assistant. Then in 1883 Hertz moved to Kiel to lecture in physics, but the lack of a proper laboratory there caused him to take up the position of professor of physics at Karlsruhe in 1885. He stayed in Karlsruhe for four years, and it was there that he carried out his most important work, discovering radio waves in 1887. In the same year, he began to suffer from toothache, the prelude to a long period of increasingly poor health due to a bone disease. Hertz moved to Bonn in 1889, succeeding Rudolf ◊Clausius as professor of physics. His physical condition steadily deteriorated and he died of blood poisoning on 1 January 1894, at the early age of 36.

In 1879 a prize was offered by the Berlin Academy for the solution of a problem concerned with James Clerk ◊Maxwell's theory of electricity, which was set by Helmholtz with Hertz particularly in mind. Hertz declined the task believing that it would take up more time and energy than he could then afford. Instead, for his doctorate he carried out a theoretical study of electromagnetic induction in rotating conductors. It was not until five years later on his move to Karlsruhe in 1885 that Hertz found the facilities and the time to work on the problem set by the Berlin Academy. He used large improved induction coils to show that dielectric polarization leads to the same electromagnetic effects as do conduction currents; and he generally confirmed the validity of Maxwell's famous equations concerning the behaviour of electricity. In 1888 he realized that Maxwell's equations implied that electric waves could be produced and would travel through air. Hertz went on to confirm his prediction by constructing an open circuit powered by an induction coil, and an open loop of wire as a receiving circuit. As a spark was produced by the induction coil, so one occurred in the open receiving loop. Electric waves travelled from the coil to the loop, creating a current in the loop and causing a spark to form across the gap. Hertz went on to determine the velocity of these waves (which were later called radio waves), and on showing that it was the same as that of light, devised experiments to show that the waves could be reflected, refracted, and diffracted. One particular experiment used prisms made of pitch to demonstrate refraction in the same way that light is refracted by a glass prism.

From about 1890 Hertz gained an interest in mechanics. He developed a system with only one law of motion – that the path of a mechanical system through space is as straight as possible and is travelled with uniform motion. When this law is developed subject to the nature of space and the constraints on matter, it can be shown that the mechanics of Isaac ◊Newton, Joseph ◊Lagrange, and William Rowan ◊Hamilton, who extended the methods of dealing with mechanical systems considerably, arise as special cases.

Heinrich Hertz made a discovery vital to the progress of technology by demonstrating the generation of radio waves. It is tragic that his early death robbed him of the opportunity to develop his achievements and to see Guglielmo ◊Marconi and others transform his discovery into a worldwide method of communication.

Further Reading

Baird, Davis; Hughes, R I; and Nordmann, Alfred (eds), *Heinrich Hertz: Classical Physicist, Modern Philosopher,* Boston Studies in the Philosophy of Science, Kluwer Academic Publishers, 1997, v 198.

Buckwald, Jed Z, *The Creation of Scientific Effects: Heinrich Hertz and Electric Waves,* University of Chicago Press, 1994.

Mulligan, Joseph F, *Heinrich Rudolph Hertz (1857–1894): A Collection of Articles and Addresses,* Garland Publishing, 1994.

Susskind, C, *Heinrich Hertz: A Short Life,* San Francisco Press, 1995.

Hertzsprung, Ejnar (1873–1967) Danish astronomer and physicist who, having proposed the concept of the absolute magnitude of a star, went on to describe for the first time the relationship between the absolute magnitude and the temperature of a star, formulating his results in the form of a graphic diagram that has since become a standard reference.

Hertzsprung was born in Frederiksberg on 8 October 1873. His father was interested in astronomy and stimulated a similar fascination for the subject in his son, but the poor financial prospects of an aspiring astronomer led Hertzsprung initially to choose chemical engineering as his career. He graduated from the Frederiksberg Polytechnic in 1898 and then went to St Petersburg, Russia, where he worked as a chemical engineer until 1901. Returning to Copenhagen via Leipzig, he studied photochemistry under Wilhelm ◊Ostwald and began to work as a private astronomer at the observatory of the University of Copenhagen and the Urania Observatory in Frederiksberg. Under the generous tutelage of H Lau, Hertzsprung rapidly acquired the skills of contemporary astronomy.

Following a correspondence with Karl ◊Schwarzschild at the University of Göttingen, Hertzsprung was invited to take up the post of assistant professor of astronomy at the Göttingen Observatory in 1909. When Schwarzschild moved on to the Potsdam Astrophysical Observatory later that year, Hertzsprung went with him. Willem ◊de Sitter was appointed director of the Leiden Observatory in the Netherlands in 1919, and he appointed Hertzsprung head of its department of astrophysics. Within a year, Leiden University appointed Hertzsprung a professor, and on the death of de Sitter he also became director of the Leiden Observatory. He retired in 1945 and returned to Denmark, but he did not cease his astronomical research until well into the 1960s.

Hertzsprung's outstanding contributions to astrophysics were recognized by his election to many prestigious scientific academies and societies and with the award of a number of honours. He died in Roskilde, Denmark, on 21 October 1967.

It was quite early in his work at the observatory in Copenhagen that Hertzsprung realized the importance of photographic techniques in astronomy. He was extremely well qualified to apply these methods and did so with great precision and energy. In 1905 he published the first of two papers (the second appearing in 1907) in a German photographic journal on the subject of stellar radiation. He proposed a standard of stellar magnitude (brightness) for scientific measurement, and defined this 'absolute magnitude' as the brightness of a star at the distance of ten parsecs (32.6 light years). As a further innovation, he described the relationship between the absolute magnitude and the colour – that is, the spectral class or temperature – of a star. During the following year (1906), Hertzsprung plotted a graph of this relationship in respect of the stars of the Pleiades. Later, he noticed that there were some stars of the same spectral class that were much brighter, and some that were much dimmer, than the Sun. He named these the red giants and the red dwarfs respectively.

Publication of his papers in a photographic journal, and refusal altogether to publish his diagrammatic material (because of diffidence in the quality of his own observations), meant that his discoveries were simply not known by Hertzsprung's fellow astronomers. And in 1913 Henry ◊Russell, a US astronomer, presented to the Royal Astronomical Society a diagram depicting the relationship that Hertzsprung had previously and independently discovered, between the temperature and absolute magnitude of stars. Credit was eventually accorded to both astronomers equally and the diagram named after both of them.

The Hertzsprung–Russell diagram, one of the most important tools of modern astronomy, consists of a log–log plot of temperature versus absolute magnitude. As plotted, the stars range themselves largely along a curve running from the upper left (the blue giant stars) to the lower right (the red dwarf stars) of the graph. This apparent arrangement is simply a reflection of the mass of each star, which is responsible for its temperature and luminosity. Approximately 90% of stars belong to this 'main sequence'; most of the rest are red giants, blue dwarfs, Cepheid variables, or novae. The blue giant stars are giant hot stars, the red dwarfs are compact cooler stars. Our Sun lies near the middle of the main sequence, and is classed as a yellow dwarf.

One of the earliest uses of the Hertzsprung–Russell diagram was devised in 1913, when Hertzsprung developed the method of 'spectroscopic parallax' (as distinct from 'trigonometric parallax') for the determination of the distances of stars from the Earth. His method relied on data for the proper motions of the nearest (galactic) Cepheids and on Henrietta ◊Leavitt's data for the periods of the Cepheids (which are variable stars) in the Small Magellanic Cloud (which is, it turned out, extragalactic). He deduced the distances of the nearest Cepheids from their proper motions and correlated them with their absolute magnitude. He then used Henrietta Leavitt's data on the length of their periods to determine their absolute magnitude and hence their distance. He found the Small Magellanic Cloud Cepheids to be at an incredible distance of 10,000 parsecs. His method was excellent, but there was a serious source of error, which led to an overestimation of the distance: he had not accounted for the effect of galactic absorption of stellar light. Nevertheless, this work earned Hertzsprung the Gold Medal of the Royal Astronomical Society.

The Hertzsprung–Russell diagram has also been essential to the development of modern theories of stellar evolution. As stars age and deplete their store of nuclear fuel, they are believed to leave the main sequence and become red giants. Eventually they radiate so much energy that they then cross the main sequence and collapse into blue dwarfs. Larger stars may follow a different pattern and explode into novae, or collapse to form black holes, at the end of their lifespans.

In 1922 Hertzsprung published a catalogue on the mean colour equivalents of nearly 750 stars of magnitude greater than 5.5. This catalogue was notable for the particularly elegant manner in which Hertzsprung managed to analyse the data to uncover a linear relationship.

Most of Hertzsprung's later work was devoted to the study of variable and of double stars. He worked on variable stars (especially Polaris) at Potsdam in 1909, and later in Johannesburg (from 1924 to 1925) and at the Harvard College Observatory (1926) and at Leiden.

Herzberg, Gerhard (1904–) German-born Canadian physicist who is best known for his work in determining – using spectrocopy – the electronic structure and geometry of molecules, especially free radicals (atoms or groups of atoms that possess a free, unbonded electron). He has received many honours for his work, including the 1971 Nobel Prize for Chemistry.

Herzberg was born on 25 December 1904 in Hamburg, where he received his early education. He then studied at the Technische Universität in Darmstadt, from which he gained his doctorate in 1928, and carried out postdoctoral work at the universities of Göttingen, 1928–29, and Bristol, 1929–30. On returning to Germany in 1930 he became a *Privatdozent* (an unsalaried lecturer) at the Darmstadt Technische Universität but in 1935, with the rise to power of Adolf Hitler, he fled to Canada, where he became research professor of physics at the University of Saskatchewan, Saskatoon, 1935–45. He spent the period 1945–48 in the USA, as professor of spectroscopy at the Yerkes Observatory (part of the University of Chicago) in Wisconsin, then returned to Canada. From 1939 until his retirement in 1969 he was director of the Division of Pure Physics for the National Research Council in Ottawa – a laboratory generally acknowledged as being one of the world's leading centres for molecular spectroscopy.

Herzberg's most important work concerned the application of spectroscopy to elucidate the properties and structure of molecules. Depending on the conditions, molecules absorb or emit electromagnetic radiation (much of it in the visible part of the spectrum) of discrete wavelengths. Moreover, the radiation spectrum is directly dependent on the electronic and geometric structure of an atom or molecule and therefore provides detailed information about molecular energies, rotations, vibrations, and electronic configurations. Herzberg, studying common molecules such as hydrogen, oxygen, nitrogen, and carbon monoxide, discovered new lines in the spectrum of molecular oxygen; called Herzberg bands, these spectral lines have been useful in analysing the upper atmosphere. He also elucidated the geometric structure of molecular oxygen, carbon monoxide, hydrogen cyanide, and acetylene (ethyne); discovered the new molecules phosphorus nitride and phosphorus carbide; proved the existence of the methyl and methylene free radicals;

and demonstrated that both neutrons and protons are part of the nucleus. His research in the field of molecular spectroscopy not only provided experimental results of fundamental importance to physical chemistry and quantum mechanics but also helped to stimulate further research into the chemical reactions of gases.

In addition, Herzberg provided much valuable information about certain aspects of astronomy. He interpreted the spectral lines of stars and comets, finding that a rare form of carbon exists in comets. He also showed that hydrogen exists in the atmospheres of some planets, and identified the spectra of certain free radicals in interstellar gas.

Herzog, Bertram (1929–) German-born US computer scientist who became one of the major pioneers in the use of computer graphics in engineering design.

Herzog was born in Offenburg, Germany, on 28 February 1929. He went to the USA and became a US citizen, and began his university education at the Case Institute of Technology, from which he graduated with a bachelor's degree in engineering in 1949. In the early 1950s he worked as a structural engineer before taking up appointments as an associate professor, first at Doanbrook and then at the University of Michigan, where he obtained a doctorate in engineering mechanics in 1961.

In 1963 Herzog joined the Ford Motor Company as engineering methods manager, where he extensively applied computers to tasks involved in planning and design. During this time he was engaged in bringing the developing field of computer graphics to the requirements of design problems in the motor industry. Herzog remained as a consultant to Ford, while returning to academic life in 1965 as professor of industrial engineering at the University of Michigan. Two years later he became director of the Computer Center and professor of electrical engineering and computer science at the University of Colorado.

Hess, Germain Henri (German Ivanovich Gess) (1802–1850) Swiss-born Russian chemist, best known for his pioneering work in thermochemistry and the law of constant heat summation named after him.

Hess was born in Geneva on 7 August 1802, the son of a Swiss artist. When he was only three years old his family moved to St Petersburg to enable his father to be a tutor in a rich Moscow family and they adopted a Russian way of life. Hess became known as German Ivanovich Gess. He qualified in medicine in 1825 at the University of Dorpat, where he also received a thorough grounding in chemistry and geology. He went to Stockholm for a short time to study chemistry under Jöns ◊Berzelius, then returned to Russia for a geological expedition to the Urals before settling in Irkutsk in a medical practice. He was elected an adjunct member of the St Petersburg Academy of Sciences in 1828 and two years later was chosen as extraordinary academician. He then settled in St Petersburg, abandoned his medical practice, and devoted the rest of his life to chemistry.

Hess then took various academic appointments: commissioner to plan the course in practical and theoretical chemistry at the St Petersburg Technological Institute (leading to a professorship there); at the Mining Institute and Chief Pedagogical Institute; and in 1838 at the Artillery School. One of his students at the Chief Pedagogical Institute was A A Voskressenskii, who was later to become the professor of chemistry and teach Dmitri ◊Mendeleyev.

In 1834 Hess was made an academician of the Russian Academy of Sciences. Between 1838 and 1843 he did the research in thermochemistry for which he is most famous, but then became less active in the field of chemical research, concentrating more on education and the seeking of due recognition for other workers. He was responsible, for example, for the granting of the prestigious Demidov Award to Karl Klaus for his discovery of ruthenium. In 1848 Hess's health failed, and he died in St Petersburg on 30 November 1850.

Hess pioneered in the field of thermochemistry, which, in a climate concerned mainly with analysis and synthesis in organic and inorganic chemistry, had largely been left to physicists. He had previously worked in various other areas, including the prevention of endemic eye disorders, the analysis of water and minerals from various parts of Russia, the analysis of natural gas from the region of Baku, the oxidation of sugars, and the chemical properties of waxes and resins. His textbook *Fundamentals of Pure Chemistry* (1831) remained the standard work in the Russian language until Mendeleyev's books of the 1860s.

Hess's first paper on thermochemistry, on 'The evolution of heat in multiple proportions', was published in 1838. Two years later he published the full text of Hess's law in both French and German, which states that the heat change in a given chemical reaction depends only on the initial and final states of the system and is independent of the path followed, provided that heat is the only form of energy to enter or leave the system. Every chemical change either absorbs or evolves heat (even if the amount is not enough to cause a measurable temperature change); reactions that absorb heat are called endothermic, those that evolve heat are termed exothermic. According to modern convention, evolved heat is negative in sign and absorbed heat is positive. The symbol ΔH denotes a change in heat content at constant pressure, the unit of heat and energy being the joule.

Suppose a substance A can be converted to substance D either directly, with a heat of reaction of w joules, or indirectly by way of substances B and C, with heats of reaction of the three stages of x, y, and z joules respectively. Then for:

$$A \rightarrow D, \Delta H = -w$$
$$A \rightarrow B, \Delta H = -x$$
$$B \rightarrow C, \Delta H = -y$$
$$C \rightarrow D, \Delta H = -z$$

According to Hess's law the heat of reaction in going from A to D is the same whether the change is achieved directly or through a series of changes, and so $w = x + y + z$. Heat changes can thus be added algebraically, allowing the calculation of heats of reaction that it would be impossible to measure directly by experiment. It is in modern terms merely an application of the law of conservation of energy, which was not formulated until 1842. In that year Hess proposed his second law, the law of thermoneutrality, which states that in exchange reactions of neutral salts in aqueous solution, no heat effect is observed. No explanation of this law was forthcoming until the announcement of Svante ◊Arrhenius's theory of electrolytic dissociation in 1887.

After Hess died no other researchers carried on his work, and thermochemistry was neglected for the next decade. Investigations of heats of reaction had to be carried out all over again (by scientists such as Pierre ◊Berthelot in France and Hans Thomsen in Denmark). Final recognition came in 1887, when Wilhelm ◊Ostwald began the section on thermochemistry in his *General Chemistry* textbook with a full account of Hess's work.

Hess, Harry Hammond (1906–1969) US geologist who played the key part in the plate tectonics revolution of the 1960s.

Born in New York on 24 May 1906, Hess first studied electrical engineering and then trained in geology at Yale and later Princeton. From 1931, he began geophysical researches into the oceans, accompanying F A Vening Meinesz on a Caribbean submarine expedition to measure gravity and take soundings. During World War II, Hess continued his oceanographic investigations, undertaking extended echo-soundings while captain of the assault transport USS *Cape Johnson.* During the war years he made studies of flat-topped sea mountains, which he called guyots (after Arnold Guyot, an earlier Princeton geologist). In the postwar years, he was one of the main advocates of the Mohole project, whose aim was to drill down through the Earth's crust to gain access to the upper mantle.

The vast increase in knowledge about the sea bed (deriving in part from wartime naval activities) led to the recognition that certain parts of the ocean floor were anomalously young. Building on Maurice ◊Ewing's discovery of the global distribution of mid-ocean ridges and their central rift valleys, Hess enunciated in 1962 the notion of deep-sea spread. This contended that convection within the Earth was continually creating new ocean floor at mid-ocean ridges. Material, Hess claimed, was incessantly rising from the Earth's mantle to create the mid-ocean ridges, which then flowed horizontally to constitute new oceanic crust. It would follow that the further from the mid-ocean ridge, the older would be the crust – an expectation confirmed by research in 1963 by D H Matthews and his student, F J Vine, into the magnetic anomalies of the sea floor. Hess envisaged that the process of sea-floor spreading would continue as far as the continental margins, where the oceanic crust would slide down beneath the lighter continental crust into a subduction zone, the entire operation thus constituting a kind of terrestrial conveyor belt. Within a few years, the plate tectonics revolution Hess had spearheaded had proved entirely successful. Hess's role in this 'revolution in the earth sciences' was largely due to his remarkable breadth as geophysicist, geologist, and oceanographer. He died in Woods Hole, Massachusetts, on 25 August 1969.

Hess, Victor Francis (1883–1964) Austrian-born US physicist who discovered cosmic rays, for which he was jointly awarded the 1936 Nobel Prize for Physics with Carl ◊Anderson.

Hess was born in Waldstein, Austria, on 24 June 1883, the son of a forester. He was educated in Graz, at the Gymnasium then at the university, obtaining his doctorate from the latter in 1906. From 1906 to 1910 he worked at the Vienna Physical Institute as a member of F Exner's research group studying radioactivity and atmospheric ionization, then from 1910 to 1920 he was an assistant to S Meyer at the Institute of Radium Research of the Viennese Academy of Sciences. In 1920 Hess was appointed extraordinary professor of experimental physics at Graz University. From 1921 to 1923, however, he was on a two-year sabbatical in the USA, as director of the research laboratory of the US Radium Corporation in Orange, New Jersey, and also as a consultant to the Department of the Interior's Bureau of Mines. In 1925, two years after his return to Graz University, he became its professor of experimental physics, a post he held until 1931, when he was appointed professor of physics at Innsbruck University and director of its newly established Institute of Radiology; while at Innsbruck he founded a cosmic-ray observatory on the Hafelekar mountain, near Innsbruck. After the Nazi occupation of Austria in 1938 Hess – a Roman Catholic himself but with a

Jewish wife – emigrated to the USA, where he was professor of physics at Fordham University, New York City, until his retirement in 1956 having become a naturalized US citizen in 1944. Hess died in Mount Vernon, New York, on 17 December 1964.

In the early 1900s it was found that gases in the atmosphere are always slightly ionized, even samples that had been enclosed in shielded containers. The theories to explain this phenomenon included radioactive contamination by the walls of the containers and the influence of gamma rays in the soil and air; these theories were later proved incorrect by Hess's findings. In 1910 Theodor Wulf, investigating atmospheric ionization using an electroscope on top of the Eiffel Tower, found that ionization at 300 m/1,000 ft above the ground was greater than at ground level, from which he concluded that the ionization was caused by extraterrestrial rays.

Continuing this line of research, Hess – with the help of the Austrian Academy of Sciences and the Austrian Aeroclub – made ten balloon ascents 1911–12 to collect data about atmospheric ionization. Ascending to altitudes of more than 5,000 m/16,000 ft, he established that the intensity of ionization decreased to a minimum at about 1,000 m/3,000 ft then increased steadily, being about four times more intense at 5,000 m/16,000 ft than at ground level. Moreover, by making ascents at night – and on one occasion, on 12 April 1912, during a nearly total solar eclipse – he proved that the ionization was not caused by the Sun. From his findings Hess concluded that radiation of great penetrating power enters the atmosphere from outer space. This discovery of cosmic rays (as this type of radiation was called by Robert ◊Millikan in 1925) led to the study of elementary particles and paved the way for Carl Anderson's discovery of the positron in 1932.

Later in his career Hess investigated the biological effects of exposure to radiation (in 1934 he had a thumb amputated following an accident with radioactive material); the gamma radiation emitted by rocks; dust pollution of the atmosphere; and the refractive indices of liquid mixtures.

Hevelius, Johannes (1611–1687) Latinized form of Jan Hewel or Hewelcke. German astronomer, most famous for his careful charting of the surface of the Moon.

Hevelius was born in Danzig (now Gdansk) in northern Poland on 28 January 1611. A wealthy brewing merchant, he had a well-equipped observatory installed on the roof of his house in 1641, and was one of the most active observers of the 17th century. During the daytime he worked in his business and some evenings he took his seat on the city council, but most of the rest of his free time he was up on his roof, observing, noting, and cataloguing. His wife Elizabeth shared his interests and assisted him greatly in the study of the Moon, his catalogue of the stars, and his work on comets. After his death, she edited and published his most famous work, *Prodromus astronomiae* (1690).

Between 1642 and 1645, Hevelius deduced a fairly accurate value for the period of the solar rotation and gave the first description of the bright ideas in the neighbourhood of sunspots. The name he gave to them, *faculae,* is still used. He also made observations of the planets, particularly of Jupiter and Saturn. On 22 November 1644 he observed the phases of Mercury, which had been predicted by Nicolaus ◊Copernicus.

In 1647, Hevelius published the first comparatively detailed map of the Moon, based on ten years' observations. It contained diagrams of the different phases for each day of lunation. He realized that the large, uniform grey regions on the lunar disc consisted of low plains, and that the bright contrasting regions represented higher, mountainous relief. He obtained better values for the heights of these lunar mountains than had ◊Galileo a generation before. His *Selenographia* also has an appendix that contains his observations of the Sun 1642–45.

Hevelius was interested in positional astronomy and planned a new star catalogue of the northern hemisphere, which was to be much more complete than that of Tycho ◊Brahe. He began in 1657, but his observatory, with some of his notes, was destroyed by fire in 1679. Nevertheless, his observations enabled him to catalogue more than 1,500 stellar positions. The resulting *Uranographia* contains an excellent celestial atlas with 54 plates, but Hevelius' practice of using only the naked eye to observe positions (despite representations by no less a man than Edmond ◊Halley) considerably reduces the value of his work. Hevelius used telescopes for details on the Moon and planets, but refused to apply them to his measuring apparatus.

Hevelius discovered four comets – he called them *pseudo-planetae* – and suggested that these bodies orbited in parabolic paths about the Sun. Many later writers have declared that this suggestion indicates that he knew the nature of comets earlier than did either Halley or ◊Newton.

A few of the names he gave to features of the Moon's surface are still in use today, particularly those that reflect geographical names on Earth. For his charting of the lunar formations, Hevelius has come to be known as the founder of lunar topography.

Hevesy, Georg von (1885–1966) Hungarian-born Swedish chemist whose main achievements were the introduction of isotopic tracers (to follow chemical reactions) and the discovery of the element hafnium.

For his work on isotopes he was awarded the 1943 Nobel Prize for Chemistry. In 1959 he received the Atoms for Peace Prize.

Hevesy was born in Budapest on 1 August 1885, into a family of industrialists. He was educated there in a Roman Catholic School and then attended the Technische Hochschule in Berlin with the intention of training as a chemical engineer. But he contracted pneumonia and moved to the more agreeable climate of Freiburg. After obtaining his doctorate in 1908 he moved again to study chemistry at the Zürich Technische Hochschule, where he worked under Richard Lorenz on the chemistry of molten salts. Following yet another move to the Karlsruhe laboratories of Fritz ⇨Haber (where he investigated the emission of electrons during the oxidation of sodium/potassium alloys), Hevesy took Haber's advice and went to Manchester in 1911 to learn some of the new research techniques being developed by Ernest ⇨Rutherford, particularly those involving radioactive elements.

In 1913 Hevesy went to join Friedrich ⇨Paneth in Vienna for a short time to continue these studies and but for the outbreak of World War I in 1914 would have carried on this work with Henry ⇨Moseley at Oxford. Instead he continued his researches in Budapest and then in 1920 went to Copenhagen to work in the Institute of Physics under the guidance of Niels ⇨Bohr and Johannes ⇨Brönsted. He returned to Freiburg in 1926, only to go back to Copenhagen eight years later. In 1943, during World War II, he escaped to Sweden from the German occupation of Denmark and became a professor at the University of Stockholm, where he remained until shortly before his death. During an academic career extending over 58 years Hevesy worked in nine major research centres in seven European countries. He died at a clinic in Freiburg on 5 July 1966.

At Manchester in 1911 Rutherford set Hevesy the task of separating radioactive radium-D from 100 kg/220 lb of lead. After a year's work Hevesy had been unable to achieve any separation – neither could he detect any chemical differences between radium-D and lead (we now know that this was an impossible task using conventional chemical techniques because radium-D is an isotope of lead, lead-210). But he turned this similarity to advantage by mixing some pure radium-D with ordinary lead and following chemical reactions of the lead by detecting its 'acquired' radioactivity. He perfected this technique while working with Paneth in Vienna from 1913, using added radium-D to study the chemistry of lead and bismuth salts. This was the beginning of the use of radioactive tracers.

During his first period at Copenhagen, 1920–26, Hevesy successfully separated isotopes of mercury using fractional distillation at low pressure. During an investigation of zirconium minerals using X-rays, he and Dirk Coster in 1922 discovered hafnium (so-called after Hafnia, the Latin name for Copenhagen). Continuing his interest in isotopes, he effected a method of isotopic enrichment of chlorine and potassium.

On his return to Freiburg in 1926, he studied the relative abundances of elements on Earth and in the universe, basing his calculations on chemical analyses by means of X-ray fluorescence. During the early 1930s he commenced experiments with his radioactive tracer technique on biological specimens, noting, for example, the take-up of radioactive lead by plants. The production of an unstable isotope of phosphorus in 1934 enabled the first tracer studies to be made on animals. Hevesy used the isotope to trace the movement of phosphorus in the tissues of the human body. During his second stay at Copenhagen, 1934–43, he took with him a sample of heavy water (donated by Harold ⇨Urey in the USA, who in 1923 had spent a year at Bohr's laboratory). Hevesy used the sample to study the mechanism of water exchange between goldfish and their surroundings and also within the human body. He then extended his experiments using potassium-42, sodium-34, and chlorine-38.

During his later years in Stockholm he continued to work on the transfer of radioactive isotopes within living material. Using radioactive calcium to label families of mice he showed that of the calcium atoms present at birth about 1 in 300 are passed on to the next generation.

Hewish, Antony (1924–) English astronomer and physicist whose research into radio scintillation resulted in the discovery of pulsars. For this achievement, and for his continued work in the field, he was awarded the 1974 Nobel Prize for Physics (jointly with Martin ⇨Ryle).

Hewish was born on 11 May 1924 in Fowey, Cornwall, and attended King's College, Taunton, before going to Gonville and Caius College at Cambridge. Graduating in 1948, Hewish worked at the Telecommunications Research Establishment in Malvern, where he met Martin Ryle, who was later to become Astronomer Royal. With Ryle, Hewish then became part of a team undertaking solar and interstellar research by radio at the Cavendish Laboratory, Cambridge, and carrying out a series of intensive surveys known respectively, when published, as the *First, Second, Third,* and *Fourth Cambridge Catalogue.* In the later 1960s, he initiated research into radio scintillation at the Mullard Observatory; the discovery of the first pulsar – and then another three – followed. Made reader at Cambridge University in 1969, he became professor of radio astronomy there in 1971. Hewish became head of the Mullard Radio Astronomy Observatory 1982–88. He retired in 1989.

Before 1950, Hewish's experimental work using radio telescopes was directed in the main to the study of solar atmosphere. He used simple corner reflectors to examine the Sun's outer corona in order to discover the electron density in its atmosphere, and to study the irregular hot gaseous clouds of plasma surrounding the Sun. After 1950, when – mainly through the efforts of Ryle – new instruments became available, radio observations were extended to sources other than the Sun. In particular Hewish examined the fluctuation in such sources of the intensity of the radiation (the scintillation) resulting from disturbances in ionized gas in the Earth's atmosphere, within the Solar System, and in interstellar space. Engaged in this research at Mullard Observatory one day in 1967, Hewish's attention was drawn by a research student, Jocelyn Bell (later ⟡Bell Burnell), to some curiously fluctuating signals being received at regular intervals during the sidereal day. Installation of a more sensitive high-speed recorder revealed, on 28 November 1967, the first indication of a pulsed emission whose fixed celestial direction ruled out the possibility of an artificial source. The November results were confirmed in early December, and verified that Hewish and Bell had discovered pulsating radio stars, or pulsars.

The results of a more detailed investigation by Hewish were published in February 1968, by which time three more pulsars had been identified. This investigation showed that pulsars are sources of radio emission in our galaxy that give out radiation in brief pulses. Although each pulsar emits radiation with nearly constant pulsation periods, the rate of emission differs between pulsars, as does the 'shape' of the pulse. Pulses can be single-, double-, or even triplepeaked. Shapes can differ within each pulsar, but the mean pulse shape changes only very slightly – generally to decrease the interval – over many months. Comparative studies of successive shapes have shown that each pulse itself is made up of two pulsatory constituents: a regular (class 1) one, and another (class 2) pulsating at irregular frequency within the first one.

It is now generally accepted – as originally proposed by Thomas ⟡Gold and others – that pulsars are rotating neutron stars (stars that are nearing the end of their stellar life, having practically exhausted their nuclear energy). It remains less clear how rotational energy is converted to emission of such shapes and such intensity as have been discovered, although many hypotheses have been put forward.

Hewish's initial discovery of four pulsars began a period of intensive research, with more than 500 pulsars having been found in our Galaxy since 1967 (although a million or so may exist). In the meanwhile, Hewish has patented a system of space navigation using three pulsars as reference points, that would provide 'fixes' in outer space accurate up to a few hundred kilometres.

Hey, James Stanley (1909–) English physicist whose work in radar led to pioneering research in radio astronomy.

Hey was born on 3 May 1909 in the Lake District. Reading physics at Manchester University, he gained his master's degree in X-ray crystallography in 1931. From 1940 to 1952 he was on the staff of an Army Operational Research Group, for the last three years as the head of the establishment. He then became a research scientist at the Royal Radar Establishment, being promoted in 1966 to chief scientific officer. He retired in 1969, although he continued to write about astronomy from his home in Sussex. Hey was made a fellow of the Royal Society in 1978.

Between 26 and 28 February 1942, during World War II, the British early-warning coastal defence radar became severely jammed. At first the jamming was attributed to enemy counter-measures but Hey, noting that the interference began as the Sun rose and ceased as it set, concluded that the spurious radio radiation emanated from the Sun and that it was related to solar activity. And he further proposed that the radiation – in strength about 10^5 of the calculated black-body radiation – was associated with a large solar flare that had just been reported.

At the end of World War II, research began in earnest at the Royal Radar Establishment in Malvern. For some years, Grote ⟡Reber had been working along the same lines in the USA, and he had published his discovery of intense radio sources located in the Milky Way, notably in the constellations Cygnus, Taurus, and Cassiopeia. The announcement in 1946 by Hey and his colleagues that they had narrowed down the location of the radiation source in Cygnus to 'a small number of discrete sources' (in fact, to Cygnus A) stimulated a search for other discrete sources around the world. Attempts were also made to devise methods to achieve better resolution in the locating of radio sources so that, eventually, it would be possible to identify such sources as optically observed objects.

Hey and his team also returned to a study of the Sun as a radio source. They discovered that large sunspots were powerful ultra-shortwave radio transmitters and that, although the Sun was constantly emitting radio waves, they were of unexpected strength.

Using radio, the team noted that they could detect and follow meteors more accurately than ever before.

Heyrovský, Jaroslav (1890–1967) Czech chemist who was awarded the 1959 Nobel Prize for Chemistry for his invention and development of polarography, an electrochemical technique of chemical analysis.

Heyrovský was born in Prague on 20 December 1890, the son of a professor of law at Charles University, Prague. He studied chemistry, physics, and mathematics at his father's university and graduated from there in 1910. He then went to University College, London, to pursue postgraduate research under William ◊Ramsay and Frederick Donnan (1870–1956), who aroused his interest in electrochemistry. He became a demonstrator in the chemistry department in 1913. He served in a military hospital during World War I and in 1920 was appointed an assistant in the Institute of Analytical Chemistry at Prague. He subsequently became a lecturer in 1922, assistant professor in 1924, and professor of physical chemistry in 1926. He remained at the institute until 1950, when he became director of the newly founded Polarographic Institute of the Czechoslovak Academy of Sciences. He revisited London in November 1955 to deliver his presidential address to the Polarographic Society. He died in Prague on 27 March 1967.

Heyrovský began the work that was to lead to the invention of polarography during his student days in London while he was investigating the electrode potential of aluminium. The technique was perfected in 1922, soon after his return to Prague. It depends on detecting the discharge of ions during electrolysis of aqueous solutions. The solution to be analysed is placed in a glass cell containing two electrodes, one above the other. The lower electrode is simply a pool of mercury. The upper electrode, called a dropping mercury electrode, consists of a fine capillary tube through which mercury flows and falls away as a series of droplets. The growing mercury droplet at the tip of the capillary tube constitutes the actual electrode.

With the dropping mercury electrode made, say, the cathode (to analyse for cations in solution), the voltage between the electrodes is slowly increased and the associated current observed on a galvanometer. When a cation is discharged (reduced) at the cathode, the current increases rapidly and then remains at a constant value. At that time, the voltage is characteristic of the cation concerned and the magnitude of the current is a measure of its concentration.

In 1925, together with M Shikita, Heyrovský developed the polarograph, an instrument that automatically applies the steadily increasing voltage and traces the resulting voltage–current curve on a chart recorder. Such curves are called polarograms.

Polarography can be used to analyse for several substances at once, because the polarogram records a separate limiting current for each. It is also extremely sensitive, being capable of detecting concentrations as little as 1 part per million. Most chemical elements can be determined by the method (as long as they form ionic species) in compounds, mixtures, or alloys. The technique has been extended to organic analysis and to the study of chemical equilibria and the rates of reactions in solutions. It can also be used for endpoint detection in titrations, a type of volumetric analysis sometimes called voltammetry.

Hilbert, David (1862–1943) German mathematician, philosopher, and physicist whose work in all three disciplines was brilliantly innovative and fundamental to further development. An excellent teacher, unsurpassed in lucidity of exposition, his influence on 20th-century mathematics has been enormous. He is particularly remembered for his research on the theory of algebraic invariants, on the theory of algebraic numbers, on the formulation of abstract axiomatic principles in geometry, on analysis and topology (in which he derived what is now called Hilbert's theory of spaces), on theoretical physics, and finally on the philosophical foundations of mathematics.

Hilbert was born on 23 January 1862 in Königsberg, then in German Prussia, now Kaliningrad in Russia. He grew up and was educated there, attending Königsberg University – an ancient and venerable seat of learning – 1880–85, when he received his PhD. He then studied further in Leipzig and in Paris before returning to Königsberg University to become an unsalaried lecturer. Six years later he was appointed professor, and three years later still (in 1895) he was offered the highly prestigious post of professor of mathematics at Göttingen University. He accepted, and held the position until he retired in 1930. Although he developed pernicious anaemia in 1925, he recovered, and died in Göttingen on 14 February 1943.

Hilbert's first period of research, 1885–92), was on algebraic invariants; he tried to find a connection between invariants and fields of algebraic functions and algebraic varieties. Representing the rational function in terms of a square, he eventually arrived at what is known as Hilbert's irreducibility theorem, which states that, in general, irreducibility is preserved if, in a polynomial of several variables with integral coefficients, some of the variables are replaced by integers. He later investigated ninth-degree equations, solving them by using algebraic functions of only four variables. In this way, Hilbert had by the end of the period not only solved all the known central problems of this branch of mathematics, he had in his methodology introduced sweeping developments and new areas for research (particularly in algebraic topology, which he himself returned to later).

In 1897, with some help from his colleague and friend Hermann ◊Minkowski, Hilbert produced *Der*

Zahlbericht, in which he gathered together all the relevant knowledge of algebraic number theory, reorganized it, and laid the basis for the developing class-field theory. He abandoned this work, however, when there was still much to be done.

Two years later, having moved to another area of study, Hilbert published his classic work, *Grundlagen der Geometrie.* In it, he gave a full account of the development of geometry in the 19th century, and although on this occasion his innovations were (for him) relatively few, his use of geometry, and of algebra within geometry, to devise systems incorporating abstract yet rigorously axiomatic principles, was important both to the further development of the subject and to Hilbert's own later work in logic and consistency proofs. In the related field of topology, he referred back to his previous work on invariants in order to derive his theory of spaces in an infinite number of dimensions.

In 1900, attending the International Congress of Mathematicians in Paris, Hilbert set the congress a total of 23 hitherto unsolved problems. Many have since been solved – but solved or not, the problems stimulated considerable scientific debate, research, and fruitful development.

In a study of mathematical analysis some years afterwards, Hilbert used a new approach in tackling Dirichlet's problem (see Lejeune ◊Dirichlet), and made other contributions to the calculus of variations. In 1909 he provided proof of Waring's hypothesis (of a century earlier) of the representation of integers as the sums of powers. From that time forward, Hilbert worked on problems of physics, such as the kinetic theory of gases, and the theory of relativity – problems, he said, too difficult to be confined to physics and physicists. His deep research led finally to his critical work on the foundations of mathematical logic, in which his contribution to proof theory was extremely important by itself.

Hill, Archibald Vivian (1886–1977) English physiologist who studied muscle action in great detail. For this work he received the 1922 Nobel Prize for Physiology or Medicine, which he shared with Otto ◊Meyerhof.

Hill was born on 26 September 1886 in Bristol. He was educated at Blundell's School, Tiverton, and then went to Trinity College, Cambridge. There he excelled at mathematics and was greatly influenced by his tutor Morley Fletcher (1873–1933), who had collaborated with Frederick Gowland ◊Hopkins in the discovery of the role of lactic acid in muscle contraction. Graduating with a medical degree in 1907 he remained at Trinity until 1914 when World War I broke out, in which he served with distinction. He became professor of physiology at Manchester University in 1920 after obtaining his doctorate from Trinity earlier that year. He joined the staff of University College, London, in 1923 and three years later took up a professorship at the Royal Society, a position he held until 1951. During that period, 1935–46, he was secretary to the Royal Society. He also served as scientific advisor to India 1943–44 and was active in various scientific organizations, including the Marine Biological Association, of which he was president from 1955. He was a member of the War Cabinet Scientific Advisory Committee during World War II. He died in Cambridge on 4 June 1977.

Influenced by his mentor Fletcher, at Cambridge, Hill researched into the workings of muscles as early as 1911. He was not concerned with the chemical details of muscle action, but wanted rather to ascertain the amount of heat produced during muscle activity. To do this he used delicate thermocouples, and discovered that contracting muscle fibres produce heat in two phases. Heat is first produced quickly as the muscle contracts. Then after the initial contraction, further heat is evolved more slowly but often in greater amounts. The thermocouples that Hill used recorded heat changes quickly and minutely in the form of tiny electric currents. He had to modify this apparatus for his particular purpose and was able to measure a rise in temperature of as little as 0.003° over a few hundredths of a second.

By 1913 Hill was aware that heat is produced after the muscles have contracted and he showed that molecular oxygen is consumed after the work of the muscles is over and not during muscular contraction. This discovery was made by proving that if muscle fibre is made to contract in an atmosphere of pure nitrogen, the first phase of heat production is not affected but the second phase does not take place at all. He realized that oxygen is not necessary in the first phase – that is, the chemical reactions immediately involved in the contraction of muscles do not require oxygen. But in phase two, when muscle contraction has taken place, oxygen is needed to produce further energy for subsequent muscle contraction.

Mammals and birds maintain a constant body temperature, which in human beings is 37 °C/98.6 °F. The maintenance of such a constant temperature is under involuntary control, but can be partly attributed to muscular activity. When a warm-blooded animal is cold, it is necessary for the body to produce heat. One of the ways in which it does so is by shivering, an involuntary contraction of muscles resulting in heat generation by the process elucidated by Hill.

Hill, Robert (1899–1991) English biochemist who contributed greatly to modern knowledge of photosynthesis.

Hill was born on 2 April 1899. After serving in World War I in the Royal Engineers Pioneer Antigas

Department 1917–18, he studied at Emmanuel College, Cambridge. He remained researching there until 1938, by which time he had become a senior research fellow. From 1943 to 1966 he was a member of the scientific staff of the Agricultural Research Council. Hill died in Cambridge on 15 March 1991.

The process of photosynthesis has been shown to occur in two separate sets of reactions: those that require sunlight (the light reactions) and those that do not (the dark reactions). Both sets of reactions are dependent on one another. In the light reactions some of the energy of sunlight is trapped within the plant and in the dark reactions this energy is used to produce potentially energy-generating chemicals, such as sugar.

In 1894 Engelman first showed that the light reactions of photosynthesis occur within the chloroplasts of leaves. Hill's experiments in 1937 confirmed the localization of the light reactions within the chloroplast, as well as elucidating in part the mechanism of the light reactions. He isolated chloroplasts from leaves and then illuminated them in the presence of an artificial electron-acceptor. The electron-acceptor he used was a ferric salt (Fe^{3+}) which was reduced to the ferrous form (Fe^{2+}) during the reaction. He showed that during the reaction, oxygen is produced and that this derived oxygen comes from water. (He also demonstrated the evolution of oxygen in human blood cells by the conversion of haemoglobin to oxyhaemoglobin.) This process, known as the Hill reaction, can be summed up by the following equation:

$$4Fe^{3+} + 2H_2O \xrightarrow[\text{chloroplasts}]{\text{light}} 4Fe^{2+} + 4H+ + O_2\uparrow$$

Much more is now known about the mechanism of the Hill reaction in plants, and the equation has become more complicated:

$$2NADP + ADP + P_1 + 2H_2O \xrightarrow[\text{chloroplasts}]{\text{light}} 2NADPH_2 + ATP + O_2\uparrow$$

The electron-acceptor in the plant is now known to be NADP (nicotinamide adenine dinucleotide phosphate), and is reduced in the reaction to $NADPH_2$. ADP (adenosine diphosphate) is phosphorylated to ATP (adenosine triphosphate) in the reaction, a process known as photophosphorylation. In the plant the energy from sunlight is first stored as ATP and later as sugar, the final product of the dark reaction of photosynthesis which uses energy from ATP. Hill's findings were of great significance in further investigations into photosynthesis.

Hinshelwood, Cyril Norman (1897–1967) English physical chemist who made fundamental studies of the kinetics and mechanisms of chemical reactions. He also investigated bacterial growth. He shared the 1956 Nobel Prize for Chemistry with the Russian scientist Nikolai ◊Semenov for his work on chain reaction mechanisms.

Hinshelwood was born in London on 19 June 1897, the son of an accountant. His family emigrated to Canada when he was a child, but returned to England after his father's death. He was educated at the Westminster City School, London, from where he won a scholarship to Balliol College, Oxford. The outbreak of World War I in 1914 interrupted his studies and from 1916 to 1918 he worked in the Department of Explosives at the Royal Ordnance Factory at Queensferry, Scotland. He returned to Balliol in 1919 and took the shortened postwar chemistry course, graduating in 1920 and being immediately elected a fellow of the college. Subsequent academic appointments were as fellow of Trinity College, Oxford, 1921–37; professor of chemistry in the University of Oxford 1937–64; and, on his retirement, senior research fellow at Imperial College, London. Hinshelwood was knighted in 1948; he died in London on 9 October 1967.

While he was at the explosive works in Scotland, Hinshelwood tried to measure the slow rate of decomposition of solid explosives by monitoring the gases they evolved. His first researches at Balliol pursued this line, and he studied the decomposition of solid substances in the presence and absence of catalysts. He then investigated homogeneous gas reactions. He found initially that the thermal decomposition of the vapours of substances such as acetone (propanone) and aliphatic aldehydes occur by means of first or second order processes. If A and B are the reactants, then the rate of a chemical reaction can be expressed as

$$R = k[A]^x[B]^y$$

where R is the rate, k is the velocity constant, [A] and [B] are the concentrations of the reactants, and x and y are powers. If $x = 1$, the reaction is said to be first order; if $x = 2$, it is second order; and so on. At that time the activation of chemical reactions was considered only in terms of collision mechanisms – easily applied to bimolecular reactions but not capable of explaining unimolecular reactions. Hinshelwood showed that even apparently simple decomposition reactions usually occur in stages. Thus for acetone (propanone):

a) $$\underset{\text{acetone (propanone)}}{(CH_3)_2CO} \rightarrow CH_2 = CO + CH_4$$

b) $$CH_2 = CO \rightarrow \tfrac{1}{2}C_2H_4 + CO$$

This early work centred on the relationship between temperature, concentration, and the influence of second and third components.

Hinshelwood went on to demonstrate that many reactions can be explained in terms of a series – a chain – of interdependent stages. At low temperatures the reaction between hydrogen and oxygen, or hydrogen and chlorine, for example, is comparatively slow because the chain reactions involved terminate at the walls of the vessel. But at high temperatures the chain reactions accelerate the process to explosion point. He provided experimental evidence for the role of activated molecules in initiating the chain reaction.

He also investigated reaction kinetics in aqueous and nonaqueous solutions, together with hydrolytic processes, esterification, and acylation of amines. He published his classic work on reaction kinetics, *Kinetics of Chemical Change,* in 1926.

In 1938 Hinshelwood published with S Daglay the first of more than a hundred papers on bacterial growth. Initially he investigated the effects of various nutrients such as carbohydrates and amino acids on simple nonpathogenic organisms. Later he studied the effects of trace elements and toxic substances such as sulphonamides, proflavine, and streptomycin. He considered that all the various chemical reactions that occurred in his bacterial growth experiments were interconnected and mutually dependent, the product of one reaction becoming the reactant for the next – a process he termed 'total integration'.

Hinshelwood's contributions to both these fields won universal acknowledgement during his lifetime, and he was the recipient of many honours. He was president of the Chemical Society during its centenary year (1947) and president of the Royal Society at its tercentenary in 1960. He was a capable linguist and artist, and a year after his death his paintings went on exhibition in Goldsmiths Hall, London.

Hinton, William Augustus (1883–1959) US bacteriologist and pathologist who achieved renown through his work on syphilis, in particular the development of the Hinton test.

William Hinton was born in Chicago on 15 December 1883. Between 1900 and 1902 he studied at the University of Kansas and went on to Harvard College, receiving a BS in 1905. Over the next four years he taught at Walden University, Nashville, Tennessee and in Langston, Oklahoma, and continued his studies in bacteriology and physiology during the summers at the University of Chicago. In 1909 he started at Harvard Medical School and graduated in 1912, having been awarded the Wigglesworth and Hayden scholarships. For three years he worked at the Wasserman Laboratory and as a volunteer assistant in the department of pathology at the Massachusetts General Hospital and in 1915 he was made chief of the Wasserman Laboratory. In 1918 he was appointed instructor in preventive medicine and hygiene at the Harvard Medical School and between 1921 and 1946 he was instructor in bacteriology and immunology and a lecturer until 1949. In 1949 he was promoted to clinical professor – the first black professor in the university's history – a post he held until he retired as professor emeritus in 1950. Hinton died in 1959. He had married in 1919 and had two daughters.

Hinton became internationally known in the field of syphilology. He developed a serological test for syphilis based on flocculation, reducing the number of false positive diagnoses of the disease. In 1934 the US Public Health Service showed the Hinton test for syphilis to be the best. He wrote many scientific papers and his book *Syphilis and its Treatment* (1926) was the first medical textbook by a black American to be published. Hinton was also the discoverer of the Davies–Hinton tests of blood and spinal fluid. Hinton was a member of many societies including the American Medical Association and the American Association of Bacteriologists. In 1938 he turned down the Spingarn Medal as he was afraid that his work would not be accepted on its merit if it were known that he was black. In his will, Hinton left $75,000 to establish the Eisenhower Scholarship Fund for graduate students at Harvard. Fifteen years after his death the Serology Laboratory of the State Laboratory Institute Building of the Massachusetts Department of Public Health was named in his honour.

Hipparchus (lived *c.* 146 BC) Greek astronomer and mathematician whose careful research and brilliant deductions led him to many discoveries that were to be of importance and relevance even 2,000 years later.

Hipparchus was born in Nicaea, in Bithynia (now in Turkey), in about 146 BC. What little is known about him and his life is contained in the writings of Strabo of Amasya and ◊Ptolemy of Alexandria, both of whom were writing well over a hundred years after Hipparchus' death. But it is recorded that Hipparchus carried out his astronomical observations in Bithynia, on the island of Rhodes, and in Alexandria.

In 134 BC Hipparchus noticed a new star in the constellation Scorpio, a discovery that inspired him to put together a star catalogue – the first of its kind ever completed. He entered his observations of stellar positions using a system of celestial latitude and longitude, making his measurements with greater accuracy than any observer before him, and taking the precaution wherever possible to state the alignments of other stars as a check on present position. His finished work, completed in 129 BC, listed about 850 stars classified not by location, but by magnitude (brightness), for which classification he devised a system very close to the modern one. The resulting catalogue was thus so excellent that it was not only plagiarized *en bloc* by Ptolemy, but was used by Edmond Halley 1,800 years later.

Hipparchus was troubled by the fact that the Babylonians had had a different number of days in their year. In trying to resolve this problem he studied the motions not only of the Sun and the Moon, but also of the Earth. In his star catalogue he noted that the star Spica was measured at 6° from the autumn equinox – whereas in the records of Timocharis of Alexandria 150 years earlier it had plainly been at 8°. He came to the conclusion that it was not Spica that was moving, but an east-to-west movement (or precession) of the Earth, and he managed to calculate a value for the annual precession of 45′ or 46′, which, considering the now accepted rate of 50.26′ per year, was an amazing feat of accuracy with the simple instruments available to him at the time. (Ptolemy credits Hipparchus with the invention of an improved kind of theodolite.) The definition of the precession of the equinoxes is usually said to be Hipparchus' most significant work. Using his knowledge of the equinoxes, Hipparchus calculated the terrestrial year to be $365^1/_4$ days, diminishing annually by a three-hundredth of a day (in which he was again astonishingly accurate), and the lunar period to be 29 days, 12 h, 44 min, and 2.5 sec (which was only 1 sec too short). From Hipparchus' time, eclipses of the Moon could be predicted to within one hour, and those of the Sun less accurately.

With such knowledge, it is surprising that Hipparchus accepted the notion of the geocentric universe, declaring that the Sun orbited the Earth in a circle of which Earth was not quite at the centre. The Moon was said to do the same, except that the centre of its orbit was itself moving (a 'moving eccentric').

Hipparchus' astronomical work led him to various areas of mathematics. He was one of the earliest formulators of trigonometry, for which he devised a table of chords. The theorem that provides a basis for the field of plane geometry was also proposed by him, although later attributed to Ptolemy (whose name is still generally attached to it).

Until Nicolaus ◊Copernicus, there was no greater astronomer than Hipparchus.

Hippocrates (*c.* 460–*c.* 377 BC) Greek physician, often known as the founder of medicine and, in ancient times, regarded as the greatest physician who had ever lived. In contrast to the general views of his time, which considered sickness to be brought about by the displeasure of the gods or by possession by demons, Hippocrates looked upon it as a purely physical phenomenon capable of rational explanation.

Hippocrates Greek physician Hippocrates was born around 460 BC on the Greek island of Kos, into a family who had been practising medicine for generations. His idea of the four 'humours' – environmental, racial, climatic, and dietetic – in causing disease remained the accepted doctrine until the 18th century. *Mary Evans Picture Library*

Hippocrates was born on the Greek island of Kos, the son of a physician (although the legends surrounding his feats ascribed him to being a member of a family of magicians). Not much is known for certain about his life as he is referred to little by his contemporaries. He travelled throughout Greece and Asia Minor where the cures he achieved, his great skill, and humanity, together with his exemplary conduct, soon made him famous. Eventually, on his return to Greece. he founded a medical school, one of the first of its kind, on Kos, where he taught.

Hippocrates concerned himself with the whole patient, regarding the body as a whole organism and not just a series of parts. He used few medicines but considered rather that it was the duty of the doctor to find out by careful observation what Nature was trying to do and to meddle as little as possible with the process of healing, relying on good diet, fresh air, rest, and cleanliness in both patient and doctor. Hippocrates did believe, however, that some diseases, such as those resulting from a poor diet, were caused by residues of undigested food that gave off vapours that seeped into the body.

The several generations of doctors who studied under Hippocrates are thought to have contributed to the 72 books known as the *Corpus Hippocraticus/*

Hippocratic Collection, the library of the medical school at Kos that was later assembled in Alexandria, in the 3rd century BC. These works comprise textbooks, research reports, lectures, essays, and clinical notebooks. They contain few correct anatomical observations, although one book contains accurately observed symptoms with the likely outcome for the patient. Another book, *About the Nature of Man,* describes a theory of 'humours', or body fluids. The theory was based on the belief in the existence of four important fluids – blood, phlegm, and yellow and black bile. If the normal levels of these fluids were unbalanced, illness ensued. This theory persisted among physicians in Europe right through the Middle Ages.

Further Reading

Byers, James M, *From Hippocrates to Virchow: Reflections on Human Disease,* ASCP Press, 1987.
Dyce, James, *Hippocrates for Today: A New Dimension to Our Calling – A Study of Basic Values,* Stress Publications, 1995.
Hippocrates; Craik, Elizabeth M (ed and transl), *Hippocrates: Places in Man,* Oxford University Press, 1998.
Jouanna, Jacques, *Hippocrates,* Medicine and Culture Series, Johns Hopkins University Press, 1998.
King, Helen, *Hippocrates and Woman: Reading the Female Body in Ancient Greece,* Routledge, 1998.
Potter, Paul, *Hippocrates,* Loeb Classical Library, Harvard University Press, 1989.
Van der Weyer, Robert, *Hippocrates: The Natural Regimen,* James (Arthur) Ltd, 1997.

Hitchings, George Herbert (1905–1998) US pharmacologist who shared the Nobel Prize for Physiology or Medicine in 1988 for his work on therapeutics, especially anti-cancer agents, and immunosuppressive drugs and antibiotics that were of vital importance in the growing surgical field of transplantation.

Hitchings was born in Hoquiam, Washington on 18 April 1905 and was educated at the University of Washington in Seattle from which he graduated in 1927 and was awarded a master's degree in 1928. He then moved to Harvard University, gaining his doctorate in biochemistry in 1933. He remained at Harvard as an instructor until 1939 when he moved to Case Western Reserve University. In 1942 he joined the biochemical research laboratory of Burroughs Wellcome, where he was joined two years later by Gertrude ◊Elion with whom he was to work closely for the rest of his career, and with whom he shared the Nobel prize. In 1955 Hitchings was appointed associate research director at Burroughs Wellcome; he became the research director of chemotherapy in 1963 and vice president of the company in 1967, from which position he retired in 1975.

Hitchings worked extensively on drug development, initially approaching problems from a basic research perspective, trying to understand fundamental biological mechanisms from which therapeutic strategies could be devised. Experiments in the 1940s on cellular metabolism revealed that some bacteria could not produce DNA and could not therefore divide and grow, in the absence of certain chemicals, particularly purines. Using this information, Hitchings and Elion developed an 'anti-metabolite' philosophy, and started to synthesise compounds that inhibited DNA synthesis, in the hope that these could be used to prevent the rapid growth of cancer cells. Their many investigations of the chemistry of purines and pyrimidines, two groups of chemicals that are involved in DNA synthesis, resulted in their jointly holding at one stage 18 pharmaceutical patents related to these two compounds. Clinical trials in patients with leukaemia revealed that these powerful drugs that rapidly inhibited the growth of cancerous cells also caused severe side effects, although further refinements of their work led to the production of clinically significant anti-cancer drugs. The basic chemical research of Hitchings and Elion provided important new information about DNA metabolism, from which several clinically important compounds were developed, against malaria, for the treatment of gout and kidney stones, and also drugs that suppressed the normal immune reactions of the body, vital tools in transplant surgery and in treating autoimmune diseases such as rheumatoid arthritis. In the 1970s Hitchings and Elion's research produced an antiviral compound, acyclovir, active against the herpes virus, which preceded the successful development by Burroughs Wellcome of AZT, the anti-AIDS compound.

Hoagland, Mahlon Bush (1921–) US biochemist who was the first to isolate transfer RNA (tRNA), which plays an essential part in intracellular protein synthesis.

Hoagland was born on 5 October 1921 in Boston, Massachusetts. He studied medicine at Harvard University Medical School, obtaining his degree in 1948, after which he worked as a research fellow in medicine in the Huntingdon Laboratory of Massachusetts General Hospital until 1951. He then spent a year at the Carlsberg Laboratory in Copenhagen as a fellow of the American Cancer Society. From 1953 to 1967 he held several positions in the Huntingdon Laboratory at Harvard Medical School; he joined the laboratory as an assistant in medicine, progressed to assistant professor of medicine, then in 1960 became associate professor of bacteriology and immunology. In 1967 he was appointed professor of biochemistry and chair of the biochemistry department at the Dartmouth Medical School. From 1970 to 1985 he was president and scientific director of the Worcester Foundation for Experimental

Biology in Shrewsbury, Massachusetts (president emeritus thereafter).

In the late 1950s Hoagland isolated various types of RNA molecules (now known as transfer RNA or tRNA) from cytoplasm and demonstrated that each type of tRNA can combine with only one specific amino acid. Within the cytoplasm a tRNA molecule and its associated amino acid combine to form a complex – amino acyl tRNA. This complex then passes to the ribosome, where it combines with a messenger RNA (mRNA) molecule in a specific way: each tRNA molecule has as part of its structure a characteristic triplet of nitrogenous bases that links to a complementary triplet on the mRNA. A number of these reactions occur on the ribosome, building up a protein one amino acid at a time.

In addition to his research on tRNA and the biosynthesis of proteins, Hoagland has also investigated the carcinogenic effects of beryllium and the biosynthesis of coenzyme A.

Hodgkin, Dorothy Mary Crowfoot (1910–1994) English chemist who used X-ray crystallographic analysis to determine the structures of numerous complex organic molecules, including penicillin and vitamin B_{12} (cyanocobalamin). She was awarded the 1964 Nobel Prize for Chemistry.

Hodgkin was born on 12 May 1910 in Cairo, where her father, John Crowfoot, was serving with the Egyptian education service. After the family returned to the UK she attended the Sir John Leman School, Beccles, before going to Somerville College, Oxford. In 1928 she indulged her interest in archaeology by accompanying her father on an expedition to Transjordan (now part of Jordan). After graduation she went to Cambridge University where between 1932 and 1934 she worked on determining the structure of sterols, at a time when X-ray analysis was limited to confirming the correct formula as predicted by organic chemical methods. She developed the technique of X-ray investigation to the point at which it became a very useful analytical method. In 1934 she returned to Oxford to take up a lecturing appointment, and three years later married Thomas Hodgkin, a noted authority on African affairs. She remained at Oxford for a further 33 years until 1970, when she became chancellor of Bristol University.

While at Cambridge, Hodgkin studied the structures of cholecalciferol (vitamin D_2) and lumisterol and, with C H Carlisle, she correctly analysed cholesterol iodide, the first complex organic molecule to be determined completely by X-ray crystallography. On her return to Oxford in 1934, she investigated various compounds of physiological importance, especially penicillin, the structure of which she and her co-workers determined before the organic chemists. This work was of national importance at the time, and was to have a lasting effect on the development of antibiotics. She later elucidated the structure of cephalosporin C, an antibiotic closely related to penicillin.

In about 1948 Hodgkin began her work on vitamin B_{12}, a compound essential to the life of red blood cells in the body; the inability to absorb sufficient vitamin B_{12} from the diet leads to pernicious anaemia. Chemical analysis had suggested that this complex compound has an approximate empirical formula of $C_{61-64}H_{86-92}O_{14}N_{14}PCo$, with a single cobalt atom, a cyanide group, and a nucleotide-like group. Hodgkin and her collaborators collected photographs and data of X-ray diffraction patterns of both wet and dry B_{12}, a hexacarboxylic acid derivative, and a derivative in which a selenium atom had been introduced. Using Fourier series the data were analysed mathematically by one of the first electronic computers, which were just becoming available. After a lengthy and painstaking step-by-step process the structure was finally worked out and announced in 1957 (the empirical formula was found to be $C_{63}H_{88}O_{14}N_{14}PCo$). Hodgkin then worked on determining the chemical structure of insulin which she completed in 1969.

In addition to the 1964 Nobel prize, Hodgkin was in 1965 admitted to the Order of Merit, becoming only the second woman to receive this honour (the first was Florence Nightingale). A committed socialist all her life, she was awarded the Lenin Peace Prize in 1987.

Further Reading

Dodson, Guy; Glusker, Jenny P; and Sayre, David, *Structural Studies on Molecules of Biological Interest : A Volume in Honour of Dorothy Hodgkin*, Clarendon, 1981.

Ferry, Georgina, *Dorothy Hodgkin: A Life*, Granta Books, 1998.

Hodgkin, Dorothy Crowfoot; Dodson, G G (et all, eds), *Collected Works of Dorothy Crowfoot Hodgkin*, Indian Academy of Sciences, 1994.

Hudson, Gill, 'Unfathering the thinkable: gender, science and pacifism in the 1930s' in: Benjamin, Marina (ed), *Science and Sensibility: Gender and Scientific Enquiry 1780–1945*, Blackwell, 1991, pp 264–286.

Hodgkin, Thomas (1798–1866) English physician who described six cases of malignant reticulosis in his paper 'On some morbid appearances of the absorbent glands and spleen', which was published in 1832. The disease is named after him.

Hodgkin was born in Tottenham, London, on 6 January 1798, and was tutored at home. Some of his education took place on the Continent, where he completed his medical training after a few years at Guy's Hospital, London. He gained his MD from

Edinburgh University in 1821. In 1825 he was selected to become one of the first fellows of the Royal College of Physicians, but he declined the honour. He became curator of the new museum at Guy's and lectured in morbid anatomy; he was the first to give regular tuition in the subject. He was lauded at home and abroad by a number of societies for his work, but in 1837 he was passed over for the post of assistant physician at Guy's and was deeply disappointed. He resigned from the hospital and gradually devoted more and more of his time to philanthropic work. He was an excellent linguist, and an active crusader in the Aborigines Protection Society. He contracted dysentery and died on 5 April 1866, while on a mercy mission to the Jewish people in Jaffa. He is buried in Israel.

Hodgkin was the first to describe a particular type of lymphoma that usually affects young adults and causes malignant inflammation of the lymph glands. The spleen and liver may also become involved. Hodgkin's disease can now be definitely diagnosed by the histological presence of Reed Sternburg cells.

Hodgkin received little recognition of his observations until 1865, when Samuel Wilks of Guy's referred to Hodgkin's account in his paper 'Cases of enlargement of the lymphatic glands and spleen (or Hodgkin's Disease)'.

Hodgkin pioneered the use of the stethoscope in the UK after being favourably impressed in France by René ◊Laênnec, who invented the instrument in 1816. He was also the first person to stress the importance of postmortem examinations.

Hodgkin and Huxley Alan Lloyd Hodgkin (1914–1998) and Andrew Fielding Huxley (1917–) English physiologists who have contributed much to the understanding of how nerve impulses are transmitted. For their work concerning the movements of ions in the excitation of nerve membranes they shared the 1963 Nobel Prize for Physiology or Medicine with the Australian physiologist John Eccles (1903–1997).

Hodgkin was born on 5 February 1914, in Banbury, Oxfordshire. He was educated at Gresham School, Holt, and then at Trinity College, Cambridge, from which he graduated in 1936. In 1937 and 1938 he worked at the Rockefeller Institute and at Woods Hole Marine Biological laboratories in Massachusetts, where he began his research on the squid. On his return to Cambridge he began his collaboration with Huxley but their work was interrupted by World War II, during which Hodgkin researched airborne radar for the Air Ministry. After the war he went back to Cambridge and served as a lecturer and assistant research director in the department of physiology 1945–52, when he became Foulerton Research Professor. In 1970 he accepted the biophysics professorship at Cambridge (retiring in 1981), and the following year took up the appointment of chancellor of the University of Leicester. Hodgkin was awarded the Royal Medal of the Royal Society in 1958 and was president of the society 1970–75. He was knighted in 1972, and was master of Trinity College, Cambridge, 1978–84. He died in Cambridge on 20 December 1998.

Andrew Huxley was born in London on 22 November 1917, the grandson of the distinguished 19th-century scientist T H ◊Huxley. He was educated at University College and Westminster schools and then at Trinity College, Cambridge, from which he graduated in 1938 and gained his MA in 1941. His researches were interrupted by World War II; he was involved in operational research for Anti-aircraft Command 1940–42, and then for the Admiralty until 1945. It was on his return to Cambridge after the war that Huxley collaborated with Hodgkin on their award-winning work. He became a demonstrator and then assistant director of research at Cambridge 1946–60. He was appointed director of studies at Cambridge 1952–60, and was elected a fellow of the Royal Society in 1955. In 1960 he moved to University College, London, to become Jodrell Professor and then, from 1969–83, Royal Society Research Professor at its department of physiology. From 1983 to 1990 he was master of Trinity College, Cambridge, becoming a fellow of the college in 1990. Huxley was president of the Royal Society 1980–85. He was knighted in 1974.

The theory concerning the nervous system developed at the end of the 19th century stated that nerve cells have as their primary function the transmission of information in the form of changes in electric potential across the cell membrane. Hodgkin first became interested in the mechanics of nerve impulses in the late 1930s and devoted most of his research to the unknown quantities and qualities of the associated electric potentials.

In 1902 Julius Bernstein, an experimental neurophysiologist, had suggested that resting potential was due to the selective permeability of nerve membrane to potassium ions. He had also suggested that the action potential was brought about by a breakdown in this selectivity so that membrane potential fell to zero. In 1945 at Cambridge, Hodgkin and Huxley attempted to measure the electrochemical behaviour of nerve membranes. They experimented on the giant axons of the squid – each axon is about 0.7 mm/0.03 in in diameter. They inserted a glass capillary tube filled with sea water into the axon to test the composition of the ions in and surrounding the cell, which also had a microelectrode inserted into it. They succeeded in measuring the potential differences between the tip of the microelectrode and the sea water. Stimulating the axon with a pair of outside electrodes, it was shown that the inside of the cell was at first negative (the resting potential) and the outside positive, and that during the conduction of the nerve impulse the membrane potential reversed so the inside became positive and the outside negative. This was the first time that electrical changes across the cell membrane had been recorded and the discovery that the membrane potential exceeds the zero level during the action potential implied that some other process than that proposed by Bernstein must be involved. Hodgkin suggested that this process was a rapid and specific increase in the permeability of the membrane to sodium ions.

Working with Bernhard ◊Katz, another cell physiologist, Hodgkin showed that when there was no current flowing through the membrane, the membrane potential could be given by a formula that acknowledges that sodium ions do play an important part in determining the membrane potential. The theory proposed that an excited membrane becomes permeable to sodium ions (which are positively charged), which on entry to a cell cause its contents also to become positively charged; it is known as the 'sodium theory' and is now accepted to be of fairly general application.

In 1947 Hodgkin concluded that during the resting phase a nerve membrane allows only potassium ions to diffuse into the cell, but when the cell is excited it allows sodium ions to enter and potassium ions to move out. By 1952 Hodgkin had demonstrated that the inside of a cell has a great concentration of potassium ions, and the surrounding solution is rich in sodium ions. In that year, Hodgkin and Katz used a special technique now known as the 'voltage clamp method' to measure the currents flowing across a nerve membrane. They found that during an action potential, the inside of the cell becomes electrically positive by 30–60 millivolts, and that the membrane recovers from an impulse within milliseconds. In 1953 Hodgkin investigated the role played by potassium ions in nerve cells and showed that the internal potassium ions are free to move in an electric field. From this he concluded that almost all the potassium in the axoplasm is effectively in a free solution and that it contributes in some way to the production of the resting potential. Two years later Hodgkin devised an apparatus to measure the extrusion of radioactive sodium from the giant axons of a squid. It was shown that there is a relationship between the efflux of sodium and the time taken for it to diffuse out and, further, that when the axon was surrounded with DNP the amount of sodium efflux fell markedly but recovered when the DNP was washed away. This experiment implied that the extrusion of sodium is probably dependent on the metabolic energy supplied either directly or indirectly in the form of ATP (adenosine triphosphate). It was also discovered that the amount of sodium flowing in equals that of the potassium flowing out.

Hodgkin, with his associates, produced more conclusive evidence of this extrusion dependence in 1960 and also showed that the sodium efflux is dependent upon the external potassium ion concentration. The action potential would not be able to revert to the resting potential inside the cell if the influx of sodium ions were not balanced by the extrusion of another positively charged ion. The decrease from the peak is followed by the exit of potassium ions, so restoring the cell interior to a negative phase. In 1959 Hodgkin had found that the inward and outward flow of sodium in frog muscle was approximately equal, implying that the greater part of the sodium efflux must be dependent upon some active transport process. Keynes has since shown that there seems to be a 'chloride pump' involved, but its function is still obscure.

Many scientists have used Hodgkin and Huxley's methods to study resting and action potentials in various excitable membranes. The 'voltage clamp' method is used to obtain information on the elements affecting nerve conduction. Investigations are also being directed to discover the mechanism that possibly involves an enzyme, which is present in the peripheral cell membrane and breaks down ATP, thus releasing energy, but only if sodium and potassium ions are present.

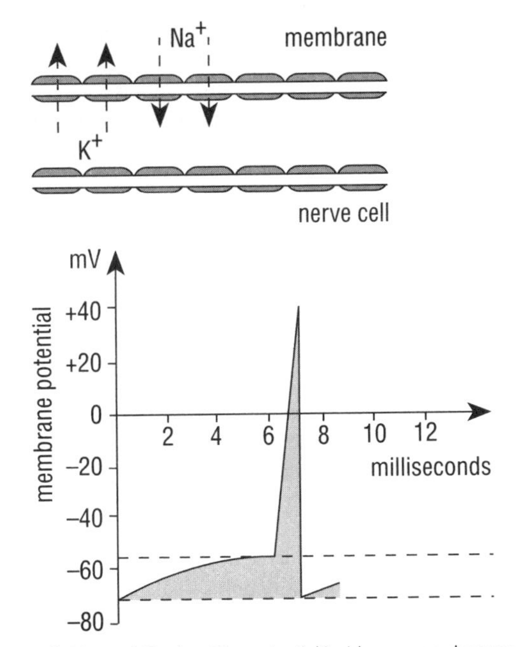

Hodgkin and Huxley The potential inside an axon changes rapidly (and reverses) with the passage of a nerve impulse, accompanied by movement of sodium and potassium ions through the cell membrane.

Hodgkinson, Easton (1789–1861) English civil engineer who worked to introduce scientific methods of measuring the strength of materials.

Hodgkinson was born on his father's farm at Anderton, Cheshire, On 26 February 1789. After showing distinct ability in mathematics during his schooldays because of the early death of his father and his mother's consequent penury he was obliged to give up hopes of a professional education and instead to assist his mother in running the family farm. He had little aptitude for this work and pursuaded his mother to invest her limited capital in a pawnbroking business in Salford, Manchester.

He also found time to develop his interest in natural science and became acquainted with the chemist John ◊Dalton and other gifted men then living in Manchester. In March 1822 he read a paper on 'The transverse strain and strength of materials' before the Literary and Philosophical Society. In this contribution he recorded a factor that became important in all his subsequent experiments, namely 'set' or the original position of a strained body and the position it assumes when the strain is removed. He fixed the exact position of the 'neutral line' in the section of rupture or fracture and made it the basis for the computation of the strength of a beam of given dimensions. His conception of the true mechanical principle by which the position of the line could be determined has long been generally accepted.

In 1828 he read before the same society an important paper on the forms of the catenary links in suspension bridges, and in 1830 one on his research into the strength of iron beams, one of the most valuable contributions ever made to the study of the strength of materials.

From a theoretical analysis of the neutral line, he devised experiments to determine the strongest beam, which resulted in the discovery of what is known as 'Hodgkinson's beam'.

Hodgkinson rendered important services to Robert ◊Stephenson in the construction of the Britannia (Menai) and Conway tubular bridges by fixing the best forms and dimensions of tubes. He edited the fourth edition of Tredgold's work on the strength of cast iron (1842) and published a volume of his own: *Experimental Researches on the Strength and other Properties of Cast Iron* (1846).

He worked 1847–49 as one of the Royal Commissioners to inquire into the application of iron to railway structures. Also in 1847 he was appointed professor of the mechanical principles of engineering at University College, London, where his lectures were somewhat impaired by his hesitancy of speech. He did not live to see an authoritative publication of all his collected papers, dying in Manchester on 18 June 1861.

Hoe, Richard March (1812–1886) US inventor and manufacturer, famous for inventing the rotary printing press.

Hoe was born in New York City on 12 September 1812. He was the eldest son of Robert Hoe (1784–1833), who, with his brothers-in-law Peter and Matthew Smith, had established in New York a firm manufacturing printing presses. Richard Hoe was educated in public schools before entering his father's firm at the age of 15. When his father retired in 1830, Richard and his cousin Matthew took over the business. Richard proved to have the same mechanical genius as his father, and the application of his ideas revolutionized printing processes. He discarded the old flatbed printing press and placed the type on a revolving cylinder. This was later developed into the Hoe rotary or 'lightning' press, patented in 1846 and first used by the *Philadephia Public Ledger* in 1847.

Under Hoe's management the company grew at a rapid rate. In 1859 he built Isaac Adams's Press Works in Boston. After the Civil War, new premises were built in Grand Street, New York, and the old buildings in Gold Street were abandoned. Between 1865 and 1870, a large manufacturing branch was built up in London, employing 600 people.

In 1871, with Stephen D Tucker as a partner, Hoe began experimenting and designed and built the Hoe web perfecting press. This press enabled publishers to satisfy the increasing circulation demands of the rapidly growing US population.

While Richard Hoe was the leading influence in the company, he spent much time and money on the welfare of his employees. Quite early in his career he started evening classes for apprentices, at which free instruction was given in those aspects of their work most likely to be of practical use to them. He was addressed by them as 'the Colonel', which dated from his early service in the National Guard.

He died suddenly while on a combined health and pleasure trip to Florence with his wife and a daughter, on 7 June 1886. He was succeeded in the business by his nephew, Robert Hoe.

At the time when Richard Hoe was made responsible for the company, it was making a single small cylinder press. Its capacity was 2,000 impressions per hour and there was demand for a greater speed of output. This prompted Hoe to concentrate on improvements to meet the demand and, in 1837, a double small cylinder press was perfected and introduced. In the next ten years, he designed and put into production a single large cylinder press. This was the first flatbed and cylinder press ever used in the USA. Hundreds of these machines were made in subsequent years and were used for book, job, and woodcut printing. During 1845 and 1846 Hoe was busily engaged in designing and inventing presses to meet the increased requirements of the newspaper publishers. The result was the construction of the revolving machine based on Hoe's patents. The basis of these inventions was a device for securely fastening the forms of type on a central cylinder placed in a horizontal position. The first of these machines, installed in the *Public Ledger* office, had four impression cylinders grouped around the central cylinder. With one boy to each cylinder to feed in blank paper, 8,000 papers could be produced per hour.

Almost immediately newspaper printing was revolutionized, and Hoe's rotary press became famous throughout the world. In 1853, he introduced the cylinder press that had been patented in France by Dutartre and improved on it in the following years for use in lithographic and letter-press work. In 1861 the curved stereotype plate was perfected, and in 1865 William Bullock succeeded in producing the first printing machine that would print on a continuous web or roll of paper. Spurred on by this latest development, Hoe and his partner began experimenting and designed and built a web press. The first of these to be used in the USA was installed in the office of the *New York Tribune.* At maximum speed, this press printed on both sides of the sheet and produced 18,000 papers per hour. Four years later, Tucker patented a rotating, folding cylinder that folded the papers as fast as they came off the press.

In 1881 the Hoe Company devised the triangular former-folder, which, when incorporated into the press, together with approximately 20 additional improvements, gave rise to the modern newspaper press. It was with the introduction of this that the 1847-type revolving press was superseded.

Hofmann, August Wilhelm von (1818–1892)

German chemist who was one of the great organic chemists of the 19th century and had enormous influence on the development of the subject in both the UK and Germany. Much of his work was connected with coal tar and its constituents, particularly aniline (phenylamine) and phenol. He was the first to explore the chemistry of the aliphatic amines (although not the first to synthesize them) and was the discoverer of the quaternary ammonium salts. His students included those (such as William ◊Perkin) who originated and developed the British synthetic dye industry, and as a chemist he discovered and patented a number of dyes of his own.

Hofmann was born on 8 April 1818 in Giessen, the son of an architect. He entered the local university in 1836 to study law and philosophy, but his interests changed when he attended some of Justus von ◊Liebig's lectures and he continued his studies by specializing in chemistry. In 1841 he obtained his doctorate for a thesis based on investigations into the constitution of coal tar. He then neglected his studies and academic advancement to look after his father, who died on 2 May 1843.

In that same year Hofmann became Liebig's assistant and in the spring of 1845 he was appointed briefly to a position in the University of Bonn. Later that year he was requested by Prince Albert, Queen Victoria's consort, to become a professor at the new Royal College of Chemistry in London. He held this position for the next 20 years, during which time the college became amalgamated with the Royal School of Mines. In 1863 he was offered chairs of chemistry in both Bonn and Berlin. He accepted the position vacated by Eilhard ◊Mitscherlich in Berlin, after designing new laboratories for both institutions, taking up the appointment in 1865 and remaining there until his death. On his seventieth birthday in 1888 he was made a baron, becoming von Hofmann. He died in Berlin on 2 May 1892.

Hofmann's first paper confirmed the presence of aniline in coal tar. He believed that phenol and aniline (phenylamine) were related, and converted the former into the latter by heating phenol with ammonia in a sealed tube for three weeks. Later he prepared nitrobenzene from the light oil distillate of coal tar and made aniline from it by reduction with nascent hydrogen.

From about 1850 Hofmann explored the behaviour and properties of amines, showing that alkyl halides (halogenoalkanes) react with aniline to give secondary and tertiary amines (although Hofmann used different terms for them). He also prepared alkyl amines by the reaction between ammonia and alkyl halides, reporting that he was replacing the hydrogen atoms of ammonia with alkyl groups and thus producing compounds of the 'ammonia type'. The theory of types held vogue in organic chemistry for a number of years and other organic compounds were classified as the 'water', 'hydrogen' and 'hydrochloric acid' types. Hofmann also set about investigating a 'phosphorus' type based on phosphine, PH_3. In 1865 he published *An Introduction to Modern Chemistry,* which was a textbook based on the theory of types.

Hofmann produced more complex amines from diamines and alkyl halides. He discovered what is now known as the Hofmann degradation, in which an amide is treated with bromine and an alkali (hypobromite) to produce an amine with one fewer carbon atoms:

$$CH_3CONH_2 + Br_2 + 4KOH \rightarrow CH_3NH_2 + 2KBr + K_2CO_3 + 2H_2O$$

acetamide (ethanamide), 2 carbon atoms; methylamine, 1 carbon atom

By the reaction of ethyl iodide (iodoethane) with triethylamine he produced tetraethyl-ammonium iodide, $(C_2H_5)_4 N^+ I^-$, the first quaternary ammonium salt. He found that the ion $(C_2H_5)_4N^+$ behaved like a sodium or potassium ion, as a 'true organic metal'. With silver oxide he was able to form from it the strong base $(C_2H_5)_4NOH$, which exists as a solid.

Hofmann was also the first to investigate the structure and properties of formaldehyde (methanal), which he called 'methyl aldehyd' and prepared by passing methanol vapour and air over heated platinum. Working with A A TF Cahours he also discovered the first unsaturated alcohol, prop-2-en-1-ol, CH_2=CH.CH_2OH.

In 1858 Hofmann obtained the dye known as fuchsine or magenta by the reaction of carbon tetrachloride (tetrachloromethane) with aniline (phenylamine). Later he isolated from it a compound that he called rosaniline and used this as a starting point for other aniline dyes, including aniline blue (triphenyl rosaniline). With alkyl iodides (iodoalkanes) he obtained a series of violet dyes, which he patented in 1863. These became known as 'Hofmann's violets' and were a considerable commercial and financial success.

On his return to Germany Hofmann founded the Deutsche Chemische Gesellschaft (German Chemical Society). The many reactions and rules that still bear his name are a testimony to his stature as an organic chemist. The term 'valence' is a contraction of his notion of 'quantivalence' and he devised much of the terminology of the paraffins (alkanes) and their derivatives that was accepted at the 1892 Geneva Conference on nomenclature. Apart from his ammonia and phosphine types he was not a great theorist but became renowned as an experimental chemist. Yet he is said to have been very clumsy with apparatus, the handling of which he usually left to his more dextrous assistants. He was married four times and had 11 children, eight of whom survived him when he died at the age of 74.

Hofstadter, Robert (1915–1990) US physicist who shared the 1961 Nobel prize for discovering that protons and neutrons contained smaller particles (now known to be quarks). Hofstadter later became active in medical physics and astrophysics.

Hofstadter was born on 5 February 1915 in New York City. He graduated from the City College of New York in 1935, and obtained a PhD from Princeton University in 1938 for research into the infrared spectroscopy of small organic molecules, particularly studies of the hydrogen bond. He remained at Princeton as a postdoctoral student, studying solid-state luminescence and photoconductivity before moving to the University of Pennsylvania where he helped build a large van de Graaff generator. It was at Pennsylvania that his interest in detectors for nuclear physics began. But World War II interrupted and Hofstadter worked on proximity fuses at the National Bureau of Standards before joining Norden Laboratory Corporation, an aerospace company, for three years. In 1946 Hofstadter returned to Princeton to pick up his studies in particle detectors and nuclear physics.

In 1948 he made what he considered his most important discovery – that thallium – activated sodium iodide (NaI(Tl)) was an excellent scintillation counter and, with John McIntyre in 1950, that it could be used to measure the energies of gamma rays. Despite being expensive and difficult to work with, NaI(Tl) has been used in gamma-ray spectrometers ever since – in medicine, chemistry, biology, and geology, in addition to nuclear, particle, and astrophysics. In 1950 Hofstadter moved to Stanford University where he began studying the elastic and inelastic scattering of electrons by nuclei. Over the next 20 years these experiments showed how electric charge (and the associated magnetic field) is distributed in the nucleus. Hofstadter also showed that although the neutron is neutral, it has a distribution of electric charge and magnetism within it. This was the first evidence that neutrons and protons were not pointlike particles, rather they were made up of other smaller particles (now known to be quarks). Hofstadter was awarded the 1961 Nobel prize for these discoveries (shared with Rudolf ◊Mössbauer).

In 1968 Hofstadter began working with Barrie Hughes on the Crystal Ball detector at the Stanford Linear Accelerator Center (SLAC). This device, which comprised 900 NaI detectors pointing at the point where beams of electrons and positrons collided, made a number of important discoveries in the spectroscopy of mesons containing charmed and bottom quarks. Hofstadter also used the synchrotron radiation given off by the electron storage rings at SLAC for coronary angiography studies in medicine. In 1970 Hofstadter also suggested putting a large gamma-ray spectrometer on a satellite to carry out gamma-ray astronomy. This device was launched aboard NASA's Gamma Ray Observatory (GRO) five months after Hofstadter's death on 17 November 1990.

Hogben, Lancelot Thomas (1895–1975) English zoologist and geneticist who, somewhat surprisingly, wrote a very successful book entitled *Mathematics for the Millions.* Although he thereafter sought to use mathematical techniques in his various posts, there were in fact few opportunities to do so until he was put in charge of the medical statistics records for the British army during World War II.

Hogben was born in Southsea, Hampshire, on 9 December 1895. Growing up, he developed a strong interest in natural history. From the Middlesex County Secondary School he won a scholarship to Trinity College, Cambridge, in 1913. Simultaneously he took an external degree course in zoology at London University. Imprisoned, however, as a conscientious objector in 1916 during World War I, he was released when his health deteriorated seriously and briefly worked as a journalist before being appointed lecturer in zoology at Birkbeck College, London, in 1917. Two years later, he transferred to Imperial College and shortly afterwards received his doctorate. He then went to teach in Edinburgh for a couple of years, to Montréal, Canada, for a similar period, and to Cape Town, South Africa, for again the same duration of time. He returned to London University in 1930 to take up the post of professor of social biology. In hospital in 1933, he wrote his famous *Mathematics for the Millions* partly as therapy, partly for self-education; he also became interested in linguistics. Following this latter interest, he was in Norway at the outbreak of World War II, and it was not until 1941 – after travelling through Sweden, the Soviet Union, and the USA – that he got back to the UK and became a colonel in the War Office. After the war he became professor of medical statistics at the University of Birmingham, where he remained until he retired in 1961. The author of many scientific papers, and the recipient of many awards and honours – a fellow of the Royal Society from 1936 – Hogben died in Glyn Ceirog, Wales, on 22 August 1975.

It was in London in the 1930s – around the time that his popular *Mathematics for the Millions* was published – that Hogben first began to try to apply mathematical principles to the study of genetics, with particular reference to his investigation of generations of the fruit fly *Drosophila* in relation to research on heredity in humans. He was especially concerned to evaluate the validity of statistical methods as applied in the biological and behavioural sciences, and was surprised at the apparent level of ignorance in such methods among the many scientists of those disciplines.

His desire to use mathematical principles was given considerably more scope on his appointment as colonel in charge of the War Office medical statistical records. Asked to reorganize the entire system, Hogben successfully did just that – quite a feat when data processing was still a matter of mechanical sorting. He was subsequently promoted to director of Army Medical Statistics. (One of his investigations into the army's clinical trials of the sulphonamide drugs proved that their indiscriminate use had favoured the selection of resistant strains of the bacteria responsible for causing gonorrhoea.)

Working at Birmingham University after the war, his attempts to reorganize civilian medical records were less successful – partly because of resistance from some senior medical practitioners reluctant to change their ways.

During his last years, Hogben spent much time revising previously published work and writing on other aspects of philosophy and mathematics.

Further Reading

Hogben, Adrian and Hogben, Anne, *Lancelot Hogben, Scientific Humanist: An Unauthorised Autobiography*, Merlin Press, 1998.

Hollerith, Herman (1860–1929) US mathematician and mechanical engineer who invented electrical tabulating machines. He was probably the first to automate the large-scale processing of information, and as such was a pioneer of the electronic calculator, particularly its application to data handling in business and commerce.

Hollerith was born on 29 February 1860 in Buffalo, New York. After his schooldays he attended the School of Mines at Columbia University, from which he graduated in 1879. The following October he became an assistant to W P Trowbridge, one of his university lecturers, in the work for the US census of 1880.

Hollerith worked with physician and librarian John Billings (1838–1913) on the statistics of manufacturing industries, especially those concerned with steam and water power in the iron and steel industries. In the course of processing the many census returns, they became aware that an automated recording process would have considerable labour-saving advantages. Whether it was Hollerith who had the idea of punching holes into cards to indicate quantities or whether Billings made the initial suggestion is not known. What is important is that later Hollerith developed the idea of punching first a continuous roll of paper, and later individual cards the same size as a dollar bill, with holes to represent information. The quantities indicated by the holes were counted when the tape or cards passed through a device in which electrical contact was made through the holes. The passage of an electric current caused electromechanical counters to advance one place for each hole. The realization of these ideas in practical terms did not, however, come about in time for the processing of the 1880 returns.

In 1882 Hollerith obtained a position as instructor in mechanical engineering at the Massachusetts Institute of Technology (MIT). He preferred experimental work to teaching, and after a year he left MIT and went to St Louis until 1884 where he worked on the development of electromechanically operated air-brakes for railways. He was then employed by the Patent Office in Washington, DC, and until the next census in 1890 he developed his recording and tabulating machines.

By 1889 he had developed not only his electrical machines for recording the information on punched cards, but also machines for punching the cards and for sorting them. But by this time he was not the only inventor of data-processing equipment, and a trial was held to decide which system was to be adopted for the 1890 census. The Hollerith system proved to be twice as fast as the best of the other two and was used to handle the returns of 63 million people. In 1891 the Hollerith system was used in the censuses in Austria, Canada, and Norway (and in the UK in 1911). The machines were later successfully adapted to the needs of government departments and businesses that handled large quantities of data, and particularly to the tabulating of railway statistics.

Hollerith soon realized that the age of large-scale data handling had begun, and in 1896 he formed the Tabulating Machine Company, to manufacture the machines and the cards they used. With the growth of business in this area. Hollerith's company was soon merged with two others into the Computing–Tabulating–Recording Company, which later became the International Business Machines (IBM) Corporation. IBM has remained one of the foremost companies in the development and manufacture of data-processing systems, and has been an important producer of electronic computers for many years.

Hollerith stayed with IBM as a consultant engineer until his retirement in 1921. He died from heart disease in Washington on 17 November 1929, aged 69.

Further Reading

Austrian, Geoffrey D, *Herman Hollerith: Forgotten Giant of Information Processing,* Columbia University Press, 1984.

Reid-Green, Keith S, 'The history of census tabulation', *Sci Amer,* 1989, v 260 (2), pp 98–103.

Holley and Nirenberg Robert William Holley (1922–1993) and Marshall Warren Nirenberg (1927–) US biochemists who shared the 1968 Nobel Prize for Physiology or Medicine (with Har Gobind ◊Khorana) for their work in deciphering the chemistry of the genetic code.

Holley was born on 28 January 1922 in Urbana, Illinois. He studied chemistry at the University of Illinois and graduated in 1942. He gained his PhD from Cornell University in 1947, having spent the latter part of World War II engaged on research into penicillin. From 1948 to 1957 he was assistant professor and then associate professor of organic chemistry at the New York State Agricultural Experimental Station at Cornell, when he moved to the Plant Soil and Nutrition Laboratory. In 1962 he became professor of biology at Cornell, and in 1968 took up a senior appointment at the Salk Institute for Biological Studies in San Diego, California, becoming an American Cancer Society Professor of Molecular Biology. He died on 11 February 1993 in Les Gatos, California.

Nirenberg was born in New York City on 10 April 1927. He studied biology at the University of Florida, where he graduated in 1948, gaining his master's degree four years later. He then went to the University of Michigan's department of biological chemistry and gained a PhD in 1957. From 1957 to 1962 he was at the National Institute of Health (Arthritic and Metabolic Diseases), becoming head of the laboratory of biochemical genetics there in 1962. He later moved to the laboratory of biochemical genetics at the National Heart, Lung and Blood Institute in Bethesda, near Washington, DC.

Holley's early work in the New York State Agricultural Experimental Station at Cornell concerned plant hormones, the volatile constituents of fruits, the nitrogen metabolism of plants, and peptide synthesis. At the Salk Institute he began to study the factors that influence growth in cultured mammalian cells. At Cornell he obtained evidence for the existence of transfer RNAs (tRNAs) and for their role as acceptors of activated amino acids. In 1958 he succeeded in isolating the alanine-, tyrosine-, and valene-specific tRNAs from baker's yeast and approximately one gram of highly purified alanine-specific tRNA (ala tRNA) was prepared and used in structural studies over the next two-and-a-half years. The technique for elucidating its structure was to break up the molecule into large 'pieces', identify the fragments and then reconstruct the original. Eventually Holley and his colleagues succeeded in solving the entire nucleotide sequence of this RNA.

Nirenberg was interested in the way in which the nitrogen bases – adenine (A), cytosine (C), guanine (G), and thymine (T) – specify a particular amino acid. To simplify the task of identifying the RNA triplet responsible for each amino acid, he used a simple synthetic RNA polymer. Using an RNA polymer containing only uracil, for example, he obtained phenylalanine, so he concluded that its code must be UUU. Similarly a cytosine RNA polymer produced only the amino acid proline.

In this way Nirenberg built up a tentative dictionary of the RNA code. He found that certain amino acids could be specified by more than one triplet, and that some triplets did not specify an amino acid at all. These 'nonsense' triplets signified the beginning or the end of a sequence. In this way he assigned values to 50 triplets. He then worked on finding the orders of the letters in the triplets. By labelling one amino acid at a time with carbon-14, and passing the experimental material through a filter that retained only the ribosomes with the tRNA and amino acids attached, he obtained unambiguous results for 60 of the possible codons. Work continued in his laboratory on the role of 'synonym' codons, codon recognition by tRNA, and the mechanics of the rate of protein synthesis during viral infection and embryonic differentiation.

Holmes, Arthur (1890–1965) English geologist who helped develop interest in the theory of continental drift.

Born in Hebburn, Newcastle upon Tyneon 14 January 1890, Holmes received his scientific education at Imperial College, London, first in physics and mathematics, then in geology. After spells teaching at Imperial College and working for oil companies, Holmes was appointed in 1924 head of the geology department at Durham University, moving in 1943 to Edinburgh University. He distinguished himself in many branches of geology, not least petrology, where his *Petrographic Methods and Calculations* (1921) became a classic, as did his more general survey, *Principles of Physical Geology* (1944).

Holmes is best remembered for his geochronological explorations, giving the first reliable modern estimates of the age of the Earth. Radioactivity had already been recognized as holding out the possibility of a technique for determining the actual ages of minerals; and the work of John William Strutt (later Lord ♢Rayleigh), in demonstrating that the profusion of radioactive minerals in the Earth's crust constituted a major heat source had scotched the thermodynamic arguments deployed by Lord ♢Kelvin against what he saw as extravagant estimates of the age of the Earth. Holmes pioneered the use of radioactive decay methods for rock-dating. Painstaking analysis of the proportions of elements formed by radioactive decay, combined with a knowledge of the rates of decay of their parent elements, would yield an absolute age. From 1913 he used the uranium–lead technique systematically to date fossils whose relative (stratigraphical) ages were established but not the absolute age. His *The Age of the Earth* (1913) summarized existing data and developed a timescale of its own. Publishing extensively, he continued work on the Phanerozoic timescale until 1959, though his original estimates did not require fundamental modification. Holmes died in London on 20 September 1965.

Almost alone in the UK, Holmes was also an early advocate of continental drift. In 1928 he proposed that convection currents within the Earth's mantle, driven by radioactive heat, might furnish the mechanism for the continental drift theory broached a few years earlier by Alfred ♢Wegener. In Holmes's view, new oceanic rocks were forming throughout the ocean ridges. Little attention was given to Holmes's ideas on this subject until the 1950s, when palaeomagnetic studies put continental drift theories on a sure footing.

Hooke, Robert (1635–1703) English physicist who was also active in many other branches of science. He is remembered mainly for the derivation of Hooke's law of elasticity, for coining the term 'cell' as used in biology, and for the invention of the hairspring regulator in timepieces and the air pump.

Hooke was born in Freshwater, Isle of Wight, on 18 July 1635. He was sickly as a child, which prevented him from studying for the church as his father had intended. Left on his own, Hooke constructed all kinds of ingenious toys, developing the great mechanical skill that he later applied to instrumentmaking. Upon the death of his father in 1648, Hooke went to London and was educated at Westminster School. There he was introduced to mathematics, mastering ♢Euclid in only a week. In 1653 he went on to Oxford University as a chorister and there became one of a group of brilliant young scientists, among them Robert ♢Boyle, to whom Hooke became an assistant.

Hooke eventually obtained his MA from Oxford in 1663, but the group broke up in 1659–60 and he and most of his colleagues moved to London, where they established the Royal Society in 1662. Hooke was appointed curator of the society, a post that entailed the demonstration of several new experiments at every weekly meeting. In 1664 he also became a lecturer in mechanics at the Royal Society, and in the following year he took up the additional post of professor of geometry at Gresham College, London. Hooke retained these positions for the rest of his life, and was also secretary of the Royal Society 1677–83. He died in London on 3 March 1703.

Hooke's post of curator at the Royal Society required him to provide a continual stream of new ideas to demonstrate before the members, and he consequently examined all fields of experimental science but made no deep or thorough investigations in any of them. His main contributions were in four main areas – mechanics, optics, geology, and instrumentmaking.

In mechanics, Hooke was the first to realize that the stress placed upon an elastic body is proportional to the strain produced. This relationship, discovered in 1678, is known as Hooke's law. But Hooke was active in mechanics long before this, and in 1658 he found that a spiral spring vibrates with a regular period in the same way as a pendulum. He began to develop this discovery to produce a watch with a spring-controlled balance wheel, but it is uncertain whether he made a working model before Christiaan ♢Huygens did so in 1674. Hooke can thus be credited with the discovery of the principle of the watch if not the invention of the device itself. Hooke himself claimed priority, however – one of many such disputes that were a feature of his life. His main antagonist in this respect was Isaac ♢Newton, with whom Hooke argued over credit for the discovery of gravitation. The idea that gravity exists between bodies was prevalent at the time, and in 1664, Hooke suggested that a body is continually pulled into an orbit around a larger body by a force of gravity directed towards the centre of the larger body. This was an important step towards an understanding of gravity, and Hooke also suggested in 1679 that the force of gravity obeys an inverse square law. Both these ideas helped Newton, with his immense analytical powers, to build on Hooke's insight and arrive at his law of universal gravitation in 1687.

Another area of contention between Hooke and Newton was in optics. In 1665 Hooke published a book called *Micrographia*. It contained superb accounts of observations that Hooke had made with the microscope and was the first important work on microscopy. From his description of the empty spaces in the structure of cork as cells came our use of the word 'cell' to mean a living unit of protoplasm. The *Micrographia* also contained Hooke's work on optics. This included the idea that light might consist of waves, which Hooke developed from his observations of spectral colours and patterns in thin films. Newton was again able to build on Hooke's work here, examining the optical effects of thin films in detail.

In geology, Hooke made an important contribution by insisting that fossils are the remains of plants and animals that existed long ago, a daring view in an age dominated by the biblical account of creation. Hooke furthermore held that the history of the Earth would be revealed by a close study of fossils.

As an instrumentmaker, Hooke made several important inventions and advances. In 1658, while working for Robert Boyle at Oxford, Hooke perfected the air pump, a development that led directly to the derivation of Boyle's law in 1662. He also made considerable advances to the microscope, and invented the wheel barometer, which registered air pressure with a moving pointer; a weather clock, which recorded such factors as air pressure and temperature on a revolving drum; and the universal joint.

Hooke is an unusual figure in the history of science. Although he made no major discoveries himself, his wide-ranging intuition and great experimental prowess sparked off important contributions in others, notably Newton and Huygens.

Further Reading

Drake, Ellen T, *Restless Genius: Robert Hooke and His Earthly Thoughts*, Oxford University Press, 1996.

Hunter, Michael and Schaffer, Simon (eds), *Robert Hooke: New Studies*, Boydell and Brewer, 1990.

Hunter, Michael Cyril William, Schaffer, Simon (eds), *Robert Hooke: New Studies*, Boydell, 1989.

Kassler, Jamie, *Inner Music: Hobbes, Hooke, and North on Internal Character*, Athlone Press, 1994.

Nichols, Richard S (ed), *The Diaries of Robert Hooke, the Leonardo of London, 1635–1703*, Book Guild, 1994.

Hooker, Joseph Dalton (1817–1911) English botanist who made many important contributions to botanical taxonomy but who is probably best known for introducing into the UK a range of previously unknown species of rhododendron and for his improvements to the Royal Botanical Gardens at Kew.

Hooker was born on 30 June 1817 in Halesworth, Suffolk. He studied medicine at Glasgow University, where his father was professor of botany, graduating in 1839. In the same year he obtained the post of assistant surgeon and naturalist on an expedition to the southern hemisphere. The expedition, which was led by Captain James Clark Ross, set out in 1839 and returned in 1843; its main aims were to locate the magnetic south Pole and to explore the Great Ice Barrier, but other places were visited, including the Falkland Islands, Tasmania, and New Zealand. On his return to England, Hooker applied for the botany chair at Edinburgh University but was not accepted and so took a job identifying fossils for a geological survey. From 1847 to 1850 he took time off to undertake a botanical exploration of northeastern India, mainly of the Himalayan state of Sikkim and eastern Nepal. In 1855 he became assistant director of Kew Gardens, where his father was by this time director. On the death of his father in 1865 Hooker became the director, a post he held until 1885, when he retired. He died on 10 December 1911 in Sunningdale, Berkshire.

While Hooker was on the expedition to the southern hemisphere he made extensive notes and sketches of the plants he saw and collected many specimens, which he pressed and mounted. On his return to the UK he produced a six-volume work (published 1844–60) of his observations and findings, with two volumes each on the flora of Antarctica, New Zealand, and Tasmania. This work combined accurate and detailed descriptions of plants with perceptive essays on plant distribution and established Hooker's reputation as a botanist of the highest calibre. The importance of this work was quickly recognized by the Royal Society, which elected Hooker a fellow in 1847.

In 1854 Hooker published a general account of his travels in the Indian subcontinent, entitled *Himalayan Journals*. He also wrote many scientific works based on his research on the Indian flora; the first of these was about rhododendrons and was published by Hooker's father while Hooker himself was still in India. In addition, Hooker sent back to England many previously unknown species of rhododendron. His first general botanical work on Indian plants was the single volume *Flora Indica* (1855), written in conjunction with Thomas Thomson. This was superseded by Hooker's monumental seven-volume *Flora of British India* (1872–97), written jointly with several other scientists. While in India, Hooker became interested in the genus *Impatiens* (a group that includes the Himalayan balsam, which has since become naturalized in the UK), and gave descriptions of about 300 species of this genus. He supplemented his Indian work by writing volumes four and five of *A Handbook to the Flora of Ceylon* 1898–1900, a work that had remained unfinished since the death in 1896 of its original author, H Trimen.

As director of Kew Gardens, Hooker introduced many improvements, with the introduction of the rock garden, the addition of new avenues, and an extension of the arboretum. Several other important developments occured during Hooker's directorship. In 1876 T J Phillips Jodrell, a friend of the Hooker family, died and left a bequest for the foundation of a botanical laboratory at Kew. The Jodrell Laboratory is now world famous for the scientific work performed there on the structure and physiology of plants. Kew Gardens also became increasingly important as a repository for collections of pressed plants and as a centre for the propagation and distribution of many crop plants, including rubber, coffee, and the oil palm. Furthermore, in 1883 the *Index Kewensis* was founded; this is a list of all scientific plant names, accompanied by descriptions, which, since the publication of the first volume in 1892, has become an invaluable aid in preventing duplication and error in the naming of plants.

As well as establishing Kew Gardens as an international centre for botanical research, Hooker also continued his own botanical work. With the botanist George Bentham he published *Genera plantarum* (1862–83), a complete catalogue of all the known genera and families of flowering plants from all parts of the world. Nevertheless, Hooker did not neglect the British flora. In 1870 he published *Student's Flora of the British Isles,* and from 1887 to 1908 he edited various editions of Bentham's *Handbook of the British Flora.* He was also interested in aspects of botany other than taxonomy, such as the dispersal of plants over large areas and the evolution of new species. After much consideration he became an evolutionist, but his belief in the theory was founded on his own rather specialized knowledge of plants and regional floras and so he contributed little to the popular debate on the subject.

Further Reading

Allen, Mea, *The Hookers of Kew 1785–1911,* Michael Joseph, 1967.

Huxley, Leonard, *Life and Letters of Sir Joseph Dalton Hooker OM GCSI,* John Murray, 1918.

Turill, W B, *Joseph Dalton Hooker: Botanist Explorer and Administrator,* Nelson and Sons, 1963.

Hooker, Stanley (1907–1984) English engineer responsible for major aircraft engine development projects, which culminated in the successful Proteus turboprop, Orpheus turbojet, Pegasus vectored-thrust turbofan, Olympus turbojet, and RB-211 turbofan.

Hooker was born in Sheerness, Kent, on 30 September 1907 and educated at Imperial College, London, and Brasenose College, Oxford. He joined the Admiralty Scientific and Research Department in 1935, working on anti-aircraft rocket development. In 1938 he moved to Rolls-Royce in charge of the Performance and Supercharger Department.

In 1940 he met Frank ◊Whittle and the following year, as chief engineer of the Rolls-Royce Barnoldswich Division, he was responsible for the development of the Whittle W2B turbojet. In the period 1944–45 the Nene and Derwent engines emerged from Barnoldswick and in 1946 the Meteor, powered by two Derwent V engines, established a world speed record of 970 kph/603 mph.

Hooker joined the Aero Engine Division of the Bristol Aeroplane Company, becoming chief engineer in 1951 and a director the following year. When Bristol Siddely Engines was formed in 1959, he became technical director of the Aero Division and, following the merger with Rolls-Royce in 1966, technical director of the Bristol Engine Division.

In January 1971 Hooker was appointed group technical director and seconded to Derby in charge of RB-211 development. The same year he took his seat on the main board as group technical director.

In his distinguished career Hooker received many honorary degrees, decorations, medals, and awards, including the OBE in 1946, CBE in 1964, and a knighthood in 1974. His fellowship of the Royal Society came in 1962 and fellowship of the Royal Aeronautical Society in 1947.

The Merlin engine, which was to power so many Allied aircraft in World War II, started life in 1935 developing under 670 kW/900 hp. As the need for greater engine power increased, particularly at high altitudes, supercharging was one of the most effective ways of meeting the demand. The Merlin supercharger, developed under Hooker, was so successful that special versions of the engine were delivering up to 1,970 kW/2,640 hp by the end of the war.

A turboprop engine uses the exhaust efflux from a turbo jet (usually called a gas generator, in this application) to drive a conventional propeller. The Proteus turboprop in its Mark 705 version entered service in the 100 Series Britainnia airliner in January 1957 at 2,900 kW/3,890 tehp (total equivalent hp, being the shaft power delivered to the propeller plus the equivalent of the residual jet).

The turbojet engine passes all the air it draws in through the combustion chambers and expels it in a hot high-velocity jet; it uses the energy of the jet alone to drive the aircraft. It is inefficient at low speeds, but as the aircraft passes through about 1500 kph/900 mph its rating improves and it is the preferred power unit for military and supersonic applications. The Orpheus turbojet, for which Hooker had overall responsibility, was chosen to power the Fiat G91, selected under the NATO mutual weapons development programme, entering service in May 1958 as the Mark 801 at 18 kN/4,050 lb thrust.

The turbofan engine passes only a proportion of the air through the combustion chambers. This air emerges as a hot high-velocity jet and is mixed with the air which by-passed the core of the engine. The amount of by-pass air varies in different engines.

The Pegasus vectored thrust engine was conceived after discussions between Hooker and Sydney Camm. It is the power unit for the Harrier V/STOL (vertical/short take-off and landing) combat aircraft and is a unique design of turbofan. Its by-pass air is ducted through two forwards nozzles and the gas coming from the core of the engine is ducted through two rear nozzles. All four nozzles can be rotated downwards to provide vertical thrust and rearwards to provide forwards thrust. Intermediate settings give the aircraft its short take-off capability. Rotating the nozzles in flight, even forwards to give reverse thrust, makes the Harrier a most manoeuvrable combat aircraft.

The Olympus turbojet, eventually to power the supersonic Concorde airliner, started life in 1946 when Bristol submitted designs to the Ministry of Supply in answer to a tender for an engine of about 36 kN/8,000 lb thrust. The design was then unique, because it was the first time the two-spool concept had been proposed. This layout uses two independent compressors driven by two independent turbines, giving the engine high compression (important for fuel economy) and great adaptability.

The engine underwent extensive development under Hooker and then flew in the prototype and preproduction aircraft, designated 593B, with a thrust of 156 kN/35,080 lb thrust. In the series aircraft designated 600, it flies with 170 kn/38,000 lb thrust.

The RB-211 turbofan was designed with one of the highest by-pass ratios of any engine in the early 1980s. This and other design features makes it the quietest and most economical in airline service. Economy encompasses every aspect of operating an engine, not just the fuel consumption. One of the most important aspects of the RB-211 operating economics is its modular design. In older concept engines, changing a turbine at the end of its service life needed a complete engine stripdown; the RB-211 has a turbine module, a compressor module, and so on, so that replacement is comparatively simple and swift.

From the Merlin supercharger to the Olympus and RB-211, Hooker led design teams that introduced some of the most advanced aeroengine concepts in the world, operating to the rigid requirements of aviation.

Hopkins, Frederick Gowland (1861–1947) English biochemist who was jointly awarded (with Christian Eijkman) the 1929 Nobel Prize for Physiology or Medicine for his work showing the necessity of certain dietary components – now known as vitamins – for the maintenance of health. He received several other honours for his work, including a knighthood in 1925.

Hopkins was born on 20 June 1861 in Eastbourne, Sussex. He showed no remarkable distinction at school, except in chemistry, and after leaving school he was articled for three years to a consulting analyst in London. He then became an analytical assistant at Guy's Hospital in London, simultaneously studying for an external degree in chemistry at the University of London. In 1888 he became a medical student at Guy's Hospital Medical School, from which he graduated in 1894. He remained at Guy's Hospital until 1898, when he was invited by Michael Foster, professor of physiology at Cambridge University, to become a lecturer in chemical physiology at Cambridge. This was an extremely taxing job, the strain of which adversely affected Hopkins's health, and it left little time for original research. In 1914, however, Hopkins was appointed professor of biochemistry at Cambridge and was able to devote more time to his own investigations; he held this position until he retired in 1943. He died in Cambridge on 16 May 1947.

Hopkins began his Nobel prizewinning work in 1906, when he realized that animals cannot survive on a diet containing only proteins, fats, and carbohydrates. Experimenting on the growth rates of rats fed on diets of artificial milk, he noticed that they failed to grow unless a small quantity of cow's milk was added to the artificial milk. From this he concluded that the cow's milk contained accessory food factors that are required in only trace amounts but which are essential for normal growth. But he failed to isolate these substances (vitamins) and his hypothesis remained controversial for many years, although it had been proved correct by the time he was awarded a Nobel prize in 1929.

Hopkins also made several other important contributions to biological knowledge. He discovered the amino acid tryptophan when one of his students – John Mellanby, who later became professor of physiology at Oxford University – failed to obtain the Adamkiewicz colour reaction for proteins (this involves adding acetic acid then strong sulphuric acid to the test solution). This led Hopkins, with the assistance of S W Cole, to investigate the reaction, which they found is the result of a reaction between glyoxylic acid – a common contaminant of acetic acid – and tryptophan. This discovery meant that tryptophan is an important constituent of proteins, which then led to the concept of essential amino acids. Hopkins also showed that tryptophan and certain other amino acids cannot be manufactured by the body and must therefore be supplied in the diet. In addition, he helped to lay the foundation for the modern understanding of muscle contraction with his demonstration – in collaboration with Morley Fletcher (1873–1933) – that contracting muscle accumulates lactic acid. Hopkins also discovered the tripeptide glutathione, which is important as a hydrogen carrier in the intracellular utilization of oxygen.

Hopper Grace, born Grace Brewster Murray, (1906–1992) US computer pioneer whose most significant contributions were in the field of software. She created the first compiler and helped invent the computer language COBOL. She was also the first person to isolate a computer 'bug' and successfully debug a computer.

After doing postgraduate work at Yale, Hopper returned to her original university, Vassar, as a member of the mathematics faculty. She volunteered for duty in World War II with the Naval Ordinance Computation Project. This was the beginning of a long association with the navy, resulting in her being appointed to the rank of rear admiral in 1983. She was then the oldest officer on active duty in the US armed forces. In 1945 Hopper was ordered to Harvard University to assist Howard ◊Aiken in building a computer. In those days, computers had to be rewired for each new task and Hopper frequently found herself entwined in the wiring of the computer. One day a breakdown of the machine was found to be due to a moth that had flown into the computer. Aiken came into the laboratory as Hopper was dealing with the insect. 'Why aren't you making numbers, Hopper?' he asked. Hopper replied: 'I am debugging the machine!' This first computer 'bug' was removed with tweezers and is preserved at the Naval Museum in Virginia in the laboratory logbook, glued beside the entry for 15.45 on 9 Sept 1945.

After the war, Hopper joined John ◊Mauchly and Presper ◊Eckert, who had set up a firm, eventually to become the Unisys Corporation, to manufacture a commercial computer. Her main contribution was to create the first computer language, together with the compiler needed to translate the instructions into a form that the computer could work with. In 1959 she was invited to join a Pentagon team attempting to create and standardize a single computer language for commercial use. This led to the development of the COBOL language, still one of the most widely used languages. Hopper died in Arlington, Virginia, on 1 January 1992.

Further Reading

Billings, Charlene, *Grace Hopper: Navy Admiral and Computer Pioneer,* Contemporary Women series, Enslow Publishers, 1990.

Hopper, Grace M and Mandell, Steven L, *Understanding Computers,* West Publishing, 1990.

Horrocks, Jeremiah (*c.* 1617–1641) English astronomer and clergyman who made the first recorded observations of a transit of Venus in 1639, which he had predicted.

Horrocks was born in Toxteth Park, Liverpool, the son of a farmer. In 1632 he entered Emmanuel College, Cambridge, as a 'sizar' (a student who paid low fees and acted as a servant to other students to pay his debt). While he was there he taught himself astronomy. In 1635 he returned to Toxteth and became a tutor, continuing his astronomical studies in his spare time. He used Johannes ◊Kepler's laws of planetary motion to prove that the Moon orbits the Earth elliptically. This made a partial basis for Isaac ◊Newton's later work. Following Kepler's prediction that Venus would transit the Sun in 1631, Horrocks calculated that the transits occurred in pairs, eight years apart, predicting that the next transit would take place in 1639.

In 1639 Horrocks was ordained as a curate in Hoole, Lancashire, and he observed the predicted transit of Venus on Sunday, 24 November, between church services. To do this he used a simple telescope set on a wooden beam, which projected a solar image onto a piece of paper marked with a 15-cm/6-in graduated circle.

Horrocks also studied tides and the mutual perturbation of Jupiter and Saturn. He derived a smaller than previously recorded value for the solar parallax, concluding that the Sun was further from the Earth than had been thought. In 1640, he returned to Toxteth, wrote his book *Venus in sole visa* and started work on a new treatise on solar dimensions. He died there suddenly on 3 January 1641. His work was commemorated in the 19th century by a plaque in Westminster Abbey and two stained-glass windows in St Michael's Church, Hoole.

The last transit of Venus was in 1882 and the next pair will occur in 2004 and 2012.

Hounsfield, Godfrey Newbold (1919–) English research scientist whose part in the invention and development of computerized axial tomography (the

CAT scanner) was recognized by the award of the 1979 Nobel Prize for Physiology or Medicine.

Hounsfield was born on 28 August 1919 and educated at the Magnus Grammar School, Newark, and later at the City and Guilds College, London, and the Faraday House Electrical Engineering College.

After serving with the Royal Air Force during World War II, he joined the Medical Systems Section of EMI in 1951 and was actively engaged in research with the company until his retirement. He was made professorial fellow of Manchester University (for imaging sciences) in 1978 and has been awarded some of the most coveted prizes in the scientific world including the Nobel prize, the Price Phillip Medical Award in 1975, the Lasker Award in 1975, and the Churchill Gold Medal in 1976. He was knighted in 1981.

Hounsfield's long career in medical research and engineering has culminated in the invention of the CAT scanner, formerly a computerized transverse axial tomography system for X-ray examination. Its development, using all the most advanced technology available, has led to major improvements in the field of X-ray radiology, enabling the whole body to be screened at one time. The X-ray crystal detectors, more sensitive than film, are rotated round the body and can distinguish between, for example, tumours and healthy tissue.

Howe, Elias (1819–1867) US engineer who invented a sewing machine, one of the first products of the Industrial Revolution that ultimately eased the burden of domestic work.

Howe was born on 9 July 1819 in Spencer, Massachusetts. As a boy, he worked on the farm his father ran, and in his grist-mills and saw-mills where he took a particular interest in the machinery. He trained to be a machinist, and having conceived the idea of a sewing machine, spent much of his life developing and patenting one. He died in Brooklyn, New York, on 3 October 1867.

Howe began work on the design of a sewing machine in about 1843. Within a year he had a rough working model, and in September 1846 he was granted a US patent for a practical machine. He was the first to patent a lock-stitch mechanism, and his machine had two other important features: a curved needle with the eye (for the thread) at the point, and an under-thread shuttle (invented by Walter Hunt in 1834).

Howe immediately went to the UK and sold the invention for £250 to a corset manufacturer named William Thomas of Cheapside, London. In December 1846 Thomas secured the English patent in his own name, and engaged Howe (on weekly wages) to adapt the machine for his needs. Howe worked with Thomas 1847–49, but his career in London was unsuccessful. He pawned his US patent rights, and returned in poverty to the USA, where he found his wife dying.

While Howe had been away, the sewing machine was beginning to arouse public curiosity, and he found that various people were making machines that infringed his patent. The most prominent of these – an inventor in his own right – was Isaac ◊Singer, who in 1851 secured a US patent for his own machine. Howe now became aware of his rights, redeemed his pawned patent, and took out law suits against the infringers. After much litigation, the courts found in his favour and from that time (about 1852) until his patent expired in 1867, Howe received royalties on all sewing machines made in the USA. When he died, he left an estate worth $2 million, an enormous sum in 1867.

The basic invention in machine sewing was the double-pointed needle, with the eye in the centre, patented by C F Weisenthal in 1755, which enabled the sewing or embroidery stitch to be made with the needle being inverted. Many of the features of the sewing machine are distinctly specified in a patent secured in England in 1790 by Thomas Saint. The machine he described was for stitching, quilting, or sewing but seems to have been chiefly intended for leatherwork. If Saint had hit upon the eye-pointed needle, his machine would have completely anticipated the modern chain-stitch machine. A real working machine was invented in France by a poor tailor, Barthélemy Thimmonier, and a patent was obtained in 1830. By 1841 about 40 of these machines were being used in Paris to make army clothing, although the machines were clumsy, being largely made of wood. An ignorant crowd wrecked his establishment and his machines, but undeterred he patented vast improvements on it. The troubles of 1848, however, blasted his prospects and his patent rights for the UK were sold. A machine of his shown at the Great Exhibition of 1851 did not attract any interest and Thimmenier died in 1857, unrewarded.

In about 1832, Walter Hunt of New York constructed a machine with a vibrating arm, at the extremity of which he fixed a curved needle with an eye near its point. The needle formed a loop of thread under the cloth to be sewn, and an oscillating shuttle passed a thread through the loop, thus making the lock-stitch of all ordinary two-thread machines. Howe was apparently unaware of Hunt's invention.

Since that time, thousands of patents have been issued in the USA and in Europe, covering improvements in the sewing machine. Although these numerous attachments and accessories have improved the machine's efficiency and usefulness, the main principles are still the same as they were in the basic machine. There are well over 2,000 types of modern sewing machine, designed for making up garments, boots and shoes, hats, for working embroidery, for edging lace curtains, and for sewing buckles on shoes. Most machines are now powered by electricity, although treadle-operated machines are still seen in specialized fields such as shoe-mending. Microelectronics are now providing push-button controls on home sewing machines, but the basic mechanism remains the same as that devised by Howe.

Hoyle, Fred (1915–) English cosmologist and astrophysicist, distinguished for his work on the evolution of stars, the development of the steady-state theory of the universe, and a new theory on gravitation.

Hoyle was born in Bingley, Yorkshire, on 24 June 1915. He attended the local grammar school and then went to Emmanuel College, Cambridge. In 1939 he was elected a fellow of St John's College, Cambridge, and in 1945 he became a lecturer in mathematics at the university. Three years later he developed the steady-state theory as a cosmological model to explain the structure and properties of the universe. In 1956 he left the UK for the USA to join the staff of the Mount Wilson and Mount Palomar observatories. He returned to the UK ten years later to become director of the Institute of Theoretical Astronomy at Cambridge 1967–73. Since 1972 he has been an honorary research professor at Manchester University. He became a fellow of the Royal Society in 1957 and was knighted in 1972.

According to Hoyle's new theory on gravitation, matter is not evenly distributed throughout space, but forms self-gravitating systems. These systems may range in diameter from a few kilometres to a million light years and they vary greatly in density. They include galaxy clusters, single galaxies, star clusters, stars, planets, and planetary satellites. Hoyle argues that this variety need not imply that the self-gravitating systems were formed in diverse ways, but rather that there is no significant intrinsic difference between one place in the universe and another or between one time and another.

To explain this structure of a universe composed of clusters of matter of different size and to explain its formation, Hoyle calculated the theoretical thermal conditions under which a large cloud of hydrogen gas would contract under the influence of its own gravitation. He found that a contracting cloud whose temperature is less than that required for it to exist in equilibrium will break up into self-gravitating fragments small enough to be at equilibrium in the new, increased density of the cloud fragments. Moreover, such fragmentation will continue until the formation of opaque fragments dense enough for gravitational contraction to offset radiation loss and to maintain the temperature at the necessary equilibrium.

To account for the origin of the elements in the universe, Hoyle, in collaboration with his colleague, William ◊Fowler, proposed that all the elements may be synthesized from hydrogen in eight separate processes that occur in different stages of the continual process by which hydrogen is converted to helium by successive fusions of hydrogen with hydrogen. The second stage occurs when the supply of hydrogen is exhausted and the cloud of gas heats up to allow the helium-burning stage, in which helium nuclei interact, to proceed. The third stage is the alpha process, whereby the cloud contracts further to reach temperatures of 10^9K (10 billion °C/18 billion °F) until neon (^{20}Ne) nuclei built up by the helium-burning stage can interact, releasing particles that in turn are used to build up nuclei of new elements, and so on, until only the element iron is left. Although alternative approaches continue to be explored, there is considerable evidence confirming such an account for the abundant distribution of the elements.

The steady-state theory was expounded, in collaboration with Thomas ◊Gold and Hermann ◊Bondi, as a model to explain the structure and properties of the universe. It postulated that the universe is expanding, but that its density remains constant at all times and places, because matter is being created at a rate fast enough to keep it so. According to this model the creation of matter and the expansion of the universe are interdependent. It calls for no new theory of space–time. Hoyle was able to propose a mathematical basis for his theory that could be reconciled with the theory of relativity.

New observations of distant galaxies have, however, led Hoyle to alter some of his initial conclusions. According to his theory, because no intrinsic feature of the universe depends on its distance from the observer, the distant parts of the universe are the same in nature as those parts that are near. But the detection of some radio galaxies indicates that in fact the universe is very different at different distances, evidence which directly contradicts the fundamental hypotheses of the steady-state theory.

Space isn't remote at all. It's only an hour's drive away if your car could go straight upwards.

Fred Hoyle Observer *Sept 1979*

Hoyle is a prolific writer of science fiction as well as popular books on science; these latter include *Of Man and the Galaxies* (1966), *Astronomy Today* (1975), *Diseases from Space* (1979), *Space Travellers: the Bringers of Life* (1981), and *Our Place in the Cosmos* (1993). His autobiography, *Home is Where the Wind Blows*, was published in 1993.

He has suggested that life originated in bacteria and viruses contained in the gas clouds of space, which were then delivered to the Earth by passing comets

Hubble, Edwin Powell (1889–1953) US astronomer who studied extragalactic nebulae and demonstrated them to be galaxies like our own. He found the first evidence for the expansion of the universe, in accordance with the cosmological theories of Georges ◊Lemaître and Willem ◊de Sitter, and his work led to an enormous expansion of our perception of the size of the universe.

Hubble was born in Marshfield, Missouri, on 20 November 1889. He went to high school in Chicago and then attended the University of Chicago where his interest in mathematics and astronomy was influenced by George ◊Hale and Robert ◊Millikan. After receiving his bachelor's degree in 1910, he became a Rhodes scholar at Queen's College, Oxford, where he took a degree in jurisprudence in 1912. When he returned to the USA in 1913, he was admitted to the Kentucky Bar, and he practised law for a brief period before returning to Chicago to take a research post at the Yerkes Observatory 1914–17.

In 1917 Hubble volunteered to serve in the US infantry and was sent to France at the end of World War I. He remained on active service in Germany until 1919, when he was able to return to the USA and take up the earlier offer made to him by Hale of a post as astronomer at the Mount Wilson Observatory near Pasadena, where the 2.5-m/100-in reflecting telescope had only recently been made operational. Hubble worked at Mount Wilson for the rest of his career, and it was there that he carried out his most important work. His research was interrupted by the outbreak of World War II, when he served as a ballistics expert for the US War Department. He was awarded the Gold Medal of the Royal Astronomical Society in 1940, and received the Presidential Medal for Merit in 1946. He was active in research until his last days, despite a heart condition, and died in San Marino, California, on 28 September 1953.

While Hubble was working at the Yerkes Observatory, he made a careful study of nebulae, and attempted to classify them into intra- and extragalactic varieties. At that time there was great interest in discovering what other structures, if any, lay beyond our Galaxy. The mysterious gas clouds, known as the smaller and larger Magellanic Clouds, which had first been systematically catalogued by Charles ◊Messier and called 'nebulae', were good extragalactic candidates and were of great interest to Hubble. He had been particularly inspired by Henrietta ◊Leavitt's work on the Cepheid variable stars in the Magellanic Clouds; and later work by Harlow ◊Shapley, Henry ◊Russell, and Ejnar ◊Hertzsprung on the distances of these stars from the Earth had demonstrated that the universe did not begin and end within the confines of our Galaxy. Hubble's doctoral thesis was based on his studies of nebulae, but he found it frustrating because he knew that more definite information depended upon the availability of telescopes of greater light-gathering power and with better resolution.

After World War I, with the 2.5-m/100-in reflector at Mount Wilson at his disposal, Hubble was able to make significant advances in his studies of nebulae. He found that the source of the light radiating from nebulae was either stars embedded in the nebular gas or stars that were closely associated with the system. In 1923 he discovered a Cepheid variable star in the Andromeda nebula. Within a year he had detected no fewer than 36 stars within that nebula alone, and found that 12 of these were Cepheids. These 12 stars could be used, following the method applied to the Cepheids that Leavitt had observed in the Magellanic Clouds, to determine the distance of the Andromeda nebula. It was approximately 900,000 light years away, much more distant than the outer boundary of our own Galaxy – then known to be about 100,000 light years in diameter.

Hubble discovered many gaseous nebulae and many other nebulae with stars. He found that they contained globular clusters, novae, and other stellar configurations that could also be found within our own Galaxy. In 1924 he finally proposed that these nebulae were in fact other galaxies like our own, a theory that became known as the 'island universe'. From 1925 onwards he studied the structures of the galaxies and classified them according to their morphology into regular and irregular forms. The regular nebulae comprised 97% of them and appeared either as ellipses or as spirals, and the spirals were further divided into normal and barred types. All the various shapes made up a continuous series, which Hubble saw as an integrated 'family'. The irregular forms comprised only 3% of the nebulae he studied. By the end of 1935, Hubble's work had extended the horizons of the universe to 500 million light years.

Having classified the various kinds of galaxies that he observed, Hubble began to assess their distances from us and the speeds at which they were receding. The radial velocity of galaxies had been studied by several other astronomers, in particular by Vesto ◊Slipher. Hubble analysed his data, and added some new observations. In 1929 he found, on the basis of information for 46 galaxies, that the speed at which the galaxies were receding (as determined from their spectroscopic red shifts) was directly correlated with their distance from us. He found that the more distant a galaxy was, the greater was its speed of recession – now known as

Hubble's law. This astonishing relationship inevitably led to the conclusion that the universe is expanding, as Lemaître had also deduced from Albert ⇨Einstein's general theory of relativity.

This data was used to determine the portion of the universe that we can ever come to know, the radius of which is called the Hubble radius. Beyond this limit, any matter will be travelling at the speed of light, and so communication with it will never be possible. The data on galactic recession was also used to determine the age and the diameter of the universe, although at the time both of these calculations were marred by erroneous assumptions, which were later corrected by Walter ⇨Baade. The ratio of the velocity of galactic recession to distance has been named the Hubble constant, and the modern value for the speed of galactic recession is 530 km/330 mi per sec – very close to Hubble's original value of 500 km/310 mi per sec.

During the 1930s, Hubble studied the distribution of galaxies and his results supported the idea that their distribution was isotropic. They also clarified the reason for the 'zone of avoidance' in the galactic plane. This effect was caused by the quantities of dust and diffuse interstellar matter in that plane.

Among his later studies was a report made in 1941 that the spiral arms of the galaxies probably did 'trail' as a result of galactic rotation, rather than open out. After World War II Hubble became very much an elder statesman of US astronomy. He was involved in the completion of the 5-m/200-in Hale telescope at Mount Palomar, which was opened in 1948. One of the original intentions for this telescope was the study of faint stellar objects, and Hubble used it for this purpose during his few remaining years.

Further Reading

Christianson, Gale E, *Edwin Hubble: Mariner of the Nebulae,* Farrar, Straus & Giroux, 1995.

Fox, Mary V, *Edwin Hubble: American Astronomer,* Book Report Biography series, Franklin Watts, 1997.

Goodwin, Simon, *Hubble's Universe: A Portrait of Our Cosmos,* Penguin, 1997.

Naeye, Robert, *Through the Eyes of Hubble: Birth, Life and Violent Death of Stars,* Institute of Physics Publishing, 1997.

Sharov, Alexander S and Novikov, Igor D; Kisin, Vitaly I (transl), *Edwin Hubble: The Discoverer of Big Bang Universe,* Cambridge University Press, 1993.

Syder, Diana, *Hubble,* Smith Doorstop Books, 1997.

Hubel, David Hunter (1926–) US neurophysiologist who worked with Torsten Wiesel (1924–) on the physiology of vision and the way in which the higher centres of the brain process visual information. They shared the 1981 Nobel Prize for Physiology or Medicine with Roger Sperry (1913–1994).

Hubel was born on 27 February 1926 in Ontario, Canada, to US parents and qualified in medicine from McGill University in Montréal, after which he was drafted into the US army. From 1955 to 1958 he worked in physiological research, studying the electrical activity of the brain, at the Walter Reed Army Research Institute. In 1958 he moved to the Johns Hopkins Medical School, and the following year to Harvard, working at both places with Stephen Kuffler (1913–1980) and becoming professor of neurophysiology in 1965. In 1968 he was appointed professor of neurobiology, where he remained until his retirement in 1982.

At Harvard he met Wiesel, and they began experiments implanting electrodes into the region of the brain of anaesthetized cats known to be involved in the processing of visual information, and then analysing the responses of individual brain cells to different types of visual stimulation. With painstaking work they correlated the anatomical structure of the visual cortex of the brain with their physiological responses. They built up a complex picture of how the brain analysed visual information by an increasingly sophisticated system of detection by the nerve cells. Later study of the development of the visual system in young animals suggested that eye defects should be treated and corrected immediately, and the then-routine ophthalmological practice of leaving a defect to correct itself was abandoned.

Hückel, Erich Armand Arthur Joseph (1896–1980) German physical chemist who, with Peter ⇨Debye, developed the modern theory that accounts for the electrochemical behaviour of strong electrolytes in solution. In his own right he was known for his discoveries relating to the structures of benzene and similar compounds that exhibit aromaticity.

Hückel was born in Berlin-Charlottenberg on 9 August 1896, the son of a doctor of medicine. When he was three years old the family moved to Göttingen and, after leaving school in 1914, he went to the local university to study physics. He interrupted his studies after two years to take a job concerned with aerodynamics in Göttingen University's Applied Mechanics Institute, under the direction of L Prandtl. At the end of World War I Hückel resumed his studies in mathematics and physics and in 1921 was awarded a doctorate based on a thesis (prepared under the direction of Debye) on the diffraction of X-rays by liquids.

He stayed at Göttingen for the next two years, first as a physics assistant to the mathematician David ⇨Hilbert and then in a similar role to Max ⇨Born, the Nobel prizewinning physicist. In 1922 he left to join Debye again, this time at the Zürich Technische Hochschule. He became an unpaid lecturer there, and in 1925 married Annemarie Zsigmondy, the daughter of the colloid chemist Richard ◊Zsigmondy. In 1928 Hückel was awarded a Rockefeller Foundation Fellowship and worked briefly with Frederick Donnan. He also spent some time with Niels ◊Bohr in Copenhagen, before becoming a fellow of the Notgemeinschaft der Deutschen Wissenschaft at the University of Leipzig. In 1930 he went to the Technische Hochschule at Stuttgart, where he remained until becoming the professor of theoretical physics at the University of Marburg in 1937. Hückel died in Marburg on 16 February 1980.

When Hückel started working with Debye in 1922, the prevailing theory of electrolyte solutions was that of Svante ◊Arrhenius, who held that an equilibrium exists in the solution between undissociated solute molecules and the ions from its dissociated molecules. In strong electrolytes, particularly those formed from strong acids and bases, dissociation was considered to be nearly complete. There were, however, many instances which this theory did not adequately explain.

Debye and Hückel suggested that strong electrolytes dissociate completely into ions, explaining the deviations from expected behaviour in terms of attraction and repulsion between the ions. In a series of highly mathematical investigations they found formulae for calculating the electrical and thermodynamic properties of electrolytic solutions, principally dilute solutions of strong electrolytes. But even taking into account the sizes of the ions concerned, they did not find a complete theory for concentrated solutions.

In 1930 Hückel began his work on aromaticity, the basis of the chemical behaviour of benzene, pyridine and similar compounds. The benzene molecule is held together by electrons that are delocalized and contribute to one large hexagonal bond above and below the plane of the molecule. This accounts for its planar shape and its ability to preserve its structure through chemical reactions. Hückel developed a mathematical approximation for the evaluation of certain integrals in the calculations concerned with the exact nature of the bonding in benzene.

Later in the 1930s he extended his research to other chemical systems that appear to possess the same kind of aromatic nature as benzene (although usually to a lower degree). From this study emerged the Hückel rule for monocyclic systems, which states that for aromaticity to occur the number of electrons contributing to the correct type of bonding (π-bonding) must be $4n + 2$, where n is a whole number. When $n = 1$, for example, there are six π-electrons, as exemplified by benzene. When $n = 2$, the ten π-electrons occur in derivatives of [10]annulene.

There seem to be exceptions to the rule in large ring systems, where the predicted aromaticity does not occur. In general the degree to which a molecule resembles benzene in aromaticity decreases with increasing ring size.

Hückel also carried out research into unsaturated compounds (those with double or triple bonds) and into the chemistry of free radical compounds (those with a free, non-bonding electron).

Huggins, William (1824–1910) English astronomer and pioneer of astrophysics. He revolutionized astronomy by using spectroscopy to determine the chemical make-up of stars and by using photography in stellar spectroscopy. With these techniques he investigated the visible spectra of the Sun, stars, planets, comets, and meteors and discovered the true nature of the so-called 'unresolved' nebulae.

Huggins was born in London on 7 February 1824. As a young boy he had been given a microscope, which helped to develop his first interest in physiology. Then when he was about 18 he bought himself a telescope and became increasingly interested in astronomy, although London, even at that time, was a poor place for making observations. It was intended that he should go to Cambridge University, but when the time came his family persuaded him to take charge of their drapery business in the City of London. He diligently followed this trade 1842–54, when he sold the business and moved with his parents to a new home south of London in Tulse Hill, where he built his own private observatory. Although he had continued to make observations while in business, it was only after he moved that he devoted his time entirely to science.

Huggins was elected a fellow of the Royal Society in 1865 and was president 1900–05. He also served terms as president of the British Association for the Advancement of Science and the Royal Astronomical Society. His country honoured him by making him a Knight Commander of the Order of Bath and he was also one of the original members of the Order of Merit in 1902. From 1890 he received a civil-list pension (of £150). Huggins died at his home in Tulse Hill on 12 May 1910.

Huggins spent his first two years of research 1858–60 observing the planets with an 20-cm/8-in refracting telescope. He was looking for a new line of action when, in 1859, Gustav ◊Kirchhoff and Robert ◊Bunsen published their findings on the use of spectroscopy to determine the chemical composition of the Sun. Huggins, together with his friend W A Miller (who was professor of chemistry at King's College, London), designed a spectroscope and attached it to the telescope. They began to observe the spectra of the Sun, Moon, planets, and brighter stars. They compared the spectral lines that they observed with those produced by various substances in the laboratory, and in this way they were able to draw conclusions about the composition of the stars and planets. They published their results in 1863 in a paper presented to the Royal Society: 'Lines of the spectra of some of the fixed stars'.

It showed that the brightest stars had elements in common with those of the Earth and the Sun, although there was a diversity in their proportions. In the same paper they suggested that starlight originated in the central part of a star and then passed through the hot gaseous envelope that surrounds it.

Following this initial success, other great achievements were to come. Some of the tiny indistinct objects known as nebulae had been observed to be faint clusters of stars. Others could not be resolved but it was believed that eventually more powerful telescopes would show up the individual stars and confirm that they were similar to those already resolved. Huggins realized that those nebulae of stellar origin would give a characteristic stellar spectrum. When he observed the unresolved nebulae in the constellation of Draco in 1864 this was not the case. Only a single bright line was observed. Seeing this, he understood the nature of these objects – they were clouds of luminous gas and not clusters of stars. In the same year he observed the great nebula in Orion and saw two unknown lines in its spectrum. Huggins postulated the existence of a previously undiscovered element, 'nebulium', but in 1927 Ira Bowden (1898–1973) showed that the lines were caused by ionized oxygen and nitrogen. Although Huggins had discovered the gaseous nature of nebulae, which suggested that they may be the 'parents' of stars, his cautious nature prevented him from stating it as a fact. Like other scientists of his day, he believed that chemical elements were unchangeable.

On 18 May 1866 Huggins made his first spectroscopic observation of a nova and found that, superimposed on a solar-type spectrum, there were a number of bright hydrogen lines that suggested that the outburst occurred with an emission of gas with a higher temperature than that of the star's surface. He then showed that the bands of this gas were coincident with those obtained from a candle flame in the laboratory and therefore arose from carbon vapours.

In 1868 Huggins spectroscopically measured the velocity of the star Sirius by observing the Doppler shift of the F line of hydrogen towards the red end of the spectrum. Since the measurements that he was making were tiny, his result (of 47.1 km/29.4 mi per second) was high compared with the modern figure, but it was as accurate as his purely visual observations would allow. In the same year he examined the spectrum of a comet and found that its light was largely due to luminous hydrocarbon vapour. He went on to more observational successes with a pair of large telescopes lent to him by the Royal Society in 1870 and he made the first ultraviolet spectrograph. In 1899 he and his wife jointly published an *Atlas of Representative Stellar Spectra.*

Besides his interest in astronomy, Huggins was also a keen violinist and fisherman. He worked briefly with the chemist and physicist William ◊Crookes on the investigation of spiritualism, but became disillusioned as his suspicion of trickery increased.

Huggins was a pioneer, and as such he was often hampered by the inadequacy of his instruments and perhaps frustrated because he knew that more accurate observations and measurements could be made when better and more powerful equipment became available.

Part of Huggins's talent lay in his ability to foresee the possibilities in new ideas that were being initiated by his contemporaries. The three fundamental discoveries that made him famous within various fields of astronomy were made with his modest 20-cm/8-in refracting telescope. He established that elements such as hydrogen, calcium, sodium, and iron were to be found in the stars and that the universe was made up of well-known elements. He resolved the long debate as to whether all nebulae were clusters of stars by proving that some, like that in Orion, were gaseous and he used the spectroscope to detect the motion of stars and to measure their compositions and velocities.

Humason, Milton Lasell (1891–1972) US astronomer famous for his investigations, at Mount Wilson Observatory, into distant galaxies.

Humason was born in Dodge Centre, Minnesota on 19 August 1891. There is no great list of educational achievements for Humason because he entered astronomy by an unusual route. When he was 14, his parents sent him to a summer camp on Mount Wilson and he so enjoyed it that when he had been back at high school for only a few days, he obtained permission from his parents to take a year off school and return there. He extended his 'year' well beyond twelve months and became one of astronomy's most notable educational dropouts. Sometime between 1908 and 1910, after his voluntary withdrawal from higher education, he became a mule driver for the packtrains that travelled the trail between the Sierra Madre and Mount Wilson during construction work on the observatory. He brought up much of the timber and other building materials for the telescope's supporting structure, the local cottages, and the scientists' quarters.

He became engaged to the daughter of the observatory's engineer and married her in 1911. In the same year he gave up his job as driver of the packtrains and went to be foreman on a relative's range in La Verne. But he still loved Mount Wilson and in 1917, when a janitor was leaving, his father-in-law suggested that this might be an opening for better opportunities. So Humason initially joined the staff of Mount Wilson Observatory as a janitor. His position was soon elevated to night assistant. George ⇨Hale, the director of the observatory at the time, recognized Humason's unusual ability as an observer and in 1919 he was appointed to the scientific staff. There was considerable opposition to this appointment, partly because Humason had had no formal education after the age of 14 and partly because he was a relative of the observatory's engineer.

Humason was assistant astronomer 1919–54 and then astronomer at both the Mount Wilson and Palomar observatories. In 1947 he was appointed secretary of the observatories, a position that involved him in handling public relations and administrative duties. In 1950 he was awarded an honorary PhD by the University of Lund in Sweden. He was a quiet, friendly man who was often consulted on administrative and personal problems. He died suddenly at his home near Mendocino, California, on 18 June 1972.

At Mount Wilson Observatory Humason took part in an extensive study of the properties of galaxies that had been initiated by Edwin Hubble. The research programme began in 1928 and consisted of making a series of systematic spectroscopic observations to test and extend the relationship that Hubble had found between the red shifts and the apparent magnitudes of galaxies. But because of the low surface brightness of galaxies there were severe technical difficulties. Special spectroscopic equipment, including the Rayton lens and the solid-block Schmidt camera, was designed. With these instruments it became possible to obtain a spectrum of a galaxy too faint to be picked up visually with the telescope used. The method was to photograph the field containing the galaxy and accurately measure the position of its image with respect to two or more bright stars. Guide microscopes were then set up at the telescope with exactly the same offsets from the slit of the spectrograph. It was a tedious task and often took several nights of exposure to produce a spectrum on a 13-mm/0.5-in plate.

Humason undertook this exacting programme. He personally developed the technique and made most of the exposures and plate measurements. During the period from 1930 to his retirement in 1957, the velocities of 620 galaxies were measured. He used the 2.5-m/100-in telescope until the 5-m/200-in Hale telescope was completed and the programme transferred to that instrument. Humason's results were published jointly with those of N U Mayall and A R Sandage in 'Redshifts and magnitudes of extragalactic nebulae', which appeared in the *Astronomical Journal* of 1956. These data still represent the majority of known values of radial velocities for normal galaxies, including most of the large values.

Humason applied the techniques he had developed for recording spectra of faint objects to the study of supernovae, old novae that were well past peak brightness, and faint blue stars (including white dwarfs). One by-product of his studies on galaxies was his discovery of Comet 1961e, which is notable for its large perihelion distance and its four-year period of visibility with remarkable changes in form.

Humason is remembered for his ability to handle instruments with meticulous care and great skill. He provided criteria for studying various models of the universe. He became internationally famous for his work on galaxies and, in spite of his lack of formal training, won a leading role in US astronomy.

Humboldt, Friedrich Heinrich Alexander, Baron von (1769–1859) German geophysicist, botanist, geologist, and writer; a founder of ecology.

Born in Berlin on 14 September 1769, Humboldt was the son of a Prussian soldier who wanted him to pursue a political career. Humboldt had other ideas. Preferring science, he studied at Göttingen University, proceeding to Abraham ⇨Werner's Freiberg mining school and spending two years as a mines engineer. In the 1790s he travelled widely in Europe, before setting out, in 1799, with the French botanist A Bonpland, on a pioneering and immensely productive expedition across Latin America. Studying physical geography above all, but also collecting vast quantities of geological, botanical, and zoological material, Humboldt covered some 9,600 km/6,000 mi. His expedition included travelling up the Orinoco and the Magdalena, passing a year in Mexico and some time in Cuba, visiting the sources of the Amazon, and cutting across the continent to the Cordilleras and on to Quito and Lima.

He returned to Europe in 1804 laden with scientific specimens, and spent the next 20 years writing up his results, not least in his *Narrative of Travels* (1818–19). Humboldt aimed to erect a new science, a 'physics of the globe', analysing the deep physical interconnectedness of all terrestrial phenomena. He believed physical relief should be grasped in terms of Earth history, and likewise that geological phenomena were to be understood in terms of more basic physical causes (for example, terrestrial magnetism or rotation). Showing a phenomenal ability to keep abreast of developments in geophysics, meteorology, and geography, he patiently amassed evidence of comparable geological phenomena from every continent.

Humboldt arrived at numerous important findings. On the basis of studies of Pacific coastal currents, he was one of the first to propose a Panama canal. In meteorology, he introduced isobars and isotherms on weathermaps, made a general study of global temperature and pressure, and finally instituted a worldwide programme for compiling magnetic and weather observations. His studies of US volcanoes demonstrated they corresponded to underlying geological faults; on that basis he deduced that volcanic action had been pivotal in geological history and that many rocks were igneous in origin. In 1804, he discovered that the Earth's magnetic field decreased from the poles to the equator.

His most popular work, *Cosmos,* begun in 1845 at the age of 76, is a profound and moving statement of our relationship with the Earth, and of the relations between physical environment and flora and fauna. A gracious polymath held in universal respect, Humboldt proved enormously influential in study of the globe, both as an explorer and as a theorist. As one who set out to chart the history of human interrelations with planet Earth, he may be called the founder of ecology. He died in Berlin on 6 May 1859.

Further Reading

Gaines, Ann, *Humboldt,* Chelsea House Publishers, 1990.

Gaines, Ann; Goetzmann, William H (ed); Collins, Michael (introduction), *Alexander von Humboldt: Colossus of Exploration,* World Explorers series, Chelsea House Publishers, 1991.

De Mauro, Tullio and Formigari, Lia (eds), *Leibniz, Humboldt, and the Origins of Comparativism: Proceedings of the International Conference, Rome, 25–28 September 1986,* Studies in the History of the Language Sciences, Benjamins (John) Publishing Co, 1990, v 49.

Malin, S R C, Barraclough, 'D R, Humboldt and the Earth's magnetic field', *Quart J Roy Astron Soc,* 1991, v 32, pp 279–293.

Sweet, Paul, *Wilhelm von Humboldt: A Biography,* Ohio State University Press, 1986, 2 vols.

Hume-Rothery, William (1899–1968) English chemist who spent his entire scientific life working on the structures of metals and their alloys. When he began his studies, metallurgy was considered to be a branch of physical chemistry; when he retired, it had come to be regarded as a distinct discipline in its own right.

Hume-Rothery was born on 15 May 1899 at Worcester Park, Surrey. He was educated at Cheltenham College before going to the Royal Military Academy, in anticipation of a career in the army. But his training was brought to a sudden stop by a serious illness that left him totally deaf. He then went to Oxford to study natural sciences, specializing in chemistry and graduating in 1926. He carried out postgraduate research and gained his PhD at the Royal School of Mines, where he became interested in metallurgy. He soon returned to Oxford and continued his research into metal alloys, first in the Dyson Perrins Laboratory and later in the chemistry department. For most of this period his research was financed by outside organizations such as the Armourers and Braziers Company, and he did not gain a formal university post until 1938. During World War II he supervised many contracts from the Ministry of Supply and Aircraft Production for work on complex aluminium and magnesium alloys.

In 1955, with the offer to Hume-Rothery of the George Kelley Readership, metallurgy finally became recognized as a discipline within Oxford University. Three years later he received the first professorship in the new department of metallurgy. He died in Oxford on 27 September 1968.

While he was working in the Dyson Perrins laboratory Hume-Rothery complained bitterly that vapours from the experiments of neighbouring organic chemists spoiled the finish of the specimens he had prepared for study with a microscope. The accepted method of preparing metals in order to study their structure was to smooth the surface by a series of grinding and polishing operations until it had a mirror finish. The surface was then carefully etched in an acid that attacked certain areas preferentially. The specimen was then viewed under a microscope using indirect reflected light. In the laboratory atmosphere the carefully prepared specimens became corroded, interfering with the study. Hume-Rothery was allowed to move.

Many of Hume-Rothery's early ideas about alloys came from interpreting their microstructures. If metals mix to form an alloy, the melting point, physical strength, and microstructure vary with the composition of the alloy. During the later part of the 1920s and the 1930s he and his continuous stream of research students established that the microstructure of an alloy depends on the different sizes of the component atoms, the valency electron concentration, and electrochemical differences. They showed that in a binary (two-metal) alloy, a common crystal lattice is possible if the two types of atoms present have a similar size. And if a common lattice forms, the alloy has a uniform structure. With atoms of widely different sizes, at least two types of lattices may form, one rich in one metal and one rich in the other. The presence of two types of structures can increase the strength of an alloy.

Hume-Rothery and his researchers discovered that the solid solubility of one element in another is extremely restricted if the atomic diameters of the two elements concerned differ by more than 15%. Within the 15% limit, solutions are formed provided that other factors are favourable. In such a solution, the solute (alloying) element replaces some of the atoms of the solvent (primary) metal. Nickel and copper have very similar atomic radii and so form a continuous range of solid solutions – no doubt assisted by the fact that they both have face-centred cubic crystal structures. Zinc crystallizes with a hexagonal structure, so that although zinc and copper alloy over the whole range of compositions, most of the alloys involve two crystal structures – making some brasses much stronger than their component metals. If the two elements differ considerably in electronegativity, a definite chemical compound is formed. Thus steel, an 'alloy' of iron and carbon, contains various iron carbides.

All this theory was backed up by experimental work, and Hume-Rothery and his team constructed the equilibrium diagrams for a great number of alloy systems. The experimental procedures became standard practice and gave rise to work that was noted for its accuracy and a stern appraisal of the source of errors. He went on to make fundamental studies of solid-state transformations and deformation, while continuing his original research, particularly his interest in equilibrium phase diagrams. Today metallurgists can produce 'tailor-made' alloys to suit particular and exacting requirements. Without Hume-Rothery's painstaking, systematic studies this would not have become possible.

Hunter, John (1728–1793) Scottish surgeon known for his occasional use of unorthodox methods. He built up a collection of 14,000 anatomical specimens and his memorial brass in Westminster Abbey records his 'services to mankind as the Founder of Scientific Surgery'.

Hunter was born in Long Calderwood, Lanarkshire, on 13 February 1728. He worked on the family farm after the death of his father and at the age of 20 went to London to join his brother, William, who was a distinguished surgeon and obstetrician. John Hunter had no formal university qualification; he assisted in the preparation of anatomical specimens for his brother's lectures and engaged in investigations of his own. He also attended surgical classes at various London hospitals. He was appointed a master of anatomy of the Surgeons' Corporation in 1753 and three years later served as house surgeon at St George's Hospital, London, (during which time he taught Edward ◊Jenner). He joined the British army in 1759 and in 1760 worked in France and Portugal as an army surgeon. He returned to London in 1763 and set up a private practice, and continued with his research. During the late 1760s he took up a senior surgical post at St George's Hospital and was appointed physician extraordinary to George III. Ten years later he became deputy surgeon to the army and in 1790 became inspector general of hospitals. Hunter collapsed and died at a meeting of the board of governors at St George's Hospital on 16 October 1793. He was buried at St Martin-in-the-Fields, but his remains were later taken to Westminster Abbey on 28 March 1859.

While Hunter was working with his brother, he made a detailed study of the structure and function of the lymphatic vessels, and the growth and structure of bone. He made these investigations by collecting a great deal of material from postmortem examinations. Many of the corpses that he dissected he obtained from 'resurrectionists', who raided graveyards at night to sell newly buried corpses to surgeons for dissection.

Hunter also kept a number of animal specimens in his garden for dissection, which at one time included a bull and the carcass of a whale. The knowledge he gained performing these dissections allowed him to improve further the embalming technique by arterial injection, as developed by William ◊Harvey. One of the most interesting of these experiments was the case of the late Mrs Martin van Butchell, whose husband took her body to Hunter for embalming in 1775. Mrs van Butchell had stated in her will that her wealth remained her husband's, as long as her body remained above the ground. Hunter embalmed the body, and Mr van Butchell clothed it and placed it in a glass case for visitors to view – and to meet the conditions of the will.

Hunter's keenness to collect specimens was well known. One of his most prized specimens was the skeleton of Charles Byrne, an Irishman who was 2.4 m/8 ft tall and who, determined not to fall into Hunter's clutches, had arranged to be buried at sea. But Hunter, on hearing of his death, arranged for his body to be seized by the resurrectionists, for a fee of £500.

Surgical training was not easily available in Hunter's day, so when he moved nearer to central London he gave lectures from his own house. Later he had a lecture room and museum built in his garden where he held meetings of the Lyceum Medicum Londinense (London Medical Academy). He had helped to found this student society and encouraged each student to prepare a paper on a medical topic and read it to his fellow students.

Hunter's experiments were wide-ranging, including studies of lymph and blood circulation, the sense of smell, the structure of teeth and bone, tissue grafting, and various diseases. He often carried out experiments on himself, such as the occasion when, trying to prove that syphilis and gonorrhoea are types of the same disease, he inoculated himself and later developed syphilis. When Hunter's health began to suffer, he took an army appointment on the surgical staff concerned with the Seven Years' War. While dealing with casualties, he gained the knowledge for his treatise on gunshot wounds.

During his lifetime Hunter published an impressive number of papers on a wide variety of medical and biological subjects such as his 'Treatise on the natural history of the human teeth' (1771), in which he describes his experiments on the transplantation of tissues – the best known of these being that of a human tooth fixed into a cock's comb. In 1786 he published his 'Treatise on the venereal disease', and his 'Treatise on the blood, inflammation and gun-shot wounds' was published posthumously in 1794. Hunter's collection was handed to the Royal College of Surgeons in 1795 and later formed the basis of the Hunterian Museum.

Further Reading

Hunter, J A, *Hunter,* Buccaneer Books, 1993.

Hunter, John; Allen, Elizabeth, Turk, J L and Murley, Sir Reginald (eds), *The Case Book of John Hunter FRS,* Royal Society of Medicine Services, 1993.

Palmer, Richard and Taylor, Jean (compilers); Findlay, David W (ed), *The Hunterian Society: A Catalogue of its records and Collections Relating to John Hunter and the Hunterian Tradition with a History of the Society,* The Hunterian Society, 1990.

Hussey, Obed (1792–1860) US inventor who developed one of the first successful reaping machines and various other agricultural machinery.

Hussey was born into a Quaker family in Maine. The family moved to Nantucket, Massachusetts, and like most boys living there at that time, the young Hussey wanted to go to sea when he grew up. He was a quiet and studious boy, always thoughtful and modest. He liked studying intricate mechanisms and became a skilled draughtsman and inventor. In addition to the reaper, he invented a steam plough, a machine for making hooks and eyes, a grinding mill for maize, a horse-powered husking machine, a sugar-cane crusher and an ice-making machine. On 4 August 1860, while travelling on a train from Boston to Portland, Maine, he got off at a station to fetch a drink of water for a child. The train started as he was getting back on, and he fell beneath the wheels and was killed.

Hussey began work on a reaping machine early in 1833 in a room at the factory of Richard Chenoweth, a manufacturer of agricultural implements in Baltimore, Maryland. The finished prototype was tested later that year near Cincinnati, Ohio, where it so impressed a local businessman, Jarvis Reynolds, that he provided the finance and a factory for manufacturing the reaper. It was patented in December 1833 and pictured in the *Mechanics Magazine* of April 1834 as 'Hussey's Grain Cutter'.

At first Hussey used a reel to gather the grain up to the cutter and throw it on to the platform, but after trials he decided that the reel was an encumbrance and discarded it. The main frame containing the gearing was suspended on two wheels about a 1 m/3 ft in diameter; the platform was attached to the rear of the frame and extended nearly 2 m/6 ft on each side. A team of horses attached to the front of the frame walked along the side of the standing grain. The cutting knife consisted of steel plates 7.5 cm/3 in wide, tapered towards the front, riveted to a flat iron bar to form a sort of saw with very coarse teeth sharpened on both edges. The knife was supported on what Hussey called guards attached to the front of the platform every 7.5 cm/3 in across the whole width of the machine. They projected forwards and had long horizontal slits in which the cutter oscillated from side to side. They thus supported the grain while it was being cut and protected the blades from damage by large stones and other obstructions. The cutter was attached by means of a Pitman rod to a crank activated by gearing, connected to one or both or the ground wheels. This arrangement gave a quick vibrating motion to the cutter as the machine was pulled along.

The point of each blade oscillated from the centre of one guard, through the space between, to the centre of the next, thus cutting equally both ways.

In his next modification to the machine, Hussey used one large ground wheel instead of two and moved the platform to a position alongside the frame, providing a seat for an operator who faced forward and raked the cut grain off the back of the platform.

During the harvest of 1834 Hussey successfully demonstrated his machine to hundreds of farmers and began to sell the reapers for $150 each. The fame of the machine spread, slowly at first, and it was eventually used in the far west. In September 1851 he went to the UK and demonstrated the reaper at Hull and Barnardscastle. He was invited to show it to Prince Albert who bought two (at £21 each), one for the estate at Windsor and one for Osborne House on the Isle of Wight.

An earlier rival design of reaper had been developed by Cyrus ◊McCormick in 1831, although Hussey's was patented first. Both machines used the principle of a reciprocating knife cutting against stationary guards or figures, although McCormick's design employed a division between the cut and uncut grain and a reel to topple the cut grain on to the platform. Hussey's contribution was summed up succinctly in the title of a book that relates the story: *Obed Hussey: Who of All Inventors, Made Bread Cheap.*

Hutton, James (1726–1797) Scottish natural philosopher who pioneered uniformitarian geology.

He was born in Edinburgh on 3 June 1726, the son of an Edinburgh merchant, and studied at Edinburgh University, Paris, and Leiden, training first for the law but taking his doctorate in medicine in 1749 (though he never practised). He spent the next two decades travelling and farming in the southeast of Scotland. During this time he cultivated a love of science and

Hutton Scottish natural philosopher James Hutton whose work is the basis for the science of geology. Much of his work was carried out around Edinburgh, where he settled in 1768. *Mary Evans Picture Library*

philosophy, developing a special taste for geology. In about 1768 he returned to his native Edinburgh. A friend of Joseph ◊Black, William Cullen, and James ◊Watt, Hutton shone as a leading member of the scientific and literary establishment, playing a large role in early history of the Royal Society of Edinburgh and in the Scottish Enlightenment. He died in Edinburgh on 26 March 1797.

Hutton wrote widely on many areas of natural science, including chemistry (where he opposed Antoine ◊Lavoisier), but he is best known for his geology, set out in his *Theory of the Earth,* of which a short version appeared in 1788, followed by the definitive statement in 1795. In that work, Hutton attempted (on the basis both of theoretical considerations and of personal fieldwork) to demonstrate that the Earth formed a steady-state system, in which terrestrial causes had always been of the same kind as at present, acting with comparable intensity (the principle later known as uniformitarianism). In the Earth's economy, in the imperceptible creation and devastation of landforms, there was no vestige of a beginning, nor prospect of an end. Continents were continually being gradually eroded by rivers and weather. Denuded debris accumulated on the sea bed, to be consolidated into strata and subsequently thrust upwards to form new continents thanks to the action of the Earth's central heat. Non-stratified rocks such as granite were of igneous origin. All the Earth's processes were exceptionally leisurely, and hence the Earth must be incalculably old. Though supported by the experimental findings of James ◊Hall, Hutton's theory was vehemently attacked in its day, partly because it appeared to point to an eternal Earth and hence to atheism. It found more favour when popularized by Hutton's friend, John Playfair, and later by Charles ◊Lyell. The notion of uniformitarianism still forms the groundwork for much geological reasoning.

Further Reading

Dean, Dennis R, *James Hutton and the History of Geology,* Cornell University Press, 1992.

Dean, Dennis R, *James Hutton in the Field and in the Study,* Scholars' Facsimiles and Reprints, 1997.

Geological Society, *James Hutton – Present and Future,* Geological Society Special Publication Series, Geological Society Publishing House, 1999 v 150.

Jones, Elizabeth Jean, *James Hutton: Founder of Modern Geology,* Scotland's Cultural Heritage, 1984.

McIntyre, D B and McKirdy, Alan, *James Hutton: The Founder of Modern Geology,* Stationery Office, 1997.

Huxley, Andrew Fielding English physiologist; see ◊Hodgkin and Huxley.

Huxley, Hugh Esmor (1924–) English physiologist whose contribution to science has been concerned with the study of muscle cells.

Huxley was born in Birkenhead, Cheshire, on 25 February 1924 (he is not a member of the Huxley family descended from the 19th-century scientist T H ◊Huxley). He was educated at Park High School, Birkenhead, and from 1914 at Christ's College, Cambridge. He read natural science there and graduated in 1943. He worked on radar research until the end of World War II and then returned to Cambridge in 1948, where he finally gained his PhD in 1952. He spent five years 1956–61 in the University of London biophysics department and was a fellow of King's College, Cambridge, 1961–67; when he became a fellow of Churchill College, Cambridge. In 1987 he moved to the USA to become professor of biology at the Rosenstiel Basic Medical Sciences Research Center, Brandeis University, Boston.

Prior to Huxley's research, the physiology of muscle structure had been under scientific scrutiny for a long time and it had been thought that muscular contraction involved a process of coiling and contraction, rather like the shortening of a helical spring. But investigations had been held up because adequate equipment, such as the electron microscope, was not available. Once this instrument was built in 1933, studies into the chemical reactions that produce muscle contraction progressed, but the real breakthrough came in the 1950s when Huxley, using the electron microscope and thin slicing techniques, established the detailed structural basis of muscle contraction. It had been ascertained that muscle fibres contain a large number of longitudinally arranged myofibrils, and that two main bands alternately cross the fibrils: the dark anisotropic (A) and the lighter isotropic (I) bands. Each A band is divided by a lighter region, the H zone (which in turn is bisected by an M line), and each I band by a dark line, the Z line.

Huxley demonstrated that the myofibrils are composed of interdigitating rows of thick and thin filaments. He found that the thin filaments are attached to the Z lines and extend through the I bands into the A bands, and that the M line is caused by a further thickening in the middle of the thick filaments. In addition he proved that the helical spring assumption was wrong. The A band does not change its length when the muscle is stretched or when it is shortened. Huxley suggested that contraction is brought about by sliding movements of the I filaments between the A filaments, and that the sliding is caused by a series of cyclic reactions between the myosin filaments and active sites on the actin filaments. This proposal is known as the sliding filament theory.

From previous research, Huxley knew that the contractile machinery of muscle cells consists of a small number of different proteins, and that actin and myosin are the two major ones involved. He proceeded to investigate the exact location of the actin and myosin. Using a solution of potassium chloride, pyrophosphate, and magnesium chloride, he found that the A filaments are composed of myosin. He then discovered, using a solution of potassium iodide, that the I filaments are composed of actin. The Z lines were unaffected by the solutions, which indicated that they are composed of some other substance. Huxley also noticed that the myosin-extracted fibres still retained their elasticity and he suggested that there is yet another set of filaments crossing the H zone. He called these hypothetical structures S filaments.

In 1953 Huxley had observed a series of projections which formed a repetitive pattern along the line of the A filament. The projections appeared to emerge in pairs from opposite sides of the myosin filaments at regular intervals, and he discovered that they are able to aggregate under suitable conditions to form 'artificial' filaments of varying lengths. He proposed that the 'tails' of the myosin molecules become attached to each other to form a filament with the heads projecting from the body of the filament, and that these projecting filaments play an important part in the sliding effect described in the sliding filament theory.

By coincidence, Andrew Huxley (see ◊Hodgkin and Huxley), working separately, came to the same conclusions at about the same time. Both Huxleys subscribe to the sliding filament theory although their interpretation of it is slightly different. Hugh Huxley suggests that the movement can be likened to that of a ratchet device or a cog-type operation, whereas Andrew Huxley proposes that each myosin filament has side pieces that can slide along the main backbone of the filament and that the slides can combine temporarily with sites on adjacent filaments. Many problems remain surrounding this discovery because although the sliding filament theory has been accepted by many scientists when applied to striated muscles, it has not yet been shown to apply to smooth muscles.

Huxley, T(homas) H(enry) (1825–1895) English biologist who helped to break down the great barrier of traditional resistance to scientific advance during the mid-19th century, and did a great deal to popularize science.

Huxley was born in Ealing, Middlesex, on 4 May 1825. He went to Ealing School in 1833, where his father taught mathematics, but his schooldays were limited to two years because his father was dismissed from his teaching post in 1835. Despite the lack of a formal education, he received some tuition in medicine from a brother-in-law, and obtained an apprenticeship in 1840 to a medical practitioner in London's East End. Huxley attended botany lectures in his spare time and in 1842, on entering a public competition, won a scholarship to study medicine at Charing Cross Hospital; he graduated in 1845. From 1846 to 1850 he was the assistant ship's surgeon on HMS *Rattlesnake* on its voyage around the South Seas. A year after his return, in 1851, he was elected to the Royal Society. Three years later he took up the appointment of professor of natural history at the Royal School of Mines (which in time became the Royal College of Science). From 1881 to 1885 he was president of the Royal Society. He then retired to continue his research and died in Eastbourne, of influenza and bronchitis, on 29 June 1895. His grandchildren include the Nobel prizewinning physiologist Andrew ◊Huxley, the biologist Julian Huxley (1887–1975), and the author Aldous Huxley.

Huxley's original intention was to be a mechanical engineer but his success in the field of medicine dissuaded him. For instance, when he was 19 years old he discovered the structure at the base of the human hair known today as Huxley's layer.

Huxley established his reputation in the scientific world during the four-year voyage to the South Seas. Each day he dissected, drew, and observed, with his microscope tied down against the lurching of the ship. Most naturalists of the time were interested only in

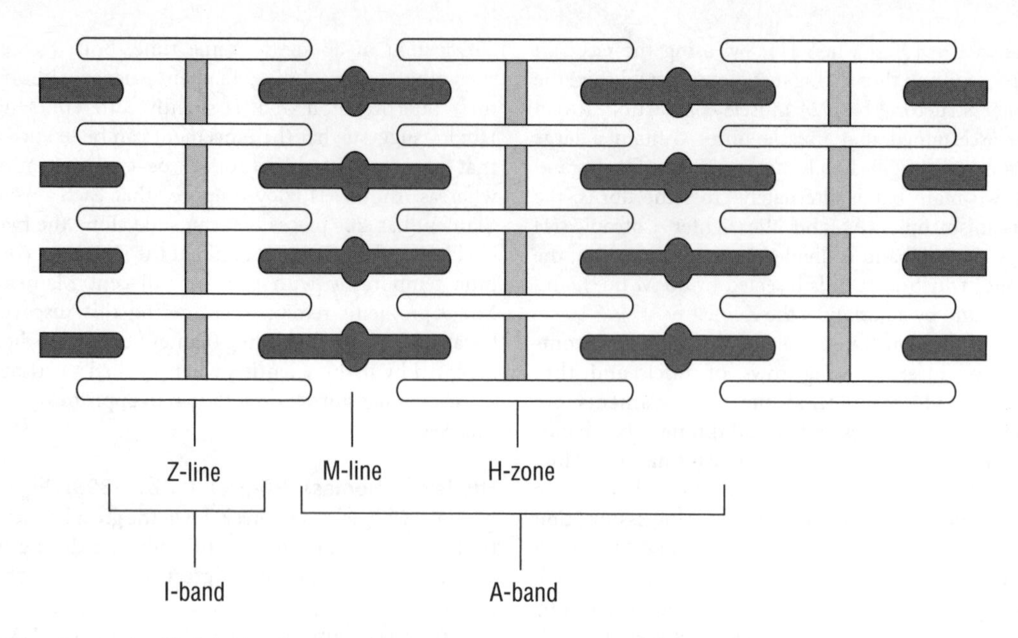

Huxley Banding in a fibril of striated muscle.

collecting specimens that could be dried, stuffed, and mounted, but Huxley concentrated on delicate invertebrates, which were difficult to preserve and liable to disintegrate. He sent his detailed recordings back to England from each port the ship docked at, and these observations were published in various influential scientific journals. On his return to England in October 1850, Huxley was acclaimed by scientists such as Joseph ◊Hooker and Charles ◊Lyell. From being an unknown assistant surgeon, he found himself at the forefront of British science.

European zoology had long been under the shadow of the French anatomist Georges ◊Cuvier. In 1817 Cuvier had formulated a system of classification that placed quadrupeds, birds, amphibians, and fish into the class Vertebrata, and insects among the Articulata. Although he recognized the molluscs as a distinct group, he had lumped most other animals together as Radiata. While carrying out his examinations on the ship Huxley realized that Cuvier's classification was not good enough and that there was a vast range of distinctions in minute anatomy of which the great man had not been aware. Huxley therefore reclassified the animal kingdom into Annuloida, Annulosa, Infusoria, Coelenterata, Mollusca, Molluscoida, Protozoa, and Vertebrata.

One of the illuminating suggestions Huxley was able to make from his detailed observations was that the inner and outer layers of the Medusae correspond with the two embryonic layers of higher animals. This suggestion provided the base to all later embryological thinking. Huxley also started a fundamental revision of the Mollusca. At that time a wide variety of non-segmented and non-radiate soft-bellied animals (sea squirts, sea-mats, and lamp shells) were grouped together indiscriminately with the true molluscs. Although his intended project to produce a regular monograph of Mollusca never really materialized, he was able to 'construct' an archetypal cephalous mollusc that was remarkably similar to the evolutionary ancestor of the molluscs deduced by zoologists more than 80 years later.

Huxley became very much embroiled in the great controversy that raged when Charles ◊Darwin published his *On the Origin of Species,* and he was one of Darwin's most outspoken champions. A famous example of his involvement in Darwinism is this extract from a public debate held in 1860 at Oxford University with Bishop Samuel Wilberforce. The bishop had asked whether Huxley traced his descent from the apes through his mother or his father, to which Huxley replied: 'If ... the question is put to me, would I rather have a miserable ape for a grandfather or a man highly endowed by nature and possessed of great means of influence, and yet who employs these faculties and that influence for the mere purpose of introducing ridicule into a grave scientific discussion – I unhesitatingly affirm my preference for the ape.'

Huxley is also credited with the founding of craniology. In his investigations into the true origins of the newly discovered Neanderthal skull, he devised a series of quantitive indices and the first real rationale of the measurement of skulls. He also produced a new system of classification of birds based mainly on the palate and other bony structures. Previously they had been classified according to their feeding habits, foot-webbing, and beaks. Huxley raised bird classification to a science and his own classification of birds is the foundation of the modern system.

Although he was never a great experimenter, Huxley's scientific work was distinguished by its critical assessment of both pre-existing and newly acquired knowledge. He produced more than 150 research papers on subjects as varied as the morphology of the molluscs, the hybridization of gentians, the taxonomy of crayfish, and the physical anthropology of the Patagonians. His dissections and observations filled numerous gaps in the knowledge of the animal kingdom, and his attitudes did much to establish the idea that science and the scientific method are the only means by which the ultimate truths of the animal world can be found.

Further Reading

Barr, Alan P (ed), *Thomas Henry Huxley's Place in Science and Letters: Centenary Essays,* University of Georgia Press, 1997.

Bibby, Cyril, *Scientist Extraordinary: The Life and Scientific Works of Thomas Henry Huxley,* Pergamon Press, 1972.

Collie, Michael, *Huxley at Work: With the Scientific Correspondence of T H Huxley and the Rev George Gordon of Birnie, near Elgin,* Macmillan, 1991.

Desmond, Adrian J, *Huxley: The Devil's Disciple,* Penguin 1994.

Desmond, Adrian J, *Huxley: From Devil's Principle to Evolution's High Priest,* Addison Wesley Longman, 1997.

Paradis, James G, *T H Huxley: Man's Place in Nature,* University of Nebraska Press, 1978.

Huygens, Christiaan (1629–1695) Dutch physicist and astronomer. He is best known in physics for his explanation of the pendulum and invention of the pendulum clock, and for the first exposition of a wave theory of light. In astronomy, Huygens was the first to recognize the rings of Saturn and discovered its satellite Titan.

Huygens was born on 14 April 1629 at The Hague into a prominent Dutch family with a tradition of diplomatic service to the ruling house of Orange, and a strong inclination towards education and culture. René ◊Descartes was a frequent guest at the house of Huygens's father, Constantijn, who was a versatile and multi-talented man, a diplomat, a prominent Dutch and Latin poet, and a composer. So it was natural for the young Christiaan to be educated at home to the highest standard. In 1645 he was sent to the University of Leiden to study mathematics and law, followed in 1647 by two years studying law at Breda.

But, eschewing his expected career in diplomacy, Huygens returned to his home to live for 16 years on an allowance from his father that enabled him to devote himself to his chosen task, the scientific study of nature. This long period of near seclusion was to be the most fruitful period of his career.

Huygens Engraving of Dutch physicist and astronomer Christiaan Huygens, notable for his contributions to the development of telescopes and to research into the nature of light. His first major scientific discovery, in 1655, was that of the rings of Saturn. *Mary Evans Picture Library*

In 1666 on the foundation of the Académie Royale des Sciences, Huygens was invited to Paris and lived and worked at the Bibliothèque Royale for 15 years, until his delicate health and the changing political climate took him back to his home. Here he continued to experiment, only occasionally venturing abroad to meet the other great scientists of the time as in 1689, when he visited London and met Isaac ◊Newton. During his stay in Paris, Huygens twice had to return home for several months because of his health and in 1694 he again fell ill. This time he did not recover, and he died in The Hague on 8 July 1695.

Huygens worked in different areas of science in a way almost impossible in our modern age of vast knowledge and increasing specialization. He made important advances in pure mathematics, applied mathematics, and mechanics, (which he virtually founded); optics and astronomy, both practical and theoretical; and mechanistic philosophy. He is also credited with the invention of the pendulum clock, enormously important for its use in navigation, since accurate timekeeping is necessary to find longitude at sea. In a seafaring nation such as Holland, this was of particular importance. Typically of Huygens, he developed this work into a thorough study of pendulum systems and harmonic oscillation in general.

At first Huygens concentrated on mathematics. In the age of such revolutionary figures as Descartes, Newton and Gottfried ◊Leibniz, his career may be termed conservative for it contained nothing completely new except the theory of evolutes and Huygens's study of probability including game theory, which originated our modern concept of the expectation of a variable. His suspicion of the new methods may have been partly due to the secrecy of scientists of his time, and partly to Huygens's fastidiousness, which often led him to delay publishing his observations and theories. The importance of his mathematical work is in its improving on available techniques, and Huygens's application of them to find solutions to problems in science.

The first of his many studies in applied mathematics was a paper on hydrostatics, including much mathematical analysis, published in 1650. Fascinating work on impact and collision followed, motivated by Huygens' disbelief in Descartes's laws of impact. Huygens used the idea of relative frames of reference, considering the motion of one body relative to the other. He discovered the law of conservation of momentum, but as yet the vectorial quantity had little intuitive meaning for him and he did not proceed beyond stating the law as the conservation of the centre of gravity of a system of bodies under impact. In *De motu corporum* (1656) he was also able to show that the quantity $^1/_2mv^2$ is conserved in an elastic collision, though again this concept had little intuitive sense for him.

Huygens also studied centrifugal force and showed, in 1659, its similarity to gravitational force, though he lacked the Newtonian concept of acceleration. Both in the early and the later part of his career he considered projectiles and gravity, developing the mathematically primitive ideas of ◊Galileo, and finding in 1659 a remarkably accurate experimental value for the distance covered by a falling body in one second. In fact his gravitational theories successfully deal with several difficult points that Newton carefully avoided. Later, in the 1670s, Huygens studied motion in resisting media, becoming convinced by experiment that the resistance in such media as air is proportional to the square of the velocity. Without calculus, however, he could not find the velocity–time curve.

Early in 1657, Huygens developed a clock regulated by a pendulum. The idea was patented and published in the same year, and was a great success; by 1658 major towns in Holland had pendulum tower clocks. Huygens worked at the theory first of the simple pendulum and then of harmonically oscillating systems throughout the rest of his life, publishing the *Horologium oscillatorium* in 1673. He made many technical and theoretical advances, including the derivation of the relationship of the period T of a simple pendulum to its length l as $T = 2\pi\sqrt{(l/g)}$. Huygens made use of his previously worked out theory of evolutes in this work.

Huygens is perhaps best known for his wave theory of light. This was the result of much optical work, begun in 1655 when Huygens and his brother began to make telescopes of high technical quality.

Through working with his brother, Constantijn, Huygens became skilful in grinding and polishing lenses, and the telescopes that the two brothers constructed were the best of their time. Huygens's comprehensive study of geometric optics led to the invention of a telescope eyepiece that reduced chromatic aberration. It consisted of two thin plano-convex lenses, rather than one fat lens, with the field lens having a focal length three times greater than that of the eyepiece lens. Its main disadvantage was that cross-wires could not be fitted to measure the size of an image. To overcome this problem Huygens developed a micrometer, which he used to measure the angular diameter of celestial objects.

Having built (with his brother) his first telescope, which had a focal length of 3.5 m/11.5 ft, Huygens discovered Titan, one of Saturn's moons, in 1655. Later that year he observed that its period of revolution was about 16 days and that it moved in the same plane as the so-called 'arms' of Saturn. This phenomenon had been somewhat of an enigma to many earlier astronomers, but because of Huygens's superior telescope, and a piece of sound if not brilliant deduction, he partially unravelled the detail of Saturn's rings. In 1659 he published a Latin anagram that, when interpreted, read 'It (Saturn) is surrounded by a thin flat ring, nowhere touching and inclined to the ecliptic'. The theory behind Huygens's hypothesis followed later in *Systema Saturnium,* which included observations on the planets, their satellites, the Orion nebula and the determination of the period of Mars. The content of this work amounted to an impressive defence of the Copernican view of the Solar System.

The *Traité de la lumière,* containing Huygen's famous wave, or pulse, theory of light, was published in 1678. Huygens had been able two years earlier to use his principle of secondary wave fronts to explain reflection and refraction, showing that refraction is related to differing velocities of light in media, and his publication was partly a counter to Newton's particle theory of light. The essence of his theory was that light is transmitted as a pulse with a 'tendency to move' through the ether by setting up a whole train of vibrations in the ether in a sort of serial displacement. The thoroughness of Huygens's analysis of this model is impressive, but although he observed the effects due to polarization, he could not yet use his ideas to explain this phenomenon.

The impact of Huygens's work in his own time, and in the 18th century, was much less than his genius deserved. Essentially a solitary man, he did not attract students or disciples and he was also slow to publish his findings. Nevertheless, after Galileo and until Newton, he was supreme in mechanics, and his other scientific work has had a significant effect on the development of physics.

Further Reading

Baker, Bevan B and Copson, E T, *The Mathematical Theory of Huygens Principle,* Chelsea Publishing Company, 1987.

Hall, A Rupert, *Newton, His Friends and His Foes,* Blackwell, 1993.

Yoder, Joella G, *Unrolling Time: Christiaan Huygens and the Mathematization of Nature,* Cambridge University Press, 1989.

Hyatt, John Wesley (1837–1920) US inventor who became famous for his invention of Celluloid, the first artificial plastic.

Hyatt was born in Starkey, New York, on 28 November 1837. As a young man he worked in Illinois as a printer, printing boards and playpieces for draughts and dominoes at his plant in Albany. Probably with a view to making such playpieces, he became interested in the material pyroxylin, a partly nitrated cellulose developed in the UK by Alexander Parkes (1813–1890) and D Spill. Then in the early 1860s the New York company of Phelan and Collender, which manufactured billiard balls, offered a prize of $10,000 for a satisfactory substitute for ivory for making the balls. Hyatt remembered pyroxylin and from it he and his brother Isaac developed and, in 1869, patented Celluloid (the US trade name; it was called Xylonite in the UK).

Celluloid consisted of a mouldable mixture of nitrated cellulose and camphor. Its chief disadvantage was its inflammability, but nevertheless for several years it was the favoured material for making a wide range of products from shirt collars and combs to dolls and babies' rattles. It was used as a flexible substrate for photographic film and, as a substitute for ivory, for making piano keys – and billiard balls. Celluloid was also used as the central 'filling' in sandwich-type safety glass for car windscreens. It has gradually been superseded in nearly all uses by other synthetic materials, but in one application it continued to be used: the manufacture of table-tennis balls.

Hyatt was never awarded the Phelan and Collender prize money. He continued to patent his inventions – more than 200 of them, including roller bearings and a multiple-stitch sewing machine. He died in Short Hills, New Jersey, on 10 May 1920.

Hyman, Libbie Henrietta (1888–1969) US zoologist renowned for her studies on the taxonomy (classification) and anatomy of invertebrates.

Hyman was born on 6 December 1888 in Des Moines, Iowa, the third in a family of four. Unhappy home circumstances encouraged her to indulge in long walks and rambles and she began to collect and classify flowers and butterflies. After graduation with honours from high school in 1905, she passed state examinations as a teacher, but was refused a position because she was considered too young. She therefore worked in a factory, until her former English and German teacher helped her gain a scholarship to the University of Chicago in 1906. She graduated in zoology in 1910 and gained her PhD in 1915, working under the supervision of Charles Manning ◊Child, for whom she then worked as a research assistant until 1930, also teaching undergraduate classes in comparative anatomy.

As a young instructor in zoology Hyman had produced a *Laboratory Manual for Elementary Zoology* (1919) and *Laboratory Manual for Comparative Vertebrate Anatomy* (1922). However her main research interests were with invertebrate animals, especially their taxonomy. Initially she worked on flatworms, but soon extended her investigations to a wide spread of invertebrates, and was frequently asked to help colleagues to identify specimens. She recognized the need for a comprehensive reference book on invertebrates, and decided to remedy the lack. A small income generated by the royalties of her books enabled her to resign her university position in 1931 to embark on an independent career. She travelled to several European laboratories, working for a period at the Stazione Zoologica, Naples, before returning to New York City to begin to write a major volume on the invertebrates, for which she was given office and laboratory space, but no salary, by the American Museum of Natural History. She read extensively in the English and foreign-language literature, and combined comparative anatomy, histology, and physiology in her assessments. Her major work, *The Invertebrates,* was published in six volumes 1940–68, and provided an encyclopedic account of most phyla of invertebrates. She received several honours and accolades for this achievement, including honorary degrees, the Gold Medal of the Linnaean Society of London, and membership of the National Academy of Sciences. Severely handicapped by Parkinson's disease towards the end of her life, she was unable to produce volumes on the molluscs or arthropods, and she died in New York City on 3 August 1969.

Hypatia (c. 370–415) Greek natural philosopher who is credited with being the first female astronomer and mathematician of note. She was regarded for centuries as being the only woman scientist of the ancient world (apart perhaps from Maria the Jewess, the 1st-century Alexandrian alchemist). She was also an inventor.

Hypatia was the daughter of the mathematician Theon, who was the director of the great school (university) of the Alexandrine Library. He paid great attention to her education and encouraged her not to accept dogma and to question everything she was taught. She travelled widely in the Mediterranean and studied in Athens under Plutarch the Younger and his daughter Asclepegeneia, gaining a reputation as a mathematician. Hypatia corresponded with many people and most of what is known about her comes from her letters.

After her studies with Plutarch, Hypatia returned to Alexandria and was offered a position as a teacher of mathematics and philosophy at the university. She became popular with her students as a teacher of geometry, astronomy, and algebra and was known as 'the philosopher'. Her philosophy was an attempt to combine neo-Platonism with Aristotelianism. In 400 she was the head of the neo-Platonic school in Alexandria. Hypatia wrote a number of books, none of which has survived. With her father she wrote treatises on ◊Euclid; she also wrote commentaries on ◊Ptolemy's *Almagest* and on ◊Diophantus' *Arithmetica*, and her commentary on Apollonius' *Conics* was later to attract the interest of 17th-century mathematicians. Some information about her comes from the letters of one of her pupils, Synesius of Cyrene, in which he asks her for scientific advice and from which it emerges that she had designed a planisphere, an astrolabe, an instrument for distilling water, and a hydrometer for measuring specific gravity.

Hypatia's learning, wisdom, and character, as well as her renowned beauty, drew students to her from all over the Greek world, both Christian and pagan. The scientifically rational neo-Platonic school of thought to which Hypatia belonged was to come into conflict with doctrinaire Christianity and was violently opposed by Cyril, Bishop of Alexandria, who was resentful of her influence. In spite of the admiration and reverence she had gained for her fine mind, she incited envy among Christians who blamed her for preventing her friend Orestes, the Roman prefect of the city, from forming an alliance with the bishop. In March 415 she was dragged from her chariot by a mob, taken into a church, stripped naked and slashed and pelted to death with oyster shells. Research suggests that this was more likely a political assassination than a random act of violence.

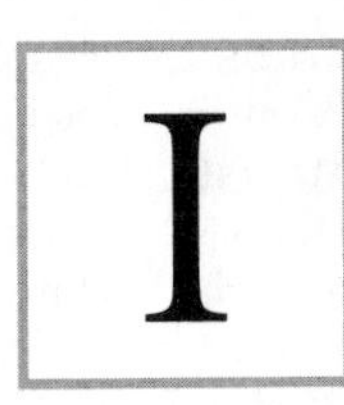

Ingenhousz, Jan (1730–1799) Dutch biologist and physiologist who discovered photosynthesis and plant respiration.

Ingenhousz was born in Breda, in the Netherlands, on 8 December 1730, the son of a pharmacist. He was educated locally and received his training in medicine and chemistry at the University of Louvain, graduating in 1753; he then studied at the University of Leyden during the following year. He went to universities in Paris and Edinburgh for short periods, after which he set up a private medical practice in Breda. On his father's death in 1765, he left for England, encouraged by a physician from the British army who had befriended his family during the War of the Austrian Succession. In 1766 he worked at the Foundling Hospital, London, where he was responsible for inoculating patients against smallpox (using the hazardous live virus). His methods were reasonably successful and in 1768 he was sent to the Austrian court in Vienna, by George III, to inoculate the royal family. He took up the appointment of court physician there 1772–79. In that year he returned to England, where he continued his research until he died on 7 September 1799 in Bowood, Wiltshire.

In 1771 Joseph ⇨Priestley had found that the flame of a candle burning in a closed space eventually goes out and that a small animal confined under similar conditions soon dies. He also found that plants can restore the capacity of the air to support life or the burning of a candle. Later Priestley discovered 'dephlogisticated air' (oxygen). It seems likely that Priestley's work inspired Ingenhousz to carry out similar experiments of his own.

Ingenhousz discovered in 1779 that only the green parts of plants are able to 'revitalize' the air, and that they are capable of doing so only in sunlight. His investigations also showed that the active part of the Sun's radiation is not in the heat generated but rather in the visible light. He found that plants, like animals, respire all the time and that respiration occurs in all the parts of plants.

In the following years Ingenhousz demonstrated that the amount of oxygen released by a plant during photosynthesis is greater than that absorbed in respiration. He suggested that green plants take in carbon dioxide and produce oxygen, whereas animals do the reverse, and therefore that animals and plants are totally dependent on each other.

Ingenhousz believed that this discovery would help to distinguish between animals and plants among the lower orders of life. At that time a controversy existed over the origin of carbon in plants, some scientists believing that it is absorbed in some form by the roots – this belief was termed the humus theory. Ingenhousz, however, was of the opinion that carbon comes from the carbon dioxide absorbed by a plant. This idea explained the disappearance of the gas, and the presence of carbon in plants. Whereas he was right about the source of carbon, he was mistaken about that of oxygen, and it is now known that oxygen given off by plants comes from the water they take in.

Apart from the life of plants, Ingenhousz had various other interests, which led him in 1776 to develop an improved apparatus for generating large amounts of electricity; he also invented a hydrogen-fuelled lighter to replace the tinderbox, and investigated the use of a mixture of air and ether as a propellant for an electrically fired pistol.

In 1779, Ingenhousz published his work *Experiments On Vegetables, Discovering their Great Power of Purifying the Common Air in Sunshine, and of Injuring it in the Shade or at Night.* His discovery laid the foundations for the study of photosynthesis, the process upon which most animals ultimately depend for their food.

Ingold, Christopher (1893–1970) English organic chemist who made a fundamental contribution to the theoretical aspects of the subject with his explanation for the mechanisms of organic reactions in terms of the behaviour of electrons in the molecules concerned.

Ingold was born in London on 28 October 1893 but moved to Shanklin, Isle of Wight, when he was a few years old because of his father's ill health. He was educated at Sandown Grammar School and then went to study chemistry at Hartley University College, Southampton (later the University of Southampton), where he graduated in 1913. He began research under J F Thorpe at Imperial College, London, investigating spiro-compounds of cyclohexane, but left in 1918 to spend two years as a research chemist with the Cassel Cyanide Company, Glasgow. He returned to Imperial College as a lecturer in 1920, where he met and married Edith Usherwood, a promising young chemist.

From 1924 to 1930 he was professor of organic chemistry at the University of Leeds, when he succeeded Robert ◊Robinson as professor of chemistry at University College, London. He remained there for the rest of his career, retiring officially in 1961 (although continuing as an active contributor to the work of the department). Ingold was knighted in 1958. He died in London on 8 December 1970.

Much of Ingold's work at University College was carried out in collaboration with E D Hughes. For 30 years he specialized in the concepts, classification, and terminology of theoretical organic chemistry. In 1926, for example, he put forward the concept of mesomerism, which allows a molecule to exist as a hybrid of a pair of equally possible structures. This work culminated in his classic reference book *Structure and Mechanisms in Organic Chemistry* (1953), whose second edition (1969) ran to 1,266 pages. His ideas, first published in 1932, are still fundamental to reaction mechanisms taught today. They concerned the role of electrons in elimination and nucleophilic aliphatic substitution reactions, which he interpreted in terms of ionic organic species.

Ingold's work removed much of the 'art' from organic chemistry and replaced it with scientific methodology. As he is reputed to have said: 'One could no longer just mix things; sophistication in physical chemistry was the base from which all chemists – including the organic – must start.'

Ipatieff, Vladimir Nikolayevich (1867–1952) Russian-born US organic chemist who is best known for his development of catalysis in organic chemistry, particularly in reactions involving hydrocarbons.

Ipatieff was born in Moscow on 21 November 1867, the son of an architect. It was intended that he should have a career in the army so he attended a military school, became an officer in the Imperial Russian Army in 1887, and in 1889 won a scholarship to continue his higher education at the Mikhail Artillery Academy in St Petersburg. From 1892, the year in which he married, he gave lectures in chemistry at the academy, and in 1897 he was given permission to go to the University of Munich for a year to study under Johann ◊Baeyer, where one of his fellow students was Richard ◊Willstätter. After a brief period in France studying explosives, he returned to Russia in 1899 as professor of chemistry and explosives at the Mikhail Artillery Academy. In 1908 he gained his PhD from the University of St Petersburg.

Ipatieff's research work was interrupted by World War I and the Russian Revolution, during which he held various administrative and advisory appointments. He was head of the Chemical Committee during World War I, and increased the monthly output of explosives from 60 tonnes to 3,300 tonnes. But he was not a Communist and when, at the age of 64, he went to Berlin in 1930 to attend a chemical conference he accepted the offer of a post in the USA and did not return to the Soviet Union. He was immediately condemned as a traitor by the Soviet authorities and expelled from the Soviet Academy of Sciences. From 1931 to 1935 he was professor of chemistry at Northwestern University, Illinois, and acted as a consultant to the Universal Oil Products Company, Chicago. In 1938 this company funded the building of the Ipatieff High Pressure Laboratory at Northwestern. He died in Chicago on 29 November 1952, just after his birthday and ten days before the death of his wife. In 1965 he was posthumously reinstated to the Soviet Academy of Sciences.

While working as a student in Munich in 1897 Ipatieff synthesized the hydrocarbon isoprene, the basis of the rubber molecule. Back in Russia in 1900 he discovered the specific nature of catalysis in high-temperature organic gas reactions, and how, by using high pressures, the method could be extended to liquids. He developed an autoclave called the Ipatieff bomb – for heating liquid compounds to above their boiling points under high pressure. He synthesized methane and iso-octane (2-methylheptane), and produced polyethylene by polymerizing ethylene (ethene).

In Chicago after 1931 Ipatieff began to apply his high-temperature catalysis reactions to petrol with low octane ratings (which produce 'knock' or pre-ignition in car engines). The result of the catalytic cracking (or 'cat cracking') is petrol with a higher octane rating, and the method became particularly important for the production of aviation fuel during World War II; it is still widely used.

Isaacs, Alick (1921–1967) Scottish virologist who discovered interferon, an antibody produced by cells when infected by viruses.

Isaacs was born on 17 July 1921 in Glasgow, the first of four sons in a family of Russian origin (his grandparents were Russian Jews who had emigrated to Scotland in about 1880). He had a conventional Jewish upbringing and was educated at Pollockshields Secondary School in Glasgow, attending classes in Judaism every day after school. His family moved to Kilmarnock in 1939 but Isaacs stayed in Glasgow and enrolled at the university there to study medicine. He was an able student, graduating in 1944 and winning several prizes, but clinical medicine did not greatly interest him and in 1945 he became a McCann research scholar in the department of bacteriology at Glasgow University, where he came under the influence of Carl Browning, the professor there. In 1947 Isaacs was awarded a Medical Research Council studentship to research into influenza viruses under Stuart Harris at Sheffield University, and in the following year he went to Australia, having won a Rockefeller Travelling Fellowship to work under Macfarlane ◊Burnet at the Walter and Eliza Hall Institute for Medical Research in Melbourne. Isaacs returned to the UK in 1951 and went to work in the laboratory of the World Influenza Centre at Mill Hill, London, where he remained for the rest of his life. His work on influenza gained him his medical degree from Glasgow University and a Bellahousten Gold Medal. In 1958–59 he suffered a three-month depression but seemed to recover, and in 1961 he took over the directorship of the World Influenza Centre. In 1964, however, he suffered a subarachnoid haemorrhage and died two years later, on 26 January, only 45 years old.

Although Isaacs began investigating influenza in 1947, it was not until 1956 that he discovered interferon. Working with a Swiss colleague, Jean Lindenmann, Isaacs found that chick embryos injected with influenza virus produce minute amounts of a protein that destroys the invading virus and also makes the embryos resistant to other viral infections. Isaacs and Lindenmann named this protein interferon. Further research demonstrated that most living creatures can make interferon, and that even plants react to viral infection in a similar way. When a virus invades a cell, the cell produces interferon, which then induces uninfected cells to make a protein that prevents the virus from multiplying. Almost any cell in the body can make interferon, which seems to act as the first line of defence against viral pathogens, because it is produced very quickly (interferon production starts within hours of infection whereas antibody production takes several days) and is thought to trigger other defence mechanisms.

In the 1950s there was no treatment or cure for viral infections (antibodies are effective only against bacteria and at that time antiviral vaccines were in their early stages of development) and so the discovery of interferon was thought to be a major breakthrough. As a result, the Medical Research Council took out a patent on interferon and established a scientific committee to undertake further research into it. This initial enthusiasm waned, however, when it was found that interferon was species specific, and that it was very difficult and costly to produce. By the time Isaacs died, research into the substance had come to a virtual standstill.

Then in the late 1960s, after Isaacs's death, interest revived as a result of a chance discovery made by Ion Gresser, a US scientist then working in Paris. He found that interferon inhibits the growth of virus-induced tumours in mice and also that it stimulates the production of special cells that attack tumours. In addition, he later showed that interferon can be made in relatively large amounts from human blood cells. These findings stimulated further research, particularly into the use of interferon against cancer, leukaemia and certain viral diseases, such as hepatitis, rabies, measles, and shingles. In early 1980 Charles Weissmann, a Swiss scientist, produced by genetic manipulation a strain of bacteria that can make human interferon, the effectiveness of which is still being investigated.

Issigonis, Alec (Alexander Arnold Constantine) (1906–1988) Turkish-born British automotive engineer and the first person to exploit scientific component packaging in the design of small volume-produced motorcars. His designs gave much greater space for the occupants together with greatly increased dynamic handling stability and improved small-car ride.

Issigonis was born on 18 November 1906 in Smyrna (now Izmir), and came to the UK with his widowed mother after the 1922 war between Turkey and Greece. He studied engineering at Battersea Polytechnic, and began his career working for a small engineering firm that was developing an automatic gearchange.

This work introduced Issigonis to the Humber division of the Rootes Group in 1934, and in 1936 he joined Morris Motors to work on suspension design. During this pre-war period he built the ingenious Lightweight Special hill-climb-and-sprint single-seater, which demonstrated the potential of all-independent suspension with rubber springs. His first complete production motorcar was the Morris Minor, launched in 1948, which brought new standards of steering and stability to small motorcars – and went on to become the first British motor-car to pass the one million sales mark (in 1961). After a spell 1952–56 working at Avis on an experimental 3.5-litre car with hydrolastic suspension, he returned to what had now become BMC to face, within a short while, his greatest challenge.

Leonard Lord (later to become Lord Lambury), the chairman of BMC, asked him to design and produce a small and economical car to counteract the flood of 'bubble cars' that had followed the Suez crisis. A period of intensive design and development led to the launch of the Mini in 1959. His other major designs were the 1100 in 1962, the 1800 in 1964, and the Maxi in 1969. He was made a CBE in 1964, became a fellow of the Royal Society in 1967, and received a knighthood in 1969.

The main significance of his work was in taking on car design as a 'vehicle architect', overseeing the separate approaches of styling, interior packaging, body engineering, and chassis layout. In this way he conceived the overall package from his knowledge and experience of the major factors affecting the product; specialists in his team then designed and engineered the subsystems of the vehicle. The approach was to make the human factor paramount in selecting design criteria.

This approach was particularly recognizable in the 'wheel at each corner' layout on the Morris Minor and its effect on handling. The vehicle's polar moment of inertia (a measure of directional stability), was made small compared with the magnitude of the tyre's cornering forces. This improved the speed of response to change in direction, and the designed-in nose-heaviness allowed quick correction of the car after a side-gust disturbance.

With the layout of the Mini a degree of interior spaciousness was achieved beyond that available in previous cars of similar exterior size. The technique included repositioning the dash panel to follow the projected line of the curved lower edge of the windscreen, using single-skinned doors and rear quarter panels, with large open lockers on the inside. By adopting a transverse engine-over-gearbox/final-drive layout with front-wheel drive, and independent suspension of all wheels, considerable gains were made in front knee-room and space at the rear of the seat base.

The use of compact wheel-location lever-arms acting on rubber suspension springs also gave a substantial packaging advantage, besides giving a well-damped ride, free of static friction. In arranging for fluid correction of the springs, the suspension could be tuned to separate motions of pitch and bounce of the vehicle; tuning virtually eliminated pitch as a prime factor in ride discomfort.

The Mini, for which Issigonis is best known, reigned supreme among small cars until the late 1970s. At the beginning of its life the car attracted a considerable cult following and all through its life it was the basis for a great diversity of modification by specialist firms.

Further Reading

Barker, Ronald and Harding, Anthony, *Automobile Design: Twelve Great Designers and Their Work,* Society of Automotive Designers, 1992.

Downs, Diarmuid, 'Alexander Arnold Constantine Issigonis', 18 November 1906–2 October 1988, *Biogr Mem Fellows Roy Soc,* 1994, v 39, pp 201–211.

Nahum, Andrew, *Alec Issigonis,* State Mutual Book and Periodical Service, 1988.

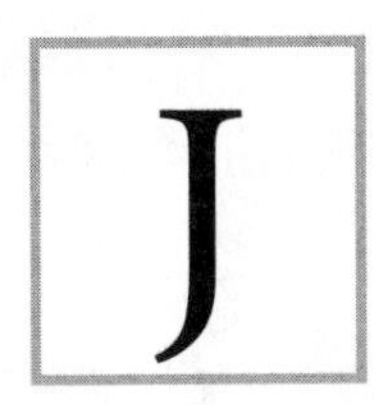

Jabir ibn Hayyan (*c.* 721–*c.* 815) Also known as Geber. Arabian alchemist, reputed to be the author of a large body of Arabic writings mainly on alchemical knowledge but also on a number of other subjects. He is regarded by some as the founder of molecular chemistry. He believed that everything is composed of a combination of earth, water, fire, and air and that these elements in the correct proportions form mercury and sulphur, the basis of all metals.

Jabir was born in Tus near Meshed, Persia (now Iran), the son of an Arabian alchemist, Abu Musa Jabir ibn Hayyan from Kufa. His father, a supporter of the Abbassid family, was on a political mission to depose the Caliph at the time of Jabir's birth in Tus. Shortly afterwards he was beheaded by the Caliph's agents.

The fatherless Jabir was sent to Arabia, where he studied most branches of contemporary Eastern knowledge including medicine, and when the Abbassids were successful in their revolution he became physician in the court of Caliph Haroun-al-Rashid. He fell out of favour in 803 and was expelled from the court. He fled to Kufa and it is thought that he remained there for the rest of his life, although it is possible that he joined the court of Caliph al-Mamun in 813. He died in Persia in about 815.

Books on alchemy that have been attributed to Jabir number over a hundred, some of which are in Latin. Scholars are still translating and investigating their origins, but it is clear that he was not the author of all the works bearing his name. The corpus of Arabic texts include volumes whose titles are related to the number of pieces of writing they contain – the largest volumes being entitled *Seventy, One Hundred Twelve,* and *Five Hundred.* Not all of the volumes deal solely with alchemy: linguistics, philosophy, astrology, cosmology, theology, medicine, agriculture, and technology are also covered. Some of these texts may have been written by Jabir's disciples and some are possibly western forgeries.

Jabir developed a theory of alchemy in which the ingredients are not entirely mineral – some are animal and vegetable. According to his theory the basic qualities in all natural objects can be combined in different quantities to arrive at a mixture suitable for the proposition at hand. Jabir, like many of his contemporaries, firmly believed in the transmutation of metals, especially in the transmutation of base metals into gold. He considered that all metals were derived from sulphur and mercury, which he believed would yield gold when combined in the right proportions, and silver when the proportions were less accurate. Jabir mastered many other chemical transformations in his drive for transmutation. He prepared a large range of simple chemicals including ammonia which he prepared using blood, hair, and urine as its material bases.

During his lifetime, Jabir was respected as a great scientist who had remarkable medical knowledge as is shown in his book on poisons. He correctly identified the ability of elements to form new compounds at a molecular level without losing their own structure. He is thought to have developed noncombustible paper, florescent ink, anti-rust coatings, and water repellents and he recommended that chemical laboratories be built away from towns to avoid polluting the population.

Jacob, François (1920–) French cellular geneticist who was jointly awarded the 1965 Nobel Prize for Physiology or Medicine with André ◊Lwoff and Jacques ◊Monod for their collaborative work on the control of gene action in bacteria.

Jacob was born on 17 June 1920 in Nancy. He was educated at the Lycée Carnot and at the University of Paris, from which he gained his medical degree in 1947 – his studies having been interrupted by military service during World War II – and his doctor of science degree in 1954. In 1950 he joined the Pasteur Institute in Paris as a research assistant, becoming head of laboratory there in 1956. From 1960 to 1991 he was head of its department of cellular genetics, and from 1965 to 1992 was professor of cellular genetics at the Collège de France.

Jacob began his Nobel prizewinning work on the control of gene action in 1958. Previous work by Francis ◊Crick, James ◊Watson and Maurice ◊Wilkins had shown that the types of proteins produced in an organism are controlled by DNA, but it was Jacob – working with Lwoff and Monod – who demonstrated how an organism controls the amount of protein produced. Jacob performed a series of experiments in which he cultured the bacterium *Escherichia coli* in various mediums to discover the effect of the medium on enzyme production. He found that *E. coli* grown on a medium containing only glucose produced very little of the enzyme ß-galactosidase, whereas *E. coli* grown on a

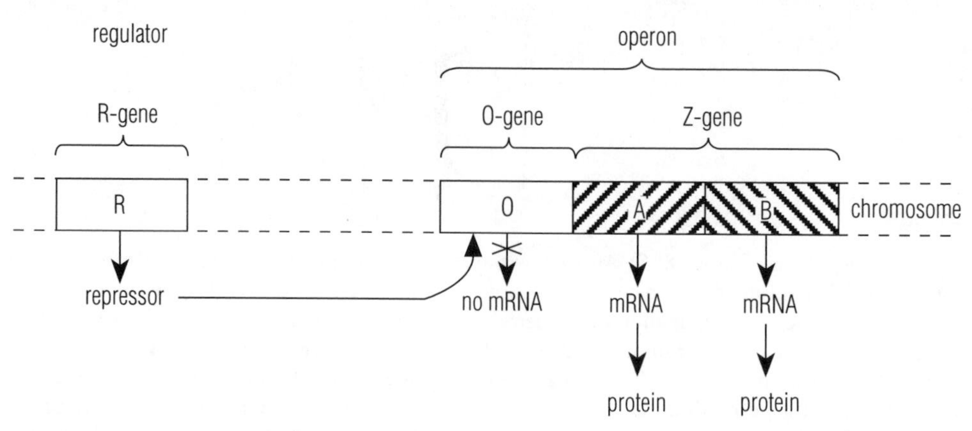

Jacob A regulator gene (R-gene) produces a repressor substance that prevents an operator (O-gene) from providing messenger RNA, blocking the production of protein.

lactose-only medium produced much greater amounts of this enzyme. This phenomenon, which is called enzyme induction, occurs because *E. coli* can metabolize glucose easily but requires ß-galactosidase to metabolize lactose. From these findings, Jacob, Lwoff, and Monod concluded that increased production of an enzyme (ß-galactosidase in this case) occurs when an organism needs that particular enzyme, and they also proposed a theory to explain the mechanisms involved in this increased enzyme production. According to their theory there are three types of genes concerned with the production of each specific protein: a structural or Z-gene, which controls the type of protein produced, and a regulator or R-gene, which produces a repressor substance that binds to the third type of gene, the operator or O-gene. The binding of the repressor substance to the O-gene prevents the production of messenger RNA (mRNA) by the appropriate Z-gene and therefore also prevents the production of the specific protein coded for by that Z-gene. In Jacob's experiments, there was minimal production of ß-galactinosidase in *E. coli* grown on glucose because the repressor substance produced by the R-gene was specifically bound to the O-gene, thereby preventing ß-galactosidase production by the Z-gene. *E. coli* grown on lactose produced large amounts of ß-galactosidase because lactose binds to the repressor substance produced by the R-gene, which alters the molecular conformation of the repressor substance so that it can no longer bind to the O-gene; as a result, the Z-gene can produce mRNA for ß-galactosidase production. This theory has since been shown to be correct in the control of protein synthesis in *E. coli* and other bacteria.

Jacobi, Karl Gustav Jacob (1804–1851) German mathematician and mathematical physicist, much of whose work was on the theory of elliptical functions, mathematical analysis, number theory, geometry, and mechanics. An influential teacher, he corresponded with many of the other great mathematicians of his time.

Jacobi was born in Potsdam on 10 December 1804, the second son of a wealthy and well-educated Jewish banker. A child prodigy, he rose to the top class of his local school within months of first entering, and was ready to go to university at the age of 12 – but that the authorities would not permit. He went to Berlin University in 1821 and graduated in the same year, with excellent results in Greek, Latin, history, and mathematics. Continuing at the university he studied philosophy and the classics, reading mathematics privately because, apart from Karl ◊Gauss at Göttingen – with whom he corresponded – there was simply no one who could teach him. At the age of 19 he qualified as a teacher. The following year he received his doctorate, underwent conversion to Christianity (as may have been professionally advisable), and became an unsalaried lecturer at Berlin University. After only one year there, however, in 1826 he joined the staff at Königsberg (now Kaliningrad) University, coming into contact there with people such as the physicists Franz Neumann (1798–1895) and Heinrich Dove (1803–1897), and the astronomer and mathematician Friedrich ◊Bessel. In 1832 Jacobi was made professor, a post he retained for 18 years. In 1843, however, he fell ill and went to Italy to recover, returning to Berlin at the end of 1844. Four years later, after an unwise and unfortunate excursion into politics, he lost the royal pension on which he had been living and his family was reduced to near-destitution. Luckily, the Prussian government was not prepared to see a man of his talent – and reputation – starve. In 1851, however, he again fell ill, first with influenza and then with smallpox. He died, in Berlin, on 18 February of that year.

There are very few areas in mathematics that Jacobi's researches at one time or another did not cover fully. He is particularly remembered for his work on elliptic functions, in which to some extent he was competing against a rival in the contemporary Norwegian mathematician Niels ◊Abel. His principal published work on elliptic function theory was *Fundamenta nova theoriae functionum ellipticarum* (1829), incorporating some of Abel's ideas and introducing his own concept of hyperelliptic functions. In analysis, Jacobi studied differential equations (the results of which he applied to problems in dynamics) and the theory of determinants. In the latter field he invented a functional determinant – now called the Jacobian determinant – which has since been of considerable use in later analytical investigations, and was even supportive in the development of quantum mechanics. He advanced the theory of the configurations of rotating liquid masses by showing that the ellipsoids now known as Jacobi's ellipsoids are figures of equilibrium.

He had immense knowledge in every branch of mathematics – Jacobi's PhD thesis was an analytical discussion on the theory of partial fractions; he was at the same time writing to Gauss on cubic residues in number theory; his first lecture, nevertheless, was on the theory of curves and surfaces in three-dimensional space. He was also always trying to link together different mathematical disciplines. For instance, he introduced elliptic functions into number theory and into the theory of integration, which in turn connected with the theory of differential equations and his own principle of the last multiplier.

Jacobi's work, and particularly his researches in collaboration with Franz Neumann and Friedrich Bessel, revived an interest in mathematics generally in Germany. He therefore became a highly respected and honoured man. After his death, his great friend Lejeune ◊Dirichlet delivered a memorial lecture at the Berlin Academy of Sciences on 1 July 1852, in which he described Jacobi as 'the greatest mathematician among members of the Academy since Joseph Lagrange'.

Jacquard, Joseph Marie (1752–1834) French engineer who developed a loom (originally for carpets), whose complicated patterns were 'programmed' on to punched cards.

Jacquard was born in Lyon on 7 July 1752, the son of a weaver. He was apprenticed when he was quite young, first to a bookbinder and later to a cutler. Throughout his time in these trades he considered ways of improving the crafts and reducing the labour needed. When his parents died, they left him their small weaving business and Jacquard took up that trade. He attempted to weave patterned fabrics, which brought a relatively high profit, but he was not successful at this intricate craft, which required long hours of hard, patient toil to finish even the most modest piece of material. He returned to the cutlery trade after becoming bankrupt as a weaver.

He continued his efforts to devise an improved weaving machine, but his first loom was not made until 1801. It combined a number of innovations that had been used before on various other machines with some ideas of his own. In 1804 he went to Paris to demonstrate a new machine for weaving net. This was welcomed and, as well as receiving a patent, he was allocated a small pension, which enabled him to work on improving looms and weaving at the Conservatoire des Arts et Métiers.

Jacquard's attachment for pattern weaving, which was developed during this time and later improved by others, allowed patterns to be woven without the intervention of the weaver. Weavers had always had to plan the pattern they wished to weave before they began their task. This planning now became the essential feature of the weaver's job. Part of Jacquard's invention was a series of cards with holes punched in them through which tools could come up and pull the warp threads down, so that the shuttle could pass over them. If there was no hole the shuttle went underneath the warp. In this way the correct threads were brought into conjunction to weave an intricate pattern.

During the first decade of the 19th century many Jacquard looms were installed in weaving mills. As Jacquard intended, his loom saved a great deal of physical labour, and because of this it was not kindly received by the workers in the weaving trade. In Lyon and elsewhere his machines were smashed, and Jacquard himself suffered violence at the hands of the angry weavers. Even angry workers, however, could not halt mechanical progress and by 1812 it is recorded that there were 11,000 Jacquard looms working in France. During the next few years they were introduced into many other countries, including England.

One pattern, when punched onto cards, could be used over and over again, which gave the new machine a tremendous advantage over the traditional one, but the preparation of the cards was a difficult and tedious task and it was only with the coming of the electronic computer that this labour was reduced.

The idea of the Jacquard cards was to feed a machine with instructions, and it was applicable in principle, at least, to many kinds of machine. One direct application of the cards was in the mechanical computing machine, called the analytical engine, designed and partly constructed by Charles ◊Babbage in the years following 1833. In this machine the cards directed a sequence of arithmetical operations and the transfer of numbers to and from the store. Other pioneers of automatic computing have also used punched cards and punched tape as a means of storing, sorting, and transferring information. Herman ◊Hollerith, for example, used a vast mechanical tabulatory system that relied on punched cards with holes in 288 positions for his 1890 census of the USA.

Only now with the rapid modern development of electronic computers and magnetic storage systems are the descendants of the Jacquard cards declining in use as representations and stores of information. The principle of the Jacquard loom still survives in weaving, even though the machinery and programming are somewhat different.

Jacquard was highly honoured, and eventually even the weavers appreciated his invention. He survived the violence of his times and died in Oullines on 7 July 1834.

Jansky, Karl Guthe (1905–1950) US radio engineer whose discovery of radio waves of extraterrestrial origin led to the development of radio astronomy.

Jansky was born in Norman, Oklahoma, on 22 October 1905. He was educated at the University of Wisconsin, where his father was a member of the faculty. He spent a year as an instructor before beginning his career as an engineer with the Bell Telephone Laboratories in 1928. His fundamental discovery of the existence of extraterrestrial radio waves was made and published by 1932. Thereafter Jansky did not continue this kind of scientific work. He became more involved with engineering and left the research in astronomy to others. Jansky died at an early age, of a heart complaint, at Red Bank, New Jersey, on 14 February 1950.

When Jansky joined the Bell Telephone Company as a research engineer, he was assigned the task of tracking down and identifying the various types of interference from which radio telephony and radio reception were suffering. The company was particularly concerned with the interference that occurred at short wavelengths of around 15 m/50 ft (then being used for ship-to-shore radio communications). It was well known that some of the static interference was caused by thunderstorms, nearby electrical equipment, or aircraft, but, by building a high quality receiver and aerial system that was set on wheels and could be rotated in various directions, Jansky was able to detect a new and unidentifiable kind of static. After months of study Jansky associated the source of the unidentified radio interference with the stars. He had noticed that the background hiss on a loudspeaker attached to the receiver and antenna system reached a maximum intensity every 24 hours. From overhead, it seemed to move steadily with the Sun but gained on the Sun by four minutes per day. This amount of time correlates with the difference of apparent motion, as seen on Earth, between the Sun and the stars, and so Jansky surmised that the source must lie beyond the Solar System. By the spring of 1932, he had concluded that the source lay in the direction of Sagittarius. This was the direction that the US astronomer Harlow ◊Shapley and the Dutch astronomer Jan ◊Oort had confirmed as being the direction of the centre of our Galaxy. Jansky published his results in December 1932. That same month, Bell Telephone Laboratories issued a press release on Jansky's discovery to the *New York Times* and it made front-page news. This was the birth of radio astronomy. There was now a second region of the electromagnetic spectrum in which it was possible to study the stars – a radio window on the universe. There was no immediate follow-up to Jansky's discovery, however, and it was left to an amateur US astronomer, Grote ◊Reber, to pursue the matter. It was not until the end of World War II that the importance of radio astronomy began to be recognized.

Although Jansky died at an early age, he did live long enough to see radio astronomy develop into an important tool of modern astronomical research. By their ability to penetrate the Earth's atmosphere and dust clouds in space, radio telescopes enable the study of celestial objects that are impossible to detect by optical means. In honour of Karl Jansky the International Astronomical Union named the unit of strength of a radio-wave emission a jansky (symbolized Jy) in 1973.

Janssen, (Pierre) Jules César (1824–1907) French astronomer, famous for his work in physical astronomy and spectroscopy.

Janssen was born in Paris on 22 February 1824. His father was a musician of Belgian descent and his maternal grandfather was a well-known architect. In his early life he had an accident that left him permanently lame, and as a result he never went to school. Because of financial difficulties he went out to work at an early age. He worked in a bank 1840–1848, but educated himself at the same time and took his Baccalaureat at the age of 25. He studied mathematics and physics at the Faculty of Sciences in Paris and gained his Licence des Sciences in 1852. He worked for a while as a substitute teacher in a lycée, before being sent on the first of his scientific missions – to Peru – in 1857. He returned with a bad attack of dysentery and became a tutor to the wealthy Schneider family who owned iron and steel mills in Le Creusot. He was awarded his doctorate in 1860, and in 1862 began work with E Follin of the Faculty of Medicine. In 1865 he was made professor of physics at the Ecole Spéciale d'Architecture. He was elected to the Academy of Sciences in 1873 and to the Bureau of Longitudes in 1875. He was appointed director of the new astronomical observatory, established by the French government at Meudon in 1875, a position he held until his death in Paris on 23 December 1907.

Although Janssen travelled to Peru to measure the magnetic equator, his first real research work was in the field of ophthalmology. He went on to work on the construction of an ophthalmoscope at the Faculty of Medicine in Paris. But after hearing that Gustav ◊Kirchhoff had demonstrated the presence of terrestrial elements in the make-up of the Sun in 1859, Janssen had decided that his real passion was for physical astronomy. He built himself an observatory on the flat roof of his wife's house in Montmartre and began to work on the nature of the dark bands in the solar spectrum. He found that these bands were most noticeable at sunrise and sunset. For his research he constructed a special spectroscope with a high dispersive power, to which he attached a device for regulating luminous intensity. By 1862 he was able to show that these bands can be resolved into rays and that they are always present. In 1864 he travelled to Italy and there he was able to show that the intensity of the rays varied during the course of a day in relation to the terrestrial atmosphere. He showed that the origin of the phenomenon was terrestrial and called the rays 'telluric rays'. To prove his point that the intensity of telluric rays would be less in thinner air, he travelled to the Bernese Alps to measure the rays at a height of 2,700 m/8,900 ft. They were even weaker than he expected, and so he attributed the extra effect to the dryness of the air, and went to the shores of Lake Geneva to study the effect of humidity on the strength of the rays. In 1867 he spent time in the Azores, carrying out optical and magnetic experiments that concluded that water vapour was present in the atmosphere of Mars.

In 1868 Janssen went to India to observe the total eclipse of the Sun, which occurred on 18 August. Together with other scientists, he noted that there were a number of bright lines in the spectrum of the solar chromosphere. He demonstrated the gaseous nature of the red prominences, but he was unable to correlate the exact positions of the lines in the solar spectrum with wavelengths of any known elements. Janssen continued to observe the unobscured Sun for 17 days after the eclipse and he reported his findings to the French Academy of Sciences by telegram. In October of the same year, Norman ◊Lockyer, an English astronomer and physicist, observed the chromosphere through a special telescope designed by him in 1866. He noted, as Janssen had, that there was a third yellow line in the spectrum, which did not correspond to either of the two known lines of sodium. He immediately reported his findings to the French Academy of Sciences and a letter from Janssen, also reporting that a new element must be responsible for the yellow line, arrived at the same time as the letter from Lockyer. The two scientists became firm friends, because of the coincidence of their discovery, and a medallion was struck by the French Academy of Sciences bearing their profiles and names. Lockyer went on to discover that the new line represented a substance that was later named helium.

In the same year, 1868, Janssen developed a spectrohelioscope so that he could observe the Sun and solar prominences spectroscopically in daylight conditions. With it he attempted to ascertain whether or not the Sun contains oxygen. He realized it would help his research if he could eliminate some of the obscuring effects of the Earth's atmosphere, and for this reason he established an observatory on Mont Blanc. By this time his health was not good, and to overcome his disability he invented a device that enabled him to be carried up Mont Blanc; even so, the journey in this conveyance took 13 hours. Janssen went to any lengths to continue his observations. In order to observe the eclipse on 22 December 1870, he travelled to Algeria by balloon, despite the Franco-Prussian War. He later devised an aeronautical compass that was capable of instantly indicating the direction and speed of flight.

From 1876 to 1903 Janssen summarized the history of the surface of the Sun by means of solar photographs. These were made at the astrophysical observatory in Meudon and were collected together in his *Atlas de photographies solaires* (1904). In order to observe the transit of Venus in 1882, he took a series of photographs in rapid succession that enabled him to measure successive positions of the planet. He obtained a series of separate images, laid out in a circle on the photographic plate. In so doing, Janssen had anticipated one of the fundamentals of cinematography, which was not invented for another 20 years. He also invented another device of historical interest, the photographic revolver, in 1873.

Jeans, James Hopwood (1877–1946) English mathematician and astrophysicist who made important contributions to cosmogony – particularly his theory of continuous creation of matter – and became known to a wide public through his popular books and broadcasts on astronomy.

Jeans was born in Ormskirk, Lancashire, on 11 September 1877, but moved at the age of three to Tulse Hill, London, where he spent his childhood. His interests in science and his literary bent disclosed themselves at an early age. When he was only nine years old he wrote a handbook on clocks, which included an explanation of the escapement principle and a description of how to make a clock from pieces of tin. He was educated at Merchant Taylors' School, London, 1890–96 and then went to Cambridge to study mathematics at Trinity College. He graduated in 1900 and was awarded a Smith's Prize. In 1901 Jeans was appointed a fellow of Trinity and from 1905 to

1909 he was professor of applied mathematics at Princeton University in the USA. From 1910 to 1912 he was Stokes Lecturer in Applied Mathematics at Cambridge. Thereafter he held no university post, but devoted himself to private research and writing, although he was a research associate at the Mount Wilson Observatory, California, 1923–44. In 1928 Jeans stated his belief that matter was continuously being created in the universe. He never developed this idea (a forerunner of the steady-state theory), however, for in that year his career as a research scientist came to an end. After that time he concentrated on broadcasting and writing popular books on science. He was awarded the Royal Medal of the Royal Society in 1919 and was knighted in 1928. He died on 1 September 1946 at Dorking, Surrey.

The first years of Jeans's research career were spent on problems in molecular physics, in particular on the foundations of the kinetic molecular theory. His *Dynamical Theory of Gases* (1904), which contained his treatment of the persistence of molecular velocities after collisions (that is, the tendency for molecules to retain some motion in the direction in which they were travelling before collision), became a standard text.

Jeans next worked on the problems of equipartition of energy in its application to specific heat capacities and black-body radiation. In 1905 he corrected a numerical error in the derivation of the classical distribution of black-body radiation made by John ◊Rayleigh and formulated the relationship known as the Rayleigh–Jeans law, which describes the spectral distribution of black-body radiation in terms of wavelength and temperature. For some time thereafter Jeans investigated various problems in quantum theory, but in about 1912 he turned his attention to astrophysics.

Jeans had been interested in astrophysics since his student days. In an undergraduate prizewinning essay he had treated compressible and incompressible fluids at a deeper level than had Henri ◊Poincaré, and this work led directly to a consideration of the origin of the universe, or cosmogony. Jeans distinguished two extremes: an incompressible mass of fluid and a gas of negligible mass surrounding a mass concentrated at its centre. He argued that if an incompressible mass were to contract or be subjected to an increase in angular momentum, it would evolve into an unstable pear-shaped configuration and then split in two. Double stars could be formed in this way. At the other extreme, gas of a negligible mass could evolve through ellipsoidal figures to a lenticular shape and then eject matter from its edge; spiral nebulae could be formed in this way. Jeans concluded that the rotation of a contracting mass could not give rise to the formation of a planetary system. He himself favoured a tidal theory of the Solar System's origins, in which planetary systems were created during the close passage of two stars. This theory is no longer held in much favour, but Jeans did point out the errors in the nebular theory by which Pierre ◊Laplace attempted to explain the origin of the Solar System.

Jenner, Edward (1749–1823) English biologist who was the first to prove by scientific experiment that cowpox gives immunity against smallpox. He was the founder of virology and was one of the pioneers of vaccination.

Jenner English biologist Edward Jenner, the pioneer of vaccination. In 1796 he discovered that innoculation with cowpox gave immunity to smallpox, and by 1800 more than 100,000 people had been vaccinated against the disease. As a result the death rate due to smallpox dropped dramatically, from around 3500 per million in the 18th century to 90 per million after 1872 when compulsory vaccination was enforced. *Mary Evans Picture Library*

Jenner was born in Berkeley, Gloucestershire, on 17 May 1749, the son of a vicar. He was educated locally and in 1761 apprenticed to a surgeon in Sodbury. In 1770 he went to London to study anatomy and surgery under John ◊Hunter, who took Jenner as his first boarding pupil at St George's Hospital in London. Returning to Gloucestershire in 1773, he set up in private practice and remained at Berkeley until he died there of a stroke on or about 26 January 1823.

In 1788, an epidemic of smallpox swept Gloucestershire and inoculations with live vaccine were used in spite of the tragedies that had previously accompanied this practice. The method used was well known in eastern countries and had been brought to England in 1721 by Lady Mary Wortley Montagu, the wife of the British ambassador to Turkey. It consisted of scratching a vein in the arm of a healthy person and working into it a small amount of matter from a smallpox pustule taken from a person with a mild attack of the disease. This treatment often resulted in the patient fatally contracting smallpox, despite the successes of Jan ◊Ingenhousz.

In the course of his inoculations Jenner noticed that people who had suffered from cowpox, a disease affecting the teats of cows and later the hands of their milkers, remained quite unaffected by the smallpox inoculation, and did not even produce the symptoms of a mild attack of smallpox (as did other patients). Over the course of 25 years he observed that he was unable to infect previous cowpox victims with smallpox, and that where whole families succumbed to smallpox, a previous cowpox victim remained healthy. He also noticed that whereas inoculation with cowpox appeared to protect the patient from smallpox, it did not give immunity against cowpox itself. In his study of cowpox, Jenner was the first to coin the word 'virus'.

In 1796 Jenner carried out an experiment on one of his patients, James Phipps, a healthy eight-year-old boy. Jenner made two small cuts in the boy's arm and worked a speck of cowpox into them. A week later, the patient had a slight fever (the usual reaction) but quickly recovered. Some weeks later, Jenner repeated the inoculation, this time using smallpox matter. The boy remained healthy – vaccination was born (named after *vaccinia*, the medical name for cowpox).

Jenner continued with his experiments and reported his findings to the Royal Society but the fellows considered that he should not risk his reputation by presenting anything 'so much at variance with established knowledge', so in 1798 Jenner published his work privately. Within a few years vaccination was a widespread practice and Jenner not only improved his technique but also found a way of preserving active vaccine.

Jenner was also interested in natural history, one of his favourite hobbies being birdwatching. He studied the habits of the cuckoo, which often lays its eggs in the nest of the hedge sparrow. It had been thought that the hen hedge sparrow threw out her young from the nest to make room for the developing cuckoo, but Jenner's patient observations revealed that it was the young cuckoo itself that heaved its competitors out of the nest. In 1788 he reported these findings to the Royal Society, who published them.

Jenner's work led to an immediate reduction in mortality from smallpox and, nearly 200 years later, the worldwide eradication of the disease. He may be considered to be one of the pioneers of immunology.

Further Reading

Fisher, Richard, *Edward Jenner: A Biography*, Deutsch (Andre) Classics, 1991.

Kerns, Thomas A, *Jenner on Trial: An Ethical Examination of Vaccine Research in the Age of Smallpox and the Age of AIDS*, University Press of America, 1997.

Miller, Genevieve (ed), *Letters of Edward Jenner and Other Documents Concerning the Early History of Vaccination*, The Henry E Sigerist Supplements to the Bulletin of the History of Medicine, New Series, Johns Hopkins University Press, 1983.

Plotkin, S A and Fantini, B (eds), *Vaccinia, Vaccination, Vaccinology: Jenner, Pasteur and Their Successors*, Elsevier Science, 1996.

Jensen, (Johannes) Hans Daniel (1907–1973)

German physicist who shared the 1963 Nobel Prize for Physics with Maria ◊Goeppert-Mayer and Eugene Wigner (1902–) for work on the detailed characteristics of atomic nuclei.

Jensen was born in Hamburg on 25 June 1907. He studied at the universities of Hamburg and Freiburg, gaining a PhD in 1932. From 1932 to 1936 he was an assistant in science at the University of Hamburg, and from 1937 to 1941 in charge of courses at the university. In 1941, he joined the Institute of Technology in Hanover as ordinary professor, staying until 1948. In 1949 he became professor at the University of Heidelberg.

With Goeppert-Mayer and Wigner, Jensen proposed the shell theory of nuclear structure in 1949, and explained it in *Elementary Theory Of Nuclear Shell Structure*, written with Goeppert-Mayer in 1955. According to this theory, a nucleus could not be thought of as a random motion of neutrons and protons about a point, but as a structure of shells, or spherical layers, each with a different radius and each filled with neutrons and protons.

Careful studies of nuclear energy levels had shown that not only do even numbers of neutrons or protons lead to more stable nuclei than odd numbers, but that nuclei containing certain definite numbers of neutrons or protons, or both, are especially stable (like the electron structures of inert gas atoms). In 1948 Goeppert-Mayer published evidence of the special stability of the following numbers of protons and neutrons: 2, 8, 20, 50, 82, and 126. These are commonly called magic numbers. Similar conclusions were reached by Jensen and Wigner. It is not only the energy levels that show particularly tight binding of nuclei with a magic number of protons or neutrons, or particularly both; the angular momentum also shows properties similar to those observed in closed shells of electrons in atoms. The value of angular momentum is zero for closed shells, but has a value where there are small numbers of neutrons or protons outside the closed shells. It seems likely that what is being dealt with here is something like the periodic table of the elements.

The three Nobel prizewinners tried to establish the significance of the magic numbers and why they have the values that they do. They concluded that the orbital and spin moments of the particles within the nucleus could be treated to a first approximation by methods used for atomic structure. One had to assume that the central field was nothing like that met in an atom, but consisted of a square well. In addition, the potential was constant throughout the interior of the sphere but had high barriers at the surface, and the energy levels were in a different order from the order of electronic levels in an atom. For example, the two 1*s* levels came lowest and then the six 2*p* levels, explaining the magic number 8. If the ten levels of the 3*d* orbitals were next, followed by two 2*s* levels, we could add 2 and 10 to 8 and get the next magic number, 20. Jensen and his colleagues then supplemented this theory with a second hypothesis – that there is a very strong spin–orbit interaction between the spin and orbit of a particle and that the lower of two states is always the one with angular momentum parallel rather than antiparallel. In this way, magic numbers seem to fall into place, although research is still going on.

Jessop, William (1745–1814) English civil engineer, a builder of canals and early railways.

Jessop was born in January 1745 in Devonport, Devon, the son of a foreman and shipwright at the local naval dockyard. His father had been associated with John ◊Smeaton in the building of the Eddystone lighthouse. When his father died in 1761, William Jessop, aged 16, became a pupil of Smeaton, who was appointed his guardian.

Jessop worked with Smeaton on the Calder and Hebble, and the Aire and Calder navigations in Yorkshire. Jessop later became England's greatest builder of large waterways; he was responsible for the Barnsley, Rochdale, and Trent navigation, and the Nottingam and the Grand Junction (Grand Union) canals. The only narrow one he worked on was the Ellesmere canal. In 1773 Smeaton went to Ireland, taking his pupil with him, and the Grand Canal of 50 km/80 mi from Dublin to the Shannon (which had been started in 1753) was completed in 1805, just two months after the Grand Junction Canal in England.

It was his work on the building of the Cromford Canal to link Richard ◊Arkwright's mill at Cromford with the Derbyshire coalfield and Nottingham that led Jessop – together with Benjamin Outram, John Wright, and Francis Beresford – to found the Butterley Iron Works Company in 1790. This company later became responsible for many iron bridges. This was the main reason why the canal engineers Jessop and Outram became much in demand for the development of the iron railways in the Midlands, especially in Nottinghamshire and Derbyshire. The first all-iron rails was laid in the 1790s. In 1792, fishbellied cast-iron rails were designed by Jessop. He was also involved in some spectacular iron bridges, particularly the Pontcysyllte aqueduct, completed in 1805. His son Josias was trained by him and became an important engineer in his own right. William Jessop was held in very high repute and for 16 years from 1774 served as secretary of the Society of Civil Engineers. He died at Butterley Hall, Derbyshire, on 18 November 1814.

Jessop was the chief engineer, appointed in 1793, on the construction of the Grand Union Canal (which was then called the Grand Junction Canal). This canal was 149.5 km/93.5 mi long from the Oxford Canal at Braunston, via Wolverton, Leighton Buzzard, Tring, and King's Langley, to the Thames at Brentford. The canal was completed in 1800, apart from the tunnel at Blisworth, which was not finished until 25 March 1805. While this tunnel was being completed, Jessop built a railway over the hill – the first in Northamptonshire – to enable traffic to operate on the canal in 1799. The canal provided a vital link between London and the Midlands.

Jessop's first tunnel was the 2.8-km/1.7-mi-long Butterley Tunnel on the Cromford Canal, and led to the forming of the Butterley Iron Works. The Cromford and High Peak Railway, engineered by Jessop's son Josias in 1825, was built to connect the Cromford Canal with the Peak Forest Canal at Whaley Bridge.

The Surrey Iron Railway grew out of a proposal of 1799 to open a part railway/part canal route from London to Portsmouth. The section from the Thames at Wandsworth to Croydon was to have been canal, but Jessop and Rennie, retained as consultants, decided that a canal would be harmful to the industries of the Wandle Valley. The railway was incorporated by an act of Parliament on 21 May 1801, with Jessop its chief engineer, and was opened in 1802.

The first all-iron rails were laid in the 1790s and spread over the country during the first years of the 19th century, coinciding with the Revolutionary and Napoleonic Wars. These wars caused the cost of wood to soar and the price of iron to fall with the intensification of industry.

The fishbellied iron rails that Jessop designed had a broad head on a thin web, deepest in the centre. One end had a flat foot, nailed with a peg onto a stone block. The other end had a round lug that was fitted into the the foot of the next rail. It was the true ancestor of the modern rail.

Jessop is not given enough credit for his involvement with Thomas ◊Telford in the construction of the Pontcysyllte aqueduct. This crossed the River Dee near Llangollen in north Wales and was completed in November 1805. In 1795, Jessop and Telford decided on a construction based on a 300-m/1,000-ft iron trough, standing on a series of slender stone piers, each solid at the base but hollow from about 21 m/70 ft upwards. The work involved building 19 arches, each with a span of 13.5 m/45 ft. The work must have been extremely dangerous, carried out high above the swirling River Dee, with huge blocks of stone. A mixture of ox blood, water, and lime was used as mortar. The flanged iron plates to make the trough were cast at Plaskynaston Foundry (within view of the aqueduct) and were bolted together, the joints being made watertight with Welsh flannel and lead dipped in boiling sugar. In November 1805, 8,000 people watched the opening. The bridge cost £47,000 with labourers paid 8–12 shillings (40–60p) per week. When Telford wrote later about the achievement, he gave himself all the credit, with no mention of Jessop's contribution, Pontcysyllte is the finest aqueduct in the UK and is certainly one of the most spectacular pieces of aerial navigation in the world. Jessop also built the Derwent aqueduct with 24-m/80-ft span, and a less successful three-arch masonry aqueduct near Cosgrove.

He worked on the construction of a large wetdock area on the Avon at Bristol, on the West India Docks and the Isle of Dogs Canal in London, on the harbours at Shoreham and Littlehampton, and on many other projects.

By 1799 or 1800, he had abandoned the fishbellied rails in favour of cast-iron sockets fixed to the sleepers in which the rails were supported in the upright position. He produced what is equivalent to the flanged wheel of today, with flanges inside the rails.

Jessop achieved an incredible amount in a relatively short lifetime and was always ready to help his fellow professionals, for many of whom he acted as a consulting engineer on their projects, and his opinion was frequently sought because of his vast experience.

Further Reading

Hadfield, Charles and Skempton, S, *William Jessop, Engineer,* David & Charles plc, 1989.

Johannsen, Wilhelm Ludvig (1857–1927) Danish botanist and one of the founders of the modern science of genetics. He coined the term 'gene' as the unit of heredity and introduced the concept of an organism having a set of variable genes – a 'genotype' and characteristics produced by certain genes – 'phenotypes'. Johannsen was hugely influential on the development of modern genetics in the early twentieth century – his major contribution being his 'pure line' theory.

Born in Copenhagen, Denmark, on 3 February 1857, Johannsen was the son of a Danish army officer, Otto Julius Georg Johannsen and Anna Margrethe Dorothea Ebbeson. He attended elementary school in Copenhagen but his father's income would not stretch to a university education and he was apprenticed to a pharmacist in 1872. He taught himself chemistry while working in pharmacies in Denmark and in Germany where he developed an interest in botany. Returning to Denmark in 1879 he passed the pharmacist's examination and continued his botanical and chemical studies, getting a job as an assistant chemist at the new Carlsberg laboratories under Danish chemist Johan Kjeldahl (1849–1900). Here he was given the freedom to do his own research on ripening, dormancy, and germination in seeds, tubers, and buds. He developed rigour and clarity in his research methods. He was also well read in several languages, philosophy, and aesthetics.

Johannsen left the Carlsberg laboratory in 1887 and continued his own research, leading to the discovery of a method of breaking the dormancy period of winter buds. He travelled to Zurich, Darmstadt, and Tübingen to study plant physiology before becoming a lecturer in 1892 and professor in 1903 at Copenhagen Agricultural College. He was influenced by Darwin and was impressed by the 'theory of heredity' of Francis ◊Galton that described his experiments with the sweet pea and demonstrated the ineffectiveness of selection applied to the progeny of self-fertilizing plants. Johannsen repeated this work with princess beans, finding that selection did work in a mixed group of self-fertilizing beans and that it was only in the offspring of a single parent that selection was ineffectual. The descendants of a single parent, which he called the 'pure line' were genetically identical, variations being due to chance and environmental factors. Johannsen dedicated his 1903 paper on this discovery to Galton in 'respect and gratitude' even though he had proved Galton's application was wrong. In 1905 Johannsen coined the terms 'genotype', which describes the genetic constitution of an individual, and 'phenotype' to describe the visible result of the interaction of a genotype with the environment.

Johannsen's proof of the existence of heritable and nonheritable variability, set out in his book *Om arvelighed og variablilitet/On Heredity and Variation* (1896), allowed naturalists to stop trying to claim (as Darwin had done) that the inheritance of acquired characteristics was a process of evolution. In the light of the rediscovery of the laws of Austrian monk Gregor ◊Mendel, he enlarged and reissued his book as *Arvelighedlaerens elementer/Elements of Heredity* in 1905. The enlarged and rewritten (by Johannsen himself) German version brought out in 1909 was to become the most influential textbook on genetics in Europe.

In spite of his lack of formal education Johannsen was made professor of plant physiology at the University of Copenhagen in 1905 and rector in 1917. He was awarded several honorary degrees and membership of the Royal Danish Academy of Sciences. Johannsen spent his later years writing on the history of science and his book *Arvelighed I historik og experimentel Belysning/Heredity in the Light of History and Experimental Study* (1923) is a survey of ideas in genetics from the ancient Greeks to US geneticist T H ⇨Morgan that went to four editions even though it was published only in Danish.

Johannsen's work played a major role in the transformation of nineteenth century ideas to the modern science of genetics and has stood the test of time in the changes brought about by the discovery of chromosomes and the structure of nucleic acids.

Joliot-Curie, Irène (1897–1956) and Frédéric (1900–1958) French chemists who received the Nobel Prize for Chemistry in 1935 for their discovery of artificial, or induced, radioactivity, in light elements. Irène was the daughter of Pierre and Marie ⇨Curie, who shared the 1903 physics prize with Henri ⇨Becquerel. (Marie also won the 1911 chemistry prize outright). Both mother and daughter died of leukaemia caused by overexposure to radioactivity during their research.

Irène Curie was born on 12 September 1897 in Paris and educated privately by tutors including Paul ⇨Langevin and Jean ⇨Perrin. She obtained her undergraduate degree at Collège Sévigné. During World War I she worked as a nurse, helping her mother operate radiography equipment, and then studied physics and mathematics at the Sorbonne, gaining a doctorate for studying the range of alpha particles. She then went to work for her mother at the Radium Institute. There she met Frédéric Joliot whom she married in 1926. In 1937 she was elected a professor at the Sorbonne and in 1946 she became director of the Radium Institute. She died in Paris on 17 March 1956.

Frédéric Joliot was born on 19 March 1900 in Paris. He was the son of a tradesman and graduated from the Ecole Supérieure de Physique et de Chimie Industrielle in engineering. After a brief job as a production engineer at a steelworks, he joined the Radium Institute in 1925 and obtained his PhD in 1930 for a thesis on the electrochemistry of radioactive elements. He taught at the Sorbonne for two years before becoming professor of nuclear chemistry at the Collège de France. In 1944 he was appointed director of the atomic synthesis laboratory at the Centre Nationale des Recherches Scientifiques (NRS) in Paris, putting him effectively in charge of the nation's atomic energy programme. From 1946 he was high commissioner for atomic energy, but in 1950 he returned to the Collège de France and then, after his wife's death, took up her professorship at the Sorbonne in 1956. He died in Paris on 14 August 1958.

Together the Joliot-Curies worked on radioactivity and the transmutation of elements. Twice they just missed major discoveries: in 1932 when James ⇨Chadwick beat them to the neutron, and in 1933 when Carl ⇨Anderson discovered the positron. However, early in 1934, while bombarding aluminium with alpha particles (helium nuclei) from polonium, they observed that positrons continued to be emitted after the alpha-particle source was removed. They had made a short-lived (half-life about three minutes) radioactive isotope of phosphorus. Similarly, starting with boron and magnesium they produced new radioisotopes of nitrogen and aluminium. Many other groups of workers, including John ⇨Cockcroft in the UK and Enrico ⇨Fermi in Italy, took up the method and by the end of 1935 about a hundred artificial radioactive elements were known. For this discovery, the Joliot-Curies received the Nobel Prize for Chemistry in 1935 – one year after Marie Curie's death.

The Joliot-Curies, thereafter, continued to work on fission in heavy elements like uranium. During World War II Frédéric worked on the application of isotopes to the study of biological processes, and was involved in the French Resistance movement. After 1945 he became active in various international peace movements and in the founding of UNESCO.

The Joliot-Curies were noted for their left-wing politics. Irène was refused entry to the American Chemical Society because of her political beliefs, and in 1950 Frédéric was replaced as high commissioner for atomic energy, despite having been appointed by Charles de Gaulle and having built a nuclear reactor without any outside help.

Jones, Harold Spencer (1890–1960) English astronomer, the tenth Astronomer Royal. He is noted for his study of the speed of rotation of the Earth; the motions of the Sun, Moon, and planets; and his determination of solar parallax.

Jones was born in London on 29 March 1890, the third child of an accountant. He attended Hammersmith Grammar School and Latymer Upper School, where he excelled at mathematics. He won a scholarship to Jesus College, Cambridge. Following his graduation in 1913 he was awarded a research fellowship to Jesus College. In the same year he was appointed chief assistant at the Royal Observatory, Greenwich, to succeed Arthur ⇨Eddington, who had gone to Cambridge as Plumian Professor. Ten years later, in 1923, Jones was appointed His Majesty's Astronomer on the Cape of Good Hope. In 1933 he returned to England to become Astronomer Royal. He held this position until his retirement in 1955. He was president of the International Astronomical Union 1944–48 and secretary general of the Scientific Union 1955–58. He had been president of the Royal Astronomical Society 1938–39 and was knighted in 1943. Jones died in London on 3 November 1960.

Jones's first period of research began when he went to Russia in 1914 to observe an eclipse. He wrote several papers on the variation of latitude, as observed using an instrument known as the Cookson floating telescope. During this time he also made determinations of the photographic magnitude scale of the North Polar Sequence (the stars located near the Celestial North Pole). While at the Cape of Good Hope, he published an important catalogue containing the radial velocities of the southern stars, calculated the orbits of a number of spectroscopic binary stars, and made a spectroscopic determination of the constant of aberration. In 1924 he made extensive observations of Mars, using an 18-cm/7-in heliometer – a refracting telescope for measuring the angular diameter of celestial objects – and these observations were later used to obtain the value for solar parallax. In 1925 he obtained and described a long series of spectra of a nova that had appeared in the constellation of Pictor.

While Jones was working at the Cape of Good Hope, he had collected more than 1,200 photographic observations towards the solar parallax programme and as a result, by international agreement in 1931, he was entrusted with dealing with all results of the observations of Eros. Eros, an asteroid whose orbit brings it to within 24 million km/15 million mi of Earth, was discovered in 1898. From photographic observations in both hemispheres, Jones derived a figure for the solar parallax that corresponded to a distance of 149,670,000 km/93,005,000 mi and published his results in 1941. By 1967 the distance of the Earth from the Sun had been obtained using direct measurements by radar. The result, a distance of 149,597,890 km/92,960,128 mi, is ten times as accurate (though the difference is only 0.05%) as that which Jones had originally derived using parallax. Any improvements on this value hinges on the limitations of our present knowledge of the velocity of light. It was unfortunate that Jones's painstaking and time-consuming work on determining the mean distance of the Earth from the Sun came just before the use of automatic computing equipment made his methods and instruments redundant.

Besides his work on solar parallax, Jones's principal contributions to astronomy were his work on the motions and secular acceleration of the Sun, Moon, and planets and the rotation of the Earth. He proved that fluctuations in the observed longitudes of these celestial bodies are due not to any peculiarities in their motion, but to fluctuations in the angular velocity of rotation of the Earth. He also successfully investigated geophysical phenomena such as the rotation of the Earth and its magnetism and oblateness, and he estimated the mass of the Moon. As Astronomer Royal, Jones campaigned for the Royal Observatory to be moved from Greenwich, because the increasing smoke and lights of London hindered observation. It was not until after World War II, however, that a new site was procured at Herstmonceux Castle, Sussex, and the new observatory was not completed until 1958, three years after Jones's retirement.

Jordan, (Marie Ennemond) Camille (1838–1922)
French mathematician originally trained as an engineer. An influential teacher, he was himself strongly influenced by the work (one generation earlier) of Evariste ◊Galois and, accordingly, concentrated on research in topology, analysis, and (particularly) group theory. He nevertheless also made further contributions to a wide range of mathematical topics.

Jordan was born in Lyon on 5 January 1838. Completing his education, he entered the Ecole Polytechnique in Paris and studied engineering. He devoted most of his spare time to mathematics, however, and although he qualified as an engineer – and even began to work professionally in that capacity – it was as a mathematician that he joined the staff of the Ecole Polytechnique at the age of 35, in 1873. He also gave lectures at the Collège de France. Eight years later he was elected to the Paris Academy of Sciences. In 1912 he retired. In 1919 – at the age of 81 – he became a foreign member of the Royal Society. He died in Paris on 20 January 1922.

Before he had even begun to teach mathematics, Jordan was already acknowledged as the greatest exponent of algebra in his day. Pursuing his interest in the work of the ill-fated Evariste Galois, Jordan systematically developed the theory of finite groups and arrived at the concept of infinite groups. An early result of this was the related concept of composition series, and what is now known as the Jordan–Holder theorem (which deals with the invariance of the system of indices of consecutive groups). He investigated Galois's study of permutation groups, in which Galois had considered the consequences of permutating the roots (solutions) of equations, and linked this with the problem of the solution of polynomial equations.

In 1870 Jordan published his famous *Traité des substitutions et des équations algébriques,* which, for the next three decades, was to be the standard work in group theory. Following this major achievement,

Jordan concentrated on his theorems of finiteness. In all he developed three, the first of which dealt with symmetrical groups. The second had its origin in the theory of linear differential equations; Immanuel ◊Fuchs had completed work on such equations of the second order; Jordan reduced the similar problem for equations of the order n to a problem in group theory. Jordan's last finiteness theorem generalized Charles ◊Hermite's results on the theory of quadratic forms with integral coefficients.

In topology, Jordan developed an entirely new approach to what is now known as homological, or combinatorial, topology by investigating symmetries in polyhedra from an exclusively combinatorial viewpoint. His other great contribution to this branch of mathematics was his formulation of the proof for the 'decomposition' of a plane into two regions by a simple closed curve.

Jordan's work in analysis was published in 1882 in the *Cours d'analyse de l'Ecole Polytechnique,* a spiritedly modern book that had a widespread influence, particularly because of Jordan's insistence on what a really rigorous proof should comprise.

Much of Jordan's later work was concerned with the theory of functions, and he applied the theory of functions of bounded variation to the particular curve that bears his name.

Josephson, Brian David (1940–) Welsh physicist who discovered the Josephson tunnelling effect in superconductivity. He shared the 1973 Nobel Prize for Physics with Leo ◊Esaki and Ivar Giaever (1929–) for their work on tunnelling in semiconductors and superconductors.

Josephson was born in Cardiff on 4 January 1940. He was educated at Cardiff High School and at Cambridge University, where, as an undergraduate, he published an important paper showing that an application of the Mössbauer effect to verify gravitational changes in the energy of photons had failed to take account of Doppler shifts associated with temperature changes.

Josephson became a fellow of Trinity College, Cambridge, in 1962 and has remained at Cambridge ever since apart from the years 1965 and 1966, when he had an assistant professorship at the University of Illinois. In 1974 he became professor of physics at Cambridge.

In 1962 Josephson saw some novel connections between solid-state theory and his own experimental problems on superconductivity. He then set out to calculate the current due to quantum mechanical tunnelling across a thin strip of insulator between two superconductors, and the current–voltage characteristics of such junctions are now known as the Josephson effect.

Superconductivity, the property of zero resistivity of some metals below a critical temperature, was discovered by Heike ◊Kamerlingh Onnes in 1911, and in 1933 it was recognized that superconductors are perfect diamagnets, that is, they totally repel magnetic field lines. Although the effect could be described by phenomenological macroscopic equations, no clear understanding of the microscopic mechanism emerged until the BCS theory developed by John ◊Bardeen and co-workers in 1957. This explained the absence of electron scattering by electrons of opposite spin pairing up so that, via the response of the atoms in the metal, the electrons had an attractive rather than repulsive influence on one another.

Josephson's contribution was to solve the problem of how such superconductivity pairs would behave when confronted by an insulating barrier, typically of a thickness of 1–2 nanometres (1–2 billionths of a metre). He recognized that the electron pairs could tunnel through barriers in a manner analogous to the behaviour of single particles in alpha decay. One observable prediction is that an alternating current (AC) occurs in the barrier when a steady external voltage is applied to a system comprising a superconductor, such as lead, with a thin film of oxide at its surface and a superconducting film evaporated on the oxide. The effect is known as the AC Josephson effect. Conversely radiation, particularly in the far infrared or microwave region, will excite an extra potential across the junction when absorbed.

In addition, when a steady magnetic field is applied across an insulating barrier, a steady current flows. This is the DC Josephson effect. A circuit containing two such barriers in parallel shows quantum interference, analogous to the two-slit experiment in optics, when a magnetic flux is applied to the system.

Josephson's predictions were soon verified by J M Rowell at the Bell Telephone Laboratories (the direct current–magnetic field characteristic) and by S Shapiro, who applied an AC voltage to a superconductor–insulator–superconductor junction and observed resonance coupling to the internal AC Josephson supercurrent.

The Josephson effect has the following applications. The frequency of the AC current is very precisely related to fundamental constants of physics, and this has led to the most accurate method of determining h/e, which in turn establishes a voltage standard. The effect may be used as a generator of radiation, particularly in the microwave and far infrared region. Quantum interference effects are used in squids (superconducting quantum interference devices), which may act as ultra-sensitive magnetometers capable of detecting tiny geophysical anomalies in the Earth's magnetic field or even the anomaly caused by the presence of a submarine. In addition, the fast switching properties of such devices can be exploited in logic elements or binary memories, and this has great potential for future generations of computers. Josephson's discovery may thus have important consequences in the development of artificial intelligence, and he is now absorbed in the mathematical modelling of intelligence.

Joule, James Prescott (1818–1889) English physicist who verified the principle of conservation of energy by making the first accurate determination of the mechanical equivalent of heat. He also discovered Joule's law, which defines the relation between heat and electricity, and – with William Thomson (Lord ◊Kelvin) – the Joule–Thomson effect. In recognition of Joule's pioneering work on energy, the SI unit of energy is named the joule.

Joule was born in Salford on 24 December 1818 into a wealthy brewing family. He and his brother were educated at home between 1833 and 1837 in elementary mathematics, natural philosophy, and chemistry, partly by the chemist John ◊Dalton. Joule was a delicate child and very shy, and apart from his early education he was entirely self-taught in science. He does not seem to have played any part in the family brewing business, although some of his first experiments were done in a laboratory at the brewery.

Joule English physicist James Prescott Joule, who established the mechanical theory of heat and discovered Joule's law, which defines the relation between heat and electricity. In 1848 he made the first estimate of the velocity of gas molecules. *Mary Evans Picture Library*

Joule had great dexterity as an experimenter, and was able to measure temperatures very exactly indeed. At first, other scientists could not credit such accuracy and were disinclined to believe the theories that Joule developed to explain his results. The encouragement of Lord Kelvin from 1847 changed these attitudes, however, and Kelvin subsequently used Joule's practical ability to great advantage. By 1850, Joule was highly thought of among scientists and became a fellow of the Royal Society. He was awarded the Society's Copley Medal in 1866 and was president of the British Association for the Advancement of Science in 1872 and again in 1887. Joule's own wealth was able to fund his scientific career, and he never took an academic post. His funds eventually ran out, however. He was awarded a pension in 1878 by Queen Victoria, but by that time his mental powers were declining. He suffered a long illness and died in Sale, Cheshire, on 11 October 1889.

Joule realized the importance of accurate measurement very early on and exact quantitative data became his hallmark. His most active research period was between 1837 and 1847 and led to the establishment of the principle of conservation of energy and the equivalence of heat and other forms of energy. In a long series of experiments, he studied the quantitative relationship between electrical, mechanical, and chemical effects and heat, and in 1843 he was able to announce his determination of the amount of work required to produce a unit of heat. This is called the mechanical equivalent of heat (currently accepted value 4.1868 joules per calorie).

Joule's first experiments related the chemical and electrical energy expended to the heat produced in metallic conductors and voltaic and electrolytic cells. These results were published between 1840 and 1843. He proved the relationship, known as Joule's law, that the heat produced in a conductor of resistance R by a current I is proportional to I^2R per second. He went on to discuss the relationship between heat and mechanical power in 1843. Joule first measured the rise in temperature and the current and the mechanical work involved when a small electromagnet rotated in water between the poles of another magnet, his training for these experiments having been provided by early research with William ◊Sturgeon, a pioneer of electromagnetism. Joule then checked the rise in temperature by a more accurate experiment, forcing water through capillary tubes. The third method depended on the compression of air and the fourth produced heat from friction in water using paddles that rotated under the action of a falling weight. This has become the best-known method for the determination of the mechanical equivalent. Joule showed that the results obtained using different liquids (water, mercury, and sperm oil) were the same. In the case of water, 772 ft lb of work produced a rise of 1 °F in 472 cu cm/29 cu in of water. This value was universally accepted as the mechanical equivalent of heat. It now has no validity, however, because as both heat and work are considered to be forms of energy, they are measured in the same units – in joules. A joule is basically defined as the energy expended when a force of 1 newton moves 1 metre.

The great value of Joule's work in the establishment of the conservation of energy lay in the variety and completeness of his experimental evidence. He showed that the same relationship held in all cases that could be examined experimentally and that the ratio of equivalence of the different forms of energy did not depend on how one form was converted into another or on the materials involved. The principle that Joule had established is in fact the first law of thermodynamics – that energy cannot be created nor destroyed but only transformed.

Because he had not received any formal mathematical training, Joule was unable to keep up with the new science of thermodynamics to which he had made such a fundamental and important contribution. However, the presentation of his final work on the mechanical equivalent of heat in 1847 attracted great interest and support from William Thomson, then only 22 and later to become Lord Kelvin. Much of Joule's later work was carried out with him, for Kelvin had need of Joule's experimental prowess to put his ideas on thermodynamics into practice. This led in 1852 to the discovery of the Joule–Thomson effect, which produces cooling in a gas when the gas expands freely. The effect is caused by the conversion of heat into work done by the molecules in overcoming attractive forces between them as they move apart. It was to prove vital to techniques in the liquefaction of gases and low-temperature physics.

Joule lives on in the use of his name to measure energy, supplanting earlier units such as the erg and calorie. It is an appropriate reflection of his great experimental ability and his tenacity in establishing a basic law of science.

Joy, James Harrison (1882–1973) US astronomer, most famous for his work on stellar distances, the radial motions of stars, and variable stars.

Joy was born on 23 September 1882 in Greenville, Illinois, the son of a merchant of New England ancestry. He was educated locally and then attended Greenville College, obtaining a PhD in 1903. From 1904 to 1914 he worked at the American University of Beirut, Lebanon, first as a teacher and then as professor of astronomy and director of the observatory. He returned to the USA in 1914, worked at the Yerkes Observatory as an instructor for a year, and then joined the staff of Mount Wilson Observatory, where he remained until 1952. He was vice president of the American Astronomical Society in 1946 and president in 1949.

When Joy joined the staff of the Yerkes Observatory in 1914, he took part in a programme of measuring stellar distances, using the 1-m/40-in Yerkes refractor, by direct photography and parallax measurements, an extremely tedious and out-of-date method. After 1916, he began to make spectroscopic observations at Mount Wilson Observatory and he was subsequently invited to take part in their research programme to obtain stellar distances.

The new method of finding the distance to stars was to compare the apparent magnitude of the star with its absolute magnitude. Joy continued with this programme, in collaboration with his colleagues Walter ◊Adams and Milton ◊Humason, for more than 20 years and their results enabled them to ascertain the spectral type, absolute magnitude, and stellar distance of more than 5,000 stars.

Joy and his colleagues also studied the Doppler displacement of the spectral lines of some stars to determine their radial velocities. By noting the variations in radial velocity, Joy and his team were able to show that many stars are spectroscopic binary stars and that their period and orbit could therefore be elucidated. From their observations of eclipsing binary stars they deduced the absolute dimensions, masses, and orbital elements of some specific stars within eclipsing binary systems. Using radial velocity data of 130 Cepheid variable stars, Joy determined the distance and direction of the centre of the Galaxy and calculated the rotation period for bodies moving in circular orbits at the distance of the Sun, with a view to calculating the rotation period of the Galaxy.

Joy also spent many years of his life observing variable stars and classifying them according to their characteristic spectra. While studying the long-period variable star, Mira Ceti, Joy observed the spectrum of a small hot companion object. It was later named Mira B and shown to be a white dwarf; it can be observed visually today. Joy later became interested in the parts of the Galaxy where dark, absorbing clouds of gas and dust exist, and by carefully observing these areas he found examples of a particular kind of variable star, called a T-Tauri star, which is strongly associated with these areas. T-Tauri stars have a wide range of spectral types combined with characteristically low magnitudes. Joy showed that these characteristics indicated that they were very young stars in an early stage of their evolutionary history.

Jung, Carl Gustav (1875–1961) Swiss psychologist who founded analytical psychology as a deliberate alternative to the psychoanalysis of Sigmund ◊Freud.

Jung was born in Kesswil, near Basel, on 26 July 1875, the son of a Protestant clergyman. Despite an early interest in archaeology – and a strong family background in religion and theology – he went to Basel University in 1895 to study medicine, graduating in 1900. He then attended Zürich University and obtained his MD in 1902, at the same time turning to psychiatry. For the next seven years he worked at the Burghölzi Psychiatric Clinic in Zürich under Eugen Bleuler, an expert on schizophrenia; also from 1905 to 1913 he lectured in psychiatry at the university. In 1907 he met Freud and for five years became his chief disciple, accepting the appointment as the first president of the International Psycho-Analytical Association on its foundation in 1911. But in 1913 following publication of his *Wandlungen und Symbole de Libido* (1912), translated as *The Psychology of the Unconscious* (1916), he broke with Freud, resigned from the association, and set up his own practice in Zürich. In 1933 he became professor of psychology at the Zürich Federal Institute of Technology, a post he held for eight years. In 1943 he resigned almost immediately after being appointed professor of medical psychology at Basel University (when he was 68 years old) because his health began to fail. But he continued to practise until he was over 80, and he died in Küsnacht, near Zürich, on 6 June 1961.

While Jung was at the Psychiatric Clinic in the early 1900s he devised the word-association test as a psychoanalytical technique for penetrating a subject's unconscious mind. He also developed his theory concerning emotional, partly repressed ideas, which he termed 'complexes'. The chief reason for his split with Freud – like Alfred ◊Adler's before him – was Freud's emphasis on infantile sexuality. Jung introduced the alternative idea of 'collective unconscious', which is made up of many archetypes or 'congenital conditions of intuition'.

The separation of psychology from the premises of biology is purely artificial, because the human psyche lives in indissoluble union with the body.

CARL JUNG FACTORS DETERMINING HUMAN BEHAVIOUR

Each person is born with access to these archetypes, which Jung tried to identify by studying cultures such as those of the North American Pueblo Indians. Mythology and folklore, alchemical writings, religious texts, and even dreams were also analysed for archetypes.

Jung Swiss psychologist Carl Gustav Jung, the founder of analytical psychology. Jung devised the technique of word association to study the unconscious mind, used dream analysis, and distinguished between the characteristics of introversion and extroversion. *Mary Evans Picture Library*

Jung also studied personality and its importance in human behaviour and in 1921 introduced the concept of 'introverts' and 'extroverts' in his book *Psychologische Typen*. This work also contained his theory that the mind has four basic functions: thinking, feeling, sensations, and intuition. Any particular person's personality can be ascribed to the predominance of one of these functions.

Further Reading

Hanah, Barbara, *Jung: His Life and Work*, Chiron Publications, 1998.

Hyde, Maggie and McGuinness, Michael, *Jung for Beginners*, Icon Books, 1992.

McLynn, Frank J, *Carl Gustav Jung*, Bantam Press, 1996.

Shamdasani, Sonu, *Cult Fictions: C G Jung and the Founding of Analytical Psychology*, Routledge, 1998.

Stevens, Anthony, *Jung*, Past Masters series, Oxford University Press, 1994.

Storr, Anthony, *Jung*, Fontana Modern Masters series, Fontana Press, 1995.

Young-Eisendrath, Polly and Dawson, Terence (eds), *The Cambridge Companion to Jung*, Cambridge University Press, 1997.

Just, Ernest Everett (1833–1941) US biologist who became internationally known for his pioneering research into fertilization, cell physiology, and experimental embryology.

Just was born on 14 August 1883 in Charlston, South Carolina. He was educated at Kimball Union Academy, Vermont, graduating in 1903 with honours and went on to Dartmouth College, New Hampshire, to study English and classics, later switching to biology; he received his AB degree in 1907 as the only *magna cum laude* in his class. Just then went as a teacher to Howard University where he remained until 1929. In 1912 he was made professor and head of the department of zoology, and in 1916 he took leave of absence to complete his PhD in zoology at the University of Chicago. From 1909 to 1929 Just spent his summers conducting research at the Marine Biological Laboratories at Woods Hole, Massachusetts. In 1929, frustrated with the limitations imposed on him by his race, Just went to Europe to conduct research in German laboratories and at French and Italian marine stations. With the German occupation of France, Just returned to his former post at Howard in 1940, where he died of pancreatic cancer on 27 October 1941. He was married twice and had two daughters and a son.

Just is chiefly known for his research into fertilization and experimental parthenogenesis in marine eggs, on which he published over 60 papers. Through his research at Woods Hole he became the leading authority on the embryological resources of the marine group of animals. His focus of attention was the cell and in particular the cytoplasm, which, contrary to popular belief, he stated was just as important as the nucleus, and was primarily responsible for the individuality and development of the cell. He was a co-author of *General Cytology* (1924) and while in Europe he published *Biology of the Cell Surface* (1939) and *Basic Methods for Experiments in Eggs of Marine Animals* (1939).

Just also worked on the philosophical implications of his research, believing that there was a need for the integration of biological science and social philosophy. He was an associate editor of the *Biological Bulletin, Physiological Zoology, Protoplasma,* and the *Journal of Morphology.* For his work he was awarded the first Spingarn Medal in 1915, he was vice president of the American Society of Zoologists in 1930, and was elected a member of the Washington Academy of Scientists in 1936.

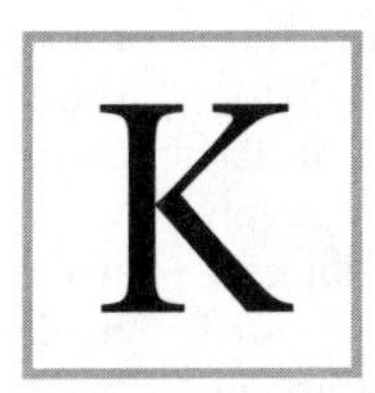

Kamerlingh Onnes, Heike (1853–1926) Dutch physicist who is particularly remembered for the contributions he made to the study of the properties of matter at low temperatures. He was the first to liquefy helium and later discovered superconductivity, gaining the 1913 Nobel Prize for Physics in recognition of his work.

Kamerlingh Onnes was born in Groningen on 21 September 1853 and attended local schools. In 1870 he entered the University of Groningen to study physics and mathematics. He won two prizes at the university during his early years there for his studies on the nature of the chemical bond. In 1871 Kamerlingh Onnes travelled to Heidelberg to study under Gustav ◊Kirchhoff and Robert ◊Bunsen. He then returned to Groningen and in 1876 completed the research for his PhD, on new proofs for the Earth's rotation, but the degree was not awarded until 1879. In the meantime, Kamerlingh Onnes served as an assistant at the polytechnic in Delft and later lectured there. In 1882, he was appointed professor of experimental physics at the University of Leiden, a post he held for 42 years. In 1894, he founded the famous cryogenic laboratories at Leiden, which became a world centre of low-temperature physics. Kamerlingh Onnes died in Leiden on 21 February 1926.

During his years at Delft, Kamerlingh Onnes met Johannes ◊van der Waals, who was then professor of physics in Amsterdam. Their discussions led Kamerlingh Onnes to become interested in the properties of matter over a wide temperature range, and in particular at the lower end of the scale. He felt that he might thereby be able to obtain experimental evidence to support van der Waals's theories concerning the equations of state for gases.

Kamerlingh Onnes applied the cascade method for cooling gases that had been developed by James ◊Dewar, and in 1908 succeeded in liquefying helium, a feat that had eluded Dewar. He was not content merely to report the technical achievements involved in these processes, he also wanted to investigate how matter behaves at very low temperatures. In 1910, Kamerlingh Onnes succeeded in lowering the temperature of liquid helium to 0.8K (–272.4 °C/–458.2 °F). Lord ◊Kelvin had postulated in 1902 that as the temperature approached absolute zero, electrical resistance would increase. In 1911, Kamerlingh Onnes found the reverse to be the case, showing that the electrical resistance of a conductor tends to decrease and vanish as temperature approaches absolute zero. He called this phenomenon supraconductivity, later renamed superconductivity.

Kamerlingh Onnes made a particular study of the effects of low temperature on the conductivity of mercury, lead, nickel, and manganese–iron alloys. He found that the imposition of a magnetic field eliminated superconductivity even at low temperatures. Special alloys have now been produced that do not behave in this manner.

Kamerlingh Onnes was also interested in the magneto-optical properties of metals at low temperature, and in the technical application of low-temperature physics to the industrial and commercial uses of refrigeration.

Kant, Immanuel (1724–1804) German philosopher whose theoretical work in astronomy inspired cosmological theories and many of his conjectures have been confirmed by observational evidence.

Kant was born in Königsberg (now Kaliningrad) on 27 April 1724 and he lived there or nearby for all of his life. His grandfather was an immigrant from Scotland, called Cant, whose name was Germanized. Kant's father was a saddlemaker of modest means and conventional religious beliefs and his mother was a member of a pietist Protestant sect.

At the age of ten Kant entered the Collegium Fridericianum, where he was expected to study theology. Instead, he concentrated on the classics and then went to the University of Königsberg to study mathematics and physics. Six years after he entered university, his father died and he was forced to work as a private tutor to three of Königsberg's leading families in order to support his brothers and sisters. He did not return to the university until 1755, when he gained his PhD and became an unpaid lecturer. Fifteen years later, Kant was offered the chair in logic and metaphysics, a post he held for 27 years. He died in Königsberg on 12 February 1804.

Kant's most important work was a philosophical text, the *Critique of Pure Reason*, published in 1781 when he was in his late fifties. In his early pre-*Critique* work, however, he dealt with various scientific subjects, including a consideration of the origin and natural history of the universe. Kant was not an experimental scientist in any sense and he has sometimes been accused of being merely an 'armchair scientist'. Yet he was not ignorant of the scientific advances of his day and his theoretical works on astronomy were deeply affected by his studies of Newtonian and Leibnizian physics. The hypotheses Kant put forward inspired other theories, notably those propounded by Carl von ◊Weizsäcker in the Gifford Lectures of 1959–60 and by Gerard ◊Kuiper of the Yerkes Observatory in the USA. They also coincided with another more empirical work on the nature of the universe that later became known as the Kant–Laplacian theory.

Besides being indebted to Isaac ◊Newton, Kant's thought on astronomy and cosmology was also influenced by a work entitled *An Original Theory or New Hypothesis of the Universe* published by Thomas ◊Wright in 1758. Kant's pre-*Critique Universal Natural History and Theory of the Heavens* was based on the theories of Newton and Wright. It also anticipated facts of astronomy that were later confirmed only by advanced observational techniques. Kant proposed that the Solar System was part of a system of stars constituting a galaxy and that there were many such galaxies making up the whole universe. His conjecture that nebulous stars were also separate galaxies similar to our own was confirmed only in the 20th century.

Although Kant accepted the religious claim that God created the world, in the *Natural History* he concerned himself with explaining how the universe evolved once it had been created. In contrast with the idea that planets were formed as a result of tidal forces set off in the Sun when some large celestial body passed close by, he argued, in the theory that came to be associated with one put forward by Pierre ◊Laplace, that planetary bodies in the Solar System were formed by the condensation of nebulous, diffuse primordial matter that was previously widely distributed in space. He used the Newtonian theory of the attraction and repulsion of materials to account for the fact that the Solar System, like other celestial systems, is flattened out like a disc. He also supposed that action at a distance was possible, explaining the gravitational attraction that held the moons and planets in their orbits. Moreover, Kant argued that the universe continues to develop: its parts come together and disperse with the condensation and diffusion of matter in the infinity of time and space.

Kant's most famous work, *Critique of Pure Reason,* inaugurated the 'Copernican revolution' in philosophy, turning attention to the mind's role in constructing our knowledge of the objective world. He examined the legitimacy and objectivity of cognition and science, an investigation that led him to doubt a number of the presuppositions necessary to the theories put forward in *Natural History.* For example, one of the self-contradictory conclusions considered in the *Critique* showed that we cannot, with certainty, make assertions concerning the finite or infinite nature of space and time. But even with this qualification it is amazing that, based only on Newtonian principles and information gleaned from reading an abstract of the book by Wright, Kant's hypotheses coincide so well with later cosmological theories.

Further Reading

Engstrom, Stephen and Whiting, Jennifer, *Aristotle, Kant, and the Stoics: Rethinking Happiness and Duty,* Cambridge University Press, 1998.

Gram, Moltke (ed), *Interpreting Kant,* University of Iowa Press, 1982.

Hoffe, Otfried; Farrier, Marshall (transl), *Immanuel Kant,* SUNY Series in Ethical Theory, State University of New York Press, 1994.

Ross, George M and McWalter, Tony (eds), *Kant and His Influence,* Thoemmes Antiquarian Books, 1990.

Scruton, Roger, *Kant,* Past Masters series, Oxford University Press, 1983.

Kapitza, Pyotr Leonidovich (1894–1984) Russian physicist best known for his work on the superfluidity of liquid helium. He also achieved the first high-intensity magnetic fields. For his achievements in low-temperature physics, Kapitza was awarded the 1978 Nobel Prize for Physics. He shared the prize with two US scientists, Arno ◊Penzias and Robert ◊Wilson, who discovered the cosmic microwave background radiation, predicted by George ◊Gamow.

Kapitza was born in Kronstadt on 8 July 1894. He graduated in 1919 from Petrograd Polytechnical Institute and then travelled to the UK and worked with Ernest ◊Rutherford in Cambridge, becoming deputy director of magnetic research at the Cavendish Laboratory in 1924. In 1930 Kapitza became director of the Mond Laboratory at Cambridge, which had been built for him. Four years later he went to the USSR for a professional meeting as he had done before, but this time he did not return. His passport was seized and he was held on Stalin's orders. In 1936 he was made director of the S I Vavilov Institute of Physical Problems of the Soviet Academy of Sciences in Moscow. The Mond Laboratory was sold to the Soviet government at cost and transported to the institute for Kapitza's use. In 1946 he refused to work on the development of nuclear weapons and was put under house arrest until after Stalin's death in 1953. He was restored as director of the institute in 1955. He died in Moscow on 8 April 1984.

For his graduate work, Kapitza went to Rutherford's laboratory and pioneered the production of strong (though temporary) magnetic fields. In 1924 he designed apparatus that achieved magnetic fields of 500,000 gauss, an intensity that was not surpassed for more than 30 years. In his earliest experiments, Kapitza, using especially constructed accumulator batteries and switch gear, passed currents of up to 8,000 amperes through a coil for short intervals of time. In a coil of 1 mm/0.04 in internal diameter, fields of the order of 500,000 gauss for 0.003 second were obtained in this way. By 1927, he had produced stronger currents by short-circuiting an alternating current generator of special construction, and a field of 320,000 gauss in a volume of 2 cu cm/0.12 cu in was obtained for 0.01 second.

Kapitza's most renowned work is in connection with liquid helium, which was first produced by Heike ◊Kamerlingh Onnes in 1908. Kapitza was one of the first to study the unusual properties of helium II – the form of liquid helium that exists below 2.2K (–270.8 °C/–455.4 °F). Helium II conducts heat far more rapidly than copper, which is the best conductor at ordinary temperatures, and Kapitza showed that this is because it flows so easily. This property of helium is known as superfluidity, and it has far less viscosity than any other liquid or gas. In 1937 Kapitza persuaded the Soviet physicist Lev ◊Landau to move to Moscow to head the theory division of his institute. Kapitza published his findings on the superfluidity of helium II in 1941. Landau then continued the work on superfluidity, resulting in the award of the 1962 Nobel Prize for Physics to him for his theory of superfluidity.

In 1939 Kapitza built apparatus for producing large quantities of liquid oxygen. This work and its application was of great importance, particularly to Soviet steel production. He invented a turbine for producing liquid air cheaply in large quantities. His technique for liquefying helium by adiabatic expansion eliminated the presence of liquid hydrogen characteristic of previous methods.

In the 1950s, Kapitza turned his attention partly to ball lightning, a puzzling phenomenon in which high-energy plasma maintains itself for a much longer period than seems likely. His analysis involves the formation of standing waves, and this research turned his interests to controlled thermonuclear fusion, upon which he published a paper in 1969.

Further Reading

Badash, Lawrence, *Kapitza, Rutherford, and the Kremlin*, Books on Demand, 1994.

Boag, J W; Rubinin, P R; and Shoenberg, D (eds), *Kapitza in Cabridge and Moscow: The Life and Letters of a Russian Physicist*, North-Holland Publishing Co, 1990.

Kedrov, F, *Kapitza: Life and Discoveries*, Collets, 1984.

Ter Haar, D (ed), *Collected Papers of P L Kapitza*, Pergamon Brassey's International Defence Publishers, 1986.

Kaplan, Viktor (1876–1934) Austrian engineer famous for inventing the water turbine with adjustable rotor-blades that now bears his name.

Kaplan was born on 27 November 1876 at Murz, Austria. He was educated at the Realschule in Vienna and then at the Technische Hochschule, where he studied machine construction and gained his engineer's diploma in 1900. After a year's voluntary service in the navy, he became a constructor of diesel engines for the firm of Ganz and Company of Leobersdorf. In 1903 he was appointed to the chair of kinematics, theoretical machine studies, and machine construction at the Technische Hochscule in Brunn. In 1918, he became professor of water turbine construction. He became debilitated and retired early in 1931. He died on 23 August 1934 at Unterach on the Attersee, Austria.

Kaplan published his first paper on turbines in 1908, writing about the Francis turbine (see James Bicheno ◊Francis) and basing it on work that he had done for his doctorate at the Technische Hochschule in Vienna. In connection with this paper, he set up a propeller turbine for the lowest possible fall of water. In 1913 the first prototype of the turbine was completed. The Kaplan turbine was patented in 1920 as a water turbine with adjustable rotor-blades; the runner blades can be varied to match the correct angle for a given flow of water. (The Francis turbine corresponds to a centrifugal pump, whereas the Kaplan turbine resembles a propeller pump.)

The Kaplan turbine in its traditional form has a vertical shaft, although one power-generator design uses a horizontal shaft and an alternator mounted in a 'nacelle' (metal shell) in the water flow. This is particularly suitable for lower-output machines operating on very low heads of water, in which an almost straight water passage is possible. This type of turbine was designed for a French project on the River Rance estuary on the Gulf of St Malo, Brittany, France. Construction was begun in January 1961, on the world's first large-scale tidal plant, and completed late in 1967. It was designed to be made up of 24 hydroelectric power units. Each consisted of an ogive-shaped shell of metal containing an alternator and a Kaplan turbine. These units were installed in apertures in a dam through which the water flows. Each turbine acts both as a turbine and a pump, in both directions of tidal flow and in both directions of rotation, with each hydroelectric power unit generating 20,000 kW.

The physical size and the operating head for which a Kaplan turbine can be used have increased considerably since 1920 when it was first patented. Runner diameters of 9 m/30 ft have been employed. British manufacturers have developed designs operating at heads of 58 m/190 ft, but it is likely that further progress in increased heads will be limited due to the advent of the Deriaz turbine.

Kapteyn, Jacobus Cornelius (1851–1922) Dutch astronomer who analysed the structure of the universe by studying the distribution of stars using photographic techniques. To achieve more accurate star counts in selected sample areas of the sky he introduced the technique of statistical astronomy. He also encouraged fruitful international collaboration among astronomers.

Kapteyn was born in Barneveld, the Netherlands, on 19 January 1851. He was born into a large and talented family and displayed great academic abilities early in his life. He began his studies at the University of Utrecht in 1868 when he was 17, although he had already satisfied the university's entrance requirements a year earlier. Kapteyn concentrated his studies on mathematics and physics and earned his doctorate for a thesis on vibration.

Kapteyn's career in astronomy began when he was employed by the astronomical observatory at Leiden in 1875. Three years later he was appointed as the first professor of astronomy and theoretical mechanics at the University of Groningen. He was a member of the French Academy of Science and a fellow of the Royal Society. He was active in the organizational work that eventually led to the establishment of the International Union of Astronomers. He retired from his post at Groningen in 1921, but continued to work and publish his results. He died in Amsterdam on 18 June 1922.

Initially there was a lack of appropriate facilities at the University of Groningen, but this did not deter Kapteyn. He proposed a cooperative arrangement with David ◊Gill at Cape Town Observatory in South Africa, whereby photographs of the stars in the southern hemisphere were analysed at Groningen. This early collaborative work with Gill resulted in the publication during the years 1896–1900 of the *Cape Photographic Durchmusterung*. It was welcomed as an essential complement to Friedrich ◊Argelander's monumental *Bonner Durchmusterung* (1859–62), because it presented accurate data on the brightness and positions of nearly 455,000 stars in the less-studied southern hemisphere. The catalogue included all stars of magnitude greater than 10 which lay within 19 ° of the south pole, and it encouraged Edward ◊Pickering's team of astronomers from the Harvard Observatory, who set up a research station in Chile, to expand their observations of stars in the southern skies.

Kapteyn's next project was to improve the technique of trigonometric parallax so that he could obtain data of a higher quality on the proper motions of stars. Although proper motion had been observed for many years, its explanation was unknown. It had been assumed by most astronomers that there was no pattern in the proper motion of stars, that they resembled the random Brownian motions of gas molecules. In 1904 Kapteyn reported that there was indeed a pattern. He found that stars could be divided into two streams, moving in nearly opposite directions. The notion that there was this element of order in the universe was to have a considerable impact, although its significance was not realized at the time.

Arthur ◊Eddington and Karl ◊Schwarzschild extended Kapteyn's work and confirmed his results, but they missed the underlying importance of their observations. It was not until 1928 that Jan ◊Oort and Bertil ◊Lindblad, greatly aided by recent discoveries on the characteristics of extragalactic nebulae, proposed that Kapteyn's results could be readily understood if they were considered in the contest of a rotating spiral galaxy. Kapteyn's data had been the first evidence of the rotation of our Galaxy (which at that time was not understood to be in any way distinct from the rest of the universe), although it was not recognized as such at the time.

William ◊Herschel had investigated the structure of stellar systems by counting the number of stars in different directions. Kapteyn decided to repeat this analysis with the aid of modern telescopes. He selected 206 specific stellar zones, and in 1906 he began a vast international programme to study them. His aim was to ascertain the magnitudes of all the stars within these zones, as well as to collect data on the spectral type, radial velocity, proper motion of the stars, and other astronomical parameters. This enormous project was the first coordinated statistical analysis in astronomy and it involved the cooperation of over 40 different observatories. The project was extended by Edward Pickering to include 46 additional stellar zones of particular interest.

The analysis of data collected for the programme was a colossal task and took many years to complete. In 1922, only weeks before his death, Kapteyn published his conclusions based on the analysis he had completed. His overall result was strikingly similar to that obtained by Herschel. He envisaged an island universe, a lens-shaped aggregation of stars that were densely packed at the centre and thinned away as empty space outside the Galaxy was reached. The Sun lay only about 2,000 light years from the centre of a structure which spanned some 40,000 light years along its main axis.

A fundamental error in the construction of 'Kapteyn's universe' was his neglect of the possibility that interstellar material would cause stellar light from a distance to be dimmed. This placed a limit on the maximum distance from which data could be obtained. Kapteyn had, in fact, considered this possibility, and he tried unsuccessfully to detect its effect.

The effect Kapteyn had sought was observed by Robert ⇨Trumpler nearly ten years after Kapteyn's death. Harlow ⇨Shapley's model for the structure of our Galaxy soon replaced that proposed by Kapteyn. Today the Galaxy is thought to span some 100,000 light years and to be shaped like a pancake that thickens considerably towards the centre. Our Solar System lies quite a long way out along one of the spiral arms, 30,000 light years from the centre of the Galaxy – which lies in the direction of the constellation Sagittarius.

Further Reading

Robert, Paul E, 'J C Kapteyn and the early 20th-century universe', *J Hist Astron,* 1986, v 17, pp 155–182.

Robert, Paul E and Hertzsprung-Kapteyn, Henrietta, *The Life and Works of J C Kapteyn: An Annotated Translation,* Kluwer Academic Publishers, 1994.

Kármán, Theodore von (1881–1963) Hungarian-born US aerodynamicist who made a fundamental contribution to rocket research, enabling the USA to acquire a lead in this technology. He possessed remarkable teaching and organizational abilities, and coordinated many famous research institutions.

Kármán was born on 11 May 1881, in Budapest, of intellectual parents. He would have been a mathematical prodigy if his father, Mór (Maurice), a professor at Budapest University and an authority on education, had not guided him towards engineering. He completed his undergraduate studies at the Budapest Royal Polytechnic University in 1902 and the following year was appointed assistant professor there.

Three years later he accepted a two-year fellowship at Gottingen to study for his doctorate. In 1908 he went to Paris and witnessed a flight by Henry Farman (1874–1958), which inspired him to concentrate his efforts on aeronautical engineering. Returning to Gottingen he worked on a wind tunnel for airship research with Ludwig Prandetl (1875–1953) and received his doctorate there in 1909.

Between 1913 and 1930, with the exception of the war years when he directed aeronautical research in the Austro-Hungarian Army, Kármán built the newly founded Aachen Institute into a world-recognized research establishment. In 1926 he went to the USA and in 1928 decided to divide his time between the Aachen Institute and the Guggenheim Aeronautical Laboratory of the Califonia Institute of Technology (GALCIT).

As the political climate in Germany deteriorated, he became the first director of GALCIT and professor emeritus at the California Institute of Technology. At the peak of his career, he became the leading figure in a number of international bodies, such as the International Union of Theoretical and Applied Mechanics, the Advisory Group for Aeronautical Research and Development, the International Council of Aeronautical Sciences, and the International Academy of Astronautics.

Kármán never married and his sister, Pipö, first with the help of their mother, Helen (born Konn), managed his households in Aachen and in Pasadena. He received many honorary degrees and medals including the US Medal for Merit 1946, the Franklin Gold Medal 1948, and the first National Medal for Science 1963. He died on 7 May 1963 at the Aachen home of his close friend Bärbel Talbot, while on holiday. His body was flown by the US Air Force to California to be buried beside those of his mother and sister.

Kármán's work concerning the flow of fluids round a cylinder is among his best-known. He discovered that the wake separates into two rows of vortices that alternate like street lights. This phenomenon is called the Kármán vortex street (or Kármán vortices) and it can build up destructive vibrations – the Tacoma Narrows suspension bridge was destroyed in 1940 by vortices that induced huge vibrations when the wind speed was exactly 67.6 kph/42 mph. The spiral flutes on factory chimneys are there to prevent such vortices forming.

In 1938 the US Army Air Corps asked the National Academy of Sciences to devise some means of helping heavily laden aircraft to take off. Kármán played a prominent part in the first work on jet assisted take-off (JATO) and in 1943, authored with H S Tsien and F J Molina, a memorandum that resulted in the first research and development programme on long-range rocket-propelled missiles. The programme called for the design and building of prototype test vehicles and basic research on propellants and construction materials, missile guidance systems, and missile aerodynamics.

In aerodynamics and flight mechanics he developed a method of generating aerofoil profiles and produced papers on flight at high subsonic, transonic, and supersonic speeds. He also worked on boundary layer and compressibility effects, stability and control at high speeds, propeller design, helicopters, and gliders.

Kármán was one of the world's great theoretical aerodynamicists. He laid the foundations for the rocket and aircraft programmes that placed the USA in the lead in these fields. Many of the establishments he brought together, such as the Jet Propulsion Laboratory at Pasadena, show every sign of carrying on his work for an indefinite time.

Karrer, Paul (1889–1971) Swiss organic chemist, famous for his work on vitamins and vegetable dyestuffs. He determined the structural formulae and carried out syntheses of various vitamins, in recognition of which achievement he shared the 1937 Nobel Prize for Chemistry with Norman ⇨Haworth.

Karrer was born in Moscow, of Swiss parents, on 21 April 1889, and when he was three years old his family moved to Switzerland. He studied in the University of Zürich under Alfred ◊Werner, gaining his doctorate in 1911. After a year at the Zürich Chemical Institute he went to Frankfurt to work with Paul ◊Ehrlich at the Georg Speyer Haus. In 1918 he returned to Zürich to be professor of chemistry and a year later succeeded Werner as director of the Institute. Karrer remained there until he retired in 1959. He died in Zürich on 18 June 1971.

Karrer's early work concerned vitamin A and its chief precursor, carotene. Wackenroder had first isolated carotene (from carrots) in 1831 and in 1907 Richard ◊Willstätter determined its molecular formula to be $C_{40}H_{56}$. Karrer worked out its correct constitutional formula in 1930 (although he was not to achieve a total synthesis until 1950). He showed in 1931 that vitamin A (molecular formula $C_{20}H_{30}O$) is related to the carotenoids, the substances that give the yellow, orange, or red colour to foodstuffs such as sweet potatoes, egg yolk, carrots, and tomatoes and to non-edible substances such as lobster shells and human skin. There are, in fact, two A vitamins, compounds known chemically as diterpenoids; vitamin A_1 influences growth in animals and its deficiency leads to night blindness and a hardening of the cornea. Karrer proved that there are several isomers of carotene, and that vitamin A_1 is equivalent to half a molecule of its precursor ß-carotene.

Karrer later confirmed Albert ◊Szent-Györgyi's constitution of vitamin C (ascorbic acid, so-called by Haworth who synthesized it and shared Karrer's Nobel prize). In 1935 he solved the structure of vitamin B_2 (riboflavin), the water-soluble thermostable vitamin in the B complex that is necessary for growth and health. It occurs in green vegetables, yeast, meat, and milk (from which it was first isolated and hence is also known as lactoflavin). Karrer made extensive studies of the chemistry of the flavins.

He also investigated vitamin E (tocopherol), the group of closely related compounds that act as antisterility factors. In 1938 he solved the structure of α-tocopherol, the most biologically active component, obtained from wheatgerm oil.

For their time, these syntheses were remarkable, being the most advanced so far undertaken. They led to a better understanding of vitamins in metabolism, and acted as a spur to the work of others. In 1930 Karrer published an organic chemistry textbook, *Lehrbuch der organischen Chemie,* which became the standard work for many years, being reprinted in many editions in several languages during the 1940s and 1950s.

Katz, Bernard (1911–) German-born British physiologist who is renowned for his research into the physiology of the nervous system. He has received many honours for his work, including a knighthood in 1969 and the 1970 Nobel Prize for Physiology or Medicine, jointly with Ulf von Euler (1905–1983) and Julius Axelrod (1912–).

Katz was born on 26 March 1911 in Leipzig and studied medicine at the university there. After graduation in 1934 he did postgraduate work at University College, London, from which he obtained his PhD in 1938 and his doctor of science degree in 1943. He was a Beit Memorial Research Fellow at the Sydney Hospital 1939–42. He then served in the Royal Australian Air Force until the end of World War II, after which he returned to England. He spent the rest of his academic career at University College, London – as assistant director of research at the Biophysics Research Unit and Henry Head Research Fellow 1946–50, reader in physiology 1950–51, and professor and head of biophysics from 1952 until his retirement in 1978.

During the 1950s, Katz found that minute amounts of acetylcholine – previously demonstrated to be a neurotransmitter by Henry ◊Dale and Otto Loewi (1873–1961) – were randomly released by nerve endings at the neuromuscular junction, giving rise to very small electrical potentials at the end plate; he also found that the size of the potential was always a multiple of a certain minimum value. These findings led him to suggest that acetylcholine was released in discrete 'packets' (analogous to quanta) of a few thousand molecules each, and that these packets were released relatively infrequently while a nerve was at rest but very rapidly when an impulse arrived at the neuromuscular junction. Electron microscopy later revealed small vesicles in the nerve endings and these are thought to be the containers of the packets of acetylcholine suggested by Katz.

Kay, John (1704–*c.* 1780) English inventor of improved textile machinery, the most important of which was the flying shuttle of 1733.

Kay was born in July 1704 at Walmersley, near Bury in Lancashire. Little is known of his early life, although he is thought to have been educated in France. It is thought that on finishing his schooling he was put in charge of a wool factory owned by his father in Colchester, Essex. By 1730 he was established in his home town of Bury as a reed-maker (for looms). In that year he was granted a patent for an 'engine' for twisting and carding mohair, and for twining and dressing thread. At about the same time he improved the reeds for looms by manufacturing 'darts' of thin polished metal, instead of cane, which were more durable and better suited to weaving finer fabrics.

In 1733 Kay patented his flying shuttle, which was probably the most important improvement that had been made to the loom. Up to that time the shuttle had been passed by hand from side to side through alternate warp threads. In weaving broadcloth two workers had to be employed to throw the shuttle from one end to the other. The weft was closed up after each 'pick' or throw of the shuttle by a 'layer' extending across the piece being woven. Kay added to the layer a grooved guide, called a 'race-board', in which the shuttle was rapidly thrown from side to side by means of a 'picker' or shuttle driver. One hand only was required, the other being employed in beating or closing up the weft. Kay's improvement allowed the shuttle to work with such speed that it became known as the flying shuttle. The amount of work a weaver could do was more than doubled, and the quality of the cloth was also improved. A powerful stimulus was thus given to improve spinning techniques, notably the inventions of James ◊Hargreaves.

The patent of 1733 also included a batting machine for removing dust from wool by beating it with sticks. Kay's next patent (granted in 1738) was for a windmill for working pumps and for an improved chain-pump, but neither of these proved to be of any practical importance. In this last patent Kay described himself as an engineer.

Kay's biographer, Woodcroft, states that he moved to Leeds in 1738. The new shuttle was widely adopted by the woollen manufacturers of Yorkshire but since they were unwilling to pay royalties an association called the Shuttle Club was formed to defray the costs of legal proceedings for infringement of the patent. Kay found himself involved in numerous law suits and, although he was successful in the courts, the expenses of prosecuting these claims nearly ruined him.

In 1745 Kay was again in Bury and in that year he obtained a patent (in conjuction with Joseph Stell of Keighley) for a small loom worked by mechanical power instead of manual labour. This attempt at a power loom does not seem to have been successful, probably because of his financial difficulties and the hostility of the workers.

In 1753 a mob broke into Kay's house at Bury and destroyed everything they could find; Kay himself barely escaped with his life. Not surprisingly he lost his enthusiasm for working in England under such trying conditions, and left for France where he remained for the next ten years, introducing the flying shuttle there with fair success.

In 1765 Kay invented a machine for making the card-cloth used in carding wool and cotton. The last years of his life are poorly documented. He is said to have been ruined by suits to protect his patents and is believed to have returned to France in 1774 and to have received a pension from the French court in 1778. After this nothing more is known of him.

Keeler, James Edward (1857–1900) US astrophysicist noted for his work on the rings of Saturn and on the abundance and structure of nebulae.

Keeler was born in La Salle, Illinois, on 10 September 1857. He did not attend school between the ages of 12 and 20, but during these years he developed a keen interest in astronomy. He made a variety of astronomical instruments and spent long hours studying the Solar System. A benefactor enabled him to study at Johns Hopkins University, where he earned his bachelor's degree in 1881. While a student he participated in an expedition to Colorado to study a solar eclipse. In 1881 he began his career as an assistant to Samuel Langley (1834–1906), director of the Allegheny Observatory, and he took part in that year's expedition to the Rocky Mountains to measure solar infrared radiation.

In 1883 Keeler went to Germany to study at the universities of Heidelberg and Berlin. A year later he returned to the Allegheny Observatory and in 1886 he went to Mount Witney, the future site of the Lick Observatory. He was appointed astronomer at the Lick Observatory upon its completion in 1888. Keeler became professor of astrophysics and director of the Allegheny Observatory in 1891, but returned to the Lick Observatory as director in 1898. He was elected fellow of the Royal Astronomical Society of London in 1898 and made a member of the National Academy of Sciences in 1900. He died suddenly in San Francisco on 12 August 1900.

Keeler's earliest work was a spectroscopic demonstration of the similarity between the Orion nebula and stars. In 1888, using the 90-cm/35-in refracting telescope at Lick Observatory and an improved spectroscopic grating, he demonstrated that nebulae resembled stars in their pattern of movement. He also studied the planet Mars, but was unable to confirm Giovanni ◊Schiaparelli's observation that the Martian surface was etched with a pattern of 'canals'. In 1895 Keeler made a spectroscopic study of Saturn and its rings, in order to examine the planet's period of rotation. He found that the rings did not rotate at a uniform rate, thus proving for the first time that they could not be solid and confirming James Clerk ◊Maxwell's theory that the rings consist of meteoritic particles.

After 1898 Keeler devoted himself to a study of all the nebulae that William ◊Herschel had catalogued a hundred years earlier. He succeeded in photographing half of them, and in the course of his work he discovered many thousands of new nebulae and showed their close relationship to stars. Keeler was not only a keen and successful observer of astronomical phenomena; he was also skilled at designing and constructing instruments. These included modifications to the Crossley reflecting telescope and a spectrograph in which spectral lines were recorded with the aid of a camera.

Kekulé von Stradonitz, Friedrich August (1829–1896) German organic chemist who founded structural organic chemistry and is best known for his 'ring' formula for benzene.

Kekulé was born in Darmstadt on 7 September 1829, a descendant of a Bohemian noble family from Stradonic, near Prague. At first he studied architecture at Giessen, but came under the influence of Justus von ◊Liebig and changed to chemistry. He left Giessen in 1851 and went to Paris, where he gained his doctorate a year later. After working in Switzerland for a while, he went to London in 1854 where he assisted John Stenhouse (1809–1890) and met many leading chemists of the day. When he returned to Germany in the following year he opened a small private laboratory at Heidelberg and became an unpaid lecturer at the university there under Johann von ◊Baeyer. In 1858 he became a professor at Ghent; in 1865 took over the chair at Bonn vacated by August ◊Hofmann, and remained there until he died. While at Ghent he got married, but his wife died in childbirth. He had three more children by his second marriage of 1876, the same year in which he suffered an attack of measles that left him in poor health for the rest of his life. He was raised to the nobility by William II of Prussia in 1895 and took the name Kekulé von Stradonitz. He died in Bonn on 13 July 1896.

In 1858 Kekulé published a paper in which, after giving reasons why carbon should be regarded as a four-valent element, he set out the essential features of his theory of the linking of atoms. (A similar paper was presented three months later to the French Academy by the Scottish chemist Archibald Couper, but went largely unnoticed.) He postulated that carbon atoms can combine with each other to form chains, and that radicals are not necessarily indestructible, being capable of persisting through one set of reactions only to disintegrate in others. He explained that compounds behave differently under different sets of conditions, although he continued to subscribe to the idea of 'types', which he thought were not inflexible but served to emphasize one or more of their properties. He pictured atoms grouping themselves in space and linking with each other to form compounds.

In 1865 Kekulé announced his theory of the structure of benzene, which he envisaged as a hexagonal ring of six interconnected carbon atoms. To make the structure compatible with carbon's valency of four, he postulated alternate single and double bonds in the same ring:

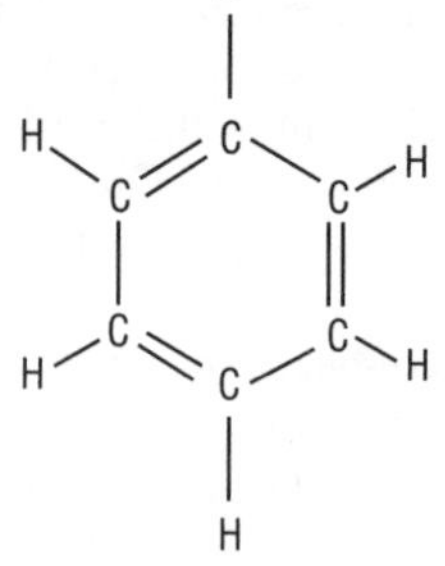

In a dissertation of 1867 outlining his views on molecular diagrams and models, he proposed the tetrahedral carbon atom (that the four valence bonds from a saturated carbon atom are directed towards the corners of a regular tetrahedron), which was to become the cornerstone of modern structural organic chemistry. Kekulé cautiously made it clear that he did not intend all the possible physical implications of his tetrahedral model to be accepted literally. It later became clear, however, that this model fulfilled strictly chemical necessities that he had not foreseen. In 1890 Baeyer said that Kekulé's molecular models were 'even cleverer than their inventor' and modern analytical methods have shown that they are true indicators of how the atoms are arranged in space.

There were two fundamental objections to Kekulé's original ring structure for benzene: it does not behave chemically like a double-bonded substance, and there are not as many di-substituted isomers as the formula would predict. Kekulé acknowledged these shortcomings and in 1872 introduced the idea of oscillation or resonance between two isomeric structures:

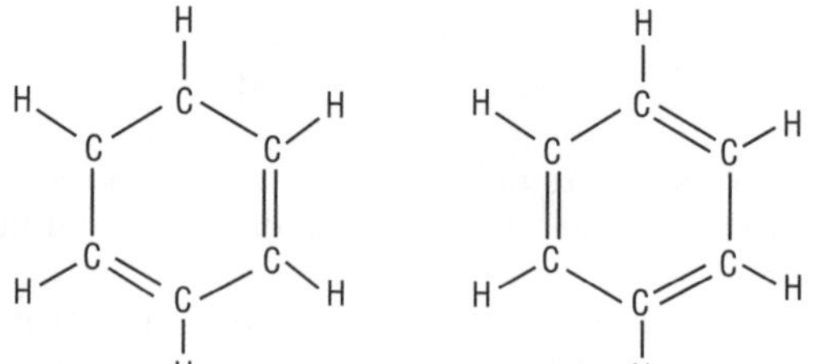

His assistant W Körner (1839–1925) developed in great detail the implications of these structures in the chemistry of benzene and its compounds. The modern view, worked out by Linus ◊Pauling in 1933 by applying quantum theory to Kekulé's resonant structures, postulates a hybrid (mesomeric) state for the benzene molecule with a carbon–carbon bond length in the ring intermediate between the normal single- and double-bond values.

Kelly, William (1811–1888) US metallurgist, arguably the original inventor of the 'air-boiling process' for making steel, known universally as the Bessemer process. Kelly remained in relative obscurity in spite of his invention, whereas his rival (Henry ◊Bessemer) received $10 million in the USA and was knighted.

Kelly was born in Pittsburgh, Pennsylvania, on 21 August 1811, the son of a wealthy landowner said to have built the first two brick houses in Pittsburgh. Kelly was educated at local schools. He was inventive and fond of metallurgy, but at the age of 35 found himself working for a dry-goods business in Philadelphia as a junior member of the firm McShane and Kelly. He was sent out collecting debts in Nashville, Tennessee, and it was there that he met his wife and her father, who was a wealthy tobacco merchant. Kelly settled there and with his brother bought nearby iron-ore lands and a furnace known as the Cobb furnace in Eddyville.

Kelly patented his steelmaking process in 1857, and in 1871 had his patent renewed for seven years. Steel under the Kelly patent was first blown commercially in the autumn of 1864 at the Wyandotte Iron Works near Detroit.

Kelly then moved to Johnstown, Pennsylvania, where he was acclaimed as a genius, and after five years he moved to Louisville, where he founded an axe-making business. He retired from active business at the age of 70 and died in Louisville on 11 February 1888. His axe-manufacturing business was carried on by his son in Charlestown, Western Virginia. On 5 October the American Society for Steel Treating erected a bronze tablet to Kelly's memory at the site of the Wyandotte Iron Works.

In Eddyville, Kelly developed the Suwanee Iron Works and the Union Forge. He manufactured sugar kettles, which were much in demand by farmers in the area. To do this he used wrought iron made from pig iron, which involved the burning out of the excess carbon. This needed a lot of charcoal, and the local supply soon ran low.

One day he noticed that although the air-blast in his 'finery fire' furnace was blowing on molten iron with no charcoal covering, the iron was still becoming white hot. When he experimented further, he found that contrary to all iron-makers' beliefs, molten iron containing sufficient carbon became much hotter when air was blown on to it. Some 3–5% of carbon contained in molten cast iron can be burnt out by the air-blast. Here the carbon itself is acting as a fuel, and makes the molten mass much hotter.

Kelly became so obsessed with his discovery that his wife called the doctor, thinking he was ill. Unfortunately, his customers could not be convinced that the steel made by this new, cheaper process was as good and so he had to revert to using charcoal. Meanwhile, with two other ironmakers, he started building a converter secretly. His first attempt was a 1.2-m/4-ft-high brick kettle. Air was blown through holes in the bottom into and through molten pig iron. The method was only partly successful. Between 1851 and 1856 he secretly built seven of these converters. In 1856, he heard that Henry Bessemer of England had been granted a US patent on the same process. Kelly immediately applied for a patent and managed to convince the Patent Office of his priority. On 23 June 1857 he was granted a patent and declared to be the original inventor. Four years later, his patent was renewed for a further seven years, whereas Bessemer's patent renewal was refused.

The financial panic of 1857 had bankrupted him, however, and he sold his patent to his father for $1,000. His father, thinking his son to be an incompetent businessman, bequeathed the patent to his daughters and would not return it to Kelly. Kelly therefore moved away to Johnstown, where a Daniel J Morrell of the Cambria Iron Works listened to his plight. Kelly built his eighth converter there (the first of a tilting type), and it is still preserved in the office of the Cambria Works, now part of the Bethlehem Steel company. His second attempt at Johnsville was a success and it produced soft steel cheaply for the first time and in large quantities. It was used for rails, bars and structural shapes and marked a milestone in the Steel Age that was just beginning.

Kelly's patent eventually came under the control of Kelly Pneumatic Process Company, with which Kelly had nothing to do. This company later merged with the Company of Alex L Holley, which was Bessemer's sole US licensee. Bessemer's name came into exclusive use for the Kelly–Bessemer steelmaking process, and Kelly remained in relative obscurity.

Kelvin, William Thomson (1824–1907) 1st Baron Kelvin. Irish physicist who first proposed the use of the absolute scale of temperature, in which the degree of temperature is now called the kelvin in his honour. Thomson also made other substantial contributions to thermodynamics and the theory of electricity and magnetism, and he was largely responsible for the first successful transatlantic telegraph cable.

Kelvin was born William Thomson in Belfast on 26 June 1824. His father was professor of mathematics at Belfast University and both he and his older brother James ◊Thomson, who also became a prominent physicist, were educated at home by their father. In

Kelvin Irish physicist William Thomson Kelvin, 1st Baron Kelvin, who introduced the absolute scale of temperature. He entered Glasgow University at the age of 10 years, and in 1846 at the age of 22 was made professor of natural philosophy in Glasgow, a position he held for the next 53 years. *Mary Evans Picture Library*

1832 Thomson's father took up the post of professor of mathematics at Glasgow University, and Thomson himself entered the university two years later at the age of ten to study natural philosophy (science). In 1841, Thomson went on to Cambridge University, graduating in 1845. He then travelled to Paris to work with Henri ⟡Regnault and in 1846 took up the position of professor of natural philosophy at Glasgow, where he created the first physics laboratory in a British university. Among his many honours were a knighthood in 1866, the Royal Society's Copley Medal in 1883, the presidency of the Royal Society 1890–94, and a peerage in 1892, when he took the title Baron Kelvin of Largs. Kelvin retired from his chair at Glasgow in 1899 and died in Largs, Ayrshire, on 17 December 1907.

Thomson's early work, begun in 1842 while he was still at Cambridge, was a comparison of the distribution of electrostatic force in a region with the distribution of heat through a solid. He found that they are mathematically equivalent, leading him in 1847 to conclude that electrical and magnetic fields are distributed in a manner analogous to the transfer of energy through an elastic solid. James Clerk ⟡Maxwell later developed this idea into a comprehensive explanation of the electromagnetic field. From 1849–59, Kelvin also developed the discoveries and theories of paramagnetism and diamagnetism made by Michael ⟡Faraday into a full theory of magnetism, developing the terms magnetic permeability and susceptibility, and arriving at an expression for the total energy of a system of magnets. In electricity, Kelvin obtained an expression for the energy possessed by a circuit carrying a current and in 1853 developed a theory of oscillating circuits that was experimentally verified in 1857 and was later used in the production of radio waves.

In Paris in 1845, Kelvin was introduced to the classic work of Sadi ⟡Carnot on the motive power of heat. From a consideration of this theory, which explains that the amount of work produced by an ideal engine is governed only by the temperature at which it operates, Kelvin developed the idea of an absolute temperature scale in which the temperature represents the total energy in a body. He proposed such a scale in 1848 and set absolute zero at –273 °C/–459 °F, showing that Carnot's theory followed if absolute temperatures were used. However, Kelvin could not accept the idea then still prevalent and accepted by Carnot that heat is a fluid, preferring to see heat as a form of motion. This followed Kelvin's championing of the work of James ⟡Joule, whom Kelvin had met in 1847, on the determination of the mechanical equivalent of heat. In 1851 Kelvin announced that Carnot's theory and the mechanical theory of heat were compatible provided that it was accepted that heat cannot pass spontaneously from a colder to a hotter body. This is now known as the second law of thermodynamics, which had been advanced independently in 1850 by Rudolf ⟡Clausius. In 1852 Kelvin also produced the idea that mechanical energy tends to dissipate as heat, which Clausius later developed into the concept of entropy.

Kelvin and Joule collaborated for several years following their meeting, Joule's experimental prowess matching Kelvin's theoretical ability. In 1852 they discovered the Joule–Thomson effect, which causes gases to undergo a fall in temperature as they expand through a nozzle. The effect is caused by the work done as the gas molecules move apart, and it proved to be of great importance in the liquefaction of gases.

Kelvin was also interested in the debate then taking place about the age of the Earth. Hermann von ⟡Helmholtz in 1854 gave a value of 25 million years for the Earth's lifetime, assuming that the Sun gained its energy by gravitational contraction. Kelvin came to a similar conclusion in 1862, basing his estimate on the rate of cooling that would have occurred from the time the Earth formed, and reaching an age of 20 million to 400 million years with 100 million years as the most likely figure. Furthermore, the cooling would have produced volcanic upheavals that would have limited the time available for the evolution of life. Although both estimates were far too low, neither scientists knowing of the nuclear processes that fuel the Sun and the radioactivity that warms the Earth, their figures were taken seriously and helped to bring about theories of mutation to explain evolution.

Kelvin's knowledge of electrical theory was applied with great practical value to the laying of the first transatlantic telegraph cable. Kelvin pointed out that a fast rate of signalling could only be achieved by using low voltages, and that these would require very sensitive detection equipment such as the mirror galvanometer that he had invented. The first cable laid in 1857 broke and high voltages were used in the second cable laid a year later as Kelvin's predictions were not believed. The cable did not work, but a third cable laid in 1866 using Kelvin's ideas was successful, and it was for this achievement that he received a knighthood.

Kelvin was also very concerned with the accurate measurement of electricity, and developed an absolute electrometer in 1870. He was instrumental in achieving the international adoption of many of our present-day electrical units in 1881.

Kelvin was one of the greatest physicists of the 19th century. His pioneering work on heat consolidated thermodynamics and his understanding of electricity and magnetism paved the way for the explanation of the electromagnetic field later achieved by Maxwell.

Further Reading

Burchfield, Joe D, *Lord Kelvin and the Age of the Earth,* University of Chicago Press, 1990.

Kargon, Robert and Achinstein, Peter, *Kelvin's Baltimore Lectures and Modern Theoretical Physics: Historical and Philosophical Perspectives,* Studies from the Johns Hopkins Center for the History and Philosophy of Science, MIT Press, 1987.

Kelvin, William Thomson and Stokes, George Gabriel; Wilson, David B (ed), *The Correspondence Between Sir George Gabriel Stokes and Sir William Thomson, Baron Kelvin of Largs,* Cambridge University Press, 1990, 2 vols.

Smith, Crosbie W and Wise, M Norton, *Energy and Empire: A Biographical Study of Lord Kelvin,* Cambridge University Press, 1989.

Tunbridge, Paul, *Lord Kelvin: His Influence on Electrical Measurements and Units,* IEE History of Technology Series, Peter Peregrinus on behalf of the Institution of Electrical Engineers, 1992, v 18.

Kendall, Edward Calvin (1886–1972) US endocrinologist who shared the Nobel Prize for Physiology or Medicine in 1950. His work on understanding the physiological chemistry of the secretions of the thyroid and adrenal glands led him to isolate some of their active constitutents. Of these cortisone from the adrenal gland has been found to offer symptomatic relief for sufferers of rheumatoid arthritis.

Kendall was born on 8 March 1886 in South Norwalk, Connecticut and received his university education at Columbia, gaining bachelor's (1908), master's (1909), and doctoral (1910) degrees in chemistry. For a few months he worked for the pharmaceutical firm Parke Davis and Company, but disliked the lack of research opportunity and regimented atmosphere there, and returned to New York to work in a hospital laboratory, before moving to the Mayo Clinic in Rochester, Minnesota in 1914. In 1921 he became professor of physiological chemistry at the Mayo, and retired in 1951. He was then appointed to a visiting professorship at Princeton University and continued to be associated with laboratory work until his death on 4 May 1972 at Princeton.

Kendall's earliest research in endocrinology was undertaken at St Luke's Hospital, New York, where he began to investigate the properties of thyroid gland secretions. In 1913 he isolated a very potent extract and after his move to Rochester he succeeded in crystallizing the substance, which he called thyroxine. For several years he continued to study the chemistry of thyroxine, although it was C R Harington (1897–1972) from London who was succesful in first determining its chemical structure and its synthetic pathway. Kendall also started investigating the physiological chemistry of the adrenal glands, directing an interdisciplinary research team that involved physicians, surgeons, and scientists. By the beginning of the 1930s it was becoming accepted that the adrenal gland produced more than one hormone, and research teams around the world, especially in Germany, turned their attentions towards adrenal chemistry. It was not until the late 1940s that Kendall and his team succeeded in isolating and synthesising cortisone. Unfortunately it seemed to have only limited therapeutic use, and did not attract much attention at the time. However Kendall had started to work with Phillip Hench (1896–1965) a physician interested in arthritis who was familiar with the experience that in some situations, such as during pregnancy, patients with arthritis improved. The two men discussed whether cortisone, which Kendall's work had shown to have important metabolic effects, was involved in these temporary improvements. They discovered by giving a severely incapicitated patient cortisone that it was an effective treatment for rheumatoid arthritis, for which they shared the Nobel prize.

Kendrew, John Cowdery (1917–1997) English biochemist who shared the 1962 Nobel Prize for Chemistry with Max ◊Perutz for his determination of the structure of the muscle protein myoglobin.

Kendrew was born in Oxford on 24 March 1917. He won a scholarship to Trinity College, Cambridge, in

1936 and graduated from there in 1939, just before the outbreak of World War II. During the war he worked for the Ministry of Aircraft Production, returning to Cambridge in 1945. A year later he was appointed departmental chairman of the Medical Council's Laboratory for Molecular Biology, Cambridge, where he remained until 1975. In that year he became director general of the European Molecular Biology Laboratory at Heidelburg. He was knighted in 1974.

In the late 1940s at Cambridge Kendrew began working with Perutz, whose other molecular biologists at that time included Francis ◊Crick. Their research centred on an investigation of the fine structure of various protein molecules; Kendrew was assigned the task of studying myoglobin, the globular protein resembling haemoglobin that occurs in muscle fibres (where it stores oxygen). He used X-ray diffraction techniques to elucidate the amino-acid sequence in the peptide chains that form the myoglobin molecule (similar to the work of Maurice Wilkins on DNA). Hundreds of X-ray diffraction photographs of the crystallized protein were analysed, using electronic computers that were becoming available in the 1950s. By 1960 Kendrew had determined the spatial arrangement of all 1,200 atoms in the molecule, showing it to be a folded helical chain of amino acids with an amino ($-NH_2$) group at one end and a carboxylic ($-COOH$) group at the other. It involves an iron-containing haem group, which allows the molecule to absorb oxygen. In the same year, Perutz determined the structure of haemoglobin.

Kenyon, Joseph (1885–1961) English organic chemist, best known for his studies of optical activity, particularly of secondary alcohols.

Kenyon was born in Blackburn, Lancashire, the eldest of seven children. He was educated locally and worked as a laboratory assistant at Blackburn Technical College 1900–03. In 1903 he won a scholarship that enabled him to study for a degree, and he graduated two years later. In 1904 he became personal assistant to R H Pickard, an industrial chemistry consultant and technical college instructor; in 1906 he became an assistant lecturer and demonstrator at Blackburn Technical College, and in 1907 a full lecturer, commencing research under Pickard. He submitted his doctoral thesis to London University in 1914 and in the following year was appointed research chemist to the Medical Research Council. From 1915 to 1916, during World War I, he worked at Leeds University and then until 1920 with William Perkin Jr (1860–1929) at the British Dyestuffs Corporation, Oxford. Finally in 1920 Kenyon became head of the chemistry department at Battersea Polytechnic, London, a year after Pickard had become its principal; he remained there until he retired in 1950. In 1951 and 1952 he was visiting professor at the universities of Alexandria and Kansas. He died in Petersham on 11 November 1961.

Kenyon published his research on secondary alcohols in 1911 while still working as an undergraduate with Pickard. The problem was to resolve the optically active stereoisomers of secondary octyl alcohol (octan-2-ol). Pickard and Kenyon converted it to secondary octyl hydrogen phthalate by heating it with phthalic anhydide, and showed that the phthalate could easily be resolved using the alkaloids brucine and cinchonidine. They went on to obtain an optically pure series of secondary alcohols and were able to relate rotatory power to chemical constitution. The technique was later used to distinguish between inter- and intramolecular rearrangements.

While at Leeds 1915–16 Kenyon worked on antidotes for gas gangrene, research that led to the development of chloramine-T for this and other purposes. With William Perkin 1916–20, he studied photographic developers and dyes, although most of their results were published only in confidential reports.

At Battersea Polytechnic Kenyon put forward the 'obstacle' theory for the cause of optical activity in certain substituted diphenic acids. He synthesized and attempted to resolve some selenoxides, confirming differences between these and sulphoxides, and investigated the geometric and optical isomerism of the methylcyclohexanols. He pointed out that when the toluene-*para*-sulphonic acids (methylbenzene-4-sulphonic acids) are prepared by the action of the sulphonic chloride on the corresponding alcohol, the four bonds of the asymmetric carbon atom remain undisturbed – there is no change in configuration. But when the toluene-4-sulphonic esters are converted to the carboxylic esters (by heating them with the alcoholic solutions of the alkali salts of the carboxylic acids), there is an almost 100% inversion of configuration.

He also discovered that the hydroxyl compounds can be converted to the sulphinic esters without inversion (by treatment with toluene-4-sulphinyl chloride) but that treatment of the esters with hypochlorous acid reconverts them into the enantiomers (mirror-images) of the original hydroxyl compounds.

Perhaps Kenyon's most important work in stereochemistry was to prove that in the Beckman, Curtius, Hofmann, Lossen and Schmidt reactions – in which a group migrates from one part of the molecule to another – in no case is the migrating fragment ever kinetically free.

Kepler, Johannes (1571–1630) German astronomer who combined great mathematical skills with patience

Kepler German astronomer Johannes Kepler who set out the laws of planetary motion early in the 17th century. One of the first followers of Copernican thought, Kepler wrote many books on optics, astronomy, and mathematics. *Mary Evans Picture Library*

and an almost mystical sense of universal harmony. He is particularly remembered for what are now known as Kepler's laws of motion. These had a profound influence on Isaac ⇨Newton and hence on all modern science. Kepler was also absorbed with the forces that govern the whole universe and he was one of the first and most powerful advocates of Copernican heliocentric (Sun-centred) cosmology.

Kepler was born on 27 December 1571 in Weil der Statt near Stuttgart, Germany. He was not a healthy child and since it was apparently thought that he was capable only of a career in the ministry, he was sent for religious training in Leonberg, Adelberg, and Maulbronn. One event that impressed him deeply during his early years was the viewing of the 'great' comet of 1577 and his interest in astronomy probably dates from that time. He passed his baccalaureate at the University of Tübingen in 1588 and then returned to Maulbronn for a year. From 1589 to 1591 he studied philosophy, mathematics, and astronomy under Michael ⇨Mästlin at the University of Tübingen; yet although he showed great aptitude and promise and obtained his MA in 1591, he then embarked on a three-year programme of theological training. This was interrupted in the last year when he was nominated for a teaching post in mathematics and astronomy in Graz. It was during this teaching period that he abandoned his plans for a career in the ministry and concentrated on astronomy. He wrote his first major paper while at Graz and attracted the attention of other notable astronomers of the time, particularly Tycho ⇨Brahe and ⇨Galileo.

Kepler was a Lutheran and so was frequently caught up in the religious troubles of his age. In 1598 a purge forced him to leave Graz. He travelled to Prague and spent a year there before he returned to Graz; he was expelled again and arrived back in Prague, where he became Tycho Brahe's assistant in 1600. Brahe died a few months later, in 1601, and Kepler succeeded him as Imperial Mathematician to Emperor Rudolph II. On his deathbed Brahe requested that Kepler complete the Rudolphine Astronomical Tables, a task that Kepler finished in 1627.

Kepler lived in Prague until 1612 and produced what was perhaps his best work during those years. He was given a telescope in 1610 by Elector Ernest of Cologne and studied optics, telescope design, and astronomy. In 1611 Rudolph II was deposed, but Kepler was retained as Imperial Mathematician. In 1612, upon Rudolph II's death, Kepler became district mathematician for the states of Upper Austria and moved to Linz. But personal problems plagued him for the next ten years – the arrest and trial of his mother, who was accused of witchcraft in 1615 but exonerated in 1621, being particularly distressing.

It was in Linz that Kepler published three of his major works. In 1628 he became the private mathematician to Wallenstein, Duke of Friedland, partly because of the duke's promise to pay the debt owed Kepler by the deposed Rudolph II. In 1630 religious persecution forced Kepler to move once again. He fell sick with an acute fever, and died on the way to Regensburg, Bavaria, on 15 November 1630.

Kepler's work in astronomy falls into three main periods of activity, at Graz, Prague, and Linz. In Graz Kepler did some work with Mästlin on optics and planetary orbits, but he devoted most of his energy to teaching. He also produced a calendar of predictions for the year 1595, which proved so uncanny in its accuracy that he gained a degree of local fame. Kepler found the production of astrological calendars a useful way of supplementing his income in later years, but he had little respect for the art. More importantly, in 1596 he published his *Mysterium cosmographicum,* in which he demonstrated that the five Platonic solids, the only five regular polyhedrons, could be fitted alternately inside a series of spheres to form a 'nest' that described quite accurately (within 5%) the distances of the planets from the Sun. Kepler regarded this discovery as a divine inspiration that revealed the secret of the universe. It was written in accordance with Copernican theories and it bought Kepler to the attention of all European astronomers.

Before Kepler arrived in Prague and was bequeathed all Brahe's data on planetary motion, he had already made a bet that, given Brahe's unfinished tables, he could find an accurate planetary orbit within a week. It took rather longer, however. It was five years before Kepler obtained his first planetary orbit, that of Mars. In 1604 his attention was diverted from the planets by his observation of the appearance of a new star, 'Kepler's nova', to which he attached great astrological significance.

In 1609 Kepler's first two laws of planetary motion were published in *Astronomia nova,* which is a long text and as unreadable as it is important. The first law states that planets travel in elliptical rather than circular or epicyclic orbits and that the Sun occupies one of the two foci of the ellipses. What is now known as the second law, but was in fact discovered first, states that the line joining the Sun and a planet traverses equal areas of space in equal periods of time, so that the planets move more quickly when they are nearer the Sun. This established the Sun as the main force governing the orbits of the planets. Kepler also showed that the orbital velocity of a planet is inversely proportional to the distance between the planet and the Sun. He suggested that the Sun itself rotates, a theory that was confirmed by using Galileo's observations of sunspots, and he postulated that this established some sort of 'magnetic' interaction between the planets and the Sun, driving them in orbit. This idea, although incorrect, was an important precursor of Newton's gravitational theory. The *Astronomia nova* had virtually no impact at all at the time, and so Kepler turned his attention to optics and telescope design. He published his second book on optics, the *Dioptrice,* in 1611. That year was a difficult one, because Kepler's wife and sons died. Then, in 1612, the Lutherans were thrown out of Prague so Kepler had to move on again to Linz.

Ubi materia, ibi geometria. *Where there is matter, there is geometry.*

JOHANNES KEPLER ATTRIBUTED REMARK

In Linz Kepler produced two more major works. The first of these was *De harmonices mundi,* which was almost a mystical text. The book was divided into five chapters and buried in the last was Kepler's third law.

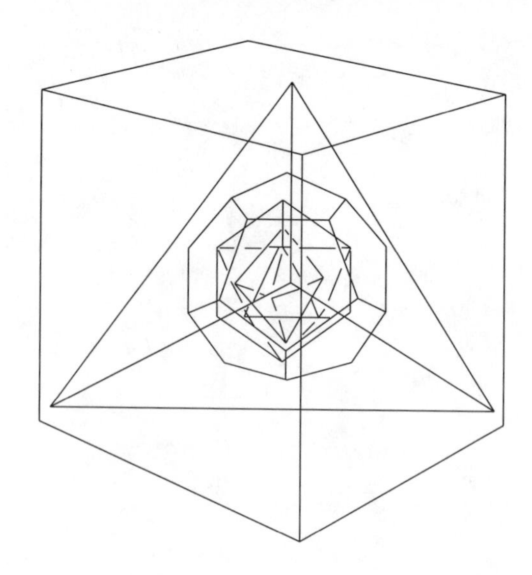

Kepler Kepler adopted Plato's idea of five nesting regular solids (from the centre; octahedron, icosahedron, dodecahedron, tetrahedron, and cube) and suggested that the planetary orbits fitted between them.

In this law he describes in precise mathematical language the link between the distances of the planets from the Sun and their velocities – a feat that afforded

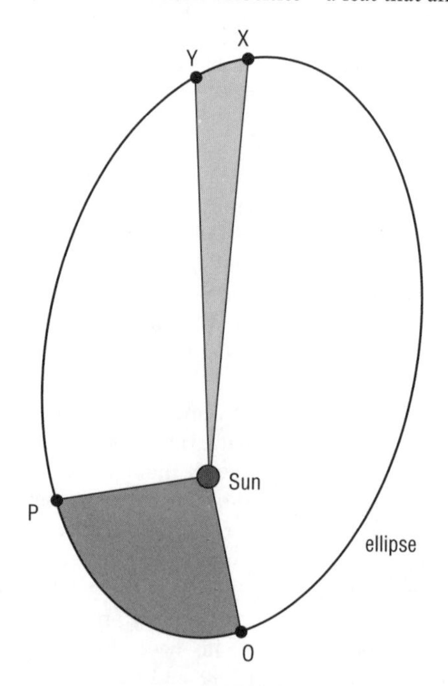

Kepler Kepler's second law states that the orange-shaded area equals the green-shaded area if the planet moves from P to O in the same time that it moves from X to Y. The law says, in effect, that a planet moves fastest when it is closest to the Sun.

him extraordinary pleasure and confirmed his belief in the harmony of the universe. The second major work to be published during his stay in Linz, the *Epitome*, intended as an introduction to Copernican astronomy, was in fact a very effective summary of Kepler's life's work in theoretical astronomy. It was a long treatise of seven books, published over a period of four years, and it had more impact than any other astronomical text of the mid-17th century.

Soon after its publication, Kepler's Lutheran background caused yet another expulsion and this time he went to Ulm, where he finally completed the Rudolphine Tables. They appeared in 1627 and brought Kepler much popular acclaim. These were the first modern astronomical tables, a vast improvement on previous attempts of this kind, and they enabled astronomers to calculate the positions of the planets at any time in the past, present, or future. The publication also included other vital information, such as a map of the world, a catalogue of stars, and the latest aid to computation, logarithms.

Kepler wrote the first science fiction story, *Solemnium*, which described a man who travelled to the Moon. It was published in 1631, a year after his death, although it had been written 20 years earlier. Kepler was a remarkable man and a brilliant scientist. He kept a steady eye on what he saw as his true vocation as a 'speculative physicist and cosmologist' and, despite living in times of political unrest and religious turmoil, was never swayed by religious bigotry or political pressures. His new astronomy provided the basis on which Newton and others were to build, and to this day his three laws of motion are considered to be the basis of our understanding of the Solar System. His work and his strong support of Copernican cosmology mark a fundamental divide between two eras – that of the Ptolemaic Earth-centred view of the universe that had been accepted for the previous 15 centuries, and the new age given birth by the Copernican heliocentric, or Sun-centred, view of the Solar System.

Further Reading

Casper, Max; Hellman, Clarisse Doris; and Gingerich, Owen, *Kepler*, Dover Publications, 1993.

Field, J V, *Kepler's Geometrical Cosmology*, Athlone Press, 1987.

Hallyn, Fernand and Leslie, Donald M, *The Poetic Structure of the World: Copernicus and Kepler*, Zone, 1990.

Kozhamthadam, Job, *The Discovery of Kepler's Laws: The Interaction of Science, Philosophy, and Religion*, University of Notre Dame Press, 1994.

Stephenson, Bruce, *The Music of the Heavens: Kepler's Harmonic Astronomy*, Princeton University Press, 1994.

Stephenson, Bruce, *Kepler's Physical Astronomy*, Studies in the History of Mathematics and Physical Sciences series, Springer-Verlag, 1987, v 13.

Kerr, John (1824–1907) Scottish physicist who discovered the Kerr effect, which produces double refraction in certain media on the application of an electric field.

Kerr was born in Ardrossan, Ayrshire, on 17 December 1824 and received his early education in Skye. He entered Glasgow University in 1941, gaining an MA in physical science in 1849. At Glasgow, Kerr became one of the first research students of Lord ◊Kelvin, and received special prizes for work in magnetism and electricity. He next became a divinity student but, although he completed his course, he did not subsequently pursue a career in the church. Kerr instead took up the post of lecturer in mathematics at the Free Church Training College for Teachers in Glasgow in 1857. He stayed in this post for 44 years, even though research facilities were virtually non-existent. Nevertheless Kerr set up a modest laboratory and although his publications were small in number, his work was of the highest quality. Kelvin must have had Kerr in mind when in a speech many years later he remarked that theology students made some of the best researchers and became all the better clergymen for their scientific experiences. Kerr's election to the Royal Society in 1890 and the award of its Royal Medal in 1898 were fitting recognition from a wider community of a life of modest excellence. He died in Glasgow on 18 August 1907.

Kerr's first important discovery was in 1875 when he demonstrated that birefringence or double refraction occurs in glass and other insulators when subjected to an intense electric field. A 5-cm/2-in block of glass had collinear holes drilled from each end until they were separated by 6 mm/0.25 in. A beam of polarized light was passed through this narrow band and at the same time an electric field was applied via electrodes inserted into the holes. Nicol prisms were arranged on either side to give extinction of the beam when the glass was free from strain. When the coil producing the field was activated, it was observed that the polarization became elliptical. The effect was strongest when the plane of polarization was 45 ° to the field, and zero when perpendicular or parallel. Kerr extended the work to other materials including amber and constructed, with great delicacy and manipulative skill, cells in which he could study liquids such as carbon disulphide and paraffin oil. He showed that the extent of the effect, which is more precisely called the electro-optical Kerr effect, is proportional to the square of the field strength.

In 1876 Kerr caused a considerable stir at the Glasgow meeting of the British Association when he demonstrated another remarkable phenomenon of polarized light that we know as the magneto-optical Kerr effect. In this a beam of plane-polarized light was reflected from the polished pole of an electromagnet. When the magnet was switched on the beam became elliptically polarized, the major axis being rotated from the direction of the original plane. The effect depended on the position of the reflecting surface with respect to the direction of magnetization and to the plane of incidence of the light. When both of these were perpendicular, the plane of polarization was turned through a small angle. The mathematical analysis of the effect and its relation to the electromagnetic theory of light was worked out some years later by George ◊FitzGerald.

The electro-optical Kerr effect is now applied in the Kerr cell, which is an electrically activated optical shutter having no moving parts. The cell contains a tube of liquid and crossed polarizers so that it normally transmits no light. Switching on the cell causes an electric field to produce double refraction in the liquid, allowing light to pass through the polarizers. The effect occurs because the field orientates the molecules of liquid to give double refraction, in which incident light is split into two beams that are polarized at right angles. Very fast shutter speeds are obtainable and Kerr cells have been used to make precise determinations of the speed of light.

Kettlewell, Henry Bernard David (1907–1979) English geneticist and lepidopterist who carried out important research into the influence of industrial melanism on natural selection in moths.

Kettlewell was born in Howden, Yorkshire, on 24 February 1907 and was educated at Charterhouse School, at Godalming in Surrey, and in Paris. He studied medicine at Gonville and Caius College, Cambridge, and at St Bartholomew's Hospital, London, graduating in 1933. After graduation he held several appointments in various London hospitals, including St Bartholomew's, and was an anaesthetist at St Luke's Hospital in Guildford, Surrey. He served as an anaesthetist during World War II. From 1949 to 1952 he investigated methods of locust control at Cape Town University, South Africa, also going on expeditions to the Kalahari Desert, the Belgian Congo, Mozambique and the Knysna Forest. After his return to England, he was awarded in 1952 a Nuffield Research Fellowship in the genetics unit of the zoology department at Oxford University. In the following year he was appointed senior research officer in genetics at Oxford, a post he held until he retired in 1974, when he became an emeritus fellow of Wolfson College, Oxford. He died in Oxford on 11 May 1979.

Kettlewell's best-known research involved the influence of industrial melanism on the survival of the peppered moth (*Biston betularia*). Until 1845, all known specimens of this moth were light-coloured, but in that year one dark specimen was found in Manchester, then an expanding industrial centre. The proportion of dark-coloured peppered moths increased rapidly, until by 1895 they comprised about 99% of Manchester's entire peppered moth population. This change from light to dark moths – a phenomenon called industrial melanism – corresponds with the increase in industrial activity (and therefore pollution, especially of the atmosphere) since the Industrial Revolution, and today only a few populations of the original light-coloured variant exist in England, being found in unindustrialized rural areas. Kettlewell performed several experiments under natural conditions to demonstrate the significance of colour in protecting the peppered moths from birds, their only predators. He released a known number of light- and dark-coloured moths – specially marked so that they could be identified – in an industrial area of Birmingham (where 90% of the indigenous peppered moths are dark-coloured) and in an unpolluted rural area of Dorset (where there are no dark-coloured moths). After an interval he collected the surviving moths and found that, in Birmingham, birds had eaten a high proportion of the light-coloured moths but a much lower proportion of the dark-coloured ones, whereas in Dorset the reverse had occurred. Thus Kettlewell demonstrated the efficiency of natural selection as an evolutionary force: the light-coloured moths are more conspicuous than the dark-coloured ones in industrial areas – where the vegetation is darkened by pollution – and are therefore easier prey for birds, but are less conspicuous in unpolluted rural areas – where the vegetation is lighter in colour – and therefore survive predation better.

In addition to his research into coloration and natural selection, Kettlewell cofounded the Rothschild–Cockayne–Kettlewell (RCK) Collection of British Lepidoptera, which is now called the National Collection (RCK) and is housed in the Natural History Museum in London.

Khorana, Har Gobind (1922–) Indian-born US chemist who has worked extensively on the chemistry of nucleic acids, for which he shared the Nobel Prize for Physiology or Medicine in 1968.

Khorana was born on 9 January 1922 in Raipur in the Punjab, now in Pakistan. His family was very poor, but believed in educating their children, and Khorana was encouraged to study, graduating from Punjab University with bachelor's degree in chemistry in 1943 and a master's degree in 1945. He was awarded a government scholarship to work for his doctorate at Liverpool University, and then moved to Switzerland for postdoctoral work in Zürich, before returning to the UK in 1950 to work in Cambridge where he came under the influence of the future Nobel laureate Alexander ◊Todd and became interested in the chemistry of nucleic acids. In 1952 he accepted a position in the division of organic chemistry at the University of British Columbia in Vancouver, Canada, and in 1960 moved to the University of Wisconsin, USA, and was professor of biochemistry there 1962–70. Since then he has been Alfred P Sloan Professor at the Massachusetts Institute of Technology. He is a member of many national academies, including the National Academy of Sciences and the Royal Society of London, and is the receipient of numerous honorary degrees and international awards.

Khorana's work has been in unravelling the chemistry of the genetic code and in establishing the steps by which proteins are synthesized. It was known that four nucleotides are involved in determining the genetic code, and Khorana systematically synthesized each possible combination of the genetic signals from these four nucleotides. He showed that a pattern of three nucleotides, called a triplet, specify a particular amino acid, which are the building blocks of proteins. He further discovered that some of the triplets provided 'punctuation' marks in the code, marking the beginning and end points of protein synthesis. For this major discovery Khorana was awarded the Nobel prize.

In 1970 he achieved the synthesis of the first artificial gene, and in more recent years has synthesized the gene for bovine rhodopsin, the pigment in the retina that is responsible for converting light energy into electrical energy. His work provides much of the basic knowledge that is important for gene therapy and biotechnology.

Kimura, Motoo (1924–) Japanese biologist who, as a result of his work on population genetics and molecular evolution, has developed a theory of neutral evolution that opposes the conventional neo-Darwinistic theory of evolution by natural selection. He has received many honours for his work, including Japan's highest cultural award, the Order of Culture, and the International Prize for Biology in 1988.

Kimura was born on 13 November 1924 in Okazaki. He studied botany at Kyoto University, from which he gained his master of science degree in 1947, then worked as an assistant there 1947–49. He spent most of his subsequent career at the National Institute of Genetics in Mishima – as a research member 1949–57, as laboratory head of the department of population genetics 1957–64, and as overall head of the same department from 1964. In 1953, however, he went to the USA as a graduate student; he spent nine months studying under Dr Lush at Iowa State College, then moved to the University of Wisconsin (where he worked under Dr Crow in the genetics department), from which Kimura gained his doctorate in 1956. Although he returned to Japan in that year, Kimura continued to collaborate with Crow, jointly writing a book on population genetics, for example. Formerly head of the division of population genetics of the National Institute of Genetics, Mishima, he is now professor emeritus there.

While a student, Kimura became interested in genetics, particularly the mathematical aspects of genetics and evolution. Stimulated by the work of J B S ◊Haldane and Sewall Wright on population genetics, Kimura began original work in this field in 1949, teaching himself the necessary mathematics. He then began to investigate the fate of mutant genes, how the genetic constitution of living organisms adapts to environmental changes, and the role of sexual reproduction in evolution.

Extending this early research, Kimura then began the work that was to lead to his postulating the theory of neutral evolution in 1968. According to the neo-Darwinistic view, evolution results from the interaction between variation and natural selection; species evolve by accumulating adaptive mutant genes, but these mutant genes increase in the population only if they confer advantageous traits on its individual members. With the advent of molecular genetics, it became possible to compare individual RNA molecules and proteins in related organisms and to assess the rate at which allelic genes (those that occupy the same relative positions on homologous chromosomes) are substituted in evolution. It also became possible to study the variability of genes within a species. Using these techniques, Kimura found that, for a given protein, the rate at which amino acids are substituted for each other is approximately the same in many diverse lineages, and that the substitutions seem to be random. Comparing the amino-acid compositions of the alpha and beta chains of the haemoglobin molecules in humans with those in carp, he found that the alpha chains have evolved in two distinct lineages, accumulating mutations independently and at about the same rate over a period of some 400 million years. Moreover, the rate of amino acid substitution observed in the carp–human comparison is very similar to the rates observed in comparisons of the alpha chains in various other animals.

These findings led Kimura to his theory of neutral evolution. According to this theory, evolutionary rates are determined by the structure and function of molecules and, at the molecular level, most intraspecific variability and evolutionary change is caused by the random drift of mutant genes that are all selectively equivalent and selectively neutral. Thus Kimura's theory directly opposes the neo-Darwinistic theory of evolution by denying that the environment influences evolution and, as a concomitant of this, also denying that mutant genes confer either advantageous or disadvantageous traits.

Kipping, Frederick Stanley (1863–1949) English chemist who pioneered the study of the organic compounds of silicon; he invented the term 'silicone', which is now applied to an entire class of silicon and oxygen polymers.

Kipping was born in Higher Broughton, near Manchester, on 16 August 1863, the son of a bank employee. He was educated at Manchester Grammar School and in 1882 graduated in chemistry from Owens College, Manchester (later Manchester University), with an external degree from the University of London. After four years as a chemist with the Manchester Gas Department, he went to Johann von ◊Baeyer's laboratory in Munich to study under William Perkin Jr (1860–1929), who was to become his close friend and collaborator. Kipping received his doctorate in 1887 and was awarded a DSc degree by the University of London in the same year – the university's first award of this qualification solely for research work.

He then followed Perkin to the Heriot-Watt College, Edinburgh, where he worked as a demonstrator. It was at Edinburgh that the two chemists began work on their classic textbook *Organic Chemistry* (1894), the first to be devoted entirely to this subject and a standard work for the next 50 years. In 1890 Kipping became chief demonstrator at the City and Guilds Institute, London, and seven years later he took up the appointment of professor of chemistry at University College, Nottingham (later Nottingham University), where he remained until he retired in 1936. He died in Criccieth, north Wales, on 1 May 1949.

In his early research Kipping investigated the preparation and properties of optically active camphor derivatives and nitrogen compounds. This interest in stereoisomerism led him in 1899 to look for such isomerism among the organic compounds of silicon, preparing them using the newly available Grignard reagents. He prepared condensation products – the first organosilicon polymers – which he called silicones. He also tried to make silicon analogues of simple carbon compounds, particularly those containing double bonds, although in this he was not successful.

$$—Si—O—Si—$$

siloxane structure

$$R—SiR_2—[—O—Si—]_n—O—SiR_2—R$$

structure of a silicone fluid

To Kipping the silicon compounds were mere chemical curiosities and as late as 1937 he could not see any practical applications for them. Yet within a very few years, spurred on by the outbreak of World War II, silicones were being used as substitutes for oils and greases. Their chemical inertness and unusual stability at high temperatures make them useful as lubricants, hydraulic fluids, waterlogging compounds, varnishes, greases, synthetic rubbers, and various other hydrocarbon substitutes.

Kirchhoff, Gustav Robert (1824–1887) German physicist who founded the science of spectroscopy. He also discovered laws that govern the flow of electricity in electrical networks and the absorption and emission of radiation in material bodies.

Kirchhoff German physicist Gustav Robert Kirchhoff who, together with R W von Bunsen, developed the technique of spectroscopic analysis. In 1860 they used this technique to discover the alkaline earth metal caesium. *Mary Evans Picture Library*

Kirchhoff was born in Königsberg, Germany (now Kaliningrad) on 12 March 1824. He studied at the University of Königsberg, graduating in 1847. In the following year, he became a lecturer at Berlin and in 1850 was appointed extraordinary professor of physics at

Breslau. Robert ◊Bunsen went to Breslau the following year and began a fruitful collaboration with Kirchhoff. In 1852 Bunsen moved to Heidelberg and Kirchhoff followed him in 1854, becoming ordinary professor of physics there. Kirchhoff stayed at Heidelberg until 1875, when he moved to Berlin as professor of mathematical physics. Illness forced him to retire in 1886 and he died in Berlin on 17 October 1887.

Kirchhoff made his first important contribution to physics while still a student. In 1845 and 1846 he extended Ohm's law to networks of conductors and derived the laws known as Kirchhoff's laws that determine the value of the current and potential at any point in a network. He went on to consider electrostatic charge and in 1849 showed that electrostatic potential is identical to tension, thus unifying static and current electricity. Kirchhoff made another fundamental discovery in electricity in 1857 by showing theoretically that an oscillating current is propagated in a conductor of zero resistance at the velocity of light. This was important in the development in the 1860s of the electromagnetic theory of light by James Clerk ◊Maxwell and Ludwig ◊Lorenz.

In the 1850s Bunsen developed his famous gas burner, which gave a colourless flame, and used it to investigate the distinctive colours that metals and their salts produce in a flame. Bunsen used coloured solutions and glass filters to distinguish the colours in a partly successful attempt to identify the substances by the colours they produced. Kirchhoff pointed out to Bunsen that sure identification could be achieved by using a prism to produce spectra of the coloured flames. They developed the spectroscope, and in 1860 discovered that the elements present in the substances each give a characteristic set of spectral lines and set about classifying elements by spectral analysis. In this way, Bunsen discovered two new elements – caesium in 1860 and rubidium in 1861.

Kirchhoff also made another important discovery in 1859 while investigating spectroscopy as an analytical tool. He noticed that certain dark lines in the Sun's spectrum, which had been discovered by Joseph von ◊Fraunhofer, were intensified if the sunlight passed through a sodium flame. This observation had in fact been made by Léon ◊Foucault ten years earlier, but he had not followed it up. Kirchhoff immediately came to the correct conclusion that the sodium flame was absorbing light from the sunlight of the same colour that it emitted, and explained that the Fraunhofer lines are due to the absorption of light by sodium and other elements present in the Sun's atmosphere.

Kirchhoff went on to identify other elements in the Sun's spectrum in this way, and also developed the theoretical aspects of this work. In 1859 he announced another important law which states that the ratio of the emission and absorption powers of all material bodies is the same at a given temperature and a given wavelength of radiation produced. From this, Kirchhoff went on in 1862 to derive the concept of a perfect black body – one that would absorb and emit radiation at all wavelengths. Balfour Stewart (1828–1887) had reached similar conclusions in 1858 by a consideration of the theory of heat exchanges discovered by Pierre ◊Prévost, but Kirchhoff presented the discovery much more cogently.

Kirchhoff's contributions to physics had far-reaching practical and theoretical consequences. The discovery of spectroscopy led to several new elements and to methods of determining the composition of stars and the structure of the atom. The study of black-body radiation led directly to the quantum theory and a radical new view of the nature of matter.

Kirkwood, Daniel (1814–1895) US astronomer who is known for his work on asteroids, meteors, and the evolution of the Solar System.

Kirkwood was born in Hartford County, Maryland, on 27 September 1814. He became a teacher in 1833 and rose to become principal of Lancaster High School 1843–49 and then of the Pottsville Academy 1849–51. He went on to become professor of mathematics at the University of Indiana. During this period of tenure, he had a two-year break which he spent as professor of mathematics and astronomy at Jefferson College in Pennsylvania. In retirement Kirkwood moved to California and in 1891 he became a lecturer in astronomy at the University of Stanford. He died in Riverside, California, on 11 June 1895.

Kirkwood's first astronomical paper was published in 1849. It consisted of a dubious mathematical description of the rotational periods of the planets. He was more interested, however, in the distribution of the asteroids in the Solar System. As early as 1857 he had noticed that three regions of the asteroid zone, sited at 2.5, 2.95, and 3.3 astronomical units from the Sun, lacked asteroids completely. In 1866 he published his analysis of this observation and proposed that the gaps, subsequently named 'Kirkwood gaps', arose as a consequence of perturbations caused by the planet Jupiter. The effect of Jupiter's mass would be to force any asteroid that appeared in one of the asteroid-free zones into another orbit, with the result that it would immediately leave the zone. It has since been proposed that non-gravitational effects, such as collisions, may also play a role in the maintenance of these gaps. Since Kirkwood's time, several more such gaps have been recognized.

Kirkwood used the same theory to explain the non-uniform distribution of particles in the ring system of Saturn. He suggested that Cassini's division was maintained by the perturbing effect that Saturn's satellites – Mimas, Enceladus, Thetys, Dione, and, to a lesser extent, Rhea and Titan – had on the orbits of the particulate material making up the rings. The gravitational forces of these satellites would prevent the ring material from entering the Cassini division, Encke's division, and the gap between the Crêpe Ring and Ring B in a manner similar to the way in which Jupiter affected the asteroids.

Kirkwood is best known for his work on the 'Kirkwood gaps', but he also carried out research into comets and meteors, made a fundamental critique of Pierre ◊Laplace's work on the evolution of the Solar System, and carried out preliminary studies on families of asteroids.

Kitasato, Shibasaburo (1852–1931) Japanese bacteriologist who is generally credited with the discovery of the bacillus that causes bubonic plague. He also did much important work on other diseases, such as tetanus and anthrax. His work gained him many international honours, and in 1923 Japan made him a baron.

Kitasato was born on 20 December 1852 in a small mountain village on the island of Kyushu in Japan. Keenly interested in science, he studied medicine at the Kumamoto Medical School and later at Tokyo University, from which he graduated in 1883. After graduating he joined the Central Bureau of the Public Health Department, but in 1885 he was sent by the government to study new developments in bacteriology in Germany and went to work in the laboratory of Robert ◊Koch in Berlin. Under Koch's guidance, Kitasato quickly mastered the new techniques and began his own research, which was so successful that he was made an honorary professor of Berlin University before he returned to Japan in 1891. On his return, Kitasato set up a small private institute of bacteriology, the first of its kind in Japan. Later the institute received financial assistance from the government, but in 1915 it was incorporated into Tokyo University against Kitasato's wishes and he resigned as its director. In the same year he founded another establishment, the Kitasato Institute, which he headed for the rest of his life. Kitasato died on 13 June 1931 in Nakanocho.

Kitasato did his first important work in Koch's laboratory in Germany where, in 1889, he became the first to obtain a pure culture of *Clostridium tetani,* the causative bacillus of tetanus. In the following year, working with Emil von ◊Behring, Kitasato discovered that animals can be protected against tetanus by inoculating them with serum containing inactive tetanus toxin. This was the very important discovery of antitoxic immunity, and Kitasato and von Behring rapidly developed a serum for treating anthrax. In the same year they published a paper on their combined work, giving details of their success with tetanus and similar results with diphtheria, on which Behring, helped by Paul Ehrlich, had concentrated.

After returning to Japan, Kitasato was sent by his government to Hong Kong in 1894 to investigate an epidemic of bubonic plague. France also sent a small research team led by Alexandre ◊Yersin, a former pupil of Louis ◊Pasteur. The two teams did not collaborate because of language difficulties and there is some doubt as to whether it was Kitasato or Yersin who first isolated *Pasteurella pestis,* the bacillus that causes bubonic plague. Kitasato published his discovery of the bacillus several weeks before Yersin announced his findings, though the bacillus has been renamed *Yersinia pestis* after Yersin.

Kitasato also isolated the causative organism of dysentery in 1898 and studied the method of infection in tuberculosis.

Klaproth, Martin Heinrich (1743–1817) German chemist famous for his discovery of several new elements and for pioneering analytical chemistry.

Klaproth was born in Wernigerode, Saxony, on 1 December 1743. His home was destroyed by fire when he was eight years old, leaving the family in poverty. He was apprenticed to an apothecary when he was 16, and after moving from employer to employer he finally, in 1771 became manager of a pharmacy in Berlin. Nine years later he set up on his own, doing chemical research. He took an appointment as a chemistry lecturer at the Berlin School of Artillery in 1792, and when the University of Berlin was founded in 1810 it was to Klaproth that the authorities offered the first chair in chemistry, which he held until his death. He died in Berlin in 1 January 1817.

Klaproth's contributions to the discovery and isolation of new elements began in 1789. From the semiprecious gemstone zircon he prepared an oxide ('earth') containing the new metallic element zirconium. In the same year he investigated pitchblende and from this black ore obtained a yellow oxide containing another new metallic element, which he named uranium after the planet Uranus, discovered by William ◊Herschel in 1781. He distinguished strontia (strontium oxide) from baryta (barium oxide) and in 1795 rediscovered and named titanium, acknowledging the prior isolation of the element four years earlier by William Gregor (1761–1817). He isolated chromium in 1797 independently of Louis ◊Vauquelin, but credited Franz Müller (1740–1825) with the priority for the discovery of tellurium, which Klaproth extracted in 1798 and named after *tellus,* the Latin for earth; he refuted, however, the claim to this discovery by the Hungarian chemist P Kitaibel (1757–1817).

In 1803 Klaproth identified cerium oxide and confirmed the existence of cerium, discovered by Jöns ◊Berzelius in the same year and named after the newly found asteroid Ceres. He also studied the rare earth minerals researched by Johan Gadolin (1760–1852), confirming that they are complex mixtures of very similar substances.

All of Klaproth's work on minerals and new elements hinged on his outstanding ability as a quantitative analytical inorganic chemist, a branch of the science which he can be credited with helping to found. He was a champion of Antoine ◊Lavoisier's antiphlogiston theory of combustion. But he was not reluctant (unlike some of his contemporaries) to publish anomalous results; he did not 'modify' his findings to make them suit some preconceived theory. He also applied his analytical skills to various archaeological finds and antiquities, such as metal artefacts, glassware, and coins.

Klein, (Christian) Felix (1849–1925) German mathematician and mathematical physicist whose unification of the various Euclidean and non-Euclidean geometries was crucial to the future development of that branch of mathematics. Equally skilled, however, in almost every branch of mathematics, Klein had as perhaps his greatest talent the ability to discover relationships between different areas of research, rather than in carrying out detailed calculations. He also possessed great organizational skills, and initiated and supervised the writing of an encyclopedia on mathematics, its teaching, and its applications. He became widely known for his lectures and his books on the historical development of mathematics in the 19th century. It was to a great extent due to him that Göttingen became the main centre for all the exact sciences – not just mathematics – in Germany.

Klein was born on 25 April 1849 in Düsseldorf, where he grew up and was educated. In 1865, at the age of 16, he went to the University of Bonn to read mathematics and physics; only three years later he was awarded his doctorate there. He decided to further his education by spending a few months in different European universities, and went to Göttingen, to Berlin, and then to Paris in 1869. At the outbreak of the Franco-Prussian War he was obliged to leave Paris so he entered the military service as a medical orderly. In 1871 he qualified as a lecturer at Göttingen University; the following year he became full professor of mathematics at Erlangen University. In his inaugural address there he introduced what he called his 'Erlangen *Programm*' on geometries. From 1875 to 1880 he was professor at the Technische Hochschule in Munich, and then took up a similar post 1880–86 at Leipzig. He spent the rest of his active research life at Göttingen, retiring as professor in 1913 because of ill health. During and after World War I, however, he gave lectures at his home. Klein died in Göttingen on 22 June 1925.

Klein's Erlangen programme of 1872 comprised a proposal for the unification of Euclidean geometry with the geometries that had been devised during the 19th century by mathematicians such as Karl ◊Gauss, Nikolai ◊Lobachevsky, János ◊Bolyai, and Bernhard ◊Riemann. He showed that the different geometries are each associated with a separate 'collection' or 'group' of tranformations. Seen in this way, the geometries could all be treated as individual members of one overall family, and from this very connection conclusions and inferences could be drawn. In the next two years, Klein developed the 'programme' and published papers that demonstrated that every individual geometry could be constructed purely projectively; he produced projective models for Euclidean, elliptic, and hyperbolic geometries. (Such projective models are now called Klein models.) Much later in his life, Klein returned to the Erlangen programme to apply it to problems in theoretical physics, with special reference to the theory of relativity.

Klein's early research was on topics in line geometry. In 1870, with Sophus ◊Lie, who was later to succeed him at Göttingen, he discovered fundamental properties of the asymptotic lines of the 'Kummer' surface, which became famous in algebraic line geometry. In his work on number theory, group theory, and the theory of differential equations, Klein was greatly influenced by Bernhard Riemann. It was Klein who redefined a Riemann surface so that it came to be regarded as an essential part of function theory, not just a valuable way of representing multi-valued functions. Borrowing concepts from physics (and especially fluid dynamics), Klein revised and developed the

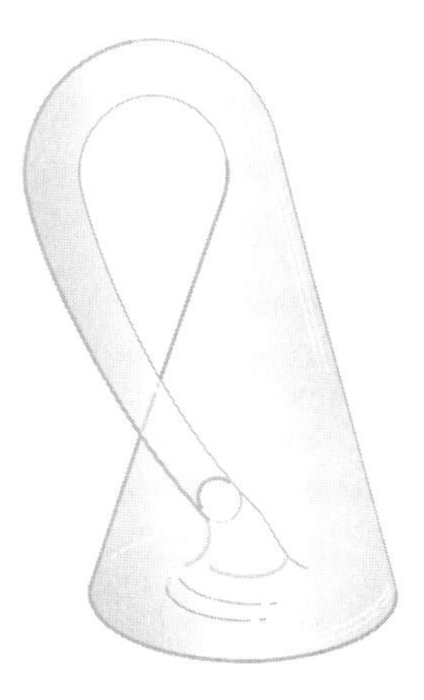

Klein Some properties of a geometric figure remain unchanged when it is distorted; for example the two lines still intersect.

theory in such a way that to him it seemed the most important work he ever accomplished.

He was also interested in fifth-degree (quintic) equations, and succeeded in solving the general algebraic equation of the fifth degree by considering the icosahedron. This work led him on to elliptic modular functions.

In the 1890s he worked on mathematical physics and engineering, and wrote a textbook with Arnold ◊Sommerfeld on the theory of the gyroscope, which is still an important work in this field of mechanics.

In 1900 he became interested in school mathematics, and recommended the introduction of differential and integral calculus and the function concept into school syllabuses. Eight years later, at the International Congress of Mathematicians in Rome, he was elected chair of the International Commission on Mathematical Instruction.

Klug, Aaron (1926–) South African molecular biologist who improved the quality of electron micrographs by using laser lighting. He used electron microscopy, X-ray diffraction, and structural modelling to study the structures of different viruses (including polio) that are too small to be visible with a light microscope or to be trapped by filters. He received the Nobel Prize for Chemistry in 1982.

Klug was born in Zelvas, Lithuania on 11 August 1926 to Lazar and Bella Klug and the family emigrated to Durban, South Africa where members of his mother's family had settled when he was two years old. Attending Durban High School the young Aaron was not interested in any subject in particular but he read widely. Inspired by Paul de Kruif's *Microbe Hunters* he decided to take up medicine at university. He studied medicine for a year at Witwatersrand University, Johannesburg, taking biochemisty in the second year and finally taking a science degree.

Klug won a scholarship at the University of Cape Town to do research in physics where he was taught by the X-ray crystallographer, R W James, attending his undergraduate lectures as well as his MSc course. He acquired a knowledge of Fourier theory and developed an interest in external and internal conical refraction. Staying on after achieving his MSc in 1947 he worked on the analysis of some small organic compounds and developed a method of using molecular structure factors to solve crystal structures. He also taught himself quantum chemistry and became interested in the organization of the structure of matter.

In 1949 Klug went to the Cavendish Laboratory, Cambridge, England, on another scholarship, where he learnt about computing, solid-state physics, and the notion of nucleation and growth in a phase change. Moving on after his PhD to the Colloid Science department in Cambridge, he showed how the reactions of haemoglobin in contact with oxygen or carbon dioxide could be simulated on a computer. He then obtained a Nuffield Fellowship to Birkbeck College in 1953 starting work in the laboratory of Irish physicist J D Bernal (1901–1971), on the polio virus, becoming a professor in 1954. Shortly after this he met Rosalind ◊Franklin and, inspired by her X-ray photographs, he took up the study of the tobacco mosaic virus with her. Four years later, with research students Kenneth Holmes and John Finch, he had mapped out the structure of the virus. He also published a paper with Francis Crick on diffraction by helical structures.

After Franklin's death in 1958, Klug continued working with Holmes and Finch on spherical viruses and in 1962 he moved as professor to the new MRC Laboratory of Molecular Biology in Cambridge working on the formation of the protein units in helical viruses. He was appointed director of the laboratory in 1986.

Klug was elected fellow of the Royal Society in 1969, was awarded the Nobel prize for his work in molecular biology, and was knighted in 1988. His many awards include the Heineken Prize, Royal Netherlands Academy of Science (1979); Louisa Gross Horowitz Prize, Columbia University (1981); the Copley Medal (1985) and the Baly Medal (1987). He married Liebe Bobrow, a choreographer he met in Cape Town in 1948 and they have two sons.

Knopf, Eleanora Frances Bliss (1883–1974) US geologist who is known for her laboratory work on metamorphic rocks and for bringing the technique of petrofabrics to the USA.

Knopf was born in Rosemont, Pennsylvania, on 15 July 1883. She was educated at the Florence Baldwin School and from 1900 at Bryn Mawr College from which she graduated in 1904 with an AB in chemistry and an AM in geology. Between 1904 and 1909 she worked as an assistant curator in the geological museum and as a demonstrator in the geological laboratory at Bryn Mawr. In 1912, having spent a year at Berkeley (1910–11), she received a PhD in geology from Bryn Mawr and after passing the civil service examinations she went to Washington, DC as a geologic aide to the US Geological Survey (USGS); she was promoted to assistant geologist in 1917 and for three years worked with the federal survey and the Maryland State Geological Survey. In 1920 Knopf married the geologist Adolf Knopf and moved to New Haven where he became professor at Yale. Knopf continued work with the USGS, in the early 1920s studying the Pennsylvania and Maryland piedmont and from 1925 onwards the mountainous region along the New York–Connecticut border. During the 1930s she was also a visiting lecturer at Yale and at Harvard. In 1951

Knopf and her husband moved to Stanford University where she was appointed research associate in the geology department. She retired in 1955 and died of arteriosclerosis on 21 January 1974 in Menlo Park, California.

Knopf's early work on the geology of the area around Bryn Mawr led to several publications including a report in the *American Museum of Natural History Bulletin* in 1913 on her discovery of the mineral glaucophane in Pennsylvania, previously unsighted in North America east of the Pacific. It was from her work on the geologically complex Stissing Mountain area on the New York–Connecticut border and her laboratory work on metamorphic rocks, however, that gained her reputation. She brought to the USA the technique of petrofabrics, developed by Bruno Sander of Innsbruck University, and applied it over the next 40 years to the study of metamorphic rocks. Her major work on the subject, *Structural Petrology,* was published in 1938.

Koch, Robert (1843–1910) German bacteriologist who, with Louis ♢Pasteur, is generally considered to be one of the two founders of modern bacteriology. He developed techniques for culturing, staining, and observing microorganisms and discovered the causative pathogens of several diseases – including tuberculosis, for which discovery he was awarded the 1905 Nobel Prize for Physiology or Medicine.

Koch German bacteriologist Robert Koch whose set out the conditions known as 'Koch's postulates' that need to be present for a microorganism to be the cause of a disease. He discovered the tubercule bacillus, which causes tuberculosis, and cholera vibrio, which causes cholera. *Mary Evans Picture Library*

Koch was born on 11 December 1843 in Klausthal, Germany, one of 13 children of a mining official. He studied natural sciences and then medicine at Göttingen University, where he was taught by Friedrich ♢Wöhler and Friedrich Henle (1809–1885), obtaining his medical degree in 1866. After serving as an army surgeon (on the Prussian side) in the Franco-Prussian War, Koch became in 1872 district medical officer in Wollstein where, despite having few research facilities, he began important investigations into anthrax. For a brief period in 1879 he was town medical officer in Breslau, before being appointed to the Imperial Health Office in Berlin to advise on hygiene and public health. By 1881 his work was becoming well known and he was invited to speak at the Seventh International Medical Congress in London. In 1882 he announced to the Berlin Physiological Society his discovery of the bacillus that causes tuberculosis, and in the following year, while investigating an outbreak of cholera in the Nile delta, he identified the cholera bacillus. In 1885 he was appointed professor of hygiene at Berlin University and director of the Institute of Hygiene. The Tenth International Medical Congress was held in Berlin in 1890 and Koch was persuaded to announce the discovery of an anti-tuberculosis vaccine; this proved to be premature, however, as the vaccine was ineffective. In 1891 he was appointed director of the newly established Institute for Infectious Diseases, but he resigned his directorship in 1904 and spent much of the rest of his life advising foreign countries on ways to combat various diseases. Koch died on 27 May 1910 in Baden-Baden, Germany.

Koch started his bacteriological research in the 1870s with the gift of a microscope from his wife, and built up a primitive laboratory in part of his consulting room. Out of necessity Koch devised simple and original methods for growing and examining bacteria. For three years he worked on anthrax in his spare time, developing techniques for culturing the bacteria in cattle blood and in aqueous humour from the eye. He trapped a small smear of blood from an anthrax victim with a drop of aqueous humour between two microscope slides and observed the bacteria grow and divide under the microscope, finding that the bacteria were short-lived but that they formed spores that were resistant to desiccation. He then inoculated animals with the spores and found that they developed anthrax, thus proving that the spores remained infective; this was the first time a bacterium cultured outside a living organism had been shown to cause disease. Koch published his findings, but only after Pasteur's demonstration of an anthrax vaccine in 1882 were Koch's findings accepted.

Koch experimented with various dyes and found some that stain bacteria and make them more visible under the microscope. He also devised an ingenious method of separating a mixture of bacteria, which involved inoculating an animal with the bacteria and passing the resulting infection from one animal to another until, at the end of the experimental chain, only one type of bacterium remained. Using this method he identified the bacteria responsible for several disorders, including septicaemia.

On joining the Imperial Health Office in 1879, Koch was provided with two assistants and for the first time had adequate laboratory facilities. Here he developed the technique of culturing bacteria on gelatin. Using this technique Koch and his assistants isolated several microorganisms and showed that they cause disease. They also investigated the effects of various disinfectants on different bacteria and showed that steam is more effective than dry heat in killing bacteria, a discovery that revolutionized hospital operating-theatre practice.

Koch then set out to discover the causative agent of tuberculosis, a common and frequently fatal disease at that time. Initially he was unable to find any microorganisms that might cause the disease, but after developing a special staining technique he identified the bacterium responsible and, despite the difficulties caused by the bacterium's small size and slow rate of growth, managed to culture it in 1882.

In 1883 Koch went to the Nile delta to investigate a cholera epidemic. Finding bacteria in the intestinal walls of dead cholera victims and the same bacteria in the excreta of cholera patients, he succeeded in isolating the causative organism. On a later visit to Calcutta, where cholera was rife, he found similar bacteria in excreta and in supplies of drinking water. On returning to Berlin he advised regular checks on the water supply, made recommendations regarding sewage disposal, and organized courses in the recognition of cholera. And when the disease occurred in Hamburg in 1892, he recommended that the victims should be isolated, all excretory matter should be disinfected and that a special check should be made on the water supply.

Koch made several other important contributions. As a result of his investigations into a bubonic plague epidemic in Calcutta in 1897, he showed that rats are vectors of the disease (although there is no evidence that he knew that the rat flea was the actual vector). He also demonstrated that sleeping sickness is transmitted by the tsetse fly. His isolation of many disease-causing organisms eventually led to the development of vaccines and to the realization of the importance of the public health measures he recommended. Furthermore, many bacteriologists received their training as his assistants, including Shibasaburo ◊Kitasato, and the Nobel prizewinners Emil von ◊Behring and Paul ◊Ehrlich. Perhaps most importantly, however, Koch formulated a systematic method for bacteriological research, including various rules – still observed today – for identifying pathogens. According to these rules (called Koch's postulates), the suspected pathogen must be identified in all of the cases examined; the pathogen must then be cultured through several generations; these later generations must be capable of causing the disease in a healthy animal; and the newly infected animal must yield the same pathogen as that found in the original victim.

Further Reading

Brock, Thomas D, *Robert Koch: A Life in Medicine and Bacteriology*, American Society of Microbiology, 1998.

Carter, K Codell (transl), *Essays of Robert Koch*, Contributions in Medical Studies, 1987, v 20.

Knight, David C, *Robert Koch: Father of Bacteriology*, Chatto & Windus, 1963.

Mielke, Martin E; Ehlers, Stefan; Hahn, Helmut; and Kaufman, Stefan H, *Robert-Koch-Symposium 1993 on 'Progress in Tuberculosis Research'*, Lubrecht and Cramer, 1994.

Kolbe, (Adolf Wilhelm) Hermann (1818–1884)

German organic chemist, generally credited as the founder of modern organic chemistry with his synthesis of acetic acid (ethanoic acid) – an organic compound – from inorganic starting materials. (Previously organic chemistry had been considered as the branch of the science devoted to compounds that occur only in living organisms.)

Kolbe was born in Elliehausen, near Göttingen, on 27 September 1818, the eldest of 15 children of a Lutheran pastor; his mother was the daughter of A F Hempel, professor of anatomy at Göttingen University. He was educated at the Gymnasium at Göttingen, where he was introduced to chemistry by a student who had studied under Robert ◊Bunsen. In 1838 he entered Göttingen University, where he attended lectures by Friedrich ◊Wöhler (who ten years earlier had synthesized urea from ammonium cyanate, arguably the first inorganic–organic transition) and became a great admirer of Jöns ◊Berzelius. He became an assistant to Bunsen at Marburg University in 1842 and three years later accepted an invitation from Lyon Playfair to work with him at the London School of Mines.

During his two-year stay in London, Kolbe met many leading chemists of the day, including Edward Frankland who was at that time developing his theory of valency. In 1847 Kolbe moved to Brunswick to join the editorial team on the *Handwortenbuch der Chemie* (founded by Justus von ◊Liebig, Wöhler, and Poggendorf), and in 1851 he was appointed Bunsen's successor as professor of chemistry at Marburg – a rapid promotion that did not meet with the full approval of the establishment. By 1865 he had moved to Leipzig and had begun to set up the largest and best equipped laboratory of the time, which had its full complement of students within three years. From 1869 he was editor of the *Journal für practische Chemie* and became notorious for his very personal and often violent criticism of the work of his contemporaries. He continued his theoretical and literary work until he died in Leipzig on 25 November 1884.

Kolbe correctly realized that organic compounds can be derived from inorganic materials by simple substitution. He introduced a modified idea of structural radicals, which contributed to the development of the structure theory, and he predicted the existence of secondary and tertiary alcohols. He is best known for his work on the electrolysis of the fatty (alkanoic) acids, for his important preparation of salicylic acid (2-hydroxybenzenecarboxylic acid) from phenol – called the Kolbe reaction, which was to lead to an easy synthesis of the drug aspirin – and for his discovery of nitromethane.

His early work, with Frankland, was on the conversion of nitriles into fatty acids. He then investigated the action of electric currents on organic compounds. But he became an extremely conservative influence on organic chemistry largely because of his adherence to the ideas of Berzelius. His method of representing molecular structures eventually gave way to the much simpler structural theory based on the work of Friedrich ◊Kekulé von Stradonitz, although his unorthodox formulae actually embodied many of the ideas that were developed by Kekulé.

One of the greatest drawbacks of Kolbe's formulae resulted from his refusal to abandon equivalent weights in favour of atomic weights (relative atomic masses). Until 1869, for example, he still followed Berzelius's contention that C = 6 and O = 8, so that he had to double the number of atoms of these elements in his formulae. He therefore wrote the methyl group as C_2H_3 – and assumed that in acetic (ethanoic) acid the methyl radical was joined to oxalic (ethanedioic) acid and water, which he expressed as $C_2H_3 + C_2O_3 + HO$. He became convinced that methyl groups existed in compounds and could be isolated from them. By electrolysing potassium acetate (ethanoate) he obtained a gas that he thought was 'methyl' (really ethane). Frankland had obtained 'free ethyl' (really butane) by the action of zinc on ethyl iodide (iodoethane) and both chemists were now certain that they had proved the existence of radicals in organic compounds. Their formulae continued to represent these misapprehensions.

Kolbe explained the relationship between aldehydes and ketones and identified a new group – the carbonyl group. He was able to predict their behaviour on oxidation, and in so doing anticipated the existence of secondary and tertiary alcohols. His later work involved the nitroparaffins (nitroalkanes) and the Kolbe synthesis for salicylic acid. But he never could see the similarity of his formula system and that of Kekulé. In 1858 he bitterly opposed the whole idea of structural formulae and ridiculed the theory of structural isomerism put forward by Jacobus ◊van't Hoff and Joseph le Bel, which are now regarded as fundamental to structural organic chemistry.

Kolmogorov, Andrey Nikolayevich (1903–1987) Russian mathematician who worked on the theory of functions of a real variable functional analysis, topology, and mathematical logic. He also worked on the theory of turbulence, information theory, and cybernetics. He is best known for developing the axiomatic theory of probability and for his work on Markov processes. He received the Lenin Prize in 1965 and the Order of Lenin on six occasions.

Kolmogorov was born in Tambov, Russia, and educated at Moscow State University, graduating in 1925. He became a research associate at the university and then a professor in 1931. In 1933 he was appointed director of the Institute of Mathematics and in 1939 he was elected an academician of the Soviet Academy of Sciences.

Kolmogorov made contributions to many mathematical projects. He took an interest in the intuitionism of Dutch mathematician Luitzen ◊Brouwer and proved that intuitionistic arithmetic is only consistent if classical arithmetic is. He settled a key problem in Fourier analysis in 1936 by constructing a function that is integrable, but whose Fourier series diverges at every point. In 1939, he published a paper on the extrapolation of time series, a project later advanced by US statistician Norbert ◊Wiener to become known as 'single-series prediction'. He died in Moscow on 20 October 1987.

In 1933 Kolmogorov wrote a major treatise on the probability theory (translated into English in 1950 as *The Foundations of the Theory of Probability*). He rigorously built up probability theory from fundamental axioms providing a definition of conditional expectation. He went on to study the motion of the planets and the turbulence of the flow of air from a jet engine, publishing two papers on turbulence in 1941. He was also interested in cybernetics in relation to algorithms and computing.

With Russian mathematician Aleksandr Khinchin (1894–1959) he founded the Soviet school of probability theory and explored Markov processes where the probability of a variable depends on its immediately previous value but not on any values before that. He formulated the partial differential equations that bear his name.

Kornberg, Arthur (1918–) US biochemist who in 1957 made the first synthetic molecules of DNA. For this achievement he shared the 1959 Nobel Prize for Physiology or Medicine with Severo ◊Ochoa. By 1967 he had synthesized a biologically active artificial viral DNA.

Kornberg was born in Brooklyn, New York City, on 3 March 1918. He was educated at local schools and in 1933 graduated from the Abraham Lincoln High School. On a state scholarship he took a pre-medical course at the College of the City of New York, obtaining a BS degree in 1937. A further scholarship enabled him to go on to the University of Rochester School of Medicine from which he gained his medical degree in 1941. Following a year as an intern at the Strong Memorial Hospital in Rochester, he joined the US Coast Guard for a short time. From 1942 to 1945 he worked in the nutritional section of the physiology department of the National Institute of Health in Bethesda, Maryland, becoming chief of the enzyme and metabolic section 1947–52. In 1953 he took up a senior appointment at the Washington University School of Medicine, and in 1959 moved to the Stanford University School of Medicine, Palo Alto, to become head of its biochemistry department 1959–69 and a professor within that department 1959–88 (thereafter emeritus).

From the beginning of his career, Kornberg was interested in enzymes – not merely what they are but what they do. For many years the fundamental genetic mystery had concerned the ability of a cell to produce one particular enzyme and no other. In 1941 George Beadle and Edward Tatum (see ◊Beadle, Tatum, and Lederberg) demonstrated that genes control the processes of life by chemical means and in 1944 Oswald ◊Avery isolated the chemical responsible – the nucleic acid DNA, whose structure was elucidated by Francis ◊Crick and James ◊Watson in 1953. It was known that DNA consists of sugar, phosphate, and nucleotides, the 'letters' of a genetic alphabet that spell out the 'recipe' of a particular genetic trait by controlling the production of the appropriate protein. Another nucleic acid, RNA, translates the DNA code in this complex chemical process.

At Washington University Kornberg set himself the task of producing a giant molecule of artificial DNA. To do this he needed a pre-existing DNA molecule as a template to be copied, the four nucleotides adenine, thymine, guanine, and cytosine – known as A, T, G, and C – and an enzyme to select and arrange the nucleotides according to the directions from the template and link them together to form the DNA chain. In 1956 he isolated the enzyme DNA polymerase and a year later made an artificial DNA that had all the physical and chemical properties of its natural counterpart, but lacked its genetic activity. One cause of this partial failure was the impurity of the enzyme (which had resulted in errors in the arrangement of nucleotides).

He then took a new and simpler template, the DNA of the virus known as Phi X174, which is single-stranded and in the form of a ring; its activity (infectivity) is lost if the ring is broken. Another enzyme was needed to close the ring and in 1966 it was discovered, and called ligase. Using the natural DNA of Phi X174 as a template, Kornberg and his co-workers mixed the enzyme DNA polymerase, the enzyme ligase, and the four nucleotides. The DNA polymerase ordered the nucleotides into the arrangement dictated by the template and the ligase closed the ring of the artificial DNA so produced.

When the synthetic DNA was added to a culture of bacteria cells (*Escherichia coli*), it infected them and usurped their genetic machinery. Within minutes the cells had abandoned their normal activity and had started to produce Phi X174 viruses. The synthetic DNA had now become the template for a second generation of synthetic viruses identical to the original: the sequence of 6,000 'building blocks' and the arrangement of the 35 atoms within each was precisely the same in the artificial ones as in the natural ones.

By this achievement, Kornberg has opened the way to future progress in the study of genetics, the possibility of curing hereditary defects, and controlling virus infections and cancer.

Kornberg, Hans Leo (1928–) German-born British biochemist who has made important contributions to the understanding of metabolic pathways and their regulation, especially in microorganisms.

Kornberg was born in Herford, Germany, on 14 January 1928, and went to the UK in 1939 as a refugee from Nazi persecution. He attended schools in southern England and in Wakefield, Yorkshire, before entering Sheffield University in 1946. He gained his BSc in 1949 and his PhD in 1953, studying under Hans ◊Krebs who in 1937 had determined the tricarboxylic acid cycle (Krebs cycle), the sequence of energy-generating biological reactions in which glucose is converted into carbon dioxide and water. From 1953 to 1955 Kornberg travelled in the USA, studying first the pentose phosphate pathway (an alternative to the Krebs cycle) under E Racker at Yale and then helping Melvin ◊Calvin at the University of California to resolve a controversy concerning one step of the 'dark reaction' of photosynthesis in plants.

On his return to England he joined the Medical Research Council Cell Metabolism Unit headed by Krebs in Oxford. From 1961 to 1975 he was professor and head of the biochemistry department at the University of Leicester, and then he became the Sir William Dunn Professor of Biochemistry at Cambridge University. In 1995 returned to the USA to become university professor and professor of biology at Boston University. He was knighted in 1978.

The tricarboxylic acid (Krebs) cycle has two apparently conflicting roles. One is the complete oxidation (breakdown) of glucose to provide the energy required in all cellular processes – a catabolic role. The other is to provide carbon 'skeletons' (backbones of organic molecules) for the various complex compounds essential to cellular function – an anabolic role. Certain simple organisms can survive with only acetate (ethanoate, a two-carbon molecule) as their source of carbon. Kornberg directed his research to seek an answer to the question of how an organism uses acetate to build up larger molecules while at the same time needing to break it down to provide the very energy required for these anabolic functions.

His study led in 1957 to the discovery of the 'glyoxylate cycle' in plants, microorganisms, and some worms (glyoxal is ethan-1, 2-dial). The cycle's basic function is to convert the two-carbon molecules of acetate into a four-carbon molecule of succinate (buten-1, 4-dioate). It is achieved by means of a by-pass reaction in which the decarboxylation reactions of the normal tricarboxylic acid cycle (in which enzymes remove carbon that ultimately is converted to carbon dioxide) are skipped over.

The glyoxylate cycle is localized in subcellular organelles called glyoxysomes in the germinating seeds of higher plants, which use it to convert stored fatty acids into carbohydrates for the production of the energy needed for rapid growth.

Two years later, in 1959, Kornberg discovered the glycerate pathway and in 1960 the dicarboxylic acid cycle, which play important roles in enabling microorganisms to grow on certain types of nutrients. He investigated the control mechanisms essential to the regulations of these metabolic processes, studying the way in which the cellular economy is balanced to prevent overproduction or waste. He introduced the concept of 'anaplerotic reactions', whereby metabolic processes are maintained by special enzymes that replenish materials syphoned off for anabolic purposes.

Kornberg studied the regulation of enzyme activity at the genetic level, where the production of enzymes is largely controlled, and at the cytoplasmic level, where the reaction rates of existing enzymes can be modified. His later research concentrated on the very first step in the processing of food materials, the selective uptake of compounds across cellular membranes. He used mutant bacterial strains that can be grown on specific, defined substrates rigorously to control the nature of the materials entering their cells. This combination of genetic and biochemical approaches provides a unique way of studying both the nature of the transport process and the way in which it is regulated.

Korolev, Sergei Pavlovich (1906–1966) Russian engineer who designed the first crewed spacecraft, interplantetary probes, and soft lunar landings. As such, he was one of the founders of practical space flight.

Korolev was born on 30 December 1906 at Zhitomir in the Ukraine, the son of a Russian schoolmaster. His early technical training was received at a building-trades school in Odessa; he graduated from there in 1924. Three years later he entered the aviation industry, but continued with his studies at the Moscow School of Aviation. He graduated from the Baumann Higher Technical School in 1930 and, in June of that year, he became a senior engineer at the Central Aerodyamics Institute. After successfully developing a number of gliders he became interested in rocket-type aircraft. In 1931, with Tsander, he formed a group for the study of jet propulsion. This became an official body, and in 1932 Korolev became its head. The group eventually trained many of the people responsible for the subsequent Soviet triumphs in space. It was this group that built the first Soviet liquid-fuel rocket, which was launched in 1933. Korolev was appointed deputy director of the new Institute for Jet Research, which was formed by combining the original Korolev group and the Gas Dynamics Laboratory in 1933. In 1934 he was appointed head of the rocket vehicle department.

Korolev was responsible for designing a number of rockets and, from 1924 to 1946, he worked as deputy chief engine designer in the Specialized Design Bureau. Later, Korolev was appointed head of the large team of scientists who developed high-power rocket systems.

Korolev received many honours. In 1953 he was made corresponding member of the Academy of Sciences of the USSR and a member of the CPSU. He was twice named Hero of Socialist Labour, the first time in 1956 and the second time in 1961. In 1958 he was made full member of the Academy of Sciences, having been awarded the Lenin Prize in 1958. In addition, his name is commemorated for all time: the largest formation on the hidden side of the moon is named after him. On his death on 14 January 1966 his body was buried in the Kremlin Wall in Red Square, Moscow – one of the highest honours that could be conferred on a Soviet citizen. Korolev published his first paper on jet propulsion in 1934. By 1939 he had designed and successfully launched the Soviet 212 guided wing rocket. This was followed in 1940 by the RIP-318–1 rocket glider, which made its first piloted flight in 1940. From then on, Korolev was responsible for a large number of historical first achievements in space exploration.

He was responsible, in an executive and scientific capacity, for directing the early research and design work which later led to vast Soviet achievements in space flight. Korolev initiated many scientific and technological ideas which have since been widely used in rocket and space technology, including ballistic missiles, rockets for geophysical research, launch vehicles, and crewed spacecraft. The most famous of these is the Vostok series of Soviet single-seater spaceships designed for close-Earth orbit in which Soviet cosmonauts made their first flights.

Vostok 1, crewed by Yuri Gagarin, made the first crewed space flight in history, which took place on 12 April 1961. Launched from the Baikonur space-launch area, Gagarin and *Vostok 1* completed one orbit of the Earth in just under two hours. This flight was followed by *Vostoks 2, 3, 4, 5,* and *6* (which was crewed by Valentina Tereshkova, the first female cosmonaut). The technological improvements made by Korolev and his team were such that Tereshkova and *Vostok 6* were able to make 48 orbits in a total time of just over 70 hours.

These close-Earth flights were followed by the launching of Sun satellites and the flights of the uncrewed interplanetary probes into the Solar System – including the Elecktron and Molniia series. Korolev was also responsible for the *Voskhod* spaceship, from which the first space walks were made.

Further Reading

Harford, James J, *Korolev: How One Man Masterminded the Soviet Drive to Beat America to the Moon,* Wiley-Liss, 1997.

Kovalevskaya, Sofia Vasiyevna, born Krukovskya (1850–1891) Russian mathematician who made great contributions to the theory of differential equations. Her professors at the University of Heidelberg saw her as an extraordinary phenomenon. She is known through the Cauchy–Kovalevsky theorem of partial differential equations.

Kovalevskaya was born in Moscow on 15 January 1850 into an aristocratic family whose social circle included Dostoyevsky. Her father, Vasily Korvin-Krukovsky, was an artillery general and her mother, Velizaveta Shubert, was a well-educated woman of noble birth. Sofia was educated at home by tutors and governesses, and from an early age enjoyed writing dramas and fictional works. Her uncle often spoke about mathematics and was the source of her interest in the subject. The family tutor, Y I Malevich began teaching her mathematics and she enjoyed it so much

Kovalevskaya Russian mathematician Sofia Kovalevskaya. *AKG*

she neglected her other studies. Her father stopped her mathematics lessons but she got hold of a copy of an algebra book, which she read secretly at night.

The family was given a physics textbook by its author, Professor Tyrtov, and he recognized Sofia's talent by the way she worked out the complicated trigonometric formulas for herself. Sofia's father rejected Tyrtov's suggestion that he should encourage her to study mathematics. He would not let her leave home to study. Women in Russia could not live apart from their families without written permission from their fathers or husbands – so she married Vladimir Kovalesky, a palaeontologist, when she was just 18.

Kovalevskaya moved with Vladimir to Heidelberg and studied unofficially, since women could not matriculate there, under Konigsberger, Gustav ◊Kirchhoff, and Hermann von ◊Helmholtz, who all enthused about her gifts. Konigsberger sent her to study with his teacher, Karl ◊Weierstrass, in Berlin but, as a woman, she was refused admission to lectures at the university, so Weierstrass taught her privately for four years. Three papers – on partial differential equations, Abelian integrals, and Saturn's rings – qualified her for a doctorate in absentia from the Göttingen University in 1874, but she was unable to get an academic position, mainly because of her gender.

Kovalevskaya felt her rejection bitterly and for six years did no research and did not reply to Weierstrass's letters. She moved back to Russia with her husband and gave birth to a daughter, 'Foufie', in 1878. In 1880 she returned to study and in 1882 wrote three articles on the refraction of light, one of which contains a valuable exposition of Weierstrass's theory for integrating partial differential equations.

During all this time, Sofia and Vasily had endured a quarrelsome marriage and they had separated in 1881. Sofia was shocked and guilt-ridden over his suicide in 1883 and immersed herself in work. She at last obtained a lectureship in Stockholm in 1884, becoming an extraordinary professor the same year. In 1889 she became the first woman in Europe since Laura ◊Bassi and Maria Gaetana ◊Agnesi (over a hundred years before) to hold a chair at a university. During this time she also published a description of her early life disguised as a novel and called *Recollections of Childhood* (or *The Sisters Rajevsky*).

In Stockholm she taught courses on analysis and became editor of the journal *Acta mathematica.* Her most important work was done here and in 1888 she won the Prix Bordin of the French Academie with *On the Rotation of a Solid Body about a Fixed Point.* In recognition of the quality of her work, the usual prize of 3,000 French francs was raised to 5,000. Further research won a prize for the Swedish Academy of Sciences in 1889 and, although she had always been denied an academic position in her home country, Kovalevskaya was finally elected to membership of the Russian Academy of Sciences.

On 15 January 1891, at the height of her career, she died of influenza complicated by pneumonia.

Krebs, Hans Adolf (1900–1981) German-born British biochemist famous for his outstanding work in elucidating the cyclical pathway involved in the intracellular metabolism of foodstuffs – a pathway known as the tricarboxylic cycle, the citric acid cycle, or the Krebs cycle. He received many honours for this work, including the 1953 Nobel Prize for Physiology or Medicine – which he shared with Fritz Lipmann (1899–1986), a German-born US biochemist – a knighthood in 1958, and several medals and honorary degrees.

Krebs was born on 25 August 1900 in Hildesheim, Germany, the son of an ear, nose, and throat specialist. He was educated at the universities of Göttingen, Freiburg, Munich, Berlin, and Hamburg, gaining his medical degree from the last in 1925. From 1926 to 1930 he worked under Otto ◊Warburg at the Kaiser Wilhelm Institute in Berlin, then taught at the University of Freiburg until 1933, when he moved to England because of the rise to power of Adolf Hitler and the Nazi movement in Germany. On his arrival in England, Krebs went to Cambridge University, as a Rockefeller Research Student 1933–34, then as a

Krebs German-born British biochemist Hans Adolf Krebs, who won the Nobel Prize for Physiology or Medicine in 1953 for his work on the metabolism of food to produce energy. Krebs was born in Hildesheim, Germany, but moved to work in Cambridge, England when the Nazis came to power. *Mary Evans Picture Library*

demonstrator in biochemistry until 1935. From 1935 to 1954 he worked at Sheffield University – as lecturer in pharmacology 1935–38, lecturer in charge of the department of biochemistry 1938–45, and professor of biochemistry 1945–54. He then moved to Oxford University as Whitley Professor of Biochemistry and a fellow of Trinity College, which positions he held until his retirement in 1967. Krebs died in Oxford, on 22 November 1981.

While working under Warburg, Krebs became interested in the process by which the body degrades amino acids. He discovered that, in amino acid degradation, nitrogen atoms are the first to be removed (deamination) and are then excreted as urea in the urine. Continuing this line of research, Krebs then investigated the processes involved in the production of urea from the removed nitrogen atoms, and by 1932 he had worked out the basic steps involved in what is now known as the urea cycle. Later workers discovered the details of the cycle, but Krebs's original basic scheme was correct.

Krebs is best known, however, for his discovery of the processes involved in the tricarboxylic acid cycle, by which carbohydrates are aerobically metabolized to carbon dioxide, water, and energy. The energy yielded by this pathway is the main source of intracellular energy and therefore of the entire organism. Previous work by Otto ◊Meyerhof and Carl and Gerty ◊Cori had shown that the carbohydrate glycogen is broken down in the liver to lactic acid by a process that does not require oxygen and that yields very little energy. Krebs continued this work (while at Sheffield University) to show how the lactic acid is then metabolized to carbon dioxide, water, and energy. As a result of his investigations – performed on pigeon breast muscle – Krebs proposed the self-regenerating biochemical pathway now known as the Krebs or tricarboxylic acid (TCA) cycle.

In this cycle the two-carbon acetyl coenzyme A (acetyl CoA) – derived from lactic acid – combines with the four-carbon oxaloacetic acid to form the six-carbon citric acid, which then undergoes a series of reactions to be reconverted to oxaloacetic acid. This oxaloacetic acid combines with another molecule of acetyl CoA to form citric acid again, and the cycle is repeated. Carbon dioxide, water, and hydrogen are produced by the various reactions in the cycle; the hydrogen combines with

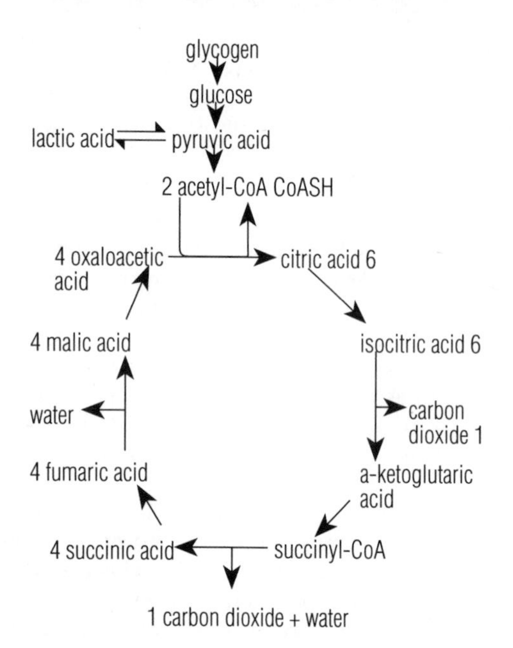

Krebs The purpose of the Krebs (or tricarboxyic acid) cycle is to complete the biochemical breakdown of food to produce energy-rich molecules, which the organism can use to fuel work. Acetyl coenzyme A (acetyl CoA) – produced by the breakdown of sugars, fatty acids, and some amino acids – reacts with oxaloacetic acid to produce citric acid, which is then converted in a series of enzyme-catalysed steps back to oxaloacetic acid. In the process, molecules of carbon dioxide and water are given off, and the precursors of the energy-rich molecules ATP are formed. (The numbers in the diagram indicate the number of carbon atoms in the principal compounds.)

oxygen in a complex series of reactions to produce energy and water. The degradation of glucose to carbon dioxide, water, and energy occurs in two stages. The first stage – in which acetyl CoA is formed – occurs in the cytoplasm; the second stage, the tricarboxylic acid cycle, takes place in mitochondria.

In addition to carbohydrate metabolism, the tricarboxylic acid cycle is also involved in the degradation of fats and amino acids, and provides substrates for the biosynthesis of other compounds, such as amino acids. Thus the cycle plays a central role in cell metabolism, and Krebs's elucidation of the steps involved in it was therefore a fundamental contribution to biochemistry.

Further Reading

Alton, Jannine and Harper, Peter, *Catalogue of the Papers and Correspondence of Sir Hans Adolf Krebs, FRS (1900–1981) Deposited in Sheffield University Library*, Reproduced for the Contemporary Scientific Archives Centre by the Royal Commission on Historical Manuscripts, 1986, 3 vols.

Holmes, Frederic L, *Hans Krebs, v 1: The Formation of a Scientific Life, 1900–1933*, Monographs in the History and Philosophy of Biology, Oxford University Press, 1991.

Holmes, Frederic L, *Hans Krebs, v 2: Architect of Intermediary Metabolism*, Monographs in the History and Philosophy of Biology, Oxford University Press, 1993.

Holmes, Frederic L, 'Antoine Lavoisier and Hans Krebs: two styles of scientific creativity' in: Doris B Wallace, Howard E Gruber (eds), *Creative People at Work*, Oxford University Press, 1989, pp 44–68.

Kronecker, Leopold (1823–1891) German mathematician, skilled in many branches of the subject but pre-eminent in none, who is remembered chiefly for the 'Kronecker delta'.

Kronecker was born in Liegnitz (now Legnica, Poland) on 7 December 1823 and attended secondary school there, where he was taught by the outstanding mathematician Ernst ◊Kummer. Kummer remained a lifelong friend and exercised great influence upon Kronecker, especially in interesting him in number theory. In 1841 Kronecker entered the University of Berlin, where he studied mathematics, chemistry, astronomy, and philosophy. In 1843 he spent some time at the universities of Bonn and Breslau (where he had been appointed professor of mathematics), before returning to Berlin. In 1845 he received his doctorate from Berlin for a thesis on complex units.

Kronecker was then diverted from mathematical research by pressing family duties – the management of an estate near Liegnitz and the winding-up of an uncle's banking business. Not until 1855, when he took up residence again in Berlin, did he begin a professional career in mathematics. Financially independent, he was able to devote himself to research without taking employment, but his election to the Berlin Academy in 1861 entitled him to give lectures at the university, an occupation that gave him such satisfaction that he turned down an invitation to take the chair of mathematics in Göttingen in 1868. By then his reputation had grown to such an extent that he was elected to the French Academy. In 1883 he succeeded Ernst Kummer as professor of mathematics at Berlin, a chair he held until his death, in Berlin on 29 December 1891.

God made the integers, man made the rest.

LEOPOLD KRONECKER *JAHRESBERICHTE DER DEUTSCHEN MATHEMATIKER VEREINIGUNG BK 2. IN F CAJORI* A HISTORY OF MATHEMATICS *1919*

Kronecker was a gifted mathematician who never quite made the kind of discovery to place him in the very front rank of mathematical geniuses. His thesis on complex units, for example, came close to hitting upon Kummer's idea of 'ideal numbers', but the actual discovery belongs to his mentor. Much of his life he was engaged in somewhat disputatious and cranky controversy. After 1855 he worked chiefly in the fields of number theory, algebra, and elliptical functions, and he achieved a great deal in his endeavour to unify analysis, algebra, and elliptical functions. But his progress, and his influence, were somewhat blocked by his obsession with the idea that all branches of mathematics (apart from geometry and mechanics) should be treated as parts of arithmetic. He also believed that whole numbers were sufficient for the study of mathematics – 'The whole numbers has the Dear God made; all else is man's work.' His refusal to grant irrational numbers equal validity with whole numbers, although it placed him in that tradition of 19th-century mathematical thought that attempted to define processes such as differentiation and integration in terms of simple arithmetic, led him into what we can now see as time-wasting debate.

One other controversy was more fruitful. His debate with Georg ◊Cantor and Karl ◊Weierstrass about the foundations of mathematics has been carried into the 20th century and beyond in the form of the debate between the formalists and the intuitionists. His system of axioms, published in 1870, has also been useful. His axioms were later shown to govern finite Abelian groups and have proved to be important in the modern development of algebra.

Kronecker remains, however, most famous for his work in linear algebra, especially for the delta named after him. (It is denoted by δrs, for which $\delta rs = 0$ when $r \neq s$, $\delta rs = 1$ when $r = 1$, where r, s are 1, 2, 3, ...) Kronecker found this a useful delta in the evaluation of determinants, the r and s being concerned with a row and a column of the determinant. The delta is a fairly simple example of a tensor and Kronecker's pupil, Hensel, who edited and published Kronecker's work 1895–1931, was largely responsible for popularizing the notation.

Kroto, Harold Walter (1939–) English chemist who with colleagues Richard ◊Smalley and Robert ◊Curl discovered a new form of carbon that they named buckminsterfullerene (carbon 60) in 1985. They shared the Nobel Prize for Chemistry in 1996 for their discovery.

Born in Wisbech, Cambridgeshire, on 7 October 1939 and brought up in Bolton, Lancashire, England, Harold Kroto was educated at the University of Sheffield graduating in chemistry in 1961 and gaining a PhD in 1964 for his research with R N Dixon on high resolution electronic spectra of free radicals produced by flash photolysis. He then took up a postdoctoral fellowship with the National Research Council of Canada, moving to the Bell Laboratories in New Jersey for a year, studying liquid phase interactions by Raman spectroscopy and making studies in quantum chemistry. In 1967 he took up a lectureship at the University of Sussex, becoming a professor there in 1985 and in 1991 being made a Royal Society Research Professor.

Kroto was interested in the carbon chains identified by radio astronomers in space in which molecules containing carbon atoms were linked by alternate triple and single bonds. He had been working on them for some time with David Walton, a colleague at Sussex, when he heard about Richard Smalley's work in Houston, Texas, in 1984. Smalley had devised a procedure in 1981 to produce microclusters of a hundred or so atoms using a new laser bombardment technique. Kroto suspected that he might be able to produce the chains he was interested in using this technique and paid a visit to Houston in 1985 to work with Smalley and his colleague Robert Curl. He persuaded Smalley to direct his laser beam at a graphite target producing the small chains he was looking for, but also, more interestingly, they found a mass-spectrum signal for a molecule of exactly 60 carbon atoms (C60) in a perfect sphere in which 12 pentagons and 20 hexagons are arranged like the panels on a modern football. The structure also resembles architect Buckminster Fuller's geodesic dome, so Kroto came up with the name buckminsterfullerene, subsequently shortened to fullerene, the molecules often being known as buckyballs. Their discovery added another allotrope to the diamond and graphite forms of carbon.

Since the discovery of buckminsterfullerene, other fullerenes have been identified with 28, 32, 50, 70, and

76 carbon atoms. New molecules have been made based on the buckyball enclosing a metal atom and buckytubes have been made consisting of cylinders of carbon atoms arranged in hexagons. Possible uses for these new molecules are as lubricants, superconductors and as a starting point for new drugs. Research in C60 has continued at Sussex focussing on the implications of the discovery for several areas of fundamental chemistry including the way it has revolutionized perspectives on carbon based materials. The research is interdisciplinary involving Curl and Smalley among others, and covers the basic chemistry of fullerenes, fundamental studies of carbon and metal clusters, carbon microparticles and nanotubes, and the study of interstellar and circumstellar molecules and dust.

Since 1990 Kroto has been chair of the editorial board of the Chemical Society Review. His awards and honours include being elected Fellow of the Royal Society in 1990, a Royal Society Research Fellowship in 1991, the Longstaff Medal from the Royal Society of Chemistry in 1993, the Hewlett Packard Europhysics Prize in 1994, honorary degrees at the Universities of Sheffield and Kingston in 1995, and the Nobel prize. He was knighted in 1996.

Krupp, Alfred (1812–1887) German metallurgist who became known as the Cannon King because of his success in manufacturing cast-steel guns. He also invented the first weldless steel tyre for railway vehicles.

Krupp was born in Essen on 26 April 1812. At 14 years of age he was obliged to leave school to help his mother run the small family steel works. The firm remained small until Krupp made his first gun in 1847. In 1851, at the Great Exhibition in London, he exhibited a solid flawless ingot of cast steel weighing 2 tonnes. Soon afterwards Krupp produced the first weldless steel tyre for railway vehicles. The works expanded its operations in the manufacture of artillery, eventually becoming the largest armaments firm in the world. Krupp was also humanely interested in the social welfare of his workers, and introduced low-cost housing and pension schemes for all his employees. He died in Essen on 14 July 1887.

In 1811 Alfred Krupp's father, Frederick, had founded a firm in Essen for the purpose of making cast steel of a superior quality. His experiments had produced a formula for making a high-quality product and, when he died prematurely in 1826, the secret was left to his son. Four years after taking charge of his father's company, Alfred Krupp extended production to include the manufacture of steel rolls. He also designed and developed new machines. He invented the spoon roll for making spoons and forks, and manufactured rolling mills for use in government mints. In 1847 the firm cast its first steel cannon, a small three-pounder muzzle-loading gun, but it still specialized in making fine steel suitable for dies, rolls, and machine-building.

At the Great Exhibition of 1851 in London, Krupp exhibited his solid flawless ingot of cast steel. In 1852 he manufactured the first seamless steel railway tyre and was the first to introduce the Bessemer and open-hearth steelmaking processes in Europe. The advent of the railways caused the firm's fortunes to increase. At first, railway axles and springs were the only cast steel products in use, but Krupp's seamless tyre greatly contributed to the speed and safety of railway travel and became extensively used. Then mass production of rails began and the company expanded still further to become a vast organization with collieries, ore mines, blast furnaces, steel plants, and manufacturing shops.

In 1863 Krupp established a physical laboratory for the testing of steels. To prove the quality of his own steel, Krupp turned to making guns and was the first to produce a successful all-steel gun. Drilled out of a single block of cast metal, it was the marvel of its day, and in the Franco-German war of 1870–1871 the performance of Krupp guns won him world renown.

By this time, the firm employed 16,000 workers and had developed into a vast integrated industrialized empire – one of the first in Germany. Although Krupp's reputation is mainly built on the manufacture of weapons of war, the output of goods from his works – even in wartime – was predominantly for peaceful purposes.

Krupp was also a philanthropist. Recognizing early the human problems of industrialization, he created a comprehensive welfare scheme for his workers. As early as 1836 he instituted a sickness fund and, in 1855, he established a pension fund for retired and incapacitated workers. Low-cost housing, medical care, and consumer cooperatives all enjoyed company backing. These institutions proved so successful that they acted as models for the social legislation enacted in Germany under Bismark 1883–1889.

Krupp built housing settlements, hospitals, schools, and churches for his employers. His social paternalism minimized serious labour troubles and fostered the Krupp employee's pride in being Kruppianer. Krupp at one time had only seven employees in his steel plant. At his death in Essen on 14 July 1887, the enterprise employed 21,000 persons.

Further Reading

Krupp, Alfred; Berdrow, Wilhelm (ed), *The Letters of Alfred Krupp, 1826–1887*, Victor Gollancz, 1930.

Kuhn, Richard (1900–1967) Austrian-born German organic chemist who worked mainly with carbohydrates and was awarded (but not allowed immediately to accept) the 1938 Nobel Prize for Chemistry for his research on the synthesis of vitamins.

Kuhn was born in Vienna on 3 December 1900, the son of a hydraulics engineer. He was first educated at home by his mother, who was a schoolteacher, and at the age of only eight began to attend the Döblinger Gymnasium, where one of his fellow pupils was Wolfgang ◊Pauli who was later to become a Nobel prizewinning physicist. He was introduced to chemistry by Ernst Ludwig, who was professor of medical chemistry at Vienna University. Kuhn was conscripted into the Austrian army in 1917, and went to Vienna University at the end of World War I in 1918. He soon moved to Germany, and completed his university education at the University of Munich where he studied under Richard ◊Willstätter, graduating in 1921 and gaining his PhD a year later with a thesis on enzymes. Kuhn then became Willstätter's assistant, before going in 1926 to teach at the Eidenössische Technische Hochschule in Zürich as professor of general and analytical chemistry. He married one of his students in 1928, and they had six children. In 1929 he became professor of organic chemistry at the University of Heidelberg and director of the Kaiser Wilhelm (later Max Planck) Institute for Medical Research. He remained there until the late 1930s when he was caught in a Nazi roundup of Jews and imprisoned in a concentration camp. The award to him of a Nobel prize came after the introduction of Adolf Hitler's policy forbidding any German to accept such an award. But after the end of World War II in 1945 Kuhn received his prize and returned to work in Heidelberg. He became editor of the chemical journal *Annalen der Chemie* in 1948. His health began to fail in 1965 and he died of cancer in Heidelberg on 31 July 1967.

Kuhn's early research concerned the carotenoids – the fat-soluble yellow pigments found in plants, which are precursors of vitamin A. In the early 1930s Kuhn and his co-workers determined the structures of vitamin A and vitamin B_2 (riboflavin) and isolated them from cow's milk, at about the same time that Paul ◊Karrer was doing similar work. Then in 1938 they took about 70,000 l/123,200 pt (148,000 US pt) of skimmed milk and after a painstaking series of extractions isolated from it 1 g/0.035 oz of vitamin B_6 (pyridoxine).

In the 1940s Kuhn continued to carry out research on carbohydrates, studying alkaloid glycosides such as those that occur in tomatoes, potatoes, and other plants of the genus *Solanum.* In 1952 he returned to experiments with milk, extracting carbohydrates from thousands of litres of milk using chromatography. This work led in the 1960s to the investigation of similar sugar-type substances in the human brain; many of these chemicals were synthesized for the first time by Kuhn and his co-workers.

Kuiper, Gerard Peter (1905–1973) Dutch-born US astronomer best known for his studies of lunar and planetary surface features and his theoretical work on the origin of the planets in the Solar System.

Kuiper was born on 7 December 1905 at Harenkarspel, the Netherlands. After completing his education in his own country, he emigrated to the USA in 1933 and four years later became a naturalized US citizen. He joined the staff of the Yerkes Observatory, which is affiliated to the University of Chicago. Between 1947 and 1949 he was director of the observatory, returning for a second term in this office 1957–60. From 1960 until his death he held a similar position at the Lunar and Planetary Laboratory at the University of Arizona, and he was closely linked with the US space programme. He died in Mexico City on 24 December 1973.

Kuiper's work on the origin of the planets stemmed from the theoretical discrepancies that arose from new 20th-century hypotheses on galactic evolution. One of the more favoured of these theories is that stars, and presumably planets, are formed from the condensation products of interstellar gas clouds. For this condensation to take place, the gravitational effects pulling the cloud together must exceed the expansive effect of the gas pressure of the cloud. However, calculations of this hypothetical process showed that under given conditions of temperature and density, there was a lower limit to the size of the condensation products. In fact, it was found that this condensation theory was inadequate to account for the temperature or the amount of material that make up the bulk of the planets in the Solar System. To compensate for this, Kuiper and his colleagues proposed that the mass of the cloud from which the planets were formed was much greater than the present mass of the planets, and suggested that the mass of the original interstellar gas cloud was approximately one-tenth the mass of the Sun, or 70 times the total mass of the planets. The results of condensation according to these new conditions would produce 'proto-planets'. But the idea of the formation of proto-planets still appears to be unworkable, partly because it involves the condensation of only 1.5% of the original interstellar gas cloud, thus leaving the rest of the material unaccounted for.

Kuiper's work on planetary features proved to be far more fruitful. In 1948 he predicted that carbon dioxide was one of the chief constituents of the Martian atmosphere – a theory he was able to see confirmed when the era of research into that planet began in 1965 with the *Mariner 1* and *2* space probes to Mars. He also discovered the fifth moon of Uranus in 1948, which he called Miranda; and in 1949 he discovered the second moon of Neptune, Nereid. Compared with Neptune's first moon, Triton, which has a diameter of 3,000 km/1,864 mi, Nereid is a dwarf and has an eccentric orbit; its distance from Neptune varies by several million kilometres. Kuiper's spectroscopic studies of the planets Uranus and Neptune led to the discovery of features, subsequently named 'Kuiper bands', found at wavelengths of 7,500 Å (7.5×10^{-7} m), which have been identified as being due to the presence of methane.

During his working life, Kuiper instigated many planetary research programmes and he played a vital role in the US space-probe programme during the late 1960s and early 1970s. In recognition of his work, the International Astronomical Union has named a ray-crater on the planet Mercury after him.

Kummer, Ernst Eduard (1810–1893) German mathematician, famous for his work in higher arithmetic and geometry, who introduced 'ideal numbers' in the attempt to prove Fermat's last theorem.

Kummer was born in Sorau (now Zary, Poland) on 29 January 1810 and was educated there before entering the University of Halle in 1828. Although his original intention was to study theology, he soon changed to mathematics and proved himself so brilliant in the subject that in 1831 the university awarded him a doctorate for his prizewinning essay on sines and cosines. From 1832 to 1842 he taught at a school at Liegnitz (now Legnica, Poland), where Leopold ◊Kronecker was one of his pupils. He was professor of mathematics at the University of Breslau 1842–55, when he was appointed to professorships at the University of Berlin and the Berlin War College. In 1857 he was awarded the Grand Prix of the French Academy of Sciences and in 1868 he was elected rector for one year of the University of Berlin. In 1882 he astonished his Berlin colleagues by announcing that failing powers of memory were weakening his ability to carry out logical, coherent thought; he therefore resigned his chair. He died in Berlin on 14 May 1893.

Kummer made an important contribution to function theory by his investigations – which surpassed those of Karl ◊Gauss – into hypergeometric series. But his two outstanding achievements were his introduction of 'ideal number' to attempt to solve Fermat's last theorem and his research into systems of rays, which led to the discovery of what is now known as the Kummer surface.

Fermat's last theorem is one of the most famous in the history of mathematics. It states that 'there do not exist integers x, y, z, none of which being zero, that satisfy

$$x^n + y^n = z^n$$

Where n is any given integer greater than 2'. Pierre de ◊Fermat left merely the statement of the theorem, without proof (except for $n = 4$), in his 1637 marginal note to his copy of ◊Diophantus' *Arithmetica*. Leonhard ◊Euler later proved the theorem for $n = 3$, but no one was able to provide a generalized proof. In general what was needed was a proof that

$$x^l + y^l + z^l = 0$$

Is impossible in non-zero integers x, y, and z for any odd prime $l > 3$. In the early 19th century Adrien ◊Legendre proved this for $l = 5$ and Henri ◊Lebesgue for $l = 7$, but later attempts to find a proof for other values of l came to naught.

Kummer studied previous attempted solutions carefully and came up with one of the most creative and influential ideas in the history of mathematics, the idea of ideal numbers. With their aid he was able to prove, in 1850, that the equation

$$x^l + y^l + z^l = 0$$

Was impossible in non-zero integers for all regular prime numbers (a special type of prime numbers related to Bernoulli numbers). He was then able, by extremely complicated calculations, to determine that the only primes less than 100 that are not regular are 37, 59, and 67. For many years Kummer continued to work on the problem and he was eventually able to prove that the equation is impossible for all primes $l < 100$. It was for his work in this field, especially the introduction of ideal numbers, an invention that some mathematicians rank in importance with the invention of non-Euclidean geometry, that Kummer received the Grand Prix of the French Academy. Much work has since been done on the problem and it is now known that the equation is impossible for all primes $l < 619$.

Kummer's other great achievement was his discovery of the fourth-order surface known as the Kummer surface. It can be described as the quartic which is the singular surface of the quadratic line complex and involves the very sophisticated and complicated concept of this surface as the wave surface in space of four dimensions.

Kundt, August Adolph (1839–1894) German physicist who is best known for Kundt's tube, a simple device for measuring the velocity of sound in gases and solids.

Kundt was born in Schwerin, Mecklenburg, on 18 November 1839 and educated at the universities of Leipzig and Berlin, from which he gained his doctorate in 1864. He began his scientific career as a lecturer in 1867, becoming professor of physics at the polytechnic in Zürich the following year. He subsequently took up the chair in physics at the University of Würzburg in 1869, and was one of the founders of the Strasbourg Physical Institute, established in 1872. He ultimately succeeded Hermann von ◊Helmholtz as professor of experimental physics and director of the Berlin Physical Institute in 1888. Kundt died in Israelsdorf, near Lübeck, on 21 May 1894.

Kundt devised his classic method of determining the velocity of sound in 1866. A Kundt's tube is a glass tube containing some dry powder and closed at one end. Into the open end, a disc attached to a rod is inserted. When the rod is sounded, the vibration of the disc sets up sound waves in the air in the tube and a position is found in which standing waves occur, causing the dust to collect at the nodes of the waves. By measuring the length of the rod and the positions of the nodes in the tube, the velocity of sound in either the rod or air can be found, provided one of these quantities is known. By using rods made of various materials and different gases in the tube, the velocity of sound in a range of solids and gases can be determined. It is also possible using the tube to find the ratio of the molar heat capacities of a gas (and hence the atomicity) and also Young's modulus of elasticity of the material in the rod.

Kundt's later work entailed the demonstration of the dispersion of light in liquids, vapours and metals, and with Wilhelm ◊Röntgen he experimented with the Faraday effect, showing that the magnetic rotation of the plane of polarizing light occurs in gases. Kundt also published observations on the spectra of lightning and on the electrical properties of crystals.

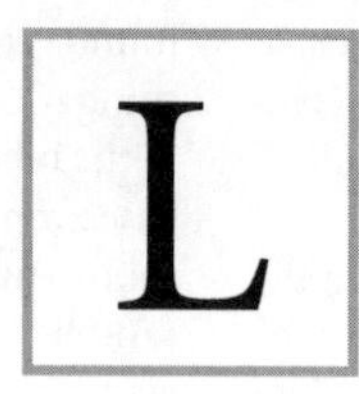

Lacaille, Nicolas Louis de (1713–1762) French scientist, a positional astronomer who also contributed to advances in geodesy. His best-known work grew out of his four-year expedition to South Africa.

Lacaille was born in Rumigny, France, on 15 March 1713. Being destined for a career in the ministry, he entered the Collège Lisieux, Paris, in 1729, but by chance he became fascinated in books on mathematics and astronomy and decided to change the direction of his career. He contacted the French Academy of Sciences in 1736 and began to make astronomical observations in 1737. He participated in two academy projects, in 1738 and 1739, and established his reputation as a thorough and talented scientist. This led to his election to the French Academy of Sciences and his appointment, at the age of only 26, as professor of mathematics at the Collège Mazarin, now the Institut de France, Paris. He was subsequently given an observatory for his personal research.

The stars of the southern hemisphere were then relatively unknown compared with those of the northern hemisphere, and so Lacaille proposed to the French Academy that a study be made of the southern skies. His suggestion was accepted, the government provided funding, and Lacaille was put in charge of the project. He left Paris in 1750 and sailed for the Cape of Good Hope. But the conditions under which he worked were poor, and he suffered from overwork and ill-health. He had completed his programme by 1753, and travelled back to France, via Mauritius and Réunion, where he continued his studies.

On his return to Paris in 1754, Lacaille quickly returned to his observatory to analyse the enormous quantity of data which he had collected. He continued his work schedule at a merciless pace and as a result died, prematurely, in Paris on 21 March 1762.

Lacaille's first serious scientific work was the task of obtaining an accurate plot of the French coastline between Nantes and Bayonne. It was assigned him by the French Academy of Sciences and was undertaken with the help of his colleague Giacoma Maraldi (1665–1729), Giovanni ◊Cassini's nephew, in 1738. The next year Lacaille was appointed to the team of scientists who were trying to resolve a dispute which had been raging for decades. In opposition to Christiaan ◊Huygens, Isaac ◊Newton, and their followers, there was a school of thought, headed by Jacques Cassini (1677–1756) that insisted that the Earth was a distorted sphere that was flattened at the Equator rather than at the poles. Lacaille and his team sought to confirm Cassini's measurement of the meridian of France which ran from Perpignan to Dunkirk. They discovered that Cassini had in fact been in error, and that his conclusions on the shape of the Earth were therefore invalid.

Upon assuming his post as professor at the Collège Mazarin, Lacaille's primary interest was to collect a huge quantity of positional data for stars in the northern skies. He also devoted considerable energy to writing popular textbooks on mathematics and applied science. The turning point in his career came in October 1750, when he began his expedition to South Africa. Despite the lack of equipment and poor research conditions Lacaille achieved an extraordinary number of observations. He determined the positions of nearly 10,000 stars, including a large number of seventh-magnitude stars that are invisible to the naked eye. He charted and named 14 new constellations, abandoning the traditional mythological naming system and instead choosing names of contemporary scientific and astronomical instruments. Lacaille measured the lunar parallax, with the aid of Joseph de ◊Lalande in Berlin. Their simultaneous observations provided them with a base-line longer than the radius of the Earth and gave an accurate measurement of the lunar parallax. It indicated that the average distance from the Earth to the Moon was approximately 60 times the radius of the Earth. Lacaille's determination of the solar parallax was less accurate, being underestimated by about 10%.

Lacaille's other astronomical work during his stay in South Africa included the compilation of the first list of 42 'nebulous stars', estimates for the distances between the Earth and both Mars and Venus, and the observation that Alpha Centauri is a double star. He also performed a number of geodetic investigations, in particular he made the first measurement of the arc of meridian in the southern hemisphere.

Soon after his return from Cape Town, Lacaille published his preliminary star catalogue, which included a star map and a description of 2,000 of the stars he had studied in South Africa. Although this catalogue contained nearly five times as many southern stars as its most complete predecessor, it suffered from the inevitable inaccuracy resulting from Lacaille's reliance on small telescopes. The catalogue of all Lacaille's data, *Coelum australe stelliferum,* was published in 1763, a year after his death.

Lacaille's later work included a brief catalogue of 400 of the brightest stars and, in 1761, the first accurate estimate of the distance between the Earth and the Moon in which the fact that the Earth is not a perfect sphere was taken into consideration. His major work during the last years of his life was the analysis of the data he had collected during his visit to South Africa, but he also performed many long computations for other purposes (including the publication of tables of logarithms in 1760).

Laênnec, René Théophile Hyacinthe (1781–1826) French physician who invented the stethoscope in 1816.

Laênnec was born in Quimper, Brittany, on 17 February 1781. Because his mother died early, he was sent to live with his uncle who was a physician in Nantes, and there he was introduced to medical work. The French Revolution struck the town with some ferocity and Laênnec, only just in his teens, worked in the city hospitals. In 1795 he was briefly appointed third surgeon at the Hôpital de la Paix, and soon afterwards at the Hospice de la Fraternité – he was then just 14 years old. It was there that Laênnec acquired his knowledge of practical medicine, including the dressing of wounds and the clinical management of large numbers of medical cases.

Because his own health was not good, he abandoned medicine until 1799, when he returned to his medical studies and became a pupil of J N Corvisart. He was appointed surgeon at the Hôtel Dieu in Nantes, from where he went on to take up the study of dissection in the in Depuytren's laboratory at the Ecole Pratique in Paris.

Between June 1802 and July 1804. Laênnec published a number of papers on anatomy. His 1804 thesis 'Propositions sur la doctrine d'Hippocrate relativement à la médecin-pratique' resulted in the award of a doctor's degree.

Laênnec became particularly interested in pathology. An associate of the Société de l'Ecole de Médecine from 1808, he established the Athenée Médical, which subsequently became incorporated in the Société Académique de Médecine de Paris. Laênnec was appointed personal physician to Cardinal Fesch, uncle of Napoleon I. In 1812–13, with France at war, Laênnec assumed responsibility for the wards in the Salpetrière, which cared solely for Breton troops. When the monarchy was restored in 1814, Laênnec took the post of physician to the Necker Hospital but, by February 1818, health problems forced him to retire to his estate in Brittany, where he lived the life of a country gentleman. At the end of that year, however, he returned to Paris, but in 1819 he was again forced to retreat to his estates at Kerbourarnec, where he resumed his existence as a country squire.

In 1822 Laênnec was appointed to the chair and a lectureship at the Collège de France. The Académie de Médecine granted him full membership from 1823 and, in 1824, he was bestowed with the order of Chevalier of the Légion d'Honneur. He gained international repute as a lecturer. However, his health failed him again and he left Paris for the last time in May, 1826. He died on 13 August of the same year.

Laênnec's important contribution to medical science was his invention of the stethoscope. In 1816, while walking in a courtyard in the Louvre, he noticed two small children with their ears close to the ends of a long stick. They were amusing themselves by tapping lightly on the stick and listening to the transmitted sound. He immediately realized the diagnostic possibilities and quickly developed an instrument based on the children's activities. It was essentially a wooden tube some 30 cm/12 in long and he used it to listen to sounds in the human body. As it was particularly useful for sounding the lungs and heart, he called it a stethoscope (from the Greek *stethos,* meaning chest).

Laênnec's main interest now lay in the diseases emphysema and tuberculosis, and the physical symptoms of chest diseases. Although auscultation (listening to internal bodily sounds) had been used as a diagnostic procedure since the time of Hippocrates, the usual 'direct' method frequently proved difficult or inconvenient. Laênnec introduced the 'mediate' method by using his stethoscopic instruments – a hollow tube for the lungs, and a solid wooden baton to listen to the heart.

In February 1818 he presented a paper on the mediate method of auscultation to the Académie de Médecine. His book, *De l'Auscultation médiate,* published in August 1819, encompassed much more than the use of the stethoscope; it was an important treatise on diseases of the heart, lungs, and liver, and was acknowledged to represent a great advance in the understanding of chest diseases, including the classification of the physical signs of egophony, rales, rhonchi, and crepitations.

Lagrange, Joseph Louis (Giuseppe Lodovico) (1736–1813) Italian-born French mathematician who revolutionized the study of mechanics.

Lagrange was born, of a French father and Italian mother, at Turin on 25 January 1736. Almost entirely by teaching himself the subject he made himself a formidable mathematician by the age of 16, and in 1755, when he was just 19, he was appointed professor of mathematics at the Royal Artillery School at Turin. In

1759 he organized a mathematical society of his students and colleagues and this ultimately became the Turin Academy of Sciences. He remained at Turin until 1766 when, on the recommendation of Leonhard ◊Euler and Jean ◊d'Alembert, Frederick the Great of Prussia invited him to Berlin to succeed Euler as the director of the Berlin Academy of Sciences. In 1787 he accepted the invitation of Louis XVI to go to Paris as a member of the French Royal Academy and in 1797 he was appointed professor of mathematics at the Ecole Polytechnique.

Like so many brilliant mathematicians, Lagrange had his best ideas as a very young man. By the time that he began to teach at Turin in 1755, he had already conceived the heart of his great work, the *Mécanique analytique,* and was near to his first important discovery, the solution to the isoperimetrical problem, on which the calculus of variations partly depends. In 1759 he sent his solution to Euler, who had been baffled and held up by the problem for years; Euler wrote back to express his delight at the young man's solution and delayed publishing his own work on the subject until Lagrange – whose proof had greater generality than Euler's – had published his discovery. In that year also Lagrange wrote a paper on minima and maxima and, advancing beyond Isaac ◊Newton's theory of sound, settled a fierce dispute on the nature of a vibrating string.

In 1764 Lagrange addressed himself to the famous three-body problem on the mutual gravitational attraction of the Earth, Sun, and Moon, and for his work on it was awarded the French Academy's Grand Prix. Two years later he won the award again for his work on the same problem with regard to Jupiter and its four (as were then known) satellites. When he moved to Berlin he diverted himself for a time with problems in number theory. He proved some of Pierre de ◊Fermat's theorems, which had remained unproven for a century; and he solved a problem known since the time of the ancients, namely, to find an integer x so that $(nx^2 + 1)$ is a square where n is a positive integer but not a square.

All the while, however, his chief concern was mechanics. By 1782 he had completed his *Mécanique analytique,* although it was not published until 1788, by which time Lagrange had become so uninterested in mathematics that the manuscript had lain untouched and unconsulted for two years. In his epoch-making treatise Lagrange, by considering a four-dimensional space with the aid of calculus, succeeded in reducing the theory of solid and fluid mechanics to an analytical principle. Breaking with conventional methods used since the time of the Greeks. Lagrange published his work without the aid of a single diagram or construction: his discussion of mechanics, that is to say, was entirely algebraic. Of this achievement Lagrange was justly proud, especially since he was able to reflect that the event that had, more than anything else, drawn him into mathematics was his reading, as a young boy, the work of English astronomer Edmond ◊Halley in the use of algebra in optics.

Lagrange never again did anything of such significance, although his lectures at the Ecole Polytechnique, published as the *Théorie des fonctions analytiques* (1797) and the *Leçons sur le calcul des fonctions* (1806), were of much use to Augustin ◊Cauchy and others. But he had still one contribution to make to the world. In 1793 he was appointed president of the commission established to standardize French weights and measures and by persuading the commission to adopt 10, not 12, as the basic unit, was the father of the metric system in use throughout most of the world today. For that, perhaps, but above all for his work on mechanics he was deservedly given a final resting place in the Panthéon.

Laithwaite, Eric Robert (1921–) English electrical engineer, most famous for his development of the linear motor.

Laithwaite was born on 14 June 1921 at Atherton, Yorkshire. He was educated at Kirkham Grammar School and then at Regent Street Polytechnic, London. He read for his BSc at the University of Manchester, but his studies were interrupted by World War II. From 1941 to 1946 he served in the Royal Air Force; for four years he worked at the Royal Aircraft Establishment, Farnborough, researching into automatic pilots for aircraft. He gained his BSc in 1949 and an MSc in the following year. He remained at the University of Manchester until 1964, holding positions of assistant lecturer, lecturer, and senior lecturer. During that time he was awarded a PhD in 1957, and a DSc in 1964. In that same year he was appointed professor of heavy electrical engineering at the Imperial College of Science and Technology at the University of London (becoming professor emeritus in 1986).

Laithwaite is unusually able to put his ideas across to nonspecialists including children, and was one of the most popular contributors to the Christmas Lectures at the Royal Institution. In 1966 the title of his lectures was 'The Engineer in Wonderland', which he followed in 1974 with 'The Engineer Through the Looking Glass'. He has also participated in radio and television broadcasts on general science including: *Young Scientist of the Year, It's Patently Obvious,* and *The Engineer's World.* He has also made a number of films, several of which have won major awards.

The idea of a linear induction motor had been suggested in 1895, but a new principle was discovered and worked on by Laithwaite: that it is possible to arrange two linear motors, back-to-back, so as to produce continuous oscillation without the use of any switching device. The important feature of the linear motor is that it is a means of propulsion without the need for wheels. This in itself is nothing remarkable. Yachts, rowing boats, surfboards, ice-skates, skis, canoes, gliders – all are examples of propulsion methods without wheels. Even swimming and walking are mechanisms requiring no rotary movement in a pure sense. Our early propulsion systems were designed to use either muscular power or natural power sources such as air currents or the wave energy of a sailing yacht in a steady wind. It is possible continually to extract power from a source, although often continuous motion has to be made up from reciprocating motion – achieved by the use of a ratchet, for example.

Linear motion is very common in industrial processes. Planing machines, looms, saws, knitting machines, and pumps all make use of motion in a straight line. In 1947 Laithwaite began research into electric linear induction motors as the shuttle drives in weaving looms. Further investigations were carried out into the use of linear motors for conveyers and as propulsion units for railway vehicles. By 1961, his research had aroused sufficient interest for it to be included in a BBC television programme, which added to public interest in the linear induction motor, and development was further stimulated by a lecture that Laithwaite gave to the Royal Institution in March 1962 entitled 'Electrical Machines of the Future'

Probably the project that aroused the greatest interest in the media was the high-speed transport project, in which hovering vehicles, moving on air cushions or jets, are powered by linear induction motors, used as high-speed propulsion units. These systems are in use in the USA, Germany, Japan, and the former Soviet Union for high-speed railways.

Lakatos, Imre (1922–1974) Born Imre Lipschitz. Hungarian philosopher of science and mathematics whose work aimed to show that mathematical discovery involves more than finding an abstract truth. His work built upon the philosophy of Austrian-born philosopher of science Karl Popper (1902–1994) and provided an alternative to Popper's theory of theory of falsificationism.

Lakatos was born in Debrecen, Hungary, on 9 November 1922, the son of Jacob Lipschitz and Margit Herzfeld. He was educated at the Gymnasium of Debrecen 1932–40, where he won prizes for mathematics. He began a law degree at the University of Debrecen in 1940 but switched in 1941 to mathematics and physics, also attending philosophy lectures. He took degrees in mathematics and physics in 1946, moving on to Budapest and receiving a PhD in 1947. During this time he was also a member of the anti-Nazi resistance, narrowly avoiding arrest. To avoid detection he changed his name first to Molnar and after the war to Lakatos.

Lakatos was a communist and served under the Stalinists in the Hungarian ministry of education as a secretary. He spent the year 1949 in Moscow. He was distrusted by the authorities however and was arrested as a dissident in 1950. He was incarcerated for three years during which time he was tortured and interrogated. On his release in 1953 he survived by translating the work of Hungarian-born US mathematician George ◊Pólya and others into Hungarian. He also spent some time as an assistant to the Hungarian Marxist philosopher Georg Lukacs from whom he is thought to have acquired some of the methods and terminology he was later to use to formulate his philosophy of science and mathematics.

Disillusioned with Marxism by his experiences of incarceration Lakatos fled to Vienna in 1956 after the Hungarian uprising, making his way eventually to Cambridge where, supported by the Rockefeller Foundation, he gained a PhD in 1958. There he met Pólya, who inspired him to study the history of mathematics at the London School of Economics (LSE). His unpublished thesis later evolved into a four-part article 'Proofs and refutations' (1963–64) and into a book, *Proofs and Refutations* (1976), which was a collection of articles in dialogue form, illustrating the creativity and informality of real mathematical discovery. The work was based on the formula

$$V - E + F = 2$$

Which had first been noticed by Swiss mathematician Leonhard ◊Euler in 1758, to be true of regular polyhedra, with V as the number of vertices, E edges, and F faces.

Lakatos taught at the LSE for the rest of his life, succeeding Karl Popper in 1969 as professor of logic and scientific method. Originally Lakatos had seen science in the same terms as Popper, but he later developed his own notion of a 'methodology of scientific research programmes', which, unlike Popper's work, did not take falsification as its central principle and was an alternative to the theories of both Popper and Thomas Kuhn (1922–). His programme rather consisted of a set of rules which indicated which research paths to avoid (negative heuristic) and which to follow (positive heuristic). In Lakatos's model research programmes can either be 'progressive' and lead to new discoveries, or 'degenerating', leading to a change of programme. He was still developing these views when he died suddenly at the age of 51 on 2 February 1974. His posthumously published *Philosophical Papers* (1977) in two volumes, contains much of his later work.

Lalande, Joseph Jérome le François de (1732–1807) French astronomer noted for his planetary tables, his account of the transit of Venus, and numerous writings on astronomy.

Lalande was born in Bourg-en-Bresse on 11 July 1732. He was the only child of relatively wealthy parents, his father being the director of both the local post office and a tobacco warehouse. Lalande was educated at the Jesuit College in Lyon in preparation for his joining the order. Instead he went to Paris to continue his studies and it was during this time that he first became intrigued by astronomy. He was influenced by Joseph-Nicolas Delisle (1688–1768) and Charles le Monnier, both leading figures in the fields of astronomy and mathematical physics. Lalande was appointed professor of astronomy at the Collège de France in 1762 and during his tenure there he published his *Treatise of Astronomy.*

In 1795 he was made director of the Paris Observatory, where he concentrated on compiling a catalogue of stars in which he noted 47,000 stars in all, including the planet Neptune, without realizing that it was not a star; thus he had observed Neptune 50 years before it was formally discovered. Lalande died in Paris on 4 April 1807.

It was Lalande's involvement in calculating the distance to the Moon that earned him his initial acclaim. In those days the method used to calculate the distance to celestial objects was parallax. This meant that measurements had to be made simultaneously by two people from observatories sited on opposite sides of the Earth. Lalande's mentor, Monnier, generously suggested that Lalande should participate in the parallax programme and so in 1751 he was despatched to Berlin, where he became a guest of the Prussian Observatory. At the same time, in collaboration with Lalande, another French venture led by Nicolas de ◊Lacaille was mounted at the Cape of Good Hope. Its purpose was to note the position of the Moon at the time of its transit from different positions on the same meridian. The combined results of these two ventures greatly improved the accuracy of the value of the Moon's parallax. Having published the results of these observations, Lalande and Monnier disagreed over the allowance to be given to the flattening of the Earth's surface. It was eventually decided, by consensus, that Lalande's results were valid, but Monnier never quite recovered from the slight and chose to remain somewhat distant from his former pupil from that time onwards.

Lalande was at the height of his profession during the 1760s and so was fortunate enough to be able to participate in the observations of the two transits of Venus that occurred in 1761 and 1769. Such transits are a rare phenomenon that occur twice within a period of eight years only every 113 years, and the event offered 18th-century astronomers the chance to establish accurately the size of the Solar System. During the transit, which takes approximately five hours, Venus can be seen silhouetted across the face of the Sun; the distance of the Earth from the Sun can be deduced by measuring the different times that the planet takes to cross the face of the Sun when seen from different latitudes on Earth. A host of expeditions was organized with worldwide collaboration, and Lalande was made personally responsible for collecting the results of observations and coordinating expeditions to all corners of the world.

Lalande was a controversial person throughout his long life, often falling into arguments with fellow astronomers. Besides his particularly well-documented account of the transit of Venus, Lalande wrote numerous textbooks on astronomy.

Lamarck, Jean Baptiste Pierre Antoine de Monet, Chevalier de (1744–1829) French naturalist best known for his alternative to Charles ◊Darwin's theory of evolution and the distinction he made between vertebrate and invertebrate animals.

Lamarck was born in Bazantin, Picardy, on 1 August 1744, the 11th child of a large family of poor aristocrats. This noble background restricted his career and

Lamarck Engraving of French naturalist Jean Baptiste de Lamarck whose theory of evolution stated that acquired characteristics are inherited. In 1781 Lamarck was made botanist to King Louis XVI, and subsequently travelled widely in Europe on botanical excursions. *Mary Evans Picture Library*

he was expected to join either the army or the church. The first intention was that he should enter the church, but when his father died in 1760 he joined the army. After serving in the Seven Years' War he left the army in 1766, when his health failed, and eventually went into medicine. He soon became interested, however, in meteorology and botany. He was admitted to the Academy of Sciences in 1778 for his book *Flore française,* written while he was posted in southern France, and attracted the interest of Georges ◊Buffon, the famous French naturalist. With Buffon's assistance he travelled across Europe, having been appointed botanist to King Louis XVI in 1781. He returned in 1785 and three years later took up a botanical appointment at the Jardin du Roi. In 1793 he was made a professor of zoology at the Museum of Natural History, in Paris, where he lectured on the 'Insecta' and 'Vermes' of ◊Linnaeus, which he later called the Invertebrata. He continued his writings, although his sight was failing, with the help of his eldest daughter and Pierre-André Latreille (1762–1833). He died blind and in great poverty, in Paris, on 18 December 1829.

Lamarck's own experiences in both botany and zoology made him realize the need to study living things as a whole, and he coined the word 'biology'. In 1785 he pointed out the necessity of natural orders in botany by an attempt at the classification of plants. He widely propagated the idea, as Buffon had done before him, that species are not unalterable and that the more complex ones have developed from pre-existent simpler forms.

Lamarck's most important contribution was his detailed investigation of living and fossil invertebrates. At the age of 50, while lecturing on Linnaeus' classifications, he began the study of invertebrates. So little was known about them at this time that some scientists grouped snakes and crocodiles with insects. Lamarck was the first to distinguish vertebrate from invertebrate animals by the presence of a bony spinal column. He was also the first to establish the crustaceans, arachnids, and annelids among the invertebrates.

In studying and classifying fossils Lamarck was led to wonder about the effect of environment on development. In 1809 he published one of his most important essays, 'Zoological philosophy', which tried to show that various parts of the body developed because they were necessary, or disappeared because of disuse when variations in the environment caused a change in habit. He believed that these body changes are inherited by the offspring, and that if this process continued for a long time, new species would eventually be produced.

Lamarck thought that it ought to be possible to arrange all living things in a branching series showing how some species gradually change into others. Unfortunately he chose poor examples to illustrate his ideas and he did not attempt to show how one species might gradually change into another on the basis of the inheritance of acquired characteristics. One of the well-known examples that constitutes the evidence on which Lamarck based his theory is the giraffe which he assumed, as an antelope-type animal, sought to browse higher and higher on the leaves of the trees on which it feeds, and 'stretched' its neck. As a result of this habit continued for a long time in all the individuals of the species, the 'antelope's' front limbs and neck gradually grew longer, and it became a giraffe. Another example is that of birds that rest on the water and which, perhaps to chase and find food, spread out their feet when they wish to swim. The skin on their feet becomes accustomed to being stretched and forms a web between the toes. These examples were regarded as rather naïve, even to his contemporaries, which could explain why Lamarck's ideas were not accepted.

A contemporary of Lamarck, Georges ◊Cuvier, proposed his own theory of evolution which, although inferior to Lamarck's, achieved considerable acclaim. The work of Lamarck received little recognition in his own lifetime and some of his successors, including Darwin, regarded it with ridicule and contempt. There is no doubt, however, that his zoological classifications have aided further study in that field and that he greatly influenced later evolutionists.

Among the many works he produced, one of Lamarck's greatest is the seven-volume *Natural History of Invertebrates,* produced between 1815 and 1822.

Further Reading

Barsanti, Giulio, 'Lamarck and the birth of biology 1740–1810' in: Poggi, Stefano and Bossi, Maurizio (eds), *Romanticism in Science: Science in Europe, 1790–1840,* Kluwer Academic, 1994, pp 47–74.

Barthelemy-Madaule, Madeleine; Shank, Michael (transl), *Lamarck the Mythical Precursor: A Study of the Relations Between Science and Technology,* MIT Press, 1982.

Corsi, Pietro, *The Age of Lamarck: Evolutionary Theories in France, 1790–1830,* University of California Press, 1988.

Jordanova, L J, *Lamarck,* Past Masters series, Oxford University Press, 1984.

Jordanova, L J, Nature's powers: a reading of Lamarck's distinction between creation and production in: Moore, James R (ed), *History, Humanity, and Evolution: Essays for John C Greene,* Cambridge University Press, 1989, pp 71–98.

Lamb, Horace (1849–1934) English applied mathematician, noted for his many books on hydrodynamics, elasticity, sound, and mechanics.

Lamb was born in Stockport, Cheshire, on 27 November 1849. His father was the foreman in a cotton mill; his mother died when he was young and he was brought up by his aunt, who was largely responsible for enrolling him at Stockport Grammar School. There Lamb excelled at Latin and Greek and at the age of 17 he won a classical scholarship to Queen's College, Cambridge. He spent a year at Owens College, Manchester, before going to the university, however, and having gained a scholarship to Trinity College entered there in the autumn of 1868. In 1872 he gained his BA in mathematics and in the same year he was elected a fellow and lecturer at Trinity.

Cambridge still imposed celibacy upon its fellows and in 1875, having married, Lamb went to Australia to take up the chair of mathematics at the University of Adelaide. He remained there until 1885, when he returned to Manchester as professor of mathematics at Owens College, a post he held until his retirement in 1920. His last years were spent in Cambridge. The university established an honorary lectureship for him, called the Rayleigh Lectureship, and Trinity elected him an honorary fellow. From 1921 to 1927 he was also a member of the Aeronautical Research Committee attached to the Admiralty and until his death he served on an advisory panel that was concerned with the fluid motion set up by aircraft.

Lamb was elected to the Royal Society in 1884 and was awarded the society's Royal Medal in 1902 and the Copley Medal, its highest honour, in 1923. He was president of the British Association in 1925. He was knighted in 1931, and he died in Cambridge on 4 December 1934.

Lamb was born a generation after physicists such as George ◊Stokes, James Clerk ◊Maxwell, and John ◊Rayleigh had begun to revive interest in the mathematical aspect of problems in physics – heat, electricity, magnetism, and elasticity. His was an age of brilliant achievement in applied mathematics and he made his own mark in many fields: electricity and magnetism, fluid mechanics, elasticity, acoustics, vibrations and wave motions, seismology, and the theory of tides and terrestrial magnetism. He was particularly adept at applying the solution of a problem in one field to problems in another. His greatest achievement was in the field of fluid mechanics. In 1879 he published his best-known work, *A Treatise on the Motion of Fluids,* issued in a new edition as *Hydrodynamics* in 1895, and re-issued in five subsequent editions (the last in 1932), each of which incorporated the latest developments in the subject. The book is rightly regarded as one of the most clearly written treatises in the whole of applied mathematics.

Lamb's contributions ranged wide over the field of applied mathematics. A paper of 1882, which analysed the modes of oscillation of an elastic sphere, achieved its true recognition in 1960, when free Earth oscillations during a Chilean earthquake behaved in the way he had described. Another paper of 1904 gave an analytical account of propagation over the surface of an elastic solid of waves generated by given initial disturbances, and the analysis he provided is now regarded as one of the seminal contributions to theoretical seismology. In 1915, with Lorna Swain, he produced the first creditable account of the marked phase differences of the tides evident in varying parts of the Earth's oceans, and thus resolved an issue which had been a matter of controversy since Isaac ◊Newton's time.

Lamb made numerous important discoveries; he published widely; he was, in the words of Ernest ◊Rutherford, 'more nearly my ideal of a university professor than anyone I have known'. By many people he is regarded as the finest applied mathematician of the century.

Lanchester, Frederick William (1868–1946) English engineer who created the first true British motorcar of reliable design and produced the first comprehensive theory of flight.

Lanchester was born in Lewisham, London, on 28 October 1868. Two year later his family moved to Hove, in Sussex, where his father set up business as an architect. In 1883 Lanchester went to the Hartley Institution (which later became University College, Southampton). From here he won a scholarship to what is now Imperial College, London.

After a brief period as a Patent Office draughtsman, he joined the Forward Gas Engine Company of Birmingham as assistant works manager in 1889, becoming works manager within a year.

It was in 1893, following his return from an unsuccessful trip to the USA to sell his gas-engine patents, that Lanchester set up a workshop next to the Forward Company. By 1896 his first motorcar had emerged, with a single cylinder 4-kW/5-hp engine and chain drive. He rebuilt it the following year with a 6-kW/8-hp flat twin engine and a worm gear transmission for which he designed both the worm form and the machine tool to produce it. His second car, built at his Ladywood Road factory, was a 6-kW/8-hp Phaeton with flywheel ignition, roller bearings, and cantilever springing.

The car won a gold medal for its performance and design at the Royal Automobile Club trials at Richmond in 1899, and the following year it completed a 1,600-km/1,000-mi tour. The company then went into production with 7.5–9-kW/10–12-hp motor-cars, of which some 350 were built between 1900 and 1905. A 15-kW/20-hp four-cylinder and 21-kW/28-hp six-cylinder models were produced in 1904 and 1905 respectively.

As early as 1891 Lanchester had acquired an interest in crewed flight and soon began experimenting with powered and gliding models. He showed that a well-designed aircraft could be inherently stable, a concept that underpins modern airworthiness requirements. On 19 June 1894 he marshalled his theories of flight in a paper of fundamental importance: 'The soaring of birds and the possibilities of mechanical flight', which was given at a meeting of the Birmingham Natural History and Philosophical Society.

Lanchester revised and amplified his paper, and submitted it to the Physical Society, only to have it rejected in September 1897. Nothing daunted, he resurrected his theories and incorporated them in his great work, *Aerial Flight* which was published in two volumes: *Aerodynamics* (1907) and *Aerodonetics* (1908). The first volume won Lanchester instant acclaim.

Lanchester and his collaborator Norman A Thompson designed an experimental aircraft whose wings were covered by an aluminium skin and a sheet steel fuselage. Sadly, the undercarriage collapsed on the first trial flight in 1911, and Lanchester decided to quit the practical side of aviation. A second aircraft, with broadly similar design, did fly successfully in the summer of 1913, and in the following year a seaplane, also incorporating many of Lanchester's ideas, took to the air. This machine was the prototype for a batch of six built for the Admiralty and used during World War I to counter German submarines.

In 1900 Lanchester acquired the Armourer Mills in Montgomery Street, Sparbrook, which were to house the Lanchester Engine Company until 1931. However, the company went into receivership in 1904 and was reconstructed the following year as the Lanchester Motor Company. This was a financial success.

The publication of *Aerial Flight* led to Lanchester's joining Prime Minister Asquith's advisory committee for aeronautics at its formation on 30 April 1909. The same year he was appointed consulting engineer and technical advisor to the Daimler company.

With the outbreak of World War I, Lanchester's involvement in military work became so great that he moved to London. On 3 September 1919 he married Dorothea Cooper, daughter of the vicar of Field Broughton, near Windermere.

After the war he left the committee, and his work in the capital tailed off. Having become increasingly busy in the Daimler business in Coventry, Lanchester decided to move back to Birmingham. He founded Lanchester's Laboratories Ltd in 1925, intending to provide research and development services, but the economic depression meant there was little call for his expertise.

Lanchester was also interested in radio and, having patented a loudspeaker and other audio equipment he went into production. Again he hit difficulties, and these, combined with his failing health, forced him finally to close down his laboratories in January 1934. His health improved towards the end of the year, and he began to write prolifically both on technical and non-technical subjects. His works included a collection of poems (under the pseudonym of Paul Netherton-Herries), an explanation of relativity, and a treatise on the theory of dimensions.

When World War II broke out, the symptons of Parkinson's disease had become clear. Although his mind was still lively, Lanchester was forced to retire. Finally, plagued by financial worries and losing his sight, Lanchester died on 8 March 1946.

Lanchester's work with motorcars is remarkable not only because of the number of new ideas that were incorporated into them, but also because he produced them with interchangeable parts. This was the concept later adopted so successfully by mass-production manufacturers. In his study of flight he was the first engineer to depart from the empirical approach and to formulate a comprehensive theory of lift and drag. His work on stability was fundamental to aviation.

Lanchester received many honours and awards, including the presidency of the Institute of Automobile Engineers in 1910, fellowship of the Royal Aeronautical Society in 1917 and of the Royal Society in 1922, and the Gold Medal of the RAS in 1926. Birmingham University made him an honorary doctor of law in 1919.

Further Reading

Clark, Chris, *Lanchester Legacy: A Trilogy of Lanchester Works, v 1: 1895 to 1931*, Coventry University, Centre for Business History, 1995.

Fletcher, John (ed), *Lanchester Legacy: A Trilogy of Lanchester Works, v 3: A Celebration of Genius*, Coventry University, Centre for Business History, 1996.

Landau, Lev Davidovich (1908–1968) Russian-Azerbaijani theoretical physicist who developed a theory of condensed matter and explained the properties of helium.

Landau was born in Baku, now in Azerbaijan on 22 January 1908. His father was a petroleum engineer who had worked in the Baku oilfields and his mother was a physician. Landau finished school in Baku at the age of 13 and wanted to go straight to university. His parents thought him too young and sent him for a year to Baku Economic Technical School before allowing him to enter Baku University in 1922 where he studied physics, mathematics, and chemistry. In 1924 at the age of 16 he transferred to the physics department at Leningrad University (now St Petersburg) publishing his first scientific work three years later, on quantum mechanics. Graduating the same year, he became a graduate student at the Leningrad Institute of Physics and Technology.

In 1929 Landau won a fellowship to visit foreign scientific institutions. He visited Germany, Switzerland, the Netherlands, England, Belgium, and in Denmark he met Neils ◊Bohr who was his greatest inspiration in the field of theoretical physics. In 1930 Landau and Rudolf ◊Peierls investigated subtle problems in quantum mechanics, and Landau formulated his first major contribution to science – the theory of diamagnetism in metals (known as Landau diamagnetism).

Landau returned to the institute in Leningrad in 1931 and in 1932 transferred to Kharkov where he became the head of the new Ukrainian Institute of Mechanical Engineering and began working on ferromagnetism, phase transitions of the second kind (subtle changes not involving heat exchange), and the superfluidity of helium. At the same time he was also chair of theoretical physics at the Kharkov Institute of Mechanical Engineering, and from 1935 he held the chair of general physics at Kharkov University. In 1934 he was awarded the degree of doctor of physical and mathematical sciences and in 1935 was given the title professor.

In 1937 Landau became the director of the section of theoretical physics at the Institute of Physical Problems of the USSR Academy of Sciences in Moscow and in 1943 he became professor of physics at Moscow State University. His work at this time included the theory of electron showers and the intermediated state of superconductors, and his interest in the physics of elementary particles and nuclear interactions began to come to the fore. He worked on the theory of explosions and the scattering of protons by protons, and the theory of ionization losses of fast particles in a medium. Working with Pyotr ◊Kapitza 1940–45, he developed a quantum mechanical explanation for the oscillatory behaviour of helium for which he was awarded the Nobel prize in 1962. In 1962 Landau was involved in a car accident receiving injuries that effectively put an end to his career, although doctors managed to keep him alive for another six years. Landau died in Moscow on 1 April 1968.

In the 1920s Landau had been an ardent Communist but disillusionment set in the 1930s as he saw the effects of military and technical directives on the freedom of scientific research. He was imprisoned in 1938 for a year for his anti-Stalinist activities as the head of a counter-revolutionary organization. In spite of this he received Stalin prizes in 1949 and 1953 for his work on the hydrogen and atomic bombs, and in 1954 was awarded the title of Hero of Socialist Labour.

Landau was an excellent teacher and a prolific and lucid writer. He published numerous books and papers among which is a nine-volume textbook *Course of Theoretical Physics* written with one of his students, Evgenni Livshitz, which became a standard work and remained in print until the 1980s. He created an important scientific school of theoretical physics that produced a number of distinguished scientists in various fields.

Landé, Alfred (1888–1976) German-born US physicist known for the Landé splitting factor in quantum theory.

Landé was born in Elberfeld on 18 December 1888. He studied at the universities of Marburg, Göttingen, and Munich, gaining his DPhil from Munich in 1914. In 1919 he became a lecturer at the University of Frankfurt am Main until 1922, when he went to Tübingen as associate professor of theoretical physics. He visited Columbus, Ohio, USA, in 1929 and again in 1930; in 1931, because of the rise of the Nazi regime, he left Germany forever to settle in Columbus as professor of theoretical physics at Ohio State University. He died on 30 October 1976.

Landé had just gained his doctorate under Arnold ◊Sommerfeld when World War I began in 1914. He joined the Red Cross and served for a few years in a hospital until Max ◊Born placed him as his assistant at the Artillerie-Prüfungs-Kommission in Berlin to work on sound-detection methods. He collaborated with Born to conclude that Niels ◊Bohr's model of co-planar electronic orbits must be wrong, and that they must be inclined to each other. When they published this paper in 1918, it caused quite a stir in the scientific world.

Several more papers were written, and in one of them Landé strongly advocated a tetrahedral arrangement for the four valence electrons of carbon. While at Frankfurt, Landé had the opportunity to visit Bohr in Copenhagen to discuss the Zeeman effect (the splitting of spectral lines in a magnetic field). Then on moving to Tübingen, he found himself at the most important centre of atomic spectroscopy in Germany. In 1923 Landé published a formula expressing a factor known as the Landé splitting factor for all multiplicities as a function of the quantum numbers of the stationary state of the atom. The Landé 'splitting' factor is the ratio of an elementary magnetic moment to its causative angular momentum when the angular momentum is measured in quantized units. It determines fine as well as hyperfine structures of optical and X-ray spectra.

Landé then collaborated with Friedrich Paschen (1865–1947) and others to analyse in great detail the fine structure of the line spectra and the further splitting of the lines under the action of magnetic fields of increasing strength. He formulated the regularities obeyed by the frequencies and intensities of the lines in terms of the sets of quantum numbers attached to the spectroscopic terms and taking integral or half-integral values.

Landsteiner, Karl (1868–1943) Austrian-born US immunologist who discovered the ABO blood groups, and the M and N factors and rhesus factor in human blood. For this work he received the 1930 Nobel Prize for Physiology or Medicine.

Landsteiner was born in Vienna on 14 June 1868, the son of a journalist. He attended the University of Vienna 1885–91, when he graduated with a doctorate in medicine. He spent the next five years studying at various universities in Europe, including a time with the German chemist Emil ⇨Fischer. In 1898 he became a research assistant at the Vienna Pathological Institute and remained there until 1908, when he was appointed professor of pathology at the Royal Imperial Wilhelminen Hospital in Vienna. Unsatisfactory working conditions there caused him to leave in 1919, and he went to the RK Hospital in The Hague. He was no happier there, and moved to the USA where he joined the Rockefeller Institute in New York in 1922. He became a US citizen in 1929 and died in New York on 26 June 1943.

While at the Vienna Pathology Institute, Landsteiner published a paper in 1900 that included information on the interagglutination of human blood cells by serum from a different human being. He explained this feature as occurring because of antigen differences in the blood between individuals. In the following year Landsteiner described a technique that enables human blood to be categorized into three groups: A, B, and O (and later, AB). His sorting method, which is still used, was to mix suspensions of blood cells with each of the sera, anti-A and anti-B. Landsteiner discovered in this way that serum contains only those antibodies that do not cause agglutination with its own blood cells. It was not until 1907, however, that a blood-match was first carried out and blood transfusions did not become a widespread practice until after 1914 when Richard Lewisohn found that the addition of citrates to blood prevents it from coagulating, enabling it to be preserved.

In 1927 Landsteiner and Philip Levine found that, in addition to antigens A and B, human blood cells contain one or other or both of two heritable antigens, M and N. These are of no importance in transfusions, because human serum does not contain the corresponding antibodies, but being inherited they are of value in resolving paternity disputes.

In 1940, Landsteiner and Levine, together with Wiener, injected blood from a rhesus monkey into rabbits and guinea pigs. The resulting antibodies agglutinated the rhesus blood cells because they contain an antigen, named Rh, which is identical or very similar to an antigen present in the rhesus monkeys. It was shown that the Rh factor is inheritable; those people possessing it are termed Rh positive, and those without it, Rh negative. The connection between this factor and a blood disorder of newborn babies (erythroblastosis foetalis) was revealed. An Rh-negative mother pregnant with an Rh-positive foetus forms antibodies against the Rh factor. In a subsequent similar pregnancy the oxygen-carrying capacity of the fetus' red cells is impaired and brain damage may result. Diagnostic tests can now enable the necessary preventive measures to be taken.

Landsteiner investigated the Wassermann test (a blood test for syphilis involving adding antigen to detect the presence of antibodies in a patient's blood). Together with Ernest Finger he discovered that the antigen, previously extracted from a syphilitic human being, could be replaced with an extract prepared from ox hearts.

Between 1908 and 1922, Landsteiner researched into poliomyelitis. He injected a preparation of brain and spinal cord tissue, obtained from a polio victim, into a Rhesus monkey, which then developed paralysis. Landsteiner could find no bacteria in the monkey's nervous system and concluded that a virus must be responsible for the disease. He later developed a procedure for the diagnosis of poliomyelitis.

Langevin, Paul (1872–1946) French physicist who invented the method of generating ultrasonic waves that is the basis of modern echolocation techniques. He was also the first to explain paramagnetism and diamagnetism, by which substances are either attracted to or repulsed by a magnetic field. Langevin was the leading mathematical physicist of his time in France, and contributed greatly to the dissemination and development of relativity and modern physics in general in his country.

Langevin was born in Paris on 23 January 1872. He attended the Ecole Lavoisier and the Ecole de Physique et de Chimie Industrielles, where he was supervised by Pierre ⇨Curie during his laboratory classes. In 1891 he entered the Sorbonne, but his studies were interrupted for a year in 1893 by his military service. In 1894 Langevin entered the Ecole Normale Supérieure, where he studied under Jean ⇨Perrin. Langevin won an academic competition which in 1897 enabled him to go to the Cavendish Laboratories at the University of Cambridge for a year. There he studied under J J ⇨Thomson, and met scientists such as Ernest ⇨Rutherford, Charles ⇨Wilson, and John ⇨Townsend.

Langevin received his PhD in 1902 for work done partly at Cambridge and partly under Curie on gaseous ionization. He spent a great deal of time in Perrin's laboratory, and was caught up in the excitement accompanying the early years of study on radioactivity and ionizing radiation. Langevin joined the faculty of the Collège de France in 1902, and was made professor of physics there in 1904, a post he held until 1909 when he was offered a similar position at the Sorbonne.

During World War I Langevin contributed to war research, improving a technique for the accurate detection and location of submarines, which continued to be developed for other purposes after the war. He became a member of the Solvay International Physics Institute in 1921, and was elected president of that organization in 1928. In 1940, after the start of World War II and the German occupation of France, Langevin became director of the Ecole Municipale de Physique et de Chimie Industrielles, where he had been teaching since 1902, but he was soon arrested by the Nazis for his outspoken anti-fascist views. He was first imprisoned in Fresnes, and later placed under house arrest in Troyes. The execution of his son-in-law and the deportation of his daughter to Auschwitz (which she survived) forced Langevin to escape to Switzerland in 1944. He returned to Paris later that year and was restored to the directorship of his old school, but died soon after in Paris on 19 December 1946.

Langevin's early work at the Cavendish Laboratories and at the Sorbonne concerned the analysis of secondary emission of X-rays from metals exposed to radiation, and Langevin discovered the emission of secondary electrons from irradiated metals independently of Georges Sagnac (1869–1928). He also studied the behaviour of ionized gases, being interested in the mobility of positive and negative ions, and in 1903 he published a theory for their recombination at different pressures. He then turned to paramagnetic (weak attractive) and diamagnetic (weak repulsive) phenomena in gases. Curie had demonstrated experimentally in 1895 that the susceptibility of a paramagnetic substance to an external magnetic field varies inversely with temperature. Langevin produced a model based on statistical mechanics to explain this in 1905. He suggested that the alignment of molecular moments in a paramagnetic substance would be random in the absence of an externally applied magnetic field, but would be non-random in its presence. The greater the temperature, however, the greater the thermal motion of the molecules and thus the greater the disturbance to their alignment by the magnetic field.

Langevin further postulated that the magnetic properties of a substance are determined by the valence electrons, a suggestion which influenced Niels ◊Bohr in the construction of his classic model describing the structure of the atom. He was able to extend his description of magnetism in terms of electron theory to account for diamagnetism. He showed how a magnetic field would affect the motion of electrons in the molecules to produce a moment that is opposed to the field. This enabled predictions to be made concerning the temperature-independence of this phenomenon and furthermore to allow estimates to be made of the size of electron orbits.

Langevin became increasingly involved with the study of Albert ◊Einstein's work on Brownian motion and on space and time. He was a firm supporter of the theory of the equivalence of energy and mass. Einstein later wrote that Langevin had all the tools for the development of the special theory of relativity at his disposal before Einstein proposed it himself and that if he had not proposed the theory, Langevin would have done so.

During World War I, Langevin took up the suggestion that had been made by several scientists that the reflection of ultrasonic waves from objects could be used to locate them. He used high-frequency radio circuitry to oscillate piezoelectric crystals and thus obtain ultrasonic waves at high intensity, and within a few years had a practical system for the echo location of submarines. This method has become the basis of modern sonar and is used for scientific as well as military purposes.

Langmuir, Irving (1881–1957) US physical chemist who is best remembered for his studies of adsorption at surfaces and for his investigations of thermionic emission. For his work on surface chemistry he was awarded the 1932 Nobel Prize for Chemistry. Unlike most scientists of world renown he did most of his research in a commercial environment, not in an academic institution.

Langmuir was born in Brooklyn, New York City, on 31 January 1881. He attended local elementary schools before his family moved to Paris for three years, where he was a boarder at a school in the suburbs. His interests in the practical aspects of science were fostered by his brother Arthur. In 1895 the family returned to the USA to Philadelphia, and Langmuir went to the Chapel Hill Academy and later to the Pratt Institute in Brooklyn. After high school he entered the School of Mines at Columbia University and graduated in 1903 with a degree in metallurgical engineering. His postgraduate studies were undertaken at the University of Göttingen under the guidance of Hermann ◊Nernst; he gained his PhD for a thesis on the recombination of dissociated gases. After a brief period as a teacher in New Jersey, he joined the General Electric Company in 1909 at their research laboratories at Schenectady and remained there until he retired in 1950; he was its assistant director 1932–50. He died in Falmouth, Massachusetts, on 16 August 1957.

Langmuir's work at Columbia concerned the dissociation of water vapour and carbon dioxide around red-hot platinum wires. His first studies at General Electric involved the thermal conduction and convection of gases around tungsten filaments. Langmuir showed that the blackening on the inside of 'vacuum-filled' electric lamps was caused by evaporation of tungsten from the filament. The introduction of nitrogen into the glass bulb prevented evaporation and blackening, but increased heat losses which were overcome by making the tungsten filament in the form of a coil.

At the same time, Langmuir was carrying out research on electric discharges in gases at very low pressures, which led to the discovery of the Child–Langmuir space–charge effect: the electron current between electrodes of any shape in vacuum is proportional to the 3/2 power of the potential difference between the electrodes. He also studied the mechanical and electrical properties of tungsten lamp filaments to which thorium oxide had been added. He showed that high thermal emissions were caused by diffusion of thorium to form a monolayer on the surface. One consequence was the development of an improved vacuum pump based on the condensation of mercury vapour. This work also initiated his 1934 patent for an atomic welding torch, in which the recombination of hydrogen atoms produced by an electric arc between tungsten electrodes generated heat at temperatures in the order of 6,000 °C/11,000 °F.

For three years Langmuir considered the problems of atomic structure. Building on Gilbert ⇨Lewis's atomic theory and valency proposals, Langmuir suggested that chemical reactions occur as a consequence of a desire by an atom to achieve a full shell of eight outer electrons. He was the first to use the terms electrovalency (for ionic bonds between metals and nonmetals) and covalency (for shared-electron bonds between nonmetals).

During the 1920s Langmuir became particularly interested in the properties of liquid surfaces. He went on to propose his general adsorption theory for the effect of a solid surface during a chemical reaction. He made the following assumptions:

(a) the surface has a fixed number of adsorption sites; at equilibrium at any temperature and gas pressure, a fraction θ of the sites are occupied by adsorbed particles and a fraction $1 - \theta$ are unoccupied;

(b) the heat of adsorption is the same for all sites and is independent of the number of sites occupied;

(c) there is no interaction between molecules on different sites, and each site can hold only one adsorbed molecule.

From this model he formulated Langmuir's adsorption isotherm, which can be expressed as

$$1/\theta = 1 + 1/bP$$

Where θ = fraction of sites occupied, b is a constant based on rate constants of evaporation and condensation, and P = pressure.

During World War II Langmuir was responsible for work that led to the generation of improved smoke-screens using smoke particles of an optimum size. He later applied this knowledge to particles of solid carbon dioxide and silver iodide which were scattered from aircraft to seed water droplets for cloud formation, in an attempt to make rain.

Laplace, Pierre Simon Marquis de (1749–1827)
French astronomer and mathematician who contributed to the fields of celestial mechanics, probability, applied mathematics, and physics.

Laplace was born on 28 March 1749 at Beaumont-en-Auge, in Normandy, France. The family were comfortably well-off – his father was a magistrate and an administrative official of the local parish and was probably also involved in the cider business. When he was seven Laplace was enrolled at the Collège in Beaumont-en-Auge run by the Benedictines and he attended as a day pupil until the age of 16. Typically, a young man would go either into the church or into the army after such an education. His father intended him to go into the church, but Laplace entered Caen University in 1766 and matriculated in the Faculty of Arts. Rather than take his MA, he went to Paris in 1768 with a letter of recommendation to Jean ⇨d'Alembert. The story is that d'Alembert gave him a problem and told him to return with the solution a week later. It is said that Laplace returned with the solution the next day and that d'Alembert immediately employed him at the Ecole Militaire.

Laplace was elected to the Academy of Sciences in 1773 and was fairly well known by the end of the 1770s. By the late 1780s, he was recognized as one of the leading people at the academy. In 1784, the Government appointed him to succeed E Bezout (1730–1783), the French mathematician, as examiner of cadets for the Royal Artillery. At the same time Gaspard ⇨Monge, the French mathematician who invented descriptive geometry, was appointed to examine naval cadets. When Napoleon became first consul in 1799, he appointed Laplace as Minister of the Interior, but dismissed him shortly afterwards and elevated him to the Senate. In 1814 Laplace voted for the overthrow of Napoleon in favour of a restored Bourbon monarchy, and so after 1815 he became an increasingly isolated figure in the scientific community. He remained loyal to the Bourbons until he died. In 1826 he was very unpopular with the Liberals for refusing to sign the declaration of the French Academy supporting the freedom of the press. After the restoration of the Bourbons, he was made a marquis. Laplace died on 5 March 1827.

Laplace's first important discovery came soon after he was appointed to the Ecole Militaire. He discovered that 'any determinant is equal to the sum of all the minors that can be formed from any selected set of its rows, each minor being described by its algebraic complement'. This is what is now known as the Laplace theorem.

He then turned his mind to problems in celestial mechanics, beginning in 1773 by examining the unexplained variations in the orbits of Jupiter and Saturn. Jupiter's orbit appeared to be continually shrinking, while Saturn's appeared to be continually expanding. No one had succeeded in explaining the phenomenon within the framework of Newtonian gravitation. In a brilliant three-part paper presented to the academy 1784–86, Laplace demonstrated that the phenomenon had a period of 929 years and that arose because the average motions of the two planets are nearly commensurable. The variations, therefore, were not at odds with, but compatible with, Isaac ◊Newton's law.

While working on that problem Laplace also published his famous paper, 'Théorie des attractions des sphéroïdes et de la figure des planètes' (1785), in which he introduced the potential function and the Laplace coefficients, both of them useful as a means of applying analysis to problems in physics. Two years later he submitted a paper to the academy in which he dealt with the average angular velocity of the Moon about the Earth, a problem with which Leonhard ◊Euler and Joseph ◊Lagrange had grappled unsuccessfully. Laplace found that, although the mean motion of the Moon around the Earth depends principally on the mutual gravitational attraction between them, it is slightly reduced by the Sun's pull on the Moon, this action of the Sun depending, in turn, upon the changes in eccentricity of the Earth's orbit due to the perturbations caused by the other planets. The result (although the inequality has a period of some millions of years) is that the Moon's mean motion is accelerated, while that of the Earth slows down and tends to become more circular.

Between 1799 and 1825, Laplace wrote volumes of *Traité de mécanique céleste*. In the first part he gave the general principles of the equilibrium and motion of bodies. By applying these principles to the motion of the planetary bodies, he obtained the law of universal attraction – the law of gravity as applied to the Earth. By analysis he was able to obtain a general expression for the motions and shapes of the planets and for the oscillations of the fluids with which they are covered. From these expressions, he was able to explain and calculate the ebb and flow of tides, the variations in the length of a degree of latitude and the associated change in gravitational force, the precession of the equinoxes, the way the Moon turns slightly to each side alternately (so that over a period slightly more than half of its surface is visible), and the rotation of Saturn's rings. He tried to show why the rings remain permanently in Saturn's equatorial plane. From the same gravitational theory he deduced equations for the principal motions of the planets, especially Jupiter and Saturn.

In 1796, Laplace published his *Exposition du système du monde*, a popular work that was widely read because it gave a theory designed to explain the origin of the Solar System. It suggested that the Solar System resulted from a primitive nebula that rotated about the Sun and condensed into successive zones or rings from which the planets themselves are supposed to have been formed. This theory was generally accepted throughout the 19th century. Since then, however, it has been shown that it is impossible for planets to form in this way because new eccentricities were later discovered that could not be explained by Laplace's concept. Laplace suggested that, as the nebula rotated and contracted, its speed of rotation would increase until the outer parts were moving too quickly for the nebula's gravitational pull to hold them. The remainder of the nebula would then shrink further, rotate faster and another planet would form. This would be repeated until only enough material to form a stable Sun was left at the centre. This theory of planetary formation does not account for the retrograde spin of Venus and the eccentric orbit of Pluto. Also, the theory fails to explain how the energy from the Sun is being continuously created.

Laplace was one of the most influential of 18th-century scientists, whose work confirmed earlier Newtonian theories and finally affirmed the permanence and stability of the Solar System. He also compiled considerably improved astronomical tables of the planetary motions, tables that remained in use until the mid-19th century. Together with Lagrange, he was largely responsible for the development of positional astronomy in the 18th and 19th centuries. Moreover, the methods by which he achieved his results established celestial mechanics as a branch of analysis and through his work potential theory, the Laplace theorem, the Laplace coefficients, orthogonal functions, and the Laplace transform (introduced in his papers on probability) became fundamental tools of analysis.

Further Reading

Gillispie, Charles Coulston; Fox, Robert; and Grattan-Guinness, I, *Pierre-Simon Laplace, 1749–1827: A Life in Exact Science*, Princeton University Press, 1997.

Hahn, Roger, *Calendar of the Correspondence of Pierre Simon Laplace*, Berkeley Papers in History of Science series, Office for History of Science and Technology, University of California, 1982, v 8.

Hahn, Roger, *The New Calendar of the Correspondence of Pierre Simon Laplace*, Berkeley Papers in History of Science series, Office for History of Science and Technology, University of California, 1994 v 16.

Lapworth, Arthur (1872–1941) Scottish organic chemist whose most important work was the enunciation of the electronic theory of organic reactions (independently of Robert ◊Robinson).

Lapworth was born in Galashiels, Scotland, on 10 October 1872, the son of the geologist Charles Lapworth who was teaching there at the time. The family moved to Birmingham on his father's appointment as professor of geology at Mason College, which Lapworth also attended to study chemistry. He graduated in 1893 and went to the City and Guilds College, London, to do research, first under H E Armstrong on the chemistry of naphthalene and then with Frederick ◊Kipping in the study of camphor. Lapworth, Kipping, and their contemporary, the chemist William Perkin Jr (1860–1929), were related through marriage into the same family. Lapworth was a demonstrator at the School of Pharmacy 1895–1900, at which time he went on to become head of the chemistry department at Goldsmiths College, London. In 1909 he moved to Manchester and became senior lecturer in organic and physical chemistry at the university. In 1922 he took up the senior professorship in the department. Spending the remainder of his life in Manchester, he suffered painful ill health during his last years. He died in Manchester on 5 April 1941.

Lapworth's early research into camphor exposed evidence of intramolecular variations, related to the pinacol–pinacolone rearrangement, a discovery that facilitated general acceptance of Bredt's structure for camphor.

His subsequent study of reaction mechanisms, in particular the formation of cyanohydrin from carbonyl compounds and the benzoin condensation reaction of benzaldehyde, brought Lapworth later recognition as a founder of the modern branch of physical-organic chemistry. He was among the first to bring attention to the ionization of organic compounds, showing the behaviour of different parts of organic molecules as following patterns of behaviour for bearing electrical charges, either at the moment of their reaction or habitually.

The development of the electronic theory of valency enabled Lapworth to classify 'alternative polarities' in organic compounds into either anionoid or cationoid reaction centres, with changes influenced by a key atom such as oxygen. At the time of Lapworth's collaboration with Robert Robinson in the mid-1920s, a controversy was sparked over these concepts when Christopher ◊Ingold and his school used a similar approach to the problem but employed a different terminology (nucleophilic for anionoid and electrophilic for cationoid). Ingold's terminology eventually gained general acceptance and the debate, if not always without friction, was beneficial to the field. In 1931 Lapworth and Ingold were listed as co-authors in a paper that was to be Lapworth's last.

Latimer, Louis Howard (1848–1928) US inventor who is best known for his work on electric lighting. He worked for a long time with Thomas ◊Edison and was the only black member of the 'Edison pioneers', an organization of scientists who worked with Edison in his pioneering work in the field of electricity.

The son of an escaped slave, Latimer was born on 4 September 1848 in Chelsea, Massachusetts. Escaping from the farm school to which he was sent, Latimer went to Boston to help support his family by selling copies of Garrison's paper *The Liberator*. During the Civil War he enlisted in the US navy from which he was honourably discharged in 1865. Latimer went to work as an office boy at a firm of patent solicitors, Crosby and Gould, where he learned mechanical drawing and went on to become chief draughtsman of the company. It was here that he became acquainted with Alexander Graham ◊Bell with whom he worked, executing the drawings and assisting in preparing descriptions for the application for the telephone patent that was issued in 1876. In 1880 Latimer was employed as a draughtsman by Hiram ◊Maxim at the US Electric Lighting Company, Connecticut. In 1883 he went to serve as an engineer and chief draughtsman to Thomas Edison at the Edison Electric Light Company (later the General Electric Company), New York City. In 1890 he transferred from the engineering department to the legal department defending Edison's patents in court as an expert witness. Between 1896 and 1911 he served as chief draughtsman and expert witness for the Board of Patent Control formed by the General Electric and Westinghouse Companies to protect their patents. Latimer died following a long illness on 11 December 1928; he was married and had two daughters.

Latimer is best known for work on electric lighting. In 1881 he and Joseph V Nichols were given a patent for their 'electric lamp'; he went on to invent a cheap method for producing long-lasting carbon filaments for the maxim electric incandescent lamp, obtaining a patent for them in 1882; these were used in railway stations both nationally and abroad. Also in 1882 he was awarded a patent for his 'globe supporter for electric lamps'. His book *Incandescent Electric Lighting: A Practical Description of the Edison System* (1890) was a pioneer on the subject. He also supervised the installation of electric lights in New York, Philadelphia, Canada, and London.

It was not only for lighting that Latimer obtained patents but also for a 'Water Closet for Railroad Cars' (1874), 'Apparatus for Cooling and Disinfecting' (1886), 'Locking Rack for Hats, Coats and Umbrellas' (1896), and 'Book Supports' (1905). Latimer also wrote poetry and his *Poems of Love and Life* was privately published in 1925 to commemorate his 77th birthday. He was honoured in May 1968 when the Lewis Howard Latimer School in Brooklyn was named after him.

Laue, Max Theodor Felix von (1879–1960)
German physicist who established that X-rays are electromagnetic waves by producing X-ray diffraction in crystals. In recognition of this achievement he was awarded the 1914 Nobel Prize for Physics.

Laue was born in Pfaffendorf, near Koblenz, on 9 October 1879. He entered the University of Strasbourg in 1899, but soon transferred to the University of Göttingen, where he chose to specialize in theoretical physics. He then attended the University of Berlin, where he obtained his DPhil for a dissertation supervised by Max ◊Planck on interference theory. In 1903 Laue returned to Göttingen, where he spent two years studying art and obtained a teaching certificate. He then became an assistant to Planck at the Institute of Theoretical Physics in Berlin, cementing a long and fruitful partnership between the two scientists. Laue remained in Berlin for four years, during which time he qualified as a lecturer at the university, and became an early adherent of ◊Einstein's special theory of relativity. He published the first monograph on the subject in 1909, which was expanded in 1919 to include the general theory of relativity and subsequently ran to several editions.

In 1909 Laue moved to the University of Munich, and became a lecturer at the Institute of Theoretical Physics, which was directed by Arnold ◊Sommerfeld. It was here that he began his work on the nature of X-rays and crystal structure for which he was awarded the Nobel prize in 1914, only two years after the research was carried out, a remarkable testimony to the importance of the work.

Laue became a full professor at the University of Frankfurt in 1914, but spent the years of World War I at the University of Würzburg, working on the development of amplifying equipment to improve communications for the military. After the end of the war he exchanged teaching posts with Max ◊Born and went to Berlin, where he became professor of theoretical physics and director of the Institute of Theoretical Physics.

In the 1920s and early 1930s, through the German Research Association, Laue directed the provision of financial resources for physics in Germany, thus helping to bring about the important discoveries in theoretical physics with which Germany led the world at that time. In 1932 he was awarded the Max Planck Medal by the German Physical Society. But the rise to power of Adolf Hitler in 1933 and the consequent persecution of Jewish scientists (such as Einstein) and manipulation of German science by the National Socialist Party was intolerable to Laue. His protests led to a gradual trimming of his influence and he lost his position in the German Research Association. He did, however, continue to act as professor at the University of Berlin until 1943, when he resigned in protest. Although Laue refused to participate in the German atomic energy project, he was nevertheless interned in the UK by the allies after the war, along with other German atomic physicists.

In 1946 Laue returned to Germany where he helped to rebuild German science. He served as deputy director of the Kaiser Wilhelm Institute for Physics, and in 1951 became the director of the Max Planck Institute for Research in Physical Chemistry. He died on 23 April 1960 as the result of a car accident.

Laue's early work in Berlin under Max Planck was chiefly concerned with interference phenomena. He also worked on the entropy of beams of rays, and developed a proof for Einstein's special theory of relativity based on optics. This used the verification in 1851 by Armand ◊Fizeau of a theory developed even earlier by Augustin ◊Fresnel that the velocity of light is affected by the motion of water. This seemed to be at odds with special relativity, but Laue was able to show that Fresnel's theory could be derived from the conclusions of relativity. Fizeau's verification thus also became an experimental confirmation of relativity, contributing greatly to its rapid acceptance.

At the time of Laue's move to Munich in 1909, the subjects of crystal structure and the nature of X-radiation were ill-defined. It had been proposed that crystals consist of orderly arrays of atoms, and that X-rays are a type of electro-magnetic radiation like light, but with very short wavelengths. The wavelength of normal monochromatic light was determined by the use of artificial diffraction gratings, but if X-rays did in fact have such short wavelengths, no grating sufficiently fine could possibly be produced with which to diffract them. Laue realized that if crystals are indeed such orderly atomic arrays, then they could be used as superfine gratings.

To test Laue's ideas, Sommerfeld's assistant Walter Friedrich and Paul Knipping, a postgraduate student, began experiments on 21 April 1912 using copper sulphate. They bombarded the crystal with X-rays and produced a photographic plate with a dark central patch representing X-rays that had penetrated straight through the crystal surrounded by a multilayered halo of regularly spaced spots representing diffracted

X-rays. The results were announced to the Bavarian Academy of Science less than two weeks later along with Laue's mathematical formulation called the 'theory of diffraction in a three-dimensional grating'. The experimental demonstration of the nature of crystal structure and of X-rays at one strike was widely acclaimed.

Following World War I, Laue expanded his original theory of X-ray diffraction. He had initially considered only the interaction between the atoms in the crystal and the radiation waves, but he now included a correction for the forces acting between the atoms. This correction accounted for the slight deviations which had already been noticed. Laue did not participate in the development of quantum theory, however, being rather cautious in his approach to it, but he did incorporate the discovery of electron interference into his diffraction theory. He also did some important work on superconductivity, including the effect of magnetic fields on this phenomenon, publishing papers on the subject 1937–47.

Laue's achievement proved to be one of the most significant discoveries of the 20th century. Einstein described it as one of the most beautiful discoveries in physics, and it has had a profound influence on physics, chemistry, and biology. Two entirely new branches of science arose out of Laue's prizewinning work, although he had no more than a passing interest in them. These were X-ray crystallography, which was developed by William and Lawrence ◊Bragg and led to the determination of the structure of DNA in 1953, and X-ray spectroscopy, which was developed by Maurice de Broglie (1875–1960) and others for the exact measurement of the wavelength of X-rays and for advances in atomic theory.

Laveran, Charles Louis Alphonse (1845–1922) French physician who discovered that the cause of malaria is a protozoan, the first time that a protozoan had been shown to be a cause of disease. For this work and later discoveries of protozoan diseases he was awarded the 1907 Nobel Prize for Physiology or Medicine.

Laveran was born in Paris on 18 June 1845, the son of an army surgeon. The family moved to Algeria 1850–55 and then returned to Paris, where he continued his education. He studied medicine in Strasbourg and graduated in 1867. When the Franco-Prussian war broke out he became an army surgeon, and was stationed at Metz 1870–71. In 1874 he was appointed professor of military medicine at the Ecole du Val-de-Grâce. Between 1878 and 1883 he was posted to Algeria, first to Bône and then Constantine. In 1884 he returned to Val-de-Grâce and served as professor of military hygiene, until 1894, when temporary posts took him to Lille and Nantes. In 1896 he left the army to join the Pasteur Institute in Paris and in 1907 he used the money from his Nobel prize to open the Laboratory of Tropical Diseases at the institute. The following year he founded the Societé de Pathologie Exotique and was its president until 1920. He died in Paris on 18 May 1922.

While in Algeria, Laveran found malaria rife there and he became interested in the disease, although his early research had concentrated on nerve regeneration. Malaria has affected humans for centuries, and at that time was explained by a number of hypotheses, one of the most common being that it was due to bad air, especially over swamps and marshes (*mal aria* are the Italian words for bad air). Scientists were beginning, however, to regard it as a bacterial disease.

It was known that malarial blood contains tiny black granules which are sometimes located in hyaline cysts inside the red blood cells. In 1880 Laveran examined blood samples from malarial patients and while investigating a cyst, saw flagellae pushed out from it, revealing that it was not a granule of pigment (as had been believed), but an amoeba-like organism. Laveran took blood samples from these patients at regular intervals and found that the organisms increased in size until they almost filled the blood cell, when they divided and formed spores. When these spores were liberated from the destroyed blood cell they invaded unaffected blood cells. He noted that the spores were released in each affected red cell at the same time and corresponded with a fresh attack of fever in the patient. The point at which the fever subsided coincided with the invasion of new blood cells. Laveran also found crescent-shaped bodies with black granules lying in the plasma.

Laveran's studies of protozoan diseases included leishmaniasis and trypanosomiasis. His investigations concerning malaria allowed Ronald ◊Ross, after him, to discover that the malarial parasite is transmitted to human beings by the *Anopheles* mosquito and these findings enabled preventive measures to be taken against the disease.

Laveran's publications included *Traité des maladies et épidemies des armées/Treatise on Army Sicknesses and Epidemics,* (1875) and *Trypanosomes et Trypanosomiasis/Trypanosomes and Trypanosomiasis* (1904).

Lavoisier, Antoine Laurent (1743–1794) French chemist, universally regarded as the founder of modern chemistry. His contributions to the science were wide ranging, but perhaps his most significant achievement was his discrediting and disproof of the phlogiston theory of combustion, which for so long had been a stumbling block to a true understanding of chemistry.

Lavoisier French chemist Antoine Lavoisier, regarded as the founder of modern chemistry. In 1775 he was appointed Inspector of Gunpowder for the French government. He and his wife, who was also his scientific assistant, lived in a house at the Arsenal. *Mary Evans Picture Library*

Lavoisier was born in Paris on 26 August 1743 into a well-off family. His mother died when he was young and he was brought up by an aunt. He received a good education at the Collège Mazarin, where he studied astronomy, botany, chemistry, and mathematics. In 1768 he was elected an associate chemist to the Academy of Sciences; he eventually became its director in 1785 and Treasurer in 1791.

Also in 1768 Lavoisier became an assistant to Baudon, one of the farmers general of the revenue, and later he became a full member of the *ferme générale*, employed by the government as tax collectors. He married 14-year-old Marie Paulze, daughter of a tax farmer, in 1771 and the following year his father bought him a title. In 1775 he was made *régisseur des poudres* and improved the method of preparing saltpetre (potassium nitrate) for the manufacture of gunpowder. A model farm he set up at Frénchines in 1778 applied scientific principles to agriculture, and he drew up various agricultural schemes as secretary to the committee on agriculture, to which he was appointed in 1785. Two years later he became a member of the provincial assembly of Orléans, in which position he initiated many improvements for the community, such as workhouses, savings banks, and canals. He was also a member of various other committees and commissions, including that formed in 1790 to rationalize the system of weights and measures throughout France, which ultimately led to the founding of the metric system.

During the French Revolution, Lavoisier came under suspicion because of his membership of the *ferme générale* (from which he derived a considerable income) and because of his marriage to one of its senior executives, although his wife acted as his scientific assistant, taking notes and even illustrating some of his books. Jean-Paul Marat, an extremist revolutionary whose membership of the Academy of Sciences had been blocked by Lavoisier, accused him of imprisoning Paris and preventing air circulation because of the wall he had built round the city in 1787. He fled from his home and laboratory in August 1792 but was arrested in the following November and sent for trial by the revolutionary tribunal in May 1794. At a cursory trial Lavoisier was one of 28 unfortunates sentenced to death. He was guillotined on 8 May 1794 and buried in a common grave. His widow later, in 1805, married the US physicist Benjamin Thompson (Count ◊Rumford).

Among Lavoisier's early scientific work were papers of the analysis of the mineral gypsum (hydrated calcium sulphate), on thunder, and a refutation that water changes into 'earth' if it is distilled repeatedly. He helped the geologist Jean-Etienne ◊Guettard to compile a mineralogical atlas of France.

But his most significant experiments concerned combustion. He found that sulphur and phosphorus increased in weight when they burned because they absorbed 'air', and reported these results in a note he left with the Academy of Sciences in 1772. He also discovered that when litharge (lead(II) oxide) was reduced to metallic lead by heating with charcoal it lost weight because it had lost 'air'. Then in 1774 Joseph ◊Priestley produced 'dephlogisticated air' and Lavoisier grasped the true explanation of combustion, inventing the name oxygen (acid-maker) for the substance that combined with caloric and formed 'oxygen gas'. He coined the word azote for the 'non-vital air' (nitrogen) that remained after the oxygen in normal air had been used up in combustion. In June 1783 he published his finding that the combustion of hydrogen in oxygen produces water, although, unknown to him, this fact had already been announced by the English chemist Henry ◊Cavendish. Lavoisier burned various organic compounds in oxygen and determined their composition by weighing the carbon dioxide and water produced – the first experiments in quantitative organic analysis. Lavoisier's findings were universally accepted after the publication of his clear and logical *Traité élémentaire de chimie* (1789), in which he listed all the chemical elements then known (although some of these were in fact oxides).

After establishing that organic compounds contain carbon, hydrogen, and oxygen, Lavoisier showed by weighing that matter is conserved during fermentation as with more conventional chemical reactions. From quantitative measurements of the changes during breathing, he discovered the composition of respired air and showed that carbon dioxide and water are both normal products of respiration.

Lavoisier also made many studies outside the field of chemistry. With Pierre ◊Laplace he experimented with calorimetry and other aspects of heat. He began to use solar energy for scientific purposes as early as 1772 and observed that 'the fire of ordinary furnaces seems less pure than that of the Sun'. He anticipated later theories about the interdependence of sequential processes in plant and animal life forms, as described in one of his papers discovered only many years after his death. His great contribution to science was summed up by Joseph ◊Lagrange who said, on the day after Lavoisier was guillotined at the Place de la Revolution, 'It required only a moment to sever that head, and perhaps a century will not be sufficient to produce another like it.'

Further Reading

Donovan, Arthur, *Antoine Lavoisier: Science, Administration, and Revolution,* Cambridge Science Biographies series, Cambridge University Press, 1996.
Holmes, Frederic L, *Lavoisier and the Chemistry of Life: An Exploration of Scientific Creativity,* Wisconsin Publications in the History of Science and Medicine, University of Wisconsin Press, 1985, v 4.
Holmes, Frederic L, *Antoine Lavoisier: The Next Crucial Year, or the Sources of His Quantitative Method in Chemistry,* Princeton University Press, 1998.
McKie, Douglas, *Antoine Lavoisier: Scientist, Economist, Social Reformer,* Constable, 1952.
Poirrier, Jean-Pierre, *Lavoisier: Chemist, Biologist, Economist,* Chemical Sciences in Society series, University of Pennsylvania Press, 1996.

Lawrence, Ernest Orlando (1901–1958) US physicist who was responsible for the concept and development of the cyclotron. For this achievement he was awarded the 1939 Nobel Prize for Physics. As director of the Radiation Laboratories at Berkeley, California, he and his co-workers then produced a remarkable sequence of discoveries, which included the creation of radioactive isotopes and the synthesis of new transuranic elements.

Lawrence was born in Canton, South Dakota, on 8 August 1901. He went to the University of South Dakota, where he was awarded a BS degree in 1922, and then continued his studies at the University of Minnesota, specializing in physics and gaining an MA the following year. After a further two years spent at the University of Chicago, he went to Yale University where he carried out research on photoelectricity and gained his PhD in 1925. He continued at Yale as a research fellow and later as assistant professor of physics until 1928. During this period he made a precise determination of the ionization potential (the energy required to remove an electron) of the mercury atom, developed methods for spark discharges of minute duration (3×10^{-9} sec), and discovered a new method of measuring the charge-to-mass ratio of the electron (e/m). Lawrence then returned to the University of California as associate professor of physics and in 1930 he became, at 29, the youngest full professor (of physics) at Berkeley. Lawrence retained this position until the end of his life, together with the directorship of the Radiation Laboratories, which he commenced in 1936.

Lawrence US physicist Ernest O Lawrence who invented the cyclotron particle accelerator, which he used to study nuclear reactions. He pioneered medical applications for neutrons and radioisotopes and during World War II worked on the development of the atomic bomb. *Mary Evans Picture Library*

During World War II, Lawrence was involved with the separation of uranium-235 and plutonium for the development of the atomic bomb, and he organized the Los Alamos Scientific Laboratories at which much of the work on this project was carried out. After the war, he continued as a believer in nuclear weapons and advocated the acceleration of their development. He died in Palo Alto, California, on 27 August 1958.

In 1929 Lawrence was pondering methods of attacking the atomic nucleus with particles at high energies.

The methods then developed involved the generation of very high voltages of electricity. To Lawrence, these seemed to be very limited in the long run as the electrical engineering necessary seemed likely to be technically very difficult, and their cost would eventually prove to be prohibitive. Early that same year, he came across an article in a German periodical that described a linear device for the acceleration of ions, and Lawrence realized the potential of the design for producing high-energy particles that could be directed at atomic nuclei. On carrying out the calculation to produce a machine that would produce protons with the energy of one million electronvolts, he discovered that the device would be too long for the laboratory space he had available. He therefore considered bending the accelerator so that the particles would travel in a spiral. This idea led to the cyclotron, which had two electrodes that were hollow and semicircular in shape mounted in a vacuum between the poles of a magnet. In between the two electrodes was the source of the particles. To the electrodes was connected a source of electricity whose polarity could be oscillated, positive and negative, from one electrode to the other at very high frequency. The negative field caused the particles to turn in a circular path in each electrode and each time they crossed between the electrodes they received an impetus that accelerated them further, giving them more and more energy without involving very high voltages. As this happened, the increased velocity carried the particles farther from the centre of the apparatus, producing a spiral path.

The first cyclotrons were made in 1930 and were only a few centimetres in diameter and of very crude design. They were made by Nels Edlefsen, one of Lawrence's research students, and one of them showed some indication of working. Later the same year M Stanley Livingston, another of Lawrence's research students, constructed a further small but improved device that accelerated protons to 80,000 electronvolts with 1,000 volts used on the electrodes. Once the principle had been seen to work well, larger and improved devices were made. Each new design produced particles of higher energy than its predecessor and new results were obtained from the use of the accelerated particles in nuclear transformations. One of these was the disintegration of the lithium nucleus to produce helium nuclei, confirming the first artificial transformation made by John ◊Cockcroft and Ernest ◊Walton in 1932, and a 68-cm/27-in model was used to produce artificial radioactivity. With increasing size, the engineering problems associated with power supplies and the production of large magnetic fields increased, and in 1936 the Radiation Laboratory with the necessary engineering and administrative facilities was opened to house the new equipment.

From his laboratory, a multitude of discoveries in chemistry, physics, and biology were made under Lawrence's direction. Hundreds of radioactive isotopes were produced, including carbon-14, iodine-131, and uranium-233. Furthermore, over the years, most of the transuranium elements were synthesized. Mesons, which were then little-known particles, were produced and studied within a laboratory for the first time at the Radiation Laboratory, and anti-particles were also found. Many of these discoveries owe their origin to other famous scientists who worked within the laboratory, but Lawrence worked tirelessly advising researchers and in making the experiments possible. Fairly early in the development of the cyclotron it became apparent that as particles were accelerated to high speeds their masses increased, as predicted by the special theory of relativity. This increase in mass caused them to lose synchronization with the oscillating electric field. The solution of this problem by Vladimir Veksler (1907–1966) and Edwin McMillan (1907–1991) independently led to the development of the synchrotron and particles of even higher energies.

Lawrence made an important contribution to physics with the invention of cyclotron, for the subsequent development of high-energy physics and the elucidation of the subatomic structure of matter was highly dependent on this machine and its successors. These have also provided radioisotopes of great value in medicine. Appropriately, Lawrence's name is remembered in the naming of element 103 as lawrencium, for it was discovered in his laboratory at Berkeley.

Leakey, Louis Seymour Bazett (1903–1972) Kenyan archaeologist, anthropologist, and palaeontologist who became famous for his discoveries of early hominid fossils in East Africa, indicating that humans probably evolved in this part of the world.

Leakey was born in Kabete on 7 August 1903, the son of a British missionary. There were few European settlers in Kenya at that time and he spent his boyhood with the local Kikuyu children, learning their language but receiving little formal schooling. He was sent to the UK when he was 16 years old and entered St John's College, Cambridge, in 1922 to study French and Kikuyu, and later archaeology and anthropology. In 1926 he went to East Africa leading the first of a series of archaeological research expeditions, which continued until 1937. From 1937 until the outbreak of World War II in 1939 he studied and recorded the customs of the Kikuyu people. During the war he was in charge of the African section at the British Special Branch Headquarters; he was a handwriting expert and remained available as a consultant in this role after the end of the war. During the war he gave freely of his spare time to the Coryndon Memorial Museum in Nairobi,

Kenya, and in 1945 he became its curator. He built up a research centre at the museum and at the same time became one of the founder trustees of the Kenya National Parks and Reserves. He resigned from the museum in 1961 and founded the National Museum Centre for Prehistory and Palaeontology. During the 1960s his health began to fail and he spent less time in the field and more time lecturing and fundraising. He died in London on 1 October 1972.

Leakey began excavations at Olduvai Gorge, now in Tanzania, in 1931 and it was to become the site of some of his most important finds. During wartime leave he and his wife Mary ◊Leakey, an expert on palaeolithic stone implements, discovered a significant Acheulian site at Olorgesailie in the Rift Valley. The Acheulian culture is characterized by stone hand axes and flourished between 1 million and 100,000 years ago. On another occasion the Leakeys found the remains of 20-million-year-old Miocene apes on the island of Rusinga in Lake Victoria. From 1947 they initiated in Nairobi a Pan-African Congress on Prehistory and Palaeontology, which was a great success and proved to be the first of many.

In 1959 the Leakeys returned to their excavations at Olduvai. In that year they found a skull of *Australopithecus boisei* (Nutcracker Man) and a year later they discovered the remains of *Homo habilis,* established by potassium/argon dating to be 1.7 million years old. A third exciting find was a skull of an Acheulian hand-axe user, *Homo erectus,* which Leakey maintained was an advanced hominid on the direct evolutionary line of *Homo sapiens,* the modern human. In 1961 at Fort Ternan, Kenya, he found jawbone fragments of another early primate, *Kenyapithecus wickeri,* believed to be 14 million years old.

Leakey was a man of many talents with firm convictions that our origins lie in Africa, a view that was totally opposed to contemporary opinion. He discovered sites of major archaeological importance containing vast numbers of bones and artefacts. The status of some of the finds, and Leakey's interpretation of them, is still in dispute, but whatever the eventual outcome he made an unparalleled contribution to knowledge of early humans and their contemporaries. He published a large number of scientific papers as well as some books of wider appeal, such as *Stone Age Africa* (1936) and *White African* (1937).

After Louis Leakey's death his work in Africa was continued by his wife and his sons John and Richard. Richard Leakey (1944–) later became a well-known author and broadcaster in his own right.

Further Reading

Pickford, Martin, *Louis S B Leakey: Beyond the Evidence,* Janus Publishing, 1998.

Willis, Delta, *The Leakey Family: Leaders in Search for Human Origins,* Makers of Modern Science series, Facts on File, 1992.

Leakey, Mary Douglas, born Mary Douglas Nichol (1913–1996) English anthropologist and archaeologist.

Born Mary Douglas Nicol in London on 6 February 1913, Leakey was the only child of a landscape painter from whom she inherited a talent for drawing that was of considerable importance to her later career. Much travelling during her childhood disrupted her formal education but a visit to prehistoric caves in southwest France, where her father went to paint, kindled an interest in archaeology. A chance meeting with the archaeologist Dorothy Liddell in the late 1920s convinced her that a career in the subject was possible and she became Liddell's assistant, chiefly as the illustrator at a major dig at a Neolithic site in Devon. She met the anthropologist Louis ◊Leakey at a dinner party and he invited her to illustrate a book he was then working on. She agreed, and the two began to work closely together. Mary Nicol travelled to join him in Kenya at the Olduvai Gorge, and after Louis Leakey's subsequent divorce, they married in 1936. Mary Leakey immediately began excavating a Late Stone Age site north of Nairobi, and with her husband discovered the remains of an important Neolithic settlement and many important artefacts.

During World War II Mary continued to develop research projects, and after the war she devoted much of her time to organizing a major Pan-African Congress of Prehistory and Palaeontology in 1947 with her husband. The success of that conference brought the work of the Leakeys to a wide audience. The attention continued in 1948 when Mary discovered, on Rusinga Island, Lake Victoria, the prehistoric ape skull known as *Proconsul,* about 20 million years old, and human footprints at Laetoli, to the south, about 3.75 million years old.

Over the following 15 years the Leakeys together and separately continued major excavations in Kenya, and accumulated evidence that East Africa was the possible cradle of the human race. By the middle of the 1960s Mary was living almost continuously at the permanent camp they had established at the Olduvai Gorge, whilst Louis was based in Nairobi and increasingly travelling around the world on lecture tours, and although they never officially separated, they lived apart until Louis's death in 1972. That event precipitated Mary onto the international stage and she became an accomplished lecturer and writer, while continuing to organize excavation work. She received several international awards and honorary doctorates, and published an account of her work

Olduvai Gorge: My Search for Early Man (1979) and an autobiography *Disclosing the Past* (1984).

Further Reading

Morell, Virginia, *Ancestral Passions: The Leakey Family and the Quest for Humankind,* Simon and Schuster, 1995.

Willis, Delta, *The Leakey Family: Leaders in Search for Human Origins,* Makers of Modern Science series, Facts on File, 1992.

Leavitt, Henrietta Swan (1868–1921) US astronomer, an expert in the photographic analysis of the magnitudes of variable stars. Her greatest achievement was the discovery of the period–luminosity relationship for variable stars, which enabled Ejnar ◊Hertzsprung and Harlow ◊Shapley to devise a method of determining stellar distance.

Leavitt was born in Lancaster, Massachusetts, on 5 July 1868. She studied at the Society for the Collegiate Instruction of Women, which later became Radcliffe College and associated with Harvard. She earned her bachelor's degree in 1892, but continued her studies there for an additional year. Her first work in astronomy was in a voluntary capacity at the Harvard College Observatory, where she was an assistant in the programme which Edward ◊Pickering had initiated on the measurement of stellar magnitudes. In 1902 Leavitt was given a permanent appointment to the staff of the observatory, and was ultimately appointed head of the department of photographic stellar photometry. She died prematurely of cancer on 12 December 1921, in Cambridge, Massachusetts.

Leavitt's work was a direct outgrowth of Pickering's overall research programme at the observatory. She was extensively involved with the establishment of a standard photographic sequence of stellar magnitudes, and besides discovering a total of 2,400 new variable stars, she also discovered four novae. Her most far-reaching discovery, however, was that the periods of some variable stars is directly related to their magnitudes.

Leavitt spent many years studying a kind of variable star called the Cepheid variable, named after the first of the type to be discovered, Delta Cephei in the Magellanic Clouds. These star collections were photographed by the astronomers at the Harvard Observatory in Arequipa, Peru. As early as 1908, when she published a preliminary report of her findings, Leavitt suspected that there was a relationship between the length of the period of variation in brightness and the average apparent magnitude of the Cepheids. In 1912 she published a table of the length of the periods, which varied from 1.253 to 127 days, with an average of around five days, and the apparent magnitudes of 25 Cepheids. There was a direct linear relationship between the average apparent magnitude and the logarithm of the period of these variable stars.

The reason Leavitt was able to notice this relationship for the variable stars in the Magellanic Clouds, while it had remained unnoticed for other variable stars, lies in the close grouping of these Cepheids in terms of their distances from each other in contrast to their enormous distance from the Earth. A nearby, albeit dim, variable star might seem to be brighter than a more distant star of greater magnitude. It has since become apparent that the period–luminosity relationship discovered by Leavitt applies only for Cepheids that lie in 'dust-free' space.

At the time this undiscovered factor did not prevent Hertzsprung and Shapley from independently applying the period–luminosity relationship to the determination of stellar distances. All of the Cepheids were too far from the Earth for their distances to be determined using the standard parallax method, pioneered by Friedrich ◊Bessel. But Hertzsprung and Shapley were able to convert Leavitt's period–luminosity curve so that it could be read in terms of absolute rather than apparent magntude. Then, once they had determined the period of any Cepheid, they could return to the curve and read off its absolute magnitude. By comparing a Cepheid's apparent magnitude with its absolute magnitude, the distance of the star from the Earth could be deduced.

Their results were nothing short of astonishing for contemporary astronomers. They found that the Magellanic Clouds were approximately 100,000 light years away, a distance that was almost beyond the comprehension of their colleagues. Shapley revised these figures in 1918, finding the Small Magellanic Cloud to be 94,000 light years away.

When Leavitt first began studying the stars in the two hazy areas of the sky, the Large and Small Magellanic Clouds, she did not realize that they were extragalactic. But, due to her work, it is now known that the Magellanic Clouds are in fact two irregular galaxies, companions of our own Galaxy. Thus Leavitt's vital contribution to astronomy was to provide the critical impetus for the discovery of the first technique capable of measuring large stellar distances.